**Mathematical methods for physics and engineering**

*A comprehensive guide*

All physical scientists and engineers need mathematical tools in their everyday work. This book presents the mathematics appropriate to undergraduate courses in a friendly and approachable way. The material covered is comprehensive, and takes the student all the way from the starting level of college or university to the end of the most mathematical of science courses. It also provides a valuable reference for active researchers in physics, engineering, chemistry, applied mathematics and earth sciences.

Throughout the text, the physical relevance of the mathematics is constantly reinforced, and numerous physical worked examples are included. In this way, it provides scientists who need to use the tools of mathematics for practical purposes with a single, comprehensive textbook.

**Ken Riley** read Mathematics at the University of Cambridge and proceeded to a Ph.D. there in theoretical and experimental nuclear physics. He became a Research Associate in elementary particle physics at Brookhaven, and then, having taken up a lectureship at the Cavendish Laboratory, Cambridge, continued this research at the Rutherford Laboratory and Stanford; in particular he was involved in the discovery of a number of the early baryonic resonances. As well as being Senior Tutor at Clare College, where he has taught physics and mathematics for over thirty years, he has served on many committees concerned with the teaching and examining of these subjects at all levels of tertiary and undergraduate education. He is also the author of *Problems for Physics Students.*

**Michael Hobson** read Natural Sciences at the University of Cambridge, specialising in theoretical physics, and remained at the Cavendish Laboratory to complete a Ph.D. in the physics of star-formation. As a Research Fellow at Trinity Hall, Cambridge, he developed an interest in cosmology, particularly the study of fluctuations in the cosmic microwave background, and was involved in the first detection of these fluctuations using a ground-based interferometer. He is currently a Director of Studies in Natural Sciences at Trinity Hall and enjoys an active role in the teaching of undergraduate physics and mathematics.

**Stephen Bence** obtained both his undergraduate degree in Natural Sciences and his Ph.D. in Astrophysics from the University of Cambridge. As a Research Associate in the Cavendish Laboratory, he is interested in the star-formation process and the structure of star-forming regions. In particular, his research has concentrated on the physics of jets and outflows from young stars. He is experienced in teaching mathematics and physics to undergraduate and pre-university students.

To our families

# Mathematical methods for physics and engineering

## *A comprehensive guide*

K. F. Riley, M. P. Hobson and S. J. Bence

PUBLISHED BY THE PRESS SYNDICATE OF THE UNIVERSITY OF CAMBRIDGE
The Pitt Building, Trumpington Street, Cambridge, United Kingdom

CAMBRIDGE UNIVERSITY PRESS
The Edinburgh Building, Cambridge CB2 2RU, UK http://www.cup.cam.ac.uk
40 West 20th Street, New York, NY 10011-4211, USA http://www.cup.org
10 Stamford Road, Oakleigh, Melbourne 3166, Australia
Ruiz de Alarcón 13, 28014 Madrid, Spain



First published 1997
Reprinted 1998 (with minor corrections), 2000

Printed in the United Kingdom at the University Press, Cambridge

Typeset in TEX Monotype Times

*A catalogue record for this book is available from the British Library*

*Library of Congress Cataloguing in Publication data*
Riley, K. F. (Kenneth Franklin), 1936–
Mathematical methods for physics and engineering / K. F. Riley,
M. P. Hobson, and S. J. Bence.
p. cm.
ISBN 0 521 55506 X (hc.) – ISBN 0 521 55529 9 (pbk.)
1. Mathematical analysis. I. Hobson, M. P. (Michael Paul), 1967– .
II. Bence, S. J. (Stephen John), 1972– . III. Title.
QA401.R537 1997
515′.1–dc21 96-52942 CIP

ISBN 0 521 55506 X hardback
ISBN 0 521 55529 9 paperback

# Contents

# Preface

A knowledge of mathematical methods is important for an increasing number of university and college courses, particularly in physics, engineering and chemistry, but also in more general science. Students embarking on such courses come from diverse mathematical backgrounds, and their core knowledge varies considerably. We have therefore decided to write a textbook that assumes knowledge only of material that can be expected to be familiar to all the current generation of students starting physical science courses at university. In the United Kingdom this corresponds to the standard of Mathematics A-level, whereas in the United States the material assumed is that which would normally be covered at junior college.

Starting from this level, the first six chapters cover a collection of topics with which the reader may already be familiar, but which are here extended and applied to typical problems encountered by first-year university students. They are aimed at providing a common base of general techniques used in the development of the remaining chapters. Students who have had additional preparation, such as Further Mathematics at A-level, will find much of this material straightforward.

Following these opening chapters, the remainder of the book is intended to cover at least that mathematical material which an undergraduate in the physical sciences might encounter up to the end of his or her course. The book is also appropriate for those beginning graduate study with a mathematical content, and naturally much of the material forms parts of courses for mathematics students. Furthermore, the text should provide a useful reference for research workers.

The general aim of the book is to present a topic in three stages. The first stage is a qualitative introduction, wherever possible from a physical point of view. The second is a more formal presentation, although we have deliberately avoided strictly mathematical questions such as the existence of limits, uniform convergence, the interchanging of integration and summation orders, etc. on the grounds that 'this is the real world; it must behave reasonably'. Finally a worked example is presented, often drawn from familiar situations in physical science and engineering. These examples have generally been fully worked, since, in the authors' experience, partially worked examples are unpopular with students. Only in a few cases, where trivial algebraic manipulation is involved, or where repetition of the main text would result, has an example been left as an exercise for the reader. Nevertheless, a number of exercises also appear at the end of each chapter, and these should give the reader ample opportunity to test his or her understanding. Hints and answers to these exercises are also provided.

With regard to the presentation of the mathematics, it has to be accepted that many equations (especially partial differential equations) can be written more compactly by using subscripts, e.g. $u_{xy}$ for a second partial derivative, instead of the more familiar $\partial^2 u/\partial x \partial y$, and that this certainly saves typographical space. However, for many students, the labour of mentally unpacking such equations is sufficiently great that it is not possible to think of an equation's physical interpretation at the same time. Consequently, wherever possible we have decided to write out such expressions in their more obvious but longer form.

During the writing of this book we have received much help and encouragement from various colleagues at the Cavendish Laboratory, Clare College, Trinity Hall and Peterhouse. In particular, we would like to thank Peter Scheuer, whose comments and general enthusiasm proved invaluable in the early stages. For reading sections of the manuscript, for pointing out misprints and for numerous useful comments, we thank many of our students and colleagues at the University of Cambridge. We are especially grateful to Chris Doran, John Huber, Garth Leder, Tom Körner and, not least, Mike Stobbs, who, sadly, died before the book was completed. We also extend our thanks to the University of Cambridge and the Cavendish teaching staff, whose examination questions and lecture hand-outs have collectively provided the basis for some of the examples included. Of course, any errors and ambiguities remaining are entirely the responsibility of the authors, and we would be most grateful to have them brought to our attention.

We are indebted to Dave Green for a great deal of advice concerning typesetting in LaTeX and to Andrew Lovatt for various other computing tips. Our thanks also go to Anja Visser and Graça Rocha for enduring many hours of (sometimes heated) debate. At Cambridge University Press, we are very grateful to our editor Adam Black for his help and patience and to Alison Woollatt for her expert typesetting of such a complicated text. We also thank our copy-editor Susan Parkinson for many useful suggestions that have undoubtedly improved the style of the book.

Finally, on a personal note, KFR wishes to thank his wife Penny, not only for a long and happy marriage, but also for her support and understanding during his recent illness – and when things have not gone too well at the bridge table! MPH is indebted both to Rebecca Morris and to his parents for their tireless support and patience, and for their unending supplies of tea. SJB is grateful to Anthony Gritten for numerous relaxing discussions about J. S. Bach, to Susannah Ticciati for her patience and understanding, and to Kate Isaak for her calming late-night e-mails from the USA.

Ken Riley, Michael Hobson and Stephen Bence
Cambridge, 1997

# 1

# Preliminary calculus

The opening chapter of this book is concerned with the formalism of probably the most widely used mathematical technique in the physical sciences, namely the calculus. The chapter divides into two sections. The first deals with the process of differentiation and the second with its inverse process, integration. The material covered is essential for the remainder of the book and serves as a reference. Readers who have previously studied these topics should ensure familiarity by looking at the worked examples in the main text, and by attempting the exercises at the end of the chapter.

## 1.1 Differentiation

Differentiation is the process of determining how quickly or slowly a function varies, as the quantity on which it depends, the argument, is changed. More specifically it is the procedure for obtaining an expression (numerical or algebraic) for the rate of change of the function with respect to its argument. Familiar examples of rates of change include acceleration (the rate of change of velocity) and the rate of a chemical reaction (the rate of change of chemical composition). Both acceleration and reaction rate give a measure of the change of a quantity with respect to time. However, differentiation may also be applied to changes with respect to other quantities, for example a change in pressure with respect to a change in temperature.

Although it will not be apparent from what we have said so far, differentiation is in fact a limiting process, that is, it deals only with the infinitesimal change in one quantity resulting from an infinitesimal change in another.

### *1.1.1 Differentiation from first principles*

Let us consider a function $f(x)$ that depends on only one variable $x$, together with numerical constants (for example, $f(x) = 3x^2$ or $f(x) = \sin x$ or $f(x) = 2 + 3/x$).

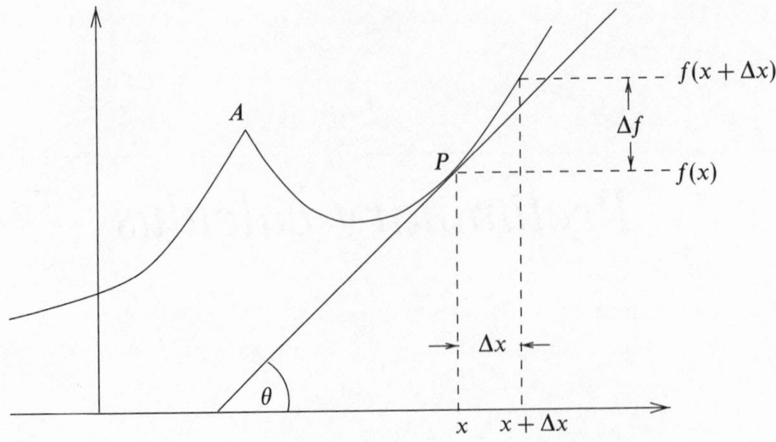


Figure 1.1 The graph of a function $f(x)$ showing that the gradient of the function at $P$, given by $\tan\theta$, is approximately equal to $\Delta f/\Delta x$.

Figure 1.1 shows an example of such a function. Near any particular point, $P$, the value of the function changes by an amount $\Delta f$, say, as $x$ changes by a small amount $\Delta x$. The slope of the tangent to the graph of $f(x)$ at $P$ is then approximately $\Delta f/\Delta x$, and the change in the value of the function is $\Delta f = f(x+\Delta x) - f(x)$. In order to calculate the true value of the gradient, or *first derivative*, of the function at $P$, we must let $\Delta x$ become infinitesimally small. We therefore define the first derivative of $f(x)$ by

$$f'(x) \equiv \frac{df(x)}{dx} \equiv \lim_{\Delta x \to 0} \frac{f(x+\Delta x) - f(x)}{\Delta x}, \tag{1.1}$$

provided that the limit exists. The limit will depend in almost all cases on the value of $x$. If the limit does exist at a point $x = a$, then the function is said to be differentiable at $a$; otherwise it is said to be non-differentiable at $a$. The formal concept of a limit and its existence or non-existence is discussed in chapter 3; for present purposes we will adopt an intuitive approach.

In the definition (1.1), we allow $\Delta x$ to tend to zero from either positive or negative values and require the same limit to be obtained in both cases. A function that is differentiable at $a$ is necessarily continuous at $a$ (there must be no jump in the value of the function at $a$) though the converse is not necessarily true. This latter assertion is illustrated in figure 1.1: the function is continuous at the 'kink' $A$ but the two limits of the gradient as $\Delta x$ tends to zero from positive or negative values are different and so the function is not differentiable at $A$.

It should be clear from the above discussion that near the point $P$ we may approximate the change in the value of the function, $\Delta f$, that results from a small change $\Delta x$ in $x$ by

$$\Delta f \approx \frac{df(x)}{dx}\Delta x. \tag{1.2}$$

As one would expect, the approximation improves as the value of $\Delta x$ is reduced. In the limit in which the change $\Delta x$ becomes infinitesimally small, we denote it by the *differential* $dx$, and (1.2) reads

$$df = \frac{df(x)}{dx}dx. \tag{1.3}$$

This *equality* relates the infinitesimal change in the function, $df$, to the infinitesimal change $dx$ in its argument that causes it.

So far we have discussed only the first derivative of a function. However, we can also define the *second derivative* as the gradient of the gradient of a function. Again we use the definition (1.1) but now with $f(x)$ replaced by $f'(x)$. Hence the second derivative is defined by

$$f''(x) \equiv \lim_{\Delta x \to 0} \frac{f'(x+\Delta x) - f'(x)}{\Delta x}, \tag{1.4}$$

provided that the limit exists. A physical example of a second derivative is the second derivative of the distance travelled by a particle with respect to time. Since the first derivative of distance travelled gives the particle's velocity, the second derivative gives its acceleration.

We can continue in this manner, the $n$th derivative of the function $f(x)$ being defined by

$$f^{(n)}(x) \equiv \lim_{\Delta x \to 0} \frac{f^{(n-1)}(x+\Delta x) - f^{(n-1)}(x)}{\Delta x}. \tag{1.5}$$

It should be noted that with this notation $f'(x) \equiv f^{(1)}(x)$, $f''(x) \equiv f^{(2)}(x)$, etc., and that formally $f^{(0)}(x) \equiv f(x)$.

All this should be familiar to the reader, though perhaps not with such formal definitions. The following example shows the differentiation of $f(x) = x^2$ from first principles. In practice, however, the reader will normally simply remember the derivatives of standard functions and apply the techniques given in the remainder of this section to find more complicated derivatives.

►*Find from first principles the derivative with respect to x of* $f(x) = x^2$.

Using the definition (1.1),

$$\begin{aligned} f'(x) &= \lim_{\Delta x \to 0} \frac{f(x+\Delta x) - f(x)}{\Delta x} \\ &= \lim_{\Delta x \to 0} \frac{(x+\Delta x)^2 - x^2}{\Delta x} \\ &= \lim_{\Delta x \to 0} \frac{2x\Delta x + (\Delta x)^2}{\Delta x} \\ &= \lim_{\Delta x \to 0} (2x + \Delta x). \end{aligned}$$

As $\Delta x$ tends to zero, $2x + \Delta x$ tends towards $2x$, hence

$$f'(x) = 2x. \blacktriangleleft$$

The derivatives of some other simple functions that may be derived in a similar manner are listed below (note that $a$ is a constant).

$$\frac{d}{dx}(x^n) = nx^{n-1}, \qquad \frac{d}{dx}(e^{ax}) = ae^{ax}, \qquad \frac{d}{dx}(\ln ax) = \frac{1}{x},$$

$$\frac{d}{dx}(\sin ax) = a\cos ax, \qquad \frac{d}{dx}(\cos ax) = -a\sin ax, \qquad \frac{d}{dx}(\sec ax) = a\sec ax\tan ax,$$

$$\frac{d}{dx}(\tan ax) = a\sec^2 ax, \qquad \frac{d}{dx}(\operatorname{cosec} ax) = -a\operatorname{cosec} ax\cot ax,$$

$$\frac{d}{dx}(\cot ax) = -a\operatorname{cosec}^2 ax, \qquad \frac{d}{dx}\left(\sin^{-1}\frac{x}{a}\right) = \frac{1}{\sqrt{a^2 - x^2}},$$

$$\frac{d}{dx}\left(\cos^{-1}\frac{x}{a}\right) = \frac{-1}{\sqrt{a^2 - x^2}}, \qquad \frac{d}{dx}\left(\tan^{-1}\frac{x}{a}\right) = \frac{a}{a^2 + x^2}.$$

Differentiation from first principles emphasises the definition of a derivative as the gradient of a function. However, for most practical purposes, returning to the definition is time consuming and does not aid our understanding. Instead we employ a number of techniques, which use the derivatives listed above as 'building blocks', to evaluate the derivatives of more complicated functions than hitherto encountered. Subsections 1.1.2–1.1.7 develop the methods required.

### *1.1.2 Differentiation of products*

As a first example of the differentiation of a more complicated function, we consider finding the derivative of a function $f(x)$ that can be written as the product of two other functions of $x$, namely $f(x) = u(x)v(x)$. For example, if $f(x) = x^3\sin x$, then we might take $u(x) = x^3$ and $v(x) = \sin x$. Clearly the separation is not unique. (In the given example possible alternative break-ups would be $u(x) = x^2$, $v(x) = x \sin x$, or even $u(x) = x^4 \tan x$, $v(x) = x^{-1} \cos x$.)

The purpose of the separation is to split the function into two (or more) parts of which we know the derivatives (or at least we can evaluate the derivatives more easily than that of the whole). We would however gain little if we did not know the relationship between the derivative of $f$ and those of $u$ and $v$. Fortunately, they are very simply related, as we shall now show.

Since $f(x)$ is written as the product $u(x)v(x)$, it follows that

$$\begin{aligned} f(x+\Delta x) - f(x) &= u(x+\Delta x)v(x+\Delta x) - u(x)v(x) \\ &= u(x+\Delta x)[v(x+\Delta x) - v(x)] + [u(x+\Delta x) - u(x)]v(x). \end{aligned}$$

From the definition of a derivative (1.1),

$$\begin{aligned} \frac{df}{dx} &= \lim_{\Delta x \to 0} \frac{f(x+\Delta x) - f(x)}{\Delta x} \\ &= \lim_{\Delta x \to 0} \left\{ u(x+\Delta x) \left[ \frac{v(x+\Delta x) - v(x)}{\Delta x} \right] + \left[ \frac{u(x+\Delta x) - u(x)}{\Delta x} \right] v(x) \right\}. \end{aligned}$$

In the limit $\Delta x \to 0$, the factors in square brackets become $dv/dx$ and $du/dx$ (by the definitions of these quantities) and $u(x + \Delta x)$ simply becomes $u(x)$. Consequently we obtain

$$\frac{df}{dx} = \frac{d}{dx}[u(x)v(x)] = u(x)\frac{dv(x)}{dx} + \frac{du(x)}{dx}v(x). \tag{1.6}$$

In primed notation, without the argument $x$ written explicitly each time, this is

$$f' = (uv)' = uv' + u'v. \tag{1.7}$$

This is a general result obtained without making any assumptions about the specific forms $f$, $u$ and $v$, other than that $f(x) = u(x)v(x)$. In words, the result reads as follows. *The derivative of the product of two functions is equal to the first function times the derivative of the second plus the second function times the derivative of the first.*

► *Find the derivative with respect to $x$ of $f(x) = x^3 \sin x$.*

Using the product rule, (1.6),

$$\begin{aligned} \frac{d}{dx}(x^3 \sin x) &= x^3 \frac{d}{dx}(\sin x) + \frac{d}{dx}(x^3) \sin x \\ &= x^3 \cos x + 3x^2 \sin x. \; ◄ \end{aligned}$$

The product rule may readily be extended to the product of three or more functions. Considering the function

$$f(x) = u(x)v(x)w(x), \tag{1.8}$$

and using (1.6), we obtain, as before omitting the argument,

$$\frac{df}{dx} = u\frac{d}{dx}(vw) + \frac{du}{dx}vw.$$

Using (1.6) again to expand the first term on the RHS gives the complete result

$$\frac{d}{dx}(uvw) = uv\frac{dw}{dx} + u\frac{dv}{dx}w + \frac{du}{dx}vw. \tag{1.9}$$

or

$$(uvw)' = uvw' + uv'w + u'vw. \tag{1.10}$$

It is readily apparent that this can be extended to products containing any number of factors, $n$, and that the expression for the derivative will consist of $n$ terms with the prime appearing in successive terms on each of the $n$ factors in turn. This is probably the easiest way to recall the product rule.

### 1.1.3 The chain rule

Products are just one type of complicated function that we may encounter in differentiation. Another is a function of a function, e.g. $f(x) = (3 + x^2)^3 = u(x)^3$, where $u(x) = (3 + x^2)$. Since $\Delta f$, $\Delta u$ and $\Delta x$ are small finite quantities, it follows that

$$\frac{\Delta f}{\Delta x} = \frac{\Delta f}{\Delta u}\frac{\Delta u}{\Delta x}.$$

Hence as the quantities become infinitesimally small we obtain

$$\frac{df}{dx} = \frac{df}{du}\frac{du}{dx}. \tag{1.11}$$

This is the *chain rule*, which we must apply when differentiating a function of a function.

►*Find the derivative with respect to $x$ of $f(x) = (3 + x^2)^3$.*

Rewriting the function as $f(x) = u^3$, where $u(x) = (3 + x^2)$, and applying (1.11) we find

$$\frac{df}{dx} = 3u^2\frac{du}{dx} = 3u^2\frac{d}{dx}(3 + x^2) = 3u^2 \times 2x = 6x(3 + x^2)^2. \blacktriangleleft$$

Similarly, the derivative with respect to $x$ of $f(x) = 1/v(x)$ may be obtained by rewriting the function as $f(x) = v^{-1}$ and applying (1.11),

$$\frac{df}{dx} = -v^{-2}\frac{dv}{dx} = -\frac{1}{v^2}\frac{dv}{dx}. \tag{1.12}$$

The chain rule is also useful for calculating the derivative of a function $f$ with respect to $x$ when both $x$ and $f$ are written in terms of a variable (or parameter), say $t$.

►*Find the derivative with respect to $x$ of $f(t) = 2at$, where $x = at^2$.*

We could of course substitute for $t$ and then differentiate $f$ as a function of $x$, but in this case it is quicker to use

$$\frac{df}{dx} = \frac{df}{dt}\frac{dt}{dx} = 2a\frac{1}{2at} = \frac{1}{t},$$

where we have used the fact that

$$\frac{dt}{dx} = \left(\frac{dx}{dt}\right)^{-1}. \blacktriangleleft$$

### *1.1.4 Differentiation of quotients*

Applying (1.6) for the derivative of a product to a function $f(x) = u(x)[1/v(x)]$, we may obtain the derivative of the quotient of two factors. Thus

$$f' = \left(\frac{u}{v}\right)' = u\left(\frac{1}{v}\right)' + u'\left(\frac{1}{v}\right) = u\left(-\frac{v'}{v^2}\right) + \frac{u'}{v},$$

where (1.12) has been used to evaluate $(1/v)'$. This can now be rearranged into the more convenient and memorisable form

$$f' = \left(\frac{u}{v}\right)' = \frac{vu' - uv'}{v^2}. \qquad (1.13)$$

This can be expressed in words as *the derivative of a quotient is equal to the bottom times the derivative of the top minus the top times the derivative of the bottom, all over the bottom squared.*

►*Find the derivative with respect to $x$ of $f(x) = \sin x/x$.*

Using (1.13) with $u(x) = \sin x$, $v(x) = x$ and hence $u'(x) = \cos x$, $v'(x) = 1$, we find

$$f'(x) = \frac{x\cos x - \sin x}{x^2} = \frac{\cos x}{x} - \frac{\sin x}{x^2}. \blacktriangleleft$$

### *1.1.5 Implicit differentiation*

So far we have only differentiated functions written in the form $y = f(x)$. We may not, however, always be presented with a relationship in this form. As an example consider the relation $x^3 - 3xy + y^3 = 2$. In this case it is not possible to rearrange the equation to give $y$ as a function of $x$. However, by differentiating term by term with respect to $x$, we can find the derivative of the implicit function.

►*Find $dy/dx$ if $x^3 - 3xy + y^3 = 2$.*

Differentiating each term in the equation with respect to $x$ we obtain

$$\frac{d}{dx}(x^3) - \frac{d}{dx}(3xy) + \frac{d}{dx}(y^3) = \frac{d}{dx}(2),$$

$$\Rightarrow \quad 3x^2 - \left(3x\frac{dy}{dx} + 3y\right) + 3y^2\frac{dy}{dx} = 0,$$

where the derivative of $3xy$ has been found using the product rule. Hence, rearranging for $dy/dx$,

$$\frac{dy}{dx} = \frac{y - x^2}{y^2 - x}.$$

Note that $dy/dx$ is a function of both $x$ and $y$ and cannot be expressed as a function of $x$ only. ◄

### 1.1.6 Logarithmic differentiation

In circumstances in which the variable with respect to which we are differentiating is an exponent, taking logarithms and then differentiating implicitly is the simplest way to find the derivative.

►*Find the derivative with respect to $x$ of $y = a^x$.*

To find the required derivative we first take logarithms, and then differentiate implicitly:

$$\ln y = \ln a^x = x \ln a \quad \Rightarrow \quad \frac{1}{y}\frac{dy}{dx} = \ln a.$$

Now, rearranging and substituting for $y$, we find

$$\frac{dy}{dx} = y \ln a = a^x \ln a. \; ◄$$

### 1.1.7 Leibnitz' theorem

We have already discussed finding the derivative of the product of two or more functions. We now consider *Leibnitz' theorem*, which gives the corresponding results for the higher derivatives of products.

Consider again the function $f(x) = u(x)v(x)$. We know from the product rule that $f' = uv' + u'v$. Using the rule once more for each of the products,

$$\begin{aligned} f'' &= (uv'' + u'v') + (u'v' + u''v) \\ &= uv'' + 2u'v' + u''v. \end{aligned}$$

Similarly differentiating two further times gives

$$f''' = uv''' + 3u'v'' + 3u''v' + u'''v$$
$$f^{(4)} = uv^{(4)} + 4u'v''' + 6u''v'' + 4u'''v' + u^{(4)}v.$$

The pattern emerging is clear and strongly suggests that the results generalise to

$$f^{(n)} = \sum_{r=0}^{n} \frac{n!}{r!(n-r)!} u^{(r)} v^{(n-r)}. \qquad (1.14)$$

The fraction $n!/[r!(n-r)!]$ is the binomial coefficient ${}^nC_r$ (see chapter 24). This result may be *proved* straightforwardly using the method of induction, but the proof is not presented here.

►*Find the third derivative of the function* $f(x) = x^3 \sin x$.

Using (1.14) we immediately find

$$f'''(x) = 6\sin x + 3(6x)\cos x + 3(3x^2)(-\sin x) + x^3(-\cos x)$$
$$= 3(2 - 3x^2)\sin x + x(18 - x^2)\cos x. \;◄$$

### *1.1.8 Special points of a function*

We have interpreted the derivative of a function as the gradient of the function at the relevant point (figure 1.1). If the gradient is zero at some point then the function is said to have a *stationary point* there. Clearly, in graphical terms, this corresponds to a horizontal tangent to the graph at that point.

Stationary points may be divided into three categories and an example of each is shown in figure 1.2. Point $B$ is said to be a *minimum* since the function *increases* in value in both directions away from it. Point $Q$ is said to be a *maximum* since the function *decreases* in both directions away from it. Note that $B$ is not the overall minimum value of the function and $Q$ is not the overall maximum, but, rather, they are a local minimum and a local maximum. The third type of stationary point is the *stationary point of inflection, S*. In this case the function falls in the positive $x$-direction and rises in the negative $x$-direction so $S$ is neither a maximum nor a minimum. Nevertheless, the gradient of the function is zero at $S$, i.e. the graph of the function is flat there, and this justifies our calling it a stationary point. Of course, a point at which the gradient of the function is zero but the function rises in the positive $x$-direction and falls in the negative $x$-direction is also a stationary point of inflection.

The above distinction between the three types of stationary point has been made rather descriptively. However, it is possible to define and distinguish stationary points mathematically. From their definition as points of zero gradient,

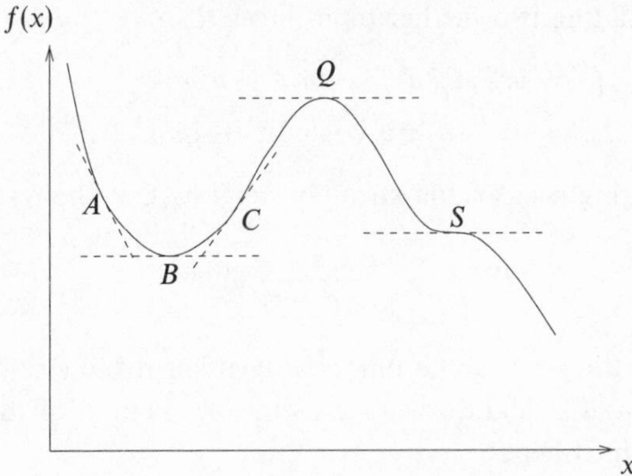


Figure 1.2 A graph of a function, $f(x)$, showing how differentiation corresponds to finding the gradient of the function at a particular point. Points $B$, $Q$ and $S$ are stationary points (see text).

all stationary points must be characterised by $df/dx = 0$. In the case of the minimum, $B$, the slope, i.e. $df/dx$, changes from negative at $A$ to positive at $C$ through zero at $B$. Thus $df/dx$ is increasing and so the second derivative $d^2f/dx^2$ must be positive. Conversely, at the maximum, $Q$, we must have that $d^2f/dx^2$ is negative.

It is less obvious, but intuitively reasonable, that at $S$, $d^2f/dx^2$ is zero. This may be inferred from the following observations. To the left of $S$ the curve is concave upwards so $df/dx$ is increasing and hence $d^2f/dx^2 > 0$. To the right of $S$, the curve is concave downwards so $df/dx$ is decreasing and hence $d^2f/dx^2 < 0$.

In summary, at a stationary point $df/dx = 0$ and

(i) for a minimum, $d^2f/dx^2 > 0$,

(ii) for a maximum, $d^2f/dx^2 < 0$,

(iii) for a stationary point of inflection, $d^2f/dx^2 = 0$ and $d^2f/dx^2$ changes sign through the point.

In case (iii) for a stationary point of inflection, in order that $d^2f/dx^2$ changes sign through the point we normally require $d^3f/dx^3 \neq 0$ at that point. This simple rule can fail for some functions, however, and in general if the first non-vanishing derivative of $f(x)$ at the stationary point is $f^{(n)}$, then if $n$ is even the point is a maximum or minimum, and if $n$ is odd the point is a stationary point of inflection. This may be seen from the Taylor expansion (see equation (3.17)) of the function about the stationary point, but it is not proved here.

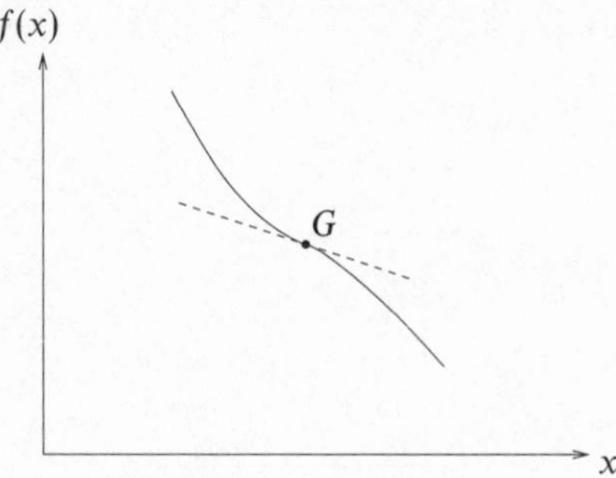


Figure 1.3 The graph of a function $f(x)$ that has a general point of inflection at the point $G$.

►*Find the positions and natures of the stationary points of the function* $f(x) = 2x^3 - 3x^2 - 36x + 2$.

The first criterion for a stationary point is that $df/dx = 0$, and hence

$$\frac{df}{dx} = 6x^2 - 6x - 36 = 0,$$

from which we obtain

$$(x-3)(x+2) = 0.$$

Hence the stationary points are at $x = 3$ and $x = -2$. To determine the nature of the stationary point we must evaluate $d^2f/dx^2$:

$$\frac{d^2f}{dx^2} = 12x - 6.$$

Now, we examine each stationary point in turn. For $x = 3$, $d^2f/dx^2 = 30$. Since this is positive, we conclude that $x = 3$ is a minimum. Similarly for $x = -2$, $d^2f/dx^2 = -30$ and so $x = -2$ is a maximum. ◄

So far we have concentrated on stationary points, which are defined to have $df/dx = 0$. We find that at a stationary point of inflection $d^2f/dx^2$ is also zero and changes sign. This naturally leads us to consider points at which $d^2f/dx^2$ is zero and changes sign, but at which $df/dx$ is *not*, in general, zero. Such points are called *general points of inflection* or simply *points of inflection*. Clearly, a stationary point of inflection is a special case for which $df/dx$ is also zero. At a general point of inflection the graph of the function changes from being concave upwards to concave downwards (or vice versa), but the tangent to the curve at this point need not be horizontal. A typical example of a general point of inflection is shown in figure 1.3.

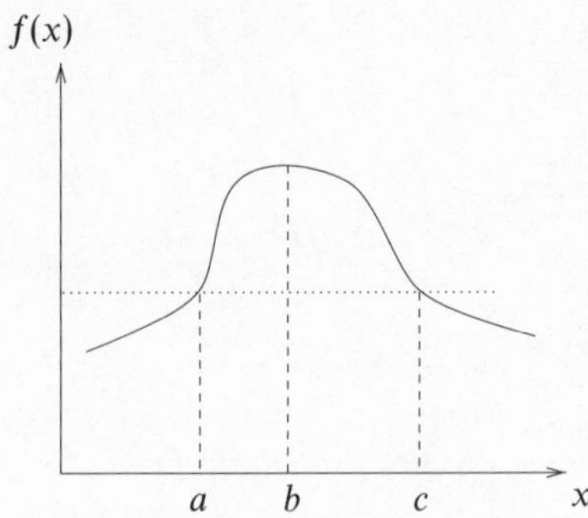


Figure 1.4 The graph of a function $f(x)$, showing that if $f(a) = f(c)$ then at one point at least between $x = a$ and $x = c$ the graph has zero gradient.

### *1.1.9 Theorems of differentiation*

*Rolle's theorem*

Rolle's theorem (figure 1.4) states that if a function $f(x)$ is continuous in the range $a \leq x \leq c$, is differentiable in the range $a < x < c$ and satisfies $f(a) = f(c)$ then for at least one point $x = b$, where $a < b < c$, $f'(b) = 0$. Thus Rolle's theorem states that for a well-behaved function (i.e. a continuous and differentiable one) that has the same value at two points, either there is at least one stationary point between those points or the function is a constant between them. The validity of the theorem is immediately apparent from figure 1.4 and a full analytic proof will not be given. The theorem is used in deriving the mean value theorem, which we now discuss.

*Mean value theorem*

The mean value theorem (figure 1.5) states that if a function $f(x)$ is continuous in the range $a \leq x \leq c$ and differentiable in the range $a < x < c$, then

$$f'(b) = \frac{f(c) - f(a)}{c - a}, \tag{1.15}$$

for at least one value $b$ where $a < b < c$. Thus the mean value theorem states that for a well-behaved function the gradient of the line joining two points on the curve is equal to the slope of the tangent to the curve for at least one intervening point.

The proof of the mean value theorem is found by examination of figure 1.5, as follows. The equation of the line $AC$ is

$$g(x) = f(a) + (x - a)\frac{f(c) - f(a)}{c - a},$$

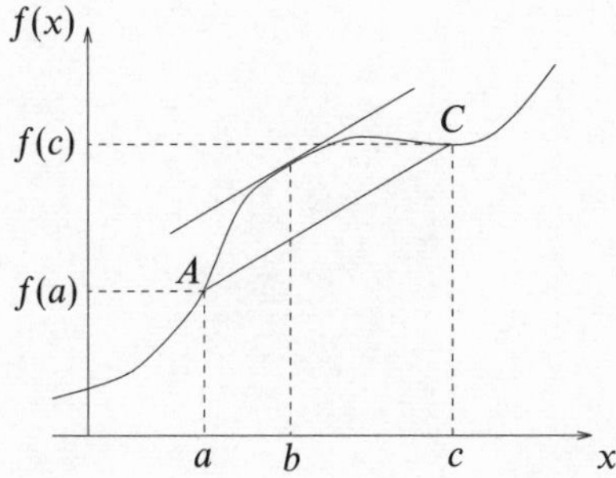


Figure 1.5 The graph of a function $f(x)$; at a point $x = b$, it has the same gradient as the line $AC$.

and hence the difference between the curve and the line is

$$h(x) = f(x) - g(x) = f(x) - f(a) - (x - a)\frac{f(c) - f(a)}{c - a}.$$

Since the curve and the line intersect at $A$ and $C$, $h(x) = 0$ at both of these points. Hence, by an application of Rolle's theorem, $h'(x) = 0$ for at least one point $b$ between $A$ and $C$. Differentiating our expression for $h(x)$, we find

$$h'(x) = f'(x) - \frac{f(c) - f(a)}{c - a},$$

and hence at $b$, where $h'(x) = 0$,

$$f'(b) = \frac{f(c) - f(a)}{c - a}.$$

## 1.2 Integration

The notion of an integral as the area under a curve will be familiar to the reader. In figure 1.6, in which the solid line is a plot of a function $f(x)$, the shaded area represents the quantity denoted by

$$I = \int_a^b f(x)\, dx. \qquad (1.16)$$

This expression is known as the *definite integral* of $f(x)$ between the *lower limit* $x = a$ and the *upper limit* $x = b$, and $f(x)$ is called the *integrand.*

### *1.2.1 Integration from first principles*

The definition of an integral as the area under a curve is not a formal definition, but one that can be readily visualised. The formal definition of $I$ involves

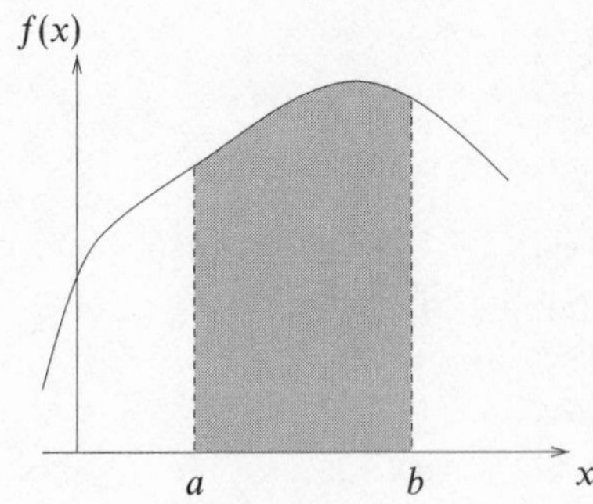


Figure 1.6 An integral as the area under a curve.

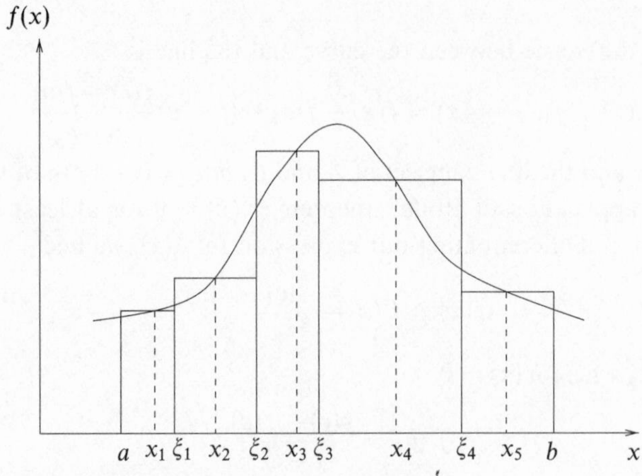


Figure 1.7 The evaluation of a definite integral by subdividing the interval $a \leq x \leq b$ into subintervals.

subdividing the interval $a \leq x \leq b$ into a large number of subintervals by defining intermediate points $\xi_i$ such that $a = \xi_0 < \xi_1 < \xi_2 < \cdots < \xi_n = b$, and then forming the sum

$$S = \sum_{i=1}^{n} f(x_i)(\xi_i - \xi_{i-1}), \tag{1.17}$$

where $x_i$ is an arbitrary point that lies in the range $\xi_{i-1} \leq x_i \leq \xi_i$ (see figure 1.7). Then, if $n$ is allowed to tend to infinity in any way whatsoever, subject only to the restriction that the length of every subinterval $\xi_{i-1}$ to $\xi_i$ tends to zero, $S$ might, or might not, tend to a unique limit, $I$. If it does, then the definite integral of $f(x)$ between $a$ and $b$ is defined as having the value $I$. If no unique limit exists the integral is undefined.

►*Evaluate from first principles the integral* $I = \int_0^b x^2\,dx$.

We first divide the area under the curve $y = x^2$ between 0 and $b$ into $n$ rectangles of equal width $h$. If we take the value at the lower end of each subinterval (in the limit of an infinite number of subintervals we could equally well have chosen the value at the upper end) to give the height of the corresponding rectangle, the area of the $k$th rectangle will be $(kh)^2 h = k^2 h^3$. The total area is thus

$$A = \sum_{k=0}^{n-1} k^2 h^3 = (h^3)\tfrac{1}{6}n(n-1)(2n-1),$$

where we have used the expression for the sum of the squares of the natural numbers derived in subsection 3.2.5. Now $h = b/n$ and so

$$A = \left(\frac{b^3}{n^3}\right)\frac{n}{6}(n-1)(2n-1) = \frac{b^3}{6}\left(1-\frac{1}{n}\right)\left(2-\frac{1}{n}\right).$$

As $n \to \infty$, $A \to b^3/3$. ◄

Some straightforward properties of definite integrals which are almost self-evident are:

$$\int_a^b 0\,dx = 0, \tag{1.18}$$

$$\int_a^b f(x)\,dx = -\int_b^a f(x)\,dx, \tag{1.19}$$

$$\int_a^c f(x)\,dx = \int_a^b f(x)\,dx + \int_b^c f(x)\,dx, \tag{1.20}$$

$$\int_a^b [f(x)+g(x)]\,dx = \int_a^b f(x)\,dx + \int_a^b g(x)\,dx. \tag{1.21}$$

### *1.2.2 Integration as the inverse of differentiation*

The definite integral has been defined as the area under a curve between two fixed limits. Let us now consider the integral

$$F(x) = \int_a^x f(u)\,du \tag{1.22}$$

in which the lower limit $a$ remains fixed but the upper limit $x$ is now variable. It will be noticed, as compared with (1.16), that in the integrand the variable $x$ has been replaced by a new variable $u$. It is conventional to rename the *dummy variable* in the integrand in this way in order that the same variable does not appear in both the integrand and the integration limits.

It is apparent from (1.22) that $F(x)$ is a continuous function of $x$, but at first glance the definition of an integral as the area under a curve does not connect with our assertion that integration is the inverse process to differentiation. However, by considering the integral (1.22) and using the elementary property (1.20), we obtain

$$\begin{aligned} F(x+\Delta x) &= \int_a^{x+\Delta x} f(u)\,du \\ &= \int_a^x f(u)\,du + \int_x^{x+\Delta x} f(u)\,du \\ &= F(x) + \int_x^{x+\Delta x} f(u)\,du. \end{aligned}$$

Rearranging and dividing through by $\Delta x$ yields

$$\frac{F(x+\Delta x)-F(x)}{\Delta x} = \frac{1}{\Delta x}\int_x^{x+\Delta x} f(u)\,du.$$

Letting $\Delta x \to 0$ and using (1.1) we find that the LHS becomes $dF/dx$, whereas the RHS becomes $f(x)$. The latter conclusion follows because when $\Delta x$ is small the value of the integral on the RHS is approximately $f(x)\Delta x$, and in the limit $\Delta x \to 0$ no approximation is involved. Thus

$$\frac{dF(x)}{dx} = f(x), \tag{1.23}$$

or, substituting for $F(x)$ from (1.22),

$$\frac{d}{dx}\left[\int_a^x f(u)\,du\right] = f(x).$$

From the last two equations it is clear that integration can be considered the inverse of differentiation. However, we see from the above analysis that the lower limit $a$ is arbitrary and so differentiation does not have a *unique* inverse. Any function $F(x)$ obeying (1.23) is called an *indefinite integral* of $f(x)$, though any two such functions can differ only by an arbitrary additive constant. Since the lower limit is arbitrary, it is usual to write

$$F(x) = \int^x f(u)\,du \tag{1.24}$$

and explicitly include the arbitrary constant when evaluating $F(x)$. The evaluation is conventionally written in the form

$$\int f(x)\,dx = F(x) + c \tag{1.25}$$

where $c$ is called the *constant of integration*. It will be noticed that, in the absence of any integration limits, we use the same symbol for the arguments of both $f$ and $F$. This can be confusing, but is sufficiently common practice that the reader needs to become familiar with it.

We also note that the definite integral of $f(x)$ between the fixed limits $x = a$ and $x = b$ can be written in terms of $F(x)$. From (1.22) we have

$$\int_a^b f(x)\,dx = \int_{x_0}^b f(x)\,dx - \int_{x_0}^a f(x)\,dx$$
$$= F(b) - F(a), \qquad (1.26)$$

where $x_0$ is any third fixed point. Using the notation $F'(x) = dF/dx$, we may rewrite (1.23) as $F'(x) = f(x)$, and so express (1.26) as

$$\int_a^b F'(x)\,dx = F(b) - F(a) \equiv [F]_a^b.$$

Unlike in differentiation, where repeated applications of the product rule and/or the chain rule will always give the derivative required, it is not always possible to find the integral of an arbitrary function. Indeed, in most real physical problems exact integration cannot be performed and we have to revert instead to numerical approximations. Despite this cautionary note, it is in fact possible to integrate many simple functions and the following subsections introduce the most common types. Many of the techniques will be familiar to the reader and so are summarised by example.

### *1.2.3 Integration by inspection*

The simplest method of integrating a function is by inspection. Some of the simplest functions have well-known integrals that should be remembered. The reader will notice that the integrals are precisely the converses of the derivatives found near the end of subsection 1.1.1. A few are presented below, using (1.25).

$$\int a\,dx = ax + c, \qquad \int ax^n\,dx = \frac{ax^{n+1}}{n+1} + c,$$

$$\int e^{ax}\,dx = \frac{e^{ax}}{a} + c, \qquad \int \frac{a}{x}\,dx = a\ln x + c,$$

$$\int a\cos bx\,dx = \frac{a\sin bx}{b} + c, \qquad \int a\sin bx\,dx = \frac{-a\cos bx}{b} + c,$$

$$\int a\tan bx\,dx = \frac{-a\ln(\cos bx)}{b} + c, \qquad \int a\cos bx\sin^n bx\,dx = \frac{a\sin^{n+1} bx}{b(n+1)} + c,$$

$$\int \frac{a}{a^2+x^2}\,dx = \tan^{-1}\left(\frac{x}{a}\right) + c, \qquad \int a\sin bx\cos^n bx\,dx = \frac{-a\cos^{n+1} bx}{b(n+1)} + c,$$

$$\int \frac{-1}{\sqrt{a^2-x^2}}\,dx = \cos^{-1}\left(\frac{x}{a}\right) + c, \qquad \int \frac{1}{\sqrt{a^2-x^2}}\,dx = \sin^{-1}\left(\frac{x}{a}\right) + c,$$

where the integrals that depend on $n$ are valid for all $n \neq -1$ and where $a$ and $b$ are constants. In the two final results $|x| \leq a$.

### 1.2.4 Integration of sinusoidal functions

Integrals of the type $\int \sin^n x\, dx$ and $\int \cos^n x\, dx$ may be found by using trigonometrical expansions. Two methods are applicable, one for odd $n$ and the other for even $n$. They are best illustrated by example.

►*Evaluate the integral* $I = \int \sin^5 x\, dx$.

Rewriting the integral as a product of $\sin x$ and an even power of $\sin x$, and then using the relation $\sin^2 x = 1 - \cos^2 x$ yields

$$\begin{aligned} I &= \int \sin^4 x \sin x\, dx \\ &= \int (1 - \cos^2 x)^2 \sin x\, dx \\ &= \int (1 - 2\cos^2 x + \cos^4 x) \sin x\, dx \\ &= \int (\sin x - 2 \sin x \cos^2 x + \sin x \cos^4 x)\, dx \\ &= -\cos x + \tfrac{2}{3}\cos^3 x - \tfrac{1}{5}\cos^5 x + c, \end{aligned}$$

where the integration has been carried out using the results of subsection 1.2.3. ◄

►*Evaluate the integral* $I = \int \cos^4 x\, dx$.

Rewriting the integral as a power of $\cos^2 x$ and using the double angle formula $\cos^2 x = \frac{1}{2}(1 + \cos 2x)$ yields

$$\begin{aligned} I &= \int (\cos^2 x)^2\, dx = \int \left(\frac{1 + \cos 2x}{2}\right)^2 dx \\ &= \int \tfrac{1}{4}(1 + 2\cos 2x + \cos^2 2x)\, dx. \end{aligned}$$

Using the double angle formula again we may write $\cos^2 2x = \frac{1}{2}(1 + \cos 4x)$, and hence

$$\begin{aligned} I &= \int \left[\tfrac{1}{4} + \tfrac{1}{2}\cos 2x + \tfrac{1}{8}(1 + \cos 4x)\right] dx \\ &= \tfrac{1}{4}x + \tfrac{1}{4}\sin 2x + \tfrac{1}{8}x + \tfrac{1}{32}\sin 4x + c \\ &= \tfrac{3}{8}x + \tfrac{1}{4}\sin 2x + \tfrac{1}{32}\sin 4x + c. \end{aligned}$$ ◄

### 1.2.5 Logarithmic integration

Integrals for which the integrand may be written as a fraction in which the numerator is the derivative of the denominator may be easily evaluated using

$$\int \frac{f'(x)}{f(x)}\,dx = \ln f(x) + c. \tag{1.27}$$

This follows directly from the differentiation of a logarithm as a function of a function (see subsection 1.1.3).

►*Evaluate the integral*

$$I = \int \frac{6x^2 + 2\cos x}{x^3 + \sin x}\,dx.$$

We firstly note that the numerator can be factorised to give $2(3x^2 + \cos x)$, and then that the quantity in brackets is the derivative of the denominator. Hence

$$I = 2\int \frac{3x^2 + \cos x}{x^3 + \sin x}\,dx = 2\ln(x^3 + \sin x) + c. \blacktriangleleft$$

### 1.2.6 Integration using partial fractions

The method of partial fractions involves no more than a simple manipulation of the integrand in such a way that a complicated fraction is written as the sum of two or more simpler fractions. Again, this is best illustrated by example.

►*Evaluate the integral*

$$I = \int \frac{1}{x^2 + x}\,dx.$$

Firstly we note that the denominator factorises to give $x(x+1)$. Hence

$$I = \int \frac{1}{x(x+1)}\,dx.$$

We now separate the fraction into the difference of two fractions, called *partial fractions*, and integrate directly, i.e.

$$I = \int \left(\frac{1}{x} - \frac{1}{x+1}\right) dx = \ln x - \ln(x+1) + c = \ln\left(\frac{x}{x+1}\right) + c. \blacktriangleleft$$

### 1.2.7 Integration by substitution

Sometimes a substitution of variables may be made that turns a complicated integral into a simpler one, which can then be integrated by a standard method.

There are many useful substitutions and knowing which to use is a matter of experience. We now present a few examples of particularly useful substitutions.

►*Evaluate the integral*

$$I = \int \frac{1}{\sqrt{1-x^2}}\,dx.$$

Making the substitution $x = \sin u$, we note that $dx = \cos u\,du$, and hence

$$I = \int \frac{1}{\sqrt{1-\sin^2 u}}\cos u\,du = \int \frac{1}{\sqrt{\cos^2 u}}\cos u\,du = \int du = u + c.$$

Now substituting back for $u$,

$$I = \sin^{-1} x + c. ◄$$

Another particular example of integration by substitution is afforded by integrals of the form

$$I = \int \frac{1}{a + b\cos x}\,dx \qquad \text{or} \qquad I = \int \frac{1}{a + b\sin x}\,dx. \tag{1.28}$$

In these cases making the substitution $t = \tan(\frac{1}{2}x)$ yields integrals which can be solved more easily than the originals.

However, before proceeding to a specific example we need to derive some important relationships. Firstly we must relate $dx$ and $dt$. Since

$$\frac{dt}{dx} = \tfrac{1}{2}\sec^2\left(\tfrac{1}{2}x\right) = \tfrac{1}{2}\left[1 + \tan^2\left(\tfrac{1}{2}x\right)\right] = \frac{1+t^2}{2},$$

we have

$$dx = \frac{2}{1+t^2}\,dt. \tag{1.29}$$

Furthermore, since $1 + t^2 = \sec^2(\frac{1}{2}x)$, it follows that $\cos^2(\frac{1}{2}x) = 1/(1+t^2)$. Thus, since $\cos x = 2\cos^2(\frac{1}{2}x) - 1$,

$$\cos x = \frac{2}{1+t^2} - 1 = \frac{1-t^2}{1+t^2}. \tag{1.30}$$

Similarly we can show that

$$\sin x = \frac{2t}{1+t^2}. \tag{1.31}$$

►*Evaluate the integral*

$$I = \int \frac{2}{1 + 3\cos x}\, dx.$$

Making the substitution $t = \tan(\frac{1}{2}x)$, and using (1.29) and (1.30) the integral becomes

$$\begin{aligned}
I &= \int \frac{2}{1 + 3\left[(1 - t^2)(1 + t^2)^{-1}\right]} \left(\frac{2}{1 + t^2}\right) dt \\
&= \int \frac{2(1 + t^2)}{1 + t^2 + 3(1 - t^2)} \left(\frac{2}{1 + t^2}\right) dt \\
&= \int \frac{2}{2 - t^2}\, dt = \int \frac{2}{(\sqrt{2} - t)(\sqrt{2} + t)}\, dt \\
&= \int \frac{1}{\sqrt{2}} \left(\frac{1}{\sqrt{2} - t} + \frac{1}{\sqrt{2} + t}\right) dt \\
&= -\frac{1}{\sqrt{2}} \ln(\sqrt{2} - t) + \frac{1}{\sqrt{2}} \ln(\sqrt{2} + t) + c \\
&= \frac{1}{\sqrt{2}} \ln\left[\frac{\sqrt{2} + \tan(\frac{1}{2}x)}{\sqrt{2} - \tan(\frac{1}{2}x)}\right] + c. \blacktriangleleft
\end{aligned}$$

Integrals of a similar form to (1.28), but involving $\sin 2x$, $\cos 2x$, $\tan 2x$, $\sin^2 x$, $\cos^2 x$ or $\tan^2 x$ instead of $\cos x$ and $\sin x$, should be evaluated by using the substitution $t = \tan x$. In this case

$$\sin x = \frac{t}{\sqrt{1 + t^2}}, \qquad \cos x = \frac{1}{\sqrt{1 + t^2}} \qquad \text{and} \qquad dx = \frac{dt}{1 + t^2}. \tag{1.32}$$

A final example of the evaluation of integrals using substitution is the method of completing the square.

►*Evaluate the integral*

$$I = \int \frac{1}{x^2 + 4x + 7}\, dx.$$

We can write the integral in the form

$$I = \int \frac{1}{(x + 2)^2 + 3}\, dx.$$

Substituting $y = x + 2$, we find $dy = dx$ and hence

$$I = \int \frac{1}{y^2 + 3}\, dy,$$

Hence, by comparison with the table of standard integrals (see subsection 1.2.3)

$$I = \frac{\sqrt{3}}{3}\tan^{-1}\left(\frac{y}{\sqrt{3}}\right) + c = \frac{\sqrt{3}}{3}\tan^{-1}\left(\frac{x+2}{\sqrt{3}}\right) + c. \blacktriangleleft$$

### *1.2.8 Integration by parts*

Integration by parts is the integration analogy of product differentiation. The principle is to break down a complicated function into two functions, at least one of which can be integrated by inspection. The method in fact relies on the result for the differentiation of a product. Recalling from (1.6) that

$$\frac{d}{dx}(uv) = u\frac{dv}{dx} + \frac{du}{dx}v,$$

where $u$ and $v$ are functions of $x$, we now integrate to find

$$uv = \int u\frac{dv}{dx}\,dx + \int \frac{du}{dx}\,v\,dx.$$

Rearranging into the standard form gives

$$\int u\frac{dv}{dx}\,dx = uv - \int \frac{du}{dx}\,v\,dx. \qquad (1.33)$$

Integration by parts is often remembered for practical purposes in the form *the integral of a product of two functions is equal to {the first times the integral of the second} minus the integral of {the derivative of the first times the integral of the second}*. Here, $u$ is 'the first' and $dv/dx$ is 'the second'; clearly the integral of 'the second' must be determinable by inspection.

►*Evaluate the integral* $I = \int x \sin x\, dx$.

In the notation given above, we identify $x$ with $u$ and $\sin x$ with $dv/dx$. Hence $v = -\cos x$ and $du/dx = 1$ and using (1.33)

$$I = x(-\cos x) - \int (1)(-\cos x)\,dx = -x\cos x + \sin x + c. \blacktriangleleft$$

The separation of the functions is not always so apparent, as is illustrated by the following example.

►*Evaluate the integral* $I = \int x^3 e^{-x^2}\, dx$.

Firstly we rewrite the integral as

$$I = \int x^2 \left(xe^{-x^2}\right)\,dx.$$

Now, using the notation given above, we identify $x^2$ with $u$ and $xe^{-x^2}$ with $dv/dx$. Hence $v = -\frac{1}{2}e^{-x^2}$ and $du/dx = 2x$, so that

$$I = -\tfrac{1}{2}x^2 e^{-x^2} - \int -xe^{-x^2}\,dx = -\tfrac{1}{2}x^2 e^{-x^2} - \tfrac{1}{2}e^{-x^2} + c. \blacktriangleleft$$

A trick that is sometimes useful is to take '1' as one factor of the product, as is illustrated by the following example.

►*Evaluate the integral* $I = \int \ln x\,dx$.

Firstly we rewrite the integral as

$$I = \int (\ln x)\,1\,dx.$$

Now, using the notation above, we identify $\ln x$ with $u$ and 1 with $dv/dx$. Hence $v = x$ and $du/dx = 1/x$,

$$I = (\ln x)(x) - \int \left(\frac{1}{x}\right) x\,dx = x\ln x - x + c. \blacktriangleleft$$

It is sometimes necessary to integrate by parts more than once. In doing so we may occasionally re-encounter the original integral $I$. In such cases we can obtain a linear algebraic equation for $I$ that can be solved to obtain its value.

►*Evaluate the integral* $I = \int e^{ax}\cos bx\,dx$.

Integrating by parts, taking $e^{ax}$ as the first function, we find

$$I = e^{ax}\left(\frac{\sin bx}{b}\right) - \int ae^{ax}\left(\frac{\sin bx}{b}\right)dx,$$

where, for convenience, we have omitted the constant of integration. Integrating by parts a second time,

$$I = e^{ax}\left(\frac{\sin bx}{b}\right) - ae^{ax}\left(\frac{-\cos bx}{b^2}\right) + \int a^2 e^{ax}\left(\frac{-\cos bx}{b^2}\right)dx.$$

But the integral on the RHS is just $-a^2/b^2$ times the original integral $I$. Thus

$$I = e^{ax}\left(\frac{1}{b}\sin bx + \frac{a}{b^2}\cos bx\right) - \frac{a^2}{b^2}I.$$

Rearranging this expression to obtain $I$ explicitly and including the constant of integration we find

$$I = \frac{e^{ax}}{a^2+b^2}(b\sin bx + a\cos bx) + c. \qquad (1.34)$$

Another method of evaluating this integral, using the exponential of a complex number, is given in section 2.6. ◄

### *1.2.9 Reduction formulae*

Integration using reduction formulae is a process that involves first evaluating a simple integral and then, in stages, using it to find a more complicated integral.

►*Using integration by parts, find a relationship between $I_n$ and $I_{n-1}$ where*

$$I_n = \int_0^1 (1-x^3)^n \, dx$$

*and n is any positive integer. Hence evaluate $I_2 = \int_0^1 (1-x^3)^2 \, dx$.*

Writing $I_n$ as a product and separating the integral we find

$$\begin{aligned} I_n &= \int_0^1 (1-x^3)(1-x^3)^{n-1} \, dx \\ &= \int_0^1 (1-x^3)^{n-1} \, dx - \int_0^1 x^3(1-x^3)^{n-1} \, dx. \end{aligned}$$

The first term on the RHS is clearly $I_{n-1}$ and so, writing the second term on the RHS as a product,

$$I_n = I_{n-1} - \int_0^1 (x)x^2(1-x^3)^{n-1} \, dx.$$

Integrating by parts we find

$$\begin{aligned} I_n &= I_{n-1} + \left[\frac{x}{3n}(1-x^3)^n\right]_0^1 - \int_0^1 \frac{1}{3n}(1-x^3)^n \, dx \\ &= I_{n-1} + 0 - \frac{1}{3n} I_n, \end{aligned}$$

which on rearranging gives

$$I_n = \frac{3n}{3n+1} I_{n-1}.$$

We now have a relation connecting successive integrals. Hence, if we can evaluate $I_0$, we can find $I_1$, $I_2$ etc. Evaluating $I_0$ is trivial:

$$I_0 = \int_0^1 (1-x^3)^0 \, dx = \int_0^1 dx = [x]_0^1 = 1.$$

Hence

$$I_1 = \frac{(3 \times 1)}{(3 \times 1)+1} \times 1 = \frac{3}{4}, \qquad I_2 = \frac{(3 \times 2)}{(3 \times 2)+1} \times \frac{3}{4} = \frac{9}{14}.$$

Although the first few $I_n$ could be evaluated by direct multiplication, this becomes tedious for integrals containing higher values of $n$; these are therefore best evaluated using the reduction formula. ◄

### *1.2.10 Infinite and improper integrals*

The definition of an integral given previously does not allow for cases in which either of the limits of integration are infinite (an *infinite integral*), nor for cases in which $f(x)$ is infinite in some part of the range (an *improper integral*), e.g. $f(x) = (2-x)^{-1/4}$ near the point $x = 2$. Nevertheless, modification of the definition of an integral gives infinite and improper integrals each a meaning.

In the case of an integral $I = \int_a^b f(x)\,dx$, the infinite integral, in which $b$ tends to $\infty$, is defined by

$$I = \int_a^\infty f(x)\,dx = \lim_{b\to\infty} \int_a^b f(x)\,dx = \lim_{b\to\infty} F(b) - F(a).$$

As previously, $F(x)$ is the indefinite integral of $f(x)$, and $\lim_{b\to\infty} F(b)$ means the limit (or value) that $F(b)$ approaches as $b \to \infty$; it is evaluated after calculating the integral. The formal concept of a limit will be introduced in chapter 3.

►*Evaluate the integral*

$$I = \int_0^\infty \frac{x}{(x^2+a^2)^2}\,dx.$$

Integrating, we find $F(x) = -\frac{1}{2}(x^2+a^2)^{-1} + c$ and so

$$I = \lim_{b\to\infty}\left[\frac{-1}{2(b^2+a^2)}\right] - \left(\frac{-1}{2a^2}\right) = \frac{1}{2a^2}. \blacktriangleleft$$

For the case of improper integrals, we adopt the approach of excluding the unbounded range from the integral. For example, if the integrand $f(x)$ is infinite at $x = c$ (say), $a \le c \le b$, then

$$\int_a^b f(x)\,dx = \lim_{\delta\to 0}\int_a^{c-\delta} f(x)\,dx + \lim_{\epsilon\to 0}\int_{c+\epsilon}^b f(x)\,dx.$$

►*Evaluate the integral* $I = \int_0^2 (2-x)^{-1/4}\,dx$.

Integrating directly,

$$I = \lim_{\epsilon\to 0}\left[-\tfrac{4}{3}(2-x)^{3/4}\right]_0^{2-\epsilon} = \lim_{\epsilon\to 0}\left[-\tfrac{4}{3}\epsilon^{3/4}\right] + \tfrac{4}{3}2^{3/4} = \left(\tfrac{4}{3}\right)2^{3/4}. \blacktriangleleft$$

### *1.2.11 Integration in plane polar coordinates*

In plane polar coordinates $\rho, \phi$ a curve is defined by its distance $\rho$ from the origin as a function of the angle $\phi$ between the line joining a point on the curve to the origin and the $x$-axis, i.e. $\rho = \rho(\phi)$. The area of an element is given by

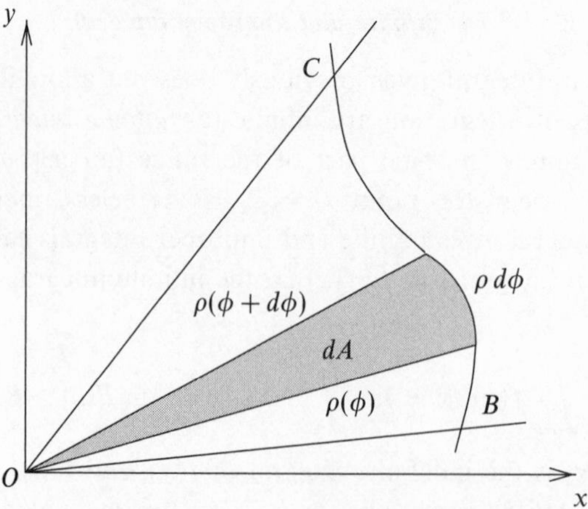


Figure 1.8 Finding the area of a sector $OBC$ defined by the curve $\rho(\phi)$ and the radii $OB$, $OC$, at angles to the $x$-axis $\phi_1$, $\phi_2$ respectively.

$dA = \frac{1}{2}\rho^2\, d\phi$, as illustrated in figure 1.8, and hence the total area between two angles $\phi_1$ and $\phi_2$ is given by

$$A = \int_{\phi_1}^{\phi_2} \tfrac{1}{2}\rho^2\, d\phi. \tag{1.35}$$

An immediate observation is that the area of a circle of radius $a$ is given by

$$A = \int_0^{2\pi} \tfrac{1}{2}a^2\, d\phi = \left[\tfrac{1}{2}a^2\phi\right]_0^{2\pi} = \pi a^2.$$

►*The equation in polar coordinates of an ellipse with semi-axes a and b is*

$$\frac{1}{\rho^2} = \frac{\cos^2\phi}{a^2} + \frac{\sin^2\phi}{b^2}.$$

*Find the area A of the ellipse.*

Using (1.35) and symmetry, we have

$$A = \frac{1}{2}\int_0^{2\pi} \frac{a^2b^2}{b^2\cos^2\phi + a^2\sin^2\phi}\, d\phi = 2a^2b^2 \int_0^{\pi/2} \frac{1}{b^2\cos^2\phi + a^2\sin^2\phi}\, d\phi.$$

To evaluate this integral we write $t = \tan\phi$ and use (1.32):

$$A = 2a^2b^2 \int_0^{\infty} \frac{1}{b^2 + a^2t^2}\, dt = 2b^2 \int_0^{\infty} \frac{1}{(b/a)^2 + t^2}\, dt.$$

Finally, from the list of standard integrals (see subsection 1.2.3),

$$A = 2b^2 \left[ \frac{1}{(b/a)} \tan^{-1} \frac{t}{(b/a)} \right]_0^\infty = 2ab \left( \frac{\pi}{2} - 0 \right) = \pi ab. \blacktriangleleft$$

### *1.2.12 Integral inequalities*

Consider the functions $f(x)$, $\phi_1(x)$ and $\phi_2(x)$ such that $\phi_1(x) \leq f(x) \leq \phi_2(x)$ for all $x$ in the range $a \leq x \leq b$. It immediately follows that

$$\int_a^b \phi_1(x)\,dx \leq \int_a^b f(x)\,dx \leq \int_a^b \phi_2(x)\,dx, \tag{1.36}$$

which gives us a way of estimating an integral that is difficult to evaluate explicitly.

►*Show that the value of the integral*

$$I = \int_0^1 \frac{1}{(1 + x^2 + x^3)^{1/2}}\,dx$$

*lies between* 0.810 *and* 0.882.

We note that for $x$ in the range $0 \leq x \leq 1$, $0 \leq x^3 \leq x^2$. Hence

$$(1 + x^2)^{1/2} \leq (1 + x^2 + x^3)^{1/2} \leq (1 + 2x^2)^{1/2},$$

and so

$$\frac{1}{(1 + x^2)^{1/2}} \geq \frac{1}{(1 + x^2 + x^3)^{1/2}} \geq \frac{1}{(1 + 2x^2)^{1/2}}.$$

Consequently,

$$\int_0^1 \frac{1}{(1 + x^2)^{1/2}}\,dx \geq \int_0^1 \frac{1}{(1 + x^2 + x^3)^{1/2}}\,dx \geq \int_0^1 \frac{1}{(1 + 2x^2)^{1/2}}\,dx,$$

from which we obtain

$$\left[ \ln(x + \sqrt{1 + x^2}) \right]_0^1 \geq I \geq \left[ \tfrac{1}{\sqrt{2}} \ln \left( x + \sqrt{\tfrac{1}{2} + x^2} \right) \right]_0^1$$

$$0.8814 \geq I \geq 0.8105$$

$$0.882 \geq I \geq 0.810. \blacktriangleleft$$

### *1.2.13 Applications of integration*

*Mean value of a function*

The mean value $m$ of a function between two limits $a$ and $b$ is defined by

$$m = \frac{1}{b - a} \int_a^b f(x)\,dx. \tag{1.37}$$

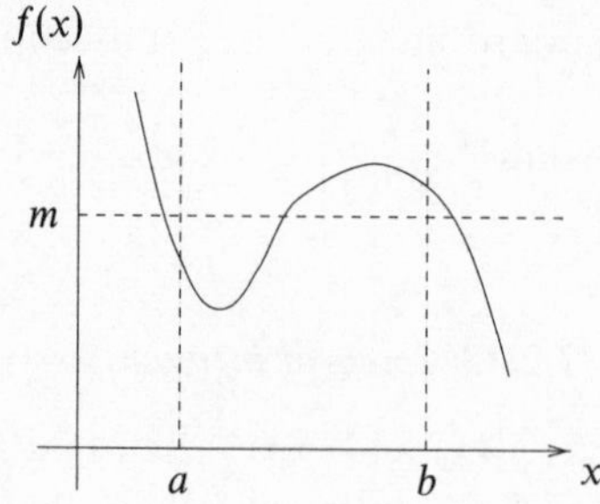


Figure 1.9 The mean value $m$ of a function.

The mean value may be thought of as the height of the rectangle that has the same area (over the same interval) as the area under the curve $f(x)$. This is illustrated in figure 1.9.

►*Find the mean value $m$ of the function $f(x) = x^2$ between the limits $x = 2$ and $x = 4$.*

Using (1.37),

$$m = \frac{1}{4-2}\int_2^4 x^2\,dx = \frac{1}{2}\left[\frac{x^3}{3}\right]_2^4 = \frac{1}{2}\left(\frac{4^3}{3} - \frac{2^3}{3}\right) = \frac{28}{3}. \blacktriangleleft$$

*Finding the length of a curve*

The definition of integration as finding the area under a curve provides only one example of its use. Another is in finding the length of a curve. If a curve is defined by $y = f(x)$ then the distance along the curve, $\Delta s$, that corresponds to small changes $\Delta x$ and $\Delta y$ in $x$ and $y$ is given by

$$\Delta s \approx \sqrt{(\Delta x)^2 + (\Delta y)^2}; \qquad (1.38)$$

this follows directly from Pythagoras' theorem (see figure 1.10). Dividing (1.38) through by $\Delta x$ and letting $\Delta x \to 0$ we obtain†

$$\frac{ds}{dx} = \sqrt{1 + \left(\frac{dy}{dx}\right)^2}.$$

† Instead of considering small changes $\Delta x$ and $\Delta y$ and letting these tend to zero, we could have derived (1.38) by considering infinitesimal changes $dx$ and $dy$ from the start. After writing $(ds)^2 = (dx)^2 + (dy)^2$, (1.38) may be deduced by using the formal device of dividing through by $dx$. Although not mathematically rigorous, this method is often used and generally leads to the correct result.

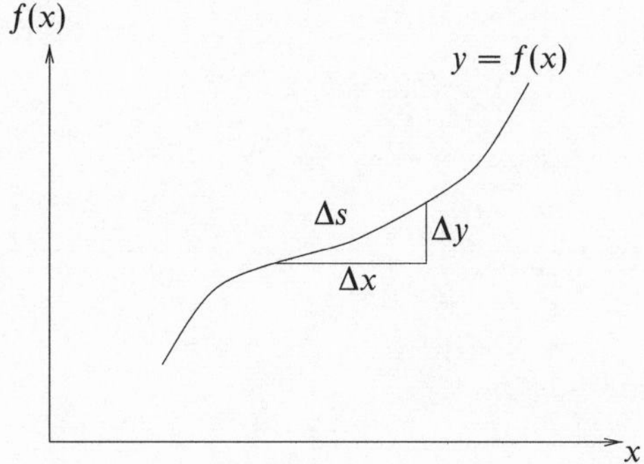


Figure 1.10 The distance moved along a curve, $\Delta s$, corresponding to the small changes $\Delta x$ and $\Delta y$.

Clearly the total length $s$ of the curve between the points $x = a$ and $x = b$ is then given by integrating both sides of the equation, to yield

$$s = \int_a^b \sqrt{1 + \left(\frac{dy}{dx}\right)^2}\, dx. \tag{1.39}$$

In plane polar coordinates,

$$ds = \sqrt{(dr)^2 + (r\,d\phi)^2} \quad \Rightarrow \quad s = \int_{r_1}^{r_2} \sqrt{1 + r^2 \left(\frac{d\phi}{dr}\right)^2}\, dr. \tag{1.40}$$

►*Find the length of the curve $y = x^{3/2}$ from $x = 0$ to $x = 2$.*

Using (1.39) and noting that $dy/dx = \frac{3}{2}\sqrt{x}$, the length of the curve is given by

$$\begin{aligned} s &= \int_0^2 \sqrt{1 + \tfrac{9}{4}x}\, dx \\ &= \left[\tfrac{2}{3}\left(\tfrac{4}{9}\right)\left(1 + \tfrac{9}{4}x\right)^{3/2}\right]_0^2 = \tfrac{8}{27}\left[\left(1 + \tfrac{9}{4}x\right)^{3/2}\right]_0^2 \\ &= \tfrac{8}{27}\left[\left(\tfrac{11}{2}\right)^{3/2} - 1\right]. \end{aligned}$$ ◄

*Surfaces of revolution*

Consider the surface $S$ formed by rotating the curve $y = f(x)$ about the $x$-axis (see figure 1.11). The surface area of the 'collar' formed by rotating an element of the curve, $ds$, about the $x$-axis is $2\pi y\, ds$, and hence the total surface area is

$$S = \int_a^b 2\pi y\, ds.$$

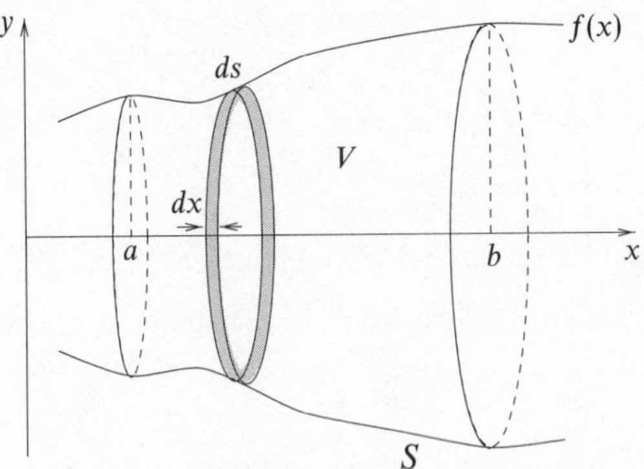


Figure 1.11 The surface and volume of revolution for the curve $y = f(x)$.

Since $(ds)^2 = (dx)^2 + (dy)^2$ from (1.38), the total surface area between the planes $x = a$ and $x = b$ is

$$S = \int_a^b 2\pi y \sqrt{1 + \left(\frac{dy}{dx}\right)^2}\, dx. \tag{1.41}$$

►*Find the surface area of a cone formed by rotating about the x-axis the line $y = 2x$ between $x = 0$ and $x = h$.*

Using (1.41), the surface area is given by

$$\begin{aligned} S &= \int_0^h (2\pi)2x\sqrt{1 + \left[\frac{d}{dx}(2x)\right]^2}\, dx \\ &= \int_0^h 4\pi x \left(1 + 2^2\right)^{1/2}\, dx = \int_0^h 4\sqrt{5}\pi x\, dx \\ &= \left[2\sqrt{5}\pi x^2\right]_0^h = 2\sqrt{5}\pi(h^2 - 0) = 2\sqrt{5}\pi h^2. \end{aligned}$$ ◄

We note that a surface of revolution may also be formed by rotating a line about the $y$-axis. In this case the surface area between $y = a$ and $y = b$ is

$$S = \int_a^b 2\pi x \sqrt{1 + \left(\frac{dx}{dy}\right)^2}\, dy. \tag{1.42}$$

*Volumes of revolution*

The volume $V$ enclosed by rotating the curve $y = f(x)$ about the $x$-axis can also be found (see figure 1.11). The volume of the disc between $x$ and $x + dx$ is given by $dV = \pi y^2\, dx$. Hence the total volume between $x = a$ and $x = b$ is

$$V = \int_a^b \pi y^2\, dx. \tag{1.43}$$

►*Find the volume of a cone enclosed by the surface formed by rotating about the x-axis the line $y = 2x$ between $x = 0$ and $x = h$.*

Using (1.43), the volume is given by

$$V = \int_0^h \pi(2x)^2\, dx = \int_0^h 4\pi x^2\, dx$$
$$= \left[\tfrac{4}{3}\pi x^3\right]_0^h = \tfrac{4}{3}\pi(h^3 - 0) = \tfrac{4}{3}\pi h^3. \blacktriangleleft$$

As before, it is also possible to form a volume of revolution by rotating a curve about the $y$-axis. In this case the volume enclosed between $y = a$ and $y = b$ is

$$V = \int_a^b \pi x^2\, dy. \tag{1.44}$$

## 1.3 Exercises

1.1 Obtain the following derivatives from first principles:

(a) the first derivative of $3x + 4$;
(b) the first, second and third derivatives of $x^2 + x$;
(c) the first derivative of $\sin x$.

1.2 Find the first derivatives of

(a) $x^2 \exp x$, (b) $2 \sin x \cos x$, (c) $\sin 2x$, (d) $x \sin ax$,
(e) $(\exp ax)(\sin ax) \tan^{-1} ax$, (f) $\ln(x^a + x^{-a})$,
(g) $\ln(a^x + a^{-x})$, (h) $x^x$.

1.3 Find the first derivatives of

(a) $x/(a + x)^2$, (b) $x/(1 - x)^{1/2}$, (c) $\tan x$ as $\sin x/ \cos x$,
(d) $(3x^2 + 2x + 1)/(8x^2 - 4x + 2)$.

1.4 Use result (1.12) to find the first derivatives of

(a) $(2x + 3)^{-3}$, (b) $\sec^2 x$, (c) $\operatorname{cosech}^3 3x$, (d) $1/ \ln x$, (e) $1/[\sin^{-1}(x/a)]$.

1.5 Find $dy/dx$ if $x = (t-2)/(t+2)$ and $y = 2t/(t+1)$ for $-\infty < t < \infty$. Show that it is always non-negative, and make use of this result in sketching the curve of $y$ as a function of $x$.

1.6 If $2y + \sin y + 5 = x^4 + 4x^3 + 2\pi$, show that $dy/dx = 16$ when $x = 1$.

1.7 Find the positions and natures of the stationary points of the following functions:

(a) $x^3 - 3x + 3$; (b) $x^3 - 3x^2 + 3x$; (c) $x^3 + 3x + 3$;
(d) $\sin ax$ with $a \neq 0$; (e) $x^5 + x^3$; (f) $x^5 - x^3$.

1.8 Show that the lowest value taken by the function $3x^4 + 4x^3 - 12x^2 + 6$ is $-26$.

1.9 Use Leibnitz' theorem to find:

(a) the second derivative of $\cos x \sin 2x$;
(b) the third derivative of $\sin x \ln x$;
(c) the fourth derivative of $(2x^3 + 3x^2 + x + 2) \exp 2x$.

1.10 If $y = \exp(-x^2)$ show that $dy/dx = -2xy$ and hence, by applying Leibnitz' theorem, prove that for $n \geq 1$

$$y^{(n+1)} + 2xy^{(n)} + 2ny^{(n-1)} = 0.$$

1.11 Find the following indefinite integrals:

(a) $\int (4 + x^2)^{-1}\, dx$; (b) $\int (8 + 2x - x^2)^{-1/2}\, dx$ for $-2 \leq x \leq 4$;
(c) $\int (1 + \sin\theta)^{-1}\, d\theta$; (d) $\int (x\sqrt{1-x})^{-1}\, dx$ for $0 < x \leq 1$.

1.12 Use integration by parts to evaluate the following:

(a) $\int_0^y x^2 \sin x\, dx$; (b) $\int_1^y x \ln x\, dx$;
(c) $\int_0^y \sin^{-1} x\, dx$; (d) $\int_1^y \ln(a^2 + x^2)/x^2\, dx$.

1.13 By integrating by parts twice, prove that $I_n$ as defined below for positive integers $n$ has the value shown.

$$I_n = \int_0^{\pi/2} \sin n\theta \cos\theta\, d\theta = \frac{n - \sin(n\pi/2)}{n^2 - 1}.$$

1.14 Evaluate the following definite integrals:

(a) $\int_0^\infty xe^{-x}\, dx$; (b) $\int_0^1 (x^3 + 1)/(x^4 + 4x + 1)\, dx$;
(c) $\int_0^{\pi/2} [a + (a-1)\cos\theta]^{-1}\, d\theta$ with $a > \frac{1}{2}$; (d) $\int_{-\infty}^{\infty} (x^2 + 6x + 18)^{-1}\, dx$.

1.15 If $J_r$ is the integral

$$\int_0^\infty x^r \exp(-x^2)\, dx$$

show that

(a) $J_{2r+1} = (r!)/2$; (b) $J_{2r} = 2^{-r}(2r-1)(2r-3)\cdots(5)(3)(1)\, J_0$.

1.16 (a) Find positive constants $a$, $b$ such that $ax \leq \sin x \leq bx$ for $0 \leq x \leq \pi/2$. Use this inequality to find (to two significant figures) upper and lower bounds for the integral

$$I = \int_0^{\pi/2} (1 + \sin x)^{1/2}\, dx.$$

(b) Use the substitution $t = \tan(\frac{1}{2}x)$ to evaluate $I$ exactly.

1.17 By noting that for $0 \leq \eta \leq 1$, $\eta^{1/2} \geq \eta^{3/4} \geq \eta$, prove that

$$\frac{2}{3} \leq \frac{1}{a^{5/2}} \int_0^a (a^2 - x^2)^{3/4}\, dx \leq \frac{\pi}{4}.$$

## 1.4 Hints and answers

1.1 (a) 3; (b) $2x + 1$, 2, 0; (c) $\cos x$.

1.2 (a) $(x^2 + 2x)\exp x$; (b) $2(\cos^2 x - \sin^2 x) = 2\cos 2x$; (c) $2\cos 2x$; (d) $\sin ax + ax\cos ax$; (e) $(a\exp ax)[(\sin ax + \cos ax)\tan^{-1} ax + (\sin ax)(1 + a^2x^2)^{-1}]$; (f) $[a(x^a - x^{-a})]/[x(x^a + x^{-a})]$; (g) $[(a^x - a^{-x})\ln a]/(a^x + a^{-x})$; (h) $(1 + \ln x)x^x$.

1.3 (a) $(a - x)(a + x)^{-3}$; (b) $(1 - x/2)(1 - x)^{-3/2}$; (c) $\sec^2 x$; (d) $(-7x^2 - x + 2)(4x^2 - 2x + 1)^{-2}$.

1.4 (a) $-6(2x + 3)^{-4}$; (b) $2\sec^2 x \tan x$; (c) $-9\,\text{cosech}^3 3x \coth 3x$; (d) $-x^{-1}(\ln x)^{-2}$; (e) $-(a^2 - x^2)^{-1/2}[\sin^{-1}(x/a)]^{-2}$.

1.5 $(t + 2)^2/[2(t + 1)^2]$.

1.6 $y = \pi$ at $x = 1$.

1.7 (a) Min. at $x = 1$, max. at $x = -1$; (b) inflection at $x = 1$; (c) no stationary points; (d) $x = (n + \frac{1}{2})\pi/a$, max. for $n$ even, min. for $n$ odd; (e) inflection at $x = 0$; (f) inflection at $x = 0$, max at $x = -(\frac{3}{5})^{1/2}$, min. at $(\frac{3}{5})^{1/2}$.

1.8 $-26$ at $x = -2$; other stationary values are 6 at $x = 0$, and 1 at $x = 1$.

1.9 (a) $2(2 - 9\cos^2 x)\sin x$; (b) $(2x^{-3} - 3x^{-1})\sin x - (3x^{-2} + \ln x)\cos x$; (c) $8(4x^3 + 30x^2 + 62x + 38)\exp 2x$.

1.11 (a) $[\tan^{-1}(x/2)]/2$; (b) $\sin^{-1}[(x - 1)/3]$; (c) $-2[1 + \tan(\theta/2)]^{-1}$; (d) put $y = (1 - x)^{1/2}$, $\ln\{[1 - (1 - x)^{1/2}]/[1 + (1 - x)^{1/2}\}$.

1.12 (a) $(2 - y^2)\cos y + 2y\sin y - 2$; (b) $[(y^2 \ln y)/2] + [(1 - y^2)/4]$; (c) $y\sin^{-1} y + (1 - y^2)^{1/2} - 1$; (d) $\ln(a^2 + 1) - (1/y)\ln(a^2 + y^2) + (2/a)[\tan^{-1}(y/a) - \tan^{-1}(1/a)]$.

1.14 (a) 1; (b) $(\ln 6)/4$; (c) $\{2\tan^{-1}[(2a - 1)^{-1/2}]\}/(2a - 1)^{1/2}$; (d) $\pi/3$.

1.16 (a) $a = 2/\pi$, $b = 1$; $\frac{2}{3}[(1 + \frac{\pi}{2})^{3/2} - 1] > I > \frac{\pi}{3}(2^{3/2} - 1)$, $2.08 > I > 1.91$; (b) $I = 2$.

1.17 Set $\eta = 1 - (x/a)^2$.

# 2

# Complex numbers and hyperbolic functions

This chapter is concerned with the representation and manipulation of complex numbers. Complex numbers pervade this book, underscoring their wide application in the mathematics of the physical sciences. The application of complex numbers to the description of physical systems is left until later chapters and only the basic tools are presented here.

## 2.1 The need for complex numbers

Although complex numbers occur in many branches of mathematics, they arise most directly out of solving polynomial equations. We examine a specific quadratic equation as an example.

Consider the quadratic equation

$$z^2 - 4z + 5 = 0. \tag{2.1}$$

Equation (2.1) has two solutions, $z_1$ and $z_2$, such that

$$(z - z_1)(z - z_2) = 0. \tag{2.2}$$

Using the familiar formula for the roots of a quadratic equation, the solutions $z_1$ and $z_2$, written in brief as $z_{1,2}$, are

$$\begin{aligned} z_{1,2} &= \frac{4 \pm \sqrt{(-4)^2 - 4(1 \times 5)}}{2} \\ &= 2 \pm \frac{\sqrt{-4}}{2}. \end{aligned} \tag{2.3}$$

Both solutions contain the square root of a negative number. However, it is not true to say that there are no solutions to the quadratic equation. The *fundamental theorem of algebra* states that a quadratic equation will always have two solutions and these are in fact given by (2.3). The second term on the RHS of (2.3) is called an *imaginary* term since it contains the square root of a negative number;

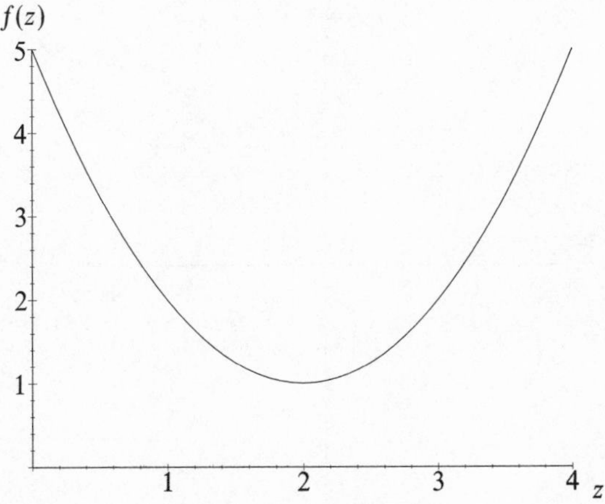


Figure 2.1 The function $f(z) = z^2 - 4z + 5$.

the first term is called a *real* term. The full solution is the sum of a real term and an imaginary term and is called a *complex number.* A plot of the function $f(z) = z^2 - 4z + 5$ is shown in figure 2.1. It will be seen that the plot does not intersect the $z$-axis, corresponding to the fact that the equation $f(z) = 0$ has no purely real solutions.

The choice of the symbol $z$ for the quadratic variable was not arbitrary; the conventional representation of a complex number is $z$, where $z$ is the sum of a real part $x$ and $i$ times an imaginary part $y$, i.e.

$$z = x + iy,$$

where $i$ is used to denote the square root of $-1$. The real part $x$ and the imaginary part $y$ are usually denoted by $\text{Re}\, z$ and $\text{Im}\, z$ respectively. We note at this point that some physical scientists, engineers in particular, use $j$ instead of $i$. However, for consistency, we will use $i$ throughout this book.

In our particular example $\sqrt{-4} = 2\sqrt{-1} = 2i$, and hence the two solutions of (2.1) are

$$z_{1,2} = 2 \pm \frac{2i}{2} = 2 \pm i.$$

Thus here $x = 2$ and $y = \pm 1$.

For compactness a complex number is sometimes written in the form

$$z = (x, y),$$

where the components of $z$ may be thought of as coordinates in an $xy$-plot. Such a plot is called an *Argand diagram* and is a common representation of a complex number; an example is shown in figure 2.2.

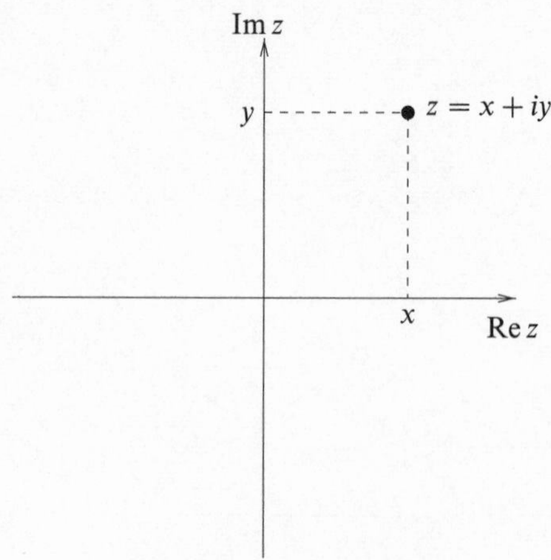


Figure 2.2 The Argand diagram.

Our particular example of a quadratic equation may readily be generalised to polynomials whose highest power (order) is greater than 2, e.g. cubic equations (order 3), quartic equations (order 4) and so on. For a general polynomial $f(z)$, of order $n$, the fundamental theorem of algebra states that the equation $f(z) = 0$ will have exactly $n$ solutions. We will examine cases of higher-order equations in subsection 2.4.3.

The remainder of this chapter deals with the algebra and manipulation of complex numbers; their polar representation, which has advantages in many circumstances; complex exponentials and logarithms; the use of complex numbers in finding the roots of polynomial equations; and hyperbolic functions.

## 2.2 Manipulation of complex numbers

This section considers basic complex number manipulation. Some analogy may be drawn with vector manipulation (see chapter 6) but this section stands alone as an introduction.

### *2.2.1 Addition and subtraction*

The addition of two complex numbers, $z_1$ and $z_2$, in general gives another complex number. The real components and the imaginary components are added separately and in a like manner to the familiar addition of real numbers:

$$z_1 + z_2 = (x_1 + iy_1) + (x_2 + iy_2) = (x_1 + x_2) + i(y_1 + y_2),$$

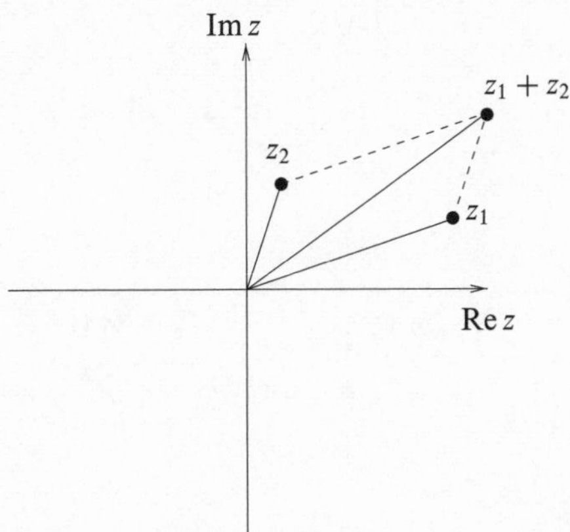


Figure 2.3 The addition of two complex numbers.

or in component notation

$$z_1 + z_2 = (x_1, y_1) + (x_2, y_2) = (x_1 + x_2, y_1 + y_2).$$

The Argand representation of the addition of two complex numbers is shown in figure 2.3.

By straightforward application of the commutativity and associativity of the real and imaginary parts separately, we can show that the addition of complex numbers is itself commutative and associative, i.e.

$$z_1 + z_2 = z_2 + z_1,$$
$$z_1 + (z_2 + z_3) = (z_1 + z_2) + z_3.$$

Thus it is immaterial in what order complex numbers are added.

►*Sum the complex numbers* $1 + 2i$, $3 - 4i$, $-2 + i$.

Summing the real terms we obtain

$$1 + 3 - 2 = 2,$$

and summing the imaginary terms we obtain

$$2i - 4i + i = -i.$$

Hence

$$(1 + 2i) + (3 - 4i) + (-2 + i) = 2 - i. \blacktriangleleft$$

The subtraction of complex numbers is very similar to their addition. As in the case of real numbers, if two identical complex numbers are subtracted then the result is zero.

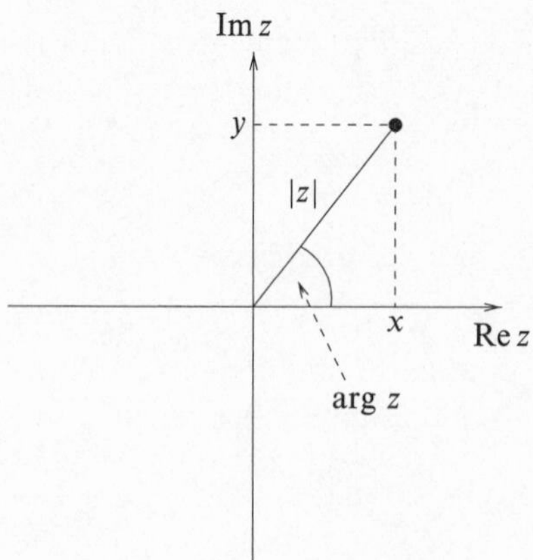


Figure 2.4 The modulus and argument of a complex number.

### 2.2.2 Modulus and argument

The modulus of the complex number $z$ is denoted by $|z|$ and is defined as

$$|z| = \sqrt{x^2 + y^2}. \tag{2.4}$$

Hence the modulus of the complex number is the distance of the corresponding point from the origin in the Argand diagram, as may be seen in figure 2.4.

The argument of the complex number $z$ is denoted by arg $z$ and is defined as

$$\arg z = \tan^{-1}\left(\frac{y}{x}\right). \tag{2.5}$$

It is the angle that the line joining the origin to $z$ on the Argand diagram makes with the positive $x$-axis. The anticlockwise direction is taken to be positive by convention. The angle arg $z$ is shown in figure 2.4. Account must be taken of the signs of $x$ and $y$ individually in determining in which quadrant arg $z$ lies. Thus, for example, if $x$ and $y$ are both negative, arg $z$ lies in the range $-\pi < \arg z < -\pi/2$, rather than in the first quadrant ($0 < \arg z < \pi/2$), though both cases give the same value for the argument of the inverse tangent function.

►*Find the modulus and the argument of the complex number $z = 2 - 3i$.*

Using (2.4), the modulus is given by

$$|z| = \sqrt{2^2 + (-3)^2} = \sqrt{13}.$$

Using (2.5), the argument is given by

$$\arg z = \tan^{-1}\left(-\tfrac{3}{2}\right).$$

The two angles whose tangents equal $-1.5$ are $-0.9828$ rad and $2.1588$ rad. Since $z$ clearly lies in the fourth quadrant in this case, $\arg z = -0.9828$ is the appropriate answer. ◄

### *2.2.3 Multiplication*

Complex numbers may be multiplied together and in general give a complex number as the result. The product of two complex numbers $z_1$ and $z_2$ is found by multiplying them out in full and remembering that $i^2 = -1$, i.e.

$$\begin{aligned} z_1 z_2 &= (x_1 + iy_1)(x_2 + iy_2) \\ &= x_1x_2 + ix_1y_2 + iy_1x_2 + i^2y_1y_2 \\ &= (x_1x_2 - y_1y_2) + i(x_1y_2 + y_1x_2). \end{aligned} \tag{2.6}$$

►*Multiply the complex numbers* $z_1 = 3 + 2i$ *and* $z_2 = -1 - 4i$.

By direct multiplication we find

$$\begin{aligned} z_1 z_2 &= (3 + 2i)(-1 - 4i) \\ &= -3 - 2i - 12i - 8i^2 \\ &= 5 - 14i. \blacktriangleleft \end{aligned} \tag{2.7}$$

The multiplication of complex numbers is both commutative and associative, i.e.

$$z_1 z_2 = z_2 z_1, \tag{2.8}$$

$$(z_1 z_2) z_3 = z_1 (z_2 z_3). \tag{2.9}$$

The product of two complex numbers also has the simple properties

$$|z_1 z_2| = |z_1||z_2|, \tag{2.10}$$

$$\arg(z_1 z_2) = \arg z_1 + \arg z_2. \tag{2.11}$$

These relationships are derived in subsection 2.3.1.

►*Verify that (2.10) holds for the product of* $z_1 = 3 + 2i$ *and* $z_2 = -1 - 4i$.

From (2.7)

$$|z_1 z_2| = |5 - 14i| = \sqrt{5^2 + (-14)^2} = \sqrt{221}.$$

We also find

$$|z_1| = \sqrt{3^2 + 2^2} = \sqrt{13},$$

$$|z_2| = \sqrt{(-1)^2 + (-4)^2} = \sqrt{17},$$

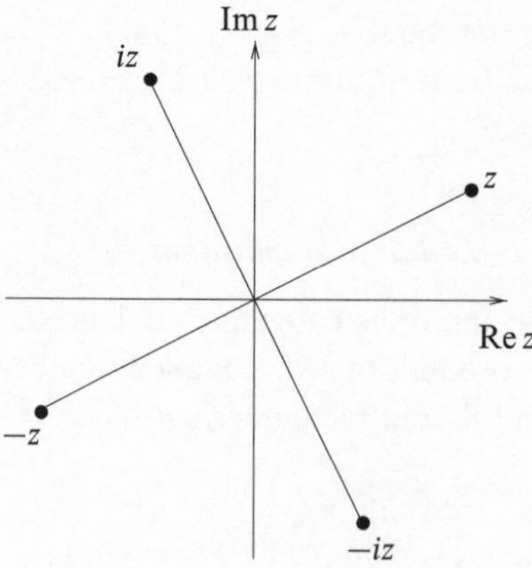


Figure 2.5 Multiplication of a complex number by $\pm 1$ and $\pm i$.

and hence

$$|z_1||z_2| = \sqrt{13}\sqrt{17} = \sqrt{221} = |z_1 z_2|. \blacktriangleleft$$

We now examine the effect on a complex number $z$ of multiplying it by $\pm 1$ and $\pm i$. These four multipliers have modulus unity and we can immediately see from (2.10) that multiplying $z$ by another complex number of unit modulus leaves the modulus of $z$ unchanged. We can also see from (2.11) that if we multiply by a complex number, the argument of the product is the sum of the argument of $z$ and the argument of the multiplier. Hence multiplying $z$ by unity (which has an argument of zero) leaves $z$ unchanged in both modulus and argument, i.e. $z$ is completely unaltered by the operation. Multiplying by $-1$ (which has argument $\pi$) leads to a rotation, through an angle $\pi$, of the line joining the origin to $z$ in the Argand diagram. Similarly, multiplication by $i$ and $-i$ lead to corresponding rotations of $\pi/2$ and $-\pi/2$ respectively. This geometrical interpretation of multiplication is shown in figure 2.5.

►*Using the geometrical interpretation of multiplication by i, find the product $i(1-i)$.*

The complex number $1-i$ has argument $-\pi/4$ and modulus $\sqrt{2}$. Thus, using (2.10) and (2.11), its product with $i$ has argument $+\pi/4$ and unchanged modulus $\sqrt{2}$. The complex number with modulus $\sqrt{2}$ and argument $+\pi/4$ is $1+i$ and so

$$i(1-i) = 1 + i,$$

as is easily verified by direct multiplication. ◄

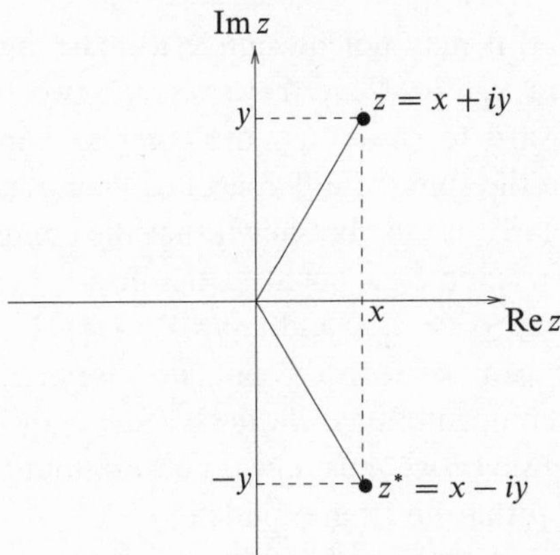


Figure 2.6 The complex conjugate as a mirror image in the real axis.

The division of two complex numbers is similar to their multiplication but requires the notion of the complex conjugate (see the following subsection) and so discussion is postponed until subsection 2.2.5.

### *2.2.4 Complex conjugate*

If $z$ has the convenient form $x + iy$ then the complex conjugate, denoted by $z^*$, may be found simply by changing the sign of the imaginary part, i.e. if $z = x + iy$ then $z^* = x - iy$. More generally, we may define the complex conjugate of $z$ as the (complex) number having the same magnitude as $z$ that when multiplied by $z$ leaves a real result, i.e. there is no imaginary component in the product.

In the case where $z$ can be written in the form $x + iy$ it is easily verified, by direct multiplication of the components, that the product $zz^*$ gives a real result:

$$zz^* = (x + iy)(x - iy) = x^2 - ixy + ixy - i^2y^2 = x^2 + y^2 = |z|^2.$$

Complex conjugation corresponds to a reflection of $z$ in the real axis of the Argand diagram, as may be seen in figure 2.6.

►*Find the complex conjugate of $z = a + 2i + 3ib$.*

The complex number is written in the standard form

$$z = a + i(2 + 3b),$$

and then replacing $i$ by $-i$ we obtain

$$z^* = a - i(2 + 3b). \blacktriangleleft$$

In some cases, however, it may not be simple to rearrange the expression for $z$ into the standard form $x + iy$. Nevertheless, given two complex numbers $z_1$ and $z_2$, it is straightforward to show that the complex conjugate of their sum (or difference) is equal to the sum (or difference) of their complex conjugates, i.e. $(z_1 \pm z_2)^* = z_1^* \pm z_2^*$. Similarly, it may be shown that the complex conjugate of the product (or quotient) of $z_1$ and $z_2$ is equal to the product (or quotient) of their complex conjugates, i.e. $(z_1 z_2)^* = z_1^* z_2^*$ and $(z_1/z_2)^* = z_1^*/z_2^*$.

Using these results, it can be deduced that, no matter how complicated the expression, its complex conjugate may *always* be found by replacing every $i$ by $-i$. To apply this rule, however, we must always ensure that all complex parts are first written out in full, so that no $i$'s are hidden.

►*Find the complex conjugate of the complex number* $z = w^{(3y+2ix)}$ *where* $w = x + 5i$.

Although we do not discuss complex powers until section 2.5, the simple rule given above still enables us to find the complex conjugate of $z$.

In this case $w$ itself contains real and imaginary components and so must be written out in full, i.e.

$$z = w^{3y+2ix} = (x + 5i)^{3y+2ix}.$$

Now we can replace each $i$ by $-i$ to obtain

$$z^* = (x - 5i)^{(3y-2ix)}.$$

It can be shown that the product $zz^*$ is real, as required. ◄

The following properties of the complex conjugate are easily proved and others may be derived from them. If $z = x + iy$ then

$$(z^*)^* = z, \tag{2.12}$$

$$z + z^* = 2\,\mathrm{Re}\,z = 2x, \tag{2.13}$$

$$z - z^* = 2i\,\mathrm{Im}\,z = 2iy, \tag{2.14}$$

$$\frac{z}{z^*} = \left(\frac{x^2 - y^2}{x^2 + y^2}\right) + i\left(\frac{2xy}{x^2 + y^2}\right). \tag{2.15}$$

The derivation of this last relation relies on the results of the following subsection.

### 2.2.5 *Division*

The division of two complex numbers $z_1$ and $z_2$ bears some similarity to their multiplication. Writing the quotient in component form we obtain

$$\frac{z_1}{z_2} = \frac{x_1 + iy_1}{x_2 + iy_2}. \tag{2.16}$$

In order to proceed, we wish to separate the real and imaginary components of the quotient. This may be achieved by multiplying both numerator and denominator by the complex conjugate of the denominator. This process will, by definition, leave the denominator as a real quantity. Equation (2.16) gives

$$\frac{z_1}{z_2} = \frac{(x_1 + iy_1)(x_2 - iy_2)}{(x_2 + iy_2)(x_2 - iy_2)} = \frac{(x_1x_2 + y_1y_2) + i(x_2y_1 - x_1y_2)}{x_2^2 + y_2^2}$$
$$= \frac{x_1x_2 + y_1y_2}{x_2^2 + y_2^2} + i\frac{x_2y_1 - x_1y_2}{x_2^2 + y_2^2}.$$

Hence we have separated the quotient into real and imaginary components.

In the special case where $z_2 = z_1^*$, so that $x_2 = x_1$ and $y_2 = -y_1$, the general result reduces to (2.15).

►*Find the quotient*

$$z = \frac{3 - 2i}{-1 + 4i}.$$

Multiplying numerator and denominator by the complex conjugate of the denominator we obtain

$$z = \frac{(3 - 2i)(-1 - 4i)}{(-1 + 4i)(-1 - 4i)} = \frac{-11 - 10i}{17}$$
$$= -\frac{11}{17} - \frac{10}{17}i. \blacktriangleleft$$

In analogy to (2.10) and (2.11), which describe the multiplication of two complex numbers, the following relations apply to division:

$$\left|\frac{z_1}{z_2}\right| = \frac{|z_1|}{|z_2|}, \qquad (2.17)$$

$$\arg\left(\frac{z_1}{z_2}\right) = \arg z_1 - \arg z_2. \qquad (2.18)$$

The proof of these relations is left until subsection 2.3.1.

## 2.3 Polar representation of complex numbers

Although considering a complex number as the sum of a real and an imaginary part is often useful, sometimes the *polar representation* proves easier to manipulate. This makes use of the complex exponential function, which is defined by

$$e^z = \exp z \equiv 1 + z + \frac{z^2}{2!} + \frac{z^3}{3!} + \cdots. \qquad (2.19)$$

Strictly speaking it is the function $\exp z$ that is defined by (2.19). The number $e$ is the value of $\exp(1)$, i.e. it is just a number. However, it may be shown that $e^z$

and $\exp z$ are equivalent when $z$ is real and rational and mathematicians then *define* their equivalence for irrational and complex $z$. For the purposes of this book we will not concern ourselves further with this mathematical nicety but, rather, assume that (2.19) is valid for all $z$. We also note that, using (2.19), by multiplying together the appropriate series we may show that (see chapter 18)

$$e^{z_1}e^{z_2} = e^{z_1+z_2}, \tag{2.20}$$

which is analogous to the familiar result for exponentials of real numbers.

From (2.19), it immediately follows that for $z = i\theta$, $\theta$ real,

$$e^{i\theta} = 1 + i\theta - \frac{\theta^2}{2!} - \frac{i\theta^3}{3!} + \cdots \tag{2.21}$$

$$= 1 - \frac{\theta^2}{2!} + \frac{\theta^4}{4!} - \cdots + i\left(\theta - \frac{\theta^3}{3!} + \frac{\theta^5}{5!} - \cdots\right), \tag{2.22}$$

and hence

$$e^{i\theta} = \cos\theta + i\sin\theta, \tag{2.23}$$

where the last equality follows from the series expansion of trigonometrical functions (see subsection 3.6.3). This last relationship is called *Euler's equation.* It also follows from (2.23) that

$$e^{in\theta} = \cos n\theta + i\sin n\theta$$

for all $n$. From Euler's equation (2.23), and using figure 2.7, we deduce that

$$\begin{aligned} re^{i\theta} &= r(\cos\theta + i\sin\theta) \\ &= x + iy. \end{aligned}$$

Thus a complex number may be represented in the polar form

$$z = re^{i\theta}. \tag{2.24}$$

Referring to figure 2.7, we can identify $r$ with $|z|$ and $\theta$ with $\arg z$. The simplicity of the representation of the modulus and the argument is one of the main reasons for using the polar representation. The angle $\theta$ is conventionally in the range $-\pi < \theta \leq \pi$, but, since rotation by $\theta$ is the same as rotation by $2n\pi + \theta$, where $n$ is any integer,

$$re^{i\theta} \equiv re^{i(\theta+2n\pi)}.$$

The algebra of the polar representation is different from that of the real and imaginary component representation, though, of course, the results are identical. Some operations prove much easier in the polar representation, others much more complicated. The best representation for a particular problem must be determined by the manipulation required.

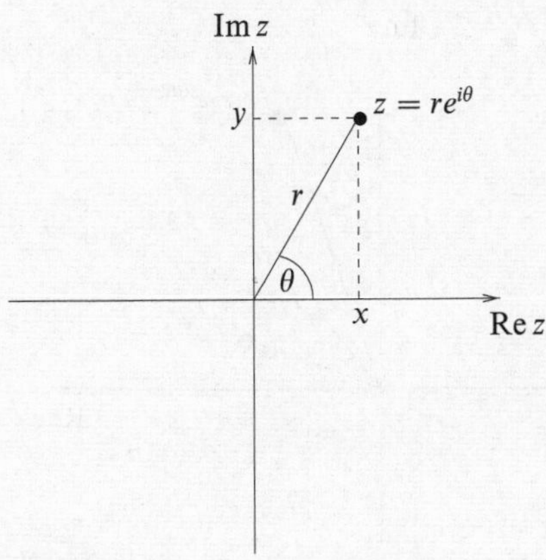


Figure 2.7 The polar representation of a complex number.

### 2.3.1 *Multiplication and division in polar form*

Multiplication and division in polar form are particularly simple. The product of $z_1 = r_1 e^{i\theta_1}$ and $z_2 = r_2 e^{i\theta_2}$ is given by

$$\begin{aligned} z_1 z_2 &= r_1 e^{i\theta_1} r_2 e^{i\theta_2} \\ &= r_1 r_2 e^{i(\theta_1+\theta_2)}. \end{aligned} \tag{2.25}$$

The relations $|z_1 z_2| = |z_1||z_2|$ and $\arg(z_1 z_2) = \arg z_1 + \arg z_2$ follow immediately. An example of the multiplication of two complex numbers is shown in figure 2.8.

Division is equally simple in polar form; the quotient of $z_1$ and $z_2$ is given by

$$\frac{z_1}{z_2} = \frac{r_1 e^{i\theta_1}}{r_2 e^{i\theta_2}} = \frac{r_1}{r_2} e^{i(\theta_1-\theta_2)}. \tag{2.26}$$

The relations $|z_1/z_2| = |z_1|/|z_2|$ and $\arg(z_1/z_2) = \arg z_1 - \arg z_2$ are again immediately apparent. The division of two complex numbers in polar form is shown in figure 2.9.

## 2.4 de Moivre's theorem

We now derive an extremely important theorem. Since $\left(e^{i\theta}\right)^n = e^{in\theta}$, we have

$$(\cos\theta + i\sin\theta)^n = \cos n\theta + i\sin n\theta, \tag{2.27}$$

where the identity $e^{in\theta} = \cos n\theta + i\sin n\theta$ follows from the series definition of $e^{in\theta}$ (see (2.21)). This result is called *de Moivre's theorem* and is often used in the manipulation of complex numbers. The theorem is valid for all $n$ whether real, imaginary or complex.

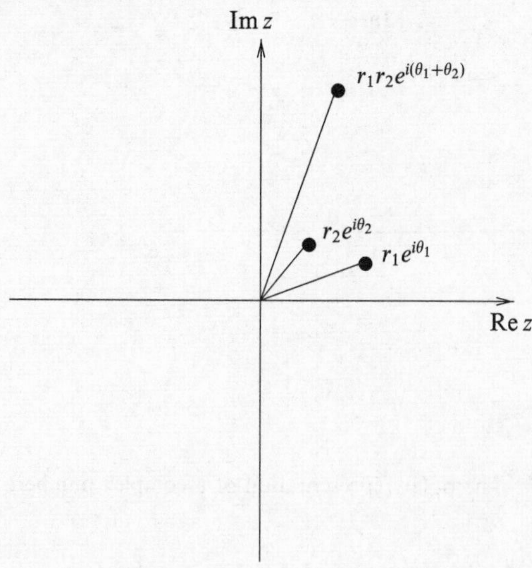


Figure 2.8 The multiplication of two complex numbers. In this case $r_1$ and $r_2$ are both greater than unity.

There are numerous applications of de Moivre's theorem but this section examines just three: proofs of trigonometrical identities; finding the $n$th roots of unity; and solving complex equations.

### 2.4.1 Trigonometrical identities

The use of de Moivre's theorem in finding trigonometrical identities is best illustrated by example. We consider the expression of a multiple-angle function in terms of a polynomial in the single-angle function, and its converse.

►*Express* $\sin 3\theta$ *and* $\cos 3\theta$ *in terms of powers of* $\cos\theta$ *and* $\sin\theta$.

Using de Moivre's theorem,

$$\begin{aligned}\cos 3\theta + i\sin 3\theta &= (\cos\theta + i\sin\theta)^3\\ &= (\cos^3\theta - 3\cos\theta\sin^2\theta) + i(3\sin\theta\cos^2\theta - \sin^3\theta). \qquad (2.28)\end{aligned}$$

We can equate the real and imaginary coefficients separately, i.e.

$$\begin{aligned}\cos 3\theta &= \cos^3\theta - 3\cos\theta\sin^2\theta\\ &= 4\cos^3\theta - 3\cos\theta \qquad (2.29)\end{aligned}$$

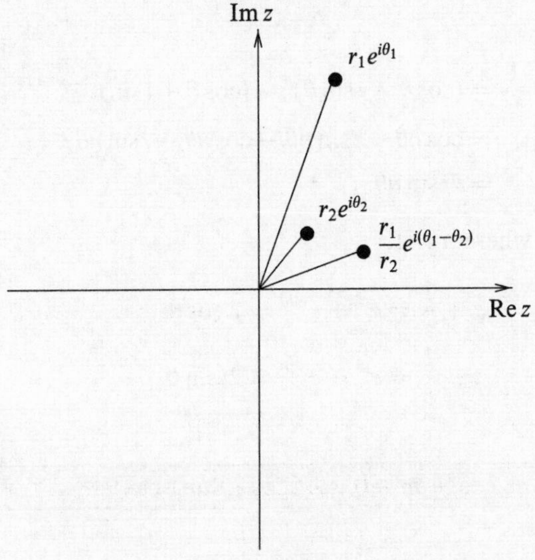


Figure 2.9 The division of two complex numbers. As in the previous figure, $r_1$ and $r_2$ are both greater than unity.

and

$$\begin{aligned}\sin 3\theta &= 3\sin\theta\cos^2\theta - \sin^3\theta \\ &= 3\sin\theta - 4\sin^3\theta. \blacktriangleleft\end{aligned}$$

This method can clearly be applied to finding power expansions of $\cos n\theta$ and $\sin n\theta$ for any positive integer $n$.

The converse process uses the following properties of $z = e^{i\theta}$,

$$z^n + \frac{1}{z^n} = 2\cos n\theta, \tag{2.30}$$

$$z^n - \frac{1}{z^n} = 2i\sin n\theta. \tag{2.31}$$

These equalities follow from simple applications of de Moivre's theorem, i.e.

$$\begin{aligned}z^n + \frac{1}{z^n} &= (\cos\theta + i\sin\theta)^n + (\cos\theta + i\sin\theta)^{-n} \\ &= \cos n\theta + i\sin n\theta + \cos(-n\theta) + i\sin(-n\theta) \\ &= \cos n\theta + i\sin n\theta + \cos n\theta - i\sin n\theta \\ &= 2\cos n\theta,\end{aligned}$$

and

$$\begin{aligned} z^n - \frac{1}{z^n} &= (\cos\theta + i\sin\theta)^n - (\cos\theta + i\sin\theta)^{-n} \\ &= \cos n\theta + i\sin n\theta - \cos n\theta + i\sin n\theta \\ &= 2i\sin n\theta. \end{aligned}$$

In the particular case where $n = 1$,

$$z + \frac{1}{z} = e^{i\theta} + e^{-i\theta} = 2\cos\theta, \tag{2.32}$$

$$z - \frac{1}{z} = e^{i\theta} - e^{-i\theta} = 2i\sin\theta. \tag{2.33}$$

►*Find an expression for* $\cos^3\theta$ *in terms of* $\cos 3\theta$ *and* $\cos\theta$.

Using (2.32),

$$\begin{aligned} \cos^3\theta &= \frac{1}{2^3}\left(z + \frac{1}{z}\right)^3 \\ &= \frac{1}{8}\left(z^3 + 3z + \frac{3}{z} + \frac{1}{z^3}\right) \\ &= \frac{1}{8}\left(z^3 + \frac{1}{z^3}\right) + \frac{3}{8}\left(z + \frac{1}{z}\right). \end{aligned}$$

Now using (2.30) and (2.32), we find

$$\cos^3\theta = \tfrac{1}{4}\cos 3\theta + \tfrac{3}{4}\cos\theta. \blacktriangleleft$$

This result happens to be a simple rearrangement of (2.29), but cases involving larger values of $n$ are better handled using this direct method than by rearranging polynomial expansions of multiple-angle functions.

### 2.4.2 Finding the nth roots of unity

The equation $z^2 = 1$ has the familiar solutions $z = \pm 1$. However, now that we have introduced the concept of complex numbers, we can solve the general equation $z^n = 1$. Recalling the fundamental theorem of algebra, we know that the equation has $n$ solutions. In order to proceed we rewrite the equation as

$$z^n = e^{2ik\pi},$$

where $k$ is any integer. Now taking the $n$th root of each side of the equation we find

$$z = e^{2ik\pi/n}.$$

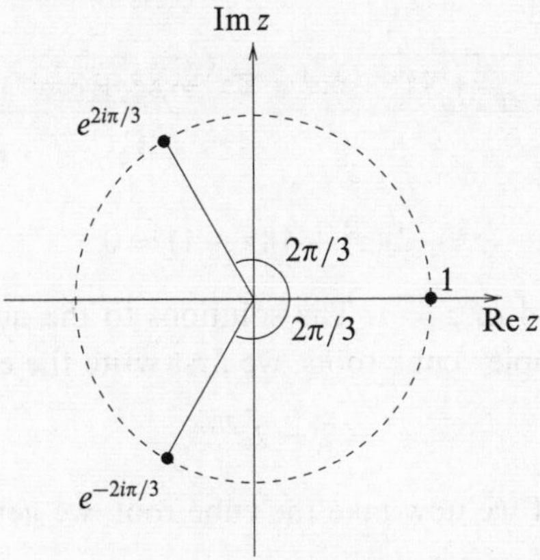


Figure 2.10 The solutions of $z^3 = 1$.

Hence, the solutions of $z^n = 1$ are

$$z_{1,2,\ldots,n} = 1, \quad e^{2i\pi/n}, \quad \ldots, \quad e^{2i(n-1)\pi/n},$$

corresponding to the values $0, 1, 2, \ldots, (n-1)$ for $k$. Larger integer values of $k$ do not give new solutions, since the listed roots are simply cyclically repeated for $k = n, n+1, n+2$, etc.

►*Find the solutions to the equation $z^3 = 1$.*

By applying the above method we find

$$z = e^{2ik\pi/3}.$$

Hence the three solutions are $z_1 = e^{0i} = 1$, $z_2 = e^{2i\pi/3}$, $z_3 = e^{4i\pi/3}$. We note that, as expected, the next solution, for which $k = 3$, gives $z_4 = e^{6i\pi/3} = 1 = z_1$, so that there are only three separate solutions.◄

Not surprisingly, given that $|z^3| = |z|^3$ from (2.10), all the roots of unity have unit modulus, i.e. they all lie on a circle in the Argand diagram of unit radius. The three roots are shown in figure 2.10.

The cube roots of unity are often written 1, $\omega$ and $\omega^2$. The properties $\omega^3 = 1$ and $1 + \omega + \omega^2 = 0$ are easily proved.

### *2.4.3 Solving polynomial equations*

A third application of de Moivre's theorem is in solving polynomial equations. Complex equations in the form of a polynomial relationship must first be solved for $z$ in a similar fashion to the method for finding the roots of real polynomial equations. Then the complex roots of $z$ may be found.

►*Solve the equation* $z^6 - z^5 + 4z^4 - 6z^3 + 2z^2 - 8z + 8 = 0$.

We first factorise to give

$$(z^3 - 2)(z^2 + 4)(z - 1) = 0.$$

Hence $z^3 = 2$ or $z^2 = -4$ or $z = 1$. The solutions to the quadratic equation are $z = \pm 2i$; to find the complex cube roots, we first write the equation in the form

$$z^3 = 2 = 2e^{2ik\pi},$$

where $k$ is any integer. If we now take the cube root, we get

$$z = 2^{1/3}e^{2ik\pi/3}.$$

To avoid the duplication of solutions, we use the fact that $-\pi < \arg z \leq \pi$, and find

$$z_1 = 2^{1/3},$$

$$z_2 = 2^{1/3}e^{2\pi i/3} = 2^{1/3}\left(-\frac{1}{2} + \frac{\sqrt{3}}{2}i\right),$$

$$z_3 = 2^{1/3}e^{-2\pi i/3} = 2^{1/3}\left(-\frac{1}{2} - \frac{\sqrt{3}}{2}i\right).$$

The complex numbers $z_1$, $z_2$ and $z_3$, together with $z_4 = 2i$, $z_5 = -2i$ and $z_6 = 1$ are the solutions to the original polynomial equation.

As expected from the fundamental theorem of algebra, we find that the total number of complex roots (six, in this case) is equal to the largest power of $z$ in the polynomial. ◄

A useful result is that the roots of a polynomial with real coefficients occur in conjugate pairs (i.e. if $z_1$ is a root, then $z_1^*$ is a second distinct root, unless $z_1$ is real). This may be proved as follows. Let the polynomial equation of which $z$ is a root be

$$a_n z^n + a_{n-1}z^{n-1} + \cdots + a_1 z + a_0 = 0.$$

Taking the complex conjugate of this equation,

$$a_n^*(z^*)^n + a_{n-1}^*(z^*)^{n-1} + \cdots + a_1^* z^* + a_0^* = 0.$$

But the $a_n$ are real, and so $z^*$ satisfies

$$a_n(z^*)^n + a_{n-1}(z^*)^{n-1} + \cdots + a_1 z^* + a_0 = 0,$$

and is also a root of the original equation.

## 2.5 Complex logarithms and complex powers

The concept of a complex exponential has already been introduced in section 2.3 where it was assumed that the definition of an exponential as a series was valid for complex numbers as well as for real numbers. Similarly we can define the logarithm of a complex number and we can use complex numbers as exponents.

Let us denote the natural logarithm of a complex number $z$ by $w = \operatorname{Ln} z$, where the notation Ln will be explained shortly. Thus, $w$ must satisfy

$$z = e^w.$$

Using (2.20), we see that

$$z_1 z_2 = e^{w_1} e^{w_2} = e^{w_1 + w_2},$$

and taking logarithms of both sides we find

$$\operatorname{Ln}(z_1 z_2) = w_1 + w_2 = \operatorname{Ln} z_1 + \operatorname{Ln} z_2, \tag{2.34}$$

which shows that the familiar rule for the logarithm of the product of two real numbers also holds for complex numbers.

We may use (2.34) to investigate further the properties of $\operatorname{Ln} z$. We have already noted that the argument of a complex number is multivalued, i.e. $\arg z = \theta + 2n\pi$, where $n$ is any integer. Thus, in polar form, the complex number $z$ should strictly be written as

$$z = re^{i(\theta + 2n\pi)}.$$

Taking the logarithm of both sides, and using (2.34), we find

$$\operatorname{Ln} z = \ln r + i(\theta + 2n\pi), \tag{2.35}$$

where $\ln r$ is the natural logarithm of the real positive quantity $r$ and so is written normally. Thus from (2.35) we see that $\operatorname{Ln} z$ is itself multivalued. To avoid this multivalued behaviour it is conventional to define another function $\ln z$, the *principal value* of $\operatorname{Ln} z$, which is obtained from $\operatorname{Ln} z$ by restricting the argument of $z$ to lie in the range $-\pi < \theta \leq \pi$.

►*Evaluate* $\operatorname{Ln}(-i)$.

By rewriting $-i$ as a complex exponential, we find

$$\operatorname{Ln}(-i) = \operatorname{Ln}\left[e^{i(-\pi/2 + 2n\pi)}\right] = i(-\pi/2 + 2n\pi),$$

where $n$ is any integer. Hence $\operatorname{Ln}(-i) = -i\pi/2,\ 3i\pi/2,\ \ldots$. We note that $\ln z$, the principal value of $\operatorname{Ln} z$, is given by $\ln z = -i\pi/2$. ◄

If $z$ and $t$ are both complex numbers then the $z$th power of $t$ is defined by

$$t^z = e^{z \operatorname{Ln} t}.$$

Since $\operatorname{Ln} t$ is multivalued, so too is this definition.

►*Simplify the expression* $z = i^{-2i}$.

Firstly we take the logarithm of both sides of the equation to give

$$\mathrm{Ln}\, z = -2i\, \mathrm{Ln}\, i.$$

Now inverting the process we find

$$e^{\mathrm{Ln}\, z} = z = e^{-2i \mathrm{Ln}\, i}.$$

We can write $i = e^{i(\pi/2+2n\pi)}$, where $n$ is any integer, and hence

$$\begin{aligned}\mathrm{Ln}\, i &= \mathrm{Ln}\left[e^{i(\pi/2+2n\pi)}\right] \\ &= i\left(\pi/2 + 2n\pi\right).\end{aligned}$$

We can now simplify $z$ to give

$$\begin{aligned}i^{-2i} &= e^{-2i\times i(\pi/2+2n\pi)} \\ &= e^{(\pi+4n\pi)},\end{aligned}$$

which, perhaps surprisingly, is a real quantity rather than a complex one.◄

Complex powers and the logarithms of complex numbers are discussed further in chapter 18.

## 2.6 Applications to differentiation and integration

We can use the exponential form of a complex number together with de Moivre's theorem (see section 2.4) to simplify the differentiation of trigonometric functions.

►*Find the derivative with respect to* $x$ *of* $e^{3x}\cos 4x$.

We could differentiate this function straightforwardly using the product rule (see subsection 1.1.2). However, an alternative method in this case is to use the complex exponential. Let us consider the complex number

$$z = e^{3x}(\cos 4x + i\sin 4x) = e^{3x}e^{4ix} = e^{(3+4i)x},$$

where we have used de Moivre's theorem to rewrite the trigonometric functions as a complex exponential. This complex number has $e^{3x}\cos 4x$ as its real part. Now, differentiating $z$ with respect to $x$ we obtain

$$\frac{dz}{dx} = (3+4i)e^{(3+4i)x} = (3+4i)e^{3x}(\cos 4x + i\sin 4x), \qquad (2.36)$$

where we have again used de Moivre's theorem. Equating real parts we then find

$$\frac{d}{dx}\left(e^{3x}\cos 4x\right) = e^{3x}(3\cos 4x - 4\sin 4x).$$

By equating the imaginary parts of (2.36), we also obtain, as a bonus,

$$\frac{d}{dx}\left(e^{3x}\sin 4x\right) = e^{3x}(4\cos 4x + 3\sin 4x). \blacktriangleleft$$

In a similar way the complex exponential can be used to evaluate integrals containing trigonometric and exponential functions.

►*Evaluate the integral* $I = \int e^{ax}\cos bx\,dx$.

Let us consider the integrand as the real part of the complex number

$$e^{ax}(\cos bx + i\sin bx) = e^{ax}e^{ibx} = e^{(a+ib)x},$$

where we use de Moivre's theorem to rewrite the trigonometrical functions as a complex exponential. Integrating we find

$$\begin{aligned}\int e^{(a+ib)x}\,dx &= \frac{e^{(a+ib)x}}{a+ib} + c\\ &= \frac{(a-ib)e^{(a+ib)x}}{(a-ib)(a+ib)} + c\\ &= \frac{e^{ax}}{a^2+b^2}\left(ae^{ibx} - ibe^{ibx}\right) + c, \qquad (2.37)\end{aligned}$$

where the constant of integration $c$ is in general complex. Denoting this constant by $c = c_1 + ic_2$ and equating real parts in (2.37) we obtain

$$I = \int e^{ax}\cos bx\,dx = \frac{e^{ax}}{a^2+b^2}(a\cos bx + b\sin bx) + c_1,$$

which agrees with result (1.34) found using integration by parts. Equating imaginary parts in (2.37) we obtain, as a bonus,

$$J = \int e^{ax}\sin bx\,dx = \frac{e^{ax}}{a^2+b^2}(a\sin bx - b\cos bx) + c_2. \blacktriangleleft$$

## 2.7 Hyperbolic functions

The *hyperbolic functions* are the complex analogues of the trigonometrical functions. The analogy may not be immediately apparent and their definitions may appear at first to be somewhat arbitrary. However, careful examination of their properties reveals the purpose of the definitions. For instance, their close relationship with the trigonometrical functions, both in their identities and in their calculus, means that many of the familiar properties of trigonometrical functions can also be applied to the hyperbolic functions. Further, hyperbolic functions occur regularly, and so giving them a special name is a notational convenience.

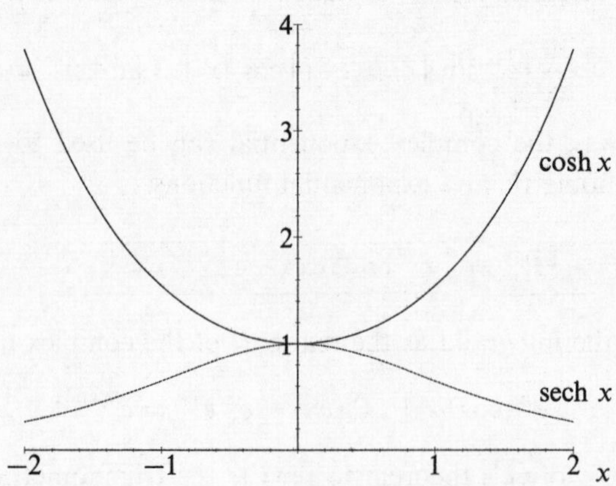


Figure 2.11 Graphs of $\cosh x$ and $\operatorname{sech} x$.

### 2.7.1 Definitions

The two fundamental hyperbolic functions are $\cosh x$ and $\sinh x$, which, as their names suggest, are the hyperbolic equivalents of $\cos x$ and $\sin x$. They are defined by the following relations:

$$\cosh x = \tfrac{1}{2}(e^x + e^{-x}), \tag{2.38}$$

$$\sinh x = \tfrac{1}{2}(e^x - e^{-x}). \tag{2.39}$$

Note that $\cosh x$ is an even function and $\sinh x$ is an odd function. By analogy with the trigonometrical functions, the remaining hyperbolic functions are

$$\tanh x = \frac{\sinh x}{\cosh x} = \frac{e^x - e^{-x}}{e^x + e^{-x}}, \tag{2.40}$$

$$\operatorname{sech} x = \frac{1}{\cosh x} = \frac{2}{e^x + e^{-x}}, \tag{2.41}$$

$$\operatorname{cosech} x = \frac{1}{\sinh x} = \frac{2}{e^x - e^{-x}}, \tag{2.42}$$

$$\coth x = \frac{1}{\tanh x} = \frac{e^x + e^{-x}}{e^x - e^{-x}}. \tag{2.43}$$

All the hyperbolic functions above have been defined in terms of the real variable $x$. However, this was simply so that they may be plotted (see figures 2.11–2.13) and the definitions are equally valid for any complex number $z$.

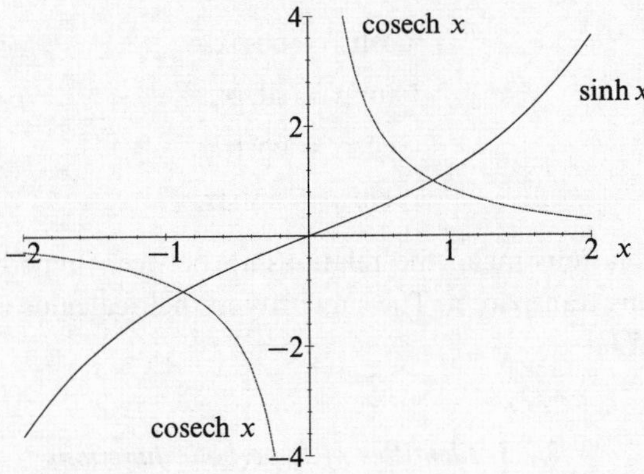


Figure 2.12 Graphs of $\sinh x$ and $\operatorname{cosech} x$.

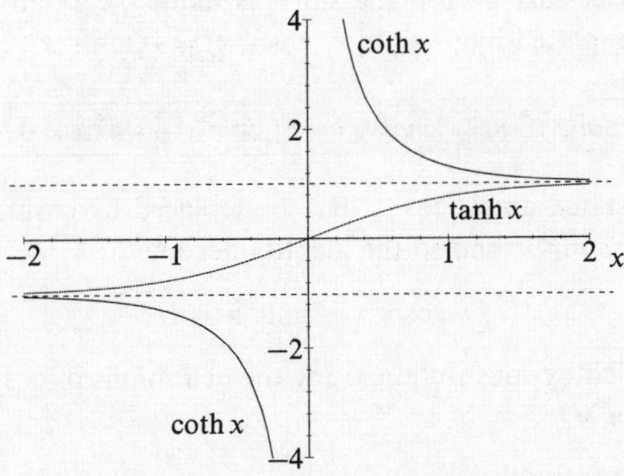


Figure 2.13 Graphs of $\tanh x$ and $\coth x$.

### 2.7.2 Hyperbolic–trigonometric analogies

In the previous subsections we have alluded to the analogy between trigonometrical and hyperbolic functions. Here, we discuss the close relationship between the two groups of functions.

Recalling (2.32) and (2.33) we find

$$\cos ix = \tfrac{1}{2}(e^x + e^{-x}),$$
$$\sin ix = \tfrac{1}{2}i(e^x - e^{-x}).$$

Hence, by the definitions given in the previous subsection,

$$\cosh x = \cos ix, \tag{2.44}$$

$$i \sinh x = \sin ix, \tag{2.45}$$

$$\cos x = \cosh ix, \tag{2.46}$$

$$i \sin x = \sinh ix. \tag{2.47}$$

These useful equations make the relationship between hyperbolic and trigonometrical functions transparent. The similarity in their calculus is discussed further in subsection 2.7.6.

### *2.7.3 Identities of hyperbolic functions*

The analogies between trigonometrical function and hyperbolic functions having been established, we should not be surprised that all the trigonometrical identities also hold for hyperbolic functions, with the following modification. Wherever $\sin^2 x$ occurs, it must be replaced by $-\sinh^2 x$ and vice versa. Note that this replacement is necessary even if the $\sin^2 x$ is hidden, e.g. $\tan^2 x = \sin^2 x/\cos^2 x$ and so must be replaced by $(-\sinh^2 x/\cosh^2 x) = -\tanh^2 x$.

►*Find the hyperbolic identity analogous to* $\cos^2 x + \sin^2 x = 1$.

Using the rules stated above $\cos^2 x$ may be replaced by $\cosh^2 x$, and $\sin^2 x$ may be replaced by $-\sinh^2 x$, and so the identity becomes

$$\cosh^2 x - \sinh^2 x = 1.$$

This can be verified by substitution, using the definitions of $\cosh x$ and $\sinh x$ (see (2.38) and (2.39)). ◄

Some other identities that can be proved in a similar way are

$$\operatorname{sech}^2 x = 1 - \tanh^2 x, \tag{2.48}$$

$$\operatorname{cosech}^2 x = \coth^2 x - 1, \tag{2.49}$$

$$\sinh 2x = 2 \sinh x \cosh x, \tag{2.50}$$

$$\cosh 2x = \cosh^2 x + \sinh^2 x. \tag{2.51}$$

### *2.7.4 Solving hyperbolic equations*

When we are presented with a hyperbolic equation to solve, we may proceed by analogy with the solution of trigonometrical equations. However, it is almost always easier to express the equation directly in terms of exponentials.

►*Solve the hyperbolic equation* $\cosh x - 5\sinh x - 5 = 0$.

Substituting the definitions of the hyperbolic functions we obtain

$$\tfrac{1}{2}(e^x + e^{-x}) - \tfrac{5}{2}(e^x - e^{-x}) - 5 = 0.$$

Rearranging, and then multiplying through by $-e^x$, gives in turn

$$-2e^x + 3e^{-x} - 5 = 0$$

and

$$2e^{2x} + 5e^x - 3 = 0.$$

Now we can factorise and solve:

$$(2e^x - 1)(e^x + 3) = 0.$$

Thus $e^x = 1/2$ or $e^x = -3$. Hence $x = -\ln 2$ or $x = \ln(-3)$. The interpretation of the logarithm of a negative number has been discussed in section 2.5. ◄

### 2.7.5 *Inverses of hyperbolic functions*

Just like trigonometrical functions, hyperbolic functions have inverses. If $y = \cosh x$ then $x = \cosh^{-1} y$, which serves as a definition of the inverse. By using the fundamental definitions of hyperbolic functions, we can find a closed-form expression for the inverse of hyperbolic functions. This is best illustrated by example.

►*Find a closed-form expression for the inverse hyperbolic function* $y = \sinh^{-1} x$.

First we write $x$ as a function of $y$, i.e.

$$y = \sinh^{-1} x \quad \Rightarrow \quad x = \sinh y.$$

Now, since $\cosh y = \frac{1}{2}(e^y + e^{-y})$ and $\sinh y = \frac{1}{2}(e^y - e^{-y})$,

$$\begin{aligned} e^y &= \cosh y + \sinh y \\ &= \sqrt{1 + \sinh^2 y} + \sinh y \\ e^y &= \sqrt{1 + x^2} + x, \end{aligned}$$

and hence

$$y = \ln(\sqrt{1 + x^2} + x). \quad ◄$$

In a similar fashion it can be shown that

$$\cosh^{-1} x = \ln(\sqrt{x^2 - 1} + x).$$

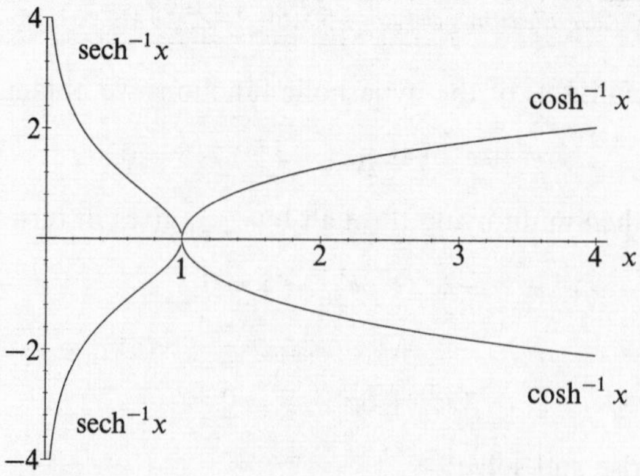


Figure 2.14 Graphs of $\cosh^{-1} x$ and $\text{sech}^{-1}x$.

►*Find a closed-form expression for the inverse hyperbolic function* $y = \tanh^{-1} x$.

First we write $x$ as a function of $y$, i.e.

$$y = \tanh^{-1} x \quad \Rightarrow \quad x = \tanh y.$$

Now, using the definition of $\tanh y$ and rearranging, we find

$$x = \frac{e^y - e^{-y}}{e^y + e^{-y}} \quad \Rightarrow \quad (x+1)e^{-y} = (1-x)e^y.$$

Thus

$$e^{2y} = \frac{1+x}{1-x} \;\Rightarrow\; e^y = \sqrt{\frac{1+x}{1-x}},$$
$$y = \ln\sqrt{\frac{1+x}{1-x}},$$
$$\tanh^{-1} x = \frac{1}{2}\ln\left(\frac{1+x}{1-x}\right). \blacktriangleleft$$

Graphs of the inverse hyperbolic functions are given in figures 2.14–2.16.

### 2.7.6 Calculus of hyperbolic functions

Just as the identities of hyperbolic functions closely follow those of their trigonometrical counterparts, so their calculus is similar. The derivatives of the two basic

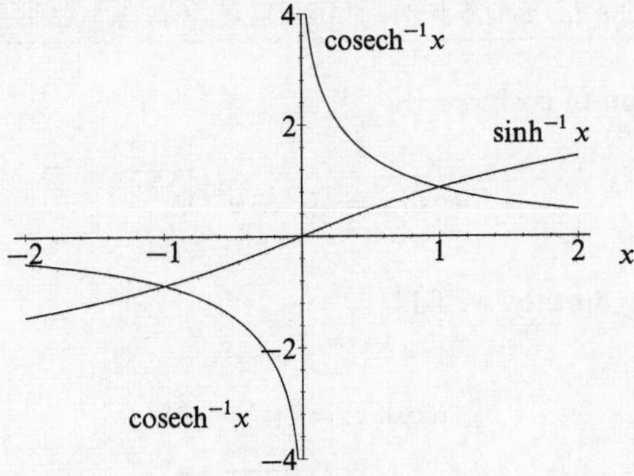


Figure 2.15 Graphs of $\sinh^{-1} x$ and $\operatorname{cosech}^{-1} x$.

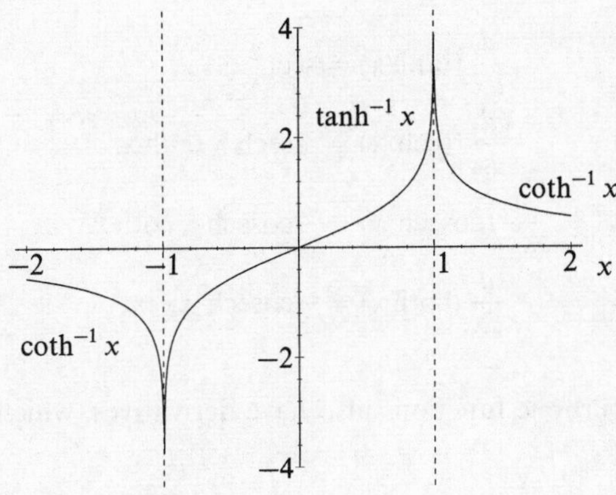


Figure 2.16 Graphs of $\tanh^{-1} x$ and $\coth^{-1} x$.

hyperbolic functions are

$$\frac{d}{dx}(\cosh x) = \sinh x, \tag{2.52}$$

$$\frac{d}{dx}(\sinh x) = \cosh x. \tag{2.53}$$

These may be deduced by considering the definitions.

►*Verify the relation* $(d/dx)\cosh x = \sinh x$.

Using the definition of $\cosh x$,

$$\cosh x = \tfrac{1}{2}(e^x + e^{-x}),$$

and differentiating directly, we find

$$\begin{aligned}\frac{d}{dx}(\cosh x) &= \tfrac{1}{2}(e^x - e^{-x})\\ &= \sinh x.\end{aligned}$$ ◄

Clearly the integrals of the fundamental hyperbolic functions are also defined by these relationships. The derivatives of the remaining hyperbolic functions can be derived by product differentiation and are presented below only for completeness.

$$\frac{d}{dx}(\tanh x) = \text{sech}^2 x, \tag{2.54}$$

$$\frac{d}{dx}(\text{sech}\, x) = -\text{sech}\, x \tanh x, \tag{2.55}$$

$$\frac{d}{dx}(\text{cosech}\, x) = -\text{cosech}\, x \coth x, \tag{2.56}$$

$$\frac{d}{dx}(\coth x) = -\text{cosech}^2 x. \tag{2.57}$$

The inverse hyperbolic functions also have derivatives, which are given by the following:

$$\frac{d}{dx}\left(\cosh^{-1}\frac{x}{a}\right) = \frac{1}{\sqrt{x^2 - a^2}}, \tag{2.58}$$

$$\frac{d}{dx}\left(\sinh^{-1}\frac{x}{a}\right) = \frac{1}{\sqrt{x^2 + a^2}}, \tag{2.59}$$

$$\frac{d}{dx}\left(\tanh^{-1}\frac{x}{a}\right) = \frac{a}{a^2 - x^2}, \quad \text{for } x^2 < a^2, \tag{2.60}$$

$$\frac{d}{dx}\left(\coth^{-1}\frac{x}{a}\right) = \frac{-a}{x^2 - a^2}, \quad \text{for } x^2 > a^2. \tag{2.61}$$

These may be derived from the logarithmic form of the inverse (see subsection 2.7.5).

►*Evaluate $(d/dx)\sinh^{-1} x$ using the logarithmic form of the inverse.*

From the results of section 2.7.5,

$$\begin{aligned}\frac{d}{dx}\left(\sinh^{-1} x\right) &= \frac{d}{dx}\left[\ln\left(x+\sqrt{x^2+1}\right)\right] \\ &= \frac{1}{x+\sqrt{x^2+1}}\left(1+\frac{x}{\sqrt{x^2+1}}\right) \\ &= \frac{1}{x+\sqrt{x^2+1}}\left(\frac{\sqrt{x^2+1}+x}{\sqrt{x^2+1}}\right) \\ &= \frac{1}{\sqrt{x^2+1}}. \blacktriangleleft\end{aligned}$$

## 2.8 Exercises

2.1 Two complex numbers $z$ and $w$ are given by $z = 3 + 4i$ and $w = 2 - i$. On an Argand diagram plot

(a) $z + w$, (b) $w - z$, (c) $wz$, (d) $z/w$,
(e) $z^*w + w^*z$, (f) $w^2$, (g) $\ln z$, (h) $(1 + z + w)^{1/2}$.

2.2 By considering the real and imaginary parts of the product $e^{i\theta}e^{i\phi}$ prove the standard formulae for $\cos(\theta + \phi)$ and $\sin(\theta + \phi)$.

2.3 Evaluate

(a) $\text{Re}(\exp 2iz)$, (b) $\text{Im}(\cosh^2 z)$, (c) $(-1 + \sqrt{3}i)^{1/2}$,
(d) $|\exp(i^{1/2})|$, (e) $\exp(i^3)$, (f) $\text{Im}(2^{i+3})$, (g) $i^i$, (h) $\ln[(\sqrt{3} + i)^3]$.

2.4 Sketch the parts of the Argand diagram in which

(a) $\text{Re}(z^2) < 0$, $|z^{1/2}| \leq 2$,
(b) $0 \leq \arg z^* \leq \pi/2$,
(c) $|\exp(z^3)| \to 0$ as $|z| \to \infty$.

What is the area of the region in which all three conditions are satisfied?

2.5 Solve the equation

$$z^7 - 4z^6 + 6z^5 - 6z^4 + 6z^3 - 12z^2 + 8z + 4 = 0,$$

(a) by examining the effect of setting $z^3$ equal to 2, and
(b) by factorising and using the binomial expansion of $(z + a)^4$.

Plot the seven roots of the equation on an Argand plot, verifying that complex roots occur in conjugate pairs.

2.6 Use de Moivre's theorem with $n = 4$ to prove that

$$\cos 4\theta = 8\cos^4\theta - 8\cos^2\theta + 1,$$

and deduce that

$$\cos\frac{\pi}{8} = \left(\frac{2+\sqrt{2}}{4}\right)^{1/2}.$$

2.7 Express $\sin^4\theta$ entirely in terms of the trigonometric functions of multiple angles and deduce that its average value over a complete cycle is $3/8$.

2.8 (a) Prove that

$$\cosh x - \cosh y = 2\sinh\left(\frac{x+y}{2}\right)\sinh\left(\frac{x-y}{2}\right).$$

(b) Prove that, if $y = \sinh^{-1} x$,

$$(x^2+1)\frac{d^2y}{dx^2} + x\frac{dy}{dx} = 0.$$

2.9 In the theory of special relativity, the position coordinate $x$ and time coordinate $t$ of an event as measured in one frame of reference are related to the coordinates $x'$ and $t'$ measured in another frame by equations of the form

$$x' = x\cosh\phi - ct\sinh\phi,$$
$$ct' = -x\sinh\phi + ct\cosh\phi.$$

Express $x$ and $ct$ in terms of $x'$, $ct'$ and $\phi$ and show that

$$x^2 - (ct)^2 = (x')^2 - (ct')^2.$$

2.10 A closed barrel has as its curved surface that obtained by rotating about the $x$-axis the part of the curve

$$y = a[2 - \cosh(x/a)]$$

lying in the range $-b \leq x \leq b$. Show that the total surface area $A$ of the barrel is given by

$$A = \pi a[9a - 8a\exp(-b/a) + a\exp(-2b/a) - 2b].$$

2.11 The principal value of the logarithmic function of a complex variable is defined to have its argument in the range $-\pi < \arg z \leq \pi$. By writing $z = \tan w$ in terms of exponentials show that

$$\tan^{-1} z = \frac{1}{2i}\ln\left(\frac{1+iz}{1-iz}\right).$$

Use this result to evaluate

$$\tan^{-1}\left(\frac{2\sqrt{3}-3i}{7}\right).$$

## 2.9 Hints and answers

2.1 (a) $5+3i$; (b) $-1-5i$; (c) $10+5i$; (d) $2/5+11i/5$; (e) 4; (f) $3-4i$; (g) $5+i[\tan^{-1}(4/3)+2n\pi]$; (h) $\pm(2.521+0.595i)$.

2.3 (a) $\exp(-2y)\cos 2x$; (b) $(\sin 2y \sinh 2x)/2$; (c) $\sqrt{2}\exp(\pi i/3)$ or $\sqrt{2}\exp(4\pi i/3)$; (d) $\exp(1/\sqrt{2})$ or $\exp(-1/\sqrt{2})$; (e) $0.540-0.841i$; (f) $8\sin(\ln 2)=5.11$; (g) $\exp(-\pi/2-2\pi n)$; (h) $\ln 8+i(2n+1/2)\pi$.

2.4 All three conditions are satisfied in $3\pi/2 \le \theta \le 7\pi/4$, $|z| \le 4$; area $=2\pi$.

2.5 $2^{1/3}\exp(2\pi ni/3)$ for $n=0,1,2$; $1\pm 3^{1/4}$; $1\pm 3^{1/4}i$.

2.7 $(\cos 4\theta)/8-(\cos 2\theta)/2+3/8$.

2.9 The same expressions but with $\phi$ replaced by $-\phi$ are obtained.

2.10 Show that $ds=(\cosh x/a)\,dx$; curved surface area $=\pi a^2[8\sinh(b/a)-\sinh(2b/a)]-2\pi ab$.

2.11 $\pi/6-i\ln\sqrt{2}$.

# 3

# *Series and limits*

## 3.1 Series

Many examples exist in the physical sciences of situations where we are presented with a *sum of terms* to evaluate. For example, we may wish to add the contributions from successive slits in a diffraction grating to find the total light intensity at a particular point behind the grating.

A series may have either a finite or infinite number of terms. In either case, the sum of the first $N$ terms of a series (often called a partial sum) is written

$$S_N = u_1 + u_2 + u_3 + \cdots + u_N,$$

where the terms of the series $u_n$, $n = 1, 2, \ldots, N$ are numbers, that may in general be complex. If the terms are complex then $S_N$ will in general be complex also, and we can write $S_N = X_N + iY_N$, where $X_N$ and $Y_N$ are the partial sums of the real and imaginary parts of each term separately, and are therefore real. If a series has only $N$ terms, then the partial sum $S_N$ is of course the sum of the series. Sometimes we may encounter series where each term depends on some variable, $x$, say. In this case the partial sum of the series will depend on the value assumed by $x$. For example, consider the infinite series

$$S(x) = 1 + x + \frac{x^2}{2!} + \frac{x^3}{3!} + \cdots .$$

This is an example of a power series; these are discussed in more detail in section 3.5. It is in fact the Maclaurin expansion of $\exp x$ (see subsection 3.6.3). Therefore $S(x) = \exp x$ and, of course, varies according to the value of the variable $x$. A series might just as easily depend on a complex variable $z$.

A general, random sequence of numbers can be described as a series and a sum of the terms found. However, for practical cases of interest, there will usually be some sort of relationship between successive terms. For example, if the $n$th term of a series is given by

$$u_n = \frac{1}{2^n},$$

for $n = 1, 2, 3, \ldots, N$, then the sum of the first $N$ terms will be

$$S_N = \sum_{n=1}^{N} u_n = \frac{1}{2} + \frac{1}{4} + \frac{1}{8} + \cdots + \frac{1}{2^N}. \tag{3.1}$$

It is clear that the sum of a finite number of terms is always finite, provided that each term is itself finite. It is often of practical interest, however, to consider the sum of a series with an infinite number of finite terms. The sum of an infinite number of terms is best defined by first considering the partial sum of the first $N$ terms, $S_N$. If the value of the partial sum $S_N$ tends to a finite limit, $S$, as $N$ tends to infinity, then the series is said to converge, and its sum is given by the limit $S$. In other words, the sum of an infinite series is given by

$$S = \lim_{N\to\infty} S_N,$$

provided the limit exists. For complex infinite series, if $S_N$ approaches a limit $S = X + iY$ as $N \to \infty$, this means that $X_N \to X$ and $Y_N \to Y$ separately, i.e. the real and imaginary parts of the series are each convergent series with sums $X$ and $Y$ respectively.

However, not all infinite series have finite sums. As $N \to \infty$, the value of the partial sum $S_N$ may diverge: it may approach $+\infty$ or $-\infty$, or oscillate finitely or infinitely. Moreover, for a series where each term depends on some variable, its convergence can depend on the value assumed by the variable. Whether an infinite series converges, diverges or oscillates has important implications when describing physical systems. Methods for determining whether a series converges are discussed in section 3.3.

## 3.2 Summation of series

It is often necessary to find the sum of a finite series or a convergent infinite series. We now describe arithmetic, geometric and arithmetico-geometric series, which are particularly common, and for which the sums are easily found. Other methods that may sometimes be used to sum more complicated series are also discussed below.

### 3.2.1 Arithmetic series

An *arithmetic series* has the characteristic that the difference between successive terms is constant. The sum of an arithmetic series is in general written

$$S_N = a + (a+d) + (a+2d) + \cdots + [a+(N-1)d] = \sum_{n=0}^{N-1}(a+nd).$$

Rewriting the series in the opposite order and adding this term by term to the original expression for $S_N$, we find

$$S_N = \frac{N}{2}\,[a + a + (N-1)d] = \frac{N}{2}(\text{first term} + \text{last term}). \qquad (3.2)$$

If an infinite number of terms are added the series will increase (or decrease) indefinitely; that is to say, it diverges.

►*Sum the integers between* 1 *and* 1000 *inclusive.*

This is an arithmetic series with $a = 1$, $d = 1$ and $N = 1000$. Therefore, using (3.2) we find

$$S_N = \frac{1000}{2}(1 + 1000) = 500500,$$

which can be checked directly only with considerable effort. ◄

### 3.2.2 Geometric series

Equation (3.1) is a particular example of a *geometric series*, which has the characteristic that the ratio of successive terms is a constant (one-half in this case). The sum of a geometric series is in general written

$$S_N = a + ar + ar^2 + \cdots + ar^{N-1} = \sum_{n=0}^{N-1} ar^n,$$

where $a$ is a constant and $r$ is the ratio of successive terms, the *common ratio*. The sum may be evaluated by considering $S_N$ and $rS_N$:

$$S_N = a + ar + ar^2 + ar^3 + \cdots + ar^{N-1},$$
$$rS_N = ar + ar^2 + ar^3 + ar^4 + \cdots + ar^N.$$

If we now subtract the second equation from the first we obtain

$$(1-r)S_N = a - ar^N,$$

and hence

$$S_N = \frac{a(1-r^N)}{1-r}. \qquad (3.3)$$

For a series with an infinite number of terms and $|r| < 1$, we have $\lim_{N\to\infty} r^N = 0$, and the sum tends to the limit

$$S = \frac{a}{1-r}. \tag{3.4}$$

In (3.1), $r = \frac{1}{2}$, $a = \frac{1}{2}$, and so $S = 1$. For $|r| \geq 1$, however, the series either diverges or oscillates.

►*Consider a ball that drops from a height of* 27 m *and on each bounce retains only a third of its kinetic energy. Thus after one bounce it will return to a height of* 9 m*, after two bounces to* 3 m*, and so on. Find the total distance travelled between the first bounce and the Mth bounce.*

The total distance travelled between the first bounce and the $M$th bounce is given by the sum of $M-1$ terms:

$$S_{M-1} = 2\,(9 + 3 + 1 + \cdots) = 2\sum_{m=0}^{M-2} \frac{9}{3^m}$$

for $M > 1$, where the factor of two is included to allow for both the upward and downward journey. Inside the parentheses we clearly have a geometric series with first term 9 and common ratio $1/3$ and hence the distance is given by (3.3), i.e.

$$S_{M-1} = 2 \times \frac{9\left[1 - \left(\frac{1}{3}\right)^{M-1}\right]}{1 - \left(\frac{1}{3}\right)} = 27\left[1 - \left(\frac{1}{3}\right)^{M-1}\right],$$

where the number of terms $N$ in (3.3) has been replaced by $M-1$. ◄

### 3.2.3 Arithmetico-geometric series

An arithmetico-geometric series, as its name suggests, is a combined arithmetic and geometric series. It has the general form

$$S_N = a + (a+d)r + (a+2d)r^2 + \cdots + [a + (N-1)d]\,r^{N-1} = \sum_{n=0}^{N-1} (a + nd)r^n,$$

and can be summed, in a similar way to a pure geometric series, by multiplying by $r$ and subtracting the result from the original series to obtain

$$(1-r)S_N = a + rd + r^2 d + \cdots + r^{N-1} d - [a + (N-1)d]\,r^N.$$

Using the expression for the sum of a geometric series (3.3) and rearranging, we find

$$S_N = \frac{a - [a + (N-1)d]\,r^N}{1-r} + \frac{rd(1 - r^{N-1})}{(1-r)^2}.$$

For an infinite series with $|r| < 1$, $\lim_{N\to\infty} r^N = 0$ as in the previous subsection, and the sum tends to the limit

$$S = \frac{a}{1-r} + \frac{rd}{(1-r)^2}. \tag{3.5}$$

As in the case of a geometric series, if $|r| \geq 1$ then the series either diverges or oscillates.

►*Sum the series*

$$S = 2 + \frac{5}{2} + \frac{8}{2^2} + \frac{11}{2^3} + \cdots.$$

This is an infinite arithmetico-geometric series with $a = 2$, $d = 3$ and $r = 1/2$. Therefore, from (3.5), we obtain $S = 10$. ◄

### 3.2.4 *The difference method*

The difference method is sometimes useful in summing series that are more complicated than the examples discussed above. Let us consider the general series

$$\sum_{n=1}^{N} u_n = u_1 + u_2 + \cdots + u_N.$$

If the terms of the series, $u_n$, can be expressed in the form

$$u_n = f(n) - f(n-1)$$

for some function $f(n)$, then its (partial) sum is given by

$$S_N = \sum_{n=1}^{N} u_n = f(N) - f(0).$$

This can be shown as follows. The sum is given by

$$S_N = u_1 + u_2 + \cdots + u_N$$

and since $u_n = f(n) - f(n-1)$, it may be rewritten

$$S_N = [f(1) - f(0)] + [f(2) - f(1)] + \cdots + [f(N) - f(N-1)].$$

By cancelling terms we see that

$$S_N = f(N) - f(0).$$

►*Evaluate the sum*

$$\sum_{n=1}^{N} \frac{1}{n(n+1)}.$$

Using partial fractions we find

$$u_n = -\left(\frac{1}{n+1} - \frac{1}{n}\right).$$

Hence $u_n = f(n) - f(n-1)$ with $f(n) = -1/(n+1)$, and so the sum is given by

$$S_N = f(N) - f(0) = -\frac{1}{N+1} + 1 = \frac{N}{N+1}. \blacktriangleleft$$

The difference method may be easily extended to evaluate sums in which each term can be expressed in the form

$$u_n = f(n) - f(n-m), \tag{3.6}$$

where $m$ is an integer. By writing out the sum to $N$ terms with each term expressed in this form, and cancelling terms in pairs as before, we find

$$S_N = \sum_{k=1}^{m} f(N-k+1) - \sum_{k=1}^{m} f(1-k).$$

►*Evaluate the sum*

$$\sum_{n=1}^{N} \frac{1}{n(n+2)}.$$

Using partial fractions we find

$$u_n = -\left[\frac{1}{2(n+2)} - \frac{1}{2n}\right].$$

Hence $u_n = f(n) - f(n-2)$ with $f(n) = -1/[2(n+2)]$, and so the sum is given by

$$S_N = f(N) + f(N-1) - f(0) - f(-1) = \frac{3}{4} - \frac{1}{2}\left(\frac{1}{N+2} + \frac{1}{N+1}\right). \blacktriangleleft$$

In fact the difference method is quite flexible, and may be used to evaluate sums even when each term cannot be expressed as in (3.6). The method still relies, however, on being able to write $u_n$ in terms of a single function such that most terms in the sum cancel, leaving only a few terms at the beginning and the end. This is best illustrated by an example.

► *Evaluate the sum*

$$\sum_{n=1}^{N} \frac{1}{n(n+1)(n+2)}.$$

Using partial fractions we find

$$u_n = \frac{1}{2(n+2)} - \frac{1}{n+1} + \frac{1}{2n}.$$

Hence $u_n = f(n) - 2f(n-1) + f(n-2)$ with $f(n) = 1/[2(n+2)]$. If we write out the sum, expressing each term $u_n$ in this form, we find that most terms cancel and the sum is given by

$$S_N = f(N) - f(N-1) - f(0) + f(-1) = \frac{1}{4} + \frac{1}{2}\left(\frac{1}{N+2} - \frac{1}{N+1}\right). \blacktriangleleft$$

### 3.2.5 Series involving natural numbers

Series consisting of the natural numbers 1, 2, 3, ..., or the square or cube of these numbers, occur frequently and deserve a special mention. Let us first consider the sum of the first $N$ natural numbers,

$$S_N = 1 + 2 + 3 + \cdots + N = \sum_{n=1}^{N} n.$$

This is clearly an arithmetic series with first term $a = 1$ and common difference $d = 1$. Therefore, from (3.2), $S_N = \frac{1}{2}N(N+1)$.

Next, we consider the sum of the squares of the first $N$ natural numbers:

$$S_N = 1^2 + 2^2 + 3^2 + \ldots + N^2 = \sum_{n=1}^{N} n^2,$$

which may be evaluated using the difference method. The $n$th term in the series is $u_n = n^2$, which we need to express in the form $f(n) - f(n-1)$ for some function $f(n)$. Consider the function

$$f(n) = n(n+1)(2n+1) \quad \Rightarrow \quad f(n-1) = (n-1)n(2n-1).$$

For this function $f(n) - f(n-1) = 6n^2$, and so we can write

$$u_n = \tfrac{1}{6}[f(n) - f(n-1)].$$

Therefore, by the difference method,

$$S_N = \tfrac{1}{6}[f(N) - f(0)] = \tfrac{1}{6}N(N+1)(2N+1).$$

Finally, we calculate the sum of the cubes of the first $N$ natural numbers,

$$S_N = 1^3 + 2^3 + 3^3 + \cdots + N^3 = \sum_{n=1}^{N} n^3,$$

again using the difference method. Consider the function

$$f(n) = [n(n+1)]^2 \qquad \Rightarrow \qquad f(n-1) = [(n-1)n]^2,$$

for which $f(n) - f(n-1) = 4n^3$. Therefore we can write the general $n$th term of the series as

$$u_n = \tfrac{1}{4}[f(n) - f(n-1)],$$

and using the difference method we find

$$S_N = \tfrac{1}{4}[f(N) - f(0)] = \tfrac{1}{4}N^2(N+1)^2.$$

Note that this is the square of the sum of the natural numbers, i.e.

$$\sum_{n=1}^{N} n^3 = \left(\sum_{n=1}^{N} n\right)^2.$$

►*Sum the series*

$$\sum_{n=1}^{N} (n+1)(n+3).$$

The general $n$th term in this series is

$$u_n = (n+1)(n+3) = n^2 + 4n + 3,$$

therefore we can write

$$\begin{aligned}\sum_{n=1}^{N} (n+1)(n+3) &= \sum_{n=1}^{N} (n^2 + 4n + 3) \\ &= \sum_{n=1}^{N} n^2 + 4\sum_{n=1}^{N} n + \sum_{n=1}^{N} 3 \\ &= \tfrac{1}{6}N(N+1)(2N+1) + 4 \times \tfrac{1}{2}N(N+1) + 3N \\ &= \tfrac{1}{6}N(2N^2 + 15N + 31). \text{ ◄}\end{aligned}$$

### *3.2.6 Transformation of series*

A complicated series may sometimes be summed by transforming it into a familiar series of which we already know the sum. The object is usually to transform the original series into a geometric series, or into the Maclaurin expansion of a simple function (see subsection 3.6.3). Various techniques are useful, and deciding which one to use in any given case is a matter of experience. We now discuss a few of the more common methods.

The differentiation or integration of a series is often useful in transforming an apparently intractable series into a more familiar one. If we wish to differentiate or integrate a series that already depends on some variable, then we may do so in a straightforward manner.

►*Sum the series*

$$S(x) = \frac{x^4}{3(0!)} + \frac{x^5}{4(1!)} + \frac{x^6}{5(2!)} + \cdots.$$

Dividing both sides by $x$ we obtain

$$\frac{S(x)}{x} = \frac{x^3}{3(0!)} + \frac{x^4}{4(1!)} + \frac{x^5}{5(2!)} + \cdots,$$

which is easily differentiated to give

$$\frac{d}{dx}\left[\frac{S(x)}{x}\right] = \frac{x^2}{0!} + \frac{x^3}{1!} + \frac{x^4}{2!} + \frac{x^5}{3!} + \cdots.$$

Using the Maclaurin expansion of $\exp x$ given in subsection 3.6.3, we recognise that the RHS is equal to $x^2 \exp x$. Therefore we can now integrate both sides to obtain

$$S(x)/x = \int x^2 \exp x \, dx.$$

Integrating the RHS by parts we find

$$S(x)/x = x^2 \exp x - 2x \exp x + 2 \exp x + c,$$

where the value of the constant of integration $c$ can be fixed by the requirement that $S(x)/x = 0$ at $x = 0$. Therefore we find $c = -2$, and the sum is given by

$$S(x) = x^3 \exp x - 2x^2 \exp x + 2x \exp x - 2x. \blacktriangleleft$$

Often, however, we require the sum of a series that does not depend on a variable. In this case, in order that we might differentiate or integrate the series, we define a function of some variable $x$ such that the value of this function is equal to the sum of the series for some particular value of $x$ (usually at $x = 1$).

►*Sum the series*

$$S = 1 + \frac{2}{2} + \frac{3}{2^2} + \frac{4}{2^3} + \cdots.$$

Let us begin by defining the function

$$f(x) = 1 + 2x + 3x^2 + 4x^3 + \cdots,$$

so that the sum $S = f(1/2)$. Integrating this function we obtain

$$\int f(x)\, dx = x + x^2 + x^3 + \cdots,$$

which we recognise as an infinite geometric series with first term $a = x$ and common ratio $r = x$. Therefore, from (3.4), we find that the sum of this series is $x/(1-x)$. In other words

$$\int f(x)\, dx = \frac{x}{1-x},$$

so that $f(x)$ is given by

$$f(x) = \frac{d}{dx}\left(\frac{x}{1-x}\right) = \frac{1}{(1-x)^2}.$$

The sum of the original series is therefore $S = f(1/2) = 4$. ◄

Aside from differentiation and integration, an appropriate substitution can sometimes transform a series into a more familiar one. In particular, series with terms that contain trigonometrical functions can often be summed by the use of complex exponentials.

►*Sum the series*

$$S(\theta) = 1 + \cos\theta + \frac{\cos 2\theta}{2!} + \frac{\cos 3\theta}{3!} + \cdots.$$

Replacing the cosine terms with a complex exponential, we obtain

$$\begin{aligned} S(\theta) &= \mathrm{Re}\left\{1 + \exp i\theta + \frac{\exp 2i\theta}{2!} + \frac{\exp 3i\theta}{3!} + \cdots\right\} \\ &= \mathrm{Re}\left\{1 + \exp i\theta + \frac{(\exp i\theta)^2}{2!} + \frac{(\exp i\theta)^3}{3!} + \cdots\right\}. \end{aligned}$$

Again using the Maclaurin expansion of $\exp x$ given in subsection 3.6.3, we notice that

$$\begin{aligned} S(\theta) &= \mathrm{Re}[\exp(\exp i\theta)] = \mathrm{Re}[\exp(\cos\theta + i\sin\theta)] \\ &= \mathrm{Re}\{[\exp(\cos\theta)][\exp(i\sin\theta)]\} = [\exp(\cos\theta)]\mathrm{Re}[\exp(i\sin\theta)] \\ &= [\exp(\cos\theta)][\cos(\sin\theta)]. \end{aligned}$$ ◄

## 3.3 Convergence of infinite series

Although the sums of some commonly occurring infinite series may be found, the sum of a general infinite series is often difficult to calculate. Nevertheless, it is often useful to know whether the partial sum of such a series converges to a limit, even if this limit cannot be explicitly found. As mentioned at the end of section 3.1, if we allow $N$ to tend to infinity, the partial sum

$$S_N = \sum_{n=1}^{N} u_n$$

of a series may tend to a definite limit (i.e. the sum $S$ of the series), or increase or decrease without limit, or oscillate finitely or infinitely.

To investigate the convergence of any given series, it is useful to have available a number of tests and theorems of general applicability. We discuss them below; some we will merely state, since once they have been stated they become almost self-evident, but are no less useful for that.

### *3.3.1 Absolute and conditional convergence*

Let us first consider some general points concerning the convergence, or otherwise, of an infinite series. In general an infinite series $\sum u_n$ can have complex terms, and in the special case of a real series the terms can be positive or negative. From any such series, however, we can always construct another series $\sum |u_n|$ in which each term is simply the modulus of the corresponding term in the original series. Then each term in the new series will be a positive real number.

If the series $\sum |u_n|$ converges then $\sum u_n$ also converges, and $\sum u_n$ is said to be *absolutely convergent*, i.e. the series formed by the absolute values is convergent. For an absolutely convergent series, the terms may be reordered without affecting the convergence of the series. However, if $\sum |u_n|$ diverges whilst $\sum u_n$ converges then $\sum u_n$ is said to be *conditionally convergent*. For a conditionally convergent series, rearranging the order of the terms can affect the behaviour of the sum and, hence, whether the series converges or diverges. In fact, a theorem due to Riemann shows that a conditionally convergent series may be made, by a suitable rearrangement, to converge to any arbitrary limit, or to diverge, or to oscillate finitely or infinitely! Of course, if the original series $\sum u_n$ consists only of positive real terms and converges, then automatically it is absolutely convergent.

### *3.3.2 Convergence of a series containing only real positive terms*

As discussed above, in order to test for the absolute convergence of a series $\sum u_n$, we first construct the corresponding series $\sum |u_n|$ that consists only of real positive terms. Therefore in this subsection we will restrict our attention to series of this type.

We discuss below some tests that may be used to investigate the convergence of such a series. Before doing so, however, we note the following *crucial consideration.* In all the tests for, or discussions of, the convergence of a series, it is not what happens in the first ten, or the first thousand, or the first million terms (or any other finite number of terms) that matters, but what happens *ultimately*.

*Preliminary test*

A necessary *but not sufficient* condition for a series of real positive terms $\sum u_n$ to be convergent is that the term $u_n$ tends to zero as $n$ tends to infinity, i.e. we require

$$\lim_{n\to\infty} u_n = 0.$$

If this condition is not satisfied then the series must diverge. Even if it is satisfied, however, the series may still diverge, and further testing is required.

*Comparison test*

The comparison test is the most basic test for convergence. Let us consider two series $\sum u_n$ and $\sum v_n$ and suppose that we *know* the latter to be convergent (by some earlier analysis, for example). Then, if each term $u_n$ in the first series is less than or equal to the corresponding term $v_n$ in the second series (for all $n$ greater than some fixed number $N$, say, which will vary from series to series) then the original series $\sum u_n$ is also convergent. In other words, if $\sum v_n$ is convergent and

$$u_n \leq v_n \qquad \text{for } n > N,$$

then $\sum u_n$ converges.

However, if $\sum v_n$ diverges and $u_n \geq v_n$ for all $n$ greater than some fixed number, then $\sum u_n$ diverges.

►*Determine whether the following series converges:*

$$\sum_{n=1}^{\infty} \frac{1}{n!+1} = \frac{1}{2} + \frac{1}{3} + \frac{1}{7} + \frac{1}{25} + \cdots. \tag{3.7}$$

Let us compare this series with the series

$$\sum_{n=0}^{\infty} \frac{1}{n!} = \frac{1}{0!} + \frac{1}{1!} + \frac{1}{2!} + \frac{1}{3!} + \cdots = 2 + \frac{1}{2!} + \frac{1}{3!} + \cdots, \tag{3.8}$$

which is merely the series obtained by setting $x = 1$ in the Maclaurin expansion of $\exp x$ (see subsection 3.6.3), i.e.

$$\exp(1) = e = 1 + \frac{1}{1!} + \frac{1}{2!} + \frac{1}{3!} + \cdots.$$

Clearly this second series is convergent, since it consists of only positive terms and has a finite sum. Thus, since each term $u_n$ in the series (3.7) is less than the corresponding term $1/n!$ in (3.8), we conclude from the comparison test that (3.7) is also convergent. ◄

*D'Alembert's ratio test*

The ratio test determines whether a series converges by comparing the relative magnitude of successive terms. If we consider a series $\sum u_n$ and set

$$\rho = \lim_{n\to\infty}\left(\frac{u_{n+1}}{u_n}\right), \tag{3.9}$$

then if $\rho < 1$ the series is convergent; if $\rho > 1$ the series is divergent; if $\rho = 1$ then the behaviour of the series is undetermined by this test.

To prove this we observe that if the limit (3.9) is less than unity, i.e. $\rho < 1$, then we can find a value $r$ in the range $\rho < r < 1$ and a value $N$ such that

$$\frac{u_{n+1}}{u_n} < r,$$

for all $n > N$. Now the terms $u_n$ of the series that follow $u_N$ are

$$u_{N+1}, \qquad u_{N+2}, \qquad u_{N+3}, \qquad \ldots,$$

and each of these is less than the corresponding term of

$$ru_N, \qquad r^2u_N, \qquad r^3u_N, \qquad \ldots. \tag{3.10}$$

However, the terms of (3.10) are those of a geometric series with a common ratio $r$ that is less than unity. This geometric series consequently converges and therefore, by the comparison test discussed above, so must the original series $\sum u_n$. An analogous argument may be used to prove the divergent case when $\rho > 1$.

►*Determine whether the following series converges.*

$$\sum_{n=0}^{\infty}\frac{1}{n!} = \frac{1}{0!}+\frac{1}{1!}+\frac{1}{2!}+\frac{1}{3!}+\cdots = 2+\frac{1}{2!}+\frac{1}{3!}+\cdots.$$

As mentioned in the previous example, this series may be obtained by setting $x = 1$ in the Maclaurin expansion of $\exp x$, and hence we know already that it converges and has the sum $\exp(1) = e$. Nevertheless, we may use the ratio test to confirm that it converges.

Using (3.9), we have

$$\rho = \lim_{n\to\infty}\left[\frac{n!}{(n+1)!}\right] = \lim_{n\to\infty}\left(\frac{1}{n+1}\right) = 0 \tag{3.11}$$

and since $\rho < 1$, the series converges, as expected. ◄

*Ratio comparison test*

As its name suggests, the ratio comparison test is a combination of the ratio and comparison tests. Let us consider the two series $\sum u_n$ and $\sum v_n$ and assume that we know the latter to be convergent. It may be shown that if

$$\frac{u_{n+1}}{u_n} \le \frac{v_{n+1}}{v_n},$$

for all $n$ greater than some fixed value $N$, then $\sum u_n$ is also convergent.

Similarly, if

$$\frac{u_{n+1}}{u_n} \ge \frac{v_{n+1}}{v_n},$$

for all sufficiently large $n$, and $\sum v_n$ diverges, then $\sum u_n$ also diverges.

►*Determine whether the following series converges:*

$$\sum_{n=1}^{\infty} \frac{1}{(n!)^2} = 1 + \frac{1}{2^2} + \frac{1}{6^2} + \cdots .$$

In this case the ratio of successive terms, as $n$ tends to infinity, is given by

$$R = \lim_{n\to\infty} \left[\frac{n!}{(n+1)!}\right]^2 = \lim_{n\to\infty} \left(\frac{1}{n+1}\right)^2,$$

which is less than the ratio seen in (3.11). Hence, by the ratio comparison test, the series converges. (It is clear that this series could also be found to be convergent using the ratio test.) ◄

*Quotient test*

The quotient test may also be considered as a combination of the ratio and comparison tests. Let us again consider the two series $\sum u_n$ and $\sum v_n$, and consider the limit

$$\rho = \lim_{n\to\infty} \left(\frac{u_n}{v_n}\right). \tag{3.12}$$

Then, it can be shown that:

(i) if $\rho \ne 0$ but finite then $\sum u_n$ and $\sum v_n$ either both converge or both diverge;

(ii) if $\rho = 0$ and $\sum v_n$ converges, then $\sum u_n$ converges;

(iii) if $\rho = \infty$ and $\sum v_n$ diverges, then $\sum u_n$ diverges.

►*Given that the series $\sum_{n=1}^{\infty} 1/n$ diverges, determine whether the following series converges:*

$$\sum_{n=1}^{\infty} \frac{4n^2 - n - 3}{n^3 + 2n}. \tag{3.13}$$

If we set $u_n = (4n^2 - n - 3)/(n^3 + 2n)$ and $v_n = 1/n$, then the limit (3.12) becomes

$$\rho = \lim_{n\to\infty} \left[ \frac{(4n^2 - n - 3)/(n^3 + 2n)}{1/n} \right] = \lim_{n\to\infty} \left[ \frac{4n^3 - n^2 - 3n}{n^3 + 2n} \right] = 4.$$

Since $\rho$ is finite but non-zero and $\sum v_n$ diverges, from (i) above $\sum u_n$ must also diverge. ◄

*Integral test*

The integral test is an extremely powerful means of investigating the convergence of a series $\sum u_n$. Suppose that there exists a function $f(x)$ which monotonically decreases for $x$ greater than some fixed value $x_0$ and for which $f(n) = u_n$, i.e. the value of the function at integer values of $x$ is equal to the corresponding term in the series under investigation. It can then be shown that if the limit of the integral

$$\lim_{N\to\infty} \int^N f(x)\,dx$$

exists, then the series $\sum u_n$ is convergent. Otherwise the series diverges. Note that the integral has no lower limit. In fact the integral test is sometimes defined to include a lower limit of unity in the integral, but this can lead to unnecessary difficulties.

►*Determine whether the following series converges:*

$$\sum_{n=1}^{\infty} \frac{1}{(n - 3/2)^2} = 4 + 4 + \frac{4}{9} + \frac{4}{25} + \cdots.$$

Let us consider the function $f(x) = (x - 3/2)^{-2}$. Clearly $f(n) = u_n$ and $f(x)$ monotonically decreases for $x > 3/2$. Applying the integral test, we consider

$$\lim_{N\to\infty} \int^N \frac{1}{(x - 3/2)^2}\,dx = \lim_{N\to\infty} \left( \frac{-1}{N - 3/2} \right) = 0.$$

Since the limit exists the series converges. Note, however, that if we had included a lower limit of unity on the integral, we would have run into problems, since the integrand diverges at $x = 3/2$. ◄

The integral test is also useful for examining the convergence of the Riemann zeta series. This is a special series that occurs regularly and is of the form

$$\sum_{n=1}^{\infty} \frac{1}{n^p}.$$

It converges for $p > 1$ and diverges if $p \leq 1$. These convergence criteria may be derived as follows.

Using the integral test, we consider

$$\lim_{N\to\infty} \int^N \frac{1}{x^p} dx = \lim_{N\to\infty} \left( \frac{N^{1-p}}{1-p} \right),$$

and it is obvious that the limit tends to zero for $p > 1$ and to $\infty$ for $p \leq 1$.

*Cauchy's root test*

Cauchy's root test may be useful in testing for convergence, especially if the $n$th terms of the series contains an $n$th power. If we define the limit

$$\rho = \lim_{n\to\infty} (u_n)^{1/n},$$

then it may be proved that the series $\sum u_n$ converges if $\rho < 1$. If $\rho > 1$ then the series diverges. Its behaviour is undetermined if $\rho = 1$.

►*Determine whether the following series converges:*

$$\sum_{n=1}^{\infty} \left( \frac{1}{n} \right)^n = 1 + \frac{1}{4} + \frac{1}{27} + \cdots.$$

Using Cauchy's root test, we find

$$\rho = \lim_{n\to\infty} \left( \frac{1}{n} \right) = 0,$$

and hence the series converges. ◄

*Grouping terms*

We now consider the Riemann zeta series, mentioned above, with an alternative proof of its convergence that uses the method of grouping terms. In general there are better ways of determining convergence, but the grouping method may be used if it is not immediately obvious how to approach a problem by a better method.

First consider the case where $p > 1$, and group the terms in the series as follows:

$$S_N = \frac{1}{1^p} + \left( \frac{1}{2^p} + \frac{1}{3^p} \right) + \left( \frac{1}{4^p} + \cdots + \frac{1}{7^p} \right) + \cdots.$$

Now we can see that each bracket of this series is less than each term of the geometric series

$$S_N = \frac{1}{1^p} + \frac{2}{2^p} + \frac{4}{4^p} + \cdots.$$

This geometric series has common ratio $r = \left(\frac{1}{2}\right)^{p-1}$; therefore $r < 1$ since $p > 1$, and so the geometric series converges. Then the comparison test shows that the Riemann zeta series also converges for $p > 1$.

The divergence of the Riemann zeta series for $p \leq 1$ can be seen by first considering the case $p = 1$. The series is

$$S_N = 1 + \frac{1}{2} + \frac{1}{3} + \frac{1}{4} + \cdots,$$

which does *not* converge, as may be seen by bracketing the terms of the series in groups in the following way:

$$S_N = \sum_{n=1}^{N} u_n = 1 + \left(\frac{1}{2}\right) + \left(\frac{1}{3} + \frac{1}{4}\right) + \left(\frac{1}{5} + \frac{1}{6} + \frac{1}{7} + \frac{1}{8}\right) + \cdots.$$

The sum of the terms in each bracket is $\geq \frac{1}{2}$ and since as many such groupings can be made as we wish, it is clear that $S_N$ increases indefinitely as $N$ is increased.

Now returning to the case of the Riemann zeta series for $p < 1$, we note that each term in the series is greater than the corresponding one in the series for which $p = 1$. In other words $1/n^p > 1/n$ for $n > 1$, $p < 1$. The comparison test then shows us that the Riemann zeta series will diverge for all $p \leq 1$.

### 3.3.3 Alternating series test

The tests discussed in the last subsection have been concerned with determining whether the series of real positive terms $\sum |u_n|$ converges, and so whether $\sum u_n$ is absolutely convergent. Nevertheless, it is sometimes useful to consider whether a series is merely convergent rather than absolutely convergent. This is especially true for series containing an infinite number of both positive and negative terms. In particular, we will consider the convergence of series in which the positive and negative terms alternate, i.e. an *alternating series*.

An alternating series can be written as

$$\sum_{n=1}^{\infty} (-1)^{n+1} u_n = u_1 - u_2 + u_3 - u_4 + u_5 - \cdots,$$

with all $u_n \geq 0$. Such a series can be shown to converge provided (i) $u_n \to 0$ as $n \to \infty$ and (ii) $u_n < u_{n-1}$ for all $n > N$ for some finite $N$. If these conditions are not met, then the series oscillates.

To prove this, suppose for definiteness that $N$ is odd and consider the series starting at $u_N$. The sum of its first $2m$ terms is

$$S_{2m} = (u_N - u_{N+1}) + (u_{N+2} - u_{N+3}) + \cdots + (u_{N+2m-2} - u_{N+2m-1}).$$

By condition (ii) above, all the parentheses are positive, and so $S_{2m}$ increases as $m$ increases. We can also write, however,

$$S_{2m} = u_N - (u_{N+1} - u_{N+2}) - \cdots - (u_{N+2m-3} - u_{N+2m-2}) - u_{N+2m-1},$$

and since each parenthesis is positive, we must have $S_{2m} < u_N$. Thus, since $S_{2m}$ is always less than $u_N$ for all $m$ and $u_n \to 0$ as $n \to \infty$, the alternating series converges. It is clear that an analogous proof can be constructed in the case where $N$ is even.

►*Determine whether the following series converges:*

$$\sum_{n=1}^{\infty} (-1)^{n+1} \frac{1}{n} = 1 - \frac{1}{2} + \frac{1}{3} - \cdots .$$

This alternating series clearly satisfies conditions (i) and (ii) above, and hence converges. However, as shown above by the method of grouping terms, the corresponding series with all positive terms is divergent. ◄

## 3.4 Operations with series

Simple operations with series are fairly intuitive, and we discuss them here only for completeness. The following points apply to both finite and infinite series unless otherwise stated.

(i) If $\sum u_n = S$, then $\sum k u_n = kS$ where $k$ is any constant.

(ii) If $\sum u_n = S$ and $\sum v_n = T$, then $\sum (u_n + v_n) = S + T$.

(iii) If $\sum u_n = S$, then $a + \sum u_n = a + S$. A simple extension of this trivial result shows that the removal or insertion of a finite number of terms anywhere in a series does not affect its convergence.

(iv) If the infinite series $\sum u_n$ and $\sum v_n$ are both absolutely convergent, then the series $\sum w_n$, where

$$w_n = u_1 v_n + u_2 v_{n-1} + \cdots + u_n v_1$$

is also absolutely convergent. The series $\sum w_n$ is called the *Cauchy product* of the two original series. Furthermore, if $\sum u_n$ converges to the sum $S$ and $\sum v_n$ converges to the sum $T$ then $\sum w_n$ converges to the sum $ST$.

(v) It is not true in general that term-by-term differentiation or integration of a series will result in a new series with the same convergence properties.

## 3.5 Power series

A power series has the form

$$P(x) = a_0 + a_1x + a_2x^2 + a_3x^3 + \cdots,$$

where $a_0, a_1, a_2, a_3$ etc. are constants. Such series regularly occur in physics and engineering and are useful because, for $|x| < 1$, the later terms in the series may become very small and be discarded. For example the series

$$P(x) = 1 + x + x^2 + x^3 + \cdots,$$

although in principle infinitely long, in practice may be simplified if $x$ happens to have a value small compared with unity. To see this note that $P(x)$ for $x = 0.1$ has the following values: 1, if just one term is taken into account; 1.1, for two terms; 1.11, for three terms; 1.111, for four terms, etc. If the quantity that it represents can only be measured with an accuracy of two decimal places, then all but the first three terms may be ignored, i.e. when $x = 0.1$ or less

$$P(x) = 1 + x + x^2 + \mathrm{O}(x^3) \approx 1 + x + x^2.$$

This sort of approximation is often used to simplify equations into manageable forms. It may seem imprecise at first, but is perfectly acceptable insofar as it matches the experimental accuracy that can be achieved.

The symbols O and $\approx$ used above need some further explanation. They are used to compare the behaviour of two functions when a variable upon which both functions depend tends to a particular limit, usually zero or infinity (obvious from the context). If two functions of $x$ are denoted by $f(x)$ and $g(x)$, and $g$ is positive, then the formal *definitions* of the above symbols are as follows:

(i) If there exists a constant $k$ such that $|f| \leq kg$ as the limit is approached, then $f = \mathrm{O}(g)$.

(ii) If as the limit of $x$ is approached $f/g$ tends to a limit $l$, where $l \neq 0$, then $f \approx lg$. The statement $f \approx g$ means that the ratio of the two sides tends to unity.

### *3.5.1 Convergence of power series*

The convergence or otherwise of power series is a crucial consideration in practical terms. For example, if we are to use a power series as an approximation, it is clearly important that it tends to the precise answer as more and more terms of the approximation are taken. Consider the general power series

$$P(x) = a_0 + a_1x + a_2x^2 + \cdots.$$

Using d'Alembert's ratio test (see subsection 3.3.2), we see that $P(x)$ converges absolutely if

$$\rho = \lim_{n\to\infty} \left| \frac{a_{n+1}}{a_n} x \right| = |x| \lim_{n\to\infty} \left| \frac{a_{n+1}}{a_n} \right| < 1.$$

Thus the convergence of $P(x)$ depends upon the value of $x$, i.e. there is, in general, a range of values of $x$ for which $P(x)$ converges, an *interval of convergence*. Note that at the limits of this range $\rho = 1$, and so the series may converge or diverge. The convergence of the series at the end-points may be determined by substituting these values of $x$ into the power series $P(x)$, and testing the resulting series using any applicable method (discussed in section 3.3).

►*Determine the range of values of $x$ for which the following power series converges:*

$$P(x) = 1 + 2x + 4x^2 + 8x^3 + \cdots.$$

By using the interval-of-convergence method discussed above,

$$\rho = \lim_{n\to\infty} \left| \frac{2^{n+1}}{2^n} x \right| = |2x|,$$

and hence the power series will converge for $|x| < 1/2$. Examining the end-points of the interval separately, we find

$$P(1/2) = 1 + 1 + 1 + \cdots,$$
$$P(-1/2) = 1 - 1 + 1 - \cdots.$$

Obviously $P(1/2)$ diverges, while $P(-1/2)$ oscillates. Therefore $P(x)$ is not convergent at either end-point of the region, but is convergent for $-1 < x < 1$. ◄

The convergence of power series may be extended to the case of a complex parameter $z$. Considering the power series

$$P(z) = a_0 + a_1 z + a_2 z^2 + \cdots,$$

we find that $P(z)$ converges if

$$\rho = \lim_{n\to\infty} \left| \frac{a_{n+1}}{a_n} z \right| = |z| \lim_{n\to\infty} \left| \frac{a_{n+1}}{a_n} \right| < 1.$$

We therefore have a range in $|z|$ for which $P(z)$ converges, i.e. $P(z)$ converges for values of $z$ lying within a circle in the Argand diagram (in this case centred on the origin of the Argand diagram). The radius of the circle is called the *radius of convergence*: if $z$ lies inside the circle, the series will converge whereas if $z$ lies outside the circle, the series will diverge; if, though, $z$ lies on the circle, the convergence must be tested using another method. Clearly the radius of convergence $R$ is given by $1/R = \lim_{n\to\infty} |a_{n+1}/a_n|$.

►*Determine the range of values of z for which the following complex power series converges:*

$$P(z) = 1 - \frac{z}{2} + \frac{z^2}{4} - \frac{z^3}{8} + \cdots .$$

We find that $\rho = |z/2|$, which shows that $P(z)$ converges for $|z| < 2$. Therefore the circle of convergence in the Argand diagram is centred on the origin and has a radius $R = 2$. On this circle we must test the convergence by substituting the value of $z$ into $P(z)$ and considering the resulting series. On the circle of convergence we can write $z = 2\exp i\theta$. Substituting this into $P(z)$, we obtain

$$P(z) = 1 - \frac{2\exp i\theta}{2} + \frac{4\exp 2i\theta}{4} - \cdots ,$$
$$= 1 - \exp i\theta + [\exp i\theta]^2 - \cdots ,$$

which is a complex infinite geometric series with first term $a = 1$ and common ratio $r = -\exp i\theta$. Therefore, on the the circle of convergence we have

$$P(z) = \frac{1}{1 + \exp i\theta}.$$

Unless $\theta = \pi$, this is a finite complex number, and so $P(z)$ converges at all points on the circle $|z| = 2$ except at $\theta = \pi$ (i.e. $z = -2$), where it diverges. Note that $P(z)$ is just the binomial expansion of $(1 + z/2)^{-1}$, for which it is obvious that $z = -2$ is a singular point. In general, for power series expansions of complex functions about a given point in the complex plane, the circle of convergence extends as far as the nearest singular point. This is discussed further in chapter 18. ◄

Note that the centre of the circle of convergence does not necessarily lie at the origin. For example, applying the ratio test to the complex power series

$$P(z) = 1 + \frac{z-1}{2} + \frac{(z-1)^2}{4} + \frac{(z-1)^3}{8} + \cdots ,$$

we find that for it to converge we require $|(z-1)/2| < 1$. Thus the series converges for $z$ lying within a circle of radius 2 centred on the point $(1, 0)$ in the Argand diagram.

### 3.5.2 Operations with power series

The following rules are useful when manipulating power series, and apply to power series of a real or complex variable.

(i) If two power series $P(x)$ and $Q(x)$ have regions of convergence that overlap to some extent, then the series produced by taking the sum, the difference or the product of $P(x)$ and $Q(x)$ converges in the common region.

(ii) If two power series $P(x)$ and $Q(x)$ converge for all values of $x$, then one series may be substituted into the other to give a third series, which also converges for all values of $x$. For example consider the power series expansions of $\sin x$ and $e^x$ given below in subsection 3.6.3,

$$\sin x = x - \frac{x^3}{3!} + \frac{x^5}{5!} - \frac{x^7}{7!} + \cdots$$

$$e^x = 1 + x + \frac{x^2}{2!} + \frac{x^3}{3!} + \frac{x^4}{4!} + \cdots,$$

both of which converge for all values of $x$. Substituting the series for $\sin x$ into that for $e^x$ we obtain

$$e^{\sin x} = 1 + x + \frac{x^2}{2!} - \frac{3x^4}{4!} - \frac{8x^5}{5!} + \cdots,$$

which also converges for all values of $x$.

If, however, either of the power series $P(x)$ and $Q(x)$ has only a limited region of convergence, or if they both do so, then further care must be taken when substituting one series into the other. For example, suppose $Q(x)$ converges for all $x$, but $P(x)$ only converges for $x$ within a finite range. We may substitute $Q(x)$ into $P(x)$ to obtain $P(Q(x))$, but we must be careful since the value of $Q(x)$ may lie outside the region of convergence for $P(x)$, with the consequence that the resulting series $P(Q(x))$ does not converge.

(iii) If a power series $P(x)$ converges for a particular range of $x$ then the series obtained by differentiating every term and the series obtained by integrating every term also converge in this range.

This is easily seen for the power series

$$P(x) = a_0 + a_1 x + a_2 x^2 + \cdots,$$

which converges if $|x| < \lim_{n\to\infty} |a_n/a_{n+1}| \equiv k$. The series obtained by differentiating $P(x)$ with respect to $x$ is given by

$$\frac{dP}{dx} = a_1 + 2a_2 x + 3a_3 x^2 + \cdots,$$

and converges if

$$|x| < \lim_{n\to\infty} \left| \frac{na_n}{(n+1)a_{n+1}} \right| = k.$$

Similarly the series obtained by integrating $P(x)$ term by term,

$$\int P(x)\, dx = a_0 x + \frac{a_1 x^2}{2} + \frac{a_2 x^3}{3} + \cdots,$$

converges if

$$|x| < \lim_{n\to\infty} \left| \frac{(n+2)a_n}{(n+1)a_{n+1}} \right| = k.$$

So, series resulting from differentiation or integration have the same interval of convergence as the original series. However, even if the original series converged at either end-point of the interval, then it is not necessarily the case that the new series will do so. These new series must be tested separately at the end-points in order to determine whether they converge there. Note that although power series may be integrated or differentiated without altering their interval of convergence, this is not true for series in general.

It is also worth noting that differentiating or integrating a power series term by term within its interval of convergence is equivalent to differentiating or integrating the function it represents. For example, consider the power series expansion of $\sin x$,

$$\sin x = x - \frac{x^3}{3!} + \frac{x^5}{5!} - \frac{x^7}{7!} + \cdots, \tag{3.14}$$

which converges for all values of $x$. If we differentiate term by term, the series becomes

$$1 - \frac{x^2}{2!} + \frac{x^4}{4!} - \frac{x^6}{6!} + \cdots,$$

which is the series expansion of $\cos x$ as we expect.

## 3.6 Taylor series

Taylor's theorem provides a way of expressing a function as a power series in $x$, known as a *Taylor series*, but it can be applied only to those functions that are continuous and differentiable within the $x$-range of interest.

### *3.6.1 Taylor's theorem*

Suppose that we have a function $f(x)$ that we wish to express as a power series in $x-a$ about the point $x=a$. We shall assume that $f(x)$ is a continuous, single-valued function of $x$ having continuous derivatives with respect to $x$, denoted by $f'(x)$, $f''(x)$ and so on, up to and including $f^{(n-1)}(x)$, in a given $x$-range, and that $f^{(n)}(x)$ exists in this range.

From the equation following (1.26) we may write

$$\int_a^{a+h} f'(x)\,dx = f(a+h) - f(a),$$

where $a$, $a+h$ are neighbouring values of $x$. Rearranging this equation, we may express the value of the function at $x = a+h$ in terms of its value at $a$ by

$$f(a+h) = f(a) + \int_a^{a+h} f'(x)\,dx. \tag{3.15}$$

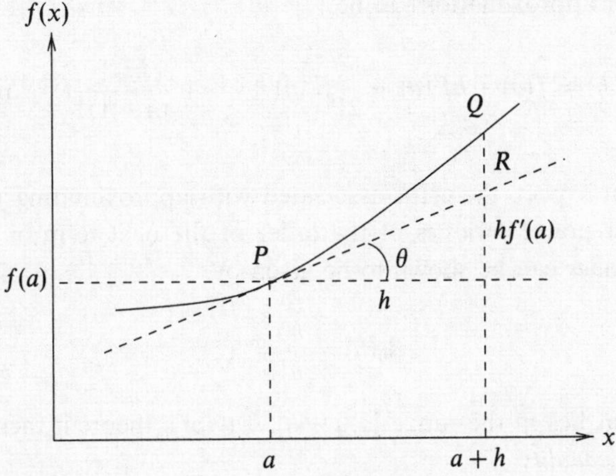


Figure 3.1 The first-order Taylor series approximation to a function $f(x)$. The slope of the function at $P$, i.e. $\tan\theta$, equals $f'(a)$. Thus the value of the function at $Q$, $f(a+h)$, is approximated by the ordinate of $R$, $f(a)+hf'(a)$.

A *first approximation* for $f(a+h)$ may be obtained by substituting $f'(a)$ for $f'(x)$ in (3.15), to obtain

$$f(a+h) \approx f(a) + hf'(a).$$

This approximation is shown graphically in figure 3.1. We may write this first approximation in terms of $x$ and $a$ as

$$f(x) \approx f(a) + (x-a)f'(a),$$

and, in a similar way,

$$\begin{aligned} f'(x) &\approx f'(a) + (x-a)f''(a) \\ f''(x) &\approx f''(a) + (x-a)f'''(a), \end{aligned}$$

and so on. Substituting for $f'(x)$ in (3.15), we obtain the *second approximation*:

$$\begin{aligned} f(a+h) &\approx f(a) + \int_a^{a+h} [f'(a) + (x-a)f''(a)]\,dx \\ &\approx f(a) + hf'(a) + \frac{h^2}{2}f''(a). \end{aligned}$$

We may repeat this procedure as often as we like (so long as the derivatives of $f(x)$ exist) to obtain higher-order approximations to $f(a+h)$; we find the

$(n-1)$th-order approximation† to be

$$f(a+h) \approx f(a) + hf'(a) + \frac{h^2}{2!}f''(a) + \cdots + \frac{h^{n-1}}{(n-1)!}f^{(n-1)}(a). \qquad (3.16)$$

As we might expect, the error associated with approximating $f(a+h)$ by this $(n-1)$th-order power series is of the order of the next term in the series. This error or *remainder* can be shown to be given by

$$R_n(h) = \frac{h^n}{n!}f^{(n)}(\xi),$$

for some $\xi$ that lies in the range $[a, a+h]$. Taylor's theorem then states that we may write the *equality*

$$f(a+h) = f(a) + hf'(a) + \frac{h^2}{2!}f''(a) + \cdots + \frac{h^{(n-1)}}{(n-1)!}f^{(n-1)}(a) + R_n(h). \qquad (3.17)$$

The theorem may also be written in a form suitable for finding $f(x)$ given the value of the function and its relevant derivatives at $x = a$, by substituting $x = a + h$ in the above expression. It then reads

$$f(x) = f(a) + (x-a)f'(a) + \frac{(x-a)^2}{2!}f''(a) + \cdots + \frac{(x-a)^{n-1}}{(n-1)!}f^{(n-1)}(a) + R_n(x), \qquad (3.18)$$

where the remainder now takes the form

$$R_n(x) = \frac{(x-a)^n}{n!}f^{(n)}(\xi),$$

and $\xi$ lies in the range $[a, x]$. Each of the formulae (3.17), (3.18) gives us the *Taylor expansion* of the function about the point $x = a$. A special case occurs when $a = 0$. Such Taylor expansions, about $x = 0$, are called *Maclaurin series.*

Taylor's theorem is also valid without modification for functions of a complex variable (see chapter 18). The extension of Taylor's theorem to functions of more than one variable is given in chapter 4.

For a function to be expressible as an infinite power series we require it to be infinitely differentiable and the remainder term $R_n$ to tend to zero as $n$ tends to infinity, i.e. $\lim_{n\to\infty} R_n = 0$. In this case the infinite power series will represent the function within the interval of convergence of the series.

† The order of the approximation is simply the highest power of $h$ in the series. Note, though, that the $(n-1)$th-order approximation contains $n$ terms.

►*Expand $f(x) = \sin x$ as a Maclaurin series, i.e. about $x = 0$.*

We must first verify that $\sin x$ may indeed be represented by an infinite power series. It is easily shown that the $n$th derivative of $f(x)$ is given by

$$f^{(n)}(x) = \sin\left(x + \frac{n\pi}{2}\right).$$

Therefore the remainder after expanding $f(x)$ as an $(n-1)$th-order polynomial about $x = 0$ is given by

$$R_n(x) = \frac{x^n}{n!}\sin\left(\xi + \frac{n\pi}{2}\right),$$

where $\xi$ lies in the range $[0, x]$. Since the modulus of the sine term is always less than or equal to unity, we can write $|R_n(x)| < |x^n|/n!$. For any particular value of $x$, say $x = c$, $R_n(c) \to 0$ as $n \to \infty$. Hence $\lim_{n\to\infty} R_n(x) = 0$, and so $\sin x$ can be represented by an infinite Maclaurin series.

Evaluating the function and its derivatives at $x = 0$ we obtain

$$\begin{aligned} f(0) &= \sin 0 = 0, \\ f'(0) &= \sin(\pi/2) = 1, \\ f''(0) &= \sin \pi = 0, \\ f'''(0) &= \sin(3\pi/2) = -1, \end{aligned}$$

and so on. Therefore, the Maclaurin series expansion of $\sin x$ is given by

$$\sin x = x - \frac{x^3}{3!} + \frac{x^5}{5!} - \cdots.$$

Note that, as expected, since $\sin x$ is an odd function, its power series expansion contains only odd powers of $x$. ◄

We may follow a similar procedure to obtain a Taylor series about an arbitrary point $x = a$.

►*Expand $f(x) = \cos x$ as a Taylor series about $x = \pi/3$.*

As in the above example, it is easily shown that the $n$th derivative of $f(x)$ is given by

$$f^{(n)}(x) = \cos\left(x + \frac{n\pi}{2}\right).$$

Therefore the remainder after expanding $f(x)$ as an $(n-1)$th-order polynomial about $x = \pi/3$ is given by

$$R_n(x) = \frac{(x - \pi/3)^n}{n!}\cos\left(\xi + \frac{n\pi}{2}\right),$$

where $\xi$ lies in the range $[\pi/3, x]$. The modulus of the cosine term is again always less than or equal to unity, and so $|R_n(x)| < |(x - \pi/3)^n|/n!$. Again, as in the previous example, $\lim_{n\to\infty} R_n(x) = 0$ for any particular value of $x$, and so $\cos x$ can be represented by an infinite Taylor series about $x = \pi/3$.

Evaluating the function and its derivatives at $x = \pi/3$ we obtain

$$\begin{aligned} f(\pi/3) &= \cos(\pi/3) = 1/2, \\ f'(\pi/3) &= \cos(5\pi/6) = -\sqrt{3}/2, \\ f''(\pi/3) &= \cos(4\pi/3) = -1/2, \end{aligned}$$

and so on. Thus the Taylor series expansion of $\cos x$ about $x = \pi/3$ is given by

$$\cos x \approx \frac{1}{2} - \frac{\sqrt{3}}{2}\left(x - \pi/3\right) - \frac{1}{2}\frac{\left(x - \pi/3\right)^2}{2!} + \cdots. \blacktriangleleft$$

### 3.6.2 *Approximation errors in Taylor series*

In the previous subsection we saw how to represent a function $f(x)$ by an infinite power series, which is exactly equal to $f(x)$ for all $x$ within the interval of convergence of the series. However, in physical problems we usually do not want to have to sum an infinite number of terms, but prefer to use only a finite number of terms in the Taylor series to *approximate* the function in some given range of $x$. In this case it is desirable to know what is the maximum possible error associated with the approximation.

As given in (3.18), a function $f(x)$ can be represented by a finite $(n-1)$th-order power series together with a remainder term such that

$$f(x) = f(a) + (x-a)f'(a) + \frac{(x-a)^2}{2!}f''(a) + \cdots + \frac{(x-a)^{n-1}}{(n-1)!}f^{(n-1)}(a) + R_n(x),$$

where

$$R_n(x) = \frac{(x-a)^n}{n!}f^{(n)}(\xi)$$

and $\xi$ lies in the range $[a, x]$. $R_n(x)$ is the remainder term, and represents the error in approximating $f(x)$ by the above $(n-1)$th-order power series. Since the exact value of $\xi$ that satisfies the expression for $R_n(x)$ is not known, an upper limit on the error may be found by differentiating $R_n(x)$ with respect to $\xi$ and equating the derivative to zero in the usual way for finding maxima.

►*Expand $f(x) = \cos x$ as a Taylor series about $x = 0$ and find the error associated with using the approximation to evaluate* $\cos(0.5)$ *if only the first two non-vanishing terms are taken. (Note that the Taylor expansions of trigonometrical functions are only valid for angles measured in radians.)*

Evaluating the function and its derivatives at $x = 0$, we find

$$f(0) = \cos 0 = 1,$$
$$f'(0) = -\sin 0 = 0,$$
$$f''(0) = -\cos 0 = -1,$$
$$f'''(0) = \sin 0 = 0.$$

So, for small $|x|$, we find from (3.18)

$$\cos x \approx 1 - \frac{x^2}{2}.$$

Note that since $\cos x$ is an even function, its power series expansion contains only even powers of $x$. Therefore, in order to estimate the error in this approximation, we must consider the term in $x^4$, which is the next in the series. The required derivative is $f^{(4)}(x)$ and this is (by chance) equal to $\cos x$. Thus, adding in the remainder term $R_4(x)$, we find

$$\cos x = 1 - \frac{x^2}{2} + \frac{x^4}{4!}\cos\xi,$$

where $\xi$ lies in the range $[0, x]$. Thus, the maximum possible error is $x^4/4!$, since $\cos\xi$ cannot exceed unity. If $x = 0.5$, taking just the first two terms yields $\cos(0.5) \approx 0.875$ with a predicted error of less than 0.00260. In fact $\cos(0.5) = 0.87758$ to 5 decimal places. Thus, to this accuracy, the true error is 0.00258, an error of about 0.3%. ◄

### *3.6.3 Standard Maclaurin series*

It is often useful to have a readily available table of Maclaurin series for standard elementary functions, and therefore these are listed below.

$$\sin x = x - \frac{x^3}{3!} + \frac{x^5}{5!} - \frac{x^7}{7!} + \cdots \quad \text{for } -\infty < x < \infty,$$
$$\cos x = 1 - \frac{x^2}{2!} + \frac{x^4}{4!} - \frac{x^6}{6!} + \cdots \quad \text{for } -\infty < x < \infty,$$
$$\tan^{-1} x = x - \frac{x^3}{3} + \frac{x^5}{5} - \frac{x^7}{7} + \cdots \quad \text{for } -1 < x < 1,$$

$$e^x = 1 + x + \frac{x^2}{2!} + \frac{x^3}{3!} + \frac{x^4}{4!} + \cdots \quad \text{for} \quad -\infty < x < \infty,$$

$$\ln(1+x) = x - \frac{x^2}{2} + \frac{x^3}{3} - \frac{x^4}{4} + \cdots \quad \text{for} \quad -1 < x \leq 1,$$

$$(1+x)^n = 1 + nx + n(n-1)\frac{x^2}{2!} + n(n-1)(n-2)\frac{x^3}{3!} + \cdots \quad \text{for} \quad -\infty < x < \infty.$$

These can all be derived by straightforward application of Taylor's theorem to the expansion of a function about $x = 0$.

## 3.7 Evaluation of limits

The idea of the limit of a function $f(x)$ as $x$ approaches a value $a$ is fairly intuitive though a strict definition exists and is stated below. In many cases, the limit of the function as $x$ approaches $a$ will be simply the value $f(a)$, but in some cases this is not so. Firstly, the function may be undefined at $x = a$ as, for example, when

$$f(x) = \frac{\sin x}{x},$$

which takes the value $0/0$ at $x = 0$. However, here the limit as $x$ approaches zero still exists and can be evaluated as unity using l'Hôpital's rule below. Another possibility is that even if $f(x)$ is defined at $x = a$ its value may not be equal to the limiting value $\lim_{x\to a} f(x)$. This can occur for a discontinuous function at a point of discontinuity. The strict definition of a limit is that *if* $\lim_{x\to a} f(x) = l$ *then for any number* $\epsilon$ *however small, it must be possible to find a number* $\eta$ *such that* $|f(x) - l| < \epsilon$ *whenever* $|x - a| < \eta$. In other words, as $x$ becomes arbitrarily close to $a$, $f(x)$ becomes arbitrarily close to its limit, $l$.

The following observations are often useful in finding the limit of a function.

(i) A limit may be $\pm\infty$. For example as $x \to 0$, $1/x^2 \to \infty$.

(ii) A limit may be approached from below or above and the value may be different in each case. For example consider the function $f(x) = \tan x$. As $x$ tends to $\pi/2$ from below, $f(x) \to \infty$, but if the limit is approached from above, $f(x) \to -\infty$. Another way of writing this is

$$\lim_{x\to\frac{\pi}{2}^-} \tan x = \infty, \qquad \lim_{x\to\frac{\pi}{2}^+} \tan x = -\infty.$$

(iii) It may ease the evaluation of limits if the function under consideration is split into a sum, product or quotient. Provided each of the limits exists, the rules for evaluating such limits follow.

$$\text{(a)} \lim_{x\to a} \{f(x) + g(x)\} = \lim_{x\to a} f(x) + \lim_{x\to a} g(x).$$

$$\text{(b)} \lim_{x\to a} \{f(x)g(x)\} = \lim_{x\to a} f(x) \lim_{x\to a} g(x).$$

$$\text{(c)} \lim_{x\to a} \frac{f(x)}{g(x)} = \frac{\lim_{x\to a} f(x)}{\lim_{x\to a} g(x)}, \text{ provided that}$$

numerator and denominator are not both equal to zero or infinity.

Examples of cases (a)–(c) are discussed below.

►*Evaluate the limits*

$$\lim_{x\to 1}(x^2 + 2x^3), \qquad \lim_{x\to 0}(x\cos x), \qquad \lim_{x\to\pi/2} \frac{\sin x}{x}.$$

Using (a) above,

$$\lim_{x\to 1}(x^2 + 2x^3) = \lim_{x\to 1} x^2 + \lim_{x\to 1} 2x^3 = 3.$$

Using (b),

$$\lim_{x\to 0}(x\cos x) = \lim_{x\to 0} x \lim_{x\to 0} \cos x = 0 \times 1 = 0.$$

Using (c),

$$\lim_{x\to\pi/2} \frac{\sin x}{x} = \frac{\lim_{x\to\pi/2} \sin x}{\lim_{x\to\pi/2} x} = \frac{2}{\pi}. \blacktriangleleft$$

(iv) Limits of functions of $x$ which contain exponents that also depend on $x$ can often be found by taking logarithms.

►*Evaluate the limit*

$$\lim_{x\to\infty}\left(1 - \frac{a^2}{x^2}\right)^{x^2}.$$

Let us define

$$y = \left(1 - \frac{a^2}{x^2}\right)^{x^2}$$

and consider the logarithm of the required limit, i.e.

$$\lim_{x\to\infty} \ln y = \lim_{x\to\infty}\left[x^2 \ln\left(1 - \frac{a^2}{x^2}\right)\right].$$

Using the Maclaurin series for $\ln(1 + x)$ given in subsection 3.6.3, we can expand the logarithm as a series, which gives

$$\lim_{x\to\infty} \ln y = \lim_{x\to\infty}\left[x^2\left(-\frac{a^2}{x^2} - \frac{a^4}{2x^4} + \cdots\right)\right]$$
$$= -a^2$$

Therefore, since $\lim_{x\to\infty} \ln y = -a^2$ it follows that $\lim_{x\to\infty} y = \exp(-a^2)$. ◄

(v) L'Hôpital's rule may be used; it is an extension of (iii)(c) above. In cases where both numerator and denominator are zero or both are infinite, further consideration of the limit must follow. Let us first consider $\lim_{x\to a} f(x)/g(x)$, where $f(a) = g(a) = 0$. Expanding the numerator and denominator as Taylor series we obtain

$$\frac{f(x)}{g(x)} = \frac{f(a) + (x-a)f'(a) + [(x-a)^2/2!]f''(a) + \cdots}{g(a) + (x-a)g'(a) + [(x-a)^2/2!]g''(a) + \cdots}.$$

However, $f(a) = g(a) = 0$ so

$$\frac{f(x)}{g(x)} = \frac{f'(a) + [(x-a)/2!]f''(a) + \cdots}{g'(a) + [(x-a)/2!]g''(a) + \cdots}.$$

Therefore we find

$$\lim_{x\to a} \frac{f(x)}{g(x)} = \frac{f'(a)}{g'(a)},$$

provided $f'(a)$ and $g'(a)$ are not themselves both equal to zero. If, however, $f'(a)$ and $g'(a)$ are both zero then the same process can be applied to the ratio $f'(x)/g'(x)$ to yield

$$\lim_{x\to a} \frac{f(x)}{g(x)} = \frac{f''(a)}{g''(a)},$$

provided at least one of $f''(a)$ and $g''(a)$ is non-zero. If the limit exists, then the process can be repeated as many times as is necessary for the ratio of the corresponding derivatives to be defined, i.e.

$$\lim_{x\to a} \frac{f(x)}{g(x)} = \frac{f^{(n)}(a)}{g^{(n)}(a)}.$$

►*Evaluate the limit*

$$\lim_{x\to 0} \frac{\sin x}{x}.$$

We first note that if $x = 0$, both numerator and denominator are zero. Thus we apply l'Hôpital's rule: differentiating, we obtain

$$\lim_{x\to 0}(\sin x/x) = \lim_{x\to 0}(\cos x/1) = 1. \blacktriangleleft$$

So far we have only considered the case where $f(a) = g(a) = 0$. For the case where $f(a) = g(a) = \infty$ we may still apply l'Hôpital's rule by writing

$$\lim_{x\to a} \frac{f(x)}{g(x)} = \lim_{x\to a} \frac{1/g(x)}{1/f(x)},$$

which is now of the form $0/0$ at $x = a$. Note also that l'Hôpital's rule is still valid for finding limits as $x \to \infty$, i.e. when $a = \infty$. This is easily shown by letting $y = 1/x$ as follows:

$$\begin{aligned}
\lim_{x\to\infty} \frac{f(x)}{g(x)} &= \lim_{y\to 0} \frac{f(1/y)}{g(1/y)} \\
&= \lim_{y\to 0} \frac{-f'(1/y)/y^2}{-g'(1/y)/y^2} \\
&= \lim_{y\to 0} \frac{f'(1/y)}{g'(1/y)} \\
&= \lim_{x\to\infty} \frac{f'(x)}{g'(x)}.
\end{aligned}$$

**Solution method.** *To find the limit of a continuous function $f(x)$ at a point $x = a$, simply substitute the value $a$ into the function noting that $\frac{0}{\infty} = 0$ and that $\frac{\infty}{0} = \infty$. The only difficulty occurs when either of the expressions $\frac{0}{0}$ or $\frac{\infty}{\infty}$ results. In this case differentiate top and bottom and try again. Continue differentiating until the top and bottom limits are no longer both zero or both infinity. If the undetermined form $0 \times \infty$ occurs then it can always be rewritten as $\frac{0}{0}$ or $\frac{\infty}{\infty}$.*

## 3.8 Exercises

3.1 Sum the even numbers between 1000 and 2000 inclusive.

3.2 If you invest £1000 on the first day of each year, and interest is paid at 5% on your balance at the end of each year, how much money do you have after 25 years?

3.3 Use the difference method to sum the series

$$\sum_{n=2}^{N} \frac{2n-1}{2n^2(n-1)^2}.$$

3.4 Determine whether the following series converge ($\theta$ and $p$ are positive real numbers):

$$\text{(a)} \sum_{n=1}^{\infty} \frac{2\sin n\theta}{n(n+1)}; \quad \text{(b)} \sum_{n=1}^{\infty} \frac{2}{n^2}; \quad \text{(c)} \sum_{n=1}^{\infty} \frac{1}{2n^{1/2}};$$

$$\text{(d)} \sum_{n=2}^{\infty} \frac{(-1)^n(n^2+1)^{1/2}}{n\ln n}; \quad \text{(e)} \sum_{n=1}^{\infty} \frac{n^p}{n!}.$$

3.5 For what real values of $x$ are the following series convergent?:

$$\text{(a)} \sum_{n=1}^{\infty} \frac{x^n}{n+1}; \quad \text{(b)} \sum_{n=1}^{\infty} (\sin x)^n; \quad \text{(c)} \sum_{n=1}^{\infty} n^x;$$

$$\text{(d)} \sum_{n=1}^{\infty} e^{nx}; \quad \text{(e)} \sum_{n=2}^{\infty} (\ln n)^x.$$

3.6 A Fabry–Pérot interferometer consists of two parallel heavily silvered glass plates; light enters normally to the plates, and undergoes repeated reflections between them, with a small transmitted fraction emerging at each reflection. Find the intensity $|B|^2$ of the emerging wave, where

$$B = A(1-r)\sum_{n=0}^{\infty} r^n e^{in\phi},$$

with $r$ and $\phi$ real.

3.7 Find the Maclaurin series for

$$(a)\ \ln\left(\frac{1+x}{1-x}\right); \quad (b)\ (x^2+4)^{-1}; \quad (c)\ \sin^2 x.$$

3.8 By using the logarithmic series, prove that if $a$ and $b$ are positive and nearly equal,

$$\ln\frac{a}{b} \simeq \frac{2(a-b)}{a+b}.$$

Show that the error in this approximation is about $2(a-b)^3/[3(a+b)^3]$.

3.9 Find the limits of the following functions:

(a) $\dfrac{x^3+x^2-5x-2}{2x^3-7x^2+4x+4}$, as $x \to 0$, $x \to \infty$ and $x \to 2$.

(b) $\dfrac{\sin x - x\cosh x}{\sinh x - x}$, as $x \to 0$.

(c) $\displaystyle\int_x^{\pi/2} \left(\frac{y\cos y - \sin y}{y^2}\right) dy$, as $x \to 0$.

3.10 Evaluate

$$\lim_{x\to 0}\left[\frac{1}{x^3}\left(\operatorname{cosec} x - \frac{1}{x} - \frac{x}{6}\right)\right].$$

3.11 In quantum theory, a system of oscillators, each of fundamental frequency $\nu$, interacting at temperature $T$ has an average energy $\bar{E}$ given by

$$\bar{E} = \frac{\sum_{n=0}^{\infty} nh\nu e^{-nx}}{\sum_{n=0}^{\infty} e^{-nx}},$$

where $x = h\nu/kT$, $h$ and $k$ being the Planck and Boltzmann constants respectively. Prove that both series converge, evaluate their sums, and show that at high temperatures $\bar{E} \approx kT$ whilst at low temperatures $\bar{E} \approx h\nu \exp(-h\nu/kT)$.

3.12 In a very simple model of a crystal, point-like atomic ions are regularly spaced along an infinite one-dimensional row with spacing $R$. Alternate ions carry equal and opposite charges $\pm e$. The potential energy of the $i$th ion in the electric field due to the $j$th ion is

$$\frac{q_i q_j}{4\pi\epsilon_0 r_{ij}},$$

where $q_j$ is the charge on the $j$th ion and $r_{ij}$ is the distance between the $i$th and $j$th ions.

Write down a series giving the total potential energy $V_i$ of the $i$th ion. Show that the series converges, and, if $V_i$ is written as

$$V_i = \frac{\alpha e^2}{4\pi\epsilon_0 R},$$

find a closed form expression for $\alpha$, the Madelung constant for this (unrealistic) lattice.

3.13 One of the factors contributing to the high relative permittivity of water to static electric fields is the permanent electric dipole moment $p$ of the water molecule. In an external field $E$ the dipoles tend to line up with the field but do not do so completely because of thermal agitation at the temperature $T$ of the water. A classical (non-quantum) calculation using the Boltzmann distribution shows that the average polarisability per molecule $\alpha$ is given by

$$\alpha = \frac{p}{E}(\coth x - x^{-1}),$$

where $x = pE/kT$ and $k$ is the Boltzmann constant.

At ordinary temperatures, even with high field strengths ($10^4$ Vm$^{-1}$ or more), $x \ll 1$. By making suitable series expansions of the hyperbolic functions involved, show that $\alpha = p^2/3kT$ to an accuracy of about one part in $15x^{-2}$.

3.14 In quantum theory a certain method (the Born approximation) gives the (so-called) amplitude $f(\theta)$ for the scattering of a particle of mass $m$ through an angle $\theta$ by a uniform potential well of depth $V_0$ and radius $b$ (i.e. the potential energy of the particle is $-V_0$ within a sphere of radius $b$, and zero elsewhere) as

$$f(\theta) = \frac{2mV_0}{\hbar^2 K^3}(\sin Kb - Kb\cos Kb).$$

Here $\hbar$ is the Planck constant divided by $2\pi$, the energy of the particle is $\hbar^2 k^2/2m$ and $K$ is given by $K = 2k\sin(\theta/2)$.

Use l'Hôpital's rule to evaluate the amplitude at low energies, i.e.

when $k$, and hence $K$, tends to zero, and so determine the low-energy total cross-section. (Note: the differential cross-section is given by $|f(\theta)|^2$ and the total cross-section by $2\pi \int_0^\pi |f(\theta)|^2 \sin\theta \, d\theta$.)

### 3.9 Hints and answers

3.1 $2\sum_{500}^{1000} n = 751500$.

3.2 $Ar(r^n - 1)/(r-1) = £50{,}113$.

3.3 $(1 - N^{-2})/2$.

3.4 (a) Convergent, compare with $\sum n^{-1}(n+1)^{-1}$; (b) convergent, ratio test; (c) divergent, compare with $\sum n^{-1}$; (d) convergent, alternating signs; (e) convergent, ratio test.

3.5 (a) $-1 \le x < 1$; (b) all $x$ except $x = (2n \pm 1)\pi/2$; (c) $x < -1$; (d) $x < 0$; (e) always divergent. Clearly divergent for $x > -1$. For $-X = x < -1$, consider

$$\sum_{k=1}^{\infty} \sum_{M_{k-1}+1}^{M_k} \frac{1}{(\ln M_k)^X},$$

where $\ln M_k = k$ and note that $M_k - M_{k-1} = e^{-1}(e-1)M_k$; hence show that the series diverges.

3.6 $|A|^2(1-r)^2/(1 + r^2 - 2r\cos\phi)$.

3.7

$$\text{(a)} \sum_{n \text{ odd}} \frac{2x^n}{n}; \ \text{(b)} \sum_{n=0}^{\infty} \frac{(-1)^n}{4}\left(\frac{x}{2}\right)^{2n}; \ \text{(c)} \sum_{n=1}^{\infty} \frac{(-1)^{n+1}(2x)^{2n}}{2(2n)!}.$$

3.8 Set $a = D + \delta$ and $b = D - \delta$ and use the expansion for $\ln(1 \pm \delta/D)$.

3.9 (a) $-1/2$, $1/2$, $\infty$; (b) $-4$; (c) $-1 + 2/\pi$.

3.10 $7/360$.

3.11 $E = h\nu[\exp(h\nu/kT) - 1]^{-1}$.

3.12 $\alpha = -2\ln 2$.

3.14 $f(\theta) = 2mV_0b^3/3\hbar^2$ (i.e. independent of $\theta$); $4\pi(2mV_0b^3/3\hbar^2)^2$.

# 4

# *Partial differentiation*

In chapter 1, we discussed functions $f$ of only one variable $x$, which were usually written $f(x)$. Certain constants and parameters may also have appeared in the definition of $f$, e.g. $f(x) = ax+2$ contains the constant 2 and the parameter $a$, but only $x$ was considered as a variable and only the derivatives $f^{(n)}(x) = d^n f/dx^n$ were defined.

However, we may equally well consider functions that depend on more than one variable, for example the function $f(x, y) = x^2 + 3xy$, which depends on the two variables $x$ and $y$ (and the constant 3). For any pair of values $x, y$, the function $f(x, y)$ has a well-defined value, e.g. $f(2, 3) = 22$. This notion can clearly be extended to functions dependent on more than two variables. For the $n$-variable case, we write $f(x_1, x_2, \ldots, x_n)$ for a function that depends on the variables $x_1, x_2, \ldots, x_n$. When $n = 2$, $x_1$ and $x_2$ correspond to the variables $x$ and $y$ used above.

Functions of one variable, like $f(x)$, can be represented by a graph on a plane sheet of paper, and it is apparent that functions of two variables can, with little effort, be represented by a surface in three-dimensional space. We may also picture $f(x, y)$ as describing the variation of height with position in a mountainous landscape. Functions of many variables, however, are usually very difficult to visualise and so the preliminary discussion in this chapter will concentrate on functions of just two variables.

## 4.1 Definition of the partial derivative

It is clear that a function $f(x, y)$ of two variables will have a gradient in all directions in the $xy$-plane. A general expression for this rate of change can be found and will be discussed the next section. However, we first consider the simpler case of finding the rate of change of $f(x, y)$ in the positive $x$- and $y$-directions. These rates of change are called the *partial derivatives* with respect to $x$ and $y$ respectively, and they are extremely important in a wide range of physical applications.

For a function of two variables $f(x, y)$ we may define the derivative with respect to $x$, for example, by saying that it is that for a one-variable function when $y$ is held fixed and treated as a constant. To signify that a derivative is with respect to $x$ but at the same time to recognize that a derivative with respect to $y$ also exists, the former is denoted $\partial f/\partial x$ and is the *partial derivative of* $f(x, y)$ *with respect to* $x$. Similarly, the partial derivative of $f$ with respect to $y$ is denoted by $\partial f/\partial y$.

To define formally the partial derivative of $f$ with respect to $x$, we have

$$\frac{\partial f}{\partial x} = \lim_{\Delta x \to 0} \frac{f(x + \Delta x, y) - f(x, y)}{\Delta x}, \tag{4.1}$$

provided that the limit exists. This is much the same as for the derivative of a one-variable function. The other partial derivative of $f(x, y)$ is similarly defined as a limit (provided it exists):

$$\frac{\partial f}{\partial y} = \lim_{\Delta y \to 0} \frac{f(x, y + \Delta y) - f(x, y)}{\Delta y}. \tag{4.2}$$

It is common practice in connection with partial derivatives of functions involving more than one variable to indicate those variables that are held constant by writing them as subscripts to the derivative symbol. Thus, the partial derivatives defined in (4.1), (4.2) would be written respectively as

$$\left(\frac{\partial f}{\partial x}\right)_y \quad \text{and} \quad \left(\frac{\partial f}{\partial y}\right)_x.$$

In this form, the subscript explicitly shows which variable is to be kept constant. A more compact notation for these partial derivatives is $f_x$ and $f_y$. However, it is extremely important when using partial derivatives to remember which variables are being held constant and it is wise to write out the partial derivative in explicit form if there is any possibility of confusion.

The extension of the definitions (4.1), (4.2) to the general $n$-variable case is straightforward and can be formally written as

$$\frac{\partial f(x_1, x_2, \ldots, x_n)}{\partial x_i} = \lim_{\Delta x_i \to 0} \frac{1}{\Delta x_i}[f(x_1, x_2, \ldots, x_i + \Delta x_i, \ldots, x_n) \\ - f(x_1, x_2, \ldots, x_i, \ldots, x_n)],$$

provided that the limit exists.

Just as for one-variable functions, second (and higher) partial derivatives may be defined in a similar way. For a two-variable function $f(x, y)$ they are

$$\frac{\partial}{\partial x}\left(\frac{\partial f}{\partial x}\right) = \frac{\partial^2 f}{\partial x^2} = f_{xx}, \qquad \frac{\partial}{\partial y}\left(\frac{\partial f}{\partial y}\right) = \frac{\partial^2 f}{\partial y^2} = f_{yy},$$

$$\frac{\partial}{\partial x}\left(\frac{\partial f}{\partial y}\right) = \frac{\partial^2 f}{\partial x \partial y} = f_{xy}, \qquad \frac{\partial}{\partial y}\left(\frac{\partial f}{\partial x}\right) = \frac{\partial^2 f}{\partial y \partial x} = f_{yx}.$$

Only three of the second derivatives are independent since the relation

$$\frac{\partial^2 f}{\partial x \partial y} = \frac{\partial^2 f}{\partial y \partial x},$$

is always obeyed, provided the second partial derivatives are continuous at the point in question. This relation often proves extremely useful as a labour-saving device when evaluating second partial derivatives. It can also be shown that for a function of $n$-variables, $f(x_1, x_2, \ldots, x_n)$, under the same conditions,

$$\frac{\partial^2 f}{\partial x_i \partial x_j} = \frac{\partial^2 f}{\partial x_j \partial x_i}.$$

►*Find the first and second partial derivatives of the function*

$$f(x, y) = 2x^3y^2 + y^3.$$

The first partial derivatives are

$$\frac{\partial f}{\partial x} = 6x^2y^2, \qquad \frac{\partial f}{\partial y} = 4x^3y + 3y^2,$$

and the second partial derivatives are

$$\frac{\partial^2 f}{\partial x^2} = 12xy^2, \qquad \frac{\partial^2 f}{\partial y^2} = 4x^3 + 6y, \qquad \frac{\partial^2 f}{\partial x \partial y} = 12x^2y, \qquad \frac{\partial^2 f}{\partial y \partial x} = 12x^2y,$$

with the last two equal, as expected. ◄

## 4.2 The total differential and total derivative

Having defined the partial derivatives of a function $f(x, y)$, which give the rate of change of $f$ along the positive $x$- and $y$-axes, we must next consider the rate of change of $f(x, y)$ in an arbitrary direction. Suppose that we simultaneously make the small changes $\Delta x$ in $x$ and $\Delta y$ in $y$ and that, as a result, $f$ changes to $f + \Delta f$. Then we must have

$$\begin{aligned}
\Delta f &= f(x + \Delta x, y + \Delta y) - f(x, y) \\
&= f(x + \Delta x, y + \Delta y) - f(x, y + \Delta y) + f(x, y + \Delta y) - f(x, y) \\
&= \left[\frac{f(x + \Delta x, y + \Delta y) - f(x, y + \Delta y)}{\Delta x}\right] \Delta x + \left[\frac{f(x, y + \Delta y) - f(x, y)}{\Delta y}\right] \Delta y.
\end{aligned} \tag{4.3}$$

In the last line we note that the quantities in brackets are very similar to those involved in the definitions of partial derivatives (4.1), (4.2). For them to be strictly equal to the partial derivatives, $\Delta x$ and $\Delta y$ would need to be infinitesimally small. But even for finite (but not too large) $\Delta x$ and $\Delta y$ the approximate formula

$$\Delta f \approx \frac{\partial f(x,y)}{\partial x}\Delta x + \frac{\partial f(x,y)}{\partial y}\Delta y, \tag{4.4}$$

can be obtained. It will be noticed that the first bracket in (4.3) actually approximates to $\partial f(x, y+\Delta y)/\partial x$ but that this has been replaced by $\partial f(x,y)/\partial x$ in (4.4). This approximation clearly has the same degree of validity as that which replaces the bracket by the partial derivative.

How valid an approximation (4.4) is to (4.3) depends not only on how small $\Delta x$ and $\Delta y$ are but also on the magnitudes of higher partial derivatives; this is discussed further in section 4.7 in the context of Taylor series for functions of more than one variable. Nevertheless, letting the small changes $\Delta x$ and $\Delta y$ in (4.4) become infinitesimal, we can define the *total differential* $df$ of the function $f(x,y)$, without any approximation, as

$$df = \frac{\partial f}{\partial x}dx + \frac{\partial f}{\partial y}dy. \tag{4.5}$$

Equation (4.5) can easily be extended to the case of a function of $n$ variables, $f(x_1, x_2, \ldots, x_n)$, in which case

$$df = \frac{\partial f}{\partial x_1}dx_1 + \frac{\partial f}{\partial x_2}dx_2 + \cdots + \frac{\partial f}{\partial x_n}dx_n. \tag{4.6}$$

►*Find the total differential of the function $f(x,y) = y\exp(x+y)$.*

Evaluating the first partial derivatives, we find

$$\frac{\partial f}{\partial x} = y\exp(x+y), \quad \frac{\partial f}{\partial y} = \exp(x+y) + y\exp(x+y).$$

Applying (4.5), we then find that the total differential is given by

$$df = [y\exp(x+y)]dx + [(1+y)\exp(x+y)]dy. \blacktriangleleft$$

In some situations, despite the fact that several variables $x_i$, $i = 1, 2, \ldots, n$, appear to be involved, effectively only one of them is. This occurs if there are subsidiary relationships constraining all the $x_i$ to have values dependent on the value of one of them, say $x_1$. These relationships may be represented by equations, typically of the form

$$x_i = x_i(x_1), \qquad i = 2, \ldots, n. \tag{4.7}$$

In principle $f$ can then be expressed as a function of $x_1$ alone by substituting from (4.7) for $x_2, x_3, \ldots, x_n$, and then the *total derivative* (or simply derivative) of $f$ with respect to $x_1$ obtained by ordinary differentiation.

Alternatively, (4.6) can be used to give

$$\frac{df}{dx_1} = \frac{\partial f}{\partial x_1} + \left(\frac{\partial f}{\partial x_2}\right)\frac{dx_2}{dx_1} + \cdots + \left(\frac{\partial f}{\partial x_n}\right)\frac{dx_n}{dx_1}. \tag{4.8}$$

It should be noticed that the LHS of this equation is the total derivative $df/dx_1$, whilst the partial derivative $\partial f/\partial x_1$ forms only a part of the RHS. In evaluating this partial derivative only *explicit* appearances of $x_1$ in the function $f$ must be taken account of, and *no* allowance must be made for the knowledge that as $x_1$ is changed, this necessarily changes $x_2, x_3, \ldots, x_n$ and these variables also appear in $f$. This latter contribution is precisely the remaining terms on the RHS of (4.8). Naturally, what has been shown using $x_1$ in the above argument applies equally well to any other of the $x_i$, with the appropriate consequent changes.

►*Find the total derivative of* $f(x, y) = x^2 + 3xy$ *with respect to* $x$*, given that* $y = \sin^{-1} x$.

We can see immediately that

$$\frac{\partial f}{\partial x} = 2x + 3y, \qquad \frac{\partial f}{\partial y} = 3x, \qquad \frac{dy}{dx} = \frac{1}{(1-x^2)^{1/2}}$$

and so

$$\begin{aligned}\frac{df}{dx} &= 2x + 3y + 3x\frac{1}{(1-x^2)^{1/2}} \\ &= 2x + 3\sin^{-1} x + \frac{3x}{(1-x^2)^{1/2}}.\end{aligned}$$

Obviously the same expression would have resulted if we had substituted for $y$ from the start, but the above method often produces results with reduced calculation, particularly in more complicated examples. ◄

## 4.3 Exact and inexact differentials

In the last section we discussed how to find the total differential of a function, i.e. its infinitesimal change in an arbitrary direction, in terms of its gradients $\partial f/\partial x$ and $\partial f/\partial y$ in the $x$- and $y$- directions (see (4.5)). Sometimes, however, we may wish to reverse the process and find the function $f$ that differentiates to give a known differential. Usually, finding such functions relies on inspection and experience.

As an example, it is easy to see that the function whose differential $df = x\,dy + y\,dx$ is simply $f(x, y) = xy + c$, where $c$ is a constant. Differentials such as this one, which integrate directly, are called *exact differentials*, whereas those that do not are *inexact differentials*. For example, $x\,dy + 3y\,dx$ is not the straightforward differential of any function (see below). Inexact differentials can be made exact, however, by multiplying through by a suitable function called an integrating factor. This is discussed further in chapter 12.

►*Show that the differential* $x\,dy + 3y\,dx$ *is inexact.*

On the one hand, if we integrate with respect to $x$ we conclude that $f(x, y) = 3xy + g(y)$, where $g(y)$ is any function of $y$. On the other hand, if we integrate with respect to $y$ we conclude that $f(x, y) = xy + h(x)$ where $h(x)$ is any function of $x$. These conclusions are inconsistent for any and every choice of $g(y)$ and $h(x)$, and therefore the differential is inexact. ◄

It is naturally of interest to investigate which properties of a differential make it exact. Consider the general differential containing two variables,

$$df = A(x, y)\,dx + B(x, y)\,dy.$$

We see that

$$\frac{\partial f}{\partial x} = A(x, y), \qquad \frac{\partial f}{\partial y} = B(x, y)$$

and, using the property $f_{xy} = f_{yx}$, we therefore require

$$\frac{\partial A}{\partial y} = \frac{\partial B}{\partial x}. \tag{4.9}$$

This is in fact both a necessary and a sufficient condition for the differential to be exact.

►*Using (4.9) show that* $x\,dy + 3y\,dx$ *is inexact.*

In the above notation, $A(x, y) = 3y$ and $B(x, y) = x$ and so

$$\frac{\partial A}{\partial y} = 3, \qquad \frac{\partial B}{\partial x} = 1.$$

As these are not equal it follows that the differential is inexact. ◄

Determining whether a differential containing many variable $x_1, x_2, \ldots, x_n$ is exact is a simple extension of the above. A differential containing many variables can in general be written as

$$df = \sum_{i=1}^{n} g_i(x_1, x_2, \ldots, x_n)\,dx_i$$

and will be exact if

$$\frac{\partial g_i}{\partial x_j} = \frac{\partial g_j}{\partial x_i} \quad \text{for all pairs } i, j. \tag{4.10}$$

There will be $\frac{1}{2}n(n-1)$ such relationships to be satisfied.

►*Show that*

$$(y+z)\,dx + x\,dy + x\,dz,$$

*is an exact differential.*

In this case, $g_1(x,y,z) = y+z$, $g_2(x,y,z) = x$, $g_3(x,y,z) = x$ and hence $\partial g_1/\partial y = 1 = \partial g_2/\partial x$, $\partial g_3/\partial x = 1 = \partial g_1/\partial z$, $\partial g_2/\partial z = 0 = \partial g_3/\partial y$; therefore, from (4.10), the differential is exact. As mentioned above, it is sometimes possible to show that a differential is exact simply by finding by inspection the function from which it originates. In this example, it can easily be seen that $f(x,y,z) = x(y+z)+c$. ◄

## 4.4 Useful theorems of partial differentiation

So far our discussion has centred on a function $f(x,y)$ dependent on two variables, $x$ and $y$. We could equally, however, have expressed $x$ as a function of $f$ and $y$, or $y$ as a function of $f$ and $x$. To emphasise the point that all the variables are of equal standing, we now replace $f$ by $z$. This does not imply that $x$, $y$ and $z$ are coordinate positions (though they might be). Since $x$ is a function of $y$ and $z$, it follows that

$$dx = \left(\frac{\partial x}{\partial y}\right)_z dy + \left(\frac{\partial x}{\partial z}\right)_y dz \tag{4.11}$$

and similarly, since $y = y(x,z)$,

$$dy = \left(\frac{\partial y}{\partial x}\right)_z dx + \left(\frac{\partial y}{\partial z}\right)_x dz. \tag{4.12}$$

We may now substitute (4.12) into (4.11) to obtain

$$dx = \left(\frac{\partial x}{\partial y}\right)_z \left(\frac{\partial y}{\partial x}\right)_z dx + \left[\left(\frac{\partial x}{\partial y}\right)_z \left(\frac{\partial y}{\partial z}\right)_x + \left(\frac{\partial x}{\partial z}\right)_y\right] dz. \tag{4.13}$$

Now if we hold $z$ constant, so that $dz = 0$, we obtain the *reciprocity relation*

$$\left(\frac{\partial x}{\partial y}\right)_z = \left(\frac{\partial y}{\partial x}\right)_z^{-1},$$

which holds provided both partial derivatives exist and neither is equal to zero. Note, further, that this relationship only holds when the variable being kept constant, in this case $z$, is the same on both sides of the equation.

Alternatively we can put $dx = 0$ in (4.13). Then the contents of the square brackets also equal zero, and we obtain the *cyclic relation*

$$\left(\frac{\partial y}{\partial z}\right)_x \left(\frac{\partial z}{\partial x}\right)_y \left(\frac{\partial x}{\partial y}\right)_z = -1,$$

which holds unless any of the derivatives vanish. In deriving this result we have used the reciprocity relation to replace $(\partial x/\partial z)_y^{-1}$ by $(\partial z/\partial x)_y$.

## 4.5 The chain rule

So far we have discussed the differentiation of a function $f(x, y)$ with respect to its variables $x$ and $y$. We now consider the case where $x$ and $y$ are themselves functions of another variable, say $u$. If we now wish to find the derivative $df/du$, we could simply substitute in $f(x, y)$ the expressions for $x(u)$ and $y(u)$, and then differentiate the resulting function of $u$. Such substitution will quickly give the desired answer in simple cases, but in more complicated examples it is easier to make use of the total differentials described in the previous section.

From the previous section the total differential of $f(x, y)$ is given by

$$df = \frac{\partial f}{\partial x}dx + \frac{\partial f}{\partial y}dy,$$

but we now note that by using the formal device of dividing through by $du$ this immediately implies

$$\frac{df}{du} = \frac{\partial f}{\partial x}\frac{dx}{du} + \frac{\partial f}{\partial y}\frac{dy}{du}, \tag{4.14}$$

which is called the *chain rule* for partial differentiation. This expression provides a direct method for calculating the derivative of $f$ with respect to $u$ and is particularly useful when an equation is expressed in a parametric form.

►*Given that $x(u) = 1+au$ and $y(u) = bu^3$, find the rate of change of $f(x, y) = xe^{-y}$ with respect to $u$.*

As discussed above, this problem could be addressed by substituting for $x$ and $y$ to obtain $f$ as a function only of $u$, and then differentiating with respect to $u$. However, using (4.14) directly we obtain

$$\frac{df}{du} = (e^{-y})a + (-xe^{-y})3bu^2,$$

which on substituting for $x$ and $y$ gives

$$\frac{df}{du} = e^{-bu^3}(a - 3bu^2 - 3bau^3). \blacktriangleleft$$

Equation (4.14) is an example of the chain rule for a function of two variables each of which depends on a single variable. The chain rule may be extended to functions of many variables, each of which is itself a function of a variable $u$, i.e. $f(x_1, x_2, x_3, \ldots, x_n)$, with $x_i = x_i(u)$. In this case the chain rule gives

$$\frac{df}{du} = \sum_{i=1}^{n} \frac{\partial f}{\partial x_i}\frac{dx_i}{du} = \frac{\partial f}{\partial x_1}\frac{dx_1}{du} + \frac{\partial f}{\partial x_2}\frac{dx_2}{du} + \cdots + \frac{\partial f}{\partial x_n}\frac{dx_n}{du}. \tag{4.15}$$

## 4.6 Change of variables

It is sometimes necessary or desirable to make a change of variables during the course of an analysis, and consequently to have to change an equation expressed in one set of variables into an equation using another set. The same situation arises if a function $f$ depends on one set of variables $x_i$, so that $f = f(x_1, x_2, \ldots, x_n)$, but the $x_i$ are given in terms of a further set of variables $u_i$ by the equations

$$x_i = x_i(u_1, u_2, \ldots, u_m). \tag{4.16}$$

For each value of $i$, the function $x_i$ on the right of this equation will be a different function of the $u_i$. In this case the chain rule (4.15) becomes

$$\frac{\partial f}{\partial u_j} = \sum_{i=1}^{n} \frac{\partial f}{\partial x_i}\frac{\partial x_i}{\partial u_j}, \qquad j = 1, 2, \ldots, m, \tag{4.17}$$

and is said to express a *change of variables*. In general the number of variables in each set need not be equal, i.e. $m$ need not equal $n$, but if both the $x_i$ and the $u_i$ are sets of independent variables then $m = n$.

►*Plane polar coordinates, $\rho$ and $\phi$, and Cartesian coordinates, $x$ and $y$, are related by the expressions*

$$x = \rho\cos\phi, \qquad y = \rho\sin\phi,$$

*as can be seen from figure 4.1. An arbitrary function $f(x, y)$ can be re-expressed as a function $g(\rho, \phi)$. Transform the expression*

$$\frac{\partial^2 f}{\partial x^2} + \frac{\partial^2 f}{\partial y^2},$$

*into one in $\rho$ and $\phi$.*

We first note that $\rho^2 = x^2 + y^2$, $\phi = \tan^{-1}(y/x)$. We can now write down the four partial derivatives

$$\frac{\partial \rho}{\partial x} = \frac{x}{(x^2+y^2)^{1/2}} = \cos\phi, \qquad \frac{\partial \phi}{\partial x} = \frac{-(y/x^2)}{1+(y/x)^2} = -\frac{\sin\phi}{\rho},$$

$$\frac{\partial \rho}{\partial y} = \frac{y}{(x^2+y^2)^{1/2}} = \sin\phi, \qquad \frac{\partial \phi}{\partial y} = \frac{1/x}{1+(y/x)^2} = \frac{\cos\phi}{\rho}.$$

Thus, from (4.17), we may write

$$\frac{\partial}{\partial x} = \cos\phi\frac{\partial}{\partial \rho} - \frac{\sin\phi}{\rho}\frac{\partial}{\partial \phi}, \qquad \frac{\partial}{\partial y} = \sin\phi\frac{\partial}{\partial \rho} + \frac{\cos\phi}{\rho}\frac{\partial}{\partial \phi}.$$

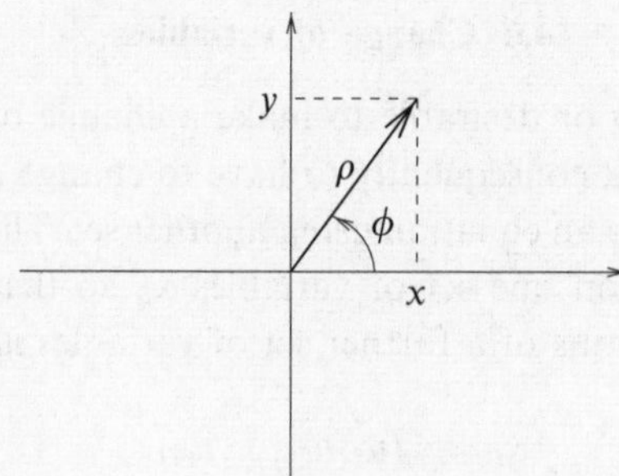


Figure 4.1 The relationship between Cartesian and plane cylindrical polar coordinates.

Now it is only a matter of writing

$$\begin{aligned}\frac{\partial^2 f}{\partial x^2} &= \frac{\partial}{\partial x}\left(\frac{\partial f}{\partial x}\right) = \frac{\partial}{\partial x}\left(\frac{\partial}{\partial x}\right) f \\ &= \left(\cos\phi\frac{\partial}{\partial\rho} - \frac{\sin\phi}{\rho}\frac{\partial}{\partial\phi}\right)\left(\cos\phi\frac{\partial}{\partial\rho} - \frac{\sin\phi}{\rho}\frac{\partial}{\partial\phi}\right) g \\ &= \left(\cos\phi\frac{\partial}{\partial\rho} - \frac{\sin\phi}{\rho}\frac{\partial}{\partial\phi}\right)\left(\cos\phi\frac{\partial g}{\partial\rho} - \frac{\sin\phi}{\rho}\frac{\partial g}{\partial\phi}\right) \\ &= \cos^2\phi\frac{\partial^2 g}{\partial\rho^2} + \frac{2\cos\phi\sin\phi}{\rho^2}\frac{\partial g}{\partial\phi} - \frac{2\cos\phi\sin\phi}{\rho}\frac{\partial^2 g}{\partial\phi\partial\rho} \\ &\quad + \frac{\sin^2\phi}{\rho}\frac{\partial g}{\partial\rho} + \frac{\sin^2\phi}{\rho^2}\frac{\partial^2 g}{\partial\phi^2}\end{aligned}$$

and a similar expression for $\partial^2 f/\partial y^2$,

$$\begin{aligned}\frac{\partial^2 f}{\partial y^2} &= \left(\sin\phi\frac{\partial}{\partial\rho} + \frac{\cos\phi}{\rho}\frac{\partial}{\partial\phi}\right)\left(\sin\phi\frac{\partial}{\partial\rho} + \frac{\cos\phi}{\rho}\frac{\partial}{\partial\phi}\right) g \\ &= \sin^2\phi\frac{\partial^2 g}{\partial\rho^2} - \frac{2\cos\phi\sin\phi}{\rho^2}\frac{\partial g}{\partial\phi} + \frac{2\cos\phi\sin\phi}{\rho}\frac{\partial^2 g}{\partial\phi\partial\rho} \\ &\quad + \frac{\cos^2\phi}{\rho}\frac{\partial g}{\partial\rho} + \frac{\cos^2\phi}{\rho^2}\frac{\partial^2 g}{\partial\phi^2}.\end{aligned}$$

When these two expressions are added together the change of variables is complete and we obtain

$$\frac{\partial^2 f}{\partial x^2} + \frac{\partial^2 f}{\partial y^2} = \frac{\partial^2 g}{\partial\rho^2} + \frac{1}{\rho}\frac{\partial g}{\partial\rho} + \frac{1}{\rho^2}\frac{\partial^2 g}{\partial\phi^2}. \blacktriangleleft$$

## 4.7 Taylor's theorem for many-variable functions

We have already introduced Taylor's theorem for a function $f(x)$ of one variable, in section 3.6. In an analogous way, the Taylor expansion of a function $f(x, y)$ of two variables is given by

$$\begin{aligned} f(x, y) = f(x_0, y_0) &+ \frac{\partial f}{\partial x}\Delta x + \frac{\partial f}{\partial y}\Delta y \\ &+ \frac{1}{2!}\left[\frac{\partial^2 f}{\partial x^2}(\Delta x)^2 + 2\frac{\partial^2 f}{\partial x \partial y}\Delta x \Delta y + \frac{\partial^2 f}{\partial y^2}(\Delta y)^2\right] + \cdots, \end{aligned} \tag{4.18}$$

where $\Delta x = x - x_0$ and $\Delta y = y - y_0$, and all the derivatives are to be evaluated at $(x_0, y_0)$.

►*Find a Taylor expansion, up to quadratic terms in $x - 2$ and $y - 3$, of $f(x, y) = y \exp xy$ about the point $x = 2$, $y = 3$.*

We first evaluate the required partial derivatives of the function i.e.

$$\frac{\partial f}{\partial x} = y^2 \exp xy, \qquad \frac{\partial f}{\partial y} = \exp xy + xy \exp xy,$$

$$\frac{\partial^2 f}{\partial x^2} = y^3 \exp xy, \qquad \frac{\partial^2 f}{\partial y^2} = 2x \exp xy + x^2 y \exp xy,$$

$$\frac{\partial^2 f}{\partial x \partial y} = 2y \exp xy + xy^2 \exp xy.$$

Using (4.18), the Taylor expansion of a two-variable function, we find

$$\begin{aligned} f(x, y) \approx e^6 \Big\{&3 + 9(x-2) + 7(y-3) \\ &+ (2!)^{-1}\left[27(x-2)^2 + 48(x-2)(y-3) + 16(y-3)^2\right]\Big\}. \end{aligned}$$ ◄

It will be noticed that the terms in (4.18) containing first derivatives can be written as

$$\frac{\partial f}{\partial x}\Delta x + \frac{\partial f}{\partial y}\Delta y = \left(\Delta x \frac{\partial}{\partial x} + \Delta y \frac{\partial}{\partial y}\right) f(x, y),$$

where both sides of this relation should be evaluated at the point $(x_0, y_0)$. Similarly the terms in (4.18) containing second derivatives can be written as

$$\frac{1}{2!}\left[\frac{\partial^2 f}{\partial x^2}(\Delta x)^2 + 2\frac{\partial^2 f}{\partial x \partial y}\Delta x \Delta y + \frac{\partial^2 f}{\partial y^2}(\Delta y)^2\right] = \frac{1}{2!}\left(\Delta x \frac{\partial}{\partial x} + \Delta y \frac{\partial}{\partial y}\right)^2 f(x, y), \tag{4.19}$$

where it is understood that the partial derivatives resulting from squaring the expression in parentheses act only on $f(x, y)$ and its derivatives, and not on $\Delta x$

or $\Delta y$; again both sides of (4.19) should be evaluated at $(x_0, y_0)$. It can be shown that the higher-order terms of the Taylor expansion of $f(x, y)$ can be written in an analogous way, and that we may write the full Taylor series as

$$f(x, y) = \sum_{n=0}^{\infty} \frac{1}{n!} \left[ \left( \Delta x \frac{\partial}{\partial x} + \Delta y \frac{\partial}{\partial y} \right)^n f(x, y) \right]_{x_0, y_0}$$

where, as indicated, all the terms on the RHS are to be evaluated at $(x_0, y_0)$.

The most general form of Taylor's theorem, for a function of many variables $f(x_1, \ldots, x_n)$, is a simple extension of the above. Although it is not neccessary to do so, we may think of the $x_i$ as coordinates in $n$-dimensional space and write the function as $f(\mathbf{x})$, where $\mathbf{x}$ is a vector from the origin to $(x_1, x_2, \ldots, x_n)$. Taylor's theorem then becomes

$$f(\mathbf{x}) = f(\mathbf{x}_0) + \sum_i \frac{\partial f}{\partial x_i} \Delta x_i + \frac{1}{2!} \sum_i \sum_j \frac{\partial^2 f}{\partial x_i \partial x_j} \Delta x_i \Delta x_j + \cdots, \tag{4.20}$$

where $\Delta x_i = x_i - x_{i_0}$ and the partial derivatives are evaluated at $(x_{1_0}, x_{2_0}, \ldots, x_{n_0})$. For completeness, we note that in this case the full Taylor series can be written in the form

$$f(\mathbf{x}) = \sum_{n=0}^{\infty} \frac{1}{n!} \left[ (\Delta \mathbf{x} \cdot \nabla)^n f(\mathbf{x}) \right]_{\mathbf{x}=\mathbf{x}_0},$$

where $\nabla$ is the vector differential operator del, to be discussed in chapter 8.

## 4.8 Stationary values of many-variable functions

The idea of the *stationary points* of a function of just one variable has already been discussed in subsection 1.1.8. We recall that the function $f(x)$ has a stationary point at $x = x_0$ if its gradient $df/dx$ is zero at that point. A function may have any number of stationary points, and their nature, i.e. whether they are maxima, minima or stationary points of inflection, is determined by the value of the second derivative at the point. A stationary point is

(i) a minimum if $d^2f/dx^2 > 0$;
(ii) a maximum if $d^2f/dx^2 < 0$;
(iii) a stationary point of inflection if $d^2f/dx^2 = 0$ and changes sign through the point.

We now consider the stationary points of functions of more than one variable; we will see that partial differential analysis is ideally suited to the determination of the position and nature of such points. It is helpful to consider first the case of a function of just two variables but, even in this case, the general situation is more complex than that for a function of one variable, as can be seen from figure 4.2.

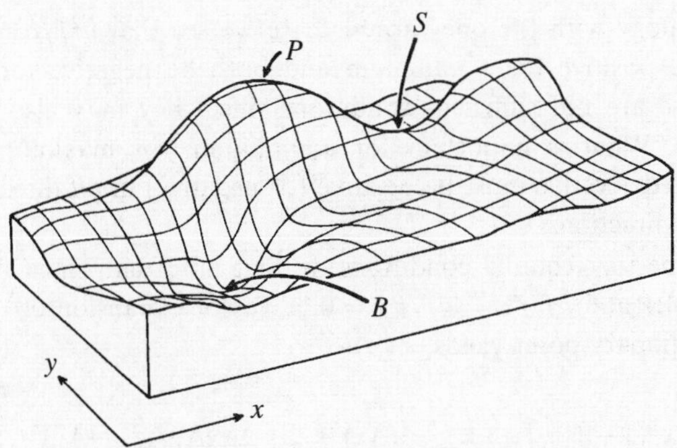


Figure 4.2 Stationary points of a function of two variables. A minimum occurs at $B$, a maximum at $P$ and a saddle point at $S$.

This figure shows part of a three-dimensional model of a function $f(x, y)$. At positions $P$ and $B$ there are a peak and a bowl respectively or, more mathematically, a local maximum and a local minimum. At position $S$ the gradient in any direction is zero but the situation is complicated, since a section parallel to the plane $x = 0$ would show a maximum, but one parallel to the plane $y = 0$ would show a minimum. A point such as $S$ is known as a *saddle point*. The orientation of the 'saddle' in the $xy$-plane is irrelevant; it is as shown in the figure solely for ease of discussion. For a typical *saddle point*, such as $S$, the function increases in some directions away from the point, but decreases in other directions.

For functions of two variables, such as the one shown, it should be clear that a necessary condition for a stationary point (maximum, minimum or saddle point) to occur is that

$$\frac{\partial f}{\partial x} = 0 \qquad \text{and} \qquad \frac{\partial f}{\partial y} = 0. \tag{4.21}$$

The vanishing of the partial derivatives in directions parallel to the axes is enough to ensure that the partial derivative in any other direction is also zero, since any such derivative can be resolved along the directions of the $x$- and $y$-axes. This may be made clearer by considering the total differential,

$$df = \frac{\partial f}{\partial x}dx + \frac{\partial f}{\partial y}dy.$$

Using (4.21) we see that although the infinitesimal changes $dx$ and $dy$ can be chosen independently the change in the value of the function $df$ is always zero.

We now turn our attention to determining the nature of a stationary point of a function of two variables, i.e. whether it is a maximum, a minimum or a saddle point. By analogy with the one-variable case we see that $\partial^2 f/\partial x^2$ and $\partial^2 f/\partial y^2$ must both be positive for a minimum and both be negative for a maximum. However these are not sufficient conditions since they may also be obeyed at saddle points. What is important for a minimum (or maximum) is that the second partial derivative must be positive (or negative) in *all* directions, not just the $x$- and $y$- directions.

To determine the required conditions we first note that, since $f$ is a function of two variables and $\partial f/\partial x = \partial f/\partial y = 0$, a Taylor expansion of the type (4.18) about the stationary point yields

$$f(x, y) - f(x_0, y_0) \approx \frac{1}{2!}\left[(\Delta x)^2 f_{xx} + 2\Delta x \Delta y f_{xy} + (\Delta y)^2 f_{yy}\right],$$

where $\Delta x = x - x_0$ and $\Delta y = y - y_0$ and where the partial derivatives have been written in more compact notation. Rearranging the contents of the square bracket as the weighted sum of two squares, we find

$$f(x, y) - f(x_0, y_0) \approx \frac{1}{2}\left[f_{xx}\left(\Delta x + \frac{f_{xy}\Delta y}{f_{xx}}\right)^2 + (\Delta y)^2\left(f_{yy} - \frac{f_{xy}^2}{f_{xx}}\right)\right]. \tag{4.22}$$

For a minimum, we require (4.22) to be positive for all $\Delta x$ and $\Delta y$, and hence $f_{xx} > 0$ and $f_{yy} - (f_{xy}^2/f_{xx}) > 0$. Given the first constraint, the second can be written $f_{xx}f_{yy} > f_{xy}^2$. Similarly for a maximum we require (4.22) to be negative, and hence $f_{xx} < 0$ and $f_{xx}f_{yy} > f_{xy}^2$. For minima and maxima, symmetry requires that $f_{yy}$ obeys the same criteria as $f_{xx}$. When (4.22) is negative (or zero) for some values of $\Delta x$ and $\Delta y$ but positive (or zero) for others, we have a saddle point. In this case $f_{xx}f_{yy} < f_{xy}^2$. In summary, all stationary points have $f_x = f_y = 0$ and they may be classified further as

(i) minima if $f_{xy}^2 < f_{xx}f_{yy}$, $f_{xx}$ is positive and $f_{yy}$ is positive,

(ii) maxima if $f_{xy}^2 < f_{xx}f_{yy}$, $f_{xx}$ is negative and $f_{yy}$ is negative,

(iii) saddle points if $f_{xy}^2 > f_{xx}f_{yy}$.

Note, however, that if

$$\frac{\partial^2 f}{\partial x^2} = 0 \qquad \text{or} \qquad \frac{\partial^2 f}{\partial y^2} = 0$$

then further investigation is required. For example, for the function $f(x, y) = x^4 + y^4$ at the point $(0, 0)$, $f_{xx} = f_{yy} = 0$ but, nevertheless, $f(x, y)$ has a minimum there.

►*Show that the function $f(x, y) = x^3 \exp(-x^2 - y^2)$ has a maximum at the point $(\sqrt{3/2}, 0)$, a minimum at $(-\sqrt{3/2}, 0)$ and a stationary point at the origin whose nature cannot be determined by the above procedures.*

Setting the first two partial derivatives to zero to locate the stationary points, we find

$$\frac{\partial f}{\partial x} = (3x^2 - 2x^4) \exp(-x^2 - y^2) = 0, \tag{4.23}$$

$$\frac{\partial f}{\partial y} = -2yx^3 \exp(-x^2 - y^2) = 0. \tag{4.24}$$

For (4.24) to be satisfied we require $x = 0$ or $y = 0$ and for (4.23) to be satisfied we require $x = 0$ or $x = \pm\sqrt{3/2}$. Hence the stationary points are at $(0, 0)$, $(\sqrt{3/2}, 0)$ and $(-\sqrt{3/2}, 0)$. We now find the second partial derivatives:

$$f_{xx} = (4x^5 - 14x^3 + 6x) \exp(-x^2 - y^2)$$
$$f_{yy} = x^3(4y^2 - 2) \exp(-x^2 - y^2)$$
$$f_{xy} = 2x^2y(2x^2 - 3) \exp(-x^2 - y^2).$$

We then substitute the pairs of values of $x$ and $y$ for each stationary point and find that at $(0, 0)$

$$f_{xx} = 0, \qquad f_{yy} = 0, \qquad f_{xy} = 0$$

and at $(\pm\sqrt{3/2}, 0)$

$$f_{xx} = \mp 6\sqrt{3/2} \exp(-3/2), \qquad f_{yy} = \mp 3\sqrt{3/2} \exp(-3/2), \qquad f_{xy} = 0.$$

Hence, applying the determination equations above, we find that $(0, 0)$ is an undetermined stationary point, $(\sqrt{3/2}, 0)$ is a maximum and $(-\sqrt{3/2}, 0)$ is a minimum. The function is shown in figure 4.3. ◄

Determining the nature of stationary points for functions of a general number of variables is considerably more difficult and requires a knowledge of the eigenvectors and eigenvalues of matrices. Although these are not discussed until chapter 7, we present the analysis here for completeness. The remainder of this section can therefore be omitted on a first reading.

For a function of $n$ real variables, $f(x_1, x_2, \ldots, x_n)$, we require that, at all stationary points,

$$\frac{\partial f}{\partial x_i} = 0 \qquad \text{for all } x_i.$$

In order to determine the nature of a stationary point, we must expand the function as a Taylor series about it. Recalling the Taylor expansion (4.20) for a

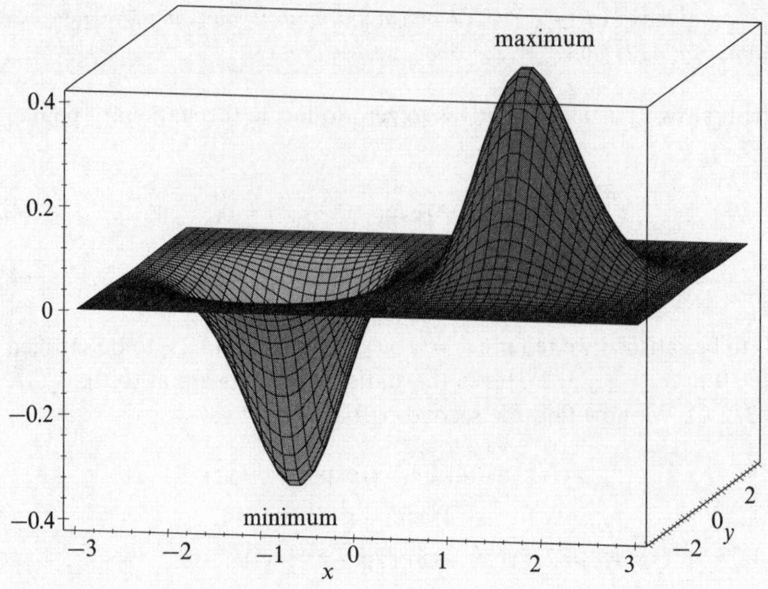


Figure 4.3 The function $f(x, y) = x^3 \exp(-x^2 - y^2)$.

function of $n$ variables, we see that

$$\Delta f = f(\mathbf{x}) - f(\mathbf{x}_0) \approx \frac{1}{2} \sum_i \sum_j \frac{\partial^2 f}{\partial x_i \partial x_j} \Delta x_i \Delta x_j. \tag{4.25}$$

If we define the matrix $\mathsf{M}$ to have elements given by

$$M_{ij} = \frac{\partial^2 f}{\partial x_i \partial x_j},$$

then we can rewrite (4.25) as

$$\Delta f = \tfrac{1}{2} \Delta \mathsf{x}^{\mathrm{T}} \mathsf{M} \Delta \mathsf{x}, \tag{4.26}$$

where $\Delta \mathsf{x}$ is the column vector with the $\Delta x_i$ as its components and $\Delta \mathsf{x}^{\mathrm{T}}$ is its transpose. Since $\mathsf{M}$ is real and symmetric it has $n$ real eigenvalues $\lambda_r$ and $n$ orthogonal eigenvectors $\mathsf{e}_r$, which after suitable normalisation satisfy

$$\mathsf{M}\mathsf{e}_r = \lambda_r \mathsf{e}_r, \qquad \mathsf{e}_r^{\mathrm{T}} \mathsf{e}_s = \delta_{rs},$$

where the *Kronecker delta* symbol $\delta_{rs}$ equals unity for $r = s$ and equals zero otherwise. These eigenvectors form a basis set for the $n$-dimensional space and we can therefore expand $\Delta\mathsf{x}$ in terms of them, obtaining

$$\Delta\mathsf{x} = \sum_r a_r \mathbf{e}_r,$$

where the $a_r$ are coefficients dependent upon $\Delta\mathsf{x}$. Substituting this into (4.26), we find

$$\Delta f = \tfrac{1}{2}\Delta\mathsf{x}^{\mathrm{T}}\mathsf{M}\Delta\mathsf{x} = \tfrac{1}{2}\sum_r \lambda_r a_r^2.$$

Now, for the stationary point to be a minimum, we require $\Delta f = \frac{1}{2}\sum_r \lambda_r a_r^2 > 0$ for all sets of values of the $a_r$, and therefore all the eigenvalues of $\mathsf{M}$ to be greater than zero. Conversely, for a maximum we require $\Delta f = \frac{1}{2}\sum_r \lambda_r a_r^2 < 0$, and therefore all the eigenvalues of $\mathsf{M}$ to be less than zero. If the eigenvalues have mixed signs, then we have a saddle point. Note that the test may fail if the eigenvalues are equal to zero.

►*Derive the conditions for maxima, minima and saddle points for a function of two real variables, using the above analysis.*

For a two-variable function the matrix $\mathsf{M}$ is given by

$$\mathsf{M} = \begin{pmatrix} f_{xx} & f_{xy} \\ f_{yx} & f_{yy} \end{pmatrix}.$$

Therefore its eigenvalues satisfy the equation

$$\begin{vmatrix} f_{xx} - \lambda & f_{xy} \\ f_{xy} & f_{yy} - \lambda \end{vmatrix} = 0.$$

Hence

$$(f_{xx} - \lambda)(f_{yy} - \lambda) - f_{xy}^2 = 0$$

$$\Rightarrow \quad f_{xx}f_{yy} - (f_{xx} + f_{yy})\lambda + \lambda^2 - f_{xy}^2 = 0$$

$$\Rightarrow \quad 2\lambda = (f_{xx} + f_{yy}) \pm \sqrt{(f_{xx} + f_{yy})^2 - 4(f_{xx}f_{yy} - f_{xy}^2)},$$

which by rearrangement of the terms under the square root gives

$$2\lambda = (f_{xx} + f_{yy}) \pm \sqrt{(f_{xx} - f_{yy})^2 + 4f_{xy}^2}.$$

Now, that $\mathsf{M}$ is real and symmetric implies that its eigenvalues are real, and so for both eigenvalues to be positive (corresponding to a minimum), we require $f_{xx}$ and $f_{yy}$ positive and also

$$f_{xx} + f_{yy} > \sqrt{(f_{xx} + f_{yy})^2 - 4(f_{xx}f_{yy} - f_{xy}^2)},$$

$$\Rightarrow \quad f_{xx}f_{yy} - f_{xy}^2 > 0.$$

A similar procedure will find the criteria for maxima and saddle points. ◄

### 4.9 Stationary values under constraints

In the previous section we looked at the problem of finding stationary values of a function of two or more variables when all the variables may be independently varied. However, it is often the case in physical problems that not all the variables used to describe a situation are in fact independent, i.e. some relationship between the variables must be satisfied. For example, if we walk through a hilly landscape and we are constrained to walk along a path, we will never reach the highest peak on the landscape, unless the path happens to take us to it. Nevertheless, we can still find the highest point that we have reached during our journey.

We first discuss the case of a function of just two variables. Let us consider finding the maximum value of $f(x,y)$ subject to the constraint $g(x,y) = c$, where $c$ is a constant. In the above analogy, $f(x,y)$ might represent the height of the land above sea-level in some hilly region, whilst $g(x,y) = c$ is the equation of the path along which we walk.

We could, of course, use the constraint $g(x,y) = c$ to substitute for $x$ or $y$ in $f(x,y)$, thereby obtaining a new function of only one variable whose stationary points could be found using the methods discussed in subsection 1.1.8. However, such a procedure can involve a lot of algebra and becomes very tedious for functions of more than two variables. A more direct method for solving such problems is the *method of Lagrange undetermined multipliers*, which we now discuss.

To maximise $f$ we require

$$df = \frac{\partial f}{\partial x}dx + \frac{\partial f}{\partial y}dy = 0.$$

If $dx$ and $dy$ were independent, we could conclude $f_x = 0 = f_y$. However, here they are not independent, but constrained because $g$ is constant:

$$dg = \frac{\partial g}{\partial x}dx + \frac{\partial g}{\partial y}dy = 0.$$

Multiplying $dg$ by an as yet unknown number $\lambda$ and adding it to $df$ we obtain

$$d(f + \lambda g) = \left(\frac{\partial f}{\partial x} + \lambda\frac{\partial g}{\partial x}\right)dx + \left(\frac{\partial f}{\partial y} + \lambda\frac{\partial g}{\partial y}\right)dy = 0,$$

where $\lambda$ is called a *Lagrange undetermined multiplier*. This introduction of an additional unknown is intended to render $dx$ and $dy$ independent; to achieve this we must therefore choose $\lambda$ such that

$$\frac{\partial f}{\partial x} + \lambda\frac{\partial g}{\partial x} = 0, \tag{4.27}$$

$$\frac{\partial f}{\partial y} + \lambda\frac{\partial g}{\partial y} = 0. \tag{4.28}$$

These equations, together with the constraint $g(x,y) = c$, are sufficient to find the three unknowns, $\lambda$ and the values of $x$, $y$ at the required stationary point.

►*The temperature of a point $(x, y)$ on a unit circle is given by $T(x, y) = xy$. Find the temperature of the two hottest points on the circle.*

We need to maximise $T(x, y)$ subject to the constraint $x^2 + y^2 = 1$. Applying (4.27) and (4.28), we obtain

$$y + 2\lambda x = 0, \tag{4.29}$$
$$x + 2\lambda y = 0. \tag{4.30}$$

These results, together with the original constraint $x^2 + y^2 = 1$, provide three simultaneous equations that may be solved for $\lambda$, $x$ and $y$.

From (4.29) and (4.30) we find $\lambda = \pm 1/2$ which in turn imply $y = \mp x$. Remembering that $x^2 + y^2 = 1$, we find that

$$y = x \quad \Rightarrow \quad x = \pm\frac{1}{\sqrt{2}}, \qquad y = \pm\frac{1}{\sqrt{2}}$$
$$y = -x \quad \Rightarrow \quad x = \mp\frac{1}{\sqrt{2}}, \qquad y = \pm\frac{1}{\sqrt{2}}.$$

We have not yet, however, determined which of these stationary points are maxima and which are minima. In this simple case, we can simply substitute the four pairs of $x$- and $y$- values into $T(x, y) = xy$, from which we find the maximum temperature on the unit circle is $T_{\rm max} = 1/2$ at the points $y = x = \pm 1/\sqrt{2}$. ◄

The method of Lagrange multipliers can be used to find the stationary points of functions of more than two variables, subject to several constraints, provided that the number of constraints is smaller than the number of variables. For example, if we wish to find the stationary points of $f(x, y, z)$ subject to the constraints $g(x, y, z) = c_1$ and $h(x, y, z) = c_2$, where $c_1$ and $c_2$ are constants, then we proceed as above, obtaining

$$\begin{aligned}
\frac{\partial}{\partial x}(f + \lambda g + \mu h) &= \frac{\partial f}{\partial x} + \lambda\frac{\partial g}{\partial x} + \mu\frac{\partial h}{\partial x} = 0, \\
\frac{\partial}{\partial y}(f + \lambda g + \mu h) &= \frac{\partial f}{\partial y} + \lambda\frac{\partial g}{\partial y} + \mu\frac{\partial h}{\partial y} = 0, \\
\frac{\partial}{\partial z}(f + \lambda g + \mu h) &= \frac{\partial f}{\partial z} + \lambda\frac{\partial g}{\partial z} + \mu\frac{\partial h}{\partial z} = 0.
\end{aligned} \tag{4.31}$$

We may now solve these three equations, together with the two constraints, to give $\lambda$, $\mu$, $x$, $y$ and $z$.

►*Find the stationary points of* $f(x,y,z) = x^3 + y^3 + z^3$ *subject to the following constraints:*

(*i*) $g(x,y,z) = x^2 + y^2 + z^2 = 1$;
(*ii*) $g(x,y,z) = x^2 + y^2 + z^2 = 1$ *and* $h(x,y,z) = x + y + z = 0$.

*Case (i).* Since there is only one constraint in this case, we need only introduce a single Lagrange multiplier to obtain

$$\begin{aligned} \frac{\partial}{\partial x}(f + \lambda g) &= 3x^2 + 2\lambda x = 0, \\ \frac{\partial}{\partial y}(f + \lambda g) &= 3y^2 + 2\lambda y = 0, \\ \frac{\partial}{\partial z}(f + \lambda g) &= 3z^2 + 2\lambda z = 0. \end{aligned} \tag{4.32}$$

These equations are highly symmetrical and clearly have the solution $x = y = z = -2\lambda/3$. Using the constraint $x^2 + y^2 + z^2 = 1$ we find $\lambda = \pm\sqrt{3}/2$ and so stationary points occur at

$$x = y = z = \pm\frac{1}{\sqrt{3}}. \tag{4.33}$$

In solving the three equations (4.32) in this way, however, we have implicitly assumed that $x$, $y$ and $z$ are non-zero. But, it is clear from (4.32) that any of these values can equal zero, with the exception of the case where $x = y = z = 0$ since this is prohibited by the constraint $x^2 + y^2 + z^2 = 1$. We must consider the other cases separately.

If $x = 0$, for example, we require

$$\begin{aligned} 3y^2 + 2\lambda y &= 0, \\ 3z^2 + 2\lambda z &= 0, \\ y^2 + z^2 &= 1. \end{aligned}$$

Clearly, we require $\lambda \neq 0$, otherwise these equations are inconsistent. If neither $y$ nor $z$ is zero we find $y = -2\lambda/3 = z$ and from the third equation we require $y = z = \pm 1/\sqrt{2}$. If $y = 0$, however, then $z = \pm 1$ and, similarly, if $z = 0$ then $y = \pm 1$. Thus the stationary points having $x = 0$ are $(0, 0, \pm 1)$, $(0, \pm 1, 0)$ and $(0, \pm 1/\sqrt{2}, \pm 1/\sqrt{2})$. A similar procedure can be followed for the cases $y = 0$ and $z = 0$ respectively and, in addition to those already obtained, we find the stationary points $(\pm 1, 0, 0)$, $(\pm 1/\sqrt{2}, 0, \pm 1/\sqrt{2})$ and $(\pm 1/\sqrt{2}, \pm 1/\sqrt{2}, 0)$.

*Case (ii).* We now have two constraints and must therefore introduce two

Lagrange multipliers to obtain (cf. (4.31))

$$\frac{\partial}{\partial x}(f + \lambda g + \mu h) = 3x^2 + 2\lambda x + \mu = 0, \tag{4.34}$$

$$\frac{\partial}{\partial y}(f + \lambda g + \mu h) = 3y^2 + 2\lambda y + \mu = 0, \tag{4.35}$$

$$\frac{\partial}{\partial z}(f + \lambda g + \mu h) = 3z^2 + 2\lambda z + \mu = 0. \tag{4.36}$$

These equations are again highly symmetrical and the simplest way to proceed is to subtract (4.35) from (4.34) to obtain

$$3(x^2 - y^2) + 2\lambda(x - y) = 0$$

$$\Rightarrow \qquad 3(x + y)(x - y) + 2\lambda(x - y) = 0. \tag{4.37}$$

This equation is clearly satisfied if $x = y$ and from the second constraint, $x + y + z = 0$, we find $z = -2x$. Substituting these values into the first constraint, $x^2 + y^2 + z^2 = 1$, we therefore obtain

$$x = \pm\frac{1}{\sqrt{6}}, \qquad y = \pm\frac{1}{\sqrt{6}}, \qquad z = \mp\frac{2}{\sqrt{6}}. \tag{4.38}$$

Because of the high degree of symmetry amongst the equations (4.34)–(4.36), we may obtain by inspection two further relations analogous to (4.37), one containing the variables $y, z$ and the other the variables $x, z$. Assuming $y = z$ in the first relation and $x = z$ in the second, we find the stationary points

$$x = \pm\frac{1}{\sqrt{6}}, \qquad y = \mp\frac{2}{\sqrt{6}}, \qquad z = \pm\frac{1}{\sqrt{6}} \tag{4.39}$$

and

$$x = \mp\frac{2}{\sqrt{6}}, \qquad y = \pm\frac{1}{\sqrt{6}}, \qquad z = \pm\frac{1}{\sqrt{6}}. \tag{4.40}$$

We note that in finding the stationary points (4.38)–(4.40) we did not need to evaluate the Lagrange multipliers $\lambda$ and $\mu$ explicitly. This is not always the case, however, and in some problems it may be simpler to begin by finding the values of these multipliers.

Returning to (4.37) we must now consider the case where $x \neq y$ and we find

$$3(x + y) + 2\lambda = 0. \tag{4.41}$$

However, in obtaining the stationary points (4.39), (4.40), we did *not* assume $x = y$, but only required $y = z$ and $x = z$ respectively. It is clear that $x \neq y$ at these stationary points, and it can be shown that they do indeed satisfy (4.41). Similarly, several stationary points for which $x \neq z$ or $y \neq z$ have already been found.

Thus we need only consider further the two cases where (i) $x = y = z$ and (ii) $x$, $y$ and $z$ are all different. The first is clearly prohibited by the constraint

$x+y+z=0$. For the second case, (4.41) must be satisfied, together with the analogous equations containing $y, z$ and $x, z$ respectively, i.e.

$$3(x+y)+2\lambda=0,$$
$$3(y+z)+2\lambda=0,$$
$$3(x+z)+2\lambda=0.$$

Adding these three equations together and using the constraint $x+y+z=0$ we find $\lambda=0$. However, for $\lambda=0$, the equations are inconsistent for non-zero $x$, $y$ and $z$.

Therefore all the stationary points have already been found and are given by (4.38)–(4.40). ◄

The method may be extended to functions of any number $n$ of variables subject to any smaller number $m$ of constraints. This means that effectively there are $n-m$ independent variables and, as mentioned above, we could solve by substitution and then by the methods of the previous section. However, for large $n$ this becomes cumbersome and Lagrange undetermined multipliers are a useful simplification.

►*A system contains a very large number $N$ of particles, each of which can be in any of $R$ energy levels with a corresponding energy $E_i$, $i=1,2,\ldots,R$. The number of particles in the ith level is $n_i$ and the total energy of the system is a constant, $E$. Find the distribution of particles amongst the energy levels that maximises the expression*

$$P=\frac{N!}{n_1!n_2!\cdots n_R!},$$

*subject to the constraints that both the number of particles and the total energy remain constant, i.e.*

$$g=N-\sum_{i=1}^{R}n_i=0 \quad \text{and} \quad h=E-\sum_{i=1}^{R}n_iE_i=0.$$

The way in which we proceed is as follows. In order to maximise $P$, we must minimise its denominator (since the numerator is fixed). Minimising the denominator is the same as minimising the logarithm of the denominator, i.e.

$$f=\ln(n_1!n_2!\cdots n_R!)=\ln(n_1!)+\ln(n_2!)+\cdots+\ln(n_R!).$$

Using Stirling's approximation, $\ln(n!)\approx n\ln n-n$, we find that

$$\begin{aligned}f&=n_1\ln n_1+n_2\ln n_2+\cdots+n_R\ln n_R-(n_1+n_2+\cdots+n_R)\\&=\left(\sum_{i=1}^{R}n_i\ln n_i\right)-N.\end{aligned}$$

It has been assumed here that, for the desired distribution, all the $n_i$ are large. Thus, we now have a function $f$ subject to two constraints, $g = 0$ and $h = 0$, and we can apply the Lagrange method, obtaining (cf. (4.31))

$$\frac{\partial f}{\partial n_1} + \lambda \frac{\partial g}{\partial n_1} + \mu \frac{\partial h}{\partial n_1} = 0,$$
$$\frac{\partial f}{\partial n_2} + \lambda \frac{\partial g}{\partial n_2} + \mu \frac{\partial h}{\partial n_2} = 0,$$
$$\vdots$$
$$\frac{\partial f}{\partial n_R} + \lambda \frac{\partial g}{\partial n_R} + \mu \frac{\partial h}{\partial n_R} = 0.$$

Since all these equations are alike, we consider the general case

$$\frac{\partial f}{\partial n_k} + \lambda \frac{\partial g}{\partial n_k} + \mu \frac{\partial h}{\partial n_k} = 0,$$

for $k = 1, 2, \ldots, R$. Substituting the functions $f$, $g$ and $h$ into this relation we find

$$\frac{n_k}{n_k} + \ln n_k + \lambda(-1) + \mu(-E_k) = 0,$$

which can be rearranged to give

$$\ln n_k = \mu E_k + \lambda - 1,$$

and hence

$$n_k = C \exp \mu E_k.$$

We now have the general form for the distribution of particles amongst energy levels, but in order to determine the two constants $\mu$, $C$ we recall that

$$\sum_{k=1}^{R} C \exp \mu E_k = N,$$

and

$$\sum_{k=1}^{R} C E_k \exp \mu E_k = E.$$

This is known as the Boltzmann distribution and is a well-known result from statistical mechanics. ◀

## 4.10 Thermodynamic relations

Thermodynamic relations provide a useful set of physical examples of partial differentiation. The relations we will derive are called *Maxwell's thermodynamic relations.* They express relationships between four thermodynamic quantities describing a unit mass of a substance. The quantities are the pressure $P$, the volume $V$, the thermodynamic temperature $T$ and the entropy $S$ of the substance. These four quantities are not independent, since only two of them are independently variable.

The first law of thermodynamics may be expressed as

$$dU = T\,dS - P\,dV, \tag{4.42}$$

where $U$ is the internal energy of the substance. Essentially this is a conservation of energy equation, but we shall concern ourselves not with the physics but rather with the use of partial differentials to relate the four basic quantities discussed above. The method involves writing $dU = T\,dS - P\,dV$ in terms of the differentials of the variables, say $X$ and $Y$, required to remain constant, equating this to the total differential

$$dU = \left(\frac{\partial U}{\partial X}\right)_Y dX + \left(\frac{\partial U}{\partial Y}\right)_X dY \tag{4.43}$$

and then using the relationship

$$\frac{\partial^2 U}{\partial X \partial Y} = \frac{\partial^2 U}{\partial Y \partial X}$$

to obtain the required Maxwell relation.

►*Show that* $(\partial T/\partial V)_S = -(\partial P/\partial S)_V$.

Here the variables to stay constant are just those corresponding to the differentials on the RHS of (4.42). We have

$$T\,dS - P\,dV = dU = \left(\frac{\partial U}{\partial S}\right)_V dS + \left(\frac{\partial U}{\partial V}\right)_S dV,$$

and so we find directly

$$\left(\frac{\partial U}{\partial S}\right)_V = T \quad \text{and} \quad \left(\frac{\partial U}{\partial V}\right)_S = -P.$$

Differentiating the first expression with respect to $V$ and the second with respect to $S$, and using

$$\frac{\partial^2 U}{\partial V \partial S} = \frac{\partial^2 U}{\partial S \partial V},$$

we find the Maxwell relation

$$\left(\frac{\partial T}{\partial V}\right)_S = -\left(\frac{\partial P}{\partial S}\right)_V. \blacktriangleleft$$

►*Show that* $(\partial S/\partial V)_T = (\partial P/\partial T)_V$.

Applying (4.43) to $dS$, with variables $V$ and $T$, we find

$$dU = T\,dS - P\,dV = T\left[\left(\frac{\partial S}{\partial V}\right)_T dV + \left(\frac{\partial S}{\partial T}\right)_V dT\right] - P\,dV.$$

Similarly applying (4.43) to $dU$, we find

$$dU = \left(\frac{\partial U}{\partial V}\right)_T dV + \left(\frac{\partial U}{\partial T}\right)_V dT.$$

Thus, equating partial derivatives,

$$\left(\frac{\partial U}{\partial V}\right)_T = T\left(\frac{\partial S}{\partial V}\right)_T - P \quad \text{and} \quad \left(\frac{\partial U}{\partial T}\right)_V = T\left(\frac{\partial S}{\partial T}\right)_V$$

but since

$$\frac{\partial^2 U}{\partial T \partial V} = \frac{\partial^2 U}{\partial V \partial T},$$

it follows that

$$\left(\frac{\partial S}{\partial V}\right)_T + T\frac{\partial^2 S}{\partial T \partial V} - \left(\frac{\partial P}{\partial T}\right)_V = \frac{\partial}{\partial V}\left[T\left(\frac{\partial S}{\partial T}\right)_V\right]_T = T\frac{\partial^2 S}{\partial V \partial T}.$$

Thus finally we get the Maxwell relation

$$\left(\frac{\partial S}{\partial V}\right)_T = \left(\frac{\partial P}{\partial T}\right)_V. \blacktriangleleft$$

The above derivation is rather cumbersome, however, and a useful trick that can simplify the working is to define a new function, called a *potential.* The internal energy $U$ discussed above is one example of a potential but three others are commonly defined and they are described below.

►*Show that* $(\partial S/\partial V)_T = (\partial P/\partial T)_V$ *by considering the potential* $U - ST$.

We first consider the differential $d(U - ST)$. From (4.5), we obtain

$$d(U - ST) = dU - SdT - TdS = -SdT - PdV$$

when use is made of (4.42). We rewrite $U - ST$ as $F$ for convenience of notation; $F$ is called the *Helmholtz potential.* Thus

$$dF = -SdT - PdV,$$

and it follows that

$$\left(\frac{\partial F}{\partial T}\right)_V = -S \quad \text{and} \quad \left(\frac{\partial F}{\partial V}\right)_T = -P.$$

Using these results together with

$$\frac{\partial^2 F}{\partial T \partial V} = \frac{\partial^2 F}{\partial V \partial T},$$

we can see immediately that

$$\left(\frac{\partial S}{\partial V}\right)_T = \left(\frac{\partial P}{\partial T}\right)_V,$$

which is the same Maxwell relation as before. ◄

Although the Helmholtz potential has other uses, in this context it has simply provided a means for a quick derivation of the Maxwell relation. The other Maxwell relations can be derived similarly by using two other potentials, the *enthalpy*, $H = U + PV$, and the *Gibbs free energy*, $G = U + PV - ST$ (see exercise 4.8).

## 4.11 Differentiation of integrals

We conclude this chapter with a discussion of the differentiation of integrals. Let us consider the indefinite integral (cf. equation (1.25))

$$F(x,t) = \int f(x,t)\,dt,$$

from which it follows immediately that

$$\frac{\partial F(x,t)}{\partial t} = f(x,t).$$

Assuming that the second partial derivatives of $F(x,t)$ are continuous, we have

$$\frac{\partial^2 F(x,t)}{\partial t \partial x} = \frac{\partial^2 F(x,t)}{\partial x \partial t},$$

so we can write

$$\frac{\partial}{\partial t}\left[\frac{\partial F(x,t)}{\partial x}\right] = \frac{\partial}{\partial x}\left[\frac{\partial F(x,t)}{\partial t}\right] = \frac{\partial f(x,t)}{\partial x}.$$

Integrating this equation with respect to $t$ then gives

$$\frac{\partial F(x,t)}{\partial x} = \int \frac{\partial f(x,t)}{\partial x}\,dt. \qquad (4.44)$$

Now consider the definite integral

$$\begin{aligned} I(x) &= \int_{t=u}^{t=v} f(x,t)\,dt \\ &= F(x,v) - F(x,u), \end{aligned}$$

where $u$ and $v$ are constants. Differentiating this integral with respect to $x$, and using (4.44), we see that

$$\begin{aligned}\frac{dI(x)}{dx} &= \frac{\partial F(x,v)}{\partial x} - \frac{\partial F(x,u)}{\partial x}\\ &= \int^{v} \frac{\partial f(x,t)}{\partial x} dt - \int^{u} \frac{\partial f(x,t)}{\partial x} dt\\ &= \int_{u}^{v} \frac{\partial f(x,t)}{\partial x} dt.\end{aligned}$$

This is *Leibnitz' rule* for differentiating integrals, and it basically states that for constant limits of integration the order of integration and differentiation can be reversed.

In the more general case where the limits of the integral are themselves functions of $x$, it follows immediately that

$$\begin{aligned}I(x) &= \int_{t=u(x)}^{t=v(x)} f(x,t)\, dt\\ &= F(x, v(x)) - F(x, u(x)),\end{aligned}$$

which yields the partial derivatives

$$\frac{\partial I}{\partial v} = f(x, v(x)), \qquad \frac{\partial I}{\partial u} = -f(x, u(x)).$$

Consequently

$$\begin{aligned}\frac{dI}{dx} &= \left(\frac{\partial I}{\partial v}\right)\frac{dv}{dx} + \left(\frac{\partial I}{\partial u}\right)\frac{du}{dx} + \frac{\partial I}{\partial x}\\ &= f(x, v(x))\frac{dv}{dx} - f(x, u(x))\frac{du}{dx} + \frac{\partial}{\partial x}\int_{u(x)}^{v(x)} f(x,t)dt\\ &= f(x, v(x))\frac{dv}{dx} - f(x, u(x))\frac{du}{dx} + \int_{u(x)}^{v(x)} \frac{\partial f(x,t)}{\partial x} dt, \qquad (4.45)\end{aligned}$$

where the partial derivative with respect to $x$ in the last term has been taken inside the integral sign using (4.44). This procedure is valid because $u(x)$ and $v(x)$ are being held constant in this term.

►*Find the derivative with respect to x of the integral*

$$I(x) = \int_x^{x^2} \frac{\sin xt}{t}\, dt.$$

Applying (4.45), we see that

$$\begin{aligned}
\frac{dI}{dx} &= \frac{\sin x^3}{x^2}(2x) - \frac{\sin x^2}{x}(1) + \int_x^{x^2} \frac{t\cos xt}{t} dt \\
&= \frac{2\sin x^3}{x} - \frac{\sin x^2}{x} + \left[\frac{\sin xt}{x}\right]_x^{x^2} \\
&= 3\frac{\sin x^3}{x} - 2\frac{\sin x^2}{x} \\
&= \frac{1}{x}(3\sin x^3 - 2\sin x^2). \blacktriangleleft
\end{aligned}$$

## 4.12 Exercises

4.1 The equation $3y = z^3 + 3xz$ defines $z$ implicitly as a function of $x$ and $y$. Evaluate all three second partial derivatives of $z$ with respect to $x$ and/or $y$. Verify that $z$ is a solution of

$$x\frac{\partial^2 z}{\partial y^2} + \frac{\partial^2 z}{\partial x^2} = 0.$$

4.2 If $x = e^u \cos\theta$ and $y = e^u \sin\theta$, show that

$$\frac{\partial^2 \phi}{\partial u^2} + \frac{\partial^2 \phi}{\partial \theta^2} = (x^2 + y^2)\left(\frac{\partial^2 f}{\partial x^2} + \frac{\partial^2 f}{\partial y^2}\right),$$

where $f(x, y) = \phi(u, \theta)$.

4.3 Find and evaluate the maxima, minima and saddle points of the function

$$f(x, y) = xy(x^2 + y^2 - 1).$$

4.4 Locate the stationary points of the function

$$f(x, y) = (x^2 - 2y^2)\exp[-(x^2 + y^2)/a^2],$$

where $a$ is a non-zero constant.

Sketch the function along the $x$- and $y$- axes and hence identify the nature and values of the stationary points.

4.5 Find the stationary points of the function

$$f(x, y) = x^3 + xy^2 - 12x - y^2$$

and identify their nature.

4.6 The temperature of a point $(x, y, z)$ on the unit sphere is given by

$$T(x, y, z) = xy + yz.$$

By using the method of Lagrange multipliers find the temperature of the hottest point on the sphere.

4.7 A rectangular parallelepiped has all eight vertices on the ellipsoid

$$x^2 + 3y^2 + 3z^2 = 1.$$

Using the symmetry of the parallelepiped about each of the planes $x = 0$, $y = 0$, $z = 0$, write down the surface area of the parallelepiped in terms of the coordinates of the vertex that lies in the octant $x, y, z \geq 0$. Hence find the maximum value of the surface area of such a parallelepiped.

4.8 By considering the differential

$$dG = d(U + PV - ST),$$

where $G$ is the Gibbs free energy, $P$ the pressure, $V$ the volume, $S$ the entropy and $T$ the temperature of a system, and given further that

$$dU = TdS - PdV,$$

derive a Maxwell relation connecting $(\partial V/\partial T)_P$ and $(\partial S/\partial P)_T$.

4.9 Functions $P(V, T)$, $U(V, T)$ and $S(V, T)$ are related by

$$TdS = dU + PdV,$$

where the symbols have the same meaning as in the previous question. $P$ is known from experiment to have the form

$$P = \tfrac{1}{3}T^4 + T/V,$$

in appropriate units. If

$$U = \alpha V T^4 + \beta T,$$

where $\alpha$, $\beta$, are constants (or at least do not depend on $T$, $V$), deduce that $\alpha$ must have a specific value but $\beta$ may have any value. Find the corresponding form of $S$.

4.10 The entropy $S(H, T)$, the magnetisation $M(H, T)$, and the internal energy $U(H, T)$ of a magnetic salt placed in a magnetic field of strength $H$ at temperature $T$ are connected by the equation

$$TdS = dU - HdM.$$

By considering $d(U - TS - HM)$, or otherwise, prove that

$$\left(\frac{\partial M}{\partial T}\right)_H = \left(\frac{\partial S}{\partial H}\right)_T.$$

For a particular salt

$$M(H,T) = M_0[1 - \exp(-\alpha H/T)].$$

Show that, at a fixed temperature, if the applied field is increased from zero to a strength such that the magnetization of the salt is $\frac{3}{4}M_0$, then the salt's entropy *decreases* by an amount

$$\frac{M_0}{4\alpha}(3 - \ln 4).$$

4.11 If

$$I(\alpha) = \int_0^1 \frac{x^\alpha - 1}{\ln x}\,dx, \qquad \alpha > -1,$$

what is the value of $I(0)$? Show that

$$\frac{d}{d\alpha}x^\alpha = x^\alpha \ln x,$$

and deduce that

$$\frac{d}{d\alpha}I(\alpha) = \frac{1}{\alpha+1}.$$

Hence prove that $I(\alpha) = \ln(1+\alpha)$.

4.12 Find the derivative with respect to $x$ of the integral

$$I(x) = \int_x^{3x} \exp xt\, dt.$$

## 4.13 Hints and answers

4.1 $\partial^2 z/\partial x^2 = 2xz(z^2+x)^{-3}$, $\partial^2 z/\partial x\partial y = (z^2-x)(z^2+x)^{-3}$, $\partial^2 z/\partial y^2 = -2z(z^2+x)^{-3}$.

4.2 Write $\partial/\partial u$ and $\partial/\partial\theta$ in terms of $x$, $y$, $\partial/\partial x$ and $\partial/\partial y$ using (4.17). The terms that cancel when $\partial^2\phi/\partial u^2$ and $\partial^2\phi/\partial\theta^2$ are added are $\pm[x(\partial f/\partial x) + y(\partial f/\partial y) + 2xy(\partial^2 f/\partial x\partial y)]$.

4.3 Maxima equal to 1/8 at $\pm(1/2, -1/2)$, minima equal to $-1/8$ at $\pm(1/2, 1/2)$, saddle points equalling 0 at $(0,0)$, $(0,\pm 1)$, $(\pm 1, 0)$.

4.4 Maxima equal to $a^2e^{-1}$ at $(\pm a, 0)$, minima equal to $-2a^2e^{-1}$ at $(0, \pm a)$, saddle point equalling 0 at $(0,0)$.

4.5 Maximum equal to 16 at $(-2,0)$, minimum equal to $-16$ at $(2,0)$, saddle points equalling $-11$ at $(1,\pm 3)$.

4.6 $1/\sqrt{2}$ at $\pm(1/2, 1/\sqrt{2}, 1/2)$.

4.7 Lagrange multiplier method gives $z = y = x/2$ for maximal area of 4.

4.8 Show that $(\partial G/\partial P)_T = V$ and $(\partial G/\partial T)_P = -S$. From each result obtain an expression for $\partial^2 G/\partial T\partial P$ and equate these, giving $(\partial V/\partial T)_P = -(\partial S/\partial P)_T$.

4.9 Establish that $(\partial U/\partial V)_T = T(\partial S/\partial V)_T - P$ and $(\partial U/\partial T)_V = T(\partial S/\partial T)_V$. Equate expressions for $\partial^2 S/\partial T \partial V$ and hence show $\alpha = 1$. Integrate $(\partial S/\partial V)_T$ and $(\partial S/\partial T)_V$ to show that $S = 4T^3V/3 + \ln V + \beta \ln T + c$.

4.10 Show $dF = d(U - TS - HM) = -S\,dT - M\,dH$ and find two expressions for $\partial^2 F/\partial H \partial T$. Establish that $(\partial S/\partial H)_T = -M_0 \alpha H T^{-2} \exp(-\alpha H/T)$ and integrate with respect to $H$.

4.11 $I(0) = 0$; use Leibnitz' rule.

4.12 $(6 - x^{-2}) \exp(3x^2) - (2 - x^{-2}) \exp x^2$.

# 5

# *Multiple integrals*

For functions of several variables, just as we may consider derivatives with respect to two or more of them, so may the integral of the function with respect to more than one variable be formed. The formal definitions of such multiple integrals are extensions of the case of a single variable, discussed in chapter 1. We first discuss double and triple integrals and illustrate some of their applications. We then consider changing the variables in multiple integrals and discuss some general properties of Jacobians.

## 5.1 Double integrals

For an integral involving two variables – a double integral – we have a function, $f(x, y)$ say, to be integrated with respect to $x$ and $y$ between certain limits. These limits can usually be represented by a closed curve $C$ bounding a region $R$ in the $xy$-plane. Following the discussion of single integrals given in chapter 1, let us divide the region $R$ into $N$ subregions $\Delta R_p$ of area $\Delta A_p$, $p = 1, \ldots, N$, and let $(x_p, y_p)$ be any point in the subregion $\Delta R_p$. Now consider the sum

$$S = \sum_{p=1}^{N} f(x_p, y_p) \Delta A_p,$$

and let $N \to \infty$ as each of the areas $\Delta A_p \to 0$. If the sum $S$ tends to a unique limit, $I$, then this is called the *double integral of* $f(x, y)$ *over the region* $R$ and is written

$$I = \int_R f(x, y)\, dA, \tag{5.1}$$

where $dA$ stands for the element of area in the $xy$-plane. By choosing the subregions to be small rectangles each of area $\Delta A = \Delta x \Delta y$, and letting both $\Delta x$

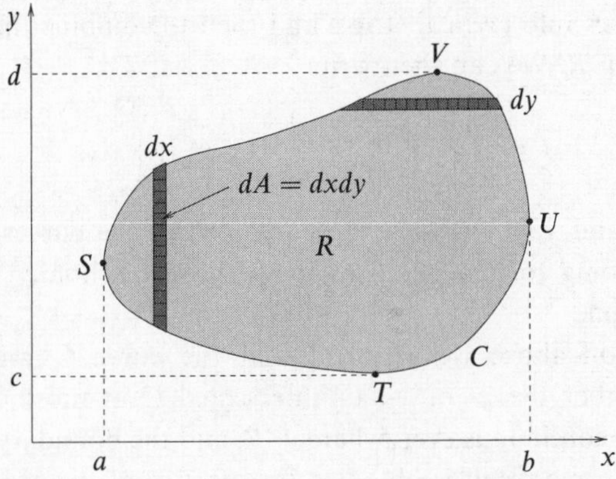


Figure 5.1 A simple curve in the $xy$-plane, enclosing the region $R$.

and $\Delta y \to 0$, we can also write the integral as

$$I = \iint_R f(x, y)\, dx\, dy, \tag{5.2}$$

where we have written out the element of area explicitly as the product of the two coordinate differentials (see figure 5.1).

Some authors use a single integration symbol whatever the dimension of the integral; others use as many symbols as the dimension. In different circumstances both have their advantages. We will adopt the convention used in (5.1) and (5.2), that as many integration symbols will be used as differentials *explicitly* written.

The form (5.2) gives us a clue as to how we may proceed in the evaluation of a double integral. Referring to figure 5.1, the limits on the integration may be written as an equation $c(x, y) = 0$ giving the boundary curve $C$. However, an explicit statement of the limits can be written in two separate ways.

One way of evaluating the integral is first to sum up the small rectangular elemental areas into horizontal strips (as shown in the figure), and then to combine these horizontal strips to cover the region $R$. In this case, we can write

$$I = \int_{y=c}^{y=d} \left\{ \int_{x=x_1(y)}^{x=x_2(y)} f(x, y)\, dx \right\} dy, \tag{5.3}$$

where $x = x_1(y)$ and $x = x_2(y)$ are the equations of the curves $TSV$ and $TUV$ respectively. This expression indicates that first $f(x, y)$ is to be integrated with respect to $x$ (treating $y$ as a constant) between the values $x = x_1(y)$ and $x = x_2(y)$, and then the result (considered as a function of $y$) is to be integrated between the limits $y = c$ and $y = d$. Thus the double integral is evaluated by expressing it in terms of two single integrals called *iterated* (or *repeated*) integrals.

An alternative way of evaluating the integral, however, is first to sum up the elemental rectangles into *vertical* strips, and then to combine these vertical strips to cover the region $R$. We can then write

$$I = \int_{x=a}^{x=b} \left\{ \int_{y=y_1(x)}^{y=y_2(x)} f(x,y)\,dy \right\} dx, \qquad (5.4)$$

where $y = y_1(x)$ and $y = y_2(x)$ are the equations of the curves $STU$ and $SVU$ respectively. In going to (5.4) from (5.3), we have essentially interchanged the order of integration.

In the discussion above we assumed that the curve $C$ was such that any line parallel to either the $x$- or $y$-axis intersected $C$ at most twice. In general, provided $f(x,y)$ is continuous everywhere in $R$, and the boundary curve $C$ has this simple shape, the same result is obtained irrespective of the order of integration. In cases where the region $R$ has a more complicated shape, it can usually be subdivided into smaller simpler regions $R_1$, $R_2$ etc. that satisfy this criterion. The double integral over $R$ is then merely the sum of the double integrals over the subregions.

►*Evaluate the double integral*

$$I = \iint_R x^2 y\,dx\,dy,$$

*where $R$ is the triangular area bounded by the lines $x = 0$, $y = 0$ and $x + y = 1$. Reverse the order of integration and demonstrate that the same result is obtained.*

The area of integration is shown in figure 5.2. Suppose we choose to carry out the integration with respect to $y$ first. With $x$ fixed, the range in $y$ is 0 to $1 - x$. We can therefore write

$$\begin{aligned} I &= \int_{x=0}^{x=1} \left\{ \int_{y=0}^{y=1-x} x^2 y\,dy \right\} dx \\ &= \int_{x=0}^{x=1} \left[ \frac{x^2 y^2}{2} \right]_{y=0}^{y=1-x} dx = \int_0^1 \frac{x^2(1-x)^2}{2}\,dx = \frac{1}{60}. \end{aligned}$$

Alternatively, we may choose to perform the integration with respect to $x$ first. With $y$ fixed, the range of $x$ is 0 to $1 - y$, so we have

$$\begin{aligned} I &= \int_{y=0}^{y=1} \left\{ \int_{x=0}^{x=1-y} x^2 y\,dx \right\} dy \\ &= \int_{y=0}^{y=1} \left[ \frac{x^3 y}{3} \right]_{x=0}^{x=1-y} dx = \int_0^1 \frac{(1-y)^3 y}{3}\,dy = \frac{1}{60}. \end{aligned}$$

As expected, we obtain the same result irrespective of the order of integration. ◄

We may avoid the use of braces in expressions such as (5.3) and (5.4) by writing

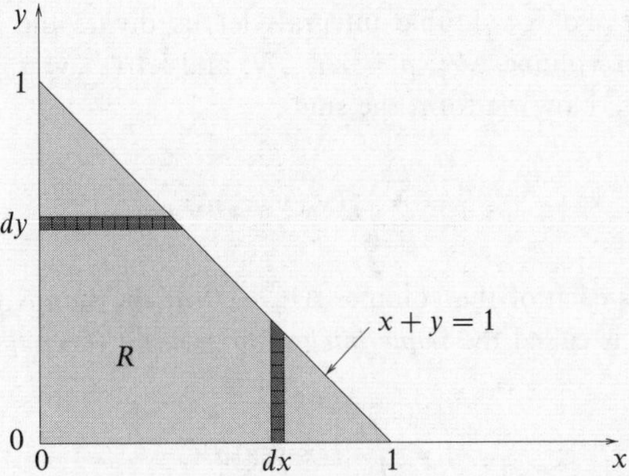


Figure 5.2 The triangular region whose sides are the lines $x = 0$, $y = 0$ and $x + y = 1$.

(5.4), for example, as

$$I = \int_a^b dx \int_{y_1(x)}^{y_2(x)} dy\, f(x, y),$$

where it is understood that each integral symbol acts on everything to its right, and that the order of integration is from right to left. So, in this example, the integrand $f(x, y)$ is first to be integrated with respect to $y$, and then with respect to $x$. With the double integral expressed in this way, we will also no longer write the independent variables explicitly in the limits of integration, since the differential of the variable with respect to which we are integrating is always adjacent to the relevant integral sign.

Using the order of integration in (5.3), we could also write the double integral as

$$I = \int_c^d dy \int_{x_1(y)}^{x_2(y)} dx\, f(x, y).$$

Occasionally, however, the interchange of the order of integration in a double integral is not permissible, as it yields a different result. Such difficulties might arise if (amongst other things) the region $R$ were unbounded, so that some of the limits are infinite. However, in many cases involving infinite limits the same result is obtained whichever order of integration is used. Difficulties can also occur if the integrand $f(x, y)$ has any discontinuities in the region $R$ or on its boundary $C$.

## 5.2 Triple integrals

The above discussion for double integrals can easily be extended to triple integrals. Consider the function $f(x, y, z)$ defined in a closed three-dimensional region $R$.

Proceeding as we did for double integrals let us divide the region $R$ into $N$ subregions $\Delta R_p$ of volume $\Delta V_p$, $p = 1, \ldots, N$, and let $(x_p, y_p, z_p)$ be any point in the subregion $\Delta R_p$. Now we form the sum

$$S = \sum_{p=1}^{N} f(x_p, y_p, z_p) \Delta V_p,$$

and let $N \to \infty$ as each of the volumes $\Delta V_p \to 0$. If the sum $S$ tends to a unique limit, $I$, then this is called the *triple integral of $f(x, y, z)$ over the region $R$* and is written

$$I = \int_R f(x, y, z)\, dV, \tag{5.5}$$

where $dV$ stands for the element of volume. By choosing the subregions to be small cuboids, each of volume $\Delta V = \Delta x \Delta y \Delta z$, and proceeding to the limit, we can also write the integral as

$$I = \iiint_R f(x, y, z)\, dx\, dy\, dz, \tag{5.6}$$

where we have written out the element of volume explicitly as the product of the three coordinate differentials. Extending the discussion of double integrals, we may write triple integrals as three iterated integrals, for example,

$$I = \int_{x_1}^{x_2} dx \int_{y_1(x)}^{y_2(x)} dy \int_{z_1(x,y)}^{z_2(x,y)} dz\, f(x, y, z),$$

where the limits on each of the integrals describe the values that $x$, $y$ and $z$ take on the boundary of the region $R$. As for double integrals, in most cases the order of integration does not affect the value of the integral.

We can extend these ideas to define multiple integrals of higher dimensionality in a similar way.

## 5.3 Applications of multiple integrals

Multiple integrals have many uses in the physical sciences, since there are numerous physical quantities which can be written in terms of them. We now discuss a few of the more common examples.

### *5.3.1 Areas and volumes*

Multiple integrals are often used in finding areas and volumes. For example, the integral

$$A = \int_R dA = \iint_R dx\, dy$$

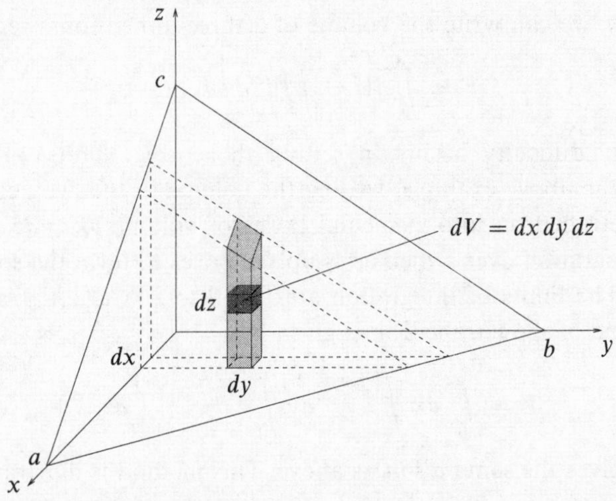


Figure 5.3 The tetrahedron bounded by the coordinate surfaces and the plane $x/a + y/b + z/c = 1$ is divided up into vertical slabs, the slabs into columns and the columns into small boxes.

is simply equal to the area of the region $R$. Similarly, if we consider the surface $z = f(x, y)$ in three-dimensional Cartesian coordinates, then the volume under this surface that stands vertically above the region $R$ is given by the integral

$$V = \int_R z\, dA = \iint_R f(x, y)\, dx\, dy,$$

where volumes above the $xy$-plane are counted as positive, and those below as negative.

►*Find the volume of the tetrahedron bounded by the coordinate surfaces ($x = 0$, $y = 0$ and $z = 0$) and the plane $x/a + y/b + z/c = 1$.*

Referring to figure 5.3, the elemental volume of the shaded region is given by $dV = z\, dx\, dy$, and we must integrate over the triangular region $R$ in the $xy$-plane whose sides are $x = 0$, $y = 0$ and $y = b - bx/a$. The total volume of the tetrahedron is therefore given by

$$\begin{aligned} V = \iint_R z\, dx\, dy &= \int_0^a dx \int_0^{b-bx/a} dy\, c\left(1 - \frac{y}{b} - \frac{x}{a}\right) \\ &= c\int_0^a dx \left[y - \frac{y^2}{2b} - \frac{xy}{a}\right]_{y=0}^{y=b-bx/a} \\ &= c\int_0^a dx \left(\frac{bx^2}{2a^2} - \frac{bx}{a} + \frac{b}{2}\right) = \frac{abc}{6}. \blacktriangleleft \end{aligned}$$

Alternatively, we can write the volume of a three-dimensional region $R$ as

$$V = \int_R dV = \iiint_R dx\,dy\,dz, \tag{5.7}$$

where the only difficulty occurs in setting the correct limits on each of the integrals. For the above example, writing the volume in this way corresponds to dividing the tetrahedron into elemental boxes of volume $dx\,dy\,dz$ (as shown in figure 5.3); integration over $z$ then adds up the boxes to form the shaded column in the figure. The limits of integration are $z = 0$ to $z = c\left(1 - y/b - x/a\right)$, and the total volume of the tetrahedron is given by

$$V = \int_0^a dx \int_0^{b-bx/a} dy \int_0^{c(1-y/b-x/a)} dz, \tag{5.8}$$

which clearly gives the same result as above. This method is illustrated further in the following example.

►*Find the volume of the region bounded by the paraboloid $z = x^2 + y^2$ and the plane $z = 2y$.*

The required region is shown in figure 5.4. In order to write the volume of the region in the form (5.7), we must deduce the limits on each of the integrals. Since the integrations can be performed in any order, let us first divide the region into vertical slabs of thickness $dy$ perpendicular to the $y$-axis, and then as shown in the figure we cut each slab into horizontal strips of height $dz$, and each strip into elemental boxes of volume $dV = dx\,dy\,dz$. Integrating first with respect to $x$ (adding up the elemental boxes to get a horizontal strip), the limits on $x$ are $x = -\sqrt{z-y^2}$ to $x = \sqrt{z-y^2}$. Now integrating with respect to $z$ (adding up the strips to form a vertical slab) the limits on $z$ are $z = y^2$ to $z = 2y$. Finally, integrating with respect to $y$ (adding up the slabs to obtain the required region), the limits on $y$ are $y = 0$ and $y = 2$ (the solutions of the simultaneous equations $z = 0^2 + y^2$ and $z = 2y$). So the volume of the region is

$$\begin{aligned} V &= \int_0^2 dy \int_{y^2}^{2y} dz \int_{-\sqrt{z-y^2}}^{\sqrt{z-y^2}} dx = \int_0^2 dy \int_{y^2}^{2y} dz\, 2\sqrt{z-y^2} \\ &= \int_0^2 dy \left[\tfrac{4}{3}(z-y^2)^{3/2}\right]_{z=y^2}^{z=2y} = \int_0^2 dy\, \tfrac{4}{3}(2y-y^2)^{3/2}. \end{aligned}$$

The integral over $y$ may be straightforwardly evaluated by making the substitution $y = 1 + \sin u$, and gives $V = \pi/2$. ◄

In general, when calculating the volume (area) of a region, the volume (area) elements need not be small boxes as in the previous example, but may be of any convenient shape. They are usually chosen to make the evaluation of the integral as simple as possible.

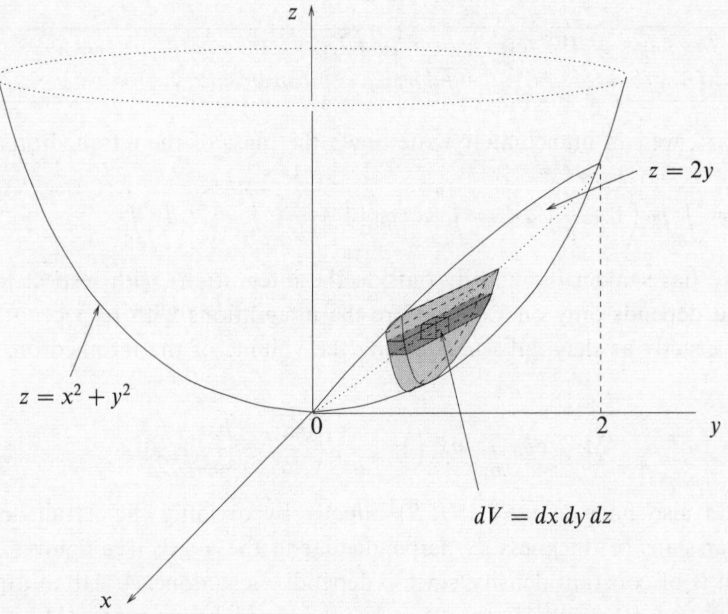


Figure 5.4 The region bounded by the paraboloid $z = x^2 + y^2$ and the plane $z = 2y$ is divided into vertical slabs, the slabs into horizontal strips and the strips into boxes.

### 5.3.2 Masses, centres of mass and centroids

It is sometimes necessary to calculate the mass of a given object having a non-uniform density. Symbolically, this mass is given simply by

$$M = \int dM,$$

where $dM$ is the element of mass, and the integral is taken over the extent of the object. For a solid three-dimensional body the element of mass is just $dM = \rho\, dV$, where $dV$ is an element of volume and $\rho$ is the variable density. For a laminar body (i.e. a sheet of material that is uniform in one dimension) the element of mass $dM = \sigma\, dA$, where $\sigma$ is the mass per unit area of the object and $dA$ is an element of area of the lamina. Finally, for a body in the form of a thin wire, $dM = \lambda\, ds$, where $\lambda$ is the mass per unit length and $ds$ is an element of arc length along the wire. When evaluating the required integral, we are free to divide up the body into those mass elements that are most convenient, provided that over each mass element the density is approximately constant.

► *Find the mass of the tetrahedron bounded by the coordinate surfaces and the plane $x/a + y/b + z/c = 1$, if its density is given by $\rho(x, y, z) = \rho_0(1 + x/a)$.*

From (5.8), we can immediately write down the mass of the tetrahedron as

$$M = \int_R \rho_0 \left(1 + \frac{x}{a}\right) dV = \int_0^a dx\, \rho_0 \left(1 + \frac{x}{a}\right) \int_0^{b-bx/a} dy \int_0^{c(1-y/b-x/a)} dz,$$

where we have taken the density outside the integrations with respect to $z$ and $y$ since it depends only on $x$. Therefore the integrations with respect to $z$ and $y$ proceed exactly as they did when finding the volume of the tetrahedron, and we have

$$M = c\rho_0 \int_0^a dx \left(1 + \frac{x}{a}\right) \left(\frac{bx^2}{2a^2} - \frac{bx}{a} + \frac{b}{2}\right). \tag{5.9}$$

We could also have arrived at (5.9) directly by dividing the tetrahedron into triangular slabs of thickness $dx$ perpendicular to the $x$-axis (see figure 5.3), each of which is of constant density, since $\rho$ depends on $x$ alone. A slab at a position $x$ has volume $dV = \frac{1}{2}c(1 - x/a)(b - bx/a)\, dx$, and hence mass $dM = \rho\, dV = \rho_0(1 + x/a)\, dV$. Integrating over $x$ we again obtain (5.9). This integral is easily evaluated and gives $M = \frac{5}{24}abc\rho_0$. ◄

The coordinates of the centre of mass of a solid or laminar body may also be written as multiple integrals. The centre of mass of a body has coordinates $(\bar{x}, \bar{y}, \bar{z})$ given by the three equations

$$\bar{x} \int dM = \int x\, dM$$
$$\bar{y} \int dM = \int y\, dM$$
$$\bar{z} \int dM = \int z\, dM,$$

where again $dM$ is an element of mass as described above, $(x, y, z)$ are the coordinates of the centre of mass of the element $dM$, and the integrals are taken over the entire body. Obviously, for any body that lies entirely in, or is symmetrical about, the $xy$-plane (say), we immediately have $\bar{z} = 0$. For completeness, we note that the three equations above can be written in the single vector equation (see chapter 6)

$$\bar{\mathbf{r}} = \frac{1}{M} \int \mathbf{r}\, dM,$$

where $\bar{\mathbf{r}}$ is the position vector of the body's centre of mass with respect to the origin, $\mathbf{r}$ is the position vector of the centre of mass of the element $dM$ and $M = \int dM$ is the total mass of the body. As above, we may divide the body

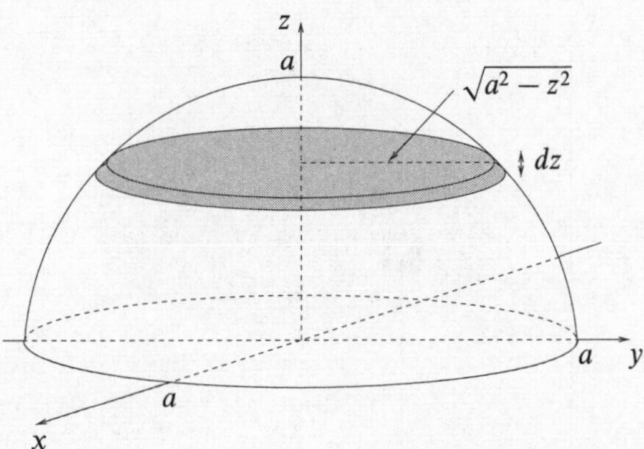


Figure 5.5 The solid hemisphere bounded by the surfaces $x^2 + y^2 + z^2 = a^2$ and the $xy$-plane.

into the most convenient mass elements for evaluating the necessary integrals, provided each mass element is of constant density.

We further note that the coordinates of the *centroid* of a body are simply those that would give the centre of mass of the body if it had a uniform density.

►*Find the centre of mass of the solid hemisphere bounded by the surfaces $x^2 + y^2 + z^2 = a^2$ and the xy-plane, assuming that it has a uniform density $\rho$.*

Referring to figure 5.5, we know from symmetry that the centre of mass must lie on the $z$-axis. Let us divide the hemisphere into circular slab volume elements of thickness $dz$ parallel to the $xy$-plane. For a slab at a height $z$, the mass of the element is $dM = \rho\, dV = \rho\pi(a^2 - z^2)\, dz$. Integrating over $z$, we find that the $z$-coordinate of the centre of mass of the hemisphere is given by

$$\bar{z} \int_0^a \rho\pi(a^2 - z^2)\, dz = \int_0^a z\rho\pi(a^2 - z^2)\, dz.$$

The integrals are easily evaluated and give $\bar{z} = 3a/8$. Since the hemisphere is of uniform density, this is also the position of its centroid. ◄

### *5.3.3 Pappus' theorems*

The theorems of Pappus (which are about seventeen centuries old) relate centroids to volumes of revolution and areas of surfaces (discussed in chapter 1), and can be useful in finding one quantity given another which may be calculated more easily.

*Pappus' first theorem* states that if a plane area is rotated about an axis that does not intersect it, then the volume of revolution generated is given by the plane area $A$ multiplied by the distance moved by its centroid (see figure 5.6).

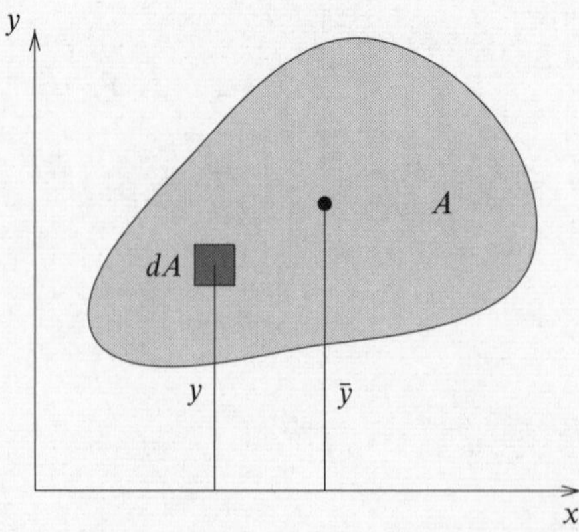


Figure 5.6 An area $A$ in the $xy$-plane, which may be rotated about the $x$-axis to form a volume of revolution.

This may be proved by considering the definition of the centroid of the plane area as the position of the centre of mass if the density is uniform, so that

$$\bar{y} = \frac{1}{A}\int y\, dA.$$

Now the volume of revolution generated by rotating the plane area about the $x$-axis is given by

$$V = \int 2\pi y\, dA = 2\pi \bar{y} A,$$

which is the area multiplied by the distance moved by the centroid.

*Pappus' second theorem* states that if a plane curve is rotated about a coplanar axis that does not intersect it, then the area of revolution generated is given by the length of the curve $L$ multiplied by the distance moved by its centroid (see figure 5.7). This may be proved in a similar manner to the first theorem by considering the definition of the centroid of a plane curve,

$$\bar{y} = \frac{1}{L}\int y\, ds,$$

and noting that the surface area of revolution is given by

$$S = \int 2\pi y\, ds = 2\pi \bar{y} L,$$

which is equal to the length of the curve multiplied by the distance moved by its centroid.

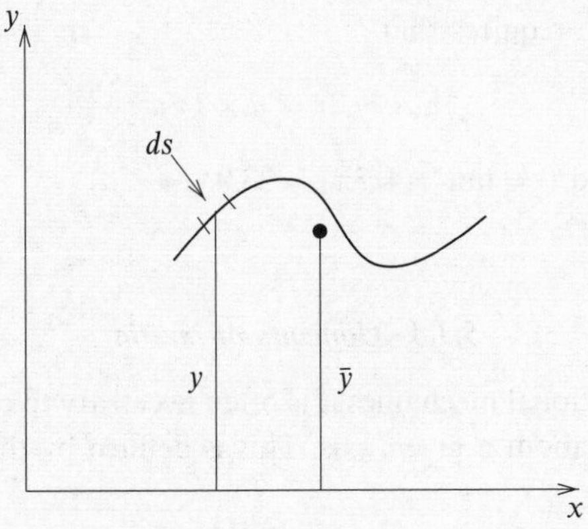


Figure 5.7 A curve in the $xy$-plane, which may be rotated about the $x$-axis to form a surface of revolution.

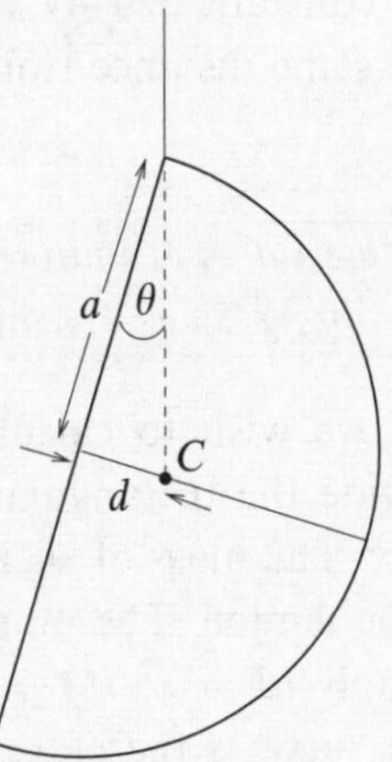


Figure 5.8 Suspending a semicircular lamina from one of its corners.

► *A semicircular uniform lamina is freely suspended from one of its corners. Show that its straight edge makes an angle of* 23.0° *with the vertical.*

Referring to figure 5.8, the suspended lamina will have its centre of gravity $C$ vertically below the suspension point and its straight edge will make an angle $\theta = \tan^{-1}(d/a)$ with the vertical, where $2a$ is the diameter of the semicircle and $d$ is the distance of its centre of mass from the diameter.

Since rotating the lamina about the diameter generates a sphere of volume $\frac{4}{3}\pi a^3$,

Pappus' first theorem requires that

$$\tfrac{4}{3}\pi a^3 = 2\pi \times d \times \tfrac{1}{2}\pi a^2.$$

Hence $d = 4a/3\pi$ and $\theta = \tan^{-1}(4/3\pi) = 23.0°$. ◄

### 5.3.4 Moments of inertia

For problems in rotational mechanics it is often necessary to calculate the moment of inertia of a body about a given axis. This is defined by the multiple integral

$$I = \int l^2 \, dM,$$

where $l$ is the distance of a mass element $dM$ from the axis. We may again choose mass elements convenient for evaluating the integral. In this case, however, in addition to elements of constant density we require all parts of each element to be at approximately the same distance from the axis about which the moment of inertia is required.

► *Find the moment of inertia of a uniform rectangular lamina of mass $M$ with sides $a$ and $b$ about one of the sides of length $b$.*

Referring to figure 5.9, we wish to calculate the moment of inertia about the $y$-axis. We therefore divide the rectangular lamina into elemental strips parallel to the $y$-axis of width $dx$. The mass of such a strip is $dM = \sigma b \, dx$, where $\sigma$ is the mass per unit area of the lamina. The moment of inertia of a strip at a distance $x$ from the $y$-axis is simply $dI = x^2 \, dM = \sigma b x^2 \, dx$. The total moment of inertia of the lamina about the $y$-axis is therefore

$$I = \int_0^a \sigma b x^2 \, dx = \frac{\sigma b a^3}{3}.$$

Since the total mass of the lamina is $M = \sigma ab$, we can write $I = \tfrac{1}{3}Ma^2$. ◄

### 5.3.5 Mean values of functions

In chapter 1 we discussed average values for functions of a single variable. This is easily extended to functions of several variables. Let us consider, for example, a function $f(x, y)$ defined in some region $R$ of the $xy$-plane. Then the average value $\bar{f}$ of the function is given by

$$\bar{f} \int_R dA = \int_R f(x, y) \, dA. \qquad (5.10)$$

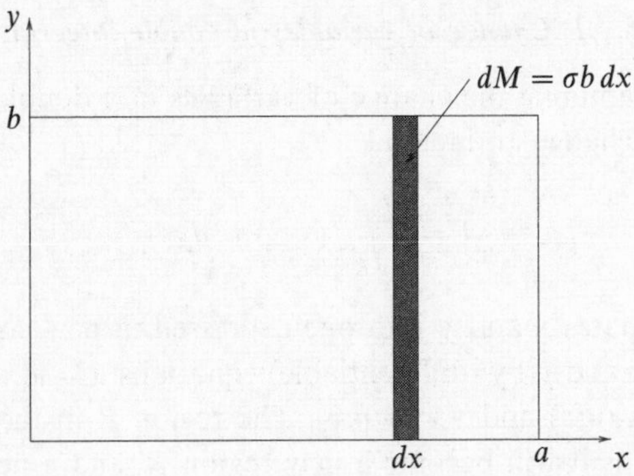


Figure 5.9 A uniform rectangular lamina of mass $M$ with sides $a$ and $b$ can be divided into vertical strips.

This definition is easily extended to three (and higher) dimensions, so that if a function $f(x, y, z)$ is defined in some three-dimensional region of space $R$, then the average value $\bar{f}$ of the function is given by

$$\bar{f} \int_R dV = \int_R f(x, y, z)\, dV. \tag{5.11}$$

► *A tetrahedron is bounded by the coordinate surfaces and the plane $x/a + y/b + z/c = 1$, and has a density $\rho(x, y, z) = \rho_0(1 + x/a)$. Find the average value of the density.*

From (5.11), the average value of the density is given by

$$\bar{\rho} \int_R dV = \int_R \rho(x, y, z)\, dV.$$

Now the integral on the LHS is just the volume of the tetrahedron, which we found in subsection 5.3.1 to be $V = \frac{1}{6}abc$, and the integral of the RHS is its mass $M = \frac{5}{24}abc\rho_0$, calculated in subsection 5.3.2. Therefore $\bar{\rho} = M/V = \frac{5}{4}\rho_0$. ◄

## 5.4 Change of variables in multiple integrals

It often happens that, either because of the form of the integrand involved or because of the shape of the boundary of the region of integration, it is desirable to express a multiple integral in terms of a new set of variables.

### 5.4.1 Change of variables in double integrals

Let us begin by examining the change of variables in a double integral. Suppose that we require to change an integral

$$I = \iint_R f(x, y)\, dx\, dy,$$

in terms of coordinates $x$ and $y$ into one expressed in new coordinates $u$ and $v$, given in terms of $x$ and $y$ by (differentiable) equations $u = u(x, y)$ and $v = v(x, y)$ with inverses $x = x(u, v)$ and $y = y(u, v)$. The region $R$ in the $xy$-plane and the curve $C$ that bounds it will become a new region $R'$ and a new boundary $C'$ in the $uv$-plane, and so we must change the limits of integration accordingly. Also, the function $f(x, y)$ becomes a new function $g(u, v)$ of the new coordinates.

Now the part of the integral that requires most consideration is the area element. In the $xy$-plane the element is the rectangular area $dA_{xy} = dx\, dy$ generated by constructing a grid of straight lines parallel to the $x$- and $y$- axes respectively. Our task is to determine the corresponding element of area in the $uv$-coordinates. In general the corresponding element $dA_{uv}$ will not be the same shape as $dA_{xy}$, but this does not matter since all elements are infinitesimally small and the value of the integrand is considered constant over them. Since the sides of the area element are infinitesimal, $dA_{uv}$ will in general have the shape of a parallelogram. We can find the connection between $dA_{xy}$ and $dA_{uv}$ by considering the grid formed by the family of curves $u =$ constant and $v =$ constant as shown in figure 5.10. Since $v$ is constant along the line element $KL$, the latter has components in the directions of the $x$- and $y$-axes of $(\partial x/\partial u)\, du$ and $(\partial y/\partial u)\, du$ respectively. Similarly, since $u$ is constant along the line element $KN$, the latter has the components $(\partial x/\partial v)\, dv$ and $(\partial y/\partial v)\, dv$. Using the result for the area of a parallelogram given in chapter 6, we find that the area of the parallelogram $KLMN$ is given by

$$\begin{aligned} dA_{uv} &= \left| \frac{\partial x}{\partial u} du \frac{\partial y}{\partial v} dv - \frac{\partial x}{\partial v} dv \frac{\partial y}{\partial u} du \right| \\ &= \left| \frac{\partial x}{\partial u}\frac{\partial y}{\partial v} - \frac{\partial x}{\partial v}\frac{\partial y}{\partial u} \right| du\, dv. \end{aligned}$$

Defining the *Jacobian* of $x$, $y$ with respect to $u$, $v$ as

$$J = \frac{\partial(x, y)}{\partial(u, v)} = \frac{\partial x}{\partial u}\frac{\partial y}{\partial v} - \frac{\partial x}{\partial v}\frac{\partial y}{\partial u},$$

we have

$$dA_{uv} = \left| \frac{\partial(x, y)}{\partial(u, v)} \right| du\, dv.$$

The reader acquainted with determinants will notice that the Jacobian can also

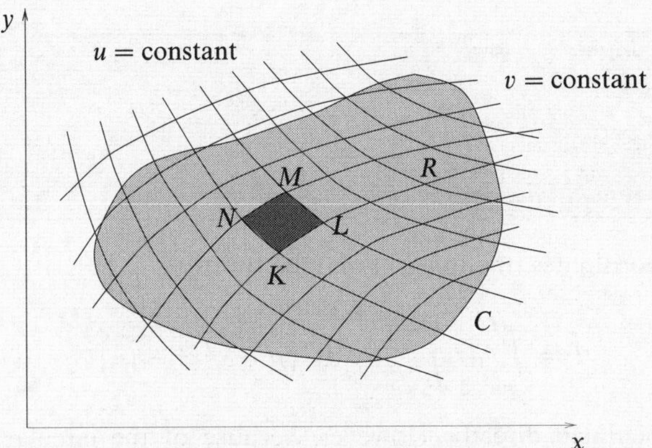


Figure 5.10 A region of integration $R$ overlaid with a grid formed by the family of curves $u$ = constant and $v$ = constant. The parallelogram $KLMN$ defines the area element $dA_{uv}$.

be written as the $2 \times 2$ determinant

$$J = \frac{\partial(x,y)}{\partial(u,v)} = \begin{vmatrix} \dfrac{\partial x}{\partial u} & \dfrac{\partial y}{\partial u} \\ \dfrac{\partial x}{\partial v} & \dfrac{\partial y}{\partial v} \end{vmatrix}.$$

Such determinants can in general be evaluated using the methods of chapter 7.

So, in summary, the relationship between the size of the area element generated by $dx$, $dy$ and the size of the corresponding area element generated by $du$, $dv$ is

$$dx\,dy = \left|\frac{\partial(x,y)}{\partial(u,v)}\right| du\,dv.$$

This equality should be taken as meaning that when transforming from coordinates $x, y$ to coordinates $u, v$, the area element $dx\,dy$ should be replaced by the expression on the RHS of the above equality. Of course, the Jacobian can, and in general will, vary over the region of integration. We may express the double integral in either coordinate system as

$$I = \iint_R f(x,y)\,dx\,dy = \iint_{R'} g(u,v) \left|\frac{\partial(x,y)}{\partial(u,v)}\right| du\,dv. \tag{5.12}$$

When evaluating the integral in the new coordinate system, it is often advisable to sketch the region of integration $R'$ in the $uv$-plane.

►*Evaluate the double integral*

$$I = \iint_R \left(a + \sqrt{x^2 + y^2}\right) dx\, dy,$$

*where R is the region bounded by the circle* $x^2 + y^2 = a^2$.

In Cartesian coordinates, the integral may be written

$$I = \int_{-a}^{a} dx \int_{-\sqrt{a^2-x^2}}^{\sqrt{a^2-x^2}} dy \left(a + \sqrt{x^2 + y^2}\right),$$

and can be calculated directly. However, because of the circular boundary of the integration region, a change of variables to plane polar coordinates $\rho$, $\phi$ is indicated. The relationship between Cartesian and plane polar coordinates is given by $x = \rho \cos\phi$ and $y = \rho \sin\phi$. Using (5.12) we can therefore write

$$I = \iint_{R'} (a + \rho) \left| \frac{\partial(x, y)}{\partial(\rho, \phi)} \right| d\rho\, d\phi,$$

where $R'$ is the rectangular region in the $\rho\phi$-plane whose sides are $\rho = 0$, $\rho = a$, $\phi = 0$ and $\phi = 2\pi$. The Jacobian is easily calculated, and we obtain

$$J = \frac{\partial(x, y)}{\partial(\rho, \phi)} = \begin{vmatrix} \cos\phi & \sin\phi \\ -\rho \sin\phi & \rho \cos\phi \end{vmatrix} = \rho(\cos^2\phi + \sin^2\phi) = \rho.$$

So the relationship between the area element in Cartesian and in plane polar coordinates is

$$dx\, dy = \rho\, d\rho\, d\phi.$$

Therefore, when expressed in plane polar coordinates, the integral is given by

$$\begin{aligned} I &= \iint_{R'} (a + \rho)\rho\, d\rho\, d\phi \\ &= \int_0^{2\pi} d\phi \int_0^a d\rho\, (a + \rho)\rho = 2\pi \left[ \frac{a\rho^2}{2} + \frac{\rho^3}{3} \right]_0^a = \frac{5\pi a^3}{3}. \end{aligned}$$ ◄

### 5.4.2 Evaluation of the integral $I = \int_{-\infty}^{\infty} e^{-x^2}\, dx$

By making a judicious change of variables, it is sometimes possible to evaluate an integral that would be intractable otherwise. An important example of this method is provided by the evaluation of the integral

$$I = \int_{-\infty}^{\infty} e^{-x^2}\, dx.$$

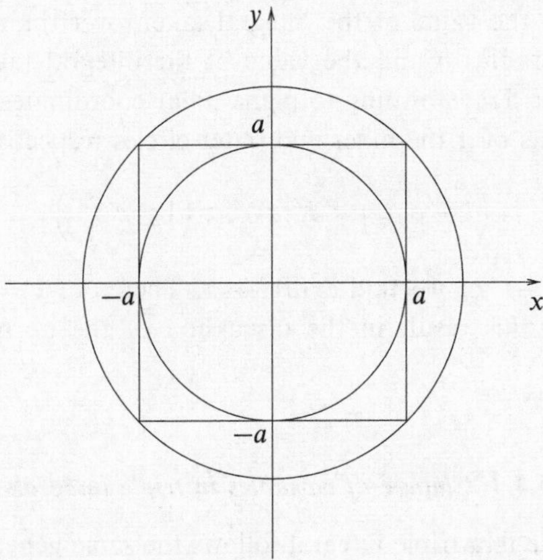


Figure 5.11 The regions used to illustrate the convergence properties of the integral $I(a) = \int_{-a}^{a} e^{-x^2}\,dx$ as $a \to \infty$.

Its value may be found by first constructing $I^2$, as follows:

$$I^2 = \int_{-\infty}^{\infty} e^{-x^2}\,dx \int_{-\infty}^{\infty} e^{-y^2}\,dy = \int_{-\infty}^{\infty} dx \int_{-\infty}^{\infty} dy\; e^{-(x^2+y^2)}$$
$$= \iint_R e^{-(x^2+y^2)}\,dx\,dy,$$

where the region $R$ is the whole $xy$-plane. Then, transforming to plane polar coordinates, we find

$$I^2 = \iint_{R'} e^{-\rho^2}\rho\,d\rho\,d\phi = \int_0^{2\pi} d\phi \int_0^{\infty} d\rho\,\rho e^{-\rho^2} = 2\pi\left[-\tfrac{1}{2}e^{-\rho^2}\right]_0^{\infty} = \pi.$$

Therefore the original integral is given by $I = \sqrt{\pi}$. We also note that since the integrand is even, the value of the integral from 0 to $\infty$ is simply $\sqrt{\pi}/2$.

We note, however, that unlike in all the previous examples, the regions of integration $R$ and $R'$ are both infinite in extent (i.e. unbounded). It is therefore wise to derive this result more rigorously, by instead considering the integral

$$I(a) = \int_{-a}^{a} e^{-x^2}\,dx.$$

We then have

$$I^2(a) = \iint_R e^{-(x^2+y^2)}\,dx\,dy,$$

where $R$ is the square of side $2a$ centred on the origin. Referring to figure 5.11, since the integrand is always positive the value of the integral taken over the square lies between the value of the integral taken over the region bounded by the inner circle of radius $a$ and the value of the integral taken over the outer circle of radius $\sqrt{2}a$. Transforming to plane polar coordinates as above, we may evaluate the integrals over the inner and outer circles respectively, and we find

$$\pi\left(1-e^{-a^2}\right) < I^2(a) < \pi\left(1-e^{-2a^2}\right).$$

Taking the limit $a \to \infty$, we find $I^2(a) \to \pi$. Therefore $I = \sqrt{\pi}$ as we found previously. We use this result in the discussion of the normal distribution in chapter 24.

### 5.4.3 *Change of variables in triple integrals*

A change of variable in a triple integral follows the same general lines as that for a double integral. Suppose we wish to change variables from $x$, $y$, $z$ to $u$, $v$, $w$. In the $x$, $y$, $z$ coordinates the element of volume is a cuboid of sides $dx$, $dy$, $dz$ and volume $dV_{xyz} = dx\,dy\,dz$. If, however, we divide up the total volume into infinitesimal elements by constructing a grid formed from the coordinate surfaces $u =$ constant, $v =$ constant and $w =$ constant, then the element of volume $dV_{uvw}$ in the new coordinates will in general have the shape of a parallelepiped whose faces are the coordinate surfaces, and whose edges are the curves formed by the intersections of these surfaces (see figure 5.12). Along the line element $PQ$ the coordinates $v$ and $w$ are constant, and so $PQ$ has components in the direction of the $x$-, $y$- and $z$- axes of $(\partial x/\partial u)\,du$, $(\partial y/\partial u)\,du$ and $(\partial z/\partial u)\,du$ respectively. The components of the line elements $PS$ and $ST$ are found by replacing $u$ by $v$ and $w$ respectively.

The expression for the volume of a parallelepiped in terms of the components of its edges with respect to the $x$-, $y$- and $z$-axes is given in chapter 6. Using this, we find that the element of volume in $u, v, w$ coordinates is given by

$$dV_{uvw} = \left|\frac{\partial(x,y,z)}{\partial(u,v,w)}\right| du\,dv\,dw,$$

where the Jacobian of $x$, $y$, $z$ with respect to $u$, $v$, $w$ is a short-hand for a $3 \times 3$ determinant:

$$\frac{\partial(x,y,z)}{\partial(u,v,w)} = \begin{vmatrix} \dfrac{\partial x}{\partial u} & \dfrac{\partial y}{\partial u} & \dfrac{\partial z}{\partial u} \\ \dfrac{\partial x}{\partial v} & \dfrac{\partial y}{\partial v} & \dfrac{\partial z}{\partial v} \\ \dfrac{\partial x}{\partial w} & \dfrac{\partial y}{\partial w} & \dfrac{\partial z}{\partial w} \end{vmatrix}.$$

So, in summary, the relationship between the elemental volumes in multiple

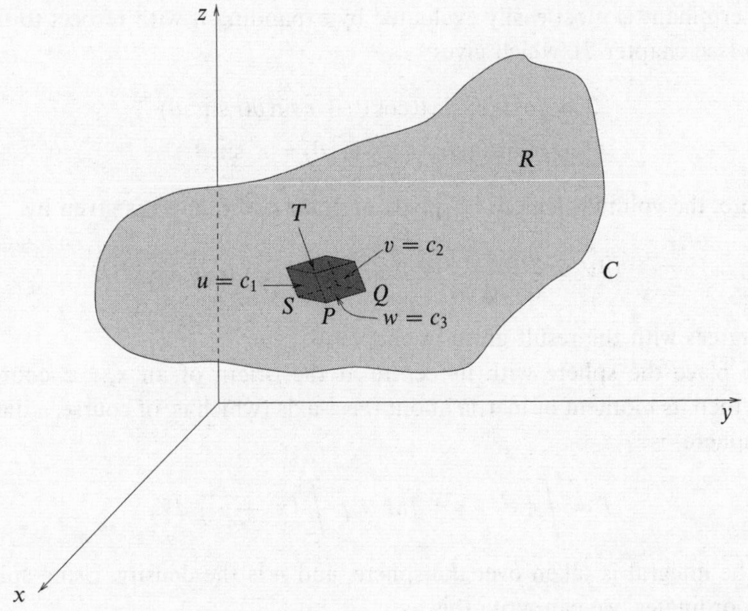


Figure 5.12 A three-dimensional region of integration $R$ showing an element of volume in $u, v, w$ coordinates formed by the coordinate surfaces $u =$ constant, $v =$ constant, $w =$ constant.

integrals formulated in the two coordinate systems is given in Jacobian form by

$$dx\,dy\,dz = \left| \frac{\partial(x, y, z)}{\partial(u, v, w)} \right| du\,dv\,dw,$$

and we can write a triple integral in either set of coordinates as

$$I = \iiint_R f(x, y, z)\,dx\,dy\,dz = \iiint_{R'} g(u, v, w) \left| \frac{\partial(x, y, z)}{\partial(u, v, w)} \right| du\,dv\,dw.$$

► *Find an expression for a volume element in spherical polar coordinates, and hence calculate the moment of inertia about a diameter of a uniform sphere of radius $a$ and mass $M$.*

Spherical polar coordinates $r, \theta, \phi$ are defined by

$$x = r \sin\theta \cos\phi, \quad y = r \sin\theta \sin\phi, \quad z = r \cos\theta$$

(and are discussed fully in chapter 8). The required Jacobian is therefore

$$J = \frac{\partial(x, y, z)}{\partial(r, \theta, \phi)} = \begin{vmatrix} \sin\theta\cos\phi & \sin\theta\sin\phi & \cos\theta \\ r\cos\theta\cos\phi & r\cos\theta\sin\phi & -r\sin\theta \\ -r\sin\theta\sin\phi & r\sin\theta\cos\phi & 0 \end{vmatrix}.$$

The determinant is most easily evaluated by expanding it with respect to the last column (see chapter 7), which gives

$$\begin{aligned} J &= \cos\theta(r^2\sin\theta\cos\theta) + r\sin\theta(r\sin^2\theta) \\ &= r^2\sin\theta(\cos^2\theta + \sin^2\theta) = r^2\sin\theta. \end{aligned}$$

Therefore, the volume element in spherical polar coordinates is given by

$$dV = \frac{\partial(x,y,z)}{\partial(r,\theta,\phi)}\,dr\,d\theta\,d\phi = r^2\sin\theta\,dr\,d\theta\,d\phi,$$

which agrees with the result given in chapter 8.

If we place the sphere with its centre at the origin of an $x$, $y$, $z$ coordinate system, then its moment of inertia about the $z$-axis (which is, of course, a diameter of the sphere) is

$$I = \int \left(x^2 + y^2\right) dM = \rho \int \left(x^2 + y^2\right) dV,$$

where the integral is taken over the sphere, and $\rho$ is the density. Using spherical polar coordinates, we can write this as

$$\begin{aligned} I &= \rho \iiint_V \left(r^2\sin^2\theta\right) r^2\sin\theta\,dr\,d\theta\,d\phi \\ &= \rho\int_0^{2\pi} d\phi \int_0^{\pi} d\theta\,\sin^3\theta \int_0^a dr\,r^4 \\ &= \rho\times 2\pi\times\tfrac{4}{3}\times\tfrac{1}{5}a^5 = \tfrac{8}{15}\pi a^5\rho. \end{aligned}$$

Since the mass of the sphere is $M = \frac{4}{3}\pi a^3\rho$, the moment of inertia can also be written as $I = \frac{2}{5}Ma^2$. ◀

### 5.4.4 General properties of Jacobians

Although we will not prove it, the general result for a change of coordinates in an $n$-dimensional integral from a set $x_i$ to a set $y_j$ (where $i$ and $j$ both run from 1 to $n$) is

$$dx_1\,dx_2\cdots dx_n = \left|\frac{\partial(x_1,x_2,\ldots,x_n)}{\partial(y_1,y_2,\ldots,y_n)}\right| dy_1\,dy_2\cdots dy_n,$$

where the $n$-dimensional Jacobian can be written as an $n\times n$ determinant (see chapter 7) in an analogous way to the two- and three-dimensional cases.

For readers who already have sufficient familiarity with matrices (see chapter 7) and their properties, a fairly compact proof of some useful general properties of Jacobians can be given as follows. Other readers should turn straight to the results (5.16) and (5.17) and return to the proof at some later time.

Consider three sets of variables $x_i$, $y_i$ and $z_i$, with $i$ running from 1 to $n$ for each set. From the chain rule in partial differentiation (see (4.17)), we know that

$$\frac{\partial x_i}{\partial z_j} = \sum_{k=1}^{n} \frac{\partial x_i}{\partial y_k}\frac{\partial y_k}{\partial z_j}. \tag{5.13}$$

Now let A, B and C be the matrices whose $(i, j)$th elements are $\partial x_i/\partial y_j$, $\partial y_i/\partial z_j$ and $\partial x_i/\partial z_j$ respectively. We can then write (5.13) as the matrix product

$$c_{ij} = \sum_{k=1}^{n} a_{ik}b_{kj} \qquad \text{or} \qquad \mathsf{C} = \mathsf{AB}. \tag{5.14}$$

We may now use the general result for the determinant of the product of two matrices, namely $|\mathsf{AB}| = |\mathsf{A}||\mathsf{B}|$, and recall that the Jacobian

$$J_{xy} = \frac{\partial(x_1, \ldots, x_n)}{\partial(y_1, \ldots, y_n)} = |\mathsf{A}|, \tag{5.15}$$

and similarly for $J_{yz}$ and $J_{xz}$. On taking the determinant of (5.14), we therefore obtain

$$J_{xz} = J_{xy}J_{yz},$$

or in the usual notation

$$\frac{\partial(x_1, \ldots, x_n)}{\partial(z_1, \ldots, z_n)} = \frac{\partial(x_1, \ldots, x_n)}{\partial(y_1, \ldots, y_n)}\frac{\partial(y_1, \ldots, y_n)}{\partial(z_1, \ldots, z_n)}. \tag{5.16}$$

As a special case, if the set $z_i$ is taken to be identical to the set $x_i$, and the obvious result $J_{xx} = 1$ is used, we obtain

$$J_{xy}J_{yx} = 1,$$

or in the usual notation

$$\frac{\partial(x_1, \ldots, x_n)}{\partial(y_1, \ldots, y_n)} = \left[\frac{\partial(y_1, \ldots, y_n)}{\partial(x_1, \ldots, x_n)}\right]^{-1}. \tag{5.17}$$

The similarity between the properties of Jacobians and those of derivatives is apparent, and to some extent is suggested by the notation. We further note from (5.15) that since $|\mathsf{A}| = |\mathsf{A}^{\mathrm{T}}|$, where $\mathsf{A}^{\mathrm{T}}$ is the transpose of A, we can interchange the rows and columns in the determinantal form of the Jacobian without changing its value.

## 5.5 Exercises

5.1 Evaluate the volume integral of $x^2 + y^2 + z^2$ over the rectangular parallelepiped bounded by the six surfaces $x = \pm a$, $y = \pm b$, $z = \pm c$.

5.2 Find the volume integral of $x^2y$ over the tetrahedral volume bounded by the planes $x = 0$, $y = 0$, $z = 0$, and $x + y + z = 1$.

5.3 Evaluate the surface integral of $f(x, y)$ over the rectangle $0 \leq x \leq a$, $0 \leq y \leq b$ for the functions

$$\text{(a) } f(x, y) = \frac{x}{x^2 + y^2}, \qquad \text{(b) } f(x, y) = (b - y + x)^{-3/2}.$$

5.4 (a) Prove that the area of the ellipse

$$\frac{x^2}{a^2} + \frac{y^2}{b^2} = 1$$

is $\pi ab$.

(b) Use this result to obtain an expression for the volume of a slice of thickness $dz$ of the ellipsoid

$$\frac{x^2}{a^2} + \frac{y^2}{b^2} + \frac{z^2}{c^2} = 1.$$

Hence show that the volume of the ellipsoid is $4\pi abc/3$.

5.5 In quantum mechanics the electron in a hydrogen atom in some particular state is described by a wavefunction $\Psi$, which is such that $|\Psi|^2\, dV$ is the probability of finding the electron in the infinitesimal volume $dV$. In spherical polar coordinates $r, \theta, \phi$, $\Psi = \Psi(r, \theta, \phi)$ and $dV = r^2 \sin\theta\, dr\, d\theta\, d\phi$. Two such states are described by

$$\Psi_1 = \left(\frac{1}{4\pi}\right)^{1/2} \left(\frac{1}{a_0}\right)^{3/2} 2e^{-r/a_0},$$

$$\Psi_2 = -\left(\frac{3}{8\pi}\right)^{1/2} \sin\theta\, e^{i\phi} \left(\frac{1}{2a_0}\right)^{3/2} \frac{re^{-r/2a_0}}{a_0\sqrt{3}}.$$

(a) Show that each $\Psi_i$ is normalised, i.e. the integral over all space $\int |\Psi|^2\, dV$ is equal to unity – physically, this means that the electron must be somewhere.

(b) The (so-called) dipole matrix element between the states 1 and 2 is given by the integral

$$p_x = \int \Psi_1^* qr \sin\theta \cos\phi\, \Psi_2\, dV,$$

where $q$ is the charge on the electron. Prove that $p_x$ has the value $-2^7 qa_0/3^5$.

5.6 A thin uniform circular disc has mass $M$ and radius $a$.

(a) Prove that its moment of inertia about an axis perpendicular to its plane and passing through its centre is $\frac{1}{2}Ma^2$.

(b) Prove that the moment of inertia of the same disc about a diameter is $\frac{1}{4}Ma^2$.

This is an example of a general result for planar bodies that the moment of inertia of the body about an axis perpendicular to the plane is equal to the sum of the moments of inertia about two perpendicular axes lying in the plane: in an obvious notation

$$I_z = \int r^2\, dm = \int (x^2 + y^2)\, dm = \int x^2\, dm + \int y^2\, dm = I_y + I_x.$$

5.7 In some applications in mechanics the moment of inertia of a body about a single point (as opposed to about an axis) is needed. The moment of inertia $I$ about the origin of a uniform solid body of density $\rho$ is given by the volume integral

$$I = \int_V (x^2 + y^2 + z^2)\rho\, dV.$$

Show that the moment of inertia of a right circular cylinder of radius $a$, length $2b$, and mass $M$ about its centre is

$$M\left(\frac{a^2}{2} + \frac{b^2}{3}\right).$$

5.8 In spherical polar coordinates $r, \theta, \phi$ the element of volume for a body symmetrical about the polar axis is $dV = 2\pi r^2 \sin\theta\, dr\, d\theta$, whilst its element of surface area is $2\pi r \sin\theta[(dr)^2 + r^2(d\theta)^2]^{1/2}$. A particular surface is defined by $r = 2a\cos\theta$, where $a$ is a constant, and $0 \leq \theta \leq \pi/2$. Find its total surface area and the volume it encloses, and hence identify the surface.

5.9 Sketch the two families of curves

$$y^2 = 4u(u - x), \qquad y^2 = 4v(v + x),$$

where $u$ and $v$ are parameters.

By transforming to the $uv$-plane evaluate the integral of $y/(x^2 + y^2)^{1/2}$ over that part of the quadrant $x > 0$, $y > 0$ bounded by the lines $x = 0$, $y = 0$ and the curve $y^2 = 4a(a - x)$.

5.10 This is a more difficult question about 'volumes' in an increasing number of dimensions.

(a) Let $R$ be a real positive number and define $K_m$ by

$$K_m = \int_{-R}^{R} \left(R^2 - x^2\right)^m dx.$$

Show, by integrating by parts, that $K_m$ satisfies the recurrence relation

$$(2m + 1)K_m = 2mR^2 K_{m-1}.$$

(b) For integer $n$, define $I_n = K_n$ and $J_n = K_{n+1/2}$. Evaluate $I_0$ and $J_0$ directly and hence prove that

$$I_n = \frac{2^{2n+1}(n!)^2 R^{2n+1}}{(2n+1)!} \qquad \text{and} \qquad J_n = \frac{\pi(2n+1)! R^{2n+2}}{2^{2n+1} n!(n+1)!}.$$

(c) A sequence of functions $V_n(R)$ is defined by

$$V_0(R) = 1,$$
$$V_n(R) = \int_{-R}^{R} V_{n-1}\left(\sqrt{R^2 - x^2}\right) dx, \qquad n \geq 1.$$

Prove by induction that

$$V_{2n}(R) = \frac{\pi^n R^{2n}}{n!}, \qquad V_{2n+1}(R) = \frac{\pi^n 2^{2n+1} n! R^{2n+1}}{(2n+1)!}.$$

(d) For interest

(i) show that $V_{2n+2}(1) < V_{2n}(1)$ and $V_{2n+1}(1) < V_{2n-1}(1)$ for all $n \geq 3$;
(ii) hence, by explicitly writing out $V_k(R)$ for $1 \leq k \leq 8$ (say), show that the 'volume' of the totally symmetric solid of unit radius is a maximum in five dimensions.

## 5.6 Hints and answers

5.1 $8abc(a^2 + b^2 + c^2)/3$.

5.2 $1/360$.

5.3 (a) Integrate by parts to obtain $(b/2)\ln[1 + (a/b)^2] + a\tan^{-1}(b/a)$;
(b) $4[a^{1/2} + b^{1/2} - (a+b)^{1/2}]$.

5.4 (a) Evaluate $\int 2b[1 - (x/a)^2]^{1/2}\, dx$ by setting $x = a\cos\phi$;
(b) $dV = \pi \times a[1 - (z/c)^2]^{1/2} \times b[1 - (z/c)^2]^{1/2}\, dz$.

5.6 (b) Evaluate $\int 2(a^2 - x^2)^{1/2} x^2 (M/\pi a^2)\, dx$ by setting $x = a\cos\phi$.

5.7 Transform to cylindrical polar coordinates.

5.8 $4\pi a^2$, $4\pi a^3/3$, a sphere.

5.9 Jacobian $= (u/v)^{1/2} + (v/u)^{1/2}$; area in $uv$-plane is the triangle bounded by $v = 0$, $u = v$, $u = a$; integral $= a^2$.

5.10 (d) $2$, $\pi$, $4\pi/3$, $\pi^2/2$, $8\pi^2/15$, $\pi^3/6$, $16\pi^3/105$, $\pi^4/24$.

# 6

# *Vector algebra*

This chapter introduces space vectors and their manipulation. Firstly we deal with the description and algebra of vectors, then consider how vectors may be used to describe lines and planes and finally we look at the practical use of vectors in finding distances. Much use of vectors will be made in subsequent chapters but this chapter gives only some basic rules.

## 6.1 Scalars and vectors

The simplest kind of physical quantity is one which can be completely specified by its magnitude, a single number, together with the units in which it is measured. Such a quantity is called a *scalar* and examples include temperature, time and density.

A *vector* is a quantity that requires both a magnitude ($\geq 0$) and a direction in space to specify it completely, and we may think of it as an arrow in space. A familiar example is force which has a magnitude (strength), measured in newtons, and a direction of application. The large number of vectors that are used to describe the physical world include velocity, displacement, momentum and electric field. Vectors are also used to describe quantities such as angular momentum and surface elements (a surface element has an area and a direction defined by the normal to its tangent plane). In such cases their definitions are somewhat arbitrary, though standard, and not as physically intuitive as for vectors such as force. A vector is notated by bold type, the convention of this book, or by underlining, particularly in handwritten work.

This chapter considers basic vector algebra and illustrates how powerful vector analysis can be. All the techniques are presented for three-dimensional space but most can be readily extended to more dimensions.

Throughout the book we will represent vectors in diagrams as a line together with an arrowhead. We will make no distinction between an arrowhead at the

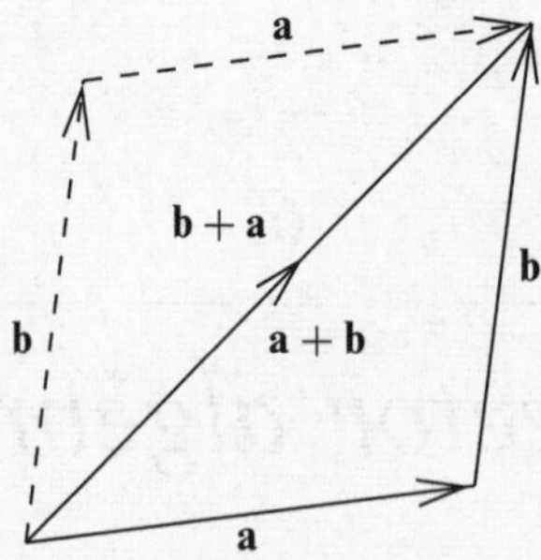


Figure 6.1 Addition of two vectors showing the commutation relation. We make no distinction between an arrowhead at the end of the line and one along the line's length, but rather use that which gives the clearer diagram.

end of the line or one along the line's length, but rather use that which gives the clearer diagram. Furthermore, even though we are considering three-dimensional vectors, we have to draw them in the plane of the paper. It should not be assumed that vectors drawn thus are coplanar, unless this is explicitly stated.

## 6.2 Addition and subtraction of vectors

The *resultant* of two displacement vectors is the displacement vector that results from performing first one and then the other displacement, as shown in figure 6.1; this process is known as vector addition. However, the principle of addition has physical meaning for vector quantities other than displacements; for example, if two forces act on the same body, the resultant force acting on the body is the vector sum of the two. The addition of vectors only makes physical sense if they are of a like kind, for example if they are both forces acting in three dimensions. It may be seen from figure 6.1 that vector addition is commutative, i.e.

$$\mathbf{a} + \mathbf{b} = \mathbf{b} + \mathbf{a}. \qquad (6.1)$$

The generalisation of this procedure to the addition of three (or more) vectors is clear and leads to the associativity property of addition (see figure 6.2), e.g.

$$\mathbf{a} + (\mathbf{b} + \mathbf{c}) = (\mathbf{a} + \mathbf{b}) + \mathbf{c}. \qquad (6.2)$$

Thus, it is immaterial in what order any number of vectors are added.

The subtraction of two vectors is very similar to their addition (see figure 6.3), that is,

$$\mathbf{a} - \mathbf{b} = \mathbf{a} + (-\mathbf{b})$$

where $-\mathbf{b}$ is a vector of equal magnitude but exactly opposite direction to vector $\mathbf{b}$. The subtraction of two equal vectors yields the zero vector, $\mathbf{0}$, which has zero magnitude and no associated direction.

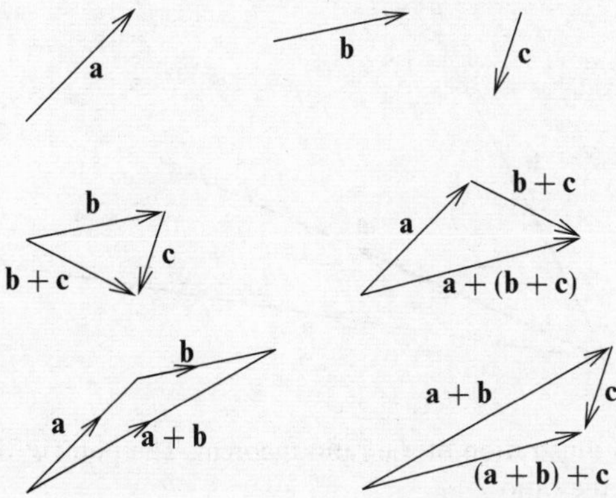


Figure 6.2 Addition of three vectors showing the associativity relation.

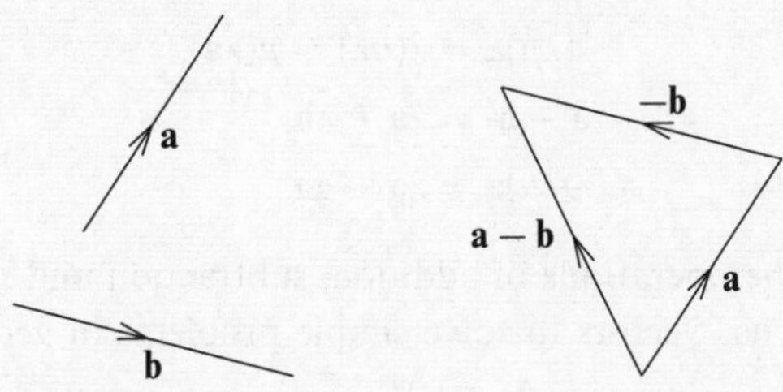


Figure 6.3 Subtraction of two vectors.

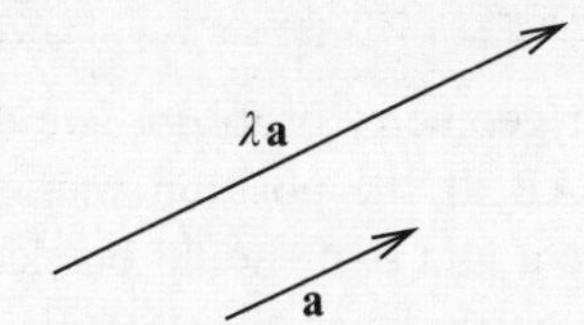


Figure 6.4 Scalar multiplication of a vector (for $\lambda > 1$).

## 6.3 Multiplication by a scalar

Multiplication of a vector by a scalar (not to be confused with the 'scalar product', to be discussed in subsection 6.6.1) gives a vector in the same direction as the original but of a proportional magnitude. This can be seen in figure 6.4. The scalar may be positive, negative or zero. It can also be complex in some applications. Clearly, when the scalar is negative we obtain a vector pointing in the opposite direction to the original vector.

Multiplication by a scalar is associative, commutative and distributive over addition. These properties may be summarised for arbitrary vectors **a** and **b** and

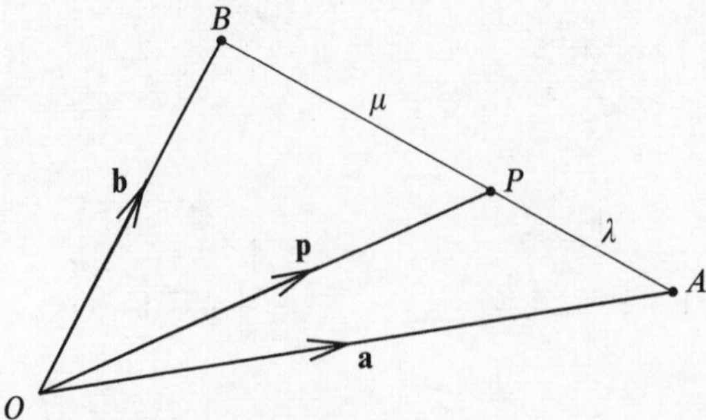


Figure 6.5 An illustration of the ratio theorem. The point $P$ divides the line segment $AB$ in the ratio $\lambda : \mu$.

arbitrary scalars $\lambda$ and $\mu$ by

$$(\lambda\mu)\mathbf{a} = \lambda(\mu\mathbf{a}) = \mu(\lambda\mathbf{a}), \tag{6.3}$$

$$\lambda(\mathbf{a}+\mathbf{b}) = \lambda\mathbf{a} + \lambda\mathbf{b}, \tag{6.4}$$

$$(\lambda+\mu)\mathbf{a} = \lambda\mathbf{a} + \mu\mathbf{a}. \tag{6.5}$$

Having defined the operations of addition, subtraction and multiplication by a scalar, we can now use vectors to solve simple problems in geometry.

►*A point $P$ divides a line segment $AB$ in the ratio $\lambda : \mu$ (see figure 6.5). If the position vectors of the points $A$ and $B$ are* **a** *and* **b** *respectively, find the position vector of the point $P$.*

As is conventional for vector geometry problems, we denote the vector from the point $A$ to the point $B$ by **AB**. If the position vectors of the points $A$ and $B$, relative to some origin $O$, are **a** and **b**, it should be clear that $\mathbf{AB} = \mathbf{b} - \mathbf{a}$.

Now, from figure 6.5 we see that one possible way of reaching the point $P$ from $O$ is first to go from $O$ to $A$ and to go along the line $AB$ for a distance equal to the the fraction $\lambda/(\lambda+\mu)$ of its total length. We may express this in terms of vectors as

$$\begin{aligned}
\mathbf{OP} = \mathbf{p} &= \mathbf{a} + \frac{\lambda}{\lambda+\mu}\mathbf{AB} \\
&= \mathbf{a} + \frac{\lambda}{\lambda+\mu}(\mathbf{b}-\mathbf{a}) \\
&= \left(1 - \frac{\lambda}{\lambda+\mu}\right)\mathbf{a} + \frac{\lambda}{\lambda+\mu}\mathbf{b} \\
&= \frac{\mu}{\lambda+\mu}\mathbf{a} + \frac{\lambda}{\lambda+\mu}\mathbf{b}, \qquad (6.6)
\end{aligned}$$

which expresses the position vector of the point $P$ in terms of those of $A$ and $B$.

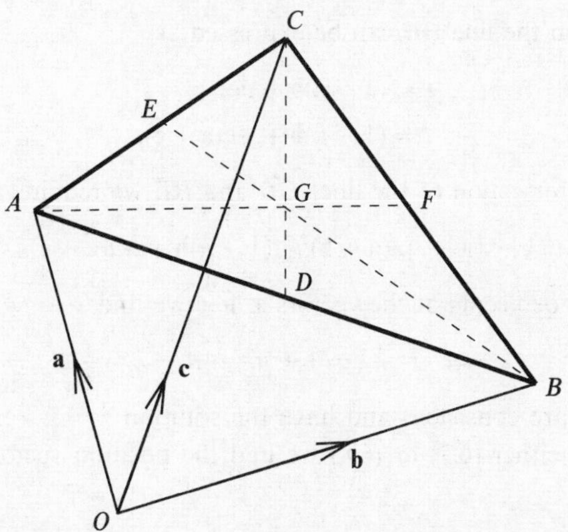


Figure 6.6 The centroid of a triangle. The triangle is defined by the points $A$, $B$ and $C$ that have position vectors $\mathbf{a}$, $\mathbf{b}$ and $\mathbf{c}$. The lines $CD$, $BE$, $AF$ connect the vertices of the triangle to the mid-points of the opposite sides; these lines intersect at the centroid $G$ of the triangle.

We would, of course, obtain the same result by considering the path from $O$ to $B$ and then to $P$. ◄

The result (6.6) is a version of the *ratio theorem* and we may use it in solving more complicated problems.

►*The vertices of triangle ABC have position vectors* $\mathbf{a}$, $\mathbf{b}$ *and* $\mathbf{c}$ *relative to some origin O (see figure 6.6). Find the position vector of the centroid G of the triangle.*

From figure 6.6, the points $D$ and $E$ bisect the lines $AB$ and $AC$ respectively. Thus from the ratio theorem (6.6), with $\lambda = \mu = 1/2$, the position vectors of $D$ and $E$ relative to the origin are

$$\mathbf{d} = \tfrac{1}{2}\mathbf{a} + \tfrac{1}{2}\mathbf{b},$$
$$\mathbf{e} = \tfrac{1}{2}\mathbf{a} + \tfrac{1}{2}\mathbf{c}.$$

Using the ratio theorem again, we may write the position vector of a general point on the line $CD$ that divides the line in the ratio $\lambda : (1-\lambda)$ as

$$\begin{aligned} \mathbf{r} &= (1-\lambda)\mathbf{c} + \lambda\mathbf{d}, \\ &= (1-\lambda)\mathbf{c} + \tfrac{1}{2}\lambda(\mathbf{a}+\mathbf{b}), \end{aligned} \qquad (6.7)$$

where we have expressed $\mathbf{d}$ in terms of $\mathbf{a}$ and $\mathbf{b}$. Similarly, the position vector of a general point on the line $BE$ can be expressed as

$$\begin{aligned} \mathbf{r} &= (1-\mu)\mathbf{b} + \mu\mathbf{e}, \\ &= (1-\mu)\mathbf{b} + \tfrac{1}{2}\mu(\mathbf{a}+\mathbf{c}). \end{aligned} \tag{6.8}$$

Thus, at the intersection of the lines $CD$ and $BE$ we require, from (6.7), (6.8),

$$(1-\lambda)\mathbf{c} + \tfrac{1}{2}\lambda(\mathbf{a}+\mathbf{b}) = (1-\mu)\mathbf{b} + \tfrac{1}{2}\mu(\mathbf{a}+\mathbf{c}).$$

By equating the coefficents of the vectors **a**, **b**, **c** we find

$$\lambda = \mu, \qquad \tfrac{1}{2}\lambda = 1-\mu, \qquad 1-\lambda = \tfrac{1}{2}\mu.$$

These equations are consistent and have the solution $\lambda = \mu = 2/3$. Substituting these values into either (6.7) or (6.8) we find the position vector of the centroid $G$ is given by

$$\mathbf{g} = \tfrac{1}{3}(\mathbf{a}+\mathbf{b}+\mathbf{c}). \blacktriangleleft$$

## 6.4 Basis vectors and components

Given any three different vectors $\mathbf{e}_1$, $\mathbf{e}_2$ and $\mathbf{e}_3$, which do not all lie in a plane, it is possible, in three-dimensional space, to write any other vector in terms of scalar multiples of them:

$$\mathbf{a} = a_1\mathbf{e}_1 + a_2\mathbf{e}_2 + a_3\mathbf{e}_3. \tag{6.9}$$

The three vectors $\mathbf{e}_1$, $\mathbf{e}_2$ and $\mathbf{e}_3$ are said to form a *basis* (for the three-dimensional space), and the scalars $a_1$, $a_2$ and $a_3$, which may be positive, negative or zero, are called the *components* of the vector **a** with respect to this basis. We say that the vector has been *resolved* into components.

Most often we shall use basis vectors that are mutually perpendicular, for ease of manipulation, though this is not necessary. In general, a basis set must

(i) have as many basis vectors as the number of dimensions (in more formal language, the basis vectors must span the space), and

(ii) be such that no basis vector may be described as a sum of the others, or more formally, the basis vectors must be *linearly independent*. Putting this mathematically, in $N$ dimensions, we require

$$c_1\mathbf{e}_1 + c_2\mathbf{e}_2 + \cdots + c_N\mathbf{e}_N \neq \mathbf{0},$$

for any set of coefficients $c_1, c_2, \ldots, c_N$ except $c_1 = c_2 = \cdots = c_N = 0$.

In this chapter we will only consider vectors in three dimensions; higher dimensionality can be achieved by simple extension.

If we wish to label points in space using a Cartesian coordinate system $(x, y, z)$, we may introduce the unit vectors **i**, **j** and **k**, which point along the positive $x$-,

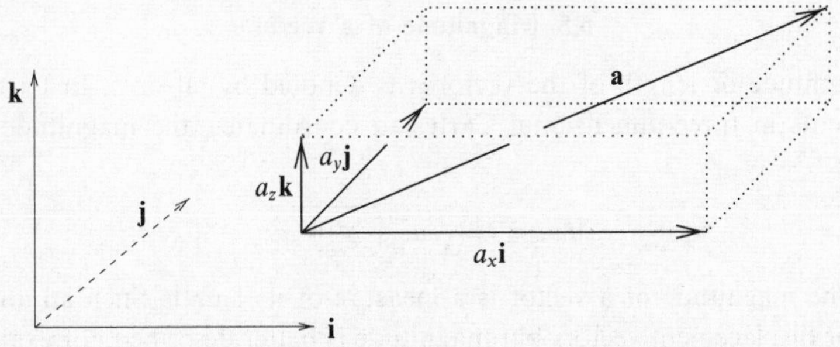


Figure 6.7 A Cartesian basis set. The vector **a** is the sum of $a_x\mathbf{i}$, $a_y\mathbf{j}$ and $a_z\mathbf{k}$.

$y$-, and $z$- axes respectively. A vector **a** may then be written as a sum of three vectors, each parallel to a different coordinate axis:

$$\mathbf{a} = a_x\mathbf{i} + a_y\mathbf{j} + a_z\mathbf{k}. \tag{6.10}$$

A vector in three-dimensional space thus requires three components to describe fully both its direction and its magnitude. A displacement in space may be thought of as the sum of displacements along the $x$-, $y$- and $z$- directions (see figure 6.7). For brevity, the components of a vector **a** with respect to a particular coordinate system are sometimes written in the form $(a_x, a_y, a_z)$. Note that the basis vectors **i**, **j** and **k** may themselves be represented by $(1, 0, 0)$, $(0, 1, 0)$ and $(0, 0, 1)$ respectively.

We can consider the addition and subtraction of vectors in terms of their components. The sum of two vectors **a** and **b** is found by simply adding their components, i.e.

$$\begin{aligned}\mathbf{a} + \mathbf{b} &= a_x\mathbf{i} + a_y\mathbf{j} + a_z\mathbf{k} + b_x\mathbf{i} + b_y\mathbf{j} + b_z\mathbf{k} \\ &= (a_x + b_x)\mathbf{i} + (a_y + b_y)\mathbf{j} + (a_z + b_z)\mathbf{k}\end{aligned} \tag{6.11}$$

and their difference by subtracting their components,

$$\begin{aligned}\mathbf{a} - \mathbf{b} &= a_x\mathbf{i} + a_y\mathbf{j} + a_z\mathbf{k} - (b_x\mathbf{i} + b_y\mathbf{j} + b_z\mathbf{k}) \\ &= (a_x - b_x)\mathbf{i} + (a_y - b_y)\mathbf{j} + (a_z - b_z)\mathbf{k}.\end{aligned} \tag{6.12}$$

► *Two particles have velocities* $\mathbf{v}_1 = \mathbf{i} + 3\mathbf{j} + 6\mathbf{k}$ *and* $\mathbf{v}_2 = \mathbf{i} - 2\mathbf{k}$ *respectively. Find the velocity* **u** *of the second particle relative to the first.*

The required relative velocity is given by

$$\begin{aligned}\mathbf{u} = \mathbf{v}_2 - \mathbf{v}_1 &= (1 - 1)\mathbf{i} + (0 - 3)\mathbf{j} + (-2 - 6)\mathbf{k} \\ &= -3\mathbf{j} - 8\mathbf{k}. \blacktriangleleft\end{aligned}$$

## 6.5 Magnitude of a vector

The magnitude or length of the vector **a** is denoted by $|\mathbf{a}|$ or $a$. In terms of its components in three-dimensional Cartesian coordinates, the magnitude of **a** is given by

$$a \equiv |\mathbf{a}| = \sqrt{a_x^2 + a_y^2 + a_z^2}. \tag{6.13}$$

Hence, the magnitude of a vector is a measure of its length. Such an analogy is useful for displacement vectors but magnitude is better described, for example, by 'strength' for vectors such as force or by 'speed' for velocity vectors. For instance, in the previous example, the speed of the second particle relative to the first is given by

$$u = |\mathbf{u}| = \sqrt{(-3)^2 + (-8)^2} = \sqrt{73}.$$

A vector whose magnitude equals unity is called a *unit vector*. The unit vector in the direction **a** is usually notated $\hat{\mathbf{a}}$ and may be evaluated as

$$\hat{\mathbf{a}} = \frac{\mathbf{a}}{|\mathbf{a}|}. \tag{6.14}$$

The unit vector is a useful concept because a vector written as $\lambda\hat{\mathbf{a}}$ then has magnitude $\lambda$ and direction $\hat{\mathbf{a}}$. Thus magnitude and direction are explicitly separated.

## 6.6 Multiplication of vectors

We have already considered multiplying a vector by a scalar. Now we consider the concept of multiplying one vector by another vector. It is not immediately obvious what the product of two vectors represents and in fact two products are commonly defined, the *scalar product* and the *vector product*. As their names imply, the scalar product of two vectors is just a number, whereas the vector product is itself a vector. Although neither the scalar nor the vector product is what we might normally think of as a product, their use is widespread and numerous examples will be described elsewhere in this book.

### 6.6.1 Scalar product

The scalar product (or dot product) of two vectors **a** and **b** is denoted by $\mathbf{a}\cdot\mathbf{b}$ and is given by

$$\mathbf{a}\cdot\mathbf{b} \equiv |\mathbf{a}||\mathbf{b}|\cos\theta, \qquad 0 \le \theta \le \pi, \tag{6.15}$$

where $\theta$ is the angle between the two vectors, placed 'tail to tail' or 'head to head'. Thus, the value of the scalar product $\mathbf{a}\cdot\mathbf{b}$ equals the magnitude of **a** multiplied by the projection of **b** onto **a** (see figure 6.8).

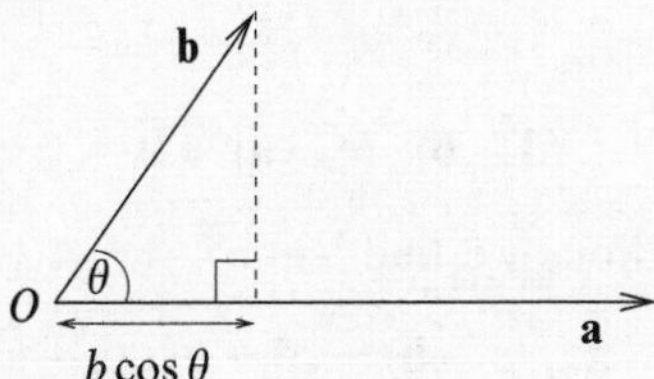


Figure 6.8 The projection of **b** onto the direction of **a** is $b\cos\theta$. The scalar product of **a** and **b** is $ab\cos\theta$.

From (6.15) we see that the scalar product has the particularly useful property that

$$\mathbf{a}\cdot\mathbf{b} = 0 \tag{6.16}$$

is a necessary and sufficient condition for **a** to be perpendicular to **b** (unless either of them is zero). It should be noted in particular that the Cartesian basis vectors **i**, **j** and **k**, being mutually orthogonal unit vectors, satisfy the equations

$$\mathbf{i}\cdot\mathbf{i} = \mathbf{j}\cdot\mathbf{j} = \mathbf{k}\cdot\mathbf{k} = 1, \tag{6.17}$$

$$\mathbf{i}\cdot\mathbf{j} = \mathbf{j}\cdot\mathbf{k} = \mathbf{k}\cdot\mathbf{i} = 0. \tag{6.18}$$

Examples of scalar products arise naturally throughout physics and in particular in connection with energy. Perhaps the simplest is the work done in moving the point of application of a constant force **F** through a displacement **r**. The work done is then given by $\mathbf{F}\cdot\mathbf{r}$; notice that, as expected, if the displacement is perpendicular to the direction of the force then $\mathbf{F}\cdot\mathbf{r} = 0$ and no work is done. A second simple example is afforded by the potential energy $-\mathbf{m}\cdot\mathbf{B}$ of a magnetic dipole, represented in strength and orientation by a vector **m**, placed in an external magnetic field **B**.

As the name implies, the scalar product has a magnitude but no direction. The scalar product is commutative, and distributive over addition:

$$\mathbf{a}\cdot\mathbf{b} = \mathbf{b}\cdot\mathbf{a} \tag{6.19}$$

$$\mathbf{a}\cdot(\mathbf{b}+\mathbf{c}) = \mathbf{a}\cdot\mathbf{b} + \mathbf{a}\cdot\mathbf{c}. \tag{6.20}$$

►*Four points $A, B, C, D$ are positioned such that the line $AD$ is perpendicular to $BC$ and $BD$ is perpendicular to $AC$. Show that $CD$ is perpendicular to $AB$.*

Let us denote the position vectors of the points $A$, $B$, $C$, $D$ by **a**, **b**, **c**, **d** respectively. Thus, since $AD \perp BC$ we have from (6.16) that

$$(\mathbf{d}-\mathbf{a})\cdot(\mathbf{c}-\mathbf{b}) = 0.$$

Similarly, since $BD \perp AC$,

$$(\mathbf{d}-\mathbf{b})\cdot(\mathbf{c}-\mathbf{a})=0.$$

Combining these two equations we find

$$(\mathbf{d}-\mathbf{a})\cdot(\mathbf{c}-\mathbf{b})=(\mathbf{d}-\mathbf{b})\cdot(\mathbf{c}-\mathbf{a}),$$

which on mutliplying out the parentheses gives

$$\mathbf{d}\cdot\mathbf{c}-\mathbf{a}\cdot\mathbf{c}-\mathbf{d}\cdot\mathbf{b}+\mathbf{a}\cdot\mathbf{b}=\mathbf{d}\cdot\mathbf{c}-\mathbf{b}\cdot\mathbf{c}-\mathbf{d}\cdot\mathbf{a}+\mathbf{b}\cdot\mathbf{a}.$$

Cancelling terms that appear on both sides and rearranging yields

$$\mathbf{d}\cdot\mathbf{b}-\mathbf{d}\cdot\mathbf{a}-\mathbf{c}\cdot\mathbf{b}+\mathbf{c}\cdot\mathbf{a}=0,$$

which simplifies to give

$$(\mathbf{d}-\mathbf{c})\cdot(\mathbf{b}-\mathbf{a})=0.$$

From (6.16), we see that this implies $CD$ is perpendicular to $AB$. ◄

If we introduce a set of basis vectors that are mutually orthogonal, such as **i**, **j**, **k**, we can write the components of a vector **a**, with respect to the basis, in terms of the scalar product of **a** with each of the basis vectors, i.e. $a_x=\mathbf{a}\cdot\mathbf{i}$, $a_y=\mathbf{a}\cdot\mathbf{j}$ and $a_z=\mathbf{a}\cdot\mathbf{k}$. In terms of the components $a_x$, $a_y$ and $a_z$ the scalar product is given by

$$\mathbf{a}\cdot\mathbf{b}=(a_x\mathbf{i}+a_y\mathbf{j}+a_z\mathbf{k})\cdot(b_x\mathbf{i}+b_y\mathbf{j}+b_z\mathbf{k})=a_xb_x+a_yb_y+a_zb_z, \qquad (6.21)$$

where the cross terms such as $a_x\mathbf{i}\cdot b_y\mathbf{j}$ are zero because the basis vectors are mutually perpendicular (see equation (6.18)). It should be clear from (6.15) that the value of $\mathbf{a}\cdot\mathbf{b}$ has a geometrical definition, and its value is independent of the actual basis vectors used.

►*Find the angle between the vectors* $\mathbf{a}=\mathbf{i}+2\mathbf{j}+3\mathbf{k}$ *and* $\mathbf{b}=2\mathbf{i}+3\mathbf{j}+4\mathbf{k}$.

From (6.15) the cosine of the angle $\theta$ between **a** and **b** is given by

$$\cos\theta=\frac{\mathbf{a}\cdot\mathbf{b}}{|\mathbf{a}||\mathbf{b}|}.$$

The scalar product $\mathbf{a}\cdot\mathbf{b}$ has the value

$$\mathbf{a}\cdot\mathbf{b}=1\times 2+2\times 3+3\times 4=20,$$

and the lengths of the vectors are

$$|\mathbf{a}|=\sqrt{1^2+2^2+3^2}=\sqrt{14} \qquad \text{and} \qquad |\mathbf{b}|=\sqrt{2^2+3^2+4^2}=\sqrt{29}.$$

Thus,

$$\cos\theta=\frac{20}{\sqrt{14}\sqrt{29}}\approx 0.9926 \qquad \Rightarrow \qquad \theta=0.12 \text{ rad.} \;◄$$

We can see from the expressions (6.15), (6.21) for the scalar product that if $\theta$ is the angle between $\mathbf{a}$ and $\mathbf{b}$ then

$$\cos\theta = \frac{a_x}{a}\frac{b_x}{b} + \frac{a_y}{a}\frac{b_y}{b} + \frac{a_z}{a}\frac{b_z}{b}$$

where $a_x/a$, $a_y/a$ and $a_z/a$ are called the *direction cosines* of $\mathbf{a}$, since they give the cosine of the angle made by $\mathbf{a}$ with each of the basis vectors. Similarly $b_x/b$, $b_y/b$ and $b_z/b$ are the direction cosines of $\mathbf{b}$.

If we take the scalar product of a vector $\mathbf{a}$ with itself then clearly $\theta = 0$ and from (6.15) we have

$$\mathbf{a}\cdot\mathbf{a} = |\mathbf{a}|^2.$$

Thus the magnitude of a vector can be written in a coordinate-independent form as $|\mathbf{a}| = \sqrt{\mathbf{a}\cdot\mathbf{a}}$.

Finally, we note that the scalar product may be extended to vectors with complex components if it is defined as

$$\mathbf{a}\cdot\mathbf{b} = a_x^* b_x + a_y^* b_y + a_z^* b_z,$$

where the asterisk represents the operation of complex conjugation. To accommodate this extension the commutation property (6.19) must be modified to read

$$\mathbf{a}\cdot\mathbf{b} = (\mathbf{b}\cdot\mathbf{a})^*. \qquad (6.22)$$

In particular it should be noted that $(\lambda\mathbf{a})\cdot\mathbf{b} = \lambda^*\mathbf{a}\cdot\mathbf{b}$ whereas $\mathbf{a}\cdot(\lambda\mathbf{b}) = \lambda\mathbf{a}\cdot\mathbf{b}$. However, the magnitude of a complex vector is still given by $|\mathbf{a}| = \sqrt{\mathbf{a}\cdot\mathbf{a}}$, since $\mathbf{a}\cdot\mathbf{a}$ is always real.

### 6.6.2 Vector product

The vector product (or cross product) of two vectors $\mathbf{a}$ and $\mathbf{b}$ is denoted by $\mathbf{a}\times\mathbf{b}$ and is a vector of magnitude $|\mathbf{a}||\mathbf{b}|\sin\theta$ in a direction perpendicular to both $\mathbf{a}$ and $\mathbf{b}$;

$$|\mathbf{a}\times\mathbf{b}| = |\mathbf{a}||\mathbf{b}|\sin\theta.$$

The direction is found by 'rotating' $\mathbf{a}$ into $\mathbf{b}$ through the smallest possible angle as a right-hand screw which then points in the direction of $\mathbf{a}\times\mathbf{b}$ (see figure 6.9). Again, $\theta$ is the angle between the two vectors placed 'tail to tail' or 'head to head'. With this definition $\mathbf{a}$, $\mathbf{b}$ and $\mathbf{a}\times\mathbf{b}$ form a right-hand set.

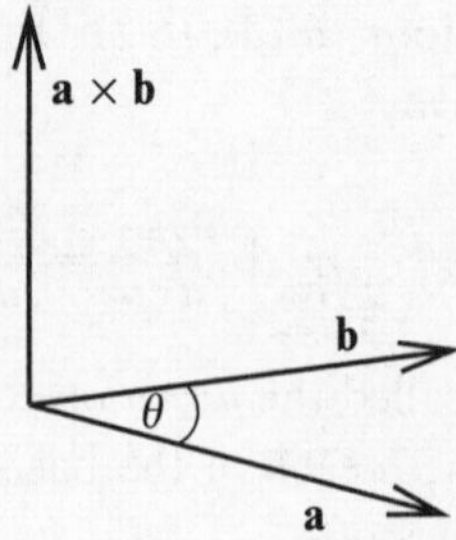


Figure 6.9 The vector product. The vectors **a**, **b** and **a** × **b** form a right-hand set.

The vector product is distributive over addition, but *anticommutative* and *non-associative*:

$$(\mathbf{a}+\mathbf{b})\times\mathbf{c} = (\mathbf{a}\times\mathbf{c}) + (\mathbf{b}\times\mathbf{c}), \tag{6.23}$$

$$\mathbf{b}\times\mathbf{a} = -(\mathbf{a}\times\mathbf{b}), \tag{6.24}$$

$$(\mathbf{a}\times\mathbf{b})\times\mathbf{c} \neq \mathbf{a}\times(\mathbf{b}\times\mathbf{c}). \tag{6.25}$$

From its definition, we see that the vector product has the very useful property that if $\mathbf{a}\times\mathbf{b} = \mathbf{0}$ then **a** is parallel or antiparallel to **b** (unless either of them is zero). We also note that

$$\mathbf{a}\times\mathbf{a} = \mathbf{0}. \tag{6.26}$$

►*Show that if* $\mathbf{a} = \mathbf{b} + \lambda\mathbf{c}$, *for some scalar* $\lambda$, *then* $\mathbf{a}\times\mathbf{c} = \mathbf{b}\times\mathbf{c}$.

From (6.23) we have

$$\mathbf{a}\times\mathbf{c} = (\mathbf{b}+\lambda\mathbf{c})\times\mathbf{c} = \mathbf{b}\times\mathbf{c} + \lambda\mathbf{c}\times\mathbf{c}.$$

However, from (6.26), $\mathbf{c}\times\mathbf{c} = \mathbf{0}$ and so

$$\mathbf{a}\times\mathbf{c} = \mathbf{b}\times\mathbf{c}. \tag{6.27}$$

We note in passing that the fact that (6.27) is satisfied does *not* in general imply that $\mathbf{a} = \mathbf{b}$. ◄

An example of the use of the vector product is that of finding the area, $A$, of a parallelogram with sides **a** and **b**, using the formula

$$A = |\mathbf{a}\times\mathbf{b}|. \tag{6.28}$$

Another example is afforded by considering a force **F** acting through a point $R$, whose vector position relative to the origin $O$ is **r** (see figure 6.10). Its *moment* or *torque* about $O$ is the strength of the force times the perpendicular distance

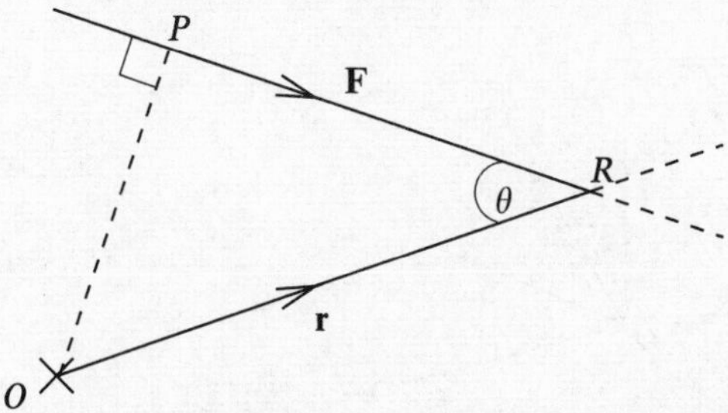


Figure 6.10 The moment of the force $\mathbf{F}$ about $O$ is $\mathbf{r}\times\mathbf{F}$. The cross represents the direction of $\mathbf{r}\times\mathbf{F}$, which is perpendicularly into the plane of the paper.

$OP$, which numerically is just $Fr\sin\theta$, i.e. the magnitude of $\mathbf{r}\times\mathbf{F}$. Furthermore, the sense of the moment is clockwise about an axis through $O$ that points perpendicularly into the plane of the paper (the axis is represented by a cross in the figure). Thus the moment is completely represented by the vector $\mathbf{r}\times\mathbf{F}$, in both magnitude and spatial sense.

Similarly, if a solid body is rotating about some axis that passes through the origin, with an angular velocity $\omega$, then we can describe this rotation by a vector $\boldsymbol{\omega}$ that has magnitude $\omega$ and points along the axis of rotation. The direction of $\boldsymbol{\omega}$ is the forward direction of a right-hand screw rotating in the same sense as the body. The velocity of any point in the body with position vector $\mathbf{r}$ is then given by $\mathbf{v}=\boldsymbol{\omega}\times\mathbf{r}$.

Since the basis vectors $\mathbf{i}$, $\mathbf{j}$, $\mathbf{k}$ are mutually perpendicular unit vectors, forming a right-handed set, their vector products are easily seen to be

$$\mathbf{i}\times\mathbf{i}=\mathbf{j}\times\mathbf{j}=\mathbf{k}\times\mathbf{k}=\mathbf{0}, \tag{6.29}$$

$$\mathbf{i}\times\mathbf{j}=-\mathbf{j}\times\mathbf{i}=\mathbf{k}, \tag{6.30}$$
$$\mathbf{j}\times\mathbf{k}=-\mathbf{k}\times\mathbf{j}=\mathbf{i}, \tag{6.31}$$
$$\mathbf{k}\times\mathbf{i}=-\mathbf{i}\times\mathbf{k}=\mathbf{j}. \tag{6.32}$$

Using these relations, it is straightforward to show that the vector product of two general vectors $\mathbf{a}$ and $\mathbf{b}$ is given in terms of their components with respect to the basis set $\mathbf{i}$, $\mathbf{j}$, $\mathbf{k}$, by

$$\mathbf{a}\times\mathbf{b}=(a_yb_z-a_zb_y)\mathbf{i}+(a_zb_x-a_xb_z)\mathbf{j}+(a_xb_y-a_yb_x)\mathbf{k}. \tag{6.33}$$

For the reader who is familiar with determinants (see chapter 7), we record that this can also be written as

$$\mathbf{a}\times\mathbf{b}=\begin{vmatrix}\mathbf{i} & \mathbf{j} & \mathbf{k}\\ a_x & a_y & a_z\\ b_x & b_y & b_z\end{vmatrix}.$$

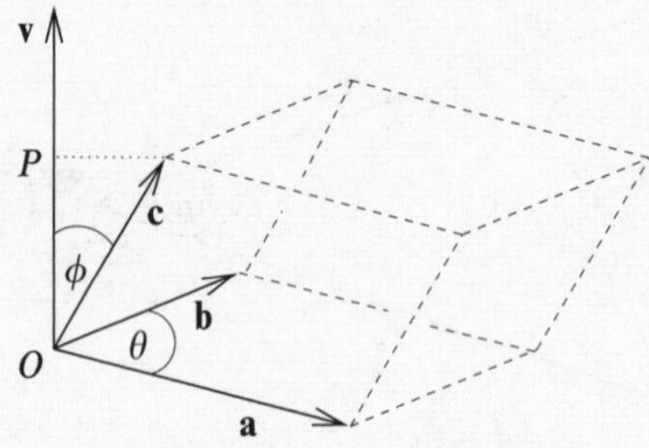


Figure 6.11 The triple scalar product gives the volume of a parallelepiped.

That the cross product $\mathbf{a} \times \mathbf{b}$ is perpendicular to both $\mathbf{a}$ and $\mathbf{b}$ can be verified in component form by forming its dot products with each of the two vectors and showing that is zero in both cases.

►*Find the area A of the parallelogram with sides* $\mathbf{a} = \mathbf{i}+2\mathbf{j}+3\mathbf{k}$ *and* $\mathbf{b} = 4\mathbf{i}+5\mathbf{j}+6\mathbf{k}$.

The vector product $\mathbf{a} \times \mathbf{b}$ is given in component form by

$$\begin{aligned}\mathbf{a} \times \mathbf{b} &= (2 \times 6 - 3 \times 5)\mathbf{i} + (3 \times 4 - 1 \times 6)\mathbf{j} + (1 \times 5 - 2 \times 4)\mathbf{k} \\ &= -3\mathbf{i} + 6\mathbf{j} - 3\mathbf{k}.\end{aligned}$$

Thus the area of the parallelogram is

$$A = |\mathbf{a} \times \mathbf{b}| = \sqrt{(-3)^2 + 6^2 + (-3)^2} = \sqrt{54}. \blacktriangleleft$$

### 6.6.3 Scalar triple product

Now that we have defined the scalar and vector products, we can extend our discussion to define products of three vectors. Again, there are two possibilities, the *scalar triple product* and the *vector triple product*.

The scalar triple product is denoted by

$$[\mathbf{a}, \mathbf{b}, \mathbf{c}] \equiv \mathbf{a} \cdot (\mathbf{b} \times \mathbf{c})$$

and, as its name suggests, it is just a number. It is most simply interpreted as the volume of a parallelepiped whose edges are given by $\mathbf{a}$, $\mathbf{b}$ and $\mathbf{c}$ (see figure 6.11). The vector $\mathbf{v} = \mathbf{a} \times \mathbf{b}$ is perpendicular to the base of the solid and has magnitude $v = ab \sin\theta$, i.e. the area of the base. Further, $\mathbf{v} \cdot \mathbf{c} = vc \cos\phi$. Thus, since $c \cos\phi = OP =$ vertical height of the parallelepiped, it is clear that $(\mathbf{a} \times \mathbf{b}) \cdot \mathbf{c} =$ area of the base $\times$ perpendicular height $=$ volume. It follows that, if the vectors $\mathbf{a}$, $\mathbf{b}$ and $\mathbf{c}$ are coplanar, $\mathbf{a} \cdot (\mathbf{b} \times \mathbf{c}) = 0$.

In terms of the components of each vector with respect to the Cartesian basis set **i**, **j**, **k** the scalar triple product is given by

$$\mathbf{a}\cdot(\mathbf{b}\times\mathbf{c}) = a_x(b_yc_z - b_zc_y) + a_y(b_zc_x - b_xc_z) + a_z(b_xc_y - b_yc_x), \tag{6.34}$$

which can also be written as a determinant:

$$\mathbf{a}\cdot(\mathbf{b}\times\mathbf{c}) = \begin{vmatrix} a_x & a_y & a_z \\ b_x & b_y & b_z \\ c_x & c_y & c_z \end{vmatrix}.$$

By writing the vectors in component form, it can be shown that

$$\mathbf{a}\cdot(\mathbf{b}\times\mathbf{c}) = (\mathbf{a}\times\mathbf{b})\cdot\mathbf{c},$$

so that the dot and cross symbols can be interchanged without changing the result. More generally, the triple scalar product is unchanged under cyclic permutation of the vectors $\mathbf{a}, \mathbf{b}, \mathbf{c}$. Other permutations simply give the negative of the original triple scalar product. These results can be summarised by

$$[\mathbf{a},\mathbf{b},\mathbf{c}] = [\mathbf{b},\mathbf{c},\mathbf{a}] = [\mathbf{c},\mathbf{a},\mathbf{b}] = -[\mathbf{a},\mathbf{c},\mathbf{b}] = -[\mathbf{b},\mathbf{a},\mathbf{c}] = -[\mathbf{c},\mathbf{b},\mathbf{a}]. \tag{6.35}$$

►*Find the volume $V$ of the parallelepiped with sides* $\mathbf{a} = \mathbf{i}+2\mathbf{j}+3\mathbf{k}$, $\mathbf{b} = 4\mathbf{i}+5\mathbf{j}+6\mathbf{k}$ *and* $\mathbf{c} = 7\mathbf{i}+8\mathbf{j}+10\mathbf{k}$.

We have already found $\mathbf{a}\times\mathbf{b} = -3\mathbf{i}+6\mathbf{j}-3\mathbf{k}$ in subsection 6.6.2. Hence the volume of the parallelepiped is given by

$$\begin{aligned} V = |\mathbf{a}\cdot(\mathbf{b}\times\mathbf{c})| &= |(\mathbf{a}\times\mathbf{b})\cdot\mathbf{c}| \\ &= |(-3\mathbf{i}+6\mathbf{j}-3\mathbf{k})\cdot(7\mathbf{i}+8\mathbf{j}+10\mathbf{k})| \\ &= |(-3)(7)+(6)(8)+(-3)(10)| = 3. \blacktriangleleft \end{aligned}$$

Another useful formula involving both the scalar and vector products is Lagrange's identity (see exercise 6.6), i.e.

$$(\mathbf{a}\times\mathbf{b})\cdot(\mathbf{c}\times\mathbf{d}) \equiv (\mathbf{a}\cdot\mathbf{c})(\mathbf{b}\cdot\mathbf{d}) - (\mathbf{a}\cdot\mathbf{d})(\mathbf{b}\cdot\mathbf{c}). \tag{6.36}$$

### *6.6.4 Vector triple product*

By the vector triple product of three vectors **a**, **b**, **c** we mean the vector $\mathbf{a}\times(\mathbf{b}\times\mathbf{c})$. Clearly, $\mathbf{a}\times(\mathbf{b}\times\mathbf{c})$ is perpendicular to **a** and lies in the plane of **b** and **c** and so can be expressed in terms of them (see (6.37) below). We note, from (6.25), that the vector triple product is not associative, i.e. $\mathbf{a}\times(\mathbf{b}\times\mathbf{c}) \neq (\mathbf{a}\times\mathbf{b})\times\mathbf{c}$.

Two useful formulae involving the vector triple product are

$$\mathbf{a}\times(\mathbf{b}\times\mathbf{c}) = (\mathbf{a}\cdot\mathbf{c})\mathbf{b} - (\mathbf{a}\cdot\mathbf{b})\mathbf{c}, \tag{6.37}$$

$$(\mathbf{a}\times\mathbf{b})\times\mathbf{c} = (\mathbf{a}\cdot\mathbf{c})\mathbf{b} - (\mathbf{b}\cdot\mathbf{c})\mathbf{a}, \tag{6.38}$$

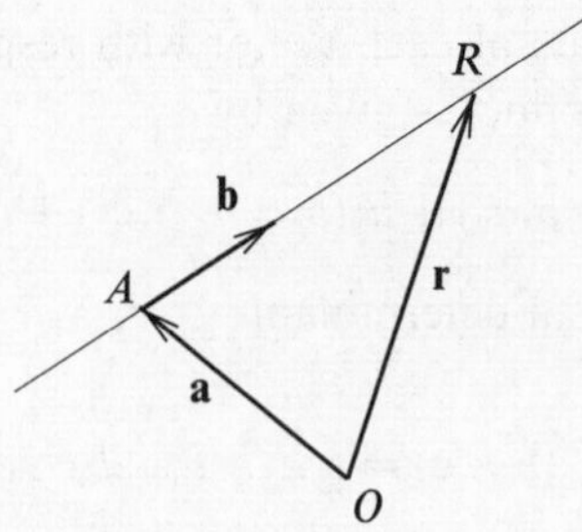


Figure 6.12 The equation of a line. The vector **b** is in the direction $AR$, and $\lambda\mathbf{b}$ is the vector from $A$ to $R$.

which may be derived by writing each vector in component form (see exercise 6.5). It can also be shown that for any three vectors **a**, **b**, **c**,

$$\mathbf{a} \times (\mathbf{b} \times \mathbf{c}) + \mathbf{b} \times (\mathbf{c} \times \mathbf{a}) + \mathbf{c} \times (\mathbf{a} \times \mathbf{b}) = \mathbf{0}.$$

## 6.7 Equations of lines and planes

Now that we have described the basic algebra of vectors, we can apply the results to a variety of problems, the first of which is to find the equations of lines and planes in vector form.

### *6.7.1 Equation of a line*

Consider the line passing through the fixed point $A$ with position vector **a** and having a direction **b** (see figure 6.12). It is clear that the position vector **r** of a general point $R$ on the line can be written as

$$\mathbf{r} = \mathbf{a} + \lambda\mathbf{b}, \tag{6.39}$$

since $R$ can be reached by starting from $O$, going by a translation vector **a** to the point $A$ on the line and then adding some multiple $\lambda\mathbf{b}$ of the vector **b**. Different values of $\lambda$ give different points $R$ on the line.

Taking the components of (6.39), we see that the equation of the line can also be written in the form

$$\frac{x - a_x}{b_x} = \frac{y - a_y}{b_y} = \frac{z - a_z}{b_z} = \text{constant}. \tag{6.40}$$

Furthermore, by taking the vector product of (6.39) with **b** and remembering that $\mathbf{b} \times \mathbf{b} = 0$, the equation of the line becomes

$$(\mathbf{r} - \mathbf{a}) \times \mathbf{b} = \mathbf{0}.$$

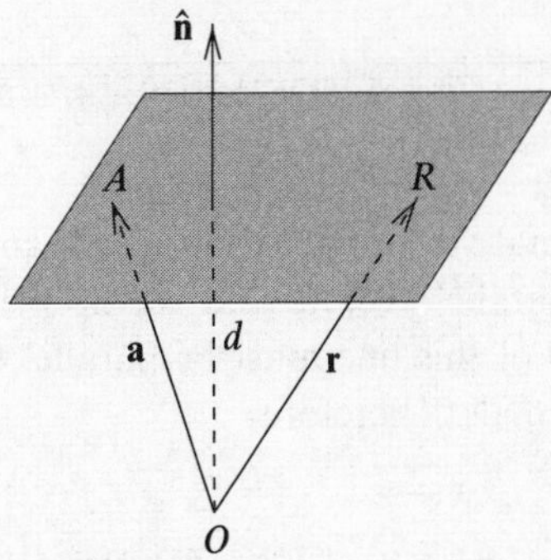


Figure 6.13 The equation of the plane is $(\mathbf{r}-\mathbf{a})\cdot\hat{\mathbf{n}}=0$.

We may also find the equation of the line that passes through two fixed points with position vectors $\mathbf{a}$ and $\mathbf{c}$. Since its direction is given simply by $\mathbf{c}-\mathbf{a}$, the position vector of a general point on the line is

$$\mathbf{r}=\mathbf{a}+\lambda(\mathbf{c}-\mathbf{a}).$$

### *6.7.2 Equation of a plane*

The equation of a plane through a point $A$ with position vector $\mathbf{a}$ and perpendicular to a unit position vector $\hat{\mathbf{n}}$ (see figure 6.13) is

$$(\mathbf{r}-\mathbf{a})\cdot\hat{\mathbf{n}}=0; \tag{6.41}$$

this follows since the vector joining $A$ to a general point $R$ with position vector $\mathbf{r}$ is $\mathbf{r}-\mathbf{a}$, and $\mathbf{r}$ will lie in the plane if this vector is perpendicular to the normal to the plane. Rewriting the above equation as $\mathbf{r}\cdot\hat{\mathbf{n}}=\mathbf{a}\cdot\hat{\mathbf{n}}$, we see that (6.41) may also be expressed in the form $\mathbf{r}\cdot\hat{\mathbf{n}}=d$, or in component form as

$$lx+my+nz=d, \tag{6.42}$$

where the unit normal to the plane is $\hat{\mathbf{n}}=l\mathbf{i}+m\mathbf{j}+n\mathbf{k}$ and $d=\mathbf{a}\cdot\hat{\mathbf{n}}$ is the perpendicular distance of the plane from the origin.

The equation of a plane containing points $\mathbf{a}$, $\mathbf{b}$ and $\mathbf{c}$ is

$$\mathbf{r}=\mathbf{a}+\lambda(\mathbf{b}-\mathbf{a})+\mu(\mathbf{c}-\mathbf{a}).$$

This is apparent because starting from the point $\mathbf{a}$ in the plane, all other points may be reached by moving a distance along each of two (non-parallel) directions in the plane. Two such directions are given by $\mathbf{b}-\mathbf{a}$ and $\mathbf{c}-\mathbf{a}$. It can be shown that the equation of this plane may also be written in the more symmetrical form

$$\mathbf{r}=\alpha\mathbf{a}+\beta\mathbf{b}+\gamma\mathbf{c},$$

where $\alpha+\beta+\gamma=1$.

►*Find the direction,* $\mathbf{p}$, *of the line of intersection of the planes* $x + 3y - z = 5$ *and* $2x - 2y + 4z = 3$.

The two planes have normal vectors $\mathbf{n}_1 = \mathbf{i} + 3\mathbf{j} - \mathbf{k}$ and $\mathbf{n}_2 = 2\mathbf{i} - 2\mathbf{j} + 4\mathbf{k}$. It is clear that these are not parallel vectors and so the planes must intersect along some line. The direction $\mathbf{p}$ of this line must be parallel to both planes and hence perpendicular to both normals. Therefore

$$\begin{aligned}\mathbf{p} &= \mathbf{n}_1 \times \mathbf{n}_2 \\ &= [(3)(4) - (-2)(-1)]\,\mathbf{i} + [(-1)(2) - (1)(4)]\,\mathbf{j} + [(1)(-2) - (3)(2)]\,\mathbf{k} \\ &= 10\mathbf{i} - 6\mathbf{j} - 8\mathbf{k}. \blacktriangleleft\end{aligned}$$

## 6.8 Using vectors to find distances

This section deals with the practical application of vectors to finding distances. Some of these problems are extremely cumbersome in component form, but all reduce to neat solutions when general vectors, with no explicit basis set, are used. These examples show the power of vectors in simplifying geometrical problems.

### *6.8.1 Distance from a point to a line*

Figure 6.14 shows a line of direction $\mathbf{b}$ that passes through a point $A$ whose position vector is $\mathbf{a}$. To find the *minimum distance* $d$ of the line from a point $P$ whose position vector is $\mathbf{p}$, we must solve the right-angled triangle shown. We see that $d = |\mathbf{p} - \mathbf{a}| \sin\theta$. So, from the definition of the vector product it follows that

$$d = |(\mathbf{p} - \mathbf{a}) \times \hat{\mathbf{b}}|.$$

►*Find the minimum distance from the point* $P$ *with coordinates* $(1, 2, 1)$ *to the line* $\mathbf{r} = \mathbf{a} + \lambda\mathbf{b}$ *where* $\mathbf{a} = \mathbf{i} + \mathbf{j} + \mathbf{k}$ *and* $\mathbf{b} = 2\mathbf{i} - \mathbf{j} + 3\mathbf{k}$.

Comparison with (6.39) shows that the line passes through the point $(1, 1, 1)$ and has direction $2\mathbf{i} - \mathbf{j} + 3\mathbf{k}$. The unit vector in this direction is

$$\hat{\mathbf{b}} = \frac{1}{\sqrt{14}}(2\mathbf{i} - \mathbf{j} + 3\mathbf{k}).$$

The position vector of $P$ is $\mathbf{p} = \mathbf{i} + 2\mathbf{j} + \mathbf{k}$ and we find

$$\begin{aligned}(\mathbf{p} - \mathbf{a}) \times \hat{\mathbf{b}} &= \frac{1}{\sqrt{14}}[\mathbf{j} \times (2\mathbf{i} - 3\mathbf{j} + 3\mathbf{k})] \\ &= \frac{1}{\sqrt{14}}(3\mathbf{i} - 2\mathbf{k}).\end{aligned}$$

Thus the minimum distance from the line to the point $P$ is $d = \sqrt{13/14}$. ◄

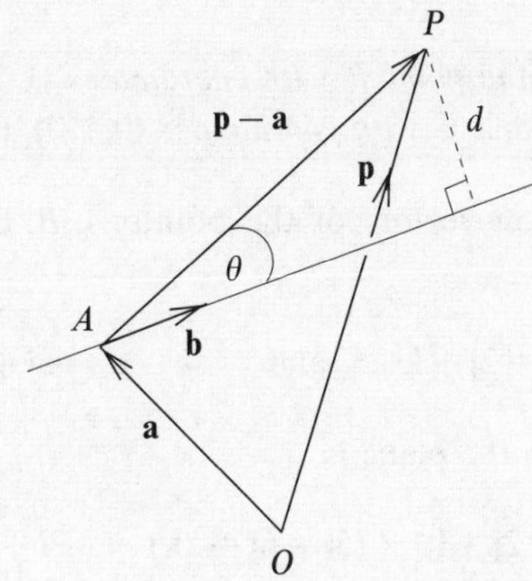


Figure 6.14 The minimum distance from a point to a line.

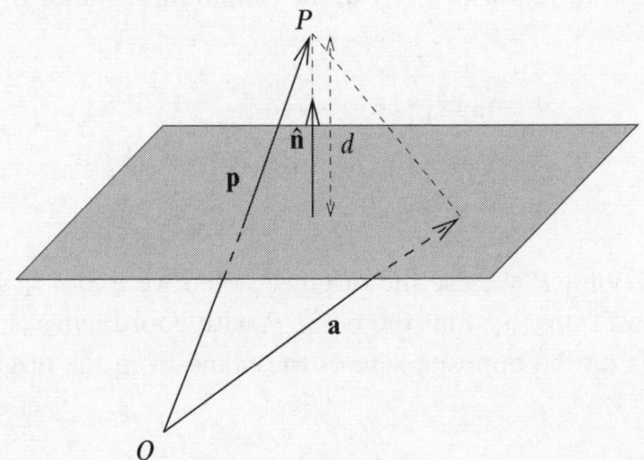


Figure 6.15 The minimum distance $d$ from a point to a plane.

### *6.8.2 Distance from a point to a plane*

The minimum distance $d$ from a point $P$ whose position vector is $\mathbf{p}$ to a plane, $(\mathbf{r}-\mathbf{a})\cdot\hat{\mathbf{n}}=0$, may be deduced by finding any vector from $P$ to the plane and then determining its component in the normal direction. This is shown in figure 6.15. Consider the vector $\mathbf{a}-\mathbf{p}$, which is a particular vector from $P$ to the plane. Its component normal to the plane, and hence its distance from the plane, is given by

$$d = (\mathbf{a}-\mathbf{p})\cdot\hat{\mathbf{n}}, \tag{6.43}$$

where the sign of $d$ depends on which side of the plane $P$ is situated.

►*Find the distance from the point P with coordinates* $(1,2,3)$ *to the plane which contains the points A, B and C with coordinates* $(0,1,0)$, $(2,3,1)$ *and* $(5,7,2)$.

Let us denote the position vectors of the points $A$, $B$, $C$ by $\mathbf{a}$, $\mathbf{b}$, $\mathbf{c}$. Thus, two vectors in the plane are

$$\mathbf{b}-\mathbf{a} = 2\mathbf{i}+2\mathbf{j}+\mathbf{k} \qquad \text{and} \qquad \mathbf{c}-\mathbf{a} = 5\mathbf{i}+6\mathbf{j}+2\mathbf{k}.$$

Hence a vector normal to the plane is

$$\mathbf{n} = (2\mathbf{i}+2\mathbf{j}+\mathbf{k}) \times (5\mathbf{i}+6\mathbf{j}+2\mathbf{k}) = -2\mathbf{i}+\mathbf{j}+2\mathbf{k},$$

and the unit normal is

$$\hat{\mathbf{n}} = \frac{\mathbf{n}}{|\mathbf{n}|} = \tfrac{1}{3}(-2\mathbf{i}+\mathbf{j}+2\mathbf{k}).$$

Denoting the position vector of $P$ by $\mathbf{p}$, the minimum distance from the plane to $P$ is given by

$$\begin{aligned} d &= (\mathbf{a}-\mathbf{p})\cdot\hat{\mathbf{n}} \\ &= (-\mathbf{i}-\mathbf{j}-3\mathbf{k})\cdot\tfrac{1}{3}(-2\mathbf{i}+\mathbf{j}+2\mathbf{k}) \\ &= \tfrac{2}{3}-\tfrac{1}{3}-2 = -\tfrac{5}{3}. \end{aligned}$$

If instead of the point $P$ we use the origin $O$, then we find $d = \frac{1}{3}$, i.e. a positive quantity. It follows from this that the point $P$ with coordinates $(1,2,3)$, for which $d$ was negative, is on the opposite side of the plane from the origin. ◄

### *6.8.3 Distance from a line to a line*

Consider two lines in the directions $\mathbf{a}$ and $\mathbf{b}$ as shown in figure 6.16. Since $\mathbf{a}\times\mathbf{b}$ is by definition perpendicular to both $\mathbf{a}$ and $\mathbf{b}$, the unit vector normal to both these lines is

$$\hat{\mathbf{n}} = \frac{\mathbf{a}\times\mathbf{b}}{|\mathbf{a}\times\mathbf{b}|}.$$

If $\mathbf{p}$ and $\mathbf{q}$ are the position vectors of any two points $P$ and $Q$ on different lines then the vector connecting them is $\mathbf{p}-\mathbf{q}$. Thus, the minimum distance $d$ between the lines is this vector's component along the unit normal, i.e.

$$d = |(\mathbf{p}-\mathbf{q})\cdot\hat{\mathbf{n}}|.$$

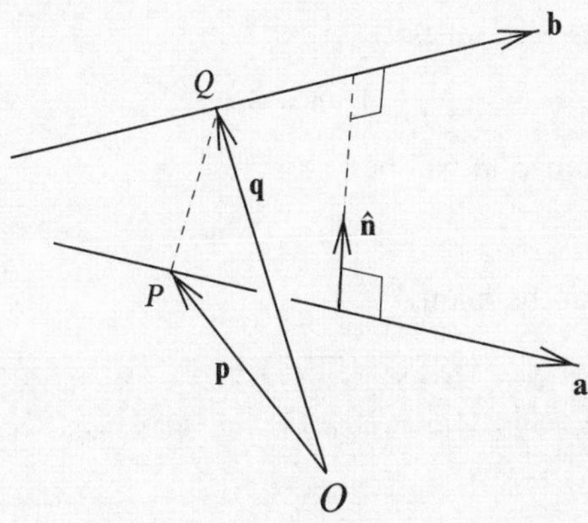


Figure 6.16 The minimum distance from one line to another.

►*A line is inclined at equal angles to the x-, y- and z- axes, and passes through the origin. Another line passes through the points* $(1, 2, 4)$ *and* $(0, 0, 1)$*. Find the minimum distance between the two lines.*

The first line is given by

$$\mathbf{r}_1 = \lambda(\mathbf{i} + \mathbf{j} + \mathbf{k}),$$

and the second by

$$\mathbf{r}_2 = \mathbf{k} + \mu(\mathbf{i} + 2\mathbf{j} + 3\mathbf{k}).$$

Hence a vector normal to both lines is

$$\mathbf{n} = (\mathbf{i} + \mathbf{j} + \mathbf{k}) \times (\mathbf{i} + 2\mathbf{j} + 3\mathbf{k}) = \mathbf{i} - 2\mathbf{j} + \mathbf{k},$$

and the unit normal is

$$\hat{\mathbf{n}} = \frac{1}{\sqrt{6}}(\mathbf{i} - 2\mathbf{j} + \mathbf{k}).$$

A vector between the two lines is, for example, the one connecting the points $(0, 0, 0)$ and $(0, 0, 1)$, which is simply $\mathbf{k}$. Thus it follows that the minimum distance between the two lines is

$$d = \frac{1}{\sqrt{6}}|\mathbf{k} \cdot (\mathbf{i} - 2\mathbf{j} + \mathbf{k})| = \frac{1}{\sqrt{6}}. \blacktriangleleft$$

### *6.8.4 Distance from a line to a plane*

Let us consider the line $\mathbf{r} = \mathbf{a} + \lambda\mathbf{b}$. This line will intersect any plane to which it is not parallel. Thus if a plane has a normal $\hat{\mathbf{n}}$, then the minimum distance from the line to the plane is zero unless

$$\mathbf{b} \cdot \hat{\mathbf{n}} = 0,$$

in which case the distance, $d$, will be

$$d = |(\mathbf{a} - \mathbf{r}) \cdot \hat{\mathbf{n}}|,$$

where $\mathbf{r}$ is any point in the plane.

►*A line is given by* $\mathbf{r} = \mathbf{a} + \lambda\mathbf{b}$, *where* $\mathbf{a} = \mathbf{i} + 2\mathbf{j} + 3\mathbf{k}$ *and* $\mathbf{b} = 4\mathbf{i} + 5\mathbf{j} + 6\mathbf{k}$. *Find the coordinates of the point P at which the line intersects the plane*

$$x + 2y + 3z = 6.$$

A vector normal to the plane is

$$\mathbf{n} = \mathbf{i} + 2\mathbf{j} + 3\mathbf{k},$$

from which we find $\mathbf{b} \cdot \mathbf{n} \neq 0$. Thus the line does indeed intersect the plane. To find the point of intersection we merely substitute the $x$-, $y$- and $z$- values of the line into the equation of the plane to obtain

$$1 + 4\lambda + 2(2 + 5\lambda) + 3(3 + 6\lambda) = 6 \quad \Rightarrow \quad 14 + 32\lambda = 6.$$

This gives $\lambda = -\frac{1}{4}$, which we may substitute into the equation for the line to obtain $x = 1 - \frac{1}{4}(4) = 0$, $y = 2 - \frac{1}{4}(5) = \frac{3}{4}$ and $z = 3 - \frac{1}{4}(6) = \frac{3}{2}$. Thus the point of intersection is $(0, \frac{3}{4}, \frac{3}{2})$. ◄

## 6.9 Reciprocal vectors

The final section of this chapter introduces the concept of reciprocal vectors, which have particular uses in crystallography.

The two sets of vectors $\mathbf{a}$, $\mathbf{b}$, $\mathbf{c}$ and $\mathbf{a}'$, $\mathbf{b}'$, $\mathbf{c}'$ are called *reciprocal sets* if

$$\mathbf{a} \cdot \mathbf{a}' = \mathbf{b} \cdot \mathbf{b}' = \mathbf{c} \cdot \mathbf{c}' = 1 \tag{6.44}$$

and

$$\mathbf{a}' \cdot \mathbf{b} = \mathbf{a}' \cdot \mathbf{c} = \mathbf{b}' \cdot \mathbf{a} = \mathbf{b}' \cdot \mathbf{c} = \mathbf{c}' \cdot \mathbf{a} = \mathbf{c}' \cdot \mathbf{b} = 0. \tag{6.45}$$

It can be verified (see exercise 6.10) that the reciprocal vectors of $\mathbf{a}$, $\mathbf{b}$ and $\mathbf{c}$ are given by

$$\mathbf{a}' = \frac{\mathbf{b} \times \mathbf{c}}{\mathbf{a} \cdot (\mathbf{b} \times \mathbf{c})}, \tag{6.46}$$

$$\mathbf{b}' = \frac{\mathbf{c} \times \mathbf{a}}{\mathbf{a} \cdot (\mathbf{b} \times \mathbf{c})}, \tag{6.47}$$

$$\mathbf{c}' = \frac{\mathbf{a} \times \mathbf{b}}{\mathbf{a} \cdot (\mathbf{b} \times \mathbf{c})}, \tag{6.48}$$

where $\mathbf{a}\cdot(\mathbf{b}\times\mathbf{c})\neq 0$. In other words, reciprocal vectors only exist if **a**, **b** and **c** are not coplanar. Moreover, if **a**, **b**, **c** are mutually orthogonal unit vectors, then $\mathbf{a}'=\mathbf{a}$, $\mathbf{b}'=\mathbf{b}$ and $\mathbf{c}'=\mathbf{c}$, so that the two systems of vectors are identical.

►*Construct the reciprocal vectors of* $\mathbf{a}=2\mathbf{i}$, $\mathbf{b}=\mathbf{j}+\mathbf{k}$, $\mathbf{c}=\mathbf{i}+\mathbf{k}$.

First we evaluate the triple scalar product:

$$\begin{aligned}\mathbf{a}\cdot(\mathbf{b}\times\mathbf{c})&=2\mathbf{i}\cdot[(\mathbf{j}+\mathbf{k})\times(\mathbf{i}+\mathbf{k})]\\&=2\mathbf{i}\cdot(\mathbf{i}+\mathbf{j}-\mathbf{k})\;=\;2.\end{aligned}$$

Now we find the reciprocal vectors:

$$\begin{aligned}\mathbf{a}'&=\tfrac{1}{2}(\mathbf{j}+\mathbf{k})\times(\mathbf{i}+\mathbf{k})\;=\tfrac{1}{2}(\mathbf{i}+\mathbf{j}-\mathbf{k}),\\\mathbf{b}'&=\tfrac{1}{2}(\mathbf{i}+\mathbf{k})\times 2\mathbf{i}\;=\;\mathbf{j},\\\mathbf{c}'&=\tfrac{1}{2}(2\mathbf{i})\times(\mathbf{j}+\mathbf{k})\;=\;-\mathbf{j}+\mathbf{k}.\end{aligned}$$

It is easily verified that these reciprocal vectors satisfy their defining properties (6.44), (6.45). ◄

We may also use the concept of reciprocal vectors to define the components of a vector **a** with respect to basis vectors $\mathbf{e}_1$, $\mathbf{e}_2$, $\mathbf{e}_3$ that are not mutually orthogonal. If the basis vectors are of unit length and mutually orthogonal, such as the Cartesian basis vectors **i**, **j**, **k**, we found in section 6.4 that

$$\mathbf{a}=(\mathbf{a}\cdot\mathbf{i})\mathbf{i}+(\mathbf{a}\cdot\mathbf{j})\mathbf{j}+(\mathbf{a}\cdot\mathbf{k})\mathbf{k}.$$

If the basis is not orthonormal, however, then this is no longer true. Nevertheless, we may write the components of **a** with respect to a non-orthonormal basis $\mathbf{e}_1$, $\mathbf{e}_2$, $\mathbf{e}_3$ in terms of its reciprocal basis vectors $\mathbf{e}_1'$, $\mathbf{e}_2'$, $\mathbf{e}_3'$, which are defined as in (6.46)–(6.48). If we let

$$\mathbf{a}=a_1\mathbf{e}_1+a_2\mathbf{e}_2+a_3\mathbf{e}_3,$$

then the scalar product $\mathbf{a}\cdot\mathbf{e}_1'$ is given by

$$\mathbf{a}\cdot\mathbf{e}_1'=a_1\mathbf{e}_1\cdot\mathbf{e}_1'+a_2\mathbf{e}_2\cdot\mathbf{e}_1'+a_3\mathbf{e}_3\cdot\mathbf{e}_1'=a_1,$$

where we have used the relations (6.45). Similarly, $a_2=\mathbf{a}\cdot\mathbf{e}_2'$ and $a_3=\mathbf{a}\cdot\mathbf{e}_3'$; so now

$$\mathbf{a}=(\mathbf{a}\cdot\mathbf{e}_1')\mathbf{e}_1+(\mathbf{a}\cdot\mathbf{e}_2')\mathbf{e}_2+(\mathbf{a}\cdot\mathbf{e}_3')\mathbf{e}_3. \qquad (6.49)$$

## 6.10 Exercises

6.1 A unit cell of a diamond is a cube of side $A$ with carbon atoms at each corner and at the centre of each face. In addition, there are carbon atoms displaced by $\frac{1}{4}A(\mathbf{i}+\mathbf{j}+\mathbf{k})$ from each of the previously mentioned ones, where **i**, **j**, **k** are unit vectors along the cube axes. One corner of the cube is taken as the origin of coordinates. What are the vectors joining the atom at $\frac{1}{4}A(\mathbf{i}+\mathbf{j}+\mathbf{k})$ to its four nearest neighbours? Determine the angle between the carbon bonds in diamond.

6.2 Identify the following surfaces:

(a) $|\mathbf{r}| = k$; (b) $\mathbf{r} \cdot \mathbf{u} = l$; (c) $\mathbf{r} \cdot \mathbf{u} = m|\mathbf{r}|$ for $-1 \leq m \leq +1$;
(d) $|\mathbf{r} - (\mathbf{r} \cdot \mathbf{u})\mathbf{u}| = n$.

Here $k$, $l$, $m$ and $n$ are fixed scalars and **u** is a fixed unit vector.

6.3 Find the angle between the position vectors to the points $(3, -4, 0)$ and $(-2, 1, 0)$ and find the direction cosines of a vector perpendicular to both.

6.4 The edges $OP$, $OQ$ and $OR$ of a tetrahedron $OPQR$ are vectors **p**, **q** and **r** respectively, where $\mathbf{p} = 2\mathbf{i} + 4\mathbf{j}$, $\mathbf{q} = 2\mathbf{i} - \mathbf{j} + 3\mathbf{k}$ and $\mathbf{r} = 4\mathbf{i} - 2\mathbf{j} + 5\mathbf{k}$. Show that $OP$ is perpendicular to the plane containing $OQR$. Express the volume of the tetrahedron in terms of **p**, **q** and **r** and hence calculate the volume.

6.5 Show that

$$(\mathbf{a} \times \mathbf{b}) \times \mathbf{c} = (\mathbf{a} \cdot \mathbf{c})\mathbf{b} - (\mathbf{b} \cdot \mathbf{c})\mathbf{a}.$$

6.6 Prove Lagrange's identity, i.e.

$$(\mathbf{a} \times \mathbf{b}) \cdot (\mathbf{c} \times \mathbf{d}) = (\mathbf{a} \cdot \mathbf{c})(\mathbf{b} \cdot \mathbf{d}) - (\mathbf{a} \cdot \mathbf{d})(\mathbf{b} \cdot \mathbf{c}).$$

6.7 Show that the points $(1, 0, 1)$, $(1, 1, 0)$ and $(1, -3, 4)$ lie on a straight line. Give the equation of the line in the form

$$\mathbf{r} = \mathbf{a} + \lambda\mathbf{b}.$$

6.8 The vectors **a**, **b** and **c** are coplanar and related by

$$\lambda\mathbf{a} + \mu\mathbf{b} + \nu\mathbf{c} = 0,$$

where $\lambda, \mu, \nu$ are not all zero. Show that the condition for the points with position vectors $\alpha\mathbf{a}, \beta\mathbf{b}$ and $\gamma\mathbf{c}$ to be collinear is

$$\frac{\lambda}{\alpha} + \frac{\mu}{\beta} + \frac{\nu}{\gamma} = 0.$$

6.9 (a) Show that the line of intersection of the planes $x + 2y + 3z = 0$ and $3x + 2y + z = 0$ is equally inclined to the $x$- and $z$- axes and makes an angle $\cos^{-1}(-2/\sqrt{6})$ with the $y$-axis.
(b) Find the perpendicular distance of one corner of a unit cube from the major diagonal not passing through it.

6.10 The vectors **a**, **b** and **c** are not coplanar. The vectors $\mathbf{a}'$, $\mathbf{b}'$ and $\mathbf{c}'$ are the associated reciprocal vectors. Verify that the expressions (6.46)–(6.48)

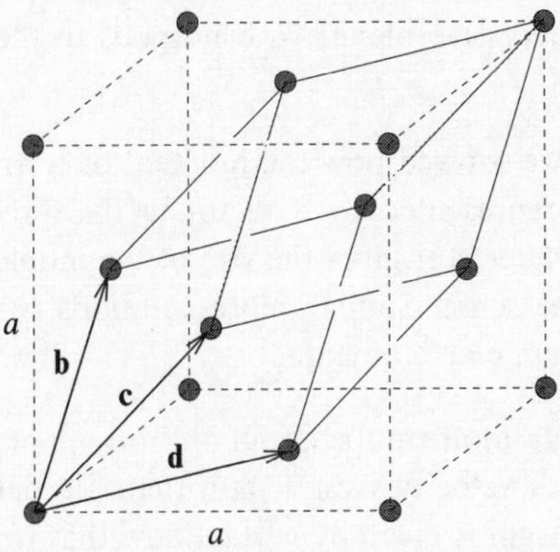


Figure 6.17 A face-centred cubic crystal.

define a set of reciprocal vectors $\mathbf{a}'$, $\mathbf{b}'$ and $\mathbf{c}'$ with the following properties:

(a) $\mathbf{a}' \cdot \mathbf{a} = \mathbf{b}' \cdot \mathbf{b} = \mathbf{c}' \cdot \mathbf{c} = 1$;
(b) $\mathbf{a}' \cdot \mathbf{b} = \mathbf{a}' \cdot \mathbf{c} = \mathbf{b}' \cdot \mathbf{a}$ etc $= 0$;
(c) $[\mathbf{a}', \mathbf{b}', \mathbf{c}'] = 1/[\mathbf{a}, \mathbf{b}, \mathbf{c}]$;
(d) $\mathbf{a} = (\mathbf{b}' \times \mathbf{c}')/[\mathbf{a}', \mathbf{b}', \mathbf{c}']$.

6.11 In a crystal with a face-centred cubic structure, the basic cell can be taken as a cube of edge $a$ with its centre at the origin of coordinates and its edges parallel to the Cartesian coordinate axes; atoms are sited at the eight corners and at the centre of each face. However, other basic cells are possible. One is the rhomboid shown in figure 6.17, which has the three vectors **b**, **c** and **d** as edges.

(a) Show that the volume of the rhomboid is one-quarter that of the cube.
(b) Show that the angles between pairs of edges of the rhomboid are 60°, and that the corresponding angles between pairs of edges of the rhomboid defined by the reciprocal vectors to **b**, **c**, **d** are each 109.5°. (This rhomboid can be used as the basic cell of a body-centred cubic structure, more easily visualised as a cube with an atom at each corner and one at its centre.)
(c) In order to use the Bragg formula, $2d \sin\theta = n\lambda$, for the scattering of x-rays by a crystal, it is necessary to know the perpendicular distance $d$ between successive planes of atoms; for a given crystal structure, this varies as different directions for the normal to the plane are selected. For the face-centred cubic structure find the distance

between successive planes with normals in the $\mathbf{k}$, $\mathbf{i}+\mathbf{j}$ and $\mathbf{i}+\mathbf{j}+\mathbf{k}$ directions.

6.12 In section 6.6 we showed how the moment or torque of a force about an axis could be represented by a vector in the direction of the axis. The magnitude of the vector gives the size of the moment and the sign of the vector gives the sense. Similar representations can be used for angular velocities and angular momenta.

(a) The angular momentum about the origin of a particle of mass $m$ moving with velocity $\mathbf{v}$ on a path that is a perpendicular distance $d$ from the origin is given by $m|\mathbf{v}|d$. Show that if $\mathbf{r}$ is the position of the particle, the vector $\mathbf{J} = \mathbf{r} \times m\mathbf{v}$ represents the angular momentum.

(b) Now consider a rigid collection of particles (or a solid body) rotating about an axis through the origin, the angular velocity of the collection being represented by $\boldsymbol{\omega}$.

(i) Show that the velocity of the $i$th particle is

$$\mathbf{v}_i = \boldsymbol{\omega} \times \mathbf{r}_i,$$

and that the total angular momentum $\mathbf{J}$ is

$$\mathbf{J} = \sum_i m_i[r_i^2\boldsymbol{\omega} - (\mathbf{r}_i \cdot \boldsymbol{\omega})\mathbf{r}_i].$$

(ii) Show further that the component of $\mathbf{J}$ along the axis of rotation can be written as $I\omega$, where $I$, the moment of inertia of the collection about the axis or rotation, is given by

$$I = \sum_i m_i \rho_i^2.$$

Interpret $\rho_i$ geometrically.

(iii) Prove that the total kinetic energy of the particles is $I\omega^2/2$.

6.13 Systems that can be modelled as damped harmonic oscillators are widespread; pendulum clocks, car shock absorbers, tuning circuits in television sets and radios, and collective electron motions in plasmas and metals are just a few examples.

In all these cases, one or more variables describing the system obey(s) an equation of the form

$$\ddot{x} + 2\gamma\dot{x} + \omega_0^2 x = P\cos\omega t,$$

where $\dot{x} = dx/dt$, etc., and the inclusion of the factor 2 is conventional.

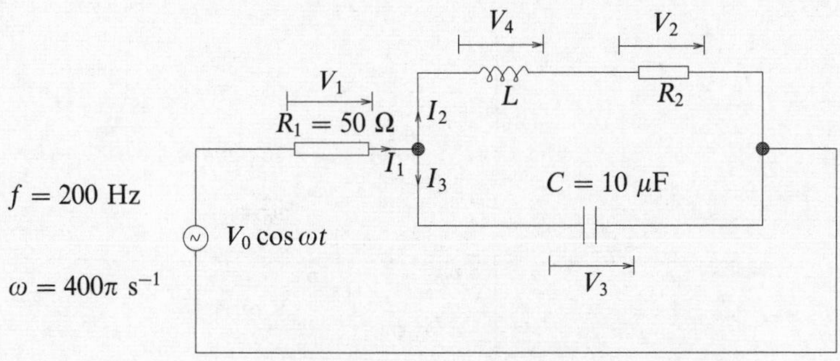


Figure 6.18 An oscillatory electric circuit.

In the steady state (i.e. after any starting transients have died away) the solution of the equation takes the form

$$x(t) = A\cos(\omega t + \phi).$$

By expressing each term in the form $B\cos(\omega t + \epsilon)$ and representing it by a vector of magnitude $B$ making an angle $\epsilon$ with the $x$-axis, draw a closed vector diagram (at $t = 0$ say) that is equivalent to the equation.

(a) Convince yourself that whatever the value of $\omega$ ($> 0$) $\phi$ must be negative ($-\pi < \phi \leq 0$) and that

$$\phi = \tan^{-1}\left(\frac{-2\gamma\omega}{\omega_0^2 - \omega^2}\right).$$

(b) Obtain an expression for $A$ in terms of $P$, $\omega_0$ and $\omega$.

6.14 According to alternating current theory, the currents and voltages in the components of the circuit shown in figure 6.18 are determined by Kirchhoff's laws and the relationships

$$I_1 = \frac{V_1}{R_1}, \qquad I_2 = \frac{V_2}{R_2}, \qquad I_3 = i\omega C V_3, \qquad V_4 = i\omega L I_2.$$

The factor $i = \sqrt{-1}$ in the expression for $I_3$ indicates that the phase of $I_3$ is 90° ahead of $V_3$. Similarly the phase of $V_4$ is 90° ahead of $I_2$.

Measurement shows that $V_3$ has an amplitude of $0.661V_0$ and a phase of $+13.4°$ relative to that of the power supply. Using a series of vector plots for voltages and currents (they could all be on the same plot if suitable scales were chosen), determine all unknown currents and voltages and find values for the inductance of $L$ and the resistance of $R_2$. (Scales of 1 cm $= 0.1V_0$ for voltages, and 1 cm $= 1$ mA $\times|V_0|$ for currents are convenient.)

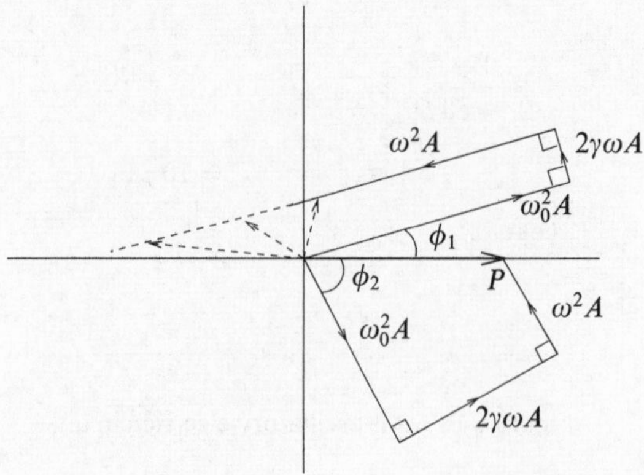


Figure 6.19 The vector diagram for the equation in exercise 6.13.

## 6.11 Hints and answers

6.1 In units of $\frac{1}{4}A$ the vectors are $-\mathbf{i}-\mathbf{j}-\mathbf{k}$, $\mathbf{i}+\mathbf{j}-\mathbf{k}$, $\mathbf{i}-\mathbf{j}+\mathbf{k}$, $-\mathbf{i}+\mathbf{j}+\mathbf{k}$; $\cos^{-1}(-1/3) = 109.5°$.

6.2 (a) Sphere of radius $k$ centred on the origin; (b) plane with its normal in the direction of $\mathbf{u}$ and a distance $l$ from the origin; (c) cone with its axis parallel to $\mathbf{u}$ and semiangle $\cos^{-1} m$; (d) circular cylinder of radius $n$ with its axis parallel to $\mathbf{u}$.

6.3 $\cos^{-1}(-2/\sqrt{5}) = 153.4°$; 0, 0, 1.

6.4 Show that $\mathbf{q}\times\mathbf{r}$ is parallel to $\mathbf{p}$; volume $= \frac{1}{3}\left[\frac{1}{2}(\mathbf{q}\times\mathbf{r})\cdot\mathbf{p}\right] = \frac{5}{3}$.

6.6 Note that $(\mathbf{a}\times\mathbf{b})\cdot(\mathbf{c}\times\mathbf{d}) = \mathbf{d}\cdot[(\mathbf{a}\times\mathbf{b})\times\mathbf{c}]$ and use the result from the previous question.

6.7 Show vectors are linearly dependent; $\mathbf{r} = \mathbf{a} + \lambda\mathbf{b}$ where $\mathbf{a} = \mathbf{i}+\mathbf{k}$ and $\mathbf{b} = -\mathbf{j}+\mathbf{k}$.

6.8 For collinearity, $\gamma\mathbf{c} = \theta\alpha\mathbf{a} + (1-\theta)\beta\mathbf{b}$ for some $\theta$.

6.9 (a) Find two points on both planes, say $(0,0,0)$ and $(1,-2,1)$, and hence determine the direction cosines of the line of intersection; (b) $(2/3)^{1/2}$.

6.10 (c) and (d) Use the result of exercise 6.5 to evaluate $(\mathbf{c}\times\mathbf{a})\times(\mathbf{a}\times\mathbf{b})$.

6.11 (b) $\mathbf{b}' = a^{-1}(-\mathbf{i}+\mathbf{j}+\mathbf{k})$, $\mathbf{c}' = a^{-1}(\mathbf{i}-\mathbf{j}+\mathbf{k})$, $\mathbf{d}' = a^{-1}(\mathbf{i}+\mathbf{j}-\mathbf{k})$; (c) $a/2$ for direction $\mathbf{k}$; successive planes through $(0,0,0)$ and $(a/2,0,0)$ give a spacing of $a/\sqrt{8}$ for direction $\mathbf{i}+\mathbf{j}$; successive planes through $(-a/2,0,0)$ and $(a/2,0,0)$ give a spacing of $a/\sqrt{3}$ for direction $\mathbf{i}+\mathbf{j}+\mathbf{k}$.

6.12 (a) Check both magnitude and rotational sense. (b) Use the result of exercise 6.5 to evaluate $\mathbf{r}_i \times m_i(\boldsymbol{\omega}\times\mathbf{r}_i)$. (c) Form $(\mathbf{J}\cdot\boldsymbol{\omega})/\omega$; $\rho_i$ is the distance of the $i$th particle from the axis of rotation; (d) use Lagrange's identity to evaluate $(\boldsymbol{\omega}\times\mathbf{r}_i)\cdot(\boldsymbol{\omega}\times\mathbf{r}_i)$.

6.13 See figure 6.19 and recall that $-\cos\theta = \cos(\theta + \pi)$ and that $-\sin\theta = \cos(\theta + \pi/2)$.
(a) With $\phi_1 > 0$, no matter what the value of $\omega$ the possible resultants (dashed arrows) can never equal $P$. With $\phi_2 < 0$, closure of the quadrilateral is possible. (b) $A = P[(\omega_0^2 - \omega^2)^2 + 4\gamma^2\omega^2]^{-1/2}$.

6.14 Currents in units of mA$/|V_0|$. Voltages in units of $V_0$.
$I_1 = (7.76, -23.2°)$, $I_2 = (14.36, -50.8°)$, $I_3 = (8.30, 103.4°)$; $V_1 = (0.388, -23.2°)$, $V_2 = (0.287, -50.8°)$, $V_4 = (0.596, 39.2°)$; $L = 33$ mH, $R_2 = 20\ \Omega$.

# 7

# Matrices and vector spaces

In the previous chapter we defined a *vector* as a geometrical object which has both a magnitude and a direction and which may be thought of as an arrow in our familiar three-dimensional space. This geometrical definition of a vector is both useful and important since it is *independent* of any coordinate system with which we choose to label points in space. If we do, however, choose a particular Cartesian coordinate system (for example), then we may express a vector **a** in terms of its components along the $x$-, $y$-, and $z$- axes (or more correctly its components in the directions of the unit base vectors **i**, **j** and **k**) as

$$\mathbf{a} = a_x\mathbf{i} + a_y\mathbf{j} + a_z\mathbf{k}. \tag{7.1}$$

Although we have so far considered only real three-dimensional space, we may extend our notion of a vector to more abstract spaces, which can in general have an arbitrary number of dimensions $N$. We may still think of such a vector as an 'arrow' in this abstract space, so that it is again *independent* of any ($N$-dimensional) coordinate system with which we choose to label the space.

In this chapter we first discuss general *vector spaces* and their properties. We then go on to discuss the transformation of one vector into another by a linear operator. By choosing a coordinate system in a given vector space, this leads naturally to the concept of a *matrix*, a two-dimensional array of numbers. The properties of matrices are then discussed and we conclude with examples of their application, for instance in describing small oscillations of mechanical systems.

## 7.1 Vector spaces

A set of objects (vectors) **a**, **b**, **c**, …, are said to form a *linear vector space* $V$ if:

(i) the set is closed under commutative and associative addition, so that

$$\mathbf{a} + \mathbf{b} = \mathbf{b} + \mathbf{a}, \tag{7.2}$$

$$(\mathbf{a} + \mathbf{b}) + \mathbf{c} = \mathbf{a} + (\mathbf{b} + \mathbf{c}); \tag{7.3}$$

(ii) the set is closed under multiplication by a scalar (any complex number) to form a new vector $\lambda\mathbf{a}$, the operation being both distributive and associative so that

$$\lambda(\mathbf{a}+\mathbf{b}) = \lambda\mathbf{a} + \lambda\mathbf{b}, \tag{7.4}$$

$$(\lambda+\mu)\mathbf{a} = \lambda\mathbf{a} + \mu\mathbf{a}, \tag{7.5}$$

$$\lambda(\mu\mathbf{a}) = (\lambda\mu)\mathbf{a}, \tag{7.6}$$

where $\lambda$ and $\mu$ are arbitrary scalars;

(iii) there exists a *null vector* $\mathbf{0}$ such that $\mathbf{a}+\mathbf{0}=\mathbf{a}$;

(iv) multiplication by unity leaves any vector unchanged, i.e. $1\times\mathbf{a}=\mathbf{a}$;

(v) all vectors have a corresponding *negative vector* $-\mathbf{a}=(-1)\times\mathbf{a}$ such that $\mathbf{a}+(-\mathbf{a})=\mathbf{0}$.

We note that if we restrict all scalars to be real we obtain a *real vector space* (an example of which is our familiar three-dimensional space), otherwise in general we obtain a *complex vector space*. We note that it is common to use the terms 'vector space' and 'space', instead of the more formal 'linear vector space'.

The *span* of a set of vectors $\mathbf{a}$, $\mathbf{b}$, ..., $\mathbf{s}$, is defined as the set of all vectors that may be written as a linear sum of the original set, i.e. all vectors

$$\mathbf{x} = \alpha\mathbf{a} + \beta\mathbf{b} + \cdots + \sigma\mathbf{s} \tag{7.7}$$

that result from the (infinitely many) possible values of the (in general complex) scalars $\alpha$, $\beta$, ..., $\sigma$. If (7.7) is equal to $\mathbf{0}$ for some choice of $\alpha$, $\beta$, ..., $\sigma$ (not *all* zero), i.e. if

$$\alpha\mathbf{a} + \beta\mathbf{b} + \cdots + \sigma\mathbf{s} = \mathbf{0}, \tag{7.8}$$

then the set of vectors is said to be *linearly dependent*. In such a set at least one of the original set of vectors $\mathbf{a}$, $\mathbf{b}$, ..., $\mathbf{s}$, is redundant, since it can be expressed as a linear sum of the others. If, however, (7.8) is not satisfied by *any* set of coefficients (other than the trivial case in which all the coefficients are zero), then the vectors are *linearly independent*, and no vector in the set can be expressed as a linear sum of the others.

If, in a given vector space, there exist sets of $N$ linearly independent vectors, but no set of $N+1$ linearly independent vectors, then the vector space is said to be $N$-dimensional. (In this chapter we will limit our discussion to vector spaces of finite dimensionality; spaces of infinite dimensionality are discussed in chapter 15.)

### 7.1.1 Basis vectors

If $V$ is an $N$-dimensional vector space then *any* set of $N$ *linearly independent* vectors $\mathbf{e}_1$, $\mathbf{e}_2$, ..., $\mathbf{e}_N$ forms a *basis* for $V$. If $\mathbf{x}$ is an arbitrary vector lying in $V$

then the set of $N+1$ vectors $\mathbf{x}$, $\mathbf{e}_1$, $\mathbf{e}_2$, ..., $\mathbf{e}_N$, must be *linearly dependent* and such that

$$\alpha\mathbf{e}_1 + \beta\mathbf{e}_2 + \cdots + \sigma\mathbf{e}_N + \chi\mathbf{x} = \mathbf{0}, \tag{7.9}$$

where the coefficients $\alpha$, $\beta$, ..., $\chi$ are not all equal to 0, and in particular $\chi \neq 0$. Rearranging (7.9) we may write $\mathbf{x}$ as a linear sum of the vectors $\mathbf{e}_i$ as follows:

$$\mathbf{x} = x_1\mathbf{e}_1 + x_2\mathbf{e}_2 + \cdots + x_N\mathbf{e}_N = \sum_{i=1}^{N} x_i\mathbf{e}_i, \tag{7.10}$$

for some set of coefficients $x_i$ which are simply related to the original coefficients, e.g. $x_1 = -\alpha/\chi$, $x_2 = -\beta/\chi$, etc. Since any $\mathbf{x}$ lying in the span of $V$ can be expressed in terms of the *basis* or *base vectors* $\mathbf{e}_i$, they are said to be *complete.* The coefficients $x_i$ are the *components* of $\mathbf{x}$ with respect to the $\mathbf{e}_i$-basis. These components are *unique*, since if both

$$\mathbf{x} = \sum_{i=1}^{N} x_i\mathbf{e}_i \qquad \text{and} \qquad \mathbf{x} = \sum_{i=1}^{N} y_i\mathbf{e}_i,$$

then

$$\sum_{i=1}^{N}(x_i - y_i)\mathbf{e}_i = \mathbf{0}, \tag{7.11}$$

which, since the $\mathbf{e}_i$ are linearly independent, has only the solution $x_i = y_i$ for all $i = 1, \ldots, N$.

From the above discussion we see that *any* set of $N$ linearly independent vectors can form a basis for an $N$-dimensional space. If we choose a different set $\mathbf{e}'_i$, $i = 1, \ldots, N$, then we can write $\mathbf{x}$ as

$$\mathbf{x} = x'_1\mathbf{e}'_1 + x'_2\mathbf{e}'_2 + \cdots + x'_N\mathbf{e}'_N = \sum_{i=1}^{N} x'_i\mathbf{e}'_i. \tag{7.12}$$

We reiterate that the vector $\mathbf{x}$ (a geometrical entity) is independent of the basis – it is only the components of $\mathbf{x}$ that depend on the basis. We note, however, that given a set of vectors $\mathbf{u}_1$, $\mathbf{u}_2$, ..., $\mathbf{u}_M$, where $M \neq N$, in an $N$-dimensional vector space, then *either* there exists a vector that cannot be expressed as a linear combination of the $\mathbf{u}_i$ *or*, for some vector that can be so expressed, the components are not unique.

### *7.1.2 The inner product*

We may introduce additional structure in a vector space by defining the *inner product* of two vectors, denoted by $\langle\mathbf{a}|\mathbf{b}\rangle$, which is a scalar function of $\mathbf{a}$ and $\mathbf{b}$. The scalar (dot) product, $\mathbf{a}\cdot\mathbf{b} \equiv |\mathbf{a}||\mathbf{b}|\cos\theta$, of vectors in real three-dimensional space (where $\theta$ is the angle between the vectors), was introduced in the last chapter and is an example of an inner product. In effect the notion of an inner product $\langle \mathbf{a}|\mathbf{b}\rangle$ is a generalisation of the dot product to more abstract vector spaces and has the following properties:

(i) $\langle \mathbf{a}|\mathbf{b}\rangle = \langle \mathbf{b}|\mathbf{a}\rangle^*$,
(ii) $\langle \mathbf{a}|\lambda\mathbf{b} + \mu\mathbf{c}\rangle = \lambda\langle \mathbf{a}|\mathbf{b}\rangle + \mu\langle \mathbf{a}|\mathbf{c}\rangle$.

The inner product is also often denoted by $(\mathbf{a}, \mathbf{b})$, or simply by $\mathbf{a}\cdot\mathbf{b}$. We note that in general, for a complex vector space, (i) and (ii) imply that

$$\langle \lambda\mathbf{a} + \mu\mathbf{b}|\mathbf{c}\rangle = \lambda^*\langle \mathbf{a}|\mathbf{c}\rangle + \mu^*\langle \mathbf{b}|\mathbf{c}\rangle, \tag{7.13}$$

$$\langle \lambda\mathbf{a}|\mu\mathbf{b}\rangle = \lambda^*\mu\langle \mathbf{a}|\mathbf{b}\rangle. \tag{7.14}$$

Following the analogy with the dot product in three-dimensional real space, two vectors in a general vector space are defined to be *orthogonal* if $\langle \mathbf{a}|\mathbf{b}\rangle = 0$. Similarly, the *norm* of a vector $\mathbf{a}$ is given by $\|\mathbf{a}\| = \langle \mathbf{a}|\mathbf{a}\rangle^{1/2}$, and is clearly a generalisation of the length (modulus) $|\mathbf{a}|$ of a vector $\mathbf{a}$ in three-dimensional space. In a general vector space $\langle \mathbf{a}|\mathbf{a}\rangle$ can be positive or negative; however, we shall be primarily concerned with spaces in which $\langle \mathbf{a}|\mathbf{a}\rangle \geq 0$ and which are thus said to have a *positive semi-definite norm*. In such a space $\langle \mathbf{a}|\mathbf{a}\rangle = 0$ implies $\mathbf{a} = \mathbf{0}$.

Let us now introduce into our $N$-dimensional vector space a basis $\hat{\mathbf{e}}_1$, $\hat{\mathbf{e}}_2$, ..., $\hat{\mathbf{e}}_N$, which has the desirable property of being *orthonormal* (the basis vectors are mutually orthogonal and each has unit norm), i.e. a basis which has the property

$$\langle \hat{\mathbf{e}}_i|\hat{\mathbf{e}}_j\rangle = \delta_{ij}. \tag{7.15}$$

Here $\delta_{ij}$ is the *Kronecker delta* symbol (of which we say more in chapter 19) and has the properties

$$\delta_{ij} = \begin{cases} 1 & \text{for } i = j, \\ 0 & \text{for } i \neq j. \end{cases}$$

In the above basis we may express any two vectors $\mathbf{a}$ and $\mathbf{b}$:

$$\mathbf{a} = \sum_{i=1}^{N} a_i\hat{\mathbf{e}}_i \qquad \text{and} \qquad \mathbf{b} = \sum_{i=1}^{N} b_i\hat{\mathbf{e}}_i.$$

Furthermore, *in such an orthonormal basis* we have (for example)

$$\langle \hat{\mathbf{e}}_j|\mathbf{a}\rangle = \sum_{i=1}^{N}\langle \hat{\mathbf{e}}_j|a_i\hat{\mathbf{e}}_i\rangle = \sum_{i=1}^{N} a_i\langle \hat{\mathbf{e}}_j|\hat{\mathbf{e}}_i\rangle = a_j. \tag{7.16}$$

Thus the components of $\mathbf{a}$ are given by $a_i = \langle \hat{\mathbf{e}}_i|\mathbf{a}\rangle$. Note that this is *not* true unless the basis is orthonormal. We can write the inner product of $\mathbf{a}$ and $\mathbf{b}$ in terms of their components in an orthonormal basis as

$$\begin{aligned}\langle \mathbf{a}|\mathbf{b}\rangle &= \langle a_1\hat{\mathbf{e}}_1 + \cdots + a_N\hat{\mathbf{e}}_N | b_1\hat{\mathbf{e}}_1 + \cdots + b_N\hat{\mathbf{e}}_N\rangle \\ &= \sum_{i=1}^{N} a_i^* b_i \langle \hat{\mathbf{e}}_i|\hat{\mathbf{e}}_i\rangle + \sum_i \sum_{j\neq i} a_i^* b_j \langle \hat{\mathbf{e}}_i|\hat{\mathbf{e}}_j\rangle \\ &= \sum_{i=1}^{N} a_i^* b_i,\end{aligned}$$

where the second equality follows from (7.14), and the third follows from (7.15). This is clearly a generalisation of the expression (6.21) for the dot product of vectors in three-dimensional space.

We may generalise the above to the case where the base vectors $\mathbf{e}_1, \mathbf{e}_2, \ldots, \mathbf{e}_N$ are *not* orthonormal (or orthogonal). In general we can define the $N^2$ numbers

$$G_{ij} = \langle \mathbf{e}_i|\mathbf{e}_j\rangle. \tag{7.17}$$

Then if $\mathbf{a} = \sum_{i=1}^{N} a_i\mathbf{e}_i$ and $\mathbf{b} = \sum_{i=1}^{N} b_i\mathbf{e}_i$, the inner product of $\mathbf{a}$ and $\mathbf{b}$ is given by

$$\begin{aligned}\langle \mathbf{a}|\mathbf{b}\rangle &= \left\langle \sum_{i=1}^{N} a_i\mathbf{e}_i \,\middle|\, \sum_{j=1}^{N} b_j\mathbf{e}_j \right\rangle \\ &= \sum_{i=1}^{N}\sum_{j=1}^{N} a_i^* b_j \langle \mathbf{e}_i|\mathbf{e}_j\rangle \\ &= \sum_{i=1}^{N}\sum_{j=1}^{N} a_i^* G_{ij} b_j.\end{aligned} \tag{7.18}$$

We further note that from (7.17) and the properties of the inner product we require $G_{ij} = G_{ji}^*$. This in turn ensures that $\|\mathbf{a}\| = \langle \mathbf{a}|\mathbf{a}\rangle$ is real, since then

$$\langle \mathbf{a}|\mathbf{a}\rangle^* = \sum_{i=1}^{N}\sum_{j=1}^{N} a_i G_{ij}^* a_j^* = \sum_{j=1}^{N}\sum_{i=1}^{N} a_j^* G_{ji} a_i = \langle \mathbf{a}|\mathbf{a}\rangle.$$

### *7.1.3 Some useful inequalities*

For a set of objects (vectors) forming a linear vector space in which $\langle \mathbf{a}|\mathbf{a}\rangle \geq 0$ for all $\mathbf{a}$, the following inequalities are often useful.

(i) *Schwarz's inequality* is the most basic result and states that

$$|\langle \mathbf{a}|\mathbf{b}\rangle| \leq \|\mathbf{a}\|\|\mathbf{b}\|, \tag{7.19}$$

where equality holds when $\mathbf{a}$ is a scalar multiple of $\mathbf{b}$, i.e. when $\mathbf{a} = \lambda\mathbf{b}$. It is important here to distinguish between the *absolute value* of a scalar

$|\lambda|$, and the *norm* of a vector, $\|\mathbf{a}\|$. Schwarz's inequality may be proved by considering

$$\begin{aligned}\|\mathbf{a}+\lambda\mathbf{b}\|^2 &= \langle\mathbf{a}+\lambda\mathbf{b}|\mathbf{a}+\lambda\mathbf{b}\rangle \\ &= \langle\mathbf{a}|\mathbf{a}\rangle+\lambda\langle\mathbf{a}|\mathbf{b}\rangle+\lambda^*\langle\mathbf{b}|\mathbf{a}\rangle+\lambda\lambda^*\langle\mathbf{b}|\mathbf{b}\rangle.\end{aligned}$$

If we write $\langle\mathbf{a}|\mathbf{b}\rangle$ as $|\langle\mathbf{a}|\mathbf{b}\rangle|e^{i\alpha}$ then

$$\|\mathbf{a}+\lambda\mathbf{b}\|^2 = \|\mathbf{a}\|^2+|\lambda|^2\|\mathbf{b}\|^2+\lambda|\langle\mathbf{a}|\mathbf{b}\rangle|e^{i\alpha}+\lambda^*|\langle\mathbf{a}|\mathbf{b}\rangle|e^{-i\alpha}.$$

But $\|\mathbf{a}+\lambda\mathbf{b}\|^2 \geq 0$ for all $\lambda$, so we may choose $\lambda = re^{-i\alpha}$ and require that, for all $r$,

$$0 \leq \|\mathbf{a}+\lambda\mathbf{b}\|^2 = \|\mathbf{a}\|^2+r^2\|\mathbf{b}\|^2+2r|\langle\mathbf{a}|\mathbf{b}\rangle|.$$

This, in turn, implies

$$4r^2|\langle\mathbf{a}|\mathbf{b}\rangle|^2 \leq 4\|\mathbf{a}\|^2r^2\|\mathbf{b}\|^2,$$

which, on cancelling $4r^2$ and taking the square root (all factors are necessarily positive) of both sides, gives Schwarz's inequality.

(ii) The *triangle inequality* states

$$\|\mathbf{a}+\mathbf{b}\| \leq \|\mathbf{a}\|+\|\mathbf{b}\|, \tag{7.20}$$

and may be derived from the properties of the inner product and Schwarz's inequality as follows. Let us first consider

$$\|\mathbf{a}+\mathbf{b}\|^2 = \|\mathbf{a}\|^2+\|\mathbf{b}\|^2+2\,\mathrm{Re}\,\langle\mathbf{a}|\mathbf{b}\rangle \leq \|\mathbf{a}\|^2+\|\mathbf{b}\|^2+2|\langle\mathbf{a}|\mathbf{b}\rangle|.$$

Using Schwarz's inequality we then have

$$\|\mathbf{a}+\mathbf{b}\|^2 \leq \|\mathbf{a}\|^2+\|\mathbf{b}\|^2+2\|\mathbf{a}\|\,\|\mathbf{b}\| = (\|\mathbf{a}\|+\|\mathbf{b}\|)^2,$$

which, on taking the square root, gives the triangle inequality (7.20).

(iii) *Bessel's inequality* requires the introduction of an orthonormal basis $\hat{\mathbf{e}}_i$, $i = 1,\ldots,N$ in the $N$-dimensional vector space, and states

$$\|\mathbf{a}\|^2 \geq \sum_i |\langle\hat{\mathbf{e}}_i|\mathbf{a}\rangle|^2, \tag{7.21}$$

where the equality holds if the sum includes all $N$ basis vectors. If only some of the basis vectors are included in the sum then the inequality results (though of course the equality remains if those basis vectors omitted all have $a_i = 0$). Bessel's inequality can also be written

$$\langle\mathbf{a}|\mathbf{a}\rangle \geq \sum_i |a_i|^2,$$

where the $a_i$ are the components of $\mathbf{a}$ in the orthonormal basis. From (7.16) these are given by $a_i = \langle \hat{\mathbf{e}}_i|\mathbf{a}\rangle$. The above may be proved by considering

$$\left\|\mathbf{a} - \sum_i \langle \hat{\mathbf{e}}_i|\mathbf{a}\rangle \hat{\mathbf{e}}_i\right\|^2 = \left\langle \mathbf{a} - \sum_i \langle \hat{\mathbf{e}}_i|\mathbf{a}\rangle \hat{\mathbf{e}}_i \,\middle|\, \mathbf{a} - \sum_j \langle \hat{\mathbf{e}}_j|\mathbf{a}\rangle \hat{\mathbf{e}}_j \right\rangle.$$

Expanding out the inner product and using $\langle \hat{\mathbf{e}}_i|\mathbf{a}\rangle^* = \langle \mathbf{a}|\hat{\mathbf{e}}_i\rangle$, we obtain

$$\left\|\mathbf{a} - \sum_i \langle \hat{\mathbf{e}}_i|\mathbf{a}\rangle \hat{\mathbf{e}}_i\right\|^2 = \langle \mathbf{a}|\mathbf{a}\rangle - 2\sum_i \langle \mathbf{a}|\hat{\mathbf{e}}_i\rangle\langle \hat{\mathbf{e}}_i|\mathbf{a}\rangle + \sum_i\sum_j \langle \mathbf{a}|\hat{\mathbf{e}}_i\rangle\langle \hat{\mathbf{e}}_j|\mathbf{a}\rangle\langle \hat{\mathbf{e}}_i|\hat{\mathbf{e}}_j\rangle.$$

Now $\langle \hat{\mathbf{e}}_i|\hat{\mathbf{e}}_j\rangle = \delta_{ij}$, since the basis is orthonormal, and so we find

$$0 \le \left\|\mathbf{a} - \sum_i \langle \hat{\mathbf{e}}_i|\mathbf{a}\rangle \hat{\mathbf{e}}_i\right\|^2 = \|\mathbf{a}\|^2 - \sum_i |\langle \hat{\mathbf{e}}_i|\mathbf{a}\rangle|^2,$$

which is Bessel's inequality.

We take this opportunity to mention also

(iv) the *parallelogram equality*

$$\|\mathbf{a}+\mathbf{b}\|^2 + \|\mathbf{a}-\mathbf{b}\|^2 = 2\left(\|\mathbf{a}\|^2 + \|\mathbf{b}\|^2\right), \qquad (7.22)$$

which may be proved straightforwardly from the properties of the inner product.

## 7.2 Linear operators

We now discuss the action of *linear operators* on vectors in a vector space. A linear operator $\mathcal{A}$ associates with every vector $\mathbf{x}$ another vector

$$\mathbf{y} = \mathcal{A}\,\mathbf{x},$$

in such a way that, for two vectors $\mathbf{a}$ and $\mathbf{b}$,

$$\mathcal{A}\,(\lambda\mathbf{a} + \mu\mathbf{b}) = \lambda\mathcal{A}\,\mathbf{a} + \mu\mathcal{A}\,\mathbf{b},$$

where $\lambda$, $\mu$ are scalars. We say that $\mathcal{A}$ 'operates' on $\mathbf{x}$ to give the vector $\mathbf{y}$. We note that the action of $\mathcal{A}$ is *independent* of any basis or coordinate system, and may be thought of as 'transforming' one geometrical entity (i.e. a vector) into another.

If we now introduce a basis $\mathbf{e}_i$, $i = 1, \ldots, N$, into our vector space, then the action of $\mathcal{A}$ on each of the basis vectors is to produce a linear combination of the basis set; this may be written as

$$\mathcal{A}\,\mathbf{e}_j = \sum_{i=1}^{N} A_{ij}\mathbf{e}_i, \qquad (7.23)$$

where $A_{ij}$ is the $i$th component of the vector $\mathcal{A}\,\mathbf{e}_j$ in this basis; collectively the numbers $A_{ij}$ are called the components of the linear operator in the $\mathbf{e}_i$-basis. *In this basis* we can express the relation $\mathbf{y} = \mathcal{A}\,\mathbf{x}$ in component form as

$$\mathbf{y} = \sum_{i=1}^{N} y_i \mathbf{e}_i = \mathcal{A}\left(\sum_{j=1}^{N} x_j \mathbf{e}_j\right) = \sum_{j=1}^{N} x_j \sum_{i=1}^{N} A_{ij}\mathbf{e}_i,$$

and hence, in purely component form, in this basis we have

$$y_i = \sum_{j=1}^{N} A_{ij} x_j. \qquad (7.24)$$

If we had chosen a different basis $\mathbf{e}'_i$, in which the components of $\mathbf{x}$, $\mathbf{y}$ and $\mathcal{A}$ are $x'_i$, $y'_i$ and $A'_{ij}$ respectively, then the geometrical relationship $\mathbf{y} = \mathcal{A}\,\mathbf{x}$ would be represented in this new basis by

$$y'_i = \sum_{j=1}^{N} A'_{ij} x'_j.$$

We have so far assumed that the vector $\mathbf{y}$ is in the same vector space as $\mathbf{x}$. If, however, $\mathbf{y}$ belongs to a different vector space, which may in general be $M$-dimensional ($M \neq N$), then the above analysis needs a slight modification. By introducing a basis set $\mathbf{f}_i$, $i = 1, \ldots, M$, into the vector space to which $\mathbf{y}$ belongs we may generalise (7.23) as

$$\mathcal{A}\,\mathbf{e}_j = \sum_{i=1}^{M} A_{ij}\mathbf{f}_i,$$

where the components $A_{ij}$ of the linear operator $\mathcal{A}$ relate to both of the bases $\mathbf{e}_j$ and $\mathbf{f}_i$.

### *7.2.1 Properties of linear operators*

If $\mathbf{x}$ is a vector and $\mathcal{A}$ and $\mathcal{B}$ are two linear operators, then it follows that

$$(\mathcal{A} + \mathcal{B})\mathbf{x} = \mathcal{A}\,\mathbf{x} + \mathcal{B}\,\mathbf{x},$$
$$(\lambda\mathcal{A})\mathbf{x} = \lambda(\mathcal{A}\,\mathbf{x}),$$
$$(\mathcal{A}\mathcal{B})\mathbf{x} = \mathcal{A}(\mathcal{B}\,\mathbf{x}),$$

where in the last equality we see that the action of two linear operators in succession is associative. The product of two linear operators is not in general commutative however, so that in general $\mathcal{A}\mathcal{B}\,\mathbf{x} \neq \mathcal{B}\mathcal{A}\,\mathbf{x}$. In an obvious way we define the null (or zero) and identity operators by

$$\mathcal{O}\,\mathbf{x} = \mathbf{0} \quad \text{and} \quad \mathcal{I}\,\mathbf{x} = \mathbf{x},$$

for any vector $\mathbf{x}$ in our vector space. Two operators $\mathcal{A}$ and $\mathcal{B}$ are equal if $\mathcal{A}\mathbf{x} = \mathcal{B}\mathbf{x}$ for all vectors $\mathbf{x}$. Finally, if there exists an operator $\mathcal{A}^{-1}$ such that

$$\mathcal{A}\mathcal{A}^{-1} = \mathcal{A}^{-1}\mathcal{A} = \mathcal{I},$$

then $\mathcal{A}^{-1}$ is the *inverse* of $\mathcal{A}$. Some linear operators do not possess an inverse and are called *singular*, whilst those operators which do have an inverse are termed *non-singular*.

## 7.3 Matrices

We have seen that in a particular basis $\mathbf{e}_i$ both vectors and linear operators can be described in terms of their components with respect to the basis. These components may be displayed as an array of numbers called a *matrix*. In general, if a linear operator $\mathcal{A}$ transforms vectors from an $N$-dimensional vector space, for which we choose a basis $\mathbf{e}_j$, $j = 1, \ldots, N$, into vectors belonging to an $M$-dimensional vector space, with basis $\mathbf{f}_i$, $i = 1, \ldots, M$, then we may represent the operator $\mathcal{A}$ by the matrix

$$\mathsf{A} = \begin{pmatrix} A_{11} & A_{12} & \ldots & A_{1N} \\ A_{21} & A_{22} & \ldots & A_{2N} \\ \vdots & \vdots & \ddots & \vdots \\ A_{M1} & A_{M2} & \ldots & A_{MN} \end{pmatrix}, \tag{7.25}$$

where the $A_{ij}$ are the components of the linear operator with respect to these bases. This array has $M$ rows and $N$ columns and is thus called an $M \times N$ matrix. If the dimensions of the two vector spaces are the same, i.e. $M = N$ (for example, if they are the same vector space) then we may represent $\mathcal{A}$ by an $N \times N$ or *square* matrix of *order* $N$. The components $A_{ij}$ of the linear operator appear in the $i$th row and $j$th column of the matrix and are called *matrix elements*. The component $A_{ij}$ is also denoted by $(\mathsf{A})_{ij}$.

In a similar way we may denote a vector $\mathbf{x}$ in terms of its components $x_i$ in a basis $\mathbf{e}_i$, $i = 1, \ldots, N$, by the array

$$\mathsf{x} = \begin{pmatrix} x_1 \\ \vdots \\ x_N \end{pmatrix},$$

which is a special case of (7.25) and is called a *column matrix* (or conventionally, and slightly confusingly, a *column vector* – strictly speaking the term 'vector' refers to the geometrical entity $\mathbf{x}$). The column matrix $\mathsf{x}$ can also be written as

$$\mathsf{x} = (x_1 \; x_2 \; \cdots \; x_N)^{\mathrm{T}},$$

which is the *transpose* of a *row matrix* (see section 7.5).

We note that in a different basis $\mathbf{e}'_i$ the vector $\mathbf{x}$ would be represented by a *different* column matrix containing the components $x'_i$ in the new basis, i.e.

$$\mathsf{x}' = \begin{pmatrix} x'_1 \\ \vdots \\ x'_N \end{pmatrix}.$$

Thus, we use $\mathsf{x}$ and $\mathsf{x}'$ to denote different column matrices which, in different bases $\mathbf{e}_i$ and $\mathbf{e}'_i$, represent the *same* vector $\mathbf{x}$. In many texts, however, this distinction is not made and $\mathbf{x}$ (rather than $\mathsf{x}$) is equated to the corresponding column matrix – if we regard $\mathbf{x}$ as the geometrical entity, however, this can be misleading and so we explicitly make the distinction. A similar argument follows for linear operators – the same linear operator $\mathcal{A}$ can be described in different bases by different matrices $\mathsf{A}$ and $\mathsf{A}'$, containing different matrix elements.

The algebra of matrices may be deduced from the properties of the linear operators that they represent. In a given basis the action of two linear operators $\mathcal{A}$ and $\mathcal{B}$ on an arbitrary vector $\mathbf{x}$ (see the beginning of subsection 7.2.1) can be written in terms of components, using (7.24), as

$$\sum_j (\mathsf{A}+\mathsf{B})_{ij} x_j = \sum_j A_{ij} x_j + \sum_j B_{ij} x_j,$$

$$\sum_j (\lambda\mathsf{A})_{ij} x_j = \lambda \sum_j A_{ij} x_j,$$

$$\sum_j (\mathsf{AB})_{ij} x_j = \sum_k A_{ik} (\mathsf{Bx})_k = \sum_j \sum_k A_{ik} B_{kj} x_j.$$

Now, since $\mathbf{x}$ is arbitrary, we can immediately deduce the way in which matrices are added or multiplied, i.e.

$$(\mathsf{A}+\mathsf{B})_{ij} = A_{ij} + B_{ij}, \tag{7.26}$$

$$(\lambda\mathsf{A})_{ij} = \lambda A_{ij}, \tag{7.27}$$

$$(\mathsf{AB})_{ij} = \sum_k A_{ik} B_{kj}. \tag{7.28}$$

We now discuss matrix addition and multiplication in more detail.

### *7.3.1 Matrix addition and multiplication by a scalar*

From (7.26) we see that the sum of two matrices, $\mathsf{S} = \mathsf{A} + \mathsf{B}$, is the matrix whose elements are given by

$$S_{ij} = A_{ij} + B_{ij}$$

for every pair of subscripts $i, j$, with $i = 1, 2, \ldots, M$ and $j = 1, 2, \ldots, N$. For example, if $\mathsf{A}$ and $\mathsf{B}$ are $2 \times 3$ matrices then $\mathsf{S} = \mathsf{A} + \mathsf{B}$ is given by

$$\begin{aligned} \begin{pmatrix} S_{11} & S_{12} & S_{13} \\ S_{21} & S_{22} & S_{23} \end{pmatrix} &= \begin{pmatrix} A_{11} & A_{12} & A_{13} \\ A_{21} & A_{22} & A_{23} \end{pmatrix} + \begin{pmatrix} B_{11} & B_{12} & B_{13} \\ B_{21} & B_{22} & B_{23} \end{pmatrix} \\ &= \begin{pmatrix} A_{11} + B_{11} & A_{12} + B_{12} & A_{13} + B_{13} \\ A_{21} + B_{21} & A_{22} + B_{22} & A_{23} + B_{23} \end{pmatrix}. \end{aligned} \tag{7.29}$$

Clearly, for the sum of two matrices to have any meaning, the matrices must have the same dimensions, i.e. both be $M \times N$ matrices.

From definition (7.29) it follows that $\mathsf{A} + \mathsf{B} = \mathsf{B} + \mathsf{A}$, and that the sum of a number of matrices can be written unambiguously without bracketing, i.e. matrix addition is *commutative* and *associative*.

The difference of two matrices is defined by direct analogy with addition. The matrix $\mathsf{D} = \mathsf{A} - \mathsf{B}$ has elements

$$D_{ij} = A_{ij} - B_{ij}, \qquad \text{for } i = 1, 2, \ldots, M,\ j = 1, 2, \ldots, N. \tag{7.30}$$

From (7.27) the product of a matrix $\mathsf{A}$ with a scalar $\lambda$ is the matrix with elements $\lambda A_{ij}$, for example

$$\lambda \begin{pmatrix} A_{11} & A_{12} & A_{13} \\ A_{21} & A_{22} & A_{23} \end{pmatrix} = \begin{pmatrix} \lambda A_{11} & \lambda A_{12} & \lambda A_{13} \\ \lambda A_{21} & \lambda A_{22} & \lambda A_{23} \end{pmatrix}. \tag{7.31}$$

Multiplication by a scalar is distributive and associative.

From the above considerations we see that the set of all (in general complex) $M \times N$ matrices (with fixed $M$ and $N$) form a linear vector space of dimension $MN$. One basis for the space is the set of $M \times N$ matrices $\mathsf{E}^{(pq)}$ with the property that $E_{ij}^{(pq)} = 1$ if $i = p$ and $j = q$, but $E_{ij}^{(pq)} = 0$ for all other values of $i$ and $j$.

►*The matrices* $\mathsf{A}$, $\mathsf{B}$ *and* $\mathsf{C}$ *are given by*

$$\mathsf{A} = \begin{pmatrix} 2 & -1 \\ 3 & 1 \end{pmatrix}, \quad \mathsf{B} = \begin{pmatrix} 1 & 0 \\ 0 & -2 \end{pmatrix}, \quad \mathsf{C} = \begin{pmatrix} -2 & 1 \\ -1 & 1 \end{pmatrix}.$$

*Find the matrix* $\mathsf{D} = \mathsf{A} + 2\mathsf{B} - \mathsf{C}$.

$$\begin{aligned} \mathsf{D} &= \begin{pmatrix} 2 & -1 \\ 3 & 1 \end{pmatrix} + 2\begin{pmatrix} 1 & 0 \\ 0 & -2 \end{pmatrix} - \begin{pmatrix} -2 & 1 \\ -1 & 1 \end{pmatrix} \\ &= \begin{pmatrix} 2 + 2 \times 1 - (-2) & -1 + 2 \times 0 - 1 \\ 3 + 2 \times 0 - (-1) & 1 + 2 \times (-2) - 1 \end{pmatrix} = \begin{pmatrix} 6 & -2 \\ 4 & -4 \end{pmatrix}. \blacktriangleleft \end{aligned}$$

### *7.3.2 Multiplication of matrices*

Let us again consider the 'transformation' of one vector into another, $\mathbf{y} = \mathcal{A}\,\mathbf{x}$, which, from (7.24), may be described in terms of components with respect to a particular basis as

$$y_i = \sum_{j=1}^{N} A_{ij} x_j \quad \text{for } i = 1, \ldots, M. \tag{7.32}$$

Writing this in matrix form $\mathsf{y} = \mathsf{Ax}$ we have

$$\begin{pmatrix} y_1 \\ \boxed{y_2} \\ \vdots \\ y_M \end{pmatrix} = \begin{pmatrix} A_{11} & A_{12} & \ldots & A_{1N} \\ \boxed{A_{21} \quad A_{22} \quad \ldots \quad A_{2N}} \\ \vdots & \vdots & \ddots & \vdots \\ A_{M1} & A_{M2} & \ldots & A_{MN} \end{pmatrix} \begin{pmatrix} \boxed{\begin{matrix} x_1 \\ x_2 \\ \vdots \\ x_N \end{matrix}} \end{pmatrix} \tag{7.33}$$

where we have highlighted with boxes the components used to calculate the element $y_2$: using (7.32) for $i = 2$,

$$y_2 = A_{21}x_1 + A_{22}x_2 + \cdots + A_{2N}x_N.$$

All the other components $y_i$ are calculated similarly.

If instead we operate with $\mathcal{A}$ on a basis vector $\mathbf{e}_j$ having all components zero except for the $j$th, which equals unity, then we find

$$\mathsf{Ae}_j = \begin{pmatrix} A_{11} & A_{12} & \ldots & A_{1N} \\ \boxed{A_{21} \quad A_{22} \quad \ldots \quad A_{2N}} \\ \vdots & \vdots & \ddots & \vdots \\ A_{M1} & A_{M2} & \ldots & A_{MN} \end{pmatrix} \begin{pmatrix} \boxed{\begin{matrix} 0 \\ 0 \\ \vdots \\ 1 \\ \vdots \\ 0 \end{matrix}} \end{pmatrix} = \begin{pmatrix} A_{1j} \\ \boxed{A_{2j}} \\ \vdots \\ A_{Mj} \end{pmatrix},$$

and so confirm our identification of the matrix element $A_{ij}$ as the $i$th component of $\mathsf{Ae}_j$ in this basis.

From (7.28) we can extend our discussion to the product of two matrices $\mathsf{P} = \mathsf{AB}$, where $\mathsf{P}$ is the matrix of the quantities formed by the operation of the rows of $\mathsf{A}$ on the columns of $\mathsf{B}$, treating each column of $\mathsf{B}$ in turn as the vector $\mathbf{x}$ represented in component form in (7.32). It is clear that, for this to be a meaningful definition, the number of columns in $\mathsf{A}$ must equal the number of rows in $\mathsf{B}$. Thus the product $\mathsf{AB}$ of an $M \times N$ matrix $\mathsf{A}$ with an $N \times R$ matrix $\mathsf{B}$ is itself an $M \times R$ matrix $\mathsf{P}$, where

$$P_{ij} = \sum_{k=1}^{N} A_{ik} B_{kj} \quad \text{for } i = 1, \ldots, M, \quad j = 1, \ldots, R.$$

For example, $\mathsf{P} = \mathsf{AB}$ may be written in matrix form

$$\begin{pmatrix} \boxed{P_{11}} & P_{12} \\ P_{21} & P_{22} \end{pmatrix} = \begin{pmatrix} \boxed{A_{11} \quad A_{12} \quad A_{13}} \\ A_{21} \quad A_{22} \quad A_{23} \end{pmatrix} \begin{pmatrix} \boxed{\begin{matrix} B_{11} \\ B_{21} \\ B_{31} \end{matrix}} & \begin{matrix} B_{12} \\ B_{22} \\ B_{32} \end{matrix} \end{pmatrix}$$

where

$$\begin{aligned} P_{11} &= A_{11}B_{11} + A_{12}B_{21} + A_{13}B_{31}, \\ P_{21} &= A_{21}B_{11} + A_{22}B_{21} + A_{23}B_{31}, \\ P_{12} &= A_{11}B_{12} + A_{12}B_{22} + A_{13}B_{32}, \\ P_{22} &= A_{21}B_{12} + A_{22}B_{22} + A_{23}B_{32}. \end{aligned}$$

Multiplication of more than two matrices follows naturally and is associative. So, for example,

$$\mathsf{A}(\mathsf{BC}) \equiv (\mathsf{AB})\mathsf{C}, \tag{7.34}$$

provided, of course, that all the products are defined.

As mentioned above, if $\mathsf{A}$ is an $M \times N$ matrix and $\mathsf{B}$ is an $N \times M$ matrix, then two product matrices are possible, i.e.

$$\mathsf{P} = \mathsf{AB} \qquad \text{and} \qquad \mathsf{Q} = \mathsf{BA}.$$

These are clearly not the same, since $\mathsf{P}$ is an $M \times M$ matrix whilst $\mathsf{Q}$ is an $N \times N$ matrix. Particular care must thus be taken to write matrix products in the intended order; $\mathsf{P} = \mathsf{AB}$ but $\mathsf{Q} = \mathsf{BA}$. We note in passing that the notation $\mathsf{A}^2$ is used to mean $\mathsf{AA}$ and $\mathsf{A}^3 = \mathsf{A}(\mathsf{AA}) = (\mathsf{AA})\mathsf{A}$, etc. Even if both $\mathsf{A}$ and $\mathsf{B}$ are square, in general

$$\mathsf{AB} \neq \mathsf{BA}, \tag{7.35}$$

i.e. multiplication of matrices is not, in general, commutative.

►*Evaluate* $\mathsf{P} = \mathsf{AB}$ *and* $\mathsf{Q} = \mathsf{BA}$ *where*

$$\mathsf{A} = \begin{pmatrix} 3 & 2 & -1 \\ 0 & 3 & 2 \\ 1 & -3 & 4 \end{pmatrix}, \qquad \mathsf{B} = \begin{pmatrix} 2 & -2 & 3 \\ 1 & 1 & 0 \\ 3 & 2 & 1 \end{pmatrix}.$$

As we saw for the $2 \times 2$ matrix above, the elements ($P_{ij}$ say) of the resultant matrices are found by mentally taking the 'scalar product' of the $i$th row of the first matrix with the $j$th column of the second one. For example, $P_{11} = 3 \times 2 + 2 \times 1 + (-1) \times 3 = 5$, $P_{12} = 3 \times (-2) + 2 \times 1 + (-1) \times 2 = -6$, etc. Thus

$$\mathsf{P} = \mathsf{AB} = \begin{pmatrix} 3 & 2 & -1 \\ 0 & 3 & 2 \\ 1 & -3 & 4 \end{pmatrix} \begin{pmatrix} 2 & -2 & 3 \\ 1 & 1 & 0 \\ 3 & 2 & 1 \end{pmatrix} = \begin{pmatrix} 5 & -6 & 8 \\ 9 & 7 & 2 \\ 11 & 3 & 7 \end{pmatrix},$$

and, similarly,

$$\mathsf{Q} = \mathsf{BA} = \begin{pmatrix} 2 & -2 & 3 \\ 1 & 1 & 0 \\ 3 & 2 & 1 \end{pmatrix} \begin{pmatrix} 3 & 2 & -1 \\ 0 & 3 & 2 \\ 1 & -3 & 4 \end{pmatrix} = \begin{pmatrix} 9 & -11 & 6 \\ 3 & 5 & 1 \\ 10 & 9 & 5 \end{pmatrix}.$$

These results illustrate that, in general, $\mathsf{AB} \neq \mathsf{BA}$. ◄

The property that matrix multiplication is distributive over addition, i.e.

$$(\mathsf{A} + \mathsf{B})\mathsf{C} = \mathsf{AC} + \mathsf{BC}, \tag{7.36}$$

and

$$\mathsf{C}(\mathsf{A} + \mathsf{B}) = \mathsf{CA} + \mathsf{CB}, \tag{7.37}$$

follows directly from its definition.

## 7.4 The null and identity matrices

Both the null matrix and the identity matrix are frequently encountered, and we take this opportunity to introduce them briefly, leaving their uses until later. The *null* or *zero* matrix $\mathsf{0}$ has all elements equal to zero, and the properties

$$\mathsf{A0} = \mathsf{0} = \mathsf{0A},$$

$$\mathsf{A} + \mathsf{0} = \mathsf{0} + \mathsf{A} = \mathsf{A}.$$

The *identity* matrix $\mathsf{I}$ has the property

$$\mathsf{AI} = \mathsf{IA} = \mathsf{A}.$$

It is clear that, in order for the above products to be defined, the identity matrix must be square. The $N \times N$ identity matrix (often denoted by $\mathsf{I}_N$) has the form

$$\mathsf{I}_N = \begin{pmatrix} 1 & & 0 \\ & \ddots & \\ 0 & & 1 \end{pmatrix}.$$

## 7.5 The transpose of a matrix

We have seen that the components of a linear operator in a given coordinate system can be written in the form of a matrix $\mathsf{A}$. We will, however, also find it useful to consider the different (but clearly related) matrix formed by interchanging the rows and columns of $\mathsf{A}$. The matrix is called the *transpose* of $\mathsf{A}$ and is denoted by $\mathsf{A}^{\mathrm{T}}$.

►*Find the transpose of the matrix*

$$\mathsf{A} = \begin{pmatrix} 3 & 1 & 2 \\ 0 & 4 & 1 \end{pmatrix}.$$

By interchanging the rows and columns of A we immediately obtain

$$\mathsf{A}^{\mathrm{T}} = \begin{pmatrix} 3 & 0 \\ 1 & 4 \\ 2 & 1 \end{pmatrix}. \blacktriangleleft$$

It is obvious that if A is an $M \times N$ matrix, then its transpose $\mathsf{A}^{\mathrm{T}}$ is a $N \times M$ matrix. As mentioned in section 7.3 the transpose of a column matrix is a row matrix and vice versa. An important use of column and row matrices is in the representation of the inner product of two real vectors in terms of their components in a given basis. This notion is discussed fully in the next section, where it is extended to complex vectors.

The transpose of the product of two matrices $(\mathsf{AB})^{\mathrm{T}}$ is given by the product of their transposes taken in the reverse order, i.e.

$$(\mathsf{AB})^{\mathrm{T}} = \mathsf{B}^{\mathrm{T}}\mathsf{A}^{\mathrm{T}}.$$

This is easily proved as follows:

$$\begin{aligned}(\mathsf{AB})^{\mathrm{T}}_{ij} = (\mathsf{AB})_{ji} &= \sum_k A_{jk}B_{ki} \\ &= \sum_k (\mathsf{A}^{\mathrm{T}})_{kj}(\mathsf{B}^{\mathrm{T}})_{ik} = \sum_k (\mathsf{B}^{\mathrm{T}})_{ik}(\mathsf{A}^{\mathrm{T}})_{kj} = (\mathsf{B}^{\mathrm{T}}\mathsf{A}^{\mathrm{T}})_{ij},\end{aligned}$$

and the proof can be extended to the product of several matrices to give

$$(\mathsf{ABC}\cdots\mathsf{G})^{\mathrm{T}} = \mathsf{G}^{\mathrm{T}}\cdots\mathsf{C}^{\mathrm{T}}\mathsf{B}^{\mathrm{T}}\mathsf{A}^{\mathrm{T}}.$$

## 7.6 The complex and Hermitian conjugates of a matrix

Two further matrices that can be derived from a given general $M \times N$ matrix are the *complex conjugate*, denoted by $\mathsf{A}^*$, and the *Hermitian conjugate*, denoted by $\mathsf{A}^\dagger$.

The complex conjugate of a matrix A is the matrix obtained by taking the complex conjugate of each of the elements of A, i.e.

$$(\mathsf{A}^*)_{ij} = (A_{ij})^*.$$

Obviously if a matrix is *real* (i.e. it contains only real elements) then $\mathsf{A}^* = \mathsf{A}$.

►*Find the complex conjugate of the matrix*

$$\mathsf{A} = \begin{pmatrix} 1 & 2 & 3i \\ 1+i & 1 & 0 \end{pmatrix}.$$

By taking the complex conjugate of each element we immediately obtain

$$\mathsf{A}^* = \begin{pmatrix} 1 & 2 & -3i \\ 1-i & 1 & 0 \end{pmatrix}. \blacktriangleleft$$

The Hermitian conjugate (or *adjoint*) of a matrix A is simply the transpose of its complex conjugate, or equivalently, the complex conjugate of its transpose, i.e.

$$\mathsf{A}^\dagger = (\mathsf{A}^*)^{\mathrm{T}} = (\mathsf{A}^{\mathrm{T}})^*.$$

Following the previous line of argument for the transpose of the product of several matrices, the Hermitian conjugate of such a product can be shown to be given by

$$(\mathsf{AB}\cdots\mathsf{G})^\dagger = \mathsf{G}^\dagger\cdots\mathsf{B}^\dagger\mathsf{A}^\dagger.$$

►*Find the Hermitian conjugate of the matrix*

$$\mathsf{A} = \begin{pmatrix} 1 & 2 & 3i \\ 1+i & 1 & 0 \end{pmatrix}.$$

Taking the complex conjugate of A and then forming the transpose we find

$$\mathsf{A}^\dagger = \begin{pmatrix} 1 & 1-i \\ 2 & 1 \\ -3i & 0 \end{pmatrix}.$$

We obtain the same result, of course, if we first take the transpose of A and then take the complex conjugate. ◄

We note that if A is real (so that $\mathsf{A}^* = \mathsf{A}$) then $\mathsf{A}^\dagger = \mathsf{A}^{\mathrm{T}}$, and so the Hermitian conjugate may be considered as a generalisation of the transpose to complex matrices.

An important use of the Hermitian conjugate (or transpose in the real case) is in connection with the inner product of two vectors. Suppose that in a given orthonormal basis the vectors **a** and **b** may be represented by the column matrices

$$\mathsf{a} = \begin{pmatrix} a_1 \\ \vdots \\ a_N \end{pmatrix} \quad \text{and} \quad \mathsf{b} = \begin{pmatrix} b_1 \\ \vdots \\ b_N \end{pmatrix}. \tag{7.38}$$

Taking the Hermitian conjugate of a, to give a row matrix, and multiplying (on the right) by b we obtain

$$\mathsf{a}^\dagger\mathsf{b} = (a_1^* \ a_2^* \ \cdots \ a_N^*)\begin{pmatrix} b_1 \\ \vdots \\ b_N \end{pmatrix} = \sum_{i=1}^{N} a_i^* b_i, \tag{7.39}$$

which is the expression for the inner product $\langle \mathbf{a}|\mathbf{b}\rangle$ in that basis. We note that for real vectors (7.39) reduces to $\mathsf{a}^\mathrm{T}\mathsf{b} = \sum_{i=1}^{N} a_i b_i$.

If the basis $\mathbf{e}_i$ is *not* orthonormal, so that, in general,

$$\langle \mathbf{e}_i|\mathbf{e}_j\rangle = G_{ij} \neq \delta_{ij},$$

then, from (7.18), the scalar product of **a** and **b** in terms of their components with respect to this basis is given by

$$\langle \mathbf{a}|\mathbf{b}\rangle = \sum_{i=1}^{N}\sum_{j=1}^{N} a_i^* G_{ij} b_j = \mathsf{a}^\dagger \mathsf{G} \mathsf{b},$$

where G is the $N \times N$ matrix with elements $G_{ij}$.

## 7.7 The determinant of a matrix

Much of the subsequent material in this chapter relies on the theory of determinants. The development work needed to establish and explain the required results is quite extensive and, so as not to break up the later material too much, we present it here.

The *determinant*, det A, of a matrix, A, is also denoted by |A|, and is a single number (or algebraic expression) that depends upon the elements of A. It is defined only for square matrices. If, for example, A is a $3 \times 3$ matrix, then its determinant, of *order* 3, is written

$$\det \mathsf{A} = |\mathsf{A}| = \begin{vmatrix} A_{11} & A_{12} & A_{13} \\ A_{21} & A_{22} & A_{23} \\ A_{31} & A_{32} & A_{33} \end{vmatrix}. \tag{7.40}$$

In order to calculate the value of a determinant, we first need to introduce the notions of the *minor* and the *cofactor* of an element of a matrix. (We shall see that we can use the cofactors to write the order-3 determinant as the weighted sum of three order-2 determinants, thereby simplifying its evaluation.) The minor $M_{ij}$ of the element $A_{ij}$ of an $N \times N$ matrix A is the determinant of the $(N-1) \times (N-1)$ matrix obtained by removing all the elements of the $i$th row and $j$th column of A; the associated cofactor, $C_{ij}$, is found by multiplying the minor by $(-1)^{i+j}$.

►*Find the cofactor of the element $A_{23}$ of the matrix*

$$\mathsf{A} = \begin{pmatrix} A_{11} & A_{12} & A_{13} \\ A_{21} & A_{22} & A_{23} \\ A_{31} & A_{32} & A_{33} \end{pmatrix}.$$

Removing all the elements of the second row and third column of A and forming the determinant of the remaining terms gives the minor

$$M_{23} = \begin{vmatrix} A_{11} & A_{12} \\ A_{31} & A_{32} \end{vmatrix}.$$

Multiplying the minor by $(-1)^{2+3} = (-1)^5 = -1$ gives

$$C_{23} = -\begin{vmatrix} A_{11} & A_{12} \\ A_{31} & A_{32} \end{vmatrix}. \blacktriangleleft$$

We now define a determinant as *the sum of the products of the elements of any row or column and their corresponding cofactors*, e.g. $A_{21}C_{21} + A_{22}C_{22} + A_{23}C_{23}$ or $A_{13}C_{13} + A_{23}C_{23} + A_{33}C_{33}$. Such a sum is called a *Laplace expansion*. We will see later that the value of the determinant is independent of the row or column chosen.

►*By considering the elements of the second row of the determinant defined by (7.40) and their corresponding cofactors, write $|\mathsf{A}|$ as a Laplace expansion.*

$$\begin{aligned} |\mathsf{A}| &= A_{21}(-1)^{(2+1)}M_{21} + A_{22}(-1)^{(2+2)}M_{22} + A_{23}(-1)^{(2+3)}M_{23} \\ &= -A_{21}\begin{vmatrix} A_{12} & A_{13} \\ A_{32} & A_{33} \end{vmatrix} + A_{22}\begin{vmatrix} A_{11} & A_{13} \\ A_{31} & A_{33} \end{vmatrix} - A_{23}\begin{vmatrix} A_{11} & A_{12} \\ A_{31} & A_{32} \end{vmatrix}. \blacktriangleleft \end{aligned}$$

Of course, we have not yet determined the value of $|\mathsf{A}|$, but rather written it as the weighted sum of three determinants of order 2. However, applying again the definition of a determinant, we can evaluate each of the order-2 determinants.

►*Evaluate the determinant*

$$\begin{vmatrix} A_{12} & A_{13} \\ A_{32} & A_{33} \end{vmatrix}.$$

By considering the products of the elements of the first row in the determinant, and their corresponding cofactors, we find

$$\begin{aligned} \begin{vmatrix} A_{12} & A_{13} \\ A_{32} & A_{33} \end{vmatrix} &= A_{12}(-1)^{(1+1)}|A_{33}| + A_{13}(-1)^{(1+2)}|A_{32}| \\ &= A_{12}A_{33} - A_{13}A_{32}, \end{aligned}$$

where the values of the order-1 determinants $|A_{33}|$ and $|A_{32}|$ are defined to be $A_{33}$ and $A_{32}$ respectively. It must be remembered that the determinant is *not* the same as the modulus, e.g. det $(-2) = |-2| = -2$, not 2. ◄

We can now combine all the above results to show that the value of the determinant (7.40) is given by

$$|\mathsf{A}| = -A_{21}(A_{12}A_{33} - A_{13}A_{32}) + A_{22}(A_{11}A_{33} - A_{13}A_{31}) \\ -A_{23}(A_{11}A_{32} - A_{12}A_{31}) \qquad (7.41)$$

$$= A_{11}(A_{22}A_{33} - A_{23}A_{32}) + A_{12}(A_{23}A_{31} - A_{21}A_{33}) \\ +A_{13}(A_{21}A_{32} - A_{22}A_{31}), \qquad (7.42)$$

where the final equality gives the form in which the determinant is usually remembered and is the form that is immediately obtained by considering the Laplace expansion of the first row of the determinant. The last equality supports our assertion that the value of the determinant is unaffected by which row or column is chosen for the Laplace expansion.

The evaluation of determinants of order greater than 3 follows the same method as presented above in that it relies on successively reducing the order of the determinant by writing it as a Laplace expansion. Thus, a determinant of order 4 is first written as a sum of four determinants of order 3, which are then evaluated using the above method. Another method for evaluating determinants, which makes use of the Levi–Civita symbol $\epsilon_{ijk}$, is discussed in chapter 19.

### *7.7.1 Properties of determinants*

A number of properties of determinants follow straightforwardly from the definition of det A; their use will often reduce the labour of evaluating a determinant. We present them without specific proofs.

(i) *Determinant of the transpose.* The transpose matrix $\mathsf{A}^{\mathrm{T}}$ (which, we recall, is obtained by interchanging the rows and columns of A) has the same determinant as A itself. It follows that *any* theorem established for the rows of A will apply to the columns as well, and vice versa.

(ii) *Interchanging two rows or two columns.* If two rows (columns) of A are interchanged, its determinant changes sign, but is unaltered in magnitude.

(iii) *Removing factors.* If all the elements of a single row (column) of A have a common factor, $\lambda$, then this factor may be removed; the value of the determinant is given by the product of the remaining determinant and $\lambda$. Clearly this implies that if all the elements of any row (column) are zero, then $|\mathsf{A}| = 0$.

(iv) *Identical rows or columns.* If any two rows (columns) of A are identical, or are multiples of one another, then it can be shown that $|\mathsf{A}| = 0$.

(v) *Adding a constant multiple of one row (column) to another.* The determinant of a matrix is unchanged in value by the addition to the elements of one row (column), of any fixed multiple of the elements of another row (column).

(vi) *Determinant of a product.* If $\mathsf{A}$ and $\mathsf{B}$ are square matrices of the same order, then $|\mathsf{AB}| = |\mathsf{A}||\mathsf{B}| = |\mathsf{BA}|$. A simple extension of this property gives $|\mathsf{AB}\cdots\mathsf{G}| = |\mathsf{A}||\mathsf{B}|\cdots|\mathsf{G}| = |\mathsf{A}||\mathsf{G}|\cdots|\mathsf{B}| = |\mathsf{A}\cdots\mathsf{GB}|$, etc.

There is no explicit procedure for using the above results in the evaluation of any given determinant, and judging the quickest route to an answer is a matter of experience. A general guide is to try to reduce all terms but one in a row or column to zero and hence in effect to obtain a determinant of smaller size. The steps taken in evaluating the determinant in the example below are certainly not the fastest, but have been chosen in order to illustrate the use of most of the properties listed above.

►*Evaluate the determinant*

$$|\mathsf{A}| = \begin{vmatrix} 1 & 0 & 2 & 3 \\ 0 & 1 & -2 & 1 \\ 3 & -3 & 4 & -2 \\ -2 & 1 & -2 & -1 \end{vmatrix}.$$

Taking a factor 2 out of the third column and then adding the second column to the third gives

$$|\mathsf{A}| = 2\begin{vmatrix} 1 & 0 & 1 & 3 \\ 0 & 1 & -1 & 1 \\ 3 & -3 & 2 & -2 \\ -2 & 1 & -1 & -1 \end{vmatrix} = 2\begin{vmatrix} 1 & 0 & 1 & 3 \\ 0 & 1 & 0 & 1 \\ 3 & -3 & -1 & -2 \\ -2 & 1 & 0 & -1 \end{vmatrix}.$$

Subtracting the second column from the fourth gives

$$|\mathsf{A}| = 2\begin{vmatrix} 1 & 0 & 1 & 3 \\ 0 & 1 & 0 & 0 \\ 3 & -3 & -1 & 1 \\ -2 & 1 & 0 & -2 \end{vmatrix}.$$

We now note that the second row has only one non-zero element and so the determinant may conveniently be written as a Laplace expansion, i.e.

$$|\mathsf{A}| = 2\times 1\times(-1)^{2+2}\begin{vmatrix} 1 & 1 & 3 \\ 3 & -1 & 1 \\ -2 & 0 & -2 \end{vmatrix} = 2\begin{vmatrix} 4 & 0 & 4 \\ 3 & -1 & 1 \\ -2 & 0 & -2 \end{vmatrix},$$

where the last equality follows by adding the second row to the first. It can now be seen that the first row is minus twice the third, and so the value of the determinant is zero, by property (iv) above. ◄

## 7.8 The inverse of a matrix

Our first use of determinants will be in defining the *inverse* of a matrix. If we were dealing with ordinary numbers we would consider the relation $\mathsf{P} = \mathsf{AB}$ as equivalent to $\mathsf{B} = \mathsf{P}/\mathsf{A}$ provided that $\mathsf{A} \neq 0$. However, if A, B and P are matrices this notation does not have an obvious meaning. What we really want to know is whether an explicit formula for B can be obtained in terms of A and P. It will be shown that this is possible for those cases where $|\mathsf{A}| \neq 0$. A square matrix whose determinant is zero is called a *singular* matrix; otherwise it is *non-singular*. We will show that if A is non-singular we can define a matrix, denoted by $\mathsf{A}^{-1}$, called the *inverse* of A, which has the property that if $\mathsf{AB} = \mathsf{P}$ then $\mathsf{B} = \mathsf{A}^{-1}\mathsf{P}$. In words, B can be obtained by multiplying P from the left by $\mathsf{A}^{-1}$. Analogously, if B is non-singular then, by multiplication from the right, $\mathsf{A} = \mathsf{PB}^{-1}$.

It is clear that

$$\mathsf{AI} = \mathsf{A} \quad \Rightarrow \quad \mathsf{I} = \mathsf{A}^{-1}\mathsf{A}, \tag{7.43}$$

where I is the unit matrix, and so $\mathsf{A}^{-1}\mathsf{A} = \mathsf{I} = \mathsf{AA}^{-1}$. These statements are equivalent to saying that if we first multiply a matrix, B say, by A and then multiply by the inverse $\mathsf{A}^{-1}$, we end up with the matrix we started with, i.e.

$$\mathsf{A}^{-1}\mathsf{AB} = \mathsf{B}. \tag{7.44}$$

This justifies our use of the term inverse. It is also clear that the inverse is only defined for square matrices.

So far we have only defined what we mean by the inverse of a matrix. Actually finding the inverse of a matrix A may be carried out in a number of ways. We will show that one method is to construct first the matrix C containing the cofactors of the elements of A, as discussed in the last subsection. Then, the required inverse $\mathsf{A}^{-1}$ can be found by forming the transpose of C and dividing by the determinant of A. Thus the elements of the inverse $\mathsf{A}^{-1}$ are given by

$$(\mathsf{A}^{-1})_{ik} = \frac{(\mathsf{C})^{\mathrm{T}}_{ik}}{|\mathsf{A}|} = \frac{C_{ki}}{|\mathsf{A}|}. \tag{7.45}$$

That this procedure does indeed result in the inverse may be seen by considering the components of $\mathsf{A}^{-1}\mathsf{A}$, i.e.

$$(\mathsf{A}^{-1}\mathsf{A})_{ij} = \sum_k (\mathsf{A}^{-1})_{ik}(\mathsf{A})_{kj} = \sum_k \frac{(C_{ki})}{|\mathsf{A}|} A_{kj} = \frac{|\mathsf{A}|}{|\mathsf{A}|}\delta_{ij}. \tag{7.46}$$

The last equality in (7.46) relies on the property

$$\sum_k C_{ki}A_{kj} = |\mathsf{A}|\delta_{ij}; \tag{7.47}$$

this can be proved by considering the matrix $\mathsf{A}'$ obtained from the original matrix A when the $i$th column of A is replaced by one of the other columns, say the $j$th.

Thus $\mathsf{A}'$ is a matrix with two identical columns and so has zero determinant. However, replacing the $i$th column by another does not change the cofactors $C_{ki}$ of the elements in the $i$th column, which are therefore the same in $\mathsf{A}$ and $\mathsf{A}'$. Recalling the Laplace expansion of a determinant, i.e.

$$|\mathsf{A}| = \sum_k a_{ki} C_{ki},$$

we obtain

$$0 = |\mathsf{A}'| = \sum_k a'_{ki} C'_{ki} = \sum_k a_{kj} C_{ki}, \qquad i \neq j,$$

which may be combined with the Laplace expansion itself to give (7.47).

It is immediately obvious from (7.45) that the inverse of a matrix is not defined if it is singular (i.e. if $|\mathsf{A}| = 0$.)

►*Find the inverse of the matrix*

$$\mathsf{A} = \begin{pmatrix} 2 & 4 & 3 \\ 1 & -2 & -2 \\ -3 & 3 & 2 \end{pmatrix}.$$

We first determine $|\mathsf{A}|$:

$$\begin{aligned} |\mathsf{A}| &= 2[-2(2) - (-2)3] + 4[(-2)(-3) - (1)(2)] + 3[(1)(3) - (-2)(-3)] \\ &= 11. \end{aligned} \tag{7.48}$$

This is non-zero and so an inverse matrix can be constructed. To do this we need the matrix of the cofactors, $\mathsf{C}$, and hence $\mathsf{C}^{\mathrm{T}}$. We find

$$\mathsf{C} = \begin{pmatrix} 2 & 4 & -3 \\ 1 & 13 & -18 \\ -2 & 7 & -8 \end{pmatrix} \quad \text{and} \quad \mathsf{C}^{\mathrm{T}} = \begin{pmatrix} 2 & 1 & -2 \\ 4 & 13 & 7 \\ -3 & -18 & -8 \end{pmatrix},$$

and hence

$$\mathsf{A}^{-1} = \frac{\mathsf{C}^{\mathrm{T}}}{|\mathsf{A}|} = \frac{1}{11}\begin{pmatrix} 2 & 1 & -2 \\ 4 & 13 & 7 \\ -3 & -18 & -8 \end{pmatrix}. \tag{7.49}$$ ◄

For a $2 \times 2$ matrix there is a particularly simple form of the inverse. If we consider the matrix

$$\mathsf{A} = \begin{pmatrix} A_{11} & A_{12} \\ A_{21} & A_{22} \end{pmatrix},$$

its determinant $|\mathsf{A}|$ is given by $|\mathsf{A}| = A_{11}A_{22} - A_{12}A_{21}$, and the matrix of cofactors is

$$\mathsf{C} = \begin{pmatrix} A_{22} & -A_{21} \\ -A_{12} & A_{11} \end{pmatrix}.$$

Thus the inverse of A is given by

$$\mathsf{A}^{-1} = \frac{\mathsf{C}^{\mathrm{T}}}{|\mathsf{A}|} = \frac{1}{A_{11}A_{22} - A_{12}A_{21}} \begin{pmatrix} A_{22} & -A_{12} \\ -A_{21} & A_{11} \end{pmatrix}. \qquad (7.50)$$

It can be seen that the transposed matrix of cofactors for a $2 \times 2$ matrix is the same as the matrix formed by swapping the elements on the leading diagonal ($A_{11}$ and $A_{22}$) and changing the signs of the other two elements ($A_{12}$ and $A_{21}$). This is completely general for a $2 \times 2$ matrix and is easy to remember.

The following are some further useful properties related to the inverse matrix, and may be straightforwardly derived.

(i) $(\mathsf{A}^{-1})^{-1} = \mathsf{A}$.
(ii) $(\mathsf{A}^{\mathrm{T}})^{-1} = (\mathsf{A}^{-1})^{\mathrm{T}}$.
(iii) $(\mathsf{AB})^{-1} = \mathsf{B}^{-1}\mathsf{A}^{-1}$.
(iv) $(\mathsf{AB}\cdots\mathsf{G})^{-1} = \mathsf{G}^{-1}\cdots\mathsf{B}^{-1}\mathsf{A}^{-1}$.

## 7.9 The rank of a matrix

The *rank* of a general $M \times N$ matrix is an important concept, particularly in the solution of sets of simultaneous linear equations, to be discussed in the next section, and we now discuss this concept in some detail. As we shall see, there are two *equivalent* definitions of the rank of a general matrix.

Firstly, the rank of a matrix may be defined in terms of the *linear independence* of vectors. Suppose that the columns of an $M \times N$ matrix are interpreted as the components in a given basis of $N$ ($M$-component) vectors $\mathbf{v}_1, \mathbf{v}_2, \ldots, \mathbf{v}_N$, as follows:

$$\mathsf{A} = \begin{pmatrix} \uparrow & & \uparrow \\ \mathsf{v}_1 & \ldots & \mathsf{v}_N \\ \downarrow & & \downarrow \end{pmatrix}.$$

Then the *rank* of A is defined as the number of *linearly independent* vectors in the set $\mathbf{v}_1, \mathbf{v}_2, \ldots, \mathbf{v}_N$, and equals the dimension of the vector space spanned by those vectors. Alternatively, we may consider the rows of A to contain the components in a given basis of the $M$ ($N$-component) vectors $\mathbf{w}_1, \mathbf{w}_2, \ldots, \mathbf{w}_M$ as follows,

$$\mathsf{A} = \begin{pmatrix} \leftarrow & \mathsf{w}_1 & \rightarrow \\ & \vdots & \\ \leftarrow & \mathsf{w}_M & \rightarrow \end{pmatrix},$$

and it may be shown that the rank of A is also equal to the number of linearly independent vectors in the set $\mathbf{w}_1, \mathbf{w}_2, \ldots, \mathbf{w}_M$.

From this definition it is should be clear that the rank of A is unchanged by exchanging two rows (or two columns), or by multiplying a row (or column) by a constant. Furthermore, suppose that a constant multiple of one row (column) is added to another row (column): for example, we might replace the row $\mathsf{w}_i$ by $\mathsf{w}_i + c\mathsf{w}_j$. This also has no effect on the number of linearly independent rows and so leaves the rank of A unchanged. We may use these properties to evaluate the rank of a given matrix.

As mentioned above, a second (equivalent) definition may be given of the rank of a matrix and uses the concept of *submatrices*. A submatrix of A is any matrix that can be formed from the elements of A by ignoring one or more rows or columns. It may be shown that the rank $r$ of a general $M \times N$ matrix is equal to the size of the largest square submatrix of A whose determinant is non-zero. Therefore, if a matrix A has an $r \times r$ submatrix S with $|\mathsf{S}| \neq 0$, but no $(r+1) \times (r+1)$ submatrix with non-zero determinant, then the rank of A is $r$. From either definition it is clear that the rank of A is less than or equal to the smaller of $M$ and $N$.

►*Determine the rank of the matrix*

$$\mathsf{A} = \begin{pmatrix} 1 & 1 & 0 & -2 \\ 2 & 0 & 2 & 2 \\ 4 & 1 & 3 & 1 \end{pmatrix}.$$

The largest possible square submatrices of A must be of dimension $3 \times 3$. Clearly, A possesses four such submatrices, the determinants of which are given by

$$\begin{vmatrix} 1 & 1 & 0 \\ 2 & 0 & 2 \\ 4 & 1 & 3 \end{vmatrix} = 0, \qquad \begin{vmatrix} 1 & 1 & -2 \\ 2 & 0 & 2 \\ 4 & 1 & 1 \end{vmatrix} = 0,$$

$$\begin{vmatrix} 1 & 0 & -2 \\ 2 & 2 & 2 \\ 4 & 3 & 1 \end{vmatrix} = 0, \qquad \begin{vmatrix} 1 & 0 & -2 \\ 0 & 2 & 2 \\ 1 & 3 & 1 \end{vmatrix} = 0.$$

(In each case the determinant may be evaluated as described in subsection 7.7.1.)

The next largest square submatrices of A are of dimension $2 \times 2$. Considering, for example, the $2 \times 2$ submatrix formed by ignoring the third row and the third and fourth columns of A, this has the determinant

$$\begin{vmatrix} 1 & 1 \\ 2 & 0 \end{vmatrix} = 1 \times 0 - 2 \times 1 = -2.$$

Since this determinant is non-zero, A is of rank 2 and we need not consider any other $2 \times 2$ submatrices. ◀

## 7.10 Simultaneous linear equations

In physical applications we often encounter sets of simultaneous linear equations. In general we may have $M$ equations in $N$ unknowns $x_1, x_2, \ldots, x_N$ of the form

$$\begin{aligned} A_{11}x_1 + A_{12}x_2 + \cdots + A_{1N}x_N &= b_1, \\ A_{21}x_1 + A_{22}x_2 + \cdots + A_{2N}x_N &= b_2, \\ &\ \ \vdots \\ A_{M1}x_1 + A_{M2}x_2 + \cdots + A_{MN}x_N &= b_M, \end{aligned} \tag{7.51}$$

where the $A_{ij}$ and $b_i$ have known values. If all the $b_i$ are zero then the system of equations is called *homogeneous*, otherwise it is *inhomogeneous*. Depending on the given values, this set of equations for the $N$ unknowns $x_1, x_2, \ldots, x_N$ may have either a unique solution, no solution or infinitely many solutions. Matrix analysis may be used to distinguish between the possibilities. The set of equations may be expressed as a single matrix equation $\mathsf{Ax} = \mathsf{b}$, or, written out in full, as

$$\begin{pmatrix} A_{11} & A_{12} & \ldots & A_{1N} \\ A_{21} & A_{22} & \ldots & A_{2N} \\ \vdots & \vdots & \ddots & \vdots \\ A_{M1} & A_{M2} & \ldots & A_{MN} \end{pmatrix} \begin{pmatrix} x_1 \\ x_2 \\ \vdots \\ x_N \end{pmatrix} = \begin{pmatrix} b_1 \\ b_2 \\ \vdots \\ b_M \end{pmatrix}.$$

### 7.10.1 *N simultaneous linear equations in N unknowns*

A special case of (7.51) occurs when $M = N$ and we have the same number of equations as unknowns. In this case A is square and possesses an inverse provided $|\mathsf{A}| \neq 0$. If A is non-singular, we immediately obtain

$$\mathsf{x} = \mathsf{A}^{-1}\mathsf{b} \tag{7.52}$$

as the unique solution to the set of equations. However, if $\mathsf{b} = 0$, making the system of equations homogeneous, and $|\mathsf{A}| \neq 0$, the set of equations possesses only the trivial solution $\mathsf{x} = 0$.

►*Show that the set of simultaneous equations*

$$\begin{aligned} 2x_1 + 4x_2 + 3x_3 &= 4, \\ x_1 - 2x_2 - 2x_3 &= 0, \\ -3x_1 + 3x_2 + 2x_3 &= -7, \end{aligned} \qquad (7.53)$$

*has a unique solution, and find that solution.*

The simultaneous equations can be represented by the matrix equation $\mathsf{Ax} = \mathsf{b}$, i.e.

$$\begin{pmatrix} 2 & 4 & 3 \\ 1 & -2 & -2 \\ -3 & 3 & 2 \end{pmatrix} \begin{pmatrix} x_1 \\ x_2 \\ x_3 \end{pmatrix} = \begin{pmatrix} 4 \\ 0 \\ -7 \end{pmatrix}.$$

As we have already shown that $\mathsf{A}^{-1}$ exists and have calculated it, see (7.49), it follows that $\mathsf{x} = \mathsf{A}^{-1}\mathsf{b}$ or, more explicitly,

$$\begin{pmatrix} x_1 \\ x_2 \\ x_3 \end{pmatrix} = \frac{1}{11} \begin{pmatrix} 2 & 1 & -2 \\ 4 & 13 & 7 \\ -3 & -18 & -8 \end{pmatrix} \begin{pmatrix} 4 \\ 0 \\ -7 \end{pmatrix} = \begin{pmatrix} 2 \\ -3 \\ 4 \end{pmatrix}. \qquad (7.54)$$

Thus the unique solution is $x_1 = 2$, $x_2 = -3$, $x_3 = 4$. ◄

Although conceptually simple, finding the solution by calculating $\mathsf{A}^{-1}$ can be computationally tedious. An alternative method of solution is to use *Cramer's rule.* To illustrate this method let us consider a set of three equations in three unknowns,

$$\begin{aligned} A_{11}x_1 + A_{12}x_2 + A_{13}x_3 &= b_1, \\ A_{21}x_1 + A_{22}x_2 + A_{23}x_3 &= b_2, \\ A_{31}x_1 + A_{32}x_2 + A_{33}x_3 &= b_3, \end{aligned} \qquad (7.55)$$

which may be represented by the matrix equation $\mathsf{Ax} = \mathsf{b}$. We wish either to find the solution(s) $\mathsf{x}$ to these equations, or establish that there are no solutions. From result (v) of subsection 7.7.1, the determinant $|\mathsf{A}|$ is unchanged by adding to its first column the combination

$$\frac{x_2}{x_1} \times (\text{second column of } |\mathsf{A}|) + \frac{x_3}{x_1} \times (\text{third column of } |\mathsf{A}|).$$

We thus obtain

$$|\mathsf{A}| = \begin{vmatrix} A_{11} & A_{12} & A_{13} \\ A_{21} & A_{22} & A_{23} \\ A_{31} & A_{32} & A_{33} \end{vmatrix} = \begin{vmatrix} A_{11} + (x_2/x_1)A_{12} + (x_3/x_1)A_{13} & A_{12} & A_{13} \\ A_{21} + (x_2/x_1)A_{22} + (x_3/x_1)A_{23} & A_{22} & A_{23} \\ A_{31} + (x_2/x_1)A_{32} + (x_3/x_1)A_{33} & A_{32} & A_{33} \end{vmatrix},$$

which, on substituting $b_i/x_1$ for the $i$th entry in the first column, yields

$$|\mathsf{A}| = \frac{1}{x_1}\begin{vmatrix} b_1 & A_{12} & A_{13} \\ b_2 & A_{22} & A_{23} \\ b_3 & A_{32} & A_{33} \end{vmatrix} = \frac{1}{x_1}\Delta_1.$$

The determinant $\Delta_1$ is known as a *Cramer determinant*. Similar manipulations of the second and third rows of $|\mathsf{A}|$ yield $x_2$ and $x_3$, and so the full set of results reads

$$x_1 = \frac{\Delta_1}{|\mathsf{A}|}, \qquad x_2 = \frac{\Delta_2}{|\mathsf{A}|}, \qquad x_3 = \frac{\Delta_3}{|\mathsf{A}|}, \tag{7.56}$$

where

$$\Delta_1 = \begin{vmatrix} b_1 & A_{12} & A_{13} \\ b_2 & A_{22} & A_{23} \\ b_3 & A_{32} & A_{33} \end{vmatrix}, \quad \Delta_2 = \begin{vmatrix} A_{11} & b_1 & A_{13} \\ A_{21} & b_2 & A_{23} \\ A_{31} & b_3 & A_{33} \end{vmatrix}, \quad \Delta_3 = \begin{vmatrix} A_{11} & A_{12} & b_1 \\ A_{21} & A_{22} & b_2 \\ A_{31} & A_{32} & b_3 \end{vmatrix}.$$

It can easily be seen that each Cramer determinant is simply $|\mathsf{A}|$ but with the relevant column replaced by the RHS of the original set of simultaneous equations. If $|\mathsf{A}| \neq 0$, then (7.56) gives the unique solution. The proof given here appears to fail if any of the solutions $x_i$ is zero, but it can be shown that the result (7.56) is valid even in such a case.

►*Use Cramer's rule to solve the set of simultaneous equations (7.53).*

Let us again represent these simultaneous equations by the matrix equation $\mathsf{Ax} = \mathsf{b}$, i.e.

$$\begin{pmatrix} 2 & 4 & 3 \\ 1 & -2 & -2 \\ -3 & 3 & 2 \end{pmatrix}\begin{pmatrix} x_1 \\ x_2 \\ x_3 \end{pmatrix} = \begin{pmatrix} 4 \\ 0 \\ -7 \end{pmatrix}.$$

From (7.48), the determinant of A is given by $|\mathsf{A}| = 11$. Following the discussion given above, the three Cramer determinants are

$$\Delta_1 = \begin{vmatrix} 4 & 4 & 3 \\ 0 & -2 & -2 \\ -7 & 3 & 2 \end{vmatrix}, \quad \Delta_2 = \begin{vmatrix} 2 & 4 & 3 \\ 1 & 0 & -2 \\ -3 & -7 & 2 \end{vmatrix}, \quad \Delta_3 = \begin{vmatrix} 2 & 4 & 4 \\ 1 & -2 & 0 \\ -3 & 3 & -7 \end{vmatrix}.$$

These may be evaluated using the properties of determinants listed in subsection 7.7.1 and we find $\Delta_1 = 22$, $\Delta_2 = -33$ and $\Delta_3 = 44$. From (7.56) the solution to the equations (7.53) is given by

$$x_1 = \frac{22}{11} = 2, \qquad x_2 = \frac{-33}{11} = -3, \qquad x_3 = \frac{44}{11} = 4,$$

which agrees with the solution found in the previous example. ◄

At this point it is useful to consider each of the three equations (7.55) as representing a plane in three-dimensional Cartesian coordinates. Using result (6.42) of chapter 6, the components of the vectors normal to the planes are $(A_{11}, A_{12}, A_{13})$, $(A_{21}, A_{22}, A_{23})$ and $(A_{31}, A_{32}, A_{33})$, and using (6.43) the perpendicular distances of the planes from the origin are given by

$$d_i = \frac{b_i}{\left(A_{i1}^2 + A_{i2}^2 + A_{i3}^2\right)^{1/2}}, \qquad \text{for } i = 1, 2, 3.$$

Finding the solution(s) to the simultaneous equations above corresponds to finding the point(s) of intersection of the planes.

If there is a unique solution the planes intersect only at a single point. This occurs if their normals are linearly independent vectors. Since the rows of $\mathsf{A}$ represent the directions of these normals, we therefore require $|\mathsf{A}| \neq 0$. If $\mathsf{b} = (0 \;\; 0 \;\; 0)^{\mathrm{T}} = \mathsf{0}$ then all the planes pass through the origin and, since there is only a single solution to the equations the origin is that solution.

Let us now turn to the cases where $|\mathsf{A}| = 0$. The simplest such case is that in which all three planes are parallel; this implies that the normals are all parallel and so $\mathsf{A}$ is of rank 1. Two possibilities exist:

(i) the planes are coincident, i.e. $d_1 = d_2 = d_3$, in which case there is an infinity of solutions;

(ii) the planes are not all coincident, i.e. $d_1 \neq d_2$ and/or $d_1 \neq d_3$ and/or $d_2 \neq d_3$, in which case there are no solutions.

It is apparent from (7.56) that case (i) occurs when all the Cramer determinants are zero and case (ii) occurs when at least one Cramer determinant is non-zero.

The most complicated cases with $|\mathsf{A}| = 0$ are those in which the normals to the planes themselves lie in a plane, but are not parallel. In this case $\mathsf{A}$ has rank 2. Again two possibilities exist and are shown in figure 7.1. Just as in the rank-1 case, if all the Cramer determinants are zero we get an infinity of solutions (this time on a line). Of course, in the special case in which $\mathsf{b} = \mathsf{0}$ (and the system of equations is homogeneous), the planes all pass through the origin and so they must intersect on a line through it. If at least one of the Cramer determinants is non-zero, we get no solution.

These rules may be summarised as follows.

(i) $|\mathsf{A}| \neq 0$, $\mathsf{b} \neq \mathsf{0}$: the three planes intersect at a single point that is not the origin, and so there is only one solution, given by both (7.52) and (7.56).

(ii) $|\mathsf{A}| \neq 0$, $\mathsf{b} = \mathsf{0}$: the three planes intersect at the origin only and there is only the trivial solution, $\mathsf{x} = \mathsf{0}$.

(iii) $|\mathsf{A}| = 0$, $\mathsf{b} \neq \mathsf{0}$, Cramer determinants all zero: there is an infinity of solutions either on a line (if $\mathsf{A}$ is rank 2, i.e. the cofactors are not all zero) or on a plane (if $\mathsf{A}$ is rank 1, i.e. the cofactors are all zero).

(iv) $|\mathsf{A}| = 0$, $\mathsf{b} \neq \mathsf{0}$, Cramer determinants not all zero: no solutions.

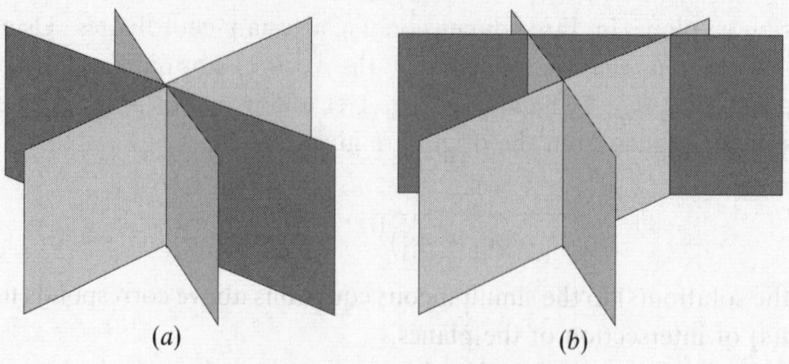

Figure 7.1 The two possible cases where A is of rank 2. In both cases all the normals lie in a horizontal plane but in (*a*) the planes all intersect on a single line (corresponding to an infinite number of solutions) whilst in (*b*) there are no common intersection points (no solutions).

(v) $|\mathsf{A}| = 0$, $\mathsf{b} = \mathsf{0}$: the three planes intersect on a line through the origin giving an infinity of solutions.

## 7.11 Special square matrices

Matrices that are square, i.e. $N \times N$, are very common in physical applications. We now consider some special forms of square matrix that are of particular importance.

### *7.11.1 Diagonal matrices*

The unit matrix, which we have already encountered, is an example of a *diagonal* matrix, in which any non-zero elements lie only on the *leading diagonal*, i.e. only elements $A_{ij}$ with $i = j$ are non-zero. For example,

$$\mathsf{A} = \begin{pmatrix} 1 & 0 & 0 \\ 0 & 2 & 0 \\ 0 & 0 & -3 \end{pmatrix},$$

is a $3 \times 3$ diagonal matrix. Such a matrix is often denoted by $\mathsf{A} = \text{diag}\,(1, 2, -3)$.

### *7.11.2 Symmetric and antisymmetric matrices*

A square matrix A of order $N$ that is equal to its transpose is said to be *symmetric*; a matrix that has a transpose equal to minus itself is said to be *anti-* or *skew*-symmetric and its diagonal elements $a_{11}, a_{22}, \ldots, a_{NN}$ are necessarily zero.

Any $N \times N$ real matrix $\mathsf{A}$ can be written as the sum of a symmetric and an antisymmetric matrix, since we may write

$$\mathsf{A} = \tfrac{1}{2}(\mathsf{A} + \mathsf{A}^{\mathrm{T}}) + \tfrac{1}{2}(\mathsf{A} - \mathsf{A}^{\mathrm{T}}) = \mathsf{B} + \mathsf{C},$$

where clearly $\mathsf{B} = \mathsf{B}^{\mathrm{T}}$ and $\mathsf{C} = -\mathsf{C}^{\mathrm{T}}$. The matrix $\mathsf{B}$ is therefore called the symmetric part of $\mathsf{A}$ and $\mathsf{C}$ the antisymmetric part.

### *7.11.3 Orthogonal matrices*

A real matrix with the property that its transpose is also its inverse,

$$\mathsf{A}^{\mathrm{T}} = \mathsf{A}^{-1}, \tag{7.57}$$

is called an *orthogonal matrix*. This clearly requires that $\mathsf{A}$ is non-singular; but further, since

$$\mathsf{A}^{\mathrm{T}}\mathsf{A} = \mathsf{I}$$

for such a matrix, $|\mathsf{A}|^2 = |\mathsf{A}^{\mathrm{T}}||\mathsf{A}| = |\mathsf{I}|$ and so $|\mathsf{A}| = \pm 1$.

An orthogonal matrix represents (in a particular basis) a linear operator that leaves the norms (lengths) of vectors unchanged, as we will now show. Suppose that $\mathbf{y} = \mathcal{A}\mathbf{x}$ is represented in some coordinate system by the matrix equation $\mathsf{y} = \mathsf{A}\mathsf{x}$; then $\langle \mathbf{y}|\mathbf{y}\rangle$ is given in this coordinate system by

$$\mathsf{y}^{\mathrm{T}}\mathsf{y} = \mathsf{x}^{\mathrm{T}}\mathsf{A}^{\mathrm{T}}\mathsf{A}\mathsf{x} = \mathsf{x}^{\mathrm{T}}\mathsf{x}.$$

Hence $\langle \mathbf{y}|\mathbf{y}\rangle = \langle \mathbf{x}|\mathbf{x}\rangle$, showing that the action of the linear operator $\mathcal{A}$ does not change the norm of vectors.

### *7.11.4 Hermitian matrices*

An *Hermitian* matrix is the complex analogue of a real symmetric matrix and satisfies $\mathsf{A} = \mathsf{A}^{\dagger}$, where $\mathsf{A}^{\dagger}$ is the Hermitian conjugate discussed in section 7.6. Clearly a real symmetric matrix is a special case of an Hermitian matrix, in which all the elements of the matrix are real. Similarly if $\mathsf{A}^{\dagger} = -\mathsf{A}$, then $\mathsf{A}$ is called *anti-Hermitian* and is the complex analogue of a real skew-symmetric matrix.

Any complex $N \times N$ matrix $\mathsf{A}$ can be written as the sum of an Hermitian matrix and an anti-Hermitian matrix, since

$$\mathsf{A} = \tfrac{1}{2}(\mathsf{A} + \mathsf{A}^{\dagger}) + \tfrac{1}{2}(\mathsf{A} - \mathsf{A}^{\dagger}) = \mathsf{B} + \mathsf{C},$$

where clearly $\mathsf{B} = \mathsf{B}^{\dagger}$ and $\mathsf{C} = -\mathsf{C}^{\dagger}$. The matrix $\mathsf{B}$ is called the Hermitian part of $\mathsf{A}$, and $\mathsf{C}$ is called the anti-Hermitian part.

### 7.11.5 *Unitary matrices*

The complex analogue of an orthogonal matrix is a *unitary* matrix. A unitary matrix $\mathsf{A}$ is defined as one for which

$$\mathsf{A}^\dagger = \mathsf{A}^{-1}; \tag{7.58}$$

it represents a linear operator that leaves the norm of a (complex) vector unchanged, since, if $\mathsf{y} = \mathsf{A}\mathsf{x}$,

$$\mathsf{y}^\dagger \mathsf{y} = \mathsf{x}^\dagger \mathsf{A}^\dagger \mathsf{A}\mathsf{x} = \mathsf{x}^\dagger \mathsf{x}.$$

Its action parallels that of an orthogonal matrix acting on a real vector.

### 7.11.6 *Normal matrices*

A final important set of special matrices consists of the *normal* matrices for which

$$\mathsf{A}\mathsf{A}^\dagger = \mathsf{A}^\dagger\mathsf{A},$$

i.e. a normal matrix is one that commutes with its Hermitian conjugate.

We can easily show that Hermitian matrices and unitary matrices (or symmetric matrices and orthogonal matrices in the real case) are examples of normal matrices. For an Hermitian matrix, $\mathsf{A} = \mathsf{A}^\dagger$, and so

$$\mathsf{A}\mathsf{A}^\dagger = \mathsf{A}\mathsf{A} = \mathsf{A}^\dagger\mathsf{A}.$$

Similarly for a unitary matrix, $\mathsf{A}^{-1} = \mathsf{A}^\dagger$ and so

$$\mathsf{A}\mathsf{A}^\dagger = \mathsf{A}\mathsf{A}^{-1} = \mathsf{A}^{-1}\mathsf{A} = \mathsf{A}^\dagger\mathsf{A}.$$

This broad class of matrices is important in the discussion of eigenvectors and eigenvalues in the next section.

## 7.12 Eigenvectors and eigenvalues

Suppose a linear operator $\mathcal{A}$ transforms vectors $\mathbf{x}$ in an $N$-dimensional vector space into other vectors $\mathcal{A}\mathbf{x}$ in the same space. The possibility then arises that there will exist vectors $\mathbf{x}$ each of which is transformed by $\mathcal{A}$ simply into a multiple of itself. Such a vector $\mathbf{x}$ must therefore satisfy

$$\mathcal{A}\mathbf{x} = \lambda\mathbf{x}. \tag{7.59}$$

Any non-zero vector $\mathbf{x}$ that satisfies (7.59) for some value of $\lambda$ is called an *eigenvector* of the linear operator $\mathcal{A}$, and $\lambda$ is called the corresponding *eigenvalue.* As will be discussed below, in general the operator $\mathcal{A}$ has $N$ independent eigenvectors $\mathbf{x}^i$, with eigenvalues $\lambda_i$. The $\lambda_i$ are not necessarily all distinct.

If we choose a particular basis in the vector space, we can write (7.59) in terms of the components of $\mathcal{A}$ and $\mathbf{x}$ with respect to this basis as the matrix equation

$$\mathsf{Ax} = \lambda\mathsf{x}, \tag{7.60}$$

where $\mathsf{A}$ is an $N \times N$ matrix. The column matrices $\mathsf{x}$ that satisfy (7.60) obviously represent the eigenvectors $\mathbf{x}$ of $\mathcal{A}$ in our chosen coordinate system. Conventionally, these column matrices are also referred to as the *eigenvectors of the matrix* $\mathsf{A}$. Clearly, if $\mathsf{x}$ is an eigenvector of $\mathsf{A}$ (with some eigenvalue $\lambda$) then any scalar multiple $\mu\mathsf{x}$ is also an eigenvector with the same eigenvalue. We therefore often use *normalised* eigenvectors, for which

$$\mathsf{x}^\dagger\mathsf{x} = 1$$

(note that $\mathsf{x}^\dagger\mathsf{x}$ corresponds to the inner product $\langle\mathbf{x}|\mathbf{x}\rangle$ in our basis). Any eigenvector $\mathsf{x}$ can be normalised by dividing all components by $(\mathsf{x}^\dagger\mathsf{x})^{1/2}$.

As will be seen, the problem of finding the eigenvalues and corresponding eigenvectors of a square matrix $\mathsf{A}$ plays an important role in many physical investigations. Throughout this chapter we denote the $i$th eigenvector of a square matrix $\mathsf{A}$ by $\mathsf{x}^i$ and the corresponding eigenvalue by $\lambda_i$. The superscript notation for eigenvectors is used to avoid any confusion with components.

In the remainder of this section we will discuss some useful results concerning the eigenvectors and eigenvalues of certain special (though commonly occurring) square matrices. The results will be established for matrices whose elements may be complex; the corresponding properties for real matrices may be obtained as special cases.

### *7.12.1 Eigenvectors and eigenvalues of a normal matrix*

In subsection 7.11.6 we defined a normal matrix $\mathsf{A}$ as one that commutes with its Hermitian conjugate, so that

$$\mathsf{A}^\dagger\mathsf{A} = \mathsf{A}\mathsf{A}^\dagger.$$

We also showed that both Hermitian and unitary matrices (or symmetric and orthogonal matrices in the real case) are examples of normal matrices. We now discuss the properties of the eigenvectors and eigenvalues of a normal matrix.

If $\mathsf{x}$ is an eigenvector of a normal matrix $\mathsf{A}$ with corresponding eigenvalue $\lambda$, then $\mathsf{Ax} = \lambda\mathsf{x}$, or equivalently

$$(\mathsf{A} - \lambda\mathsf{I})\mathsf{x} = 0. \tag{7.61}$$

Denoting $\mathsf{B} = \mathsf{A} - \lambda\mathsf{I}$, (7.61) becomes $\mathsf{Bx} = 0$, and taking the Hermitian conjugate we also have

$$(\mathsf{Bx})^\dagger = \mathsf{x}^\dagger\mathsf{B}^\dagger = 0. \tag{7.62}$$

From (7.61) and (7.62) we then have

$$\mathsf{x}^\dagger \mathsf{B}^\dagger \mathsf{B}\mathsf{x} = 0. \tag{7.63}$$

However, the product $\mathsf{B}^\dagger\mathsf{B}$ is given by

$$\mathsf{B}^\dagger\mathsf{B} = (\mathsf{A} - \lambda\mathsf{I})^\dagger(\mathsf{A} - \lambda\mathsf{I}) = (\mathsf{A}^\dagger - \lambda^*\mathsf{I})(\mathsf{A} - \lambda\mathsf{I}) = \mathsf{A}^\dagger\mathsf{A} - \lambda^*\mathsf{A} - \lambda\mathsf{A}^\dagger + \lambda\lambda^*.$$

Now since $\mathsf{A}$ is normal, $\mathsf{A}\mathsf{A}^\dagger = \mathsf{A}^\dagger\mathsf{A}$ (see subsection 7.11.6) and so

$$\mathsf{B}^\dagger\mathsf{B} = \mathsf{A}\mathsf{A}^\dagger - \lambda^*\mathsf{A} - \lambda\mathsf{A}^\dagger + \lambda\lambda^* = (\mathsf{A} - \lambda\mathsf{I})(\mathsf{A} - \lambda\mathsf{I})^\dagger = \mathsf{B}\mathsf{B}^\dagger,$$

and hence $\mathsf{B}$ is also normal. From (7.63) we then find

$$\mathsf{x}^\dagger\mathsf{B}^\dagger\mathsf{B}\mathsf{x} = \mathsf{x}^\dagger\mathsf{B}\mathsf{B}^\dagger\mathsf{x} = (\mathsf{B}^\dagger\mathsf{x})^\dagger\mathsf{B}^\dagger\mathsf{x} = 0,$$

from which we obtain

$$\mathsf{B}^\dagger\mathsf{x} = (\mathsf{A}^\dagger - \lambda^*\mathsf{I})\mathsf{x} = 0.$$

Therefore, for a normal matrix $\mathsf{A}$, *the eigenvalues of* $\mathsf{A}^\dagger$ *are the complex conjugates of the eigenvalues of* $\mathsf{A}$.

Let us now consider two eigenvectors $\mathsf{x}^i$ and $\mathsf{x}^j$ of a normal matrix $\mathsf{A}$ corresponding to two *different* eigenvalues $\lambda_i$ and $\lambda_j$. We then have

$$\mathsf{A}\mathsf{x}^i = \lambda_i\mathsf{x}^i, \tag{7.64}$$

$$\mathsf{A}\mathsf{x}^j = \lambda_j\mathsf{x}^j. \tag{7.65}$$

Multiplying (7.65) on the left by $(\mathsf{x}^i)^\dagger$ we obtain

$$(\mathsf{x}^i)^\dagger\mathsf{A}\mathsf{x}^j = \lambda_j(\mathsf{x}^i)^\dagger\mathsf{x}^j. \tag{7.66}$$

However, on the LHS of (7.66) we have

$$(\mathsf{x}^i)^\dagger\mathsf{A} = (\mathsf{A}^\dagger\mathsf{x}^i)^\dagger = (\lambda_i^*\mathsf{x}^i)^\dagger = \lambda_i(\mathsf{x}^i)^\dagger, \tag{7.67}$$

where we have used (7.64) and the property just proved for a normal matrix to write $\mathsf{A}^\dagger\mathsf{x}^i = \lambda_i^*\mathsf{x}^i$. From (7.66) and (7.67) we have

$$(\lambda_i - \lambda_j)(\mathsf{x}^i)^\dagger\mathsf{x}^j = 0. \tag{7.68}$$

Thus, *if* $\lambda_i \neq \lambda_j$ *the eigenvectors* $\mathsf{x}^i$ *and* $\mathsf{x}^j$ *must be orthogonal*, i.e. $(\mathsf{x}^i)^\dagger\mathsf{x}^j = 0$.

It follows immediately from (7.68) that if all $N$ eigenvalues of a normal matrix $\mathsf{A}$ are distinct then all $N$ eigenvectors of $\mathsf{A}$ are mutually orthogonal. If, however, two or more eigenvalues are the same then further consideration is required. An eigenvalue corresponding to two (or more) different eigenvectors (one not simply a multiple of the other) is said to be *degenerate*. Suppose that $\lambda_1$ is $k$-fold degenerate, i.e.

$$\mathsf{A}\mathsf{x}^i = \lambda_1\mathsf{x}^i \quad \text{for } i = 1, 2, \ldots, k, \tag{7.69}$$

but that $\lambda_1$ is different from any of $\lambda_{k+1}$, $\lambda_{k+2}$, etc. Then any linear combination of these $\mathsf{x}^i$ is also an eigenvector with eigenvalue $\lambda_1$ since, for $\mathsf{z} = \sum_{i=1}^k c_i \mathsf{x}^i$,

$$\mathsf{Az} \equiv \mathsf{A} \sum_{i=1}^{k} c_i \mathsf{x}^i = \sum_{i=1}^{k} c_i \mathsf{A}\mathsf{x}^i = \sum_{i=1}^{k} c_i \lambda_1 \mathsf{x}^i = \lambda_1 \mathsf{z}. \tag{7.70}$$

If the $\mathsf{x}^i$ defined in (7.69) are not already mutually orthogonal, consider the new eigenvectors $\mathsf{z}^i$ constructed to be orthogonal by the following procedure, in which each of the new vectors is normalised to give $\hat{\mathsf{z}}^i$ before proceeding to the construction of the next one (the normalisation is carried out by dividing each element of the vector $\mathsf{z}^i$ by $[(\mathsf{z}^i)^\dagger \mathsf{z}^i]^{1/2}$):

$$\begin{aligned}
\mathsf{z}^1 &= \mathsf{x}^1, \\
\mathsf{z}^2 &= \mathsf{x}^2 - \left[(\hat{\mathsf{z}}^1)^\dagger \mathsf{x}^2\right] \hat{\mathsf{z}}^1, \\
\mathsf{z}^3 &= \mathsf{x}^3 - \left[(\hat{\mathsf{z}}^2)^\dagger \mathsf{x}^3\right] \hat{\mathsf{z}}^2 - \left[(\hat{\mathsf{z}}^1)^\dagger \mathsf{x}^3\right] \hat{\mathsf{z}}^1, \\
&\vdots \\
\mathsf{z}^k &= \mathsf{x}^k - \left[(\hat{\mathsf{z}}^{k-1})^\dagger \mathsf{x}^k\right] \hat{\mathsf{z}}^{k-1} - \cdots - \left[(\hat{\mathsf{z}}^1)^\dagger \mathsf{x}^k\right] \hat{\mathsf{z}}^1.
\end{aligned}$$

Each of the factors in brackets $(\hat{\mathsf{z}}^m)^\dagger \mathsf{x}^n$ is a scalar product and thus only a number; hence each new vector $\mathsf{z}^i$ is, as shown in (7.70), an eigenvector of $\mathsf{A}$ with eigenvalue $\lambda_1$ and will remain so on normalisation. It is straightforward to check that, provided the previous new eigenvectors have been normalised as prescribed, each $\mathsf{z}^i$ is orthogonal to all its predecessors. This method is called *Gram–Schmidt orthogonalisation*; however, in practice it is laborious and the example in subsection 7.13.1 gives a less rigorous but considerably quicker way.

Therefore, even if $\mathsf{A}$ has some degenerate eigenvalues we can *by construction* obtain a set of $N$ mutually orthogonal eigenvectors. Moreover, it may be shown (although the proof is beyond the scope of this book) that these eigenvectors are *complete* in that they form a basis for the $N$-dimensional vector space. As a result any arbitrary vector $\mathsf{y}$ can be expressed as a linear combination of the eigenvectors $\mathsf{x}^i$,

$$\mathsf{y} = \sum_{i=1}^{N} a_i \mathsf{x}^i,$$

i.e. the eigenvectors form an orthogonal basis for the vector space. By normalising the eigenvectors so that $(\mathsf{x}^i)^\dagger \mathsf{x}^i = 1$ this basis is made orthonormal.

### 7.12.2 *Eigenvectors and eigenvalues of Hermitian and anti-Hermitian matrices*

For a normal matrix we showed that if $\mathsf{A}\mathsf{x} = \lambda\mathsf{x}$ then $\mathsf{A}^\dagger\mathsf{x} = \lambda^*\mathsf{x}$. However, if A is also Hermitian, $\mathsf{A} = \mathsf{A}^\dagger$, it follows necessarily that $\lambda = \lambda^*$. Thus, the eigenvalues of an Hermitian matrix are real, a result which may be proved directly.

►*Prove that the eigenvalues of an Hermitian matrix are real.*

For any particular eigenvector $\mathsf{x}^i$, we take the Hermitian conjugate of $\mathsf{A}\mathsf{x}^i = \lambda_i\mathsf{x}^i$ to give

$$(\mathsf{x}^i)^\dagger\mathsf{A}^\dagger = \lambda_i^*(\mathsf{x}^i)^\dagger. \tag{7.71}$$

Using $\mathsf{A}^\dagger = \mathsf{A}$, since A is Hermitian, and multiplying on the right by $\mathsf{x}^i$, we obtain

$$(\mathsf{x}^i)^\dagger\mathsf{A}\mathsf{x}^i = \lambda_i^*(\mathsf{x}^i)^\dagger\mathsf{x}^i. \tag{7.72}$$

But multiplying $\mathsf{A}\mathsf{x}^i = \lambda_i\mathsf{x}^i$ through on the left by $(\mathsf{x}^i)^\dagger$ gives

$$(\mathsf{x}^i)^\dagger\mathsf{A}\mathsf{x}^i = \lambda_i(\mathsf{x}^i)^\dagger\mathsf{x}^i.$$

Subtracting this from (7.72) yields

$$0 = (\lambda_i^* - \lambda_i)(\mathsf{x}^i)^\dagger\mathsf{x}^i.$$

But $(\mathsf{x}^i)^\dagger\mathsf{x}^i$ is the modulus squared of the non-zero vector $\mathsf{x}^i$ and is thus non-zero. Hence $\lambda_i^*$ must equal $\lambda_i$ and thus be real. The same argument can be used to show that the eigenvalues of a real symmetric matrix are themselves real. ◄

The importance of the above result will be apparent to any student of quantum mechanics. In quantum mechanics the eigenvalues of operators correspond to measured values of observable quantities, e.g. energy, angular momentum, parity and so on, and these clearly must be real. If we use Hermitian operators to formulate the theories of quantum mechanics, the above property guarantees physically meaningful results.

Since an Hermitian matrix is also a normal matrix, its eigenvectors are orthogonal (or can be made so using the Gram–Schmidt orthogonalisation procedure). Alternatively we can prove the orthogonality of the eigenvectors directly.

►*Prove that the eigenvectors of an Hermitian matrix corresponding to different eigenvalues are orthogonal.*

Consider two unequal eigenvalues $\lambda_i$ and $\lambda_j$. It follows that

$$\mathsf{A}\mathsf{x}^i = \lambda_i\mathsf{x}^i, \tag{7.73}$$

$$\mathsf{A}\mathsf{x}^j = \lambda_j\mathsf{x}^j. \tag{7.74}$$

Taking the Hermitian conjugate of (7.73) we find $(\mathsf{x}^i)^\dagger \mathsf{A}^\dagger = \lambda_i^*(\mathsf{x}^i)^\dagger$. Multiplying this on the right by $\mathsf{x}^j$ we obtain

$$(\mathsf{x}^i)^\dagger \mathsf{A}^\dagger \mathsf{x}^j = \lambda_i^*(\mathsf{x}^i)^\dagger \mathsf{x}^j,$$

and similarly multiplying (7.74) through on the left by $(\mathsf{x}^i)^\dagger$ we find

$$(\mathsf{x}^i)^\dagger \mathsf{A} \mathsf{x}^j = \lambda_j(\mathsf{x}^i)^\dagger \mathsf{x}^j.$$

Then, using the fact that $\mathsf{A}^\dagger = \mathsf{A}$, the two left-hand sides are equal and, since the $\lambda_i$ are real, on subtraction we obtain

$$0 = (\lambda_i - \lambda_j)(\mathsf{x}^i)^\dagger \mathsf{x}^j.$$

Finally we note that $\lambda_i \neq \lambda_j$ and so $(\mathsf{x}^i)^\dagger \mathsf{x}^j = 0$, i.e. the eigenvectors $\mathsf{x}^i$ and $\mathsf{x}^j$ are orthogonal. ◀

In the case where some of the eigenvalues are equal, further justification of the orthogonality of the eigenvectors is needed. The Gram–Schmidt orthogonalisation procedure discussed above provides a proof of, and a means of achieving, orthogonality. The general method has already been described and we shall not repeat it here.

We may also consider the properties of the eigenvalues and eigenvectors of an anti-Hermitian matrix, for which $\mathsf{A}^\dagger = -\mathsf{A}$ and so

$$\mathsf{A}\mathsf{A}^\dagger = \mathsf{A}(-\mathsf{A}) = (-\mathsf{A})\mathsf{A} = \mathsf{A}^\dagger\mathsf{A}.$$

Therefore matrices that are anti-Hermitian are also normal and so have mutually orthogonal eigenvectors. The properties of the eigenvalues are also simply deduced, since if $\mathsf{A}\mathsf{x} = \lambda\mathsf{x}$ then

$$\lambda^*\mathsf{x} = \mathsf{A}^\dagger\mathsf{x} = -\mathsf{A}\mathsf{x} = -\lambda\mathsf{x}.$$

Hence $\lambda^* = -\lambda$ and so $\lambda$ must be *pure imaginary* (or *zero*). In a similar manner to that used for Hermitian matrices, these properties may be proved directly.

### *7.12.3 Eigenvectors and eigenvalues of a unitary matrix*

A unitary matrix satisfies $\mathsf{A}^\dagger = \mathsf{A}^{-1}$ and is also a normal matrix, with mutually orthogonal eigenvectors. Regarding the eigenvalues of a unitary matrix, if $\mathsf{A}\mathsf{x} = \lambda\mathsf{x}$ then

$$\mathsf{x}^\dagger\mathsf{x} = \mathsf{x}^\dagger\mathsf{A}^\dagger\mathsf{A}\mathsf{x} = \lambda^*\lambda\mathsf{x}^\dagger\mathsf{x},$$

and we deduce that $\lambda\lambda^* = |\lambda|^2 = 1$. Thus, the eigenvalues of a unitary matrix have unit modulus.

### 7.12.4 *Eigenvectors and eigenvalues of a general square matrix*

When an $N \times N$ matrix is not normal there are no general properties of its eigenvalues and eigenvectors. In general it is not possible to find any orthogonal set of $N$ eigenvectors, or even *pairs* of orthogonal eigenvectors (except by chance in some cases). While the $N$ non-orthogonal eigenvectors are usually linearly independent, and hence form a basis for the $N$-dimensional vector space, this is not necessarily so. It may be shown (although we shall not prove it) that any $N \times N$ matrix with *distinct* eigenvalues has $N$ linearly independent eigenvectors, which therefore form a basis for the $N$-dimensional vector space. If a general square matrix has degenerate eigenvalues, however, then it may or may not have $N$ linearly independent eigenvectors. A matrix whose eigenvectors are not linearly independent is said to be *defective*.

## 7.13 Determination of eigenvalues and eigenvectors

The next step is to show how the eigenvalues and eigenvectors of a given $N \times N$ matrix $\mathsf{A}$ are found. To do this we refer to the definition (7.60) and as in (7.61) rewrite it as

$$\mathsf{A}\mathsf{x} - \lambda\mathsf{I}\mathsf{x} = (\mathsf{A} - \lambda\mathsf{I})\mathsf{x} = 0. \tag{7.75}$$

The slight rearrangement used here is to write $\mathsf{I}\mathsf{x}$ instead of $\mathsf{x}$ where $\mathsf{I}$ is the unit matrix of order $N$. The point of doing this is immediate since (7.75) now has the form of a homogeneous set of simultaneous equations, the theory of which was developed in section 7.10. What was proved there was that the equation $\mathsf{B}\mathsf{x} = 0$ only has a non-trivial solution $\mathsf{x}$ if $|\mathsf{B}| = 0$. Correspondingly, therefore, we must have in the present case that

$$|\mathsf{A} - \lambda\mathsf{I}| = 0, \tag{7.76}$$

if there are to be non-zero solutions $\mathsf{x}$ to (7.75).

Equation (7.76) is known as the *characteristic equation* for $\mathsf{A}$ and its LHS as the *characteristic* or *secular determinant* of $\mathsf{A}$. The equation is a polynomial (of degree $N$) in the quantity $\lambda$. The $N$ roots of this equation $\lambda_i$, $i = 1, 2, \ldots, N$, give the eigenvalues of $\mathsf{A}$. Corresponding to each $\lambda_i$ there will be a column vector $\mathsf{x}^i$, which is the $i$th eigenvector of $\mathsf{A}$ and can be found by using (7.60).

It will be observed that when (7.76) is written out as a polynomial equation in $\lambda$, the coefficient of $-\lambda^{N-1}$ in the equation will be simply $A_{11} + A_{22} + \cdots + A_{NN}$ relative to the coefficient of $\lambda^N$. The quantity $\sum_{i=1}^{N} A_{ii}$ is called the *trace* or *spur* of $\mathsf{A}$ and, from the ordinary theory of polynomial equations, will be equal to the sum of the roots of (7.76):

$$\sum_{i=1}^{N} \lambda_i = \mathrm{Tr}\,\mathsf{A}. \tag{7.77}$$

This can be used as one check that a computation of the eigenvalues has been done correctly. Another proof of (7.77) is given in section 7.15.

►*Find the eigenvalues and normalised eigenvectors of the real symmetric matrix*

$$\mathsf{A} = \begin{pmatrix} 1 & 1 & 3 \\ 1 & 1 & -3 \\ 3 & -3 & -3 \end{pmatrix}.$$

Using (7.76),

$$\begin{vmatrix} 1-\lambda & 1 & 3 \\ 1 & 1-\lambda & -3 \\ 3 & -3 & -3-\lambda \end{vmatrix} = 0.$$

Expanding out this determinant gives

$$(1-\lambda)\,[(1-\lambda)(-3-\lambda)-(-3)(-3)] + 1\,[(-3)(3)-1(-3-\lambda)] \\ + 3\,[1(-3)-(1-\lambda)(3)] = 0,$$

which simplifies to give

$$(1-\lambda)(\lambda^2+2\lambda-12)+(\lambda-6)+3(3\lambda-6)=0,$$
$$\Rightarrow \quad (\lambda-2)(\lambda-3)(\lambda+6)=0.$$

Hence the roots of the characteristic equation are $\lambda_1 = 2$, $\lambda_2 = 3$, $\lambda_3 = -6$, which are thus the eigenvalues of A. We note that, as expected,

$$\lambda_1+\lambda_2+\lambda_3 = -1 = 1+1-3 = A_{11}+A_{22}+A_{33} = \mathrm{Tr}\,\mathsf{A}.$$

For the first root, $\lambda_1 = 2$, a suitable eigenvector $\mathsf{x}^1$, with elements $x_1$, $x_2$, $x_3$, must satisfy $\mathsf{A}\mathsf{x}^1 = 2\mathsf{x}^1$, or equivalently

$$\begin{aligned} x_1+x_2+3x_3 &= 2x_1, \\ x_1+x_2-3x_3 &= 2x_2, \\ 3x_1-3x_2-3x_3 &= 2x_3. \end{aligned} \tag{7.78}$$

These three equations are consistent (to ensure this was the purpose in finding the particular values of $\lambda$) and yield $x_3 = 0$, $x_1 = x_2 = k$, where $k$ is any non-zero number. A suitable eigenvector would thus be

$$\mathsf{x}^1 = (k \;\; k \;\; 0)^{\mathrm{T}}.$$

If we apply the normalisation condition, we require $k^2+k^2+0^2 = 1$ or $k = \frac{1}{\sqrt{2}}$. Hence

$$\mathsf{x}^1 = \left(\tfrac{1}{\sqrt{2}} \;\; \tfrac{1}{\sqrt{2}} \;\; 0\right)^{\mathrm{T}} = \tfrac{1}{\sqrt{2}}(1 \;\; 1 \;\; 0)^{\mathrm{T}}.$$

Repeating the last paragraph, but with the factor 2 on the RHS of (7.78) replaced successively by $\lambda_2 = 3$ and $\lambda_3 = -6$, gives two further normalised eigenvectors

$$\mathsf{x}^2 = \tfrac{1}{\sqrt{3}}(1 \quad -1 \quad 1)^{\mathrm{T}}, \qquad \mathsf{x}^3 = \tfrac{1}{\sqrt{6}}(1 \quad -1 \quad -2)^{\mathrm{T}}. \blacktriangleleft$$

In the above example, the three values of $\lambda$ are all different and $\mathsf{A}$ is a real symmetric matrix. We thus expect, and it is easily checked, that the three eigenvectors are mutually orthogonal, i.e.

$$\left(\mathsf{x}^1\right)^{\mathrm{T}}\mathsf{x}^2 = \left(\mathsf{x}^1\right)^{\mathrm{T}}\mathsf{x}^3 = \left(\mathsf{x}^2\right)^{\mathrm{T}}\mathsf{x}^3 = 0.$$

It will also be apparent that the normalisation of the eigenvectors has no effect on their orthogonality, as expected.

### *7.13.1 Degenerate eigenvalues*

We now return to the case of degenerate eigenvalues, i.e. those that have two or more associated eigenvectors. We have already shown that it is always possible to construct an orthogonal set of eigenvectors for a normal matrix, see subsection 7.12.1, and the following example illustrates one method for constructing such a set.

►*Construct an orthonormal set of eigenvectors for the matrix*

$$\mathsf{A} = \begin{pmatrix} 1 & 0 & 3 \\ 0 & -2 & 0 \\ 3 & 0 & 1 \end{pmatrix}.$$

We first evaluate the eigenvalues using $|\mathsf{A} - \lambda\mathsf{I}| = 0$:

$$0 = \begin{vmatrix} 1-\lambda & 0 & 3 \\ 0 & -2-\lambda & 0 \\ 3 & 0 & 1-\lambda \end{vmatrix} = -(1-\lambda)^2(2+\lambda) + 3(3)(2+\lambda)$$

$$= (4-\lambda)(\lambda+2)^2.$$

Thus $\lambda_1 = 4$, $\lambda_2 = -2 = \lambda_3$. The eigenvector $\mathsf{x}^1 = (x_1 \quad x_2 \quad x_3)^{\mathrm{T}}$ is determined by

$$\begin{pmatrix} 1 & 0 & 3 \\ 0 & -2 & 0 \\ 3 & 0 & 1 \end{pmatrix}\begin{pmatrix} x_1 \\ x_2 \\ x_3 \end{pmatrix} = 4\begin{pmatrix} x_1 \\ x_2 \\ x_3 \end{pmatrix} \quad \Rightarrow \quad \mathsf{x}^1 = \frac{1}{\sqrt{2}}\begin{pmatrix} 1 \\ 0 \\ 1 \end{pmatrix}.$$

A general column vector that is orthogonal to $\mathsf{x}^1$ is

$$\mathsf{x} = (a \quad b \quad -a)^{\mathrm{T}}, \tag{7.79}$$

and it is easily shown that

$$\mathsf{Ax} = \begin{pmatrix} 1 & 0 & 3 \\ 0 & -2 & 0 \\ 3 & 0 & 1 \end{pmatrix} \begin{pmatrix} a \\ b \\ -a \end{pmatrix} = -2 \begin{pmatrix} a \\ b \\ -a \end{pmatrix} = -2\mathsf{x}.$$

Thus $\mathsf{x}$ is a eigenvector of $\mathsf{A}$ with associated eigenvalue $-2$. It is clear, however, that there is an infinite set of eigenvectors $\mathsf{x}$ all possessing the required property; the geometrical analogue is that there are an infinite number of corresponding vectors $\mathbf{x}$ in the plane having $\mathbf{x}^1$ as its normal. We do require that the two remaining eigenvectors are orthogonal to one another, but this still leaves an infinite number of possibilities. For $\mathsf{x}^2$, therefore, let us choose a simple form of (7.79), suitably normalised, say,

$$\mathsf{x}^2 = (0 \quad 1 \quad 0)^{\mathrm{T}}.$$

The third eigenvector is then also specified (to within an arbitrary multiplicative constant), since it must be orthogonal to $\mathsf{x}^1$ and $\mathsf{x}^2$; thus $\mathsf{x}^3$ may be found by evaluating the vector product of $\mathsf{x}^1$ and $\mathsf{x}^2$ and normalising the result. This gives

$$\mathsf{x}^3 = \frac{1}{\sqrt{2}}(-1 \quad 0 \quad 1)^{\mathrm{T}},$$

to complete the construction of an orthonormal set of eigenvectors. ◄

## 7.14 Change of basis and similarity transformations

Throughout this chapter we have considered the vector $\mathbf{x}$ as a geometrical quantity which is independent of any basis (or coordinate system). If we introduce a basis $\mathbf{e}_i$, $i = 1, \ldots, N$, into our $N$-dimensional vector space then we may write

$$\mathbf{x} = x_1\mathbf{e}_1 + \cdots + x_N\mathbf{e}_N,$$

and represent $\mathbf{x}$ in this basis by the column matrix

$$\mathsf{x} = (x_1 \quad \cdots \quad x_n)^{\mathrm{T}},$$

containing the components $x_i$. We now consider how these components change as a result of a prescribed change of basis. Let us introduce a new basis $\mathbf{e}'_i$, $i = 1, \ldots, N$, which is related to the old basis by

$$\mathbf{e}'_j = \sum_{i=1}^{N} S_{ij}\mathbf{e}_i, \tag{7.80}$$

the coefficient $S_{ij}$ being the $i$th component of $\mathbf{e}'_j$ with respect to the old (unprimed) basis. For an arbitrary vector $\mathbf{x}$ it follows that

$$\mathbf{x} = \sum_{i=1}^{N} x_i\mathbf{e}_i = \sum_{j=1}^{N} x'_j\mathbf{e}'_j = \sum_{j=1}^{N} x'_j \sum_{i=1}^{N} S_{ij}\mathbf{e}_i.$$

From this we derive the relation between the components of $\mathbf{x}$ in the two coordinate systems as

$$x_i = \sum_{j=1}^{N} S_{ij} x'_j,$$

which we can write in matrix form as

$$\mathsf{x} = \mathsf{S}\mathsf{x}' \tag{7.81}$$

where $\mathsf{S}$ is the *transformation matrix* associated with the change of basis.

Furthermore, since the vectors $\mathbf{e}'_j$ are linearly independent, the matrix $\mathsf{S}$ is non-singular and so possesses an inverse $\mathsf{S}^{-1}$. Multiplying (7.81) on the left by $\mathsf{S}^{-1}$ we find

$$\mathsf{x}' = \mathsf{S}^{-1}\mathsf{x}, \tag{7.82}$$

which relates the components of $\mathbf{x}$ in the new basis with those in the old basis. Comparing (7.82) and (7.80) we note that the components of $\mathbf{x}$ transform inversely to the way the basis vectors $\mathbf{e}_i$ themselves transform. This occurs because $\mathbf{x}$ itself is unchanged.

We may also find the transformation law for the components of a linear operator under the same change of basis. Now, the operator equation $\mathbf{y} = \mathcal{A}\mathbf{x}$ (which is basis independent) can be written as a matrix equation in each of the two bases as

$$\mathsf{y} = \mathsf{A}\mathsf{x}, \qquad \mathsf{y}' = \mathsf{A}'\mathsf{x}'. \tag{7.83}$$

But, using (7.81), we may rewrite the first equation as

$$\mathsf{S}\mathsf{y}' = \mathsf{A}\mathsf{S}\mathsf{x}' \quad \Rightarrow \quad \mathsf{y}' = \mathsf{S}^{-1}\mathsf{A}\mathsf{S}\mathsf{x}'.$$

Comparing this with the second equation in (7.83) we find that the components of the linear operator $\mathcal{A}$ must transform as

$$\mathsf{A}' = \mathsf{S}^{-1}\mathsf{A}\mathsf{S}. \tag{7.84}$$

Equation (7.84) is an example of a *similarity transformation* – a transformation that can be particularly useful in converting matrices into convenient forms for computation.

Given a square matrix $\mathsf{A}$, we may interpret it as representing the linear operator $\mathcal{A}$ in a given basis $\mathbf{e}_i$. From (7.84), however, we may also consider the matrix $\mathsf{A}' = \mathsf{S}^{-1}\mathsf{A}\mathsf{S}$, for any non-singular matrix $\mathsf{S}$, as representing the same linear operator $\mathcal{A}$ but in a new basis $\mathbf{e}'_j$ related to the old basis by

$$\mathbf{e}'_j = \sum_i S_{ij}\mathbf{e}_i,$$

Therefore we would expect that any property of the matrix $\mathsf{A}$ which represents some (basis-independent) property of the linear operator it describes will also be shared by $\mathsf{A}'$. We list these properties below.

(i) If $\mathsf{A} = \mathsf{I}$ then $\mathsf{A}' = \mathsf{I}$, since, from (7.84),

$$\mathsf{A}' = \mathsf{S}^{-1}\mathsf{I}\mathsf{S} = \mathsf{S}^{-1}\mathsf{S} = \mathsf{I}. \tag{7.85}$$

(ii) The value of the determinant is unchanged:

$$|\mathsf{A}'| = |\mathsf{S}^{-1}\mathsf{A}\mathsf{S}| = |\mathsf{S}^{-1}||\mathsf{A}||\mathsf{S}| = |\mathsf{A}||\mathsf{S}^{-1}||\mathsf{S}| = |\mathsf{A}||\mathsf{S}^{-1}\mathsf{S}| = |\mathsf{A}|. \tag{7.86}$$

(iii) The characteristic determinant and hence the eigenvalues of $\mathsf{A}'$ are the same as those of $\mathsf{A}$: from (7.76),

$$\begin{aligned}|\mathsf{A}' - \lambda\mathsf{I}| &= |\mathsf{S}^{-1}\mathsf{A}\mathsf{S} - \lambda\mathsf{I}| = |\mathsf{S}^{-1}(\mathsf{A} - \lambda\mathsf{I})\mathsf{S}| \\ &= |\mathsf{S}^{-1}||\mathsf{S}||\mathsf{A} - \lambda\mathsf{I}| = |\mathsf{A} - \lambda\mathsf{I}|.\end{aligned} \tag{7.87}$$

(iv) The value of the trace is unchanged: from (7.77),

$$\begin{aligned}\mathrm{Tr}\,\mathsf{A}' &= \sum_i A'_{ii} = \sum_i\sum_j\sum_k (\mathsf{S}^{-1})_{ij}A_{jk}S_{ki} \\ &= \sum_i\sum_j\sum_k S_{ki}(\mathsf{S}^{-1})_{ij}A_{jk} = \sum_j\sum_k \delta_{kj}A_{jk} = \sum_j A_{jj} \\ &= \mathrm{Tr}\,\mathsf{A}.\end{aligned} \tag{7.88}$$

An important class of similarity transformations is that for which $\mathsf{S}$ is a unitary matrix and so $\mathsf{A}' = \mathsf{S}^{-1}\mathsf{A}\mathsf{S} = \mathsf{S}^\dagger\mathsf{A}\mathsf{S}$. Unitary transformation matrices are particularly important for the following reason. If the original basis $\mathbf{e}_i$ is orthonormal and the transformation matrix $\mathsf{S}$ is unitary then

$$\begin{aligned}\langle \mathbf{e}'_i|\mathbf{e}'_j\rangle &= \left\langle \sum_k S_{ki}\mathbf{e}_k \middle| \sum_r S_{rj}\mathbf{e}_r\right\rangle \\ &= \sum_k S^*_{ki}\sum_r S_{rj}\langle \mathbf{e}_k|\mathbf{e}_r\rangle \\ &= \sum_k S^*_{ki}\sum_r S_{rj}\delta_{kr} = \sum_k S^*_{ki}S_{kj} = (\mathsf{S}^\dagger\mathsf{S})_{ij} = \delta_{ij},\end{aligned}$$

showing that the new basis is also orthonormal.

Furthermore, in addition to the properties of general similarity transformations, for unitary transformations the following also hold.

(i) If $\mathsf{A}$ is Hermitian (anti-Hermitian) then $\mathsf{A}'$ is Hermitian (anti-Hermitian), i.e. if $\mathsf{A}^\dagger = \pm\mathsf{A}$, then

$$(\mathsf{A}')^\dagger = (\mathsf{S}^\dagger\mathsf{A}\mathsf{S})^\dagger = \mathsf{S}^\dagger\mathsf{A}^\dagger\mathsf{S} = \pm\mathsf{S}^\dagger\mathsf{A}\mathsf{S} = \pm\mathsf{A}'. \tag{7.89}$$

(ii) If $\mathsf{A}$ is unitary (so that $\mathsf{A}^\dagger = \mathsf{A}^{-1}$), then $\mathsf{A}'$ is unitary, since

$$(\mathsf{A}')^\dagger \mathsf{A}' = (\mathsf{S}^\dagger \mathsf{A} \mathsf{S})^\dagger (\mathsf{S}^\dagger \mathsf{A} \mathsf{S}) = \mathsf{S}^\dagger \mathsf{A}^\dagger \mathsf{S} \mathsf{S}^\dagger \mathsf{A} \mathsf{S} = \mathsf{S}^\dagger \mathsf{A}^\dagger \mathsf{A} \mathsf{S}$$
$$= \mathsf{S}^\dagger \mathsf{I} \mathsf{S} = \mathsf{I}. \qquad (7.90)$$

### 7.15 Diagonalisation of matrices

Suppose a linear operator $\mathcal{A}$ is represented in some basis $\mathbf{e}_i$, $i = 1, \ldots, N$, by the matrix $\mathsf{A}$, which contains the components of the linear operator with respect to this basis. Let us consider a new basis $\mathbf{x}^j$ given by

$$\mathbf{x}^j = \sum_{i=1}^{N} S_{ij} \mathbf{e}_i,$$

where the $\mathbf{x}^j$ are chosen to be the eigenvectors of the linear operator $\mathcal{A}$, i.e.

$$\mathcal{A} \mathbf{x}^j = \lambda_j \mathbf{x}^j. \qquad (7.91)$$

In the new basis, $\mathcal{A}$ is represented by the matrix $\mathsf{A}' = \mathsf{S}^{-1} \mathsf{A} \mathsf{S}$, which has a particularly simple form. The element $S_{ij}$ is the $i$th component, in the old (unprimed) basis, of the $j$th eigenvector $\mathsf{x}^j$, i.e. the columns of $\mathsf{S}$ are the eigenvectors of the matrix $\mathsf{A}$:

$$\mathsf{S} = \begin{pmatrix} \uparrow & \uparrow & & \uparrow \\ \mathsf{x}^1 & \mathsf{x}^2 & \cdots & \mathsf{x}^N \\ \downarrow & \downarrow & & \downarrow \end{pmatrix},$$

or $S_{ij} = (\mathsf{x}^j)_i$. Therefore $\mathsf{A}'$ is given by

$$\begin{aligned}
(\mathsf{S}^{-1} \mathsf{A} \mathsf{S})_{ij} &= \sum_k \sum_l (\mathsf{S}^{-1})_{ik} A_{kl} S_{lj} \\
&= \sum_k \sum_l (\mathsf{S}^{-1})_{ik} A_{kl} (\mathsf{x}^j)_l \\
&= \sum_k (\mathsf{S}^{-1})_{ik} \lambda_j (\mathsf{x}^j)_k \\
&= \sum_k \lambda_j (\mathsf{S}^{-1})_{ik} S_{kj} = \lambda_j \delta_{ij}.
\end{aligned}$$

So the matrix $\mathsf{A}'$ is diagonal with the eigenvalues of $\mathcal{A}$ as the diagonal elements, i.e.

$$\mathsf{A}' = \begin{pmatrix} \lambda_1 & & \mathbf{0} \\ & \ddots & \\ \mathbf{0} & & \lambda_N \end{pmatrix}.$$

Therefore, given a matrix $\mathsf{A}$, if we construct the matrix $\mathsf{S}$ that has the eigenvectors of $\mathsf{A}$ as its columns, then the matrix $\mathsf{A}' = \mathsf{S}^{-1} \mathsf{A} \mathsf{S}$ is diagonal with the eigenvalues of A as the diagonal elements. Since we require S to be non-singular ($|\mathsf{S}| \neq 0$), the $N$ eigenvectors of A must be linearly independent and form a basis for the $N$-dimensional vector space. It may be shown that *any matrix with distinct eigenvalues* may be diagonalised by this procedure. If, however, a general square matrix has degenerate eigenvalues then it may, or may not, have $N$ linearly independent eigenvectors. If it does not then it *cannot* be diagonalised.

For normal matrices (which include Hermitian, anti-Hermitian and unitary matrices) the $N$ eigenvectors are indeed linearly independent. Moreover, when normalised, these eigenvectors form an *orthonormal* set (or can be made to do so). Therefore the matrix S with these normalised eigenvectors as columns, so that its elements are $S_{ij} = (\mathsf{x}^j)_i$, has the property

$$(\mathsf{S}^\dagger\mathsf{S})_{ij} = \sum_k (\mathsf{S}^\dagger)_{ik}(\mathsf{S})_{kj} = \sum_k S^*_{ki}S_{kj} = \sum_k (\mathsf{x}^i)^*_k(\mathsf{x}^j)_k = (\mathsf{x}^i)^\dagger\mathsf{x}^j = \delta_{ij}.$$

Hence S is unitary ($\mathsf{S}^{-1} = \mathsf{S}^\dagger$) and the original matrix A can be diagonalised by

$$\mathsf{A}' = \mathsf{S}^{-1}\mathsf{A}\mathsf{S} = \mathsf{S}^\dagger\mathsf{A}\mathsf{S}.$$

Therefore, any normal matrix A can be diagonalised by a similarity transformation using a *unitary* transformation matrix S.

►*Diagonalise the matrix*

$$\mathsf{A} = \begin{pmatrix} 1 & 0 & 3 \\ 0 & -2 & 0 \\ 3 & 0 & 1 \end{pmatrix}.$$

The matrix A is symmetric and so may be diagonalised by a transformation of the form $\mathsf{A}' = \mathsf{S}^\dagger\mathsf{A}\mathsf{S}$, where S has the normalised eigenvectors of A as its columns. We have already found these eigenvectors (see subsection 7.13.1) and so

$$\mathsf{S} = \frac{1}{\sqrt{2}}\begin{pmatrix} 1 & 0 & -1 \\ 0 & \sqrt{2} & 0 \\ 1 & 0 & 1 \end{pmatrix}.$$

We note that although the eigenvalues of A are degenerate, its three eigenvectors are linearly independent and so A can still be diagonalised. Thus, calculating $\mathsf{S}^\dagger\mathsf{A}\mathsf{S}$ we obtain

$$\mathsf{S}^\dagger\mathsf{A}\mathsf{S} = \frac{1}{2}\begin{pmatrix} 1 & 0 & 1 \\ 0 & \sqrt{2} & 0 \\ -1 & 0 & 1 \end{pmatrix}\begin{pmatrix} 1 & 0 & 3 \\ 0 & -2 & 0 \\ 3 & 0 & 1 \end{pmatrix}\begin{pmatrix} 1 & 0 & -1 \\ 0 & \sqrt{2} & 0 \\ 1 & 0 & 1 \end{pmatrix}$$

$$= \begin{pmatrix} 4 & 0 & 0 \\ 0 & -2 & 0 \\ 0 & 0 & -2 \end{pmatrix},$$

which is diagonal, as required, and has as its diagonal elements the eigenvalues of $\mathsf{A}$. ◀

If a matrix $\mathsf{A}$ is diagonalised by the similarity transformation $\mathsf{A}' = \mathsf{S}^{-1}\mathsf{A}\mathsf{S}$, so that $\mathsf{A}' = \mathrm{diag}(\lambda_1, \ldots, \lambda_N)$, then we immediately have

$$\mathrm{Tr}\,\mathsf{A}' = \mathrm{Tr}\,\mathsf{A} = \sum_{i=1}^{N} \lambda_i,$$

$$|\mathsf{A}'| = |\mathsf{A}| = \prod_{i=1}^{N} \lambda_i,$$

since the eigenvalues of the matrix are unchanged by the transformation.

## 7.16 Quadratic and Hermitian forms

Let us now introduce the concept of quadratic forms (and their complex analogues, Hermitian forms). A quadratic form $Q$ is a scalar function of a real vector $\mathbf{x}$ given by

$$Q(\mathbf{x}) = \langle \mathbf{x} | \mathcal{A}\, \mathbf{x} \rangle, \tag{7.92}$$

for some real linear operator $\mathcal{A}$. In any given basis (coordinate system) we can write (7.92) in matrix form as

$$Q(\mathsf{x}) = \mathsf{x}^{\mathrm{T}}\mathsf{A}\mathsf{x}. \tag{7.93}$$

We need only consider the case where $\mathsf{A}$ is symmetric (see below), i.e. $\mathsf{A} = \mathsf{A}^{\mathrm{T}}$. As an example in a three-dimensional space,

$$\begin{aligned} Q = \mathsf{x}^{\mathrm{T}}\mathsf{A}\mathsf{x} &= \begin{pmatrix} x_1 & x_2 & x_3 \end{pmatrix} \begin{pmatrix} 1 & 1 & 3 \\ 1 & 1 & -3 \\ 3 & -3 & -3 \end{pmatrix} \begin{pmatrix} x_1 \\ x_2 \\ x_3 \end{pmatrix} \\ &= x_1^2 + x_2^2 - 3x_3^2 + 2x_1x_2 + 6x_1x_3 - 6x_2x_3. \end{aligned} \tag{7.94}$$

It is reasonable to ask whether a quadratic form $Q = \mathsf{x}^{\mathrm{T}}\mathsf{M}\mathsf{x}$, where $\mathsf{M}$ is any (possibly non-symmetric) real square matrix, is a more general definition. This is not the case since we may express $\mathsf{M}$ in terms of a symmetric matrix $\mathsf{A} = \frac{1}{2}(\mathsf{M}+\mathsf{M}^{\mathrm{T}})$ and an antisymmetric matrix $\mathsf{B} = \frac{1}{2}(\mathsf{M}-\mathsf{M}^{\mathrm{T}})$, such that $\mathsf{M} = \mathsf{A}+\mathsf{B}$. We then have

$$Q = \mathsf{x}^{\mathrm{T}}\mathsf{M}\mathsf{x} = \mathsf{x}^{\mathrm{T}}\mathsf{A}\mathsf{x} + \mathsf{x}^{\mathrm{T}}\mathsf{B}\mathsf{x}. \tag{7.95}$$

But $Q$ is a scalar quantity and so

$$Q = Q^{\mathrm{T}} = (\mathsf{x}^{\mathrm{T}}\mathsf{A}\mathsf{x})^{\mathrm{T}} + (\mathsf{x}^{\mathrm{T}}\mathsf{B}\mathsf{x})^{\mathrm{T}} = \mathsf{x}^{\mathrm{T}}\mathsf{A}^{\mathrm{T}}\mathsf{x} + \mathsf{x}^{\mathrm{T}}\mathsf{B}^{\mathrm{T}}\mathsf{x} = \mathsf{x}^{\mathrm{T}}\mathsf{A}\mathsf{x} - \mathsf{x}^{\mathrm{T}}\mathsf{B}\mathsf{x}. \tag{7.96}$$

Comparing (7.95) and (7.96) shows that $\mathsf{x}^{\mathrm{T}}\mathsf{B}\mathsf{x} = 0$, and hence $\mathsf{x}^{\mathrm{T}}\mathsf{M}\mathsf{x} = \mathsf{x}^{\mathrm{T}}\mathsf{A}\mathsf{x}$, i.e.

$Q$ is unchanged by considering only the symmetric part of M. Hence, with no loss of generality, we may assume $\mathsf{A} = \mathsf{A}^{\mathrm{T}}$ in (7.93).

From its definition (7.92), $Q$ is clearly a basis- (i.e. coordinate-) independent quantity. Let us therefore consider a new basis related to the old one by an orthogonal transformation matrix S, so that the components of any vector **x** in the two bases are related (as in (7.81)) by $\mathsf{x} = \mathsf{S}\mathsf{x}'$, or equivalently $\mathsf{x}' = \mathsf{S}^{-1}\mathsf{x} = \mathsf{S}^{\mathrm{T}}\mathsf{x}$. We then have

$$Q = \mathsf{x}^{\mathrm{T}}\mathsf{A}\mathsf{x} = (\mathsf{x}')^{\mathrm{T}}\mathsf{S}^{\mathrm{T}}\mathsf{A}\mathsf{S}\mathsf{x}' = (\mathsf{x}')^{\mathrm{T}}\mathsf{A}'\mathsf{x}',$$

where (as expected) the matrix describing the linear operator $\mathcal{A}$ in the new basis is given by $\mathsf{A}' = \mathsf{S}^{\mathrm{T}}\mathsf{A}\mathsf{S}$ (since $\mathsf{S}^{\mathrm{T}} = \mathsf{S}^{-1}$). But, from the last section, if we choose as S the matrix whose columns are the *normalised* eigenvectors of A, then $\mathsf{A}' = \mathsf{S}^{\mathrm{T}}\mathsf{A}\mathsf{S}$ is diagonal with the eigenvalues of A as the diagonal elements. (Since A is symmetric its normalised eigenvectors are orthogonal (or can be made so) and hence S is orthogonal with $\mathsf{S}^{-1} = \mathsf{S}^{\mathrm{T}}$.)

In this basis

$$Q = \mathsf{x}^{\mathrm{T}}\mathsf{A}\mathsf{x} = (\mathsf{x}')^{\mathrm{T}}\Lambda\mathsf{x}' = \lambda_1 x_1'^2 + \lambda_2 x_2'^2 + \cdots + \lambda_N x_N'^2,$$

where $\Lambda = \mathrm{diag}(\lambda_1, \lambda_2, \ldots, \lambda_N)$, $\lambda_i$ are the eigenvalues of A and $Q$ contains no cross-terms of the form $x_1'x_2'$, etc.

►*Find an orthogonal transformation that takes the quadratic form (7.94) into the form* $\lambda_1 x_1'^2 + \lambda_2 x_2'^2 + \lambda_3 x_3'^2$.

The required transformation matrix S has the *normalised* eigenvectors of A as its columns. We have already found these in section 7.13, and so

$$\mathsf{S} = \frac{1}{\sqrt{6}}\begin{pmatrix} \sqrt{3} & \sqrt{2} & 1 \\ \sqrt{3} & -\sqrt{2} & -1 \\ 0 & \sqrt{2} & -2 \end{pmatrix},$$

which is easily verified as being orthogonal. Since the eigenvalues of A are $\lambda = 2, 3, -6$, the general result already proved shows that the transformation $\mathsf{x} = \mathsf{S}\mathsf{x}'$ will carry (7.94) into the form $2x_1'^2 + 3x_2'^2 - 6x_3'^2$. This may be verified most easily by writing out the inverse transformation $\mathsf{x}' = \mathsf{S}^{-1}\mathsf{x} = \mathsf{S}^{\mathrm{T}}\mathsf{x}$ and substituting. The inverse equations are

$$\begin{aligned} x_1' &= (x_1 + x_2)/\sqrt{2}, \\ x_2' &= (x_1 - x_2 + x_3)/\sqrt{3}, \qquad\qquad (7.97) \\ x_3' &= (x_1 - x_2 - 2x_3)/\sqrt{6}. \end{aligned}$$

If these are substituted into the form $Q = 2x_1'^2 + 3x_2'^2 - 6x_3'^2$, the original expression (7.94) is recovered. ◄

In the definition of $Q$ it was assumed that the components $x_1$, $x_2$, $x_3$ and the matrix $\mathsf{A}$ were real. It is clear that in this case the quadratic form $Q \equiv \mathsf{x}^{\mathrm{T}}\mathsf{A}\mathsf{x}$ is real also. Another, rather more general, expression that is also real is the *Hermitian form*

$$H(\mathsf{x}) \equiv \mathsf{x}^{\dagger}\mathsf{A}\mathsf{x}, \tag{7.98}$$

where $\mathsf{A}$ is Hermitian (i.e. $\mathsf{A}^{\dagger} = \mathsf{A}$), and the components of $\mathsf{x}$ may now be complex. It is straightforward to show that $H$ is real, since

$$H^* = (H^{\mathrm{T}})^* = \mathsf{x}^{\dagger}\mathsf{A}^{\dagger}\mathsf{x} = \mathsf{x}^{\dagger}\mathsf{A}\mathsf{x} = H.$$

With suitable generalisation, the properties of quadratic forms apply also to Hermitian forms, but to keep the presentation simple we will restrict our discussion to quadratic forms.

A special case of quadratic (Hermitian) forms is that for which $Q = \mathsf{x}^{\mathrm{T}}\mathsf{A}\mathsf{x}$ is greater than zero for all column vectors $\mathsf{x}$. By choosing as basis the eigenvectors of $\mathsf{A}$ we have $Q$ in the form

$$Q = \lambda_1 x_1^2 + \lambda_2 x_2^2 + \lambda_3 x_3^2.$$

The requirement that $Q > 0$ for all $\mathsf{x}$ means that all the eigenvalues $\lambda_i$ of $\mathsf{A}$ must be positive. A symmetric (Hermitian) matrix $\mathsf{A}$ with this property is called *positive definite*. If instead $Q \geq 0$ for all $\mathsf{x}$, then it is possible that some of the eigenvalues are zero, and $\mathsf{A}$ is called *positive semi-definite*.

### *7.16.1 The stationary properties of the eigenvectors*

Consider a quadratic form such as (7.92) in a given fixed basis. As the vector $\mathbf{x}$ is varied, through changes in its three components $x_1$, $x_2$ and $x_3$, the value of the quantity $Q$ also varies. Because of the homogeneous form of $Q$ we may restrict any investigation of these variations to vectors of unit length (since multiplying any vector $\mathbf{x}$ by any scalar $k$ simply multiplies the value of $Q$ by a factor $k^2$).

Of particular interest are any vectors $\mathbf{x}$ that make the value of the quadratic form a maximum or minimum. A necessary (but not sufficient) condition for this is that $Q$ is stationary with respect to small (infinitesimal) variations $\Delta\mathbf{x}$ in $\mathbf{x}$, whilst $\langle \mathbf{x}|\mathbf{x}\rangle$ is maintained at a constant value (unity).

In our chosen basis the quadratic form is given by $Q = \mathsf{x}^{\mathrm{T}}\mathsf{A}\mathsf{x}$ and, using Lagrange undetermined multipliers to deal with the conditional variations, we are led to seek solutions of

$$\Delta[\mathsf{x}^{\mathrm{T}}\mathsf{A}\mathsf{x} - \lambda(\mathsf{x}^{\mathrm{T}}\mathsf{x} - 1)] = 0. \tag{7.99}$$

This may be used directly, together with the fact that $(\Delta\mathsf{x}^{\mathrm{T}})\mathsf{A}\mathsf{x} = \mathsf{x}^{\mathrm{T}}\mathsf{A}\,\Delta\mathsf{x}$ (since $\mathsf{A}$ is symmetric), to obtain the necessary condition

$$\mathsf{A}\mathsf{x} = \lambda\mathsf{x}, \tag{7.100}$$

that $\mathsf{x}$ must satisfy. If (7.100) is satisfied for some eigenvector $\mathsf{x}$, the value of $Q(\mathsf{x})$ is then just

$$Q = \mathsf{x}^{\mathrm{T}}\mathsf{A}\mathsf{x} = \mathsf{x}^{\mathrm{T}}\lambda\mathsf{x} = \lambda. \tag{7.101}$$

However, if $\mathsf{x}$ and $\mathsf{y}$ are eigenvectors corresponding to different eigenvalues they are (or can be chosen to be) orthogonal. Consequently the expression $\mathsf{y}^{\mathrm{T}}\mathsf{A}\mathsf{x}$ is necessarily zero, since

$$\mathsf{y}^{\mathrm{T}}\mathsf{A}\mathsf{x} = \mathsf{y}^{\mathrm{T}}\lambda\mathsf{x} = \lambda\mathsf{y}^{\mathrm{T}}\mathsf{x} = 0. \tag{7.102}$$

Summarising, those column vectors $\mathsf{x}$ of unit magnitude which make the quadratic form $Q$ stationary are eigenvectors of the matrix $\mathsf{A}$, and the stationary value of $Q$ is then equal to the corresponding eigenvalue. It is straightforward to see from the proof of (7.100) that conversely any eigenvector of $\mathsf{A}$ makes $Q$ stationary.

Instead of maximising or minimising $Q = \mathsf{x}^{\mathrm{T}}\mathsf{A}\mathsf{x}$ subject to the constraint $\mathsf{x}^{\mathrm{T}}\mathsf{x} = 1$, an equivalent procedure is to extremise the function

$$\lambda(\mathsf{x}) = \frac{\mathsf{x}^{\mathrm{T}}\mathsf{A}\mathsf{x}}{\mathsf{x}^{\mathrm{T}}\mathsf{x}}.$$

►*Show that if $\lambda(\mathsf{x})$ is stationary, then $\mathsf{x}$ is an eigenvector of $\mathsf{A}$ and $\lambda(\mathsf{x})$ is equal to the corresponding eigenvalue.*

We require $\Delta\lambda(\mathsf{x}) = 0$ with respect to small variations in $\mathsf{x}$. Now

$$\begin{aligned}\Delta\lambda &= \frac{1}{(\mathsf{x}^{\mathrm{T}}\mathsf{x})^2}\left[(\mathsf{x}^{\mathrm{T}}\mathsf{x})\left(\Delta\mathsf{x}^{\mathrm{T}}\mathsf{A}\mathsf{x} + \mathsf{x}^{\mathrm{T}}\mathsf{A}\,\Delta\mathsf{x}\right) - \mathsf{x}^{\mathrm{T}}\mathsf{A}\mathsf{x}\left(\Delta\mathsf{x}^{\mathrm{T}}\mathsf{x} + \mathsf{x}^{\mathrm{T}}\Delta\mathsf{x}\right)\right]\\ &= \frac{2\Delta\mathsf{x}^{\mathrm{T}}\mathsf{A}\mathsf{x}}{\mathsf{x}^{\mathrm{T}}\mathsf{x}} - 2\left(\frac{\mathsf{x}^{\mathrm{T}}\mathsf{A}\mathsf{x}}{\mathsf{x}^{\mathrm{T}}\mathsf{x}}\right)\frac{\Delta\mathsf{x}^{\mathrm{T}}\mathsf{x}}{\mathsf{x}^{\mathrm{T}}\mathsf{x}},\end{aligned}$$

since $\mathsf{x}^{\mathrm{T}}\mathsf{A}\,\Delta\mathsf{x} = (\Delta\mathsf{x}^{\mathrm{T}})\mathsf{A}\mathsf{x}$ and $\mathsf{x}^{\mathrm{T}}\Delta\mathsf{x} = (\Delta\mathsf{x}^{\mathrm{T}})\mathsf{x}$. Thus

$$\Delta\lambda = \frac{2}{\mathsf{x}^{\mathrm{T}}\mathsf{x}}\Delta\mathsf{x}^{\mathrm{T}}[\mathsf{A}\mathsf{x} - \lambda(\mathsf{x})\mathsf{x}].$$

Hence, if $\Delta\lambda = 0$ then $\mathsf{A}\mathsf{x} = \lambda(\mathsf{x})\mathsf{x}$, i.e. $\mathsf{x}$ is an eigenvector of $\mathsf{A}$ with eigenvalue $\lambda(\mathsf{x})$. ◄

Thus the eigenvalues of a symmetric matrix $\mathsf{A}$ are the values of the function

$$\lambda(\mathsf{x}) = \frac{\mathsf{x}^{\mathrm{T}}\mathsf{A}\mathsf{x}}{\mathsf{x}^{\mathrm{T}}\mathsf{x}}$$

at its stationary points. The eigenvectors of $\mathsf{A}$ thus lie along those directions in space for which the quadratic form $Q = \mathsf{x}^{\mathrm{T}}\mathsf{A}\mathsf{x}$ has stationary values, given a fixed magnitude for the vector $\mathsf{x}$. Similar results hold for Hermitian matrices.

### 7.16.2 *Quadratic surfaces*

The results of the previous subsection may be turned round to state that the surface given by

$$\mathsf{x}^{\mathrm{T}}\mathsf{A}\mathsf{x} = \text{constant} = 1(\text{say}), \tag{7.103}$$

called a *quadratic surface*, has stationary values of its radius (i.e. origin–surface distance) in those directions that are along the eigenvectors of A. More specifically, in three dimensions the quadratic surface $\mathsf{x}^{\mathrm{T}}\mathsf{A}\mathsf{x} = 1$ has its principal axes along the three mutually perpendicular eigenvectors of A, and the squares of the corresponding principal radii are given by $\lambda_i^{-1}$, $i = 1, 2, 3$. If any two of the eigenvalues are degenerate, then the quadratic has rotational symmetry about some axis and the choice of a pair of axes perpendicular to that axis is not uniquely defined.

►*Describe the shape of the quadratic surface*

$$x_1^2 + x_2^2 - 3x_3^2 + 2x_1x_2 + 6x_1x_3 - 6x_2x_3 = 1.$$

If, instead of expressing the quadratic surface in terms of $x_1$, $x_2$, $x_3$, we were to use the new variables $x_1'$, $x_2'$, $x_3'$ defined in (7.97), for which the coordinate axes are along the three mutually perpendicular eigenvector directions $(1, 1, 0)$, $(1, -1, 1)$ and $(1, -1, -2)$, then the equation of the surface would take the form

$$\frac{x_1'^2}{(2^{-1/2})^2} + \frac{x_2'^2}{(3^{-1/2})^2} - \frac{x_3'^2}{(6^{-1/2})^2} = 1.$$

Thus, for example, a section of the quadratic surface in the plane $x_3' = 0$, i.e. $x_1 - x_2 - 2x_3 = 0$, is an ellipse, with semi-axes $2^{-1/2}$ and $3^{-1/2}$. Similarly a section in the plane $x_1' = x_1 + x_2 = 0$ is a hyperbola. ◄

Clearly the simplest three-dimensional situation to visualise is that in which all the eigenvalues are positive, since then the quadratic surface is an ellipsoid.

## 7.17 Mechanical oscillations and normal modes

Any student of the physical sciences will encounter the subject of oscillations on many occasions and in a wide variety of applications, e.g. the oscillations in an electric circuit and the oscillations of molecules. Matrices provide a particularly simple way to approach what may, at a first glance, appear to be a difficult physical problem.

We will consider only mechanical systems for which a potential exists, i.e., the potential energy of the system in any particular configuration can depend upon the coordinates of the configuration, which need not be restricted to spatial positions, but must not depend upon the time derivative (general velocity) of these coordinates. A further restriction that we place is that the potential has a local minimum at the equilibrium point; it is physically obvious that this is a necessary and sufficient condition for stable equilibrium. By suitably defining the origin of the potential, we may take its value at the equilibrium point as zero.

We denote the coordinates chosen to describe a configuration of the system by $q_i$, $i = 1, \dots, N$. The $q_i$ need not be distances; some could be angles, for example. For convenience we can define the $q_i$ so that they are all zero at the equilibrium point. The instantaneous velocities of various parts of the system will depend on the time derivatives of the $q_i$, denoted by $\dot{q}_i$. For small oscillations the velocities will be linear in the $\dot{q}_i$ and consequently the total kinetic energy $T$ will be quadratic in them – and will include cross terms of the form $\dot{q}_i \dot{q}_j$ with $i \neq j$. The general expression can be written as the quadratic form

$$T = \sum_i \sum_j a_{ij} \dot{q}_i \dot{q}_j = \dot{\mathsf{q}}^{\mathrm{T}} \mathsf{A} \dot{\mathsf{q}}, \tag{7.104}$$

where $\dot{\mathsf{q}}^{\mathrm{T}} = (\dot{q}_1 \quad \dot{q}_2 \quad \cdots \quad \dot{q}_N)$, and the $N \times N$ matrix $\mathsf{A}$ is real and may be chosen to be symmetric. Furthermore, $\mathsf{A}$, like any matrix corresponding to a kinetic energy, is positive definite (more strictly positive semi-definite); that is, whatever real values the $\dot{q}_i$ take the quadratic form (7.104) has a value $\geq 0$.

Turning now to the potential energy, we may write its value for a configuration $\mathsf{q}$ by means of a Taylor expansion about the origin $\mathsf{q} = 0$,

$$V(\mathsf{q}) = V(0) + \sum_i \frac{\partial V(0)}{\partial q_i} q_i + \frac{1}{2} \sum_i \sum_j \frac{\partial^2 V(0)}{\partial q_i \partial q_j} q_i q_j + \cdots .$$

However, we have chosen $V(0) = 0$ and, since the origin is an equilibrium point, there is no force there and $\partial V(0)/\partial q_i = 0$. Consequently, to second order in the $q_i$ we also have a quadratic form, but in the coordinates rather than in their time derivatives:

$$V = \sum_i \sum_j b_{ij} q_i q_j = \mathsf{q}^{\mathrm{T}} \mathsf{B} \mathsf{q}, \tag{7.105}$$

where $\mathsf{B}$ is, or can be made, symmetric. In this case, and in general, the requirement that the potential is a minimum means that the potential matrix, like the kinetic energy matrix, is real and positive definite.

We will now introduce a particular example, although the results of this section are general, given the above constraints, and the reader will find it easy to apply the results to many other instances.

Consider a uniform rod of mass $M$ and length $l$, attached by a light string of the same length to a fixed point $P$, executing small oscillations in a vertical plane. We choose as coordinates the angles $\theta_1$ and $\theta_2$ shown, with exaggerated

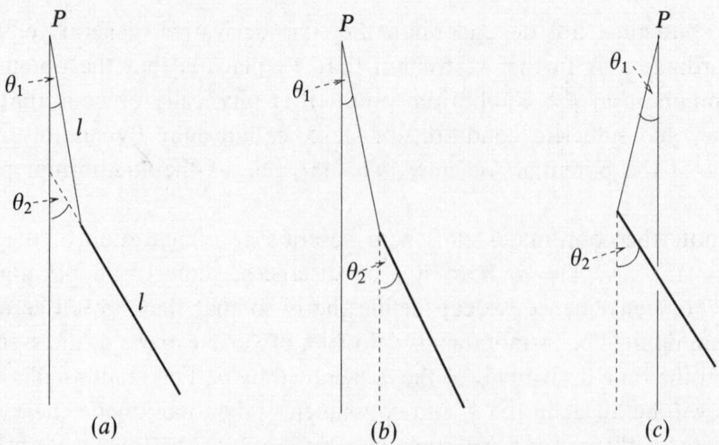


Figure 7.2 A uniform rod of length $l$ attached to the fixed point $P$ by a light string of the same length: (*a*) the general coordinate system; (*b*) approximation to the normal mode with lower frequency; (*c*) approximation to the mode with higher frequency.

magnitude, in figure 7.2. In terms of these coordinates the centre of gravity of the rod has, to *first order* in the $\theta_i$, a velocity component in the $x$-direction equal to $l\dot{\theta}_1 + \frac{1}{2}l\dot{\theta}_2$ and in the $y$-direction equal to zero. Adding in the rotational kinetic energy of the rod about its centre of gravity we obtain, to second order in the $\dot{\theta}_i$,

$$\begin{aligned} T &\approx \tfrac{1}{2}Ml^2(\dot{\theta}_1^2 + \tfrac{1}{4}\dot{\theta}_2^2 + \dot{\theta}_1\dot{\theta}_2) + \tfrac{1}{24}Ml^2\dot{\theta}_2^2 \\ &= \tfrac{1}{6}Ml^2\left(3\dot{\theta}_1^2 + 3\dot{\theta}_1\dot{\theta}_2 + \dot{\theta}_2^2\right) = \tfrac{1}{12}Ml^2\dot{\mathsf{q}}^{\mathrm{T}}\begin{pmatrix} 6 & 3 \\ 3 & 2 \end{pmatrix}\dot{\mathsf{q}}, \end{aligned} \tag{7.106}$$

where $\dot{\mathsf{q}}^{\mathrm{T}} = (\dot{\theta}_1 \quad \dot{\theta}_2)$. The potential energy is given by

$$\begin{aligned} V &= Mlg\left[(1-\cos\theta_1) + \tfrac{1}{2}(1-\cos\theta_2)\right] \\ &\approx \tfrac{1}{4}Mlg(2\theta_1^2 + \theta_2^2) = \tfrac{1}{12}Mlg\mathsf{q}^{\mathrm{T}}\begin{pmatrix} 6 & 0 \\ 0 & 3 \end{pmatrix}\mathsf{q}, \end{aligned}$$

where $g$ is the acceleration due to gravity, $\mathsf{q}^{\mathrm{T}} = (\theta_1 \quad \theta_2)$, and the expression is valid to second order in the $\theta_i$.

With these expressions for $T$ and $V$ we now apply the conservation of energy,

$$\frac{d}{dt}(T+V) = 0, \tag{7.107}$$

assuming that there are no external forces other than gravity. In matrix form

(7.107) becomes

$$\frac{d}{dt}(\dot{\mathsf{q}}^{\mathrm{T}}\mathsf{A}\dot{\mathsf{q}} + \mathsf{q}^{\mathrm{T}}\mathsf{B}\mathsf{q}) = \ddot{\mathsf{q}}^{\mathrm{T}}\mathsf{A}\dot{\mathsf{q}} + \dot{\mathsf{q}}^{\mathrm{T}}\mathsf{A}\ddot{\mathsf{q}} + \dot{\mathsf{q}}^{\mathrm{T}}\mathsf{B}\mathsf{q} + \mathsf{q}^{\mathrm{T}}\mathsf{B}\dot{\mathsf{q}} = 0,$$

which, using $\mathsf{A} = \mathsf{A}^{\mathrm{T}}$ and $\mathsf{B} = \mathsf{B}^{\mathrm{T}}$, gives

$$2\dot{\mathsf{q}}^{\mathrm{T}}(\mathsf{A}\ddot{\mathsf{q}} + \mathsf{B}\mathsf{q}) = 0.$$

Now, although it is not clear that this gives the only possible solution, we will assume that the above equation implies that the coefficient of each $\dot{q}_i$ is separately zero. Hence

$$\mathsf{A}\ddot{\mathsf{q}} + \mathsf{B}\mathsf{q} = 0. \tag{7.108}$$

For a rigorous derivation Lagrange's equations should be used, as in chapter 20.

Now we search for sets of coordinates $\mathsf{q}$ that *all* oscillate with the same period, i.e. the total motion repeats itself *exactly* after a *finite* interval. Solutions of this form will satisfy

$$\mathsf{q} = \mathsf{x}\cos\omega t, \tag{7.109}$$

where the relative values of the elements of $\mathsf{x}$ (assuming we can find such a solution) will indicate how each coordinate is involved in this special motion. In general there will be $N$ values of $\omega$ for $N \times N$ matrices $\mathsf{A}$ and $\mathsf{B}$, and the values are known as *normal frequencies* or *eigenfrequencies*.

Putting (7.109) into (7.108) yields

$$-\omega^2\mathsf{A}\mathsf{x} + \mathsf{B}\mathsf{x} = (\mathsf{B} - \omega^2\mathsf{A})\mathsf{x} = 0. \tag{7.110}$$

Our work in section 7.10 shows that this can have non-trivial solutions only if

$$|\mathsf{B} - \omega^2\mathsf{A}| = 0. \tag{7.111}$$

This is a form of characteristic equation for $\mathsf{B}$, except that $\mathsf{A}$ has replaced the unit matrix $\mathsf{I}$. It has the more familiar form if a choice of coordinates is made in which the kinetic energy $T$ is a simple sum of squared terms, i.e. it has been diagonalised, the scale of the new coordinates being chosen to make each diagonal element unity.

However, even in the present case (7.111) can be solved to yield $\omega_k^2$ for $k = 1, 2, \ldots, N$, where $N$ is the order of $\mathsf{A}$ and $\mathsf{B}$. The values of $\omega_k$ can be used with (7.110) to find the corresponding column vector $\mathsf{x}^k$ and the initial (stationary) physical configuration, which on release will execute motion with period $2\pi/\omega_k$, to be found.

Returning to the suspended rod, we find

$$\left| \frac{Mlg}{12}\begin{pmatrix} 6 & 0 \\ 0 & 3 \end{pmatrix} - \frac{\omega^2 Ml^2}{12}\begin{pmatrix} 6 & 3 \\ 3 & 2 \end{pmatrix} \right| = 0.$$

Writing $\omega^2 l/g = \lambda$, this becomes

$$\begin{vmatrix} 6-6\lambda & -3\lambda \\ -3\lambda & 3-2\lambda \end{vmatrix} = 0 \quad \Rightarrow \quad \lambda^2 - 10\lambda + 6 = 0,$$

which has the roots $\lambda = 5 \pm \sqrt{19}$. We thus find that the two normal frequencies are given by $\omega_1 = (0.641g/l)^{1/2}$ and $\omega_2 = (9.359g/l)^{1/2}$. Putting the lower of the two values for $\omega^2$, namely $(5 - \sqrt{19})g/l$, into (7.110) shows that for this mode

$$x_1 : x_2 = 3(5 - \sqrt{19}) : 6(\sqrt{19} - 4) = 1.923 : 2.153.$$

This corresponds to the case where the rod and string are almost straight out, i.e. they almost form a simple pendulum. Similarly it may be shown that the higher frequency corresponds to a solution when the string and rod are moving with opposite phase and $x_1 : x_2 = 9.359 : -16.718$. The two situations are shown in figure 7.2.

In connection with quadratic forms it has been shown how to make a change of coordinates such that the matrix for a particular form becomes diagonal. In exercise 7.18 a method is developed for simultaneously diagonalising two quadratic forms (but not in general with an orthogonal transformation). If this process is carried out for $\mathsf{A}$ and $\mathsf{B}$ in a general system undergoing stable oscillations, the kinetic and potential energies in the new variables $\eta_i$ take the forms

$$T = \sum_i \mu_i \dot{\eta}_i^2 = \dot{\eta}^{\mathrm{T}} \mathsf{M} \dot{\eta}, \qquad \mathsf{M} = \mathrm{diag}\,(\mu_1, \ldots, \mu_N), \tag{7.112}$$

$$V = \sum_i \nu_i \eta_i^2 = \eta^{\mathrm{T}} \mathsf{N} \eta, \qquad \mathsf{N} = \mathrm{diag}\,(\nu_1, \ldots, \nu_N), \tag{7.113}$$

and the equations of motion are the *uncoupled* equations

$$\mu_i \ddot{\eta}_i + \nu_i \eta_i = 0, \qquad i = 1, 2, \ldots, N. \tag{7.114}$$

Clearly a simple renormalisation of the $\eta_i$ can be made to reduce all the $\mu_i$ in (7.112) to unity. When this is done the variables so formed are called *normal coordinates*, and equations (7.114) the *normal equations*.

When a system is executing one of these truly periodic motions it is said to be in a *normal mode*, and once started in such a mode it will repeat its motion exactly after each interval of $2\pi/\omega_i$. Any arbitrary motion may be written as a superposition of the normal modes and each component will execute harmonic motion with the corresponding eigenfrequency; however, unless by chance the eigenfrequencies are in integer relationship the system will never return to its initial configuration after any finite time interval. The orthogonality properties of eigenvectors corresponding to different normal modes are discussed in the next section.

As a second example we will consider a number of masses coupled together by

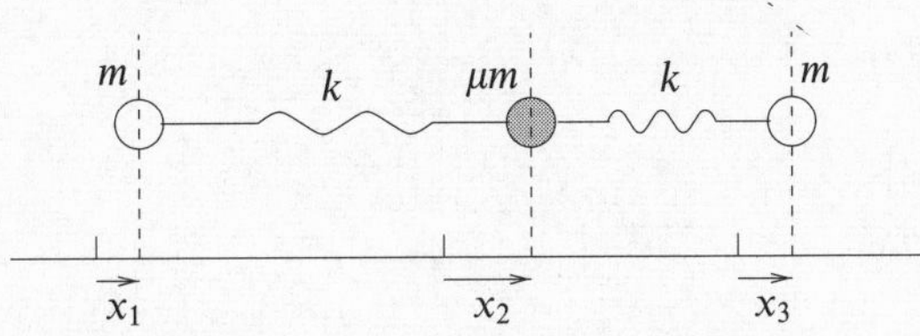


Figure 7.3 Three masses $m$, $\mu m$ and $m$ connected by two equal light springs of force constant $k$.

springs. For this type of situation the potential and kinetic energies are automatically quadratic functions of the coordinates and their derivatives (provided the elastic limits of the springs are not exceeded), and the oscillations do not have to be vanishingly small for the analysis to be valid.

►*Find the normal frequencies and modes of oscillation of three particles of masses m, μm, m connected in that order in a straight line by two equal light springs of force constant k. (This arrangement could serve as a model for some linear molecules, e.g.* $CO_2$.*)*

The situation is shown in figure 7.3; the coordinates of the particles, $x_1$, $x_2$, $x_3$, are measured from their equilibrium positions at which the springs are neither extended nor compressed.

The kinetic energy of the system is simply

$$T = \tfrac{1}{2}m\left(\dot{x}_1^2 + \mu\dot{x}_2^2 + \dot{x}_3^2\right),$$

whilst the potential energy stored in the springs is

$$V = \tfrac{1}{2}k\left[(x_2 - x_1)^2 + (x_3 - x_2)^2\right].$$

The kinetic and potential energy symmetric matrices are thus

$$\mathsf{A} = \frac{m}{2}\begin{pmatrix} 1 & 0 & 0 \\ 0 & \mu & 0 \\ 0 & 0 & 1 \end{pmatrix}, \qquad \mathsf{B} = \frac{k}{2}\begin{pmatrix} 1 & -1 & 0 \\ -1 & 2 & -1 \\ 0 & -1 & 1 \end{pmatrix}.$$

From (7.111), to find the normal frequencies we have to solve $|\mathsf{B} - \omega^2\mathsf{A}| = 0$. Thus, writing $m\omega^2/k = \lambda$ we have

$$\begin{vmatrix} 1-\lambda & -1 & 0 \\ -1 & 2-\mu\lambda & -1 \\ 0 & -1 & 1-\lambda \end{vmatrix} = 0,$$

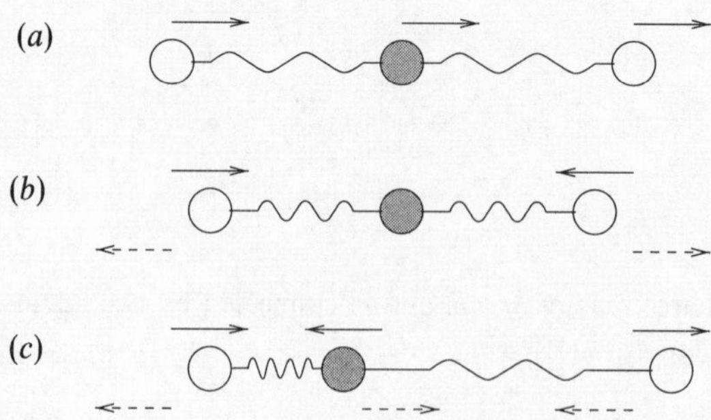


Figure 7.4 The normal modes of the masses and springs of a linear molecule such as $CO_2$. (a) $\omega^2 = 0$; (b) $\omega^2 = k/m$; (c) $\omega^2 = k(\mu+2)/(\mu m)$.

which leads to $\lambda = 0$, 1, or $1+2/\mu$. The corresponding eigenvectors are respectively

$$\mathsf{x}^1 = \frac{1}{\sqrt{3}}\begin{pmatrix} 1 \\ 1 \\ 1 \end{pmatrix}, \qquad \mathsf{x}^2 = \frac{1}{\sqrt{2}}\begin{pmatrix} 1 \\ 0 \\ -1 \end{pmatrix}, \qquad \mathsf{x}^3 = \frac{1}{\sqrt{2+(4/\mu^2)}}\begin{pmatrix} 1 \\ -2/\mu \\ 1 \end{pmatrix}.$$

The physical motions associated with these solutions are illustrated in figure 7.4. The first, with $\lambda = \omega = 0$ and all the $x_i$ equal, merely describes the bodily translation of the whole system, with no (i.e. zero-frequency) internal oscillations.

In the second solution the central particle remains stationary, $x_2 = 0$, whilst the other two oscillate with equal amplitudes in antiphase with each other. The motion of frequency $\omega = (k/m)^{1/2}$ is illustrated in figure 7.4(b).

The final and most complicated of the three normal modes has a frequency $\omega = [k(\mu+2)/(\mu m)]^{1/2}$, and involves a motion of the central particle which is in antiphase with that of the two outer ones and which has an amplitude $2/\mu$ times as great. In this motion (see figure 7.4(c)) the two springs are compressed and extended in turn. We also note that in the second and third normal modes the centre of mass of the molecule remains stationary. ◄

## 7.18 Rayleigh–Ritz method

We conclude this chapter with a discussion of the Rayleigh–Ritz method for estimating the eigenfrequencies of an oscillating system. We recall from the previous section that for a system undergoing small oscillations the potential and kinetic energy are given by

$$V = \mathsf{q}^{\mathrm{T}}\mathsf{B}\mathsf{q} \qquad \text{and} \qquad T = \dot{\mathsf{q}}^{\mathrm{T}}\mathsf{A}\dot{\mathsf{q}},$$

where the components of $\mathsf{q}$ are the coordinates chosen to represent the configuration of the system and $\mathsf{A}$ and $\mathsf{B}$ are symmetric matrices (or may be chosen to be such). We also recall that the normal modes $\mathsf{x}^i$ and the eigenfrequencies $\omega_i^2$

are from (7.110) given by

$$(\mathsf{B} - \omega_i^2 \mathsf{A})\mathsf{x}^i = 0. \tag{7.115}$$

It may be shown that the eigenvectors $\mathsf{x}^i$ corresponding to different normal modes are linearly independent and so form a complete set. Thus, any coordinate vector $\mathsf{q}$ can be written $\mathsf{q} = \sum_j c_j \mathsf{x}^j$. We now consider the value of the generalised quadratic form

$$\lambda(\mathsf{x}) = \frac{\mathsf{x}^{\mathrm{T}}\mathsf{B}\mathsf{x}}{\mathsf{x}^{\mathrm{T}}\mathsf{A}\mathsf{x}} = \frac{\sum_m (\mathsf{x}^m)^{\mathrm{T}} c_m^* \mathsf{B} \sum_i c_i \mathsf{x}^i}{\sum_j (\mathsf{x}^j)^{\mathrm{T}} c_j^* \mathsf{A} \sum_k c_k \mathsf{x}^k}$$

and use (7.115) to replace $\mathsf{B}\mathsf{x}^i$, with the result that

$$\begin{aligned}\lambda(\mathsf{x}) &= \frac{\sum_m (\mathsf{x}^m)^{\mathrm{T}} c_m^* \mathsf{A} \sum_i \omega_i^2 c_i \mathsf{x}^i}{\sum_j (\mathsf{x}^j)^{\mathrm{T}} c_j^* \mathsf{A} \sum_k c_k \mathsf{x}^k} \\ &= \frac{\sum_m (\mathsf{x}^m)^{\mathrm{T}} c_m^* \sum_i \omega_i^2 c_i \mathsf{A} \mathsf{x}^i}{\sum_j (\mathsf{x}^j)^{\mathrm{T}} c_j^* \mathsf{A} \sum_k c_k \mathsf{x}^k}.\end{aligned}$$

Now the eigenvectors obtained by solving $(\mathsf{B} - \omega^2 \mathsf{A})\mathsf{x} = \mathsf{0}$ are not mutually orthogonal unless $\mathsf{A}$ is a multiple of the unit matrix. However, from (7.115) it may be shown that the $\mathsf{x}^i$ do possess the desirable properties (see exercise 7.18)

$$(\mathsf{x}^i)^{\mathrm{T}}\mathsf{A}\mathsf{x}^j = 0 \qquad \text{and} \qquad (\mathsf{x}^i)^{\mathrm{T}}\mathsf{B}\mathsf{x}^j = 0 \quad \text{if } i \neq j.$$

Using the first of these relations we find

$$\lambda(\mathsf{x}) = \frac{\sum_i |c_i|^2 \omega_i^2 (\mathsf{x}^i)^{\mathrm{T}} \mathsf{A} \mathsf{x}^i}{\sum_k |c_k|^2 (\mathsf{x}^k)^{\mathrm{T}} \mathsf{A} \mathsf{x}^k}. \tag{7.116}$$

Now, if $\omega_0^2$ is the lowest eigenfrequency then $\omega_i^2 \geq \omega_0^2$ for all $i$ and further, since $(\mathsf{x}^i)^{\mathrm{T}}\mathsf{A}\mathsf{x}^i \geq 0$ for all $i$ the numerator of (7.116) is $\geq \omega_0^2 \sum_i |c_i|^2 (\mathsf{x}^i)^{\mathrm{T}}\mathsf{A}\mathsf{x}^i$. Hence

$$\lambda(\mathsf{x}) \equiv \frac{\mathsf{x}^{\mathrm{T}}\mathsf{B}\mathsf{x}}{\mathsf{x}^{\mathrm{T}}\mathsf{A}\mathsf{x}} \geq \omega_0^2, \tag{7.117}$$

for any $\mathsf{x}$ whatsoever (an eigenvector or not). Thus we are able to estimate the lowest eigenfrequency of the system by evaluating $\lambda$ for a variety of column vectors $\mathsf{x}$, the components of which, it will be recalled, give the ratios of the coordinate amplitudes. This is sometimes a useful approach if many coordinates are involved and direct solution for the eigenvalues is not possible.

An additional result is that the maximum eigenfrequency $\omega_m^2$ may also be estimated. It is obvious that if we replace the statement '$\omega_i^2 \geq \omega_0^2$ for all $i$' by '$\omega_i^2 \leq \omega_m^2$ for all $i$', then $\lambda(\mathsf{x}) \leq \omega_m^2$ for any $\mathsf{x}$. Thus $\lambda(\mathsf{x})$ always lies between the lowest and highest eigenfrequencies of the system. Furthermore, $\lambda(\mathsf{x})$ has a *stationary* value equal to $\omega_k^2$ when $\mathsf{x}$ is the $k$th eigenvector (see subsection 7.16.1).

►*Estimate the eigenfrequencies of the oscillating rod of section 7.17.*

We firstly recall that

$$\mathsf{A} = \frac{Ml^2}{12}\begin{pmatrix} 6 & 3 \\ 3 & 2 \end{pmatrix} \quad \text{and} \quad \mathsf{B} = \frac{Mlg}{12}\begin{pmatrix} 6 & 0 \\ 0 & 3 \end{pmatrix}.$$

Physical intuition suggests that the slower mode will have a configuration approximating that of a simple pendulum (figure 7.2), in which $\theta_1 = \theta_2$, and so we use this as a *trial vector*. Taking $\mathsf{x} = (\theta \quad \theta)^{\mathrm{T}}$,

$$\lambda(\mathsf{x}) = \frac{\mathsf{x}^{\mathrm{T}}\mathsf{B}\mathsf{x}}{\mathsf{x}^{\mathrm{T}}\mathsf{A}\mathsf{x}} = \frac{3Mlg\theta^2/4}{7Ml^2\theta^2/6} = \frac{9g}{14l} = 0.643\frac{g}{l},$$

and we conclude that the lower (angular) frequency is $\leq (0.643g/l)^{1/2}$. We have already seen that the true answer is $(0.641g/l)^{1/2}$ and so we have come very close to it.

Next we turn to the higher frequency. Here, what a typical configuration looks like is not so obvious but, rather preempting the answer, we try $\theta_2 = -2\theta_1$ and obtain $\lambda = 9g/l$ and so conclude that the higher eigenfrequency $\geq (9g/l)^{1/2}$. We have already seen that the exact answer is $(9.359g/l)^{1/2}$ and so we have again come close. ◄

A simplified version of the Rayleigh–Ritz method may be used to estimate the eigenvalues of a symmetric (or in general Hermitian) matrix $\mathsf{B}$, whose eigenvectors will be mutually orthogonal. By repeating the calculations leading to (7.116), $\mathsf{A}$ being replaced by the unit matrix $\mathsf{I}$, it is easily verified that if

$$\lambda(\mathsf{x}) = \frac{\mathsf{x}^{\mathrm{T}}\mathsf{B}\mathsf{x}}{\mathsf{x}^{\mathrm{T}}\mathsf{x}}$$

is evaluated for *any* vector $\mathsf{x}$ then

$$\lambda_1 \leq \lambda(\mathsf{x}) \leq \lambda_m,$$

where $\lambda_1, \ldots, \lambda_m$ are the eigenvalues of $\mathsf{B}$ in order of increasing size. A similar result holds for Hermitian matrices.

## 7.19 Exercises

7.1 Evaluate the determinants

$$\text{(a)} \quad \begin{vmatrix} a & h & g \\ h & b & f \\ g & f & c \end{vmatrix}, \qquad \text{(b)} \quad \begin{vmatrix} 1 & 0 & 2 & 3 \\ 0 & 1 & -2 & 1 \\ 3 & -3 & 4 & -2 \\ -2 & 1 & -2 & 1 \end{vmatrix},$$

$$\text{(c)}\quad \begin{vmatrix} gc & ge & a+ge & gb+ge \\ 0 & b & b & b \\ c & e & e & b+e \\ a & b & b+f & b+d \end{vmatrix}.$$

7.2 Do the following sets of equations have non-zero solutions? If so, find them.

(a) $3x+2y+z=0$, $x-3y+2z=0$, $2x+y+3z=0$.
(b) $2x=b(y+z)$, $x=2a(y-z)$, $x=(6a-b)y-(6a+b)z$.

7.3 Using the properties of determinants, solve with a minimum of calculation the following equations for $x$:

$$\text{(a)}\quad \begin{vmatrix} x & a & a & 1 \\ a & x & b & 1 \\ a & b & x & 1 \\ a & b & c & 1 \end{vmatrix} = 0, \qquad \text{(b)}\quad \begin{vmatrix} x+2 & x+4 & x-3 \\ x+3 & x & x+5 \\ x-2 & x-1 & x+1 \end{vmatrix} = 0.$$

7.4 Show that the following equations have solutions only if $\eta = 1$ or 2, and find them in these cases:

$$x+y+z=1,$$
$$x+2y+4z=\eta,$$
$$x+4y+10z=\eta^2.$$

7.5 Consider the matrices

$$\text{(a)}\ \mathsf{B} = \begin{pmatrix} 0 & -i & i \\ i & 0 & -i \\ -i & i & 0 \end{pmatrix}, \qquad \text{(b)}\ \mathsf{C} = \frac{1}{\sqrt{8}} \begin{pmatrix} \sqrt{3} & -\sqrt{2} & -\sqrt{3} \\ 1 & \sqrt{6} & -1 \\ 2 & 0 & 2 \end{pmatrix}.$$

Are they (i) real, (ii) diagonal, (iii) symmetric, (iv) antisymmetric, (v) singular, (vi) orthogonal, (vii) Hermitian, (viii) anti-Hermitian, (ix) unitary, (x) normal?

7.6 By considering the matrices

$$\mathsf{A} = \begin{pmatrix} 1 & 0 \\ 0 & 0 \end{pmatrix}, \qquad \mathsf{B} = \begin{pmatrix} 0 & 0 \\ 3 & 4 \end{pmatrix}$$

show that $\mathsf{AB} = \mathsf{0}$ does *not* imply that either A or B is the zero matrix, but that it does imply that at least one of them is singular.

7.7 (a) The basis vectors of the unit cell of a crystal, with the origin $O$ at one corner, are denoted by $\mathbf{e}_1$, $\mathbf{e}_2$, $\mathbf{e}_3$. The matrix G has elements $G_{ij}$ where $G_{ij} = \mathbf{e}_i \cdot \mathbf{e}_j$ and $H_{ij}$ are the elements of the matrix $\mathsf{H} \equiv \mathsf{G}^{-1}$.

Show that the vectors $\mathbf{f}_i = \sum_j H_{ij}\mathbf{e}_j$ are the reciprocal vectors and that $H_{ij} = \mathbf{f}_i \cdot \mathbf{f}_j$.

(b) If the vectors $\mathbf{u}$ and $\mathbf{v}$ are given by

$$\mathbf{u} = \sum_i u_i \mathbf{e}_i, \qquad \mathbf{v} = \sum_i v_i \mathbf{f}_i,$$

obtain expressions for $|\mathbf{u}|$, $|\mathbf{v}|$, and $\mathbf{u} \cdot \mathbf{v}$.

(c) If the basis vectors are each of length $a$ and the angle between each pair is $\pi/3$, write down $\mathsf{G}$ and hence obtain $\mathsf{H}$.

(d) Calculate (i) the length of the normal from $O$ onto the plane containing the points $p^{-1}\mathbf{e}_1$, $q^{-1}\mathbf{e}_2$, $r^{-1}\mathbf{e}_3$, and (ii) the angle between this normal and $\mathbf{e}_1$.

7.8 (a) Show that if $\mathsf{A}$ is Hermitian and $\mathsf{U}$ is unitary then $\mathsf{U}^{-1}\mathsf{A}\mathsf{U}$ is Hermitian.

(b) Show that if $\mathsf{A}$ is anti-Hermitian then $i\mathsf{A}$ is Hermitian.

(c) Prove that the product of two Hermitian matrices $\mathsf{A}$ and $\mathsf{B}$ is Hermitian if and only if $\mathsf{A}$ and $\mathsf{B}$ commute.

(d) Prove that if $\mathsf{S}$ is a real antisymmetric matrix then $\mathsf{A} = (\mathsf{I}-\mathsf{S})(\mathsf{I}+\mathsf{S})^{-1}$ is orthogonal and express the matrix

$$\begin{pmatrix} \cos\theta & \sin\theta \\ -\sin\theta & \cos\theta \end{pmatrix}$$

in this form.

7.9 $\mathsf{A}$ and $\mathsf{B}$ are real non-zero $3 \times 3$ matrices and satisfy the equation

$$(\mathsf{AB})^{\mathrm{T}} + \mathsf{B}^{-1}\mathsf{A} = 0.$$

(a) Prove that if $\mathsf{B}$ is orthogonal then $\mathsf{A}$ is antisymmetric.

(b) Without assuming that $\mathsf{B}$ is orthogonal, prove that $\mathsf{A}$ is singular.

7.10 A general triangle has angles $\alpha$, $\beta$ and $\gamma$ and corresponding opposite sides $a$, $b$ and $c$. Express the length of each side in terms of the lengths of the other two sides and the relevant cosines, writing the relationships in matrix and vector form using the vectors with components $a, b, c$ and $\cos\alpha, \cos\beta, \cos\gamma$. Invert the matrix and hence deduce the cosine-law expressions involving $\alpha$, $\beta$ and $\gamma$.

7.11 For the matrix

$$\mathsf{A} = \begin{pmatrix} 1 & \alpha & 0 \\ \beta & 1 & 0 \\ 0 & 0 & 1 \end{pmatrix},$$

where $\alpha$ and $\beta$ are non-zero complex numbers, find the eigenvalues and eigenvectors. Find the respective conditions for (a) the eigenvalues to be real, and (b) the eigenvectors to be orthogonal. Show that the conditions are jointly satisfied only if $\mathsf{A}$ is Hermitian.

7.12 Find the eigenvalues and a set of eigenvectors of the matrix

$$\begin{pmatrix} 1 & 3 & -1 \\ 3 & 4 & -2 \\ -1 & -2 & 2 \end{pmatrix}.$$

Verify that its eigenvectors are mutually orthogonal.

7.13 The equation of a conic section is

$$Q \equiv 8x_1^2 + 8x_2^2 - 6x_1x_2 = 110.$$

Determine the type of conic section this represents, the orientation of its principal axes, and relevant lengths in the directions of these axes.

7.14 Show that the quadratic surface

$$5x^2 + 11y^2 + 5z^2 - 10yz + 2xz - 10xy = 4$$

is an ellipsoid with semi-axes of 2, 1 and 0.5. Find the direction of the longest axis.

7.15 Find the direction of the axis of symmetry of the quadratic surface

$$7x^2 + 7y^2 + 7z^2 - 20yz - 20xz + 20xy = 3.$$

7.16 Three coupled pendulums swing perpendicularly to the horizontal line containing their points of suspension, and the following equations of motion are satisfied:

$$\begin{aligned} -m\ddot{x}_1 &= cmx_1 + d(x_1 - x_2), \\ -M\ddot{x}_2 &= cMx_2 + d(x_2 - x_1) + d(x_2 - x_3), \\ -m\ddot{x}_3 &= cmx_3 + d(x_3 - x_2), \end{aligned}$$

where $x_1$, $x_2$ and $x_3$ are measured from the equilibrium points, $m$ and $M$ are masses and $c$ and $d$ are positive constants. Find the normal frequencies of the system and sketch the corresponding patterns of oscillation. What happens as $d \to 0$, or $d \to \infty$?

7.17 Find an orthogonal transformation that takes the quadratic form

$$Q \equiv -x_1^2 - 2x_2^2 - x_3^2 + 8x_2x_3 + 6x_1x_3 + 8x_1x_2$$

into the form

$$\mu_1 y_1^2 + \mu_2 y_2^2 - 4y_3^2,$$

where $\mu_1$ and $\mu_2$ are to be determined (see section 7.16).

7.18 *The simultaneous reduction to diagonal form of two real symmetric quadratic forms.*

Consider the two real symmetric quadratic forms $\mathsf{u}^{\mathrm{T}}\mathsf{A}\mathsf{u}$ and $\mathsf{u}^{\mathrm{T}}\mathsf{B}\mathsf{u}$, where $\mathsf{u}^{\mathrm{T}}$ stands for the row matrix $(x \quad y \quad z)$, and denote by $\mathsf{u}^n$ those vectors which satisfy

$$\mathsf{B}\mathsf{u}^n = \lambda_n \mathsf{A}\mathsf{u}^n, \tag{E7.1}$$

in which $n$ is a label and the $\lambda_n$ are real, non-zero and all different.

(a) By multiplying (E7.1) on the left by $(\mathsf{u}^m)^{\mathrm{T}}$ and the transpose of the corresponding equation for $\mathsf{u}^m$ on the right by $\mathsf{u}^n$, show that $(\mathsf{u}^m)^{\mathrm{T}}\mathsf{A}\mathsf{u}^n = 0$ for $n \neq m$.

(b) By noting that $\mathsf{A}\mathsf{u}^n = (\lambda_n)^{-1}\mathsf{B}\mathsf{u}^n$, deduce that $(\mathsf{u}^m)^{\mathrm{T}}\mathsf{B}\mathsf{u}^n = 0$ for $m \neq n$. It can be shown that the $\mathsf{u}^n$ are linearly independent; the next step is to construct a matrix $\mathsf{P}$ whose columns are the vectors $\mathsf{u}^n$.

(c) Make a change of variables $\mathsf{u} = \mathsf{P}\mathsf{v}$ such that $\mathsf{u}^{\mathrm{T}}\mathsf{A}\mathsf{u}$ becomes $\mathsf{v}^{\mathrm{T}}\mathsf{C}\mathsf{v}$, and $\mathsf{u}^{\mathrm{T}}\mathsf{B}\mathsf{u}$ becomes $\mathsf{v}^{\mathrm{T}}\mathsf{D}\mathsf{v}$. Show that $\mathsf{C}$ and $\mathsf{D}$ are diagonal by showing $c_{ij} = 0$ if $i \neq j$ and similarly for $d_{ij}$.

Thus $\mathsf{u} = \mathsf{P}\mathsf{v}$ or $\mathsf{v} = \mathsf{P}^{-1}\mathsf{u}$ reduces both quadratics to diagonal form.

To summarise, the method is:

(a) find the $\lambda_n$ which allow (E7.1) a non-zero solution, by solving $|\mathsf{B} - \lambda\mathsf{A}| = 0$;

(b) for each $\lambda_n$ construct $\mathsf{u}^n$,

(c) construct the non-singular matrix $\mathsf{P}$ whose columns are the vectors $\mathsf{u}^n$,

(d) make the change of variable $\mathsf{u} = \mathsf{P}\mathsf{v}$.

7.19 (*It is recommended that the reader does not attempt this question until exercise 7.18 has been studied.*)

If, in the pendulum system studied in section 7.17 the string is replaced by a second rod identical to the first, the expressions for the kinetic energy $T$ and the potential energy $V$ become (to second order in the $\theta_i$)

$$T \approx Ml^2\left(\tfrac{8}{3}\dot\theta_1^2 + 2\dot\theta_1\dot\theta_2 + \tfrac{2}{3}\dot\theta_2^2\right),$$
$$V \approx Mgl\left(\tfrac{3}{2}\theta_1^2 + \tfrac{1}{2}\theta_2^2\right).$$

Determine the normal frequencies of the system and find new variables $\xi$ and $\eta$ that will reduce these two expressions to diagonal form, i.e. to

$$a_1\dot\xi^2 + a_2\dot\eta^2 \qquad \text{and} \qquad b_1\xi^2 + b_2\eta^2.$$

7.20 (*It is recommended that the reader does not attempt this question until exercise 7.18 has been studied.*)

Find a real linear transformation that simultaneously reduces the quadratic forms

$$3x^2 + 5y^2 + 5z^2 + 2yz + 6zx - 2xy,$$

$$5x^2 + 12y^2 + 8yz + 4zx$$

to diagonal form.

7.21 Three particles of mass $m$ are attached to a light horizontal string having fixed ends, the string being thus divided into four equal portions of length $a$ each under a tension $T$. Show that for small transverse vibrations the amplitudes $\mathsf{x}^i$ of the normal modes satisfy $\mathsf{Bx} = (ma\omega^2/T)\mathsf{x}$, where $\mathsf{B}$ is the matrix

$$\begin{pmatrix} 2 & -1 & 0 \\ -1 & 2 & -1 \\ 0 & -1 & 2 \end{pmatrix}.$$

Estimate the lowest and highest eigenfrequencies using the trial column vectors $(3 \;\; 4 \;\; 3)^{\mathrm{T}}$ and $(3 \;\; -4 \;\; 3)^{\mathrm{T}}$. Use also the exact column vectors $\begin{pmatrix}1 & \sqrt{2} & 1\end{pmatrix}^{\mathrm{T}}$ and $\begin{pmatrix}1 & -\sqrt{2} & 1\end{pmatrix}^{\mathrm{T}}$ and compare the results.

7.22 Use the Rayleigh–Ritz method to estimate the lowest oscillation frequency of a heavy chain of $N$ links, each of length $a$ $(= L/N)$, which hangs freely from one end. (Try simple calculable configurations such as all links but one vertical, or all links collinear, etc.)

## 7.20 Hints and answers

7.1 (a) $abc + 2fgh - af^2 - bg^2 - ch^2$; (b) 0; (c) $ab(ab - cd)$.

7.2 (a) No, $|\mathsf{A}| = -24 \neq 0$; Yes, $x : y : z = 4ab : 4a + b : 4a - b$.

7.3 (a) $x = a, b, c$; (b) $x = -1$, equation is linear in $x$.

7.4 $\eta = 1$, $x = 1 + 2z$, $y = -3z$; $\eta = 2$, $x = 2z$, $y = 1 - 3z$.

7.5 (a) iv, v, vii, x; (b) i, vi, ix, x.

7.7 (b) $(\sum_{ij} u_i G_{ij} u_j)^{1/2}$, $(\sum_{ij} v_i H_{ij} v_j)^{1/2}$, $\sum_i u_i v_i$;

(c) $$\mathsf{H} = \frac{1}{a^2}\begin{pmatrix} 3/2 & -1/2 & -1/2 \\ -1/2 & 3/2 & -1/2 \\ -1/2 & -1/2 & 3/2 \end{pmatrix}.$$

(d) (i) $M^{-1}$, (ii) $\cos^{-1}(p/Ma)$, where $M = a^{-1}[3(p^2 + q^2 + r^2)/2 - qr - pr - pq]^{1/2}$.

7.8 (d) $$\mathsf{S} = \begin{pmatrix} 0 & -\tan(\theta/2) \\ \tan(\theta/2) & 0 \end{pmatrix}.$$

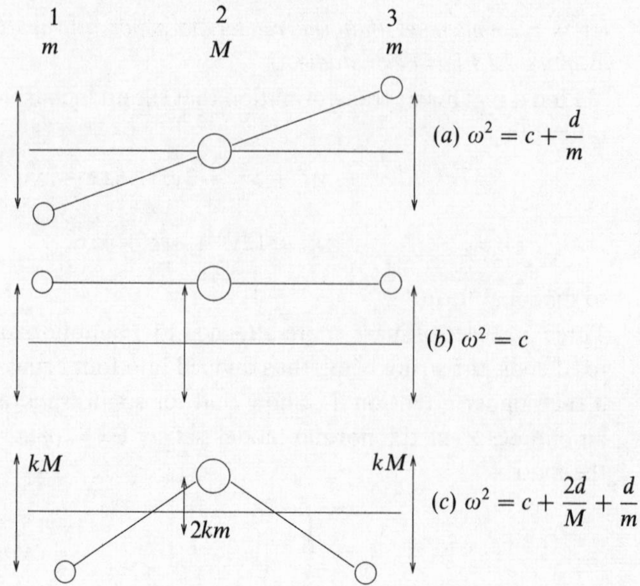


Figure 7.5 The normal modes, as viewed from above, of the coupled pendulums in example 7.16.

7.9 (b) Note that $|-\mathsf{A}| = (-1)^3|\mathsf{A}|$.

7.10 $a = b\cos\gamma + c\cos\beta$, and cyclic permutations; $a^2 = b^2 + c^2 - 2bc\cos\alpha$, and cyclic permutations.

7.11 $\lambda = 1$, $(0 \;\; 0 \;\; 1)^{\mathrm{T}}$; $\lambda = 1 + (\alpha\beta)^{1/2}$, $\left(\alpha^{1/2} \;\; \beta^{1/2} \;\; 0\right)^{\mathrm{T}}$; $\lambda = 1 - (\alpha\beta)^{1/2}$, $\left(\alpha^{1/2} \;\; -\beta^{1/2} \;\; 0\right)^{\mathrm{T}}$. (a) $\alpha\beta$ real and $> 0$; (b) $|\alpha| = |\beta|$.

7.12 $\lambda = 1$, $(1 \;\; 1 \;\; 3)^{\mathrm{T}}$; $\lambda = 3 + \sqrt{15}$, $\left(5 + \sqrt{15} \;\; 7 + 2\sqrt{15} \;\; -4 - \sqrt{15}\right)^{\mathrm{T}}$; $\lambda = 3 - \sqrt{15}$, $\left(5 - \sqrt{15} \;\; 7 - 2\sqrt{15} \;\; -4 + \sqrt{15}\right)^{\mathrm{T}}$.

7.13 Ellipse; $\theta = \pi/4$, $a = \sqrt{22}$; $\theta = 3\pi/4$, $b = \sqrt{10}$.

7.14 That of the eigenvector corresponding to the *smallest* eigenvalue is $\sqrt{3}\,(1 \;\; 1 \;\; 1)^{\mathrm{T}}$.

7.15 That of the eigenvector of the non-repeated eigenvalue is $\sqrt{3}\,(1 \;\; 1 \;\; -1)^{\mathrm{T}}$.

7.16 See figure 7.5.

7.17 $y_1 = 3^{-1/2}(x_1 + x_2 + x_3)$, $y_2 = 6^{-1/2}(x_1 - 2x_2 + x_3)$, $y_3 = 2^{-1/2}(-x_1 + x_3)$; $\mu_1 = 6$, $\mu_2 = -6$.

7.18 (a) Obtain $(\lambda^{(n)} - \lambda^{(m)})(\mathsf{u}^{(m)})^{\mathrm{T}}\mathsf{A}\mathsf{u}(n) = 0$;
(c) $c_{ij} = (\mathsf{P}^{\mathrm{T}}\mathsf{A}\mathsf{P})_{ij} = (\mathsf{P}^{\mathrm{T}})_{ik}A_{kl}P_{lj} = u_k^{(i)}A_{kl}u_l^{(j)} = (\mathsf{u}^{(i)})^{\mathrm{T}}\mathsf{A}\mathsf{u}^{(j)} = 0$ for $i \neq j$.

7.19 $\omega = (2.634g/l)^{1/2}$ and $(0.3661g/l)^{1/2}$; $\theta_1 = \xi + \eta$, $\theta_2 = 1.431\xi - 2.097\eta$.

7.20 $\lambda = -1, 2, 4$; $x = 2\xi - 2\eta + 2\chi$, $y = \xi + \eta + \chi$, $z = -3\xi + \eta - \chi$.

7.21 Estimated, $10/17 < Ma\omega^2/T < 58/17$; exact, $2 - \sqrt{2} \leq Ma\omega^2/T \leq 2 + \sqrt{2}$.

7.22 Collinear case gives the best estimate, $\omega^2 \leq 6n^2 g/4n^3 a \approx 3g/2l$.

# 8

# Vector calculus

In chapter 6 we discussed the algebra of vectors, and in chapter 7 we considered how to transform one vector into another using a linear operator. In this chapter and the next we discuss the calculus of vectors, i.e. the differentiation and integration both of vectors describing particular bodies, such as the velocity of a particle, and of vector fields, in which a vector is defined as a function of the coordinates throughout some volume (one-, two- or three-dimensional). Since the aim of this chapter is to develop methods for handling multidimensional physical situations, we will assume throughout that the functions with which we have to deal have sufficiently amenable mathematical properties, in particular that they are continuous and differentiable.

## 8.1 Differentiation of vectors

Let us consider a vector that is a function of a scalar variable $u$. By this we mean that with each value of $u$ we associate a vector $\mathbf{a}(u)$. For example, in Cartesian coordinates $\mathbf{a}(u) = a_x(u)\mathbf{i} + a_y(u)\mathbf{j} + a_z(u)\mathbf{k}$, where $a_x(u)$, $a_y(u)$ and $a_z(u)$ are scalar functions of $u$, and are the components of the vector $\mathbf{a}(u)$ in the $x$-, $y$- and $z$-directions respectively. We note that if $\mathbf{a}(u)$ is continuous at some point $u = u_0$, this implies that each of the Cartesian components $a_x(u)$, $a_y(u)$ and $a_z(u)$ are also continuous there.

Let us consider the derivative of the vector function $\mathbf{a}(u)$ with respect to $u$. The derivative of a vector function is defined in a similar manner to the ordinary derivative of a scalar function $f(x)$ given in chapter 1. The small change in the vector $\mathbf{a}(u)$ resulting from a small change $\Delta u$ in the value of $u$ is given by $\Delta\mathbf{a} = \mathbf{a}(u + \Delta u) - \mathbf{a}(u)$ (see figure 8.1). The derivative of $\mathbf{a}(u)$ with respect to $u$ is defined to be

$$\frac{d\mathbf{a}}{du} = \lim_{\Delta u \to 0} \frac{\mathbf{a}(u + \Delta u) - \mathbf{a}(u)}{\Delta u}, \tag{8.1}$$

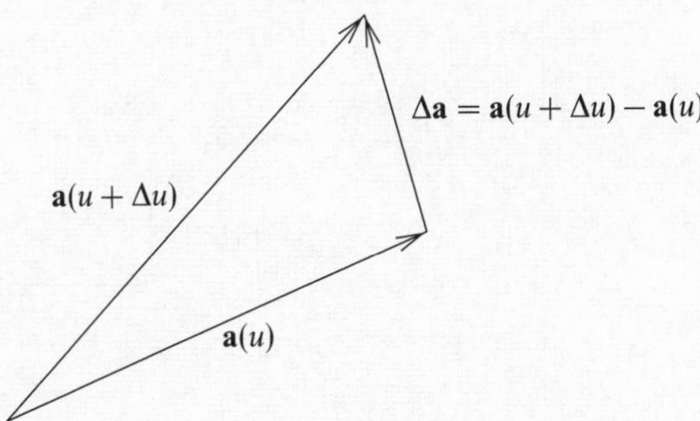


Figure 8.1 A small change in a vector $\mathbf{a}(u)$ resulting from a small change in $u$.

assuming that the limit exists, in which case $\mathbf{a}(u)$ is said to be differentiable at that point. Note that $d\mathbf{a}/du$ is also a vector, which is not, in general, parallel to $\mathbf{a}(u)$. In Cartesian coordinates, the derivative of the vector $\mathbf{a}(u) = a_x\mathbf{i} + a_y\mathbf{j} + a_z\mathbf{k}$ is given by

$$\frac{d\mathbf{a}}{du} = \frac{da_x}{du}\mathbf{i} + \frac{da_y}{du}\mathbf{j} + \frac{da_z}{du}\mathbf{k}.$$

Perhaps the simplest application of the above is to finding the velocity and acceleration of a particle in classical mechanics. If the time-dependent position vector of the particle with respect to the origin in Cartesian coordinates is given by $\mathbf{r}(t) = x(t)\mathbf{i} + y(t)\mathbf{j} + z(t)\mathbf{k}$, then the velocity of the particle is given by the vector

$$\mathbf{v}(t) = \frac{d\mathbf{r}}{dt} = \frac{dx}{dt}\mathbf{i} + \frac{dy}{dt}\mathbf{j} + \frac{dz}{dt}\mathbf{k}.$$

The direction of the velocity vector is along the tangent to the path $\mathbf{r}(t)$ at the instantaneous position of the particle, and its magnitude $|\mathbf{v}(t)|$ is equal to the speed of the particle. The acceleration of the particle is given in a similar manner by

$$\mathbf{a}(t) = \frac{d\mathbf{v}}{dt} = \frac{d^2x}{dt^2}\mathbf{i} + \frac{d^2y}{dt^2}\mathbf{j} + \frac{d^2z}{dt^2}\mathbf{k}.$$

►*The position vector of a particle at time $t$ in Cartesian coordinates is given by $\mathbf{r}(t) = 2t^2\mathbf{i} + (3t - 2)\mathbf{j} + (3t^2 - 1)\mathbf{k}$. Find the speed of the particle at $t = 1$, and the component of its acceleration in the direction $\mathbf{s} = \mathbf{i} + 2\mathbf{j} + \mathbf{k}$.*

The velocity and acceleration of the particle are given by

$$\mathbf{v}(t) = \frac{d\mathbf{r}}{dt} = 4t\mathbf{i} + 3\mathbf{j} + 6t\mathbf{k},$$

$$\mathbf{a}(t) = \frac{d\mathbf{v}}{dt} = 4\mathbf{i} + 6\mathbf{k}.$$

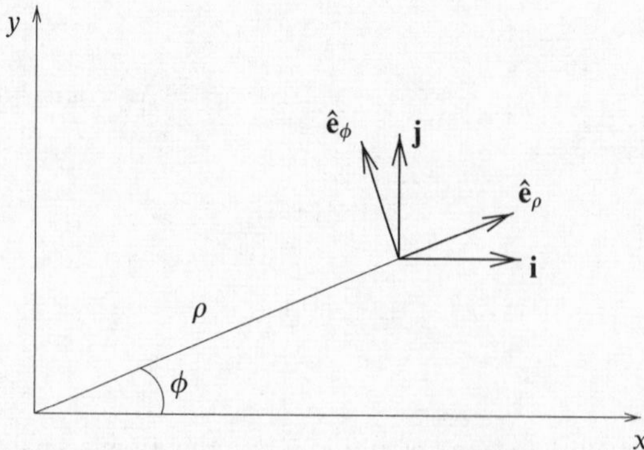


Figure 8.2 Unit basis vectors for two-dimensional Cartesian and plane polar coordinates.

The speed of the particle at $t = 1$ is simply

$$|\mathbf{v}(1)| = \sqrt{4^2 + 3^2 + 6^2} = \sqrt{61}.$$

The acceleration of the particle is constant (independent of $t$), and its component in the direction $\mathbf{s}$ is given by

$$\mathbf{a} \cdot \hat{\mathbf{s}} = \frac{(4\mathbf{i} + 6\mathbf{k}) \cdot (\mathbf{i} + 2\mathbf{j} + \mathbf{k})}{\sqrt{1^2 + 2^2 + 1^2}} = \frac{5\sqrt{6}}{3}. \blacktriangleleft$$

Note that in the case discussed above $\mathbf{i}$, $\mathbf{j}$ and $\mathbf{k}$ are fixed, time-independent basis vectors. This may not be true of basis vectors in general; when we are not using Cartesian coordinates the basis vectors themselves must also be differentiated. We discuss basis vectors for non-Cartesian coordinate systems in detail in section 8.10. Nevertheless, as a simple example, let us now consider two-dimensional plane polar coordinates $\rho, \phi$.

Referring to figure 8.2, imagine holding $\phi$ fixed and moving radially outwards, i.e. in the direction of increasing $\rho$. Let us denote the unit vector in this direction by $\hat{\mathbf{e}}_\rho$. Similarly, imagine keeping $\rho$ fixed and moving around a circle of fixed radius in the direction of increasing $\phi$. Let us denote the unit vector tangent to the circle as $\hat{\mathbf{e}}_\phi$. The two vectors $\hat{\mathbf{e}}_\rho$ and $\hat{\mathbf{e}}_\phi$ are the basis vectors for this two-dimensional coordinate system, just as $\mathbf{i}$ and $\mathbf{j}$ are basis vectors for two-dimensional Cartesian coordinates. All these basis vectors are shown in figure 8.2.

An important difference between the two sets of basis vectors is that while $\mathbf{i}$ and $\mathbf{j}$ are constant in magnitude *and direction*, the vectors $\hat{\mathbf{e}}_\rho$ and $\hat{\mathbf{e}}_\phi$ have constant magnitudes but their directions change as $\rho$ and $\phi$ vary. Therefore, when calculating the derivative of a vector written in polar coordinates, we must also differentiate the basis vectors. One way of doing this is to express $\hat{\mathbf{e}}_\rho$ and $\hat{\mathbf{e}}_\phi$

in terms of **i** and **j**. From figure 8.2, we see that

$$\hat{\mathbf{e}}_\rho = \cos\phi\,\mathbf{i} + \sin\phi\,\mathbf{j},$$
$$\hat{\mathbf{e}}_\phi = -\sin\phi\,\mathbf{i} + \cos\phi\,\mathbf{j}.$$

Since **i** and **j** are constant vectors, we find that the derivatives of the basis vectors $\hat{\mathbf{e}}_\rho$ and $\hat{\mathbf{e}}_\phi$ with respect to $t$ are given by

$$\frac{d\hat{\mathbf{e}}_\rho}{dt} = -\sin\phi\frac{d\phi}{dt}\mathbf{i} + \cos\phi\frac{d\phi}{dt}\mathbf{j} = \dot{\phi}\,\hat{\mathbf{e}}_\phi, \qquad (8.2)$$
$$\frac{d\hat{\mathbf{e}}_\phi}{dt} = -\cos\phi\frac{d\phi}{dt}\mathbf{i} - \sin\phi\frac{d\phi}{dt}\mathbf{j} = -\dot{\phi}\,\hat{\mathbf{e}}_\rho, \qquad (8.3)$$

where the overdot is the conventional notation for differentiation with respect to time.

►*The position vector of a particle in plane polar coordinates is* $\mathbf{r}(t) = \rho(t)\hat{\mathbf{e}}_\rho$. *Find expressions for the velocity and acceleration of the particle in these coordinates.*

Using result (8.4) below, the velocity of the particle is given by

$$\mathbf{v}(t) = \dot{\mathbf{r}}(t) = \dot{\rho}\,\hat{\mathbf{e}}_\rho + \rho\dot{\hat{\mathbf{e}}}_\rho = \dot{\rho}\,\hat{\mathbf{e}}_\rho + \rho\dot{\phi}\,\hat{\mathbf{e}}_\phi,$$

where we have used (8.2). In a similar way its acceleration is given by

$$\begin{aligned}\mathbf{a}(t) &= \frac{d}{dt}(\dot{\rho}\,\hat{\mathbf{e}}_\rho + \rho\dot{\phi}\,\hat{\mathbf{e}}_\phi)\\ &= \ddot{\rho}\,\hat{\mathbf{e}}_\rho + \dot{\rho}\dot{\hat{\mathbf{e}}}_\rho + \rho\dot{\phi}\dot{\hat{\mathbf{e}}}_\phi + \rho\ddot{\phi}\,\hat{\mathbf{e}}_\phi + \dot{\rho}\dot{\phi}\,\hat{\mathbf{e}}_\phi\\ &= \ddot{\rho}\,\hat{\mathbf{e}}_\rho + \dot{\rho}(\dot{\phi}\,\hat{\mathbf{e}}_\phi) + \rho\dot{\phi}(-\dot{\phi}\,\hat{\mathbf{e}}_\rho) + \rho\ddot{\phi}\,\hat{\mathbf{e}}_\phi + \dot{\rho}\dot{\phi}\,\hat{\mathbf{e}}_\phi\\ &= (\ddot{\rho} - \rho\dot{\phi}^2)\,\hat{\mathbf{e}}_\rho + (\rho\ddot{\phi} + 2\dot{\rho}\dot{\phi})\,\hat{\mathbf{e}}_\phi. \blacktriangleleft\end{aligned}$$

Here we have used (8.2) and (8.3).

### *8.1.1 Differentiation of composite vector expressions*

In composite vector expressions each of the vectors or scalars involved may be a function of some scalar variable $u$, as we have seen. The derivatives of such expressions are easily found using the definition (8.1) and the rules of ordinary differential calculus. They may be summarised by the following, assuming that **a** and **b** are differentiable vector functions of a scalar $u$, and that $\phi$ is a differentiable scalar function of $u$:

$$\frac{d}{du}(\phi\mathbf{a}) = \phi\frac{d\mathbf{a}}{du} + \frac{d\phi}{du}\mathbf{a}, \qquad (8.4)$$
$$\frac{d}{du}(\mathbf{a}\cdot\mathbf{b}) = \mathbf{a}\cdot\frac{d\mathbf{b}}{du} + \frac{d\mathbf{a}}{du}\cdot\mathbf{b}, \qquad (8.5)$$
$$\frac{d}{du}(\mathbf{a}\times\mathbf{b}) = \mathbf{a}\times\frac{d\mathbf{b}}{du} + \frac{d\mathbf{a}}{du}\times\mathbf{b}, \qquad (8.6)$$

where of course the order of the factors in the terms on the RHS of (8.6) is just as important as it is in the original vector product.

►*A particle of mass $m$ with position vector $\mathbf{r}$ relative to some origin, $O$, experiences a force $\mathbf{F}$, which produces a torque (moment) $\mathbf{T} = \mathbf{r} \times \mathbf{F}$ about $O$. The angular momentum of the particle about $O$ is given by $\mathbf{L} = \mathbf{r} \times m\mathbf{v}$, where $\mathbf{v}$ is the particle's velocity. Show that the rate of change of angular momentum is equal to the applied torque.*

The rate of change of angular momentum is given by

$$\frac{d\mathbf{L}}{dt} = \frac{d}{dt}(\mathbf{r} \times m\mathbf{v}).$$

Using (8.6) we obtain

$$\begin{aligned}\frac{d\mathbf{L}}{dt} &= \frac{d\mathbf{r}}{dt} \times m\mathbf{v} \ + \ \mathbf{r} \times \frac{d}{dt}(m\mathbf{v}) \\ &= \mathbf{v} \times m\mathbf{v} \ + \ \mathbf{r} \times \frac{d}{dt}(m\mathbf{v}) \\ &= \mathbf{0} \ + \ \mathbf{r} \times \mathbf{F} = \mathbf{T},\end{aligned}$$

where in the last line we use Newton's second law, namely $\mathbf{F} = d(m\mathbf{v})/dt$. ◄

If a vector $\mathbf{a}(s)$ is a function of the scalar variable $s$, which is itself a function of $u$ such that $s = s(u)$, then the chain rule (see subsection 1.1.3) gives

$$\frac{d\mathbf{a}(s)}{du} = \frac{ds}{du}\frac{d\mathbf{a}}{ds}. \tag{8.7}$$

The derivatives of more complicated vector expressions may be found by repeated application of the above equations.

One further useful result can be derived by considering the derivative

$$\frac{d}{du}(\mathbf{a} \cdot \mathbf{a}) = 2\mathbf{a} \cdot \frac{d\mathbf{a}}{du};$$

since $\mathbf{a} \cdot \mathbf{a} = a^2$, where $a = |\mathbf{a}|$, we see that

$$\mathbf{a} \cdot \frac{d\mathbf{a}}{du} = 0 \quad \text{if } a \text{ is constant.} \tag{8.8}$$

In other words, if a vector $\mathbf{a}(u)$ has a constant magnitude as $u$ varies, then it is perpendicular to the vector $d\mathbf{a}/du$.

### *8.1.2 Differential of a vector*

As a final note on the differentiation of vectors, we can also define the *differential* of a vector in a similar way to that of a scalar in ordinary differential calculus. In the definition of the vector derivative (8.1), we used the notion of a small change

$\Delta\mathbf{a}$ in a vector $\mathbf{a}(u)$ resulting from a small change $\Delta u$ in its argument. In the limit $\Delta u \to 0$, the change in $\mathbf{a}$ becomes infinitesimally small, and we denote it by the differential $d\mathbf{a}$. From (8.1) we see that the differential is given by

$$d\mathbf{a} = \frac{d\mathbf{a}}{du}\,du. \tag{8.9}$$

Note that the differential of a vector is also a vector. As an example, the infinitesimal change in the position vector of a particle in an infinitesimal time $dt$ is

$$d\mathbf{r} = \frac{d\mathbf{r}}{dt}\,dt = \mathbf{v}\,dt,$$

where $\mathbf{v}$ is the particle's velocity.

## 8.2 Integration of vectors

The integration of a vector (or of an expression involving vectors that may itself be either a vector or scalar) with respect to a scalar $u$ can be regarded as the inverse of differentiation. We must remember, however, that

(i) the integral has the same nature (vector or scalar) as the integrand,
(ii) the constant of integration for indefinite integrals must be of the same nature as the integral.

For example, if $\mathbf{a}(u) = d[\mathbf{A}(u)]/du$, then the indefinite integral of $\mathbf{a}(u)$ is given by

$$\int \mathbf{a}(u)\,du = \mathbf{A}(u) + \mathbf{b},$$

where $\mathbf{b}$ is a constant vector. The definite integral of $\mathbf{a}(u)$ from $u = u_1$ to $u = u_2$ is given by

$$\int_{u_1}^{u_2} \mathbf{a}(u)\,du = \mathbf{A}(u_2) - \mathbf{A}(u_1).$$

► *A small particle of mass $m$ orbits a much larger mass $M$ centred at the origin $O$. According to Newton's law of gravitation, the position vector $\mathbf{r}$ of the small mass obeys the differential equation*

$$m\frac{d^2\mathbf{r}}{dt^2} = -\frac{GMm}{r^2}\,\hat{\mathbf{r}}.$$

*Show that the vector $\mathbf{r} \times d\mathbf{r}/dt$ is a constant of the motion.*

Forming the vector product of the differential equation with $\mathbf{r}$, we obtain

$$\mathbf{r} \times \frac{d^2\mathbf{r}}{dt^2} = -\frac{GM}{r^2}\mathbf{r} \times \hat{\mathbf{r}}.$$

Since $\mathbf{r}$ and $\hat{\mathbf{r}}$ are collinear $\mathbf{r} \times \hat{\mathbf{r}} = \mathbf{0}$, and therefore we have

$$\mathbf{r} \times \frac{d^2\mathbf{r}}{dt^2} = \mathbf{0}. \tag{8.10}$$

However,

$$\frac{d}{dt}\left(\mathbf{r} \times \frac{d\mathbf{r}}{dt}\right) = \mathbf{r} \times \frac{d^2\mathbf{r}}{dt^2} + \frac{d\mathbf{r}}{dt} \times \frac{d\mathbf{r}}{dt} = \mathbf{0},$$

since the first term is zero by (8.10), and the second because it is the vector product of two parallel (in this case identical) vectors. Integrating, we obtain the required result

$$\mathbf{r} \times \frac{d\mathbf{r}}{dt} = \mathbf{c}, \tag{8.11}$$

where $\mathbf{c}$ is a constant vector.

As a further point of interest we may note that in an infinitesimal time $dt$ the change in the position vector of the small mass is $d\mathbf{r}$ and the element of area swept out by the position vector of the particle is simply $dA = \frac{1}{2}|\mathbf{r} \times d\mathbf{r}|$. Dividing both sides of this equation by $dt$, we conclude that

$$\frac{dA}{dt} = \frac{1}{2}\left|\mathbf{r} \times \frac{d\mathbf{r}}{dt}\right| = \frac{|\mathbf{c}|}{2},$$

and that the physical interpretation of the above result (8.11) is that the position vector $\mathbf{r}$ of the small mass sweeps out equal areas in equal times. This result is in fact valid for motion under any force that acts along the line joining the two particles. ◄

## 8.3 Space curves

In the previous section we mentioned that the velocity vector of a particle is a tangent to the curve in space along which the particle moves. We now give a more complete discussion of curves in space, and the geometrical interpretation of the vector derivative.

A curve $C$ in space can be described by the vector $\mathbf{r}(u)$ joining the origin $O$ of a coordinate system to a point on the curve (see figure 8.3). As the parameter $u$ varies, the end-point of the vector moves along the curve. In Cartesian coordinates,

$$\mathbf{r}(u) = x(u)\mathbf{i} + y(u)\mathbf{j} + z(u)\mathbf{k},$$

where $x = x(u)$, $y = y(u)$ and $z = z(u)$ are the *parametric* equations of the curve. This parametric representation can be very useful, particularly in mechanics when the parameter may be the time $t$. We can, however, also represent a space curve by $y = f(x)$, $z = g(x)$, which can be easily converted into the parametric form by setting $u = x$, so that

$$\mathbf{r}(u) = u\mathbf{i} + f(u)\mathbf{j} + g(u)\mathbf{k}.$$

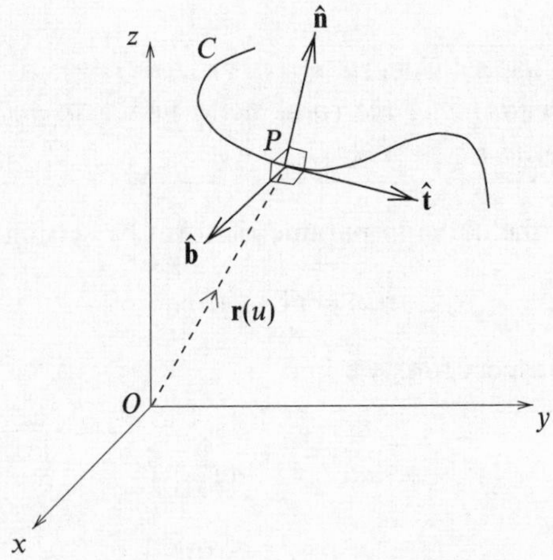


Figure 8.3 The unit tangent $\hat{\mathbf{t}}$, normal $\hat{\mathbf{n}}$ and binormal $\hat{\mathbf{b}}$ to the space curve $C$ at a particular point $P$.

Alternatively, a space curve can be represented in the form $F(x, y, z) = 0$, $G(x, y, z) = 0$, where each equation represents a surface and the curve is the intersection of the two surfaces.

A curve may sometimes be described in parametric form by the vector $\mathbf{r}(s)$, where the parameter $s$ is the *arc length* along the curve measured from a fixed point. Even when the curve is expressed in terms of some other parameter, it is straightforward to find the arc length between any two points on the curve. For the curve described by $\mathbf{r}(u)$, let us consider an infinitesimal vector displacement

$$d\mathbf{r} = dx\,\mathbf{i} + dy\,\mathbf{j} + dz\,\mathbf{k}$$

along the curve. The square of the infinitesimal distance moved is then given by

$$(ds)^2 = d\mathbf{r}\cdot d\mathbf{r} = (dx)^2 + (dy)^2 + (dz)^2,$$

from which it can be shown that

$$\left(\frac{ds}{du}\right)^2 = \frac{d\mathbf{r}}{du}\cdot\frac{d\mathbf{r}}{du}.$$

Therefore, the arc length between two points on the curve $\mathbf{r}(u)$, given by $u = u_1$ and $u = u_2$, is

$$s = \int_{u_1}^{u_2} \sqrt{\frac{d\mathbf{r}}{du}\cdot\frac{d\mathbf{r}}{du}}\,du. \tag{8.12}$$

► *A curve lying in the xy-plane is given by $y = y(x)$, $z = 0$. Using (8.12), show that the arc length along the curve between $x = a$ and $x = b$ is given by $s = \int_a^b \sqrt{1+y'^2}\,dx$, where $y' = dy/dx$.*

Let us first represent the curve in parametric form by setting $u = x$, so that

$$\mathbf{r}(u) = u\mathbf{i} + y(u)\mathbf{j}.$$

Differentiating with respect to $u$, we find

$$\frac{d\mathbf{r}}{du} = \mathbf{i} + \frac{dy}{du}\mathbf{j},$$

from which we obtain

$$\frac{d\mathbf{r}}{du}\cdot\frac{d\mathbf{r}}{du} = 1 + \left(\frac{dy}{du}\right)^2.$$

Therefore, remembering that $u = x$, from (8.12) the arc length between $x = a$ and $x = b$ is given by

$$s = \int_a^b \sqrt{\frac{d\mathbf{r}}{du}\cdot\frac{d\mathbf{r}}{du}}\,du = \int_a^b \sqrt{1+\left(\frac{dy}{dx}\right)^2}\,dx.$$

This result was derived using more elementary methods in chapter 1. ◄

If a curve $C$ is described by $\mathbf{r}(u)$, then by considering figures 8.1 and 8.3, we see that, at any given point on the curve, $d\mathbf{r}/du$ is a vector tangent to $C$ at that point, in the direction of increasing $u$. In the special case where the parameter $u$ is the arc length $s$ along the curve then $d\mathbf{r}/ds$ is a *unit* tangent vector to $C$ and is denoted by $\hat{\mathbf{t}}$.

The rate at which the unit tangent $\hat{\mathbf{t}}$ changes with respect to $s$ is given by $d\hat{\mathbf{t}}/ds$, and its magnitude is defined as the *curvature* $\kappa$ of the curve $C$ at a given point,

$$\kappa = \left|\frac{d\hat{\mathbf{t}}}{ds}\right| = \left|\frac{d^2\hat{\mathbf{r}}}{ds^2}\right|.$$

We can also define the quantity $\rho = 1/\kappa$, which is called the *radius of curvature*.

Since $\hat{\mathbf{t}}$ is of constant (unit) magnitude, it follows from (8.8) that it is perpendicular to $d\hat{\mathbf{t}}/ds$. The unit vector in the direction perpendicular to $\hat{\mathbf{t}}$ is denoted by $\hat{\mathbf{n}}$, and is called the *principal normal* at the point. We therefore have

$$\frac{d\hat{\mathbf{t}}}{ds} = \kappa\,\hat{\mathbf{n}}. \tag{8.13}$$

The unit vector $\hat{\mathbf{b}} = \hat{\mathbf{t}} \times \hat{\mathbf{n}}$, which is perpendicular to the plane containing $\hat{\mathbf{t}}$ and $\hat{\mathbf{n}}$, is called the *binormal* to $C$. The vectors $\hat{\mathbf{t}}$, $\hat{\mathbf{n}}$ and $\hat{\mathbf{b}}$ form a right-handed rectangular cooordinate system (or *triad*) at any given point on $C$ (see figure 8.3). As $s$ changes so that the point of interest moves along $C$, the triad of vectors also changes.

The rate at which $\hat{\mathbf{b}}$ changes with respect to $s$ is given by $d\hat{\mathbf{b}}/ds$, and is a measure of the *torsion* $\tau$ of the curve at any given point. Since $\hat{\mathbf{b}}$ is of constant magnitude, from (8.8) it is perpendicular to $d\hat{\mathbf{b}}/ds$. We may further show that $d\hat{\mathbf{b}}/ds$ is also perpendicular to $\hat{\mathbf{t}}$, as follows. By definition $\hat{\mathbf{b}}\cdot\hat{\mathbf{t}} = 0$, which on differentiating yields

$$\begin{aligned} 0 = \frac{d}{ds}\left(\hat{\mathbf{b}}\cdot\hat{\mathbf{t}}\right) &= \frac{d\hat{\mathbf{b}}}{ds}\cdot\hat{\mathbf{t}} + \hat{\mathbf{b}}\cdot\frac{d\hat{\mathbf{t}}}{ds} \\ &= \frac{d\hat{\mathbf{b}}}{ds}\cdot\hat{\mathbf{t}} + \hat{\mathbf{b}}\cdot\kappa\hat{\mathbf{n}} \\ &= \frac{d\hat{\mathbf{b}}}{ds}\cdot\hat{\mathbf{t}}, \end{aligned}$$

where we have used the fact that $\hat{\mathbf{b}}\cdot\hat{\mathbf{n}} = 0$. Hence, since $d\hat{\mathbf{b}}/ds$ is perpendicular to both $\hat{\mathbf{b}}$ and $\hat{\mathbf{t}}$, we must have $d\hat{\mathbf{b}}/ds \propto \hat{\mathbf{n}}$. The constant of proportionality is $-\tau$, so we finally obtain

$$\frac{d\hat{\mathbf{b}}}{ds} = -\tau\,\hat{\mathbf{n}}. \tag{8.14}$$

Taking the dot product of each side with $\hat{\mathbf{n}}$, we see that the torsion of a curve is given by

$$\tau = -\hat{\mathbf{n}}\cdot\frac{d\hat{\mathbf{b}}}{ds}.$$

We may also define the quantity $\sigma = 1/\tau$, which is called the *radius of torsion*.

Finally, we consider the derivative $d\hat{\mathbf{n}}/ds$. Since $\hat{\mathbf{n}} = \hat{\mathbf{b}}\times\hat{\mathbf{t}}$ we have

$$\begin{aligned} \frac{d\hat{\mathbf{n}}}{ds} &= \frac{d\hat{\mathbf{b}}}{ds}\times\hat{\mathbf{t}} + \hat{\mathbf{b}}\times\frac{d\hat{\mathbf{t}}}{ds} \\ &= -\tau\,\hat{\mathbf{n}}\times\hat{\mathbf{t}} + \hat{\mathbf{b}}\times\kappa\,\hat{\mathbf{n}} \\ &= \tau\,\hat{\mathbf{b}} - \kappa\,\hat{\mathbf{t}}. \end{aligned} \tag{8.15}$$

In summary, $\hat{\mathbf{t}}$, $\hat{\mathbf{n}}$ and $\hat{\mathbf{b}}$ and their derivatives with respect to $s$ are related to one another by the relations (8.13), (8.14) and (8.15), the *Frenet-Serret formulae*,

$$\frac{d\hat{\mathbf{t}}}{ds} = \kappa\,\hat{\mathbf{n}}, \qquad \frac{d\hat{\mathbf{n}}}{ds} = \tau\,\hat{\mathbf{b}} - \kappa\,\hat{\mathbf{t}}, \qquad \frac{d\hat{\mathbf{b}}}{ds} = -\tau\,\hat{\mathbf{n}}. \tag{8.16}$$

►*Show that the acceleration of a particle travelling along a trajectory* $\mathbf{r}(t)$ *is given by*

$$\mathbf{a}(t) = \frac{dv}{dt}\,\hat{\mathbf{t}} + \frac{v^2}{\rho}\,\hat{\mathbf{n}},$$

*where* $v$ *is the speed of the particle,* $\hat{\mathbf{t}}$ *is the unit tangent to the trajectory,* $\hat{\mathbf{n}}$ *is its principal normal and* $\rho$ *is its radius of curvature.*

The velocity of the particle is given by

$$\mathbf{v}(t) = \frac{d\mathbf{r}}{dt} = \frac{d\mathbf{r}}{ds}\frac{ds}{dt} = \frac{ds}{dt}\,\hat{\mathbf{t}},$$

where $ds/dt$ is the speed of the particle, which we denote by $v$, and $\hat{\mathbf{t}}$ is the unit vector tangent to the trajectory. Writing the velocity as $\mathbf{v} = v\,\hat{\mathbf{t}}$, and differentiating once more with respect to time $t$, we obtain

$$\mathbf{a}(t) = \frac{d\mathbf{v}}{dt} = \frac{dv}{dt}\,\hat{\mathbf{t}} + v\frac{d\hat{\mathbf{t}}}{dt};$$

but we note that

$$\frac{d\hat{\mathbf{t}}}{dt} = \frac{ds}{dt}\frac{d\hat{\mathbf{t}}}{ds} = v\kappa\,\hat{\mathbf{n}} = \frac{v}{\rho}\,\hat{\mathbf{n}}.$$

Therefore, we have

$$\mathbf{a}(t) = \frac{dv}{dt}\,\hat{\mathbf{t}} + \frac{v^2}{\rho}\,\hat{\mathbf{n}}.$$

This shows that in addition to an acceleration $dv/dt$ along the tangent to the particle's trajectory, there is also an acceleration $v^2/\rho$ in the direction of the principal normal. The latter is often called the *centripetal* acceleration. ◄

Finally, we note that a curve $\mathbf{r}(u)$ representing the trajectory of a particle may sometimes be given in terms of some parameter $u$ that is not necessarily equal to the time $t$ but is functionally related to it in some way. In this case the velocity of the particle is given by

$$\mathbf{v} = \frac{d\mathbf{r}}{dt} = \frac{d\mathbf{r}}{du}\frac{du}{dt}.$$

Differentiating again with respect to time gives the acceleration as

$$\mathbf{a} = \frac{d\mathbf{v}}{dt} = \frac{d}{dt}\left(\frac{d\mathbf{r}}{du}\frac{du}{dt}\right) = \frac{d^2\mathbf{r}}{du^2}\left(\frac{du}{dt}\right)^2 + \frac{d\mathbf{r}}{du}\frac{d^2u}{dt^2}.$$

## 8.4 Vector functions of several arguments

The concept of the derivative of a vector is easily extended to cases where the vectors (or scalars) are functions of more than one independent scalar variable,

$u_1, u_2, \ldots, u_n$. In this case, the results of subsection 8.1.1 are still valid, except that the derivatives become partial derivatives $\partial\mathbf{a}/\partial u_i$ defined as in ordinary differential calculus. For example, in Cartesian coordinates,

$$\frac{\partial\mathbf{a}}{\partial u} = \frac{\partial a_x}{\partial u}\mathbf{i} + \frac{\partial a_y}{\partial u}\mathbf{j} + \frac{\partial a_z}{\partial u}\mathbf{k}.$$

In particular, (8.7) generalises to the chain rule of partial differentiation discussed in section 4.5. If $\mathbf{a} = \mathbf{a}(u_1, u_2, \ldots, u_n)$ and each of the $u_i$ is also a function $u_i(v_1, v_2, \ldots, v_n)$ of the variables $v_i$ then, generalising (4.17),

$$\frac{\partial\mathbf{a}}{\partial v_i} = \frac{\partial\mathbf{a}}{\partial u_1}\frac{\partial u_1}{\partial v_i} + \frac{\partial\mathbf{a}}{\partial u_2}\frac{\partial u_2}{\partial v_i} + \cdots + \frac{\partial\mathbf{a}}{\partial u_n}\frac{\partial u_n}{\partial v_i} = \sum_{j=1}^{n}\frac{\partial\mathbf{a}}{\partial u_j}\frac{\partial u_j}{\partial v_i}. \tag{8.17}$$

A special case of this rule arises when $\mathbf{a}$ is an explicit function of some variable $v$, as well as of scalars $u_1, u_2, \ldots, u_n$ that are themselves functions of $v$. Then we have

$$\frac{d\mathbf{a}}{dv} = \frac{\partial\mathbf{a}}{\partial v} + \sum_{j=1}^{n}\frac{\partial\mathbf{a}}{\partial u_j}\frac{\partial u_j}{\partial v}. \tag{8.18}$$

We may also extend the concept of the differential of a vector given in (8.9) to vectors dependent on several variables $u_1, u_2, \ldots, u_n$:

$$d\mathbf{a} = \frac{\partial\mathbf{a}}{\partial u_1}\,du_1 + \frac{\partial\mathbf{a}}{\partial u_2}\,du_2 + \cdots + \frac{\partial\mathbf{a}}{\partial u_n}\,du_n = \sum_{j=1}^{n}\frac{\partial\mathbf{a}}{\partial u_j}\,du_j. \tag{8.19}$$

As an example, the infinitesimal change in an electric field $\mathbf{E}$ in moving from a position $\mathbf{r}$ to a neighbouring one $\mathbf{r} + d\mathbf{r}$ is given by

$$d\mathbf{E} = \frac{\partial\mathbf{E}}{\partial x}\,dx + \frac{\partial\mathbf{E}}{\partial y}\,dy + \frac{\partial\mathbf{E}}{\partial z}\,dz. \tag{8.20}$$

## 8.5 Surfaces

A surface $S$ in space can be described by the vector $\mathbf{r}(u, v)$ joining the origin, $O$, of a coordinate system to a point on the surface (see figure 8.4). As the parameters $u$ and $v$ vary, the end-point of the vector moves over the surface. This is very similar to the parametric representation $\mathbf{r}(u)$ of a curve discussed in section 8.3, but with the important difference that we require *two* parameters to describe a surface, whereas we needed only one to describe a curve.

In Cartesian coordinates the surface is given by

$$\mathbf{r}(u, v) = x(u, v)\mathbf{i} + y(u, v)\mathbf{j} + z(u, v)\mathbf{k},$$

where $x = x(u, v)$, $y = y(u, v)$ and $z = z(u, v)$ are the parametric equations of the surface. We can also represent a surface by $z = f(x, y)$ or $g(x, y, z) = 0$. Either of these representations can be converted into the parametric form in a similar

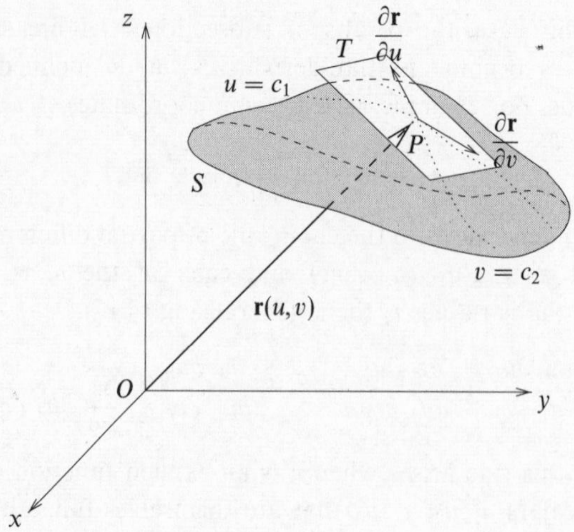


Figure 8.4 The tangent plane $T$ to a surface $S$ at a particular point $P$; $u = c_1$ and $v = c_2$ are the coordinate curves.

manner to that used for equations of curves. For example, if $z = f(x, y)$ then by setting $u = x$ and $v = y$ the surface can be represented in parametric form by

$$\mathbf{r}(u, v) = u\mathbf{i} + v\mathbf{j} + f(u, v)\mathbf{k}.$$

Any curve $\mathbf{r}(\lambda)$, where $\lambda$ is a parameter, on the surface $S$ can be represented by a pair of equations relating the parameters $u$ and $v$, for example $u = f(\lambda)$ and $v = g(\lambda)$. A parametric representation of the curve can easily be found by straightforward substitution, i.e. $\mathbf{r}(\lambda) = \mathbf{r}(u(\lambda), v(\lambda))$. Using (8.17), for the case where the vector is a function of a single variable $\lambda$ so that the LHS becomes a total derivative, the tangent to the curve $\mathbf{r}(\lambda)$ at any point is given by

$$\frac{d\mathbf{r}}{d\lambda} = \frac{\partial \mathbf{r}}{\partial u}\frac{du}{d\lambda} + \frac{\partial \mathbf{r}}{\partial v}\frac{dv}{d\lambda}. \tag{8.21}$$

The two curves $u =$ constant and $v =$ constant passing through any point $P$ on $S$ are called *coordinate curves*. For the curve $u =$ constant, for example, we have $du/d\lambda = 0$, and so from (8.21) its tangent vector is in the direction $\partial\mathbf{r}/\partial v$. Similarly, the tangent vector to the curve $v =$ constant is in the direction $\partial\mathbf{r}/\partial u$.

If the surface is smooth, then at any point $P$ on $S$ the vectors $\partial\mathbf{r}/\partial u$ and $\partial\mathbf{r}/\partial v$ are linearly independent, and define the *tangent plane* $T$ at the point $P$ (see figure 8.4). A vector normal to the surface at $P$ is given by

$$\mathbf{n} = \frac{\partial \mathbf{r}}{\partial u} \times \frac{\partial \mathbf{r}}{\partial v}. \tag{8.22}$$

In the neighbourhood of $P$, an infinitesimal vector displacement $d\mathbf{r}$ is written

$$d\mathbf{r} = \frac{\partial \mathbf{r}}{\partial u}\,du + \frac{\partial \mathbf{r}}{\partial v}\,dv.$$

If we consider an infinitesimal parallelogram near $P$, whose sides are the coordinate curves, then the *element of area* at $P$ is

$$dS = \left|\frac{\partial \mathbf{r}}{\partial u}\,du \times \frac{\partial \mathbf{r}}{\partial v}\,dv\right| = \left|\frac{\partial \mathbf{r}}{\partial u} \times \frac{\partial \mathbf{r}}{\partial v}\right| du\,dv. \qquad (8.23)$$

Therefore the total area of the surface is simply

$$A = \iint_R \left|\frac{\partial \mathbf{r}}{\partial u} \times \frac{\partial \mathbf{r}}{\partial v}\right| du\,dv, \qquad (8.24)$$

where $R$ is the region in the $uv$-plane corresponding to the range of parameter values that define the surface.

► *Find the element of area on the surface of a sphere of radius $a$, and hence calculate its total surface area.*

We can represent a point $\mathbf{r}$ on the surface of the sphere in terms of the two parameters $\theta$ and $\phi$:

$$\mathbf{r}(\theta, \phi) = a\sin\theta\cos\phi\,\mathbf{i} + a\sin\theta\sin\phi\,\mathbf{j} + a\cos\theta\,\mathbf{k},$$

where $\theta$ and $\phi$ are the polar and azimuthal angles respectively. At any point $P$, vectors tangent to the coordinate curves $\theta$ = constant and $\phi$ = constant are

$$\frac{\partial \mathbf{r}}{\partial \theta} = a\cos\theta\cos\phi\,\mathbf{i} + a\cos\theta\sin\phi\,\mathbf{j} - a\sin\theta\,\mathbf{k},$$
$$\frac{\partial \mathbf{r}}{\partial \phi} = -a\sin\theta\sin\phi\,\mathbf{i} + a\sin\theta\cos\phi\,\mathbf{j}.$$

A normal $\mathbf{n}$ to the surface at this point is then given by

$$\mathbf{n} = \frac{\partial \mathbf{r}}{\partial \theta} \times \frac{\partial \mathbf{r}}{\partial \phi} = \begin{vmatrix} \mathbf{i} & \mathbf{j} & \mathbf{k} \\ a\cos\theta\cos\phi & a\cos\theta\sin\phi & -a\sin\theta \\ -a\sin\theta\sin\phi & a\sin\theta\cos\phi & 0 \end{vmatrix}$$
$$= a^2\sin\theta(\sin\theta\cos\phi\,\mathbf{i} + \sin\theta\sin\phi\,\mathbf{j} + \cos\theta\,\mathbf{k}),$$

which has a magnitude of $a^2\sin\theta$. Therefore, the element of area at $P$ is, from (8.23),

$$dS = a^2\sin\theta\,d\theta\,d\phi,$$

and the total surface area of the sphere is given by

$$A = \int_0^{\pi} d\theta \int_0^{2\pi} d\phi \; a^2 \sin\theta = 4\pi a^2.$$

This familiar result can, of course, be proved by much simpler methods! ◀

## 8.6 Scalar and vector fields

We now turn to the case where a particular scalar or vector quantity is defined not just at a point in space, but continuously as a *field* throughout some region of space $R$ (which is often the whole space). Although the concept of a field is valid for spaces with an arbitrary number of dimensions, in the remainder of this chapter we will restrict our attention to the familiar three-dimensional case. A *scalar field* $\phi(x, y, z)$ associates a scalar with each point in $R$, while a *vector field* $\mathbf{a}(x, y, z)$ associates a vector with each point. In what follows, we will assume that the variation of the scalar or vector field from point to point is both continuous and differentiable in $R$.

Simple examples of scalar fields include the pressure at each point in a fluid and the electrostatic potential at each point in space in the presence of an electric charge. Vector fields drawn from the same physical systems are the velocity vector in a fluid (giving the local speed and direction of the flow) and the electric field.

With the study of continuously varying scalar and vector fields there arises the need to consider their derivatives, and also the integration of field quantities along lines, over surfaces and throughout volumes in the field. We defer the discussion of line, surface and volume integrals until the next chapter, and in the remainder of this chapter we concentrate on the definition of vector differential operators and their properties.

## 8.7 Vector operators

Certain differential operations may be performed on scalar and vector fields, and have wide-ranging applications in the physical sciences. The most important operations are those of finding the *gradient* of a scalar field and the *divergence* and *curl* of a vector field. It is usual to define these operators from a strictly mathematical point of view, as we do below. In the following chapter, however, we will discuss their geometrical definitions, which rely on the concept of integrating vector quantities along lines and over surfaces.

Central to all these differential operations is the vector operator $\nabla$, which is called *del* (or sometimes *nabla*), and in Cartesian coordinates is defined by

$$\nabla \equiv \mathbf{i}\frac{\partial}{\partial x} + \mathbf{j}\frac{\partial}{\partial y} + \mathbf{k}\frac{\partial}{\partial z}. \tag{8.25}$$

The form of this operator in non-Cartesian coordinate systems is discussed in sections 8.9 and 8.10.

### *8.7.1 Gradient of a scalar field*

The *gradient* of a scalar field $\phi(x, y, z)$ is defined by

$$\text{grad}\ \phi = \nabla\phi = \mathbf{i}\frac{\partial\phi}{\partial x} + \mathbf{j}\frac{\partial\phi}{\partial y} + \mathbf{k}\frac{\partial\phi}{\partial z}. \tag{8.26}$$

Clearly, $\nabla\phi$ is a vector field whose $x$-, $y$- and $z$- components are the first partial derivatives of $\phi(x, y, z)$ with respect to $x$, $y$ and $z$ respectively. Also note that the vector field $\nabla\phi$ should not be confused with $\phi\nabla$, which has components $(\phi\,\partial/\partial x, \phi\,\partial/\partial y, \phi\,\partial/\partial z)$ and is still a vector operator.

►*Find the gradient of the scalar field* $\phi = xy^2z^3$.

From (8.26) the gradient of $\phi$ is given by

$$\nabla\phi = y^2z^3\mathbf{i} + 2xyz^3\mathbf{j} + 3xy^2z^2\mathbf{k}. \blacktriangleleft$$

The gradient of a scalar field $\phi$ has some interesting geometrical properties. Let us first consider the problem of *calculating the rate of change of $\phi$ in some particular direction.* For an infinitesimal vector displacement $d\mathbf{r}$, forming its scalar product with $\nabla\phi$ we obtain

$$\begin{aligned}\nabla\phi \cdot d\mathbf{r} &= \left(\mathbf{i}\frac{\partial\phi}{\partial x} + \mathbf{j}\frac{\partial\phi}{\partial y} + \mathbf{k}\frac{\partial\phi}{\partial z}\right) \cdot (\mathbf{i}\,dx + \mathbf{j}\,dy + \mathbf{k}\,dx),\\ &= \frac{\partial\phi}{\partial x}\,dx + \frac{\partial\phi}{\partial y}\,dy + \frac{\partial\phi}{\partial z}\,dz,\\ &= d\phi,\end{aligned} \tag{8.27}$$

which is the infinitesimal change in $\phi$ in going from position $\mathbf{r}$ to $\mathbf{r} + d\mathbf{r}$. In particular, if $\mathbf{r}$ depends on some parameter $u$ such that $\mathbf{r}(u)$ defines a space curve, then the total derivative of $\phi$ with respect to $u$ along the curve is simply

$$\frac{d\phi}{du} = \nabla\phi \cdot \frac{d\mathbf{r}}{du}. \tag{8.28}$$

In the particular case where the parameter $u$ is the arc length $s$ along the curve, the total derivative of $\phi$ with respect to $s$ along the curve is given by

$$\frac{d\phi}{ds} = \nabla\phi \cdot \hat{\mathbf{t}}, \tag{8.29}$$

where $\hat{\mathbf{t}}$ is the unit tangent to the curve at the given point, as discussed in section 8.3.

In general, the rate of change of $\phi$ with respect to the distance $s$ in a particular direction $\mathbf{a}$ is given by

$$\frac{d\phi}{ds} = \nabla\phi \cdot \hat{\mathbf{a}}, \tag{8.30}$$

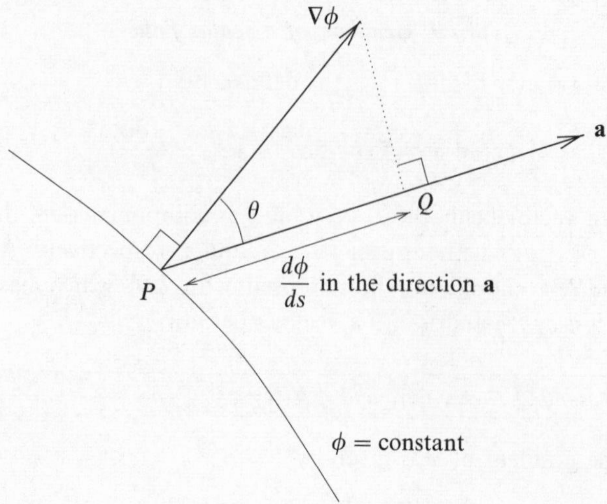


Figure 8.5 Geometrical properties of $\nabla\phi$. $PQ$ gives the value of $d\phi/ds$ in the direction $\mathbf{a}$.

and is called the directional derivative. Since $\hat{\mathbf{a}}$ is a unit vector we have

$$\frac{d\phi}{ds} = |\nabla\phi| \cos\theta,$$

where $\theta$ is the angle between $\hat{\mathbf{a}}$ and $\nabla\phi$ as shown in figure 8.5. Clearly $\nabla\phi$ lies in the direction of the fastest increase in $\phi$, and $|\nabla\phi|$ is the largest possible value of $d\phi/ds$. Similarly, the largest rate of decrease of $\phi$ is $d\phi/ds = -|\nabla\phi|$ in the direction of $-\nabla\phi$.

►*Find the rate of change with respect to distance s of* $\phi = x^2y + yz$ *at the point* $(1, 2, -1)$ *in the direction* $\mathbf{a} = \mathbf{i} + 2\mathbf{j} + 3\mathbf{k}$. *In which direction is the rate of change of* $\phi$ *greatest at this point, and what is its value in this direction?*

The gradient of $\phi$ is given by (8.26):

$$\begin{aligned}\nabla\phi &= 2xy\mathbf{i} + (x^2 + z)\mathbf{j} + y\mathbf{k},\\ &= 4\mathbf{i} + 2\mathbf{k} \qquad \text{at the point } (1, 2, -1).\end{aligned}$$

The unit vector in the direction of $\mathbf{a}$ is $\hat{\mathbf{a}} = \frac{1}{\sqrt{14}}(\mathbf{i} + 2\mathbf{j} + 3\mathbf{k})$, so the rate of change of $\phi$ with distance in this direction is, using (8.30),

$$\frac{d\phi}{ds} = \nabla\phi \cdot \hat{\mathbf{a}} = \frac{1}{\sqrt{14}}(4 + 6) = \frac{10}{\sqrt{14}}.$$

From the above discussion, $d\phi/ds$ will be greatest in the direction of $\nabla\phi = 4\mathbf{i} + 2\mathbf{k}$ at the point $(1, 2, -1)$, and has the value $|\nabla\phi| = \sqrt{20}$ at this point. ◄

We can extend the above analysis to find the rate of change of a vector field (rather than a scalar field as above) in a particular direction. The scalar differential operator $\hat{\mathbf{a}} \cdot \nabla$ can be shown to give the rate of change with distance in the direction $\hat{\mathbf{a}}$ of the quantity (vector or scalar) on which it acts. In Cartesian coordinates it may be written as

$$\hat{\mathbf{a}} \cdot \nabla = a_x \frac{\partial}{\partial x} + a_y \frac{\partial}{\partial y} + a_z \frac{\partial}{\partial z}. \tag{8.31}$$

Thus we can write the infinitesimal change in an electric field in moving from $\mathbf{r}$ to $\mathbf{r} + d\mathbf{r}$ given in (8.20) as $d\mathbf{E} = (d\mathbf{r} \cdot \nabla)\mathbf{E}$.

A second interesting geometrical property of $\nabla\phi$ may be found by considering the surface defined by $\phi(x, y, z) = c$, where $c$ is some constant. If $\hat{\mathbf{t}}$ is a unit tangent to this surface at some point, then clearly $d\phi/ds = 0$ in this direction, and from (8.29) we have $\nabla\phi \cdot \hat{\mathbf{t}} = 0$. In other words, *$\nabla\phi$ is a vector normal to the surface $\phi(x, y, z) = c$ at every point*, as shown in figure 8.5. If $\hat{\mathbf{n}}$ is a unit normal to the surface in the direction of increasing $\phi(x, y, z)$, then the gradient is sometimes written

$$\nabla\phi \equiv \frac{\partial\phi}{\partial n}\hat{\mathbf{n}}, \tag{8.32}$$

where $\partial\phi/\partial n \equiv |\nabla\phi|$ is the rate of change of $\phi$ in the direction $\hat{\mathbf{n}}$, and is called the *normal derivative*.

►*Find expressions for the equations of the tangent plane and the line normal to the surface $\phi(x, y, z) = c$ at the point $P$ with coordinates $(x_0, y_0, z_0)$. Use the results to find the equations of the tangent plane and the line normal to the surface of the sphere $\phi = x^2 + y^2 + z^2 = a^2$ at the point $(0, 0, a)$.*

A vector normal to the surface $\phi(x, y, z) = c$ at the point $P$ is simply $\nabla\phi$ evaluated at that point; we denote it by $\mathbf{n}_0$. If $\mathbf{r}_0$ is the position vector of the point $P$ relative to the origin, and $\mathbf{r}$ is the position vector of any point on the tangent plane, then the vector equation of the tangent plane is, from (6.41),

$$(\mathbf{r} - \mathbf{r}_0) \cdot \mathbf{n}_0 = 0.$$

Similarly, if $\mathbf{r}$ is the position vector of any point on the straight line passing through $P$ (with position vector $\mathbf{r}_0$) in the direction of the normal $\mathbf{n}_0$, then the vector equation of this line is, from subsection 6.7.1,

$$(\mathbf{r} - \mathbf{r}_0) \times \mathbf{n}_0 = \mathbf{0}.$$

For the surface of the sphere $\phi = x^2 + y^2 + z^2 = a^2$,

$$\begin{aligned} \nabla\phi &= 2x\mathbf{i} + 2y\mathbf{j} + 2z\mathbf{k} \\ &= 2a\mathbf{k} \quad \text{at the point } (0, 0, a). \end{aligned}$$

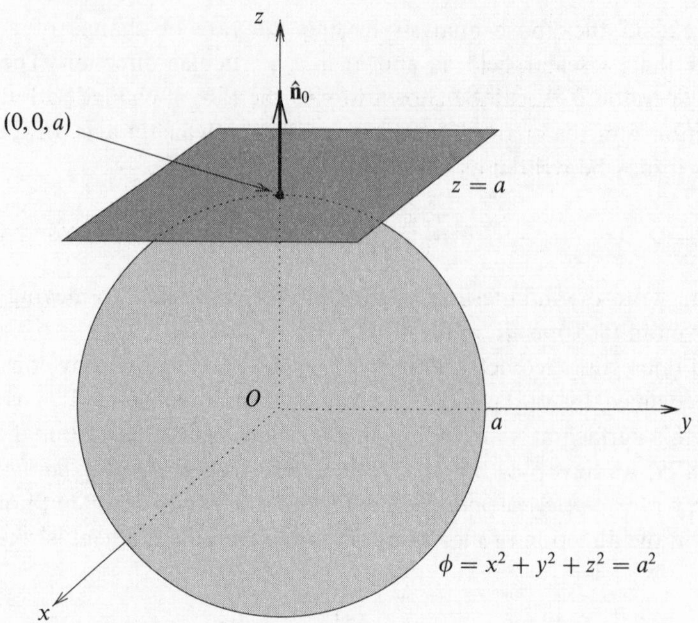


Figure 8.6 The tangent plane and the normal to the surface of the sphere $\phi = x^2 + y^2 + z^2 = a^2$ at the point $\mathbf{r}_0$ with coordinates $(0, 0, a)$.

Therefore the equation of the tangent plane to the sphere at this point is

$$(\mathbf{r} - \mathbf{r}_0) \cdot 2a\mathbf{k} = 0.$$

This gives $2a(z - a) = 0$ or $z = a$, as expected. The equation of the line normal to the sphere at the point $(0, 0, a)$ is

$$(\mathbf{r} - \mathbf{r}_0) \times 2a\mathbf{k} = \mathbf{0},$$

which gives $2ay\mathbf{i} - 2ax\mathbf{j} = \mathbf{0}$ or $x = y = 0$, i.e. the $z$-axis, as expected. The tangent plane and normal to the surface of the sphere at this point are shown in figure 8.6. ◀

Further properties of the gradient operation, which are analogous to those of the ordinary derivative, are listed in subsection 8.8.1 and may be easily proved. In addition to these, we note that the gradient operation also obeys the chain rule as in ordinary differential calculus, i.e. if $\phi$ and $\psi$ are scalar fields in some region $R$, then

$$\nabla [\phi(\psi)] = \frac{\partial \phi}{\partial \psi} \nabla \psi.$$

### *8.7.2 Divergence of a vector field*

The *divergence* of a vector field $\mathbf{a}(x, y, z)$ is defined by

$$\operatorname{div} \mathbf{a} = \nabla \cdot \mathbf{a} = \frac{\partial a_x}{\partial x} + \frac{\partial a_y}{\partial y} + \frac{\partial a_z}{\partial z}, \tag{8.33}$$

where $a_x$, $a_y$ and $a_z$ are the $x$-, $y$- and $z$- components of $\mathbf{a}$. Clearly, $\nabla \cdot \mathbf{a}$ is a scalar field. Any vector field $\mathbf{a}$ for which $\nabla \cdot \mathbf{a} = 0$ is said to be *solenoidal.*

►*Find the divergence of the vector field* $\mathbf{a} = x^2y^2\mathbf{i} + y^2z^2\mathbf{j} + x^2z^2\mathbf{k}$.

From (8.33) the divergence of $\mathbf{a}$ is given by

$$\nabla \cdot \mathbf{a} = 2xy^2 + 2yz^2 + 2x^2z = 2(xy^2 + yz^2 + x^2z). \blacktriangleleft$$

We will discuss fully the geometric definition of divergence and its physical meaning in the next chapter. For the moment, we merely note that the divergence can be considered as a quantitative measure of how much a vector field diverges (spreads out) or converges at any given point. For example, if we consider the vector field $\mathbf{v}(x, y, z)$ describing the local velocity at any point in a fluid, then $\nabla \cdot \mathbf{v}$ is equal to the net rate of outflow of fluid per unit volume evaluated at a point (by letting the infinitesimal volume at that point tend to zero).

Now if some vector field $\mathbf{a}$ is itself derived from a scalar field via $\mathbf{a} = \nabla\phi$ then $\nabla \cdot \mathbf{a}$ has the form $\nabla \cdot \nabla\phi$ or, as it is usually written, $\nabla^2\phi$, where $\nabla^2$ (del squared) is the scalar differential operator

$$\nabla^2 \equiv \frac{\partial^2}{\partial x^2} + \frac{\partial^2}{\partial y^2} + \frac{\partial^2}{\partial z^2}. \tag{8.34}$$

$\nabla^2\phi$ is called the *Laplacian* of $\phi$, and appears in several important partial differential equations of mathematical physics, discussed in chapters 16 and 17.

►*Find the Laplacian of the scalar field* $\phi = xy^2z^3$.

From (8.34) the Laplacian of $\phi$ is given by

$$\nabla^2\phi = \frac{\partial^2\phi}{\partial x^2} + \frac{\partial^2\phi}{\partial y^2} + \frac{\partial^2\phi}{\partial z^2} = 2xz^3 + 6xy^2z. \blacktriangleleft$$

### *8.7.3 Curl of a vector field*

The *curl* of a vector field $\mathbf{a}(x, y, z)$ is defined by

$$\operatorname{curl} \mathbf{a} = \nabla \times \mathbf{a} = \left(\frac{\partial a_z}{\partial y} - \frac{\partial a_y}{\partial z}\right)\mathbf{i} + \left(\frac{\partial a_x}{\partial z} - \frac{\partial a_z}{\partial x}\right)\mathbf{j} + \left(\frac{\partial a_y}{\partial x} - \frac{\partial a_x}{\partial y}\right)\mathbf{k},$$

where $a_x$, $a_y$ and $a_z$ are the $x$-, $y$- and $z$- components of $\mathbf{a}$. The RHS can be written in a more memorable form as a determinant:

$$\nabla \times \mathbf{a} = \begin{vmatrix} \mathbf{i} & \mathbf{j} & \mathbf{k} \\ \dfrac{\partial}{\partial x} & \dfrac{\partial}{\partial y} & \dfrac{\partial}{\partial z} \\ a_x & a_y & a_z \end{vmatrix}, \tag{8.35}$$

where it is understood that, on expanding the determinant, the partial derivatives in the second row act on the components of $\mathbf{a}$ in the third row. Clearly, $\nabla \times \mathbf{a}$ is itself a vector field. Any vector field $\mathbf{a}$ for which $\nabla \times \mathbf{a} = \mathbf{0}$ is said to be *irrotational.*

►*Find the curl of the vector field* $\mathbf{a} = x^2y^2z^2\mathbf{i} + y^2z^2\mathbf{j} + x^2z^2\mathbf{k}$.

The curl of $\mathbf{a}$ is given by

$$\nabla\phi = \begin{vmatrix} \mathbf{i} & \mathbf{j} & \mathbf{k} \\ \dfrac{\partial}{\partial x} & \dfrac{\partial}{\partial y} & \dfrac{\partial}{\partial z} \\ x^2y^2z^2 & y^2z^2 & x^2z^2 \end{vmatrix} = -2\left[y^2z\mathbf{i} + (xz^2 - x^2y^2z)\mathbf{j} + x^2yz^2\mathbf{k}\right]. \blacktriangleleft$$

For a vector field $\mathbf{v}(x, y, z)$ describing the local velocity at any point in a fluid, $\nabla \times \mathbf{v}$ is a measure of the angular velocity of the fluid in the neighbourhood of that point. If a small paddle wheel were placed at various points in the fluid, then it would tend to rotate in regions where $\nabla \times \mathbf{v} \neq \mathbf{0}$, while it would not rotate in regions where $\nabla \times \mathbf{v} = \mathbf{0}$.

Another insight into the physical interpretation of the curl operator is gained by considering the vector field $\mathbf{v}$ describing the velocity at any point in a rigid body rotating about some axis with angular velocity $\boldsymbol{\omega}$. If $\mathbf{r}$ is the position vector of the point with respect to some origin on the axis of rotation, then the velocity of the point is given by $\mathbf{v} = \boldsymbol{\omega} \times \mathbf{r}$. Without any loss of generality, we may take $\boldsymbol{\omega}$ to lie along the $z$-axis of our coordinate system, so that $\boldsymbol{\omega} = \omega\,\mathbf{k}$. The velocity field is then $\mathbf{v} = -\omega y\,\mathbf{i} + \omega x\,\mathbf{j}$. The curl of this vector field is easily found to be

$$\nabla \times \mathbf{v} = \begin{vmatrix} \mathbf{i} & \mathbf{j} & \mathbf{k} \\ \dfrac{\partial}{\partial x} & \dfrac{\partial}{\partial y} & \dfrac{\partial}{\partial z} \\ -\omega y & \omega x & 0 \end{vmatrix} = 2\omega\,\mathbf{k} = 2\boldsymbol{\omega}. \tag{8.36}$$

Therefore the curl of the velocity field is a vector equal to twice the angular velocity vector of the rigid body about its axis of rotation. We give a full geometrical discussion of the curl of a vector in the next chapter.

## 8.8 Vector operator formulae

As in chapter 6 for ordinary vectors, certain identities exist for vector operators. In addition, we must consider various relations involving the action of vector operators on sums and products of scalar and vector fields. Some of these relations have been mentioned above, but we list all the most important ones here for convenience. The validity of these relations may be easily verified by direct calculation (a quick method of deriving them using tensor notation is given in chapter 19).

Although some of the following vector relations are expressed in Cartesian coordinates, it may be proved that they are all independent of the choice of coordinate system. This is to be expected since grad, div and curl all have clear geometrical definitions, which are discussed more fully in the next chapter and which do not rely on any particular choice of coordinate system.

### *8.8.1 Vector operators acting on sums and products*

Let $\phi$ and $\psi$ be scalar fields and $\mathbf{a}$ and $\mathbf{b}$ be vector fields. Assuming these fields are differentiable, the action of grad, div and curl on various sums and products of them is presented in table 8.1.

These relations can be proved by direct calculation.

►*Show that*

$$\nabla \times (\phi\mathbf{a}) = \nabla\phi \times \mathbf{a} + \phi\nabla \times \mathbf{a}.$$

The $x$-component of the LHS is

$$\begin{aligned}\frac{\partial}{\partial y}(\phi a_z) - \frac{\partial}{\partial z}(\phi a_y) &= \phi\frac{\partial a_z}{\partial y} + \frac{\partial \phi}{\partial y}a_z - \phi\frac{\partial a_y}{\partial z} - \frac{\partial \phi}{\partial z}a_y,\\ &= \phi\left(\frac{\partial a_z}{\partial y} - \frac{\partial a_y}{\partial z}\right) + \left(\frac{\partial \phi}{\partial y}a_z - \frac{\partial \phi}{\partial z}a_y\right),\\ &= \phi(\nabla \times \mathbf{a})_x + (\nabla\phi \times \mathbf{a})_x,\end{aligned}$$

where, for example, $(\nabla\phi \times \mathbf{a})_x$ denotes the $x$-component of the vector $\nabla\phi \times \mathbf{a}$. Incorporating the $y$- and $z$- components, which can be similarly found, we obtain the stated result. ◄

Some useful special cases of the relations in table 8.1 are worth noting. If $\mathbf{r}$ is

$$\begin{aligned}
\nabla(\phi+\psi) &= \nabla\phi+\nabla\psi \\
\nabla\cdot(\mathbf{a}+\mathbf{b}) &= \nabla\cdot\mathbf{a}+\nabla\cdot\mathbf{b} \\
\nabla\times(\mathbf{a}+\mathbf{b}) &= \nabla\times\mathbf{a}+\nabla\times\mathbf{b} \\
\nabla(\phi\psi) &= \phi\nabla\psi+\psi\nabla\phi \\
\nabla(\mathbf{a}\cdot\mathbf{b}) &= \mathbf{a}\times(\nabla\times\mathbf{b})+\mathbf{b}\times(\nabla\times\mathbf{a})+(\mathbf{a}\cdot\nabla)\mathbf{b}+(\mathbf{b}\cdot\nabla)\mathbf{a} \\
\nabla\cdot(\phi\mathbf{a}) &= \phi\nabla\cdot\mathbf{a}+\mathbf{a}\cdot\nabla\phi \\
\nabla\cdot(\mathbf{a}\times\mathbf{b}) &= \mathbf{b}\cdot(\nabla\times\mathbf{a})-\mathbf{a}\cdot(\nabla\times\mathbf{b}) \\
\nabla\times(\phi\mathbf{a}) &= \nabla\phi\times\mathbf{a}+\phi\nabla\times\mathbf{a} \\
\nabla\times(\mathbf{a}\times\mathbf{b}) &= \mathbf{a}(\nabla\cdot\mathbf{b})-\mathbf{b}(\nabla\cdot\mathbf{a})+(\mathbf{b}\cdot\nabla)\mathbf{a}-(\mathbf{a}\cdot\nabla)\mathbf{b}
\end{aligned}$$

Table 8.1 Vector operators acting on sums and products. The operator $\nabla$ is defined in (8.25); $\phi$ and $\psi$ are scalar fields, **a** and **b** are vector fields.

the position vector relative to some origin, and $r = |\mathbf{r}|$, then

$$\begin{aligned}
\nabla\phi(r) &= \frac{d\phi}{dr}\,\hat{\mathbf{r}}, \\
\nabla\cdot[\phi(r)\mathbf{r}] &= 3\phi(r)+r\frac{d\phi(r)}{dr}, \\
\nabla^2\phi(r) &= \frac{d^2\phi(r)}{dr^2}+\frac{2}{r}\frac{d\phi(r)}{dr}, \\
\nabla\times[\phi(r)\mathbf{r}] &= \mathbf{0}.
\end{aligned}$$

These results may be proved straightforwardly using Cartesian coordinates, but far more simply using spherical polar coordinates, which are discussed in subsection 8.9.2. Particular cases of these results are

$$\nabla r = \hat{\mathbf{r}}, \qquad \nabla\cdot\mathbf{r} = 3, \qquad \nabla\times\mathbf{r} = \mathbf{0},$$

together with

$$\begin{aligned}
\nabla\left(\frac{1}{r}\right) &= -\frac{\hat{\mathbf{r}}}{r^2}, \\
\nabla\cdot\left(\frac{\hat{\mathbf{r}}}{r^2}\right) &= -\nabla^2\left(\frac{1}{r}\right) = 4\pi\delta(\mathbf{r}),
\end{aligned}$$

where $\delta(\mathbf{r})$ is the Dirac delta function, which is discussed in chapter 11. The last equation is important in the solution of certain partial differential equations and is discussed further in chapter 16.

### *8.8.2 Combinations of grad, div and curl*

We now consider the action of two vector operators in succession on a scalar or vector field. We can immediately discard four of the nine obvious combinations of grad, div and curl, since they clearly do not make sense. If $\phi$ is a scalar field and

**a** is a vector field, these four combinations are grad(grad $\phi$), div(div **a**), curl(div **a**) and grad(curl **a**). In each case the second (outer) vector operator is acting on the wrong type of field, i.e. scalar instead of vector or vice versa. In grad(grad $\phi$), for example, grad acts on grad $\phi$, which is a vector field, but we know that grad only acts on scalar fields (although in fact we will see in chapter 19 that we can form the *outer product* of the del operator with a vector to form a tensor, but that need not concern us here).

Of the five valid combinations of grad, div and curl, two are identically zero, namely

$$\text{curl grad}\,\phi = \nabla\times\nabla\phi = \mathbf{0}, \tag{8.37}$$

$$\text{div curl}\,\mathbf{a} = \nabla\cdot(\nabla\times\mathbf{a}) = 0. \tag{8.38}$$

From (8.37), we see that if **a** is derived from the gradient of some scalar function such that $\mathbf{a} = \nabla\phi$, then it is necessarily irrotational ($\nabla\times\mathbf{a} = 0$). We also note that if **a** is an irrotational vector field, then another irrotational vector field is $\mathbf{a} + \nabla\phi + \mathbf{c}$, where $\phi$ is any scalar field and **c** is a constant vector. This follows since

$$\nabla\times(\mathbf{a}+\nabla\phi+\mathbf{c}) = \nabla\times\mathbf{a} + \nabla\times\nabla\phi = \mathbf{0}.$$

Similarly, from (8.38) we may infer that if **b** is the curl of some vector field **a** such that $\mathbf{b} = \nabla\times\mathbf{a}$ then **b** is solenoidal ($\nabla\cdot\mathbf{b} = 0$). Obviously, if **b** is solenoidal and **c** is any constant vector, then $\mathbf{b}+\mathbf{c}$ is also solenoidal.

The three remaining combinations of grad, div and curl are

$$\text{div grad}\,\phi = \nabla\cdot\nabla\phi = \nabla^2\phi = \frac{\partial^2\phi}{\partial x^2} + \frac{\partial^2\phi}{\partial y^2} + \frac{\partial^2\phi}{\partial z^2}, \tag{8.39}$$

$$\begin{aligned}\text{grad div}\,\mathbf{a} &= \nabla(\nabla\cdot\mathbf{a}),\\ &= \left(\frac{\partial^2 a_x}{\partial x^2} + \frac{\partial^2 a_y}{\partial x\partial y} + \frac{\partial^2 a_z}{\partial x\partial z}\right)\mathbf{i} + \left(\frac{\partial^2 a_x}{\partial y\partial x} + \frac{\partial^2 a_y}{\partial y^2} + \frac{\partial^2 a_z}{\partial y\partial z}\right)\mathbf{j}\\ &\quad + \left(\frac{\partial^2 a_x}{\partial z\partial x} + \frac{\partial^2 a_y}{\partial z\partial y} + \frac{\partial^2 a_z}{\partial z^2}\right)\mathbf{k},\end{aligned} \tag{8.40}$$

$$\text{curl curl}\,\mathbf{a} = \nabla\times(\nabla\times\mathbf{a}) = \nabla(\nabla\cdot\mathbf{a}) - \nabla^2\mathbf{a}, \tag{8.41}$$

where (8.39) and (8.40) are expressed in Cartesian coordinates. In (8.41), the term $\nabla^2\mathbf{a}$ has the linear differential operator $\nabla^2$ acting on a vector (as opposed to a scalar as in (8.39)), which of course consists of a sum of unit vectors multiplied by components. Two cases arise.

(i) If the unit vectors are constants (independent of the values of the coordinates) the differential operator gives a non-zero contribution only when acting upon the coordinates (with the unit vectors merely as multipliers).

(ii) If the unit vectors vary as the values of the coordinates change (i.e. are not constant in direction throughout the whole space) then the derivatives of these vectors appear as contributions to $\nabla^2\mathbf{a}$.

Cartesian coordinates are an example of the first case in which each component satisfies $(\nabla^2\mathbf{a})_i = \nabla^2 a_i$. In this case (8.41) can be applied to each component separately:

$$[\nabla \times (\nabla \times \mathbf{a})]_i = [\nabla(\nabla \cdot \mathbf{a})]_i - \nabla^2 a_i. \tag{8.42}$$

However, cylindrical and spherical polar coordinates come in the second class. For them (8.41) is still true, but the further step to (8.42) cannot be made.

More complicated vector operator relations may be proved using the relations given above.

►*Show that*

$$\nabla \cdot (\nabla\phi \times \nabla\psi) = 0,$$

*where $\phi$ and $\psi$ are scalar fields.*

From the previous section we have

$$\nabla \cdot (\mathbf{a} \times \mathbf{b}) = \mathbf{b} \cdot (\nabla \times \mathbf{a}) - \mathbf{a} \cdot (\nabla \times \mathbf{b}).$$

If we let $\mathbf{a} = \nabla\phi$ and $\mathbf{b} = \nabla\psi$ then we obtain

$$\nabla \cdot (\nabla\phi \times \nabla\psi) = \nabla\psi \cdot (\nabla \times \nabla\phi) - \nabla\phi \cdot (\nabla \times \nabla\psi) = 0, \tag{8.43}$$

since $\nabla \times \nabla\phi = 0 = \nabla \times \nabla\psi$, from (8.37). ◄

## 8.9 Cylindrical and spherical polar coordinates

The operators we have discussed in this chapter, i.e. grad, div, curl and $\nabla^2$, were all defined in terms of Cartesian coordinates, but for many physical situations other coordinate systems are more natural. For example, many systems, such as an isolated charge in space, have spherical symmetry and spherical polar coordinates would be the obvious choice. For axisymmetric systems, such as fluid flow in a pipe, cylindrical polar coordinates are the natural choice. The physical laws governing the behaviour of the systems are often expressed in terms of the vector operators we have been discussing, and so it is necessary to be able to express these operators in these other, non-Cartesian, coordinates. We first consider the two most common non-Cartesian coordinate systems, i.e. cylindrical and spherical polars, and go on to discuss general curvilinear coordinates in the next section.

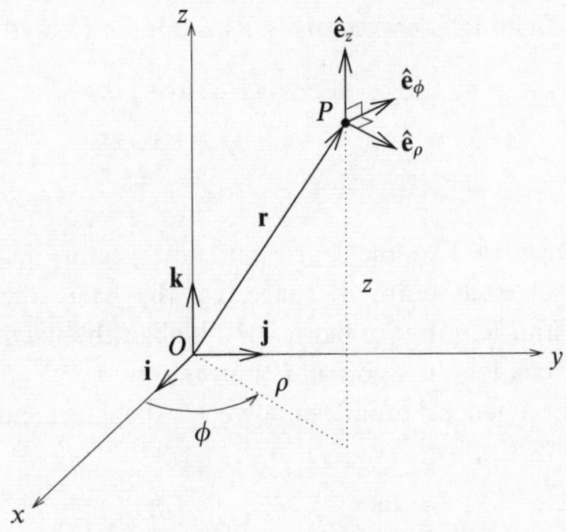


Figure 8.7 Cylindrical polar coordinates $\rho, \phi, z$.

### *8.9.1 Cylindrical polar coordinates*

As shown in figure 8.7, the position of a point in space $P$ having Cartesian coordinates $x, y, z$ may be expressed in terms of cylindrical polar coordinates $\rho, \phi, z$, where

$$x = \rho \cos \phi, \qquad y = \rho \sin \phi, \qquad z = z, \tag{8.44}$$

and $\rho \geq 0$, $0 \leq \phi < 2\pi$ and $-\infty < z < \infty$. The position vector of $P$ may therefore be written

$$\mathbf{r} = \rho \cos \phi \, \mathbf{i} + \rho \sin \phi \, \mathbf{j} + z \, \mathbf{k}. \tag{8.45}$$

If we take the partial derivatives of $\mathbf{r}$ with respect to $\rho$, $\phi$ and $z$ respectively, we obtain the three vectors

$$\mathbf{e}_\rho = \frac{\partial \mathbf{r}}{\partial \rho} = \cos \phi \, \mathbf{i} + \sin \phi \, \mathbf{j}, \tag{8.46}$$

$$\mathbf{e}_\phi = \frac{\partial \mathbf{r}}{\partial \phi} = -\rho \sin \phi \, \mathbf{i} + \rho \cos \phi \, \mathbf{j}, \tag{8.47}$$

$$\mathbf{e}_z = \frac{\partial \mathbf{r}}{\partial z} = \mathbf{k}. \tag{8.48}$$

These vectors lie in the directions of increasing $\rho$, $\phi$ and $z$ respectively, but are not all of unit length. Although $\mathbf{e}_\rho$, $\mathbf{e}_\phi$ and $\mathbf{e}_z$ form a useful set of basis vectors in their own right (we will see in section 8.10 that such a basis is sometimes the *most* useful), it is usual to work with the corresponding *unit* vectors, which are easily obtained by dividing each vector by its modulus to give

$$\hat{\mathbf{e}}_\rho = \mathbf{e}_\rho \quad = \cos\phi\,\mathbf{i} + \sin\phi\,\mathbf{j}, \tag{8.49}$$

$$\hat{\mathbf{e}}_\phi = \tfrac{1}{\rho}\,\mathbf{e}_\phi = -\sin\phi\,\mathbf{i} + \cos\phi\,\mathbf{j}, \tag{8.50}$$

$$\hat{\mathbf{e}}_z = \mathbf{e}_z \quad = \mathbf{k}. \tag{8.51}$$

These three unit vectors, like the Cartesian unit vectors **i**, **j** and **k**, form an orthonormal triad at each point in space, i.e. the basis vectors are mutually orthogonal and of unit length (see figure 8.7). Unlike the fixed vectors **i**, **j** and **k**, however, $\hat{\mathbf{e}}_\rho$ and $\hat{\mathbf{e}}_\phi$ change direction as $P$ moves.

The expression for a general infinitesimal vector displacement $d\mathbf{r}$ in the position of $P$ is given, from (8.19), by

$$\begin{aligned} d\mathbf{r} &= \frac{\partial \mathbf{r}}{\partial \rho}\,d\rho + \frac{\partial \mathbf{r}}{\partial \phi}\,d\phi + \frac{\partial \mathbf{r}}{\partial z}\,dz \\ &= d\rho\,\mathbf{e}_\rho + d\phi\,\mathbf{e}_\phi + dz\,\mathbf{e}_z \\ &= d\rho\,\hat{\mathbf{e}}_\rho + \rho\,d\phi\,\hat{\mathbf{e}}_\phi + dz\,\hat{\mathbf{e}}_z. \end{aligned} \tag{8.52}$$

This expression illustrates an important difference between Cartesian and cylindrical polar coordinates (or non-Cartesian coordinates in general). In Cartesian coordinates, the distance moved in going from $x$ to $x + dx$, with $y$ and $z$ held constant, is simply $ds = dx$. However, in cylindrical polars, if $\phi$ changes by $d\phi$, with $\rho$ and $z$ held constant, then the distance moved is *not* $d\phi$, but $ds = \rho\,d\phi$. Factors, such as the $\rho$ in $\rho\,d\phi$, that multiply the coordinate differentials (in the orthonormal basis) to get distances, are known as *scale factors*. From (8.52), the scale factors for the $\rho$-, $\phi$- and $z$- coordinates are therefore 1, $\rho$ and 1 respectively.

The magnitude $ds$ of the displacement $d\mathbf{r}$ is given in cylindrical polar coordinates by

$$(ds)^2 = d\mathbf{r}\cdot d\mathbf{r} = (d\rho)^2 + \rho^2(d\phi)^2 + (dz)^2,$$

where in the second equality we have used the fact that the basis vectors are orthonormal. We can also find the volume element in a cylindrical polar system (see figure 8.8) by calculating the volume of the infinitesimal parallelepiped defined by the vectors $d\rho\,\hat{\mathbf{e}}_\rho$, $\rho\,d\phi\,\hat{\mathbf{e}}_\phi$ and $dz\,\hat{\mathbf{e}}_z$, which is given by

$$dV = |d\rho\,\hat{\mathbf{e}}_\rho \cdot (\rho\,d\phi\,\hat{\mathbf{e}}_\phi \times dz\,\hat{\mathbf{e}}_z)| = \rho\,d\rho\,d\phi\,dz,$$

which again uses the fact that the basis vectors are orthonormal. For a simple coordinate system such as cylindrical polars the expressions for $(ds)^2$ and $dV$ are obvious from the geometry.

We will now express the vector operators discussed in this chapter in terms of cylindrical polar coordinates. Let us consider a scalar field $\Phi(\rho, \phi, z)$ (where we use $\Phi$ for the scalar field to avoid confusion with the azimuthal angle $\phi$) and a vector field $\mathbf{a}(\rho, \phi, z)$. We must first write the vector field in terms of the basis

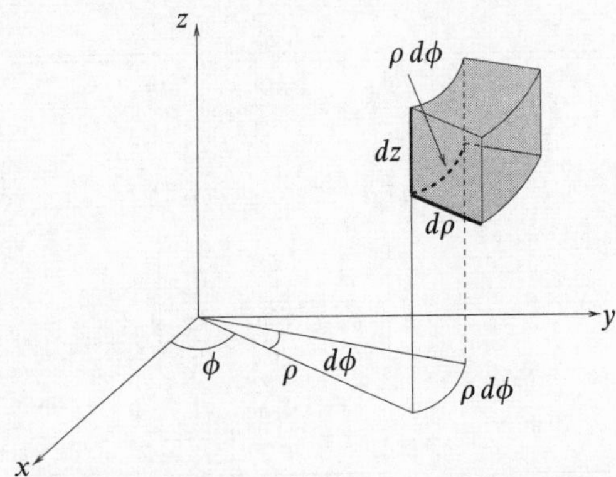


Figure 8.8 The element of volume in cylindrical polar coordinates is given by $\rho\, d\rho d\phi dz$.

vectors of the cylindrical polar coordinate system, i.e.

$$\mathbf{a} = a_\rho\,\hat{\mathbf{e}}_\rho + a_\phi\,\hat{\mathbf{e}}_\phi + a_z\,\hat{\mathbf{e}}_z,$$

where $a_\rho$, $a_\phi$ and $a_z$ are the components of $\mathbf{a}$ in the $\rho$-, $\phi$- and $z$- directions respectively. The expressions for grad, div, curl and $\nabla^2$ can then be calculated, and are given in table 8.2. Since the derivations of these expressions are rather complicated we leave them until our discussion of general curvilinear coordinates in the next section; the reader could well postpone examination of these formal proofs until some experience of using the expressions has been gained.

►*Express the vector field* $\mathbf{a} = yz\,\mathbf{i} - y\,\mathbf{j} + xz^2\,\mathbf{k}$ *in cylindrical polar coordinates, and hence calculate its divergence. Show that the same result is obtained by evaluating the divergence in Cartesian coordinates.*

The basis vectors of the cylindrical polar coordinate system are given in (8.49)–(8.51). Solving these equations simultaneously for $\mathbf{i}$, $\mathbf{j}$ and $\mathbf{k}$ we obtain

$$\begin{aligned}\mathbf{i} &= \cos\phi\,\hat{\mathbf{e}}_\rho - \sin\phi\,\hat{\mathbf{e}}_\phi\\ \mathbf{j} &= \sin\phi\,\hat{\mathbf{e}}_\rho + \cos\phi\,\hat{\mathbf{e}}_\phi\\ \mathbf{k} &= \hat{\mathbf{e}}_z.\end{aligned}$$

Substituting these relations and (8.44) into the expression for $\mathbf{a}$ we find

$$\begin{aligned}\mathbf{a} &= z\rho\sin\phi(\cos\phi\,\hat{\mathbf{e}}_\rho - \sin\phi\,\hat{\mathbf{e}}_\phi) - \rho\sin\phi(\sin\phi\,\hat{\mathbf{e}}_\rho + \cos\phi\,\hat{\mathbf{e}}_\phi) + z^2\rho\cos\phi\,\hat{\mathbf{e}}_z\\ &= (z\rho\sin\phi\cos\phi - \rho\sin^2\phi)\,\hat{\mathbf{e}}_\rho - (z\rho\sin^2\phi + \rho\sin\phi\cos\phi)\,\hat{\mathbf{e}}_\phi + z^2\rho\cos\phi\,\hat{\mathbf{e}}_z.\end{aligned}$$

$$\nabla\Phi = \frac{\partial\Phi}{\partial\rho}\hat{\mathbf{e}}_\rho + \frac{1}{\rho}\frac{\partial\Phi}{\partial\phi}\hat{\mathbf{e}}_\phi + \frac{\partial\Phi}{\partial z}\hat{\mathbf{e}}_z$$

$$\nabla\cdot\mathbf{a} = \frac{1}{\rho}\frac{\partial}{\partial\rho}(\rho a_\rho) + \frac{1}{\rho}\frac{\partial a_\phi}{\partial\phi} + \frac{\partial a_z}{\partial z}$$

$$\nabla\times\mathbf{a} = \frac{1}{\rho}\begin{vmatrix} \hat{\mathbf{e}}_\rho & \rho\hat{\mathbf{e}}_\phi & \hat{\mathbf{e}}_z \\ \dfrac{\partial}{\partial\rho} & \dfrac{\partial}{\partial\phi} & \dfrac{\partial}{\partial z} \\ a_\rho & \rho a_\phi & a_z \end{vmatrix}$$

$$\nabla^2\Phi = \frac{1}{\rho}\frac{\partial}{\partial\rho}\left(\rho\frac{\partial\Phi}{\partial\rho}\right) + \frac{1}{\rho^2}\frac{\partial^2\Phi}{\partial\phi^2} + \frac{\partial^2\Phi}{\partial z^2}$$

Table 8.2 Vector operators in cylindrical polar coordinates. $\Phi$ is a scalar field and $\mathbf{a}$ is a vector field.

Substituting into the expression for $\nabla\cdot\mathbf{a}$ given in table 8.2,

$$\begin{aligned}\nabla\cdot\mathbf{a} &= 2z\sin\phi\cos\phi - 2\sin^2\phi - 2z\sin\phi\cos\phi - \cos^2\phi + \sin^2\phi + 2z\rho\cos\phi \\ &= 2z\rho\cos\phi - 1.\end{aligned}$$

Calculating the divergence directly in Cartesian coordinates, we have

$$\nabla\cdot\mathbf{a} = \frac{\partial a_x}{\partial x} + \frac{\partial a_y}{\partial y} + \frac{\partial a_z}{\partial z} = 2zx - 1,$$

which on substituting $x = \rho\cos\phi$ yields the same result as in cylindrical polars. ◄

Finally, we note that similar results can be obtained for (two-dimensional) polar coordinates in a plane by omitting the $z$-dependence. For example, $(ds)^2 = (d\rho)^2 + \rho^2(d\phi)^2$, while the element of volume is replaced by the element of area $dA = \rho\, d\rho\, d\phi$.

### *8.9.2 Spherical polar coordinates*

As shown in figure 8.9, the position of a point in space $P$, with Cartesian coordinates $x, y, z$, may be expressed in terms of spherical polar coordinates $r, \theta, \phi$, where

$$x = r\sin\theta\cos\phi, \qquad y = r\sin\theta\sin\phi, \qquad z = r\cos\theta, \tag{8.53}$$

and $r \geq 0$, $0 \leq \theta \leq \pi$ and $0 \leq \phi < 2\pi$. The position vector of $P$ may therefore be written as

$$\mathbf{r} = r\sin\theta\cos\phi\,\mathbf{i} + r\sin\theta\sin\phi\,\mathbf{j} + r\cos\theta\,\mathbf{k}.$$

If, in a similar manner to that used in the previous section for cylindrical polars, we find the partial derivatives of $\mathbf{r}$ with respect to $r$, $\theta$ and $\phi$ respectively and

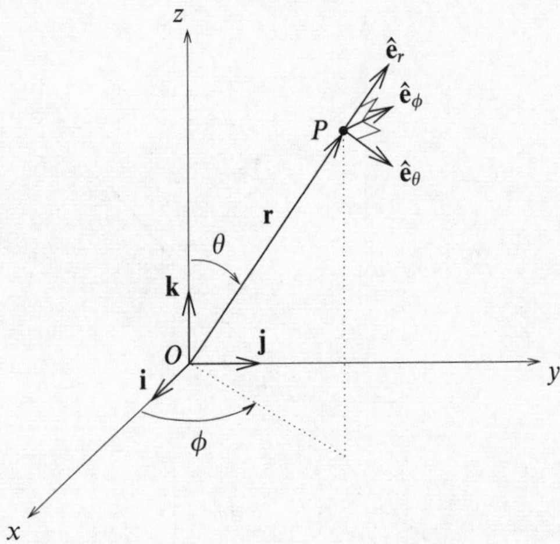


Figure 8.9 Spherical polar coordinates $r, \theta, \phi$.

divide each of the resulting vectors by its modulus, we obtain the unit basis vectors

$$\begin{aligned}\hat{\mathbf{e}}_r &= \sin\theta\cos\phi\,\mathbf{i} + \sin\theta\sin\phi\,\mathbf{j} + \cos\theta\,\mathbf{k},\\ \hat{\mathbf{e}}_\theta &= \cos\theta\cos\phi\,\mathbf{i} + \cos\theta\sin\phi\,\mathbf{j} - \sin\theta\,\mathbf{k},\\ \hat{\mathbf{e}}_\phi &= -\sin\phi\,\mathbf{i} + \cos\phi\,\mathbf{j}.\end{aligned}$$

These unit vectors are in the directions of increasing $r$, $\theta$ and $\phi$ respectively, and are the orthonormal basis set for spherical polar coordinates, as shown in figure 8.9.

A general infinitesimal vector displacement in spherical polars is, from (8.19),

$$d\mathbf{r} = dr\,\hat{\mathbf{e}}_r + r\,d\theta\,\hat{\mathbf{e}}_\theta + r\sin\theta\,d\phi\,\hat{\mathbf{e}}_\phi, \tag{8.54}$$

i.e. the scale factors for the $r$-, $\theta$- and $\phi$- coordinates are $1, r$ and $r\sin\theta$ respectively. The magnitude $ds$ of the displacement $d\mathbf{r}$ is therefore given by

$$(ds)^2 = d\mathbf{r}\cdot d\mathbf{r} = (dr)^2 + r^2(d\theta)^2 + r^2\sin^2\theta(d\phi)^2,$$

since the basis vectors form an orthonormal set. The element of volume in spherical polar coordinates (see figure 8.10) is the volume of the infinitesimal parallelepiped defined by the vectors $dr\,\hat{\mathbf{e}}_r$, $r\,d\theta\,\hat{\mathbf{e}}_\theta$ and $r\sin\theta\,d\phi\,\hat{\mathbf{e}}_\phi$, and is given by

$$dV = |dr\,\hat{\mathbf{e}}_r\cdot(r\,d\theta\,\hat{\mathbf{e}}_\theta\ \times\ r\sin\theta\,d\phi\,\hat{\mathbf{e}}_\phi)| = r^2\sin\theta\,dr\,d\theta\,d\phi,$$

where again we use the fact that the basis vectors are orthonormal. The expressions for $(ds)^2$ and $dV$ in spherical polars can be obtained from the geometry of this coordinate system.

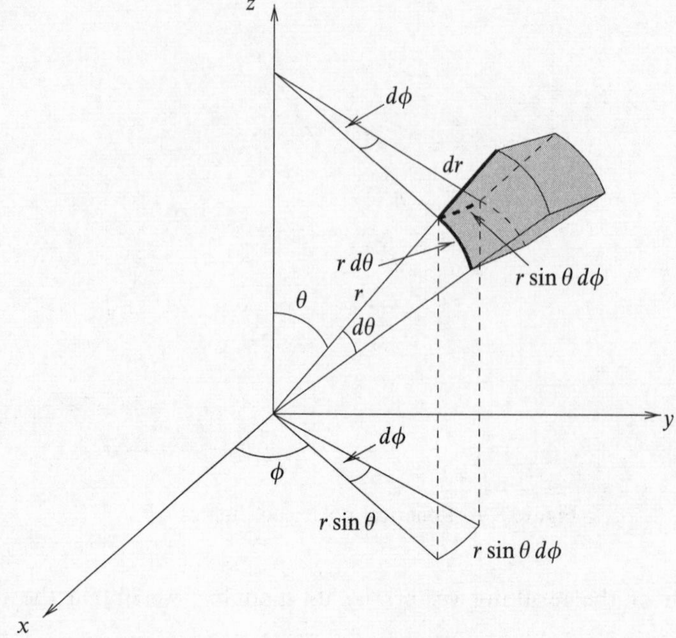


Figure 8.10 The element of volume in spherical polar coordinates is given by $r^2 \sin\theta\, dr d\theta d\phi$.

We will now express the standard vector operators in spherical polar coordinates, using the same techniques as for cylindrical polar coordinates. We consider a scalar field $\Phi(r, \theta, \phi)$ and a vector field $\mathbf{a}(r, \theta, \phi)$. The latter may be written in terms of the basis vectors of the spherical polar coordinate system as

$$\mathbf{a} = a_r\, \hat{\mathbf{e}}_r + a_\theta\, \hat{\mathbf{e}}_\theta + a_\phi\, \hat{\mathbf{e}}_\phi,$$

where $a_r$, $a_\theta$ and $a_\phi$ are the components of $\mathbf{a}$ in the $r$-, $\theta$- and $\phi$- directions respectively. The expressions for grad, div, curl and $\nabla^2$ are given in table 8.3. The derivations of these results are given in the next section.

As a final note we mention that in the expression for $\nabla^2\Phi$ given in table 8.3 we can rewrite the first term on the RHS as follows:

$$\frac{1}{r^2}\frac{\partial}{\partial r}\left(r^2\frac{\partial \Phi}{\partial r}\right) = \frac{1}{r}\frac{\partial^2}{\partial r^2}(r\Phi),$$

which can often be useful in shortening calculations.

$$\nabla\Phi = \frac{\partial\Phi}{\partial r}\hat{\mathbf{e}}_r + \frac{1}{r}\frac{\partial\Phi}{\partial\theta}\hat{\mathbf{e}}_\theta + \frac{1}{r\sin\theta}\frac{\partial\Phi}{\partial\phi}\hat{\mathbf{e}}_\phi$$

$$\nabla\cdot\mathbf{a} = \frac{1}{r^2}\frac{\partial}{\partial r}(r^2 a_r) + \frac{1}{r\sin\theta}\frac{\partial}{\partial\theta}(\sin\theta\, a_\theta) + \frac{1}{r\sin\theta}\frac{\partial a_\phi}{\partial\phi}$$

$$\nabla\times\mathbf{a} = \frac{1}{r^2\sin\theta}\begin{vmatrix} \hat{\mathbf{e}}_r & r\hat{\mathbf{e}}_\theta & r\sin\theta\,\hat{\mathbf{e}}_\phi \\ \dfrac{\partial}{\partial r} & \dfrac{\partial}{\partial\theta} & \dfrac{\partial}{\partial\phi} \\ a_r & ra_\theta & r\sin\theta\, a_\phi \end{vmatrix}$$

$$\nabla^2\Phi = \frac{1}{r^2}\frac{\partial}{\partial r}\left(r^2\frac{\partial\Phi}{\partial r}\right) + \frac{1}{r^2\sin\theta}\frac{\partial}{\partial\theta}\left(\sin\theta\frac{\partial\Phi}{\partial\theta}\right) + \frac{1}{r^2\sin^2\theta}\frac{\partial^2\Phi}{\partial\phi^2}$$

Table 8.3 Vector operators in spherical polar coordinates. $\Phi$ is a scalar field and $\mathbf{a}$ is a vector field.

## 8.10 General curvilinear coordinates

As indicated earlier, the contents of this section are more formal and technically complicated than hitherto. The section could be omitted until the reader has had some experience of using its results.

Cylindrical and spherical polars are just two examples of what are called *general curvilinear coordinates*. In the general case, the position of a point $P$ having Cartesian coordinates $x, y, z$ may be expressed in terms of the three curvilinear coordinates $u_1, u_2, u_3$, where

$$x = x(u_1, u_2, u_3), \qquad y = y(u_1, u_2, u_3), \qquad z = z(u_1, u_2, u_3),$$

and similarly

$$u_1 = u_1(x, y, z), \qquad u_2 = u_2(x, y, z), \qquad u_3 = u_3(x, y, z).$$

We assume that all these functions are continuous, are differentiable, and have a single-valued inverse, except perhaps at or on certain isolated points or lines, so that there is a one-to-one correspondence between the $x$, $y$, $z$ and $u_1$, $u_2$, $u_3$ systems. The $u_1$, $u_2$ and $u_3$ coordinate curves of a general curvilinear system are analogous to the $x$-, $y$- and $z$- axes of Cartesian coordinates. The surfaces $u_1 = c_1$, $u_2 = c_2$ and $u_3 = c_3$, where $c_1$, $c_2$, $c_3$ are constants, are called the *coordinate surfaces* and each pair of these surfaces intersect in curves called *coordinate curves or lines* (see figure 8.11). If at each point in space the three coordinate surfaces passing through the point meet at right angles, then the curvilinear coordinate system is called *orthogonal*. For example, in spherical polars $u_1 = r$, $u_2 = \theta$, $u_3 = \phi$ and the three coordinate surfaces passing through the point $(R, \Theta, \Phi)$ are the sphere $r = R$, the circular cone $\theta = \Theta$ and the plane $\phi = \Phi$, which intersect at right angles at that point. Therefore spherical polars (and cylindrical polars) form an orthogonal coordinate system.

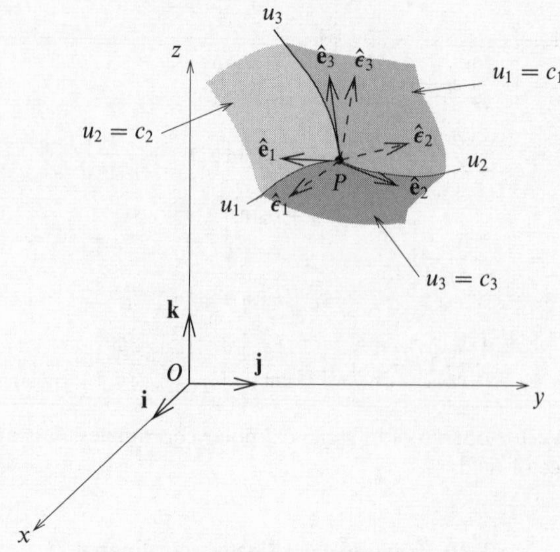


Figure 8.11 General curvilinear coordinates.

If $\mathbf{r}(u_1, u_2, u_3)$ is the position vector of the point $P$, then $\mathbf{e}_1 = \partial\mathbf{r}/\partial u_1$ is a vector tangent to the $u_1$-curve at $P$ (for which $u_2$ and $u_3$ are constants) in the direction of increasing $u_1$. Similarly, $\mathbf{e}_2 = \partial\mathbf{r}/\partial u_2$ and $\mathbf{e}_3 = \partial\mathbf{r}/\partial u_3$ are vectors tangent to the $u_2$- and $u_3$- curves at $P$ in the direction of increasing $u_2$ and $u_3$ respectively. Denoting the lengths of these vectors by $h_1$, $h_2$ and $h_3$, the *unit* vectors in each of these directions are given by

$$\hat{\mathbf{e}}_1 = \frac{1}{h_1}\frac{\partial\mathbf{r}}{\partial u_1},$$
$$\hat{\mathbf{e}}_2 = \frac{1}{h_2}\frac{\partial\mathbf{r}}{\partial u_2},$$
$$\hat{\mathbf{e}}_3 = \frac{1}{h_3}\frac{\partial\mathbf{r}}{\partial u_3},$$

where $h_1 = |\partial\mathbf{r}/\partial u_1|$, $h_2 = |\partial\mathbf{r}/\partial u_2|$ and $h_3 = |\partial\mathbf{r}/\partial u_3|$.

The quantities $h_1, h_2, h_3$ are called the *scale factors* of the curvilinear coordinate system. The element of distance associated with an infinitesimal change $du_i$ in one of the coordinates is $h_i\,du_i$. In the previous section we found that the scale factors for cylindrical and spherical polar coordinates were

$$\begin{aligned} &\text{for cylindrical polars,} \quad h_\rho = 1, \quad h_\phi = \rho, \quad h_z = 1, \\ &\text{for spherical polars,} \quad h_r = 1, \quad h_\theta = r, \quad h_\phi = r\sin\theta. \end{aligned}$$

Although the vectors $\mathbf{e}_1$, $\mathbf{e}_2$, $\mathbf{e}_3$ form a perfectly good basis for the curvilinear coordinate system, it is usual to work with the corresponding unit vectors $\hat{\mathbf{e}}_1$, $\hat{\mathbf{e}}_2$,

$\hat{\mathbf{e}}_3$. For an orthogonal curvilinear coordinate system these unit vectors form an orthonormal basis.

An infinitesimal vector displacement in general curvilinear coordinates is given by, from (8.19),

$$d\mathbf{r} = \frac{\partial \mathbf{r}}{\partial u_1}\,du_1 + \frac{\partial \mathbf{r}}{\partial u_2}\,du_2 + \frac{\partial \mathbf{r}}{\partial u_3}\,du_3 \tag{8.55}$$

$$= du_1\,\mathbf{e}_1 + du_2\,\mathbf{e}_2 + du_3\,\mathbf{e}_3 \tag{8.56}$$

$$= h_1\,du_1\,\hat{\mathbf{e}}_1 + h_2\,du_2\,\hat{\mathbf{e}}_2 + h_3\,du_3\,\hat{\mathbf{e}}_3. \tag{8.57}$$

In the case of *orthogonal* curvilinear coordinates, where the $\hat{\mathbf{e}}_i$ are mutually perpendicular, the element of arc length is given by

$$(ds)^2 = d\mathbf{r}\cdot d\mathbf{r} = h_1^2(du_1)^2 + h_2^2(du_2)^2 + h_3^2(du_3)^2. \tag{8.58}$$

The volume element for the coordinate system is the volume of the infinitesimal parallelepiped defined by the vectors $(\partial\mathbf{r}/\partial u_i)\,du_i = du_i\,\mathbf{e}_i = h_i\,du_i\,\hat{\mathbf{e}}_i$, for $i = 1, 2, 3$. For orthogonal coordinates this is given by

$$\begin{aligned} dV &= |du_1\,\mathbf{e}_1\cdot(du_2\,\mathbf{e}_2\ \times\ du_3\,\mathbf{e}_3)| \\ &= |h_1\,\hat{\mathbf{e}}_1\cdot(h_2\,\hat{\mathbf{e}}_2\ \times\ h_3\,\hat{\mathbf{e}}_3)|\,du_1\,du_2\,du_3 \\ &= h_1h_2h_3\,du_1\,du_2\,du_3. \end{aligned}$$

Now, in addition to the set $\{\hat{\mathbf{e}}_i\}$, for $i = 1, 2, 3$, there exists another useful set of three unit basis vectors at $P$. Since $\nabla u_1$ is a vector normal to the surface $u_1 = c_1$, a unit vector in this direction is $\hat{\boldsymbol{\epsilon}}_1 = \nabla u_1/|\nabla u_1|$. Similarly, $\hat{\boldsymbol{\epsilon}}_2 = \nabla u_2/|\nabla u_2|$ and $\hat{\boldsymbol{\epsilon}}_3 = \nabla u_3/|\nabla u_3|$ are unit vectors normal to the surfaces $u_2 = c_2$ and $u_3 = c_3$ respectively.

Therefore at each point $P$ in a curvilinear coordinate system, there exist, in general, two sets of unit vectors: $\{\hat{\mathbf{e}}_i\}$, tangent to the coordinate curves, and $\{\hat{\boldsymbol{\epsilon}}_i\}$, normal to the coordinate surfaces. A vector $\mathbf{a}$ can be written in terms of either set of unit vectors:

$$\mathbf{a} = a_1\hat{\mathbf{e}}_1 + a_2\hat{\mathbf{e}}_2 + a_3\hat{\mathbf{e}}_3 = A_1\hat{\boldsymbol{\epsilon}}_1 + A_2\hat{\boldsymbol{\epsilon}}_2 + A_3\hat{\boldsymbol{\epsilon}}_3,$$

where $a_1$, $a_2$, $a_3$ and $A_1$, $A_2$, $A_3$ are the components of $\mathbf{a}$ in the two systems. It may be shown that the two bases become identical if the coordinate system is orthogonal.

Instead of the *unit* vectors discussed above, we could instead work directly with the two sets of vectors $\{\mathbf{e}_i = \partial\mathbf{r}/\partial u_i\}$ and $\{\boldsymbol{\epsilon}_i = \nabla u_i\}$, which are not, in general, of unit length. We can then write a vector $\mathbf{a}$ as

$$\mathbf{a} = \alpha_1\mathbf{e}_1 + \alpha_2\mathbf{e}_2 + \alpha_3\mathbf{e}_3 = \beta_1\boldsymbol{\epsilon}_1 + \beta_2\boldsymbol{\epsilon}_2 + \beta_3\boldsymbol{\epsilon}_3,$$

or more explicitly as

$$\mathbf{a} = \alpha_1\frac{\partial\mathbf{r}}{\partial u_1} + \alpha_2\frac{\partial\mathbf{r}}{\partial u_2} + \alpha_3\frac{\partial\mathbf{r}}{\partial u_3} = \beta_1\nabla u_1 + \beta_2\nabla u_2 + \beta_3\nabla u_3,$$

where $\alpha_1$, $\alpha_2$, $\alpha_3$ and $\beta_1$, $\beta_2$, $\beta_3$ are called the *contravariant* and *covariant* components of $\mathbf{a}$ respectively. A more detailed discussion of these components, in the context of tensor analysis, is given in chapter 19. The (in general) non-unit bases $\{\mathbf{e}_i\}$ and $\{\boldsymbol{\epsilon}_i\}$ are often the most natural bases in which to express vector quantities.

►*Show that* $\{\mathbf{e}_i\}$ *and* $\{\boldsymbol{\epsilon}_i\}$ *are reciprocal systems of vectors.*

Let us consider the scalar product $\mathbf{e}_i \cdot \boldsymbol{\epsilon}_j$; using the Cartesian expressions for $\mathbf{r}$ and $\nabla$, we obtain

$$\begin{aligned}
\mathbf{e}_i \cdot \boldsymbol{\epsilon}_j &= \frac{\partial \mathbf{r}}{\partial u_i} \cdot \nabla u_j \\
&= \left( \frac{\partial x}{\partial u_i}\mathbf{i} + \frac{\partial y}{\partial u_i}\mathbf{j} + \frac{\partial z}{\partial u_i}\mathbf{k} \right) \cdot \left( \frac{\partial u_j}{\partial x}\mathbf{i} + \frac{\partial u_j}{\partial y}\mathbf{j} + \frac{\partial u_j}{\partial z}\mathbf{k} \right) \\
&= \frac{\partial x}{\partial u_i}\frac{\partial u_j}{\partial x} + \frac{\partial y}{\partial u_i}\frac{\partial u_j}{\partial y} + \frac{\partial z}{\partial u_i}\frac{\partial u_j}{\partial z} = \frac{\partial u_j}{\partial u_i},
\end{aligned}$$

where in the last step we use the chain rule for partial differentiation. Therefore $\mathbf{e}_i \cdot \boldsymbol{\epsilon}_j = 1$ if $i = j$, and $\mathbf{e}_i \cdot \boldsymbol{\epsilon}_j = 0$ otherwise. Hence $\{\mathbf{e}_i\}$ and $\{\boldsymbol{\epsilon}_j\}$ are reciprocal systems of vectors. ◄

We now derive expressions for the standard vector operators in *orthogonal* curvilinear coordinates. Despite the useful properties of the non-unit bases discussed above, the remainder of our discussion in this section will be in terms of the unit basis vectors $\{\hat{\mathbf{e}}_i\}$. The expressions for the vector operators in cylindrical and spherical polar coordinates given in tables 8.2 and 8.3 respectively can be found from those derived below by inserting the appropriate scale factors.

*Gradient.* The change $d\Phi$ in a scalar field $\Phi$ resulting from changes $du_1, du_2, du_3$ in the coordinates $u_1, u_2, u_3$ is given by, from (4.5),

$$d\Phi = \frac{\partial \Phi}{\partial u_1}\,du_1 + \frac{\partial \Phi}{\partial u_2}\,du_2 + \frac{\partial \Phi}{\partial u_3}\,du_3.$$

For orthogonal curvilinear coordinates $u_1, u_2, u_3$ we find from (8.57), and comparison with (8.27), that we can write this as

$$d\Phi = \nabla\Phi \cdot d\mathbf{r}, \tag{8.59}$$

where $\nabla\Phi$ is given by

$$\nabla\Phi = \frac{1}{h_1}\frac{\partial \Phi}{\partial u_1}\,\hat{\mathbf{e}}_1 + \frac{1}{h_2}\frac{\partial \Phi}{\partial u_2}\,\hat{\mathbf{e}}_2 + \frac{1}{h_3}\frac{\partial \Phi}{\partial u_3}\,\hat{\mathbf{e}}_3. \tag{8.60}$$

This implies that the del operator can be written

$$\nabla = \frac{\hat{\mathbf{e}}_1}{h_1}\frac{\partial}{\partial u_1} + \frac{\hat{\mathbf{e}}_2}{h_2}\frac{\partial}{\partial u_2} + \frac{\hat{\mathbf{e}}_3}{h_3}\frac{\partial}{\partial u_3}.$$

►*Show that for orthogonal curvilinear coordinates $\nabla u_i = \hat{\mathbf{e}}_i / h_i$. Hence show that the two sets of vectors $\{\hat{\mathbf{e}}_i\}$ and $\{\hat{\epsilon}_i\}$ are identical in this case.*

Letting $\Phi = u_i$ in (8.60) we find immediately that $\nabla u_i = \hat{\mathbf{e}}_i / h_i$. Therefore $|\nabla u_i| = 1/h_i$, and so $\hat{\epsilon}_i = \nabla u_i / |\nabla u_i| = h_i \nabla u_i = \hat{\mathbf{e}}_i$. ◄

*Divergence.* In order to derive the expression for the divergence of a vector field in orthogonal curvilinear coordinates, we must first write the vector field in terms of the basis vectors of the coordinate system:

$$\mathbf{a} = a_1\, \hat{\mathbf{e}}_1 + a_2\, \hat{\mathbf{e}}_2 + a_3\, \hat{\mathbf{e}}_3.$$

The divergence is then given by

$$\nabla \cdot \mathbf{a} = \frac{1}{h_1 h_2 h_3} \left[ \frac{\partial}{\partial u_1}(h_2 h_3 a_1) + \frac{\partial}{\partial u_2}(h_3 h_1 a_2) + \frac{\partial}{\partial u_3}(h_1 h_2 a_3) \right]. \tag{8.61}$$

►*Prove the expression for $\nabla \cdot \mathbf{a}$ in orthogonal curvilinear coordinates.*

Let us consider the sub-expression $\nabla \cdot (a_1 \hat{\mathbf{e}}_1)$. Now $\hat{\mathbf{e}}_1 = \hat{\mathbf{e}}_2 \times \hat{\mathbf{e}}_3 = h_2 \nabla u_2 \times h_3 \nabla u_3$. Therefore

$$\begin{aligned} \nabla \cdot (a_1 \hat{\mathbf{e}}_1) &= \nabla \cdot (a_1 h_2 h_3 \nabla u_2 \times \nabla u_3), \\ &= \nabla(a_1 h_2 h_3) \cdot (\nabla u_2 \times \nabla u_3) + a_1 h_2 h_3 \nabla \cdot (\nabla u_2 \times \nabla u_3). \end{aligned}$$

However, $\nabla \cdot (\nabla u_2 \times \nabla u_3) = 0$, from (8.43), so we obtain

$$\nabla \cdot (a_1 \hat{\mathbf{e}}_1) = \nabla(a_1 h_2 h_3) \cdot \left( \frac{\hat{\mathbf{e}}_2}{h_2} \times \frac{\hat{\mathbf{e}}_3}{h_3} \right) = \nabla(a_1 h_2 h_3) \cdot \frac{\hat{\mathbf{e}}_1}{h_2 h_3};$$

letting $\Phi = a_1 h_2 h_3$ in (8.60) and substituting into the above equation, we find

$$\nabla \cdot (a_1 \hat{\mathbf{e}}_1) = \frac{1}{h_1 h_2 h_3} \frac{\partial}{\partial u_1}(a_1 h_2 h_3).$$

Repeating the analysis for $\nabla \cdot (a_2 \hat{\mathbf{e}}_2)$ and $\nabla \cdot (a_3 \hat{\mathbf{e}}_3)$, and adding the results we obtain (8.61), as required. ◄

*Laplacian.* In the expression for the divergence (8.61), let

$$\mathbf{a} = \nabla \Phi = \frac{1}{h_1} \frac{\partial \Phi}{\partial u_1}\, \hat{\mathbf{e}}_1 + \frac{1}{h_2} \frac{\partial \Phi}{\partial u_2}\, \hat{\mathbf{e}}_2 + \frac{1}{h_3} \frac{\partial \Phi}{\partial u_3}\, \hat{\mathbf{e}}_3,$$

where we have used (8.60). We then obtain

$$\nabla^2 \Phi = \frac{1}{h_1 h_2 h_3} \left[ \frac{\partial}{\partial u_1} \left( \frac{h_2 h_3}{h_1} \frac{\partial \Phi}{\partial u_1} \right) + \frac{\partial}{\partial u_2} \left( \frac{h_3 h_1}{h_2} \frac{\partial \Phi}{\partial u_2} \right) + \frac{\partial}{\partial u_3} \left( \frac{h_1 h_2}{h_3} \frac{\partial \Phi}{\partial u_3} \right) \right],$$

which is the expression for the Laplacian in orthogonal curvilinear coordinates.

| | | |
|---|---|---|
| $\nabla\Phi$ | $=$ | $\dfrac{1}{h_1}\dfrac{\partial\Phi}{\partial u_1}\hat{\mathbf{e}}_1+\dfrac{1}{h_2}\dfrac{\partial\Phi}{\partial u_2}\hat{\mathbf{e}}_2+\dfrac{1}{h_3}\dfrac{\partial\Phi}{\partial u_3}\hat{\mathbf{e}}_3$ |
| $\nabla\cdot\mathbf{a}$ | $=$ | $\dfrac{1}{h_1h_2h_3}\left[\dfrac{\partial}{\partial u_1}(h_2h_3a_1)+\dfrac{\partial}{\partial u_2}(h_3h_1a_2)+\dfrac{\partial}{\partial u_3}(h_1h_2a_3)\right]$ |
| $\nabla\times\mathbf{a}$ | $=$ | $\dfrac{1}{h_1h_2h_3}\begin{vmatrix} h_1\hat{\mathbf{e}}_1 & h_2\hat{\mathbf{e}}_2 & h_3\hat{\mathbf{e}}_3 \\ \dfrac{\partial}{\partial u_1} & \dfrac{\partial}{\partial u_2} & \dfrac{\partial}{\partial u_3} \\ h_1a_1 & h_2a_2 & h_3a_3 \end{vmatrix}$ |
| $\nabla^2\Phi$ | $=$ | $\dfrac{1}{h_1h_2h_3}\left[\dfrac{\partial}{\partial u_1}\left(\dfrac{h_2h_3}{h_1}\dfrac{\partial\Phi}{\partial u_1}\right)+\dfrac{\partial}{\partial u_2}\left(\dfrac{h_3h_1}{h_2}\dfrac{\partial\Phi}{\partial u_2}\right)+\dfrac{\partial}{\partial u_3}\left(\dfrac{h_1h_2}{h_3}\dfrac{\partial\Phi}{\partial u_3}\right)\right]$ |

Table 8.4 Vector operators in orthogonal curvilinear coordinates $u_1, u_2, u_3$. $\Phi$ is a scalar field and $\mathbf{a}$ is a vector field.

*Curl.* The curl of a vector field $\mathbf{a} = a_1\hat{\mathbf{e}}_1 + a_2\hat{\mathbf{e}}_2 + a_3\hat{\mathbf{e}}_3$ in orthogonal curvilinear coordinates is given by

$$\nabla\times\mathbf{a} = \frac{1}{h_1h_2h_3}\begin{vmatrix} h_1\hat{\mathbf{e}}_1 & h_2\hat{\mathbf{e}}_2 & h_3\hat{\mathbf{e}}_3 \\ \dfrac{\partial}{\partial u_1} & \dfrac{\partial}{\partial u_2} & \dfrac{\partial}{\partial u_3} \\ h_1a_1 & h_2a_2 & h_3a_3 \end{vmatrix}. \qquad (8.62)$$

►*Prove the expression for* $\nabla\times\mathbf{a}$ *in orthogonal curvilinear coordinates.*

Let us consider the sub-expression $\nabla\times(a_1\hat{\mathbf{e}}_1)$. Since $\hat{\mathbf{e}}_1 = h_1\nabla u_1$ we have

$$\begin{aligned}\nabla\times(a_1\hat{\mathbf{e}}_1) &= \nabla\times(a_1h_1\nabla u_1),\\ &= \nabla(a_1h_1)\times\nabla u_1 \;+\; a_1h_1\nabla\times\nabla u_1.\end{aligned}$$

But $\nabla\times\nabla u_1 = 0$, so we obtain

$$\nabla\times(a_1\hat{\mathbf{e}}_1) = \nabla(a_1h_1)\times\frac{\hat{\mathbf{e}}_1}{h_1}.$$

Letting $\Phi = a_1h_1$ in (8.60) and substituting into the above equation, we find

$$\nabla\times(a_1\hat{\mathbf{e}}_1) = \frac{\hat{\mathbf{e}}_2}{h_3h_1}\frac{\partial}{\partial u_3}(a_1h_1) - \frac{\hat{\mathbf{e}}_3}{h_1h_2}\frac{\partial}{\partial u_2}(a_1h_1).$$

The corresponding analysis of $\nabla\times(a_2\,\hat{\mathbf{e}}_2)$ produces terms in $\hat{\mathbf{e}}_3$ and $\hat{\mathbf{e}}_1$, whilst that of $\nabla\times(a_3\,\hat{\mathbf{e}}_3)$ produces terms in $\hat{\mathbf{e}}_1$ and $\hat{\mathbf{e}}_2$. When the three results are added together, the coefficients multiplying $\hat{\mathbf{e}}_1$, $\hat{\mathbf{e}}_2$ and $\hat{\mathbf{e}}_3$ are the same as those obtained by writing out (8.62) explicitly, thus proving the stated result. ◄

The general expressions for the vector operators in orthogonal curvilinear coordinates are shown for reference in table 8.4. The explicit results for cylindrical and spherical polar coordinates, given in tables 8.2 and 8.3 respectively, are obtained by substituting the appropriate set of scale factors in each case.

A discussion of the expressions for vector operators in tensor form, which are valid even for non-orthogonal curvilinear coordinate systems, is given in chapter 19.

## 8.11 Exercises

8.1 Evaluate the integral

$$\int \left[\mathbf{a}(\dot{\mathbf{b}}\cdot\mathbf{a}+\mathbf{b}\cdot\dot{\mathbf{a}})+\dot{\mathbf{a}}(\mathbf{b}\cdot\mathbf{a})-2(\dot{\mathbf{a}}\cdot\mathbf{a})\mathbf{b}-\dot{\mathbf{b}}|\mathbf{a}|^2\right]dt$$

in which $\dot{\mathbf{a}}$, $\dot{\mathbf{b}}$ are the derivatives of $\mathbf{a}$, $\mathbf{b}$ with respect to $t$.

8.2 The general equation of motion of a (non-relativistic) particle of mass $m$ and charge $q$ when it is placed in a magnetic field $\mathbf{B}$ and an electric field $\mathbf{E}$ is

$$m\ddot{\mathbf{r}} = q(\mathbf{E} + \dot{\mathbf{r}}\times\mathbf{B}),$$

where $\mathbf{r}$ is the position of the particle at time $t$ and $\dot{\mathbf{r}} = d\mathbf{r}/dt$, etc. Write this as three separate equations in terms of the Cartesian components of the vectors involved.

For the simple case of crossed uniform fields $\mathbf{E} = E\mathbf{i}$, $\mathbf{B} = B\mathbf{j}$ in which the particle starts from the origin at $t = 0$ with $\dot{\mathbf{r}} = v_0\mathbf{k}$, find the equations of motion and show the following.

(a) If $v_0 = E/B$, the particle continues its initial motion.
(b) If $v_0 = 0$, the particle follows the space curve given in terms of the parameter $\xi$ by

$$x = \frac{mE}{B^2q}(1-\cos\xi), \qquad y = 0, \qquad z = \frac{mE}{B^2q}(\xi - \sin\xi).$$

Interpret this curve geometrically and relate $\xi$ to $t$. Show that the total distance travelled by the particle after time $t$ is

$$\frac{2E}{B}\int_0^t \left|\sin\frac{Bqt'}{2m}\right| dt'.$$

8.3 Prove that for a space curve $\mathbf{r} = \mathbf{r}(s)$, where $s$ is the arc length measured along the curve from a fixed point, the triple scalar product

$$\left(\frac{d\mathbf{r}}{ds}\times\frac{d^2\mathbf{r}}{ds^2}\right)\cdot\frac{d^3\mathbf{r}}{ds^3}$$

at any point on the curve has the value $\kappa^2\tau$, where $\kappa$ is the curvature and $\tau$ the torsion at that point.

8.4 The shape of the curving slip road joining two motorways that cross at right angles and are at vertical heights $z = 0$ and $z = h$ can be approximated by the space curve

$$\mathbf{r} = \frac{\sqrt{2}h}{\pi} \ln\cos\left(\frac{z\pi}{2h}\right)\mathbf{i} + \frac{\sqrt{2}h}{\pi} \ln\sin\left(\frac{z\pi}{2h}\right)\mathbf{j} + z\mathbf{k}.$$

Show that at height $z$ the radius of curvature $\rho$ of the curve is $(2h/\pi)\text{cosec}\,(z\pi/h)$ and that the torsion $\tau = -1/\rho$. (To shorten the algebra, set $z = 2h\theta/\pi$ and use $\theta$ as the parameter.)

8.5 (a) Parameterising the hyperboloid

$$\frac{x^2}{a^2} + \frac{y^2}{b^2} - \frac{z^2}{c^2} = 1$$

by $x = a\cos\theta\sec\phi$, $y = b\sin\theta\sec\phi$, $z = c\tan\phi$, show that an area element on its surface is

$$dS = \sec^2\phi\left[c^2\sec^2\phi\left(b^2\cos^2\theta + a^2\sin^2\theta\right) + a^2b^2\tan^2\phi\right]^{1/2} d\theta d\phi.$$

(b) Use this formula to show that the area of the curved surface $x^2 + y^2 - z^2 = a^2$ between the planes $z = 0$ and $z = 2a$ is

$$\pi a^2\left(6 + \frac{1}{\sqrt{2}}\sinh^{-1} 2\sqrt{2}\right).$$

8.6 Verify by direct calculation that

$$\nabla\cdot(\mathbf{a}\times\mathbf{b}) = \mathbf{b}\cdot(\nabla\times\mathbf{a}) - \mathbf{a}\cdot(\nabla\times\mathbf{b}).$$

8.7 (a) Simplify

$$\nabla\times\mathbf{a}(\nabla\cdot\mathbf{a}) \;+\; \mathbf{a}\times[\nabla\times(\nabla\times\mathbf{a})] \;+\; \mathbf{a}\times\nabla^2\mathbf{a}.$$

(b) By explicitly writing out the terms in Cartesian coordinates prove that

$$[\mathbf{c}\cdot(\mathbf{b}\cdot\nabla) - \mathbf{b}\cdot(\mathbf{c}\cdot\nabla)]\,\mathbf{a} = (\nabla\times\mathbf{a})\cdot(\mathbf{b}\times\mathbf{c}).$$

(c) Prove that $\mathbf{a}\times(\nabla\times\mathbf{a}) = \nabla(a^2/2) - (\mathbf{a}\cdot\nabla)\mathbf{a}$.

8.8 Verify that (8.42) is valid for each component separately when $\mathbf{a}$ is the Cartesian vector $x^2y\mathbf{i} + xyz\mathbf{j} + z^2y\mathbf{k}$, by showing that each side of the equation is equal to $z\mathbf{i} + (2x + 2z)\mathbf{j} + x\mathbf{k}$.

8.9 (a) For cylindrical polar coordinates $\rho, \phi, z$ evaluate the derivatives of the three unit vectors with respect to each of the coordinates, showing that only $\partial\hat{\mathbf{e}}_\rho/\partial\phi$ and $\partial\hat{\mathbf{e}}_\phi/\partial\phi$ are non-zero.

(i) Hence evaluate $\nabla^2\mathbf{a}$ when $\mathbf{a}$ is the vector $\hat{\mathbf{e}}_\rho$, i.e. a vector of unit magnitude everywhere directed radially outwards from the z-axis.

(ii) Note that it is trivially obvious that $\nabla \times \mathbf{a} = \mathbf{0}$, and hence that equation (8.41) requires that $\nabla(\nabla \cdot \mathbf{a}) = \nabla^2 \mathbf{a}$.

(iii) Evaluate $\nabla(\nabla \cdot \mathbf{a})$ and show that the latter equation holds, but that

$$[\nabla(\nabla \cdot \mathbf{a})]_\rho \neq \nabla^2 a_\rho.$$

(b) Rework the same problem in Cartesian coordinates (where, as it happens, the algebra is more complicated).

8.10 Maxwell's equations for electromagnetism in free space can be written

$$\text{(i)}\ \nabla \cdot \mathbf{B} = 0, \qquad \text{(ii)}\ \nabla \cdot \mathbf{E} = 0,$$
$$\text{(iii)}\ \nabla \times \mathbf{E} + \frac{1}{c}\frac{\partial \mathbf{B}}{\partial t} = \mathbf{0}, \quad \text{(iv)}\ \nabla \times \mathbf{B} - \frac{1}{c}\frac{\partial \mathbf{E}}{\partial t} = \mathbf{0}.$$

A vector $\mathbf{A}$ is defined by $\mathbf{B} = \nabla \times \mathbf{A}$, and a scalar $\phi$ by $\mathbf{E} = -\nabla\phi - (1/c)\partial\mathbf{A}/\partial t$. Show that if the condition

$$\text{(v)}\ \nabla \cdot \mathbf{A} + \frac{1}{c}\frac{\partial \phi}{\partial t} = 0$$

is imposed (this is known as choosing the Lorentz gauge), then both $\mathbf{A}$ and $\phi$ satisfy the wave equations

$$\text{(vi)}\ \nabla^2\phi - \frac{1}{c^2}\frac{\partial^2 \phi}{\partial t^2} = 0,$$
$$\text{(vii)}\ \nabla^2\mathbf{A} - \frac{1}{c^2}\frac{\partial^2 \mathbf{A}}{\partial t^2} = \mathbf{0}.$$

The reader is invited to proceed as follows.

(a) Verify that the expressions for $\mathbf{B}$ and $\mathbf{E}$ in terms of $\mathbf{A}$ and $\phi$ are consistent with (i) and (iii).

(b) Substitute for $\mathbf{E}$ in (ii) and use the derivative with respect to time of (v) to eliminate $\mathbf{A}$ from the resulting expression. Hence obtain (vi).

(c) Substitute for $\mathbf{B}$ and $\mathbf{E}$ in (iv) in terms of $\mathbf{A}$ and $\phi$. Then use the divergence of (v) to simplify the resulting equation and so obtain (vii).

8.11 Non-orthogonal curvilinear coordinates are difficult to work with and should be avoided if at all possible, but the following example is provided to illustrate the content of section 8.10.

In a new coordinate system for the region of space in which the Cartesian coordinate $z \geq 0$, the position of a point $\mathbf{r}$ is given by $(\alpha_1, \alpha_2, R)$, where $\alpha_1$ and $\alpha_2$ are respectively the cosines of the angles made by $\mathbf{r}$ with the $x$- and $y$- coordinate axes of a Cartesian system, and $R = |\mathbf{r}|$. The ranges are $-1 \leq \alpha_i \leq 1$, $0 \leq R < \infty$.

(a) Express $\mathbf{r}$ in terms of $\alpha_1$, $\alpha_2$, $R$ and the unit Cartesian vectors $\mathbf{i}$, $\mathbf{j}$, $\mathbf{k}$.

(b) Obtain expressions for the vectors $\mathbf{e}_i$ $(= \partial\mathbf{r}/\partial\alpha_1, \ldots)$ and hence show that the scale factors $h_i$ have expressions

$$h_1 = \frac{R(1-\alpha_2^2)^{1/2}}{(1-\alpha_1^2-\alpha_2^2)^{1/2}}, \qquad h_2 = \frac{R(1-\alpha_1^2)^{1/2}}{(1-\alpha_1^2-\alpha_2^2)^{1/2}}, \qquad h_3 = 1.$$

(c) Verify formally that the system is not an orthogonal one.

(d) Show that the volume element of the coordinate system is

$$dV = \frac{R^2\,d\alpha_1\,d\alpha_2\,dR}{(1-\alpha_1^2-\alpha_2^2)^{1/2}},$$

and demonstrate that this is always less than or equal to the corresponding expression for an orthogonal curvilinear system.

(e) Calculate the expression for $(ds)^2$ for the system, and show that it differs from that for an orthogonal system by

$$\frac{2\alpha_1\alpha_2 R^2}{1-\alpha_1^2-\alpha_2^2}\,d\alpha_1 d\alpha_2.$$

8.12 In a Cartesian system, $A$ and $B$ are the points $(0,0,-1)$ and $(0,0,1)$ respectively. In a new coordinate system a general point $P$ is given by $(u_1,u_2,u_3)$ with $u_1 = \frac{1}{2}(r_1+r_2)$, $u_2 = \frac{1}{2}(r_1-r_2)$, $u_3 = \phi$; here $r_1$ and $r_2$ are the distances $AP$ and $BP$ and $\phi$ is the angle between the plane $ABP$ and $y=0$.

(a) Express $z$ and the distance $\rho$ from $P$ to the $z$-axis in terms of $u_1$, $u_2$, $u_3$.

(b) Evaluate $\partial x/\partial u_i$, $\partial y/\partial u_i$, $\partial z/\partial u_i$, $i = 1,2,3$.

(c) Find the Cartesian components of $\hat{\mathbf{u}}_j$ and hence show that the new coordinates are mutually orthogonal. Evaluate the scale factors and the infinitesimal volume in the new coordinate system.

(d) Determine and sketch the forms of the surfaces $u_i =$ constant.

(e) Find the most general function $f$ of $u_1$ only that satisfies $\nabla^2 f = 0$.

## 8.12 Hints and answers

8.1 $\mathbf{a}\times(\mathbf{a}\times\mathbf{b})+\mathbf{h}$.

8.2 For crossed uniform fields $\ddot{x} + (Bq/m)^2 x = q(E - Bv_0)/m$; $\ddot{y} = 0$; $m\dot{z} = qBx + mv_0$; (b) $\xi = Bqt/m$; the path is a cycloid in the plane $y = 0$; $ds = [(dx/dt)^2 + (dz/dt)^2]^{1/2}\,dt$.

8.3 Differentiate $\hat{\mathbf{b}} = \hat{\mathbf{t}}\times\hat{\mathbf{n}}$ with respect to $s$; express the result in terms of the derivatives of $\mathbf{r}$; take the scalar product with $d^2\mathbf{r}/ds^2$.

8.4 $ds/d\theta = \sqrt{2}h/(\pi\sin\theta\cos\theta)$; $\hat{\mathbf{t}} = -\sin^2\theta\,\mathbf{i} + \cos^2\theta\,\mathbf{j} + \sqrt{2}\sin\theta\cos\theta\,\mathbf{k}$; $\hat{\mathbf{b}} = \cos^2\theta\,\mathbf{i} - \sin^2\theta\,\mathbf{j} + \sqrt{2}\sin\theta\cos\theta\,\mathbf{k}$.

8.5 (b) Put $\tan\phi = 2^{-1/2}\sinh\psi$.

8.7 (a) $(\nabla \cdot \mathbf{a})(\nabla \times \mathbf{a})$; (b) terms of the form $b_x c_x(\partial a_x/\partial x)$ cancel; (c) for the $x$-component, add and subtract $a_x(\partial a_x/\partial x)$ and regroup.

8.9 (a) $\partial \hat{\mathbf{e}}_\rho/\partial \phi = \hat{\mathbf{e}}_\phi$, $\partial \hat{\mathbf{e}}_\phi/\partial \phi = -\hat{\mathbf{e}}_\rho$; (i) $-\rho^{-2}\hat{\mathbf{e}}_\rho$.
(b) $\nabla^2 \mathbf{a} = -(x^2 + y^2)^{-3/2}(x\mathbf{i} + y\mathbf{j})$.

8.11 (c) $\mathbf{e}_1 \cdot \mathbf{e}_2 = R^2\alpha_1\alpha_2/(1 - \alpha_1^2 - \alpha_2^2) \neq 0$.

8.12 (a) $z = u_1u_2$, $\rho = u_1^2 + u_2^2 - u_1^2u_2^2 - 1$.
(b) $u_1(1-u_2^2)\cos u_3/\rho$, $u_1(1-u_2^2)\sin u_3/\rho$, $u_2$, $u_2(1-u_1^2)\cos u_3/\rho$, $u_2(1-u_1^2)\sin u_3/\rho$, $u_1$, $-\rho \sin u_3$, $\rho\cos u_3$, 0.
(c) $[(u_1^2 - u_2^2)/(u_1^2 - 1)]^{1/2}$, $[(u_2^2 - u_1^2)/(u_2^2 - 1)]^{1/2}$, $\rho$; $|u_1^2 - u_2^2|\, du_1 du_2 du_3$.
(d) Confocal ellipsoids, hyperboloids, planes containing the $z$-axis.
(e) $B \ln[(u_1 - 1)/(u_1 + 1)]$.

# 9

# *Line, surface and volume integrals*

In the previous chapter we encountered continuously varying scalar and vector fields, and discussed the action of various differential operators on them. In addition to these differential operations, the need often arises to consider the integration of field quantities along lines, over surfaces and throughout volumes. In general the integrand may be scalar or vector in nature, but the evaluation of such integrals involves their reduction to one or more scalar integrals, which are then evaluated. In the case of surface and volume integrals this requires the evaluation of double and triple integrals (see chapter 5).

## 9.1 Line integrals

In this section we discuss *line integrals*, in which some quantity related to the field is integrated between two given points in space, $A$ and $B$, along a prescribed curve $C$ that joins them. In general, we may encounter line integrals of the forms

$$\int_C \phi \, d\mathbf{r}, \qquad \int_C \mathbf{a} \cdot d\mathbf{r}, \qquad \int_C \mathbf{a} \times d\mathbf{r}, \tag{9.1}$$

where $\phi$ is a scalar field and $\mathbf{a}$ is a vector field. The three integrals themselves are respectively vector, scalar and vector in nature. As we will see below, in physical applications line integrals of the second type are by far the most common.

The formal definition of a line integral closely follows that of ordinary integrals, and can be considered as the limit of a sum. We may divide the path $C$ joining the points $A$ and $B$ into $N$ small line elements $\Delta\mathbf{r}_p$, $p = 1, \ldots, N$. If $(x_p, y_p, z_p)$ is any point on the line element $\Delta\mathbf{r}_p$, then the second type of line integral in (9.1), for example, is defined as

$$\int_C \mathbf{a} \cdot d\mathbf{r} = \lim_{N\to\infty} \sum_{p=1}^{N} \mathbf{a}(x_p, y_p, z_p) \cdot \Delta\mathbf{r}_p,$$

where it is assumed that all $|\Delta\mathbf{r}_p| \to 0$ as $N \to \infty$.

Each of the line integrals in (9.1) is evaluated over some curve $C$ that may be either open (with $A$ and $B$ being distinct points) or closed (forming a loop so that $A$ and $B$ are coincident). In the case where $C$ is closed, the line integral is written with $\int_C$ replaced by $\oint_C$ to indicate this. The curve may be given either parametrically by $\mathbf{r}(u) = x(u)\mathbf{i} + y(u)\mathbf{j} + z(u)\mathbf{k}$, or by means of simultaneous equations relating $x, y, z$ for the given path (in Cartesian coordinates). A full discussion of the different representations of space curves was given in section 8.3.

In general, the value of the line integral depends not only on the end-points $A$ and $B$ but also on the path $C$ joining them. For a closed curve we must also specify the direction around the loop in which the integral is taken. It is usually taken so that a person walking around the loop $C$ in this direction always has the region $R$ on his/her left; this is equivalent to traversing $C$ in the anticlockwise direction.

### *9.1.1 Evaluating line integrals*

The method of evaluating a line integral is to reduce it to a set of scalar integrals. It is usual to work in Cartesian coordinates, in which case $d\mathbf{r} = dx\,\mathbf{i} + dy\,\mathbf{j} + dz\,\mathbf{k}$. The first type of line integral in (9.1) then becomes simply

$$\int_C \phi\, d\mathbf{r} = \mathbf{i}\int_C \phi(x,y,z)\,dx + \mathbf{j}\int_C \phi(x,y,z)\,dy + \mathbf{k}\int_C \phi(x,y,z)\,dz.$$

The three integrals on the RHS are ordinary scalar integrals that can be evaluated in the usual way once the path of integration $C$ has been specified. Note that in the above we have used relations of the form

$$\int \phi\,\mathbf{i}\,dx = \mathbf{i}\int \phi\,dx,$$

which is allowable since the Cartesian unit vectors are of constant magnitude and direction, and hence may be taken out of the integral. If we were using a different coordinate system such as spherical polars, for example, then, as we saw in the last chapter, the unit basis vectors are not constant. In that case the basis vectors could not be factorised out of the integral.

The second and third line integrals in (9.1) can also be reduced to a set of scalar integrals by writing the vector field $\mathbf{a}$ in terms of its Cartesian components as $\mathbf{a} = a_x\mathbf{i} + a_y\mathbf{j} + a_z\mathbf{k}$, where $a_x$, $a_y$, $a_z$ are each (in general) functions of $x$, $y$, $z$. The second line integral in (9.1), for example, can then be written as

$$\begin{aligned}\int_C \mathbf{a}\cdot d\mathbf{r} &= \int_C (a_x\mathbf{i} + a_y\mathbf{j} + a_z\mathbf{k})\cdot(dx\,\mathbf{i} + dy\,\mathbf{j} + dz\,\mathbf{k})\\ &= \int_C (a_x\,dx \;+\; a_y\,dy \;+\; a_z\,dz)\\ &= \int_C a_x\,dx \;+\; \int_C a_y\,dy \;+\; \int_C a_z\,dz. \qquad (9.2)\end{aligned}$$

A similar procedure may be followed for the third type of line integral in (9.1).

Line integrals have properties that are analogous to those of ordinary integrals. In particular, the following are useful properties (which we illustrate using the second form of line integral in (9.1) but which are valid for all three types).

(i) Reversing the path of integration changes the sign of the integral. If the path $C$ along which the line integrals are evaluated has $A$ and $B$ as its end-points, then

$$\int_A^B \mathbf{a} \cdot d\mathbf{r} = -\int_B^A \mathbf{a} \cdot d\mathbf{r}.$$

This implies that if the path $C$ is a loop then integrating around the loop in the opposite direction changes the sign of the integral.

(ii) If the path of integration is subdivided into smaller segments, then the sum of the separate line integrals along each segment is equal to the line integral along the whole path. So, if $P$ is any point on the path of integration that lies between the path's end-points $A$ and $B$ then

$$\int_A^B \mathbf{a} \cdot d\mathbf{r} = \int_A^P \mathbf{a} \cdot d\mathbf{r} + \int_P^B \mathbf{a} \cdot d\mathbf{r}.$$

►*Evaluate the line integral* $I = \int_C \mathbf{a} \cdot d\mathbf{r}$, *where* $\mathbf{a} = (x+y)\mathbf{i} + (y-x)\mathbf{j}$, *along each of the paths in the xy-plane shown in figure 9.1, namely*

*(i) the parabola* $y^2 = x$ *from* $(1,1)$ *to* $(4,2)$,
*(ii) the curve* $x = 2u^2 + u + 1$, $y = 1 + u^2$ *from* $(1,1)$ *to* $(4,2)$,
*(iii) the line* $y = 1$ *from* $(1,1)$ *to* $(4,1)$, *followed by the line* $x = 4$ *from* $(4,1)$ *to* $(4,2)$.

Since each of the paths lies entirely in the $xy$-plane, we have $d\mathbf{r} = dx\,\mathbf{i} + dy\,\mathbf{j}$. We can therefore write the line integral as

$$I = \int_C \mathbf{a} \cdot d\mathbf{r} = \int_C [(x+y)\,dx + (y-x)\,dy]. \tag{9.3}$$

We must now evaluate this line integral along each of the prescribed paths.

*Case (i)*. Along the parabola $y^2 = x$ we have $2y\,dy = dx$. Substituting for $x$ in (9.3) and using just the limits on $y$, we obtain

$$I = \int_{(1,1)}^{(4,2)} [(x+y)\,dx + (y-x)\,dy] = \int_1^2 [(y^2+y)2y + (y-y^2)]\,dy = 11\tfrac{1}{3}.$$

Note that we could just as easily have substituted for $y$ and obtained an integral in $x$, which would have given the same result.

*Case (ii)*. The second path is given in terms of a parameter $u$. We could eliminate $u$ between the two equations to obtain a relationship between $x$ and $y$

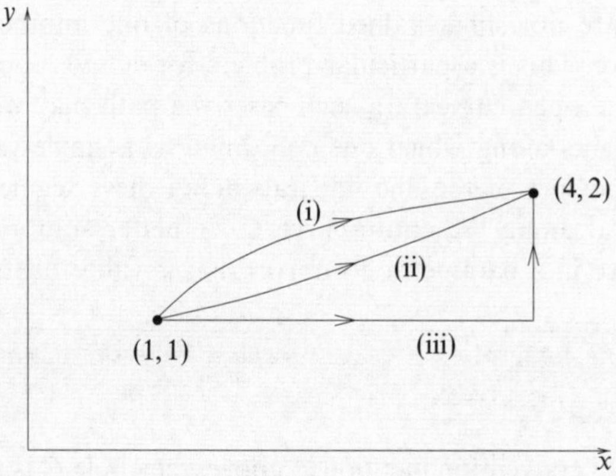


Figure 9.1 Different possible paths between the points (1,1) and (4,2).

directly, and proceed as above, but it is usually quicker to write the line integral in terms of the parameter $u$. Along the curve $x = 2u^2 + u + 1$, $y = 1 + u^2$ we have $dx = (4u + 1)\,du$ and $dy = 2u\,du$. Substituting for $x$ and $y$ in (9.3) and writing the correct limits on $u$, we obtain

$$I = \int_{(1,1)}^{(4,2)} [(x + y)\,dx + (y - x)\,dy]$$
$$= \int_0^1 [(3u^2 + u + 2)(4u + 1) - (u^2 + u)2u]\,du = 10\tfrac{2}{3}.$$

*Case (iii).* For the third path the line integral must be evaluated along the two line segments separately and the results added together. First, along the line $y = 1$ we have $dy = 0$. Substituting this into (9.3) and using just the limits on $x$ for this segment, we obtain

$$\int_{(1,1)}^{(4,1)} [(x + y)\,dx + (y - x)\,dy] = \int_1^4 (x + 1)\,dx = 10\tfrac{1}{2}.$$

Next, along the line $x = 4$ we have $dx = 0$. Substituting this into (9.3) and using just the limits on $y$ for this segment, we obtain

$$\int_{(4,1)}^{(4,2)} [(x + y)\,dx + (y - x)\,dy] = \int_1^2 (y - 4)\,dy = -2\tfrac{1}{2}.$$

The value of the line integral along the whole path is just the sum of the values of the line integrals along each segment, and is given by $I = 10\frac{1}{2} - 2\frac{1}{2} = 8$. ◄

When calculating a line integral along some curve $C$, which is given in terms of $x$, $y$ and $z$, we are sometimes faced with the problem that the curve $C$ is such that $x$, $y$ and $z$ are not single-valued functions of one another over the entire length of the curve. This is a particular problem for closed loops in the $xy$-plane (and also for some open curves). In such cases the path may be subdivided into shorter line segments along which one coordinate is a single-valued function of the other two. The sum of the line integrals along these segments is then equal to the line integral along the entire curve $C$. A better solution, however, is to represent the curve in a parametric form $\mathbf{r}(u)$ that is valid for its entire length.

►*Evaluate the line integral* $I = \oint_C x\,dy$, *where* $C$ *is the circle in the* $xy$*-plane defined by* $x^2 + y^2 = a^2$, $z = 0$.

Adopting the usual convention mentioned above, the circle $C$ is to be traversed in the anticlockwise direction. Taking the circle as a whole means $x$ is not a single-valued function of $y$. We must therefore divide the path into two parts with $x = +\sqrt{a^2 - y^2}$ for the semicircle lying to the right of $x = 0$, and $x = -\sqrt{a^2 - y^2}$ for the semicircle lying to the left of $x = 0$. The required line integral is then the sum of the integrals along the two semicircles. Substituting for $x$, it is given by

$$\begin{aligned} I = \oint_C x\,dy &= \int_{-a}^{a} \sqrt{a^2 - y^2}\,dy + \int_{a}^{-a} \left(-\sqrt{a^2 - y^2}\right) dy \\ &= 4\int_0^a \sqrt{a^2 - y^2}\,dy = \pi a^2. \end{aligned}$$

Alternatively, we can represent the entire circle parametrically, in terms of the azimuthal angle $\phi$, so that $x = a\cos\phi$ and $y = a\sin\phi$ with $\phi$ running from 0 to $2\pi$. The integral can therefore be evaluated over the whole circle at once. Noting that $dy = a\cos\phi\,d\phi$, we can rewrite the line integral completely in terms of the parameter $\phi$ and obtain

$$I = \oint_C x\,dy = a^2 \int_0^{2\pi} \cos^2\phi\,d\phi = \pi a^2. \blacktriangleleft$$

### 9.1.2 Physical examples of line integrals

There are many physical examples of line integrals, but perhaps the most common is the expression for the total work done by a force $\mathbf{F}$ when it moves its point of application from a point $A$ to a point $B$ along a given curve $C$. We allow the magnitude and direction of $\mathbf{F}$ to vary along the curve. Let the force act at a point $\mathbf{r}$ and consider a small displacement $d\mathbf{r}$ along the curve; then the small amount of work done is $dW = \mathbf{F}\cdot d\mathbf{r}$, as discussed in chapter 6 (note that $dW$ can be either positive or negative). Therefore, the total work done in traversing the path $C$ is

$$W_C = \int_C \mathbf{F}\cdot d\mathbf{r}.$$

Naturally, other physical quantities can be expressed in such a way. For example, the electrostatic potential energy gained by moving a charge $q$ along a path $C$ in an electric field $\mathbf{E}$ is $-q\int_C \mathbf{E}\cdot d\mathbf{r}$. We may also note that Ampère's law concerning the magnetic field $\mathbf{B}$ associated with a current-carrying wire can be written as

$$\oint_C \mathbf{B}\cdot d\mathbf{r} = \mu_0 I,$$

where $I$ is the current enclosed by a closed path $C$ traversed in a right-handed sense with respect to the current direction.

Magnetostatics also provides a physical example of the third type of line integral in (9.1). If a loop of wire $C$ carrying a current $I$ is placed in a magnetic field $\mathbf{B}$, then the force $d\mathbf{F}$ on a small length $d\mathbf{r}$ of the wire is given by $d\mathbf{F} = I\, d\mathbf{r}\times\mathbf{B}$, and so the total (vector) force on the loop is

$$\mathbf{F} = I\oint_C d\mathbf{r}\times\mathbf{B}.$$

### *9.1.3 Line integrals with respect to a scalar*

In addition to those listed in (9.1), we can form other types of line integral, which depend on a particular curve $C$ but for which we integrate with respect to a scalar $du$, rather than the vector differential $d\mathbf{r}$. This distinction is rather arbitrary, however, since we can always rewrite line integrals containing the vector differential $d\mathbf{r}$ as a line integral with respect to some scalar parameter. If the path $C$ along which the integral is taken is described parametrically by $\mathbf{r}(u)$, then

$$d\mathbf{r} = \frac{d\mathbf{r}}{du}\,du,$$

and the second type of line integral in (9.1), for example, can be written as

$$\int_C \mathbf{a}\cdot d\mathbf{r} = \int_C \mathbf{a}\cdot\frac{d\mathbf{r}}{du}\,du.$$

A similar procedure can be followed for the other types of line integral in (9.1).

Commonly occurring special cases of line integrals with respect to a scalar are

$$\int_C \phi\, ds, \qquad \int_C \mathbf{a}\, ds,$$

where $s$ is the arc length along the curve $C$. We can always represent $C$ parametrically by $\mathbf{r}(u)$, and from section 8.3 we have

$$ds = \sqrt{\frac{d\mathbf{r}}{du}\cdot\frac{d\mathbf{r}}{du}}\,du.$$

The line integrals can therefore be expressed entirely in terms of the parameter $u$ and thence evaluated.

►*Evaluate the line integral $I = \int_C (x-y)^2\,ds$, where $C$ is the semicircle of radius $a$ joining $(a,0)$ and $(-a,0)$ (in that sense) and for which $y \geq 0$.*

The semicircular path can be described in terms of the azimuthal angle $\phi$ (measured from the $x$-axis) by

$$\mathbf{r}(\phi) = a\cos\phi\,\mathbf{i} + a\sin\phi\,\mathbf{j},$$

where $\phi$ runs from 0 to $\pi$. Therefore the element of arc length is given by

$$ds = \sqrt{\frac{d\mathbf{r}}{d\phi}\cdot\frac{d\mathbf{r}}{d\phi}}\,d\phi = a(\cos^2\phi + \sin^2\phi)\,d\phi = a\,d\phi.$$

Since $(x-y)^2 = a^2(1-\sin 2\phi)$, the line integral becomes

$$I = \int_C (x-y)^2\,ds = \int_0^\pi a^3(1-\sin 2\phi)\,d\phi = \pi a^3. \blacktriangleleft$$

As discussed in the previous chapter, the expression (8.58) for the square of the element of arc length in three-dimensional orthogonal curvilinear coordinates $u_1, u_2, u_3$ is

$$(ds)^2 = h_1^2\,(du_1)^2 + h_2^2\,(du_2)^2 + h_3^2\,(du_3)^2,$$

where $h_1$, $h_2$, $h_3$ are the scale factors of the coordinate system. If a curve $C$ in three dimensions is given parametrically by the equations $u_i = u_i(\lambda)$ for $i = 1, 2, 3$ then the element of arc length along the curve is

$$ds = \sqrt{h_1^2\left(\frac{du_1}{d\lambda}\right)^2 + h_2^2\left(\frac{du_2}{d\lambda}\right)^2 + h_3^2\left(\frac{du_3}{d\lambda}\right)^2}\,d\lambda.$$

## 9.2 Connectivity of regions

In physical systems it is usual to define a scalar or vector field in some region $R$. In the next section (and subsequent ones) we will need the concept of the *connectivity* of such a region in both two and three dimensions.

We begin by discussing planar regions. A plane region $R$ is said to be *simply connected* if every simple closed curve within $R$ can be continuously shrunk to a point without leaving the region (see figure 9.2($a$)). If, however, the region $R$ contains a hole then there exist simple closed curves that cannot by shrunk to a point without leaving $R$ (see figure 9.2($b$)). Such a region is said to be doubly connected, since its boundary has two distinct parts. Similarly, a region with $n-1$ holes is said to be *n-fold connected*, or *multiply connected* (the region in figure 9.2($c$) is triply connected).

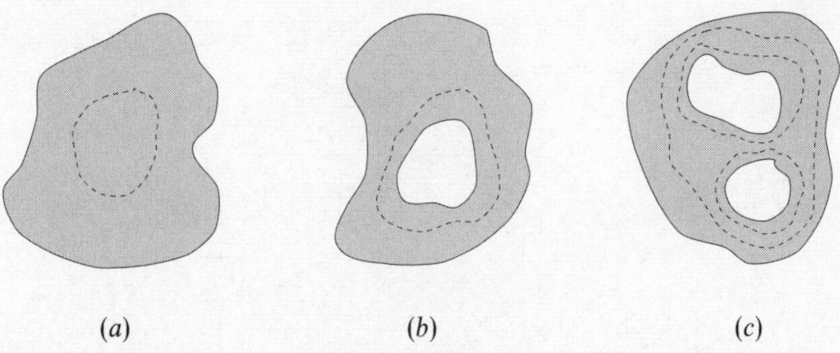


Figure 9.2 (*a*) A simply connected region; (*b*) a doubly connected region; (*c*) a triply connected region.

These ideas can be extended to regions that are not planar, such as general three-dimensional surfaces and volumes. The same criteria concerning shrinking closed curves to a point also apply when deciding the connectivity of such regions. In these cases, however, the curves must lie in the surface or volume in question. For example, the interior of a torus is not simply connected, since there exist closed curves in the interior that cannot be shrunk to a point without leaving the torus. On the other hand, the region between two concentric spheres of different radii is simply connected.

## 9.3 Green's theorem in a plane

In subsection 9.1.1 we considered (amongst other things) the evaluation of line integrals for which the path $C$ is closed and lies entirely in the $xy$-plane. Since the path is closed it will enclose a region $R$ of the plane. We now discuss how to express the line integral around the loop as a double integral over the enclosed region $R$.

Suppose the functions $P(x,y)$, $Q(x,y)$ and their partial derivatives are single-valued, finite and continuous inside and on the boundary $C$ of some simply connected region $R$ in the $xy$-plane. *Green's theorem in a plane* then states

$$\oint_C (P\,dx + Q\,dy) = \iint_R \left( \frac{\partial Q}{\partial x} - \frac{\partial P}{\partial y} \right) dx\,dy, \qquad (9.4)$$

and so relates the line integral around $C$ to a double integral over the enclosed region $R$. This theorem may be proved straightforwardly in the following way. Consider the simply connected region $R$ in figure 9.3, and let $y = y_1(x)$ and $y = y_2(x)$ be the equations of the curves $STU$ and $SVU$ respectively. We then

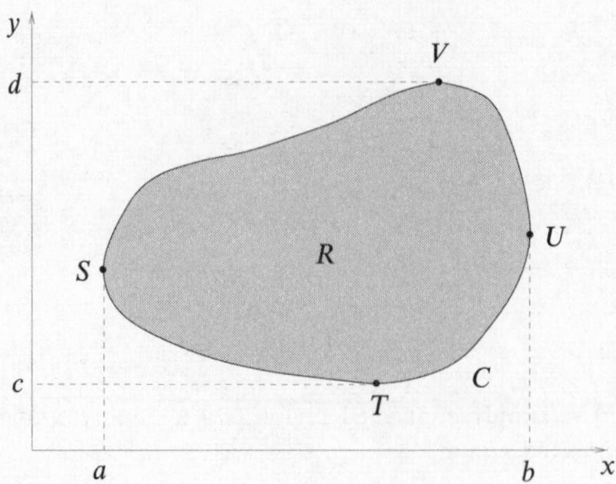


Figure 9.3 A simply connected region $R$ bounded by the curve $C$.

write

$$\iint_R \frac{\partial P}{\partial y}\,dx\,dy = \int_a^b dx \int_{y_1(x)}^{y_2(x)} dy\,\frac{\partial P}{\partial y} = \int_a^b dx \Big[P(x,y)\Big]_{y=y_1(x)}^{y=y_2(x)}$$

$$= \int_a^b \Big[P(x, y_2(x)) - P(x, y_1(x))\Big]\,dx$$

$$= -\int_a^b P(x, y_1(x))\,dx - \int_b^a P(x, y_2(x))\,dx = -\oint_C P\,dx.$$

If we now let $x = x_1(y)$ and $x = x_2(y)$ be the equations of the curves $TSV$ and $TUV$ respectively, we can similarly show that

$$\iint_R \frac{\partial Q}{\partial x}\,dx\,dy = \int_c^d dy \int_{x_1(y)}^{x_2(y)} dx\,\frac{\partial Q}{\partial x} = \int_c^d dy \Big[Q(x,y)\Big]_{x=x_1(y)}^{x=x_2(y)}$$

$$= \int_c^d \Big[Q(x_2(y), y) - Q(x_1(y), y)\Big]\,dy$$

$$= \int_d^c Q(x_1, y)\,dy + \int_c^d Q(x_2, y)\,dy = \oint_C Q\,dy.$$

Subtracting these two results gives Green's theorem in a plane.

► *Show that the area of a region $R$ enclosed by a simple closed curve $C$ is given by $A = \frac{1}{2}\oint_C (x\,dy - y\,dx) = \oint_C x\,dy = -\oint_C y\,dx$. Hence calculate the area of the ellipse $x = a\cos\phi$, $y = b\sin\phi$.*

In Green's theorem put $P = -y$ and $Q = x$; then

$$\oint_C (x\,dy - y\,dx) = \iint_R (1+1)\,dx\,dy = 2\iint_R dx\,dy = 2A.$$

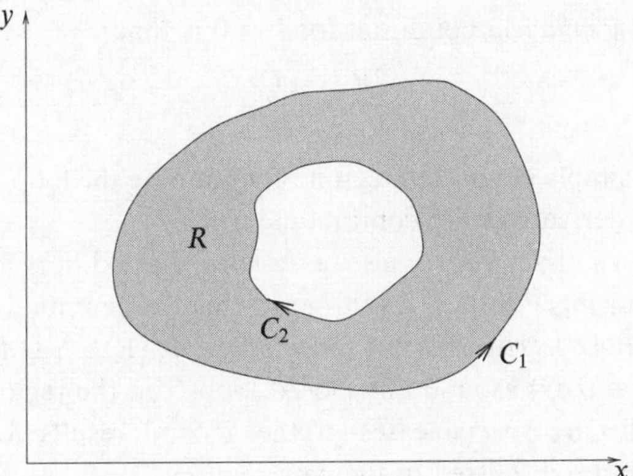


Figure 9.4 A doubly connected region $R$ bounded by the curves $C_1$ and $C_2$.

Therefore the area of the region is $A = \frac{1}{2}\oint_C (x\,dy - y\,dx)$. Alternatively, we could put $P = 0$ and $Q = x$ and obtain $A = \oint_C x\,dy$, or put $P = -y$ and $Q = 0$, which gives $A = -\oint_C y\,dx$.

The area of the ellipse $x = a\cos\phi$, $y = b\sin\phi$ is given by

$$A = \frac{1}{2}\oint_C (x\,dy - y\,dx) = \frac{1}{2}\int_0^{2\pi} ab(\cos^2\phi + \sin^2\phi)\,d\phi$$

$$= \frac{ab}{2}\int_0^{2\pi} d\phi = \pi ab. \blacktriangleleft$$

It may further be shown that Green's theorem in a plane is also valid for multiply connected regions. In this case, the line integral must be taken over all the distinct boundaries of the region. Furthermore, each boundary must be traversed in the positive direction, such that a person travelling along it in this direction always has the region $R$ on their left. In order to apply Green's theorem to the region $R$ shown in figure 9.4, the line integrals must be taken over both boundaries, $C_1$ and $C_2$ (in the directions indicated), and the results added together.

We may also use Green's theorem in a plane to investigate the path independence (or not) of line integrals when the paths lie in the $xy$-plane. Let us consider the line integral

$$I = \int_A^B (P\,dx + Q\,dy).$$

For the line integral from $A$ to $B$ to be independent of the path taken, it must have the same value along any two arbitrary paths $C_1$ and $C_2$ joining the points. Moreover, if we consider as the path the closed loop $C$ formed by $C_1 - C_2$, then the line integral around this loop must be zero. From Green's theorem in a plane,

(9.4), we see that a *sufficient* condition for $I = 0$ is that

$$\frac{\partial P}{\partial y} = \frac{\partial Q}{\partial x}, \tag{9.5}$$

throughout some simply connected region $R$ containing the loop, where we assume that these partial derivatives are continuous in $R$.

It may be shown that (9.5) is also a *necessary* condition for $I = 0$, and is equivalent to requiring $P\,dx + Q\,dy$ to be an exact differential of some function $\phi(x, y)$ such that $P\,dx + Q\,dy = d\phi$. It follows that $\int_A^B (P\,dx + Q\,dy) = \phi(B) - \phi(A)$ and that $\oint_C (P\,dx + Q\,dy)$ around any closed loop $C$ in the region $R$ is identically zero. These results are special cases of the general results for paths in three dimensions, which are discussed in the next section.

►*Evaluate the line integral*

$$I = \oint_C [(e^x y + \cos x \sin y)\,dx + (e^x + \sin x \cos y)\,dy],$$

*around the ellipse* $x^2/a^2 + y^2/b^2 = 1$.

Clearly, it is not straightforward to calculate this line integral directly. However, if we let

$$P = e^x y + \cos x \sin y \qquad \text{and} \qquad Q = e^x + \sin x \cos y,$$

then $\partial P/\partial y = e^x + \cos x \cos y = \partial Q/\partial x$, and so $P\,dx + Q\,dy$ is an exact differential (it is actually the differential of the function $f(x, y) = e^x y + \sin x \sin y$). From the above discussion, we therefore immediately conclude that $I = 0$. ◄

## 9.4 Conservative fields and potentials

So far we have made the point that, in general, the value of a line integral between two points $A$ and $B$ depends on the path $C$ taken from $A$ to $B$. In the previous section, however, we saw that for paths in the $xy$-plane line integrals whose integrands have certain properties are independent of the path taken. We now extend that discussion to the full three-dimensional case.

For line integrals of the form $\int_C \mathbf{a} \cdot d\mathbf{r}$, there exists a class of vector fields for which the line integral between two points is *independent* of the path taken. Such vector fields are called *conservative*. A vector field $\mathbf{a}$ that has continuous partial derivatives in a simply connected region $R$ is conservative if, and only if, any of the following is true.

(i) The integral $\int_A^B \mathbf{a} \cdot d\mathbf{r}$, where $A$ and $B$ lie in the region $R$, is independent of the path from $A$ to $B$. Hence the integral $\oint_C \mathbf{a} \cdot d\mathbf{r}$ around any closed loop in $R$ is zero.

(ii) There exists a single-valued function $\phi$ of position such that $\mathbf{a} = \nabla\phi$.

(iii) $\nabla \times \mathbf{a} = \mathbf{0}$.
(iv) $\mathbf{a} \cdot d\mathbf{r}$ is an exact differential.

The validity or otherwise of any of these statements implies the same for the other three, which we now show.

First, let us assume that (i) above is true. If the line integral from $A$ to $B$ is independent of the path taken between the points, then its value must be a function only of the positions of $A$ and $B$. We may therefore write

$$\int_A^B \mathbf{a} \cdot d\mathbf{r} = \phi(B) - \phi(A), \tag{9.6}$$

which defines a single-valued scalar function of position $\phi$. If the points $A$ and $B$ are separated by an infinitesimal displacement $d\mathbf{r}$ then (9.6) becomes

$$\mathbf{a} \cdot d\mathbf{r} = d\phi,$$

which shows that we require $\mathbf{a} \cdot d\mathbf{r}$ to be an exact differential: condition (iv). From (8.27) we can write $d\phi = \nabla\phi \cdot d\mathbf{r}$, and so we have

$$(\mathbf{a} - \nabla\phi) \cdot d\mathbf{r} = 0.$$

Since $d\mathbf{r}$ is arbitrary, we find that $\mathbf{a} = \nabla\phi$, which immediately implies $\nabla \times \mathbf{a} = \mathbf{0}$, condition (iii) (see (8.37)).

Alternatively, if we suppose there exists a single-valued function of position $\phi$ such that $\mathbf{a} = \nabla\phi$, then $\nabla \times \mathbf{a} = \mathbf{0}$ follows as before. The line integral around a closed loop then becomes

$$\oint_C \mathbf{a} \cdot d\mathbf{r} = \oint_C \nabla\phi \cdot d\mathbf{r} = \oint d\phi.$$

Since we defined $\phi$ to be single-valued, this integral is zero as required.

Now suppose $\nabla \times \mathbf{a} = \mathbf{0}$. From Stoke's theorem, which is discussed in section 9.9, we immediately obtain $\oint_C \mathbf{a} \cdot d\mathbf{r} = 0$; then $\mathbf{a} = \nabla\phi$ and $\mathbf{a} \cdot d\mathbf{r} = d\phi$ follow as above.

Finally, let us suppose $\mathbf{a} \cdot d\mathbf{r} = d\phi$. Then immediately we have $\mathbf{a} = \nabla\phi$, and the other results follow as above.

►*Evaluate the line integral $I = \int_A^B \mathbf{a} \cdot d\mathbf{r}$, where $\mathbf{a} = (xy^2 + z)\mathbf{i} + (x^2y + 2)\mathbf{j} + x\mathbf{k}$, $A$ is the point $(c, c, h)$ and $B$ is the point $(2c, c/2, h)$, along the different paths*

*(i) $C_1$, given by $x = cu$, $y = c/u$, $z = h$, and*
*(ii) $C_2$, given by $2y = 3c - x$, $z = h$.*

*Show that the vector field $\mathbf{a}$ is in fact conservative, and find $\phi$ such that $\mathbf{a} = \nabla\phi$.*

Expanding out the integrand, we have

$$I = \int_{(c,c,h)}^{(2c,c/2,h)} \left[(xy^2 + z)\,dx + (x^2y + 2)\,dy + x\,dz\right], \tag{9.7}$$

which we must evaluate along each of the paths $C_1$ and $C_2$.

(i) Along $C_1$ we have $dx = c\,du$, $dy = -(c/u^2)\,du$, $dz = 0$, and on substituting in (9.7) and finding the limits on $u$, we obtain

$$I = \int_1^2 c\left(h - \frac{2}{u^2}\right)\,du = c(h-1).$$

(ii) Along $C_2$ we have $2\,dy = -dx$, $dz = 0$, and on substituting in (9.7) and using the limits on $x$, we obtain

$$I = \int_c^{2c} \left(\tfrac{1}{2}x^3 - \tfrac{9}{4}cx^2 + \tfrac{9}{4}c^2x + h - 1\right)\,dx = c(h-1).$$

Hence the line integral has the same value along both paths $C_1$ and $C_2$. Taking the curl of $\mathbf{a}$ we have

$$\nabla \times \mathbf{a} = (0-0)\mathbf{i} + (1-1)\mathbf{j} + (2xy - 2xy)\mathbf{k} = \mathbf{0},$$

so $\mathbf{a}$ is a conservative vector field, and the line integral between two points must be independent of the path taken. Since $\mathbf{a}$ is conservative, we can write $\mathbf{a} = \nabla\phi$. Therefore, $\phi$ must satisfy

$$\frac{\partial \phi}{\partial x} = xy^2 + z,$$

which implies that $\phi = \frac{1}{2}x^2y^2 + zx + f(y,z)$ for some function $f$. Secondly, we require

$$\frac{\partial \phi}{\partial y} = x^2y + \frac{\partial f}{\partial y} = x^2y + 2,$$

which implies $f = 2y + g(z)$. Finally, since

$$\frac{\partial \phi}{\partial z} = x + \frac{\partial g}{\partial z} = x,$$

we have $g$ = constant = $k$. So we have explicitly constructed the function $\phi = \frac{1}{2}x^2y^2 + zx + 2y + k$. ◀

The quantity $\phi$ that figures so prominently in this section is called the *scalar potential function* of the conservative vector field $\mathbf{a}$ (which satisfies $\nabla\times\mathbf{a} = \mathbf{0}$), and is unique up to an arbitrary additive constant. Scalar potentials that are multivalued functions of position (but in simple ways) are also of value in describing some physical situations, the most obvious example being the scalar magnetic potential associated with a current-carrying wire. When the integral of a field quantity around a closed loop is considered, provided the loop does not enclose a net current, the potential is single-valued and all the above results still hold. If the loop does enclose a net current, however, our analysis is no longer valid, and extra care must be taken.

If, instead of being conservative, a vector field $\mathbf{b}$ satisfies $\nabla\cdot\mathbf{b} = 0$ (i.e. $\mathbf{b}$ is solenoidal), then it is both possible and useful (for example in the theory of electromagnetism) to define a *vector potential function* $\mathbf{a}$ such that the vector field

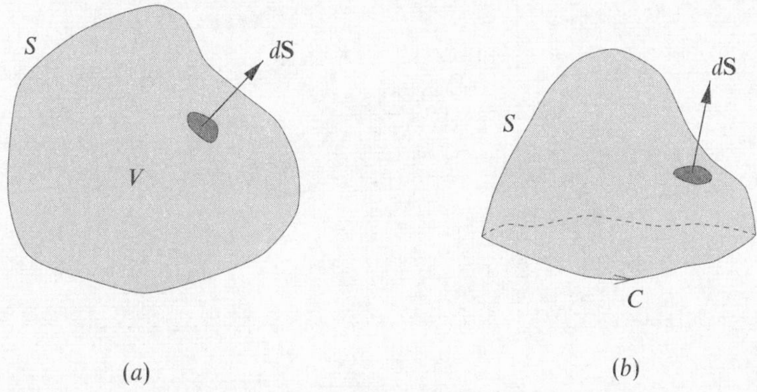


Figure 9.5 (*a*) A closed surface and (*b*) an open surface. In each case a normal to the surface is shown: $d\mathbf{S} = \hat{\mathbf{n}}\, dS$.

$\mathbf{b} = \nabla \times \mathbf{a}$. It may be shown that such a vector field $\mathbf{a}$ always exists. Further, if $\mathbf{a}$ is one such vector field, then $\mathbf{a}' = \mathbf{a} + \nabla\psi + \mathbf{c}$, where $\psi$ is any scalar function and $\mathbf{c}$ is any constant vector, is also a vector potential, such that $\mathbf{b} = \nabla \times \mathbf{a}'$. This was discussed more fully in subsection 8.8.2.

## 9.5 Surface integrals

As with line integrals, integrals over surfaces can involve vector and scalar fields, and can equally result in either a vector or a scalar. The simplest case involves entirely scalars and is of the form

$$\int_S \phi \, dS. \tag{9.8}$$

As analogues of the line integrals listed in (9.1), we may also encounter surface integrals involving vectors, namely

$$\int_S \phi \, d\mathbf{S}, \qquad \int_S \mathbf{a} \cdot d\mathbf{S}, \qquad \int_S \mathbf{a} \times d\mathbf{S}. \tag{9.9}$$

All the above integrals are taken over some surface $S$, which may be either open or closed, and are therefore, in general, double integrals. Following the notation for line integrals, for surface integrals over a closed surface $\int_S$ is replaced by $\oint_S$.

The vector differential $d\mathbf{S}$ in (9.9) represents a vector area element of the surface $S$. It may also be written $d\mathbf{S} = \hat{\mathbf{n}}\, dS$, where $\hat{\mathbf{n}}$ is a unit normal to the surface at the position of the element and $dS$ is the (scalar) area of the element used in (9.8). The convention for the direction of the normal $\hat{\mathbf{n}}$ to a surface depends on whether the surface is open or closed. A closed surface (see figure 9.5(*a*)) does not have to be simply connected (for example, the surface of a torus is not), but it does have to enclose a volume $V$, which may be of infinite extent. The direction of $\hat{\mathbf{n}}$

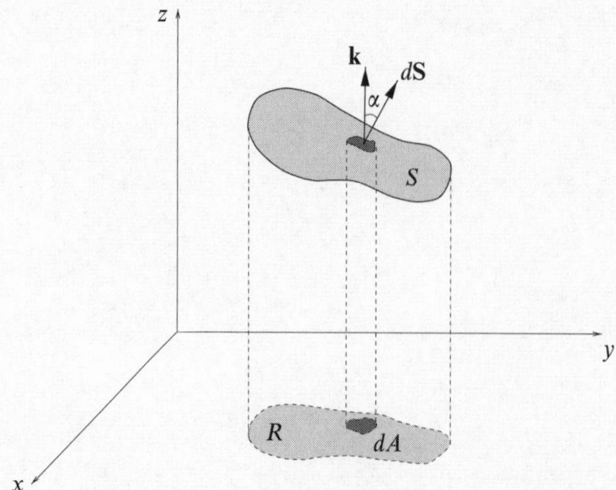


Figure 9.6 A surface $S$ (or part thereof) projected onto a region $R$ in the $xy$-plane; $d\mathbf{S}$ is the surface element at a point $P$.

is taken to point outwards from the enclosed volume as shown. An open surface (see figure 9.5($b$)) spans some perimeter curve $C$. The direction of $\hat{\mathbf{n}}$ is then given by the right-hand sense with respect to the direction in which the perimeter is traversed. An open surface does not have to be simply connected but it must be two-sided (a Möbius strip is an example of a one-sided surface).

The formal definition of a surface integral is very similar to that of a line integral. We divide the surface $S$ into $N$ elements of area $\Delta S_p$, $p = 1, \ldots, N$, each with a unit normal $\hat{\mathbf{n}}_p$. If $(x_p, y_p, z_p)$ is any point in $\Delta S_p$, then the second type of line integral in (9.9), for example, is defined as

$$\int_S \mathbf{a} \cdot d\mathbf{S} = \lim_{N\to\infty} \sum_{p=1}^{N} \mathbf{a}(x_p, y_p, z_p) \cdot \hat{\mathbf{n}}_p \Delta S_p,$$

where it is required that all $\Delta S_p \to 0$ as $N \to \infty$.

### *9.5.1 Evaluating surface integrals*

We now consider how to evaluate surface integrals over some general surface. This involves writing the scalar area element $dS$ in terms of the coordinate differentials of our chosen coordinate system. In some particularly simple cases this is very straightforward. For example, if $S$ is the surface of a sphere of radius $a$ (or some part thereof), then using spherical polar coordinates $\theta, \phi$ on the sphere we have $dS = a^2 \sin\theta\, d\theta\, d\phi$. For a general surface, however, it is not usually possible to represent the surface simply in any particular coordinate system. In such cases,

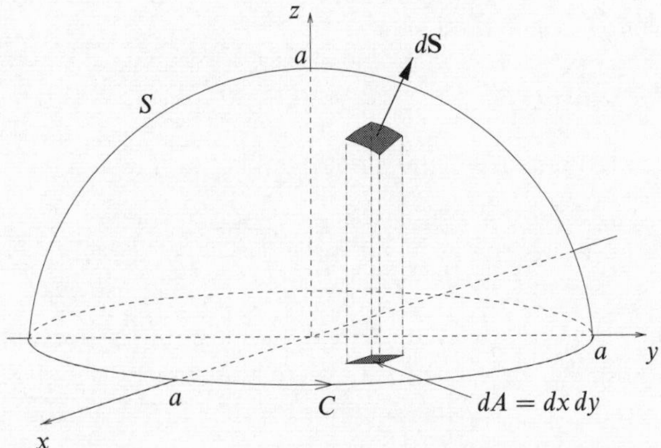


Figure 9.7 The surface of the hemisphere $x^2 + y^2 + z^2 = a^2$, $z \geq 0$.

it is usual to work in Cartesian coordinates and consider the projections of the surface onto the coordinate planes.

Consider a surface (or part of a surface) $S$ as in figure 9.6. The surface $S$ is projected onto a region $R$ of the $xy$-plane, so that an element of surface area $dS$ at point $P$ projects onto the area element $dA$. From the figure, we see that $dA = |\cos\alpha|\, dS$, where $\alpha$ is the angle between the unit vector $\mathbf{k}$ in the $z$-direction and the unit normal $\hat{\mathbf{n}}$ to the surface at $P$. So, at any given point of $S$, we have simply

$$dS = \frac{dA}{|\cos\alpha|} = \frac{dA}{|\hat{\mathbf{n}} \cdot \mathbf{k}|}.$$

Now, if the surface $S$ is given by the equation $f(x, y, z) = 0$ then, as shown in subsection 8.7.1, the unit normal at any point of the surface is simply given by $\hat{\mathbf{n}} = \nabla f / |\nabla f|$ evaluated at that point, cf. (8.32). The scalar element of surface area then becomes

$$dS = \frac{dA}{|\hat{\mathbf{n}} \cdot \mathbf{k}|} = \frac{|\nabla f|\, dA}{\nabla f \cdot \mathbf{k}} = \frac{|\nabla f|\, dA}{\partial f / \partial z}, \qquad (9.10)$$

where $|\nabla f|$ and $\partial f/\partial z$ are evaluated on the surface $S$. We can therefore express any surface integral over $S$ as a double integral over the region $R$ in the $xy$-plane.

►*Evaluate the surface integral $I = \int_S \mathbf{a} \cdot d\mathbf{S}$, where $\mathbf{a} = x\mathbf{i}$ and $S$ is the surface of the hemisphere $x^2 + y^2 + z^2 = a^2$ with $z \geq 0$.*

The surface of the hemisphere is shown in figure 9.7. In this case $dS$ may be easily expressed in spherical polar coordinates as $dS = a^2 \sin\theta\, d\theta\, d\phi$, and the unit normal to the surface at any point is simply $\hat{\mathbf{r}}$. On the surface of the hemisphere we have $x = a\sin\theta\cos\phi$ and so

$$\mathbf{a}\cdot d\mathbf{S} = x\ (\mathbf{i}\cdot\hat{\mathbf{r}})\ dS = (a\sin\theta\cos\phi)(\sin\theta\cos\phi)(a^2\sin\theta\, d\theta\, d\phi).$$

Therefore, inserting the correct limits on $\theta$ and $\phi$, we have

$$I = \int_S \mathbf{a}\cdot d\mathbf{S} = a^3 \int_0^{\pi/2} d\theta\, \sin^3\theta \int_0^{2\pi} d\phi\, \cos^2\phi = \frac{2\pi a^3}{3}.$$

We could, however, follow the general prescription above and project the hemisphere $S$ onto the region $R$ in the $xy$-plane, which is a circle of radius $a$ centred at the origin. Writing the equation of the surface of the hemisphere as $f(x, y) = x^2 + y^2 + z^2 - a^2 = 0$ and using (9.10), we have

$$I = \int_S \mathbf{a}\cdot d\mathbf{S} = \int_S x\,(\mathbf{i}\cdot\hat{\mathbf{r}})\ dS = \int_R x\,(\mathbf{i}\cdot\hat{\mathbf{r}})\,\frac{|\nabla f|\, dA}{\partial f/\partial z}.$$

Now $\nabla f = 2x\mathbf{i} + 2y\mathbf{j} + 2z\mathbf{k} = 2\mathbf{r}$, so on the surface $S$ we have $|\nabla f| = 2|\mathbf{r}| = 2a$. On $S$, we also have $\partial f/\partial z = 2z = 2\sqrt{a^2 - x^2 - y^2}$ and $\mathbf{i}\cdot\hat{\mathbf{r}} = x/a$. Therefore, the integral becomes

$$I = \iint_R \frac{x^2}{\sqrt{a^2 - x^2 - y^2}}\, dx\, dy.$$

Although this integral may be evaluated directly, it is quicker to transform to plane polar coordinates:

$$I = \iint_{R'} \frac{\rho^2\cos^2\phi}{\sqrt{a^2 - \rho^2}}\,\rho\, d\rho\, d\phi$$
$$= \int_0^{2\pi} \cos^2\phi\, d\phi \int_0^a \frac{\rho^3\, d\rho}{\sqrt{a^2 - \rho^2}}.$$

Making the substitution $\rho = a\sin u$, we finally obtain

$$I = \int_0^{2\pi} \cos^2\phi\, d\phi \int_0^{\pi/2} a^3 \sin^3 u\, du = \frac{2\pi a^3}{3}.\ ◀$$

In the above discussion we assumed that any line parallel to the $z$-axis intersects $S$ only once. If this is not the case, we must split up the surface into smaller surfaces $S_1$, $S_2$ etc. that are of this type. The surface integral over $S$ is then the sum of the surface integrals over $S_1$, $S_2$ and so on. This is always necessary for closed surfaces.

We may also sometimes wish to project a surface $S$ (or some part of it) onto the $xz$- or $yz$-plane, rather than the $xy$-plane. In such cases, the above analysis is easily modified.

### *9.5.2 Vector areas of surfaces*

The vector area of a surface $S$ is defined simply as

$$\mathbf{S} = \int_S d\mathbf{S},$$

where the surface integral may be evaluated as above.

►*Find the vector area of the surface of the hemisphere $x^2 + y^2 + z^2 = a^2$ with $z \geq 0$.*

As in the previous example, $d\mathbf{S} = a^2 \sin\theta \, d\theta \, d\phi \, \hat{\mathbf{r}}$ in spherical polar coordinates. Therefore the vector area is given by

$$\mathbf{S} = \iint_S a^2 \sin\theta \, \hat{\mathbf{r}} \, d\theta \, d\phi.$$

Now, since $\hat{\mathbf{r}}$ varies over the surface $S$, it also must be integrated. This is most easily achieved by writing $\hat{\mathbf{r}}$ in terms of the constant Cartesian basis vectors. On $S$ we have

$$\hat{\mathbf{r}} = \sin\theta\cos\phi \, \mathbf{i} + \sin\theta \sin\phi \, \mathbf{j} + \cos\theta \, \mathbf{k},$$

so the expression for the vector area becomes

$$\begin{aligned}\mathbf{S} = \mathbf{i}\left(a^2\int_0^{2\pi}\cos\phi\,d\phi\int_0^{\pi/2}\sin^2\theta\,d\theta\right) + \mathbf{j}\left(a^2\int_0^{2\pi}\sin\phi\,d\phi\int_0^{\pi/2}\sin^2\theta\,d\theta\right)\\ + \mathbf{k}\left(a^2\int_0^{2\pi}d\phi\int_0^{\pi/2}\sin\theta\cos\theta\,d\theta\right)\\ = \mathbf{0} + \mathbf{0} + \pi a^2\mathbf{k} = \pi a^2 \mathbf{k}.\end{aligned}$$

Note that the magnitude of $\mathbf{S}$ is the projected area, of the hemisphere onto the $xy$-plane, and not the surface area of the hemisphere. ◄

The hemispherical shell discussed above is an example of an open surface. For a closed surface, however, the vector area is always zero. This may be seen by projecting the surface down onto each Cartesian coordinate plane in turn. For each projection, every positive element of area on the upper surface is cancelled by the corresponding negative element on the lower surface. Therefore, each component of $\mathbf{S} = \oint_S d\mathbf{S}$ vanishes.

An important corollary of this result is that the vector area of an open surface depends only on its perimeter, or boundary curve, $C$. This may be proved as follows. If surfaces $S_1$ and $S_2$ have the same perimeter, then $S_1 - S_2$ is a closed surface for which

$$\oint d\mathbf{S} = \int_{S_1} d\mathbf{S} - \int_{S_2} d\mathbf{S} = \mathbf{0}.$$

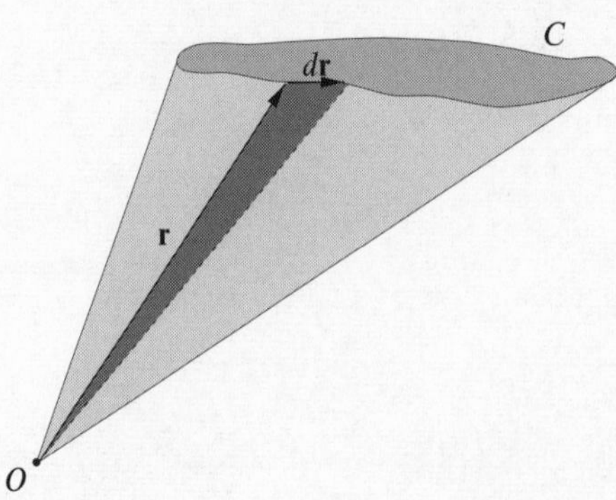


Figure 9.8 The conical surface spanning the perimeter $C$ with its vertex at the origin.

Hence $\mathbf{S}_1 = \mathbf{S}_2$. Moreover, we may derive an expression for the vector area of an open surface $S$ solely in terms of a line integral around its perimeter $C$. Since we may choose any surface with perimeter $C$, we will consider a cone with its vertex at the origin (see figure 9.8). The vector area of the elementary triangular region shown in the figure is $d\mathbf{S} = \frac{1}{2}\mathbf{r} \times d\mathbf{r}$. Therefore, the vector area of the cone, and hence of *any* open surface with perimeter $C$, is given by the line integral

$$\mathbf{S} = \frac{1}{2}\oint_C \mathbf{r} \times d\mathbf{r}.$$

For a surface confined to the $xy$-plane, $\mathbf{r} = x\mathbf{i} + y\mathbf{j}$ and $d\mathbf{r} = dx\,\mathbf{i} + dy\,\mathbf{j}$, and we obtain the special case that the area of the surface is given by $A = \frac{1}{2}\oint_C (x\,dy - y\,dx)$, as we found in section 9.3.

►*Find the vector area of the surface of the hemisphere $x^2 + y^2 + z^2 = a^2$, $z \geq 0$, by evaluating the line integral $\mathbf{S} = \frac{1}{2}\oint_C \mathbf{r} \times d\mathbf{r}$ around its perimeter.*

The perimeter $C$ of the hemisphere is the circle $x^2 + y^2 = a^2$, on which we have

$$\mathbf{r} = a\cos\phi\,\mathbf{i} + a\sin\phi\,\mathbf{j}, \qquad d\mathbf{r} = -a\sin\phi\,d\phi\,\mathbf{i} + a\cos\phi\,d\phi\,\mathbf{j}.$$

Therefore the cross product $\mathbf{r} \times d\mathbf{r}$ is given by

$$\mathbf{r} \times d\mathbf{r} = \begin{vmatrix} \mathbf{i} & \mathbf{j} & \mathbf{k} \\ a\cos\phi & a\sin\phi & 0 \\ -a\sin\phi\,d\phi & a\cos\phi\,d\phi & 0 \end{vmatrix} = a^2(\cos^2\phi + \sin^2\phi)\,d\phi\,\mathbf{k} = a^2\,d\phi\,\mathbf{k},$$

and the vector area becomes

$$\mathbf{S} = \tfrac{1}{2}a^2\mathbf{k}\int_0^{2\pi} d\phi = \pi a^2\,\mathbf{k}. \blacktriangleleft$$

### 9.5.3 *Physical examples of surface integrals*

There are many examples of surface integrals in the physical sciences. Surface integrals of the form (9.8) may occur in computing the total electric charge on a surface or the mass of a shell, $\int_S \rho(\mathbf{r})\, dS$, when the charge or mass density is known. For surface integrals involving vectors, the second form in (9.9) is the most common. For a vector field $\mathbf{a}$, the surface integral $\int_S \mathbf{a} \cdot d\mathbf{S}$ is called the *flux* of $\mathbf{a}$ through $S$. Examples of physically important flux integrals are numerous. For example, let us consider a surface $S$ in a fluid with density $\rho(\mathbf{r})$ in motion with a velocity field $\mathbf{v}(\mathbf{r})$. The mass of fluid crossing an element of surface area $d\mathbf{S}$ in time $dt$ is $dM = \rho\mathbf{v} \cdot d\mathbf{S}\, dt$. Therefore the *net* total mass flux of fluid crossing $S$ is $M = \int_S \rho(\mathbf{r})\mathbf{v}(\mathbf{r}) \cdot d\mathbf{S}$. As a another example, the electromagnetic flux of energy out of a given volume $V$ bounded by a surface $S$ is $\oint_S (\mathbf{E} \times \mathbf{H}) \cdot d\mathbf{S}$.

The solid angle, to be defined below, subtended at a point $O$ by a surface (closed or otherwise) can also be represented by an integral of this form, although it is not strictly a flux integral (except for imaginary isotropic rays radiating from $O$). The integral

$$\Omega = \int_S \frac{\mathbf{r} \cdot d\mathbf{S}}{r^3} = \int_S \frac{\hat{\mathbf{r}} \cdot d\mathbf{S}}{r^2}, \qquad (9.11)$$

gives the *solid angle subtended at O by a surface S* if $\mathbf{r}$ is the position of an element of surface measured from $O$. A little thought will show that (9.11) takes account of all three relevant factors, the size of the element of surface, its inclination to the line joining the element to $O$ and the distance from $O$. Such a general expression is often useful for computing solid angles when the three-dimensional geometry is complicated. Note that (9.11) remains valid when the surface $S$ is not convex and when a single ray from $O$ in certain directions would cut $S$ in more than one place (but we exclude multiply connected regions). In particular, when the surface is closed $\Omega = 0$ if $O$ is outside $S$, and $\Omega = 4\pi$ if $O$ is an interior point.

Surface integrals resulting in vectors occur less frequently. An example is afforded, however, by the total resultant force experienced by a body immersed in a stationary fluid, in which the hydrostatic pressure is given by $p(\mathbf{r})$. The pressure is everywhere inwardly directed and so the resultant force is $\mathbf{F} = -\oint_S p\, d\mathbf{S}$, taken over the whole surface.

## 9.6 Volume integrals

Volume integrals are defined in an obvious way, and are generally simpler than line or surface integrals since the element of volume $dV$ is a scalar quantity. We may encounter volume integrals of the form

$$\int_V \phi\, dV, \qquad \int_V \mathbf{a}\, dV. \qquad (9.12)$$

Clearly, the first form results in a scalar, whereas the second form yields a vector. Two closely related physical examples, one of each kind, are provided by the total mass of a fluid contained in a volume $V$, given by $\int_V \rho(\mathbf{r})\,dV$, and the total linear momentum of that same fluid $\int_V \rho(\mathbf{r})\mathbf{v}(\mathbf{r})\,dV$, where $\mathbf{v}(\mathbf{r})$ is the velocity field in the fluid. As a slightly more complicated example of a volume integral we may consider the following.

▶ *Find an expression for the angular momentum of a solid body rotating with angular velocity* $\boldsymbol{\omega}$ *about an axis through the origin.*

Consider a small volume element $dV$ situated at position $\mathbf{r}$; its linear momentum is $\rho\,dV\dot{\mathbf{r}}$, where $\rho = \rho(\mathbf{r})$ is the density distribution, and its angular momentum about $O$ is $\mathbf{r} \times \rho\dot{\mathbf{r}}\,dV$. Thus for the whole body the angular momentum $\mathbf{L}$ is

$$\mathbf{L} = \int_V (\mathbf{r} \times \dot{\mathbf{r}})\rho\,dV.$$

Putting $\dot{\mathbf{r}} = \boldsymbol{\omega} \times \mathbf{r}$ yields

$$\mathbf{L} = \int_V [\mathbf{r} \times (\boldsymbol{\omega} \times \mathbf{r})]\,\rho\,dV = \int_V \boldsymbol{\omega} r^2 \rho\,dV - \int_V (\mathbf{r} \cdot \boldsymbol{\omega})\mathbf{r}\rho\,dV. \blacktriangleleft$$

The evaluation of volume integrals of the first type in (9.12) has already been considered in our discussion of multiple integrals in chapter 5. The evaluation of the second type of volume integral follows directly since we can write

$$\int_V \mathbf{a}\,dV = \mathbf{i}\int_V a_x\,dV + \mathbf{j}\int_V a_y\,dV + \mathbf{k}\int_V a_z\,dV, \qquad (9.13)$$

where $a_x$, $a_y$, $a_z$ are the Cartesian components of $\mathbf{a}$. We could, of course, write $\mathbf{a}$ in terms of the basis vectors of some other coordinate system (e.g. spherical polars). Since such basis vectors are in general not constant, however, they cannot be taken out of the integral sign as in (9.13), but must also be included as part of the integrand.

### 9.6.1 Volumes of three-dimensional regions

As discussed in chapter 5, the volume of a three-dimensional region $V$ is simply $V = \int_V dV$, which may be evaluated directly once the limits of integration have been found. However, the volume of the region obviously depends only on the surface $S$ that bounds it. We should therefore be able to express the volume $V$ in terms of a surface integral over $S$. This is indeed possible, and may derived as follows. Referring to figure 9.9, let suppose that the origin $O$ is contained within $V$. Then the volume of the small shaded cone is $dV = \frac{1}{3}\mathbf{r} \cdot d\mathbf{S}$. The total volume of the region is then given by

$$V = \frac{1}{3}\oint_S \mathbf{r} \cdot d\mathbf{S},$$

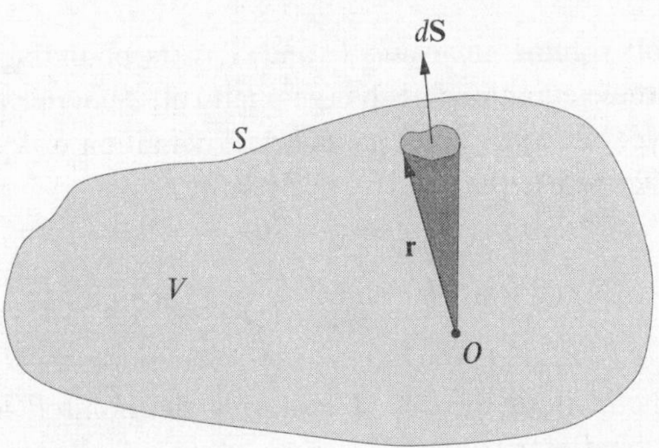


Figure 9.9 A general volume $V$ containing the origin and bounded by the closed surface $S$.

although in many cases it may be simpler to evaluate the volume integral directly. It may be shown that this expression is still valid even when $O$ is not contained in $V$.

►*Find the volume enclosed between a sphere of radius a centred on the origin, and a circular cone of half angle α with its vertex at the origin.*

The element of vector area $d\mathbf{S}$ on the surface of the sphere is given in spherical polar coordinates by $a^2 \sin\theta\, d\theta\, d\phi\, \hat{\mathbf{r}}$. Now taking the axis of the cone to lie along the $z$-axis (from which $\theta$ is measured) the required volume is given by

$$V = \frac{1}{3}\oint_S \mathbf{r}\cdot d\mathbf{S} = \frac{1}{3}\int_0^{2\pi} d\phi \int_0^{\alpha} a^2 \sin\theta\, \mathbf{r}\cdot\hat{\mathbf{r}}\, d\theta$$

$$= \frac{1}{3}\int_0^{2\pi} d\phi \int_0^{\alpha} a^3 \sin\theta\, d\theta = \frac{2\pi a^3}{3}(1-\cos\alpha). \quad ◄$$

## 9.7 Integral forms for grad, div and curl

In the previous chapter we defined the vector operators grad, div and curl in purely mathematical terms, which depended on the coordinate system in which they were expressed. An interesting application of line, surface and volume integrals is the expression of grad, div and curl in coordinate-free, geometrical terms. If $\phi$ is a scalar field and $\mathbf{a}$ is a vector field, then it may be shown that at any point $P$

$$\nabla\phi = \lim_{V\to 0}\left(\frac{1}{V}\oint_S \phi\, d\mathbf{S}\right) \tag{9.14}$$

$$\nabla\cdot\mathbf{a} = \lim_{V\to 0}\left(\frac{1}{V}\oint_S \mathbf{a}\cdot d\mathbf{S}\right) \tag{9.15}$$

$$\nabla\times\mathbf{a} = \lim_{V\to 0}\left(\frac{1}{V}\oint_S d\mathbf{S}\times\mathbf{a}\right) \tag{9.16}$$

where $V$ is a small volume enclosing $P$ and $S$ is its bounding surface. Indeed, we may consider these equations as the (geometrical) *definitions* of grad, div and curl. An alternative (but equivalent) geometrical definition of $\nabla \times \mathbf{a}$ at a point $P$, which is often easier to use than (9.16), is given by

$$(\nabla \times \mathbf{a}) \cdot \hat{\mathbf{n}} = \lim_{A \to 0} \left( \frac{1}{A} \oint_C \mathbf{a} \cdot d\mathbf{r} \right), \qquad (9.17)$$

where $C$ is a plane contour of area $A$ enclosing the point $P$ and $\hat{\mathbf{n}}$ is the unit normal to the enclosed planar area.

It may be shown that all the above equations are consistent with our definitions in the last chapter *in any coordinate system*, although the difficulty of the proofs depends on the chosen coordinate system. The most general coordinate system encountered in the last chapter was one with orthogonal curvilinear coordinates $u_1, u_2, u_3$, of which Cartesians, cylindrical polars and spherical polars are all special cases. Although it may be shown that (9.14) leads to the usual expression for grad in curvilinear coordinates, the proof requires complicated manipulations of the derivatives of the basis vectors with respect to the coordinates, and is not presented here. In Cartesian coordinates, however, the proof is quite simple.

► *Show that the geometrical definition of* grad *leads to the usual expression for* $\nabla\phi$ *in Cartesian coordinates.*

Consider a small rectangular volume element $\Delta V = \Delta x \, \Delta y \, \Delta z$ with its faces parallel to the $x$, $y$, $z$ coordinate surfaces and with the point $P$ at one corner. We must calculate the surface integral (9.14) over each of its six faces. Remembering that the normal to the surface points outwards from the volume on each face, the two faces with $x =$ constant have areas $\Delta\mathbf{S} = -\mathbf{i}\,\Delta y\,\Delta z$ and $\Delta\mathbf{S} = \mathbf{i}\,\Delta y\,\Delta z$ respectively. Furthermore, over each small surface element, we may take $\phi$ to be constant, so that the net contribution to the surface integral from these two faces is then

$$\begin{aligned}[(\phi + \Delta\phi) - \phi]\,\Delta y\,\Delta z\,\mathbf{i} &= \left(\phi + \frac{\partial\phi}{\partial x}\Delta x - \phi\right)\Delta y\,\Delta z\,\mathbf{i} \\ &= \frac{\partial\phi}{\partial x}\Delta x\,\Delta y\,\Delta z\,\mathbf{i}.\end{aligned}$$

The surface integral over the pairs of faces with $y =$ constant and $z =$ constant respectively may be found in a similar way, and we obtain

$$\oint_S \phi\, d\mathbf{S} = \left(\frac{\partial\phi}{\partial x}\mathbf{i} + \frac{\partial\phi}{\partial y}\mathbf{j} + \frac{\partial\phi}{\partial z}\mathbf{k}\right)\Delta x\,\Delta y\,\Delta z.$$

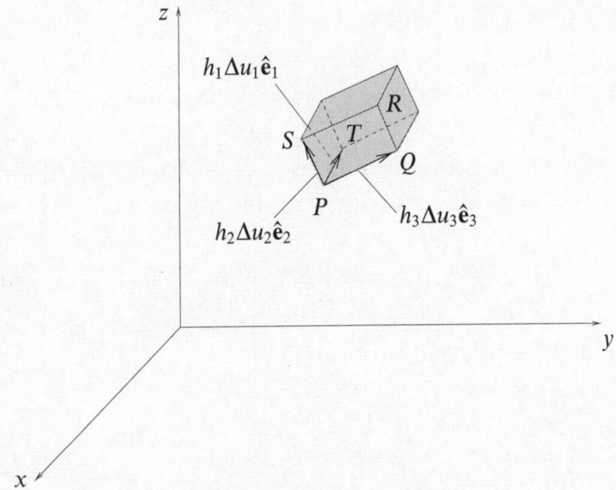


Figure 9.10 A general volume $\Delta V$ in orthogonal curvilinear coordinates $u_1, u_2, u_3$. $PT$ gives the vector $h_1\,\Delta u_1\,\hat{\mathbf{e}}_1$, $PS$ gives $h_2\,\Delta u_2\,\hat{\mathbf{e}}_2$ and $PQ$ gives $h_3\,\Delta u_3\,\hat{\mathbf{e}}_3$.

Therefore $\nabla\phi$ at the point $P$ is given by

$$\nabla\phi = \lim_{\Delta x,\Delta y,\Delta z\to 0}\left[\frac{1}{\Delta x\,\Delta y\,\Delta z}\left(\frac{\partial\phi}{\partial x}\mathbf{i}+\frac{\partial\phi}{\partial y}\mathbf{j}+\frac{\partial\phi}{\partial z}\mathbf{k}\right)\Delta x\,\Delta y\,\Delta z\right]$$
$$= \frac{\partial\phi}{\partial x}\mathbf{i}+\frac{\partial\phi}{\partial y}\mathbf{j}+\frac{\partial\phi}{\partial z}\mathbf{k}. \blacktriangleleft$$

The geometrical definitions (9.15) and (9.17) may be straightforwardly shown to lead to the usual expressions for div and curl in orthogonal curvilinear coordinates.

► *By considering the infinitesimal volume element $dV = h_1h_2h_3\,\Delta u_1\,\Delta u_2\,\Delta u_3$ shown in figure 9.10, show that (9.15) leads to the usual expression for $\nabla\cdot\mathbf{a}$ in orthogonal curvilinear coordinates.*

Let us write the vector field in terms of its components with respect to the basis vectors of the curvilinear coordinate system as $\mathbf{a} = a_1\hat{\mathbf{e}}_1+a_2\hat{\mathbf{e}}_2+a_3\hat{\mathbf{e}}_3$, and consider first the two faces with $u_1 =$ constant in figure 9.10. Now, the volume element is formed from the orthogonal vectors $h_1\,\Delta u_1\,\hat{\mathbf{e}}_1$, $h_2\,\Delta u_2\,\hat{\mathbf{e}}_2$ and $h_3\,\Delta u_3\,\hat{\mathbf{e}}_3$, at the point $P$, so for the front face $PQRS$ we have

$$\Delta\mathbf{S} = h_2h_3\,\Delta u_2\,\Delta u_3\;\hat{\mathbf{e}}_3\times\hat{\mathbf{e}}_2 = -h_2h_3\,\Delta u_2\,\Delta u_3\,\hat{\mathbf{e}}_1.$$

The contribution to the surface integral of $\mathbf{a}\cdot d\mathbf{S}$ over $PQRS$ and its opposite face taken together is therefore given by

$$\frac{\partial}{\partial u_1}(\mathbf{a}\cdot\Delta\mathbf{S})\,\Delta u_1 = \frac{\partial}{\partial u_1}(a_1h_2h_3)\,\Delta u_1\,\Delta u_2\,\Delta u_3.$$

The surface integrals over the pairs of faces with $u_2 =$ constant and $u_3 =$ constant respectively may be found in a similar way, and we obtain

$$\oint_S \mathbf{a}\cdot d\mathbf{S} = \left[\frac{\partial}{\partial u_1}(a_1h_2h_3) + \frac{\partial}{\partial u_2}(a_2h_3h_1) + \frac{\partial}{\partial u_3}(a_3h_1h_2)\right]\Delta u_1\,\Delta u_2\,\Delta u_3.$$

Therefore $\nabla\cdot\mathbf{a}$ at the point $P$ is given by

$$\begin{aligned}\nabla\cdot\mathbf{a} &= \lim_{\Delta u_1,\Delta u_2,\Delta u_3\to 0}\left[\frac{1}{h_1h_2h_3\,\Delta u_1\,\Delta u_2\,\Delta u_3}\oint_S \mathbf{a}\cdot d\mathbf{S}\right]\\ &= \frac{1}{h_1h_2h_3}\left[\frac{\partial}{\partial u_1}(a_1h_2h_3) + \frac{\partial}{\partial u_2}(a_2h_3h_1) + \frac{\partial}{\partial u_3}(a_3h_1h_2)\right]. \blacktriangleleft\end{aligned}$$

► *By considering the infinitesimal planar surface element PQRS in figure 9.10, show that (9.17) leads to the usual expression for $\nabla\times\mathbf{a}$ in orthogonal curvilinear coordinates.*

The planar surface $PQRS$ is defined by the orthogonal vectors $h_2\,\Delta u_2\,\hat{\mathbf{e}}_2$ and $h_3\,\Delta u_3\,\hat{\mathbf{e}}_3$ at the point $P$. If we traverse the loop in the direction $PSRQ$, then by the right-handed convention the unit normal to the plane is $\hat{\mathbf{e}}_1$. Writing $\mathbf{a} = a_1\hat{\mathbf{e}}_1 + a_2\hat{\mathbf{e}}_2 + a_3\hat{\mathbf{e}}_3$, the line integral around the loop in this direction is given by

$$\begin{aligned}\oint_{PSRQ}\mathbf{a}\cdot d\mathbf{r} &= a_2h_2\,\Delta u_2 + \left[a_3h_3 + \frac{\partial}{\partial u_2}(a_3h_3)\,\Delta u_2\right]\Delta u_3\\ &\quad - \left[a_2h_2 + \frac{\partial}{\partial u_3}(a_2h_2)\,\Delta u_3\right]\Delta u_2 - a_3h_3\,\Delta u_3\\ &= \left[\frac{\partial}{\partial u_2}(a_3h_3) - \frac{\partial}{\partial u_3}(a_2h_2)\right]\Delta u_2\,\Delta u_3.\end{aligned}$$

Therefore from (9.17) the component of $\nabla\times\mathbf{a}$ in the direction $\hat{\mathbf{e}}_1$ at $P$ is given by

$$\begin{aligned}(\nabla\times\mathbf{a})_1 &= \lim_{\Delta u_2,\Delta u_3\to 0}\left[\frac{1}{h_2h_3\,\Delta u_2\,\Delta u_3}\oint_{PSRQ}\mathbf{a}\cdot d\mathbf{r}\right]\\ &= \frac{1}{h_2h_3}\left[\frac{\partial}{\partial u_2}(h_3a_3) - \frac{\partial}{\partial u_3}(h_2a_2)\right].\end{aligned}$$

The other two components are found by cyclically permuting the subscripts 1, 2, 3. ◄

Finally, we note that we can also write the $\nabla^2$ operator as a surface integral by setting $\mathbf{a} = \nabla\phi$ in (9.15) to obtain

$$\nabla^2\phi = \nabla \cdot \nabla\phi = \lim_{V \to 0} \left( \frac{1}{V} \oint_S \nabla\phi \cdot d\mathbf{S} \right).$$

## 9.8 Divergence theorem and related theorems

The divergence theorem relates the total flux of a vector field out of a closed surface $S$ to the integral of the divergence of the vector field over the enclosed volume $V$, and follows almost immediately from our geometrical definition of divergence (9.15).

Imagine a volume $V$, in which a vector field $\mathbf{a}$ is continuous and differentiable, to be divided up into a large number of small volumes $V_i$. Using (9.15), we have for each small volume

$$(\nabla \cdot \mathbf{a})V_i \approx \oint_{S_i} \mathbf{a} \cdot d\mathbf{S},$$

where $S_i$ is the surface of the small volume $V_i$. Summing over $i$ we find that contributions from surface elements interior to $S$ cancel, since each surface element appears in two terms with opposite signs, the outward normals in the two terms being equal and opposite. Only contributions from surface elements which are also parts of $S$ survive. If each $V_i$ is allowed to tend to zero, we obtain the *divergence theorem*,

$$\int_V \nabla \cdot \mathbf{a}\, dV = \oint_S \mathbf{a} \cdot d\mathbf{S}. \qquad (9.18)$$

We note that the divergence theorem holds for both simply and multiply connected surfaces, provided that they are closed and enclose some non-zero volume $V$. The divergence theorem may also be extended to tensor fields (see chapter 19).

The theorem finds most use as a tool in formal manipulations, but sometimes it is of value in evaluating surface integrals of the form $\int_S \mathbf{a} \cdot d\mathbf{S}$ as volume integrals or vice versa. For example, setting $\mathbf{a} = \mathbf{r}$ we immediately obtain

$$\int_V \nabla \cdot \mathbf{r}\, dV = \int_V 3\, dV = 3V = \oint_S \mathbf{r} \cdot d\mathbf{S},$$

which gives the expression for the volume of a region found in subsection 9.6.1. The use of the divergence theorem is further illustrated in the following example.

►*Evaluate the surface integral* $I = \int_S \mathbf{a} \cdot d\mathbf{S}$, *where* $\mathbf{a} = (y - x)\,\mathbf{i} + x^2 z\,\mathbf{j} + (z + x^2)\,\mathbf{k}$, *and* $S$ *is the open surface of the hemisphere* $x^2 + y^2 + z^2 = a^2$, $z \geq 0$.

We could evaluate this surface integral directly, but the algebra is somewhat lengthy. We will therefore evaluate it by use of the divergence theorem. Since the latter only holds for closed surfaces enclosing a non-zero volume $V$, let us first consider the closed surface $S' = S + S_1$, where $S_1$ is the circular area in the $xy$-plane given by $x^2 + y^2 \leq a^2$, $z = 0$; $S'$ then encloses a hemispherical volume $V$. By the divergence theorem we have

$$\int_V \nabla \cdot \mathbf{a}\, dV = \oint_{S'} \mathbf{a} \cdot d\mathbf{S} = \int_S \mathbf{a} \cdot d\mathbf{S} + \int_{S_1} \mathbf{a} \cdot d\mathbf{S}.$$

Now $\nabla \cdot \mathbf{a} = -1 + 0 + 1 = 0$, so we can write

$$\int_S \mathbf{a} \cdot d\mathbf{S} = -\int_{S_1} \mathbf{a} \cdot d\mathbf{S}.$$

The surface integral over $S_1$ is easily evaluated. Remembering that the normal to the surface points outward from the volume the surface element on $S_1$ is simply $d\mathbf{S} = -\mathbf{k}\, dx\, dy$. On $S_1$ we also have $\mathbf{a} = (y - x)\,\mathbf{i} + x^2\,\mathbf{k}$, so that

$$I = -\int_{S_1} \mathbf{a} \cdot d\mathbf{S} = \iint_R x^2\, dx\, dy,$$

where $R$ is the circular region in the $xy$-plane given by $x^2 + y^2 \leq a^2$. Transforming to plane polar coordinates we have

$$I = \iint_{R'} \rho^2 \cos^2\phi\ \rho\, d\rho\, d\phi = \int_0^{2\pi} \cos^2\phi\, d\phi \int_0^a \rho^3\, d\rho = \frac{\pi a^4}{4}. \blacktriangleleft$$

It is also interesting to consider the two-dimensional version of the divergence theorem. As an example, let us consider a two-dimensional planar region $R$ in the $xy$-plane bounded by some closed curve $C$ (see figure 9.11). At any point on the curve the vector $d\mathbf{r} = dx\,\mathbf{i} + dy\,\mathbf{j}$ is a tangent to the curve and the vector $\hat{\mathbf{n}}\, ds = dy\,\mathbf{i} - dx\,\mathbf{j}$ is a normal pointing out of the region $R$. If the vector field $\mathbf{a}$ is continuous and differentiable in $R$, then the two-dimensional divergence theorem in Cartesian coordinates gives

$$\iint_R \left( \frac{\partial a_x}{\partial x} + \frac{\partial a_y}{\partial y} \right) dx\, dy = \oint \mathbf{a} \cdot \hat{\mathbf{n}}\, ds = \oint_C (a_x\, dy - a_y\, dx).$$

Letting $P = -a_y$ and $Q = a_x$, we recover Green's theorem in a plane, which was discussed in section 9.3.

### *9.8.1 Green's theorems*

Consider two scalar functions $\phi$ and $\psi$ that are continuous and differentiable in some volume $V$ bounded by a surface $S$. Applying the divergence theorem to the vector field $\phi\nabla\psi$ we obtain

$$\begin{aligned} \oint_S \phi\nabla\psi \cdot d\mathbf{S} &= \int_V \nabla \cdot (\phi\nabla\psi)\, dV \\ &= \int_V \left[\phi\nabla^2\psi + (\nabla\phi)\cdot(\nabla\psi)\right] dV. \end{aligned} \qquad (9.19)$$

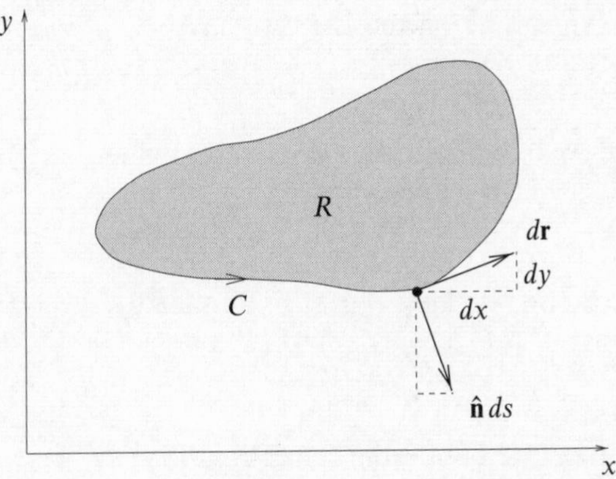


Figure 9.11 A closed curve $C$ in the $xy$-plane bounding a region $R$. Vectors tangent and normal to the curve at a given point are also shown.

Reversing the roles of $\phi$ and $\psi$ in (9.19) and subtracting the two equations gives

$$\oint_S (\phi\nabla\psi - \psi\nabla\phi) \cdot d\mathbf{S} = \int_V (\phi\nabla^2\psi - \psi\nabla^2\phi)\, dV. \qquad (9.20)$$

Equation (9.19) is usually known as Green's first theorem and (9.20) as his second. Green's second theorem is useful in the development of the Green's functions used in the solution of partial differential equations (see chapter 17).

### *9.8.2 Other related integral theorems*

There exist two other integral theorems which are closely related to the divergence theorem and which are of some use in physical applications. If $\phi$ is a scalar field and $\mathbf{b}$ is a vector field and both $\phi$ and $\mathbf{b}$ satisfy our usual differentiability conditions in some volume $V$ bounded by a closed surface $S$ then

$$\int_V \nabla\phi\, dV = \oint_S \phi\, d\mathbf{S}, \qquad (9.21)$$

$$\int_V \nabla\times\mathbf{b}\, dV = \oint_S d\mathbf{S}\times\mathbf{b}. \qquad (9.22)$$

►*Use the divergence theorem to prove (9.21).*

In the divergence theorem (9.18) let $\mathbf{a} = \phi\mathbf{c}$, where $\mathbf{c}$ is a constant vector. We then have

$$\int_V \nabla\cdot(\phi\mathbf{c})\, dV = \oint_S \phi\mathbf{c}\cdot d\mathbf{S}.$$

Expanding out the integrand on the LHS we have

$$\nabla \cdot (\phi \mathbf{c}) = \phi \nabla \cdot \mathbf{c} + \mathbf{c} \cdot \nabla \phi = \mathbf{c} \cdot \nabla \phi,$$

since $\mathbf{c}$ is constant. Also, $\phi \mathbf{c} \cdot d\mathbf{S} = \mathbf{c} \cdot \phi d\mathbf{S}$, so we obtain

$$\int_V \mathbf{c} \cdot (\nabla \phi)\, dV = \oint_S \mathbf{c} \cdot \phi\, d\mathbf{S}.$$

Since $\mathbf{c}$ is constant we may take it out of both integrals to give

$$\mathbf{c} \cdot \int_V \nabla \phi\, dV = \mathbf{c} \cdot \oint_S \phi\, d\mathbf{S},$$

and since $\mathbf{c}$ is arbitrary we obtain the stated result (9.21). ◄

Equation (9.22) may be proved in a similar way by letting $\mathbf{a} = \mathbf{b} \times \mathbf{c}$ in the divergence theorem, where $\mathbf{c}$ is again a constant vector.

### *9.8.3 Physical applications of the divergence theorem*

The divergence theorem is useful in deriving many of the most important partial differential equations in physics (see chapter 16). The basic idea is to use the divergence theorem to convert an integral form (often derived from observation) into an equivalent differential form (used in theory).

►*For a compressible fluid with time-varying position-dependent density $\rho(\mathbf{r}, t)$ and velocity field $v(\mathbf{r}, t)$, in which fluid is neither being created nor destroyed, show that*

$$\frac{\partial \rho}{\partial t} + \nabla \cdot (\rho \mathbf{v}) = 0.$$

For an arbitrary volume $V$ in the fluid, conservation of mass tells us that the rate of increase or decrease of the mass $M$ of fluid in the volume must equal the net rate at which fluid is entering or leaving the volume, i.e.

$$\frac{dM}{dt} = -\oint_S \rho \mathbf{v} \cdot d\mathbf{S},$$

where $S$ is the surface bounding $V$. But the mass of fluid in $V$ is simply $M = \int_V \rho\, dV$, so we have

$$\frac{d}{dt} \int_V \rho\, dV + \oint_S \rho \mathbf{v} \cdot d\mathbf{S} = 0.$$

Taking the derivative inside the first integral on the RHS, and using the divergence theorem to rewrite the second integral, we obtain

$$\int_V \frac{\partial \rho}{\partial t}\, dV + \int_V \nabla \cdot (\rho \mathbf{v})\, dV = \int_V \left[ \frac{\partial \rho}{\partial t} + \nabla \cdot (\rho \mathbf{v}) \right] dV = 0.$$

Since the volume $V$ is arbitrary, the integrand (which is assumed continuous) must be identically zero, so

$$\frac{\partial \rho}{\partial t} + \nabla \cdot (\rho \mathbf{v}) = 0.$$

This is known as the *continuity equation*. It can also be applied to other systems, for example those in which $\rho$ is the density of electric charge or the heat content, etc. In the flow of an incompressible fluid $\rho$ = constant and the continuity equation becomes simply $\nabla \cdot \mathbf{v} = 0$. ◀

In the previous example, we assumed that there were no sources or sinks in the volume $V$, i.e. that there was no part of $V$ in which fluid was being created or destroyed. We now consider the case where a finite number of *point* sources and/or sinks are present in an incompressible fluid. Let us first consider the simple case where a single source is located at the origin, out of which a quantity of fluid flows radially at a rate $Q$ ($\text{m}^3\ \text{s}^{-1}$). The velocity field is simply given by

$$\mathbf{v} = \frac{Q\mathbf{r}}{4\pi r^3} = \frac{Q\hat{\mathbf{r}}}{4\pi r^2}.$$

Now, for a sphere $S_1$ of radius $r$ centred on the source, the flux across $S_1$ is

$$\oint_{S_1} \mathbf{v} \cdot d\mathbf{S} = |\mathbf{v}| 4\pi r^2 = Q.$$

Since $\mathbf{v}$ has a singularity at the origin it is not differentiable there, but at all other points $\nabla \cdot \mathbf{v} = 0$, as required for an incompressible fluid. Therefore, from the divergence theorem, for any closed surface $S_2$ that does not enclose the origin we have

$$\oint_{S_2} \mathbf{v} \cdot d\mathbf{S} = \int_V \nabla \cdot \mathbf{v}\, dV = 0.$$

We thus see that the surface integral $\oint_S \mathbf{v} \cdot d\mathbf{S}$ has value $Q$ or zero depending on whether or not $S$ encloses the source at the origin. In order that the divergence theorem is valid for *all* surfaces $S$, irrespective of whether they enclose the source, we write

$$\nabla \cdot \mathbf{v} = Q\delta(\mathbf{r}),$$

where $\delta(\mathbf{r})$ is the three-dimensional Dirac delta function. The properties of this function are discussed fully in chapter 11, but for the moment we note that it is defined in such a way that

$$\delta(\mathbf{r} - \mathbf{a}) = 0 \quad \text{for } \mathbf{r} \neq \mathbf{a},$$

$$\int_V f(\mathbf{r})\delta(\mathbf{r} - \mathbf{a})\, dV = \begin{cases} f(\mathbf{a}) & \text{if } \mathbf{a} \text{ lies in } V \\ 0 & \text{otherwise} \end{cases}$$

for any well-behaved function $f(\mathbf{r})$. Therefore, for any volume $V$ containing the source at the origin, we have

$$\int_V \nabla \cdot \mathbf{v} \, dV = Q \int_V \delta(\mathbf{r}) \, dV = Q,$$

which is consistent with $\oint_S \mathbf{v} \cdot d\mathbf{S} = Q$ for a closed surface enclosing the source. Hence, by introducing the Dirac delta function the divergence theorem can be made valid even for non-differentiable point sources.

The generalisation to several sources and sinks is straightforward. For example, if a source is located at $\mathbf{r} = \mathbf{a}$ and a sink at $\mathbf{r} = \mathbf{b}$, then the velocity field is

$$\mathbf{v} = \frac{Q(\mathbf{r} - \mathbf{a})}{4\pi|\mathbf{r} - \mathbf{a}|^3} - \frac{Q(\mathbf{r} - \mathbf{b})}{4\pi|\mathbf{r} - \mathbf{b}|^3},$$

and its divergence is given by

$$\nabla \cdot \mathbf{v} = Q\delta(\mathbf{r} - \mathbf{a}) - Q\delta(\mathbf{r} - \mathbf{b}).$$

Therefore, the integral $\oint_S \mathbf{v} \cdot d\mathbf{S}$ has the value $Q$ if $S$ encloses the source, $-Q$ if $S$ encloses the sink and 0 if $S$ encloses neither the source nor sink or encloses them both. This analysis also applies to other physical systems – for example, in electrostatics we can regard the sources and sinks as positive and negative point charges respectively, and replace $\mathbf{v}$ by the electric field $\mathbf{E}$.

## 9.9 Stokes' theorem and related theorems

Stokes' theorem is the 'curl analogue' of the divergence theorem and relates the integral of the curl of a vector field over an open surface $S$ to the line integral of the vector field around the perimeter $C$ bounding the surface.

Following the same lines as for the derivation of the divergence theorem, we can divide the surface $S$ into many small areas $S_i$ with boundaries $C_i$ and with unit normals $\hat{\mathbf{n}}_i$. Using (9.17), we have for each small area

$$(\nabla \times \mathbf{a}) \cdot \hat{\mathbf{n}}_i \; S_i \approx \oint_{C_i} \mathbf{a} \cdot d\mathbf{r}.$$

Summing over $i$ we find that on the RHS all parts of all interior boundaries that are not part of $C$ are included twice, being traversed in opposite directions on each occasion and thus contributing nothing. Only contributions from line elements that are also parts of $C$ survive. If each $S_i$ is allowed to tend to zero, we obtain Stokes' theorem,

$$\int_S (\nabla \times \mathbf{a}) \cdot d\mathbf{S} = \oint_C \mathbf{a} \cdot d\mathbf{r}. \qquad (9.23)$$

We note that Stokes' theorem holds for both simply and multiply connected open surfaces, provided that they are two-sided. Stokes' theorem may also be extended to tensor fields (see chapter 19).

Just as the divergence theorem (9.18) can be used to relate volume and surface integrals, Stokes' theorem can be used in evaluating surface integrals of the form $\oint_S (\nabla \times \mathbf{a}) \cdot d\mathbf{S}$ as line integrals or vice versa.

►*Given the vector field* $\mathbf{a} = y\,\mathbf{i} - x\,\mathbf{j} + z\,\mathbf{k}$, *verify Stokes' theorem for the hemispherical surface* $x^2 + y^2 + z^2 = a^2$, $z \geq 0$.

Let us first evaluate the surface integral

$$\int_S (\nabla \times \mathbf{a}) \cdot d\mathbf{S},$$

over the hemisphere. It is easily shown that $\nabla \times \mathbf{a} = -2\,\mathbf{k}$, and the surface element is $d\mathbf{S} = a^2 \sin\theta \, d\theta \, d\phi \, \hat{\mathbf{r}}$ in spherical polar coordinates. Therefore

$$\begin{aligned}
\int_S (\nabla \times \mathbf{a}) \cdot d\mathbf{S} &= \int_0^{2\pi} d\phi \int_0^{\pi/2} d\theta \, \left(-2a^2 \sin\theta\right) \hat{\mathbf{r}} \cdot \mathbf{k} \\
&= -2a^2 \int_0^{2\pi} d\phi \int_0^{\pi/2} \sin\theta \left(\frac{z}{a}\right) d\theta \\
&= -2a^2 \int_0^{2\pi} d\phi \int_0^{\pi/2} \sin\theta \cos\theta \, d\theta = -2\pi a^2.
\end{aligned}$$

We now evaluate the line integral around the perimeter curve $C$ of the surface, which is the circle $x^2 + y^2 = a^2$ in the $xy$-plane. This is given by

$$\begin{aligned}
\oint_C \mathbf{a} \cdot d\mathbf{r} &= \oint_C (y\,\mathbf{i} - x\,\mathbf{j} + z\,\mathbf{k}) \cdot (dx\,\mathbf{i} + dy\,\mathbf{j} + dz\,\mathbf{k}) \\
&= \oint_C (y\,dx - x\,dy).
\end{aligned}$$

Using plane polar coordinates, on $C$ we have $x = a\cos\phi$, $y = a\sin\phi$ so that $dx = -a\sin\phi\,d\phi$, $dy = a\cos\phi\,d\phi$, and the line integral becomes

$$\oint_C (y\,dx - x\,dy) = -a^2 \int_0^{2\pi} (\sin^2\phi + \cos^2\phi)\,d\phi = -a^2 \int_0^{2\pi} d\phi = -2\pi a^2.$$

Since the surface and line integrals have the same value, we have verified Stokes' theorem in this case. ◄

The two-dimensional version of Stokes' theorem also yields Green's theorem in a plane. Consider the region $R$ in the $xy$-plane shown in figure 9.11, in which a vector field $\mathbf{a}$ is defined. Since $\mathbf{a} = a_x\,\mathbf{i} + a_y\,\mathbf{j}$, we have $\nabla \times \mathbf{a} = (\partial a_y/\partial x - \partial a_x/\partial y)\,\mathbf{k}$, and Stokes' theorem becomes

$$\iint_R \left(\frac{\partial a_y}{\partial x} - \frac{\partial a_x}{\partial y}\right) dx\,dy = \oint_C (a_x\,dx + a_y\,dy).$$

Letting $P = a_x$ and $Q = a_y$ we recover Green's theorem in a plane, (9.4).

### 9.9.1 Related integral theorems

As for the divergence theorem, there exist two other integral theorems that are closely related to Stokes' theorem. If $\phi$ is a scalar field and $\mathbf{b}$ is a vector field, and both $\phi$ and $\mathbf{b}$ satisfy our usual differentiability conditions on some two-sided open surface $S$ bounded by a closed perimeter curve $C$, then

$$\int_S d\mathbf{S} \times \nabla\phi = \oint_C \phi\, d\mathbf{r}, \tag{9.24}$$

$$\int_S (d\mathbf{S} \times \nabla) \times \mathbf{b} = \oint_C d\mathbf{r} \times \mathbf{b}. \tag{9.25}$$

►*Use Stokes' theorem to prove (9.24).*

In Stokes' theorem, (9.23), let $\mathbf{a} = \phi\mathbf{c}$, where $\mathbf{c}$ is a constant vector. We then have

$$\int_S [\nabla \times (\phi\mathbf{c})] \cdot d\mathbf{S} = \oint_C \phi\mathbf{c} \cdot d\mathbf{r}. \tag{9.26}$$

Expanding out the integrand on the LHS we have

$$\nabla \times (\phi\mathbf{c}) = \nabla\phi \times \mathbf{c} + \phi\nabla \times \mathbf{c} = \nabla\phi \times \mathbf{c},$$

since $\mathbf{c}$ is constant, and the triple scalar product on the LHS of (9.26) can therefore be written

$$[\nabla \times (\phi\mathbf{c})] \cdot d\mathbf{S} = (\nabla\phi \times \mathbf{c}) \cdot d\mathbf{S} = \mathbf{c} \cdot (d\mathbf{S} \times \nabla\phi).$$

Substituting this into (9.26), and taking $\mathbf{c}$ out of both integrals because it is constant, we find

$$\mathbf{c} \cdot \int_S d\mathbf{S} \times \nabla\phi = \mathbf{c} \cdot \oint_C \phi\, d\mathbf{r}.$$

Since $\mathbf{c}$ is an arbitrary constant vector we therefore obtain the stated result (9.24). ◄

Equation (9.25) may be proved in a similar way, by letting $\mathbf{a} = \mathbf{b} \times \mathbf{c}$ in Stokes' theorem, where $\mathbf{c}$ is again a constant vector. We also note that by setting $\mathbf{b} = \mathbf{r}$ in (9.25) we find

$$\int_S (d\mathbf{S} \times \nabla) \times \mathbf{r} = \oint_C d\mathbf{r} \times \mathbf{r}.$$

Expanding out the integrand on the LHS we find

$$(d\mathbf{S} \times \nabla) \times \mathbf{r} = d\mathbf{S} - d\mathbf{S}(\nabla \cdot \mathbf{r}) = d\mathbf{S} - 3\, d\mathbf{S} = -2\, d\mathbf{S}.$$

Therefore, as we found in subsection 9.5.2, the vector area of an open surface $S$ is given by

$$\mathbf{S} = \int_S d\mathbf{S} = \frac{1}{2}\oint_C \mathbf{r} \times d\mathbf{r}.$$

### 9.9.2 Physical applications of Stokes' theorem

Like the divergence theorem, Stokes' theorem is useful in converting integral equations into differential equations.

► *From Ampère's law derive Maxwell's equation in the case where the currents are steady, i.e.* $\nabla \times \mathbf{B} - \mu_0 \mathbf{J} = \mathbf{0}$.

Ampère's rule for a distributed current with current density $\mathbf{J}$ is

$$\oint_C \mathbf{B} \cdot d\mathbf{r} = \mu_0 \int_S \mathbf{J} \cdot d\mathbf{S},$$

for any circuit $C$ bounding a surface $S$. Using Stokes' theorem, the LHS can be transformed into $\int_S (\nabla \times \mathbf{B}) \cdot d\mathbf{S}$; hence

$$\int_S (\nabla \times \mathbf{B} - \mu_0 \mathbf{J}) \cdot d\mathbf{S} = 0$$

for *any* surface $S$. This can only be so if $\nabla \times \mathbf{B} - \mu_0 \mathbf{J} = \mathbf{0}$, which is the required relation. Similarly, from Faraday's law of electromagnetic induction we can derive Maxwell's equation $\nabla \times \mathbf{E} = -\partial \mathbf{B}/\partial t$. ◄

In subsection 9.8.3 we discussed the flow of an incompressible fluid in the presence of several sources and sinks. Let us now consider *vortex* flow in an incompressible fluid with a velocity field

$$\mathbf{v} = \frac{1}{\rho} \hat{\mathbf{e}}_\phi,$$

in cylindrical polar coordinates $\rho$, $\phi$, $z$. For this velocity field $\nabla \times \mathbf{v}$ equals zero everywhere except on the axis $\rho = 0$, where $\mathbf{v}$ has a singularity. Therefore $\oint_C \mathbf{v} \cdot d\mathbf{r}$ equals zero for any path $C$ that does not enclose the vortex line on the axis and $2\pi$ if $C$ does enclose the axis. In order for Stokes' theorem to be valid for all paths $C$, we therefore set

$$\nabla \times \mathbf{v} = 2\pi\delta(\rho),$$

where $\delta(\rho)$ is the Dirac delta function, to be discussed in subsection 11.1.3. Now, since $\nabla \times \mathbf{v} = \mathbf{0}$, except on the axis $\rho = 0$, there exists a scalar potential $\psi$ such that $\mathbf{v} = \nabla\psi$. It may easily be shown that $\psi = \phi$, the polar angle. Therefore, if $C$ does not enclose the axis,

$$\oint_C \mathbf{v} \cdot d\mathbf{r} = \oint d\phi = 0,$$

and if $C$ does enclose the axis,

$$\oint_C \mathbf{v} \cdot d\mathbf{r} = \Delta\phi = 2\pi n,$$

where $n$ is the number of times we traverse $C$. Thus $\phi$ is a multivalued potential.

A similar analysis is valid for other physical systems – for example, in magnetostatics we may replace the vortex lines by current-carrying wires and $\mathbf{v}$ by the magnetic field $\mathbf{B}$.

## 9.10 Exercises

9.1 $\mathbf{F}$ is a vector field $xy^2\mathbf{i} + 2\mathbf{j} + x\mathbf{k}$, and $L$ a path parameterised by $x = ct$, $y = c/t$, $z = d$ for the range $1 \leq t \leq 2$. Evaluate (a) $\int_L \mathbf{F}\, dt$, (b) $\int_L \mathbf{F}\, dy$ and (c) $\int_L \mathbf{F} \cdot d\mathbf{r}$.

9.2 A single-turn coil $C$ of arbitrary shape is placed in a magnetic field $\mathbf{B}$ and carries a current $I$. Show that the couple acting upon the coil can be written as

$$\mathbf{M} = I \int_C (\mathbf{B} \cdot \mathbf{r})\, d\mathbf{r} - I \int_C \mathbf{B}(\mathbf{r} \cdot d\mathbf{r}).$$

For a planar rectangular coil of sides $2a$ and $2b$ placed with its plane vertical and at an angle $\phi$ to a uniform horizontal field $\mathbf{B}$, show that $\mathbf{M}$ is, as expected, $4abBI \cos\phi\, \mathbf{k}$.

9.3 An axially symmetric solid body with its axis $AB$ vertical is immersed in an incompressible fluid of density $\rho_0$. Use the following method to show that, whatever the shape of the body, for $\rho = \rho(z)$ in cylindrical polars the Archimedean upthrust is, as expected, $\rho_0 g V$, where $V$ is the volume of the body.

Express the vertical component of the resultant force ($-\int p\, d\mathbf{S}$, where $p$ is the pressure) on the body in terms of an integral; note that $p = \rho_0 g(-z)$ and that for an annular surface element of width $dl$, $\mathbf{n} \cdot \mathbf{n}_z\, dl = -d\rho$. Integrate by parts and use the fact that $\rho(z_A) = \rho(z_B) = 0$.

9.4 Show that the expression below is equal to the solid angle subtended by a rectangular aperture of sides $2a$ and $2b$ at a point a distance $c$ from the aperture along the normal to its centre:

$$\Omega = 4 \int_0^b \frac{ac}{(y^2 + c^2)(y^2 + c^2 + a^2)^{1/2}}\, dy.$$

By setting $y = (a^2 + c^2)^{1/2} \tan\phi$, change this integral into the form

$$\int_0^{\phi_1} \frac{4ac \cos\phi}{c^2 + a^2 \sin^2\phi}\, d\phi,$$

where $\tan\phi_1 = b/(a^2 + c^2)^{1/2}$, and hence show that

$$\Omega = 4 \tan^{-1}\left[\frac{ab}{c(a^2 + b^2 + c^2)^{1/2}}\right].$$

9.5 A vector field $\mathbf{a} = -zxr^{-3}\mathbf{i} - zyr^{-3}\mathbf{j} + (x^2+y^2)r^{-3}\mathbf{k}$, where $r^2 = x^2+y^2+z^2$. Show that the field is conservative (a) by showing $\nabla \times \mathbf{a} = \mathbf{0}$, and (b) by constructing its potential function $\phi$.

9.6 The vector field $\mathbf{a} = (z^2 + 2xy)\mathbf{i} + (x^2 + 2yz)\mathbf{j} + (y^2 + 2zx)\mathbf{k}$. Show that $\mathbf{a}$ is conservative and that the line integral $\int \mathbf{a} \cdot d\mathbf{r}$ along any line joining $(1, 1, 1)$ and $(1, 2, 2)$ has the value 11.

9.7 A force $\mathbf{F}(\mathbf{r})$ acts on a particle at $\mathbf{r}$. In which of the following cases can $\mathbf{F}$ be represented in terms of a potential? Where it can, find the potential.

(a) $\mathbf{F} = [\mathbf{i} - \mathbf{j} - 2(x - y)\mathbf{r}] \exp(-r^2)$.
(b) $\mathbf{F} = [z\mathbf{k} + (x^2 + y^2 - 1)\mathbf{r}] \exp(-r^2)$.
(c) $\mathbf{F} = \mathbf{k} + (\mathbf{r} \times \mathbf{k})r^{-2}$.

9.8 A vector field $\mathbf{a} = f(r)\mathbf{r}$ is spherically symmetric and everywhere directed away from the origin. Show that $\mathbf{a}$ is irrotational, but that it is also solenoidal only if $f(r)$ is of the form $Ar^{-3}$.

9.9 Demonstrate the validity of the divergence theorem:

(a) by calculating the flux of the vector

$$\mathbf{F} = \frac{\mathbf{r}}{(r^2 + a^2)^{3/2}}$$

through the spherical surface $|\mathbf{r}| = \sqrt{3}a$,

(b) by showing that

$$\nabla \cdot \mathbf{F} = \frac{3a^2}{(r^2 + a^2)^{5/2}}$$

and evaluating the volume integral of $\nabla \cdot \mathbf{F}$ over the interior of the sphere $|\mathbf{r}| = \sqrt{3}a$.

(The substitution $r = a \tan\theta$ will prove useful in carrying out the integration.)

9.10 The vector field $\mathbf{F}$ is defined in cylindrical polar coordinates $\rho, \theta, z$ by

$$\mathbf{F} = (x\cos z)\mathbf{i} + (y\cos z)\mathbf{j} + (\sin z)\mathbf{k} \;\equiv\; (\rho\cos z)\mathbf{e}_r + (\sin z)\mathbf{k},$$

where $\mathbf{i}$, $\mathbf{j}$ and $\mathbf{k}$ are the unit vectors along the Cartesian axes and $\mathbf{e}_r$ is the unit vector $(x/\rho)\mathbf{i} + (y/\rho)\mathbf{j}$.

(a) Calculate, as a surface integral, the flux of $\mathbf{F}$ through the closed surface bounded by the cylinders $\rho = 1$ and $\rho = 2$ and the planes $z = \pm\pi/2$.
(b) Evaluate the same integral using the divergence theorem.

9.11 The vector field $\mathbf{F}$ is defined in Cartesian coordinates by

$$\mathbf{F} = \left(\tfrac{1}{3}y^3 + ye^{xy} + 1\right)\mathbf{i} + \left[xy^2 + (x+y)e^{xy}\right]\mathbf{j} + ze^{xy}\mathbf{k}.$$

Use Stokes' theorem to calculate

$$\oint_L \mathbf{F}\cdot d\mathbf{r},$$

where $L$ is the perimeter of the rectangle $ABCD$ given by $A = (0,1,0)$, $B = (1,1,0)$, $C = (1,3,0)$ and $D = (0,3,0)$.

## 9.11 Hints and answers

9.1 (a) $c^3 \ln 2\,\mathbf{i} + 2\mathbf{j} + (3c/2)\mathbf{k}$; (b) $(-3c^4/8)\mathbf{i} - c\mathbf{j} - c^2 \ln 2\,\mathbf{k}$; (c) $c^4 \ln 2 - c$.
9.2 $M = I\int_C \mathbf{r}\times(d\mathbf{r}\times\mathbf{B})$.
9.5 (b) $\phi = c + z/r$.
9.6 $f(x,y,z) = z^2x + x^2y + y^2z$.
9.7 (a) Yes, $(x-y)\exp(-r^2)$; (b) yes, $-[(x^2+y^2)/2]\exp(-r^2)$; (c) no, $\nabla\times\mathbf{F} \neq \mathbf{0}$.
9.8 $\nabla\times\mathbf{a} = \mathbf{0}$; $\nabla\cdot\mathbf{a} = 3f(r) + rf'(r) = 0$ if $f(r) = Ar^{-3}$.
9.9 $3\sqrt{3}\pi/2$ in each case.
9.10 $18\pi$.
9.11 $\int_0^1 dx \int_1^3 dy\, y^2e^{xy} = 2e^3 - 4$.

# 10

# Fourier series

We have already discussed in chapter 3 how complicated functions may be expressed as power series. However, this is not the only way in which a function may be represented as a series, and the subject of this chapter is the expression of functions as a sum of sine and cosine terms. Such a representation is called a *Fourier series*. Unlike Taylor series, a Fourier series can describe functions that are not everywhere continuous and/or differentiable. There are also other advantages in using trigonometrical terms. They are easy to differentiate and integrate, their moduli are easily taken and each term contains only one characteristic frequency. This last point is important because, as we shall see later, Fourier series are often used to represent the response of a system to a periodic input, and this response often depends directly on the frequency content of the input. Fourier series are used in a wide variety of such physical situations, including the vibrations of a finite string, the scattering of light by a diffraction grating and the transmission of an input signal by an electronic circuit.

## 10.1 The Dirichlet conditions

We have already mentioned that Fourier series may be used to represent some functions for which a Taylor series expansion is not possible. The particular conditions that a function $f(x)$ must fulfil in order that it may be expanded as a Fourier series are known as the *Dirichlet conditions*, and may be summarised by the following four points:

(i) the function must be periodic;
(ii) it must be single-valued and continuous, except possibly at a finite number of finite discontinuities;
(iii) it must have only a finite number of maxima and minima within one period;
(iv) the integral over one period of $|f(x)|$ must converge.

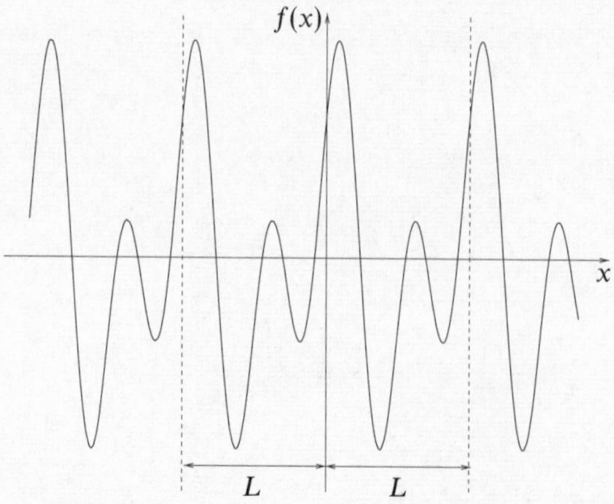


Figure 10.1 An example of a function that may, without modification, be represented as a Fourier series.

If the above conditions are satisfied, then the Fourier series converges to $f(x)$ at all points where $f(x)$ is continuous. The convergence of the Fourier series at points of discontinuity is discussed in section 10.4. The last three Dirichlet conditions are almost always met in real applications, but not all functions are periodic and hence do not fulfil the first condition. It may be possible, however, to represent a non-periodic function as a Fourier series by manipulation of the function into a periodic form. This is discussed in section 10.5. An example of a function that may, without modification, be represented as a Fourier series is shown in figure 10.1.

We have stated without proof that any function which satisfies the Dirichlet conditions may be represented as a Fourier series. Let us now show why this is a plausible statement. We require that any reasonable function (one that satisfies the Dirichlet conditions) can be expressed as a linear sum of sine and cosine terms. We first note that we cannot use just a sum of sine terms since sine, being an odd function (i.e. a function for which $f(-x) = -f(x)$), cannot represent even functions (i.e. functions for which $f(-x) = f(x)$). This is obvious when we try to express a function $f(x)$ that takes a non-zero value at $x = 0$. Clearly, since $\sin nx = 0$ for all values of $n$, we cannot represent $f(x)$ at $x = 0$ by a sine series. Similarly odd functions cannot be represented by a cosine series since cosine is an even function. Nevertheless, it is possible to represent *all* odd functions by a sine series and *all* even functions by a cosine series. Now, since all functions may be written as the sum of an odd and an even part,

$$\begin{aligned} f(x) &= \tfrac{1}{2}[f(x) + f(-x)] + \tfrac{1}{2}[f(x) - f(-x)] \\ &= f_{\text{even}}(x) + f_{\text{odd}}(x), \end{aligned}$$

we can write any function as the sum of a sine series and a cosine series.

All the terms of a Fourier series are mutually orthogonal, that is, the integrals, over one period, of the product of any two terms have the following properties:

$$\int_{x_0}^{x_0+L} \sin\left(\frac{2\pi rx}{L}\right) \cos\left(\frac{2\pi px}{L}\right) dx = 0 \quad \text{for all } r \text{ and } p, \tag{10.1}$$

$$\int_{x_0}^{x_0+L} \cos\left(\frac{2\pi rx}{L}\right) \cos\left(\frac{2\pi px}{L}\right) dx = \begin{cases} L & \text{for } r = p = 0, \\ \frac{1}{2}L & \text{for } r = p > 0, \\ 0 & \text{for } r \neq p, \end{cases} \tag{10.2}$$

$$\int_{x_0}^{x_0+L} \sin\left(\frac{2\pi rx}{L}\right) \sin\left(\frac{2\pi px}{L}\right) dx = \begin{cases} 0 & \text{for } r = p = 0, \\ \frac{1}{2}L & \text{for } r = p > 0, \\ 0 & \text{for } r \neq p, \end{cases} \tag{10.3}$$

where $r$ and $p$ are integers greater than or equal to zero; these formulae are easily derived. A full discussion of why it is possible to expand a function as a sum of mutually orthogonal functions is given in chapter 15.

The Fourier series expansion of the function, $f(x)$, is conventionally written

$$f(x) = \frac{a_0}{2} + \sum_{r=1}^{\infty} \left[ a_r \cos\left(\frac{2\pi rx}{L}\right) + b_r \sin\left(\frac{2\pi rx}{L}\right) \right], \tag{10.4}$$

where $a_0, a_r, b_r$ are constants called the *Fourier coefficients*. These coefficients are analogous to those in a power series expansion and the determination of their numerical values is the essential step in writing a function as a Fourier series.

This chapter continues with a discussion of how to find the Fourier coefficients for particular functions. We then discuss simplifications to the general Fourier series that may save considerable effort in calculations. This is followed by the alternative representation of a function as a complex Fourier series, and we conclude with a discussion of Parseval's theorem.

## 10.2 The Fourier coefficients

We have indicated that a series that satisfies the Dirichlet conditions may be written in the form (10.4). We now consider how to find the Fourier coefficients for any particular function. For a periodic function $f(x)$ of period $L$ we will find that the Fourier coefficients are given by

$$a_r = \frac{2}{L} \int_{x_0}^{x_0+L} f(x) \cos\left(\frac{2\pi rx}{L}\right) dx, \tag{10.5}$$

$$b_r = \frac{2}{L} \int_{x_0}^{x_0+L} f(x) \sin\left(\frac{2\pi rx}{L}\right) dx, \tag{10.6}$$

where $x_0$ is arbitrary but is often taken as 0 or $-L/2$. The apparently arbitrary factor $\frac{1}{2}$ which appears in the $a_0$ term in (10.4) is included so that (10.5) may apply for $r = 0$ as well as $r > 0$. The relations (10.5) and (10.6) may be derived as follows.

Suppose the Fourier series expansion of $f(x)$ can be written as in (10.4),

$$f(x) = \frac{a_0}{2} + \sum_{r=1}^{\infty} \left[ a_r \cos\left(\frac{2\pi rx}{L}\right) + b_r \sin\left(\frac{2\pi rx}{L}\right) \right].$$

Then, multiplying by $\cos(2\pi px/L)$, integrating over one full period in $x$ and changing the order of the summation and integration, we get

$$\begin{aligned}\int_{x_0}^{x_0+L} f(x)\cos\left(\frac{2\pi px}{L}\right) dx &= \frac{a_0}{2}\int_{x_0}^{x_0+L} \cos\left(\frac{2\pi px}{L}\right) dx \\ &\quad + \sum_{r=1}^{\infty} a_r \int_{x_0}^{x_0+L} \cos\left(\frac{2\pi rx}{L}\right)\cos\left(\frac{2\pi px}{L}\right) dx \\ &\quad + \sum_{r=1}^{\infty} b_r \int_{x_0}^{x_0+L} \sin\left(\frac{2\pi rx}{L}\right)\cos\left(\frac{2\pi px}{L}\right) dx.\end{aligned} \tag{10.7}$$

We can now find the Fourier coefficients by considering (10.7) as $p$ takes different values. Using the orthogonality conditions (10.1)–(10.3) of the previous section, we find that when $p = 0$ (10.7) becomes

$$\int_{x_0}^{x_0+L} f(x)dx = \frac{a_0}{2}L.$$

When $p \neq 0$ the only non-vanishing term on the RHS of (10.7) occurs when $r = p$, and so

$$\int_{x_0}^{x_0+L} f(x)\cos\left(\frac{2\pi rx}{L}\right) dx = \frac{a_r}{2}L.$$

The other Fourier coefficients $b_r$ may be found by repeating the above process but multiplying by $\sin(2\pi px/L)$ instead of $\cos(2\pi px/L)$ (see exercise 10.2).

►*Express the square-wavefunction illustrated in figure 10.2 as a Fourier series.*

Physically this might represent the input to a electrical circuit that switches between a high and a low state with a time period of $T$. The square wave may be represented by

$$f(t) = \begin{cases} -1 & \text{for } -\frac{1}{2}T \leq t < 0, \\ +1 & \text{for } 0 \leq t < \frac{1}{2}T. \end{cases}$$

In deriving the Fourier coefficients, we note firstly that the function is an odd function and so the series will contain only sine terms (this simplification is

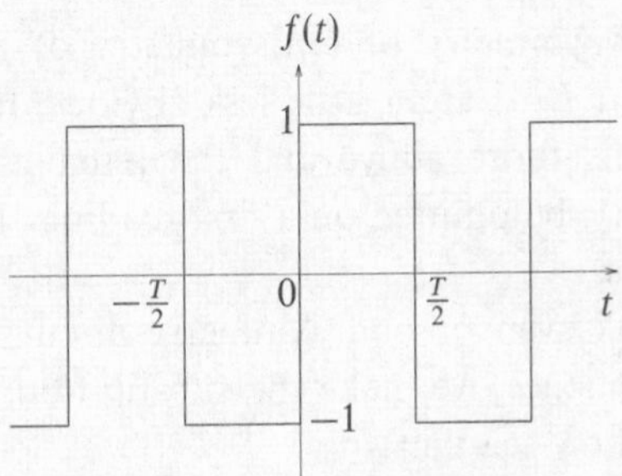


Figure 10.2 A square-wavefunction.

discussed further in the following section). To evaluate the coefficients in the sine series we use (10.6). Hence

$$\begin{aligned} b_r &= \frac{2}{T}\int_{-T/2}^{T/2} f(t)\sin\left(\frac{2\pi rt}{T}\right) dt \\ &= \frac{4}{T}\int_{0}^{T/2} \sin\left(\frac{2\pi rt}{T}\right) dt \\ &= \frac{2}{\pi r}\,[1-(-1)^r]. \end{aligned}$$

Thus the sine coefficients are zero if $r$ is even and equal to $4/\pi r$ if $r$ is odd. Hence the Fourier series for the square-wavefunction may be written as

$$f(t) = \frac{4}{\pi}\left(\sin\omega t + \frac{\sin 3\omega t}{3} + \frac{\sin 5\omega t}{5} + \cdots\right), \tag{10.8}$$

where $\omega = 2\pi/T$ is called the *angular frequency*. ◄

## 10.3 Symmetry considerations

The example of the previous section employed the useful property that since the function to be represented was odd, all the cosine terms of the Fourier series were zero. It is often the case that the function we wish to express as a Fourier series has a particular symmetry, which we can exploit to reduce the calculational labour of evaluating Fourier coefficients. Functions that are symmetric or antisymmetric about the origin (i.e. even and odd functions respectively) admit particularly useful simplifications. Functions that are odd in $x$ have no cosine terms (see section 10.1) and all the $a$-coefficients are equal to zero. Similarly, functions that are even in $x$ have no sine terms and all the $b$-coefficients are zero. Since the Fourier series of odd and even functions contain only half of the coefficients required for a general periodic function, there is a considerable reduction in the algebra needed to find a Fourier series.

The consequences of symmetry or antisymmetry of the function about the quarter period (i.e. about $L/4$) are a little less obvious. Furthermore, the results are not used as often as those above and the remainder of this section can, without loss of continuity, be omitted on a first reading. The following argument gives the required results.

Suppose that $f(x)$ has even or odd symmetry about $L/4$, i.e. $f(L/4 - x) = \pm f(x - L/4)$. For convenience, we make the substitution $s = x - L/4$ and hence $f(-s) = \pm f(s)$. We can now see that

$$b_r = \frac{2}{L}\int_{x_0}^{x_0+L} f(s)\sin\left(\frac{2\pi rs}{L} + \frac{\pi r}{2}\right) ds,$$

where the limits of integration have been left unaltered since $f$ is of course periodic in $s$ as well as in $x$. If we use the expansion

$$\sin\left(\frac{2\pi rs}{L} + \frac{\pi r}{2}\right) = \sin\left(\frac{2\pi rs}{L}\right)\cos\left(\frac{\pi r}{2}\right) + \cos\left(\frac{2\pi rs}{L}\right)\sin\left(\frac{\pi r}{2}\right),$$

we can immediately see that the trigonometrical part of the integrand is an odd function of $s$ if $r$ is even, and an even function of $s$ if $r$ is odd. Hence if $f(s)$ is even and $r$ is even then the integral is zero, and if $f(s)$ is odd and $r$ is odd then the integral is zero. Similar results can be derived for the Fourier $a$-coefficients and we conclude that

(i) if $f(x)$ is even about $L/4$, then $a_{2r+1} = 0$ and $b_{2r} = 0$,
(ii) if $f(x)$ is odd about $L/4$ then $a_{2r} = 0$ and $b_{2r+1} = 0$.

All the above results follow automatically when the Fourier coefficients are evaluated in any particular case, but prior knowledge of them will often enable some coefficients to be set equal to zero on inspection, and so substantially reduce the computational labour. As an example, the square-wave function shown in figure 10.2 is (i) an odd function of $t$, so that all $a_r = 0$, and (ii) even about the point $t = T/4$, so that $b_{2r} = 0$. Thus we can say immediately that only sine terms of odd harmonics will be present and therefore will need to be calculated; this is confirmed in the expansion (10.8).

## 10.4 Discontinuous functions

The Fourier series expansion usually works well for functions that are discontinuous in the required range. However, the series itself does not produce a discontinuous function and we state without proof that the value of the expanded $f(x)$ at a discontinuity will be half-way between the upper and lower values. Expressing this more mathematically, at a point of finite discontinuity, $x_d$, the Fourier series converges to

$$\tfrac{1}{2}\lim_{\epsilon\to 0}[f(x_d + \epsilon) + f(x_d - \epsilon)].$$

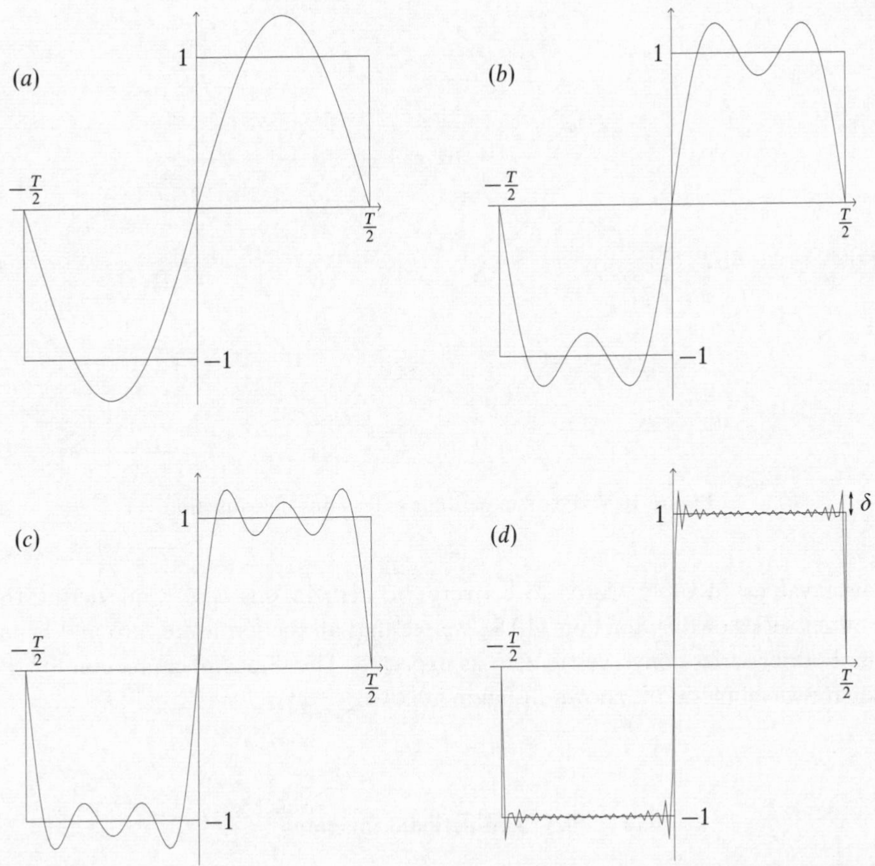


Figure 10.3 The convergence of a Fourier series expansion of a square-wave function, including (*a*) one term, (*b*) two terms, (*c*) three terms and (*d*) 20 terms. The overshoot $\delta$ is shown in (*d*).

At a discontinuity, the Fourier series representation of the function will overshoot its value. Although as more terms are included the overshoot moves in position arbitrarily close to the discontinuity, it never disappears even in the limit of an infinite number of terms. This behaviour is known as *Gibbs' phenomenon*. A full discussion is not pursued here but suffice it to say that the size of the overshoot is proportional to the magnitude of the discontinuity.

►*Find the value to which the Fourier series of the square-wave function discussed in section 10.2 converges at $t = 0$.*

It can be seen that the function is discontinuous at $t = 0$ and, by the above rule, we expect the series to converge to a value half-way between the upper and

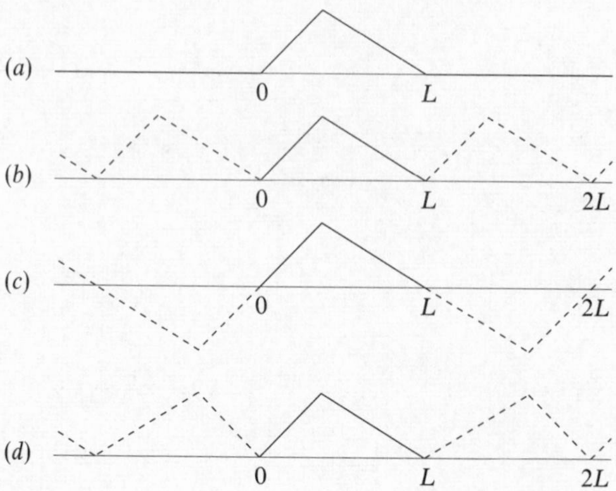


Figure 10.4 Possible periodic extensions of a function.

lower values, in other words to converge to zero in this case. Considering the Fourier series of this function, (10.8), we see that all the terms are zero and hence the Fourier series converges to zero as expected. The Gibbs phenomenon for the square-wavefunction is shown in figure 10.3. ◄

## 10.5 Non-periodic functions

We have already mentioned that a Fourier representation may sometimes be used for non-periodic functions. If we wish to find the Fourier series of a non-periodic function only within a fixed range, then we may *continue* this function outside the range so as to make it periodic. The Fourier series of this periodic function would then correctly represent the non-periodic function in the desired range. Since we are often at liberty to extend the function in a number of ways, we can sometimes make it odd or even and so reduce the calculation required. Figure 10.4($b$) shows the simplest extension to the function shown in figure 10.4($a$). However, this extension has no particular symmetry. Figures 10.4($c$), ($d$) show extensions as odd and even functions respectively with the benefit that only sine or cosine terms appear in the resulting Fourier series. We note that these last two extensions give a function of period $2L$.

In view of the result of section 10.4, it must be added that the continuation must not be discontinuous at the end-points of the interval of interest; if it is the series will not converge to the required value there. This requirement that the series converges appropriately may reduce the choice of continuations. This is discussed further at the end of the following example.

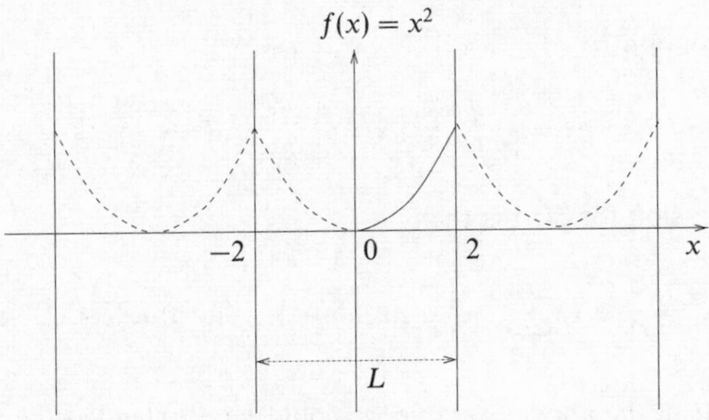


Figure 10.5 $f(x) = x^2$, $0 < x \leq 2$, with extended range and with periodicity.

► *Find the Fourier series of $f(x) = x^2$ for $0 < x \leq 2$.*

We must first make the function periodic. We do this by extending the range of interest to $-2 < x \leq 2$ in such a way that $f(x) = f(-x)$ and then letting $f(x+4k) = f(x)$ where $k$ is any integer. This is shown in figure 10.5. Now we have an even function of period 4. The Fourier series will faithfully represent $f(x)$ in the range, $-2 < x \leq 2$, although not outside it. Firstly we note that since we have made the specified function even in $x$ by extending the range, all the coefficients $b_r$ will be zero. Now we apply (10.5) and (10.6) with $L = 4$ to determine the remaining coefficients:

$$a_r = \frac{2}{4}\int_{-2}^{2} x^2 \cos\left(\frac{2\pi r x}{4}\right) dx = \frac{4}{4}\int_{0}^{2} x^2 \cos\left(\frac{\pi r x}{2}\right) dx,$$

where the second equality holds because the function is even in $x$. Thus

$$\begin{aligned}
a_r &= \left[\frac{2}{\pi r} x^2 \sin\left(\frac{\pi r x}{2}\right)\right]_0^2 - \frac{4}{\pi r}\int_0^2 x \sin\left(\frac{\pi r x}{2}\right) dx \\
&= \frac{8}{\pi^2 r^2}\left[x \cos\left(\frac{\pi r x}{2}\right)\right]_0^2 - \frac{8}{\pi^2 r^2}\int_0^2 \cos\left(\frac{\pi r x}{2}\right) dx \\
&= \frac{16}{\pi^2 r^2}\cos \pi r \\
&= \frac{16}{\pi^2 r^2}(-1)^r.
\end{aligned}$$

Since this expression for $a_r$ has $r^2$ in its denominator, to evaluate $a_0$ we must return to the original definition,

$$a_r = \frac{2}{4}\int_{-2}^{2} f(x) \cos\left(\frac{\pi r x}{2}\right) dx.$$

From this we obtain

$$a_0 = \frac{2}{4}\int_{-2}^{2} x^2\,dx = \frac{4}{4}\int_{0}^{2} x^2\,dx = \frac{8}{3}.$$

The final expression for $f(x)$ is then

$$x^2 = \frac{4}{3} + 16\sum_{r=1}^{\infty}\frac{(-1)^r}{\pi^2 r^2}\cos\left(\frac{\pi r x}{2}\right) \qquad \text{for } 0 < x \le 2. \blacktriangleleft$$

We note that in the above example we could have extended the range so as to make the function odd. In other words we could have set $f(x) = -f(-x)$, and then made $f(x)$ periodic in such a way that $f(x+4) = f(x)$. In this case the resulting Fourier series would be a series of just sine terms. However, although this will faithfully represent the function inside the required range, it does not converge to the correct values of $f(x) = \pm 4$ at $x = \pm 2$, but instead converges to zero because of the discontinuity at the end of the range which results from extending the function so as to make it odd.

## 10.6 Integration and differentiation

It is sometimes possible to find the Fourier series of a function by integration or differentiation of another Fourier series. If the Fourier series of $f(x)$ is integrated term by term, the resulting Fourier series converges to the integral of $f(x)$. Clearly, when integrating in such a way there is a constant of integration that must be found. If $f(x)$ is a continuous function of $x$ for all $x$ and $f(x)$ is also periodic, then the Fourier series that results from differentiating term by term converges to $f'(x)$, provided that $f'(x)$ itself satisfies the Dirichlet conditions. These properties of Fourier series may be useful in calculating complicated Fourier series since simple Fourier series may easily be evaluated (or found from standard tables) and the more complicated series may often be built up by integration and/or differentiation.

►*Find the Fourier series of $f(x) = x^3$ for $0 < x \le 2$.*

In the example discussed in the previous section we found the Fourier series for $f(x) = x^2$ in the required range. So, if we *integrate* this term by term, we obtain

$$\frac{x^3}{3} = \frac{4}{3}x + 32\sum_{r=1}^{\infty}\frac{(-1)^r}{\pi^3 r^3}\sin\left(\frac{\pi r x}{2}\right) + c,$$

where $c$ is, so far, an arbitrary constant. We have not yet found the Fourier series for $x^3$ because the term $\frac{4}{3}x$ appears in the expansion. However, now *differentiating* our expression for $x^2$ we obtain

$$2x = -8\sum_{r=1}^{\infty}\frac{(-1)^r}{\pi r}\sin\left(\frac{\pi r x}{2}\right).$$

We can now write the full Fourier expansion of $x^3$ as

$$x^3 = -16\sum_{r=1}^{\infty}\frac{(-1)^r}{\pi r}\sin\left(\frac{\pi r x}{2}\right) + 96\sum_{r=1}^{\infty}\frac{(-1)^r}{\pi^3 r^3}\sin\left(\frac{\pi r x}{2}\right) + c.$$

Finally, we can find the constant, c, by considering $f(0)$. At $x = 0$, our Fourier expansion gives $x^3 = c$ since all the sine terms are zero, and hence $c = 0$. ◄

## 10.7 Complex Fourier series

As a Fourier series expansion in general contains both sine and cosine parts, it may be written more compactly using a complex exponential expansion. This simplification makes use of the property that $\exp(irx) = \cos rx + i\sin rx$. The complex Fourier series expansion is written

$$f(x) = \sum_{r=-\infty}^{\infty} c_r \exp\left(\frac{2\pi i r x}{L}\right), \qquad (10.9)$$

where the Fourier coefficients are given by

$$c_r = \frac{1}{L}\int_{x_0}^{x_0+L} f(x)\exp\left(-\frac{2\pi i r x}{L}\right)dx. \qquad (10.10)$$

This relation can be derived, in a similar manner to that of section 10.2, by multiplying (10.9) by $\exp(-2\pi i p x/L)$ before integrating, and using the orthogonality relation

$$\int_{x_0}^{x_0+L}\exp\left(-\frac{2\pi i p x}{L}\right)\exp\left(\frac{2\pi i r x}{L}\right)dx = \begin{cases} L & \text{for } r = p,\\ 0 & \text{for } r \neq p.\end{cases}$$

The complex Fourier coefficients in (10.9) have the following relations with the real Fourier coefficients:

$$\begin{aligned} c_r &= \tfrac{1}{2}(a_r - i b_r),\\ c_{-r} &= \tfrac{1}{2}(a_r + i b_r). \end{aligned} \qquad (10.11)$$

Note that if $f(x)$ is real then $c_{-r} = c_r^*$, where the asterisk represents complex conjugation.

►*Find a complex Fourier series for $f(x) = x$ in the range $-2 < x < 2$.*

Using (10.10), for $r \neq 0$,

$$\begin{aligned}
c_r &= \frac{1}{4}\int_{-2}^{2} x \exp\left(-\frac{\pi i r x}{2}\right) dx \\
&= \left[-\frac{x}{2\pi i r}\exp\left(-\frac{\pi i r x}{2}\right)\right]_{-2}^{2} + \int_{-2}^{2}\frac{1}{2\pi i r}\exp\left(-\frac{\pi i r x}{2}\right) dx \\
&= -\frac{1}{\pi i r}\left[\exp(-\pi i r) + \exp(\pi i r)\right] + \left[\frac{1}{r^2\pi^2}\exp\left(-\frac{\pi i r x}{2}\right)\right]_{-2}^{2} \\
&= \frac{2i}{\pi r}\cos \pi r - \frac{2i}{r^2\pi^2}\sin \pi r = \frac{2i}{\pi r}(-1)^r. \qquad (10.12)
\end{aligned}$$

For $r = 0$, we find $c_0 = 0$ and hence

$$x = \sum_{\substack{r=-\infty \\ r\neq 0}}^{\infty} \frac{2i(-1)^r}{r\pi} \exp\left(\frac{\pi i r x}{2}\right).$$

We note that the Fourier series derived for $x$ in section 10.6 gives $a_r = 0$ for all $r$ and

$$b_r = -\frac{4(-1)^r}{\pi r},$$

and so, using (10.11), we confirm that $c_r$ and $c_{-r}$ have the forms derived above. It is also apparent that the relationship $c_r^* = c_{-r}$ holds, as we expect, since $f(x)$ is real. ◄

## 10.8 Parseval's theorem

*Parseval's theorem* gives a useful way of relating the Fourier coefficients to the function that they describe. Essentially a conservation law, it states that

$$\begin{aligned}
\frac{1}{L}\int_{x_0}^{x_0+L} |f(x)|^2 dx &= \sum_{r=-\infty}^{\infty} |c_r|^2 \\
&= \left(\tfrac{1}{2}a_0\right)^2 + \tfrac{1}{2}\sum_{r=1}^{\infty}(a_r^2 + b_r^2). \qquad (10.13)
\end{aligned}$$

In a more memorable form, this says that the sum of the moduli squared of the complex Fourier coefficients is equal to the average value of $|f(x)|^2$ over one period. Parseval's theorem can be proved straightforwardly by writing $f(x)$ as a Fourier series and evaluating the required integral, but the algebra is messy. Therefore, we shall use an alternative method, for which the algebra is simple and which in fact leads to a more general form of the theorem.

Let us consider two functions $f(x)$ and $g(x)$, which are (or can be made) periodic with period $L$, and which have Fourier series (expressed in complex form)

$$f(x) = \sum_{r=-\infty}^{\infty} c_r \exp\left(\frac{2\pi irx}{L}\right),$$

$$g(x) = \sum_{r=-\infty}^{\infty} \gamma_r \exp\left(\frac{2\pi irx}{L}\right),$$

where $c_r$ and $\gamma_r$ are the complex Fourier coefficients of $f(x)$ and $g(x)$ respectively. Thus

$$f(x)g^*(x) = \sum_{r=-\infty}^{\infty} c_r g^*(x) \exp\left(\frac{2\pi irx}{L}\right).$$

Integrating this equation with respect to $x$ over the interval $(x_0, x_0 + L)$, and dividing by $L$, we find

$$\begin{aligned}
\frac{1}{L}\int_{x_0}^{x_0+L} f(x)g^*(x)\,dx &= \sum_{r=-\infty}^{\infty} c_r \frac{1}{L}\int_{x_0}^{x_0+L} g^*(x)\exp\left(\frac{2\pi irx}{L}\right) dx \\
&= \sum_{r=-\infty}^{\infty} c_r \left[\frac{1}{L}\int_{x_0}^{x_0+L} g(x)\exp\left(\frac{-2\pi irx}{L}\right) dx\right]^* \\
&= \sum_{r=-\infty}^{\infty} c_r\gamma_r^*,
\end{aligned}$$

where the last equality uses (10.10). Finally, if we let $g(x) = f(x)$ we obtain Parseval's theorem (10.13). This proof can be performed in a similar manner using the sine and cosine form of the Fourier series, but the algebra is slightly more complicated.

Parseval's theorem is sometimes used to sum series. However, if one is presented with a series to sum, it is not usually possible to decide which Fourier series should be used to evaluate it. Instead, useful summations are sometimes found serendipitously. The following example shows the evaluation of a sum by a Fourier series method.

►*Using Parseval's theorem and the Fourier series for $f(x) = x^2$ found in section 10.5, calculate the sum $\sum_{r=1}^{\infty} r^{-4}$.*

Firstly we find the average value of $[f(x)]^2$ over the interval $-2 < x \leq 2$.

$$\frac{1}{4}\int_{-2}^{2} x^4\,dx = \frac{16}{5}.$$

Now we evaluate the right-hand side of (10.13):

$$\left(\tfrac{1}{2}a_0\right)^2 + \tfrac{1}{2}\sum_{1}^{\infty} a_r^2 + \tfrac{1}{2}\sum_{1}^{\infty} b_n^2 = \left(\tfrac{4}{3}\right)^2 + \tfrac{1}{2}\sum_{r=1}^{\infty} \frac{16^2}{\pi^4 r^4}.$$

Equating the two expression we find

$$\sum_{r=1}^{\infty} \frac{1}{r^4} = \frac{\pi^4}{90}. \blacktriangleleft$$

### 10.9 Exercises

10.1 Prove the orthogonality relations stated in section 10.1.

10.2 Derive the Fourier coefficients $b_r$ in a similar manner to the derivation of the $a_r$ in section 10.2.

10.3 Find the Fourier series of the function $f(x) = x$ in the range $-\pi < x \leq \pi$. Hence show that

$$1 - \frac{1}{3} + \frac{1}{5} - \frac{1}{7} + \cdots = \frac{\pi}{4}.$$

10.4 Find the Fourier coefficients in the expansion of $f(x) = \exp x$ over the range $-1 < x < 1$. What value will the expansion have when $x = 2$?

10.5 By integrating term by term the Fourier series found in the previous question, and using the Fourier series for $f(x) = x$ found in section 10.6, show that $\int \exp x \, dx = \exp x + c$. Why is it not possible to show that $d(\exp x)/dx = \exp x$ by differentiating the Fourier series of $f(x) = \exp x$ in a similar manner?

10.6 Consider the function $f(x) = \exp(-x^2)$ in the range $0 \leq x \leq 1$. Show how it should be continued to give as its Fourier series a series (the actual form is not wanted)

(a) with only cosine terms, (b) with only sine terms,
(c) with period 1, (d) with period 2.

Would there be any difference between the values of the last two series at (i) $x = 0$, (ii) $x = 1$?

10.7 Express the function $f(x) = x^2$ as a Fourier sine series in the range $0 < x \leq 2$, and show that it converges to zero at $x = \pm 2$.

10.8 Demonstrate explicitly for the square-wavefunction discussed in section 10.2 that Parseval's theorem (10.13) is valid. You will need to use the relationship

$$\sum_{m=0}^{\infty} \frac{1}{(2m+1)^2} = \frac{\pi^2}{8}.$$

Show that a filter that transmits frequencies only up to $8\pi/T$ will still transmit more than 90 per cent of the power in such a square-wave voltage signal.

10.9 Show that the Fourier series for $|\sin\theta|$ in the range $-\pi \le \theta \le \pi$ is

$$|\sin\theta| = \frac{2}{\pi} - \frac{4}{\pi}\sum_{m=1}^{\infty} \frac{\cos 2m\theta}{4m^2-1}.$$

By setting $\theta = 0$ and $\theta = \pi/2$, deduce values for

$$\sum_{m=1}^{\infty} \frac{1}{4m^2-1} \quad \text{and} \quad \sum_{m=1}^{\infty} \frac{1}{16m^2-1}.$$

10.10 An odd function $f(x)$ of period $2\pi$ is to be approximated by a Fourier sine series having only $m$ terms. The error in this approximation is measured by the square deviation

$$E_m = \int_{-\pi}^{\pi} \left[ f(x) - \sum_{n=1}^{m} b_n \sin nx \right]^2 dx.$$

By differentiating $E_m$ with respect to the coefficients $b_n$, find the values of $b_n$ that minimise $E_m$.

Sketch the graph of the function $f(x)$, where

$$f(x) = \begin{cases} -x(\pi + x) & \text{for } -\pi \le x < 0, \\ x(x-\pi) & \text{for } 0 \le x < \pi. \end{cases}$$

$f(x)$ is to be approximated by the first three terms of a Fourier sine series. What coefficients minimise $E_3$? What is the resulting value of $E_3$?

## 10.10 Hints and answers

10.3 $f(x) = 2\sum_1^{\infty}(-1)^{n+1}n^{-1}\sin nx$; set $x = \pi/2$.

10.4 $f(x) = \sinh(1)\left\{1 + 2\sum_1^{\infty}(-1)^n(1+n^2\pi^2)^{-1}[\cos(n\pi x) - n\pi\sin(n\pi x)]\right\}$; $f(2) = f(0) = 1$.

10.5 Combine the coefficients of the $\sin(n\pi x)$ terms from the Fourier series for $x$ and (part of) $\int \exp x\, dx$; the partial series obtained by differentiating the $\sin(n\pi x)$ terms does not converge, having coefficients of the form $(n\pi)^2/[1+(n\pi)^2]$.

10.6 See figure 10.6. (c) (i) $(1+e^{-1})/2$, (ii) $(1+e^{-1})/2$; (d) (i) $(1+e^{-4})/2$, (ii) $e^{-1}$.

10.7 Consider $f(x) = -x^2$ for $-2 < x \le 0$, to ensure a sine series; $\sum_n b_n \sin(n\pi x/2)$ with $b_n = (-1)^{n+1}8/(n\pi)$ for $n$ even, and $(-1)^{n+1}8/(n\pi) - 32/(n\pi)^3$ for $n$ odd.

10.8 $C_{\pm(2m+1)} = \mp 2i/[(2m+1)\pi]$; $\sum |C_n|^2 = (4/\pi^2) \times 2 \times (\pi^2/8)$; the values $n = \pm 1, \pm 3$ contribute $> 90\%$ of the total.

10.9 Write $\sin\theta\cos n\theta$ as $\frac{1}{2}[\sin(n+1)\theta - \sin(n-1)\theta]$; obtain $\sum_1^{\infty}(4m^2-1)^{-1} = \frac{1}{2}$ and $\sum_1^{\infty}(-1)^m(4m^2-1)^{-1} = \frac{1}{2} - \frac{\pi}{4}$ and add the two equations; $\frac{1}{2}$, $\frac{1}{2} - \frac{\pi}{8}$.

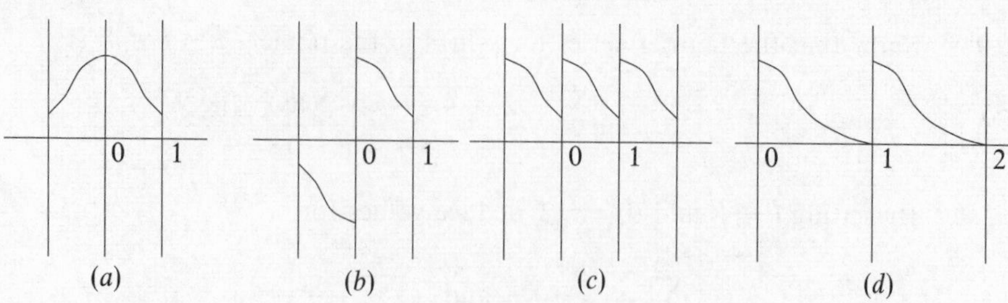


Figure 10.6 Continuations of $\exp(-x^2)$ in $0 \leq x \leq 1$ to give: (*a*) cosine terms only; (*b*) sine terms only; (*c*) period 1; (*d*) period 2.

10.10 Show that the minimising value $b_k$ is given by $0 = \int_{-\pi}^{\pi} f(x) \sin kx \, dx - \sum_{n=1}^{m} b_n \pi \delta_{kn}$, and hence that $b_k$ is equal to the normal Fourier coefficient; $b_1 = -8/\pi$, $b_2 = 0$, $b_3 = -8/(27\pi)$; $E_3 = (64/\pi) \sum_2^{\infty} (2m+1)^{-6}$.

# 11

# Integral transforms

In the previous chapter we encountered the Fourier series representation of a periodic function in a fixed interval as a superposition of sinusoidal functions. It is often desirable, however, to obtain such a representation even for functions defined over an infinite interval and with no particular periodicity. Such a representation is called a *Fourier transform*, and is one of a class of representations called *integral transforms*.

We begin by considering Fourier transforms as a generalisation of Fourier series. We then go on to discuss the properties of the Fourier transform and its applications. In the second part of the chapter we present an analogous discussion of the closely related *Laplace transform*.

## 11.1 Fourier transforms

The Fourier transform provides a representation of functions defined over an infinite interval, and having no particular periodicity, in terms of a superposition of sinusoidal functions. It may thus be considered as a generalisation of the Fourier series representation of periodic functions. Since Fourier transforms are often used to represent time-varying functions, we shall present much of our discussion in terms of $f(t)$, rather than $f(x)$, although in some spatial examples $f(x)$ will be the more natural notation and we shall use it as appropriate. Our only requirement on $f(t)$ will be that $\int_{-\infty}^{\infty} |f(t)|\, dt$ is finite.

In order to develop the transition from Fourier series to Fourier transforms, we first recall that a function of period $T$ may be represented as a complex Fourier series, cf. (10.9),

$$f(t) = \sum_{r=-\infty}^{\infty} c_r\, e^{2\pi irt/T} = \sum_{r=-\infty}^{\infty} c_r\, e^{i\omega_r t}, \tag{11.1}$$

where $\omega_r = 2\pi r/T$. As the period $T$ tends to infinity, the 'frequency quantum'

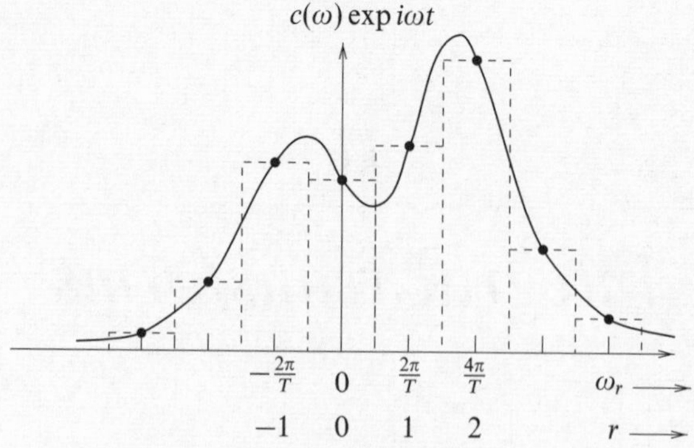


Figure 11.1 The relationship between the Fourier terms for a function of period $T$ and the Fourier integral (the area below the solid line) of the function.

$\Delta\omega = 2\pi/T$ becomes vanishingly small and the spectrum of allowed frequencies $\omega_r$ becomes a continuum. Thus, the infinite sum of terms in the Fourier series becomes an integral, and the coefficients $c_r$ become functions of the *continuous* variable $\omega$, as follows.

We recall, cf. (10.10), that the coefficients $c_r$ in (11.1) are given by

$$c_r = \frac{1}{T}\int_{-T/2}^{T/2} f(t)\, e^{-2\pi i r t/T}\, dt = \frac{\Delta\omega}{2\pi}\int_{-T/2}^{T/2} f(t)\, e^{-i\omega_r t}\, dt, \tag{11.2}$$

where we have written the integral in two alternative forms and, for convenience, made one period run from $-T/2$ to $+T/2$, rather than from 0 to $T$. Substituting from (11.2) into (11.1) gives

$$f(t) = \sum_{r=-\infty}^{\infty} \frac{\Delta\omega}{2\pi}\int_{-T/2}^{T/2} f(u)\, e^{-i\omega_r u}\, du\; e^{i\omega_r t}. \tag{11.3}$$

At this stage $\omega_r$ is still a discrete function of $r$ equal to $2\pi r/T$.

The solid points in figure 11.1 are a plot of (say, the real part of) $c_r\, e^{i\omega_r t}$ as a function of $r$ (or equivalently of $\omega_r$) and it is clear that $(2\pi/T)c_r\, e^{i\omega_r t}$ gives the area of the $r$th (broken-line) rectangle. If $T$ tends to $\infty$, $\Delta\omega$ $(= 2\pi/T)$ consequently becomes infinitesimal, the width of the rectangles tends to zero and, from the mathematical definition of an integral,

$$\sum_{r=-\infty}^{\infty} \frac{\Delta\omega}{2\pi} g(\omega_r)\, e^{i\omega_r t} \;\;\rightarrow\;\; \frac{1}{2\pi}\int_{-\infty}^{\infty} g(\omega)\, e^{i\omega t}\, d\omega.$$

In this particular case

$$g(\omega_r) = \int_{-T/2}^{T/2} f(u)\, e^{-i\omega_r u}\, du,$$

and (11.3) becomes

$$f(t) = \frac{1}{2\pi} \int_{-\infty}^{\infty} d\omega\, e^{i\omega t} \int_{-\infty}^{\infty} du\, f(u)\, e^{-i\omega u}. \tag{11.4}$$

This result is known as *Fourier's inversion theorem.*

From it we may define the *Fourier transform* of $f(t)$ by

$$\widetilde{f}(\omega) = \frac{1}{\sqrt{2\pi}} \int_{-\infty}^{\infty} f(t)\, e^{-i\omega t}\, dt, \tag{11.5}$$

and its inverse by

$$f(t) = \frac{1}{\sqrt{2\pi}} \int_{-\infty}^{\infty} \widetilde{f}(\omega)\, e^{i\omega t}\, d\omega. \tag{11.6}$$

Including the constant $1/\sqrt{2\pi}$ in the definition of $\widetilde{f}(\omega)$ (whose mathematical existence as $T \to \infty$ is assumed here without proof) is clearly arbitrary, the only requirement being that the product of the constants in (11.5) and (11.6) should equal $1/(2\pi)$. Our definition is chosen to be as symmetric as possible.

► *Find the Fourier transform of the exponential decay function $f(t) = 0$ for $t < 0$ and $f(t) = A\, e^{-\lambda t}$ for $t \geq 0$ ($\lambda > 0$).*

Using the definition (11.5) and separating the integral into two parts,

$$\begin{aligned} \widetilde{f}(\omega) &= \frac{1}{\sqrt{2\pi}} \int_{-\infty}^{0} (0)\, e^{-i\omega t}\, dt + \frac{A}{\sqrt{2\pi}} \int_{0}^{\infty} e^{-\lambda t}\, e^{-i\omega t}\, dt \\ &= 0 + \frac{A}{\sqrt{2\pi}} \left[ -\frac{e^{-(\lambda + i\omega)t}}{\lambda + i\omega} \right]_0^{\infty} \\ &= \frac{A}{\sqrt{2\pi}(\lambda + i\omega)}, \end{aligned}$$

which is the required transform. It is clear that the multiplicative constant $A$ does not affect the form of the transform, merely its amplitude. This transform may be verified by re-substitution of the above result into (11.6) to recover $f(t)$, but evaluation of the integral requires the use of complex-variable contour integration (chapter 18). ◄

### 11.1.1 The uncertainty principle

An important function that appears in many areas of physical science, either precisely or as an approximation to a physical situation, is the *Gaussian* or *normal* distribution. Its Fourier transform is of importance both in itself and because, when interpreted statistically, it readily illustrates a form of *uncertainty principle.*

►*Find the Fourier transform of the normalised Gaussian distribution*

$$f(t) = \frac{1}{\tau\sqrt{2\pi}} \exp\left(-t^2/2\tau^2\right), \qquad -\infty < t < \infty.$$

This Gaussian distribution is centred on $t = 0$ and has a root mean square deviation $\Delta t = \tau$. (Any reader who is unfamiliar with this interpretation of the distribution should refer to chapter 24.)

Using the definition (11.5), the Fourier transform of $f(t)$ is given by

$$\begin{aligned}\widetilde{f}(\omega) &= \frac{1}{\sqrt{2\pi}} \int_{-\infty}^{\infty} \frac{1}{\tau\sqrt{2\pi}} \exp(-t^2/2\tau^2) \exp(-i\omega t)\, dt \\ &= \frac{1}{\sqrt{2\pi}} \int_{-\infty}^{\infty} \frac{1}{\tau\sqrt{2\pi}} \exp\left\{-\left[t^2 + 2\tau^2 i\omega t + (\tau^2 i\omega)^2 - (\tau^2 i\omega)^2\right]/2\tau^2\right\} dt,\end{aligned}$$

where the quantity $-(\tau^2 i\omega)^2/2\tau^2$ has been both added and subtracted in the exponent in order to allow the factors involving the variable of integration $t$ to be expressed as a complete square. Hence the expression can be written

$$\widetilde{f}(\omega) = \frac{\exp\left(-\tau^2\omega^2/2\right)}{\sqrt{2\pi}} \left\{ \frac{1}{\tau\sqrt{2\pi}} \int_{-\infty}^{\infty} \exp\left[-(t + i\tau^2\omega)^2/2\tau^2\right] dt \right\}.$$

The quantity inside the braces is the normalisation integral for the Gaussian and equals unity, although in order to show this strictly needs results from complex variable theory (chapter 18). That it is equal to unity can be made plausible by changing the variable to $s = t + i\tau^2\omega$ and assuming that the imaginary parts introduced into the integration path and limits (where the integrand goes rapidly to zero anyway) make no difference.

We are left with the result that

$$\widetilde{f}(\omega) = \frac{1}{\sqrt{2\pi}} \exp\left(\frac{-\tau^2\omega^2}{2}\right), \tag{11.7}$$

which is another Gaussian distribution, centred on zero and with a root mean square deviation $\Delta\omega = 1/\tau$. It is interesting to note, and an important property, that the Fourier transform of a Gaussian is another Gaussian. ◄

In the above example the root mean square deviation in $t$ was $\tau$, and so it is seen that the deviations or 'spreads' in $t$ and in $\omega$ are inversely related:

$$\Delta\omega\Delta t = 1,$$

independently of the value of $\tau$. In physical terms, the narrower in time is, say, an electrical impulse the greater the spread of frequency components it must contain. Similar physical statements are valid for other pairs of Fourier-related variables, such as spatial position and wave number. In an obvious notation, $\Delta k\Delta x = 1$ for a Gaussian wave packet.

The uncertainty relations as usually expressed in quantum mechanics can be related to the above if the de Broglie and Einstein relationships for momentum and energy are introduced,

$$p = \hbar k \qquad \text{and} \qquad E = \hbar\omega.$$

Here $\hbar$ is Planck's constant $h$ divided by $2\pi$. In quantum mechanics $f(t)$ is a wavefunction and the distribution of the wave intensity in time is given by $|f|^2$ (also a Gaussian). Similarly, the intensity distribution in frequency is given by $|\widetilde{f}|^2$. These two distributions have root mean square deviations of $\tau/\sqrt{2}$ and $1/(\sqrt{2}\tau)$, giving, after incorporation of the above relations,

$$\Delta E \Delta t = \hbar/2 \qquad \text{and} \qquad \Delta p \Delta x = \hbar/2.$$

The factors of 1/2 that appear are specific to the Gaussian form, but any distribution $f(t)$ produces for the product $\Delta E \Delta t$ a quantity $\lambda\hbar$ in which $\lambda$ is strictly positive (in fact the value of 1/2 for a Gaussian is the minimum possible).

### *11.1.2 Fraunhofer diffraction*

We take our final example of the Fourier transform from the field of optics. The pattern of transmitted light produced by a partially opaque (or phase-changing) object upon which a coherent beam of radiation falls is called a *diffraction pattern* and, in particular, when the cross-section of the object is small compared with the distance at which the light is observed the pattern is known as a *Fraunhofer* diffraction pattern.

We will consider only the case in which the light is monochromatic with wavelength $\lambda$. The direction of the incident beam of light can then be described by the *wave vector* $\mathbf{k}$; the magnitude of this vector is given by the *wave number* $k = 2\pi/\lambda$ of the light. The essential quantity in a Fraunhofer diffraction pattern is the dependence of the observed amplitude (and hence intensity) on the angle $\theta$ between the viewing direction $\mathbf{k}'$ and the direction $\mathbf{k}$ of the incident beam. This is entirely determined by the spatial distribution of the amplitude and phase of the light at the object, the transmitted intensity in a particular direction $\mathbf{k}'$ being determined by the corresponding Fourier component of this spatial distribution.

As an example, we take as an object a simple one-dimensional screen of width $2Y$ on which light of wave number $k$ is incident normally; see figure 11.2. We suppose that at the position $(0, y)$ the amplitude of the transmitted light is $f(y)$ per unit length in the $y$-direction ($f(y)$ may be complex). The function $f(y)$ is called an *aperture function.* Both the screen and beam are assumed infinite in the $z$-direction.

Denoting the unit vectors in the $x$- and $y$- directions by $\mathbf{i}$ and $\mathbf{j}$ respectively, the total light amplitude at a position $\mathbf{r}_0 = x_0\mathbf{i} + y_0\mathbf{j}$, with $x_0 > 0$, will be the

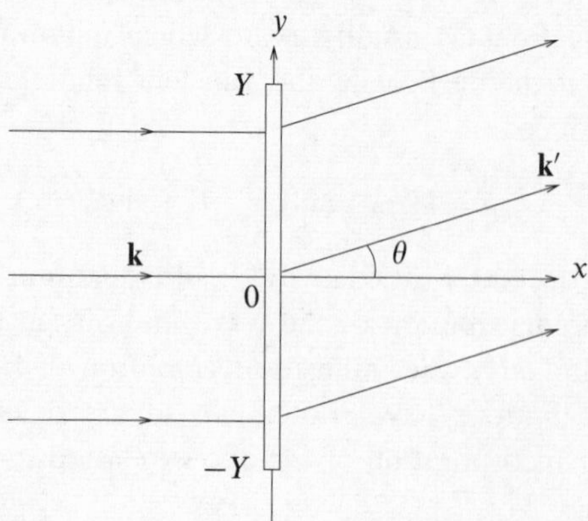


Figure 11.2 Diffraction grating of width $2Y$ with light of wavelength $2\pi/k$ being diffracted through an angle $\theta$.

superposition of all the (Huygens') wavelets originating from the various parts of the screen. For large $r_0$ $(= |\mathbf{r}_0|)$, these can be treated as plane waves to give†

$$A(\mathbf{r}_0) = \int_{-Y}^{Y} \frac{f(y)\exp[i\mathbf{k}'\cdot(\mathbf{r}_0 - y\mathbf{j})]}{|\mathbf{r}_0 - y\mathbf{j}|}\,dy, \tag{11.8}$$

The factor $\exp[i\mathbf{k}'\cdot(\mathbf{r}_0 - y\mathbf{j})]$ represents the phase change undergone by the light in travelling from the point $y\mathbf{j}$ on the screen to the point $\mathbf{r}_0$, and the denominator represents the reduction in amplitude with distance. (Recall that the system is infinite in the $z$-direction and so the 'spreading' is effectively in two dimensions only.)

If the medium is the same on both sides of the screen then $\mathbf{k}' = k\cos\theta\,\mathbf{i} + k\sin\theta\,\mathbf{j}$, and if $r_0 \gg Y$, expression (11.8) can be approximated by

$$A(\mathbf{r}_0) = \frac{\exp(i\mathbf{k}'\cdot\mathbf{r}_0)}{r_0}\int_{-\infty}^{\infty} f(y)\exp(-iky\sin\theta)\,dy. \tag{11.9}$$

We have used that $f(y) = 0$ for $|y| > Y$, to extend the integral to infinite limits. The intensity in the direction $\theta$ is then given by

$$I(\theta) = |A|^2 = \frac{2\pi}{{r_0}^2}|\tilde{f}(q)|^2, \tag{11.10}$$

where $q = k\sin\theta$.

† This is the approach first used by Fresnel. For simplicity we have omitted from the integral a multiplicative inclination factor which depends on angle $\theta$ and decreases as $\theta$ increases.

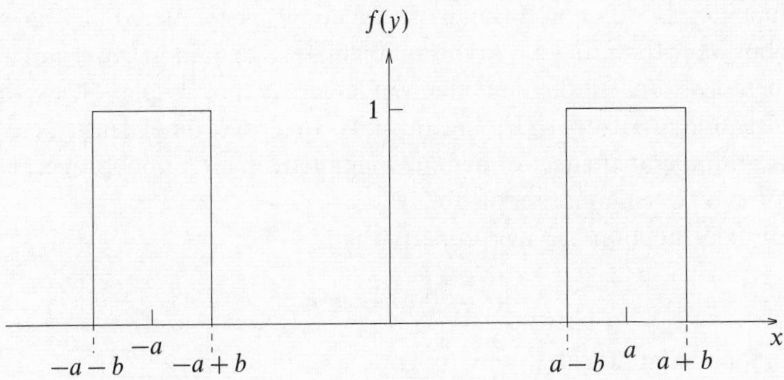


Figure 11.3 The aperture function $f(y)$ for two wide slits.

► *Evaluate $I(\theta)$ for an aperture consisting of two long slits each of width $2b$, their centres being separated by a distance $2a$, $a > b$, illuminated by light of wavelength $\lambda$.*

The aperture function is plotted in figure 11.3. We first need to find $\widetilde{f}(q)$:

$$\begin{aligned}
\widetilde{f}(q) &= \frac{1}{\sqrt{2\pi}} \int_{-a-b}^{-a+b} e^{-iqx}\, dx + \frac{1}{\sqrt{2\pi}} \int_{a-b}^{a+b} e^{-iqx}\, dx \\
&= \frac{1}{\sqrt{2\pi}} \left[ -\frac{e^{-iqx}}{iq} \right]_{-a-b}^{-a+b} + \frac{1}{\sqrt{2\pi}} \left[ -\frac{e^{-iqx}}{iq} \right]_{a-b}^{a+b} \\
&= \frac{-1}{iq\sqrt{2\pi}} \left[ e^{-iq(-a+b)} - e^{-iq(-a-b)} + e^{-iq(a+b)} - e^{-iq(a-b)} \right].
\end{aligned}$$

After some manipulation we obtain

$$\widetilde{f}(q) = \frac{4 \cos qa \sin qb}{q\sqrt{2\pi}}.$$

Now applying (11.10), and remembering that $q = (2\pi \sin\theta)/\lambda$, we find

$$I(\theta) = \frac{16 \cos^2 qa \sin^2 qb}{q^2 {r_0}^2},$$

where $r_0$ is the distance from the centre of the aperture. ◄

### *11.1.3 The Dirac δ-function*

Before going on to consider further properties of Fourier transforms we make a digression to discuss the Dirac $\delta$-function and its relation to Fourier transforms.

The $\delta$-function is different from most functions encountered in the physical sciences but we will see that a rigorous mathematical definition exists and its utility will be demonstrated throughout the remainder of this chapter. The $\delta$-function can be visualised as a very sharp narrow pulse (in space, time, density, etc.) which produces an integrated effect of definite magnitude. The formal properties of the $\delta$-function may be summarised as follows.

The Dirac $\delta$-function has the property that

$$\delta(t) = 0 \quad \text{for } t \neq 0, \tag{11.11}$$

but its fundamental defining property is

$$\int f(t)\delta(t-a)\,dt = f(a), \tag{11.12}$$

provided the range of integration includes the point $t = a$; otherwise the integral equals zero. This leads immediately to two further useful results:

$$\int_{-a}^{b} \delta(t)\,dt = 1 \quad \text{for all } a, b > 0. \tag{11.13}$$

and

$$\int \delta(t-a)\,dt = 1, \tag{11.14}$$

provided the range of integration includes $t = a$.

Equation (11.12) can be used to derive further useful properties of the Dirac $\delta$-function:

$$\delta(t) = \delta(-t), \tag{11.15}$$

$$\delta(at) = \frac{1}{|a|}\delta(t), \tag{11.16}$$

$$t\delta(t) = 0. \tag{11.17}$$

►*Prove that* $\delta(bt) = \delta(t)/|b|$.

Let us first consider the case where $b > 0$. It follows that

$$\int_{-\infty}^{\infty} f(t)\delta(bt)\,dt = \int_{-\infty}^{\infty} f\left(\frac{t'}{b}\right)\delta(t')\frac{dt'}{b} = \frac{1}{b}f(0) = \frac{1}{b}\int_{-\infty}^{\infty} f(t)\delta(t)\,dt,$$

where we have made the substitution $t' = bt$. But $f(t)$ is arbitrary and so we immediately see that $\delta(bt) = \delta(t)/b = \delta(t)/|b|$ for $b > 0$.

Now consider the case where $b = -c < 0$. It follows that

$$\int_{-\infty}^{\infty} f(t)\delta(bt)\,dt = \int_{\infty}^{-\infty} f\left(\frac{t'}{-c}\right)\delta(t')\left(\frac{dt'}{-c}\right) = \int_{-\infty}^{\infty} \frac{1}{c} f\left(\frac{t'}{-c}\right)\delta(t')\,dt'$$

$$= \frac{1}{c}f(0) = \frac{1}{|b|}f(0) = \frac{1}{|b|}\int_{-\infty}^{\infty} f(t)\delta(t)\,dt,$$

where we have made the substitution $t' = bt = -ct$. But $f(t)$ is arbitrary and so

$$\delta(bt) = \frac{1}{|b|}\delta(t),$$

for all $b$, which establishes the result. ◄

Furthermore, by considering an integral of the form

$$\int f(t)\delta(h(t))\,dt,$$

and making a change of variables to $z = h(t)$, we may show that

$$\delta(h(t)) = \sum_i \frac{\delta(t - t_i)}{|h'(t_i)|}, \tag{11.18}$$

where the $t_i$ are those values of $t$ for which $h(t) = 0$, and $h'(t)$ stands for $dh/dt$.

The derivative of the delta function, $\delta'(t)$, is defined by

$$\int_{-\infty}^{\infty} f(t)\delta'(t)\,dt = \Big[f(t)\delta(t)\Big]_{-\infty}^{\infty} - \int_{-\infty}^{\infty} f'(t)\delta(t)\,dt$$
$$= -f'(0), \tag{11.19}$$

and similarly for higher derivatives.

For many practical purposes, effects which are not strictly described by a $\delta$-function may be analysed as such, if they take place in an interval much shorter than the response interval of the system on which they act. For example, the idealised notion of an instantaneous impulse of magnitude $J$ applied at time $t_0$ can be represented by

$$j(t) = J\delta(t - t_0). \tag{11.20}$$

Many physical situations are described by a $\delta$-function in space rather than in time. Moreover, we often require the $\delta$-function to be defined in more than one dimension. For example, the charge density of a point charge $q$ at a point $\mathbf{r}_0$, may be expressed as a three-dimensional $\delta$-function

$$\rho(\mathbf{r}) = q\delta(\mathbf{r} - \mathbf{r_0}) = q\delta(x - x_0)\delta(y - y_0)\delta(z - z_0), \tag{11.21}$$

so that a discrete 'quantum' is expressed as if it were a continuous distribution.

From (11.21) we see that (as expected) the total charge enclosed in a volume $V$ is given by

$$\int_V \rho(\mathbf{r})\, dV = \int_V q\delta(\mathbf{r}-\mathbf{r}_0)\, dV = \begin{cases} q & \text{if } \mathbf{r}_0 \text{ lies in } V, \\ 0 & \text{otherwise.} \end{cases}$$

Closely related to the Dirac $\delta$-function is the *Heaviside* or *unit step function* $H(t)$, for which

$$H(t) = \begin{cases} 1 & \text{for } t > 0, \\ 0 & \text{for } t < 0. \end{cases} \tag{11.22}$$

This function is clearly discontinuous at $t = 0$ and it is usual to take $H(0) = 1/2$. The Heaviside function is related to the delta function by

$$H'(t) = \delta(t). \tag{11.23}$$

►*Prove relation (11.23).*

Considering the integral

$$\begin{aligned} \int_{-\infty}^{\infty} f(t)H'(t)\, dt &= \Big[f(t)H(t)\Big]_{-\infty}^{\infty} - \int_{-\infty}^{\infty} f'(t)H(t)\, dt \\ &= f(\infty) - \int_0^{\infty} f'(t)\, dt \\ &= f(\infty) - \Big[f(t)\Big]_0^{\infty} = f(0), \end{aligned}$$

and comparing it with (11.12) when $a = 0$ immediately shows that $H'(t) = \delta(t)$. ◄

### 11.1.4 Relation of the δ-function to Fourier transforms

In the previous section we introduced the Dirac $\delta$-function as a way of representing very sharp narrow pulses, but in no way related it to Fourier transforms. We now show that the $\delta$-function can equally well be defined in a way that more naturally relates it to the Fourier transform.

Referring back to the Fourier inversion theorem (11.4), we have

$$\begin{aligned} f(t) &= \frac{1}{2\pi}\int_{-\infty}^{\infty} d\omega\, e^{i\omega t} \int_{-\infty}^{\infty} du\, f(u)\, e^{-i\omega u} \\ &= \int_{-\infty}^{\infty} du\; f(u)\left\{\frac{1}{2\pi}\int_{-\infty}^{\infty} e^{i\omega(t-u)}\, d\omega\right\}. \end{aligned}$$

Comparison of this with (11.12) shows that we may write the $\delta$-function as

$$\delta(t-u) = \frac{1}{2\pi}\int_{-\infty}^{\infty} e^{i\omega(t-u)}\, d\omega. \tag{11.24}$$

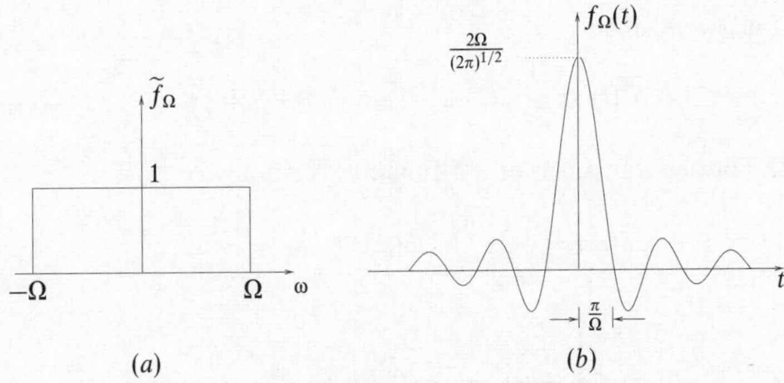


Figure 11.4 (*a*) A Fourier transform showing a rectangular distribution of frequencies between $\pm\Omega$; (*b*) the function of which it is the transform, which is proportional to $t^{-1}\sin\Omega t$.

Considered as a Fourier transform, this representation shows that a very narrow time peak at $t = u$ results from the superposition of a complete spectrum of harmonic waves, all frequencies having the same amplitude, and all waves being in phase at $t = u$. This suggests that the $\delta$-function may also be represented as the limit of the transform of a uniform distribution of unit height, as the width of this distribution becomes infinite.

Consider the rectangular distribution of frequencies shown in figure 11.4(*a*). From (11.6), taking the inverse Fourier transform,

$$\begin{aligned} f_\Omega(t) &= \frac{1}{\sqrt{2\pi}}\int_{-\Omega}^{\Omega} 1 \times e^{i\omega t}\,d\omega \\ &= \frac{2\Omega}{\sqrt{2\pi}}\frac{\sin\Omega t}{\Omega t}. \end{aligned} \tag{11.25}$$

This function is illustrated in figure 11.4 and it is apparent that, for large $\Omega$, it becomes very large at $t = 0$ and also very narrow about $t = 0$, as we qualitatively expect and require. We also note that in the limit $\Omega \to \infty$, $f_\Omega(t)$, as defined in (11.25), tends to $(2\pi)^{1/2}\delta(t)$ by virtue of (11.24). Hence we may conclude that the $\delta$-function can also be represented by

$$\delta(t) = \lim_{\Omega\to\infty}\left(\frac{\sin\Omega t}{\pi t}\right). \tag{11.26}$$

Several other function representations are equally valid, e.g. the limiting cases of rectangular, triangular or Gaussian distributions; the only essential requirements are a knowledge of the area under such a curve and that undefined operations such as dividing by zero are not inadvertently carried out on the $\delta$-function whilst some non-explicit representation is being employed.

We also note that the Fourier transform definition of the delta function, (11.24), shows that it is real since

$$\delta^*(t) = \frac{1}{2\pi}\int_{-\infty}^{\infty} e^{-i\omega t}\,d\omega = \delta(-t) = \delta(t).$$

Finally, the Fourier transform of a $\delta$-function is simply

$$\tilde{\delta}(\omega) = \frac{1}{\sqrt{2\pi}}\int_{-\infty}^{\infty} \delta(t)\,e^{-i\omega t}\,dt = \frac{1}{\sqrt{2\pi}}. \tag{11.27}$$

### *11.1.5 Properties of Fourier transforms*

Having considered the Dirac $\delta$-function, we now return to our discussion of the properties of Fourier transforms. As we would expect, Fourier transforms have many properties analogous to those of Fourier series in respect of the connection between transforms of related functions. Here we only list these properties without proof; they can be verified by working from the definition of the transform. As previously, we denote the Fourier transform of $f(t)$ by $\tilde{f}(\omega)$ or $\mathcal{F}[f(t)]$.

(i) Differentiation:

$$\mathcal{F}\left[f'(t)\right] = i\omega\tilde{f}(\omega). \tag{11.28}$$

This may be extended to higher derivatives, so that

$$\mathcal{F}\left[f''(t)\right] = i\omega\mathcal{F}\left[f'(t)\right] = -\omega^2\tilde{f}(\omega),$$

and so on.

(ii) Integration:

$$\mathcal{F}\left[\int^t f(s)\,ds\right] = \frac{1}{i\omega}\tilde{f}(\omega) + 2\pi c\delta(\omega), \tag{11.29}$$

where the term $2\pi c\delta(\omega)$ represents the Fourier transform of the constant of integration associated with the indefinite integral.

(iii) Scaling:

$$\mathcal{F}[f(at)] = \frac{1}{a}\tilde{f}\left(\frac{\omega}{a}\right). \tag{11.30}$$

(iv) Translation:

$$\mathcal{F}[f(t+a)] = e^{ia\omega}\tilde{f}(\omega). \tag{11.31}$$

(v) Exponential multiplication:

$$\mathcal{F}\left[e^{\alpha t}f(t)\right] = \tilde{f}(\omega + i\alpha), \tag{11.32}$$

where $\alpha$ may be real, imaginary or complex.

►*Prove relation (11.28).*

Calculating the Fourier transform of $f'(t)$ directly, we obtain

$$\begin{aligned}\mathcal{F}\left[f'(t)\right] &= \frac{1}{\sqrt{2\pi}}\int_{-\infty}^{\infty} f'(t)\, e^{-i\omega t}\, dt \\ &= \frac{1}{\sqrt{2\pi}}\left[e^{-i\omega t} f(t)\right]_{-\infty}^{\infty} + \frac{1}{\sqrt{2\pi}}\int_{-\infty}^{\infty} i\omega\, e^{-i\omega t} f(t)\, dt \\ &= i\omega \widetilde{f}(\omega),\end{aligned}$$

if $f(t) \to 0$ at $t = \pm\infty$ (as it must since $\int_{-\infty}^{\infty} |f(t)|\, dt$ is finite). ◄

To illustrate a use and also a proof of (11.32), let us consider an amplitude-modulated radio wave. Suppose a message to be broadcast is represented by $f(t)$. The message can be added electronically to a constant signal $a$ of such a magnitude that $a + f(t)$ is never negative, and the sum then used to modulate the amplitude of a carrier signal of frequency $\omega_c$. Using a complex exponential notation, the transmitted amplitude is now

$$g(t) = A\,[a + f(t)]\, e^{i\omega_c t}. \tag{11.33}$$

Ignoring in the present context the effect of the term $Aa\exp(i\omega_c t)$, which gives a contribution to the transmitted spectrum only at $\omega = \omega_c$, we obtain for the new spectrum

$$\begin{aligned}\widetilde{g}(\omega) &= \frac{1}{\sqrt{2\pi}} A \int_{-\infty}^{\infty} f(t)\, e^{i\omega_c t}\, e^{-i\omega t}\, dt \\ &= \frac{1}{\sqrt{2\pi}} A \int_{-\infty}^{\infty} f(t)\, e^{-i(\omega-\omega_c)t}\, dt \\ &= A\widetilde{f}(\omega - \omega_c),\end{aligned} \tag{11.34}$$

which is simply a shift of the whole spectrum by the carrier frequency. The use of different carrier frequencies enables signals to be separated.

### 11.1.6 Odd and even functions

If $f(t)$ is odd or even then we may derive alternative forms of Fourier's inversion theorem, which lead to the definition of different transform pairs. Let us first consider an odd function $f(t) = -f(-t)$, whose Fourier transform is given by

$$\begin{aligned}\widetilde{f}(\omega) &= \frac{1}{\sqrt{2\pi}}\int_{-\infty}^{\infty} f(t)\, e^{-i\omega t}\, dt \\ &= \frac{1}{\sqrt{2\pi}}\int_{-\infty}^{\infty} f(t)(\cos\omega t - i\sin\omega t)\, dt \\ &= \frac{-2i}{\sqrt{2\pi}}\int_{0}^{\infty} f(t)\sin\omega t\, dt,\end{aligned}$$

where in the last line we use the fact that $f(t)$ and $\sin \omega t$ are odd, whereas $\cos \omega t$ is even.

We note that $\tilde{f}(-\omega) = -\tilde{f}(\omega)$, i.e. $\tilde{f}(\omega)$ is an odd function of $\omega$. Hence

$$f(t) = \frac{1}{\sqrt{2\pi}} \int_{-\infty}^{\infty} \tilde{f}(\omega)\, e^{i\omega t}\, d\omega = \frac{2i}{\sqrt{2\pi}} \int_0^{\infty} \tilde{f}(\omega) \sin \omega t\, d\omega$$
$$= \frac{2}{\pi} \int_0^{\infty} d\omega\ \sin \omega t \left\{ \int_0^{\infty} f(u) \sin \omega u\, du \right\}.$$

Thus we may define the *Fourier sine transform pair* for odd functions:

$$\tilde{f}_s(\omega) = \sqrt{\frac{2}{\pi}} \int_0^{\infty} f(t)\ \sin \omega t\, dt, \qquad (11.35)$$

$$f(t) = \sqrt{\frac{2}{\pi}} \int_0^{\infty} \tilde{f}_s(\omega)\ \sin \omega t\, d\omega. \qquad (11.36)$$

Note that although the Fourier sine transform pair was derived by considering an odd function $f(t)$ defined over all $t$, the definitions (11.35) and (11.36) only require $f(t)$ and $\tilde{f}_s(\omega)$ to be defined for positive $t$ and $\omega$ respectively. For an even function, $f(t) = f(-t)$, we can define the *Fourier cosine transform pair* in a similar way, but with $\sin \omega t$ replaced by $\cos \omega t$.

### *11.1.7 Convolution and deconvolution*

It is apparent that any attempt to measure the value of a physical quantity is, to some extent, limited by the finite resolution of the measuring apparatus used. On the one hand, the physical quantity we wish to measure will in general be a function of an independent variable, $x$ say, i.e. the true function to be measured takes the form $f(x)$. On the other hand, the apparatus we are using does not give the true value of the function; a resolution function $g(y)$ is involved. By this we mean that the probability that a reading which should have been at $y = 0$ is recorded instead as being between $y$ and $y + dy$ is given by $g(y)\, dy$. Some possible resolution functions of this sort are shown in figure 11.5. To obtain good results we wish the resolution function to be as close to a $\delta$-function as possible (case (*a*)). A typical piece of apparatus has a resolution function of finite width, although the mean is centred on the true value (case (*b*)). However, some apparatus may show a bias that tends to shift observations to higher or lower values than the true ones (cases (*c*) and (*d*)).

Given that the true distribution is $f(x)$ and the resolution function of our measuring apparatus is $g(y)$, we wish to calculate what the observed distribution $h(z)$ will be. The symbols $x$, $y$ and $z$ all refer to the same physical variable (e.g. length or angle), but are denoted differently because the variable appears in the analysis in three different roles.

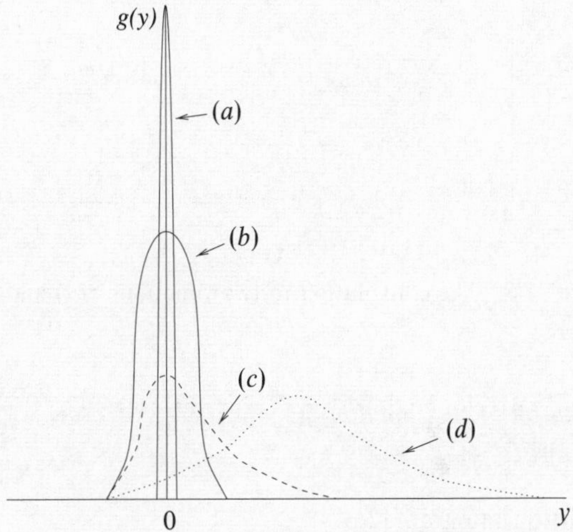


Figure 11.5 Resolution functions: (*a*) ideal $\delta$-function; (*b*) typical unbiased resolution; (*c*) and (*d*) biases tending to shift observations to higher values than the true one.

The probability that a true reading lying between $x$ and $x + dx$ (which has a probability $f(x)\,dx$ of being selected) is moved by the instrumental resolution through an interval $z - x$ into an interval of width $dz$ is $g(z - x)\,dz$. Hence the combined probability that the interval $dx$ will give rise to an observation appearing in the interval $dz$ is $f(x)\,dx\,g(z-x)\,dz$ and, adding together the contributions from all values of $x$ that can lead to an observation in the range $z$ to $z + dz$, we find that the observed distribution is given by

$$h(z) = \int_{-\infty}^{\infty} f(x)g(z - x)\,dx. \tag{11.37}$$

The integral in (11.37) is called the *convolution* of the functions $f$ and $g$ and is often written $f * g$. The convolution defined above is commutative ($f * g = g * f$), associative and distributive. The observed distribution is thus the convolution of the true distribution and the experimental resolution function. The result will be that the observed distribution is broader and smoother than the true one and, if $g(y)$ has a bias, the maxima will normally be displaced from their true positions. It is also obvious from (11.37) that if the resolution is the ideal $\delta$-function, $g(y) = \delta(y)$, then $h(z) = f(z)$ and the observed distribution is the true one.

It is interesting to note, and a very important property, that the convolution of any function $g(y)$ with a number of delta functions leaves a copy of $g(y)$ at the position of each of the delta functions.

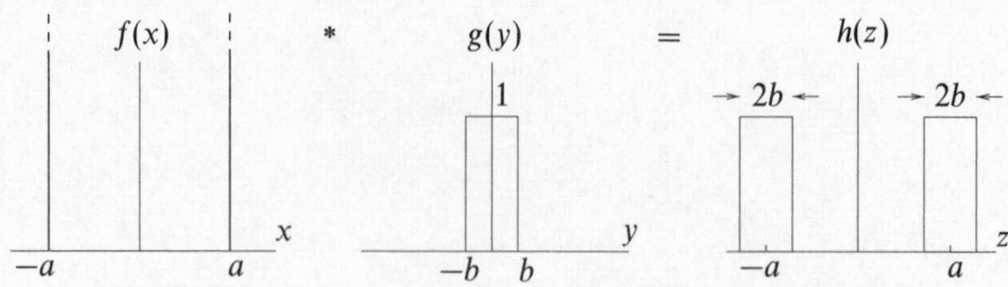


Figure 11.6 The convolution of two functions $f(x)$ and $g(y)$.

►*Find the convolution of the function $f(x) = \delta(x+a) + \delta(x-a)$ with the function $g(y)$ plotted in figure 11.6.*

Using the convolution integral (11.37)

$$h(z) = \int_{-\infty}^{\infty} f(x)g(z-x)\,dx = \int_{-\infty}^{\infty} [\delta(x+a) + \delta(x-a)]g(z-x)\,dx$$
$$= g(z+a) + g(z-a).$$

The convolution $h(z)$ is also plotted in figure 11.6. ◄

Let us now consider the Fourier transform of the convolution (11.37), which is given by

$$\widetilde{h}(k) = \frac{1}{\sqrt{2\pi}} \int_{-\infty}^{\infty} dz\, e^{-ikz} \left\{ \int_{-\infty}^{\infty} f(x)g(z-x)\,dx \right\}$$
$$= \frac{1}{\sqrt{2\pi}} \int_{-\infty}^{\infty} dx\, f(x) \left\{ \int_{-\infty}^{\infty} g(z-x)\, e^{-ikz}\,dz \right\}.$$

If we let $u = z - x$ in the second integral we have

$$\widetilde{h}(k) = \frac{1}{\sqrt{2\pi}} \int_{-\infty}^{\infty} dx\, f(x) \left\{ \int_{-\infty}^{\infty} g(u)\, e^{-ik(u+x)}\,du \right\}$$
$$= \frac{1}{\sqrt{2\pi}} \int_{-\infty}^{\infty} f(x)\, e^{-ikx}\,dx \int_{-\infty}^{\infty} g(u)\, e^{-iku}\,du$$
$$= \frac{1}{\sqrt{2\pi}} \times \sqrt{2\pi}\,\widetilde{f}(k) \times \sqrt{2\pi}\widetilde{g}(k) = \sqrt{2\pi}\,\widetilde{f}(k)\widetilde{g}(k). \qquad (11.38)$$

Hence the Fourier transform of a convolution $f * g$ is equal to the product of the separate Fourier transforms multiplied by $\sqrt{2\pi}$; this result is called the *convolution theorem*.

It may similarly be proved that the converse is also true, namely that the Fourier transform of the product $f(x)g(x)$ is given by

$$\mathscr{F}[f(x)g(x)] = \frac{1}{\sqrt{2\pi}} \widetilde{f}(k) * \widetilde{g}(k). \qquad (11.39)$$

► *Find the Fourier transform of the function representing two wide slits in figure 11.3 by considering the Fourier transforms of (i) two δ-functions, at $x = \pm a$, (ii) a rectangular function of height 1 and width 2b centred on $x = 0$.*

(i) The Fourier transform of the two δ-functions is given by

$$\begin{aligned}\widetilde{f}(q) &= \frac{1}{\sqrt{2\pi}}\int_{-\infty}^{\infty}\delta(x-a)\,e^{-iqx}\,dx + \frac{1}{\sqrt{2\pi}}\int_{-\infty}^{\infty}\delta(x+a)\,e^{-iqx}\,dx\\ &= \frac{1}{\sqrt{2\pi}}\left(e^{-iqa}+e^{iqa}\right) = \frac{2\cos qa}{\sqrt{2\pi}}.\end{aligned}$$

(ii) The Fourier transform of the broad slit is

$$\begin{aligned}\widetilde{g}(q) &= \frac{1}{\sqrt{2\pi}}\int_{-b}^{b} e^{-iqx}\,dx = \frac{1}{\sqrt{2\pi}}\left[\frac{e^{-iqx}}{-iq}\right]_{-b}^{b}\\ &= \frac{-1}{iq\sqrt{2\pi}}(e^{-iqb}-e^{iqb}) = \frac{2\sin qb}{q\sqrt{2\pi}}.\end{aligned}$$

We have already seen that the convolution of these functions is the required function respresenting two wide slits (see figure 11.6), and so, using the convolution theorem, the Fourier transform of the convolution is simply $\sqrt{2\pi}$ times the product of the individual transforms, i.e. $4\cos qa\,\sin qb/(q\sqrt{2\pi})$. This is, of course, the same result as obtained in the example in subsection 11.1.2. ◄

The inverse of convolution, called *deconvolution*, allows us to find a true distribution $f(x)$ given an observed distribution $h(z)$ and a resolution function $g(y)$.

►*An experimental quantity $f(x)$ is measured by an instrument with known resolution function $g(y)$ to give an observed distribution $h(z)$. How may $f(x)$ be extracted from the measured distribution?*

From the convolution theorem (11.38), the Fourier transform of the measured distribution is

$$\widetilde{h}(k) = \sqrt{2\pi}\,\widetilde{f}(k)\widetilde{g}(k),$$

from which we obtain

$$\widetilde{f}(k) = \frac{1}{\sqrt{2\pi}} \times \frac{\widetilde{h}(k)}{\widetilde{g}(k)}.$$

Then on inverse Fourier transforming we find

$$f(x) = \frac{1}{\sqrt{2\pi}}\mathcal{F}^{-1}\left[\frac{\widetilde{h}(k)}{\widetilde{g}(k)}\right].$$

In words, to extract the true distribution, we divide the Fourier transform of the observed distribution by that of the resolution function for each value of $k$, and then take the inverse Fourier transform of the function so generated. ◄

This explicit method of extracting true distributions is straightforward for exact functions but, in practice, because of experimental and statistical uncertainties in the experimental data or because data over only a limited range are available, it is often not very precise, involving as it does three (numerical) transforms each requiring in principle an integral over an infinite range.

### *11.1.8 Correlation functions and energy spectra*

The *cross-correlation* of two functions $f$ and $g$ is defined by

$$C(z) = \int_{-\infty}^{\infty} f^*(x)g(x+z)\,dx. \tag{11.40}$$

Despite the formal similarity between (11.40) and the definition of the convolution in (11.37), the use and interpretation of the cross-correlation and of the convolution are very different; the cross-correlation provides a quantitative measure of the similarity of two functions $f$ and $g$ as one is displaced through a distance $z$ relative to the other. The cross-correlation is often notated as $C = f \otimes g$, and like convolution it is both associative and distributive. Unlike convolution, however, it is *not* commutative, in fact

$$[f \otimes g](z) = [g \otimes f]^*(-z). \tag{11.41}$$

►*Prove the Wiener–Kinchin theorem,*

$$\widetilde{C}(k) = \sqrt{2\pi}[\widetilde{f}(k)]^*\widetilde{g}(k). \tag{11.42}$$

Following a method similar to that for the convolution of $f$ and $g$, let us consider the Fourier transform of (11.40):

$$\begin{aligned}\widetilde{C}(k) &= \frac{1}{\sqrt{2\pi}}\int_{-\infty}^{\infty} dz\, e^{-ikz}\left\{\int_{-\infty}^{\infty} f^*(x)g(x+z)\,dx\right\}\\ &= \frac{1}{\sqrt{2\pi}}\int_{-\infty}^{\infty} dx\, f^*(x)\left\{\int_{-\infty}^{\infty} g(x+z)\,e^{-ikz}\,dz\right\}.\end{aligned}$$

Making the substitution $u = x+z$ in the second integral we obtain

$$\begin{aligned}\widetilde{C}(k) &= \frac{1}{\sqrt{2\pi}}\int_{-\infty}^{\infty} dx\, f^*(x)\left\{\int_{-\infty}^{\infty} g(u)\,e^{-ik(u-x)}\,du\right\}\\ &= \frac{1}{\sqrt{2\pi}}\int_{-\infty}^{\infty} f^*(x)\,e^{ikx}\,dx\int_{-\infty}^{\infty} g(u)\,e^{-iku}\,du\\ &= \frac{1}{\sqrt{2\pi}}\times\sqrt{2\pi}\,[\widetilde{f}(k)]^*\times\sqrt{2\pi}\widetilde{g}(k) = \sqrt{2\pi}\,[\widetilde{f}(k)]^*\widetilde{g}(k).\end{aligned}$$ ◄

Thus the Fourier transform of the cross-correlation of $f$ and $g$ is equal to the product of $[\tilde{f}(k)]^*$ and $\tilde{g}(k)$ multiplied by $\sqrt{2\pi}$. This a statement of the *Wiener–Kinchin theorem.* Similarly we can derive the converse theorem

$$\mathcal{F}\left[f^*(x)g(x)\right] = \frac{1}{\sqrt{2\pi}}\tilde{f} \otimes \tilde{g}.$$

If we now consider the special case where $g$ is taken to be equal to $f$ in (11.40), then, writing the LHS as $a(z)$, we have

$$a(z) = \int_{-\infty}^{\infty} f^*(x)f(x+z)\,dx; \tag{11.43}$$

this is called the *auto-correlation function* of $f(x)$. Using the Wiener–Kinchin theorem (11.42) we see that

$$\begin{aligned} a(z) &= \frac{1}{\sqrt{2\pi}}\int_{-\infty}^{\infty} \tilde{a}(k)\,e^{ikz}\,dk \\ &= \frac{1}{\sqrt{2\pi}}\int_{-\infty}^{\infty} \sqrt{2\pi}\,[\tilde{f}(k)]^*\tilde{f}(k)\,e^{ikz}\,dk, \end{aligned}$$

so that $a(z)$ is the inverse Fourier transform of $\sqrt{2\pi}\,|\tilde{f}(k)|^2$, which is in turn called the *energy spectrum* of $f$.

### *11.1.9 Parseval's theorem*

Using the results of the previous section we can immediately obtain *Parseval's theorem.* The most general form of this (also called the *multiplication theorem*) is obtained simply by noting from (11.42) that the cross-correlation (11.40) of two functions $f$ and $g$ can be written as

$$C(z) = \int_{-\infty}^{\infty} f^*(x)g(x+z)\,dx = \int_{-\infty}^{\infty} [\tilde{f}(k)]^*\tilde{g}(k)\,e^{ikz}\,dk. \tag{11.44}$$

Then, setting $z = 0$ gives the multiplication theorem

$$\int_{-\infty}^{\infty} f^*(x)g(x)\,dx = \int [\tilde{f}(k)]^*\tilde{g}(k)\,dk. \tag{11.45}$$

Specialising further, by letting $g = f$, we derive the most common form of Parseval's theorem,

$$\int_{-\infty}^{\infty} |f(x)|^2\,dx = \int_{-\infty}^{\infty} |\tilde{f}(k)|^2\,dk. \tag{11.46}$$

When $f$ is a physical amplitude these integrals relate to the total intensity involved in some physical process. We have already met a form of Parseval's theorem for Fourier series in chapter 10; it is in fact a special case of (11.46).

► *The displacement of a damped harmonic oscillator as a function of time is given by*

$$f(t) = \begin{cases} 0 & \text{for } t < 0, \\ e^{-t/\tau} \sin \omega_0 t & \text{for } t \geq 0. \end{cases}$$

*Find the Fourier transform of this function and so give a physical interpretation of Parseval's theorem.*

Using the usual definition for the Fourier transform we find

$$\widetilde{f}(\omega) = \int_{-\infty}^{0} 0 \times e^{-i\omega t}\, dt + \int_{0}^{\infty} e^{-t/\tau} \sin \omega_0 t\, e^{-i\omega t}\, dt.$$

Writing $\sin \omega_0 t$ as $(e^{i\omega_0 t} - e^{-i\omega_0 t})/2i$ we obtain

$$\begin{aligned} \widetilde{f}(\omega) &= 0 + \frac{1}{2i} \int_{0}^{\infty} \left[ e^{-it(\omega-\omega_0-i/\tau)} - e^{-it(\omega+\omega_0-i/\tau)} \right] dt \\ &= \frac{1}{2} \left[ \frac{1}{\omega+\omega_0-i/\tau} - \frac{1}{\omega-\omega_0-i/\tau} \right], \end{aligned}$$

which is the required Fourier transform. The physical interpretation of $|\widetilde{f}(\omega)|^2$ is the energy content per unit frequency interval (i.e. the *energy spectrum*) whilst $|f(t)|^2$ is proportional to the sum of the kinetic and potential energies of the oscillator. Hence (to within a constant) Parseval's theorem shows the equivalence of these two alternative specifications for the total energy. ◄

### *11.1.10 Fourier transforms in higher dimensions*

The concept of the Fourier transform can be naturally extended to more than one dimension. For instance we may wish to find the spatial Fourier transform of two- or three-dimensional functions (of position). For example, in three dimensions we can define the Fourier transform of $f(x, y, z)$ as

$$\widetilde{f}(k_x, k_y, k_z) = \frac{1}{(2\pi)^{3/2}} \iiint f(x, y, z)\, e^{-ik_x x} e^{-ik_y y} e^{-ik_z z}\, dx\, dy\, dz, \tag{11.47}$$

and its inverse by

$$f(x, y, z) = \frac{1}{(2\pi)^{3/2}} \iiint \widetilde{f}(k_x, k_y, k_z)\, e^{ik_x x} e^{ik_y y} e^{ik_z z}\, dk_x\, dk_y\, dk_z. \tag{11.48}$$

Denoting the vector with components $k_x$, $k_y$, $k_z$ by $\mathbf{k}$ and that with components $x$, $y$, $z$ by $\mathbf{r}$, we can write the Fourier transform pair (11.47), (11.48) as

$$\widetilde{f}(\mathbf{k}) = \frac{1}{(2\pi)^{3/2}} \int f(\mathbf{r})\, e^{-i\mathbf{k}\cdot\mathbf{r}}\, d^3\mathbf{r}, \tag{11.49}$$

$$f(\mathbf{r}) = \frac{1}{(2\pi)^{3/2}} \int \widetilde{f}(\mathbf{k})\, e^{i\mathbf{k}\cdot\mathbf{r}}\, d^3\mathbf{k}. \tag{11.50}$$

From these relations we may deduce that the three-dimensional Dirac $\delta$-function can be written as

$$\delta(\mathbf{r}) = \frac{1}{(2\pi)^3} \int e^{i\mathbf{k}\cdot\mathbf{r}}\, d^3\mathbf{k}. \tag{11.51}$$

Similar relations to (11.49), (11.50) and (11.51) exist for spaces of other dimensionalities.

► *In three-dimensional space a function $f(\mathbf{r})$ possesses spherical symmetry, so that $f(\mathbf{r}) = f(r)$. Find the Fourier transform of $f(\mathbf{r})$ as a one-dimensional integral.*

Let us choose spherical polar coordinates in which the vector $\mathbf{k}$ of the Fourier transform lies along the polar axis ($\theta = 0$). This we can do since $f(\mathbf{r})$ is spherically symmetric. We then have

$$d^3\mathbf{r} = r^2 \sin\theta\, dr\, d\theta\, d\phi \qquad \text{and} \qquad \mathbf{k}\cdot\mathbf{r} = kr\cos\theta,$$

where $k = |\mathbf{k}|$. The Fourier transform is then given by

$$\begin{aligned}
\widetilde{f}(\mathbf{k}) &= \frac{1}{(2\pi)^{3/2}} \int f(\mathbf{r})\, e^{-i\mathbf{k}\cdot\mathbf{r}}\, d^3\mathbf{r} \\
&= \frac{1}{(2\pi)^{3/2}} \int_0^\infty dr \int_0^\pi d\theta \int_0^{2\pi} d\phi\, f(r) r^2 \sin\theta\, e^{-ikr\cos\theta} \\
&= \frac{1}{(2\pi)^{3/2}} \int_0^\infty dr\, 2\pi f(r) r^2 \int_0^\pi d\theta\, \sin\theta\, e^{-ikr\cos\theta}.
\end{aligned}$$

The integral over $\theta$ may be straightforwardly evaluated by noting that

$$\frac{d}{d\theta}(e^{-ikr\cos\theta}) = ikr\sin\theta\, e^{-ikr\cos\theta}.$$

Therefore

$$\begin{aligned}
\widetilde{f}(\mathbf{k}) &= \frac{1}{(2\pi)^{3/2}} \int_0^\infty dr\, 2\pi f(r) r^2 \left[\frac{e^{-ikr\cos\theta}}{ikr}\right]_{\theta=0}^{\theta=\pi} \\
&= \frac{1}{(2\pi)^{3/2}} \int_0^\infty 4\pi r^2 f(r) \left(\frac{\sin kr}{kr}\right) dr. \blacktriangleleft
\end{aligned}$$

A similar result may be obtained for two-dimensional Fourier transforms in which $f(\mathbf{r}) = f(\rho)$, i.e. it is independent of azimuthal angle $\phi$. In this case, using the integral representation of the Bessel function $J_0(x)$ given at the very end of subsection 14.7.3, we find

$$\widetilde{f}(\mathbf{k}) = \frac{1}{2\pi} \int_0^\infty 2\pi\rho f(\rho) J_0(k\rho)\, d\rho. \tag{11.52}$$

## 11.2 Laplace transforms

Often we are interested in functions $f(t)$ for which the Fourier transform does not exist because $f \not\to 0$ as $t \to \infty$, and so the integral defining $\tilde{f}$ does not converge. For example, the function $f(t) = t$ does not possess a Fourier transform. Furthermore, we are often only interested in a given function for $t > 0$, for example when we are given the value at $t = 0$ in an initial-value problem. This leads us to consider the Laplace transform $\bar{f}(s)$, or $\mathscr{L}[f(t)]$, of $f(t)$, which is defined by

$$\bar{f}(s) \equiv \int_0^\infty f(t)e^{-st}\,dt, \tag{11.53}$$

provided that the integral exists. We here assume that $s$ is real, but complex values have to be considered in a more detailed study. In practice, for a given function $f(t)$ there will be some real number $s_0$ such that the integral in (11.53) exists for $s > s_0$, but diverges for $s \le s_0$.

Through (11.53) we define a *linear* transformation that converts functions of the variable $t$ to functions of a new variable $s$, such that

$$\mathscr{L}[af_1(t) + bf_2(t)] = a\mathscr{L}[f_1(t)] + b\mathscr{L}[f_2(t)] = a\bar{f}_1(s) + b\bar{f}_2(s). \tag{11.54}$$

► *Find the Laplace transforms of the functions (i) $f(t) = 1$, (ii) $f(t) = e^{at}$, (iii) $f(t) = t^n$, for $n = 0, 1, 2, \ldots$.*

(i) By direct application of the definition of a Laplace transform (11.53), we find

$$\mathscr{L}[1] = \int_0^\infty e^{-st}\,dt = \left[\frac{-1}{s}e^{-st}\right]_0^\infty = \frac{1}{s}, \qquad \text{if } s > 0,$$

where the restriction $s > 0$ is required for the integral to exist.

(ii) Again using (11.53) directly, we find

$$\begin{aligned}\bar{f}(s) &= \int_0^\infty e^{at}e^{-st}\,dt = \int_0^\infty e^{(a-s)t}\,dt \\ &= \left[\frac{e^{(a-s)t}}{a-s}\right]_0^\infty = \frac{1}{s-a} \qquad \text{if } s > a.\end{aligned}$$

(iii) Once again using the definition (11.53) we have

$$\bar{f}_n(s) = \int_0^\infty t^n e^{-st}\,dt.$$

Integrating by parts we find

$$\begin{aligned}\bar{f}_n(s) &= \left[\frac{-t^n e^{-st}}{s}\right]_0^\infty + \frac{n}{s}\int_0^\infty t^{n-1}e^{-st}\,dt \\ &= 0 + \frac{n}{s}\bar{f}_{n-1}(s), \qquad \text{if } s > 0.\end{aligned}$$

| $f(t)$ | $\bar{f}(s)$ | $s_0$ |
|---|---|---|
| $c$ | $c/s$ | 0 |
| $ct^n$ | $cn!/s^{n+1}$ | 0 |
| $\sin bt$ | $b/(s^2+b^2)$ | 0 |
| $\cos bt$ | $s/(s^2+b^2)$ | 0 |
| $e^{at}$ | $1/(s-a)$ | $a$ |
| $t^n e^{at}$ | $n!/(s-a)^{n+1}$ | $a$ |
| $\sinh at$ | $a/(s^2-a^2)$ | $\|a\|$ |
| $\cosh at$ | $s/(s^2-a^2)$ | $\|a\|$ |
| $e^{at}\sin bt$ | $a/[(s-a)^2+b^2]$ | $a$ |
| $e^{at}\cos bt$ | $(s-a)/[(s-a)^2+b^2]$ | $a$ |
| $t^{1/2}$ | $\frac{1}{2}(\pi/s^3)^{1/2}$ | 0 |
| $t^{-1/2}$ | $(\pi/s)^{1/2}$ | 0 |
| $\delta(t-t_0)$ | $e^{-st_0}$ | 0 |
| $H(t-t_0)=\begin{cases}1 & \text{for } t\geq t_0\\ 0 & \text{for } t<t_0\end{cases}$ | $e^{-st_0}/s$ | 0 |

Table 11.1 Standard Laplace transforms. The transforms are valid for $s > s_0$.

We now have a recursion relation between successive transforms and by calculating $\bar{f}_0$ we can infer $\bar{f}_1$, $\bar{f}_2$, etc. Since $t^0 = 1$, (i) above gives

$$\bar{f}_0 = \frac{1}{s}, \qquad \text{if } s > 0, \tag{11.55}$$

and

$$\bar{f}_1(s) = \frac{1}{s^2}, \qquad \bar{f}_2(s) = \frac{2!}{s^3}, \qquad \ldots, \qquad \bar{f}_n(s) = \frac{n!}{s^{n+1}} \qquad \text{if } s > 0.$$

Thus, in each case (i)–(iii), direct application of the definition of the Laplace transform (11.53) yields the required result. ◀

Unlike that for the Fourier transform, the inversion of the Laplace transform is not an easy operation to perform, since an explicit formula for $f(t)$, given $\bar{f}(s)$, is not straightforwardly obtained from (11.53). The general method for obtaining an inverse Laplace transform makes use of complex variable theory and is not discussed until chapter 18. However, progress can be made without having to find an *explicit* inverse, since we can prepare from (11.53) a 'dictionary' of the Laplace transforms of common functions, and, when faced with an inversion to carry out, hope to find the given transform (together with its parent function) in the listing. Such a list is given in table 11.1.

When finding inverse Laplace transforms using table 11.1, it is useful to note that for all practical purposes the inverse Laplace transform is unique† and linear so that

$$\mathcal{L}^{-1}\left[a\bar{f}_1(s)+b\bar{f}_2(s)\right] = af_1(t)+bf_2(t). \tag{11.56}$$

In many practical problems the method of partial fractions can be useful in producing an expression from which the inverse Laplace transform can be found.

►*Using table 11.1 find $f(t)$ if*

$$\bar{f}(s) = \frac{s+3}{s(s+1)}.$$

Using partial fractions $\bar{f}(s)$ may be written

$$\bar{f}(s) = \frac{3}{s} - \frac{2}{s+1}.$$

Comparing this with the standard Laplace transforms in table 11.1, we find that the inverse transform of $3/s$ is 3 for $s > 0$ and the inverse transform of $2/(s+1)$ is $2e^{-t}$ for $s > -1$, and so

$$f(t) = 3 - 2e^{-t}, \quad \text{if } s > 0. ◄$$

### *11.2.1 Laplace transforms of derivatives and integrals*

One of the main uses of Laplace transforms is in solving differential equations. Differential equations are the subject of the next six chapters and we will return to the application of Laplace transforms to their solution in chapter 13. In the meantime we derive the required results, i.e. the Laplace transforms of derivatives.

The Laplace transform of the first derivative of $f(t)$ is given by

$$\begin{aligned}\mathcal{L}\left[\frac{df}{dt}\right] &= \int_0^\infty \frac{df}{dt}e^{-st}\,dt \\ &= \left[f(t)e^{-st}\right]_0^\infty + s\int_0^\infty f(t)e^{-st}\,dt \\ &= -f(0)+s\bar{f}(s), \quad \text{for } s>0. \end{aligned} \tag{11.57}$$

The evaluation relies on integration by parts and higher-order derivatives may be found in a similar manner.

† This is not strictly true, since two functions can differ from one another at a finite number of isolated points but have the *same* Laplace transform.

►*Find the Laplace transform of $d^2f/dt^2$.*

Using the definition of the Laplace transform and integrating by parts we obtain

$$\begin{aligned}\mathcal{L}\left[\frac{d^2f}{dt^2}\right] &= \int_0^\infty \frac{d^2f}{dt^2}e^{-st}\,dt \\ &= \left[\frac{df}{dt}e^{-st}\right]_0^\infty + s\int_0^\infty \frac{df}{dt}e^{-st}\,dt \\ &= -\frac{df}{dt}(0) + s[s\bar{f}(s) - f(0)], \qquad \text{for } s > 0,\end{aligned}$$

where (11.57) has been substituted for the integral. This can be written more neatly as

$$\mathcal{L}\left[\frac{d^2f}{dt^2}\right] = s^2\bar{f}(s) - sf(0) - \frac{df}{dt}(0), \qquad \text{for } s > 0. \blacktriangleleft$$

In general the Laplace transform of the $n$th derivative is given by

$$\mathcal{L}\left[\frac{d^nf}{dt^n}\right] = s^n\bar{f} - s^{n-1}f(0) - s^{n-2}\frac{df}{dt}(0) - \cdots - \frac{d^{n-1}f}{dt^{n-1}}(0), \qquad \text{for } s > 0. \tag{11.58}$$

We now turn to integration, which is much more straightforward. From the definition (11.53),

$$\begin{aligned}\mathcal{L}\left[\int_0^t f(u)\,du\right] &= \int_0^\infty dt\, e^{-st}\int_0^t f(u)\,du \\ &= \left[-\frac{1}{s}e^{-st}\int_0^t f(u)\,du\right]_0^\infty + \int_0^\infty \frac{1}{s}e^{-st}f(t)\,dt.\end{aligned}$$

The first term on the RHS vanishes at both limits, and so

$$\mathcal{L}\left[\int_0^t f(u)\,du\right] = \frac{1}{s}\mathcal{L}\left[f\right]. \tag{11.59}$$

### *11.2.2 Other properties of Laplace transforms*

From table 11.1 it will be apparent that multiplying a function $f(t)$ by $e^{at}$ has the effect on its transform that $s$ is replaced by $s-a$. This is easily proved generally:

$$\begin{aligned}\mathcal{L}\left[e^{at}f(t)\right] &= \int_0^\infty f(t)e^{at}e^{-st}\,dt \\ &= \int_0^\infty f(t)e^{-(s-a)t}\,dt \\ &= \bar{f}(s-a). \end{aligned} \tag{11.60}$$

As it were, multiplying $f(t)$ by $e^{at}$ moves the origin of $s$ by an amount $a$.

We may now consider the effect of multiplying the Laplace transform $\bar{f}(s)$ by $e^{-bs}$ $(b > 0)$. From the definition (11.53),

$$\begin{aligned} e^{-bs}\bar{f}(s) &= \int_0^\infty e^{-s(t+b)} f(t)\, dt \\ &= \int_0^\infty e^{-sz} f(z-b)\, dz, \end{aligned}$$

on putting $t + b = z$. Thus $e^{-bs}\bar{f}(s)$ is the Laplace transform of a function $g(t)$ defined by

$$g(t) = \begin{cases} 0 & \text{for } 0 < t \le b, \\ f(t-b) & \text{for } t > b. \end{cases}$$

In other words, the function $f$ has been translated to 'later' $t$ (larger values of $t$) by an amount $b$.

Further properties of Laplace transforms can be proved in similar ways and are listed below.

(i) $$\mathscr{L}\left[f(at)\right] = \frac{1}{a}\bar{f}\left(\frac{s}{a}\right), \tag{11.61}$$

(ii) $$\mathscr{L}\left[t^n f(t)\right] = (-1)^n \frac{d^n \bar{f}(s)}{ds^n}, \quad \text{for } n = 1, 2, 3, \ldots, \tag{11.62}$$

(iii) $$\mathscr{L}\left[\frac{f(t)}{t}\right] = \int_s^\infty \bar{f}(u)\, du, \tag{11.63}$$

provided $\lim_{t\to 0}[f(t)/t]$ exists.

Related results may be easily proved.

►*Find an expression for the Laplace transform of $t\, d^2f/dt^2$.*

From the definition of the Laplace transform we have

$$\begin{aligned} \mathscr{L}\left[t\frac{d^2 f}{dt^2}\right] &= \int_0^\infty e^{-st} t \frac{d^2 f}{dt^2}\, dt \\ &= -\frac{d}{ds}\int_0^\infty e^{-st}\frac{d^2 f}{dt^2}\, dt \\ &= -\frac{d}{ds}[s^2 \bar{f}(s) - sf(0) - f'(0)] \\ &= -s^2 \frac{d\bar{f}}{ds} - 2s\bar{f} + f(0). \quad ◄ \end{aligned}$$

Finally we mention the convolution theorem for Laplace transforms (which is analogous to that for Fourier transforms discussed in subsection 11.1.7). If the functions $f$ and $g$ have Laplace transforms $\bar{f}(s)$ and $\bar{g}(s)$, then

$$\mathcal{L}\left[\int_0^t f(u)g(t-u)\,du\right] = \bar{f}(s)\bar{g}(s), \tag{11.64}$$

where the integral in the brackets on the LHS is the *convolution* of $f$ and $g$, denoted by $f * g$. As in the case of Fourier transforms, the convolution defined above is commutative, i.e. $f * g = g * f$, associative and distributive. From (11.64) we also see that

$$\mathcal{L}^{-1}\left[\bar{f}(s)\bar{g}(s)\right] = \int_0^t f(u)g(t-u)\,du = f * g.$$

►*Prove the convolution theorem (11.64) for Laplace transforms.*

From the definition (11.64),

$$\begin{aligned}\bar{f}(s)\bar{g}(s) &= \int_0^\infty e^{-su} f(u)\,du \int_0^\infty e^{-sv} g(v)\,dv \\ &= \int_0^\infty du \int_0^\infty dv\, e^{-s(u+v)} f(u)g(v).\end{aligned}$$

Now letting $u+v=t$ changes the limits on the integrals with the result that

$$\bar{f}(s)\bar{g}(s) = \int_0^\infty du\, f(u) \int_u^\infty dt\, g(t-u)\, e^{-st}.$$

As shown in figure 11.7($a$) the shaded area of integration may be considered as the sum of vertical strips. However, we may instead integrate over this area by summing over horizontal strips as shown in figure 11.7($b$). Then the integral can be written as

$$\begin{aligned}\bar{f}(s)\bar{g}(s) &= \int_0^t du\, f(u) \int_0^\infty dt\, g(t-u)\, e^{-st} \\ &= \int_0^\infty dt\, e^{-st}\left\{\int_0^t f(u)g(t-u)\,du\right\} \\ &= \mathcal{L}\left[\int_0^t f(u)g(t-u)\,du\right]. \blacktriangleleft\end{aligned}$$

The properties of the Laplace transform derived in this section can sometimes be useful in finding the Laplace transforms of particular functions.

►*Find the Laplace transform of $f(t) = t\sin bt$.*

Although we could calculate the Laplace transform directly, we can use (11.62) to give

$$\bar{f}(s) = (-1)\frac{d}{ds}\mathcal{L}\left[\sin bt\right] = -\frac{d}{ds}\left(\frac{b}{s^2+b^2}\right) = \frac{2bs}{(s^2+b^2)^2}, \qquad \text{for } s>0. \blacktriangleleft$$

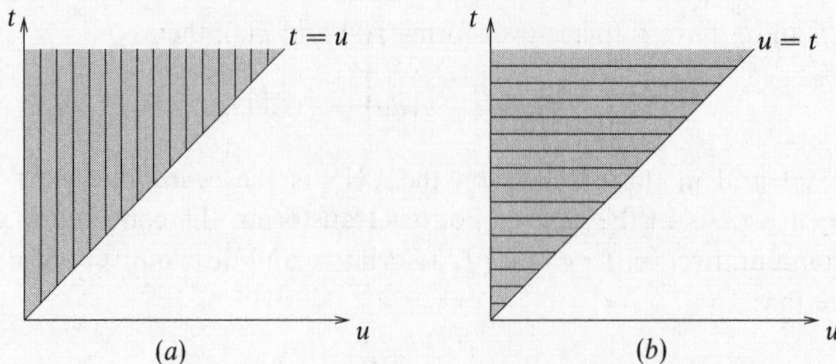


Figure 11.7 Two representations of the Laplace transform convolution (see text).

## 11.3 Concluding remarks

In this chapter we have discussed the Fourier and Laplace transforms in some detail. Both are examples of *integral transforms*, which can be considered in a more general context.

A general integral transform of a function $f(t)$ takes the form

$$F(\alpha) = \int_a^b K(\alpha, t) f(t)\, dt, \tag{11.65}$$

where $F(\alpha)$ is the transform of $f(t)$ with respect to the *kernel* $K(\alpha, t)$, and $\alpha$ is the transform variable. For example, in the Laplace transform case $K(s, t) = e^{-st}$, $a = 0$, $b = \infty$.

Very often the inverse transform can also be written straightforwardly and we obtain a transform pair similar to that encountered in Fourier transforms. Examples of such pairs are

(i) the Hankel transform

$$F(k) = \int_0^\infty f(x) J_n(kx) x\, dx,$$

$$f(x) = \int_0^\infty F(k) J_n(kx) k\, dk,$$

where the $J_n$ are Bessel functions of order $n$, and

(ii) the Mellin transform

$$F(z) = \int_0^\infty t^{z-1} f(t)\, dt,$$

$$f(t) = \frac{1}{2\pi i} \int_{-i\infty}^{i\infty} t^{-z} F(z)\, dz.$$

Although we do not have the space to discuss their general properties, the reader should at least be aware of this wider class of integral transforms.

## 11.4 Exercises

11.1 Find the Fourier transform of the function $f(t) = \exp(-|t|)$.

(a) By applying Fourier's inversion theorem prove that

$$\frac{\pi}{2}\exp(-|t|) = \int_0^\infty \frac{\cos\omega t}{1+\omega^2}\,d\omega.$$

(b) By making the substitution $\omega = \tan\theta$, demonstrate the validity of Parseval's theorem for this function.

11.2 By taking the Fourier transform of the equation

$$\frac{d^2\phi}{dx^2} - K^2\phi = f(x)$$

show that its solution $\phi(x)$ can be written as

$$\phi(x) = \frac{-1}{\sqrt{2\pi}}\int_{-\infty}^{\infty} \frac{e^{ikx}\widetilde{f}(k)}{k^2+K^2}\,dk,$$

where $\widetilde{f}(k)$ is the Fourier transform of $f(x)$.

11.3 Calculate the Fraunhofer spectrum produced by a diffraction grating, uniformly illuminated by light of wavelength $2\pi/k$, as follows. Consider a grating with $4N$ equal strips each of width $a$ and alternately opaque and transparent. The aperture function is then

$$f(y) = \begin{cases} A & \text{for } (2n+1)a \le y \le (2n+2)a, \quad -N \le n < N, \\ 0 & \text{otherwise.} \end{cases}$$

(a) Show, for diffraction at angle $\theta$ to the normal to the grating, that the required Fourier transform can be written

$$\widetilde{f}(q) = (2\pi)^{-1/2}\sum_{r=-N}^{N-1}\exp(-2iarq)\int_a^{2a} A\exp(-iqu)\,du,$$

where $q = k\sin\theta$.

(b) Evaluate the integral and sum to show that

$$\widetilde{f}(q) = (2\pi)^{-1/2}\exp(-iqa/2)\frac{A\sin(2qaN)}{q\cos(qa/2)},$$

and hence that the intensity distribution $I(\theta)$ in the spectrum is proportional to

$$\frac{\sin^2(2qaN)}{q^2\cos^2(qa/2)}.$$

(c) For large values of $N$, the numerator in the above expression has very closely spaced maxima and minima as a function of $\theta$, and effectively takes its mean value of $1/2$, giving a low-intensity background. Much more significant peaks in $I(\theta)$ occur when $\theta = 0$ or the cosine term in the denominator vanishes. Show that the corresponding values of $|\widetilde{f}(q)|$ are

$$\frac{2aNA}{(2\pi)^{1/2}} \quad \text{and} \quad \frac{4aNA}{(2\pi)^{1/2}(2m+1)\pi} \quad \text{with } m \text{ integral.}$$

Note that the constructive interference makes the maxima in $I(\theta) \propto N^2$, not $N$. Of course, observable maxima only occur for $0 \leq \theta \leq \pi/2$.

11.4 In quantum mechanics, two equal-mass particles of momenta $\mathbf{p}_j = \hbar\mathbf{k}_j$ and energies $E_j = \hbar\omega_j$ and represented by plane wavefunctions $\phi_j = \exp[i(\mathbf{k}_j \cdot \mathbf{r}_j - \omega_j t)]$, $j = 1, 2$, interact through a potential $V = V(|\mathbf{r}_1 - \mathbf{r}_2|)$. In first-order perturbation theory the probability of scattering to a state with momenta and energies $\mathbf{p}'_j$, $E'_j$ is determined by the modulus squared of the quantity

$$M = \iiint \psi_f^* V \psi_i \, d\mathbf{r}_1 \, d\mathbf{r}_2 \, dt.$$

The initial state $\psi_i$ is $\phi_1\phi_2$ and the final state $\psi_f$ is $\phi'_1\phi'_2$.

(a) By writing $\mathbf{r}_1 + \mathbf{r}_2 = 2\mathbf{R}$ and $\mathbf{r}_1 - \mathbf{r}_2 = \mathbf{r}$ and assuming that $d\mathbf{r}_1 \, d\mathbf{r}_2 = d\mathbf{R} \, d\mathbf{r}$, show that $M$ can be written as the product of three one-dimensional integrals.

(b) From two of the integrals deduce energy and momentum conservation in the form of $\delta$-functions.

(c) Show that $M$ is proportional to the Fourier transform of $V$, i.e. $\widetilde{V}(\mathbf{k})$ where $2\hbar\mathbf{k} = (\mathbf{p}_2 - \mathbf{p}_1) - (\mathbf{p}'_2 - \mathbf{p}'_1)$.

11.5 For some ion–atom scattering processes, the potential $V$ of the previous example may be approximated by $V = |\mathbf{r}_1 - \mathbf{r}_2|^{-1} \exp(-\mu|\mathbf{r}_1 - \mathbf{r}_2|)$. Show, using the result of the worked example in subsection 11.1.10, that the probability of the ion scattering from, say, $\mathbf{p}_1$ to $\mathbf{p}'_1$ is proportional to $(\mu^2 + k^2)^{-2}$ where $k = |\mathbf{k}|$ and $\mathbf{k}$ is as given in part (c) of exercise 11.4.

11.6 Prove the expressions given in table 11.1 for the Laplace transforms of $t^{-1/2}$ and $t^{1/2}$ by setting $x^2 = ts$ in the result

$$\int_0^\infty \exp(-x^2) \, dx = \tfrac{1}{2}\sqrt{\pi}.$$

11.7 Use the properties of Laplace transforms to prove the following without evaluating any Laplace integrals explicitly:

(a) $\mathcal{L}\left[t^{5/2}\right] = \frac{15}{8}\sqrt{\pi}s^{-7/2}$.

(b) $\mathcal{L}\left[(\sinh at)/t\right] = \frac{1}{2}\ln\left[(s+a)/(s-a)\right], \qquad s > |a|$.

(c) $\mathcal{L}\left[\sinh at \cos bt\right] = a(s^2 - a^2 + b^2)[(s-a)^2 + b^2]^{-1}[(s+a)^2 + b^2]^{-1}$.

11.8 By writing $f(x)$ as an integral involving the $\delta$-function $\delta(\xi - x)$ and taking the Laplace transforms of both sides, show that the transform of the solution of the equation

$$\frac{d^4y}{dx^4} - y = f(x)$$

for which $y$ and its first three derivatives vanish at $x = 0$ can be written as

$$\bar{y}(s) = \int_0^\infty f(\xi)\frac{e^{-s\xi}}{s^4 - 1}\,d\xi.$$

Use the properties of Laplace transforms and the entries in table 11.1 to show that

$$y(x) = \frac{1}{2}\int_0^x f(\xi)\left[\sinh(x-\xi) - \sin(x-\xi)\right]\,d\xi.$$

11.9 The function $f_a(x)$ is defined as unity for $0 < x < a$ and zero otherwise. Find its Laplace transform $\bar{f}_a(s)$ and deduce that the transform of $xf_a(x)$ is

$$\frac{1}{s^2}\left[1 - (1+as)e^{-sa}\right].$$

Write $f_a(x)$ in terms of Heaviside functions and hence obtain an explicit expression for

$$g_a(x) = \int_0^x f_a(y)f_a(x-y)\,dy.$$

Use the expression to write $\bar{g}_a(s)$ in terms of $\bar{f}_a(s)$, $\bar{f}_{2a}(s)$ and their derivatives, and hence show that $\bar{g}_a(s)$ is equal to the square of $\bar{f}_a(s)$, in accordance with the convolution theorem.

## 11.5 Hints and answers

11.1 $(2/\pi)^{1/2}(1+\omega^2)^{-1}$.

11.3 (c) Use l'Hôpital's rule to evaluate the expressions of the form 0/0.

11.4 (b) The $t$-integral is $\int \exp[i(E_1'+E_2'-E_1-E_2)]\,dt \propto \delta(E_1'+E_2'-(E_1+E_2))$; similarly the $\mathbf{R}$-integral yields $\delta(p_1' + p_2' - (p_1 + p_2))$.

11.5 $\widetilde{V}(\mathbf{k}) \propto (-2\pi/ik)\int\{\exp[-(\mu - ik)r] - \exp[-(\mu + ik)r]\}\,dr$.

11.6 Prove the result for $t^{1/2}$ by integrating that for $t^{-1/2}$ by parts.

11.7 (a) Use (11.62) with $n = 2$ on $\mathscr{L}\left[t^{1/2}\right]$; (b) use (11.63); (c) consider $\mathscr{L}\left[\exp(\pm at)\cos bt\right]$ and use the translation property, subsection 11.2.2.

11.8 Factorise $(s^4-1)^{-1}$ as $[(s^2-1)^{-1}-(s^2+1)^{-1}]/2$.

11.9 $s^{-1}[1-\exp(-sa)]$; $g_a(x) = x$ for $0 < x < a$, $g_a(x) = 2a - x$ for $a \le x \le 2a$, $g_a(x) = 0$ otherwise.

# 12

# First-order ordinary differential equations

Differential equations are the group of equations that contain derivatives. Chapters 12–17 discuss a variety of differential equations, starting in this chapter and the next with those ordinary differential equations (ODEs) that have closed-form solutions. As its name suggests, an ODE contains only ordinary derivatives (and not partial derivatives), and describes the relationship between these derivatives of the *dependent variable*, usually called $y$, with respect to the *independent variable*, usually called $x$. The solution to such an ODE is therefore a function of $x$, and is written $y(x)$. For an ODE to have a closed-form solution, it must be possible to express $y(x)$ in terms of the standard elementary functions such as $\exp x$, $\ln x$, $\sin x$ etc. The solutions of some differential equations may not, however, be written in closed form, but only as an infinite series; these are discussed in chapter 14.

Ordinary differential equations may conveniently be separated into different categories according to their general characteristics. The primary grouping adopted here is by the *order* of the equation. The order of an ODE is simply the order of the highest derivative it contains. Thus equations containing $dy/dx$, but no higher derivatives, are called first order, those containing $d^2y/dx^2$ are called second order and so on. In this chapter we consider first-order equations, and in the next, second- and higher-order equations.

Ordinary differential equations may be further classified according to *degree*. The degree of an ODE is the power to which the highest-order derivative is raised, after the equation has been rationalised to contain only integer powers of derivatives. Hence the ODE

$$\frac{d^3y}{dx^3} + x\left(\frac{dy}{dx}\right)^{3/2} + x^2y = 0,$$

is of third order and second degree, since after rationalisation it contains the term $(d^3y/dx^3)^2$.

The *general solution* to an ODE is the most general function $y(x)$ that satisfies the equation; it will contain *constants of integration* that may be determined by the application of some suitable *boundary conditions*. For example, we may be told that for a certain first-order differential equation, the solution $y(x)$ is equal to zero when the parameter $x$ is equal to unity. This will then allow us to determine the value of the constant of integration. The *general solutions* to $n$th-order ODEs, which are considered in detail in the next chapter, will contain $n$ (essential) arbitrary constants of integration and we therefore need $n$ boundary conditions if these constants are to be determined (see section 12.1). When the boundary conditions have been applied, and the constants found, we are left with a *particular solution* to the ODE, which obeys the given boundary conditions. Some ODEs of degree greater than one also possess *singular solutions*, which are solutions that contain no arbitrary constants and cannot be found from the general solution; singular solutions are discussed in more detail in section 12.3. When any solution to an ODE has been found, it is always possible to check its validity by substitution into the original equation, checking that any given boundary conditions are also met.

In this chapter we first discuss various types of first-degree ODE, and then go on to examine those higher-degree equations that can be solved in closed form. First, however, we discuss the general form of the solutions of ODEs; this discussion is relevant to both first- and higher-order ODEs.

## 12.1 General form of solution

It is helpful when considering the general form of the solution of an ODE to consider the inverse process, namely that of obtaining an ODE from a given group of functions each one of which is a solution of the ODE. Suppose the members of the group can be written as

$$y = f(x, a_1, a_2, \ldots, a_n), \tag{12.1}$$

each member being given by a different set of values of the parameters $a_i$. For example, consider the group of functions

$$y = a_1 \sin x + a_2 \cos x; \tag{12.2}$$

here $n = 2$.

Since an ODE is required for which *any* of the group is a solution, it clearly must not contain any of the $a_i$. As there are $n$ of the $a_i$ in expression (12.1), we must obtain $n+1$ equations involving them, in order that, by elimination, we can obtain one final equation without them.

Initially we have only (12.1), but if this is differentiated $n$ times, a total of $n+1$ equations is obtained from which (in principle) all the $a_i$ can be eliminated to give one ODE satisfied by all the group. As a result of the $n$ differentiations, $d^n y/dx^n$ will be present in one of the $n+1$ equations and hence in the final equation, which will therefore be of $n$th order.

In the case of (12.2), we have

$$\frac{dy}{dx} = a_1 \cos x - a_2 \sin x,$$
$$\frac{d^2y}{dx^2} = -a_1 \sin x - a_2 \cos x.$$

Here the elimination of $a_1$ and $a_2$ is trivial (because of the similarity of the forms of $y$ and $d^2y/dx^2$), resulting in

$$\frac{d^2y}{dx^2} + y = 0,$$

a second-order equation.

Thus, to summarise, a group of functions (12.1) with $n$ parameters satisfies an $n$th-order ODE in general (although in some degenerate cases an ODE of less than $n$th order is obtained). The intuitive converse of this, that the general solution of an $n$th-order ODE contains $n$ arbitrary parameters (constants), will, for our purposes, be assumed to be valid, although a totally general proof is difficult.

As mentioned earlier, external factors affect a system described by an ODE, by fixing the values of the dependent variables for particular values of the independent ones. These externally imposed (or *boundary*) conditions on the solution are thus the means of determining the parameters, and so of specifying precisely which function is the required solution. It is apparent that the number of boundary conditions should match the number of parameters and hence the order of the equation, if a unique solution is to be obtained. Fewer independent boundary conditions than this will lead to a number of undetermined parameters in the solution, whilst an excess will usually mean that no acceptable solution is possible.

For an $n$th-order equation the required $n$ boundary conditions can take many forms, for example the value of $y$ at $n$ different values of $x$, or the value of any $n-1$ of the $n$ derivatives $dy/dx$, $d^2y/dx^2$, ..., $d^ny/dx^n$ together with that of $y$, all for the same value of $x$, or many intermediate combinations.

## 12.2 First-degree first-order equations

First-degree first-order ODEs contain only $dy/dx$ equated to some function of $x$ and $y$, and can be written in either of two equivalent standard forms,

$$\frac{dy}{dx} = F(x, y),$$

or

$$A(x, y)\,dx + B(x, y)\,dy = 0,$$

where $F(x, y) = -A(x, y)/B(x, y)$, and $F(x, y)$, $A(x, y)$ and $B(x, y)$ are in general functions of both $x$ and $y$. Which of the two above forms is the more useful for finding a solution depends on the type of equation being considered. There are several different types of first-degree first-order ODEs that are of interest in the physical sciences. These equations and their respective solutions are discussed below.

### *12.2.1 Separable-variable equations*

A separable-variable equation is one which may be written in the conventional form

$$\frac{dy}{dx} = f(x)g(y), \tag{12.3}$$

where $f(x)$ and $g(y)$ are functions of $x$ and $y$ respectively (including cases in which $f(x)$ or $g(y)$ is simply a constant). Rearranging this equation so that the terms depending on $x$ and on $y$ appear on opposite sides (i.e. are separated), and integrating, we obtain

$$\int \frac{dy}{g(y)} = \int f(x)\, dx.$$

Finding the solution $y(x)$ that satisfies (12.3) then depends only on the ease with which the integrals in the above equation can be evaluated. It is also worth noting that ODEs that at first sight do not appear to be of the form (12.3) can sometimes be made separable by an appropriate factorisation.

►*Solve*

$$\frac{dy}{dx} = x + xy.$$

Since the RHS of this equation can be factorised to give $x(1 + y)$, the equation becomes separable and we obtain

$$\int \frac{dy}{1+y} = \int x\, dx.$$

Now integrating both sides separately, we find

$$\ln(1 + y) = \frac{x^2}{2} + c,$$

and so

$$1 + y = \exp\left(\frac{x^2}{2} + c\right) = A \exp\left(\frac{x^2}{2}\right),$$

where $c$ and hence $A$ is an arbitrary constant. ◄

**Solution method.** *Factorise the equation so that it becomes separable. After rearranging it so that the terms depending on x and those depending on y appear on opposite sides, integrate directly. Remember the constant of integration, which can be evaluated if further information is given.*

### 12.2.2 Exact equations

An *exact* first-degree first-order ODE is one of the form

$$A(x, y)\,dx + B(x, y)\,dy = 0 \quad \text{and for which} \quad \frac{\partial A}{\partial y} = \frac{\partial B}{\partial x}. \tag{12.4}$$

In this case $A(x, y)\,dx + B(x, y)\,dy$ is an exact differential, $dU(x, y)$ say (see section 4.3). In other words

$$A\,dx + B\,dy = dU = \frac{\partial U}{\partial x}\,dx + \frac{\partial U}{\partial y}\,dy,$$

from which we obtain

$$A(x, y) = \frac{\partial U}{\partial x}, \tag{12.5}$$

$$B(x, y) = \frac{\partial U}{\partial y}. \tag{12.6}$$

Since $\partial^2 U/\partial x \partial y = \partial^2 U/\partial y \partial x$ we therefore require

$$\frac{\partial A}{\partial y} = \frac{\partial B}{\partial x}. \tag{12.7}$$

If (12.7) holds then (12.4) can be written $dU(x, y) = 0$, which has the solution $U(x, y) = c$, where $c$ is a constant and from (12.5) $U(x, y)$ is given by

$$U(x, y) = \int A(x, y)\,dx + F(y). \tag{12.8}$$

The function $F(y)$ can be found from (12.6) by differentiating (12.8) with respect to $y$ and equating to $B(x, y)$.

►*Solve*

$$x\frac{dy}{dx} + 3x + y = 0.$$

Rearranging into the form (12.4) we have

$$(3x + y)\,dx + x\,dy = 0,$$

i.e. $A(x, y) = 3x + y$ and $B(x, y) = x$. Since $\partial A/\partial y = 1 = \partial B/\partial x$, the equation is exact, and by (12.8) the solution is given by

$$U(x, y) = \int (3x + y)\,dx + F(y) = c_1 \quad \Rightarrow \quad \frac{3x^2}{2} + yx + F(y) = c_1.$$

Differentiating $U(x, y)$ with respect to $y$ and equating it to $B(x, y) = x$ we obtain $dF/dy = 0$, which integrates immediately to give $F(y) = c_2$. Therefore, letting $c = c_1 - c_2$, the solution to the original ODE is

$$\frac{3x^2}{2} + xy = c. \blacktriangleleft$$

**Solution method.** *Check that the equation is an exact differential using (12.7) then solve using (12.8). Find the function $F(y)$ by differentiating (12.8) with respect to $y$ and using (12.6).*

### 12.2.3 Inexact equations: integrating factors

Equations that may be written in the form

$$A(x, y)\,dx + B(x, y)\,dy = 0 \quad \text{but for which} \quad \frac{\partial A}{\partial y} \neq \frac{\partial B}{\partial x} \tag{12.9}$$

are known as inexact equations. However, the differential $A\,dx + B\,dy$ can always be made exact by multiplying by an integrating factor $\mu(x, y)$ that obeys

$$\frac{\partial(\mu A)}{\partial y} = \frac{\partial(\mu B)}{\partial x}. \tag{12.10}$$

For an integrating factor that is a function of both $x$ and $y$, i.e. $\mu = \mu(x, y)$, there exists no general method for finding it; in such cases it may sometimes be found by inspection. If, however, an integrating factor exists that is a function of either $x$ or $y$ alone, then (12.10) can be solved to find it. For example, if we assume that the integrating factor is a function of $x$ alone, i.e. $\mu = \mu(x)$, then (12.10) reads

$$\mu\frac{\partial A}{\partial y} = \mu\frac{\partial B}{\partial x} + B\frac{d\mu}{dx}.$$

Rearranging this expression we find

$$\frac{d\mu}{\mu} = \frac{1}{B}\left(\frac{\partial A}{\partial y} - \frac{\partial B}{\partial x}\right) dx = f(x)\,dx,$$

where we require $f(x)$ also to be a function of $x$ only; indeed this provides a general method of determining whether the integrating factor $\mu$ is a function of $x$ alone. This integrating factor is then given by

$$\mu(x) = \exp\left\{\int f(x)\,dx\right\} \quad \text{where} \quad f(x) = \frac{1}{B}\left(\frac{\partial A}{\partial y} - \frac{\partial B}{\partial x}\right). \tag{12.11}$$

Similarly, if $\mu = \mu(y)$ then

$$\mu(y) = \exp\left\{\int g(y)\,dy\right\} \quad \text{where} \quad g(y) = \frac{1}{A}\left(\frac{\partial B}{\partial x} - \frac{\partial A}{\partial y}\right). \tag{12.12}$$

►*Solve*

$$\frac{dy}{dx} = -\frac{2}{y} - \frac{3y}{2x}.$$

Rearranging into the form (12.9), we have

$$(4x + 3y^2)\,dx + 2xy\,dy = 0, \qquad (12.13)$$

i.e. $A(x, y) = 4x + 3y^2$ and $B(x, y) = 2xy$. Now it is easy to show that

$$\frac{\partial A}{\partial y} = 6y, \qquad \frac{\partial B}{\partial x} = 2y,$$

so the ODE is not exact in its present form. However, we see that

$$\frac{1}{B}\left(\frac{\partial A}{\partial y} - \frac{\partial B}{\partial x}\right) = \frac{2}{x},$$

a function of $x$ alone. Therefore an integrating factor exists that is also a function of $x$ alone and, ignoring the arbitrary constant of integration, is given by

$$\mu(x) = \exp\left\{2\int \frac{dx}{x}\right\} = \exp(2\ln x) = x^2.$$

Multiplying (12.13) through by $\mu(x) = x^2$ we obtain

$$(4x^3 + 3x^2y^2)\,dx + 2x^3y\,dy = 4x^3\,dx + (3x^2y^2\,dx + 2x^3y\,dy) = 0.$$

By inspection this integrates immediately to give the solution $x^4 + y^2x^3 = c$, where $c$ is a constant. ◄

**Solution method.** *Examine whether $f(x)$ and $g(y)$ are functions of only $x$ or $y$ respectively. If so, then the required integrating factor is a function of either $x$ or $y$ only, and is given by (12.11) or (12.12) respectively. If the integrating factor is a function of both $x$ and $y$, then it may sometimes be found by inspection or by trial and error. In any case, the integrating factor $\mu$ must satisfy (12.10). Once the equation has been made exact, solve by the method of subsection 12.2.2.*

### *12.2.4 Linear equations*

Linear first-order ODEs are a special case of inexact ODEs (discussed in the previous subsection) and can be written in the conventional form

$$\frac{dy}{dx} + P(x)y = Q(x). \qquad (12.14)$$

Such equations can be made exact by multiplying through by an appropriate integrating factor in a similar manner to that discussed above. In this case, however, the integrating factor is always a function of $x$ alone, and may be expressed in a particularly simple form. An integrating factor $\mu(x)$ must be such that

$$\mu(x)\frac{dy}{dx} + \mu(x)P(x)y = \frac{d}{dx}\left[\mu(x)y\right] = \mu(x)Q(x), \tag{12.15}$$

which may then be integrated directly to give

$$\mu(x)y = \int \mu(x)Q(x)\,dx. \tag{12.16}$$

The required integrating factor $\mu(x)$ is determined by the first equality in (12.15), i.e.

$$\frac{d}{dx}(\mu y) = \mu\frac{dy}{dx} + \frac{d\mu}{dx}y = \mu\frac{dy}{dx} + \mu P y,$$

which immediately gives the simple relation

$$\frac{d\mu}{dx} = \mu(x)P(x) \qquad \Rightarrow \qquad \mu(x) = \exp\left\{\int P(x)\,dx\right\}. \tag{12.17}$$

►*Solve*

$$\frac{dy}{dx} + 2xy = 4x.$$

The integrating factor is given immediately by

$$\mu(x) = \exp\left\{\int 2x\,dx\right\} = \exp x^2.$$

Multiplying through the ODE by $\mu(x) = \exp x^2$, and integrating, we have

$$y\exp x^2 = 4\int x\exp x^2\,dx = 2\exp x^2 + c.$$

The solution to the ODE is therefore given by $y = 2 + c\exp(-x^2)$. ◄

**Solution method.** *Rearrange the equation into the form (12.14), and multiply by the integrating factor $\mu(x)$ given by (12.17). The left- and right-hand sides can then be integrated directly, giving $y$ from (12.16).*

### *12.2.5 Homogeneous equations*

Homogeneous equation are ODEs that may be written in the form

$$\frac{dy}{dx} = \frac{A(x,y)}{B(x,y)} = F\left(\frac{y}{x}\right), \tag{12.18}$$

where $A(x,y)$ and $B(x,y)$ are homogeneous functions of the same degree. A function $f(x,y)$ is homogeneous of degree $n$ if, for any $\lambda$, it obeys

$$f(\lambda x, \lambda y) = \lambda^n f(x,y).$$

For example, if $A = x^2y - xy^2$ and $B = x^3 + y^3$ then we see that $A$ and $B$ are both homogeneous functions of degree 3. In general, for functions of the form of $A$ and $B$, we see that for both to be homogeneous, and of the same degree, we require the sum of the powers in $x$ and $y$ in each term of $A$ and $B$ to be the same (in this example equal to 3). The RHS of a homogeneous ODE can be written as a function of $y/x$. The equation may then be solved by making the substitution $y = vx$, so that

$$\frac{dy}{dx} = v + x\frac{dv}{dx} = F(v).$$

This is now a separable equation and can be integrated directly to give

$$\int \frac{dv}{F(v) - v} = \int \frac{dx}{x}. \tag{12.19}$$

►*Solve*

$$\frac{dy}{dx} = \frac{y}{x} + \tan\left(\frac{y}{x}\right).$$

Substituting $y = vx$ we obtain

$$v + x\frac{dv}{dx} = v + \tan v.$$

Cancelling $v$ on both sides, rearranging and integrating gives

$$\int \cot v \, dv = \int \frac{dx}{x} = \ln x + c_1.$$

But

$$\int \cot v \, dv = \int \frac{\cos v}{\sin v} \, dv = \ln(\sin v) + c_2,$$

so the solution to the ODE is $y = x \sin^{-1} Ax$, where $A$ is a constant. ◄

**Solution method.** *Check to see whether the equation is homogeneous. If so, make the substitution $y = vx$, then separate variables as in (12.19) and integrate directly. Finally replace $v$ by $y/x$ to obtain the solution.*

### *12.2.6 Isobaric equations*

An isobaric ODE is a generalisation of the homogeneous ODE discussed in the previous section, and is of the form

$$\frac{dy}{dx} = \frac{A(x, y)}{B(x, y)}, \tag{12.20}$$

where the equation is dimensionally consistent if $y$ and $dy$ are each given a weight $m$ relative to $x$ and $dx$, i.e. if the substitution $y = vx^m$ makes it separable.

►*Solve*

$$\frac{dy}{dx} = \frac{-1}{2yx}\left(y^2 + \frac{2}{x}\right).$$

Rearranging we have

$$\left(y^2 + \frac{2}{x}\right) dx + 2yx\,dy = 0,$$

Giving $y$ and $dy$ the weight $m$ and $x$ and $dx$ the weight 1, the sums of the powers in each term on the LHS are $2m+1$, 0 and $2m+1$ respectively. These are equal if $2m+1=0$, i.e. if $m = -\frac{1}{2}$. Substituting $y = vx^m = vx^{-1/2}$, with the result that $dy = x^{-1/2}\,dv - \frac{1}{2}vx^{-3/2}\,dx$, we obtain

$$v\,dv + \frac{dx}{x} = 0,$$

which is separable and may be integrated directly to give $\frac{1}{2}v^2 + \ln x = c$. Replacing $v$ by $y\sqrt{x}$ we obtain the solution $\frac{1}{2}y^2x + \ln x = c$. ◄

**Solution method.** *Write the equation in the form $A\,dx + B\,dy = 0$. Giving $y$ and $dy$ each a weight $m$ and $x$ and $dx$ a weight 1, write down the sum of powers in each term. Then, if a value of $m$ that makes all these sums equal can be found, substitute $y = vx^m$ into the original equation to make it separable. Integrate the separated equation directly, and then replace $v$ by $yx^{-m}$ to obtain the solution.*

### *12.2.7 Bernoulli's equation*

Bernoulli's equation has the form

$$\frac{dy}{dx} + P(x)y = Q(x)y^n \qquad \text{where } n \neq 0 \text{ or } 1. \tag{12.21}$$

This equation is very similar in form to the linear equation (12.14), but is in fact non-linear due to the extra $y^n$ factor on the RHS. The equation can, however, be made linear by substituting $v = y^{1-n}$, so that

$$\frac{dy}{dx} = \left(\frac{y^n}{1-n}\right)\frac{dv}{dx}.$$

Substituting this into (12.21) and dividing through by $y^n$, we find

$$\frac{dv}{dx} + (1-n)P(x)v = (1-n)Q(x),$$

which is a linear equation, and may be solved by the method described in subsection 12.2.4.

►*Solve*

$$\frac{dy}{dx} + \frac{y}{x} = 2x^3y^4.$$

If we let $v = y^{1-4} = y^{-3}$, then

$$\frac{dy}{dx} = -\frac{y^4}{3}\frac{dv}{dx}.$$

Substituting this into the ODE and rearranging, we obtain

$$\frac{dv}{dx} - \frac{3v}{x} = -6x^3,$$

which is linear and may be solved by multiplying through by the integrating factor (see subsection 12.2.4)

$$\exp\left\{-3\int\frac{dx}{x}\right\} = \exp(-3\ln x) = \frac{1}{x^3}.$$

This yields the solution

$$\frac{v}{x^3} = -6x + c.$$

Remembering that $v = y^{-3}$, we obtain $y^{-3} = -6x^4 + cx^3$. ◄

**Solution method.** *Rearrange the equation into the form (12.21), and make the substitution $v = y^{1-n}$. This leads to a linear equation in $v$, which can be solved by the method of subsection 12.2.4. Then replace $v$ by $y^{1-n}$ to obtain the solution.*

### *12.2.8 Miscellaneous equations*

There are two further types of first-degree first-order equation that occur fairly regularly but do not fall into any of the above categories. They may, however, be reduced to one of the above equations by a suitable change of variable. We first consider

$$\frac{dy}{dx} = F(ax + by + c), \tag{12.22}$$

where $a$, $b$ and $c$ are constants, i.e. $x$ and $y$ *only* appear on the RHS in the particular combination $ax+by+c$ and not in any other combination or by themselves. This equation can be solved by making the substitution $v = ax + by + c$, in which case

$$\frac{dv}{dx} = a + b\frac{dy}{dx} = a + bF(v), \tag{12.23}$$

which is separable and may be integrated directly.

►*Solve*

$$\frac{dy}{dx} = (x + y + 1)^2.$$

Making the substitution $v = x + y + 1$, from (12.23), we obtain

$$\frac{dv}{dx} = v^2 + 1,$$

which is separable and integrates directly to give

$$\int \frac{dv}{1 + v^2} = \int dx \quad \Rightarrow \quad \tan^{-1} v = x + c_1.$$

So the solution to the original ODE is $\tan^{-1}(x + y + 1) = x + c_1$, where $c_1$ is a constant of integration. ◄

**Solution method.** *Substitute $v = ax + by + c$ to obtain a separable equation that can be integrated directly. Then replace $v$ by $ax + by + c$ to obtain the solution.*

Secondly, we discuss

$$\frac{dy}{dx} = \frac{ax + by + c}{ex + fy + g}, \tag{12.24}$$

where $a$, $b$, $c$, $e$, $f$ and $g$ are all constants. This equation may be solved by letting $x = X + \alpha$ and $y = Y + \beta$, where $\alpha$ and $\beta$ are constants found from

$$a\alpha + b\beta + c = 0 \tag{12.25}$$

$$e\alpha + f\beta + g = 0. \tag{12.26}$$

Then (12.24) can be written as

$$\frac{dY}{dX} = \frac{aX + bY}{eX + fY},$$

which is homogeneous and can be solved by the method of subsection 12.2.5. Note, however, that if $a/e = b/f$ then (12.25) and (12.26) are not independent, and so cannot be solved uniquely for $\alpha$ and $\beta$. However, in this case, (12.24) reduces to an equation of the form (12.22), which was discussed above.

►*Solve*

$$\frac{dy}{dx} = \frac{2x - 5y + 3}{2x + 4y - 6}.$$

Let $x = X + \alpha$ and $y = Y + \beta$, where $\alpha$ and $\beta$ obey the relations

$$2\alpha - 5\beta + 3 = 0$$

$$2\alpha + 4\beta - 6 = 0,$$

which solve to give $\alpha = \beta = 1$. Making these substitutions we find

$$\frac{dY}{dX} = \frac{2X - 5Y}{2X + 4Y},$$

which is a homogeneous ODE and can be solved by substituting $Y = vX$ (see subsection 12.2.5) to obtain

$$\frac{dv}{dX} = \frac{2 - 7v - 4v^2}{X(2 + 4v)}.$$

This equation is separable, and using partial fractions we find

$$\int \frac{2 + 4v}{2 - 7v - 4v^2}\, dv = -\frac{4}{3}\int \frac{dv}{4v - 1} - \frac{2}{3}\int \frac{dv}{v + 2} = \int \frac{dX}{X},$$

which integrates to give

$$\ln X + \tfrac{1}{3}\ln(4v - 1) + \tfrac{2}{3}\ln(v + 2) = c_1,$$

or

$$X^3(4v - 1)(v + 2)^2 = \exp(3c_1).$$

Remembering that $Y = vX$, $x = X + 1$ and $y = Y + 1$, the solution to the original ODE is given by $(4y - x - 3)(y + 2x - 3)^2 = c_2$, where $c_2 = \exp(3c_1)$. ◄

**Solution method.** *If $a/e \neq b/f$ then make the substitution $x = X + \alpha$, $y = Y + \beta$, where $\alpha$ and $\beta$ are given by (12.25, 12.26); the resulting equation is homogeneous and can be solved as in subsection 12.2.5. Substitute $v = Y/X$, $X = x - \alpha$, and $Y = y - \beta$ to obtain the solution. If $a/e = b/f$ then (12.24) is of the same form as (12.22), and may be solved accordingly.*

## 12.3 Higher-degree first-order equations

First-order equations of degree higher than the first do not occur often in the description of physical systems, since squared and higher powers of first-order derivatives usually arise from resistive or driving mechanisms, when an acceleration or other higher-order derivative is also present. They do, however, sometimes appear in connection with geometrical problems.

Higher-degree first-order equations can be written as $F(x, y, dy/dx) = 0$. The most general standard form is

$$p^n + a_{n-1}(x, y)p^{n-1} + \cdots + a_1(x, y)p + a_0(x, y) = 0, \qquad (12.27)$$

where for ease of notation we write $p = dy/dx$. If the equation can be solved for one of $x$, $y$ or $p$, then either an explicit or a parametric solution can sometimes be obtained. We discuss the main types of such equations below, including Clairaut's equation, which is a special case of an equation explicitly soluble for $y$.

### 12.3.1 *Equations soluble for* $p$

Sometimes the LHS of (12.27) can be factorised into the form

$$(p - F_1)(p - F_2) \cdots (p - F_n) = 0, \tag{12.28}$$

where $F_i = F_i(x, y)$. We are then left with solving the $n$ first-degree equations $p = F_i(x, y)$. Writing the solutions to these first-degree equations as $G_i(x, y) = 0$, the general solution to (12.28) is given by the product

$$G_1(x, y)G_2(x, y) \cdots G_n(x, y) = 0. \tag{12.29}$$

►*Solve*

$$(x^3 + x^2 + x + 1)p^2 - (3x^2 + 2x + 1)yp + 2xy^2 = 0. \tag{12.30}$$

This equation may be factorised to give

$$[(x + 1)p - y][(x^2 + 1)p - 2xy] = 0.$$

Taking each bracket in turn we have

$$(x + 1)\frac{dy}{dx} - y = 0,$$
$$(x^2 + 1)\frac{dy}{dx} - 2xy = 0,$$

which have the solutions $y - c(x + 1) = 0$ and $y - c(x^2 + 1) = 0$ respectively (see section 12.2 on first-degree first-order equations). Note that the arbitrary constants in these two solutions can be taken to be the same, since only one is required for a first-order equation. The general solution to (12.30) is then given by

$$[y - c(x + 1)]\,[y - c(x^2 + 1)] = 0. \blacktriangleleft$$

**Solution method.** *If the equation can be factorised into the form (12.28), then solve the first-order ODE* $p - F_i = 0$ *in each factor, and write the solution in the form* $G_i(x, y) = 0$*. The solution to the original equation is then given by the product (12.29).*

### 12.3.2 *Equations soluble for* $x$

Equations that can be solved for $x$, i.e. such that they may be written in the form

$$x = F(y, p), \tag{12.31}$$

can be reduced to first-degree first-order equations in $p$ by differentiating both sides with respect to $y$, so that

$$\frac{dx}{dy} = \frac{1}{p} = \frac{\partial F}{\partial y} + \frac{\partial F}{\partial p}\frac{dp}{dy}.$$

This results in an equation of the form $G(y, p) = 0$, which can be used together with (12.31) to eliminate $p$ and give the general solution. Note that a singular solution to the equation will often be found at the same time.

►*Solve*

$$6y^2p^2 + 3xp - y = 0. \tag{12.32}$$

This equation can be solved for $x$ explicitly to give $3x = (y/p) - 6y^2p$. Differentiating both sides with respect to $y$, we find

$$3\frac{dx}{dy} = \frac{3}{p} = \frac{1}{p} - \frac{y}{p^2}\frac{dp}{dy} - 6y^2\frac{dp}{dy} - 12yp,$$

which factorises to give

$$\left(1 + 6yp^2\right)\left(2p + y\frac{dp}{dy}\right) = 0. \tag{12.33}$$

Setting the factor containing $dp/dy$ equal to zero gives a first-degree first-order equation in $p$, which may be solved to give $py^2 = c$. Substituting for $p$ in (12.32) then yields the general solution of (12.32):

$$y^3 = 3cx + 6c^2. \tag{12.34}$$

If we now consider the first factor in (12.33), we find $6p^2y = -1$ as a possible solution. Substituting for $p$ in (12.32) we find the singular solution

$$8y^3 + 3x^2 = 0.$$

Note that the singular solution contains no arbitrary constants and cannot be found from the general solution (12.34) by any choice of the constant $c$. ◄

**Solution method.** *Write the equation in the form (12.31) and differentiate both sides with respect to y. Rearrange the resulting equation into the form $G(y, p) = 0$, which can be used together with the original ODE to eliminate p and so give the general solution. If $G(y, p)$ can be factorised, then the factor containing $dp/dy$ should be used to eliminate p and give the general solution. Using the other factors in this fashion will instead lead to singular solutions.*

### 12.3.3 Equations soluble for $y$

Equations that can be solved for $y$, i.e. such that they may be written in the form

$$y = F(x, p), \tag{12.35}$$

can be reduced to first-degree first-order equations in $p$ by differentiating both sides with respect to $x$, so that

$$\frac{dy}{dx} = p = \frac{\partial F}{\partial x} + \frac{\partial F}{\partial p}\frac{dp}{dx}.$$

This results in an equation of the form $G(x, p) = 0$, which can be used together with (12.35) to eliminate $p$ and give the general solution. An additional (singular) solution to the equation is also often found.

►*Solve*

$$xp^2 + 2xp - y = 0. \tag{12.36}$$

This equation can be solved for $y$ explicitly to give $y = xp^2 + 2xp$. Differentiating both sides with respect to $x$, we find

$$\frac{dy}{dx} = p = 2xp\frac{dp}{dx} + p^2 + 2x\frac{dp}{dx} + 2p,$$

which after factorising gives

$$(p+1)\left(p + 2x\frac{dp}{dx}\right) = 0. \tag{12.37}$$

To obtain the general solution of (12.36), we first consider the factor containing $dp/dx$. This first-degree first-order equation in $p$ has the solution $xp^2 = c$ (see subsection 12.3.1), which we then use to eliminate $p$ from (12.36). We therefore find that the general solution to (12.36) is

$$(y - c)^2 = 4cx. \tag{12.38}$$

If we now consider the first factor in (12.37), we find this has the simple solution $p = -1$. Substituting this into (12.36) then gives

$$x + y = 0,$$

which is a singular solution to (12.36). ◄

**Solution method.** *Write the equation in the form (12.35) and differentiate both sides with respect to x. Rearrange the resulting equation into the form $G(x, p) = 0$, which can be used together with the original ODE to eliminate p and so give the general solution. If $G(x, p)$ can be factorised, then the factor containing $dp/dx$ should be used to eliminate p and give the general solution. Using the other factors in this fashion will instead lead to singular solutions.*

### 12.3.4 Clairaut's equation

Finally, we consider Clairaut's equation, which has the form

$$y = px + F(p), \tag{12.39}$$

and is therefore a special case of equations soluble for $y$ (12.35). It may be solved by a similar method to that given in subsection 12.3.3, but for Clairaut's equation the form of the general solution is particularly simple. Differentiating (12.39) with respect to $x$, we find

$$\frac{dy}{dx} = p = p + x\frac{dp}{dx} + \frac{dF}{dp}\frac{dp}{dx} \quad \Rightarrow \quad \frac{dp}{dx}\left(\frac{dF}{dp} + x\right) = 0. \tag{12.40}$$

Considering first the factor containing $dp/dx$, we find

$$\frac{dp}{dx} = \frac{d^2y}{dx^2} = 0 \quad \Rightarrow \quad y = c_1x + c_2. \tag{12.41}$$

Since $p = dy/dx = c_1$, if we substitute (12.41) into (12.39) we find $c_1x + c_2 = c_1x + F(c_1)$. Therefore the constant $c_2$ is given by $F(c_1)$, and the general solution to (12.39) is

$$y = c_1x + F(c_1), \tag{12.42}$$

i.e. the general solution to Clairaut's equation can be obtained by replacing $p$ in the ODE by the arbitrary constant $c_1$. Now considering the second factor in (12.40), we also have

$$\frac{dF}{dp} + x = 0, \tag{12.43}$$

which has the form $G(x, p) = 0$. This relation may be used to eliminate $p$ from (12.39) to give a singular solution.

►*Solve*

$$y = px + p^2. \tag{12.44}$$

From (12.42) the general solution is $y = cx + c^2$. But from (12.43) we also have $2p + x = 0 \Rightarrow p = -x/2$. Substituting this into (12.44) we find the singular solution $x^2 + 4y = 0$. ◄

**Solution method.** *Write the equation in the form (12.39), then the general solution is given by replacing p by some constant c, as shown in (12.42). Using the relation $dF/dp + x = 0$ to eliminate p from the original equation yields the singular solution.*

## 12.4 Exercises

12.1 A radioactive isotope decays in such a way that the number of atoms present at a given time, $N(t)$, obeys the equation

$$\frac{dN}{dt} = -\lambda N.$$

If there are initially $N_0$ atoms present, find $N(t)$ at later times.

12.2 By finding an appropriate integrating factor solve

$$\frac{dy}{dx} = -\frac{2x^2 + y^2 + x}{xy}.$$

12.3 An electric circuit contains a resistance $R$ and a capacitor $C$ in series, and a battery supplying a time-varying electromotive force $V(t)$. The charge $q$ on the capacitor therefore obeys the equation

$$R\frac{dq}{dt} + \frac{q}{C} = V(t).$$

Assuming that there is initially no charge on the capacitor, and given that $V(t) = V_0 \sin \omega t$, find the charge on the capacitor as a function of time.

12.4 Solve

$$(y - x)\frac{dy}{dx} + 2x + 3y = 0.$$

12.5 A mass $m$ is accelerated by a time-varying force $\exp(-\beta t)v^3$, where $v$ is its velocity. It also experiences a resistive force $\eta v$, where $\eta$ is a constant, owing to its motion through the air. The equation of motion of the mass is therefore given by

$$m\frac{dv}{dt} = \exp(-\beta t)v^3 - \eta v.$$

Find an expression for the velocity $v$ of the mass as a function of time, given that it has an initial velocity $v_0$.

12.6 Solve

$$\frac{dy}{dx} = \frac{1}{x + 2y + 1}.$$

12.7 Solve

$$\frac{dy}{dx} = -\frac{x + y}{3x + 3y - 4}.$$

12.8 Find the general solutions of the following

$$\text{(a)}\ \frac{dy}{dx} + \frac{xy}{a^2 + x^2} = x; \quad \text{(b)}\ \frac{dy}{dx} = \frac{4y^2}{x^2} - y^2.$$

12.9 Solve the differential equation

$$\sin x \frac{dy}{dx} + 2y \cos x = 1$$

subject to the boundary condition $y(\pi/2) = 1$.

12.10 If $u = 1 + \tan y$, calculate $d(\ln u)/dy$; hence find the general solution of

$$\frac{dy}{dx} = \tan x \cos y \,(\cos y + \sin y).$$

12.11 By treating $y$ as the independent variable, show that the general solution of

$$(x + y^3)\frac{dy}{dx} = y$$

is $x = y(A + \frac{1}{2}y^2)$.

12.12 Solve

$$x(1 - 2x^2 y)\frac{dy}{dx} + y = 3x^2 y^2,$$

given that $y(1) = \frac{1}{2}$.

12.13 A reflecting mirror is made in the shape of the surface of revolution generated by revolving the curve $y(x)$ about the $x$-axis. In order that light rays emitted from a point source at the origin are reflected back parallel to the $x$-axis, the curve $y(x)$ must obey

$$\frac{y}{x} = \frac{2p}{1 - p^2},$$

where $p = dy/dx$. By solving this equation for $x$ find the curve $y(x)$.

12.14 Find the curve such that at each point on it the sum of the intercepts on the $x$- and $y$-axes of the tangent to the curve (taking account of sign) is equal to 1.

## 12.5 Hints and answers

12.1 $N(t) = N_0 \exp(-\lambda t)$.

12.2 Integrating factor is $x$; $3x^4 + 2x^3 + 3x^2y^2 = c$.

12.3 $q(t) = CV_0[1 + (\omega CR)^2]^{-1}\{\sin \omega t + CR\omega[\exp(-t/RC) - \cos \omega t]\}$.

12.4 Homogeneous equation, put $y = vx$ to obtain $(1 - v)(v^2 + 2v + 2)^{-1}\,dv = x^{-1}\,dx$; write $1 - v$ as $2 - (1 + v)$, and $v^2 + 2v + 2$ as $1 + (1 + v)^2$; $A[x^2 + (x + y)^2] = \exp\{4\tan^{-1}[(x + y)/x]\}$.

12.5 Bernoulli's equation; set $v = u^{-1/2}$ to obtain $m\,du/dt - 2\eta u = -2\exp(-\beta t)$; $v^{-2} = 2(m\beta + 2\eta)^{-1}[\exp(-\beta t) - \exp(2\eta t/m)] + v_0^{-2}\exp(2\eta t/m)$.

12.6 Follow subsection 12.2.8; $k + y = \ln(x + 2y + 3)$.

12.7 Equation is of the form of (12.22), set $v = x + y$; $x + 3y + 2\ln(x + y - 2) = A$.

12.8 (a) Integrating factor is $(a^2+x^2)^{1/2}$; $y=(a^2+x^2)/3+A(a^2+x^2)^{-1/2}$; (b) separable; $y=x(x^2+Ax+4)^{-1}$.

12.9 Integrating factor is $\sin x$; $y=(1+\cos x)^{-1}$.

12.10 $y=\tan^{-1}(k\sec x-1)$.

12.12 Equation is isobaric with weight $y=-2$; setting $y=vx^{-2}$ gives $v^{-1}(1-v)^{-1}(1-2v)\,dv=x^{-1}\,dx$; $4xy(1-x^2y)=1$.

12.13 Eliminate $y$ to obtain, in turn, $p(p^2-1)=2x(dp/dx)$; $p=\pm(1-Ax)^{-1/2}$; $A^2y^2=4(1-Ax)$, i.e. a parabola.

12.14 The curve must satisfy $y=(1-p^{-1})^{-1}(1-x+px)$, which has solution $x=(p-1)^{-2}$, leading to $y=(1\pm\sqrt{x})^2$ or $x=(1\pm\sqrt{y})^2$; singular solution $p'=0$ gives straight lines joining $(\theta,0)$ and $(0,1-\theta)$ for any $\theta$.

# 13

# *Higher-order ordinary differential equations*

Following on from the discussion of first-order ordinary differential equations (ODEs) given in the previous chapter, we now examine equations of second and higher order. Since a brief outline of the general properties of ODEs and their solutions was given at the beginning of the previous chapter, we will not repeat it here. Instead, we will begin with a discussion of various types of higher-order equation. This chapter is divided into three main parts. We first discuss linear equations with constant coefficients and then investigate linear equations with variable coefficients. Finally, we discuss a few methods that may be of use in solving general linear or non-linear ODEs. Let us start by considering some general points relating to *all* linear ODEs.

Linear equations are of paramount importance in the description of physical processes. Moreover, it is an empirical fact that, when put into mathematical form, many natural processes appear as higher-order linear ODEs, most often as second-order equations. Although we could restrict our attention just to these second-order equations, the generalisation to $n$th-order equations requires little extra work, and so we will consider this more general case.

A linear ODE of general order $n$ has the form

$$a_n(x)\frac{d^n y}{dx^n} + a_{n-1}(x)\frac{d^{n-1} y}{dx^{n-1}} + \cdots + a_1(x)\frac{dy}{dx} + a_0(x)y = f(x). \qquad (13.1)$$

If $f(x) = 0$ then the equation is called *homogeneous*; otherwise it is *inhomogeneous*. The first-order linear equation discussed in subsection 12.2.4 is a special case of (13.1). As discussed at the beginning of the previous chapter, the general solution to (13.1) will contain $n$ arbitrary constants, which may be determined if $n$ boundary conditions are also provided.

In order to solve any equation of the form (13.1), we must first find the general solution of the *complementary equation*, i.e. the equation formed by setting $f(x) = 0$

in (13.1):

$$a_n(x)\frac{d^n y}{dx^n} + a_{n-1}(x)\frac{d^{n-1} y}{dx^{n-1}} + \cdots + a_1(x)\frac{dy}{dx} + a_0(x)y = 0. \tag{13.2}$$

To find the general solution of (13.2), we must find $n$ linearly independent functions that satisfy it. Once we have found these solutions, the general solution is given by a linear superposition of these $n$ functions. In other words, if the $n$ solutions of (13.2) are $y_1(x)$, $y_2(x), \ldots, y_n(x)$, then the general solution is given by the linear superposition

$$y_c(x) = c_1 y_1(x) + c_2 y_2(x) + \cdots + c_n y_n(x), \tag{13.3}$$

where the $c_m$ are arbitrary constants that may be determined if $n$ boundary conditions are provided. The linear combination $y_c(x)$ is called the *complementary function* of (13.1).

The question naturally arises how we establish that any $n$ individual solutions to (13.2) are indeed linearly independent. For $n$ functions to be linearly independent over an interval, there must not exist *any* set of constants $c_1, c_2, \ldots, c_n$ such that

$$c_1 y_1(x) + c_2 y_2(x) + \cdots + c_n y_n(x) = 0 \tag{13.4}$$

over the interval in question, except for the trivial case $c_1 = c_2 = \cdots = c_n = 0$.

A statement equivalent to the above, which is perhaps more useful for the practical determination of linear independence, can be found by repeatedly differentiating (13.4), $n-1$ times in all, to obtain $n$ simultaneous equations for $c_1, c_2, \ldots, c_n$, which read

$$\begin{aligned}
c_1 y_1(x) + c_2 y_2(x) + \cdots + c_n y_n(x) &= 0 \\
c_1 y_1'(x) + c_2 y_2'(x) + \cdots + c_n y_n'(x) &= 0 \\
&\vdots \\
c_1 y_1^{(n-1)}(x) + c_2 y_2^{(n-1)} + \cdots + c_n y_n^{(n-1)}(x) &= 0,
\end{aligned} \tag{13.5}$$

where the primes denote differentiation with respect to $x$. Referring to the discussion of simultaneous linear equations given in chapter 7, if the deteminant of the coefficients of $c_1, c_2, \ldots, c_n$ is non-zero, then the only solution to equations (13.5) is the trivial solution $c_1 = c_2 = \cdots = c_n = 0$. In other words, the $n$ functions $y_1(x)$, $y_2(x), \ldots, y_n(x)$ are linearly independent over an interval if

$$W(y_1, \ldots, y_n) = \begin{vmatrix} y_1 & y_2 & \ldots & y_n \\ y_1' & & & \\ \vdots & & \ddots & \\ y_1^{(n-1)} & & & y_n^{(n-1)} \end{vmatrix} \neq 0 \tag{13.6}$$

over that interval; $W(y_1, \ldots, y_n)$ is called the *Wronskian* of the set of functions. It should be noted, however, that the vanishing of the Wronskian does not guarantee that the functions are linearly dependent.

If the original equation (13.1) has $f(x) = 0$ (i.e. it is homogeneous) then of course the complementary function $y_c(x)$ in (13.3) is already the general solution. If, however, the equation has $f(x) \neq 0$ (i.e. it is inhomogeneous), then $y_c(x)$ is only one part of the solution. The general solution of (13.1) is then given by

$$y(x) = y_c(x) + y_p(x), \tag{13.7}$$

where $y_p(x)$ is the *particular integral*, which can be *any* function that satisfies (13.1) directly, provided it is linearly independent of $y_c(x)$. It should be emphasised for practical purposes that *any* such function, no matter how simple (or complicated), is equally valid in forming the general solution (13.7).

It is important to realise that the above method for finding the general solution to an ODE by superposing particular solutions assumes crucially that the ODE is linear. For the non-linear equations discussed in section 13.3, this method cannot be used, and indeed it is often impossible to find closed-form solutions to such equations.

## 13.1 Linear equations with constant coefficients

If the $a_m$ in (13.1) are constants rather than functions of $x$, then we have

$$a_n \frac{d^n y}{dx^n} + a_{n-1} \frac{d^{n-1} y}{dx^{n-1}} + \cdots + a_1 \frac{dy}{dx} + a_0 y = f(x). \tag{13.8}$$

Equations of this sort are very common throughout the physical sciences and engineering, and the method of their solution falls into two parts as discussed in the previous section, i.e. finding the complementary function $y_c(x)$ and finding the particular integral $y_p(x)$. If $f(x) = 0$ in (13.8) then we do not have to find a particular integral, and the complementary function is by itself the general solution.

### *13.1.1 Finding the complementary function* $y_c(x)$

The complementary function must satisfy

$$a_n \frac{d^n y}{dx^n} + a_{n-1} \frac{d^{n-1} y}{dx^{n-1}} + \cdots + a_1 \frac{dy}{dx} + a_0 y = 0 \tag{13.9}$$

and contain $n$ arbitrary constants (see equation (13.3)). The standard method for finding $y_c(x)$ is to try a solution of the form $y = Ae^{\lambda x}$, substituting this into (13.9). After dividing the resulting equation through by $Ae^{\lambda x}$, we are left with a polynomial equation in $\lambda$ of order $n$; this is the *auxiliary equation* and reads

$$a_n \lambda^n + a_{n-1} \lambda^{n-1} + \cdots + a_1 \lambda + a_0 = 0. \tag{13.10}$$

In general the auxiliary equation has $n$ roots, say $\lambda_1, \ldots, \lambda_n$. In certain cases, some of these roots may be repeated and some may be complex. The three main cases are as follows.

(i) *All roots real and distinct.* In this case the $n$ solutions to (13.9) are $\exp \lambda_m x$ for $m = 1$ to $n$. It is easily shown by calculating the Wronskian (13.6) of these functions that if all the $\lambda_m$ are distinct these solutions are linearly independent. We can therefore linearly superpose them, as in (13.3), to form the complementary function

$$y_c(x) = c_1 e^{\lambda_1 x} + c_2 e^{\lambda_2 x} + \cdots + c_n e^{\lambda_n x}. \tag{13.11}$$

(ii) *Some roots complex.* For the special (but usual) case that all the coefficients $a_m$ in (13.9) are real, if one of the roots of the auxiliary equation (13.10) is complex, say $\alpha + i\beta$, then its complex conjugate $\alpha - i\beta$ is also a root. In this case we can write

$$\begin{aligned} c_1 e^{(\alpha+i\beta)x} + c_2 e^{(\alpha-i\beta)x} &= e^{\alpha x}(d_1 \cos \beta x + d_2 \sin \beta x) \\ &= A e^{\alpha x} \left\{ \begin{matrix} \sin \\ \cos \end{matrix} \right\} (\beta x + \phi), \end{aligned} \tag{13.12}$$

where $A$ and $\phi$ are arbitrary constants.

(iii) *Some roots repeated.* If, for example, $\lambda_1$ occurs $k$ times ($k > 1$) as a root of the auxiliary equation, then we have not found $n$ linearly independent solutions of (13.9); formally the Wronskian (13.6) of these solutions, having two or more identical columns, is equal to zero. We must therefore find $k - 1$ further solutions that are linearly independent of those already found and also of each other. By direct substitution into (13.9) we find that $xe^{\lambda_1 x}, x^2 e^{\lambda_1 x}, \ldots, x^{k-1} e^{\lambda_1 x}$ are also solutions, and by calculating the Wronskian it is easily shown that they, together with the solutions already found, form a linearly independent set of $n$ functions. Therefore the complementary function is given by

$$y_c(x) = (c_1 + c_2 x + \cdots + c_k x^{k-1}) e^{\lambda_1 x} + c_{k+1} e^{\lambda_{k+1} x} + \cdots + c_n e^{\lambda_n x}. \tag{13.13}$$

If more than one root is repeated the above argument is easily extended. For example, suppose as before that $\lambda_1$ is a $k$-fold root of the auxiliary equation, and, further, that $\lambda_2$ is an $l$-fold root (of course, $k > 1$ and $l > 1$). Then, from the above argument, the complementary function reads

$$\begin{aligned} y_c(x) = (c_1 + c_2 x + \cdots &+ c_k x^{k-1}) e^{\lambda_1 x} \\ &+ (c_{k+1} + c_{k+2} x + \cdots + c_{k+l} x^{l-1}) e^{\lambda_2 x} \\ &+ c_{k+l+1} e^{\lambda_{k+l+1} x} + \cdots + c_n e^{\lambda_n x}. \end{aligned} \tag{13.14}$$

►*Find the complementary function of the equation*

$$\frac{d^2y}{dx^2} - 2\frac{dy}{dx} + y = e^x. \qquad (13.15)$$

Setting the RHS to zero, substituting $y = Ae^{\lambda x}$ and dividing through by $Ae^{\lambda x}$ we obtain the auxiliary equation

$$\lambda^2 - 2\lambda + 1 = 0.$$

This equation has the repeated root $\lambda = 1$ (twice), and so although $e^x$ is a solution to (13.15), we must find a further solution to the equation that is linearly independent of $e^x$. From the above discussion, we deduce that $xe^x$ is such a solution, so that the full complementary function is given by the linear superposition

$$y_c(x) = (c_1 + c_2 x)e^x. \blacktriangleleft$$

**Solution method.** *Set the RHS of the ODE to zero (if it is not already so), and substitute $y = Ae^{\lambda x}$. After dividing through the resulting equation by $Ae^{\lambda x}$, obtain an nth-order polynomial equation in $\lambda$ (the auxiliary equation, see (13.10)). Solve the auxiliary equation to find the n roots, $\lambda_1, \ldots, \lambda_n$, say. If all these roots are real and distinct, then $y_c(x)$ is given by (13.11). If, however, some of the roots are complex or repeated, then $y_c(x)$ is given by (13.12) or (13.13) (or the extension (13.14) of the latter) respectively.*

### *13.1.2 Finding the particular integral $y_p(x)$*

There is no generally applicable method for finding the particular integral $y_p(x)$, but for linear ODEs with constant coefficients and a simple RHS, $y_p(x)$ can often be found by inspection or by assuming a parameterised form similar to $f(x)$. The latter method is sometimes called the *method of undetermined coefficients.* If $f(x)$ contains only polynomial, exponential, or sine and cosine terms, then by assuming a trial function for $y_p(x)$ of similar form but which contains a number of undetermined parameters, and substituting this trial function into (13.9), the parameters can be found and $y_p(x)$ deduced. Standard trial functions are as follows.

(i) If $f(x) = ae^{rx}$ then try $y_p(x) = be^{rx}$.

(ii) If $f(x) = a_1 \sin rx + a_2 \cos rx$ ($a_1$ or $a_2$ may be zero) then try $y_p(x) = b_1 \sin rx + b_2 \cos rx$.

(iii) If $f(x) = a_0 + a_1 x + \cdots + a_N x^N$ (some $a_m$ may be zero) then try $y_p(x) = b_0 + b_1 x + \cdots + b_N x^N$.

(iv) If $f(x)$ is the sum or product of any of the above then try $y_p(x)$ as the sum or product of the corresponding individual trial functions.

It should be noted that this method fails if any term in the assumed trial function is also contained within the complementary function $y_c(x)$. In such a case the trial function should be multiplied by the smallest integer power of $x$ such that it will then contain no term that already appears in the complementary function. The undetermined coefficients in the trial function can now be found by substitution into (13.8).

Three further methods that are useful in finding the particular integral $y_p(x)$ are Green's functions, the variation of parameters, and a change of the dependent variable using one part of the complementary function. However, since these methods are also applicable to equations with variable coefficients, a discussion of them is postponed until section 13.2.

►*Find a particular integral of the equation*

$$\frac{d^2y}{dx^2} - 2\frac{dy}{dx} + y = e^x.$$

From the above discussion our first guess at a trial particular integral would be $y_p(x) = be^x$. However, since the complementary function of this equation is $y_c(x) = (c_1 + c_2x)e^x$ (see previous subsection), we see that $e^x$ is already contained in it, as indeed is $xe^x$. Multiplying our first guess by the lowest necessary integer power of $x$ such that it does not appear in $y_c(x)$, we therefore try $y_p(x) = bx^2e^x$. Substituting this into the ODE, we find that $b = 1/2$, so the particular integral is given by $y_p(x) = x^2e^x/2$. ◄

**Solution method.** *If the RHS of the equation contains only the functions outlined above, then the appropriate trial function should be substituted into the ODE, thereby fixing the undetermined parameters. If, however, the RHS of the equation is not of this form then one of the more general methods outlined in subsections 13.2.3–13.2.5 should be used; perhaps the most straightforward of these is that of variation of parameters.*

### *13.1.3 Constructing the general solution* $y_c(x) + y_p(x)$

As stated earlier, the full solution to the ODE (13.8) is found by adding together the complementary function and any particular integral. In order to illustrate further the material discussed in the last two subsections, let us find the general solution to a new example, starting from the beginning.

►*Solve*

$$\frac{d^2y}{dx^2} + 4y = x^2 \sin 2x. \tag{13.16}$$

First we set the RHS to zero and assume the trial solution $y = Ae^{\lambda x}$. Substituting this into (13.16) leads to the auxiliary equation

$$\lambda^2 + 4 = 0 \quad \Rightarrow \quad \lambda = \pm 2i. \tag{13.17}$$

Therefore the complementary function is given by

$$y_c(x) = c_1 e^{2ix} + c_2 e^{-2ix} = d_1 \cos 2x + d_2 \sin 2x. \tag{13.18}$$

We must now turn our attention to the particular integral $y_p(x)$. Consulting the list of standard trial functions in the previous subsection, we find that a first guess at a suitable trial function for this case should be

$$(ax^2 + bx + c)(d \sin 2x + e \cos 2x). \tag{13.19}$$

However, we see that this trial function contains terms in $\sin 2x$ and $\cos 2x$, both of which already appear in the complementary function (13.18). We must therefore multiply (13.19) by the smallest integer power of $x$ that ensures that none of the resulting terms appear in $y_c(x)$. Since multiplying by $x$ will suffice, we finally assume the trial function

$$(ax^3 + bx^2 + cx)(d \sin 2x + e \cos 2x). \tag{13.20}$$

Substituting this into (13.16) to fix the constants in (13.20), we find the particular integral to be

$$y_p(x) = -\frac{x^3}{12}\cos 2x + \frac{x^2}{16}\sin 2x + \frac{x}{32}\cos 2x. \tag{13.21}$$

The general solution to (13.16) then reads

$$\begin{aligned} y(x) &= y_c(x) + y_p(x) \\ &= d_1 \cos 2x + d_2 \sin 2x - \frac{x^3}{12}\cos 2x + \frac{x^2}{16}\sin 2x + \frac{x}{32}\cos 2x. \end{aligned}$$ ◄

### *13.1.4 Laplace transform method*

The method of Laplace transforms is very useful in solving linear ODEs with constant coefficients. Taking the Laplace transform of such an equation transforms it into a purely *algebraic* equation for the Laplace transform of the required solution. Once the algebraic equation has been solved for this Laplace transform, the general solution to the original ODE can be obtained by performing an inverse Laplace transform. One advantage of this method is that, for given boundary conditions, it provides the solution in just one step, instead of having to find the complementary function and particular integral separately.

In order to apply this method we need only two results from Laplace transform theory (see section 11.2). First, the Laplace transform of a function $f(x)$ is defined by

$$\bar{f}(s) \equiv \int_0^\infty e^{-sx} f(x)\, dx, \tag{13.22}$$

from which we can derive the second useful relation concerning the Laplace transform of derivatives of $f(x)$, namely

$$\overline{f^{(n)}}(s) = s^n \bar{f}(s) - s^{n-1} f(0) - s^{n-2} f'(0) - \cdots - s f^{(n-2)}(0) - f^{(n-1)}(0), \tag{13.23}$$

where the primes and superscripts in parentheses denote differentiation with respect to $x$. Using these relations, along with the table 11.1 giving Laplace transforms of standard functions, we are now in a position to solve a linear ODE with constant coefficients by this method.

►*Solve*

$$\frac{d^2y}{dx^2} - 3\frac{dy}{dx} + 2y = 2e^{-x}, \tag{13.24}$$

*subject to the boundary conditions* $y(0) = 2$, $y'(0) = 1$.

Taking the Laplace transform of (13.24) and using the table of standard results we obtain

$$s^2\bar{y}(s) - sy(0) - y'(0) - 3\,[s\bar{y}(s) - y(0)] + 2\bar{y}(s) = \frac{2}{s+1},$$

which reduces to

$$(s^2 - 3s + 2)\bar{y}(s) - 2s + 5 = \frac{2}{s+1}. \tag{13.25}$$

Solving the algebraic equation (13.25) for $\bar{y}(s)$, the Laplace transform of the required solution to (13.24), we obtain

$$\bar{y}(s) = \frac{2s^2 - 3s - 3}{(s+1)(s-1)(s-2)} = \frac{1}{3(s+1)} + \frac{2}{s-1} - \frac{1}{3(s-2)}, \tag{13.26}$$

where in the final step we have used partial fractions. Taking the inverse Laplace transform of (13.26), again using table 11.1, we find the specific solution to (13.24) to be

$$y(x) = \tfrac{1}{3}e^{-x} + 2e^{x} - \tfrac{1}{3}e^{2x}. \blacktriangleleft$$

Note that if the boundary conditions in a problem are given as symbols, rather than just numbers, then the step involving partial fractions can often involve a considerable amount of algebra. The Laplace transform method is also useful in solving sets of *simultaneous* linear ODEs with constant coefficients.

►*Two electrical circuits, both of negligible resistance, each consist of a self-inductance $L$ and a capacitance $C = 1/G$. The mutual inductance of the two circuits is $M$. Suppose that there is no external e.m.f. applied, but that initially the second capacitance is given a charge $CV_0$ with the first one uncharged, and that at time $t = 0$ a switch in the second circuit is closed. Find the subsequent current in the first circuit.*

Subject to the initial conditions $q_1(0) = \dot{q}_1(0) = \dot{q}_2(0) = 0$ and $q_2(0) = CV_0 = V_0/G$, we have to solve

$$L\ddot{q}_1 + M\ddot{q}_2 + Gq_1 = 0$$
$$M\ddot{q}_1 + L\ddot{q}_2 + Gq_2 = 0.$$

On taking the Laplace transform of the above equations, we obtain

$$(Ls^2 + G)\bar{q}_1 + Ms^2\bar{q}_2 = sMV_0C$$
$$Ms^2\bar{q}_1 + (Ls^2 + G)\bar{q}_2 = sLV_0C.$$

Eliminating $\bar{q}_2$ and rewriting as an equation for $\bar{q}_1$, we find

$$\bar{q}_1(s) = \frac{MV_0 s}{[(L+M)s^2 + G][(L-M)s^2 + G]}$$
$$= \frac{V_0}{2G}\left[\frac{(L+M)s}{(L+M)s^2 + G} - \frac{(L-M)s}{(L-M)s^2 + G}\right].$$

Using table 11.1

$$q_1(t) = \tfrac{1}{2}V_0C(\cos\omega_1 t - \cos\omega_2 t),$$

where $\omega_1^2(L+M) = G$ and $\omega_2^2(L-M) = G$. Thus the current is given by

$$i_1(t) = \tfrac{1}{2}V_0C(\omega_2\sin\omega_2 t - \omega_1\sin\omega_1 t). \blacktriangleleft$$

**Solution method.** *Perform a Laplace transform, as defined in (13.22), on the entire equation, using (13.23) to calculate the transform of the derivatives. Then solve the resulting algebraic equation for $\bar{y}(s)$, the Laplace transform of the required solution to the ODE. By using the method of partial fractions and consulting a table of Laplace transforms of standard functions, calculate the inverse Laplace transform. The resulting function $y(x)$ is the solution of the ODE which obeys the given boundary conditions.*

## 13.2 Linear equations with variable coefficients

There is no generally applicable method of solving equations with coefficients that are functions of $x$. Nevertheless, there are certain cases in which a solution is possible. Some of the methods discussed in this section are also useful in finding the general solution or particular integral for equations with constant coefficients that have proved impenetrable by the techniques discussed above.

### 13.2.1 The Legendre and Euler linear equations

*Legendre's linear equation* has the form

$$a_n(\alpha x+\beta)^n \frac{d^n y}{dx^n} + \cdots + a_1(\alpha x+\beta)\frac{dy}{dx} + a_0 y = f(x), \tag{13.27}$$

where $\alpha$, $\beta$ and the $a_n$ are constants, and may be solved by making the substitution $\alpha x+\beta = e^t$. We then have

$$\frac{dy}{dx} = \frac{dt}{dx}\frac{dy}{dt} = \frac{\alpha}{\alpha x+\beta}\frac{dy}{dt}$$

$$\frac{d^2y}{dx^2} = \frac{d}{dx}\frac{dy}{dx} = \frac{\alpha^2}{(\alpha x+\beta)^2}\left(\frac{d^2y}{dt^2} - \frac{dy}{dt}\right),$$

and so on for higher derivatives. Therefore we can write the terms of (13.27) as

$$\begin{aligned}
(\alpha x+\beta)\frac{dy}{dx} &= \alpha\frac{dy}{dt}, \\
(\alpha x+\beta)^2\frac{d^2y}{dx^2} &= \alpha^2\frac{d}{dt}\left(\frac{d}{dt}-1\right)y, \\
&\ \vdots \\
(\alpha x+\beta)^n\frac{d^ny}{dx^n} &= \alpha^n\frac{d}{dt}\left(\frac{d}{dt}-1\right)\cdots\left(\frac{d}{dt}-n+1\right)y.
\end{aligned} \tag{13.28}$$

Substituting equations (13.28) into the original equation (13.27), the latter becomes a linear ODE with constant coefficients, i.e.

$$a_n\alpha^n\frac{d}{dt}\left(\frac{d}{dt}-1\right)\cdots\left(\frac{d}{dt}-n+1\right)y + \cdots + a_1\alpha\frac{dy}{dt} + a_0 y = f\left(\frac{e^t-\beta}{\alpha}\right),$$

which can be solved by the methods of section 13.1.

A special case of Legendre's linear equation, with $\alpha = 1$ and $\beta = 0$, is *Euler's equation*, which reads

$$a_n x^n\frac{d^ny}{dx^n} + \cdots + a_1 x\frac{dy}{dx} + a_0 y = f(x), \tag{13.29}$$

and may be solved in a similar manner to the above by substituting $x = e^t$. Alternatively, in the special case where $f(x) = 0$ in (13.29), substituting $y = x^\lambda$ leads to a simple algebraic equation in $\lambda$, which can be solved to yield the solution to (13.29). In the event that the algebraic equation for $\lambda$ has repeated roots, extra care is needed. If $\lambda_1$ is a $k$-fold root ($k > 1$), the $k$ linearly independent solutions corresponding to this root are $x^{\lambda_1}$, $x^{\lambda_1}\ln x, \ldots, x^{\lambda_1}(\ln x)^{k-1}$.

►*Solve*

$$x^2\frac{d^2y}{dx^2}+x\frac{dy}{dx}-4y=0 \tag{13.30}$$

*by both of the methods discussed above.*

First we make the substitution $x=e^t$, which, cancelling $e^t$, gives an equation with constant coefficients, i.e.

$$\frac{d}{dt}\left(\frac{d}{dt}-1\right)y+\frac{dy}{dt}-4y=0 \quad\Rightarrow\quad \frac{d^2y}{dt^2}-4y=0. \tag{13.31}$$

Using the methods of section 13.1, the general solution of (13.31), and therefore of (13.30), is given by

$$y=c_1e^{2t}+c_2e^{-2t}=c_1x^2+c_2x^{-2}.$$

Since the RHS of (13.30) is zero, we can reach the same solution by instead substituting $y=x^\lambda$ into (13.30). This gives

$$\lambda(\lambda-1)x^\lambda+\lambda x^\lambda-4x^\lambda=0,$$

which reduces to

$$(\lambda^2-4)x^\lambda=0.$$

This has the solutions $\lambda=\pm 2$, so we again obtain the general solution

$$y=c_1x^2+c_2x^{-2}.$$ ◄

**Solution method.** *If the ODE is of the Legendre form (13.27), then substitute $\alpha x+\beta=e^t$. This results in an equation of the same order but with constant coefficients, which can be solved by the methods of section 13.1. If the ODE is of the Euler form (13.29) with a non-zero RHS, then substitute $x=e^t$; this again leads to an equation of the same order but with constant coefficients. If, however, $f(x)=0$ in the Euler equation (13.29) then the equation may also be solved by substituting $y=x^\lambda$. This leads to an algebraic equation whose solution gives the allowed values of $\lambda$; the general solution is then the linear superposition of these functions.*

### *13.2.2 Exact equations*

Sometimes an ODE may be merely the derivative of another ODE of one order lower. If this is the case then the ODE is called exact. The $n$th-order linear ODE

$$a_n(x)\frac{d^ny}{dx^n}+\cdots+a_1(x)\frac{dy}{dx}+a_0(x)y=f(x), \tag{13.32}$$

is exact if the LHS can be written as a simple derivative, i.e. if

$$a_n(x)\frac{d^ny}{dx^n}+\cdots+a_0(x)y=\frac{d}{dx}\left[b_{n-1}(x)\frac{d^{n-1}y}{dx^{n-1}}+\cdots+b_0(x)y\right]. \tag{13.33}$$

It may be shown that, for (13.33) to hold, we require

$$a_0(x) - a_1'(x) + a_2''(x) - \cdots + (-1)^n a_n^{(n)}(x) = 0, \tag{13.34}$$

where the prime again denotes differentiation with respect to $x$. If (13.34) is satisfied then a straightforward integration leads to a new equation of one order lower. If this simpler equation can be solved, then a solution to the original equation is obtained. Of course, if the above process leads to an equation which is itself exact, then the analysis can be repeated to reduce the order still further.

►*Solve*

$$(1-x^2)\frac{d^2y}{dx^2} - 3x\frac{dy}{dx} - y = 1. \tag{13.35}$$

Comparing with (13.32), we have $a_2 = 1 - x^2$, $a_1 = -3x$ and $a_0 = -1$. It is easily shown that $a_0 - a_1' + a_2'' = 0$, so (13.35) is exact and can therefore be written in the form

$$\frac{d}{dx}\left[b_1(x)\frac{dy}{dx} + b_0(x)y\right] = 1. \tag{13.36}$$

Expanding the LHS of (13.36) we find

$$\frac{d}{dx}\left(b_1\frac{dy}{dx} + b_0 y\right) = b_1\frac{d^2y}{dx^2} + (b_1' + b_0)\frac{dy}{dx} + b_0' y. \tag{13.37}$$

Comparing (13.35) and (13.37) we find

$$b_1 = 1 - x^2, \qquad b_1' + b_0 = -3x, \qquad b_0' = -1.$$

These relations integrate consistently to give $b_1 = 1 - x^2$ and $b_0 = -x$, so (13.35) can be written as

$$\frac{d}{dx}\left[(1-x^2)\frac{dy}{dx} - xy\right] = 1. \tag{13.38}$$

Integrating (13.38) gives us directly the first-order linear ODE

$$\frac{dy}{dx} - \left(\frac{x}{1-x^2}\right)y = \frac{x + c_1}{1-x^2},$$

which can be solved by the method of subsection 12.2.4 and has the solution

$$y = \frac{c_1 \sin^{-1} x + c_2}{\sqrt{1-x^2}} - 1. \; ◄$$

It is worth noting that even if a higher-order ODE is not exact in its given form, it may sometimes be made exact by multiplying through by some suitable function, an *integrating factor*, cf. subsection 12.2.3. Unfortunately, no straightforward method for finding an integrating factor exists, and one often has to rely on inspection or experience.

►*Solve*

$$x(1-x^2)\frac{d^2y}{dx^2} - 3x^2\frac{dy}{dx} - xy = x. \qquad (13.39)$$

It is easily shown that (13.39) is not exact, but we also see immediately that by multiplying it through by $1/x$ we recover (13.35), which is exact and is solved above. ◄

Another important point is that an ODE need not be linear to be exact, although no simple rule such as (13.34) exists if it is not linear. Nevertheless, it is often worth exploring the possibility that a non-linear equation may be exact, since it too may then be reduced in order by one, which may lead to a soluble equation. This is discussed further in subsection 13.3.3.

**Solution method.** *For a linear ODE of the form (13.32) check whether it is exact using equation (13.34). If it is not then attempt to find an integrating factor which when multiplying the equation makes it exact. Once the equation is exact write the LHS as a derivative as in (13.33), and by expanding this derivative and comparing with the LHS of the ODE, calculate the functions $b_m(x)$ in (13.33). Integrate the resulting equation to yield another ODE but of one order lower. This may be solved or simplified further if the new ODE is itself exact, or can be made so.*

### *13.2.3 Partially known complementary function*

Suppose we wish to solve the $n$th-order linear ODE

$$a_n(x)\frac{d^ny}{dx^n} + \cdots + a_1(x)\frac{dy}{dx} + a_0(x)y = f(x), \qquad (13.40)$$

and we happen to know that $u(x)$ is a solution of (13.40) when the RHS is set to zero, i.e. $u(x)$ is one part of the complementary function. By making the substitution $y(x) = u(x)v(x)$, we can transform (13.40) into an equation of order $n-1$ in $dv/dx$. This simpler equation may prove soluble.

In particular, if the original equation is of second order, then we obtain a first-order equation in $dv/dx$, which may be soluble using the methods of section 12.2. In this way both the remaining term in the complementary function and the particular integral are found. This method therefore provides a useful way of calculating particular integrals for second-order equations with variable (or constant) coefficients.

►*Solve*

$$\frac{d^2y}{dx^2} + y = \operatorname{cosec} x. \tag{13.41}$$

We see that the RHS does not fall into any of the categories listed in subsection 13.1.2, and so we are at a loss as to how to find the particular integral. But the complementary function of (13.41) is

$$y_c(x) = c_1 \sin x + c_2 \cos x,$$

and so let us choose the solution $u(x) = \cos x$ (we could equally well choose $\sin x$) and make the substitution $y(x) = v(x)u(x) = v(x) \cos x$ into (13.41). This gives

$$\cos x \frac{d^2v}{dx^2} - 2 \sin x \frac{dv}{dx} = \operatorname{cosec} x, \tag{13.42}$$

which is a first-order linear ODE in $dv/dx$ and may be solved by multiplying through by a suitable integrating factor, as discussed in subsection 12.2.4. Writing (13.42) as

$$\frac{d^2v}{dx^2} - 2 \tan x \frac{dv}{dx} = \frac{\operatorname{cosec} x}{\cos x}, \tag{13.43}$$

we see that the required integrating factor is given by

$$\exp\left\{-2 \int \tan x \, dx\right\} = \exp\left[2 \ln(\cos x)\right] = \cos^2 x.$$

Multiplying through (13.43) by the integrating factor $\cos^2 x$ we obtain

$$\frac{d}{dx}\left(\cos^2 x \frac{dv}{dx}\right) = \cot x,$$

which integrates to give

$$\cos^2 x \frac{dv}{dx} = \ln(\sin x) + c_1.$$

After rearranging and integrating again this becomes

$$\begin{aligned} v &= \int \sec^2 x \ln(\sin x) \, dx + c_1 \int \sec^2 x \, dx \\ &= \tan x \ln(\sin x) - x + c_1 \tan x + c_2. \end{aligned}$$

Therefore the general solution to (13.41) is given by $y = uv = v \cos x$, i.e.

$$y = c_1 \sin x + c_2 \cos x + \sin x \ln(\sin x) - x \cos x,$$

which contains the full complementary function and the particular integral. ◄

**Solution method.** *If $u(x)$ is a known solution of the nth-order equation (13.40) with $f(x) = 0$, then make the substitution $y(x) = u(x)v(x)$ in (13.40). This leads to an equation of order $n-1$ in $dv/dx$, which might be soluble.*

### *13.2.4 Variation of parameters*

The variation of parameters proves useful in finding particular integrals for linear ODEs with variable (and constant) coefficients. However, this method requires knowledge of the entire complementary function, not just of one part of it as in the previous subsection.

Suppose we wish to find a particular integral of the equation

$$a_n(x)\frac{d^n y}{dx^n} + \cdots + a_1(x)\frac{dy}{dx} + a_0(x)y = f(x), \tag{13.44}$$

and the complementary function $y_c(x)$ (the general solution of (13.44) with $f(x) = 0$) is

$$y_c(x) = c_1 y_1(x) + c_2 y_2(x) + \cdots + c_n y_n(x),$$

where the functions $y_m(x)$ are known. We now assume that the particular integral of (13.44) can be expressed in a similar form to the complementary function, but with the constants $c_m$ replaced by functions of $x$, i.e. we assume a particular integral of the form

$$y_p(x) = k_1(x)y_1(x) + k_2(x)y_2(x) + \cdots + k_n(x)y_n(x). \tag{13.45}$$

This will no longer satisfy the complementary equation (i.e. (13.44) with the RHS set to zero), but might with suitable choices of the functions $k_i(x)$ be made equal to $f(x)$, thus producing not a complementary function but a particular integral.

Since we have $n$ arbitrary functions $k_1(x), \ldots, k_n(x)$, but only one restriction on them (namely the ODE), we must impose a further $n-1$ constraints. We can choose these constraints to be as convenient as possible, and the simplest choice is given by

$$\begin{aligned}
k_1'(x)y_1(x) + \cdots + k_n'(x)y_n(x) &= 0 \\
k_1'(x)y_1'(x) + \cdots + k_n'(x)y_n'(x) &= 0 \\
&\ \ \vdots \\
k_1'(x)y_1^{(n-2)}(x) + \cdots + k_n'(x)y_n^{(n-2)}(x) &= 0 \\
k_1'(x)y_1^{(n-1)}(x) + \cdots + k_n'(x)y_n^{(n-1)}(x) &= \frac{f(x)}{a_n(x)},
\end{aligned} \tag{13.46}$$

where the primes denote differentiation with respect to $x$. The last of these equations is not a freely chosen constraint, but must be satisfied given the previous $n-1$ constraints and the original ODE.

This choice of constraints is easily justified (although the algebra is quite messy). Differentiating (13.45) with respect to $x$, we obtain

$$y_p' = k_1 y_1' + \cdots + k_n y_n' + (k_1' y_1 + \cdots + k_n' y_n),$$

where, for the moment, we drop the explicit $x$-dependence of these functions. Since we are free to choose our constraints as we wish, let us define the expression in parentheses to be zero, giving the first equation in (13.46). Differentiating again we find

$$y_p'' = k_1 y_1'' + \cdots + k_n y_n'' + (k_1' y_1' + \cdots + k_n' y_n').$$

Once more we can choose the expression in brackets to be zero, giving the second equation in (13.46). We can repeat this procedure, choosing the corresponding expression in each case to be zero. This yields the first $n-1$ equations in (13.46). The $m$th derivative of $y_p$ for $m < n$ is then given by

$$y_p^{(m)} = k_1 y_1^{(m)} + \cdots + k_n y_n^{(m)}.$$

Differentiating $y_p$ once more we find its $n$th derivative is given by

$$y_p^{(n)} = k_1 y_1^{(n)} + \cdots + k_n y_n^{(n)} + (k_1' y_1^{(n-1)} + \cdots + k_n' y_n^{(n-1)}).$$

Substituting the expressions for $y_p^{(m)}$, $m = 0$ to $n$, into the original ODE (13.44), we obtain

$$\sum_{m=0}^{n} a_m (k_1 y_1^{(m)} + \cdots + k_n y_n^{(m)}) + a_n (k_1' y_1^{(n-1)} + \cdots + k_n' y_n^{(n-1)}) = f(x).$$

Rearranging the sum over $m$ on the LHS, we find

$$\sum_{m=1}^{n} k_m (a_n y_m^{(n)} + \cdots + a_1 y_m' + a_0 y_m) + a_n (k_1' y_1^{(n-1)} + \cdots + k_n' y_n^{(n-1)}) = f(x). \tag{13.47}$$

But since the functions $y_m$ are solutions of the complementary equation of (13.44) we have (for all $m$)

$$a_n y_m^{(n)} + \cdots + a_1 y_m' + a_0 y_m = 0.$$

Therefore (13.47) becomes

$$a_n (k_1' y_1^{(n-1)} + \cdots + k_n' y_n^{(n-1)}) = f(x),$$

which is the final equation given in (13.46).

Considering (13.46) to be a set of simultaneous equations in the unknowns $k_1'(x), \ldots, k_n'(x)$, we see that the determinant of the coefficients of these functions is equal to the Wronskian $W(y_1, \ldots, y_n)$, which is non-zero since the solutions $y_m(x)$ are linearly independent (see equation (13.6)). Therefore (13.46) can be solved for the functions $k_m'(x)$, which in turn can be integrated, setting all constants of integration equal to zero, to give $k_m(x)$. The general solution to (13.44) is then given by

$$y(x) = y_c(x) + y_p(x) = \sum_{m=1}^{n} [c_m + k_m(x)]y_m(x).$$

Note that if the constants of integration are included in the $k_m(x)$, then we simply recover the complementary function in addition to the particular integral.

►*Use the variation of parameters to solve*

$$\frac{d^2y}{dx^2} + y = \operatorname{cosec} x, \tag{13.48}$$

*subject to the boundary conditions* $y(0) = y(\pi/2) = 0$.

The complementary function of (13.48) is again

$$y_c(x) = c_1 \sin x + c_2 \cos x.$$

We therefore assume a particular integral of the form

$$y_p(x) = k_1(x) \sin x + k_2(x) \cos x,$$

and impose the additional constraints of (13.46), i.e.

$$k_1'(x) \sin x + k_2'(x) \cos x = 0,$$
$$k_1'(x) \cos x - k_2'(x) \sin x = \operatorname{cosec} x.$$

Solving these equations for $k_1'(x)$ and $k_2'(x)$ gives

$$k_1'(x) = \cos x \operatorname{cosec} x = \cot x,$$
$$k_2'(x) = -\sin x \operatorname{cosec} x = -1.$$

Hence, ignoring the constants of integration, $k_1(x)$ and $k_2(x)$ are given by

$$k_1(x) = \ln(\sin x),$$
$$k_2(x) = -x.$$

The general solution to the ODE (13.48) is therefore

$$y(x) = [c_1 + \ln(\sin x)] \sin x + (c_2 - x) \cos x,$$

which is identical to the solution found in the section 13.2.3. Applying the boundary conditions $y(0) = y(\pi/2) = 0$ we find $c_1 = c_2 = 0$, so that

$$y(x) = \ln(\sin x) \sin x - x \cos x. \blacktriangleleft$$

**Solution method.** *If the complementary function of (13.44) is known, then assume a particular integral of the same form but with the constants replaced by functions of x. Impose the constraints in (13.46), and solve the resulting system of equations for the unknowns $k_1'(x), \ldots, k_n'(x)$. Integrate these functions, setting constants of integration equal to zero, to obtain $k_1(x), \ldots, k_n(x)$ and hence the particular integral.*

### *13.2.5 Green's functions*

The Green's function method of solving linear ODEs bears a striking resemblance to the method of variation of parameters discussed in the previous subsection; it too requires knowledge of the entire complementary function in order to find the particular integral, and therefore the general solution. The Green's function approach differs, however, since once the Green's function for a particular LHS of (13.1) and accompanying boundary conditions has been found, then the solution for *any* RHS (i.e. $f(x)$) can be written down immediately, albeit in the form of an integral.

Although the Green's function method can be approached by considering the superposition of eigenfunctions of the equation (see chapter 15), and is also applicable to the solution of partial differential equations (see chapter 17), this section adopts a more utilitarian approach based on the properties of the Dirac delta function (see subsection 11.1.3), and deals only with the use of Green's functions in solving ODEs.

Let us again consider the equation

$$a_n(x)\frac{d^n y}{dx^n} + \cdots + a_1(x)\frac{dy}{dx} + a_0(x)y = f(x), \tag{13.49}$$

but for the sake of brevity denote the LHS by $\mathcal{L}y(x)$, i.e. a linear differential operator acting on $y(x)$. Thus (13.49) now reads

$$\mathcal{L}y(x) = f(x). \tag{13.50}$$

Let us suppose that a function $G(x,z)$ exists (the Green's function) such that the general solution to (13.50), which obeys some set of imposed boundary conditions in the range $a \le x \le b$, is given by

$$y(x) = \int_a^b G(x,z)f(z)\,dz, \tag{13.51}$$

where $z$ is the integration variable. If we apply the linear differential operator $\mathcal{L}$ to both sides of (13.51), and use (13.50), we obtain

$$\mathcal{L}y(x) = \int_a^b [\mathcal{L}G(x,z)]\, f(z)\,dz = f(x). \tag{13.52}$$

Comparison of (13.52) with a standard property of the Dirac delta function (see subsection 11.1.3), namely

$$f(x) = \int_a^b \delta(x-z)f(z)\,dz,$$

for $a \leq x \leq b$, shows that for (13.52) to hold for any function $f(x)$, we require (for $a \leq x \leq b$)

$$\mathcal{L}G(x,z) = \delta(x-z), \tag{13.53}$$

*i.e. the Green's function* $G(x,z)$ *must satisfy the original ODE with the RHS set equal to a delta function.* $G(x,z)$ may be thought of physically as the response of a system to a unit impulse at $x = z$.

In addition to (13.53), we must impose two further sets of restrictions on $G(x,z)$. The first concerns the requirement that the general solution $y(x)$ in (13.51) obeys the boundary conditions. For *homogeneous* boundary conditions, in which $y(x)$ and/or its derivatives are required to be *zero* at specified points, this is most simply arranged by demanding that $G(x,z)$ itself obeys the boundary conditions when it is considered as a function of $x$ alone: if, for example, we require $y(a) = y(b) = 0$, so we should also demand $G(a,z) = G(b,z) = 0$. Problems having inhomogeneous boundary conditions are discussed at the end of this subsection.

The second set of restrictions concerns the continuity or discontinuity of $G(x,z)$ and its derivatives at $x = z$, and can be found by integrating (13.53) with respect to $x$ over the small interval $[z-\epsilon, z+\epsilon]$ and taking the limit as $\epsilon \to 0$. We then obtain

$$\lim_{\epsilon\to 0} \sum_{m=0}^{n} \int_{z-\epsilon}^{z+\epsilon} a_m(x)\frac{d^m G(x,z)}{dx^m}\,dx = \lim_{\epsilon\to 0}\int_{z-\epsilon}^{z+\epsilon} \delta(x-z)\,dx = 1. \tag{13.54}$$

Since $d^nG/dx^n$ exists at $x = z$ but its value there is infinite, the $(n-1)$th-order derivative must have a finite discontinuity there, whereas all the lower-order derivatives, $d^mG/dx^m$ for $m < n-1$, must be continuous at this point. Therefore the terms containing these derivatives cannot contribute to the value of the integral on the LHS of (13.54). Noting that, apart from an arbitrary additive constant, $\int (d^mG/dx^m)\,dx = d^{m-1}G/dx^{m-1}$, integrating by parts we find

$$\lim_{\epsilon\to 0}\int_{z-\epsilon}^{z+\epsilon} a_m(x)\frac{d^m G(x,z)}{dx^m}\,dx = 0, \tag{13.55}$$

for $m = 0$ to $n-1$. Thus, since only the term containing $d^nG/dx^n$ contributes to the integral in (13.54), we find on performing the integration by parts that

$$\lim_{\epsilon\to 0}\left[a_n(x)\frac{d^{n-1}G(x,z)}{dx^{n-1}}\right]_{z-\epsilon}^{z+\epsilon} = 1. \tag{13.56}$$

Thus we have the further $n$ constraints that $G(x,z)$ and its derivatives up to order

$n-2$ are continuous at $x = z$, but that $d^{n-1}G/dx^{n-1}$ has a discontinuity of $1/a_n(z)$ at $x = z$.

Therefore the properties of the Green's function $G(x,z)$ for an $n$th-order linear ODE may be summarised by the following.

(i) $G(x,z)$ obeys the original ODE with the RHS set equal to a delta function $\delta(x-z)$.

(ii) When considered as a function of $x$ alone $G(x,z)$ obeys the specified (homogeneous) boundary conditions on $y(x)$.

(iii) The derivatives of $G(x,z)$ with respect to $x$ up to order $n-2$ are continuous at $x = z$, but the $(n-1)$th-order derivative has a discontinuity of $1/a_n(z)$ at this point.

►*Use Green's functions to solve*

$$\frac{d^2y}{dx^2} + y = \operatorname{cosec} x, \tag{13.57}$$

*subject to the boundary conditions* $y(0) = y(\pi/2) = 0$.

From (13.53) we see that the Green's function $G(x,z)$ must satisfy

$$\frac{d^2G(x,z)}{dx^2} + G(x,z) = \delta(x-z). \tag{13.58}$$

Now it is clear that for $x \neq z$ the RHS of (13.58) is zero, and we are left with the task of finding the general solution to the homogeneous equation, i.e. the complementary function. The complementary function of (13.58) consists of a linear superposition of $\sin x$ and $\cos x$, and *must* consist of different superpositions on either side of $x = z$ since its $(n-1)$th derivative (i.e. first derivative in this case) is required to have a discontinuity there. Therefore we assume the form of the Green's function to be

$$G(x,z) = \begin{cases} A(z)\sin x + B(z)\cos x & \text{for } x < z, \\ C(z)\sin x + D(z)\cos x & \text{for } x > z. \end{cases}$$

Note that we have performed a similar (but not identical) operation to that used in the variation of parameters, i.e. we have replaced the constants in the complementary function with functions (this time of $z$).

We must now impose the relevant restrictions on $G(x,z)$ in order to determine the functions $A(z), \ldots, D(z)$. The first of these is that $G(x,z)$ should itself obey the homogeneous boundary conditions $G(0,z) = G(\pi/2,z) = 0$. This leads to the conclusion that $B(z) = C(z) = 0$, so we now have

$$G(x,z) = \begin{cases} A(z)\sin x & \text{for } x < z, \\ D(z)\cos x & \text{for } x > z. \end{cases}$$

The second restriction is the continuity conditions given in equations (13.55),

(13.56), namely that (for this second-order equation) $G(x,z)$ is continuous at $x = z$ and that $dG/dx$ has a discontinuity of $1/a_2(z) = 1$ at this point. Applying these two constraints we have

$$D(z)\cos z - A(z)\sin z = 0$$
$$-D(z)\sin z - A(z)\cos z = 1.$$

Solving these equations for $A(z)$ and $D(z)$, we find

$$A(z) = -\cos z, \qquad D(z) = -\sin z.$$

Thus we have

$$G(x,z) = \begin{cases} -\cos z \sin x & \text{for } x < z, \\ -\sin z \cos x & \text{for } x > z. \end{cases}$$

Therefore from (13.51), the general solution to (13.57) that obeys the boundary conditions $y(0) = y(\pi/2) = 0$ is given by

$$\begin{aligned} y(x) &= \int_0^{\pi/2} G(x,z)\operatorname{cosec} z\, dz \\ &= -\cos x \int_0^x \sin z \operatorname{cosec} z\, dz - \sin x \int_x^{\pi/2} \cos z \operatorname{cosec} z\, dz \\ &= -x\cos x + \sin x \ln(\sin x), \end{aligned}$$

which agrees with the result obtained in the previous subsections. ◄

As mentioned earlier, once a Green's function has been obtained for a given LHS and boundary conditions, it can be used to find a general solution for any RHS, i.e. the solution of $d^2y/dx^2 + y = f(x)$, with $y(0) = y(\pi/2) = 0$, is given immediately by

$$\begin{aligned} y(x) &= \int_0^{\pi/2} G(x,z) f(z)\, dz \\ &= -\cos x \int_0^x \sin z\, f(z)\, dz - \sin x \int_x^{\pi/2} \cos z\, f(z)\, dz. \end{aligned} \qquad (13.59)$$

As an example, the reader may wish to verify that if $f(x) = \sin 2x$, then (13.59) gives $y(x) = -\frac{1}{3}\sin 2x$, a solution easily verified by direct substitution. Analytic integration of (13.59) for arbitrary $f(x)$ will, in general, prove intractable; then the integrals must be evaluated numerically.

Another important point is that although the Green's function method above has provided a general solution, it is also useful for finding a particular integral if the complementary function is known. This is easily seen since in (13.59) the constant integration limits 0 and $\pi/2$ lead merely to constant values by which the factors $\sin x$ and $\cos x$ are multiplied, so reconstructing the complementary function. The rest of the general solution, i.e. the particular integral, comes from the variable integration limit $x$. Therefore by changing $\int_x^{\pi/2}$ to $-\int^x$, so dropping the constant integration limits, we can find just the particular integral. For example, a particular integral of $d^2y/dx^2 + y = f(x)$ that satisfies the above boundary conditions is given by

$$y_p(x) = -\cos x \int^x \sin z \; f(z)\, dz + \sin x \int^x \cos z \; f(z)\, dz.$$

A very important point to realise about the Green's function method is that a particular $G(x,z)$ applies to a given LHS of an ODE *and* the imposed boundary conditions, i.e. *the same equation with different boundary conditions will have a different Green's function.* To illustrate this point, let us again consider the same ODE as solved above, but with different boundary conditions.

►*Use Green's functions to solve*

$$\frac{d^2y}{dx^2} + y = f(x), \tag{13.60}$$

*subject to the one-point boundary condition* $y(0) = y'(0) = 0$.

We again require (13.58) to hold and so again we assume a Green's function of the form

$$G(x,z) = \begin{cases} A(z)\sin x + B(z)\cos x & \text{for } x < z, \\ C(z)\sin x + D(z)\cos x & \text{for } x > z. \end{cases}$$

However, we now require $G(x,z)$ to obey the boundary conditions $G(0,z) = G'(0,z) = 0$, which implies $A(z) = B(z) = 0$. Therefore we have

$$G(x,z) = \begin{cases} 0 & \text{for } x < z, \\ C(z)\sin x + D(z)\cos x & \text{for } x > z. \end{cases}$$

Applying the continuity conditions on $G(x,z)$ as before now gives

$$C(z)\sin z + D(z)\cos z = 0,$$
$$C(z)\cos z - D(z)\sin z = 1,$$

which are solved to give

$$C(z) = \cos z, \qquad D(z) = -\sin z.$$

So finally the Green's function is given by

$$G(x,z) = \begin{cases} 0 & \text{for } x < z, \\ \sin(x-z) & \text{for } x > z, \end{cases}$$

and the general solution to (13.60) that obeys the boundary conditions $y(0) = y'(0) = 0$ is

$$y(x) = \int_0^\infty G(x,z)f(z)\,dz$$
$$= \int_0^x \sin(x-z)f(z)\,dz. \blacktriangleleft$$

As a final comment, we should consider how to deal with inhomogeneous boundary conditions such as $y(a) = \alpha$, $y(b) = \beta$ or $y(0) = y'(0) = \gamma$ etc., where $\alpha$, $\beta$, $\gamma$ are non-zero. The simplest way of solving such problems is to make a change of variable such that the boundary conditions in the new variable, $u$ say, are homogeneous, i.e. $u(a) = u(b) = 0$ or $u(0) = u'(0) = 0$ etc. For $n$th-order equations we generally require $n$ boundary conditions to fix the solution, but these boundary conditions can be of various types. For example we may have the $n$-point boundary conditions $y(x_m) = y_m$ for $m = 1$ to $n$, or the one-point boundary conditions $y(x_0) = y'(x_0) = \cdots = y^{(n-1)}(x_0) = y_0$, or something in between. In all cases a suitable change of variable is

$$u = y - h(x),$$

where $h(x)$ is an $(n-1)$th-order polynomial that obeys the boundary conditions. For example, if we consider the second-order case with boundary conditions $y(a) = \alpha$, $y(b) = \beta$ then a suitable change of variable is

$$u = y - (mx + c),$$

where $y = mx + c$ is the straight line through the points $(a, \alpha)$ and $(b, \beta)$; this is given by $m = (\alpha - \beta)/(a - b)$ and $c = (\beta a - \alpha b)/(a - b)$. Alternatively, if the boundary conditions for our second-order equation are $y(0) = y'(0) = \gamma$, then we would make the same change of variable, but this time $y = mx + c$ would be the straight line through $(0, \gamma)$ with slope $\gamma$, i.e. $m = c = \gamma$.

**Solution method.** *Require that the Green's function $G(x,z)$ obeys the original ODE, but with the RHS set to a delta function $\delta(x - z)$. This is equivalent to assuming that $G(x,z)$ is given by the complementary function of the original ODE, with the constants replaced by functions of $z$; these functions are different for $x < z$ and $x > z$. Now require also that $G(x,z)$ obeys the given homogeneous boundary conditions, and also impose the continuity conditions given in (13.55) and (13.56). The general solution to the original ODE is then given by (13.51). For inhomogeneous boundary conditions make the change of dependent variable $u = y - h(x)$, where $h(x)$ is a polynomial obeying the given boundary conditions.*

### *13.2.6 Canonical form for second-order equations*

In this section we specialise from $n$th-order linear ODEs with variable coefficients to those of order two. In particular we consider the equation

$$\frac{d^2y}{dx^2} + a_1(x)\frac{dy}{dx} + a_0(x)y = f(x), \qquad (13.61)$$

which has been rearranged so that the coefficient of $d^2y/dx^2$ is unity. By making the substitution $y(x) = u(x)v(x)$ we obtain

$$v'' + \left(\frac{2u'}{u} + a_1\right)v' + \left(\frac{u'' + a_1u' + a_0u}{u}\right)v = \frac{f}{u}, \qquad (13.62)$$

where the prime denotes differentiation with respect to $x$. Since (13.62) would be much simplified if there were no term in $v'$, let us choose $u(x)$ such that the first term in parentheses on the LHS of (13.62) is zero, i.e. so that

$$\frac{2u'}{u} + a_1 = 0 \quad \Rightarrow \quad u(x) = \exp\left\{-\tfrac{1}{2}\int a_1(z)\,dz\right\}. \qquad (13.63)$$

We then obtain an equation of the form

$$\frac{d^2v}{dx^2} + g(x)v = h(x), \qquad (13.64)$$

where

$$g(x) = a_0(x) - \tfrac{1}{4}[a_1(x)]^2 - \tfrac{1}{2}a_1'(x)$$
$$h(x) = f(x)\exp\left\{\tfrac{1}{2}\int a_1(z)\,dz\right\}.$$

Since (13.64) is of a simpler form than the original equation, (13.61), it may prove easier to solve.

►*Solve*

$$4x^2\frac{d^2y}{dx^2} + 4x\frac{dy}{dx} + (x^2 - 1)y = 0. \qquad (13.65)$$

Dividing (13.65) through by $4x^2$, we see that it is of the form (13.61) with $a_1(x) = 1/x$, $a_0(x) = (x^2-1)/4x^2$ and $f(x) = 0$. Therefore, making the substitution

$$y = vu = v\exp\left(-\int\frac{1}{2x}\,dx\right) = \frac{Av}{\sqrt{x}},$$

we obtain

$$\frac{d^2v}{dx^2} + \frac{v}{4} = 0. \qquad (13.66)$$

Equation (13.66) is easily solved to give

$$v = c_1 \sin \tfrac{1}{2}x + c_2 \cos \tfrac{1}{2}x,$$

so the solution of (13.65) is

$$y = \frac{v}{\sqrt{x}} = \frac{c_1 \sin \frac{1}{2}x + c_2 \cos \frac{1}{2}x}{\sqrt{x}}. \blacktriangleleft$$

As an alternative to choosing $u(x)$ such that the first bracket in (13.62) is zero, we could choose a different $u(x)$ such that the second bracket vanishes. For this to be the case, we see from (13.62) that we would require

$$u'' + a_1 u' + a_0 u = 0,$$

so $u(x)$ would have to be a solution of the original ODE with the RHS set to zero, i.e. part of the complementary function. If such a solution is known then the substitution $y = uv$ yields an equation with no term in $v$, which can be solved by two straightforward integrations. This is a special (second-order) case of the method discussed in subsection 13.2.3.

**Solution method.** *Write the equation in the form (13.61), then substitute* $y = uv$*, where* $u(x)$ *is given by (13.63). This leads to an equation of the form (13.64), in which there is no term in* $dv/dx$*, and which may be easier to solve. Alternatively, if part of the complementary function is known, then follow the method of subsection 13.2.3.*

## 13.3 General ordinary differential equations

In this section, we discuss miscellaneous methods for simplifying general ODEs. These methods are applicable to both linear and non-linear equations, and in some cases may lead to a solution. More often, however, finding a closed-form solution to a general non-linear ODE proves impossible.

### *13.3.1 Dependent variable absent*

If an ODE does not contain the dependent variable $y$ explicitly, but only its derivatives, then the change of variable $p = dy/dx$ leads to an equation of one order lower.

►*Solve*

$$\frac{d^2y}{dx^2} + 2\frac{dy}{dx} = 4x \tag{13.67}$$

This is transformed by the substitution $p = dy/dx$ to the first-order equation

$$\frac{dp}{dx} + 2p = 4x. \tag{13.68}$$

The solution to (13.68) is then found by the method of subsection 12.2.4, and reads

$$p = \frac{dy}{dx} = ae^{-2x} + 2x - 1,$$

where $a$ is a constant. Thus by direct integration the solution to the original equation, (13.67), is

$$y(x) = c_1 e^{-2x} + x^2 - x + c_2. \blacktriangleleft$$

An extension to the above method is appropriate if an ODE contains only derivatives of $y$ that are of order $m$ and greater. Then the substitution $p = d^m y/dx^m$ reduces the order of the ODE by $m$.

**Solution method.** *If the ODE contains only derivatives of y that are of order m and greater, then the substitution $p = d^m y/dx^m$ reduces the order of the equation by m.*

### *13.3.2 Independent variable absent*

If an ODE does not contain the independent variable $x$ explicitly, except in $d/dx$, $d^2/dx^2$ etc., then as in the previous subsection we make the substitution $p = dy/dx$, but also write

$$\begin{aligned}
\frac{d^2y}{dx^2} &= \frac{dp}{dx} = \frac{dy}{dx}\frac{dp}{dy} = p\frac{dp}{dy} \\
\frac{d^3y}{dx^3} &= \frac{d}{dx}\left(p\frac{dp}{dy}\right) = \frac{dy}{dx}\frac{d}{dy}\left(p\frac{dp}{dy}\right) = p^2\frac{d^2p}{dy^2} + p\left(\frac{dp}{dy}\right)^2,
\end{aligned} \tag{13.69}$$

and so on for higher-order derivatives. This leads to an equation of one order lower.

►*Solve*

$$1 + y\frac{d^2y}{dx^2} + \left(\frac{dy}{dx}\right)^2 = 0. \tag{13.70}$$

Making the substitution $dy/dx = p$ and $d^2y/dx^2 = p(dp/dy)$ we obtain the first-order ODE

$$1 + yp\frac{dp}{dy} + p^2 = 0,$$

which is separable and may be solved as in subsection 12.2.1 to obtain

$$(1 + p^2)y^2 = c_1.$$

Using $p = dy/dx$ we therefore have

$$p = \frac{dy}{dx} = \pm\sqrt{\frac{c_1^2 - y^2}{y^2}},$$

which may be integrated to give the general solution of (13.70); after squaring this reads

$$(x + c_2)^2 + y^2 = c_1^2. \blacktriangleleft$$

**Solution method.** *If the ODE does not contain x explicitly, then substitute $p = dy/dx$, along with the relations for higher derivatives given in (13.69), to obtain an equation of one order lower, which may prove easier to solve.*

### *13.3.3 Non-linear exact equations*

As discussed in subsection 13.2.2, an exact ODE is one that can be obtained by straightforward differentiation of an equation of one order lower. Moreover, the notion of exact equations is useful for both linear and non-linear equations, since an exact equation can be immediately integrated. It is, of course, possible that the resulting equation may itself be exact, so that the process can be repeated. In the non-linear case, however, there is no simple relation (such as (13.34) for the linear case) by which an equation can be shown to be exact. Nevertheless, a general procedure does exist, and is illustrated in the following example.

►*Solve*

$$2y\frac{d^3y}{dx^3} + 6\frac{dy}{dx}\frac{d^2y}{dx^2} = x. \qquad (13.71)$$

Directing our attention to the term on the LHS of (13.71) that contains the highest-order derivative, i.e. $2y\,d^3y/dx^3$, we see that it can be obtained by differentiating $2y\,d^2y/dx^2$ since

$$\frac{d}{dx}\left(2y\frac{d^2y}{dx^2}\right) = 2y\frac{d^3y}{dx^3} + 2\frac{dy}{dx}\frac{d^2y}{dx^2}. \qquad (13.72)$$

Rewriting the LHS of (13.71) using (13.72), we are left with $4(dy/dx)(d^2y/dy^2)$, which may itself be written as a derivative, i.e.

$$4\frac{dy}{dx}\frac{d^2y}{dx^2} = \frac{d}{dx}\left[2\left(\frac{dy}{dx}\right)^2\right]. \tag{13.73}$$

Since we can write the LHS of (13.71) as a sum of simple derivatives of other functions, (13.71) is exact. Integrating (13.71) with respect to $x$, and using (13.72) and (13.73), now gives

$$2y\frac{d^2y}{dx^2} + 2\left(\frac{dy}{dx}\right)^2 = \int x\,dx = \frac{x^2}{2} + c_1. \tag{13.74}$$

Now we can repeat the process to find whether (13.74) is itself exact. Considering the term on the LHS of (13.74) that contains the highest-order derivative, i.e. $2y\,d^2y/dx^2$, we note that we obtain this by differentiating $2y\,dy/dx$, as follows:

$$\frac{d}{dx}\left(2y\frac{dy}{dx}\right) = 2y\frac{d^2y}{dx^2} + 2\left(\frac{dy}{dx}\right)^2.$$

The above expression already contain all the terms on the LHS of (13.74), so we can integrate (13.74) to give

$$2y\frac{dy}{dx} = \frac{x^3}{6} + c_1x + c_2.$$

Integrating once more we obtain the solution

$$y^2 = \frac{x^4}{24} + \frac{c_1x^2}{2} + c_2x + c_3. \blacktriangleleft$$

It is worth noting that both linear equations (as discussed in subsection 13.2.2) and non-linear equations may sometimes be made exact by multiplying through by an appropriate integrating factor. Although no general method exists for finding such a factor, one may sometimes be found by inspection or inspired guesswork.

**Solution method.** *Rearrange the equation so that all the terms containing y or its derivatives are on the LHS, then check to see whether the equation is exact by attempting to write the LHS as a simple derivative. If this is possible then the equation is exact and may be integrated directly to give an equation of one order lower. If the new equation is itself exact the process can be repeated.*

### *13.3.4 Isobaric or homogeneous equations*

It is straightforward to generalise the discussion of first-order isobaric equations given in subsection 12.2.6 to equations of general order $n$. An $n$th-order isobaric equation is one in which every term can be made dimensionally consistent upon giving $y$ and $dy$ each a weight $m$, and $x$ and $dx$ each a weight 1. In this case the

$n$th derivative of $y$ with respect to $x$, for example, would have dimensions $m$ in $y$ and $-n$ in $x$. In the special case where the equation is dimensionally consistent with $m = 1$, the equation is called homogeneous (not to be confused with linear equations with a zero RHS). If an equation is isobaric or homogeneous, then the change in dependent variable $y = vx^m$ (or $y = vx$ for the homogeneous case) followed by the change in independent variable $x = e^t$ leads to an equation in which the new independent variable $t$ is absent except in the form $d/dt$.

►*Solve*

$$x^3 \frac{d^2y}{dx^2} - (x^2 + xy)\frac{dy}{dx} + (y^2 + xy) = 0. \tag{13.75}$$

Assigning $y$ and $dy$ the weight $m$, and $x$ and $dx$ the weight 1, the weights of the five terms on the LHS of (13.75) are, from left to right: $m+1$, $m+1$, $2m$, $2m$, $m+1$. For these weights all to be equal we require $m = 1$; thus (13.75) is a homogeneous equation. Since it is homogeneous we now make the substitution $y = vx$, which, after dividing the resulting equation through by $x^3$, gives

$$x\frac{d^2v}{dx^2} + (1 - v)\frac{dv}{dx} = 0. \tag{13.76}$$

Now substituting $x = e^t$ into (13.76) we obtain (after some working)

$$\frac{d^2v}{dt^2} - v\frac{dv}{dt} = 0, \tag{13.77}$$

which can be integrated directly to give

$$\frac{dv}{dt} = \tfrac{1}{2}v^2 + c_1. \tag{13.78}$$

Equation (13.78) is separable, and integrates to give

$$\begin{aligned} \tfrac{1}{2}t + d_2 &= \int \frac{dv}{v^2 + d_1^2} \\ &= \frac{1}{d_1}\tan^{-1}\left(\frac{v}{d_1}\right). \end{aligned}$$

Rearranging and using $x = e^t$ and $y = vx$ we finally obtain the solution to (13.75) as

$$y = d_1 x \tan\left(\tfrac{1}{2}d_1 \ln x + d_1 d_2\right). \blacktriangleleft$$

**Solution method.** *Assume that $y$ and $dy$ have weight $m$, and $x$ and $dx$ weight 1, and write down the combined weights of each term in the ODE. If these weights can be made equal by assuming a particular value for $m$ then the equation is isobaric (or homogeneous if $m = 1$). Making the substitution $y = vx^m$ followed by $x = e^t$ leads to an equation in which the new independent variable $t$ is absent except in the form $d/dt$.*

### 13.3.5 *Equations homogeneous in x or y alone*

It will be seen that the intermediate equation (13.76) in the example of the previous subsection was simplified by the substitution $x = e^t$, in that it led to an equation in which the new independent variable $t$ occurs only in the form $d/dt$, see (13.77). A closer examination of (13.76) reveals that it is dimensionally consistent in the independent variable $x$ *taken alone*; this is equivalent to giving the dependent variable and its differential a weight $m = 0$. For any equation that is homogeneous in $x$ alone, the substitution $x = e^t$ will lead to an equation that does not contain the new independent variable $t$, except as $d/dt$. Note that the Euler equation of subsection 13.2.1 is a special, linear example of an equation homogeneous in $x$ alone. Similarly if an equation is homogeneous in $y$ alone, then substituting $y = e^v$ leads to an equation in which the new dependent variable, $v$, occurs only in the form $d/dv$.

►*Solve*

$$x^2\frac{d^2y}{dx^2} + x\frac{dy}{dx} + \frac{2}{y^3} = 0.$$

This equation is homogeneous in $x$ alone, and on substituting $x = e^t$ we obtain

$$\frac{d^2y}{dt^2} + \frac{2}{y^3} = 0,$$

which does not contain the new independent variable $t$ except as $d/dt$. Such equations may often be solved by the method of subsection 13.3.2, but in this case we can integrate directly to obtain

$$\frac{dy}{dt} = \sqrt{2(c_1 + 1/y^2)}.$$

This equation is separable, and we find

$$\int \frac{dy}{\sqrt{2(c_1 + 1/y^2)}} = t + c_2.$$

By multiplying the numerator and denominator of the integrand on the LHS by $y$, we find the solution

$$\frac{\sqrt{c_1 y^2 + 1}}{\sqrt{2}c_1} = t + c_2.$$

Remembering that $t = \ln x$ we finally obtain

$$\frac{\sqrt{c_1 y^2 + 1}}{\sqrt{2}c_1} = \ln x + c_2. \blacktriangleleft$$

**Solution method.** *If the weight of $x$ taken alone is the same in every term in the ODE, then the substitution $x = e^t$ leads to an equation in which the new independent variable $t$ is absent. If the weight of $y$ taken alone is the same in every term then the substitution $y = e^v$ leads to an equation in which the new dependent variable $v$ is absent except in the form $d/dv$.*

### *13.3.6 Equations having $y = Ae^x$ as a solution*

As a final comment we note that if any general (linear or non-linear) $n$th-order ODE is satisfied identically by assuming

$$y = \frac{dy}{dx} = \cdots = \frac{d^n y}{dx^n}, \tag{13.79}$$

then $y = Ae^x$ is a solution of that equation. This must be so because $y = Ae^x$ is a non-zero function that satisfies (13.79).

►*Find a solution of*

$$(x^2 + x)\frac{dy}{dx}\frac{d^2y}{dx^2} - x^2 y\frac{dy}{dx} - x\left(\frac{dy}{dx}\right)^2 = 0. \tag{13.80}$$

Setting $y = dy/dx = d^2y/dx^2$ in (13.80), we obtain

$$(x^2 + x)y^2 - x^2y^2 - xy^2 = 0,$$

which is satisfied identically. Therefore $y = Ae^x$ is a solution of (13.80); this is easily verified by directly substituting $y = Ae^x$ into (13.80). ◄

**Solution method.** *If the equation is satisfied identically by assuming $y = dy/dx = \cdots = d^n y/dx^n$, then $y = Ae^x$ is a solution.*

## 13.4 Exercises

13.1 A simple harmonic oscillator, with natural frequency $\omega_0$, experiences an oscillating driving force $f(t) = \cos \omega t$. Therefore, its equation of motion is

$$\frac{d^2x}{dt^2} + \omega_0^2 x = \cos \omega t,$$

where $x$ is its position. Given that at $t = 0$ we have $x = dx/dt = 0$, find the function $x(t)$. Describe the solution if $\omega$ is approximately, but not exactly, equal to $\omega_0$.

13.2 Find the general solutions of

(a) $\dfrac{d^3y}{dx^3} - 12\dfrac{dy}{dx} + 16y = 32x - 8,$

(b) $\dfrac{d}{dx}\left(\dfrac{1}{y}\dfrac{dy}{dx}\right) + (2a\coth 2ax)\left(\dfrac{1}{y}\dfrac{dy}{dx}\right) = 2a^2,$

where $a$ is a constant.

13.3 The quantities $x(t)$, $y(t)$ satisfy the simultaneous equations

$$\ddot{x} + 2n\dot{x} + n^2x = 0$$
$$\ddot{y} + 2n\dot{y} + n^2y = \mu\dot{x},$$

where $x(0) = y(0) = \dot{y}(0) = 0$ and $\dot{x}(0) = \lambda$. Show that

$$y(t) = \tfrac{1}{2}\mu\lambda t^2\left(1 - \tfrac{1}{3}nt\right)\exp(-nt).$$

13.4 Two unstable isotopes $A$ and $B$ and a stable isotope $C$ have the following decay rates per atom present: $A \to B$, $3\text{s}^{-1}$; $A \to C$, $1\text{s}^{-1}$; $B \to C$, $2\text{s}^{-1}$. Initially a quantity $x_0$ of $A$ is present and none of the other two types. Using Laplace transforms, find the amount of $C$ present at a later time $t$.

13.5 Find the general solution of

$$x^2\frac{d^2y}{dx^2} - x\frac{dy}{dx} + y = x,$$

given that $y(1) = 1$ and $y(e) = 2e$.

13.6 Use the method of variation of parameters to find the general solutions of

(a) $\dfrac{d^2y}{dx^2} - y = x^n$, (b) $\dfrac{d^2y}{dx^2} - 2\dfrac{dy}{dx} + y = 2xe^x$.

13.7 Show that the Green's function for the equation

$$\frac{d^2y}{dx^2} + \frac{y}{4} = f(x),$$

subject to the boundary conditions $y(0) = y(\pi) = 0$, is given by

$$G(x,z) = \begin{cases} -2\cos\tfrac{1}{2}x\sin\tfrac{1}{2}z & 0 \le z \le x, \\ -2\sin\tfrac{1}{2}x\cos\tfrac{1}{2}z & x \le z \le \pi. \end{cases}$$

13.8 Find the Green's function $x = G(t, t_0)$ that solves

$$\frac{d^2x}{dt^2} + \alpha\frac{dx}{dt} = \delta(t - t_0),$$

under the initial conditions $x = dx/dt = 0$ at $t = 0$. Hence solve

$$\frac{d^2x}{dt^2} + \alpha\frac{dx}{dt} = f(t),$$

where $f(t) = 0$ for $t < 0$. Evaluate your answer explicitly for $f(t) = Ae^{-at}$ $(t > 0)$.

13.9 (a) By multiplying through by $dy/dx$, write down the solution to the equation

$$\frac{d^2y}{dx^2} + f(y) = 0,$$

where $f(y)$ can be any function.

(b) A mass $m$, initially at rest at the point $x = 0$, is accelerated by a force

$$f(x) = A(x_0 - x)\left[1 + 2\ln\left(1 - \frac{x}{x_0}\right)\right].$$

Its equation of motion is $m\,d^2x/dt^2 = f(x)$. Find $x$ as a function of time and show that ultimately the particle has travelled a distance $x_0$.

13.10 Solve

$$2y\frac{d^3y}{dx^3} + 2\left(y + 3\frac{dy}{dx}\right)\frac{d^2y}{dx^2} + 2\left(\frac{dy}{dx}\right)^2 = \sin x.$$

## 13.5 Hints and answers

13.1 $(\omega_0^2 - \omega^2)^{-1}(\cos\omega t - \cos\omega_0 t)$; for moderate $t$, $x(t)$ is a sine wave of linearly increasing amplitude $(t\sin\omega_0 t)/(2\omega_0)$; for large $t$ it shows beats of maximum amplitude $2(\omega_0^2 - \omega^2)^{-1}$.

13.2 (a) Auxiliary equation has roots 2, 2, −4; $(A+Bx)\exp 2x + C\exp(-4x) + 2x + 1$; (b) multiply through by $\sinh 2ax$ and note that $\int \operatorname{cosech} 2ax\,dx = (2a)^{-1}\ln(|\tanh ax|)$; $y = B(\sinh 2ax)^{1/2}(|\tanh ax|)^A$.

13.3 Use Laplace transforms; write $s(s+n)^{-4}$ as $(s+n)^{-3} - n(s+n)^{-4}$.

13.4 $\mathscr{L}[C(t)] = x_0(s+8)/[s(s+2)(s+4)]$ yielding $C(t) = x_0[1 + \frac{1}{2}\exp(-4t) - \frac{3}{2}\exp(-2t)]$.

13.5 Euler's equation; setting $x = \exp t$ produces $d^2z/dt^2 - 2\,dz/dt + z = \exp t$ with complementary function $(A + Bt)\exp t$ and particular integral $t^2(\exp t)/2$; $y(x) = x + [x\ln x(1 + \ln x)]/2$.

13.6 (a) Complementary function is $A\exp x + B\exp(-x)$; writing the particular integral as $k_1(x)\exp(x) + k_2(x)\exp(-x)$ gives $k_1' = x^n[\exp(-x)]/2$ and $k_2' = -x^n(\exp x)/2$. These lead to the particular integral $-(n!/2)\sum_{m=0}^{n}[1+(-1)^{n+m}]x^m/m!$. (b) Setting the particular integral equal to $k_1(x)\exp x + k_2(x)x\exp x$ gives the general solution as $y = (A + Bx + x^3/3)\exp x$.

13.7 With $y = A(x)\sin(x/2) + B(x)\cos(x/2)$, obtain $A'(z) = 2f(z)\cos(z/2)$ and $B'(z) = -2f(z)\sin(z/2)$ and hence identify $G(x,z)$.

13.8 Use continuity and the step condition on $\partial G/\partial t$ at $t = t_0$ to show $G(t,t_0) = \alpha^{-1}\{1 - \exp[\alpha(t_0 - t)]\}$ for $0 \le t_0 \le t$; $x(t) = A(\alpha - a)^{-1}\{a^{-1}[1 - \exp(-at)] - \alpha^{-1}[1 - \exp(-\alpha t)]\}$.

13.9 (a) $B + x = \int^y dz[A - 2\int^z f(u)\,du]^{-1/2}$; (b) show that the force is proportional to the derivative of $(x_0 - x)^2 \ln[x_0/(x_0 - x)]$; $x = x_0\{1 - \exp[-At^2/(2m)]\}$.

13.10 LHS of the equation is exact for two stages of integration, and then needs an integrating factor $\exp x$; $2y\,d^2y/dx^2 + 2y\,dy/dx + 2(dy/dx)^2$; $2y\,dy/dx + y^2 = d(y^2)/dx + y^2$; $y^2 = A\exp(-x) + Bx + C - (\sin x - \cos x)/2$.

# 14

# *Series solutions of ordinary differential equations*

In the previous chapter we discussed the solution of both homogeneous and non-homogeneous linear ordinary differential equations (ODEs) of order greater than or equal to two. In particular we developed methods for solving some equations in which the coefficients were not constant but functions of the independent variable $x$. In each case we were able to write the solutions to such equations in terms of elementary functions, or as integrals. In general, however, the solutions of equations with variable coefficients may not be written in this way, and we must consider alternative methods.

In this chapter we discuss a method for obtaining solutions to linear ODEs in the form of convergent series. Such series can be evaluated numerically, and those occurring most commonly are named and tabulated. There is in fact no distinct borderline between this and the previous chapter, since solutions in terms of elementary functions may equally well be written as convergent series (i.e. the relevant Taylor series). Indeed it is partly the fact that some series occur so frequently that they are given special names such as $\sin x$, $\cos x$ or $\exp x$.

Since we shall be principally concerned with second-order linear ODEs in this chapter, we begin with a discussion of these equations, and obtain some general results that will prove useful when we come to discuss series solutions.

## 14.1 Second-order linear ordinary differential equations

Any homogeneous second-order linear ODE can be written in the form

$$y'' + p(x)y' + q(x)y = 0, \qquad (14.1)$$

where $y' = dy/dx$ and $p(x)$ and $q(x)$ are given functions of $x$. From the previous chapter, we recall that the most general form of the solution to (14.1) is

$$y(x) = c_1 y_1(x) + c_2 y_2(x), \qquad (14.2)$$

where $y_1(x)$ and $y_2(x)$ are *linearly independent* solutions of (14.1), and $c_1$ and $c_2$ are constants which may be fixed by the boundary conditions (if supplied).

A full discussion of the linear independence of sets of functions was given at the beginning of the previous chapter, but for just two functions $y_1$ and $y_2$ to be linearly independent we simply require that $y_2$ is not a multiple of $y_1$. Equivalently, $y_1$ and $y_2$ must be such that the equation

$$c_1 y_1(x) + c_2 y_2(x) = 0$$

is *only* satisfied for $c_1 = c_2 = 0$. The linear independence of $y_1(x)$ and $y_2(x)$ can therefore usually be deduced by inspection, but in any case can always be verified by the evaluation of the Wronskian of the two solutions

$$W(x) = \begin{vmatrix} y_1 & y_2 \\ y_1' & y_2' \end{vmatrix} = y_1 y_2' - y_2 y_1'. \qquad (14.3)$$

If $W(x) \neq 0$ anywhere in a given interval, then $y_1$ and $y_2$ are linearly independent in that interval.

An alternative expression for $W(x)$, which we will make use of later, may be derived by differentiating (14.3) with respect to $x$ to give

$$W' = y_1 y_2'' + y_1' y_2' - y_2 y_1'' - y_2' y_1' = y_1 y_2'' - y_1'' y_2.$$

Since both $y_1$ and $y_2$ satisfy (14.1), we may substitute for $y_1''$ and $y_2''$ to obtain

$$W' = -y_1(py_2' + qy_2) + (py_1' + qy_1)y_2 = -p(y_1 y_2' - y_1' y_2) = -pW.$$

Integrating we find

$$W(x) = C \exp\left\{ -\int^x p(u)\, du \right\}, \qquad (14.4)$$

where $C$ is a constant. We further note that in the special case $p(x) \equiv 0$, we obtain $W =$ constant.

►*The functions $y_1 = \sin x$ and $y_2 = \cos x$ are both solutions of the equation $y'' + y = 0$. Evaluate the Wronskian of these two solutions, and hence show that they are linearly independent.*

The Wronskian of $y_1$ and $y_2$ is given by

$$W = y_1 y_2' - y_2 y_1' = -\sin^2 x - \cos^2 x = -1.$$

Since $W \neq 0$ the two solutions are linearly independent. We also note that $y'' + y = 0$ is a special case of (14.1) with $p(x) \equiv 0$, and $W =$ constant as expected from (14.4). ◄

From the previous chapter, we recall that, once we have obtained the general solution to the homogeneous second-order ODE (14.1) in the form (14.2), then the general solution to the *inhomogeneous* equation

$$y'' + p(x)y' + q(x)y = f(x), \tag{14.5}$$

can be written as the sum of the solution to the homogeneous equation $y_c(x)$ (the complementary function) and *any* function $y_p(x)$ (the particular integral) that satisfies (14.5) and is linearly independent of $y_c(x)$. We therefore have

$$y(x) = c_1 y_1(x) + c_2 y_2(x) + y_p(x). \tag{14.6}$$

General methods for obtaining $y_p$, which are applicable to equations with variable coefficients, such as the variation of parameters or Green's functions, were discussed in the previous chapter. An alternative description of the Green's function method for solving inhomogeneous equations is given in the next chapter. For the present, however, we will restrict our attention to the solutions of homogeneous ODEs in the form of convergent series.

### 14.1.1 Ordinary and singular points of an ODE

So far we have implicitly assumed that $y(x)$ is a *real* function of a *real* variable $x$. In general, however, this is not always the case, and in the remainder of this chapter we broaden our discussion by generalising to a *complex* function $y(z)$ of a *complex* variable $z$.

Let us therefore consider the second-order linear homogeneous ODE

$$y'' + p(z)y' + q(z) = 0. \tag{14.7}$$

where now $y' = dy/dz$; this is a straightforward generalisation of (14.1). A full discussion of complex functions and differentiation with respect to a complex variable $z$ is given in chapter 18, but for the purposes of the present chapter we need not concern ourselves with many of the subtleties that exist. In particular, we may treat differentation with respect to $z$ in an way analogous to ordinary differentiation with respect to a real variable $x$.

In (14.7), if at some point $z = z_0$ the functions $p(z)$ and $q(z)$ are finite, and can be expressed as complex power series (see section 3.5)

$$p(z) = \sum_{n=0}^{\infty} p_n (z - z_0)^n, \qquad q(z) = \sum_{n=0}^{\infty} q_n (z - z_0)^n,$$

then $p(z)$ and $q(z)$ are said to be *analytic* at $z = z_0$, and this point is called an *ordinary point* of the ODE. If, however, $p(z)$ or $q(z)$, or both, diverge at $z = z_0$, then it is called a *singular point* of the ODE.

Even if an ODE is singular at a given point $z = z_0$, it may still possess a non-singular (finite) solution at that point. In fact the necessary and sufficient condition† for such a solution to exist is that $(z-z_0)p(z)$ and $(z-z_0)^2q(z)$ are both analytic at $z = z_0$. Singular points that have this property are *regular singular points*, whereas any singular point not satisfying both these criteria is termed an *irregular* or *essential* singularity.

►*Legendre's equation has the form*

$$(1-z^2)y'' - 2zy' + \ell(\ell+1)y = 0, \tag{14.8}$$

*where $\ell$ is a constant. Show that $z = 0$ is an ordinary point and $z = \pm 1$ are regular singular points of this equation.*

We first divide through by $1 - z^2$ to put the equation into our standard form (14.7):

$$y'' - \frac{2z}{1-z^2}y' + \frac{\ell(\ell+1)}{1-z^2}y = 0.$$

Comparing with (14.7), we identify

$$p(z) = \frac{-2z}{1-z^2} = \frac{-2z}{(1+z)(1-z)}$$

$$q(z) = \frac{\ell(\ell+1)}{1-z^2} = \frac{\ell(\ell+1)}{(1+z)(1-z)}.$$

By inspection, $p(z)$ and $q(z)$ are analytic at $z = 0$, which is therefore an ordinary point, but both diverge for $z = \pm 1$, which are thus singular points. However, at $z = 1$ we see that both $(z-1)p(z)$ and $(z-1)^2q(z)$ are analytic, and hence $z = 1$ is a regular singular point. Similarly, at $z = -1$ both $(z+1)p(z)$ and $(z+1)^2q(z)$ are analytic, and it too is a regular singular point. ◄

So far we have assumed that $z_0$ is finite. However, we may sometimes wish to determine the nature of the point $|z| \to \infty$. This may be achieved straightforwardly by substituting $w = 1/z$ into the equation and investigating the behaviour at $w = 0$.

►*Show that Legendre's equation has a regular singularity at $|z| \to \infty$.*

Letting $w = 1/z$ the derivatives with respect to $z$ become

$$\frac{dy}{dz} = \frac{dy}{dw}\frac{dw}{dz} = -\frac{1}{z^2}\frac{dy}{dw} = -w^2\frac{dy}{dw},$$

$$\frac{d^2y}{dz^2} = \frac{dw}{dz}\frac{d}{dw}\left(\frac{dy}{dz}\right) = -w^2\left(-2w\frac{dy}{dw} - w^2\frac{d^2y}{dw^2}\right) = w^3\left(2\frac{dy}{dw} + w\frac{d^2y}{dw^2}\right).$$

† See, for example, Jeffreys and Jeffreys, *Mathematical Methods of Physics*, 3rd ed. (Cambridge University Press, 1966), p. 479.

| Equation | Regular singularities | Essential singularities |
|---|---|---|
| Legendre* | | |
| $(1-z^2)y''-2zy'+\ell(\ell+1)y=0$ | $-1,1,\infty$ | — |
| Chebyshev | | |
| $(1-z^2)y''-zy'+n^2y=0$ | $-1,1,\infty$ | — |
| Bessel | | |
| $z^2y''+zy+(z^2-\nu^2)y=0$ | 0 | $\infty$ |
| Laguerre* | | |
| $zy''+(1-z)y'+\alpha y=0$ | 0 | $\infty$ |
| Simple harmonic oscillator | | |
| $y''+\omega^2y=0$ | — | $\infty$ |
| Hermite | | |
| $y''-2zy'+2\alpha y=0$ | — | $\infty$ |

Table 14.1 Important ODEs in the physical sciences and engineering. The asterisks indicate that the corresponding *associated* equations (discussed in the next chapter) also have the same singular points.

If we substitute these derivatives into Legendre's equation (14.8) we obtain

$$\left(1-\frac{1}{w^2}\right)w^3\left(2\frac{dy}{dw}+w\frac{d^2y}{dw^2}\right)+2\frac{1}{w}w^2\frac{dy}{dw}+\ell(\ell+1)y=0,$$

which simplifies to give

$$w^2(w^2-1)\frac{d^2y}{dw^2}+2w^3\frac{dy}{dw}+\ell(\ell+1)y=0.$$

Dividing through by $w^2(w^2-1)$ to put the equation into standard form, and comparing with (14.7), we identify

$$p(w)=\frac{2w}{w^2-1},$$
$$q(w)=\frac{\ell(\ell+1)}{w^2(w^2-1)}.$$

At $w=0$, $p(w)$ is analytic but $q(w)$ diverges, and so the point $|z|\to\infty$ is a singular point of Legendre's equation. However, since $wp$ and $w^2q$ are both analytic at $w=0$, $|z|\to\infty$ is a regular singular point. ◄

Table 14.1 lists the singular points of several second-order linear ODEs that play important roles in the analysis of many physics and engineering problems. In sections 14.6 and 14.7 we consider the the solution of Legendre's and Bessel's equations in terms of convergent series, and discuss some useful properties of these solutions. The solutions of the remaining equations in table 14.1 may also be found in the form of convergent series, but a discussion of these solutions and their properties is left until the next chapter, where they are considered in the context of Sturm–Liouville systems. We now discuss the methods by which series solutions may be obtained.

### 14.2 Series solutions about an ordinary point

If $z = z_0$ is an ordinary point of (14.7), then it may be shown that *every* solution $y(z)$ of the equation is also analytic at $z = z_0$. In our subsequent discussion we will take $z_0$ as the origin, i.e. $z_0 = 0$. If this is not already the case, then a substitution $Z = z - z_0$ will make it so. Since every solution is analytic, $y(z)$ can be represented by a power series of the form (see section 18.13)

$$y(z) = \sum_{n=0}^{\infty} a_n z^n. \tag{14.9}$$

Moreover, it may be shown that such a power series converges for $|z| < R$, where $R$ is the radius of convergence and is equal to the distance from $z = 0$ to the nearest singular point of the ODE (see chapter 18). At the radius of convergence, however, the series may or may not converge (as shown in section 3.5).

Since every solution of (14.7) is analytic at an ordinary point, it is always possible to obtain two *independent* solutions (from which the general solution (14.2) can be constructed) of the form (14.9). The derivatives of $y$ with respect to $z$ are given by

$$y' = \sum_{n=0}^{\infty} n a_n z^{n-1} = \sum_{n=0}^{\infty} (n+1) a_{n+1} z^n, \tag{14.10}$$

$$y'' = \sum_{n=0}^{\infty} n(n-1) a_n z^{n-2} = \sum_{n=0}^{\infty} (n+2)(n+1) a_{n+2} z^n. \tag{14.11}$$

Note that, in each case, in the first equality the sum can still start at $n = 0$ since the first term in (14.10) and the first two terms in (14.11) are automatically zero. The second equality in each case is simply obtained by shifting the summation index so that the sum can be written in terms of coefficients of $z^n$. By substituting (14.9)–(14.11) into the ODE (14.7), and requiring that the coefficients of each power of $z$ sum to zero, we obtain a *recurrence relation* expressing $a_n$ as a function of the previous $a_r$ $(0 \leq r \leq n-1)$.

►*Find the series solutions, about $z = 0$, of*

$$y'' + y = 0.$$

By inspection $z = 0$ is an ordinary point of the equation and so we may obtain two independent solutions by making the substitution $y = \sum_{n=0}^{\infty} a_n z^n$. Using (14.9) and (14.11) we find

$$\sum_{n=0}^{\infty}(n+2)(n+1)a_{n+2}z^n + \sum_{n=0}^{\infty} a_n z^n = 0,$$

which may be written as

$$\sum_{n=0}^{\infty}[(n+2)(n+1)a_{n+2} + a_n]z^n = 0.$$

For this equation to be satisfied we require that the coefficient of each power of $z$ vanishes *separately*, and so we obtain the two-term recurrence relation

$$a_{n+2} = -\frac{a_n}{(n+2)(n+1)} \qquad \text{for } n \geq 0.$$

Using this relation, we can calculate, say, the even coefficients $a_2$, $a_4$, $a_6$ and so on, for a given $a_0$. Alternatively, starting with $a_1$, we may obtain the odd coefficients $a_3$, $a_5$ etc. Two independent solutions of the ODE may be obtained by setting either $a_0 = 0$ or $a_1 = 0$. If we first set $a_1 = 0$ and choose $a_0 = 1$, then we obtain the solution

$$y_1(z) = 1 - \frac{z^2}{2!} + \frac{z^4}{4!} - \cdots = \sum_{n=0}^{\infty} \frac{(-1)^n}{(2n)!} z^{2n}.$$

However, if we set $a_0 = 0$ and choose $a_1 = 1$, we obtain a second *independent* solution

$$y_2(z) = z - \frac{z^3}{3!} + \frac{z^5}{5!} - \cdots = \sum_{n=0}^{\infty} \frac{(-1)^n}{(2n+1)!} z^{2n+1}.$$

Recognising these two series as $\cos z$ and $\sin z$, we can write the general solution as

$$y(z) = c_1 \cos z + c_2 \sin z,$$

where $c_1$ and $c_2$ are arbitrary constants which may be fixed by boundary conditions (if supplied). We note that both solutions converge for all $z$, as might be expected since the ODE possesses no singular points (except $|z| \to \infty$). ◀

Solving the above example was quite straightforward and the resulting series were easily recognised and written in *closed form* (i.e. in terms of elementary functions); *this is not usually the case*. Another simplifying feature of the previous example was that we obtained a two-term recurrence relation relating $a_{n+2}$ and $a_n$, so that the odd- and even-numbered coefficients were independent of one another. In general the recurrence relation expresses $a_n$ as a function of any number of the previous $a_r$ $(0 \leq r \leq n-1)$.

►*Find the series solutions, about $z = 0$, of*

$$y'' - \frac{2}{(1-z)^2}y = 0.$$

By inspection $z = 0$ is an ordinary point, and we may therefore find two independent solutions by substituting $y = \sum_{n=0}^{\infty} a_n z^n$. Using (14.10) and (14.11), and multiplying through by $(1-z)^2$, we find

$$(1 - 2z + z^2)\sum_{n=0}^{\infty} n(n-1)a_n z^{n-2} - 2\sum_{n=0}^{\infty} a_n z^n = 0,$$

which leads to

$$\sum_{n=0}^{\infty} n(n-1)a_n z^{n-2} - 2\sum_{n=0}^{\infty} n(n-1)a_n z^{n-1} + \sum_{n=0}^{\infty} n(n-1)a_n z^n - 2\sum_{n=0}^{\infty} a_n z^n = 0.$$

In order to write all these series in terms of the coefficients of $z^n$, we must shift the summation index in the first two sums to obtain

$$\sum_{n=0}^{\infty}(n+2)(n+1)a_{n+2}z^n - 2\sum_{n=0}^{\infty}(n+1)na_{n+1}z^n + \sum_{n=0}^{\infty}(n^2 - n - 2)a_n z^n = 0,$$

which can be written as

$$\sum_{n=0}^{\infty}(n+1)[(n+2)a_{n+2} - 2na_{n+1} + (n-2)a_n]z^n = 0.$$

By demanding that the coefficients of each power of $z$ vanish separately, we obtain the three-term recurrence relation

$$(n+2)a_{n+2} - 2na_{n+1} + (n-2)a_n = 0 \qquad \text{for } n \geq 0,$$

which determines $a_n$ for $n \geq 2$ in terms of $a_0$ and $a_1$. Three-term (or more) recurrence relations are a nuisance and, in general, can be difficult to solve. This particular recurrence relation, however, has two straightforward solutions. One solution is $a_n = a_0$ for all $n$, in which case (choosing $a_0 = 1$) we find

$$y_1(z) = 1 + z + z^2 + z^3 + \cdots = \frac{1}{1-z}.$$

The other solution to the recurrence relation is $a_1 = -2a_0$, $a_2 = a_0$ and $a_n = 0$ for $n > 2$, so that (again choosing $a_0 = 1$) we obtain a *polynomial* solution to the ODE:

$$y_2(z) = 1 - 2z + z^2 = (1-z)^2.$$

The linear independence of $y_1$ and $y_2$ is obvious, but can be checked by computing the Wronskian

$$W = y_1 y_2' - y_1' y_2 = \frac{1}{1-z}[-2(1-z)] - \frac{1}{(1-z)^2}(1-z)^2 = -3.$$

Since $W \neq 0$ the two solutions $y_1$ and $y_2$ are indeed linearly independent. The general solution of the ODE is therefore

$$y(z) = \frac{c_1}{1-z} + c_2(1-z)^2.$$

We observe that $y_1$ (and hence the general solution) is singular at $z = 1$, which is the singular point of the ODE nearest to $z = 0$, but the polynomial solution $y_2$ is valid for all finite $z$. ◄

The above example illustrates the possibility that, in some cases, we may find that the recurrence relation leads to $a_n = 0$ for $n > N$, for one or both of the two solutions; we then obtain a *polynomial* solution to the equation. Polynomial solutions are discussed more fully in section 14.5, but one obvious property of such solutions is that they converge for all finite $z$, in contrast to solutions in the form of an infinite series for which (as mentioned above) the radius of convergence extends only as far as the distance to the singular point of the equation closest to the point about which the series solution is obtained.

## 14.3 Series solutions about a regular singular point

From table 14.1 we see that several of the most important second-order linear ODEs in physics and engineering have regular singular points in the finite complex plane. We must therefore extend our discussion to obtaining series solutions to ODEs about such points. In what follows we assume that the regular singular point about which the solution is required is at $z = 0$, since (as mentioned above) if this is not already the case then a substitution of the form $Z = z - z_0$ will make it so.

If $z = 0$ is a regular singular point of the equation

$$y'' + p(z)y' + q(z)y = 0,$$

then $p(z)$ and $q(z)$ are not analytic at $z = 0$, and in general we should not expect to find a power series solution of the form (14.9). We must therefore extend the method to include a more general form for the solution. In fact it may be shown (Fuch's theorem) that there exists *at least one* solution to the above equation, of the form

$$y = z^\sigma \sum_{n=0}^{\infty} a_n z^n, \tag{14.12}$$

where the exponent $\sigma$ may be any real or complex number, and where $a_0 \neq 0$ (since otherwise one could just shift $\sigma$ to $\sigma + 1$). Such a series is called a generalised power series or *Frobenius series.* As in the case of the simple power series solution, the radius of convergence of the Frobenius series is, in general, equal to the distance to the nearest singularity of the ODE.

Since $z = 0$ is a regular singularity of the ODE, $zp(z)$ and $z^2q(z)$ are analytic at $z = 0$, so that we may write

$$\begin{aligned} zp(z) &= s(z) = \sum_{n=0}^{\infty} s_n z^n \\ z^2 q(z) &= t(z) = \sum_{n=0}^{\infty} t_n z^n, \end{aligned}$$

where we have defined the analytic functions $s(z)$ and $t(z)$ for later convenience. The original ODE therefore becomes

$$y'' + \frac{s(z)}{z} y' + \frac{t(z)}{z^2} y = 0.$$

Let us substitute the Frobenius series (14.12) into this equation. The derivatives of (14.12) with respect to $x$ are given by

$$y' = \sum_{n=0}^{\infty} (n+\sigma) a_n z^{n+\sigma-1} \tag{14.13}$$

$$y'' = \sum_{n=0}^{\infty} (n+\sigma)(n+\sigma-1) a_n z^{n+\sigma-2}, \tag{14.14}$$

and we obtain

$$\sum_{n=0}^{\infty} (n+\sigma)(n+\sigma-1) a_n z^{n+\sigma-2} + s(z) \sum_{n=0}^{\infty} (n+\sigma) a_n z^{n+\sigma-2} + t(z) \sum_{n=0}^{\infty} a_n z^{n+\sigma-2} = 0.$$

Dividing this equation through by $z^{\sigma-2}$ we find

$$\sum_{n=0}^{\infty} \left[ (n+\sigma)(n+\sigma-1) + s(z)(n+\sigma) + t(z) \right] a_n z^n = 0. \tag{14.15}$$

Setting $z = 0$, all terms in the sum with $n > 0$ vanish, so that

$$[\sigma(\sigma-1) + s(0)\sigma + t(0)] a_0 = 0,$$

which, since we require $a_0 \neq 0$, yields the *indicial equation*

$$\sigma(\sigma-1) + s(0)\sigma + t(0) = 0. \tag{14.16}$$

This equation is a quadratic in $\sigma$ and in general has two roots, the nature of which determines the forms of possible series solutions.

The two roots of the indicial equation $\sigma_1$ and $\sigma_2$ are called the *indices* of the regular singular point. By substituting each of these roots into (14.15) in turn, and requiring that the coefficients of each power of $z$ vanish separately, we obtain a recurrence relation (for each root) expressing each $a_n$ as a function of the previous $a_r$ $(0 \le r \le n-1)$. Depending on the roots of the indicial equation $\sigma_1$ and $\sigma_2$, there are three possible general cases, which we now discuss.

### *14.3.1 Distinct roots not differing by an integer*

If the roots of the indicial equation $\sigma_1$ and $\sigma_2$ differ by an amount that is not an integer, then the recurrence relations corresponding to each root lead to two linearly independent solutions of the ODE,

$$y_1(z) = z^{\sigma_1} \sum_{n=0}^{\infty} a_n z^n, \qquad y_2(z) = z^{\sigma_2} \sum_{n=0}^{\infty} b_n z^n.$$

The linear independence of these two solutions follows from the fact that $y_2/y_1$ is not a constant since $\sigma_1 - \sigma_2$ is not an integer. Since $y_1$ and $y_2$ are linearly independent we may use them to construct the general solution $y = c_1 y_1 + c_2 y_2$.

We also note that this case includes complex conjugate roots where $\sigma_2 = \sigma_1^*$, since $\sigma_1 - \sigma_2 = \sigma_1 - \sigma_1^* = 2i \operatorname{Im} \sigma_1$ cannot be equal to a real integer.

►*Find the power series solutions about $z = 0$ of*

$$4zy'' + 2y' + y = 0.$$

Dividing through by $4z$ to put the equation into standard form, we obtain

$$y'' + \frac{1}{2z} y' + \frac{1}{4z} y = 0, \tag{14.17}$$

and on comparing with (14.7) we identify $p(z) = 1/(2z)$ and $q(z) = 1/(4z)$. Clearly $z = 0$ is a singular point of (14.17), but since $zp(z) = 1/2$ and $z^2 q(z) = z/4$ are finite there, it is a regular singular point. We therefore substitute the Frobenius series $y = z^\sigma \sum_{n=0}^{\infty} a_n z^n$ into (14.17). Using (14.13) and (14.14), we obtain

$$\sum_{n=0}^{\infty} (n+\sigma)(n+\sigma-1) a_n z^{n+\sigma-2} + \frac{1}{2z} \sum_{n=0}^{\infty} (n+\sigma) a_n z^{n+\sigma-1} + \frac{1}{4z} \sum_{n=0}^{\infty} a_n z^{n+\sigma} = 0,$$

which on dividing through by $z^{\sigma-2}$ gives

$$\sum_{n=0}^{\infty} \left[ (n+\sigma)(n+\sigma-1) + \tfrac{1}{2}(n+\sigma) + \tfrac{1}{4}z \right] a_n z^n = 0. \tag{14.18}$$

If we set $z = 0$ then all terms in the sum with $n > 0$ vanish, and we obtain the indicial equation

$$\sigma(\sigma - 1) + \tfrac{1}{2}\sigma = 0,$$

which has roots $\sigma = 1/2$ and $\sigma = 0$. Since these roots do not differ by an integer we expect to find two independent solutions to (14.17), in the form of Frobenius series.

Demanding that the coefficients of $z^n$ vanish separately in (14.18), we obtain the recurrence relation

$$(n+\sigma)(n+\sigma-1)a_n + \tfrac{1}{2}(n+\sigma)a_n + \tfrac{1}{4}a_{n-1} = 0. \qquad (14.19)$$

If we choose the larger root of the indicial equation, $\sigma = 1/2$, this becomes

$$(4n^2+2n)a_n + a_{n-1} = 0 \qquad \Rightarrow \qquad a_n = \frac{-a_{n-1}}{2n(2n+1)}.$$

Setting $a_0 = 1$ we find $a_n = (-1)^n/(2n+1)!$ and so the solution to (14.17) is

$$\begin{aligned} y_1(z) &= \sqrt{z}\sum_{n=0}^{\infty} \frac{(-1)^n}{(2n+1)!} z^n \\ &= \sqrt{z} - \frac{(\sqrt{z})^3}{3!} + \frac{(\sqrt{z})^5}{5!} - \cdots = \sin\sqrt{z}. \end{aligned}$$

To obtain the second solution we set $\sigma = 0$ (the smaller root of the indicial equation) in (14.19), which gives

$$(4n^2-2n)a_n + a_{n-1} = 0 \qquad \Rightarrow \qquad a_n = -\frac{a_{n-1}}{2n(2n-1)}.$$

Setting $a_0 = 1$ now gives $a_n = (-1)^n/(2n)!$, and so the second (independent) solution to (14.17) is

$$y_2(z) = \sum_{n=0}^{\infty} \frac{(-1)^n}{(2n)!} z^n = 1 - \frac{(\sqrt{z})^2}{2!} + \frac{(\sqrt{4})^4}{4!} - \cdots = \cos\sqrt{z}.$$

We may check that $y_1(z)$ and $y_2(z)$ are indeed linearly independent by computing the Wronskian

$$\begin{aligned} W &= y_1 y_2' - y_2 y_1' \\ &= \sin\sqrt{z}\left(-\frac{1}{2\sqrt{z}}\sin\sqrt{z}\right) - \cos\sqrt{z}\left(\frac{1}{2\sqrt{z}}\cos\sqrt{z}\right) \\ &= -\frac{1}{2\sqrt{z}}\left(\sin^2\sqrt{z} + \cos^2\sqrt{z}\right) = -\frac{1}{2\sqrt{z}} \neq 0. \end{aligned}$$

Since $W \neq 0$ the solutions $y_1(z)$ and $y_2(z)$ are linearly independent. Hence the general solution to (14.17) is given by

$$y(z) = c_1 \sin\sqrt{z} + c_2 \cos\sqrt{z}. \blacktriangleleft$$

### *14.3.2 Repeated root of the indicial equation*

If the indicial equation has a repeated root, so that $\sigma_1 = \sigma_2 = \sigma$, then obviously only one solution in the form of a Frobenius series (14.12) may be found as described above, i.e.

$$y_1(z) = z^\sigma \sum_{n=0}^{\infty} a_n z^n.$$

Methods for obtaining a second, linearly independent, solution are discussed in section 14.4.

### 14.3.3 Distinct roots differing by an integer

If the roots of the indicial equation differ by an integer, then the recurrence relation corresponding to the larger of the two roots leads to a solution of the ODE. However, the recurrence relation corresponding to the smaller root may or may not lead to a second linearly independent solution, depending on the ODE under consideration. Note that for complex roots of the indicial equation, the 'larger' root is taken to be the one with the larger real part.

►*Find the power series solutions about $z = 0$ of*

$$z(z-1)y'' + 3zy' + y = 0. \tag{14.20}$$

Dividing through by $z(z-1)$ to put the equation into standard form, we obtain

$$y'' + \frac{3}{(z-1)}y' + \frac{1}{z(z-1)}y = 0, \tag{14.21}$$

and on comparing with (14.7) we identify $p(z) = 3/(z-1)$ and $q(z) = 1/[z(z-1)]$. We immediately see that $z = 0$ is a singular point of (14.21), but since $zp(z) = 3z/(z-1)$ and $z^2q(z) = z/(z-1)$ are finite there, it is a regular singular point and we expect to find at least one solution in the form of a Frobenius series. We therefore substitute $y = z^\sigma \sum_{n=0}^{\infty} a_n z^n$ into (14.21), and using (14.13) and (14.14), we obtain

$$\sum_{n=0}^{\infty}(n+\sigma)(n+\sigma-1)a_n z^{n+\sigma-2} + \frac{3}{z-1}\sum_{n=0}^{\infty}(n+\sigma)a_n z^{n+\sigma-1} + \frac{1}{z(z-1)}\sum_{n=0}^{\infty} a_n z^{n+\sigma} = 0,$$

which on dividing through by $z^{\sigma-2}$ gives

$$\sum_{n=0}^{\infty}\left[(n+\sigma)(n+\sigma-1) + \frac{3z}{z-1}(n+\sigma) + \frac{z}{z-1}\right] a_n z^n = 0.$$

Although we could use this expression to find the indicial equation and recurrence relations, the working is simpler if we now multiply through by $z-1$ to give

$$\sum_{n=0}^{\infty}[(z-1)(n+\sigma)(n+\sigma-1)+3z(n+\sigma)+z]\,a_n z^n = 0. \qquad (14.22)$$

If we set $z=0$ then all terms in the sum with $n>0$ vanish, and we obtain the indicial equation

$$\sigma(\sigma-1)=0,$$

which has the roots $\sigma=1$ and $\sigma=0$. Since the roots differ by an integer (unity), it may not be possible to find two linearly independent solutions of (14.21) in the form of Frobenius series. We are, however, guaranteed to find one such solution corresponding to the larger root, $\sigma=1$.

Demanding that the coefficients of $z^n$ vanish separately in (14.22), we obtain the recurrence relation

$$(n-1+\sigma)(n-2+\sigma)a_{n-1}-(n+\sigma)(n+\sigma-1)a_n+3(n-1+\sigma)a_{n-1}+a_{n-1}=0,$$

which may be simplified to give

$$(n+\sigma-1)a_n=(n+\sigma)a_{n-1}. \qquad (14.23)$$

Substituting $\sigma=1$ into this expression, we obtain

$$a_n=\left(\frac{n+1}{n}\right)a_{n-1},$$

and setting $a_0=1$ we find $a_n=n+1$, so one solution to (14.21) is

$$y_1(z)=z\sum_{n=0}^{\infty}(n+1)z^n = z(1+2z+3z^2+\cdots)$$
$$=\frac{z}{(1-z)^2}. \qquad (14.24)$$

If we attempt to find a second solution (corresponding to the smaller root of the indicial equation) by setting $\sigma=0$ in (14.23), we find

$$a_n=\left(\frac{n}{n-1}\right)a_{n-1},$$

but we require $a_0\neq 0$, so $a_1$ is formally infinite and the method fails. We discuss how to find a second linearly independent solution in the next section. ◀

One particular case is also worth mentioning. If the point about which the solution is required, i.e. $z=0$, is in fact an ordinary point of the ODE rather than a regular singular point, then substitution of the Frobenius series (14.12) leads to an indicial equation with roots $\sigma=0$ and $\sigma=1$. Although these roots differ by an integer (unity), the recurrence relations corresponding to the two roots yield two linearly independent power series solutions (one for each root), as expected from section 14.2.

Finally, the reader may have noticed that every infinite series solution found so far has been summable in closed form or expressible in terms of known functions. This was primarily intended to show that it is always worth investigating this possibility once a series solution has been found. It could, however, give the impression that this is always possible or that, if one worked hard enough, a closed-form solution can always be found without using the series method. As mentioned earlier, *this is not the case*, and very often an infinite series solution cannot be written in closed form.

## 14.4 Obtaining a second solution

In attempting to find a solution to an ODE in the form of a Frobenius series about a regular singular point, we found in the previous section that when the indicial equation has a repeated root or roots differing by an integer, we can (in general) find only one solution of this form. In order to construct the general solution to the ODE, however, we require two linearly independent solutions $y_1$ and $y_2$. We now consider several methods for obtaining a second solution.

### *14.4.1 The Wronskian method*

If $y_1$ and $y_2$ are two linearly independent solutions of the standard equation

$$y'' + p(z)y' + q(z)y = 0,$$

then the Wronskian of these two solutions is given by $W(z) = y_1 y_2' - y_2 y_1'$. Dividing the Wronskian by $y_1^2$ we obtain

$$\frac{W}{y_1^2} = \frac{y_2'}{y_1} - \frac{y_1'}{y_1^2} y_2 = \frac{y_2'}{y_1} + \left[\frac{d}{dz}\left(\frac{1}{y_1}\right)\right] y_2 = \frac{d}{dz}\left(\frac{y_2}{y_1}\right),$$

which integrates to give

$$y_2(z) = y_1(z) \int^z \frac{W(u)}{y_1^2(u)}\, du.$$

Now using the alternative expression for $W(z)$ given in (14.4) with $C = 1$ (since we are not concerned with this normalising factor), we find

$$y_2(z) = y_1(z) \int^z \frac{1}{y_1^2(u)} \exp\left\{ -\int^u p(v)\, dv \right\} du. \tag{14.25}$$

Hence, given $y_1$, we can in principle compute $y_2$. Note that the lower limits of integration have been omitted. If constant lower limits are included they merely lead to a constant times the first solution.

►*Find a second solution to (14.21) using the Wronskian method.*

For the ODE (14.21) we have $p(z) = 3/(z-1)$, and from (14.24) we see that one solution to (14.21) is $y_1 = z/(1-z)^2$. Substituting for $p$ and $y_1$ in (14.25) we have

$$\begin{aligned} y_2(z) &= \frac{z}{(1-z)^2} \int^z \frac{(1-u)^4}{u^2} \exp\left(-\int^u \frac{3}{v-1}\, dv\right) du \\ &= \frac{z}{(1-z)^2} \int^z \frac{(1-u)^4}{u^2} \exp\left[-3\ln(u-1)\right] du \\ &= \frac{z}{(1-z)^2} \int^z \frac{u-1}{u^2}\, du \\ &= \frac{z}{(1-z)^2} \left(\ln z + \frac{1}{z}\right). \end{aligned}$$

By calculating the Wronskian of $y_1$ and $y_2$ it is easily shown that, as expected, these two solutions are linearly independent. ◄

An alternative (but equivalent) method of finding a second solution is simply to assume the second solution has the form $y_2(z) = u(z)y_1(z)$ for some function $u(z)$ to be determined (this method was discussed more fully in subsection 13.2.3 of the previous chapter). From (14.25), we see that the second solution derived from the Wronskian is indeed of this form. Substituting $y_2(z) = u(z)y_1(z)$ into the ODE leads to a first-order ODE in which $u'$ is the dependent variable; this may then be solved.

### 14.4.2 The derivative method

The derivative method of finding a second solution begins with the derivation of a recurrence relation for the coefficients $a_n$ in a Frobenius series solution, as in the previous section. However, rather than putting $\sigma = \sigma_1$ in this recurrence relation to evaluate the first series solution, we instead keep $\sigma$ as a variable parameter. This means that the computed $a_n$ are functions of $\sigma$ and the computed solution is now a function of $z$ and $\sigma$:

$$y(z, \sigma) = z^\sigma \sum_{n=0}^{\infty} a_n(\sigma) z^n. \qquad (14.26)$$

Of course, if we put $\sigma = \sigma_1$ in this, we immediately obtain the first series solution, but for the moment we leave it as a parameter.

For brevity let us denote the differential operator on the LHS of our standard ODE (14.7) by $\mathcal{L}$, so that

$$\mathcal{L} = \frac{d^2}{dz^2} + p(z)\frac{d}{dz} + q(z),$$

and examine the effect of $\mathcal{L}$ on the series $y(z,\sigma)$ in (14.26). It is clear that the series $\mathcal{L}y(z,\sigma)$ will contain only a term in $z^\sigma$, since the recurrence relation defining the $a_n(\sigma)$ is such that these coefficients vanish for higher powers of $z$. But the coefficient of $z^\sigma$ is simply the LHS of the indicial equation. Therefore, if the roots of the indicial equation are $\sigma = \sigma_1$ and $\sigma = \sigma_2$, it follows that

$$\mathcal{L}y(z,\sigma) = a_0(\sigma - \sigma_1)(\sigma - \sigma_2)z^\sigma. \qquad (14.27)$$

We therefore see (as in the previous section) that for $y(z,\sigma)$ to be a solution of the ODE $\mathcal{L}y = 0$, $\sigma$ must equal $\sigma_1$ or $\sigma_2$. For simplicity we shall set $a_0 = 1$ in the following discussion.

Let us first consider the case in which the two roots of the indicial equation are equal, i.e. $\sigma_2 = \sigma_1$. From (14.27) we then have

$$\mathcal{L}y(z,\sigma) = (\sigma - \sigma_1)^2 z^\sigma.$$

Differentiating this equation with respect to $\sigma$ we obtain

$$\frac{\partial}{\partial\sigma}\left[\mathcal{L}y(z,\sigma)\right] = (\sigma - \sigma_1)^2 z^\sigma \ln z + 2(\sigma - \sigma_1)z^\sigma,$$

which equals zero if $\sigma = \sigma_1$. But since $\partial/\partial\sigma$ and $\mathcal{L}$ are operators that differentiate with respect to different variables we can reverse their order, so that

$$\mathcal{L}\left[\frac{\partial}{\partial\sigma}y(z,\sigma)\right] = 0 \qquad \text{at } \sigma = \sigma_1.$$

Hence the function in square brackets, evaluated at $\sigma = \sigma_1$, and denoted by

$$\left[\frac{\partial}{\partial\sigma}y(z,\sigma)\right]_{\sigma=\sigma_1}, \qquad (14.28)$$

is also a solution of the original ODE $\mathcal{L}y = 0$, and is in fact the second linearly independent solution for which we were looking.

The case in which the roots of the indicial equation differ by an integer is slightly more complicated, but can be treated in a similar way. In (14.27), since $\mathcal{L}$ differentiates with respect to $z$ we may multiply (14.27) by any function of $\sigma$, say $\sigma - \sigma_2$, and take this function inside the operator $\mathcal{L}$ on the LHS to obtain

$$\mathcal{L}\left[(\sigma - \sigma_2)y(z,\sigma)\right] = (\sigma - \sigma_1)(\sigma - \sigma_2)^2 z^\sigma. \qquad (14.29)$$

Therefore the function

$$\left[(\sigma - \sigma_2)y(z,\sigma)\right]_{\sigma=\sigma_2}$$

is also a solution of the ODE $\mathcal{L}y = 0$. However, it is straightforwardly shown that this function is just a simple multiple of the first solution $y(z,\sigma_1)$, so that it is not linearly independent and we must find another solution. Differentiating

(14.29) with respect to $\sigma$ we then find

$$\frac{\partial}{\partial\sigma}\{\mathcal{L}\,[(\sigma-\sigma_2)y(z,\sigma)]\} = (\sigma-\sigma_2)^2 z^\sigma + 2(\sigma-\sigma_1)(\sigma-\sigma_2)z^\sigma + (\sigma-\sigma_1)(\sigma-\sigma_2)^2 z^\sigma \ln z,$$

which is equal to zero if $\sigma = \sigma_2$. As above, since $\partial/\partial\sigma$ and $\mathcal{L}$ are operators that differentiate with respect to different variables, we can reverse their order, to obtain

$$\mathcal{L}\left\{\frac{\partial}{\partial\sigma}\,[(\sigma-\sigma_2)y(z,\sigma)]\right\} = 0 \qquad \text{at } \sigma = \sigma_2,$$

and so the function

$$\left\{\frac{\partial}{\partial\sigma}\,[(\sigma-\sigma_2)y(z,\sigma)]\right\}_{\sigma=\sigma_2} \tag{14.30}$$

is also a solution of the original ODE $\mathcal{L}y = 0$, and is in fact the second linearly independent solution.

►*Find a second solution to (14.21) using the derivative method.*

From (14.23) the recurrence relation (with $\sigma$ as a parameter) is given by

$$(n+\sigma-1)a_n = (n+\sigma)a_{n-1}.$$

Setting $a_0 = 1$ we find that the cofficients have the particularly simple form $a_n(\sigma) = (\sigma+n)/\sigma$. We therefore consider the function

$$y(z,\sigma) = z^\sigma \sum_{n=0}^{\infty} a_n(\sigma) z^n = z^\sigma \sum_{n=0}^{\infty} \frac{\sigma+n}{\sigma} z^n.$$

The smaller root of the indicial equation for (14.21) is $\sigma_2 = 0$, and so from (14.30) a second, linearly independent, solution to the ODE is

$$\left\{\frac{\partial}{\partial\sigma}\,[\sigma y(z,\sigma)]\right\}_{\sigma=0} = \left\{\frac{\partial}{\partial\sigma}\left[z^\sigma \sum_{n=0}^{\infty}(\sigma+n)z^n\right]\right\}_{\sigma=0}.$$

The derivative with respect to $\sigma$ is given by

$$\frac{\partial}{\partial\sigma}\left[z^\sigma \sum_{n=0}^{\infty}(\sigma+n)z^n\right] = z^\sigma \ln z \sum_{n=0}^{\infty}(\sigma+n)z^n + z^\sigma \sum_{n=0}^{\infty} z^n,$$

which on setting $\sigma = 0$ gives the second solution

$$y_2(z) = \ln z \sum_{n=0}^{\infty} nz^n + \sum_{n=0}^{\infty} z^n$$
$$= \frac{z}{(1-z)^2} \ln z + \frac{1}{1-z}$$
$$= \frac{z}{(1-z)^2} \left( \ln z + \frac{1}{z} - 1 \right).$$

This second solution is the same as that obtained by the Wronskian method in the previous subsection except for the addition of some of the first solution. ◀

### 14.4.3 Series form of the second solution

Using any of the methods discussed above we can find the general form of the second solution to the ODE. This form is most easily found, however, using the derivative method. Let us first consider the case when the two solutions of the indicial equation are equal. In this case a second solution is given by (14.28), which may be written as

$$y_2(z) = \left[ \frac{\partial y(z,\sigma)}{\partial \sigma} \right]_{\sigma=\sigma_1}$$
$$= (\ln z) z^{\sigma_1} \sum_{n=0}^{\infty} a_n(\sigma_1) z^n + z^{\sigma_1} \sum_{n=1}^{\infty} \left[ \frac{da_n(\sigma)}{d\sigma} \right]_{\sigma=\sigma_1} z^n$$
$$= y_1(z) \ln z + z^{\sigma_1} \sum_{n=1}^{\infty} b_n z^n,$$

where $b_n = [da_n(\sigma)/d\sigma]_{\sigma=\sigma_1}$.

In the case where the roots of the indicial equation differ by an integer (not equal to zero), then from (14.30) a second solution is given by

$$y_2(z) = \left\{ \frac{\partial}{\partial \sigma} \left[ (\sigma - \sigma_2) y(z,\sigma) \right] \right\}_{\sigma=\sigma_2}$$
$$= \ln z \left[ (\sigma - \sigma_2) z^{\sigma} \sum_{n=0}^{\infty} a_n(\sigma) z^n \right]_{\sigma=\sigma_2} + z^{\sigma_2} \sum_{n=0}^{\infty} \left[ \frac{d}{d\sigma} (\sigma - \sigma_2) a_n(\sigma) \right]_{\sigma=\sigma_2} z^n.$$

But, as we mentioned in the previous section, $[(\sigma - \sigma_2) y(z,\sigma)]$ at $\sigma = \sigma_2$ is just a multiple of the first solution $y(z,\sigma_1)$. Therefore the second solution is of the form

$$y_2(z) = c y_1(z) \ln z + z^{\sigma_2} \sum_{n=0}^{\infty} b_n z^n,$$

where $c$ is a constant. In some cases, however, $c$ might be zero and so the second solution would not contain the term in $\ln z$ and could be written simply as a

Frobenius series. This clearly corresponds to the case in which the substitution of a Frobenius series into the original ODE yields two solutions automatically.

## 14.5 Polynomial solutions

We have seen that the evaluation of successive terms of a series solution to a differential equation is carried out by means of a recurrence relation. The form of the relation for $a_n$ depends upon $n$, the previous values of $a_r$ $(r < n)$ and the parameters of the equation. It may happen, as a result of this, that for some value of $n = N + 1$ the computed value $a_{N+1}$ is zero and that all higher $a_r$ also vanish. If this is so, and the corresponding solution of the indicial equation is a positive integer or zero, we are left with a finite polynomial of degree $N$ as a solution of the ODE

$$y(z) = \sum_{n=0}^{N} a_n z^n. \tag{14.31}$$

In many applications in theoretical physics (particularly in quantum mechanics) the termination of a potentially infinite series after a finite number of terms is of crucial importance in establishing physically acceptable descriptions and properties of systems. The condition that such a termination occurs is therefore of considerable importance.

►*Find power series solutions about $z = 0$ of*

$$y'' - 2zy' + \lambda y = 0. \tag{14.32}$$

*For what values of $\lambda$ does the equation possess a polynomial solution? Find such a solution for $\lambda = 4$.*

Clearly $z = 0$ is an ordinary point of (14.32) and so we look for solutions of the form $y = \sum_{n=0}^{\infty} a_n z^n$. Substituting this into the ODE and multiplying through by $z^2$ we find

$$\sum_{n=0}^{\infty} [n(n-1) - 2z^2 n + \lambda z^2] a_n z^n = 0.$$

By demanding that the coefficients of each power of $z$ vanish separately we derive the recurrence relation

$$n(n-1)a_n - 2(n-2)a_{n-2} + \lambda a_{n-2} = 0,$$

which may be rearranged to give

$$a_n = \frac{2(n-2) - \lambda}{n(n-1)} a_{n-2} \qquad \text{for } n \geq 2. \tag{14.33}$$

The odd and even coefficients are therefore independent of one another, and two solutions to (14.32) may be derived. We either set $a_1 = 0$ and $a_0 = 1$ to obtain

$$y_1(z) = 1 - \lambda\frac{z^2}{2!} - \lambda(4-\lambda)\frac{z^4}{4!} - \lambda(4-\lambda)(8-\lambda)\frac{z^6}{6!} - \cdots, \qquad (14.34)$$

or set $a_0 = 0$ and $a_1 = 1$ to obtain

$$y_2(z) = z + (2-\lambda)\frac{z^3}{3!} + (2-\lambda)(6-\lambda)\frac{z^5}{5!} + (2-\lambda)(6-\lambda)(10-\lambda)\frac{z^7}{7!} + \cdots.$$

Now from the recurrence relation (14.33) (or in this case from the expressions for $y_1$ and $y_2$ themselves) we see that for the ODE to possess a polynomial solution we require $\lambda = 2(n-2)$ for $n \geq 2$, or more simply $\lambda = 2n$ for $n \geq 0$, i.e. $\lambda$ must be an even positive integer. If $\lambda = 4$ then from (14.34) the ODE has the polynomial solution

$$y_1(z) = 1 - \frac{4z^2}{2!} = 1 - 2z^2. \blacktriangleleft$$

A simpler method of obtaining finite polynomial solutions is to *assume* a solution of the form (14.31), where $a_N \neq 0$. Instead of starting with the lowest power of $z$, as we have up to now, we this time start by considering the coefficient of the highest power $z^N$; such a power now exists because of our assumed form of solution.

►*By assuming a polynomial solution find the values of $\lambda$ in (14.32) for which such a solution exists.*

We assume a polynomial solution to (14.32) of the form $y = \sum_{n=0}^{N} a_n z^n$. Substituting this form into (14.32) we find

$$\sum_{n=0}^{N} \left[n(n-1)a_n z^{n-2} - 2zna_n z^{n-1} + \lambda a_n z^n\right] = 0.$$

Now, instead of starting with the lowest power of $z$ we start with the highest. Thus, demanding that the coefficient of $z^N$ vanishes, we require $-2N + \lambda = 0$, i.e. $\lambda = 2N$, as we found in the previous example. By demanding that the coefficient of a general power of $z$ is zero, the same recurrence relation as above may be derived, and the solutions found. ◄

## 14.6 Legendre's equation

In previous sections we have discussed methods for obtaining series solutions of second-order linear ODEs. In this section and the next we apply some of these methods to finding the series solutions of the two most important equations listed in table 14.1, namely Legendre's equation and Bessel's equation. As mentioned earlier, the remaining equations in table 14.1 may also be solved by the methods discussed in this chapter. These equations, and the properties of their solutions, are discussed briefly in the next chapter.

We now consider Legendre's equation

$$(1-z^2)y'' - 2zy' + \ell(\ell+1)y = 0, \tag{14.35}$$

which occurs in numerous physical applications, and particularly in problems with axial symmetry when they are expressed in spherical polar coordinates. In normal usage the variable $z$ in Legendre's equation is the cosine of the polar angle in spherical polars, and thus $-1 \leq z \leq 1$. The parameter $\ell$ is a given real number, and any solution of (14.35) is called a *Legendre function*.

In subsection 14.1.1, we showed that $z = 0$ is an ordinary point of (14.35), and so we expect to find two linearly independent solutions of the form $y = \sum_{n=0}^{\infty} a_n z^n$. Substituting we find

$$\sum_{n=0}^{\infty} \left[ n(n-1)a_n z^{n-2} - n(n-1)a_n z^n - 2na_n z^n + \ell(\ell+1)a_n z^n \right] = 0,$$

which on collecting terms gives

$$\sum_{n=0}^{\infty} \{(n+2)(n+1)a_{n+2} - [n(n+1) - \ell(\ell+1)]a_n\} \, z^n = 0.$$

The recurrence relation is therefore

$$a_{n+2} = \frac{[n(n+1) - \ell(\ell+1)]}{(n+1)(n+2)} a_n, \tag{14.36}$$

for $n = 0, 1, 2, \ldots$. If we choose $a_0 = 1$ and $a_1 = 0$ then we obtain the solution

$$y_1(z) = 1 - \ell(\ell+1)\frac{z^2}{2!} + (\ell-2)\ell(\ell+1)(\ell+3)\frac{z^4}{4!} - \cdots, \tag{14.37}$$

whereas choosing $a_0 = 0$ and $a_1 = 1$ we find a second solution

$$y_2(z) = z - (\ell-1)(\ell+2)\frac{z^3}{3!} + (\ell-3)(\ell-1)(\ell+2)(\ell+4)\frac{z^5}{5!} - \cdots. \tag{14.38}$$

By applying the ratio test to these series (see subsection 3.3.2), we find that both series converge for $|z| < 1$, so their radius of convergence is unity, which (as expected) is the distance to the nearest singular point of the equation. Since (14.37) contains only even powers of $z$ and (14.38) contains only odd powers, these two solutions cannot be proportional to one another, and are therefore linearly independent. Hence $y = c_1 y_1 + c_2 y_2$ is the general solution to (14.35) for $|z| < 1$.

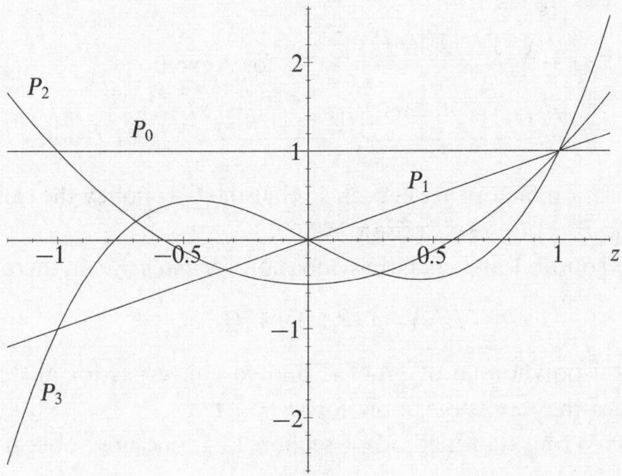


Figure 14.1 The first four Legendre polynomials.

### 14.6.1 General solution for integer $\ell$

Now, if $\ell$ is an integer in Legendre's equation (14.35), i.e. $\ell = 0, 1, 2, \ldots$, then the recurrence relation (14.36) gives

$$a_{\ell+2} = \frac{[\ell(\ell+1) - \ell(\ell+1)]}{(\ell+1)(\ell+2)} a_\ell = 0,$$

so that the series terminates and we obtain a polynomial solution of order $\ell$. These solutions (suitably normalised) are called *Legendre polynomials* of order $\ell$; they are written $P_\ell(z)$ and are valid for all finite $z$. It is conventional to normalise $P_\ell(z)$ in such a way that $P_\ell(1) = 1$, and as a consequence $P_\ell(-1) = (-1)^\ell$. The first few Legendre polynomials are easily constructed and are given by

$$\begin{aligned} P_0(z) &= 1 & P_1(z) &= z \\ P_2(z) &= \tfrac{1}{2}(3z^2 - 1) & P_3(z) &= \tfrac{1}{2}(5z^3 - 3z) \\ P_4(z) &= \tfrac{1}{8}(35z^4 - 30z^2 + 3) & P_5(z) &= \tfrac{1}{8}(63z^5 - 70z^3 + 15z). \end{aligned}$$

The first four Legendre polynomials are plotted in figure 14.1.

According to whether $\ell$ is an even or odd integer respectively, either $y_1(z)$ in (14.37) or $y_2(z)$ in (14.38) terminates to give a multiple of the corresponding Legendre polynomial $P_\ell(z)$. In either case, however, the other series does not terminate and therefore converges only for $|z| < 1$. According to whether $\ell$ is even or odd we define *Legendre functions of the second kind* as $Q_\ell(z) = \alpha_\ell y_2(z)$

or $Q_\ell(z) = \beta_\ell y_1(z)$ respectively, where the constants $\alpha_\ell$ and $\beta_\ell$ are conventionally taken to have the values

$$\alpha_\ell = \frac{(-1)^{\ell/2} 2^\ell [(\ell/2)!]^2}{\ell!} \qquad \text{for } \ell \text{ even}, \tag{14.39}$$

$$\beta_\ell = \frac{(-1)^{(\ell+1)/2} 2^{\ell-1} \{[(\ell-1)/2]!\}^2}{\ell!} \qquad \text{for } \ell \text{ odd}. \tag{14.40}$$

These normalisation factors are chosen so that the $Q_\ell(z)$ obey the same recurrence relations as the $P_\ell(z)$ (see subsection 14.6.2).

The general solution of Legendre's equation for *integer* $\ell$ is therefore

$$y(z) = c_1 P_\ell(z) + c_2 Q_\ell(z), \tag{14.41}$$

where $P_\ell(z)$ is a polynomial of order $\ell$, and so converges for all $z$, and $Q_\ell(z)$ is an infinite series that converges only for $|z| < 1$.†

By using the Wronskian method of section 14.4, one may obtain closed forms for the $Q_\ell(z)$.

►*Use the Wronskian method of section 14.4 to find a closed-form expression for* $Q_0(z)$.

From (14.25) a second solution to Legendre's equation (14.35), with $\ell = 0$, is

$$\begin{aligned} y_2(z) &= P_0(z) \int^z \frac{1}{[P_0(u)]^2} \exp\left( \int^u \frac{2v}{1-v^2}\, dv \right) du \\ &= \int^z \exp\left[ -\ln(1-u^2) \right] du \\ &= \int^z \frac{du}{(1-u^2)} = \frac{1}{2} \ln\left( \frac{1+z}{1-z} \right), \end{aligned} \tag{14.42}$$

where in the second line we used the fact that $P_0(z) = 1$.

All that remains is to fix the normalisation of this solution so that it agrees with (14.39). Expanding the logarithm in (14.42) as a Maclaurin series we obtain

$$y_2(z) = z + \frac{z^3}{3} + \frac{z^5}{5} + \cdots .$$

Comparing this with the expression for $Q_0(z)$, using (14.38) with $\ell = 0$ and the normalisation (14.39), we find that $y_2(z)$ is already correctly normalised, and so

$$Q_0(z) = \frac{1}{2} \ln\left( \frac{1+z}{1-z} \right).$$

We might, of course, have recognised the series (14.38) for $\ell = 0$, but to do so for larger $\ell$ would prove progressively more difficult. ◄

† It is in fact possible to find a second solution in terms of an infinite series of *negative* powers that is finite for $|z| > 1$.

Using the above method for $\ell = 1$, we find

$$Q_1(z) = \frac{1}{2} z \ln\left(\frac{1+z}{1-z}\right) - 1.$$

Closed forms for higher-order $Q_\ell(z)$ may now be found using the recurrence relation (14.55) derived in the next subsection.

### *14.6.2 Properties of Legendre polynomials*

As stated earlier, when encountered in physical problems the variable $z$ in Legendre's equation is usually the cosine of the polar angle $\theta$ in spherical polar coordinates, and we then require the solution $y(z)$ to be regular at $z = \pm 1$ (which corresponds to $\theta = 0$ or $\theta = \pi$). For this to occur we require the equation to have a polynomial solution, and so $\ell$ must be an integer. Furthermore, we also require the coefficient $c_2$ of the function $Q_\ell(z)$ in (14.41) to be zero (since $Q_\ell(z)$ is singular at $z = \pm 1$), with the result that the general solution is simply some multiple of the relevant Legendre polynomial $P_\ell(z)$. In this section we will study the properties of the Legendre polynomials $P_\ell(z)$ in some detail.

*Rodrigues' formula*

As an aid to establishing further properties of the Legendre polynomials we now develop Rodrigues' representation of these functions. Rodrigues' formula for the $P_\ell(z)$ is

$$P_\ell(z) = \frac{1}{2^\ell \ell!} \frac{d^\ell}{dz^\ell}(z^2 - 1)^\ell. \tag{14.43}$$

To prove that this is a representation we let $u = (z^2-1)^\ell$, so that $u' = 2\ell z(z^2-1)^{\ell-1}$ and

$$(z^2 - 1)u' - 2\ell z u = 0.$$

If we differentiate this expression $\ell + 1$ times using Leibnitz' theorem, we obtain

$$\left[(z^2-1)u^{(\ell+2)} + 2z(\ell+1)u^{(\ell+1)} + \ell(\ell+1)u^{(\ell)}\right] - 2\ell\left[zu^{(\ell+1)} + (\ell+1)u^{(\ell)}\right] = 0,$$

which reduces to

$$(z^2-1)u^{(\ell+2)} + 2zu^{(\ell+1)} - \ell(\ell+1)u^{(\ell)} = 0.$$

Changing the sign all through, and comparing the resulting expression with Legendre's equation (14.35), we see that $u^{(\ell)}$ satisfies the same equation as $P_\ell(z)$, so that

$$u^{(\ell)}(z) = c_\ell P_\ell(z), \tag{14.44}$$

for some constant $c_\ell$ that depends on $\ell$. To establish the value of $c_\ell$ we note that the only term in the expression for the $\ell$th derivative of $(z^2-1)^\ell$ that does not contain a factor $z^2-1$, and therefore does not vanish at $z=1$, is $(2z)^\ell \ell!(z^2-1)^0$. Putting $z=1$ in (14.44) therefore shows that $c_\ell = 2^\ell \ell!$, thus completing the proof of Rodrigues' formula (14.43).

►*Use Rodrigues' formula to show that*

$$I_\ell = \int_{-1}^{1} P_\ell(z)P_\ell(z)\,dz = \frac{2}{2\ell+1}. \tag{14.45}$$

The result is trivially obvious for $\ell = 0$ and so we assume $\ell \geq 1$. Then, by Rodrigues' formula,

$$I_\ell = \frac{1}{2^{2\ell}(\ell!)^2}\int_{-1}^{1}\left[\frac{d^\ell(z^2-1)^\ell}{dz^\ell}\right]\left[\frac{d^\ell(z^2-1)^\ell}{dz^\ell}\right]dz.$$

Repeated integration by parts, with all boundary terms vanishing, reduces this to

$$\begin{aligned} I_\ell &= \frac{(-1)^\ell}{2^{2\ell}(\ell!)^2}\int_{-1}^{1}(z^2-1)^\ell\frac{d^{2\ell}}{dz^{2\ell}}(z^2-1)^\ell\,dz \\ &= \frac{(2\ell)!}{2^{2\ell}(\ell!)^2}\int_{-1}^{1}(1-z^2)^\ell\,dz. \end{aligned}$$

If we write

$$K_\ell = \int_{-1}^{1}(1-z^2)^\ell\,dz,$$

then integration by parts (taking a factor 1 as the second part) gives

$$K_\ell = \int_{-1}^{1} 2\ell z^2(1-z^2)^{\ell-1}\,dz.$$

Writing $2\ell z^2$ as $2\ell - 2\ell(1-z^2)$ we obtain

$$\begin{aligned} K_\ell &= 2\ell\int_{-1}^{1}(1-z^2)^{\ell-1}\,dz - 2\ell\int_{-1}^{1}(1-z^2)^\ell\,dz, \\ &= 2\ell K_{\ell-1} - 2\ell K_\ell \end{aligned}$$

and hence the recurrence relation $(2\ell+1)K_\ell = 2\ell K_{\ell-1}$. We therefore find

$$K_\ell = \frac{2\ell}{2\ell+1}\frac{2\ell-2}{2\ell-1}\cdots\frac{2}{3}K_0 = 2^\ell \ell!\frac{2^\ell \ell!}{(2\ell+1)!}2 = \frac{2^{2\ell+1}(\ell!)^2}{(2\ell+1)!},$$

which, when substituted into the expression for $I_\ell$, establishes the required result. ◄

*Mutual orthogonality of Legendre polynomials*

Another useful property of the $P_\ell(z)$ is their mutual orthogonality, i.e. that

$$\int_{-1}^{1} P_\ell(z)P_k(z)\,dz = 0 \qquad \text{if } \ell \neq k. \tag{14.46}$$

More general considerations concerning the mutual orthogonality of solutions to various classes of second-order linear ODEs are discussed in the next chapter, but for the moment we concentrate on the specific proof of (14.46).

Since the $P_\ell(z)$ satisfy Legendre's equation we may write

$$\left[(1-z^2)P_\ell'\right]' + \ell(\ell+1)P_\ell = 0,$$

where $P_\ell' = dP_\ell/dz$. Multiplying through by $P_k$ and integrating from $z = -1$ to $z = 1$, we obtain

$$\int_{-1}^{1} P_k \left[(1-z^2)P_\ell'\right]' \, dz + \int_{-1}^{1} P_k \ell(\ell+1)P_\ell \, dz = 0.$$

Integrating the first term by parts and noting that the boundary contribution vanishes at both limits because of the factor $1 - z^2$, we find

$$-\int_{-1}^{1} P_k'(1-z^2)P_\ell' \, dz + \int_{-1}^{1} P_k \ell(\ell+1)P_\ell \, dz = 0.$$

Now, if we reverse the roles of $\ell$ and $k$ and subtract one expression from the other, we conclude that

$$[k(k+1) - \ell(\ell+1)] \int_{-1}^{1} P_k P_\ell \, dz = 0,$$

and therefore since $k \neq \ell$ we must have the result (14.46). As a particular case we note that if we put $k = 0$ we obtain

$$\int_{-1}^{1} P_\ell(z) \, dz = 0 \qquad \text{for } \ell \neq 0.$$

As will be discussed more fully in the next chapter, the mutual orthogonality of the $P_\ell(z)$ means that any reasonable function $f(z)$ (i.e. one obeying the Dirichlet conditions discussed at the start of chapter 10) can be expressed in the interval $|z| < 1$ as an infinite sum of Legendre polynomials,

$$f(z) = \sum_{\ell=0}^{\infty} a_\ell P_\ell(z), \tag{14.47}$$

where the coefficients $a_\ell$ are given by

$$a_\ell = \frac{2\ell+1}{2} \int_{-1}^{1} f(z) P_\ell(z) \, dz. \tag{14.48}$$

►*Prove the expression (14.48) for the coefficients in the Legendre polynomial expansion of a function $f(z)$.*

If we multiply (14.47) by $P_m(z)$ and integrate from $z = -1$ to $z = 1$, then we obtain

$$\begin{aligned}\int_{-1}^{1} P_m(z) f(z)\, dz &= \sum_{\ell=0}^{\infty} a_\ell \int_{-1}^{1} P_m(z) P_\ell(z)\, dz \\ &= a_m \int_{-1}^{1} P_m(z) P_m(z)\, dz = \frac{2a_m}{2m+1},\end{aligned}$$

where we have used the orthogonality property (14.46) and the normalisation property (14.45). ◄

*Generating function for Legendre polynomials*

A useful device for manipulating and studying sequences of functions or quantities labelled by an integer variable (here, the Legendre polynomials $P_\ell(z)$ labelled by $\ell$) is a *generating function*. The generating function has perhaps its greatest utility in the areas of probability theory and statistics (see chapter 24). However, it is also a great convenience in our present study.

The generating function for, say, a series of functions $f_n(z)$ for $n = 0, 1, 2, \ldots$ is a function $G(z, h)$, containing as well as $z$ a dummy variable $h$, such that

$$G(z, h) = \sum_{n=0}^{\infty} f_n(z) h^n,$$

i.e. $f_n(z)$ is the coefficient of $h^n$ in the expansion of $G$ in powers of $h$. The utility of the device lies in the fact that sometimes it is possible to find a closed form for $G(z, h)$.

For our study of Legendre polynomials let us consider the functions $P_n(z)$ defined by the equation

$$G(z, h) = (1 - 2zh + h^2)^{-1/2} = \sum_{n=0}^{\infty} P_n(z) h^n. \tag{14.49}$$

As we show below, the functions so defined are identical to the Legendre polynomials and the function $(1 - 2zh + h^2)^{-1/2}$ is in fact the generating function for them. In the process we will also deduce several useful relationships between the various polynomials and their derivatives.

In the following $dP_n(z)/dz$ will be denoted by $P_n'$. First, we differentiate the defining equation (14.49) with respect to $z$ to get

$$h(1 - 2zh + h^2)^{-3/2} = \sum P_n' h^n. \tag{14.50}$$

Also, we differentiate (14.49) with respect to $h$ to yield

$$(z-h)(1-2zh+h^2)^{-3/2} = \sum nP_n h^{n-1}. \tag{14.51}$$

Equation (14.50) can be written using (14.49) as

$$h\sum P_n h^n = (1-2zh+h^2)\sum P_n' h^n,$$

and thus equating coefficients of $h^{n+1}$ we obtain the recurrence relation

$$P_n = P_{n+1}' - 2zP_n' + P_{n-1}'. \tag{14.52}$$

Equations (14.50) and (14.51) can be combined as

$$(z-h)\sum P_n' h^n = h\sum nP_n h^{n-1},$$

from which the coefficent of $h^n$ yields a second recurrence relation

$$zP_n' - P_{n-1}' = nP_n. \tag{14.53}$$

Eliminating $P_{n-1}'$ between (14.52) and (14.53) gives the further result

$$(n+1)P_n = P_{n+1}' - zP_n'. \tag{14.54}$$

If we now take the result (14.54) with $n$ replaced by $n-1$ and add $z$ times (14.53) to it, we obtain

$$(1-z^2)P_n' = n(P_{n-1} - zP_n).$$

Finally, differentiating both sides with respect to $z$ and using (14.53) again, we find

$$\begin{aligned}(1-z^2)P_n'' - 2zP_n' &= n[(P_{n-1}' - zP_n') - P_n] \\ &= n(-nP_n - P_n) = -n(n+1)P_n,\end{aligned}$$

and so the $P_n$ defined by (14.49) do indeed satisfy Legendre's equation. It only remains to verify the normalisation. This is easily done at $z = 1$ when $G$ becomes

$$G(1,h) = [(1-h)^2]^{-1/2} = 1 + h + h^2 + \cdots,$$

and thus all the $P_n$ so defined have $P_n(1) = 1$ as required. Many other useful recurrence relations can be derived from those found above.

►*Prove the recurrence relation*

$$(n+1)P_{n+1} - (2n+1)zP_n + nP_{n-1} = 0. \qquad (14.55)$$

Substituting from (14.49) into (14.51) we find

$$(z-h)\sum P_n h^n = (1 - 2zh + h^2)\sum nP_n h^{n-1}.$$

Equating coefficients of $h^n$ we obtain

$$zP_n - P_{n-1} = (n+1)P_{n+1} - 2znP_n + (n-1)P_{n-1},$$

which on rearranging gives the stated result. ◄

Another use of the generating function (14.49) is in representing the inverse distance between two points in three-dimensional space in terms of Legendre polynomials. If two points $\mathbf{r}$ and $\mathbf{r}'$ are at distances $r$ and $r'$ respectively from the origin, with $r' < r$, then

$$\begin{aligned}\frac{1}{|\mathbf{r}-\mathbf{r}'|} &= \frac{1}{(r^2 + r'^2 - 2rr'\cos\theta)^{1/2}} \\ &= \frac{1}{r[1 - 2(r'/r)\cos\theta + (r'/r)^2]^{1/2}} \\ &= \frac{1}{r}\sum_{\ell=0}^{\infty}\left(\frac{r'}{r}\right)^{\ell} P_\ell(\cos\theta), \end{aligned} \qquad (14.56)$$

where $\theta$ is the angle between the two position vectors $\mathbf{r}$ and $\mathbf{r}'$. If $r' > r$, however, then $r$ and $r'$ must be exchanged in (14.56) or the series would not converge.

To summarise the position concerning Legendre polynomials, we now have three possible starting points, which have been shown to be equivalent: the defining equation (14.35) together with the condition $P_n(1) = 1$; Rodrigues' formula (14.43); and the generating function (14.49). In addition we have proved a variety of relationships and recurrence relations (not particularly memorable, but collectively useful) and, as will be apparent from the work of chapter 16, developed a powerful tool for use in axially symmetric situations, in which the $\nabla^2$ operator is involved and spherical polar coordinates are employed.

## 14.7 Bessel's equation

Bessel's equation arises from physical situations similar to those involving Legendre's equation but when cylindrical, rather than spherical, polar coordinates are employed. It has the form

$$z^2y'' + zy' + (z^2 - \nu^2)y = 0, \qquad (14.57)$$

where the parameter $\nu$ is a given number, which we may take as $\geq 0$ with no loss of generality. In Bessel's equation, $z$ is usually a multiple of a radial distance and therefore ranges from 0 to $\infty$.

Writing (14.57) in our standard form we have

$$y'' + \frac{1}{z}y' + \left(1 - \frac{\nu^2}{z^2}\right) y = 0. \tag{14.58}$$

By inspection $z = 0$ is a regular singular point; hence we try a solution of the form $y = z^\sigma \sum_{n=0}^{\infty} a_n z^n$. Substituting this into (14.58) and multiplying the resulting equation by $z^{2-\sigma}$, we obtain

$$\sum_{n=0}^{\infty} \left[(\sigma + n)(\sigma + n - 1) + (\sigma + n) - \nu^2\right] a_n z^n + \sum_{n=0}^{\infty} a_n z^{n+2} = 0,$$

which simplifies to

$$\sum_{n=0}^{\infty} \left[(\sigma + n)^2 - \nu^2\right] a_n z^n + \sum_{n=0}^{\infty} a_n z^{n+2} = 0.$$

Considering coefficients of $z^0$ we obtain the indicial equation

$$\sigma^2 - \nu^2 = 0,$$

so that $\sigma = \pm\nu$. For coefficients of higher powers of $z$ we find

$$\left[(\sigma + 1)^2 - \nu^2\right] a_1 = 0, \tag{14.59}$$

$$\left[(\sigma + n)^2 - \nu^2\right] a_n + a_{n-2} = 0 \quad \text{for } n \geq 2. \tag{14.60}$$

Substituting $\sigma = \pm\nu$ into (14.59) and (14.60) we obtain the recurrence relations

$$(1 \pm 2\nu)a_1 = 0, \tag{14.61}$$

$$n(n \pm 2\nu)a_n + a_{n-2} = 0 \quad \text{for } n \geq 2. \tag{14.62}$$

We now consider the form of the general solution to Bessel's equation (14.57) for two cases, the case for which $\nu$ is an integer (including zero) and the case for which it is not.

### *14.7.1 General solution for non-integer* $\nu$

If $\nu$ is a non-integer, then in general the two roots of the indicial equation, $\sigma_1 = \nu$ and $\sigma_2 = -\nu$, will not differ by an integer, and we may obtain two linearly independent solutions in the form of Frobenius series. Special cases do arise, however, when $\nu = m/2$ for $m = 1, 3, 5, \ldots$, and $\sigma_1 - \sigma_2 = 2\nu = m$ is an (odd positive) integer. For such cases, we may always obtain a solution in the form of a Frobenius series corresponding to the larger root $\sigma_1 = \nu = m/2$, as described above. For the smaller root $\sigma_2 = -\nu = -m/2$, however, we must determine whether a second Frobenius series solution is possible by examining the recurrence relation (14.62), which reads

$$n(n-m)a_n + a_{n-2} = 0 \quad \text{for } n \geq 2.$$

Since $m$ is an *odd* positive integer in this case, we can use this recurrence relation (starting with $a_0 \neq 0$) to calculate $a_2, a_4, a_6, \ldots$ in the knowledge that all these terms will remain finite. It is therefore possible in this case to find a second solution in the form of a Frobenius series corresponding to the smaller root $\sigma_2$.

Therefore, in general, for non-integer $\nu$ we have from (14.61) and (14.62)

$$\begin{aligned} a_n &= -\frac{1}{n(n \pm 2\nu)} a_{n-2} \quad &&\text{for } n = 2, 4, 6, \ldots, \\ &= 0 &&\text{for } n = 1, 3, 5, \ldots. \end{aligned}$$

Setting $a_0 = 1$ in each case, we obtain the two solutions

$$y_{\pm\nu}(z) = z^{\pm\nu}\left[1 - \frac{z^2}{2(2 \pm 2\nu)} + \frac{z^4}{2 \times 4(2 \pm 2\nu)(4 \pm 2\nu)} - \cdots\right].$$

It is customary, however, to set

$$a_0 = \frac{1}{2^{\pm\nu}\Gamma(1 \pm \nu)},$$

where $\Gamma(x)$ is the *gamma function*, described in the appendix; it may be regarded as the generalisation of the factorial function to non-integer and/or negative arguments.† The two solutions of (14.57) are then written as $J_\nu(z)$ and $J_{-\nu}(z)$, where

$$\begin{aligned} J_\nu(z) &= \frac{1}{\Gamma(\nu+1)}\left(\frac{z}{2}\right)^\nu \left[1 - \frac{1}{\nu+1}\left(\frac{z}{2}\right)^2 + \frac{1}{(\nu+1)(\nu+2)}\frac{1}{2!}\left(\frac{z}{2}\right)^4 - \cdots\right] \\ &= \sum_{n=0}^{\infty} \frac{(-1)^n}{n!\,\Gamma(\nu+n+1)}\left(\frac{z}{2}\right)^{\nu+2n}; \end{aligned} \qquad (14.63)$$

and replacing $\nu$ by $-\nu$ gives $J_{-\nu}(z)$. The functions $J_\nu(z)$ and $J_{-\nu}(z)$ are called *Bessel functions of the first kind, of order* $\nu$. Since the first term of each series is a finite non-zero multiple of $z^\nu$ and $z^{-\nu}$ respectively, if $\nu$ is not an integer $J_\nu(z)$ and $J_{-\nu}(z)$ are linearly independent. This may be confirmed by calculating the Wronskian of these two functions. Therefore, for non-integer $\nu$ the general solution of Bessel's equation (14.57) is

$$y(z) = c_1 J_\nu(z) + c_2 J_{-\nu}(z). \qquad (14.64)$$

† In particular, $\Gamma(n+1) = n!$ for $n = 0, 1, 2, \ldots$, and $\Gamma(n)$ is infinite if $n$ is any integer $\leq 0$.

►*Find the general solution of*

$$z^2y'' + zy' + (z^2 - \tfrac{1}{4})y = 0.$$

This is Bessel's equation with $\nu = 1/2$, so from (14.64) the general solution is simply

$$y(z) = c_1 J_{1/2}(z) + c_2 J_{-1/2}(z).$$

However, Bessel functions of half-integral order can be expressed in terms of trigonometric functions. To show this, we note from (14.63) that

$$J_{\pm 1/2}(z) = z^{\pm 1/2} \sum_{n=0}^{\infty} \frac{(-1)^n z^{2n}}{2^{2n \pm 1/2} n! \Gamma(1 + n \pm 1/2)}.$$

Using the fact that $\Gamma(x+1) = x\Gamma(x)$ and $\Gamma(1/2) = \sqrt{\pi}$ we find that, for $\nu = 1/2$,

$$\begin{aligned} J_{1/2}(z) &= \frac{(z/2)^{1/2}}{\Gamma(3/2)} - \frac{(z/2)^{5/2}}{1!\Gamma(5/2)} + \frac{(z/2)^{9/2}}{2!\Gamma(7/2)} - \cdots \\ &= \frac{(z/2)^{1/2}}{(1/2)\sqrt{\pi}} - \frac{(z/2)^{5/2}}{1!(3/2)(1/2)\sqrt{\pi}} + \frac{(z/2)^{9/2}}{2!(5/2)(3/2)(1/2)\sqrt{\pi}} - \cdots \\ &= \frac{(z/2)^{1/2}}{(1/2)\sqrt{\pi}} \left(1 - \frac{z^2}{3!} + \frac{z^4}{5!} - \cdots\right) = \frac{(z/2)^{1/2}}{(1/2)\sqrt{\pi}} \frac{\sin z}{z} = \sqrt{\frac{2}{\pi z}} \sin z, \end{aligned}$$

whereas for $\nu = -1/2$ we obtain

$$\begin{aligned} J_{-1/2}(z) &= \frac{(z/2)^{-1/2}}{\Gamma(1/2)} - \frac{(z/2)^{3/2}}{1!\Gamma(3/2)} + \frac{(z/2)^{7/2}}{2!\Gamma(5/2)} - \cdots \\ &= \frac{(z/2)^{-1/2}}{\sqrt{\pi}} \left(1 - \frac{z^2}{2!} + \frac{z^4}{4!} - \cdots\right) = \sqrt{\frac{2}{\pi z}} \cos z. \end{aligned}$$

Therefore the general solution we require is

$$y(z) = c_1 J_{1/2}(z) + c_2 J_{-1/2}(z) = c_1 \sqrt{\frac{2}{\pi z}} \sin z + c_2 \sqrt{\frac{2}{\pi z}} \cos z. \blacktriangleleft$$

Corresponding to the discussion in subsection 14.6.2 of the general solution of Legendre's equation, we note that when Bessel's equation is encountered in physical situations the argument $z$ is usually some multiple of a radial distance, and so takes values in the range $0 \leq z \leq \infty$. We often require that the solution is regular at $z = 0$ but, from (14.63), we see immediately that $J_{-\nu}(z)$ is singular at the origin (remember that we restricted $\nu$ to be non-negative). In such cases, the coefficient $c_2$ in (14.64) must be set to zero, and the solution is simply some multiple of $J_\nu(z)$.

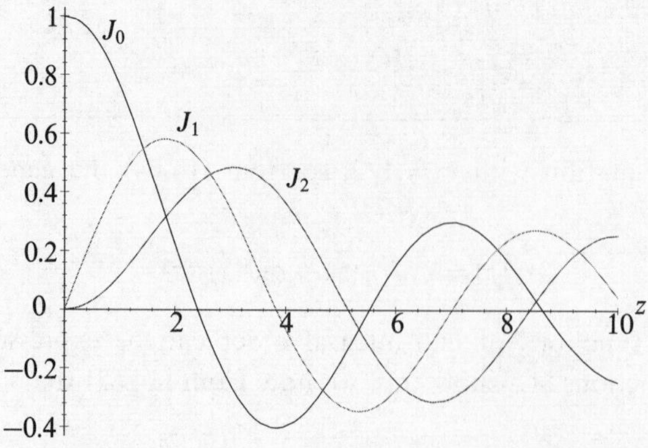


Figure 14.2 The first three integer-order Bessel functions.

### 14.7.2 General solution for integer $\nu$

The definition of the Bessel function $J_\nu(z)$ given in (14.63) is, of course, valid for all values of $\nu$ but, as we shall see, in the case of integer $\nu$ the general solution of Bessel's equation cannot be written in the form (14.64). Let us first consider the case when $\nu = 0$, so that the two solutions to the indicial equation are equal, and we clearly obtain only one solution in the form of a Frobenius series. From (14.63), this is given by

$$J_0(z) = \sum_{n=0}^{\infty} \frac{(-1)^n z^{2n}}{2^{2n} n! \Gamma(1+n)}$$

$$= 1 - \frac{z^2}{2^2} + \frac{z^4}{2^2 4^2} - \frac{z^6}{2^2 4^2 6^2} + \cdots .$$

In general, however, if $\nu$ is a positive integer then the solutions of the indicial equation differ by an integer. For the larger root, $\sigma_1 = \nu$, we may find a solution $J_\nu(z)$ for $\nu = 1, 2, 3, \ldots$, in the form of a Frobenius series given by (14.63). Graphs of $J_0(z)$, $J_1(z)$ and $J_2(z)$ are plotted in figure 14.2 for real $z$. For the smaller root $\sigma_2 = -\nu$, however, the recurrence relation (14.62) becomes

$$n(n-m)a_n + a_{n-2} = 0 \qquad \text{for } n \geq 2,$$

where $m = 2\nu$ is now an *even* positive integer, i.e. $m = 2, 4, 6, \ldots$. Starting with $a_0 \neq 0$ we may then calculate $a_2, a_4, a_6, \ldots$, but we see that when $n = m$ the coefficient $a_n$ is formally infinite, and the method fails to produce a second solution in the form of a Frobenius series.

In fact, by replacing $\nu$ by $-\nu$ in the definition of $J_\nu(z)$ given in (14.63), it can be shown that, for integer $\nu$,

$$J_{-\nu}(z) = (-1)^\nu J_\nu(z)$$

and hence that $J_\nu(z)$ and $J_{-\nu}(z)$ are linearly dependent. So, in this case, we cannot write the general solution to Bessel's equation in the form (14.64). One therefore defines the function

$$Y_\nu(z) = \frac{J_\nu(z)\cos\nu\pi - J_{-\nu}(z)}{\sin\nu\pi}, \tag{14.65}$$

which is called a Bessel's function of the *second kind* of order $\nu$. As Bessel's equation is linear, $Y_\nu(z)$ is clearly a solution, since it is just the weighted sum of Bessel functions of the first kind. Furthermore, for non-integer $\nu$ it is clear that $Y_\nu(z)$ is linearly independent of $J_\nu(z)$. It may also be shown that the Wronskian of $J_\nu(z)$ and $Y_\nu(z)$ is non-zero for *all* values of $\nu$. Hence $J_\nu(z)$ and $Y_\nu(z)$ always constitute a pair of independent solutions. The expression (14.65) does, however, become an indeterminate form $0/0$ when $\nu$ is an integer. This is so because for integer $\nu$ we have $\cos\nu\pi = (-1)^\nu$ and $J_{-\nu}(z) = (-1)^\nu J_\nu(z)$. Nevertheless, this indeterminate form can be evaluated using l'Hôpital's rule (see chapter 3). Thus for integer $\nu$ we set

$$Y_\nu(z) = \lim_{\mu\to\nu}\left[\frac{J_\mu(z)\cos\mu\pi - J_{-\mu}(z)}{\sin\mu\pi}\right], \tag{14.66}$$

which gives a linearly independent second solution for integer $\nu$. Therefore, we may write the general solution of Bessel's equation, valid for *all* $\nu$, as

$$y(z) = c_1 J_\nu(z) + c_2 Y_\nu(z). \tag{14.67}$$

As mentioned above for the case when $\nu$ is not an integer, in physical situations we often require the solution of Bessel's equation to be regular at $z = 0$. But, from its definition (14.65) or (14.66), it is clear that $Y_\nu(z)$ is singular at the origin, and so in such physical situations the coefficient $c_2$ in (14.67) must be set to zero; the solution is then simply some multiple of $J_\nu(z)$.

### *14.7.3 Properties of Bessel functions*

Bessel functions of the first and second kind, $J_\nu(z)$ and $Y_\nu(z)$, have various useful properties that are worthy of further discussion.

#### *Recurrence relations*

The recurrence relations enjoyed by Bessel functions of the first kind, $J_\nu(z)$, can be derived directly from the power series definition (14.63).

►*Prove the recurrence relation*

$$\frac{d}{dz}[z^\nu J_\nu(z)] = z^\nu J_{\nu-1}(z). \tag{14.68}$$

From the power series definition (14.63) of $J_\nu(z)$ we obtain

$$\begin{aligned}\frac{d}{dz}[z^\nu J_\nu(z)] &= \frac{d}{dz}\sum_{n=0}^{\infty}\frac{(-1)^n z^{2\nu+2n}}{2^{\nu+2n}n!\Gamma(\nu+n+1)}\\ &= \sum_{n=0}^{\infty}\frac{(-1)^n z^{2\nu+2n-1}}{2^{\nu+2n-1}n!\Gamma(\nu+n)}\\ &= z^\nu\sum_{n=0}^{\infty}\frac{(-1)^n z^{(\nu-1)+2n}}{2^{(\nu-1)+2n}n!\Gamma((\nu-1)+n+1)} = z^\nu J_{\nu-1}(z). \blacktriangleleft\end{aligned}$$

It may similarly be shown that

$$\frac{d}{dz}[z^{-\nu} J_\nu(z)] = -z^{-\nu} J_{\nu+1}(z). \tag{14.69}$$

From (14.68) and (14.69) the remaining recurrence relations may be easily derived. Expanding out the derivative on the LHS of (14.68) and dividing through by $z^{\nu-1}$ we obtain the relation

$$zJ'_\nu(z) + \nu J_\nu(z) = zJ_{\nu-1}(z). \tag{14.70}$$

Similarly, by expanding out the derivative on the LHS of (14.69), and multiplying through by $z^{\nu+1}$, we find

$$zJ'_\nu(z) - \nu J_\nu(z) = -zJ_{\nu+1}(z). \tag{14.71}$$

Adding (14.70) and (14.71) and dividing through by $z$ gives

$$J_{\nu-1}(z) - J_{\nu+1}(z) = 2J'_\nu(z). \tag{14.72}$$

Finally, subtracting (14.71) from (14.70) and dividing by $z$ gives

$$J_{\nu-1}(z) + J_{\nu+1}(z) = \frac{2\nu}{z}J_\nu(z). \tag{14.73}$$

►*Given that $J_{1/2}(z) = (2/\pi z)^{1/2}\sin z$ and $J_{-1/2}(z) = (2/\pi z)^{1/2}\cos z$, express $J_{3/2}(z)$ and $J_{-3/2}(z)$ in terms of trigonometric functions.*

From (14.71) we have

$$\begin{aligned}
J_{3/2}(z) &= \frac{1}{2z}J_{1/2}(z) - J'_{1/2}(z) \\
&= \frac{1}{2z}\left(\frac{2}{\pi z}\right)^{1/2}\sin z - \left(\frac{2}{\pi z}\right)^{1/2}\cos z + \frac{1}{2z}\left(\frac{2}{\pi z}\right)^{1/2}\sin z \\
&= \left(\frac{2}{\pi z}\right)^{1/2}\left(\frac{1}{z}\sin z - \cos z\right).
\end{aligned}$$

Similarly, from (14.70), we have

$$\begin{aligned}
J_{-3/2}(z) &= -\frac{1}{2z}J_{-1/2}(z) + J'_{-1/2}(z) \\
&= -\frac{1}{2z}\left(\frac{2}{\pi z}\right)^{1/2}\cos z - \left(\frac{2}{\pi z}\right)^{1/2}\sin z - \frac{1}{2z}\left(\frac{2}{\pi z}\right)^{1/2}\cos z \\
&= \left(\frac{2}{\pi z}\right)^{1/2}\left(-\frac{1}{z}\cos z - \sin z\right).
\end{aligned}$$

By repeated use of these recurrence relations, we see that all Bessel functions $J_\nu(z)$ of half-integer order may be expressed in terms of trigonometric functions. From their definition (14.65), Bessel functions of the second kind, $Y_\nu(z)$, of half-integer order can be similarly expressed. ◄

Finally, we note that the relations (14.68) and (14.69) may be rewritten in integral form as

$$\int z^\nu J_{\nu-1}(z)\,dz = z^\nu J_\nu(z)$$

$$\int z^{-\nu} J_{\nu+1}(z)\,dz = -z^{-\nu} J_\nu(z).$$

If $\nu$ is an integer, the recurrence relations of this section may be proved using the generating function for Bessel functions discussed below. It may also be shown that Bessel functions of the second kind, $Y_\nu(z)$, also satisfy the recurrence relations derived above.

### *Mutual orthogonality of Bessel functions*

Bessel functions of the first kind, $J_\nu(z)$, possess an orthogonality relation analogous to that of the Legendre polynomials discussed in subsection 14.6.2. A more general discussion of the mutual orthogonality of solutions to second-order linear ODEs (such as Bessel's equation) is given in chapter 15.

By definition, the function $J_\nu(z)$ satisfies Bessel's equation (14.57),

$$z^2y'' + zy' + (z^2 - \nu^2)y = 0.$$

Let us instead consider the functions $f(z) = J_\nu(\lambda z)$ and $g(z) = J_\nu(\mu z)$, which (as proved below) respectively satisfy the equations

$$z^2f'' + zf' + (\lambda^2z^2 - \nu^2)f = 0, \tag{14.74}$$

$$z^2g'' + zg' + (\mu^2z^2 - \nu^2)g = 0. \tag{14.75}$$

►*Show that $f(z) = J_\nu(\lambda z)$ satisfies (14.74).*

If $f(z) = J_\nu(\lambda z)$ and we write $w = \lambda z$, then

$$\frac{df}{dz} = \lambda\frac{dJ_\nu(w)}{dw} \qquad \text{and} \qquad \frac{d^2f}{dz^2} = \lambda^2\frac{d^2J_\nu(w)}{dw^2}.$$

When these expressions are substituted, the LHS of (14.74) becomes

$$z^2\lambda^2\frac{d^2J_\nu(w)}{dw^2} + z\lambda\frac{dJ_\nu(w)}{dw} + (\lambda^2z^2 - \nu^2)J_\nu(w)$$

$$= w^2\frac{d^2J_\nu(w)}{dw^2} + w\frac{dJ_\nu(w)}{dw} + (w^2 - \nu^2)J_\nu(w).$$

But, from Bessel's equation itself, this final expression is equal to zero, thus verifying that $f(z)$ does satisfy (14.74). ◄

Now multiplying (14.75) by $f(z)$ and (14.74) by $g(z)$ and subtracting gives

$$\frac{d}{dz}[z(fg' - gf')] = (\lambda^2 - \mu^2)zfg, \tag{14.76}$$

where we have used the fact that

$$\frac{d}{dz}[z(fg' - gf')] = z(fg'' - gf'') + (fg' - gf').$$

By integrating (14.76) over any given range $z = a$ to $z = b$ we obtain

$$\int_a^b zf(z)g(z)\,dz = \frac{1}{\lambda^2 - \mu^2}\Big[zf(z)g'(z) - zg(z)f'(z)\Big]_a^b,$$

which, on setting $f(z) = J_\nu(\lambda z)$ and $g(z) = J_\nu(\mu z)$, becomes

$$\int_a^b zJ_\nu(\lambda z)J_\nu(\mu z)\,dz = \frac{1}{\lambda^2 - \mu^2}\Big[\mu zJ_\nu(\lambda z)J'_\nu(\mu z) - \lambda zJ_\nu(\mu z)J'_\nu(\lambda z)\Big]_a^b. \tag{14.77}$$

If $\lambda \neq \mu$, and the interval $[a, b]$ is such that the expression on the RHS of (14.77) equals zero we obtain the orthogonality condition

$$\int_a^b zJ_\nu(\lambda z)J_\nu(\mu z)\,dz = 0. \tag{14.78}$$

This happens, for example, if $J_\nu(\lambda z)$ and $J_\nu(\mu z)$ vanish at $z = a$ and $z = b$, or if $J'_\nu(\lambda z)$ and $J'_\nu(\mu z)$ vanish at $z = a$ and $z = b$, or for many more general conditions.

If $\lambda = \mu$, however, then the RHS of (14.77) takes the indeterminant form 0/0. This may be evaluated using l'Hôpital's rule, or alternatively we may calculate the relevant integral directly.

►*Evaluate the integral*

$$\int_a^b J_\nu^2(\lambda z) z \, dz.$$

Ignoring the integration limits for the moment,

$$\int J_\nu^2(\lambda z) z \, dz = \frac{1}{\lambda^2} \int J_\nu^2(u) u \, du,$$

where $u = \lambda z$. Integrating by parts yields

$$I = \int J_\nu^2(u) u \, du = \tfrac{1}{2} u^2 J_\nu^2(u) - \int J_\nu(u) J'_\nu(u) u^2 \, du.$$

Now Bessel's equation (14.57) can be rearranged as

$$u^2 J_\nu(u) = \nu^2 J_\nu(u) - u J'_\nu(u) - u^2 J''_\nu(u),$$

which, on substitution into the expression for $I$, gives

$$\begin{aligned} I &= \tfrac{1}{2} u^2 J_\nu^2(u) - \int J'_\nu(u)[\nu^2 J_\nu(u) - u J'_\nu(u) - u^2 J''_\nu(u)] \, du \\ &= \tfrac{1}{2} u^2 J_\nu^2(u) - \tfrac{1}{2} \nu^2 J_\nu^2(u) + \tfrac{1}{2} u^2 [J'_\nu(u)]^2 + c. \end{aligned}$$

Since $u = \lambda z$ the required integral is given by

$$\int_a^b J_\nu^2(\lambda z) z \, dz = \frac{1}{2} \left[ \left( z^2 - \frac{\nu^2}{\lambda^2} \right) J_\nu^2(\lambda z) + z^2 [J'_\nu(\lambda z)]^2 \right]_a^b, \tag{14.79}$$

which gives the normalisation condition for Bessel functions of the first kind.◄

Since the Bessel functions $J_\nu(z)$ possess the orthogonality property (14.78) we may expand any reasonable function $f(z)$ (i.e. one obeying the Dirichlet conditions discussed in chapter 10) in the interval $0 \le z \le a$ as a sum of Bessel functions of a given order $\nu$,

$$f(z) = \sum_{n=0}^{\infty} c_n J_\nu(\lambda_n z), \tag{14.80}$$

where the $\lambda_n$ are chosen such that $J_\nu(\lambda_n a) = 0$. The coefficients $c_n$ are then given by

$$c_n = \frac{2}{a^2 J_{\nu+1}^2(\lambda_n a)} \int_0^a f(z) J_\nu(\lambda_n z) z \, dz. \tag{14.81}$$

►*Prove the expression (14.81) for the coefficients in a Bessel function expansion of a function $f(z)$.*

If we multiply (14.80) by $zJ_\nu(\lambda_m z)$ and integrate from $z = 0$ to $z = a$, then we obtain

$$\begin{aligned}\int_0^a zJ_\nu(\lambda_m z)f(z)\,dz &= \sum_{n=0}^{\infty} c_n \int_0^a zJ_\nu(\lambda_m z)J_\nu(\lambda_n z)\,dz \\ &= c_m \int_0^a J_\nu^2(\lambda_m z)z\,dz \\ &= \tfrac{1}{2}c_m a^2 J'^2_\nu(\lambda_m a) = \tfrac{1}{2}c_m a^2 J^2_{\nu+1}(\lambda_m a),\end{aligned}$$

where in the last two lines we use (14.77), (14.79), the fact that $J_\nu(\lambda_m a) = 0$ and (14.71). ◄

*Generating function for Bessel functions*

The Bessel functions $J_\nu(z)$, where $\nu$ is an integer, can be described by a generating function in a similar way to that discussed for Legendre polynomials in subsection 14.6.2. The generating function for Bessel functions of integer order is given by

$$G(z,h) = \exp\left[\frac{z}{2}\left(h - \frac{1}{h}\right)\right] = \sum_{n=-\infty}^{\infty} J_n(z)h^n. \qquad (14.82)$$

By expanding the exponential as a power series, it is straightfoward to verify that the functions $J_n(z)$ defined by (14.82) are indeed Bessel functions of the first kind.

The generating function is useful in finding for Bessel functions of integer order, properties which can often be extended to the non-integer case. In particular, the Bessel function recurrence relations may be easily proved.

►*Use the generating function (14.82) to prove, for integer $\nu$, the recurrence relation (14.73), i.e.*

$$J_{\nu-1}(z) + J_{\nu+1}(z) = \frac{2\nu}{z}J_\nu(z).$$

Differentiating $G(z,h)$ with respect to $h$ we obtain

$$\frac{\partial G(z,h)}{\partial h} = \frac{z}{2}\left(1 + \frac{1}{h^2}\right)G(z,h) = \sum_{n=-\infty}^{\infty} nJ_n(z)h^{n-1},$$

which can be written using (14.82) again as

$$\frac{z}{2}\left(1 + \frac{1}{h^2}\right)\sum_{n=-\infty}^{\infty} J_n(z)h^n = \sum_{n=-\infty}^{\infty} nJ_n(z)h^{n-1}.$$

Equating coefficients of $h^n$ we obtain

$$\frac{z}{2}[J_n(z) + J_{n+2}(z)] = (n+1)J_{n+1}(z),$$

which on replacing $n$ by $\nu - 1$ gives the required recurrence relation. ◄

The generating function (14.82) is also useful in deriving the *integral representation* of Bessel functions of integer order.

►*Show that for integer n the Bessel function $J_n(z)$ is given by*

$$J_n(z) = \frac{1}{\pi}\int_0^\pi \cos(n\theta - z\sin\theta)\, d\theta. \tag{14.83}$$

By expanding out the cosine term in the integrand in (14.83) we are led to consider the integral

$$I = \frac{1}{\pi}\int_0^\pi [\cos(z\sin\theta)\cos n\theta + \sin(z\sin\theta)\sin n\theta]\, d\theta. \tag{14.84}$$

Now we may express $\cos(z\sin\theta)$ and $\sin(z\sin\theta)$ in terms of Bessel functions by setting $h = \exp i\theta$ in (14.82) to give

$$\exp\left[\frac{z}{2}(\exp i\theta - \exp(-i\theta))\right] = \exp(iz\sin\theta) = \sum_{m=-\infty}^{\infty} J_m(z)\exp im\theta.$$

Using de Moivre's theorem $\exp i\theta = \cos\theta + i\sin\theta$ we then obtain

$$\exp(iz\sin\theta) = \cos(z\sin\theta) + i\sin(z\sin\theta) = \sum_{m=-\infty}^{\infty} J_m(z)(\cos m\theta + i\sin m\theta).$$

Equating the real and imaginary parts of this expression we find

$$\cos(z\sin\theta) = \sum_{m=-\infty}^{\infty} J_m(z)\cos m\theta$$

$$\sin(z\sin\theta) = \sum_{m=-\infty}^{\infty} J_m(z)\sin m\theta.$$

Substituting these expressions into (14.84) we find

$$I = \frac{1}{\pi}\sum_{m=-\infty}^{\infty}\int_0^\pi [J_m(z)\cos m\theta\cos n\theta + J_m(z)\sin m\theta\sin n\theta]\, d\theta.$$

However, using the orthogonality properties of the trigonometrical functions given in equations (10.1)–(10.3), we obtain

$$I = \frac{1}{\pi}\frac{\pi}{2}[J_n(z) + J_n(z)] = J_n(z),$$

which proves the integral representation (14.83). ◄

Finally, we mention the special case of the integral representation (14.83) for $n = 0$,

$$J_0(z) = \frac{1}{\pi}\int_0^{\pi} \cos(z\sin\theta)\,d\theta = \frac{1}{2\pi}\int_0^{2\pi} \cos(z\sin\theta)\,d\theta,$$

since $\cos(z\sin\theta)$ repeats itself in the range $\theta = \pi$ to $\theta = 2\pi$. However, $\sin(z\sin\theta)$ changes sign in this range and so

$$\frac{1}{2\pi}\int_0^{2\pi} \sin(z\sin\theta)\,d\theta = 0.$$

Using de Moivre's theorem, we can therefore write

$$J_0(z) = \frac{1}{2\pi}\int_0^{2\pi} \exp(iz\sin\theta)\,d\theta = \frac{1}{2\pi}\int_0^{2\pi} \exp(iz\cos\theta)\,d\theta.$$

There are in fact many other integral representations of Bessel functions, which can be derived from those given.

## 14.8 General remarks

As was our intention, in respect of infinite series solutions we have concentrated to a very marked degree on Bessel's equation and in respect of finite polynomial solutions on Legendre's equation. The techniques used are, however, applicable to many equations other than these, but since they are in all essentials the same, we do not deal with them explicitly. The solutions of the remaining equations in table 14.1 are discussed briefly in the next chapter in connection with Sturm–Liouville systems.

## 14.9 Exercises

14.1 Find solutions, as power series in $z$, of the equation

$$4zy'' + 2(1-z)y' - y = 0.$$

Identify one of the solutions and verify it by direct substitution.

14.2 Verify that $z = 0$ is a regular singular point of the equation

$$z^2y'' - \tfrac{3}{2}zy' + (1+z)y = 0,$$

and that the indicial equation has roots 2 and $\frac{1}{2}$. Show that the general solution is

$$y(z) = 6a_0z^2\sum_{n=0}^{\infty}\frac{(-1)^n(n+1)2^{2n}z^n}{(2n+3)!}$$
$$+ b_0\left(z^{1/2} + 2z^{3/2} - \frac{z^{1/2}}{4}\sum_{n=2}^{\infty}\frac{(-1)^n2^{2n}z^n}{n(n-1)(2n-3)!}\right).$$

14.3 Use the derivative method to obtain as a second solution of Bessel's equation for the case when $\nu = 0$ the following expression:

$$J_0(z)\ln z - \sum_{n=1}^{\infty} \frac{(-1)^n}{(n!)^2}\left(\sum_{r=1}^{n}\frac{1}{r}\right)\left(\frac{z}{2}\right)^{2n},$$

given that the first solution is $J_0(z)$ as specified by (14.63).

14.4 (a) Show that the indicial equation for

$$zy'' - 2y' + yz = 0$$

has roots that differ by an integer, but that the two roots generate linearly independent solutions

$$y_1(z) = 3a_0 \sum_{n=1}^{\infty} \frac{(-1)^{n+1}\, 2nz^{2n+1}}{(2n+1)!},$$

$$y_2(z) = a_0 \sum_{n=0}^{\infty} \frac{(-1)^{n+1}(2n-1)z^{2n}}{(2n)!}.$$

(b) By expanding the sinusoidal functions, show that $y_1(z)$ is equal to $3a_0(\sin z - z\cos z)$ and then, using the Wronskian method, find an expression for $y_2(z)$ in terms of sinusoidal functions. (You will need to write $z^2$ as $(z/\sin z)(z\sin z)$ and integrate by parts to evaluate the integral involved.)

(c) Confirm that the two solutions are linearly independent by showing that their Wronskian is equal to $-z^2$, in accordance with (14.4).

14.5 Find series solutions of the equation $y'' - 2zy' - 2y = 0$. Identify one of the series as $y_1(z) = \exp z^2$ and verify this by direct substitution. By setting $y_2(z) = u(z)y_1(z)$ and solving the resulting equation for $u(z)$, find an explicit form for $y_2(z)$ and deduce that

$$\int_0^x e^{-v^2}\, dv = e^{-x^2} \sum_{n=0}^{\infty} \frac{n!}{2(2n+1)!}(2x)^{2n+1}.$$

14.6 Find the radius of convergence of a series solution about the origin for the equation $(z^2 + az + b)y'' + 2y = 0$ in the following cases:

(a) $a = 5$, $b = 6$; (b) $a = 5$, $b = 7$.

Show that if $a$ and $b$ are real and $4b > a^2$, the radius of convergence is always given by $b^{1/2}$.

14.7 For the equation $y'' + z^{-3}y = 0$, show that the origin becomes a regular singular point if the independent variable is changed from $z$ to $x = 1/z$. Hence find a series solution of the form $y_1(z) = \sum_0^\infty a_n z^{-n}$. By setting

$y_2(z) = u(z)y_1(z)$ and expanding the resulting expression for $du/dz$ in powers of $z^{-1}$, show that $y_2(z)$ has the asymptotic form

$$y_2(z) = c\left[z + \ln z - \tfrac{1}{2} + \mathrm{O}\left(\frac{\ln z}{z}\right)\right],$$

where $c$ is an arbitrary constant.

14.8 Equation (14.32) was shown to have a polynomial solution provided that $\lambda = 2n$ with $n$ an integer $\geq 0$. The polynomials are known as Hermite polynomials $H_n(x)$ and are of importance in the quantum mechanical treatment of the harmonic oscillator problem. They may also be defined by

$$\Phi(x, h) = \exp(2xh - h^2) = \sum_{n=0}^{\infty} \frac{1}{n!} H_n(x)h^n.$$

Show that

$$\frac{\partial^2 \Phi}{\partial x^2} - 2x\frac{\partial \Phi}{\partial x} + 2h\frac{\partial \Phi}{\partial h} = 0,$$

and hence that the $H_n(x)$ satisfy (14.32). Use $\Phi$ to prove that

(a) $H_n'(x) = 2nH_{n-1}(x)$,
(b) $H_{n+1}(x) - 2xH_n(x) + 2nH_{n-1}(x) = 0$.

14.9 By writing $\Phi(x, h)$ of the previous example as a function of $h - x$ rather than of $h$, show that an alternative representation of the $n$th Hermite polynomial is

$$H_n(x) = (-1)^n \left(\exp x^2\right) \frac{d^n}{dx^n}[\exp(-x^2)].$$

[Note that $H_n(x) = \partial^n \Phi / \partial h^n$ at $h = 0$.]

14.10 Carry through the following procedure as an alternative proof of result (14.45).

(a) Square both sides of (14.49), giving the generating-function definition of the Legendre polynomials.
(b) Express the RHS as a sum of powers of $h$, obtaining expressions for the coefficients.
(c) Integrate the RHS from $-1$ to $1$ and use the orthogonality results (14.46).
(d) Similarly integrate the LHS and expand the result in powers of $h$.
(e) Compare coefficients.

14.11 A charge $+2q$ is situated at the origin and charges of $-q$ at distances $\pm a$ from it along the polar axis. By relating it to the generating function for the Legendre polynomials, show that the electrostatic potential $\Phi$ at a point $(r, \theta, \phi)$ with $r > a$ is

$$\Phi(r, \theta, \phi) = \frac{2q}{4\pi\epsilon_0 r} \sum_{s=1}^{\infty} \left(\frac{a}{r}\right)^{2s} P_{2s}(\cos\theta).$$

14.12 The origin is an ordinary point of the Chebyshev equation,

$$(1 - z^2)y'' - zy' + m^2 y = 0,$$

which therefore has series solutions of the form $z^\sigma \sum_0^\infty a_n z^n$ for $\sigma = 0$ and $\sigma = 1$.

(a) Find the recurrence relationships for the $a_n$ in the two cases and show that there exist polynomial solutions $T_m(z)$:

(i) for $\sigma = 0$, when $m$ is an even integer, the polynomial having $\frac{1}{2}(m + 2)$ terms;

(ii) for $\sigma = 1$, when $m$ is an odd integer, the polynomial having $\frac{1}{2}(m + 1)$ terms.

(b) $T_m(z)$ is normalised so as to have $T_m(1) = 1$. Find explicit forms for $T_m(z)$ for $m = 0, 1, 2, 3$.

(c) Show that the corresponding non-terminating series solutions $S_m(z)$ have as their first few terms

$$S_0(z) = a_0 \left( z + \frac{1}{3!}z^3 + \frac{9}{5!}z^5 + \cdots \right),$$

$$S_1(z) = a_0 \left( 1 - \frac{1}{2!}z^2 - \frac{3}{4!}z^4 - \cdots \right),$$

$$S_2(z) = a_0 \left( z - \frac{3}{3!}z^3 - \frac{15}{5!}z^5 - \cdots \right),$$

$$S_3(z) = a_0 \left( 1 - \frac{9}{2!}z^2 + \frac{45}{4!}z^4 + \cdots \right).$$

## 14.10 Hints and answers

14.1 $a_0 \exp(z/2)$; $b_0 z^{1/2} \sum_{n=0}^{\infty} (2z)^n n!/(2n + 1)!$.

14.4 (b) $\cos z + z \sin z$.

14.5 $y_2(z) = (\exp z^2) \int_0^z \exp(-x^2)\, dx$.

14.6 (a) 2; (b) $\sqrt{7}$.

14.7 Transformed equation is $xy'' + 2y' + y = 0$; $a_n = (-1)^n (n + 1)^{-1} (n!)^{-2} a_0$; $du/dz = A[y_1(z)]^{-2}$.

14.8 Consider $\partial\Phi/\partial x$; (b) differentiate result (a) and then use (a) again to replace the derivatives.

14.10 At step (d)

$$\frac{1}{h}\ln\frac{1+h}{1-h} = \sum_{h=0}^{\infty} h^{2n}\int_{-1}^{1} P_n^2(x)\,dx.$$

14.11 Using the cosine law, the distances from the charges $-q$ are of the form $r\left[1 \pm 2(a/r)\cos\theta + (a/r)^2\right]^{1/2}$.

14.12 (a) $a_{n+2} = [a_n(n^2 - m^2)]/[(n+2)(n+1)]$,
$a_{n+2} = [a_n((n+1)^2 - m^2)]/[(n+3)(n+2)]$; (b) 1, $z$, $2z^2 - 1$, $4z^3 - 3z$.

# 15

# *Eigenfunction methods for differential equations*

In the previous three chapters we have dealt with the solution of differential equations of order $n$ by two methods. In one, we found $n$ independent solutions of the equation and then combined them, weighted with coefficients determined by the boundary conditions; in the other we found solutions in terms of series whose coefficients were related by (in general) an $n$-term recurrence relation, and thence fixed by the boundary conditions. For both approaches the linearity of the equation was an important or essential factor in the utility of the method, and in this chapter our aim will be to exploit the superposition properties of linear differential equations even further.

We will be concerned with the solution of equations of the inhomogeneous form

$$\mathcal{L}y(x) = f(x), \tag{15.1}$$

where $f(x)$ is a prescribed or general function, and the boundary conditions to be satisfied by the solution $y = y(x)$, for example at the limits $x = a$ and $x = b$, are given. In this equation the expression $\mathcal{L}y(x)$ stands for a linear differential operator acting upon the function $y(x)$.

In general, unless $f(x)$ is both known and simple, it will not be possible to find particular integrals of (15.1), even if complementary functions can be found that satisfy $\mathcal{L}y = 0$. The idea is therefore to exploit the linearity of $\mathcal{L}$ by building up the required solution as a *superposition*, generally containing an infinite number of terms, of some set of functions that each individually satisfy the boundary conditions. This clearly brings in a quite considerable complication, but since, within reason, we may select the set of functions to suit ourselves, we can obtain sizeable compensation for this complication. Indeed, if the set chosen is one containing functions that, when acted upon by $\mathcal{L}$, produce particularly simple results, we can 'show a profit' on the operation. In particular, if the set consists of those functions $y_i$ for which

$$\mathcal{L}y_i(x) = \lambda_i y_i(x), \tag{15.2}$$

where $\lambda_i$ is a constant, then a distinct advantage may be obtained from the manoeuvre because all the differentiation will have disappeared from (15.1).

Equation (15.2) is clearly reminiscent of the equation satisfied by the *eigenvectors* $\mathbf{x}^i$ of a linear operator $\mathcal{A}$, namely

$$\mathcal{A}\mathbf{x}^i = \lambda_i \mathbf{x}^i, \tag{15.3}$$

where $\lambda_i$ is a constant and is called the *eigenvalue* associated with $\mathbf{x}^i$. By analogy, in the context of differential equations a function satisfying (15.2) is called an *eigenfunction* of the operator $\mathcal{L}$ and $\lambda_i$ is then called the eigenvalue associated with the particular eigenfunction $y_i(x)$.

Probably the most familiar equation of the form (15.2) is that which describes a simple harmonic oscillator, i.e.

$$\mathcal{L}y = -\frac{d^2y}{dt^2} = \omega^2 y, \qquad \text{where } \mathcal{L} = -d^2/dt^2. \tag{15.4}$$

In this case the eigenfunctions are given by $y_n(t) = A_n e^{i\omega_n t}$, where $\omega_n = 2\pi n/T$, $T$ is the period of oscillation, $n = 0, \pm 1, \pm 2, \ldots$ and the $A_n$ are constants. The eigenvalues are $\omega_n^2 = n^2\omega_1^2 = n^2(2\pi/T)^2$. (Sometimes $\omega_n$ is referred to as the eigenvalue of this equation but we will avoid this confusing terminology here.)

Another equation of the form (15.2) is Legendre's equation

$$\mathcal{L}y = -(1-x^2)\frac{d^2y}{dx^2} + 2x\frac{dy}{dx} = \ell(\ell+1)y, \tag{15.5}$$

where

$$\mathcal{L} = -(1-x^2)\frac{d^2}{dx^2} + 2x\frac{d}{dx}. \tag{15.6}$$

We found the eigenfunctions of $\mathcal{L}$ by a series method in chapter 14, and for solutions to Legendre's equation that are regular at $x = \pm 1$ these are the Legendre polynomials, which are given by

$$y_\ell(x) = P_\ell(x) = \frac{1}{2^\ell \ell!}\frac{d^\ell}{dx^\ell}(x^2-1)^\ell, \tag{15.7}$$

for $\ell = 0, 1, 2, \ldots$, and which have associated eigenvalues $\ell(\ell+1)$. (Again, $\ell$ is sometimes, confusingly, referred to as the eigenvalue of this equation.)

We may discuss a somewhat wider class of differential equations by considering a slightly more general form of (15.2), namely

$$\mathcal{L}y(x) = \lambda\rho(x)y(x), \tag{15.8}$$

where $\rho(x)$ is a *weight function.* In many applications $\rho(x)$ is unity for all $x$, in which case (15.2) is recovered; in general, though, it is a function determined by the choice of coordinate system used in describing a particular physical situation. The only requirement on $\rho(x)$ is that it is real and does not change sign in the range $a \leq x \leq b$ and can, therefore, without loss of generality, be taken to be non-negative throughout. A function $y(x)$ that satisfies (15.8) is called an eigenfunction of the operator $\mathcal{L}$ with respect to the weight function $\rho(x)$.

This chapter will not cover methods used to determine the eigenfunctions of (15.2) or (15.8), since we have already discussed these in previous chapters, but, rather, will use the properties of the eigenfunctions to solve inhomogeneous equations of the form (15.1). We shall see later that the set of eigenfunctions $y_i(x)$ of a particular class of operators called *Hermitian operators* (the operator in the simple harmonic oscillator equation and that in Legendre's equation are examples) have particularly useful properties and these will be studied in detail. For reasons that we will discuss later, many of the interesting operators met with in the physical sciences are Hermitian. Before continuing our discussion of the eigenfunctions of Hermitian operators, however, we will consider the properties of general sets of functions.

## 15.1 Sets of functions

In chapter 7 we discussed the definition of a vector space, but concentrated on spaces of finite dimensionality. We now consider the *infinite*-dimensional space of all reasonably well-behaved functions $f(x)$, $g(x)$, $h(x)$, ... on the interval $a \leq x \leq b$. That these functions form a linear vector space can be verified since the set is closed under

(i) addition, which is commutative and associative,

$$f(x) + g(x) = g(x) + f(x),$$
$$[f(x) + g(x)] + h(x) = f(x) + [g(x) + h(x)],$$

(ii) multiplication by a scalar, which is distributive and associative,

$$\lambda\,[f(x) + g(x)] = \lambda f(x) + \lambda g(x),$$
$$\lambda\,[\mu f(x)] = (\lambda\mu) f(x),$$
$$(\lambda + \mu) f(x) = \lambda f(x) + \mu f(x).$$

Furthermore, in such a space

(iii) there exists a 'null vector' 0 such that $f(x) + 0 = f(x)$,
(iv) multiplication by unity leaves any function unchanged, i.e. $1 \times f(x) = f(x)$,
(v) each function has a negative function $-f(x)$ such that $f(x) + [-f(x)] = 0$.

By analogy with finite-dimensional vector spaces we may introduce a set of linearly independent basis functions $y_n(x)$, $n = 0, 1, \ldots, \infty$, such that *any* 'reasonable' function in the interval $a \leq x \leq b$ (i.e. one that obeys the Dirichlet conditions discussed in chapter 10) can be expressed as the linear sum of these functions:

$$f(x) = \sum_{n=0}^{\infty} c_n y_n(x).$$

Clearly if a different set of linearly independent basis functions $z_n(x)$ is chosen, then the function can be expressed in terms of the new basis,

$$f(x) = \sum_{n=0}^{\infty} d_n z_n(x),$$

where the $d_n$ are a different set of coefficients. In each case, provided the basis functions are linearly independent, the coefficients are unique.

We may also define an inner product on our function space by

$$\langle f|g\rangle = \int_a^b f^*(x)g(x)\rho(x)\,dx, \tag{15.9}$$

where $\rho(x)$ is the weight function, which we require to be real and non-negative in the interval $a \leq x \leq b$. As mentioned above, $\rho(x)$ is often unity for all $x$. Two functions are said to be *orthogonal* on the interval $[a, b]$ if

$$\langle f|g\rangle = \int_a^b f^*(x)g(x)\rho(x)\,dx = 0, \tag{15.10}$$

and the *norm* of a function is defined as

$$\|f\| = \langle f|f\rangle^{1/2} = \left[\int_a^b f^*(x)f(x)\rho(x)\,dx\right]^{1/2} = \left[\int_a^b |f(x)|^2\rho(x)\,dx\right]^{1/2}. \tag{15.11}$$

An infinite-dimensional vector space of functions, for which an inner product is defined, is called a *Hilbert space*. Using the concept of the inner product we can choose a basis of linearly independent functions $\phi_n(x)$, $n = 0, 1, 2, \ldots$, that are orthonormal, i.e. such that

$$\langle \phi_i|\phi_j\rangle = \int_a^b \phi_i^*(x)\phi_j(x)\rho(x)\,dx = \delta_{ij}. \tag{15.12}$$

If $y_n(x)$, $n = 0, 1, 2, \ldots$, are a linearly independent, but not orthonormal, basis for the Hilbert space then an orthonormal set of basis functions $\phi_n$ may be produced (in a similar manner to that used in the construction of a set of orthogonal eigenvectors of an Hermitian matrix, see chapter 7) by the following procedure, in which each of the new functions $\psi_n$ is to be normalised, giving

$\phi_n = \psi_n\langle\psi_n|\psi_n\rangle^{-1/2}$, before proceeding to the construction of the next one:

$$\begin{aligned}
\psi_0 &= y_0, \\
\psi_1 &= y_1 - \phi_0\langle\phi_0|y_1\rangle, \\
\psi_2 &= y_2 - \phi_1\langle\phi_1|y_2\rangle - \phi_0\langle\phi_0|y_2\rangle, \\
&\vdots \\
\psi_n &= y_n - \phi_{n-1}\langle\phi_{n-1}|y_n\rangle - \cdots - \phi_0\langle\phi_0|y_n\rangle \\
&\vdots
\end{aligned}$$

It is straightforward to check that each $\phi_n = \psi_n\langle\psi_n|\psi_n\rangle^{-1/2}$ is orthogonal to all its predecessors $\phi_i$, $i = 0, 1, 2, \ldots, n-1$. This method is called *Gram–Schmidt orthogonalisation.* Clearly, the functions $\psi_n$ also form an orthogonal set, but do not, in general, have unit norm.

►*Starting from the linearly independent functions $y_n(x) = x^n$, $n = 0, 1, \ldots,$ construct the first three orthonormal functions over the range $-1 < x < 1$.*

The first unnormalised function $\psi_0$ is simply equal to the first of the original functions, i.e.

$$\psi_0 = 1.$$

The normalisation is carried out by dividing by

$$\langle\psi_0|\psi_0\rangle^{1/2} = \left(\int_{-1}^{1} 1 \times 1\, du\right)^{1/2} = \sqrt{2},$$

with the result that the first normalised function $\phi_0$ is given by

$$\phi_0 = \frac{\psi_0}{\sqrt{2}} = \sqrt{\frac{1}{2}}.$$

The second unnormalised function is found by applying the above Gram–Schmidt orthogonalisation procedure, i.e.

$$\psi_1 = y_1 - \phi_0\langle\phi_0|y_1\rangle.$$

It can easily be shown that $\langle\phi_0|y_1\rangle = 0$, and so $\psi_1 = x$. Normalising then gives

$$\phi_1 = \psi_1\left(\int_{-1}^{1} u \times u\, du\right)^{-1/2} = \sqrt{\tfrac{3}{2}}x.$$

The third unnormalised function is similarly given by

$$\begin{aligned}
\psi_2 &= y_2 - \phi_1\langle\phi_1|y_2\rangle - \phi_0\langle\phi_0|y_2\rangle \\
&= x^2 - 0 - \tfrac{1}{3},
\end{aligned}$$

which, on normalising, gives

$$\phi_2 = \psi_2 \left( \int_{-1}^{1} \left(u^2 - \tfrac{1}{3}\right)^2 du \right)^{-1/2} = \tfrac{1}{2}\sqrt{\tfrac{5}{2}}(3x^2 - 1).$$

By comparing the functions $\phi_0$, $\phi_1$ and $\phi_2$, with the list in subsection 14.6.1, we see that this procedure has generated (multiples of) the first three Legendre polynomials. ◀

If a function is expressed in terms of an *orthonormal* basis $\phi_n(x)$ as

$$f(x) = \sum_{n=0}^{\infty} a_n \phi_n(x), \tag{15.13}$$

then the coefficients $a_n$ are given by

$$a_n = \langle \phi_n | f \rangle = \int_a^b \phi_n^*(x) f(x) \rho(x)\, dx. \tag{15.14}$$

Note that this is true only if the basis is orthonormal.

### *15.1.1 Some useful inequalities*

Since for a Hilbert space $\langle f|f \rangle \geq 0$, the inequalities discussed in subsection 7.1.3 hold. The proofs are not repeated here, but the relationships are listed for completeness.

(i) The Schwarz inequality states that

$$|\langle f|g \rangle| \leq \langle f|f \rangle^{1/2} \langle g|g \rangle^{1/2}, \tag{15.15}$$

where the equality holds when $f(x)$ is a scalar multiple of $g(x)$, i.e. when they are linearly dependent.

(ii) The triangle inequality states that

$$\|f + g\| \leq \|f\| + \|g\|, \tag{15.16}$$

where again equality holds when $f$ is a scalar multiple of $g$.

(iii) Bessel's inequality requires the introduction of an *orthonormal* basis $\phi_n(x)$ so that any function $f(x)$ can be written as

$$f(x) = \sum_{n=0}^{\infty} c_n \phi_n(x),$$

where $c_n = \langle \phi_n | f \rangle$. Bessel's inequality then states that

$$\langle f|f \rangle \geq \sum_n |c_n|^2. \tag{15.17}$$

The equality holds if the summation is over all the basis functions. If some values of $n$ are omitted from the sum then the inequality results (unless, of course, the $c_n$ are zero for all values of $n$ omitted, in which case the equality remains).

## 15.2 Adjoint and Hermitian operators

Having discussed general sets of functions we now return to the discussion of eigenfunctions of linear operators. The *adjoint* of an operator $\mathcal{L}$, denoted by $\mathcal{L}^\dagger$, is defined by

$$\int_a^b f(x)^* \left[\mathcal{L}g(x)\right] \rho(x)\, dx = \left\{ \int_a^b g^*(x) \left[\mathcal{L}^\dagger f(x)\right] \rho(x)\, dx \right\}^*, \qquad (15.18)$$

or $\langle f|\mathcal{L}g\rangle = \langle g|\mathcal{L}^\dagger f\rangle^*$ in inner product notation. An operator is then said to be *self-adjoint* or *Hermitian* if $\mathcal{L}^\dagger = \mathcal{L}$, i.e. if

$$\int_a^b f^*(x) \left[\mathcal{L}g(x)\right] \rho(x)\, dx = \left\{ \int_a^b g^*(x) \left[\mathcal{L}f(x)\right] \rho(x)\, dx \right\}^*, \qquad (15.19)$$

or, in inner product notation, $\langle f|\mathcal{L}g\rangle = \langle g|\mathcal{L}f\rangle^*$. From (15.19) we note that for an Hermitian operator

$$\langle g|\mathcal{L}f\rangle^* = \langle \mathcal{L}f|g\rangle \quad \Rightarrow \quad \langle \mathcal{L}f|g\rangle = \langle f|\mathcal{L}g\rangle = \langle f|\mathcal{L}|g\rangle,$$

where the notation of the final equality emphasises that $\mathcal{L}$ can act on either $f$ or $g$ without changing the value of the inner product. A little careful study will reveal the similarity between the definition of an Hermitian operator and the definition of an Hermitian matrix given in chapter 7. In general, however, an operator $\mathcal{L}$ is Hermitian over an interval $a \le x \le b$ only if certain boundary conditions are met by the functions $f$ and $g$ on which it acts.

►*Find the required boundary conditions for the linear operator $\mathcal{L} = d^2/dt^2$ to be Hermitian over the interval $t_0$ to $t_0 + T$.*

Substituting into the LHS of the definition of an Hermitian operator (15.19) and integrating by parts gives

$$\int_{t_0}^{t_0+T} f^* \frac{d^2 g}{dt^2}\, dt = \left[ f^* \frac{dg}{dt} \right]_{t_0}^{t_0+T} - \int_{t_0}^{t_0+T} \frac{df^*}{dt} \frac{dg}{dt}\, dt,$$

where we have taken the weight function $\rho(x)$ to be unity. Integrating the second term on the RHS by parts yields

$$\int_{t_0}^{t_0+T} f^* \frac{d^2 g}{dt^2}\, dt = \left[ f^* \frac{dg}{dt} \right]_{t_0}^{t_0+T} + \left[ -\frac{df^*}{dt} g \right]_{t_0}^{t_0+T} + \int_{t_0}^{t_0+T} g \frac{d^2 f^*}{dt^2}\, dt.$$

Remembering that the operator is real, and taking the complex conjugate outside the integral gives

$$\int_{t_0}^{t_0+T} f^* \frac{d^2 g}{dt^2}\, dt = \left[ f^* \frac{dg}{dt} \right]_{t_0}^{t_0+T} - \left[ \frac{df^*}{dt} g \right]_{t_0}^{t_0+T} + \left( \int_{t_0}^{t_0+T} g^* \frac{d^2 f}{dt^2}\, dt \right)^*,$$

which, by comparison with (15.19), proves that $\mathcal{L}$ is Hermitian provided

$$\left[ f^* \frac{dg}{dt} \right]_{t_0}^{t_0+T} = \left[ \frac{df^*}{dt} g \right]_{t_0}^{t_0+T}. \blacktriangleleft$$

We showed in chapter 7 that the eigenvalues of Hermitian matrices are real and that their eigenvectors can be chosen to be orthogonal. Similarly, the eigenvalues of Hermitian operators are real and their eigenfunctions can be chosen to be orthogonal (we will prove these properties in the following section). Hermitian operators (or matrices) are often used in the formulation of quantum mechanics. The eigenvalues then give a measure of an observable quantity such as energy or angular momentum, and the physical requirement that such quantities must be real is ensured by the reality of these eigenvalues. Furthermore, the infinite set of eigenfunctions of an Hermitian operator form a complete basis set, so that it is possible to expand any function $y(x)$ obeying the appropriate conditions in an eigenfunction series, i.e.

$$y(x) = \sum_{n=0}^{\infty} c_n y_n(x), \qquad (15.20)$$

where the choice of suitable values for the $c_n$ will make the sum arbitrarily close to $y(x)$. † These useful properties provide the motivation for a detailed study of Hermitian operators.

## 15.3 The properties of Hermitian operators

We now provide proofs of some of the useful properties of Hermitian operators. Much of the analysis is similar to that for Hermitian matrices in chapter 7, although the present section stands alone. (Here, and throughout the remainder of this chapter, we will write out inner products in full. We note, however, that the inner product notation often provides a neat form in which to express the results.)

† The proof of the completeness of the eigenfunctions of an Hermitian operator is beyond the scope of this book. The reader should refer to e.g. Courant and Hilbert, *Methods of Mathematical Physics* (Interscience Publishers, 1953).

### 15.3.1 Reality of the eigenvalues

Consider an Hermitian operator for which (15.8) is satisfied by at least two eigenfunctions $y_i(x)$ and $y_j(x)$, which have eigenvalues $\lambda_i$ and $\lambda_j$ respectively, so that

$$\mathcal{L}y_i = \lambda_i \rho(x) y_i, \tag{15.21}$$

$$\mathcal{L}y_j = \lambda_j \rho(x) y_j. \tag{15.22}$$

Multiplying (15.21) by $y_j^*$ and (15.22) by $y_i^*$ and then integrating gives

$$\int_a^b y_j^* \mathcal{L} y_i \, dx = \lambda_i \int_a^b y_j^* y_i \rho \, dx, \tag{15.23}$$

$$\int_a^b y_i^* \mathcal{L} y_j \, dx = \lambda_j \int_a^b y_i^* y_j \rho \, dx. \tag{15.24}$$

Remembering that we have required $\rho(x)$ to be real, the complex conjugate of (15.23) becomes

$$\left[ \int_a^b y_j^* \mathcal{L} y_i \, dx \right]^* = \lambda_i^* \int_a^b y_i^* y_j \rho \, dx, \tag{15.25}$$

and using the definition of an Hermitian operator (15.19) it follows that the LHS of (15.25) is equal to the LHS of (15.24). Thus

$$(\lambda_i^* - \lambda_j) \int_a^b y_i^* y_j \rho \, dx = 0. \tag{15.26}$$

If $i = j$ then $\lambda_i = \lambda_i^*$ (since $\int_a^b y_i^* y_i \rho \, dx \neq 0$), which is a statement that the eigenvalue $\lambda_i$ is real.

### 15.3.2 Orthogonality of the eigenfunctions

From (15.26), it is immediately apparent that two eigenfunctions $y_i$ and $y_j$ that correspond to different eigenvalues, i.e. such that $\lambda_i \neq \lambda_j$, satisfy

$$\int_a^b y_i^* y_j \rho \, dx = 0, \tag{15.27}$$

which is a statement of the orthogonality of $y_i$ and $y_j$. Since the normalisation of the eigenfunctions $y_i(x)$ is arbitrary, because $\mathcal{L}$ is linear, we shall assume for definiteness that they are normalised, so that $\int_a^b y_i^* y_i \rho \, dx = 1$. Thus we can write (15.27) in the form

$$\int_a^b y_i^* y_j \rho \, dx = \delta_{ij}, \tag{15.28}$$

which is valid for all pairs of values $i, j$.

If one (or more) of the eigenvalues is degenerate, however, we have different eigenfunctions corresponding to the same eigenvalue, and the proof of orthogonality is not so straightforward. Nevertheless, an orthogonal set of eigenfunctions may be constructed in a similar manner to that used in the construction of a set of orthogonal eigenvectors of an Hermitian matrix (see chapter 7). This is the *Gram–Schmidt orthogonalisation* method, which we also met in section 15.1. We repeat the analysis here for completeness.

Suppose, for the sake of our proof, that $\lambda_0$ is $k$-fold degenerate, i.e.

$$\mathcal{L}y_i = \lambda_0 \rho y_i \quad \text{for } i = 0, 1, \ldots, k-1, \tag{15.29}$$

but that $\lambda_0$ is different from any of $\lambda_k$, $\lambda_{k+1}$, etc. Then any linear combination of these $y_i$ is also an eigenfunction with eigenvalue $\lambda_0$ since

$$\mathcal{L}z \equiv \mathcal{L}\sum_{i=0}^{k-1} c_i y_i = \sum_{i=0}^{k-1} c_i \mathcal{L}y_i = \sum_{i=0}^{k-1} c_i \lambda_0 \rho y_i = \lambda_0 \rho z. \tag{15.30}$$

If the $y_i$ defined in (15.29) are not already mutually orthogonal, consider the new eigenfunctions $z_i$ constructed by the following procedure, in which each of the new functions $w_i$ is to be normalised, to give $z_i$, before proceeding to the construction of the next one (the normalisation can be carried out by dividing the eigenfunction $w_i$ by $(\int_a^b w_i^* w_i \rho\, dx)^{1/2}$):

$$\begin{aligned}
w_0 &= y_0, \\
w_1 &= y_1 - \left(z_0 \int_a^b z_0^* y_1 \rho\, dx\right), \\
w_2 &= y_2 - \left(z_1 \int_a^b z_1^* y_2 \rho\, dx\right) - \left(z_0 \int_a^b z_0^* y_2 \rho\, dx\right), \\
&\vdots \\
w_{k-1} &= y_{k-1} - \left(z_{k-2} \int_a^b z_{k-2}^* y_{k-1} \rho\, dx\right) - \cdots - \left(z_0 \int_a^b z_0^* y_{k-1} \rho\, dx\right).
\end{aligned}$$

Each of the integrals is just a number and thus each new function $z_i = w_i(\int_a^b w_i^* w_i \rho\, dx)^{-1/2}$ is, as can be shown from (15.30), an eigenvector of $\mathcal{L}$ with eigenvalue $\lambda_0$. It is straightforward to check that each $z_i$ is orthogonal to all its predecessors. Thus, by this explicit construction we have shown that an orthogonal set of eigenfunctions of an Hermitian operator $\mathcal{L}$ can be obtained. Clearly the orthonormal set obtained, $z_i$, is not unique.

### *15.3.3 Construction of real eigenfunctions*

Recall that the eigenfunction $y_i$ satisfies

$$\mathcal{L}y_i = \lambda_i \rho y_i, \tag{15.31}$$

and that the complex conjugate of this gives

$$\mathcal{L}y_i^* = \lambda_i^* \rho y_i^* = \lambda_i \rho y_i^*, \tag{15.32}$$

where the last equality follows because the eigenvalues are real, i.e. $\lambda_i = \lambda_i^*$. Thus, $y_i$ and $y_i^*$ are eigenfunctions corresponding to the same eigenvalue and hence, because of the linearity of $\mathcal{L}$, at least one of $y_i^* + y_i$ and $i(y_i^* - y_i)$ (which are both real) is a non-zero eigenfunction corresponding to that eigenvalue. Therefore the eigenfunctions can always be made real by taking suitable linear combinations. Such linear combinations will only be necessary in cases where a particular $\lambda$ is degenerate, i.e. corresponds to more than one linearly independent eigenfunction.

## 15.4 Sturm–Liouville equations

One of the most important applications of our discussion of Hermitian operators is to the study of *Sturm–Liouville equations*, which take the general form

$$p(x)\frac{d^2y}{dx^2} + r(x)\frac{dy}{dx} + q(x)y + \lambda\rho(x)y = 0, \quad \text{where } r(x) = \frac{dp(x)}{dx} \tag{15.33}$$

and $p$, $q$ and $r$ are real functions of $x$. (We note that sign conventions vary in this expression for the general Sturm–Liouville equation, with some authors using $-\lambda\rho(x)y$ on the LHS of (15.33).) A variational approach to the Sturm–Liouville equation, which is useful in estimating the eigenvalues $\lambda$ of the equation, is discussed in chapter 20. We now, however, turn to a demonstration that the Sturm–Liouville equation can be solved by superposition methods.

It is clear that (15.33) can be written

$$\mathcal{L}y = \lambda\rho(x)y \quad \text{where } \mathcal{L} = -\left[p(x)\frac{d^2}{dx^2} + r(x)\frac{d}{dx} + q(x)\right]. \tag{15.34}$$

An example is Legendre's equation (15.5), which is a Sturm–Liouville equation with $p(x) = 1 - x^2$, $r(x) = -2x = p'(x)$, $q(x) = 0$, $\rho(x) = 1$ and eigenvalues $\ell(\ell + 1)$.

It will be seen that the general Sturm–Liouville equation (15.33) can be rewritten

$$(py')' + qy + \lambda\rho y = 0, \tag{15.35}$$

where primes denote differentiation with respect to $x$. Using (15.34) this may also be written $\mathcal{L}y = -(py')' - qy = \lambda\rho y$. We will show in the next section that, under certain boundary conditions on the solutions $y(x)$, linear operators that can be written in this form are *self-adjoint*.

Whilst it is true that Sturm–Liouville equations represent only a small fraction of the differential equations one encounters in practice, nevertheless, as we shall demonstrate in subsection 15.4.2, *any* second-order differential equation of the form

$$p(x)y'' + r(x)y' + q(x)y + \lambda\rho(x)y = 0 \tag{15.36}$$

can be converted into Sturm–Liouville form by multiplying through by a suitable factor; this is discussed in subsection 15.4.2.

### 15.4.1 Valid boundary conditions

For the linear operator of the Sturm–Liouville equation (15.34) to be Hermitian over the range $[a, b]$ requires certain boundary conditions to be satisfied, namely, any two eigenfunctions $y_i$ and $y_j$ of (15.34) must satisfy

$$\left[y_i^* p y_j'\right]_{x=a} = \left[y_i^* p y_j'\right]_{x=b} \qquad \text{for all } i, j. \tag{15.37}$$

Rearranging (15.37) we find that

$$\left[y_i^* p y_j'\right]_{x=a}^{x=b} = 0, \tag{15.38}$$

is an equivalent statement of the required boundary conditions. These boundary conditions are in fact not too restrictive and are met, for instance, by the sets $y(a) = y(b) = 0$; $y(a) = y'(b) = 0$; $p(a) = p(b) = 0$ and by many other sets. It is important to note that in order to satisfy (15.37) and (15.38) one boundary condition must be specified at each end of the range.

►*Prove that the Sturm–Liouville operator is Hermitian over the range $[a, b]$ and under the boundary conditions (15.38).*

Putting the Sturm–Liouville form $\mathcal{L}y = -(py')' - qy$ into the definition (15.19) of an Hermitian operator, the LHS may be written as a sum of two terms, i.e.

$$-\int_a^b \left[y_i^*(py_j')' + y_i^* q y_j\right] dx = -\int_a^b y_i^*(py_j')'\, dx - \int_a^b y_i^* q y_j\, dx.$$

The first term may be integrated by parts to give

$$-\left[y_i^* p y_j'\right]_a^b + \int_a^b (y_i^*)' p y_j'\, dx.$$

The first term is zero because of the boundary conditions, and thus, integrating by parts again yields

$$\left[(y_i^*)' p y_j\right]_a^b - \int_a^b ((y_i^*)' p)' y_j\, dx.$$

The first term is once again zero. Thus

$$-\int_a^b [y_i^*(py_j')' + y_i^* q y_j]\,dx = \int_a^b [-((y_i^*)'p)'y_j - y_i^* q y_j]\,dx,$$
$$= \left\{-\int_a^b [y_j^*(py_i')' + y_j^* q y_i]\,dx\right\}^*,$$

which proves that the Sturm–Liouville operator is Hermitian over the prescribed interval. ◄

### 15.4.2 Putting an equation into Sturm–Liouville form

The Sturm–Liouville equation (15.33) requires that $r(x) = p'(x)$. However, any equation of the form

$$p(x)y'' + r(x)y' + q(x)y + \lambda\rho(x)y = 0, \tag{15.39}$$

can be put into self-adjoint form by multiplying through by the integrating factor

$$F(x) = \exp\left\{\int^x \frac{r(z) - p'(z)}{p(z)}\,dz\right\}. \tag{15.40}$$

It is easily verified that (15.39) then takes on the Sturm–Liouville form

$$[F(x)p(x)y']' + F(x)q(x)y + \lambda F(x)\rho(x)y = 0, \tag{15.41}$$

with a different, but still non-negative, weight function $F(x)\rho(x)$.

►*Put the Hermite equation*

$$y'' - 2xy' + 2\alpha y = 0$$

*into Sturm–Liouville form.*

Using (15.40), with $p(z) = 1$, $p'(z) = 0$ and $r(z) = -2z$, gives the integrating factor

$$F(x) = \exp\left(\int^x -2z\,dz\right) = \exp\left(-x^2\right).$$

Thus, the Hermite equation becomes

$$e^{-x^2}y'' - 2xe^{-x^2}y' + 2\alpha e^{-x^2}y = (e^{-x^2}y')' + 2\alpha e^{-x^2}y = 0,$$

which is clearly in Sturm–Liouville form with $p(x) = e^{-x^2}$, $q(x) = 0$, $\rho(x) = e^{-x^2}$ and $\lambda = 2\alpha$. ◄

## 15.5 Examples of Sturm–Liouville equations

In order to illustrate the wide applicability of Sturm–Liouville theory, in this section we present a short catalogue of some common equations of Sturm–Liouville form. Many of them have already been discussed in chapter 14. In particular the reader should note the orthogonality properties of the various solutions, which, in each case, follow because the differential operator is self-adjoint. For completeness we also quote the associated generating functions.

### *15.5.1 Legendre's equation*

We have already met *Legendre's equation,*

$$(1-x^2)y'' - 2xy' + \ell(\ell+1)y = [(1-x^2)y']' + \ell(\ell+1)y = 0, \tag{15.42}$$

and shown that it is a Sturm–Liouville equation with $p(x) = 1 - x^2$, $q(x) = 0$, $\rho(x) = 1$ and eigenvalues $\ell(\ell+1)$. In the previous chapter we found the solutions of Legendre's equation that are regular for all finite $x$. These are the Legendre polynomials $P_\ell(x)$, which are given by a Rodrigues' formula:

$$P_\ell(x) = \frac{1}{2^\ell \ell!}\frac{d^\ell}{dx^\ell}(x^2-1)^\ell.$$

The orthogonality and normalisation of the functions in the interval $-1 \le x \le 1$ is expressed by

$$\int_{-1}^{1} P_\ell(x)P_k(x)\,dx = \frac{2}{2\ell+1}\delta_{\ell k}.$$

The generating function is

$$G(x,h) = (1-2xh+h^2)^{-1/2} = \sum_{n=0}^{\infty} P_n(x)h^n.$$

Legendre's equations appear in the analysis of physical situations involving the operator $\nabla^2$ and axial symmetry, since the linear differential operator involved has the form of the polar-angle part of $\nabla^2$, when the latter is expressed in spherical polar coordinates. Examples include the solution of Laplace's equation in axially symmetric situations and of the Schrödinger equation for a quantum mechanical system involving a central potential.

### *15.5.2 The associated Legendre equation*

Very closely related to the Legendre equation is the *associated Legendre equation*

$$[(1-x^2)y']' + \left[\ell(\ell+1) - \frac{m^2}{1-x^2}\right]y = 0, \tag{15.43}$$

which reduces to Legendre's equation when $m = 0$. In physical applications $-\ell \leq m \leq \ell$ and $m$ is restricted to integer values. If $y(x)$ is a solution of Legendre's equation, then

$$w(x) = (1 - x^2)^{|m|/2} \frac{d^{|m|} y}{dx^{|m|}}$$

is a solution of the associated equation. The solutions of the associated Legendre equation that are regular for all finite $x$ are called the *associated Legendre functions* and are therefore given by

$$P_\ell^m(x) = (1 - x^2)^{|m|/2} \frac{d^{|m|} P_\ell}{dx^{|m|}}.$$

Note also that $P_\ell^m(x) = 0$ for $m > \ell$. As for the Legendre polynomials, the associated Legendre functions $P_\ell^m(x)$ are orthogonal in the range $-1 \leq x \leq 1$. This, and their normalisation, is given by

$$\int_{-1}^{1} P_\ell^m(x) P_k^m(x)\, dx = \frac{2}{2\ell + 1} \frac{(\ell + m)!}{(\ell - m)!} \delta_{\ell k}.$$

They have the generating function

$$G(x, h) = \frac{(2m)!(1 - x^2)^{m/2}}{2^m m!(1 - 2hx + h^2)^{m+1/2}} = \sum_{n=0}^{\infty} P_{n+m}^m(x) h^n.$$

The associated Legendre equation arises in physical situations in which there is a dependence on the azimuthal angle $\phi$ of the form $e^{im\phi}$ or $\cos m\phi$.

### 15.5.3 Bessel's equation

Physical situations that when described in spherical polar coordinates give rise to Legendre and associated Legendre equations lead to Bessel's equation when cylindrical polar coordinates are used. Bessel's equation has the form

$$x^2 y'' + xy' + (x^2 - n^2) y = 0, \tag{15.44}$$

but on dividing by $x$ and changing variables to $\xi = x/a$,† it takes on the Sturm-Liouville form

$$(\xi y')' - \frac{n^2}{\xi} y + a^2 \xi y = 0, \tag{15.45}$$

where a prime now indicates differentiation with respect to $\xi$.

† This change of scale is required to give the conventional normalisation, but is not needed for the transformation into Sturm–Liouville form.

We met Bessel's equation in chapter 14, where we saw that its solutions that are regular for finite $x$ are the Bessel functions, given by

$$J_n(x) = \sum_{r=0}^{\infty} \frac{(-1)^r (x/2)^{n+2r}}{r!\Gamma(n+r+1)}, \tag{15.46}$$

where $\Gamma$ is the gamma function discussed in the Appendix. Their orthogonality and normalisation over the range $0 \le x < \infty$ have been discussed in detail in chapter 14. The generating function for the Bessel functions is

$$G(x,h) = \exp\left[\frac{x}{2}\left(h - \frac{1}{h}\right)\right] = \sum_{n=-\infty}^{\infty} J_n(x)h^n. \tag{15.47}$$

### 15.5.4 *The simple harmonic equation*

The most trivial of Sturm–Liouville equations is the simple harmonic motion equation

$$y'' + \omega^2 y = 0. \tag{15.48}$$

which has $p(x) = 1$, $q(x) = 0$, $\rho(x) = 1$ and eigenvalue $\omega^2$. We have already met the solutions of this equation in the Fourier analysis of chapter 10, and the properties of orthogonality and normalisation of the eigenfunctions given there can now be seen in the wider context of general Sturm–Liouville equations.

### 15.5.5 *Hermite's equation*

The Hermite equation appears in the description of the wavefunction of a harmonic oscillator and is given by

$$y'' - 2xy' + 2\alpha y = 0. \tag{15.49}$$

We have already seen that it can be converted to Sturm–Liouville form by multiplying by the integrating factor $\exp(-x^2)$, which yields

$$e^{-x^2}y'' - 2xe^{-x^2}y' + 2\alpha e^{-x^2}y = (e^{-x^2}y')' + 2\alpha e^{-x^2}y = 0. \tag{15.50}$$

The solutions, the Hermite polynomials $H_n(x)$, are given by a Rodrigues' formula:

$$H_n(x) = (-1)^n e^{x^2} \frac{d^n}{dx^n}\left(e^{-x^2}\right). \tag{15.51}$$

Their orthogonality over the range $-\infty < x < \infty$ and their normalisation are given by

$$\int_{-\infty}^{\infty} e^{-x^2} H_m(x) H_n(x)\, dx = 2^n n! \sqrt{\pi}\delta_{mn}, \tag{15.52}$$

and their generating function is

$$G(x,h) = e^{2hx-h^2} = \sum_{n=0}^{\infty} \frac{H_n(x)}{n!} h^n. \tag{15.53}$$

### 15.5.6 Laguerre's equation

The Laguerre equation appears in the description of the wavefunction of the hydrogen atom and is given by

$$xy'' + (1-x)y' + ny = 0. \tag{15.54}$$

It can be converted to Sturm–Liouville form by multiplying by the integrating factor $\exp(-x)$, which yields

$$xe^{-x}y'' + (1-x)e^{-x}y' + ne^{-x}y = (xe^{-x}y')' + ne^{-x}y = 0. \tag{15.55}$$

The solutions, the Laguerre polynomials $L_n(x)$, are again given by a Rodrigues' formula:

$$L_n(x) = e^x \frac{d^n}{dx^n}\left(x^n e^{-x}\right). \tag{15.56}$$

Their orthogonality over the range $0 \le x < \infty$ and their normalisation are given by

$$\int_0^{\infty} e^{-x} L_m(x) L_n(x)\, dx = (n!)^2 \delta_{mn}, \tag{15.57}$$

and their generating function is

$$G(x,h) = \frac{e^{-xh/(1-h)}}{1-h} = \sum_{n=0}^{\infty} \frac{L_n(x)}{n!} h^n. \tag{15.58}$$

### 15.5.7 Chebyshev's equation

The Chebyshev equation

$$(1-x^2)y'' - xy' + n^2 y = 0 \tag{15.59}$$

can be converted to Sturm–Liouville form by multiplying by the integrating factor $(1-x^2)^{-1/2}$. Simplifying, this yields

$$\left[(1-x^2)^{1/2} y'\right]' + n^2(1-x^2)^{-1/2} y = 0. \tag{15.60}$$

The solutions, the Chebyshev polynomials, $T_n(x)$, are once again given by a Rodrigues' formula:

$$T_n(x) = \frac{(-2)^n n!(1-x^2)^{1/2}}{(2n)!} \frac{d^n}{dx^n}(1-x^2)^{n-1/2}. \tag{15.61}$$

Their orthogonality over the range $-1 \leq x \leq 1$ and their normalisation are given by

$$\int_{-1}^{1} (1-x^2)^{-1/2} T_m(x) T_n(x)\, dx = \begin{cases} 0 & \text{for } m \neq n, \\ \pi/2 & \text{for } n = m \neq 0, \\ \pi & \text{for } n = m = 0, \end{cases} \qquad (15.62)$$

and their generating function is

$$G(x,h) = \frac{1-xh}{1-2xh+h^2} = \sum_{n=0}^{\infty} T_n(x) h^n. \qquad (15.63)$$

## 15.6 Superposition of eigenfunctions: Green's functions

We have already seen that if

$$\mathcal{L} y_n(x) = \lambda_n \rho(x) y_n(x), \qquad (15.64)$$

where $\mathcal{L}$ is an Hermitian operator, then the eigenvalues $\lambda_n$ are real and the eigenfunctions $y_n(x)$ are orthogonal (or can be made so). Let us assume that we know the eigenfunctions $y_n(x)$ of $\mathcal{L}$ that individually satisfy (15.64) and some imposed boundary conditions (for which $\mathcal{L}$ is Hermitian).

Now let us suppose we wish to solve the inhomogeneous differential equation

$$\mathcal{L} y(x) = f(x), \qquad (15.65)$$

subject to the same boundary conditions. Since the eigenfunctions of $\mathcal{L}$ form a complete set, the full solution, $y(x)$, to (15.65) may be written as a superposition of eigenfunctions, i.e.

$$y(x) = \sum_{n=0}^{\infty} c_n y_n(x), \qquad (15.66)$$

for some choice of the constants $c_n$. Now, making full use of the linearity of $\mathcal{L}$, we have

$$f(x) = \mathcal{L} y(x) = \mathcal{L}\left( \sum_{n=0}^{\infty} c_n y_n(x) \right) = \sum_{n=0}^{\infty} c_n \mathcal{L} y_n(x) = \sum_{n=0}^{\infty} c_n \lambda_n \rho(x) y_n(x). \qquad (15.67)$$

Multiplying the first and last terms of (15.67) by $y_j^*$, and integrating, we obtain

$$\int_a^b y_j^*(z) f(z)\, dz = \sum_{n=0}^{\infty} \int_a^b c_n \lambda_n y_j^*(z) y_n(z) \rho(z)\, dz, \qquad (15.68)$$

where we have used $z$ as the integration variable for later convenience. Finally, using the orthogonality condition (15.28), we see that the integrals on the RHS are zero unless $n = j$, and obtain

$$c_n = \frac{1}{\lambda_n} \frac{\int_a^b y_n^*(z) f(z)\, dz}{\int_a^b y_n^*(z) y_n(z) \rho(z)\, dz}. \tag{15.69}$$

Thus, if we can find all the eigenfunctions of a differential operator, (15.69) can be used to find the weighting coefficients for the superposition to give as the full solution

$$y(x) = \sum_{n=0}^{\infty} \frac{1}{\lambda_n} \frac{\int_a^b y_n^*(z) f(z)\, dz}{\int_a^b y_n^*(z) y_n(z) \rho(z)\, dz} y_n(x). \tag{15.70}$$

If the eigenfunctions have already been normalised, so that

$$\int_a^b y_n^*(z) y_n(z) \rho(z)\, dz = 1 \qquad \text{for all } n,$$

and we assume that we may interchange the order of summation and integration (15.70) can then be written

$$y(x) = \int_a^b \left\{ \sum_{n=0}^{\infty} \left[ \frac{1}{\lambda_n} y_n(x) y_n^*(z) \right] \right\} f(z)\, dz.$$

The quantity in braces, which is a function of $x$ and $z$ only, is usually written $G(x, z)$, and is the *Green's function* for the problem. With this notation,

$$y(x) = \int_a^b G(x, z) f(z)\, dz, \tag{15.71}$$

where

$$G(x, z) = \sum_{n=0}^{\infty} \frac{1}{\lambda_n} y_n(x) y_n^*(z). \tag{15.72}$$

We note that $G(x, z)$ is determined entirely by the boundary conditions and the eigenfunctions $y_n$, and hence by $\mathcal{L}$ itself, and $f(z)$ depends purely on the RHS of the inhomogeneous equation (15.65). Thus, for a given $\mathcal{L}$ and boundary conditions we can, once and for all, establish a function $G(x, z)$ that will enable us to solve the inhomogeneous equation for *any* RHS. From (15.72) we also note that

$$G(x, z) = G^*(z, x). \tag{15.73}$$

We have already met the Green's function in the solution of second-order differential equations in chapter 13, as the function that satisfies the equation $\mathcal{L}[G(x, z)] = \delta(x - z)$ (and the boundary conditions). The formulation given above is an alternative, though equivalent, one.

►*Find an appropriate Green's function for the equation*

$$y'' + \tfrac{1}{4}y = f(x),$$

*with boundary conditions* $y(0) = y(\pi) = 0$. *Hence, solve for (i)* $f(x) = \sin 2x$ *and (ii)* $f(x) = x/2$.

One approach to solving this problem is to use the methods of chapter 13 and find a complementary function and particular integral. However, in order to illustrate the techniques developed in the present chapter we will use the superposition of eigenfunctions, which, as may easily be checked, produces the same solution.

The operator on the LHS of this equation is already self-adjoint under the given boundary conditions, and so we seek its eigenfunctions. These satisfy the equation

$$y'' + \tfrac{1}{4}y = \lambda y.$$

This equation has the familiar solution

$$y(x) = A\sin\left(\sqrt{\tfrac{1}{4} - \lambda}\right)x + B\cos\left(\sqrt{\tfrac{1}{4} - \lambda}\right)x.$$

Now, the boundary conditions require that $B = 0$ and $\sin\left(\sqrt{\frac{1}{4} - \lambda}\right)\pi = 0$, and so

$$\sqrt{\tfrac{1}{4} - \lambda} = n, \qquad \text{where } n = 0, \pm 1, \pm 2, \ldots.$$

Therefore, the independent eigenfunctions that satisfy the boundary conditions are

$$y_n(x) = A_n \sin nx,$$

where $n$ is any non-negative integer. The normalisation condition further requires

$$\int_0^\pi A_n^2 \sin^2 nx\,dx = 1 \quad\Rightarrow\quad A_n = \left(\frac{2}{\pi}\right)^{1/2}.$$

Comparison with (15.72) shows that the appropriate Green's function is therefore given by

$$G(x,z) = \frac{2}{\pi}\sum_{n=0}^{\infty}\frac{\sin nx \sin nz}{\frac{1}{4} - n^2}.$$

*Case (i)* Using (15.71), the solution with $f(x) = \sin 2x$ is given by

$$\begin{aligned} y(x) &= \frac{2}{\pi}\int_0^\pi\left(\sum_{n=0}^{\infty}\frac{\sin nx \sin nz}{\frac{1}{4} - n^2}\right)\sin 2z\,dz \\ &= \frac{2}{\pi}\sum_{n=0}^{\infty}\frac{\sin nx}{\frac{1}{4} - n^2}\int_0^\pi \sin nz \sin 2z\,dz. \end{aligned}$$

Now the integral is zero unless $n = 2$, in which case it is

$$\int_0^\pi \sin^2 2z\, dz = \frac{\pi}{2}.$$

Thus

$$y(x) = -\frac{2}{\pi}\frac{\sin 2x}{15/4}\frac{\pi}{2} = -\frac{4}{15}\sin 2x,$$

is the full solution for $f(x) = \sin 2x$. This is, of course, exactly the solution found by using the methods of chapter 13.

*Case (ii)* The solution with $f(x) = x/2$ is given by

$$\begin{aligned} y(x) &= \int_0^\pi \left(\frac{2}{\pi}\sum_{n=0}^\infty \frac{\sin nx \sin nz}{\frac{1}{4} - n^2}\right)\frac{z}{2}\, dz \\ &= \frac{1}{\pi}\sum_{n=0}^\infty \frac{\sin nx}{\frac{1}{4} - n^2}\int_0^\pi z \sin nz\, dz. \end{aligned}$$

The integral may be evaluated by integrating by parts, i.e.

$$\begin{aligned} \int_0^\pi z \sin nz\, dz &= \left[-\frac{z\cos nz}{n}\right]_0^\pi + \int_0^\pi \frac{\cos nz}{n}\, dz \\ &= \frac{-\pi \cos n\pi}{n} + \left[\frac{\sin nz}{n^2}\right]_0^\pi \\ &= -\frac{\pi(-1)^n}{n}. \end{aligned}$$

For $n = 0$ the integral is zero, and thus

$$y(x) = \sum_{n=1}^\infty (-1)^{n+1}\frac{\sin nx}{n\left(\frac{1}{4} - n^2\right)},$$

is the full solution for $f(x) = x/2$. Using the methods of subsection 13.1.2 the solution is found to be $y(x) = 2x - 2\pi \sin(x/2)$, which may be shown to be equal to the above solution by expanding $2x - 2\pi\sin(x/2)$ as a Fourier sine series. ◀

A useful relation between the eigenfunctions of $\mathcal{L}$ is given by writing

$$\begin{aligned} f(x) &= \sum_n y_n(x)\int_a^b y_n^*(z) f(z)\rho(z)\, dz \\ &= \int_a^b f(z)\rho(z)\sum_n y_n(x)y_n^*(z)\, dz, \end{aligned}$$

and hence

$$\rho(z)\sum_n y_n(x)y_n^*(z) = \delta(x - z). \tag{15.74}$$

This is called the *completeness* or *closure* property of the eigenfunctions. It defines a complete set. If the spectrum of eigenvalues of $\mathcal{L}$ is anywhere continuous then the eigenfunction $y_n(x)$ must be treated as $y(n,x)$ and an integration carried out over $n$.

We also note that the RHS of (15.74) is a $\delta$-function and so is only non-zero when $z = x$; thus $\rho(z)$ on the LHS can be replaced by $\rho(x)$ if required, i.e.

$$\rho(z)\sum_n y_n(x)y_n^*(z) = \rho(x)\sum_n y_n(x)y_n^*(z). \tag{15.75}$$

## 15.7 A useful generalisation

We may sometimes encounter inhomogeneous equations of a slightly more general form than (15.1), given by

$$\mathcal{L}y(x) - \lambda\rho(x)y(x) = f(x) \tag{15.76}$$

for some self-adjoint operator $\mathcal{L}$, with $y$ subject to the appropriate boundary conditions and $\lambda$ a given constant. To solve this equation we expand $y(x)$ and $f(x)$ in terms of the eigenfunctions $y_n(x)$ of the operator $\mathcal{L}$, which satisfy

$$\mathcal{L}y_n(x) = \lambda_n\rho(x)y_n(x).$$

We expand $f(x)$ as follows:

$$\begin{aligned} f(x) &= \sum_{n=0}^{\infty} y_n(x)\int_a^b y_n^*(z)f(z)\rho(z)\,dz \\ &= \int_a^b \rho(z)\sum_{n=0}^{\infty} y_n(x)y_n^*(z)f(z)\,dz. \end{aligned} \tag{15.77}$$

Using (15.75) this becomes

$$\begin{aligned} f(x) &= \int_a^b \rho(x)\sum_{n=0}^{\infty} y_n(x)y_n^*(z)f(z)\,dz \\ &= \rho(x)\sum_{n=0}^{\infty} y_n(x)\int_a^b y_n^*(z)f(z)\,dz. \end{aligned} \tag{15.78}$$

Next we expand $y(x)$ as $y = \sum_{n=0}^{\infty} c_n y_n(x)$ and seek the coefficients $c_n$. Substituting this and (15.78) in (15.76) we have

$$\rho(x)\sum_{n=0}^{\infty}(\lambda_n - \lambda)c_n y_n(x) = \rho(x)\sum_{n=0}^{\infty} y_n(x)\int_a^b y_n^*(z)f(z)\,dz,$$

from which we find

$$c_n = \sum_{n=0}^{\infty}\frac{\int_a^b y_n^*(z)f(z)\,dz}{\lambda_n - \lambda}.$$

Hence the solution of (15.76) is given by

$$y = \sum_{n=0}^{\infty} c_n y_n(x) = \sum_{n=0}^{\infty} \frac{y_n(x)}{\lambda_n - \lambda} \int_a^b y_n^*(z) f(z)\, dz$$

$$= \int_a^b \sum_{n=0}^{\infty} \frac{y_n(x) y_n^*(z)}{\lambda_n - \lambda} f(z)\, dz.$$

From this we may identify the Green's function

$$G(x,z) = \sum_{n=0}^{\infty} \frac{y_n(x) y_n^*(z)}{\lambda_n - \lambda}.$$

We note that if $\lambda = \lambda_n$, i.e. if $\lambda$ equals one of the eigenvalues of $\mathcal{L}$, then $G(x,z)$ becomes infinite, and this method runs into difficulty. No solution then exists unless the RHS of (15.76) satisfies the relation

$$\int_a^b y_n^*(x) f(x)\, dx = 0.$$

If the spectrum of eigenvalues of the operator $\mathcal{L}$ is anywhere continuous, the orthogonality and closure relationships of the eigenfunctions become

$$\int_a^b y_n^*(x) y_m(x) \rho(x)\, dx = \delta(n - m)$$

$$\int_0^{\infty} y_n^*(z) y_n(x) \rho(x)\, dn = \delta(x - z).$$

Repeating the above analysis we then find that the Green's function is given by

$$G(x,z) = \int_0^{\infty} \frac{y_n(x) y_n^*(z)}{\lambda_n - \lambda}\, dn.$$

## 15.8 Exercises

15.1 Express the hypergeometric equation

$$(x^2 - x)y'' + [(1 + \alpha + \beta)x - \gamma]y' + \alpha\beta y = 0$$

in Sturm–Liouville form, determining the conditions imposed on $x$ and the parameters $\alpha$, $\beta$ and $\gamma$ by the boundary conditions and the allowed forms of weight function.

15.2 (a) Find the solution of $(1 - x^2)y'' - 2xy' + by = f(x)$ valid in the range $-1 \leq x \leq 1$ and finite at $x = 0$, in terms of Legendre polynomials.

(b) If $b = 14$ and $f(x) = 5x^3$, find the explicit solution and verify it by direct substitution.

15.3 Use the generating function for the Legendre polynomials $P_n(x)$ to show that

$$\int_0^1 P_{2n+1}(x)\,dx = (-1)^n \frac{(2n)!}{2^{2n+1} n!(n+1)!},$$

but that, except for the case $n = 0$,

$$\int_0^1 P_{2n}(x)\,dx = 0.$$

15.4 The quantum mechanical wavefunction for a one-dimensional simple harmonic oscillator in its $n$th energy level is of the form

$$\psi(x) = \exp(-x^2/2) H_n(x),$$

where $H_n(x)$ is the $n$th Hermite polynomial. The generating function for the polynomials (15.53) is

$$G(x,h) = e^{2hx-h^2} = \sum_{n=0}^{\infty} \frac{H_n(x)}{n!} h^n.$$

(a) Find $H_i(x)$ for $i = 1, 2, 3, 4$.
(b) Evaluate by direct calculation

$$\int_{-\infty}^{\infty} e^{-x^2} H_p(x) H_q(x)\,dx,$$

(i) for $p = 2$, $q = 3$; (ii) for $p = 2$, $q = 4$; (iii) for $p = q = 3$. Check your answers against equation (15.52). (You will find it convenient to use

$$\int_{-\infty}^{\infty} x^{2n} e^{-x^2}\,dx = \frac{(2n)!\sqrt{\pi}}{2^{2n} n!}$$

for integer $n \geq 0$.)

15.5 The Laguerre polynomials, which are required for the quantum mechanical description of the hydrogen atom, can be defined by the generating function (equation (15.58))

$$G(x,h) = \frac{e^{-hx/(1-h)}}{1-h} = \sum_{n=0}^{\infty} \frac{L_n(x)}{n!} h^n.$$

By differentiating the equation separately with respect to $x$ and $h$, and re-substituting for $G(x,h)$, prove that $L_n$ and $L_n'$ ($= dL_n(x)/dx$) satisfy the recurrence relations

$$L_n' - nL_{n-1}' + nL_{n-1} = 0,$$

$$L_{n+1} - (2n+1-x)L_n + n^2 L_{n-1} = 0.$$

From these two equations and others derived from them, show that $L_n(x)$ satisfies the Laguerre equation

$$xL_n'' + (1-x)L_n' + nL_n = 0.$$

15.6 The Chebyshev polynomials $T_n(x)$ can be written as

$$T_n(x) = \cos(n \cos^{-1} x).$$

(a) Verify that these functions do satisfy the Chebyshev equation.
(b) Use de Moivre's theorem to show that an alternative expression is

$$T_n(x) = \sum_{r \text{ even}}^{n} (-1)^{r/2} \frac{n!}{(n-r)!r!} x^{n-r}(1-x^2)^{r/2}.$$

15.7 A particle moves in a parabolic potential in which its natural angular frequency of oscillation is 1/2. At time $t = 0$ it passes through the origin with velocity $v$ and is suddenly subjected to an additional acceleration of $+1$ for $0 \leq t \leq \pi/2$, and then $-1$ for $\pi/2 < t \leq \pi$. At the end of this period it is at the origin again. Apply the results of the worked example in section 15.6 to show that

$$v = -\frac{8}{\pi} \sum_{m=0}^{\infty} \frac{1}{(4m+2)^2 - \frac{1}{4}} \approx -0.81.$$

15.8 By substituting $x = \exp t$ find the normalized eigenfunctions $y_n(x)$ and the eigenvalues $\lambda_n$ of the operator $\mathcal{L}$ defined by

$$\mathcal{L}y = x^2 y'' + 2xy' + \tfrac{1}{4}y, \qquad 1 \leq x \leq e,$$

with $y(1) = y(e) = 0$. Find, as a series $\sum a_n y_n(x)$, the solution of $\mathcal{L}y = x^{-1/2}$.

15.9 Express the solution of Poisson's equation in electrostatics,

$$\nabla^2 \phi(\mathbf{r}) = -\rho(\mathbf{r})/\epsilon_0,$$

where $\rho$ is the non-zero charge density over a finite part of space, in the form of an integral and hence identify the Green's function for the $\nabla^2$ operator.

15.10 In the quantum mechanical study of the scattering of a particle by a potential, a (Born-approximation) solution can be obtained in terms of a function $y(\mathbf{r})$ that satisfies an equation of the form

$$(-\nabla^2 - K^2)y(\mathbf{r}) = F(\mathbf{r}).$$

Assuming that $y_{\mathbf{k}}(\mathbf{r}) = (2\pi)^{-3/2} \exp(i\mathbf{k} \cdot \mathbf{r})$ is a suitably normalised eigenfunction of $-\nabla^2$ corresponding to eigenvalue $-k^2$, find a suitable Green's function $G_K(\mathbf{r}, \mathbf{r}')$. By taking the direction of the vector $\mathbf{r} - \mathbf{r}'$ as the polar axis for a $\mathbf{k}$-space integration, show that $G_K(\mathbf{r}, \mathbf{r}')$ can be reduced to

$$\frac{1}{4\pi|\mathbf{r} - \mathbf{r}'|} \int_{-\infty}^{\infty} \frac{w \sin w}{w^2 - w_0^2} \, dw,$$

where $w_0 = K|\mathbf{r} - \mathbf{r}'|$. (This integral can be evaluated by a contour integration (chapter 18) to give $(4\pi|\mathbf{r} - \mathbf{r}'|)^{-1} \exp(iK|\mathbf{r} - \mathbf{r}'|)$.)

## 15.9 Hints and answers

15.1 $[x^{\gamma}(1 - x)^{\alpha+\beta-\gamma+1} y']' = \alpha\beta x^{\gamma-1}(1 - x)^{\alpha+\beta-\gamma} y$; $0 \leq x \leq 1$, $\alpha + \beta > \gamma > 1$.

15.2 (a) $y = \sum a_n P_n(x)$ with

$$a_n = \frac{n + 1/2}{b - n(n + 1)} \int_{-1}^{1} f(z) P_n(z) \, dz;$$

(b) $5x^3 = 2P_3(x) + 3P_1(x)$, giving $a_1 = 1/4$ and $a_3 = 1$, leading to $y = 5(2x^3 - x)/4$.

15.4 (a) $2x$, $4x^2 - 2$, $8x^3 - 12x$, $16x^4 - 48x^2 + 12$; (b) (i) 0, (ii) 0, (iii) $48\sqrt{\pi}$.

15.8 $y_n(x) = \sqrt{2} x^{-1/2} \sin(n\pi \ln x)$ with $\lambda_n = -n^2\pi^2$;

$$a_n = \begin{cases} -(n\pi)^{-2} \int_1^e \sqrt{2} x^{-1} \sin(n\pi \ln x) = -\sqrt{8}(n\pi)^{-3} & \text{for } n \text{ odd}, \\ 0 & \text{for } n \text{ even}. \end{cases}$$

15.9 $G(\mathbf{r}, \mathbf{r}') = (4\pi|\mathbf{r} - \mathbf{r}'|)^{-1}$.

# 16

# *Partial differential equations: general and particular solutions*

In this chapter and the next the solution of differential equations of types typically encountered in the physical sciences and engineering is extended to situations involving more than one independent variable. A partial differential equation (PDE) is an equation relating an unknown function (the dependent variable) of two or more variables with one or more of its partial derivatives with respect to those variables. The most commonly occuring independent variables are those describing position and time, and so we will couch our discussion and examples in notation appropriate to them.

As in other chapters we will concentrate most of our attention on the equations that arise most often in physical situations. We shall therefore restrict our discussion to linear PDEs, i.e. those of first degree in the dependent variable. Furthermore, we shall primarily discuss second-order equations. The solution of first-order PDEs will necessarily be involved in treating these, and some of the methods discussed can be extended without difficulty to third- and higher-order equations. We shall also see that many of the ideas developed for ordinary differential equations can be carried over directly into the study of PDEs.

In this chapter we shall concentrate on general solutions of PDEs in terms of arbitrary functions, and particular solutions that may be derived from them in the presence of boundary conditions. We also discuss the existence and uniqueness of the solutions to PDEs under given boundary conditions.

In the next chapter we discuss the methods most commonly used in practice for obtaining solutions to PDEs subject to given boundary conditions. These include the separation of variables, integral transform methods and Green's functions. This division of material is rather arbitrary and really has only been made because of the general usefulness of the latter methods. In particular, it will be readily apparent that some of the results of the present chapter are in fact solutions in the form of separated variables, but arrived at by a different approach.

## 16.1 Important partial differential equations

Most of the important PDEs of physics are second-order and linear. In order to gain familiarity with their general form, some of the more important ones will now be briefly discussed. These equations apply to a wide variety of different physical systems.

Since, in general, the PDEs listed below describe three-dimensional situations, the independent variables are $\mathbf{r}$ and $t$, where $\mathbf{r}$ is the position vector and $t$ is time. The actual variables used to specify the position vector $\mathbf{r}$ are dictated by the coordinate system in use. For example, in Cartesian coordinates the independent variables of position are $x$, $y$ and $z$, whereas in spherical polar coordinates they are $r$, $\theta$ and $\phi$. The equations may, however, be written in a coordinate-independent manner by the use of the Laplacian operator $\nabla^2$.

### 16.1.1 The wave equation

The wave equation

$$\nabla^2 u = \frac{1}{c^2}\frac{\partial^2 u}{\partial t^2}, \tag{16.1}$$

describes as a function of position and time the displacement from equilibrium, $u(\mathbf{r}, t)$, of a vibrating string or membrane, or a vibrating solid, gas or liquid. The equation also occurs in electromagnetism, where $u$ may be a component of the electric or magnetic field in an elecromagnetic wave, or the current or voltage along a transmission line. The quantity $c$ is the speed of propagation of the waves.

► *Find the equation satisfied by small transverse displacements $u(x,t)$ of a uniform string of mass per unit length $\rho$ held under a uniform tension $T$, assuming that the string is initially located along the x-axis in a Cartesian coordinate system.*

Figure 16.1 shows the forces acting on an elemental length $\Delta s$ of the string. If the tension $T$ in the string is uniform along its length, the net upward vertical force on the element is

$$\Delta F = T\sin\theta_2 - T\sin\theta_1.$$

Assuming that the angles $\theta_1$ and $\theta_2$ are both small, we may make the approximation $\sin\theta \approx \tan\theta$. Since at any point on the string the slope $\tan\theta = \partial u/\partial x$, the force can be written

$$\Delta F = T\left(\frac{\partial u(x+\Delta x, t)}{\partial x} - \frac{\partial u(x,t)}{\partial x}\right) \approx T\frac{\partial^2 u(x,t)}{\partial x^2}\Delta x,$$

where we have used the definition of the partial derivative to simplify the RHS.

This upward force may be equated, by Newton's second law, to the product of the mass of the element and its upward acceleration. The element has a mass

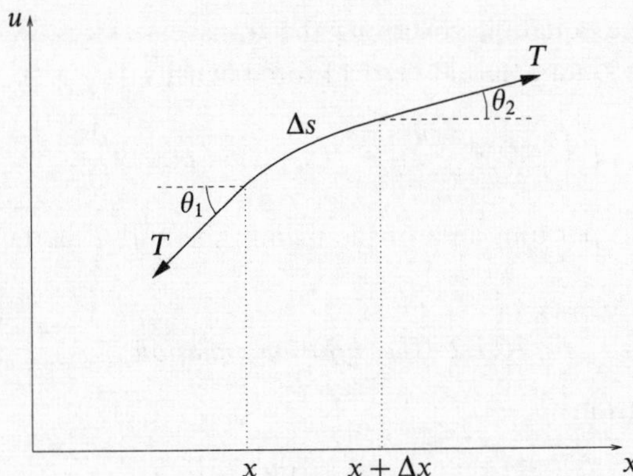


Figure 16.1 The forces acting on an element of a string under uniform tension $T$.

$\rho\,\Delta s$, which is approximately equal to $\rho\,\Delta x$ if the vibrations of the string are small, and so we have

$$\rho\,\Delta x\,\frac{\partial^2 u(x,t)}{\partial t^2} = T\frac{\partial^2 u(x,t)}{\partial x^2}\Delta x.$$

Dividing both sides by $\Delta x$ we obtain, for the vibrations of the string, the one-dimensional wave equation

$$\frac{\partial^2 u}{\partial x^2} = \frac{1}{c^2}\frac{\partial^2 u}{\partial t^2},$$

where $c^2 = T/\rho$. ◀

The longitudinal vibrations of an elastic rod obey a very similar equation to that derived in the above example, namely

$$\frac{\partial^2 u}{\partial x^2} = \frac{\rho}{E}\frac{\partial^2 u}{\partial t^2},$$

where the only difference is that here $\rho$ is the mass per unit volume and $E$ is Young's modulus.

The wave equation can be generalised slightly, for example in the case of the vibrating string, if in addition to the tension $T$ there is an external upward vertical force $f(x,t)$ per unit length acting on the string at time $t$. The transverse vibrations then satisfy the equation

$$T\frac{\partial^2 u}{\partial x^2} + f(x,t) = \rho\frac{\partial^2 u}{\partial t^2},$$

which is clearly of the form 'upward force per unit length = mass per unit length × upward acceleration'.

Similar examples, but involving two or three spatial dimensions rather than one, are provided by the equation governing the transverse vibrations of a stretched membrane subject to an external vertical force density $f(x, y, t)$,

$$T\left(\frac{\partial^2 u}{\partial x^2} + \frac{\partial^2 u}{\partial y^2}\right) + f(x, y, t) = \rho(x, y)\frac{\partial^2 u}{\partial t^2},$$

where $\rho$ is the mass per unit area of the membrane and $T$ is the tension.

### *16.1.2 The diffusion equation*

The diffusion equation

$$\kappa \nabla^2 u = \frac{\partial u}{\partial t} \tag{16.2}$$

describes the temperature $u$ in a region containing no heat sources or sinks. It also applies to the diffusion of a chemical that has a concentration $u(\mathbf{r}, t)$. The constant $\kappa$ is called the diffusivity. The equation is clearly second order in the three spatial variables, but first order in time.

►*Derive the equation satisfied by the temperature $u(\mathbf{r}, t)$ at time $t$ for a material of uniform thermal conductivity $k$, specific heat capacity $s$ and density $\rho$. Express the equation in Cartesian coordinates.*

Let us consider an arbitrary volume $V$ lying within the solid, and bounded by a surface $S$ (this may coincide with the surface of the solid if so desired). At any point in the solid the rate of heat flow per unit area in any given direction $\hat{\mathbf{r}}$ is proportional to minus the component of the temperature gradient in that direction and is given by $(-k\nabla u) \cdot \hat{\mathbf{r}}$. The total flux of heat *out* of the volume $V$ per unit time is given by

$$\begin{aligned} -\frac{dQ}{dt} &= \iint_S (-k\nabla u) \cdot \hat{\mathbf{n}}\, dS \\ &= \iiint_V \nabla \cdot (-k\nabla u)\, dV, \end{aligned} \tag{16.3}$$

where $Q$ is the total heat energy in $V$ at time $t$, and $\hat{\mathbf{n}}$ is the outward-pointing unit normal to $S$; note that we have used the divergence theorem to convert the surface integral into a volume integral.

We can also express $Q$ as a volume integral over $V$,

$$Q = \iiint_V s\rho u\, dV,$$

and so its rate of change is given by

$$\frac{dQ}{dt} = \iiint_V s\rho \frac{\partial u}{\partial t}\, dV, \tag{16.4}$$

where we have taken the derivative with respect to time inside the integral (see section 4.11).

Comparing (16.3) and (16.4), and remembering that the volume $V$ is arbitrary, we obtain the three-dimensional diffusion equation

$$\kappa \nabla^2 u = \frac{\partial u}{\partial t},$$

where the diffusion coefficient $\kappa = k/(s\rho)$. To express this equation in Cartesian coordinates, we simply write $\nabla^2$ in terms of $x$, $y$ and $z$ to obtain

$$\kappa \left( \frac{\partial^2 u}{\partial x^2} + \frac{\partial^2 u}{\partial y^2} + \frac{\partial^2 u}{\partial z^2} \right) = \frac{\partial u}{\partial t}. \blacktriangleleft$$

The diffusion equation derived above can be generalised to

$$k\nabla^2 u + f(\mathbf{r}, t) = s\rho \frac{\partial u}{\partial t}.$$

The second term $f(\mathbf{r}, t)$ represents a varying density of heat sources throughout the material, but is often not required in physical applications. In the most general case, $k$, $s$ and $\rho$ may depend on position $\mathbf{r}$, in which case the first term becomes $\nabla \cdot (k\nabla u)$. However, in the simplest application the heat flow is one-dimensional with no heat sources, and the equation becomes (in Cartesian coordinates)

$$\frac{\partial^2 u}{\partial x^2} = \frac{s\rho}{k} \frac{\partial u}{\partial t}.$$

### 16.1.3 Laplace's equation

Laplace's equation

$$\nabla^2 u = 0 \tag{16.5}$$

may be obtained by setting $\partial u/\partial t = 0$ in the diffusion equation (16.2), and describes (for example) the *steady-state* temperature of a solid in which there are no heat sources – i.e. the temperature after a long time has elapsed.

Laplace's equation also describes the gravitational potential in a region containing no matter, or the electrostatic potential in a charge-free region. It also applies to the flow of an incompressible fluid with no sources, sinks or vortices; in this case $u$ is the velocity potential, from which the velocity is given by $v = \nabla u$.

### 16.1.4 Poisson's equation

Poisson's equation

$$\nabla^2 u = \rho(\mathbf{r}), \tag{16.6}$$

describes the same physical situations as Laplace's equation, but in regions containing matter, charges, or sources of heat or fluid. The function $\rho(\mathbf{r})$ is called the source density, and in physical applications usually contains some multiplicative physical constants. For example, if $u$ is the electrostatic potential in some region of space, in which case $\rho$ is the density of electric charge, then $\nabla^2 u = -\rho(\mathbf{r})/\epsilon_0$, where $\epsilon_0$ is the permittivity of free space. Alternatively, $u$ might represent the gravitational potential in some region where the matter density is given by $\rho$; then $\nabla^2 u = 4\pi G\rho(\mathbf{r})$, where $G$ is the gravitational constant.

### 16.1.5 *Schrödinger's equation*

The Schrödinger equation

$$-\frac{\hbar^2}{2m}\nabla^2 u + V(\mathbf{r})u = i\hbar\frac{\partial u}{\partial t}, \tag{16.7}$$

describes the quantum mechanical wavefunction $u(\mathbf{r}, t)$ of a non-relativistic particle of mass $m$; $\hbar$ is Planck's constant divided by $2\pi$. Like the diffusion equation it is second order in the three spatial variables and first order in time.

## 16.2 General form of solution

Before turning to the methods by which we may hope to solve PDEs such as those listed in the previous section, it is instructive, as for ordinary differential equations in chapter 12, to study how PDEs may be formed from a set of possible solutions. Such a study can provide an indication of how equations obtained not from possible solutions but from physical arguments might be solved.

For definiteness let us suppose we have a set of functions involving two independent variables $x$ and $y$. Without further specification this is of course a very wide set of functions, and we could not expect to find a useful equation that they all satisfy. However, let us consider a type of function $u_i(x, y)$ in which $x$ and $y$ appear in a particular way, such that $u_i$ can be written as a function (however complicated) *of a single variable* $p$, itself a simple function of $x$ and $y$.

Let us illustrate this by considering the three functions

$$\begin{aligned}
u_1(x, y) &= x^4 + 4(x^2 y + y^2 + 1)\\
u_2(x, y) &= \sin x^2 \cos 2y + \cos x^2 \sin 2y\\
u_3(x, y) &= \frac{x^2 + 2y + 2}{3x^2 + 6y + 5}.
\end{aligned}$$

These are all fairly complicated functions of $x$ and $y$ and a single differential equation of which each one is a solution is not obvious. However, if we observe that in fact each is a function of the variable $p = x^2+2y$ then a great simplification takes place. Written in terms of $p$ the above equations become

$$u_1(x, y) = (x^2 + 2y)^2 + 4 = p^2 + 4 = f_1(p)$$
$$u_2(x, y) = \sin(x^2 + 2y) = \sin p = f_2(p)$$
$$u_3(x, y) = \frac{(x^2 + 2y) + 2}{3(x^2 + 2y) + 5} = \frac{p + 2}{3p + 5} = f_3(p).$$

Let us now form, for each $u_i$, the partial derivatives $\partial u_i/\partial x$ and $\partial u_i/\partial y$. In each case these are (writing both the form for general $p$ and the one appropriate to our particular case, $p = x^2 + 2y$)

$$\frac{\partial u_i}{\partial x} = \frac{df_i(p)}{dp}\frac{\partial p}{\partial x} = 2xf_i'$$
$$\frac{\partial u_i}{\partial y} = \frac{df_i(p)}{dp}\frac{\partial p}{\partial y} = 2f_i'$$

for $i = 1$, 2 or 3. All reference to the form of $f_i$ can be eliminated from these equations by cross-multiplication, obtaining

$$\frac{\partial p}{\partial y}\frac{\partial u_i}{\partial x} = \frac{\partial p}{\partial x}\frac{\partial u_i}{\partial y},$$

or, for our specific form, $p = x^2 + 2y$,

$$\frac{\partial u_i}{\partial x} = x\frac{\partial u_i}{\partial y}. \qquad (16.8)$$

It is thus apparent that not only are the three functions $u_1$, $u_2$ and $u_3$ solutions of the PDE (16.8), but so also is *any arbitrary function* $f(p)$ of which the argument $p$ has the form $x^2 + 2y$.

## 16.3 General and particular solutions

In the last section we found that the first-order PDE (16.8) has as a solution *any* function of the variable $x^2 + 2y$. This points the way for the solution of PDEs of other orders as follows. It is *not* generally true that an $n$th-order PDE can always be considered as resulting from the elimination of $n$ arbitrary *functions* from its solution (as opposed to the elimination of $n$ arbitrary *constants* for an $n$th-order ODE, see section 12.1). However, given specific PDEs we may try to solve them by seeking combinations of variables in terms of which the solutions may be expressed as arbitrary functions. Where this is possible we may expect $n$ combinations to be involved in the solution.

Naturally, the exact functional form of the solution for any particular situation must be determined by some set of boundary conditions. For instance, if the PDE contains two independent variables $x$ and $y$, then for complete determination of its solution the boundary conditions will take a form equivalent to specifying $u(x, y)$ along a suitable continuum of points in the $xy$-plane (usually along a line).

We now discuss the general and particular solutions of first- and second-order PDEs. In order to simplify the algebra, we shall restrict our discussion to equations containing just two independent variables $x$ and $y$. Nevertheless, the method presented below may be extended to equations containing several independent variables.

### 16.3.1 First-order equations

Although most of the PDEs encountered in physical contexts are second order (i.e. containing $\partial^2 u/\partial x^2$ or $\partial^2 u/\partial x\partial y$, etc.), we now discuss first-order equations to illustrate the general considerations involved in the form of the solution and in satisfying any boundary conditions on the solution.

The most general first-order linear PDE (containing two independent variables) is of the form

$$A(x,y)\frac{\partial u}{\partial x} + B(x,y)\frac{\partial u}{\partial y} + C(x,y)u = R(x,y), \tag{16.9}$$

where $A(x,y)$, $B(x,y)$, $C(x,y)$ and $R(x,y)$ are given functions. Clearly, if either $A(x,y)$ or $B(x,y)$ is zero then the PDE may be solved straightforwardly as a first-order linear ODE (as discussed in chapter 12), the only modification being that the arbitrary constant of integration becomes an *arbitrary function* of $x$ or $y$ respectively.

►*Find the general solution $u(x,y)$ of*

$$x\frac{\partial u}{\partial x} + 3u = x^2.$$

Dividing through by $x$ we obtain

$$\frac{\partial u}{\partial x} + \frac{3u}{x} = x,$$

which is a linear equation with integrating factor (see subsection 12.2.4)

$$\exp\left(\int \frac{3}{x}\,dx\right) = \exp(3\ln x) = x^3.$$

Multiplying through by this factor we find

$$\frac{\partial}{\partial x}(x^3 u) = x^4,$$

which, on integrating with respect to $x$, gives

$$x^3 u = \frac{x^5}{5} + f(y),$$

where $f(y)$ is an *arbitrary function* of $y$. Finally, dividing through by $x^3$, we obtain the solution

$$u(x, y) = \frac{x^2}{5} + \frac{f(y)}{x^3}. \blacktriangleleft$$

When the PDE contains partial derivatives with respect to both independent variables then, of course, we cannot employ the above procedure but must seek an alternative method. Let us for the moment restrict our attention to the special case in which $C(x, y) = R(x, y) = 0$ and, following the discussion of the previous section, look for solutions of the form $u(x, y) = f(p)$ where $p$ is some, at present unknown, combination of $x$ and $y$. We then have

$$\frac{\partial u}{\partial x} = \frac{df(p)}{dp}\frac{\partial p}{\partial x}$$
$$\frac{\partial u}{\partial y} = \frac{df(p)}{dp}\frac{\partial p}{\partial y},$$

which, when substituted into the PDE (16.9), gives

$$\left[A(x, y)\frac{\partial p}{\partial x} + B(x, y)\frac{\partial p}{\partial y}\right]\frac{df(p)}{dp} = 0.$$

This removes all reference to the actual form of the function $f(p)$ since for non-trivial $p$ we must have

$$A(x, y)\frac{\partial p}{\partial x} + B(x, y)\frac{\partial p}{\partial y} = 0. \qquad (16.10)$$

Let us now consider the necessary condition for $f(p)$ to remain constant as $x$ and $y$ vary; this is that $p$ itself remains constant. Thus $f$ remaining constant implies that $x$ and $y$ vary in such a way that

$$dp = \frac{\partial p}{\partial x}\,dx + \frac{\partial p}{\partial y}\,dy = 0. \qquad (16.11)$$

The forms of (16.10) and (16.11) are very alike, and become the same if we require that

$$\frac{dx}{A(x, y)} = \frac{dy}{B(x, y)}. \qquad (16.12)$$

By integrating this expression the form of $p$ can be found.

►*For*

$$x\frac{\partial u}{\partial x} - 2y\frac{\partial u}{\partial y} = 0, \qquad (16.13)$$

*find (i) the solution that takes the value $2y + 1$ on the line $x = 1$, and (ii) a solution that has the value 4 at the point $(1, 1)$.*

If we seek a solution of the form $u(x, y) = f(p)$, we deduce from (16.12) that $u(x, y)$ will be constant along lines of $(x, y)$ that satisfy

$$\frac{dx}{x} = \frac{dy}{-2y},$$

which on integrating gives $x = cy^{-1/2}$. Identifying the constant of integration $c$ with $p^{1/2}$ (to avoid fractional powers) we conclude that $p = x^2y$. Thus the general solution of the PDE (16.13) is

$$u(x, y) = f(x^2y),$$

where $f$ is an arbitrary function.

We must now find the particular solutions that obey each of the imposed boundary conditions. For boundary condition (i) a little thought shows that the particular solution required is

$$u(x, y) = 2(x^2y) + 1 = 2x^2y + 1. \qquad (16.14)$$

For boundary condition (ii) some obviously acceptable solutions are

$$u(x, y) = x^2y + 3$$
$$u(x, y) = 4x^2y$$
$$u(x, y) = 4.$$

Each is a valid solution (the freedom of choice of form arises from the fact that $u$ is specified at only one point $(1, 1)$, and not (say) along a continuum, as in boundary condition (i)) and all three are particular examples of the general solution, which may be written, for example,

$$u(x, y) = x^2y + 3 + g(x^2y),$$

where $g = g(x^2y) = g(p)$ is an arbitrary function subject only to $g(1) = 0$. For this example, the forms of $g$ corresponding to the particular solutions listed above are $g(p) = 0$, $g(p) = 3p - 3$, $g(p) = 1 - p$. ◄

As mentioned above, in order to find a solution of the form $u(x, y) = f(p)$ we require that the original PDE contains no term in $u$, but only terms containing its partial derivatives. If a term in $u$ is present, so that $C(x, y) \neq 0$ in (16.9), then the procedure needs some modification, since we cannot simply divide out the dependence on $f(p)$ to obtain (16.10). In such cases we look instead for a solution of the form $u(x, y) = h(x, y)f(p)$. We illustrate this method in the following example.

►*Find the general solution of*

$$x\frac{\partial u}{\partial x} + 2\frac{\partial u}{\partial y} - 2u = 0. \qquad (16.15)$$

We seek a solution of the form $u(x, y) = h(x, y)f(p)$, with the consequence that

$$\frac{\partial u}{\partial x} = \frac{\partial h}{\partial x}f(p) + h\frac{df(p)}{dp}\frac{\partial p}{\partial x},$$
$$\frac{\partial u}{\partial y} = \frac{\partial h}{\partial y}f(p) + h\frac{df(p)}{dp}\frac{\partial p}{\partial y}.$$

Substituting these expressions into the PDE (16.15) and rearranging, we obtain

$$\left(x\frac{\partial h}{\partial x} + 2\frac{\partial h}{\partial y} - 2h\right)f(p) + \left(x\frac{\partial p}{\partial x} + 2\frac{\partial p}{\partial y}\right)h\frac{df(p)}{dp} = 0.$$

The first term in parentheses is just the original PDE with $u$ replaced by $h$. Therefore, if $h$ is *any* solution of the PDE, *however simple*, this term will vanish, to leave

$$\left(x\frac{\partial p}{\partial x} + 2\frac{\partial p}{\partial y}\right)h\frac{df(p)}{dp} = 0,$$

from which, as in the previous case, we obtain

$$x\frac{\partial p}{\partial x} + 2\frac{\partial p}{\partial y} = 0.$$

From (16.11), (16.12) we see that $u(x, y)$ will be constant along lines of $(x, y)$ that satisfy

$$\frac{dx}{x} = \frac{dy}{2},$$

which integrates to give $x = c\exp(y/2)$. Identifying the constant of integration $c$ with $p$ we find $p = x\exp(-y/2)$. Thus the general solution of (16.15) is

$$u(x, y) = h(x, y)f(x\exp(-y/2)),$$

where $f(p)$ is any arbitrary function of $p$ and $h(x, y)$ is any solution of (16.15).

If we take, for example, $h(x, y) = \exp y$ (which clearly satisfies (16.15)), then the general solution is

$$u(x, y) = (\exp y)f(x\exp(-y/2)).$$

Alternatively, $h(x, y) = x^2$ also satisfies (16.15) and so the general solution to the equation can also be written

$$u(x, y) = x^2 g(x \exp(-y/2)),$$

where $g$ is an arbitrary function of $p$; clearly $g(p) = f(p)/p^2$. ◄

### 16.3.2 Inhomogeneous equations and problems

Let us discuss in a more general form the particular solutions of (16.13) found in the second example of the previous subsection. It is clear that, so far as this equation is concerned, if $u(x, y)$ is a solution then so is any multiple of $u(x, y)$ or any linear sum of separate solutions $u_1(x, y) + u_2(x, y)$. However, when it comes to fitting the boundary conditions this is not so.

For example, although $u(x, y)$ in (16.14) satisfies the PDE and the boundary condition $u(1, y) = 2y + 1$, the function $u_1(x, y) = 4u(x, y) = 8xy + 4$, whilst satisfying the PDE, takes the value $8y+4$ on the line $x = 1$ and so does not satisfy the required boundary condition. Likewise the function $u_2(x, y) = u(x, y)+f_1(x^2y)$, for arbitrary $f_1$, satisfies (16.13) but takes the value $u_2(1, y) = 2y + 1 + f_1(y)$ on the line $x = 1$, and thus is not of the required form unless $f_1$ is identically zero.

Thus we see that when treating the superposition of solutions of PDEs two considerations arise, one concerning the equation itself, the other connected to the boundary conditions. The *equation* is said to be *homogeneous* if the fact that $u(x, y)$ is a solution implies that $\lambda u(x, y)$, for any constant $\lambda$, is also a solution. However, the *problem* is said to be homogeneous if, in addition, the boundary conditions are such that if they are satisfied by $u(x, y)$ they are also satisfied by $\lambda u(x, y)$. The last requirement itself is referred to as that of *homogeneous boundary conditions.*

For example, the PDE (16.13) is homogeneous, but the general first-order equation (16.9) would not be homogeneous unless $R(x, y) = 0$. Furthermore, the boundary condition (i) imposed on the solution of (16.13) in the previous subsection is not homogeneous, though, in this case, the boundary condition

$$u(x, y) = 0 \qquad \text{on the line } y = 4x^{-2}$$

would be, since $u(x, y) = \lambda(x^2y - 4)$ satisfies this condition for any $\lambda$, and being a function of $x^2y$ satisfies (16.13).

The reason for discussing the homogeneity of PDEs and their boundary conditions is that in linear PDEs there is a close parallel to the complementary function and particular integral property of ODEs. The general solution of an inhomogeneous problem can be written as the sum of *any* particular solution of the problem and the general solution of the corresponding homogeneous problem (as for ODEs, we require that the particular solution is not already contained in the general solution of the homogeneous problem). Thus, for example, the general solution of

$$\frac{\partial u}{\partial x} - x\frac{\partial u}{\partial y} + au = f(x, y), \tag{16.16}$$

subject to, say, the boundary condition $u(0, y) = g(y)$, is given by

$$u(x, y) = v(x, y) + w(x, y),$$

where $v(x, y)$ is any solution (however simple) of (16.16) such that $v(0, y) = g(y)$, and $w(x, y)$ is the general solution of

$$\frac{\partial w}{\partial x} - x\frac{\partial w}{\partial y} + aw = 0, \tag{16.17}$$

with $w(0, y) = 0$. If the boundary conditions are sufficiently specified then the only possible solution of (16.17) will be $w(x, y) \equiv 0$ and $v(x, y)$ will be the complete solution by itself.

Alternatively, we may begin by finding the general solution of the inhomogeneous equation (16.16) *without* regard for any boundary conditions; it is just the sum of the general solution to the homogeneous equation and a particular integral of (16.16), both without reference to the boundary conditions. The boundary conditions can then be used to find the appropriate particular solution from the general solution.

We will not discuss at length general methods of obtaining particular integrals of PDEs, but merely note that some of those methods available for ordinary differential equations can be suitably extended.†

►*Find the general solution of*

$$y\frac{\partial u}{\partial x} - x\frac{\partial u}{\partial y} = 3x. \tag{16.18}$$

*Hence find the most general particular solution (i) which satisfies* $u(x, 0) = x^2$, *and (ii) which has the value* $u(x, y) = 2$ *at the point* $(1, 0)$.

This equation is inhomogeneous, and so let us first find the general solution of (16.18) without regard for any boundary conditions. We start by looking for a solution of the corresponding homogeneous equation (with the RHS equal to zero) of the form $u(x, y) = f(p)$. Following the same procedure as that used in the solution of (16.13) we find that $u(x, y)$ will be constant along lines of $(x, y)$ that satisfy

$$\frac{dx}{y} = \frac{dy}{-x} \quad \Rightarrow \quad \frac{x^2}{2} + \frac{y^2}{2} = c.$$

† See for example Piaggio, *Differential Equations* (Bell, 1954), p. 175 *et seq.*

Identifying the constant of integration $c$ with $p/2$, we find that the general solution of the homogeneous equation is $u(x, y) = f(x^2 + y^2)$ for arbitrary function $f$. Now by inspection a particular integral of (16.18) is $u(x, y) = -3y$, and so the general solution to (16.18) is

$$u(x, y) = f(x^2 + y^2) - 3y.$$

Now the boundary condition (i) requires $u(x, 0) = f(x^2) = x^2$, i.e. $f(z) = z$, and so the particular solution in this case is

$$u(x, y) = x^2 + y^2 - 3y.$$

Similarly, boundary condition (ii) requires $u(1, 0) = f(1) = 2$. One possibility is $f(z) = 2z$, and if we make this choice, one way of writing the most general particular solution is

$$u(x, y) = 2x^2 + 2y^2 - 3y + g(x^2 + y^2),$$

where $g$ is any arbitrary function for which $g(1) = 0$. Alternatively, a simpler choice would be $f(z) = 2$, leading to

$$u(x, y) = 2 - 3y + g(x^2 + y^2). \blacktriangleleft$$

Although we have discussed the solution of inhomogeneous problems only for first-order equations, the general considerations hold true for linear PDEs of higher order.

### *16.3.3 Second-order equations*

As noted in section 16.1, second-order linear PDEs are of great importance in describing the behaviour of many physical systems. As in our discussion of first-order equations, we shall for the moment restrict our discussion to equations with just two independent variables; extensions to a greater number of independent variables are straightforward.

The most general second-order linear PDE (containing two independent variables) has the form

$$A\frac{\partial^2 u}{\partial x^2} + B\frac{\partial^2 u}{\partial x \partial y} + C\frac{\partial^2 u}{\partial y^2} + D\frac{\partial u}{\partial x} + E\frac{\partial u}{\partial y} + Fu = R(x, y), \qquad (16.19)$$

where $A, B, \ldots, F$ and $R(x, y)$ are given functions of $x$ and $y$. Because of the nature of the solutions to such equations, they are usually divided into three classes, a division of which we will make further use in subsection 16.6.2. The equation (16.19) is called *hyperbolic* if $B^2 > 4AC$, *parabolic* if $B^2 = 4AC$ and *elliptic* if $B^2 < 4AC$. Clearly, if $A$, $B$ and $C$ are functions of $x$ and $y$ (rather than just constants) then the equation might be of a different type in different parts of the $xy$-plane.

Equation (16.19) obviously represents a very large class of PDEs, and it is usually impossible to find closed-form solutions to most of these equations. Therefore, for the moment we shall consider only homogeneous equations, with $R(x, y) = 0$, and make the further (greatly simplifying) restriction that, throughout the remainder of this section, $A, B, \ldots, F$ are not functions of $x$ and $y$, but merely constants.

We now tackle the problem of solving some types of second-order PDE with constant coefficients by seeking solutions that are arbitrary functions of particular combinations of independent variables, just as we did for first-order equations.

Following the discussion of the previous section, we can only hope to find such solutions if all the terms of the equation involve the same total number of differentiations, i.e. all terms are of the same order, although the number of differentiations with respect to the individual independent variables may be different. This means that in (16.19) we require the constants $D$, $E$ and $F$ to be identically zero (we have, of course, already assumed that $R(x, y)$ is zero), so that we are now only considering equations of the form

$$A\frac{\partial^2 u}{\partial x^2} + B\frac{\partial^2 u}{\partial x \partial y} + C\frac{\partial^2 u}{\partial y^2} = 0, \qquad (16.20)$$

where $A$, $B$ and $C$ are constants. We note that both the one-dimensional wave equation,

$$\frac{\partial^2 u}{\partial x^2} - \frac{1}{c^2}\frac{\partial^2 u}{\partial t^2} = 0,$$

and the two-dimensional Laplace equation,

$$\frac{\partial^2 u}{\partial x^2} + \frac{\partial^2 u}{\partial y^2} = 0,$$

are of this form, but that the diffusion equation,

$$\kappa\frac{\partial^2 u}{\partial x^2} - \frac{\partial u}{\partial t} = 0,$$

is not, since it contains a first-order derivative.

Since all the terms in (16.20) involve two differentiations, by assuming a solution of the form $u(x, y) = f(p)$, where $p$ is some unknown function of $x$ and $y$ (or $t$), we may be able to obtain a common factor $d^2 f(p)/dp^2$ as the only appearance of $f$ on the LHS. Then, because of the zero RHS, all reference to the form of $f$ can be cancelled out.

We can gain some guidance on the suitable forms for the combination $p = p(x, y)$ by considering $\partial u/\partial x$ when $u$ is given by $u(x, y) = f(p)$, for then

$$\frac{\partial u}{\partial x} = \frac{df(p)}{dp}\frac{\partial p}{\partial x}.$$

Clearly differentiation of this equation with respect to $x$ (or $y$) will not lead to a single term on the RHS, containing $f$ only as $d^2f(p)/dp^2$, unless the factor $\partial p/\partial x$ is a constant so that $\partial^2 p/\partial x^2$ and $\partial^2 p/\partial x \partial y$ are necessarily zero. This shows that $p$ must be a linear function of $x$. In an exactly similar way $p$ must also be a linear function of $y$, i.e. $p = ax + by$.

If we assume a solution of (16.20) of the form $u(x, y) = f(ax+by)$, and evaluate the terms ready for substitution into (16.20), we obtain

$$\frac{\partial u}{\partial x} = a\frac{df(p)}{dp}, \qquad \frac{\partial u}{\partial y} = b\frac{df(p)}{dp},$$

$$\frac{\partial^2 u}{\partial x^2} = a^2\frac{d^2f(p)}{dp^2}, \qquad \frac{\partial^2 u}{\partial x \partial y} = ab\frac{d^2f(p)}{dp^2}, \qquad \frac{\partial^2 u}{\partial y^2} = b^2\frac{d^2f(p)}{dp^2},$$

which on substituting give

$$\left(Aa^2 + Bab + Cb^2\right)\frac{d^2f(p)}{dp^2} = 0. \qquad (16.21)$$

This is the form we have been seeking, since now a solution independent of the form of $f$ can be obtained if we require that $a$ and $b$ satisfy

$$Aa^2 + Bab + Cb^2 = 0.$$

From this quadratic, two values for the ratio of the two constants $a$ and $b$ are obtained,

$$b/a = [-B \pm (B^2 - 4AC)^{1/2}]/2C.$$

If we denote these two ratios by $\lambda_1$ and $\lambda_2$, then *any* functions of the two variables

$$p_1 = x + \lambda_1 y, \qquad p_2 = x + \lambda_2 y$$

will be solutions of the original equation (16.20). The omission of the constant factor $a$ from $p_1$ and $p_2$ is of no consequence since this can always be absorbed into the particular form of any chosen function – only the *relative* weighting of $x$ and $y$ in $p$ is important.

Since $p_1$ and $p_2$ are in general different, we can thus write the general solution of (16.20) as

$$u(x, y) = f(x + \lambda_1 y) + g(x + \lambda_2 y), \qquad (16.22)$$

where $f$ and $g$ are arbitrary functions.

Finally, we note that the alternative solution $d^2f(p)/dp^2 = 0$ to (16.21) leads only to the trivial solution $u(x, y) = kx + ly + m$, for which all second derivatives are individually zero.

► *Find the general solution of the one-dimensional wave equation*

$$\frac{\partial^2 u}{\partial x^2} - \frac{1}{c^2}\frac{\partial^2 u}{\partial t^2} = 0.$$

This equation is (16.20) with $A = 1$, $B = 0$ and $C = -1/c^2$, and so the values of $\lambda_1$ and $\lambda_2$ are the solutions of

$$1 - \frac{\lambda^2}{c^2} = 0,$$

namely $\lambda_1 = -c$ and $\lambda_2 = c$. This means that arbitrary functions of the quantities

$$p_1 = x - ct, \qquad p_2 = x + ct$$

will be satisfactory solutions of the equation and that the general solution will be

$$u(x,t) = f(x - ct) + g(x + ct), \tag{16.23}$$

where $f$ and $g$ are arbitrary functions. This solution is discussed further in section 16.4. ◄

The method used to obtain the general solution of the wave equation may also be applied straightforwardly to Laplace's equation.

► *Find the general solution of the two-dimensional Laplace equation*

$$\frac{\partial^2 u}{\partial x^2} + \frac{\partial^2 u}{\partial y^2} = 0. \tag{16.24}$$

Following the established procedure, we look for a solution that is a function $f(p)$ of $p = x + \lambda y$, where from (16.24) $\lambda$ satisfies

$$1 + \lambda^2 = 0.$$

This requires that $\lambda = \pm i$, and satisfactory variables $p$ are $p = x \pm iy$. The general solution required is therefore, in terms of arbitrary functions $f$ and $g$,

$$u(x,y) = f(x + iy) + g(x - iy). \; ◄$$

It will be apparent from the last two examples that the nature of the appropriate linear combination of $x$ and $y$ depends upon whether $B^2 > 4AC$ or $B^2 < 4AC$. This is exactly the same criterion as determines whether the PDE is hyperbolic or elliptic. Hence as a general result, hyperbolic and elliptic equations of the form (16.20), with the restriction that the constants $A$, $B$ and $C$ are real, have as solutions functions whose arguments have the form $x+\alpha y$ and $x+i\beta y$ respectively, where $\alpha$ and $\beta$ themselves are real.

The one case not covered by this result is that in which $B^2 = 4AC$, i.e. a parabolic equation. In this case $\lambda_1$ and $\lambda_2$ are not different and only one suitable combination of $x$ and $y$ results, namely

$$u(x, y) = f(x - (B/2C)y).$$

To find the second part of the general solution we try, in analogy with the corresponding situation for ordinary differential equations, a solution of the form

$$u(x, y) = h(x, y)g(x - (B/2C)y).$$

Substituting this into (16.20) and using $A = B^2/4C$ results in

$$\left(A\frac{\partial^2 h}{\partial x^2} + B\frac{\partial^2 h}{\partial x \partial y} + C\frac{\partial^2 h}{\partial y^2}\right) g = 0.$$

Therefore we require $h(x, y)$ to be any solution of the original PDE. There are several simple solutions of this equation, but as only one is required we take the simplest non-trivial one, $h(x, y) = x$, to give the general solution of the parabolic equation

$$u(x, y) = f(x - (B/2C)y) + xg(x - (B/2C)y). \qquad (16.25)$$

We could, of course, have taken $h(x, y) = y$, but this in any case leads to a solution which is represented by (16.25).

►*Solve*

$$\frac{\partial^2 u}{\partial x^2} + 2\frac{\partial^2 u}{\partial x \partial y} + \frac{\partial^2 u}{\partial y^2} = 0,$$

*subject to the boundary conditions* $u(0, y) = 0$ *and* $u(x, 1) = x^2$.

From our general result, functions of $p = x + \lambda y$ will be solutions provided

$$1 + 2\lambda + \lambda^2 = 0,$$

so that $\lambda = -1$ and the equation is parabolic. The general solution is therefore

$$u(x, y) = f(x - y) + xg(x - y).$$

The boundary condition $u(0, y) = 0$ implies $f(p) \equiv 0$, whilst $u(x, 1) = x^2$ yields

$$xg(x - 1) = x^2,$$

which gives $g(p) = p + 1$, Therefore the particular solution required is

$$u(x, y) = x(p + 1) = x(x - y + 1). \text{ ◄}$$

To reinforce the material discussed above we will now give alternative derivations of the general solutions (16.22) and (16.25) by expressing the original PDE in terms of new variables before solving it. The actual solution will then become almost trivial; but, of course, it will be recognised that suitable new variables could hardly have been guessed if it were not for the work already done. This does not detract from the validity of the derivation to be described, only from the likelihood that it would be discovered by inspection.

We start again with (16.20) and change to new variables

$$\zeta = x + \lambda_1 y, \qquad \eta = x + \lambda_2 y.$$

With this change of variables, we can replace the differential operators (16.20) as follows:

$$\frac{\partial}{\partial x} = \frac{\partial}{\partial \zeta} + \frac{\partial}{\partial \eta},$$

$$\frac{\partial}{\partial y} = \lambda_1 \frac{\partial}{\partial \zeta} + \lambda_2 \frac{\partial}{\partial \eta}.$$

Carrying this out, and using the fact that

$$A + B\lambda_i + C\lambda_i^2 = 0 \quad \text{for } i = 1, 2,$$

equation (16.20) becomes

$$[2A + B(\lambda_1 + \lambda_2) + 2C\lambda_1\lambda_2] \frac{\partial^2 u}{\partial \zeta \partial \eta} = 0.$$

Then, providing the factor in square brackets does not vanish, for which the required condition is easily shown to be $B^2 \neq 4AC$, we obtain

$$\frac{\partial^2 u}{\partial \zeta \partial \eta} = 0,$$

which has the successive integrals

$$\frac{\partial u}{\partial \eta} = F(\eta), \quad u(\zeta, \eta) = f(\eta) + g(\zeta).$$

This solution is just the same as (16.22),

$$u(x, y) = f(x + \lambda_2 y) + g(x + \lambda_1 y).$$

If the equation is parabolic ($B^2 = 4AC$), we instead use the new variables

$$\zeta = x + \lambda y, \qquad \eta = x,$$

and recalling that $\lambda = -(B/2C)$ we reduce (16.20) to

$$A \frac{\partial^2 u}{\partial \eta^2} = 0.$$

Two straightforward integrations give as the general solution

$$u(\zeta, \eta) = \eta g(\zeta) + f(\zeta),$$

which in terms of $x$ and $y$ has exactly the form of (16.25),

$$u(x, y) = xg(x + \lambda y) + f(x + \lambda y).$$

Finally, as mentioned in subsection 16.3.1 with reference to first-order linear PDEs, we note that some of the methods used to find particular integrals of linear ODEs can be suitably modified to find particular integrals of PDEs of higher order. In simple cases, however, an appropriate solution may often be found by inspection.

►*Find the general solution of*

$$\frac{\partial^2 u}{\partial x^2} + \frac{\partial^2 u}{\partial y^2} = 6(x + y).$$

Following our previous methods and results, the complementary function is

$$u(x, y) = f(x + iy) + g(x - iy),$$

and only a particular integral remains to be found. By inspection a particular integral of the equation is $u(x, y) = x^3 + y^3$, and so the general solution can be written

$$u(x, y) = f(x + iy) + g(x - iy) + x^3 + y^3. \blacktriangleleft$$

## 16.4 The wave equation

We have already found that the general solution of the one-dimensional wave equation is

$$u(x, t) = f(x - ct) + g(x + ct), \qquad (16.26)$$

where $f$ and $g$ are arbitrary functions. However, the equation is of such general importance that further discussion will not be out of place.

Let us imagine that $u(x, t) = f(x-ct)$ represents the displacement of a string at time $t$ and position $x$. It is clear that all positions $x$ and times $t$ for which $x-ct =$ constant will have the same instantaneous displacement. But $x - ct =$ constant is exactly the relation between the time and position of an observer travelling with speed $c$ along the positive $x$-direction. Consequently this moving observer sees a constant displacement of the string, whereas to a stationary observer, the initial profile $u(x, 0)$ moves with speed $c$ along the $x$-axis as if it were a rigid system. Thus $f(x - ct)$ represents a wave form of constant shape travelling along the positive $x$-axis with speed $c$, the actual form of the wave depending upon the function $f$. Similarly, the term $g(x + ct)$ is a constant wave form travelling with speed $c$ in the negative $x$-direction. The general solution (16.23) represents a superposition of these.

If the functions $f$ and $g$ are the same, then the complete solution (16.23) represents identical progressive waves going in opposite directions. This may result in a wave pattern whose profile does not progress, described as a *standing wave*. As a simple example, suppose both $f(p)$ and $g(p)$ have the form†

$$f(p) = g(p) = A\cos(kp + \epsilon).$$

Then (16.23) can be written as

$$\begin{aligned} u(x,t) &= A[\cos(kx - kct + \epsilon) + \cos(kx + kct + \epsilon)] \\ &= 2A\cos(kct)\cos(kx + \epsilon). \end{aligned}$$

The important thing to notice is that the shape of the wave pattern, given by the factor in $x$, is the same at all times but that its amplitude $2A\cos(kct)$ depends upon time. At some points $x$ that satisfy

$$\cos(kx + \epsilon) = 0$$

there is no displacement at any time; such points are called *nodes*.

So far we have not imposed any boundary conditions on the solution (16.26). The problem of finding a solution to the wave equation that satisfies given boundary conditions is normally treated using the method of separation of variables discussed in the next chapter. Nevertheless, we now consider *D'Alembert's solution* to the wave equation subject to initial conditions (boundary conditions) in the following general form,

$$\text{initial displacement: } u(x,0) = \phi(x); \quad \text{initial velocity: } \frac{\partial u(x,0)}{\partial t} = \psi(x),$$

for all parts of a string whose transverse displacement is $u(x,t)$. The functions $\phi(x)$ and $\psi(x)$ are given and describe the displacement and velocity of each part of the string at the (arbitrary) time $t = 0$.

It is clear that what we need are the particular forms of the functions $f$ and $g$ in (16.26) that lead to the required values at $t = 0$. This means that

$$\phi(x) = u(x,0) = f(x - 0) + g(x + 0), \tag{16.27}$$

$$\psi(x) = \frac{\partial u(x,0)}{\partial t} = -cf'(x - 0) + cg'(x + 0), \tag{16.28}$$

where it should be noted that $f'(x - 0)$ stand for $df(p)/dp$ evaluated, after the differentiation, at $p = x - c \times 0$ and likewise with $g'(x + 0)$.

Looking on the above two left-hand sides as functions of $p = x \pm ct$, but everywhere evaluated at $t = 0$, we may integrate (16.28) between an arbitrary (and irrelevant) lower limit $p_0$ and an indefinite upper limit $p$ to obtain

$$\frac{1}{c}\int_{p_0}^{p} \psi(q)\,dq + K = -f(p) + g(p),$$

† In the usual notation, $k$ is the wave number ($= 2\pi/\text{wavelength}$) and $kc = \omega$, the angular frequency of the wave.

with the constant of integration $K$ depending on $p_0$. Comparing this equation with (16.27), with $x$ replaced by $p$, we can establish the forms of the functions $f$ and $g$ as

$$f(p) = \frac{\phi(p)}{2} - \frac{1}{2c}\int_{p_0}^{p} \psi(q)\,dq - \frac{K}{2}, \tag{16.29}$$

$$g(p) = \frac{\phi(p)}{2} + \frac{1}{2c}\int_{p_0}^{p} \psi(q)\,dq + \frac{K}{2}. \tag{16.30}$$

Adding (16.29) with $p = x - ct$ to (16.30) with $p = x + ct$ gives as the solution to the original problem

$$u(x,t) = \frac{1}{2}\left[\phi(x-ct) + \phi(x+ct)\right] + \frac{1}{2c}\int_{x-ct}^{x+ct} \psi(q)\,dq, \tag{16.31}$$

in which we notice that all dependence on $p_0$ has disappeared.

Each of the terms in (16.31) has a fairly straightforward physical interpretation. In each case the factor $1/2$ represents the fact that only half of a displacement profile that starts at any particular point on the string travels towards any other position $x$, the other half travelling away from it. The first term $\frac{1}{2}\phi(x-ct)$ arises from the initial displacement at a distance $ct$ to the left of $x$; this travels forward arriving at $x$ at time $t$. Similarly the second contribution is due to the initial displacement at a distance $ct$ to the right of $x$. The interpretation of the final term is a little less obvious. It can be viewed as representing the accumulated transverse displacement at position $x$ due to the passage past $x$ of all parts of the initial motion whose effects can reach $x$ within a time $t$, both backward and forward travelling.

The extension to the three-dimensional wave equation of solutions of the type we have so far encountered presents no serious difficulty. In Cartesian coordinates the three-dimensional wave equation is

$$\frac{\partial^2 u}{\partial x^2} + \frac{\partial^2 u}{\partial y^2} + \frac{\partial^2 u}{\partial z^2} - \frac{1}{c^2}\frac{\partial^2 u}{\partial t^2} = 0. \tag{16.32}$$

In close analogy with the one-dimensional case we try solutions that are functions of linear combinations of all four variables,

$$p = lx + my + nz + \mu t.$$

It is clear that a solution $u(x,y,z,t) = f(p)$ will be acceptable provided that

$$\left(l^2 + m^2 + n^2 - \frac{\mu^2}{c^2}\right)\frac{d^2 f(p)}{dp^2} = 0.$$

Thus, as in the one-dimensional case, $f$ can be arbitrary provided that

$$l^2 + m^2 + n^2 = \mu^2/c^2.$$

Using an obvious normalisation, we take $\mu = \pm c$ and $l$, $m$, $n$ as three numbers such that

$$l^2 + m^2 + n^2 = 1.$$

In other words $(l, m, n)$ are the Cartesian components of a unit vector $\hat{\mathbf{n}}$ that points along the direction of propagation of the wave. The quantity $p$ can be written in terms of vectors as the scalar expression $p = \hat{\mathbf{n}} \cdot \mathbf{r} \pm ct$, and the general solution of (16.32) is then

$$u(x, y, z, t) = u(\mathbf{r}, t) = f(\hat{\mathbf{n}} \cdot \mathbf{r} - ct) + g(\hat{\mathbf{n}} \cdot \mathbf{r} + ct), \qquad (16.33)$$

where $\hat{\mathbf{n}}$ is *any* unit vector. It would perhaps be more transparent to write $\hat{\mathbf{n}}$ explicitly as one of the arguments of $u$.

## 16.5 The diffusion equation

One important class of second-order PDEs, which we have not yet considered in detail, is that in which the second derivative with respect to one variable appears, but only the first derivative with respect to another (usually time). This is exemplified by the one-dimensional diffusion equation

$$\kappa \frac{\partial^2 u(x, t)}{\partial x^2} = \frac{\partial u}{\partial t}, \qquad (16.34)$$

in which $\kappa$ is a constant with the dimensions of length$^2$ $\times$ time$^{-1}$. The physical constants that go to make up $\kappa$ in a particular case depend upon the nature of the process (e.g. solute diffusion, heat flow, etc.) and the material being described.

With (16.34) we cannot hope to repeat successfully the method of subsection 16.3.3, since now $u(x, t)$ is differentiated a different number of times on the two sides of the equation. Any attempted solution in the form $u(x, t) = f(p)$ with $p = ax + bt$ will only lead, therefore, to an equation in which the form of $f$ cannot be cancelled out. Clearly we must try other methods.

Solutions may be obtained by using the standard method of separation of variables discussed in the next chapter. Alternatively, a simple solution is also given if both sides of (16.34), as it stands, are separately set equal to a constant $\alpha$ (say), so that

$$\frac{\partial^2 u}{\partial x^2} = \frac{\alpha}{\kappa}$$
$$\frac{\partial u}{\partial t} = \alpha.$$

These equations have the general solutions

$$u(x, t) = \frac{\alpha}{2\kappa} x^2 + x g(t) + h(t)$$
$$\text{and} \quad u(x, t) = \alpha t + m(x)$$

respectively, and may be made compatible with each other if $g(t)$ is taken as constant, $g(t) = g$ (or zero), $h(t) = \alpha t$, and $m(x) = (\alpha/2\kappa)x^2 + gx$. An acceptable solution is thus

$$u(x,t) = \frac{\alpha}{2\kappa}x^2 + gx + \alpha t + \text{constant}.$$

Let us now return to seeking solutions of equations by combining the independent variables in particular ways. Having seen that a linear combination of $x$ and $t$ will be of no value, we must search for other possible combinations. It has already been noted that $\kappa$ has the dimension of length$^2$ $\times$ time$^{-1}$ and so the combination of variables

$$\eta = \frac{x^2}{\kappa t},$$

will be dimensionless. Let us see if we can mathematically satisfy (16.34) with a solution of the form $u(x,t) = f(\eta)$. Evaluating the necessary derivatives we have

$$\begin{aligned}
\frac{\partial u}{\partial x} &= \frac{df(\eta)}{d\eta}\frac{\partial \eta}{\partial x} = \frac{2x}{\kappa t}\frac{df(\eta)}{d\eta}, \\
\frac{\partial^2 u}{\partial x^2} &= \frac{2}{\kappa t}\frac{df(\eta)}{d\eta} + \left(\frac{2x}{\kappa t}\right)^2 \frac{d^2 f(\eta)}{d\eta^2}, \\
\frac{\partial u}{\partial t} &= -\frac{x^2}{\kappa t^2}\frac{df(\eta)}{d\eta}.
\end{aligned}$$

Substituting these expressions into (16.34) we find the new equation can be written entirely in terms of $\eta$,

$$4\eta\frac{d^2 f(\eta)}{d\eta^2} + (2+\eta)\frac{df(\eta)}{d\eta} = 0.$$

This is a straightforward ODE, that can be solved (using a minimum of explanation) as follows. Writing $f'(\eta) = df(\eta)/d\eta$, etc., we have

$$\begin{aligned}
\frac{f''(\eta)}{f'(\eta)} &= -\frac{1}{2\eta} - \frac{1}{4} \\
\Rightarrow \quad \ln[\eta^{1/2} f'(\eta)] &= -\frac{\eta}{4} + c \\
\Rightarrow \quad f'(\eta) &= \frac{A}{\eta^{1/2}}\exp\left(\frac{-\eta}{4}\right) \\
\Rightarrow \quad f(\eta) &= A\int_{\eta_0}^{\eta} \mu^{-1/2}\exp\left(\frac{-\mu}{4}\right) d\mu.
\end{aligned}$$

If we now write this in terms of a slightly different variable

$$\zeta = \frac{\eta^{1/2}}{2} = \frac{x}{2(\kappa t)^{1/2}},$$

then $d\zeta = \frac{1}{4}\eta^{-1/2}\,d\eta$, and the solution to (16.34) is given by

$$u(x,t) = f(\eta) = g(\zeta) = B\int_{\zeta_0}^{\zeta} \exp(-v^2)\,dv. \qquad (16.35)$$

Here $B$ is a constant and it should be noticed that $x$ and $t$ appear on the RHS only in the indefinite upper limit $\zeta$, and then only in the combination $xt^{-1/2}$. If $\zeta_0$ is chosen as zero, then $u(x,t)$ is, to within a constant factor,† the error function $\text{erf}[x/2(\kappa t)^{1/2}]$, which is tabulated in many reference books. Only non-negative values of $x$ and $t$ are to be considered here, so that $\zeta \geq \zeta_0$.

Let us try to determine what kind of (say) temperature distribution and flow this represents. For definiteness we take $\zeta_0 = 0$. Firstly, since $u(x,t)$ in (16.35) only depends upon the product $xt^{-1/2}$, it is clear that all points $x$ at times $t$ such that $xt^{-1/2}$ has the same value have the same temperature. Put another way, at any specific time $t$, the region having a particular temperature has moved along the positive $x$-axis a distance proportional to the square root of $t$. This is a typical *diffusion* process.

It will also be noticed that, on the one hand, at $t = 0$, $\zeta \to \infty$ and $u$ becomes quite independent of $x$ (except perhaps at $x = 0$); the solution then represents a uniform spatial temperature distribution. On the other hand, at $x = 0$, $u(x,t)$ is identically zero for all $t$.

► *An infrared laser delivers a pulse of (heat) energy $E$ to a point $P$ on a large insulated sheet of thickness $b$, thermal conductivity $k$, specific heat $s$ and density $\rho$. The sheet is initially at a uniform temperature. If $u(r,t)$ is the excess temperature a time $t$ later, at a point a distance $r$ ($\gg b$) from $P$, then show that a suitable expression for $u$ is*

$$u(r,t) = \frac{\alpha}{t}\exp\left(-\frac{r^2}{2\beta t}\right), \qquad (16.36)$$

*where $\alpha$ and $\beta$ are constants. (Note that we use $r$ instead of $\rho$ to denote the radial coordinate in plane polars so as to avoid confusion with the density.)*

*Further show that (i) $\beta = 2k/s\rho$, (ii) the excess heat energy in the sheet is independent of $t$ and hence evaluate $\alpha$; and (iii) the total heat flow past any circle of radius $r$ is $E$.*

The equation to be solved is the heat diffusion equation

$$k\nabla^2 u(\mathbf{r},t) = s\rho\frac{\partial u(\mathbf{r},t)}{\partial t}.$$

Since we only require the solution for $r \gg b$ we can treat the problem as two-

† Take $B = 2\pi^{-1/2}$ to give the usual error function normalised such that $\text{erf}(\infty) = 1$. See the Appendix.

dimensional with obvious circular symmetry. We thus need only the $r$-derivative term in the expression for $\nabla^2$, giving

$$\frac{k}{r}\frac{\partial}{\partial r}\left(r\frac{\partial u}{\partial r}\right) = s\rho\frac{\partial u}{\partial t}, \tag{16.37}$$

where now $u(\mathbf{r},t) = u(r,t)$.

(i) Substituting the given expression (16.36) into (16.37) we obtain

$$\frac{2k\alpha}{\beta t^2}\left(\frac{r^2}{2\beta t} - 1\right)\exp\left(-\frac{r^2}{2\beta t}\right) = \frac{s\rho\alpha}{t^2}\left(\frac{r^2}{2\beta t} - 1\right)\exp\left(-\frac{r^2}{2\beta t}\right),$$

from which we find that (16.36) is a solution, provided $\beta = 2k/s\rho$.

(ii) The excess heat in the system at any time $t$ is

$$b\rho s\int_0^\infty u(r,t)2\pi r\,dr = 2\pi b\rho s\alpha\int_0^\infty \frac{r}{t}\exp\left(-\frac{r^2}{2\beta t}\right)dr$$
$$= 2\pi b\rho s\alpha\beta.$$

The excess heat is therefore independent of $t$ and must be equal to the total heat input $E$, implying that

$$\alpha = \frac{E}{2\pi b\rho s\beta} = \frac{E}{4\pi bk}.$$

(iii) The total heat flow past a circle of radius $r$ is

$$-2\pi rbk\int_0^\infty \frac{\partial u(r,t)}{\partial r}dt = -2\pi rbk\int_0^\infty \frac{E}{4\pi bkt}\left(\frac{-r}{\beta t}\right)\exp\left(-\frac{r^2}{2\beta t}\right)dt$$
$$= E\left[\exp\left(-\frac{r^2}{2\beta t}\right)\right]_0^\infty = E \quad \text{for all } r.$$

As we would expect, all the heat energy $E$ deposited by the laser will eventually flow past a circle of any given radius $r$. ◂

## 16.6 Characteristics and the existence of solutions

So far in this chapter we have discussed how to find general solutions to various types of first- and second-order linear PDE. Moreover, given a set of boundary conditions we have shown how to find the particular solution (or class of solutions) that satisfies them. For first-order equations, for example, we found that if the value of $u(x,y)$ was specified along some curve in the $xy$-plane then the solution to the PDE was in general unique, but that if $u(x,y)$ was specified at only a single point then the solution was not unique: there existed a class of particular solutions all of which satisfied the boundary condition. In this section and the next we make more rigorous the notion of the types of boundary condition that lead to a PDE having a unique solution, a class of solutions, or no solution at all.

### 16.6.1 First-order equations

Let us consider the general first-order PDE (16.9), but now write it as

$$A(x,y)\frac{\partial u}{\partial x} + B(x,y)\frac{\partial u}{\partial y} = F(x,y,u). \tag{16.38}$$

Suppose we wish to solve this PDE subject to the boundary condition that $u(x,y) = \phi(s)$ is specified along some curve $C$ in the $xy$-plane which is described parametrically by the equations $x = x(s)$ and $y = y(s)$, where $s$ is the arc length along $C$. The variation of $u$ along $C$ is therefore given by

$$\frac{du}{ds} = \frac{\partial u}{\partial x}\frac{dx}{ds} + \frac{\partial u}{\partial y}\frac{dy}{ds} = \frac{d\phi}{ds}. \tag{16.39}$$

We may then solve the two (inhomogeneous) simultaneous linear equations (16.38) and (16.39) for $\partial u/\partial x$ and $\partial u/\partial y$, *unless* the determinant of the coefficients vanishes (see section 7.10), i.e. unless

$$\begin{vmatrix} dx/ds & dy/ds \\ A & B \end{vmatrix} = 0.$$

At each point in the $xy$-plane this equation determines a set of curves called *characteristic curves* (or just *characteristics*), which thus satisfy

$$B\frac{dx}{ds} - A\frac{dy}{ds} = 0,$$

or, multiplying through by $ds/dx$ (and dividing through by $A$),

$$\frac{dy}{dx} = \frac{B(x,y)}{A(x,y)}. \tag{16.40}$$

We have, however, already met (16.40) in subsection 16.3.1 on first-order PDEs, where solutions of the form $u(x,y) = f(p)$, where $p$ is some combination of $x$ and $y$, were discussed. Comparing (16.40) with (16.12) we see that the characteristics are merely those curves along which $p$ is constant.

Since the partial derivatives $\partial u/\partial x$ and $\partial u/\partial y$ may be evaluated provided the boundary curve $C$ does *not* lie along a characteristic, defining $u(x,y) = \phi(s)$ along $C$ is sufficient to specify the solution to the original problem (equation plus boundary conditions) near the curve $C$, in terms of a Taylor expansion about $C$. Therefore the characteristics can be considered as the curves along which information about the solution $u(x,y)$ 'propagates'. This is best understood by using an example.

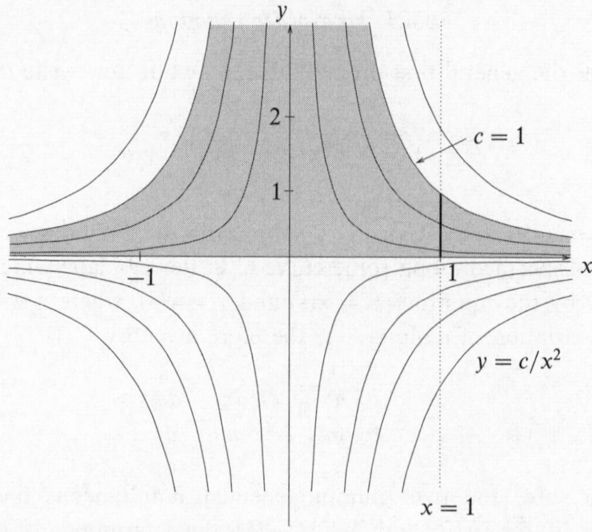


Figure 16.2 The characteristics of equation (16.41). The shaded region shows where the solution to the equation is defined, given the imposed boundary condition at $x = 1$ between $y = 0$ and $y = 1$, shown as a bold vertical line.

► *Find the general solution of*

$$x\frac{\partial u}{\partial x} - 2y\frac{\partial u}{\partial y} = 0 \qquad (16.41)$$

*that takes the value* $2y + 1$ *on the line* $x = 1$ between $y = 0$ and $y = 1$.

We have already solved this problem in subsection 16.3.1 for the case where $u(x, y)$ takes the value $2y + 1$ along the *entire* line $x = 1$. We found then that the general solution to the equation (ignoring boundary conditions) is of the form

$$u(x, y) = f(p) = f(x^2y),$$

for some arbitrary function $f$. Hence the characteristics of (16.41) are given by $x^2y = c$ where $c$ is a constant; some of these curves are plotted in figure 16.2 for various values of $c$. Furthermore, we found that the particular solution for which $u(1, y) = 2y + 1$ *for all* $y$ was given by

$$u(x, y) = 2x^2y + 1.$$

In the present case the value of $x^2y$ is fixed by the boundary conditions only between $y = 0$ and $y = 1$. However, since the characteristics are curves along which $x^2y$, and hence $f(x^2y)$, remains constant, the solution is determined everywhere along any characteristic that intersects the line segment denoting the boundary

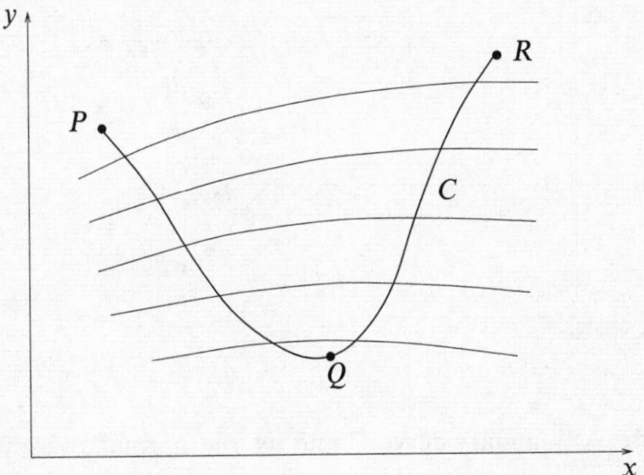


Figure 16.3 A boundary curve $C$ that crosses characteristics more than once.

conditions. Thus $u(x, y) = 2x^2y + 1$ is the particular solution which holds in the shaded region in figure 16.2 (corresponding to $0 \leq c \leq 1$).

Outside this region, however, the solution is not precisely specified, and any function of the form

$$u(x, y) = 2x^2y + 1 + g(x^2y)$$

will satisfy both the equation and the boundary condition, provided $g(p) = 0$ for $0 \leq p \leq 1$. ◄

In the above example the boundary curve was not itself a characteristic and furthermore it crossed each characteristic *once only*. For a general boundary curve $C$ this may not be the case. Firstly, if $C$ is itself a characteristic (or is just a single point), then information about the solution cannot 'propagate' away from $C$, and so the solution remains unspecified everywhere except on $C$.

The second possibility is that $C$ (although not a characteristic itself) crosses some characteristics more than once, as in figure 16.3. In this case specifying the value of $u(x, y)$ along the curve $PQ$ determines the solution along all the characteristics that intersect it. Therefore, also specifying $u(x, y)$ along $QR$ can *overdetermine* the problem solution and generally results in there being no solution.

### *16.6.2 Second-order equations*

The concept of characteristics can be naturally extended to second- (and higher-) order equations. In this case let us write the general second-order linear PDE (16.19) as

$$A(x, y)\frac{\partial^2 u}{\partial x^2} + B(x, y)\frac{\partial^2 u}{\partial x \partial y} + C(x, y)\frac{\partial^2 u}{\partial y^2} = F\left(x, y, u, \frac{\partial u}{\partial x}, \frac{\partial u}{\partial y}\right). \qquad (16.42)$$

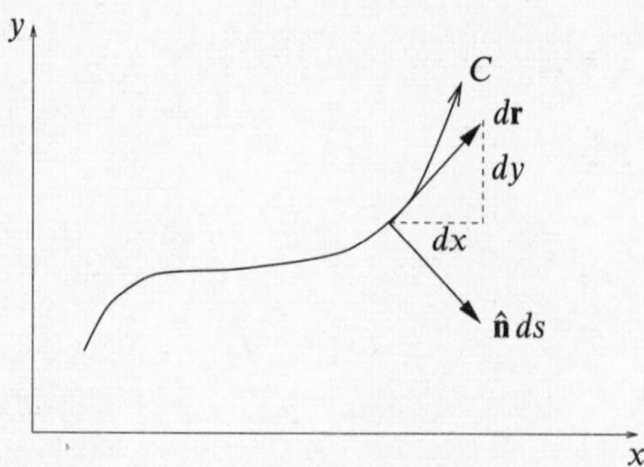


Figure 16.4 A boundary curve $C$ and its unit normal at a given point.

For second-order equations we might expect that relevant boundary conditions would involve specifying $u$, or some of its first derivatives, or both, along a suitable set of boundaries bordering or enclosing the region over which a solution is sought. Three common types of boundary condition occur and are associated with the names of Dirichlet, Neumann and Cauchy. They are as follows.

(i) *Dirichlet*: $u$ is specified at each point of the boundary.

(ii) *Neumann*: $\partial u/\partial n$, the *normal derivative* of $u$, is specified at each point of the boundary. Note that $\partial u/\partial n = \nabla u \cdot \hat{\mathbf{n}}$, where $\hat{\mathbf{n}}$ is the normal to the boundary at each point.

(iii) *Cauchy*: both $u$ and $\partial u/\partial n$ are specified at each point of the boundary.

Let us for the moment consider the solution of (16.42) subject to the Cauchy boundary conditions, i.e. $u$ and $\partial u/\partial n$ are specified along some boundary curve $C$ in the $xy$-plane defined by the parametric equations $x = x(s)$, $y = y(s)$, $s$ being the arc length along $C$ (see figure 16.4). Let us suppose that along $C$ we have $u(x, y) = \phi(s)$ and $\partial u/\partial n = \psi(s)$. At any point on $C$ the vector $d\mathbf{r} = dx\,\mathbf{i} + dy\,\mathbf{j}$ is a tangent to the curve and $\hat{\mathbf{n}}\,ds = dy\,\mathbf{i} - dx\,\mathbf{j}$ is a vector normal to the curve. Thus on $C$ we have

$$\frac{\partial u}{\partial s} \equiv \nabla u \cdot \frac{d\mathbf{r}}{ds} = \frac{\partial u}{\partial x}\frac{dx}{ds} + \frac{\partial u}{\partial y}\frac{dy}{ds} = \frac{d\phi(s)}{ds},$$

$$\frac{\partial u}{\partial n} \equiv \nabla u \cdot \hat{\mathbf{n}} = \frac{\partial u}{\partial x}\frac{dy}{ds} - \frac{\partial u}{\partial y}\frac{dx}{ds} = \psi(s).$$

These two equations may then be solved straightforwardly for the first partial derivatives $\partial u/\partial x$ and $\partial u/\partial y$ along $C$. Using the chain rule to write

$$\frac{d}{ds} = \frac{dx}{ds}\frac{\partial}{\partial x} + \frac{dy}{ds}\frac{\partial}{\partial y},$$

we may differentiate the two (known) first derivatives $\partial u/\partial x$ and $\partial u/\partial y$ along the boundary to obtain the pair of equations

$$\frac{d}{ds}\left(\frac{\partial u}{\partial x}\right) = \frac{dx}{ds}\frac{\partial^2 u}{\partial x^2} + \frac{dy}{ds}\frac{\partial^2 u}{\partial x \partial y},$$
$$\frac{d}{ds}\left(\frac{\partial u}{\partial y}\right) = \frac{dx}{ds}\frac{\partial^2 u}{\partial x \partial y} + \frac{dy}{ds}\frac{\partial^2 u}{\partial y^2}.$$

We may now solve these two equations, together with the original PDE (16.42), for the second partial derivatives of $u$, *except* where the determinant of their coefficients equals zero,

$$\begin{vmatrix} A & B & C \\ \dfrac{dx}{ds} & \dfrac{dy}{ds} & 0 \\ 0 & \dfrac{dx}{ds} & \dfrac{dy}{ds} \end{vmatrix} = 0.$$

Expanding out the determinant,

$$A\left(\frac{dy}{ds}\right)^2 - B\left(\frac{dx}{ds}\right)\left(\frac{dy}{ds}\right) + C\left(\frac{dx}{ds}\right)^2 = 0.$$

Multiplying through by $(ds/dx)^2$ we obtain

$$A\left(\frac{dy}{dx}\right)^2 - B\frac{dy}{dx} + C = 0, \tag{16.43}$$

which is the ODE for the curves in the $xy$-plane along which the second partial derivatives of $u$ *cannot* be found.

As for the first-order case, curves satisfying (16.43) are called characteristics of the original PDE. These characteristics have tangents at each point given by (when $A \neq 0$)

$$\frac{dy}{dx} = \frac{B \pm \sqrt{B^2 - 4AC}}{2A}. \tag{16.44}$$

Clearly, when the original PDE is hyperbolic ($B^2 > 4AC$), equation (16.44) defines two families of real curves in the $xy$-plane; when the equation is parabolic ($B^2 = 4AC$) it defines one family of real curves; and when the equation is elliptic ($B^2 < 4AC$) it defines two families of complex curves. Furthermore, when $A$, $B$ and $C$ are constants (rather than functions of $x$ and $y$), the equations of the characteristics will be of the form $x + \lambda y =$ constant, which is reminiscent of the form of solution discussed in subsection 16.3.3.

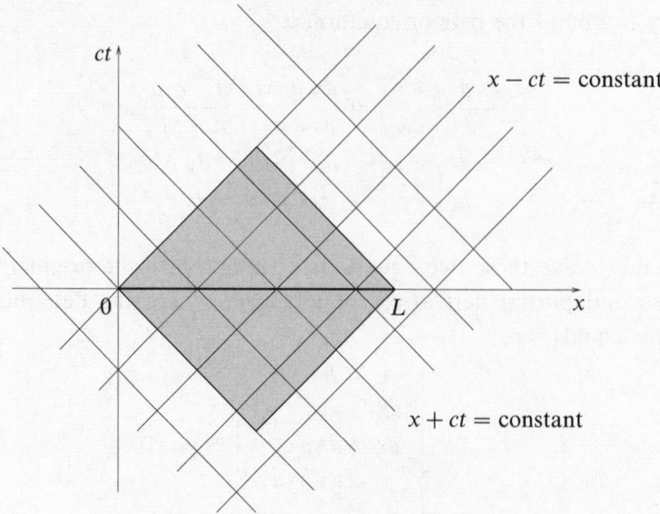


Figure 16.5 The characteristics for the one-dimensional wave equation. The shaded region indicates the region over which the solution is determined by specifying Cauchy boundary conditions at $t = 0$ on the line segment $x = 0$ to $x = L$.

►*Find the characteristics of the one-dimensional wave equation*

$$\frac{\partial^2 u}{\partial x^2} - \frac{1}{c^2}\frac{\partial^2 u}{\partial t^2} = 0.$$

This is a hyperbolic equation with $A = 1$, $B = 0$ and $C = -1/c^2$. Therefore from (16.43) the characteristics are given by

$$\left(\frac{dx}{dt}\right)^2 = c^2,$$

and so the characteristics are the straight lines $x - ct =$ constant and $x + ct =$ constant. ◄

The characteristics of second-order PDEs can be considered as the curves along which *partial* information about the solution $u(x, y)$ 'propagates'. In particular, if the equation is hyperbolic, so that we obtain two families of real characteristics in the $xy$-plane, then Cauchy boundary conditions propagate partial information concerning the solution along the characteristics, belonging to each family, that intersect the boundary curve $C$. The solution $u$ is then specified in the region common to these two families of characteristics. For instance, the characteristics of the hyperbolic one-dimensional wave equation in the last example are shown

| Equation type | Boundary | Conditions |
|---|---|---|
| hyperbolic | open | Cauchy |
| parabolic | open | Dirichlet or Neumann |
| elliptic | closed | Dirichlet or Neumann |

Table 16.1 The appropriate boundary conditions for different types of partial differential equation.

in figure 16.5. By specifying Cauchy boundary conditions $u$ and $\partial u/\partial t$ on the line segment $t = 0$, $x = 0$ to $L$, the solution is specified in the shaded region.

As in the case of first-order PDEs, however, problems can arise. For example, if for a hyperbolic equation the boundary curve intersects any characteristic more then once then Cauchy conditions along $C$ can overdetermine the problem, resulting in there being no solution. In this case either the boundary curve $C$ must be altered, or the boundary conditions on the offending parts of $C$ must be relaxed to Dirichlet or Neumann conditions.

The general considerations involved in deciding which boundary conditions are appropriate for a particular problem are complex, and we do not discuss them any further here.† We merely note that whether the various types of boundary condition are appropriate (in that they give a solution that is unique, sometimes to within a constant, and is well defined) depends upon the type of second-order equation under consideration and on whether the region of solution is bounded by a closed or an open curve (or a surface if there are more than two independent variables). Note that part of a closed boundary may be at infinity if conditions are imposed on $u$ or $\partial u/\partial n$ there.

It may be shown that the appropriate boundary condition and equation type pairings are as shown in table 16.1.

For example, Laplace's equation $\nabla^2 u = 0$ is elliptic and thus requires either Dirichlet or Neumann boundary conditions on a closed boundary which, as we have already noted, may be at infinity if the behaviour of $u$ is specified there (most often $u$ or $\partial u/\partial n \to 0$ at infinity).

## 16.7 Uniqueness of solutions

Although we have merely stated the appropriate boundary types and conditions for which, in the general case, a PDE has a unique, well-defined solution, sometimes to within an additive constant, it is often important to be able to prove that a unique solution is obtained.

† For a discussion the reader is referred, for example, to Morse and Feshbach, *Methods of Theoretical Physics, Part I* (McGraw-Hill, 1953) chapter 6.

As an extremely important example let us consider Poisson's equation in three dimensions,

$$\nabla^2 u(\mathbf{r}) = \rho(\mathbf{r}), \tag{16.45}$$

with either Dirichlet or Neumann conditions on a closed boundary appropriate to such an elliptic equation; for brevity, in (16.45), we have absorbed any physical constants into $\rho$. We aim to show that, to within an unimportant constant, the solution of (16.45) is *unique* if either the potential $u$ or its normal derivative $\partial u/\partial n$ is specified on all surfaces bounding a given region of space (including, if necessary, a hypothetical spherical surface of indefinitely large radius on which $u$ or $\partial u/\partial n$ is prescribed to have an arbitrarily small value). Stated more formally this is as follows.

**Uniqueness theorem.** *If $u$ is real and its first and second partial derivatives are continuous in a region $V$ and on its boundary $S$, and $\nabla^2 u = \rho$ in $V$ and either $u = f$ or $\partial u/\partial n = g$ on $S$, where $\rho$, $f$ and $g$ are prescribed functions, then $u$ is unique (at least to within an additive constant).*

►*Prove the uniqueness theorem for Poisson's equation.*

Let us suppose on the contrary that two solutions $u_1(\mathbf{r})$ and $u_2(\mathbf{r})$ both satisfy the conditions given above, and denote their difference by the function $w = u_1 - u_2$. We then have

$$\nabla^2 w = \nabla^2 u_1 - \nabla^2 u_2 = \rho - \rho = 0,$$

so that $w$ satisfies Laplace's equation in $V$. Furthermore, since either $u_1 = f = u_2$ or $\partial u_1/\partial n = g = \partial u_2/\partial n$ on $S$, we must have either $w = 0$ or $\partial w/\partial n = 0$ on $S$.

If we now use Green's first theorem, (9.19) for the case where both scalar functions are taken as $w$ we have

$$\int_V \left[ w\nabla^2 w + (\nabla w)\cdot(\nabla w) \right] dV = \int_S w \frac{\partial w}{\partial n}\, dS.$$

But either condition $w = 0$ or $\partial w/\partial n = 0$ makes the RHS vanish, whilst the first term on the LHS vanishes since $\nabla^2 w = 0$ in $V$. Thus we are left with

$$\int_V |\nabla w|^2 \, dV = 0.$$

Since $|\nabla w|^2$ can never be negative, this can only be satisfied if

$$\nabla w = \mathbf{0},$$

i.e. if $w$, and hence $u_1 - u_2$, is a constant in $V$.

If Dirichlet conditions are given, then $u_1 \equiv u_2$ on (some part of) $S$ and hence $u_1 = u_2$ everywhere in $V$. For Neumann conditions, however, $u_1$ and $u_2$ can differ throughout $V$ by an arbitrary (but unimportant) constant. ◄

The importance of this uniqueness theorem lies in the fact that if a solution to Poisson's (or Laplace's) equation that fits the given set of Dirichlet or Neumann conditions can be found by any means whatever, then that solution is the correct one, since only one exists. This result is the mathematical justification for the *method of images,* which is discussed more fully in the next chapter.

We also note that the same general method, used in the above example for proving the uniqueness theorem for Poisson's equation, can often be employed to prove the uniqueness (or otherwise) of solutions to other equations and boundary conditions.

## 16.8 Exercises

16.1 Find the most general solutions $u(x, y)$ of the following equations consistent with the boundary conditions stated.

(a) $y\dfrac{\partial u}{\partial x} - x\dfrac{\partial u}{\partial y} = 0; \quad u(x, 0) = 1 + \sin x.$

(b) $i\dfrac{\partial u}{\partial x} = 3\dfrac{\partial u}{\partial y}; \quad u = (4 + 3i)x^2$ on the line $x = y$.

(c) $\sin x \sin y\dfrac{\partial u}{\partial x} + \cos x \cos y\dfrac{\partial u}{\partial y} = 0; \quad u = \cos 2y$ on $x + y = \pi/2$.

(d) $\dfrac{\partial u}{\partial x} + 2x\dfrac{\partial u}{\partial y} = 0; \quad u = 2$ on the parabola $y = x^2$.

16.2 Find solutions of

$$\frac{1}{x}\frac{\partial u}{\partial x} + \frac{1}{y}\frac{\partial u}{\partial y} = 0$$

for which (a) $u(0, y) = y$, (b) $u(1, 1) = 1$.

16.3 Find the most general solutions $u(x, y)$ of the following equations consistent with the boundary conditions stated.

(a) $y\dfrac{\partial u}{\partial x} - x\dfrac{\partial u}{\partial y} = 3x; \quad u = x^2$ on the line $y = 0$.

(b) $y\dfrac{\partial u}{\partial x} - x\dfrac{\partial u}{\partial y} = 3x; \quad u(1, 0) = 2.$

(c) $y^2\dfrac{\partial u}{\partial x} + x^2\dfrac{\partial u}{\partial y} = x^2y^2(x^3 + y^3);$ no boundary conditions.

16.4 Solve

$$\sin x\frac{\partial u}{\partial x} + \cos x\frac{\partial u}{\partial y} = \cos x$$

subject to (a) $u(\pi/2, y) = 0$, (b) $u(\pi/2, y) = y(y + 1)$.

16.5 A function $u(x, y)$ satisfies

$$2\frac{\partial u}{\partial x} + 3\frac{\partial u}{\partial y} = 10,$$

and takes the value 3 on the line $y = 4x$. Evaluate $u(2, 4)$.

16.6 If $u(x, y)$ satisfies

$$\frac{\partial^2 u}{\partial x^2} - 3\frac{\partial^2 u}{\partial x \partial y} + 2\frac{\partial^2 u}{\partial y^2} = 0$$

and $u = -x^2$ and $\partial u/\partial y = 0$ for $y = 0$ and all $x$, find the value of $u(0, 1)$.

16.7 (a) Solve the previous question if the boundary condition is $u = \partial u/\partial y = 1$ when $y = 0$ for all $x$.

(b) In which region of the $xy$-plane would $u$ be determined if the boundary condition were $u = \partial u/\partial y = 1$ when $y = 0$ for all $x > 0$?

16.8 Solve

$$6\frac{\partial^2 u}{\partial x^2} - 5\frac{\partial^2 u}{\partial x \partial y} + \frac{\partial^2 u}{\partial y^2} = 14,$$

subject to $u = 2x + 1$ and $\partial u/\partial y = 4 - 6x$, both on the line $y = 0$.

16.9 Solve

$$\frac{\partial^2 u}{\partial x \partial y} + 3\frac{\partial^2 u}{\partial y^2} = x(2y + 3x).$$

16.10 Find the most general solution of $\partial^2 u/\partial x^2 + \partial^2 u/\partial y^2 = x^2 y^2$.

16.11 An incompressible fluid of density $\rho$ and negligible viscosity flows with velocity $v$ along a thin straight tube, perfectly light and flexible, of cross-section $A$ and held under tension $T$. Assume that small transverse displacements $u$ of the tube are governed by

$$\frac{\partial^2 u}{\partial t^2} + 2v\frac{\partial^2 u}{\partial x \partial t} + \left(v^2 - \frac{T}{\rho A}\right)\frac{\partial^2 u}{\partial x^2} = 0.$$

(a) Show that the general solution consists of a superposition of two waveforms travelling with different speeds.

(b) The tube initially has a small transverse displacement $u = a \cos kx$ and is suddenly released from rest. Find its subsequent motion.

16.12 In an electrical cable of resistance $R$ and capacitance $C$ per unit length, voltage signals obey the equation $\partial^2 V/\partial x^2 = RC\partial V/\partial t$. This has solutions of the form given in (16.35) and also of the form $V = Ax + D$.

(a) Find a combination of these that represents the situation after a steady voltage $V_0$ is applied at $x = 0$ at time $t = 0$.

(b) Obtain a solution describing the propagation of the voltage signal resulting from application of the signal $V = V_0$ for $0 < t < T$, $V = 0$ otherwise, to the end $x = 0$ of an infinite cable.

(c) Show that for $t \gg T$ the maximum signal occurs at a value of $x$ proportional to $t^{1/2}$ and has a magnitude proportional to $t^{-1}$.

16.13 The daily and annual variations of temperature at the surface of the earth may be represented by sine-wave oscillations with equal amplitudes and periods of 1 and 365 days respectively. Assume that for (angular) frequency $\omega$ the temperature at depth $x$ in the earth is given by $u(x,t) = A\sin(\omega t + \mu x)\exp(-\lambda x)$, where $\lambda$ and $\mu$ are constants.

(a) Use the diffusion equation to find the values of $\lambda$ and $\mu$.
(b) Find the ratio of the depths below the surface at which the amplitudes have dropped to $1/20$ of their surface values.
(c) At what time of year is the soil coldest at the greater of these depths, assuming that the smoothed total annual variation in temperature at the surface has a minimum on February 1st?

16.14 *This example gives a formal demonstration that the type of a second-order PDE (elliptic, parabolic or hyperbolic) cannot be changed by a new choice of independent variable. The algebra is somewhat lengthy, but straightforward.*

If a change of variable $\xi = \xi(x,y)$, $\eta = \eta(x,y)$ is made in (16.19) so that it reads

$$A'\frac{\partial^2 u}{\partial \xi^2} + B'\frac{\partial^2 u}{\partial \xi \partial \eta} + C'\frac{\partial^2 u}{\partial \eta^2} + D'\frac{\partial u}{\partial \xi} + E'\frac{\partial u}{\partial \eta} + F'u = R'(\xi,\eta),$$

show that

$$B'^2 - 4A'C' = (B^2 - 4AC)\left[\frac{\partial(\xi,\eta)}{\partial(x,y)}\right]^2.$$

Hence deduce the conclusion stated above.

## 16.9 Hints and answers

16.1 (a) $p = x^2 + y^2$, $\sin(x^2+y^2)^{1/2} + 1$; (b) $p = 3x + iy$, $(3x+iy)^{1/2}/2$; (c) $p = \sin x \cos y$, $2\sin x\cos y - 1$; (d) $p = y - x^2$, $y - x^2 + 2$.

16.2 (a) $(y^2 - x^2)^{1/2}$; (b) $1 + f(y^2 - x^2)$ where $f(0) = 0$.

16.3 (a) $p = x^2 + y^2$, particular integral $u = -3y$, $u = x^2 + y^2 - 3y$; (b) $u = x^2 + y^2 - 3y + 1 + g(x^2 + y^2)$ where $g(1) = 0$; (c) $(x^6 + y^6)/6 + g(x^3 - y^3)$.

16.4 $u = y + f(y - \ln(\sin x))$; (a) $u = \ln(\sin x)$; (b) $u = y + [y - \ln(\sin x)]^2$.

16.5 $u = f(3x-2y) + 2(x+y)$; $f(p) = 3 + 2p$; $u = 8x - 2y + 3$ and $u(2,4) = 11$.

16.6 General solution is $u(x,y) = f(x+y) + g(x + y/2)$. Show that $2p = -g'(p)/2$, and hence that $g(p) = k - 2p^2$, whilst $f(p) = p^2 - k$, leading to $u(x,y) = -x^2 + y^2/2$; $u(0,1) = 1/2$.

16.7 (a) $u(x, y) = 2(x + y) - 2(x + y/2) + 1 = y + 1$; $u(0, 1) = 2$; (b) in the sector $-\pi/4 \le \theta \le \pi/2 + \phi$, where $\tan\phi = 1/2$ and $\theta$ is measured from the positive $x$-axis.

16.8 $u(x, y) = f(x+2y)+g(x+3y)+x^2+y^2$ leading to $u = 1+2x+4y-6xy-8y^2$.

16.9 $u = f(y - 3x) + g(x) + x^2y^2/2$.

16.10 $u(x, y) = f(x + iy) + g(x - iy) + \frac{1}{12}x^4(y^2 - \frac{1}{15}x^2)$. In the last term, $x$ and $y$ may be interchanged.

16.11 (a) $c = v \pm \alpha$ where $\alpha^2 = T/\rho A$;
(b) $u(x, t) = a\cos[k(x - vt)]\cos(k\alpha t) - (va/\alpha)\sin[k(x - vt)]\sin(k\alpha t)$.

16.12 (a) $V_0\left[1 - (2/\sqrt{\pi})\int^{\frac{1}{2}x(CR/t)^{1/2}} \exp(-v^2)\,dv\right]$; (b) consider as $V_0$ applied at $t = 0$ and continued and $-V_0$ at $t = T$ and continued;

$$V(x, t) = \frac{2V_0}{\sqrt{\pi}}\int_{\frac{1}{2}x(CR/t)^{1/2}}^{\frac{1}{2}x[CR/(t-T)]^{1/2}} \exp\left(-v^2\right)\,dv;$$

(c) For $t \gg T$, maximum at $x = (2t/CR)^{1/2}$ with value

$$\frac{V_0 T\exp(-\frac{1}{2})}{(2\pi)^{1/2}t}.$$

16.13 (a) $\lambda = -\mu = (\omega/2\kappa)^{1/2}$, where $\kappa$ is the diffusion constant; (b) $x_A = (365)^{1/2}x_D$; (c) only the annual variation is significant at this depth and has a phase $\mu_A x_A = \ln 20$ behind the surface. Thus the coldest day is February 1st + $[(365\ln 20)/2\pi]$ days $\approx$ July 23rd.

16.14

$$A' = A\left(\frac{\partial\xi}{\partial x}\right)^2 + B\frac{\partial\xi}{\partial x}\frac{\partial\xi}{\partial y} + C\left(\frac{\partial\xi}{\partial y}\right)^2,$$

$$B' = 2A\frac{\partial\xi}{\partial x}\frac{\partial\eta}{\partial x} + B\left(\frac{\partial\xi}{\partial x}\frac{\partial\eta}{\partial y} + \frac{\partial\xi}{\partial y}\frac{\partial\eta}{\partial x}\right) + 2C\frac{\partial\xi}{\partial y}\frac{\partial\eta}{\partial y}, \qquad \text{etc.}$$

# 17

# *Partial differential equations: separation of variables and other methods*

In the previous chapter we demonstrated the methods by which general solutions of some partial differential equations (PDEs) may be obtained in terms of arbitrary functions. In particular, solutions containing the independent variables in definite combinations were sought, thus reducing the effective number of them.

In the present chapter we begin by taking the opposite approach, namely that of trying to keep the independent variables as separate as possible, using the method of separation of variables. We then discuss integral transform methods by which one of the independent variables may be eliminated, at least from differential coefficients. Finally, we discuss the use of Green's functions in solving inhomogeneous problems.

## 17.1 Separation of variables: the general method

Suppose we seek a solution $u(x, y, z, t)$ to some PDE (expressed in Cartesian coordinates). Let us attempt to obtain one which has the product form†

$$u(x, y, z, t) = X(x)Y(y)Z(z)T(t). \qquad (17.1)$$

A solution that has this form is said to be *separable* in $x$, $y$, $z$ and $t$, and seeking solutions of this form is called the method of *separation of variables.*

As simple examples we may observe that, of the functions

$$\text{(i) } xyz^2 \sin bt, \qquad \text{(ii) } xy + zt, \qquad \text{(iii) } (x^2 + y^2)z \cos \omega t,$$

(i) is completely separable, (ii) is inseparable in that no single variable can be separated out from it and written as a multiplicative factor, whilst (iii) is separable in $z$ and $t$ but not in $x$ and $y$.

† It should be noted that the conventional use here of upper-case (capital) letters to denote the functions of the corresponding lower-case variable is intended to enable an easy correspondence between a function and its argument to be made.

When seeking PDE solutions of the form (17.1), we are not requiring that there is no connection at all between the functions $X$, $Y$, $Z$ and $T$ (for example, certain parameters may appear in two or more of them), only that the function $X$ does not depend upon $y$, $z$, $t$; that $Y$ does not depend on $x$, $z$, $t$, and so on.

For a general PDE it is likely that a separable solution is impossible, but certainly some common and important equations do have useful solutions of this form and we will illustrate the method of solution by studying the three-dimensional wave equation

$$\nabla^2 u(\mathbf{r}) = \frac{1}{c^2}\frac{\partial^2 u(\mathbf{r})}{\partial t^2}. \tag{17.2}$$

We will work in Cartesian coordinates for the present and assume a solution of the form (17.1); the solutions in alternative coordinate systems, e.g. spherical or cylindrical polars, are considered in section 17.3. Expressed in Cartesian coordinates (17.2) takes the form

$$\frac{\partial^2 u}{\partial x^2} + \frac{\partial^2 u}{\partial y^2} + \frac{\partial^2 u}{\partial z^2} = \frac{1}{c^2}\frac{\partial^2 u}{\partial t^2}, \tag{17.3}$$

and substituting (17.1) gives

$$\frac{d^2 X}{dx^2}YZT + X\frac{d^2 Y}{dy^2}ZT + XY\frac{d^2 Z}{dz^2}T = \frac{1}{c^2}XYZ\frac{d^2 T}{dt^2},$$

which can also be written as

$$X''YZT + XY''ZT + XYZ''T = \frac{1}{c^2}XYZT'', \tag{17.4}$$

where in each case the primes refer to the *ordinary* derivative with respect to the independent variable upon which the function depends. This emphasises the fact that each of the functions $X$, $Y$, $Z$ and $T$ has only one independent variable and thus its only derivative is its total derivative. For the same reason, in each term in (17.4) three of the four functions are unaltered by the differentiation and behave exactly as constant multipliers.

If we now divide (17.4) throughout by $u = XYZT$ we obtain

$$\frac{X''}{X} + \frac{Y''}{Y} + \frac{Z''}{Z} = \frac{1}{c^2}\frac{T''}{T}. \tag{17.5}$$

This form shows the particular characteristic that is the basis of the method of separation of variables, namely that of the four terms the first is a function of $x$ only, the second of $y$ only, the third of $z$ only, and the RHS a function of $t$ only, and yet there is an equation connecting them. This can only be so if *each* of the terms does not in fact, despite appearances, depend upon the corresponding independent variable but *is equal to a constant*, the four constants being such that (17.5) is satisfied.

Since there is only one equation to be satisfied and four constants involved, there is considerable freedom in the values they may take. For the purposes of our illustrative example let us make the choice of $-l^2$, $-m^2$, $-n^2$, for the first three constants. The constant associated with $c^{-2}T''/T$ must then necessarily have the value $-\mu^2 = -(l^2 + m^2 + n^2)$.

Having recognised that each term of (17.5) is individually equal to a constant (or parameter), we can now replace (17.5) by four separate ordinary differential equations (ODEs),

$$\frac{X''}{X} = -l^2, \qquad \frac{Y''}{Y} = -m^2, \qquad \frac{Z''}{Z} = -n^2, \qquad \frac{1}{c^2}\frac{T''}{T} = -\mu^2. \tag{17.6}$$

The important point to notice is not the simplicity of the equations (17.6) (the corresponding ones for a general PDE are usually far from simple) but that, by the device of assuming a separable solution, a *partial* differential equation (17.3), containing derivatives with respect to the four independent variables all in one equation has been reduced to four *separate ordinary* differential equations (17.6). The ordinary equations are connected through four constant parameters that satisfy an algebraic relation. These constants are called *separation constants.*

The general solutions of the equations (17.6) can be straightforwardly deduced and are

$$\begin{aligned} X(x) &= A\exp(ilx) + B\exp(-ilx) \\ Y(y) &= C\exp(imy) + D\exp(-imy) \\ Z(z) &= E\exp(inz) + F\exp(-inz) \\ T(t) &= G\exp(ic\mu t) + H\exp(-ic\mu t), \end{aligned} \tag{17.7}$$

where $A, B, \ldots, H$ are constants, which may be determined if boundary condtions are imposed on the solution. Depending on the geometry of the problem and any boundary conditions, it is sometimes more appropriate to write the solutions (17.7) in the alternative form

$$\begin{aligned} X(x) &= A'\cos lx + B'\sin lx \\ Y(y) &= C'\cos my + D'\sin my \\ Z(z) &= E'\cos nz + F'\sin nz \\ T(t) &= G'\cos(c\mu t) + H'\sin(c\mu t), \end{aligned} \tag{17.8}$$

for some different set of constants $A', \ldots, H'$. Clearly the choice of how best to represent the solution depends on the problem being considered.

As an example, suppose that we take as particular solutions the four functions

$$X(x) = \exp(ilx), \qquad Y(y) = \exp(imy),$$
$$Z(z) = \exp(inz), \qquad T(t) = \exp(-ic\mu t).$$

This gives a particular solution of the original PDE (17.3)

$$\begin{aligned} u(x, y, z, t) &= \exp(ilx)\exp(imy)\exp(inz)\exp(-ic\mu t) \\ &= \exp[i(lx + my + nz - c\mu t)]. \end{aligned}$$

This is a special case of the solution (16.33) obtained in the previous chapter and represents a plane wave of unit amplitude propagating in a direction given by the vector with components $l, m, n$ in a Cartesian coordinate system. In the conventional notation of wave theory, $l$, $m$ and $n$ are the components of the wave number vector $\mathbf{k}$, whose magnitude $k = 2\pi/\lambda$, where $\lambda$ is the wavelength of the wave; $c\mu$ is the angular frequency $\omega$ of the wave. This gives the equation in the form

$$\begin{aligned} u(x, y, z, t) &= \exp[i(k_x x + k_y y + k_z z - \omega t)] \\ &= \exp[i(\mathbf{k} \cdot \mathbf{r} - \omega t)], \end{aligned}$$

and makes the exponent dimensionless.

The method of separation of variables can be applied to many commonly occurring PDEs encountered in physical applications.

► *Use the method of separation of variables to obtain for the one-dimensional diffusion equation*

$$\kappa \frac{\partial^2 u}{\partial x^2} = \frac{\partial u}{\partial t}, \tag{17.9}$$

*a solution that tends to zero as $t \to \infty$ for all $x$.*

Here we have only two independent variables $x$ and $t$, and therefore assume a solution of the form

$$u(x, t) = X(x)T(t).$$

Substituting this expression into (17.9) and dividing through by $u = XT$ (and also by $\kappa$) we obtain

$$\frac{X''}{X} = \frac{T'}{\kappa T}.$$

Now, arguing exactly as above that the LHS is a function of $x$ only and the RHS a function of $t$ only, we conclude that each side must equal a constant, which, anticipating the result and noting the imposed boundary condition, we will take as $-\lambda^2$. This gives us two ordinary equations,

$$X'' + \lambda^2 X = 0, \tag{17.10}$$

$$T' + \lambda^2 \kappa T = 0, \tag{17.11}$$

which have the solutions

$$X(x) = A\cos\lambda x + B\sin\lambda x,$$
$$T(t) = C\exp(-\lambda^2\kappa t).$$

Combining these to give the assumed solution $u = XT$ yields (absorbing the constant $C$ into $A$ and $B$)

$$u(x,t) = (A\cos\lambda x + B\sin\lambda x)\exp(-\lambda^2\kappa t). \qquad (17.12)$$

In order to satisfy the boundary condition $u \to 0$ as $t \to \infty$, $\lambda^2\kappa$ must be $> 0$. Since $\kappa$ is real and $> 0$, this implies that $\lambda$ is a real non-zero number and that the solution is sinusoidal in $x$, and is not a disguised hyperbolic function; this was our reason for choosing the separation constant as $-\lambda^2$. ◄

As a final example we consider Laplace's equation in Cartesian coordinates, which may be treated in a similar manner.

► *Use the method of separation of variables to obtain a solution for the two-dimensional Laplace equation,*

$$\frac{\partial^2 u}{\partial x^2} + \frac{\partial^2 u}{\partial y^2} = 0. \qquad (17.13)$$

If we assume a solution of the form $u(x,y) = X(x)Y(y)$ then, following the above method, and taking the separation constant as $\lambda^2$, we find

$$X'' = \lambda^2 X, \qquad Y'' = -\lambda^2 Y.$$

Taking $\lambda^2$ as $> 0$, the general solution becomes

$$u(x,y) = (A\cosh\lambda x + B\sinh\lambda x)(C\cos\lambda y + D\sin\lambda y), \qquad (17.14)$$

An alternative form, in which the exponentials are written explicitly, may be useful for other geometries or boundary conditions:

$$u(x,y) = [A\exp\lambda x + B\exp(-\lambda x)](C\cos\lambda y + D\sin\lambda y), \qquad (17.15)$$

with different constants $A$ and $B$.

If $\lambda^2 < 0$ then the roles of $x$ and $y$ interchange. The particular combination of sinusoidal and hyperbolic functions and the values of $\lambda$ allowed will be determined by the geometrical properties of any specific problem, together with any prescribed or necessary boundary conditions. ◄

We note here that a particular case of the solution (17.14) links up with the 'combination' result $u(x,y) = f(x+iy)$ of the previous chapter (equations (16.24) and following), namely that when $A = B$, $D = iC$ and $f(p) = AC\exp\lambda p$.

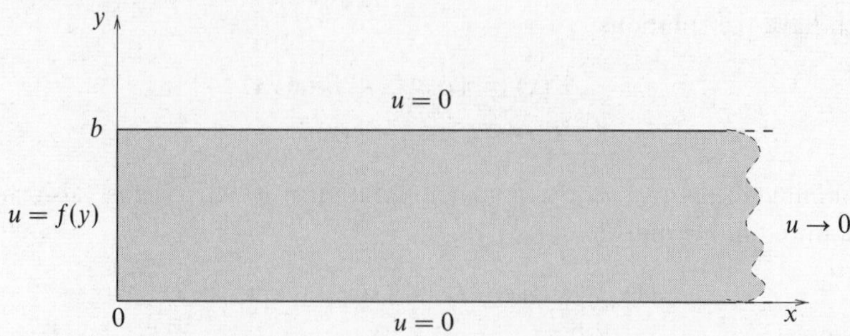


Figure 17.1 A semi-infinite metal plate whose edges are kept at fixed temperatures.

## 17.2 Superposition of separated solutions

It will be noticed in the previous two examples that there is considerable freedom in the values of the separation constant, the only essential requirement being that $\lambda$ has the *same* value in both parts of the solution, i.e. the part depending on $x$ and the part depending on $y$ (or $t$). This is a general feature for solutions in separated form, which, if the original PDE has $n$ independent variables, will contain $n - 1$ separation constants. All that is required in general is that we associate the correct function of one independent variable with the appropriate functions of the others – the correct function being the one with the same values of the separation constants.

If the original PDE is linear (as are the Laplace, Schrödinger, diffusion and wave equations) then mathematically acceptable solutions can be formed by superposing solutions corresponding to different allowed values of the separation constants. To take a two-variable example: if

$$u_{\lambda_1}(x, y) = X_{\lambda_1}(x)Y_{\lambda_1}(y)$$

is a solution of a linear PDE obtained by giving the separation constant the value $\lambda_1$, then the superposition

$$u(x, y) = a_1 X_{\lambda_1}(x)Y_{\lambda_1}(y) + a_2 X_{\lambda_2}(x)Y_{\lambda_2}(y) + \cdots = \sum_i a_i X_{\lambda_i}(x)Y_{\lambda_i}(y), \tag{17.16}$$

is also a solution for any constants $a_i$, provided that the $\lambda_i$ are the allowed values of the separation constant $\lambda$ given the imposed boundary conditions. Note that if the boundary conditions allow any of the separation constants to be zero then the form of the general solution is normally different and must be deduced by returning to the separated ordinary differential equations. We encounter this behaviour in section 17.3.

The value of the superposition approach is that a boundary condition, say that

$u(x, y)$ takes a particular form $f(x)$ when $y = 0$, might be met by choosing the constants $a_i$ such that

$$f(x) = \sum_i a_i X_{\lambda_i}(x) Y_{\lambda_i}(0).$$

This will in general be possible provided the functions $X_{\lambda_i}(x)$ form a complete set – as do the sinusoidal functions of Fourier series or the spherical harmonics that we shall discuss in subsection 17.3.2.

►*A semi-infinite rectangular metal plate occupies the region* $0 \leq x \leq \infty$ *and* $0 \leq y \leq b$ *in the xy-plane. The temperature at the far end of the plate and along its two long sides is fixed at* $0\,^\circ$C. *If the temperature of the plate at* $x = 0$ *is also fixed and is given by* $f(y)$, *find the steady-state temperature distribution* $u(x,y)$ *of the plate. Hence find the temperature distribution if* $f(y) = u_0$, *where* $u_0$ *is a constant.*

The physical situation is illustrated in figure 17.1. With the notation we have used several times before, the (two-dimensional) heat diffusion equation satisfied by the temperature $u(x, y, t)$ is

$$\kappa\left(\frac{\partial^2 u}{\partial x^2} + \frac{\partial^2 u}{\partial y^2}\right) = \frac{\partial u}{\partial t},$$

with $\kappa = k/s\rho$. In this case, however, we are asked to find the steady-state temperature, which corresponds to $\partial u/\partial t = 0$, and so we must instead consider the (two-dimensional) Laplace equation

$$\frac{\partial^2 u}{\partial x^2} + \frac{\partial^2 u}{\partial y^2} = 0.$$

We saw that assuming a separable solution of the form $u(x, y) = X(x)Y(y)$ led to solutions such as (17.14) or (17.15), or equivalent forms with $x$ and $y$ interchanged. In the current problem we have to satisfy the boundary conditions $u(x, 0) = 0 = u(x, b)$ and so a solution that is sinusoidal in $y$ seems appropriate. Furthermore, since we require $u(\infty, y) = 0$ it is best to write the $x$-dependence of the solution in terms explicitly of exponentials rather than of hyperbolic functions. We therefore write the separable solution in the form (17.15) as

$$u(x, y) = [A \exp \lambda x + B \exp(-\lambda x)](C \cos \lambda y + D \sin \lambda y).$$

Applying the boundary conditions, we see firstly that $u(\infty, y) = 0$ implies $A = 0$ if we take $\lambda > 0$. Secondly, since $u(x, 0) = 0$ we may set $C = 0$, which leaves us with (absorbing the constant $D$ into $B$)

$$u(x, y) = B \exp(-\lambda x) \sin \lambda y.$$

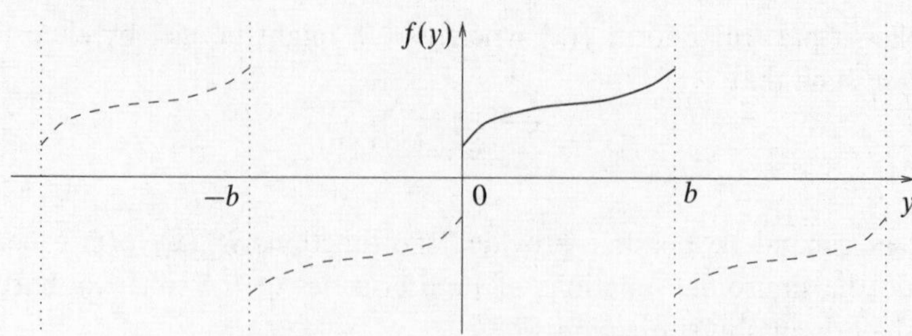


Figure 17.2 The continuation of $f(y)$ for a Fourier sine series.

But using the condition $u(x, b) = 0$ we require $\sin \lambda b = 0$, and so the constant $\lambda$ is constrained to equal $n\pi/b$, where $n$ is any positive integer.

Using the principle of superposition (17.16), the general solution can therefore be written

$$u(x, y) = \sum_{n=1}^{\infty} B_n \exp(-n\pi x/b) \sin(n\pi y/b), \tag{17.17}$$

for some constants $B_n$. Notice that in the sum in (17.17) we have omitted negative values of $n$ since they would lead to exponential terms that diverge as $x \to \infty$. The $n = 0$ term is also omitted since it is identically zero. Using the remaining boundary condition $u(0, y) = f(y)$ we see that the constants $B_n$ must satisfy

$$f(y) = \sum_{n=1}^{\infty} B_n \sin(n\pi y/b). \tag{17.18}$$

This is clearly a Fourier sine series expansion of $f(y)$ (see chapter 10). For (17.18) to hold, however, the continuation of $f(y)$ outside the region $0 \leq y \leq b$ must be an odd periodic function with period $2b$ (see figure 17.2). We also see from figure 17.2 that if the original function $f(y)$ does not equal zero at either of $y = 0$ and $y = b$ then its continuation has a discontinuity at the corresponding point(s); nevertheless, as discussed in chapter 10, the Fourier series will converge to the mid-points of these jumps, and hence tend to zero in this case. If, however, the top and bottom edges of the plate were held not at 0 °C, but at some other non-zero temperature, then the final solution would, in general, possess discontinuities at the corners $x = 0$, $y = 0$ and $x = 0$, $y = b$.

Bearing in mind these technicalities, the coefficients $B_n$ in (17.18) are given by

$$B_n = \frac{2}{b} \int_0^b f(y) \sin\left(\frac{n\pi y}{b}\right) dy. \tag{17.19}$$

Therefore, if $f(y) = u_0$ (i.e. the temperature of the side at $x = 0$ is constant along its length), (17.19) becomes

$$\begin{aligned} B_n &= \frac{2}{b}\int_0^b u_0 \sin\left(\frac{n\pi y}{b}\right) dy \\ &= \left[-\frac{2u_0}{b}\frac{b}{n\pi}\cos\left(\frac{n\pi y}{b}\right)\right]_0^b \\ &= -\frac{2u_0}{n\pi}[(-1)^n - 1] = \begin{cases} 4u_0/n\pi & \text{for } n \text{ odd} \\ 0 & \text{for } n \text{ even.} \end{cases} \end{aligned}$$

Therefore the required solution is

$$u(x,y) = \sum_{n \text{ odd}} \frac{4u_0}{n\pi} \exp\left(-\frac{n\pi x}{b}\right) \sin\left(\frac{n\pi y}{b}\right). \blacktriangleleft$$

In the above example the boundary conditions meant that one term in each part of the separable solution could be immediately discarded, making the problem much easier to solve. Sometimes, however, a little ingenuity is required in writing the separable solution in such a way that certain parts can be neglected immediately.

►*Suppose that the semi-infinite rectangular metal plate in the previous example is replaced by one that has a finite length $a$ in the $x$-direction, with the temperature of the right-hand edge fixed at 0°C and all other boundary conditions remaining as before. Find the steady-state temperature in the plate.*

As in the previous example, the boundary conditions $u(x,0) = 0 = u(x,b)$ suggest a solution that is sinusoidal in $y$. In this case, however, we require $u = 0$ on $x = a$ (rather than at infinity), and so a solution in which the $x$-dependence is written in terms of hyperbolic functions, such as (17.14), rather than exponentials is more appropriate. Moreover, since the constants in front of the hyperbolic functions are, at this stage, arbitrary, we may write the separable solution in the most convenient way that ensures that the condition $u(a,y) = 0$ is straightforwardly satisfied. We therefore write

$$u(x,y) = [A\cosh\lambda(a-x) + B\sinh\lambda(a-x)](C\cos\lambda y + D\sin\lambda y).$$

Now the condition $u(a,y) = 0$ is easily satisfied by setting $A = 0$. As before the conditions $u(x,0) = 0 = u(x,b)$ imply $C = 0$ and $\lambda = n\pi/b$ for integer $n$. Superposing the solutions for different $n$ we then obtain

$$u(x,y) = \sum_{n=1}^{\infty} B_n \sinh[n\pi(a-x)/b]\,\sin(n\pi y/b), \qquad (17.20)$$

for some constants $B_n$. We have omitted negative values of $n$ in the sum (17.20) since the relevant terms are already included in those obtained for positive $n$.

Again the $n = 0$ term is identically zero. Using the final boundary condition $u(0, y) = f(y)$ as above we find that the constants $B_n$ must satisfy

$$f(y) = \sum_{n=1}^{\infty} B_n \sinh(n\pi a/b) \sin(n\pi y/b),$$

and, remembering the caveats discussed in the previous example, the $B_n$ are therefore given by

$$B_n = \frac{2}{b \sinh(n\pi a/b)} \int_0^b f(y) \sin(n\pi y/b)\, dy. \tag{17.21}$$

For the case where $f(y) = u_0$, following the working of the previous example gives (17.21) as

$$B_n = \frac{4u_0}{n\pi \sinh(n\pi a/b)} \quad \text{for } n \text{ odd}, \qquad B_n = 0 \quad \text{for } n \text{ even}. \tag{17.22}$$

The required solution is thus

$$u(x, y) = \sum_{n \text{ odd}} \frac{4u_0}{n\pi \sinh(n\pi a/b)} \sinh[n\pi(a - x)/b] \sin\left(n\pi y/b\right).$$

We note that, as required, in the limit $a \to \infty$ this solution tends to the solution of the previous example. ◄

Often the principle of superposition can be used to write the solution to problems with more complicated boundary conditions as the sum of solutions to problems that each satisfy only some part of the boundary condition, but when added togther satisfy all the conditions.

►*Find the steady-state temperature in the (finite) rectangular plate of the previous example, subject to the boundary conditions $u(x, b) = 0$, $u(a, y) = 0$ and $u(0, y) = f(y)$ as before, but now in addition $u(x, 0) = g(x)$.*

Figure 17.3($c$) shows the imposed boundary conditions for the metal plate. Although we could find a solution to this problem using the methods presented above, we can arrive at the answer almost immediately by using the principle of superposition and the result of the previous example.

Let us suppose the required solution $u(x, y)$ is made up of two parts:

$$u(x, y) = v(x, y) + w(x, y),$$

where $v(x, y)$ is a solution to the problem satisfying the boundary conditions shown in figure 17.3($a$), whilst $w(x, y)$ is a solution satisfying the boundary conditions in figure 17.3($b$). It is clear that $v(x, y)$ is simply given by the solution to the previous example,

$$v(x, y) = \sum_{n \text{ odd}} B_n \sinh\left[\frac{n\pi(a - x)}{b}\right] \sin\left(\frac{n\pi y}{b}\right),$$

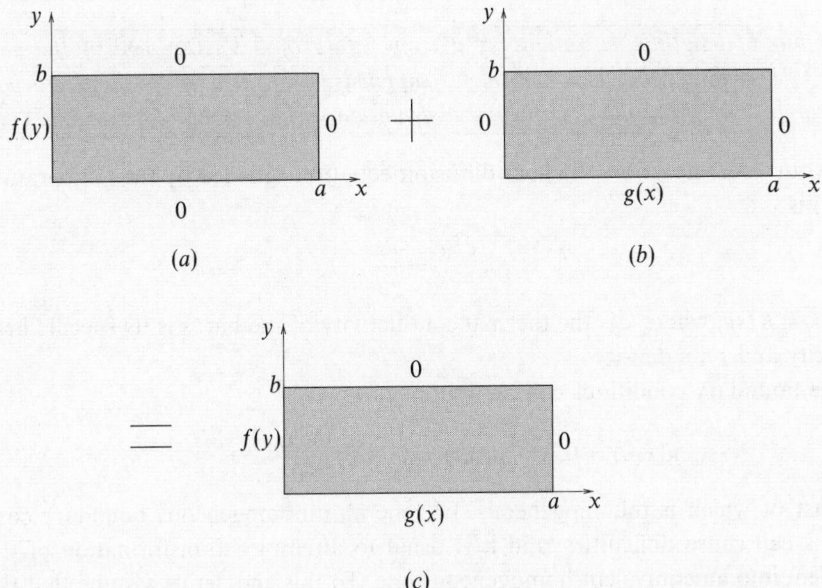


Figure 17.3 Superposition of boundary conditions for a metal plate.

where $B_n$ is given by (17.21). Moreover, by symmetry, $w(x, y)$ must be of the same form as $v(x, y)$ but with $x$ and $a$ interchanged with $y$ and $b$ respectively, and with $f(y)$ in (17.21) replaced by $g(x)$. Therefore the required solution can be written down immediately without further calculation as

$$u(x, y) = \sum_{n \text{ odd}} B_n \sinh\left[\frac{n\pi(a-x)}{b}\right] \sin\left(\frac{n\pi y}{b}\right) + \sum_{n \text{ odd}} C_n \sinh\left[\frac{n\pi(b-y)}{a}\right] \sin\left(\frac{n\pi x}{a}\right),$$

with the $B_n$ given by (17.21) and $C_n$ by

$$C_n = \frac{2}{a\sinh(n\pi b/a)} \int_0^a g(x) \sin(n\pi x/a)\, dx.$$

Clearly, this method may be extended to cases in which three or four sides of the plate have non-zero boundary conditions. ◄

As a final example of the usefulness of the principle of superposition we now consider a problem that illustrates how to deal with inhomogeneous boundary conditions by a suitable change of variables.

► *A bar of length L is initially at a temperature of 0 °C. One end of the bar (x = 0) is held at 0 °C and the other is supplied with heat at a constant rate per unit area of H. Find the temperature distribution within the bar after a time t.*

With our usual notation, the heat diffusion equation satisfied by the temperature $u(x,t)$ is

$$\kappa \frac{\partial^2 u}{\partial x^2} = \frac{\partial u}{\partial t},$$

with $\kappa = k/s\rho$, where $k$ is the thermal conductivity of the bar, $s$ is its specific heat capacity and $\rho$ its density.

The boundary conditions can be written as

$$u(x,0) = 0, \qquad u(0,t) = 0, \qquad \frac{\partial u(L,t)}{\partial x} = \frac{H}{k},$$

the last of which is inhomogeneous. In general, inhomogeneous boundary conditions can cause difficulties, and it is usual to attempt a transformation of the problem into an equivalent homogeneous one. To this end, let us assume that the solution to our problem takes the form

$$u(x,t) = v(x,t) + w(x),$$

where the function $w(x)$ is to be suitably determined. In terms of $v$ and $w$ the problem becomes

$$\begin{aligned} \kappa\left(\frac{\partial^2 v}{\partial x^2} + \frac{d^2 w}{dx^2}\right) &= \frac{\partial v}{\partial t}, \\ v(x,0) + w(x) &= 0, \\ v(0,t) + w(0) &= 0, \\ \frac{\partial v(L,t)}{\partial x} + \frac{dw(L)}{dx} &= \frac{H}{k}. \end{aligned}$$

There are several ways of choosing $w(x)$ so as to make the new problem straightforward. Using some physical insight, however, it is clear that ultimately (at $t = \infty$), when all transients have died away, the end $x = L$ will attain a temperature $u_0$ such that $ku_0/L = H$, and there will be a constant temperature gradient $u(x,\infty) = u_0 x/L$. We therefore choose

$$w(x) = \frac{Hx}{k}.$$

Since the second derivative of $w(x)$ is zero, $v$ satisfies the diffusion equation and the boundary conditions on $v$ are now

$$v(x,0) = -\frac{Hx}{k}, \qquad v(0,t) = 0, \qquad \frac{\partial v(L,t)}{\partial x} = 0,$$

which are homogeneous in $x$.

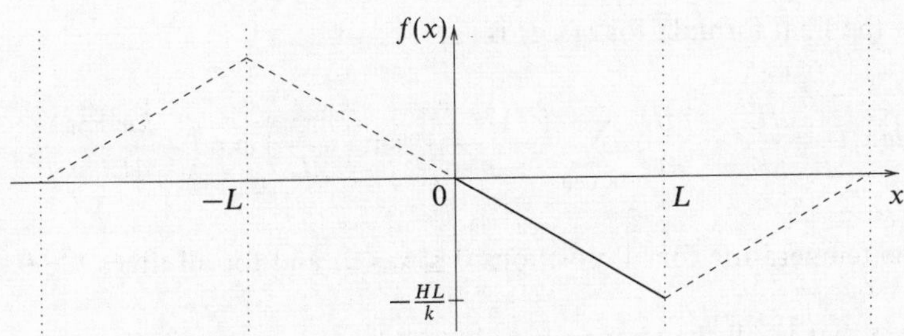


Figure 17.4 The appropriate continuation for a Fourier series containing only sine terms.

From (17.12) a separated solution for the one-dimensional diffusion equation is

$$v(x,t) = (A\cos\lambda x + B\sin\lambda x)\exp(-\lambda^2\kappa t),$$

corresponding to a separation constant $-\lambda^2$. If we restrict $\lambda$ to be real then all these solutions are transient ones decaying to zero as $t \to \infty$. These are just what is needed for adding to $w(x)$ to give the correct solution as $t \to \infty$. In order to satisfy $v(0,t) = 0$, however, we require $A = 0$. Furthermore, since

$$\frac{\partial v}{\partial x} = B\exp(-\lambda^2\kappa t)\lambda\cos\lambda x,$$

in order to satisfy $\partial v(L,t)/\partial x = 0$ we require $\cos\lambda L = 0$, and so $\lambda$ is restricted to take the values

$$\lambda = \frac{n\pi}{2L},$$

where $n$ is an odd non-negative integer, i.e. $n = 1, 3, 5, \ldots$.

Thus, to satisfy the boundary condition $v(x,0) = -Hx/k$, we must have

$$\sum_{n\ \text{odd}} B_n \sin\left(\frac{n\pi x}{2L}\right) = -\frac{Hx}{k},$$

in the range $x = 0$ to $x = L$. In this case we must be more careful about the continuation of the function $-Hx/k$ for which the Fourier sine series is needed. We want a series that is odd in $x$ (sine terms only) and continuous as $x = 0$ and $x = L$ (no discontinuities, since the series must converge at the end-points). This leads to a continuation of the function as shown in figure 17.4, with a period of $L' = 4L$. Following the discussion of section 10.3, since this continuation is odd about $x = 0$ and even about $x = L'/4 = L$ it can indeed be expressed as a Fourier sine series containing only odd-numbered terms.

The corresponding Fourier series coefficients are found to be

$$B_n = \frac{-8HL}{k\pi^2}\frac{(-1)^{(n-1)/2}}{n^2} \qquad \text{for } n \text{ odd},$$

and thus the final formula for $u(x,t)$ is

$$u(x,t) = \frac{Hx}{k} - \frac{8HL}{k\pi^2} \sum_{n \text{ odd}} \frac{(-1)^{(n-1)/2}}{n^2} \sin\left(\frac{n\pi x}{2L}\right) \exp\left(-\frac{kn^2\pi^2 t}{4L^2 s\rho}\right),$$

giving the temperature for all positions $0 \leq x \leq L$ and for all times $t \geq 0$. ◄

We note that in all the above examples the boundary conditions restricted the separation constant(s) to an infinite number of *discrete* values, usually integers. If, however, the boundary conditions allow the separation constant(s) $\lambda$ to take a *continuum* of values, then the summation in (17.16) is replaced by an integral over $\lambda$. This is discussed further in connection with integral transform methods in section 17.4.

## 17.3 Separation of variables in polar coordinates

So far we have considered the solution of PDEs only in Cartesian coordinates, but many systems in two and three dimensions are more naturally expressed in some form of polar coordinates, in which full advantage can be taken of any inherent symmetries. For example, the potential associated with an isolated point charge has a very simple expression, $q/4\pi\epsilon_0 r$, when polar coordinates are used, but involves all three coordinates, and square roots, when Cartesians are employed. For these reasons we now turn to the separation of variables in plane polar, cylindrical polar and spherical polar coordinates.

Most of the PDEs we have considered so far have involved the operator $\nabla^2$, e.g. the wave equation, the diffusion equation, Schrödinger's equation and Poisson's equation (and of course Laplace's equation). It is therefore appropriate that we recall the expressions for $\nabla^2$ when expressed in polar coordinate systems. From chapter 8, in plane polars, cylindrical polars and spherical polars respectively we have

$$\nabla^2 = \frac{1}{\rho}\frac{\partial}{\partial \rho}\left(\rho \frac{\partial}{\partial \rho}\right) + \frac{1}{\rho^2}\frac{\partial^2}{\partial \phi^2}, \tag{17.23}$$

$$\nabla^2 = \frac{1}{\rho}\frac{\partial}{\partial \rho}\left(\rho \frac{\partial}{\partial \rho}\right) + \frac{1}{\rho^2}\frac{\partial^2}{\partial \phi^2} + \frac{\partial^2}{\partial z^2}, \tag{17.24}$$

$$\nabla^2 = \frac{1}{r^2}\frac{\partial}{\partial r}\left(r^2 \frac{\partial}{\partial r}\right) + \frac{1}{r^2 \sin\theta}\frac{\partial}{\partial \theta}\left(\sin\theta \frac{\partial}{\partial \theta}\right) + \frac{1}{r^2 \sin^2\theta}\frac{\partial^2}{\partial \phi^2}. \tag{17.25}$$

Of course the first of these may be obtained from the second by taking $z$ to be identically zero.

### *17.3.1 Laplace's equation in polar coordinates*

The simplest of the equations containing $\nabla^2$ is Laplace's equation,

$$\nabla^2 u(\mathbf{r}) = 0. \tag{17.26}$$

Since it contains most of the essential features of the other more complicated equations we will consider its solution first.

*Laplace's equation in plane polars*

Suppose that we need to find a solution of (17.26) that has a prescribed behaviour on the circle $\rho = a$ (e.g. if we are finding the shape taken up by a circular drumskin when its rim is slightly deformed from being planar). Then we may seek solutions of (17.26) that are separable in $\rho$ and $\phi$ (measured from some arbitrary radius as $\phi = 0$), and hope to accommodate the boundary condition by examining the solution for $\rho = a$.

Thus, writing $u(\rho, \phi) = P(\rho)\Phi(\phi)$ and using the expression (17.23), Laplace's equation (17.26) becomes

$$\frac{\Phi}{\rho}\frac{\partial}{\partial \rho}\left(\rho \frac{\partial P}{\partial \rho}\right) + \frac{P}{\rho^2}\frac{\partial^2 \Phi}{\partial \phi^2} = 0.$$

Now, employing the same device as previously, that of dividing through by $u = P\Phi$ and multiplying through by $\rho^2$, results in the separated equation

$$\frac{\rho}{P}\frac{\partial}{\partial \rho}\left(\rho \frac{\partial P}{\partial \rho}\right) + \frac{1}{\Phi}\frac{\partial^2 \Phi}{\partial \phi^2} = 0.$$

Following our earlier argument, since the first term on the RHS is a function of $\rho$ only, whilst the second term depends only on $\phi$, we obtain the two *ordinary* equations

$$\frac{\rho}{P}\frac{d}{d\rho}\left(\rho \frac{dP}{d\rho}\right) = n^2 \tag{17.27}$$

$$\frac{1}{\Phi}\frac{d^2\Phi}{d\phi^2} = -n^2, \tag{17.28}$$

where we have taken the separation constant to have the form $n^2$ for later convenience; for the present $n$ is a general (complex) number.

Let us first consider the case where $n \neq 0$. The second equation, (17.28), then has the general solution

$$\Phi(\phi) = A\exp(in\phi) + B\exp(-in\phi). \tag{17.29}$$

Equation (17.27), on the other hand, is the homogeneous equation

$$\rho^2 P'' + \rho P' - n^2 P = 0,$$

which must be solved either by trying a power solution in $\rho$ or by making the substitution $\rho = \exp t$ as described in section 13.2.1 and so reducing it to an equation with constant coefficients. Carrying out this procedure we find

$$P(\rho) = C\rho^n + D\rho^{-n}. \tag{17.30}$$

Returning to the solution (17.29) of the azimuthal equation (17.28), we can see that if $\Phi$, and hence $u$, is to be single-valued and so not change when $\phi$ increases by $2\pi$, then $n$ must be an integer. Mathematically, other values of $n$ are permissible, but for the description of real physical situations it is clear that this limitation must be imposed. Having thus restricted the possible values of $n$ in one part of the solution, the same limitations must be carried over into the radial part (17.30). Thus we may write a particular solution of the two-dimensional Laplace equation as

$$u(\rho, \phi) = (A\cos n\phi + B\sin n\phi)(C\rho^n + D\rho^{-n}),$$

where $A$, $B$, $C$, $D$ are arbitrary constants and $n$ is any integer.

We have not yet, however, considered the solution when $n = 0$. In this case, the solutions of the separated ordinary equations (17.28) and (17.27) respectively are easily shown to be

$$\Phi(\phi) = A\phi + B,$$

$$P(\rho) = C\ln\rho + D.$$

But, in order that $u = P\Phi$ is single-valued, we require $A = 0$ and so the solution for $n = 0$ is simply (absorbing $B$ into $C$ and $D$)

$$u(\rho, \phi) = C\ln\rho + D.$$

Superposing the solutions for the different allowed values of $n$, we can write the general solution to Laplace's equation in plane polars as

$$u(\rho, \phi) = (C_0\ln\rho + D_0) + \sum_{n=1}^{\infty}(A_n\cos n\phi + B_n\sin n\phi)(C_n\rho^n + D_n\rho^{-n}), \tag{17.31}$$

where $n$ can take only integer values. Negative values of $n$ have been omitted from the sum since they are already included in the terms obtained for positive $n$. We note that, since $\ln\rho$ is singular at $\rho = 0$, whenever we solve Laplace's equation in a region containing the origin, $C_0$ must be identically zero.

►*A circular drumskin has a supporting rim at $\rho = a$. If the rim is twisted so that it is displaced vertically by a small amount $\epsilon(\sin\phi + 2\sin 2\phi)$, where $\phi$ is the azimuthal angle with respect to a given radius, find the resulting displacement $u(\rho, \phi)$ of the entire drumskin.*

The transverse displacement of a circular drumskin is usually described by the two-dimensional wave equation. In this case, however, there is no time dependence and so $u(\rho, \phi)$ solves the two-dimensional Laplace equation, subject to the imposed boundary condition.

Referring to (17.31), since we wish to find a solution that is finite everywhere inside $\rho = a$, we require $C_0 = 0$ and $D_n = 0$ for all $n > 0$. Now the boundary condition at the rim requires

$$u(a, \phi) = D_0 + \sum_{n=1}^{\infty} C_n a^n (A_n \cos n\phi + B_n \sin n\phi) = \epsilon(\sin\phi + 2\sin 2\phi).$$

Firstly we see that we require $D_0 = 0$ and $A_n = 0$ for all $n$. Furthermore, we must have $C_1 B_1 a = \epsilon$, $C_2 B_2 a^2 = 2\epsilon$ and $B_n = 0$ for $n > 2$. Hence the appropriate shape for the drumskin (valid over the whole skin, not just the rim) is

$$u(\rho, \phi) = \frac{\epsilon\rho}{a}\sin\phi + \frac{2\epsilon\rho^2}{a^2}\sin 2\phi = \frac{\epsilon\rho}{a}\left(\sin\phi + \frac{2\rho}{a}\sin 2\phi\right). \blacktriangleleft$$

*Laplace's equation in cylindrical polars*

Passing to three dimensions we now consider the solution of Laplace's equation in cylindrical polar coordinates,

$$\frac{1}{\rho}\frac{\partial}{\partial\rho}\left(\rho\frac{\partial u}{\partial\rho}\right) + \frac{1}{\rho^2}\frac{\partial^2 u}{\partial\phi^2} + \frac{\partial^2 u}{\partial z^2} = 0. \qquad (17.32)$$

We note here that even when considering a cylindrical physical system, if there is no dependence of the physical variables on $z$ (i.e. along the length of the cylinder) then the problem may be treated using two-dimensional plane polars, as discussed above.

For the more general case, we proceed as previously by trying a solution of the form

$$u(\rho, \phi, z) = P(\rho)\Phi(\phi)Z(z),$$

which on substitution into (17.32) and division through by $u = P\Phi Z$ gives

$$\frac{1}{P\rho}\frac{d}{d\rho}\left(\rho\frac{dP}{d\rho}\right) + \frac{1}{\Phi\rho^2}\frac{d^2\Phi}{d\phi^2} + \frac{1}{Z}\frac{d^2 Z}{dz^2} = 0.$$

The last term depends only on $z$ and the first and second (taken together) only on $\rho$ and $\phi$. Taking the separation constant to be $k^2$, we find

$$\frac{1}{Z}\frac{d^2 Z}{dz^2} = k^2,$$

$$\frac{1}{P\rho}\frac{d}{d\rho}\left(\rho\frac{dP}{d\rho}\right) + \frac{1}{\Phi\rho^2}\frac{d^2\Phi}{d\phi^2} + k^2 = 0.$$

The first of these equations has the straightforward solution

$$Z(z) = E\exp(-kz) + F\exp kz.$$

Multiplying the second equation through by $\rho^2$, we obtain

$$\frac{\rho}{P}\frac{d}{d\rho}\left(\rho\frac{dP}{d\rho}\right)+\frac{1}{\Phi}\frac{d^2\Phi}{d\phi^2}+k^2\rho^2=0,$$

in which the second term depends only on $\Phi$ and the other terms only on $\rho$. Taking the second separation constant to be $m^2$, we find

$$\frac{1}{\Phi}\frac{d^2\Phi}{d\phi^2}=-m^2, \tag{17.33}$$

$$\rho\frac{d}{d\rho}\left(\rho\frac{dP}{d\rho}\right)+(k^2\rho^2-m^2)P=0. \tag{17.34}$$

The equation in the azimuthal angle $\phi$ has the very familiar solution

$$\Phi(\phi)=C\cos m\phi+D\sin m\phi.$$

As in the two-dimensional case, single-valuedness of $u$ requires that $m$ is an integer. However, in the particular case $m=0$ the solution is

$$\Phi(\phi)=C\phi+D.$$

This form is appropriate to a solution with axial symmetry ($C=0$) or one that is multivalued, but manageably so, such as the magnetic scalar potential associated with a current $I$ (in which case $C=I/2\pi$ and $D$ is arbitrary).

Finally the $\rho$-equation (17.34) may be transformed into Bessel's equation of order $m$ by writing $\mu=k\rho$. This has the solution

$$P(\rho)=AJ_m(k\rho)+BY_m(k\rho).$$

The properties of these functions were investigated in chapter 14 and will not be pursued here. We merely note that $Y_m(k\rho)$ is singular at $\rho=0$, and so when seeking solutions to Laplace's equation in cylindrical coordinates within some region containing the $\rho=0$ axis, we require $B=0$.

The complete separated-variable solution in cylindrical polars of Laplace's equation $\nabla^2u=0$ is thus

$$u(\rho,\phi,z)=[AJ_m(k\rho)+BY_m(k\rho)][C\cos m\phi+D\sin m\phi][E\exp(-kz)+F\exp kz]. \tag{17.35}$$

Of course we may use the principle of superposition to build up more general solutions by adding together solutions of the form (17.35) for all allowed values of the separation constants $k$ and $m$.

►*A semi-infinite solid cylinder of radius $a$ has its curved surface held at* 0 °C *and its base held at a temperature $T_0$. Find the steady-state temperature distribution in the cylinder.*

The physical situation is shown in figure 17.5. The steady-state temperature distribution $u(\rho,\phi,z)$ must satisfy Laplace's equation subject to the imposed

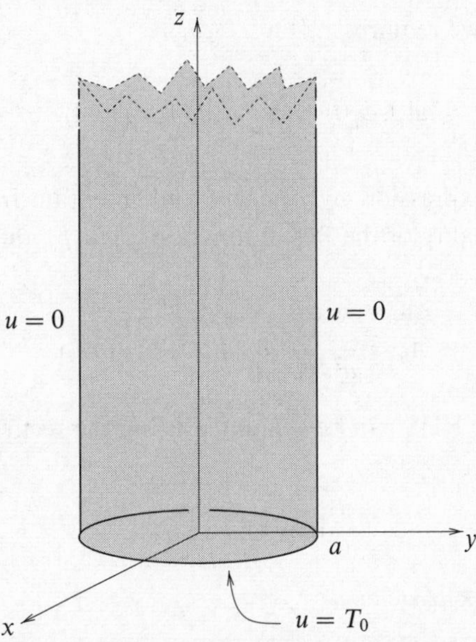


Figure 17.5 A uniform metal cylinder whose curved surface is kept at 0 °C and whose base is held at a temperature $T_0$.

boundary conditions. Let us take the cylinder to have its base in the $z = 0$ plane and to extend along the positive $z$-axis. From (17.35), in order that $u$ is finite everywhere in the cylinder we immediately require $B = 0$ and $F = 0$. Furthermore, since the boundary conditions, and hence the temperature distribution, are axially symmetric we require $m = 0$, and so the general solution must be a superposition of solutions of the form $J_0(k\rho)\exp(-kz)$ for all allowed values of the separation constant $k$.

The boundary condition $u(a, \phi, z) = 0$ restricts the allowed values of $k$ since we must have $J_0(ka) = 0$. The zeroes of Bessel functions are given in most books of mathematical tables, and we find that, to two decimal places,

$$J_0(x) = 0 \qquad \text{for } x = 2.40, 5.52, 8.65, \ldots.$$

Writing the allowed values of $k$ as $k_n$ for $n = 1, 2, 3, \ldots$ (so, for example, $k_1 = 2.40/a$), the required solution takes the form

$$u(\rho, \phi, z) = \sum_{n=1}^{\infty} A_n J_0(k_n\rho)\exp(-k_n z).$$

By imposing the remaining boundary condition $u(\rho, \phi, 0) = T_0$, the coefficients $A_n$ can be found in a similar way to Fourier coefficients but this time by exploiting the orthogonality of the Bessel functions, as discussed in chapter 14. From this boundary condition we require

$$u(\rho, \phi, 0) = \sum_{n=1}^{\infty} A_n J_0(k_n \rho) = T_0.$$

If we multiply this expression by $\rho J_0(k_r \rho)$ and integrate from $\rho = 0$ to $\rho = a$, and use the orthogonality of the Bessel functions $J_0(k_n \rho)$, then the coefficients are given by (14.81) as

$$A_n = \frac{2T_0}{a^2 J_1^2(k_n a)} \int_0^a J_0(k_n \rho) \rho \, d\rho. \tag{17.36}$$

The integral on the RHS can be evaluated using the recurrence relation (14.68) of chapter 14,

$$\frac{d}{dz}[z J_1(z)] = z J_0(z),$$

which on setting $z = k_n \rho$ yields

$$\frac{1}{k_n} \frac{d}{d\rho}[k_n \rho J_1(k_n \rho)] = k_n \rho J_0(k_n \rho).$$

Therefore the integral in (17.36) is given by

$$\int_0^a J_0(k_n \rho) \rho \, d\rho = \left[ \frac{1}{k_n} \rho J_1(k_n \rho) \right]_0^a = \frac{1}{k_n} a J_1(k_n a),$$

and the coefficients $A_n$ may be expressed as

$$A_n = \frac{2T_0}{a^2 J_1^2(k_n a)} \left[ \frac{a J_1(k_n a)}{k_n} \right] = \frac{2T_0}{k_n a J_1(k_n a)}.$$

The steady-state temperature in the cylinder is then given by

$$u(\rho, \phi, z) = \sum_{n=1}^{\infty} \frac{2T_0}{k_n a J_1(k_n a)} J_0(k_n \rho) \exp(-k_n z). \blacktriangleleft$$

We note that if, in the above example, the base of the cylinder were not kept at a uniform temperature $T_0$, but instead had some fixed temperature distribution $T(\rho, \phi)$, then the solution of the problem would become more complicated. In such a case, the required temperature distribution $u(\rho, \phi, z)$ is in general *not* axially symmetric, and so the separation constant $m$ is not restricted to be zero, but may take any integer value. The solution will then take the form

$$u(\rho, \phi, z) = \sum_{m=0}^{\infty} \sum_{n=1}^{\infty} J_m(k_{nm} \rho)(C_{nm} \cos m\phi + D_{nm} \sin m\phi) \exp(-k_{nm} z),$$

where the separation constants $k_{nm}$ are such that $J_m(k_{nm} a) = 0$, i.e. $k_{nm} a$ is the $n$th zero of the $m$th-order Bessel function. At the base of the cylinder we would then require

$$u(\rho, \phi, 0) = \sum_{m=0}^{\infty} \sum_{n=1}^{\infty} J_m(k_{nm}\rho)(C_{nm} \cos m\phi + D_{nm} \sin m\phi) = T(\rho, \phi). \tag{17.37}$$

The coefficients $C_{nm}$ can be found by multiplying (17.37) by $J_q(k_{rq}\rho) \cos q\phi$, integrating with respect to $\rho$ and $\phi$ over the base of the cylinder and exploiting the orthogonality of the Bessel functions and of the trigonometric functions. The $D_{nm}$ can be found in a similar way by multiplying (17.37) by $J_q(k_{rq}\rho) \sin q\phi$.

*Laplace's equation in spherical polars*

We now come to a very widely applicable equation in physical science, namely $\nabla^2 u = 0$ in spherical polar coordinates:

$$\frac{1}{r^2} \frac{\partial}{\partial r} \left( r^2 \frac{\partial u}{\partial r} \right) + \frac{1}{r^2 \sin\theta} \frac{\partial}{\partial \theta} \left( \sin\theta \frac{\partial u}{\partial \theta} \right) + \frac{1}{r^2 \sin^2\theta} \frac{\partial^2 u}{\partial \phi^2} = 0. \tag{17.38}$$

Our method of procedure will be as before; we try a solution of the form

$$u(r, \theta, \phi) = R(r)\Theta(\theta)\Phi(\phi).$$

Substituting this in (17.38), dividing through by $u = R\Theta\Phi$, and multiplying by $r^2$, we obtain

$$\frac{1}{R} \frac{d}{dr} \left( r^2 \frac{dR}{dr} \right) + \frac{1}{\Theta \sin\theta} \frac{d}{d\theta} \left( \sin\theta \frac{d\Theta}{d\theta} \right) + \frac{1}{\Phi \sin^2\theta} \frac{d^2\Phi}{d\phi^2} = 0. \tag{17.39}$$

The first term depends only on $r$ and the second and third terms (taken together) only on $\theta$ and $\phi$. Thus (17.39) is equivalent to the two equations

$$\frac{1}{R} \frac{d}{dr} \left( r^2 \frac{dR}{dr} \right) = \lambda, \tag{17.40}$$

$$\frac{1}{\Theta \sin\theta} \frac{d}{d\theta} \left( \sin\theta \frac{d\Theta}{d\theta} \right) + \frac{1}{\Phi \sin^2\theta} \frac{d^2\Phi}{d\phi^2} = -\lambda. \tag{17.41}$$

Equation (17.40) is a homogeneous equation,

$$r^2 \frac{d^2 R}{dr^2} + 2r \frac{dR}{dr} - \lambda R = 0,$$

which can be reduced by the substitution $r = \exp t$ (and writing $R(r) = S(t)$) to

$$\frac{d^2 S}{dt^2} + \frac{dS}{dt} - \lambda S = 0.$$

This has the straightforward solution

$$S(t) = A \exp \lambda_1 t + B \exp \lambda_2 t,$$

and so the solution to the radial equation is

$$R(r) = Ar^{\lambda_1} + Br^{\lambda_2}.$$

where $\lambda_1 + \lambda_2 = -1$ and $\lambda_1\lambda_2 = -\lambda$. We can thus take $\lambda_1$ and $\lambda_2$ as given by $\ell$ and $-(\ell+1)$; $\lambda$ then has the form $\ell(\ell+1)$. (It should be noted that at this stage nothing has been either assumed or proved about whether $\ell$ is an integer.)

Hence we have obtained some information about the first factor in the separated-variable solution, which will now have the form

$$u(r, \theta, \phi) = (Ar^{\ell} + Br^{-(\ell+1)})\Theta(\theta)\Phi(\phi), \tag{17.42}$$

where $\Theta$ and $\Phi$ must satisfy (17.41) with $\lambda = \ell(\ell+1)$.

The next step is to take (17.41) further. Multiplying it through by $\sin^2\theta$ and substituting for $\lambda$, it too takes a separated form:

$$\left[\frac{\sin\theta}{\Theta}\frac{d}{d\theta}\left(\sin\theta\frac{d\Theta}{d\theta}\right) + \ell(\ell+1)\sin^2\theta\right] + \frac{1}{\Phi}\frac{d^2\Phi}{d\phi^2} = 0. \tag{17.43}$$

Taking the separation constant as $m^2$, the equation in the azimuthal angle $\phi$ has the same solution as in cylindrical polars, namely

$$\Phi(\phi) = C\cos m\phi + D\sin m\phi.$$

As before, single-valuedness of $u$ requires that $m$ is an integer; for $m = 0$ we again have $\Phi(\phi) = C\phi + D$.

Having settled the form of $\Phi(\phi)$, we are left only with the equation satisfied by $\Theta(\theta)$, which is

$$\frac{\sin\theta}{\Theta}\frac{d}{d\theta}\left(\sin\theta\frac{d\Theta}{d\theta}\right) + \ell(\ell+1)\sin^2\theta = m^2. \tag{17.44}$$

A change of independent variable from $\theta$ to $\mu = \cos\theta$ will reduce this to a form for which solutions are known, and of which some study has been made in chapter 14. Putting

$$\mu = \cos\theta, \qquad \frac{d\mu}{d\theta} = -\sin\theta, \qquad \frac{d}{d\theta} = -(1-\mu^2)^{1/2}\frac{d}{d\mu},$$

the equation for $M(\mu) \equiv \Theta(\theta)$ reads

$$\frac{d}{d\mu}\left[(1-\mu^2)\frac{dM}{d\mu}\right] + \left[\ell(\ell+1) - \frac{m^2}{1-\mu^2}\right]M = 0. \tag{17.45}$$

This equation is the *associated Legendre equation*, which was mentioned in subsection 15.5.2 in the context of Sturm–Liouville equations.

We recall that for the case $m = 0$, (17.45) reduces to Legendre's equation, which was studied at length in chapter 14, and has the solution

$$M(\mu) = EP_\ell(\mu) + FQ_\ell(\mu). \tag{17.46}$$

We have not explicitly solved (17.45) for general $m$, but the solutions were mentioned in subsection 15.5.2, and are the associated Legendre functions $P_\ell^m(\mu)$ and $Q_\ell^m(\mu)$, where

$$P_\ell^m(\mu) = (1-\mu^2)^{|m|/2}\frac{d^{|m|}}{d\mu^{|m|}}P_\ell(\mu),$$

and similarly for $Q_\ell^m$. We then have

$$M(\mu) = EP_\ell^m(\mu) + FQ_\ell^m(\mu); \tag{17.47}$$

here $m$ must be an integer, $0 \le |m| \le \ell$. We note that if we require solutions to Laplace's equation that are finite when $\mu = \cos\theta = \pm 1$ (i.e. on the polar axis where $\theta = 0, \pi$), then we must have $F = 0$ in (17.46) and (17.47) since $Q_\ell^m(\mu)$ diverges at $\mu = \pm 1$.

It will be remembered that one of the important conditions for obtaining finite polynomial solutions of Legendre's equation is that $\ell$ is an integer $\ge 0$. This condition therefore also applies to the solutions (17.46) and (17.47), and is reflected back into the radial part of the general solution as given in (17.42).

Now that the solutions of each of the three ordinary differential equations governing $R$, $\Theta$ and $\Phi$ have been obtained, we may assemble a complete separated-variable solution of Laplace's equation in spherical polars. It is

$$u(r,\theta,\phi) = (Ar^\ell + Br^{-(\ell+1)})(C\cos m\phi + D\sin m\phi)[EP_\ell^m(\cos\theta) + FQ_\ell^m(\cos\theta)], \tag{17.48}$$

where the three bracketed factors are connected only through the *integer* parameters $\ell$ and $m$, $0 \le |m| \le \ell$. As before, a general solution may be obtained by superposing solutions of this form for the allowed values of the separation constants $\ell$ and $m$. As mentioned above, if the solution is required to be finite on the polar axis then $F = 0$ for all $\ell$ and $m$.

► *An uncharged conducting sphere of radius a is placed at the origin in an initially uniform electrostatic field E. Show that it behaves as an electric dipole.*

The uniform field, taken in the direction of the polar axis, has a electrostatic potential

$$u = -Ez = -Er\cos\theta,$$

where $u$ is arbitrarily taken as zero at $z = 0$. This satisfies Laplace's equation $\nabla^2 u = 0$, as must the potential $v$ when the sphere is present; for large $r$ the asymptotic form of $v$ must still be $-Er\cos\theta$.

Since the problem is clearly axially symmetric we have immediately that $m = 0$, and since we require $v$ to be finite on the polar axis we must have $F = 0$ in

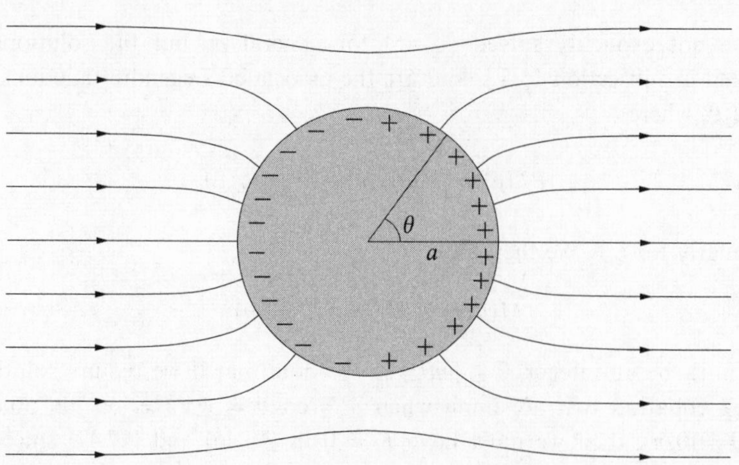


Figure 17.6 Induced charge and field lines associated with a conducting sphere placed in an initially uniform electrostatic field.

(17.48). Therefore the solution must be of the form

$$v(r,\theta,\phi) = \sum_{\ell=0}^{\infty}(A_\ell r^\ell + B_\ell r^{-(\ell+1)})P_\ell(\cos\theta).$$

Now the $\cos\theta$-dependence of $v$ for large $r$ indicates that the $(\theta,\phi)$-dependence of $v(r,\theta,\phi)$ is given by $P_1^0(\cos\theta) = \cos\theta$. Thus the $r$-dependence of $v$ must also correspond to an $\ell = 1$ solution, and the most general such solution (outside the sphere, $r \geq a$) is

$$v(r,\theta,\phi) = (A_1 r + B_1 r^{-2})P_1(\cos\theta).$$

The asymptotic form of $v$ for large $r$ immediately gives $A_1 = -E$, and so yields the solution

$$v(r,\theta,\phi) = \left(-Er + \frac{B_1}{r^2}\right)\cos\theta.$$

Since the sphere is conducting, it is an equipotential region and so $v$ must not depend on $\theta$ for $r = a$. This can only be so if $B_1/a^2 = Ea$, thus fixing $B_1$. The final solution is therefore

$$v(r,\theta,\phi) = -Er\left(1 - \frac{a^3}{r^3}\right)\cos\theta.$$

Since a dipole of moment $P$ gives rise to a potential $p/(4\pi\epsilon_0 r^2)$, this result shows that the sphere behaves as a dipole of moment $4\pi\epsilon_0 a^3 E$, because of the effect of the charge distribution induced on its surface; see figure 17.6. ◄

Often the boundary conditions are not so easily met, and it is necessary to

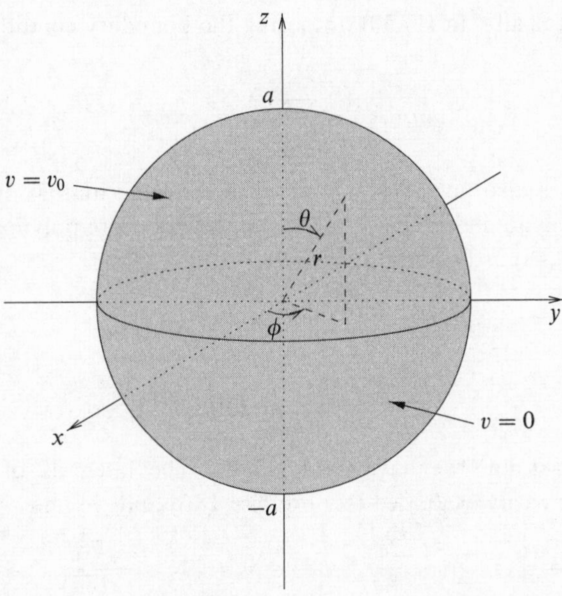


Figure 17.7 A hollow split conducting sphere with its top half charged to a potential $v_0$ and its bottom half at zero potential.

use the mutual orthogonality of the associated Legendre functions (and the trigonometric functions) to obtain the coefficients in the general solution.

► *A hollow split conducting sphere of radius a is placed at the origin. If one half of its surface is charged to a potential $v_0$ and the other half is kept at zero potential, find the potential v inside and outside the sphere.*

Let us choose the top hemisphere to be charged to $v_0$ and the bottom hemisphere to be at zero potential, with the plane in which the two hemispheres meet perpendicular to the polar axis; this is shown in figure 17.7. The boundary condition then becomes

$$v(a,\theta,\phi) = \begin{cases} v_0 & \text{for } 0 < \theta < \pi/2 \quad (\text{or } 0 < \cos\theta < 1), \\ 0 & \text{for } \pi/2 < \theta < \pi \quad (\text{or } -1 < \cos\theta < 0). \end{cases} \qquad (17.49)$$

The problem is clearly axially symmetric and so we may set $m = 0$. Also we require the solution to be finite on the polar axis and so it cannot contain $Q_\ell(\cos\theta)$. Therefore the general form of the solution to (17.38) is

$$v(r,\theta,\phi) = \sum_{\ell=0}^{\infty} (A_\ell r^\ell + B_\ell r^{-(\ell+1)}) P_\ell(\cos\theta). \qquad (17.50)$$

Inside the sphere (for $r < a$) we require the solution to be finite at the origin and so $B_\ell = 0$ for all $\ell$ in (17.50). Imposing the boundary condition at $r = a$ we must then have

$$v(a, \theta, \phi) = \sum_{\ell=0}^{\infty} A_\ell a^\ell P_\ell(\cos\theta),$$

where $v(a, \theta, \phi)$ is also given by (17.49). Exploiting the mutual orthogonality of the Legendre polynomials, the coefficients in the Legendre polynomial expansion are given by (14.48) as (writing $\mu = \cos\theta$)

$$\begin{aligned} A_\ell a^\ell &= \frac{2\ell+1}{2} \int_{-1}^{1} v(a, \theta, \phi) P_\ell(\mu)\, d\mu \\ &= \frac{2\ell+1}{2} v_0 \int_0^1 P_\ell(\mu)\, d\mu, \end{aligned}$$

where in the last line we have used (17.49). The integrals of the Legendre polynomials are easily evaluated (see exercise 15.3) and we find

$$A_0 = \frac{v_0}{2}, \qquad A_1 = \frac{3v_0}{4a}, \qquad A_2 = 0, \qquad A_3 = -\frac{7v_0}{16a^3}, \qquad \cdots,$$

so that the required solution inside the sphere is

$$v(r, \theta, \phi) = \frac{v_0}{2}\left[1 + \frac{3r}{2a} P_1(\cos\theta) - \frac{7r^3}{8a^3} P_3(\cos\theta) + \cdots\right].$$

Outside the sphere (for $r > a$) we require the solution to be bounded as $r$ tends to infinity and so in (17.50) we must have $A_\ell = 0$ for all $\ell$. In this case, by imposing the boundary condition at $r = a$ we require

$$v(a, \theta, \phi) = \sum_{\ell=0}^{\infty} B_\ell a^{-(\ell+1)} P_\ell(\cos\theta),$$

where $v(a, \theta, \phi)$ is given by (17.49). Following the above argument the coefficients in the expansion are given by

$$B_\ell a^{-(\ell+1)} = \frac{2\ell+1}{2} v_0 \int_0^1 P_\ell(\mu)\, d\mu,$$

so that the required solution outside the sphere is

$$v(r, \theta, \phi) = \frac{v_0 a}{2r}\left[1 + \frac{3a}{2r} P_1(\cos\theta) - \frac{7a^3}{8r^3} P_3(\cos\theta) + \cdots\right]. \blacktriangleleft$$

In the above example, on the equator of the sphere (i.e. at $r = a$ and $\theta = \pi/2$) the potential is given by

$$v(a, \pi/2, \phi) = v_0/2,$$

i.e. half-way between the potentials of the top and bottom hemispheres. This is so because a Legendre polynomial expansion of a function behaves in the same way as a Fourier series expansion, in that it converges to the average value of any discontinuities present in the original function.

If the potential on the surface of the sphere had been given as a function of $\theta$ *and* $\phi$, then we would have had to consider a double series summed over $\ell$ and $m$ (for $-\ell \leq m \leq \ell$), since the solution would not, in general, have been axially symmetric.

### *17.3.2 Spherical harmonics*

In obtaining solutions in spherical polar coordinates of $\nabla^2 u = 0$, we found that, for solutions which are finite on the polar axis, the angular part of the solution was given by

$$\Theta(\theta)\Phi(\phi) = P_\ell^m(\cos\theta)(C\cos m\phi + D\sin m\phi).$$

This general form is sufficiently common that particular functions of $\theta$ and $\phi$ called *spherical harmonics* are defined and tabulated. The spherical harmonics $Y_\ell^m(\theta, \phi)$ are defined for $m \geq 0$ by

$$Y_\ell^m(\theta, \phi) = (-1)^m \left[\frac{2\ell+1}{4\pi}\frac{(\ell-m)!}{(\ell+m)!}\right]^{1/2} P_\ell^m(\cos\theta)\exp(im\phi). \qquad (17.51)$$

For values of $m < 0$ the relation

$$Y_\ell^{-|m|}(\theta, \phi) = (-1)^{|m|}\left[Y_\ell^{|m|}(\theta, \phi)\right]^*$$

defines the spherical harmonic, the asterisk denoting complex conjugation. Since they contain as their $\theta$-dependent part the solution to the associated Legendre equation, which is a Sturm–Liouville equation (see chapter 15), the $Y_\ell^m$ are mutually orthogonal when integrated from $-1$ to $+1$ over $d(\cos\theta)$. Their mutual orthogonality with respect to $\phi$ ($0 \leq \phi \leq 2\pi$) is even more obvious. The numerical factor in (17.51) is chosen to make the $Y_\ell^m$ an orthonormal set, so that

$$\int_{-1}^{1}\int_0^{2\pi} \left[Y_\ell^m(\theta, \phi)\right]^* Y_{\ell'}^{m'}(\theta, \phi)\, d\phi\, d(\cos\theta) = \delta_{\ell\ell'}\delta_{mm'}.$$

In addition, the spherical harmonics form a complete set in that any reasonable function (i.e. one that is likely to be met in a physical situation) of $\theta$ and $\phi$ can be expanded as a sum of such functions,

$$f(\theta, \phi) = \sum_{\ell=0}^{\infty}\sum_{m=-\ell}^{\ell} a_{\ell m} Y_\ell^m(\theta, \phi), \qquad (17.52)$$

the constants $a_{\ell m}$ being given by

$$a_{\ell m} = \int_{-1}^{1}\int_0^{2\pi} \left[Y_\ell^m(\theta, \phi)\right]^* f(\theta, \phi)\, d\phi\, d(\cos\theta). \qquad (17.53)$$

This is in exact analogy with Fourier series and is a particular example of the general property of Sturm–Liouville solutions.

The first few spherical harmonics $Y_\ell^m(\theta, \phi) \equiv Y_\ell^m$ are as follows:

$$\begin{aligned}
Y_0^0 &= \sqrt{\tfrac{1}{4\pi}} & Y_1^0 &= \sqrt{\tfrac{3}{4\pi}}\cos\theta \\
Y_1^{\pm 1} &= \mp\sqrt{\tfrac{3}{8\pi}}\sin\theta\exp(\pm i\phi) & Y_2^0 &= \sqrt{\tfrac{5}{16\pi}}(3\cos^2\theta - 1) \\
Y_2^{\pm 1} &= \mp\sqrt{\tfrac{15}{8\pi}}\sin\theta\cos\theta\exp(\pm i\phi) & Y_2^{\pm 2} &= \sqrt{\tfrac{15}{32\pi}}\sin^2\theta\exp(\pm 2i\phi).
\end{aligned}$$

### *17.3.3 Other equations in polar coordinates*

The development of the solutions of $\nabla^2 u = 0$ carried out in the previous subsection can be readily employed in solving other equations in which the $\nabla^2$ operator appears. Since we have discussed the general method in some depth already, only an outline of the solutions will be given here.

Let us first consider the wave equation

$$\nabla^2 u = \frac{1}{c^2}\frac{\partial^2 u}{\partial t^2}, \tag{17.54}$$

and look for a separated solution of the form $u = F(\mathbf{r})T(t)$, so that initially we are separating only the spatial and time dependences. Substituting this form into (17.54) and taking the separation constants as $k^2$ we obtain

$$\nabla^2 F + k^2 F = 0, \qquad \frac{d^2 T}{dt^2} + k^2 c^2 T = 0. \tag{17.55}$$

The second equation has the simple solution

$$T(t) = A\exp i\omega t + B\exp(-i\omega t), \tag{17.56}$$

where $\omega = kc$; this may, of course, also be expressed in terms of sines and cosines. The first equation in (17.55) is referred to as *Helmholtz's equation*; we discuss it below.

We may treat the diffusion equation

$$\kappa\nabla^2 u = \frac{\partial u}{\partial t}$$

in a similar way. Separating the spatial and time dependences by assuming a solution of the form $u = F(\mathbf{r})T(t)$, and taking the separation constants as $k^2$, we find

$$\nabla^2 F + k^2 F = 0, \qquad \frac{dT}{dt} + k^2\kappa T = 0.$$

Thus the spatial part of the solution for the diffusion equation satisfies Helmholtz's equation, just as in the case of the wave equation. It only remains to consider the time dependence, which has the simple solution

$$T(t) = A \exp(-k^2 \kappa t).$$

Helmholtz's equation is clearly of great importance for the solution of the wave and diffusion equations. It can be solved in polar coordinates in much the same way as Laplace's equation, and indeed reduces to Laplace's equation when $k = 0$. We therefore only sketch its solution in each of the three polar coordinate systems.

*Helmholtz's equation in plane polars*

In two-dimensional plane polar cooordinates Helmholtz's equation takes the form

$$\frac{1}{\rho}\frac{\partial}{\partial \rho}\left(\rho \frac{\partial F}{\partial \rho}\right) + \frac{1}{\rho^2}\frac{\partial^2 F}{\partial \phi^2} + k^2 F = 0.$$

If we try a separated solution of the form $F(\mathbf{r}) = P(\rho)\Phi(\phi)$, and take the separation constants as $m^2$, we find

$$\frac{d^2\Phi}{d\phi^2} + m^2 \phi = 0,$$

$$\frac{d^2 P}{d\rho^2} + \frac{1}{\rho}\frac{dP}{d\rho} + \left(k^2 - \frac{m^2}{\rho^2}\right) P = 0.$$

As for Laplace's equation the angular part has the familiar solution (if $m \neq 0$)

$$\Phi(\phi) = A \cos m\phi + B \sin m\phi,$$

or an equivalent form in terms of complex exponentials. The radial equation differs from that found in the solution of Laplace's equation, but by making the substitution $\mu = k\rho$ it is easily transformed into Bessel's equation of order $m$ (discussed in chapter 14), and has the solution

$$P(\rho) = CJ_m(k\rho) + DY_m(k\rho),$$

where $Y_m$ is a Bessel function of the second kind, which is infinite at the origin and is not to be confused with a spherical harmonic (these are written with a superscript as well as a subscript).

Putting the two parts of the solution together we have

$$F(\rho, \phi) = [A \cos m\phi + B \sin m\phi][CJ_m(k\rho) + DY_m(k\rho)]. \qquad (17.57)$$

Clearly, for solutions of Helmholtz's equation that are required to be finite at the origin, we must set $D = 0$.

► *Find the four modes of oscillation that are lowest in frequency of a circular drumskin of radius a whose circumference is held fixed in a plane.*

The transverse displacement $u(\mathbf{r}, t)$ of the drumskin satisfies the two-dimensional wave equation

$$\nabla^2 u = \frac{1}{c^2}\frac{\partial^2 u}{\partial t^2},$$

with $c^2 = T/\sigma$, where $T$ is the tension of the drumskin and $\sigma$ is its mass per unit area. From (17.56) and (17.57) a separated solution of this equation, in plane polar coordinates, that is finite at the origin is

$$u(\rho, \phi, t) = J_m(k\rho)(A\cos m\phi + B\sin m\phi)\exp(\pm i\omega t),$$

where $\omega = kc$. Since we require the solution to be single-valued we must have $m$ as an integer. Furthermore, if the drumskin is clamped at its outer edge $\rho = a$ we also require $u(a, \phi, t) = 0$. Thus we need

$$J_m(ka) = 0,$$

which in turn restricts the allowed values of $k$. The zeroes of Bessel functions can be obtained from most books of tables, and the first few are

$$\begin{aligned} J_0(x) &= 0 \quad \text{for } x \approx 2.40, 5.52, 8.65, \ldots, \\ J_1(x) &= 0 \quad \text{for } x \approx 3.83, 7.02, 10.17, \ldots, \\ J_2(x) &= 0 \quad \text{for } x \approx 5.14, 8.42, 11.62 \ldots. \end{aligned}$$

The smallest value of $x$ for which any of the Bessel functions is zero is $x \approx 2.40$, which occurs for $J_0(x)$. Thus the lowest-frequency mode has $k = 2.40/a$ and angular frequency $\omega = 2.40c/a$. Since $m = 0$ for this mode, the shape of the drumskin in this mode of oscillation is

$$u \propto J_0\left(2.40\frac{\rho}{a}\right);$$

this is illustrated in figure 17.8.

Continuing in the same way the next three modes are given by

$$\begin{aligned} \omega &= 3.83\frac{c}{a}, \quad u \propto J_1\left(3.83\frac{\rho}{a}\right)\cos\phi, \quad J_1\left(3.83\frac{\rho}{a}\right)\sin\phi; \\ \omega &= 5.14\frac{c}{a}, \quad u \propto J_2\left(5.14\frac{\rho}{a}\right)\cos 2\phi, \quad J_2\left(5.14\frac{\rho}{a}\right)\sin 2\phi; \\ \omega &= 5.52\frac{c}{a}, \quad u \propto J_0\left(5.52\frac{\rho}{a}\right). \end{aligned}$$

These modes are also shown in figure 17.8. We note that the second and third frequencies have *two* corresponding modes of oscillation. These frequencies are therefore two-fold degenerate. ◄

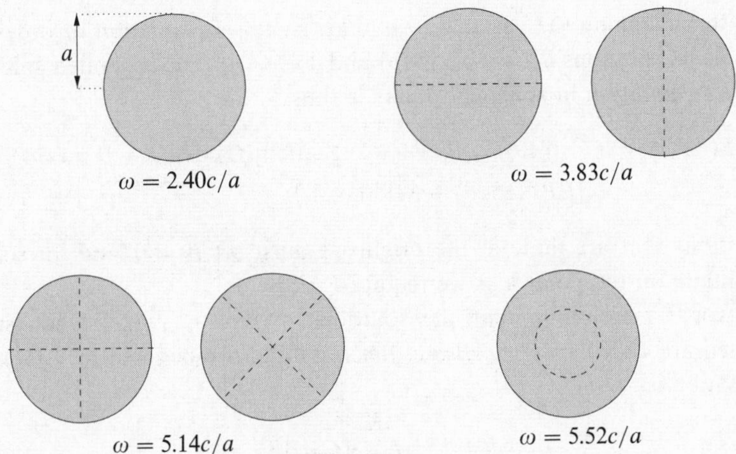


Figure 17.8 The four lowest-frequency modes of oscillation of a circular drumskin of radius $a$. The dotted lines indicate the nodes, where the displacement of the drumskin is always zero.

*Helmholtz's equation in cylindrical polars*

Generalising the above method to three-dimensional cylindrical polars is straightforward, and following a similar procedure to that used for Laplace's equation we find the separated solution of Helmholtz's equation takes the form

$$F(\rho, \phi, z) = \left[AJ_m\left(\sqrt{k^2-\alpha^2}\,\rho\right) + BY_m\left(\sqrt{k^2-\alpha^2}\,\rho\right)\right]$$
$$\times (C\cos m\phi + D\sin m\phi)[E\exp(i\alpha z) + F\exp(-i\alpha z)],$$

where $\alpha$ and $m$ are separation constants. We again note that the angular part of the solution is the same as for Laplace's equation in cylindrical polars.

*Helmholtz's equation in spherical polars*

In spherical polars, we again find that the angular parts of the solution $\Theta(\theta)\Phi(\phi)$ are identical to those of Laplace's equation in this coordinate system, i.e. they are the spherical harmonics $Y_\ell^m(\theta, \phi)$, and so we shall not discuss them further.

The radial equation in this case is given by

$$r^2R'' + 2rR' + [k^2r^2 - \ell(\ell+1)]R = 0, \tag{17.58}$$

which has an additional term $k^2r^2R$ as compared to the radial equation for the Laplace solution. The equation (17.58) looks very much like Bessel's equation and can in fact be reduced to it by writing $R(r) = r^{-1/2}S(r)$. The function $S(r)$ then satisfies

$$r^2S'' + rS' + \left[k^2r^2 - \left(\ell + \tfrac{1}{2}\right)^2\right]S = 0,$$

which, after changing the variable to $\mu = kr$, is Bessel's equation of order $\ell + \frac{1}{2}$ and has as its solutions $S(\mu) = J_{\ell+1/2}(\mu)$ and $Y_{\ell+1/2}(\mu)$. The separated solution to Helmholtz's equation in spherical polars is thus

$$F(r, \theta, \phi) = r^{-1/2}[AJ_{\ell+1/2}(kr) + BY_{\ell+1/2}(kr)](C\cos m\phi + D\sin m\phi) \\ \times [EP_\ell^m(\cos\theta) + FQ_\ell^m(\cos\theta)]. \qquad (17.59)$$

For solutions that are finite at the origin we must set $B = 0$, and for solutions that are finite on the polar axis we require $F = 0$.

It is worth mentioning that the solutions $\propto r^{-1/2}J_{\ell+1/2}(kr)$ when suitably normalised are called *spherical Bessel functions* and are denoted by $j_\ell(kr)$. Their normalisation is

$$j_\ell(\mu) = \sqrt{\frac{\pi}{2\mu}}J_{\ell+1/2}(\mu).$$

They are trigonometric functions of $\mu$ (as discussed in chapter 14), and for $\ell = 0$ and $\ell = 1$ are given by

$$j_0(\mu) = \sin\mu,$$
$$j_1(\mu) = \frac{\sin\mu}{\mu} - \cos\mu.$$

The second, linearly-independent solution of (17.58), $n_\ell(\mu)$, is derived from $Y_{\ell+1/2}(\mu)$ in a similar way.

As mentioned at the beginning of this subsection, the separated solution of the wave equation in spherical polars is the product of the time-dependent part (17.56) and a spatial part (17.59). It will be noticed that, although this solution corresponds to a definite frequency solution $\omega = kc$, except for the case $\ell = 0$ involving $j_0(kr)$, the zeroes of the radial function $j_\ell(kr)$ are not equally spaced in $r$ and so there is no precise wavelength associated with the solution.

To conclude this subsection, let us mention briefly the Schrödinger equation for the electron in a hydrogen atom, the nucleus of which is taken at the origin and is assumed massive compared to the electron. Under these circumstances the Schrödinger equation is

$$-\frac{\hbar^2}{2m}\nabla^2 u - \frac{e^2}{4\pi\epsilon_0}\frac{u}{r} = i\hbar\frac{\partial u}{\partial t}.$$

For a 'stationary-state' solution, for which the energy is a constant $E$ and the time-dependent factor $T$ in $u$ is given by $T(t) = A\exp(-iEt/\hbar)$, the above equation also becomes similar to the Helmholtz equation, except that the radial part is modified.† However, as with the wave equation, the angular parts of the

† For the solution by series of the $r$-equation in this case the reader may consult, e.g., Schiff, *Quantum Mechanics* (McGraw-Hill, 1955) p. 82.

solution are identical to those for Laplace's equation and are expressed in terms of spherical harmonics.

The important point to note is that for *any* equation involving $\nabla^2$, provided $\theta$ and $\phi$ do not appear in the equation other than as part of $\nabla^2$, a separated-variable solution in spherical polars will always lead to spherical harmonic solutions.

### 17.3.4 Solution by expansion

It is sometimes possible to use the uniqueness theorem discussed in the last chapter, together with the results of the last few subsections, in which Laplace's equation (and other equations) were considered in polar coordinates, to obtain solutions of such equations appropriate to particular physical situations.

We will illustrate the method for Laplace's equation in spherical polars and first assume that the required solution of $\nabla^2 u = 0$ can be written as a superposition in the normal way:

$$u(r,\theta,\phi) = \sum_{\ell=0}^{\infty}\sum_{m=-\ell}^{\ell}(Ar^{\ell} + Br^{-(\ell+1)})P_{\ell}^{m}(\cos\theta)(C\cos m\phi + D\sin m\phi). \tag{17.60}$$

Here, all the constants $A$, $B$, $C$, $D$ may depend upon $\ell$ and $m$, and we have assumed that the required solution is finite on the polar axis. As usual, boundary conditions of a physical nature will then fix or eliminate some of the constants; for example, $u$ finite at the origin implies all $B = 0$, or axial symmetry implies that only $m = 0$ terms are present.

The essence of the method is then to find the remaining constants by determining $u$ at values of $r, \theta, \phi$ for which it can be evaluated *by other means*, e.g. by direct calculation on an axis of symmetry. Once the remaining constants have been fixed by these special considerations to have particular values, the uniqueness theorem can be invoked to establish that they must have these values in general.

► *Calculate the gravitational potential at a general point in space due to a uniform ring of matter of radius $a$ and total mass $M$.*

Everywhere except on the ring the potential $u(\mathbf{r})$ satisfies the Laplace equation, and so if we use polar coordinates with the normal to the ring as polar axis, as in figure 17.9, a solution of the form (17.60) can be assumed.

We expect the potential $u(r,\theta,\phi)$ to tend to zero as $r \to \infty$, and also to be finite at $r = 0$. At first sight this might seem to imply that all $A$ and $B$, and hence $u$, must be identically zero – an unacceptable result. In fact, what it means is that different expressions must apply to different regions of space. On the ring itself we no longer have $\nabla^2 u = 0$ and so it is not surprising that the form of the expression for $u$ changes there. Let us therefore take two separate regions.

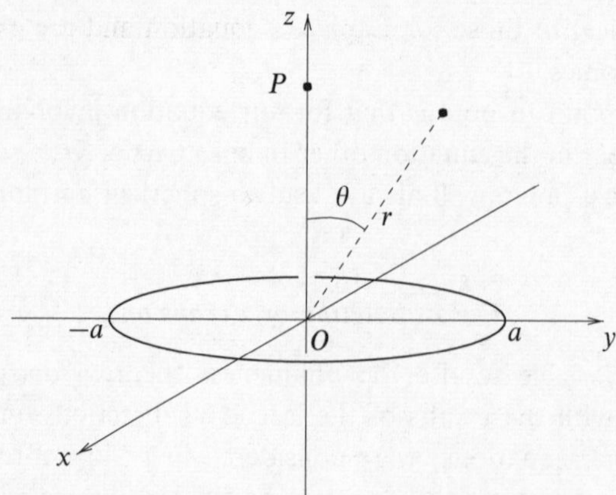


Figure 17.9 The polar axis $Oz$ is taken as normal to the plane of the ring of matter and passing through its centre.

In the region $r > a$

(i) we must have $u \to 0$ as $r \to \infty$ implying all $A = 0$,
(ii) the system is axially symmetric and so only $m = 0$ terms appear.

With these restrictions we can write as a trial form

$$u(r, \theta, \phi) = \sum_{\ell=0}^{\infty} B_\ell r^{-(\ell+1)} P_\ell^0(\cos\theta), \tag{17.61}$$

with the constants $B_\ell$ still to be determined. This we do by calculating *directly* the potential where this can be done simply – in this case, on the polar axis.

Considering a point $P$ on the polar axis at a distance $z$ $(> a)$ from the plane of the ring (taken as $\theta = \pi/2$), all parts of the ring are at a distance $(z^2 + a^2)^{1/2}$ from it. The potential at $P$ is thus straightforwardly

$$u(z, 0, \phi) = \frac{GM}{(z^2 + a^2)^{1/2}}, \tag{17.62}$$

where $G$ is the gravitational constant. This must be the same as (17.61) for the particular values $r = z$, $\theta = 0$, and $\phi$ undefined. Since $P_\ell^0(\cos\theta) = P_\ell(\cos\theta)$ with $P_\ell(1) = 1$, putting $r = z$ in (17.61) gives

$$u(z, 0, \phi) = \sum_{\ell=0}^{\infty} \frac{B_\ell}{z^{\ell+1}}. \tag{17.63}$$

However, expanding (17.62) for $z > a$ (as it applies to this region of space) we obtain

$$u(z, 0, \phi) = \frac{GM}{z}\left[1 - \frac{1}{2}\left(\frac{a}{z}\right)^2 + \frac{3}{8}\left(\frac{a}{z}\right)^4 - \cdots\right],$$

which on comparison with (17.63) gives us†

$$\begin{aligned} B_0 &= GM, \\ B_{2\ell} &= \frac{GMa^{2\ell}(-1)^\ell(2\ell-1)!!}{2^\ell \ell!} \qquad \text{for } \ell \geq 1, \\ B_{2\ell+1} &= 0. \end{aligned} \tag{17.64}$$

We now conclude the argument by saying that if a solution for a general point $(r, \theta, \phi)$ exists at all, which of course we very much expect on physical grounds, then it must be (17.61) with the $B_\ell$ given by (17.64). This is so because thus defined it is a function with no arbitrary constants and which satisfies all the boundary conditions, and the uniqueness theorem states that there is only one such function. The expression for the potential in the region $r > a$ is therefore

$$u(r,\theta,\phi) = \frac{GM}{r}\left[1 + \sum_{\ell=1}^{\infty} \frac{(-1)^\ell(2\ell-1)!!}{2^\ell \ell!}\left(\frac{a}{r}\right)^{2\ell} P_{2\ell}(\cos\theta)\right].$$

The expression for $r < a$ can be found in a similar way. The finiteness of $u$ at $r = 0$ and the axial symmetry give

$$u(r,\theta,\phi) = \sum_{\ell=0}^{\infty} A_\ell r^\ell P_\ell^0(\cos\theta).$$

Comparing this expression for $r = z$, $\theta = 0$ with the $z < a$ expansion of (17.62), which is valid for any $z$, establishes $A_{2\ell+1} = 0$, $A_0 = GM/a$ and

$$A_{2\ell} = \frac{GM}{a^{2\ell+1}} \frac{(-1)^\ell(2\ell-1)!!}{2^\ell \ell!},$$

so that the final expression valid, and convergent, for $r < a$ is thus

$$u(r,\theta,\phi) = \frac{GM}{a}\left[1 + \sum_{\ell=1}^{\infty} \frac{(-1)^\ell(2\ell-1)!!}{2^\ell \ell!}\left(\frac{r}{a}\right)^{2\ell} P_{2\ell}(\cos\theta)\right].$$

It may be easily checked that the solution obtained has the expected physical value for large $r$ and for $r = 0$, and is continuous at $r = a$. ◄

### *17.3.5 Separation of variables for inhomogeneous equations*

So far our discussion of the method of separation of variables has been limited to the solution of homogeneous equations such as the Laplace equation and the wave equation. The solutions of inhomogeneous PDEs are usually obtained using the Green's function methods discussed below in section 17.5. However, as a final illustration of the usefulness of the separation of variables, we now consider its application to the solution of inhomogeneous equations.

† $(2\ell-1)!! = 1 \times 3 \times \cdots \times (2\ell-1)$.

Because of the added complexity in dealing with inhomogeneous equations, we shall restrict our discussion to the solution of Poisson's equation,

$$\nabla^2 u = \rho(\mathbf{r}), \tag{17.65}$$

in spherical polar coordinates, although the general method can accommodate other coordinate systems and equations. In physical problems the RHS of (17.65) usually contains some multiplicative constant(s). If $u$ is the electrostatic potential in some region of space in which $\rho$ is the density of electric charge, then $\nabla^2 u = -\rho(\mathbf{r})/\epsilon_0$. Alternatively, $u$ might represent the gravitational potential in some region where the matter density is given by $\rho$, so that $\nabla^2 u = 4\pi G\rho(\mathbf{r})$.

We will simplify our discussion by assuming that the required solution $u$ is finite on the polar axis, and also that the system possesses axial symmetry about that axis – in which case $\rho$ does not depend on the azimuthal angle $\phi$. The key to the method is then to assume a separated form for both the solution $u$ *and* the density term $\rho$.

From the discussion of Laplace's equation, for systems with axial symmetry only $m = 0$ terms appear, and so the angular part of the solution can be expressed in terms of Legendre polynomials $P_\ell(\cos\theta)$. Since these functions form an orthogonal set let us expand both $u$ and $\rho$ in terms of them:

$$u = \sum_{\ell=0}^{\infty} R_\ell(r) P_\ell(\cos\theta), \tag{17.66}$$

$$\rho = \sum_{\ell=0}^{\infty} F_\ell(r) P_\ell(\cos\theta), \tag{17.67}$$

where the coefficients $R_\ell(r)$ and $F_\ell(r)$ in the Legendre polynomial expansions are functions of $r$. Since in any particular problem $\rho$ is given, we can find the coefficients $F_\ell(r)$ in the expansion in the usual way (see subsection 14.6.2). It then only remains to find the coefficients $R_\ell(r)$ in the expansion of the solution $u$.

Writing $\nabla^2$ in spherical polars and substituting (17.66) and (17.67) into (17.65) we obtain

$$\sum_{\ell=0}^{\infty} \left[ \frac{P_\ell(\cos\theta)}{r^2} \frac{d}{dr}\left( r^2 \frac{dR_\ell}{dr} \right) + \frac{R_\ell}{r^2 \sin\theta} \frac{d}{d\theta}\left( \sin\theta \frac{dP_\ell(\cos\theta)}{d\theta} \right) \right] = \sum_{\ell=0}^{\infty} F_\ell(r) P_\ell(\cos\theta). \tag{17.68}$$

But if, in equation (17.44) of our discussion of the angular part of the solution to Laplace's equation, we set $m = 0$ we conclude that

$$\frac{1}{\sin\theta} \frac{d}{d\theta}\left( \sin\theta \frac{dP_\ell(\cos\theta)}{d\theta} \right) = -\ell(\ell+1) P_\ell(\cos\theta).$$

Substituting this into (17.68), we find that the LHS is greatly simplified and we obtain

$$\sum_{\ell=0}^{\infty}\left[\frac{1}{r^2}\frac{d}{dr}\left(r^2\frac{dR_\ell}{dr}\right)-\frac{\ell(\ell+1)R_\ell}{r^2}\right]P_\ell(\cos\theta)=\sum_{\ell=0}^{\infty}F_\ell(r)P_\ell(\cos\theta).$$

This relation is most easily satisfied by equating terms on both sides for each value of $\ell$ separately, so that for $\ell = 0, 1, 2, \ldots$ we have

$$\frac{1}{r^2}\frac{d}{dr}\left(r^2\frac{dR_\ell}{dr}\right)-\frac{\ell(\ell+1)R_\ell}{r^2}=F_\ell(r). \tag{17.69}$$

This is an ODE in which $F_\ell(r)$ is given, and it can therefore be solved for $R_\ell(r)$. The solution to Poisson's equation, $u$, is then obtained by making the superposition (17.66).

► *In a certain system, the electric charge density $\rho$ is distributed as follows:*

$$\rho=\begin{cases}Ar\cos\theta & \text{for } 0\le r<a,\\ 0 & \text{for } r\ge a.\end{cases}$$

*Find the electrostatic potential inside and outside the charge distribution, given that both the potential and its radial derivative are continuous everywhere.*

The electrostatic potential $u$ satisfies

$$\nabla^2 u=\begin{cases}-(A/\epsilon_0)r\cos\theta & \text{for } 0\le r<a,\\ 0 & \text{for } r\ge a.\end{cases}$$

For $r < a$ the RHS can be written $-(A/\epsilon_0)rP_1(\cos\theta)$, and the coefficients in (17.67) are simply $F_1(r) = -(Ar/\epsilon_0)$ and $F_\ell(r) = 0$ for $\ell \neq 1$. Therefore we need only calculate $R_1(r)$, which satisfies (17.69) for $\ell = 1$:

$$\frac{1}{r^2}\frac{d}{dr}\left(r^2\frac{dR_1}{dr}\right)-\frac{2R_1}{r^2}=-\frac{Ar}{\epsilon_0}.$$

This can be rearranged to give

$$r^2R_1''+2rR_1'-2R_1=-\frac{Ar^3}{\epsilon_0},$$

where the prime denotes differentiation with respect to $r$. The LHS is homogeneous and the equation can be reduced by the substitution $r = \exp t$, and writing $R_1(r) = S(t)$, to

$$\ddot{S}+\dot{S}-2S=-\frac{A}{\epsilon_0}\exp 3t, \tag{17.70}$$

where the dots indicate differentiation with respect to $t$.

This is an inhomogeneous second-order ODE with constant coefficients, and can be straightforwardly solved by the methods of subsection 13.2.1 to give

$$S(t)=c_1\exp t+c_2\exp(-2t)-\frac{A}{10\epsilon_0}\exp 3t.$$

Recalling that $r = \exp t$ we find

$$R_1(r) = c_1 r + c_2 r^{-2} - \frac{A}{10\epsilon_0} r^3.$$

Since we are interested in the region $r < a$ we must have $c_2 = 0$ for the solution to remain finite. Thus inside the charge distribution the electrostatic potential has the form

$$u_1(r, \theta, \phi) = \left( c_1 r - \frac{A}{10\epsilon_0} r^3 \right) P_1(\cos\theta). \qquad (17.71)$$

Outside the charge distribution (for $r \geq a$), however, the electrostatic potential obeys Laplace's equation, $\nabla^2 u = 0$, and so given the symmetry of the problem and the requirement that $u \to \infty$ as $r \to \infty$ the solution must take the form

$$u_2(r, \theta, \phi) = \sum_{\ell=0}^{\infty} \frac{B_\ell}{r^{\ell+1}} P_\ell(\cos\theta). \qquad (17.72)$$

We can now use the boundary conditions at $r = a$ to fix the constants in (17.71) and (17.72). The requirement of continuity of the potential and its radial derivative at $r = a$ imply that

$$u_1(a, \theta, \phi) = u_2(a, \theta, \phi),$$
$$\frac{\partial u_1}{\partial r}(a, \theta, \phi) = \frac{\partial u_2}{\partial r}(a, \theta, \phi).$$

Clearly $B_\ell = 0$ for $\ell \neq 1$; carrying out the necessary differentiations and setting $r = a$ in (17.71) and (17.72) we obtain the simultaneous equations

$$c_1 a - \frac{A}{10\epsilon_0} a^3 = \frac{B_1}{a^2},$$
$$c_1 - \frac{3A}{10\epsilon_0} a^2 = -\frac{2B_1}{a^3}$$

which may be solved to give $c_1 = Aa^2/6\epsilon_0$ and $B_1 = Aa^5/15\epsilon_0$. Since $P_1(\cos\theta) = \cos\theta$, the electrostatic potentials inside and outside the charge distribution are given respectively by

$$u_1(r, \theta, \phi) = \frac{A}{\epsilon_0}\left( \frac{a^2 r}{6} - \frac{r^3}{10} \right) \cos\theta, \qquad u_2(r, \theta, \phi) = \frac{Aa^5}{15\epsilon_0} \frac{\cos\theta}{r^2}. \blacktriangleleft$$

## 17.4 Integral transform methods

In the method of separation of variables our aim was to keep the independent variables in a PDE as separate as possible. We now discuss the use of integral transforms in solving PDEs in which one of the independent variables can be eliminated from the differential coefficients. It will be assumed that the reader is familiar with Laplace and Fourier transforms and their properties, as discussed in chapter 11.

The method consists simply of transforming the PDE into one containing derivatives with respect to a smaller number of variables. If the original equation has just two independent variables, it is often possible to reduce the PDE into an ODE, which may then be soluble. The solution obtained can then (where possible) be transformed back to give the solution of the original PDE. As we shall see, boundary conditions can usually be incorporated in a natural way.

Which sort of transform to use, and the choice of the variable(s) with respect to which the transform is to be taken, is a matter of experience; we illustrate this in the example below. In practice, transforms can be taken with respect to each variable in turn, and the transformation that affords the greatest simplification can be pursued further.

► *A semi-infinite tube of constant cross-section initially contains pure water. At time $t = 0$, one end of the tube is put into contact with a salt solution and maintained at a concentration $u_0$. Find the total amount of salt that has diffused into the tube at time $t$, if the diffusion constant is $\kappa$.*

The concentration $u(x,t)$ at time $t$ and distance $x$ from the end of the tube satisfies the diffusion equation

$$\kappa \frac{\partial^2 u}{\partial x^2} = \frac{\partial u}{\partial t}, \tag{17.73}$$

which has to be solved subject to the boundary conditions $u(0,t) = u_0$ for all $t$ and $u(x,0) = 0$ for all $x > 0$.

Since we are interested only in $t > 0$, the use of the Laplace transform is suggested. Furthermore, it will be recalled from chapter 11 that one of the major virtues of Laplace transformations is the possibility they afford of replacing derivatives of functions by simple multiplication by a scalar. Equation (17.73), if the derivative with respect to time were so removed, would contain only differentiation with respect to a single variable. Let us therefore take the Laplace transform of (17.73) with respect to $t$:

$$\int_0^\infty \kappa \frac{\partial^2 u}{\partial x^2} \exp(-st)\, dt = \int_0^\infty \frac{\partial u}{\partial t} \exp(-st)\, dt.$$

On the LHS the (double) differentiation is with respect to $x$, whereas the integration is with respect to the independent variable $t$. Therefore the derivative can be taken outside the integral. Denoting the Laplace transform of $u(x,t)$ by $\bar{u}(x,s)$ and using result (11.57) to rewrite the transform of the derivative on the RHS (or by integrating directly by parts), we obtain

$$\kappa \frac{\partial^2 \bar{u}}{\partial x^2} = s\bar{u}(x,s) - u(x,0).$$

But from the boundary condition $u(x, 0) = 0$ the last term on the RHS vanishes, and the solution is immediate:

$$\bar{u}(x, s) = A \exp\left(\sqrt{s/\kappa}\, x\right) + B \exp\left(-\sqrt{s/\kappa}\, x\right),$$

where the constants $A$ and $B$ may depend on $s$.

We require $u(x, t) \to 0$ as $x \to \infty$ and so we must also have $\bar{u}(\infty, s) = 0$; consequently we require that $A = 0$. The value of $B$ is determined by the need for $u(0, t) = u_0$ and hence that

$$\bar{u}(0, s) = \int_0^\infty u_0 \exp(-st)\, dt = \frac{u_0}{s}.$$

We thus conclude that the appropriate expression for the Laplace transform of $u(x, t)$ is

$$\bar{u}(x, s) = \frac{u_0}{s} \exp\left(-\sqrt{\frac{s}{\kappa}}\, x\right). \qquad (17.74)$$

To obtain $u(x, t)$ from this result requires the inversion of this transform – a task that is generally difficult and requires a contour integration. This is discussed in chapter 18, but for completeness we note that the solution is

$$u(x, t) = u_0 \left[1 - \text{erf}\left(\frac{x}{\sqrt{4\kappa t}}\right)\right],$$

where $\text{erf}(x)$ is the error function discussed in the Appendix. (The more complete sets of mathematical tables list this inverse Laplace transform.)

In the present problem, however, an alternative method is available. Let $w(t)$ be the amount of salt that has diffused into the tube in time $t$; then

$$w(t) = \int_0^\infty u(x, t)\, dx,$$

and its transform is given by

$$\begin{aligned}\bar{w}(s) &= \int_0^\infty dt\, \exp(-st) \int_0^\infty u(x, t)\, dx \\ &= \int_0^\infty dx \int_0^\infty u(x, t) \exp(-st)\, dt \\ &= \int_0^\infty \bar{u}(x, s)\, dx.\end{aligned}$$

Substituting for $\bar{u}(x, s)$ from (17.74) into the last integral and integrating, we obtain

$$\bar{w}(s) = u_0 \kappa^{1/2} s^{-3/2}.$$

This expression is much simpler to invert, and referring to the table of standard

Laplace transforms (table 11.1) we find

$$w(t) = 2(\kappa/\pi)^{1/2} u_0 t^{1/2},$$

which is thus the required expression for the amount of diffused salt at time $t$. ◄

The above example shows that in some circumstances the use of a Laplace transformation can greatly simplify the solution of a PDE. However, it will have been observed that (as with ODEs) the easy elimination of some derivatives is usually paid for by the introduction of a difficult inverse transformation. This problem, although still present, is less severe for Fourier transformations.

►*An infinite metal bar has an initial temperature distribution $f(x)$ along its length. Find the temperature distribution at a later time $t$.*

We are interested in values of $x$ from $-\infty$ to $\infty$, which suggests Fourier transformation with respect to $x$. Assuming that the solution obeys the boundary conditions $u(x,t) \to 0$ and $\partial u/\partial x \to 0$ as $|x| \to \infty$, we may Fourier-transform the one-dimensional diffusion equation (17.73) to obtain

$$\frac{\kappa}{\sqrt{2\pi}} \int_{-\infty}^{\infty} \frac{\partial^2 u(x,t)}{\partial x^2} \exp(-ikx)\, dx = \frac{1}{\sqrt{2\pi}} \frac{\partial}{\partial t} \int_{-\infty}^{\infty} u(x,t) \exp(-ikx)\, dx,$$

where on the RHS we have taken the partial derivative with respect to $t$ outside the integral. Denoting the Fourier transform of $u(x,t)$ by $\widetilde{u}(k,t)$, and using equation (11.28) to rewrite the Fourier transform of the second derivative on the LHS, we then have

$$-\kappa k^2 \widetilde{u}(k,t) = \frac{\partial \widetilde{u}(k,t)}{\partial t}.$$

This first-order equation has the simple solution

$$\widetilde{u}(k,t) = \widetilde{u}(k,0) \exp(-\kappa k^2 t),$$

where the initial conditions give

$$\begin{aligned} \widetilde{u}(k,0) &= \frac{1}{\sqrt{2\pi}} \int_{-\infty}^{\infty} u(x,0) \exp(-ikx)\, dx \\ &= \frac{1}{\sqrt{2\pi}} \int_{-\infty}^{\infty} f(x) \exp(-ikx)\, dx = \widetilde{f}(k). \end{aligned}$$

Thus we may write the Fourier transform of the solution as

$$\widetilde{u}(k,t) = \widetilde{f}(k) \exp(-\kappa k^2 t) = \sqrt{2\pi}\, \widetilde{f}(k) \widetilde{G}(k,t), \tag{17.75}$$

where we have defined the function $\widetilde{G}(k,t) = (\sqrt{2\pi})^{-1} \exp(-\kappa k^2 t)$. Since $\widetilde{u}(k,t)$ can be written as the product of two Fourier transforms, we can use the convolution theorem, subsection 11.1.7, to write the solution as

$$u(x,t) = \int_{-\infty}^{\infty} G(x - x', t) f(x')\, dx',$$

where $G(x,t)$ is the Green's function for this problem (see subsection 13.2.5). This function is the inverse Fourier transform of $\widetilde{G}(k,t)$ and is thus given by

$$\begin{aligned} G(x,t) &= \frac{1}{2\pi}\int_{-\infty}^{\infty} \exp(-\kappa k^2 t)\exp(ikx)\,dk \\ &= \frac{1}{2\pi}\int_{-\infty}^{\infty} \exp\left[-\kappa t\left(k^2 - \frac{ix}{\kappa t}k\right)\right]\,dk. \end{aligned}$$

Completing the square in the integrand we find

$$\begin{aligned} G(x,t) &= \frac{1}{2\pi}\exp\left(-\frac{x^2}{4\kappa t}\right)\int_{-\infty}^{\infty} \exp\left[-\kappa t\left(k - \frac{ix}{2\kappa t}\right)^2\right]\,dk \\ &= \frac{1}{2\pi}\exp\left(-\frac{x^2}{4\kappa t}\right)\int_{-\infty}^{\infty} \exp\left(-\kappa t k'^2\right)\,dk' \\ &= \frac{1}{\sqrt{4\pi\kappa t}}\exp\left(-\frac{x^2}{4\kappa t}\right), \end{aligned}$$

where in the second line we have made the substitution $k' = k - ix/2\kappa t$, and in the last line we have used the standard result for the integral of a Gaussian, given in subsection 5.4.2. (Strictly speaking the change of variable from $k$ to $k'$ shifts the path of integration off the real axis, since $k'$ is complex for real $k$, and so results in a complex integral, as will be discussed in chapter 18. Nevertheless, in this case the path of integration can be shifted back to the real axis without affecting the value of the integral.)

Thus the temperature in the bar at a later time $t$ is given by

$$u(x,t) = \frac{1}{\sqrt{4\pi\kappa t}}\int_{-\infty}^{\infty} \exp\left[-\frac{(x-x')^2}{4\kappa t}\right] f(x')\,dx', \tag{17.76}$$

which may be evaluated (numerically if necessary) when the form of $f(x)$ is given. ◂

As we might expect from our discussion of Green's functions in chapter 13, we see from (17.76) that, if the initial temperature distribution is $f(x) = \delta(x-a)$, i.e. a 'point' source at $x = a$, then the temperature distribution at later times is simply given by

$$u(x,t) = G(x-a,t) = \frac{1}{\sqrt{4\pi\kappa t}}\exp\left[-\frac{(x-a)^2}{4\kappa t}\right].$$

The temperature at several later times is illustrated in figure 17.10, from which we see that the heat diffuses out from its initial position; the width of the Gaussian increases as $\sqrt{t}$, which is characteristic of diffusion processes.

The reader may have noticed that, in both examples using integral transforms, the solutions have been obtained in closed form – albeit in one case in the form of an integral. This differs from the infinite series solutions usually obtained via

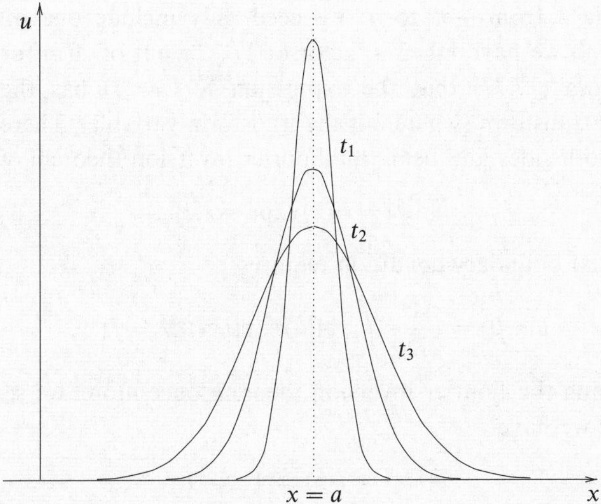


Figure 17.10 Diffusion of heat from a point source in a metal bar for different times $t_1 < t_2 < t_3$. The area under the curves remains constant, since the total heat energy is conserved.

the separation of variables. It should be noted that this behaviour is a result of the infinite range in $x$ rather than of the transform method itself. In fact the method of separation of variables would yield the same solutions, since in the infinite-range case the separation constant is not restricted to take on an infinite set of discrete values but may have any real value, with the result that the sum over $\lambda$ becomes an integral, as mentioned at the end of section 17.2.

► *An infinite metal bar has an initial temperature distribution $f(x)$ along its length. Find the temperature distribution at a later time $t$ using the method of separation of variables.*

This is the same problem as in the previous example, but we now seek a solution by separating variables. From (17.12) a separated solution for the one-dimensional diffusion equation is

$$u(x,t) = [A\exp(i\lambda x) + B\exp(-i\lambda x)]\exp(-\kappa\lambda^2 t),$$

where $-\lambda^2$ is the separation constant. Since the bar is infinite we do not require the solution to take a given form at any finite value of $x$ (for instance at $x = 0$) and so there is no restriction on $\lambda$ other than its being real. Therefore instead of the superposition of such solutions in the form of a sum over allowed values of $\lambda$ we have an integral over all $\lambda$,

$$u(x,t) = \frac{1}{\sqrt{2\pi}}\int_{-\infty}^{\infty} A(\lambda)\exp(-\kappa\lambda^2 t)\exp(i\lambda x)\,d\lambda, \qquad (17.77)$$

where in taking $\lambda$ from $-\infty$ to $\infty$ we need only include one of the complex exponentials, and we have taken a factor of $1/\sqrt{2\pi}$ out of $A(\lambda)$ for convenience. But we see from (17.77) that the expression for $u(x,t)$ has the form of an inverse Fourier transform (where $\lambda$ is the transform variable). Therefore, Fourier-transforming both sides and using the Fourier inversion theorem, we find

$$\widetilde{u}(\lambda,t) = A(\lambda)\exp(-\kappa\lambda^2 t).$$

Now the initial boundary condition requires

$$u(x,0) = \frac{1}{\sqrt{2\pi}}\int_{-\infty}^{\infty} A(\lambda)\exp(i\lambda x)\,d\lambda = f(x),$$

from which, using the Fourier inversion theorem once more, we see that $A(\lambda) = \widetilde{f}(\lambda)$. Therefore we have

$$\widetilde{u}(\lambda,t) = \widetilde{f}(\lambda)\exp(-\kappa\lambda^2 t),$$

which is identical to (17.75) in the previous example (but with $k$ replaced by $\lambda$), and hence leads to the same result. ◂

## 17.5 Inhomogeneous problems – Green's functions

In chapters 13 and 15 we encountered Green's functions as a useful tool in solving inhomogeneous linear ODEs. We now discuss their usefulness in solving inhomogeneous linear PDEs.

For the sake of brevity we shall again denote a linear PDE by

$$\mathcal{L}u(\mathbf{r}) = \rho(\mathbf{r}), \tag{17.78}$$

where $\mathcal{L}$ is a linear partial differential operator. For example in Laplace's equation we have $\mathcal{L} = \nabla^2$, whereas for Helmholtz's equation $\mathcal{L} = \nabla^2 + k^2$. Note that we have not specified the dimensionality of the problem, and (17.78) may, for example, represent Poisson's equation in two or three (or more) dimensions. The reader will also notice that for the sake of simplicity we have not included any time dependence in (17.78). Nevertheless, the following discussion can be easily generalised to include it.

As we discussed in subsection 16.3.2, a problem is inhomogeneous if the fact that $u(\mathbf{r})$ is a solution does *not* imply that any constant multiple $\lambda u(\mathbf{r})$ is also a solution. This inhomogeneity may derive from either the PDE itself or from the boundary conditions imposed on the solution.

In our discussion of Green's function solutions of inhomogeneous ODEs (see subsection 13.2.5) we dealt with inhomogeneous boundary conditions by making a suitable change of variable so that in the new variable the boundary conditions were homogeneous. In an analogous way, as illustrated in the final example of section 17.2, it is usually possible to make a change of variables in PDEs to transform between inhomogeneity of the boundary conditions and inhomogeneity of the equation. Therefore let us for the moment assume that the boundary conditions imposed on the solution $u(\mathbf{r})$ of (17.78) are homogeneous. This most commonly means that, if we seek a solution to (17.78) in some region $V$, then on the surface $S$ that bounds $V$ the solution obeys the conditions $u(\mathbf{r}) = 0$ or $\partial u/\partial n = 0$, where $\partial u/\partial n$ is the normal derivative of $u$ at the surface $S$.

We shall discuss the extension of the Green's function method to the direct solution of problems with inhomogeneous boundary conditions in subsection 17.5.2, but we first highlight how the Green's function approach to solving ODEs can be simply extended to PDEs for homogeneous boundary conditions.

### *17.5.1 Similarities with Green's functions for ODEs*

As in the discussion of ODEs in chapter 13, we may consider the Green's function for a system described by a PDE as the response of the system to a 'unit impulse' or 'point source'. Thus if we seek a solution to (17.78) that satisfies some homogeneous boundary conditions on $u(\mathbf{r})$, then the Green's function $G(\mathbf{r}, \mathbf{r}_0)$ for the problem is a solution of

$$\mathcal{L}G(\mathbf{r}, \mathbf{r}_0) = \delta(\mathbf{r} - \mathbf{r}_0), \qquad (17.79)$$

where $\mathbf{r}_0$ lies in $V$. $G(\mathbf{r}, \mathbf{r}_0)$ must also satisfy the imposed (homogeneous) boundary conditions.

It is understood that in (17.79) the $\mathcal{L}$ operator expresses differentiation with respect to $\mathbf{r}$ as opposed to $\mathbf{r}_0$. Also, $\delta(\mathbf{r} - \mathbf{r}_0)$ is the Dirac delta function (see chapter 11), of dimension appropriate for the problem, and may be thought of as representing a unit-strength point source at $\mathbf{r} = \mathbf{r}_0$.

Following an analogous argument to that given in subsection 13.2.5 for ODEs, if the boundary conditions on $u(\mathbf{r})$ are homogeneous then a solution to (17.78) that satisfies the imposed boundary conditions is given by

$$u(\mathbf{r}) = \int G(\mathbf{r}, \mathbf{r}_0)\rho(\mathbf{r}_0)\, dV(\mathbf{r}_0), \qquad (17.80)$$

where the integral on $\mathbf{r}_0$ is over some appropriate 'volume'. In two or more dimensions, however, the task of finding directly a solution to (17.79) that satisfies the imposed boundary conditions on $S$ can be a difficult one, and we return to this in the next subsection.

An alternative approach is to follow a similar argument to that presented in chapter 15 for ODEs, and construct the Green's function for (17.78) as a superposition of eigenfunctions of the operator $\mathcal{L}$, provided $\mathcal{L}$ is Hermitian. By analogy with an ordinary differential operator, a partial differential operator is

Hermitian if it satisfies

$$\int_V v^*(\mathbf{r})\mathcal{L}w(\mathbf{r})\,dV = \left[\int_V w^*(\mathbf{r})\mathcal{L}v(\mathbf{r})\,dV\right]^*,$$

where the asterisk denotes complex conjugation and $v$ and $w$ are arbitrary functions obeying the imposed (homogeneous) boundary condition on the solution of $\mathcal{L}u(\mathbf{r}) = 0$.

The eigenfunctions $u_n(\mathbf{r})$, $n = 0, 1, 2, \ldots$, of $\mathcal{L}$ satisfy

$$\mathcal{L}u_n(\mathbf{r}) = \lambda_n u_n(\mathbf{r}),$$

where $\lambda_n$ are the corresponding eigenvalues, which are all real for an Hermitian operator $\mathcal{L}$. Furthermore, each eigenfunction must obey any imposed (homogeneous) boundary conditions. Using an argument analogous to that given in chapter 15, the Green's function for the problem is given by

$$G(\mathbf{r}, \mathbf{r}_0) = \sum_{n=0}^{\infty} \frac{u_n(\mathbf{r})u_n^*(\mathbf{r}_0)}{\lambda_n}. \tag{17.81}$$

From (17.81) we immediately see that the Green's function (irrespective of how it is found) enjoys the property

$$G(\mathbf{r}, \mathbf{r}_0) = G^*(\mathbf{r}_0, \mathbf{r}).$$

Thus, if the Green's function is real, it is symmetric in its two arguments.

Once the Green's function has been obtained, the solution to (17.78) is again given by (17.80). For PDEs this approach can become very cumbersome, however, and so we shall not pursue it further here.

### *17.5.2 General boundary-value problems*

As mentioned above, inhomogeneous boundary conditions can often be dealt with by making an appropriate change of variables, such that the boundary conditions in the new variables are homogeneous although the equation itself is generally inhomogeneous. In this section, however, we extend the use of Green's functions to problems with inhomogeneous boundary conditions (and equations). This provides a more consistent and intuitive approach to the solution of such *boundary-value problems*.

For definiteness we shall consider Poisson's equation

$$\nabla^2 u(\mathbf{r}) = \rho(\mathbf{r}), \tag{17.82}$$

but the material of this section may be straightforwardly extended to other linear PDEs of the form (17.78). Clearly, Poisson's equation reduces to Laplace's equation for $\rho(\mathbf{r}) = 0$ and so our discussion is equally applicable to this case.

We wish to solve (17.82) in some region $V$ bounded by a surface $S$, which may

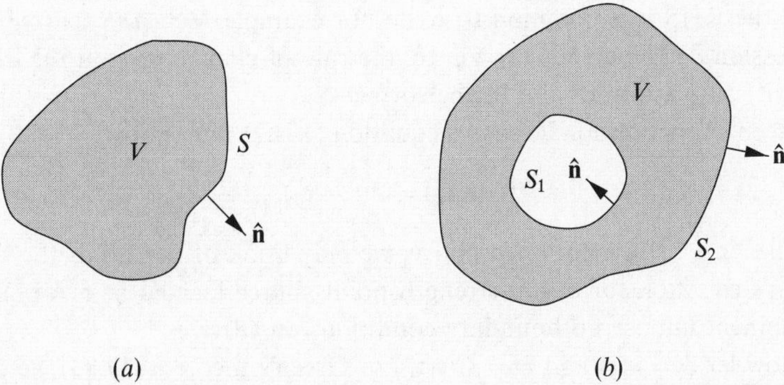


Figure 17.11 Surfaces used for solving Poisson's equation in different regions $V$.

consist of several disconnected parts. As stated above, we shall allow the possibility that the boundary conditions on the solution $u(\mathbf{r})$ may be inhomogeneous on $S$, although as we shall see this method reduces to those discussed above in the special case that the boundary conditions are in fact homogeneous.

The two common types of inhomogeneous boundary condition for Poisson's equation are (as discussed in subsection 16.6.2):

(i) Dirichlet conditions, in which $u(\mathbf{r})$ is specified on $S$, and
(ii) Neumann conditions, in which $\partial u/\partial n$ is specified on $S$.

In general, specifying *both* Dirichlet *and* Neumann conditions on $S$ overdetermines the problem and leads to there being no solution.

The specification of the surface $S$ requires some further comment, since $S$ may have several disconnected parts. If we wish to solve Poisson's equation inside some closed surface $S$, then the situation is straightforward and is shown in figure 17.11($a$). If, however, we wish to solve Poisson's equation in the gap between two closed surfaces (for example in the gap between two concentric conducting cylinders), then the volume $V$ is bounded by a surface $S$ which has two disconnected parts $S_1$ and $S_2$ as shown in figure 17.11($b$); the direction of the normal to the surface is always taken as pointing *out* of the volume $V$. A similar situation arises when we wish to solve Poisson's equation *outside* some closed surface $S_1$. In this case the volume $V$ is infinite, but is treated formally by taking the surface $S_2$ as a large sphere of radius $R$ and letting $R$ tend to infinity.

In order to solve (17.82) subject to either Dirichlet or Neumann boundary conditions on $S$, we first remind ourselves of Green's second theorem, equation (9.20), which states that for two scalar functions $\phi(\mathbf{r})$ and $\psi(\mathbf{r})$ defined in some volume $V$ bounded by a surface $S$,

$$\int_V (\phi\nabla^2\psi - \psi\nabla^2\phi)\,dV = \int_S (\phi\nabla\psi - \psi\nabla\phi)\cdot\hat{\mathbf{n}}\,dS, \qquad (17.83)$$

where on the RHS it is common to write, for example, $\nabla\psi \cdot \hat{\mathbf{n}}\, dS$ as $(\partial\psi/\partial n)\, dS$. The expression $\partial\psi/\partial n$ stands for $\nabla\psi \cdot \hat{\mathbf{n}}$, the rate of change of $\psi$ in the direction of the (unit) outward normal $\hat{\mathbf{n}}$ to the surface $S$.

The Green's function for Poisson's equation (17.82) must satisfy

$$\nabla^2 G(\mathbf{r}, \mathbf{r}_0) = \delta(\mathbf{r} - \mathbf{r}_0), \tag{17.84}$$

where $\mathbf{r}_0$ lies in $V$. (As mentioned above, we may think of $G(\mathbf{r}, \mathbf{r}_0)$ as the solution to Poisson's equation for a unit-strength point source located at $\mathbf{r} = \mathbf{r}_0$.) Let us for the moment impose no boundary conditions on $G(\mathbf{r}, \mathbf{r}_0)$.

If we now let $\phi = u(\mathbf{r})$ and $\psi = G(\mathbf{r}, \mathbf{r}_0)$ in Green's theorem (17.83) we obtain

$$\int_V \left[u(\mathbf{r})\nabla^2 G(\mathbf{r}, \mathbf{r}_0) - G(\mathbf{r}, \mathbf{r}_0)\, \nabla^2 u(\mathbf{r})\right] dV(\mathbf{r}) = \int_S \left[u(\mathbf{r})\frac{\partial G(\mathbf{r}, \mathbf{r}_0)}{\partial n} - G(\mathbf{r}, \mathbf{r}_0)\frac{\partial u(\mathbf{r})}{\partial n}\right] dS(\mathbf{r}),$$

where we have made explicit that the volume and surface integrals are with respect to $\mathbf{r}$. Using (17.82) and (17.84) the LHS can be simplified to give

$$\int_V [u(\mathbf{r})\delta(\mathbf{r} - \mathbf{r}_0) - G(\mathbf{r}, \mathbf{r}_0)\rho(\mathbf{r})]\, dV(\mathbf{r}) = \int_S \left[u(\mathbf{r})\frac{\partial G(\mathbf{r}, \mathbf{r}_0)}{\partial n} - G(\mathbf{r}, \mathbf{r}_0)\frac{\partial u(\mathbf{r})}{\partial n}\right] dS(\mathbf{r}), \tag{17.85}$$

Since $\mathbf{r}_0$ lies within the volume $V$,

$$\int_V u(\mathbf{r})\delta(\mathbf{r} - \mathbf{r}_0)\, dV(\mathbf{r}) = u(\mathbf{r}_0),$$

and thus rearranging (17.85) the solution to Poisson's equation (17.82) can be written

$$u(\mathbf{r}_0) = \int_V G(\mathbf{r}, \mathbf{r}_0)\rho(\mathbf{r})\, dV(\mathbf{r}) + \int_S \left[u(\mathbf{r})\frac{\partial G(\mathbf{r}, \mathbf{r}_0)}{\partial n} - G(\mathbf{r}, \mathbf{r}_0)\frac{\partial u(\mathbf{r})}{\partial n}\right] dS(\mathbf{r}). \tag{17.86}$$

Clearly, we can interchange the roles of $\mathbf{r}$ and $\mathbf{r}_0$ in (17.86) if we wish. (Remember also that for a real Green's function $G(\mathbf{r}, \mathbf{r}_0) = G(\mathbf{r}_0, \mathbf{r})$.)

Equation (17.86) is *central* to the extension of the Green's function method to problems with inhomogeneous boundary conditions, and we next discuss its application to both Dirichlet and Neumann boundary-value problems. But, before doing so, we also note that if the boundary condition on $S$ is in fact homogeneous, so that $u(\mathbf{r}) = 0$ or $\partial u(\mathbf{r})/\partial n = 0$ on $S$, then demanding that the Green's function $G(\mathbf{r}, \mathbf{r}_0)$ also obeys the same boundary condition causes the surface integral in (17.86) to vanish and we are left with the familiar form of solution given in (17.80). The extension of (17.86) to PDEs other than Poisson's equation is discussed in exercise 17.14.

### *17.5.3 Dirichlet problems*

In a Dirichlet problem we require the solution $u(\mathbf{r})$ of Poisson's equation (17.82) to take specific values on some surface $S$ that bounds $V$, i.e. we require $u(\mathbf{r}) = f(\mathbf{r})$ on $S$ where $f$ is a given function.

If we seek a Green's function for this problem it must clearly satisfy (17.84), but we are free to choose the boundary conditions satisfied by $G(\mathbf{r}, \mathbf{r}_0)$ in such a way as to make the solution (17.86) as simple as possible. From (17.86), we see that by choosing

$$G(\mathbf{r}, \mathbf{r}_0) = 0 \qquad \text{for } \mathbf{r} \text{ on } S, \tag{17.87}$$

the second term in the surface integral vanishes. Since $u(\mathbf{r}) = f(\mathbf{r})$ on $S$, (17.86) then becomes

$$u(\mathbf{r}_0) = \int_V G(\mathbf{r}, \mathbf{r}_0)\rho(\mathbf{r})\, dV(\mathbf{r}) + \int_S f(\mathbf{r}) \frac{\partial G(\mathbf{r}, \mathbf{r}_0)}{\partial n}\, dS(\mathbf{r}). \tag{17.88}$$

Thus we wish to find the *Dirichlet Green's function* that

(i) satisfies (17.84) and hence is singular at $\mathbf{r} = \mathbf{r}_0$, and
(ii) obeys the boundary condition $G(\mathbf{r}, \mathbf{r}_0) = 0$ for $\mathbf{r}$ on $S$.

It is, in general, difficult to obtain this function directly, and so it is useful to separate these two requirements. We therefore look for a solution of the form

$$G(\mathbf{r}, \mathbf{r}_0) = F(\mathbf{r}, \mathbf{r}_0) + H(\mathbf{r}, \mathbf{r}_0),$$

where $F(\mathbf{r}, \mathbf{r}_0)$ satisfies (17.84) and has the required singular character at $\mathbf{r} = \mathbf{r}_0$, but does not necessarily obey the boundary condition on $S$, whilst $H(\mathbf{r}, \mathbf{r}_0)$ satisfies the corresponding homogeneous equation (i.e. Laplace's equation) inside $V$, but is adjusted so that the sum $G(\mathbf{r}, \mathbf{r}_0)$ equals zero on $S$. $G(\mathbf{r}, \mathbf{r}_0)$ is still a solution of (17.84) since

$$\nabla^2 G(\mathbf{r}, \mathbf{r}_0) = \nabla^2 F(\mathbf{r}, \mathbf{r}_0) + \nabla^2 H(\mathbf{r}, \mathbf{r}_0) = \nabla^2 F(\mathbf{r}, \mathbf{r}_0) + 0 = \delta(\mathbf{r} - \mathbf{r}_0).$$

The function $F(\mathbf{r}, \mathbf{r}_0)$ is called the *fundamental solution*, and will clearly take different forms depending on the dimensionality of the problem. Let us first consider the fundamental solution to (17.84) in three dimensions.

►*Find the fundamental solution to Poisson's equation in three dimensions that tends to zero as* $|\mathbf{r}| \to \infty$.

We wish to solve

$$\nabla^2 F(\mathbf{r}, \mathbf{r}_0) = \delta(\mathbf{r} - \mathbf{r}_0) \tag{17.89}$$

in three dimensions, subject to the boundary condition $F(\mathbf{r}, \mathbf{r}_0) \to 0$ as $|\mathbf{r}| \to \infty$. Since the problem is spherically symmetric about $\mathbf{r}_0$, let us consider a large sphere

$S$ of radius $R$ centred on $\mathbf{r}_0$, and integrate (17.89) over the enclosed volume $V$. We then obtain

$$\int_V \nabla^2 F(\mathbf{r}, \mathbf{r}_0)\, dV = \int_V \delta(\mathbf{r} - \mathbf{r}_0)\, dV = 1, \qquad (17.90)$$

since $V$ encloses the point $\mathbf{r}_0$. However, using the divergence theorem,

$$\int_V \nabla^2 F(\mathbf{r}, \mathbf{r}_0)\, dV = \int_S \nabla F(\mathbf{r}, \mathbf{r}_0) \cdot \hat{\mathbf{n}}\, dS, \qquad (17.91)$$

where $\hat{\mathbf{n}}$ is the unit normal to the large sphere $S$ at any point.

Since the problem is spherically symmetric about $\mathbf{r}_0$, we expect

$$F(\mathbf{r}, \mathbf{r}_0) = F(|\mathbf{r} - \mathbf{r}_0|) = F(r),$$

i.e. it has the same value everywhere on $S$. Thus evaluating the surface integral in (17.91) and equating it to unity from (17.90) we have†

$$4\pi r^2 \left.\frac{dF}{dr}\right|_{r=R} = 1.$$

Integrating this expression we obtain

$$F(r) = -\frac{1}{4\pi r} + \text{constant},$$

but, since we require $F(\mathbf{r}, \mathbf{r}_0) \to 0$ as $|\mathbf{r}| \to \infty$, the constant must be zero. The fundamental solution in three dimensions is consequently given by

$$F(\mathbf{r}, \mathbf{r}_0) = -\frac{1}{4\pi |\mathbf{r} - \mathbf{r}_0|}. \qquad (17.92)$$

This is clearly also the full Green's function for Poisson's equation subject to the boundary condition $u(\mathbf{r}) \to 0$ as $|\mathbf{r}| \to \infty$. ◀

Using (17.92) we can write down the solution of Poisson's equation to find e.g. the electrostatic potential $u(\mathbf{r})$ due to some distribution of electric charge $\rho(\mathbf{r})$. The electrostatic potential satisfies

$$\nabla^2 u = -\frac{\rho}{\epsilon_0},$$

where $u \to 0$ as $|\mathbf{r}| \to \infty$. Since the boundary condition on the surface at infinity is homogeneous the surface integral in (17.88) vanishes, and using (17.92) we recover the familiar solution

$$u(\mathbf{r}_0) = \int \frac{\rho(\mathbf{r})}{4\pi\epsilon_0 |\mathbf{r} - \mathbf{r}_0|}\, dV(\mathbf{r}),$$

where the volume integral is over all space.

† A vertical bar to the right of an expression is a common alternative notation to enclosing the expression in square brackets; as usual, the subscript shows the value of the variable at which the expression is to be evaluated.

We can develop an analogous theory in two dimensions. As before the fundamental solution satisfies

$$\nabla^2 F(\mathbf{r}, \mathbf{r}_0) = \delta(\mathbf{r} - \mathbf{r}_0), \tag{17.93}$$

where $\delta(\mathbf{r}-\mathbf{r}_0)$ is now the two-dimensional delta function. Following an analogous method to that used in the previous example, we find the fundamental solution in two dimensions to be given by

$$F(\mathbf{r}, \mathbf{r}_0) = \frac{1}{2\pi} \ln |\mathbf{r} - \mathbf{r}_0| + \text{constant}. \tag{17.94}$$

From the form of the solution we see that in two dimensions we cannot apply the condition $F(\mathbf{r}, \mathbf{r}_0) \to 0$ as $|\mathbf{r}| \to \infty$, and in this case the constant does not necessarily vanish.

We now return to the task of constructing the full Dirichlet Green's function. To do so we wish to add to the fundamental solution a solution of the homogeneous equation (in this case Laplace's equation) such that $G(\mathbf{r}, \mathbf{r}_0) = 0$ on $S$, as required by (17.88) and its attendant conditions. The appropriate Green's function is constructed by adding to the fundamental solution 'copies' of itself that represent 'image' sources at different locations *outside* $V$. Hence this approach is called the *method of images.*

In summary, if we wish to solve Poisson's equation in some region $V$ subject to Dirichlet boundary conditions on its surface $S$, the procedure and argument are as follows.

(i) To the single source $\delta(\mathbf{r} - \mathbf{r}_0)$ inside $V$ add image sources *outside* $V$

$$\sum_{n=1}^{N} q_n \delta(\mathbf{r} - \mathbf{r}_n), \qquad \text{with } \mathbf{r}_n \text{ outside } V,$$

where the positions $\mathbf{r}_n$ and the strengths $q_n$ of the image sources are to be determined as described in step (iii) below.

(ii) Since all the image sources lie outside $V$, the fundamental solution corresponding to each source satisfies Laplace's equation *inside* $V$. Thus we may add the fundamental solutions $F(\mathbf{r}, \mathbf{r}_n)$ corresponding to each image source to that corresponding to the single source inside $V$, to obtain the Green's function

$$G(\mathbf{r}, \mathbf{r}_0) = F(\mathbf{r}, \mathbf{r}_0) + \sum_{n=1}^{N} q_n F(\mathbf{r}, \mathbf{r}_n).$$

(iii) Adjust the positions $\mathbf{r}_n$ and strengths $q_n$ of the image sources so that the required boundary conditions are satisfied on $S$. For a Dirichlet Green's function we require $G(\mathbf{r}, \mathbf{r}_0) = 0$ for $\mathbf{r}$ on $S$.

(iv) The solution to Poisson's equation subject to the Dirichlet boundary condition $u(\mathbf{r}) = f(\mathbf{r})$ on $S$ is then given by (17.88).

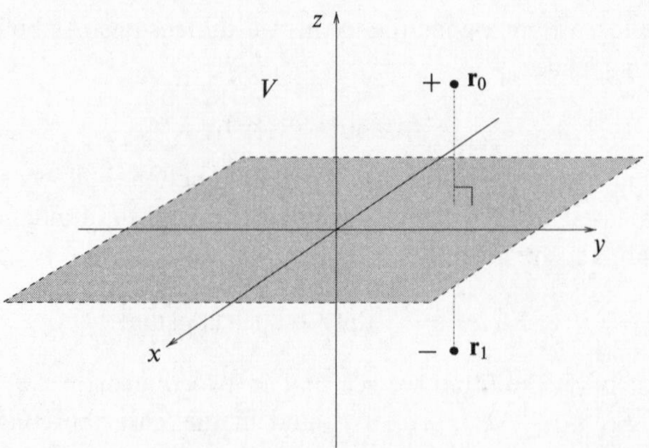


Figure 17.12 The arrangement of images for solving Laplace's equation in the half-space $z > 0$.

In general it is very difficult to find the correct positions and strengths for the images, i.e. such that the boundary conditions on $S$ are satisfied. Nevertheless, it is possible in certain problems having simple geometry. In particular, for problems in which the boundary $S$ consists of straight lines (in two dimensions) or planes (in three dimensions), positions of the image points can be deduced simply by imagining the boundary lines or planes to be mirrors in which the single source in $V$ (at $\mathbf{r}_0$) is reflected.

►*Solve Laplace's equation* $\nabla^2 u = 0$ *in three dimensions in the half-space* $z > 0$*, given that* $u(\mathbf{r}) = f(\mathbf{r})$ *on the plane* $z = 0$.

The surface $S$ bounding $V$ consists of the $xy$-plane and the surface at infinity. Therefore, the Dirichlet Green's function for this problem must satisfy $G(\mathbf{r}, \mathbf{r}_0) = 0$ on $z = 0$ and $G(\mathbf{r}, \mathbf{r}_0) \to 0$ as $|\mathbf{r}| \to \infty$. Thus it is clear in this case that we require one image source at a position $\mathbf{r}_1$ which is the reflection of $\mathbf{r}_0$ in the plane $z = 0$, as shown in figure 17.12 (so that $\mathbf{r}_1$ lies in $z < 0$, outside the region in which we wish to obtain a solution). It is also clear that the strength of this image should be $-1$.

Therefore by adding the fundamental solutions corresponding to the original source and its image we obtain the Green's function

$$G(\mathbf{r}, \mathbf{r}_0) = -\frac{1}{4\pi|\mathbf{r} - \mathbf{r}_0|} + \frac{1}{4\pi|\mathbf{r} - \mathbf{r}_1|}, \tag{17.95}$$

where $\mathbf{r}_1$ is the reflection of $\mathbf{r}_0$ in the plane $z = 0$, i.e. if $\mathbf{r}_0 = (x_0, y_0, z_0)$ then $\mathbf{r}_1 = (x_0, y_0, -z_0)$. Clearly $G(\mathbf{r}, \mathbf{r}_0) \to 0$ as $|\mathbf{r}| \to \infty$ as required. Also $G(\mathbf{r}, \mathbf{r}_0) = 0$ on $z = 0$, and so (17.95) is the desired Dirichlet Green's function.

The solution to Laplace's equation is then given by (17.88) with $\rho(\mathbf{r}) = 0$,

$$u(\mathbf{r}_0) = \int_S f(\mathbf{r}) \frac{\partial G(\mathbf{r}, \mathbf{r}_0)}{\partial n} \, dS(\mathbf{r}). \qquad (17.96)$$

Clearly the surface at infinity makes no contribution to this integral. The outward-pointing unit vector normal to the $xy$-plane is simply $\hat{\mathbf{n}} = -\mathbf{k}$ (where $\mathbf{k}$ is the unit vector in the $z$-direction), so that

$$\frac{\partial G(\mathbf{r}, \mathbf{r}_0)}{\partial n} = -\frac{\partial G(\mathbf{r}, \mathbf{r}_0)}{\partial z} = -\mathbf{k} \cdot \nabla G(\mathbf{r}, \mathbf{r}_0).$$

We may evaluate this normal derivative by writing the Green's function (17.95) explicitly in terms of $x$, $y$ and $z$ (and $x_0$, $y_0$ and $z_0$) and calculating the partial derivative with respect to $z$ directly. It is usually quicker, however, to use the fact that†

$$\nabla |\mathbf{r} - \mathbf{r}_0| = \frac{\mathbf{r} - \mathbf{r}_0}{|\mathbf{r} - \mathbf{r}_0|}, \qquad (17.97)$$

and so we find

$$\nabla G(\mathbf{r}, \mathbf{r}_0) = \frac{\mathbf{r} - \mathbf{r}_0}{4\pi |\mathbf{r} - \mathbf{r}_0|^3} - \frac{\mathbf{r} - \mathbf{r}_1}{4\pi |\mathbf{r} - \mathbf{r}_1|^3}.$$

Since $\mathbf{r}_0 = (x_0, y_0, z_0)$ and $\mathbf{r}_1 = (x_0, y_0, -z_0)$ the normal derivative is given by

$$\begin{aligned} -\frac{\partial G(\mathbf{r}, \mathbf{r}_0)}{\partial z} &= -\mathbf{k} \cdot \nabla G(\mathbf{r}, \mathbf{r}_0) \\ &= -\frac{z - z_0}{4\pi |\mathbf{r} - \mathbf{r}_0|^3} + \frac{z + z_0}{4\pi |\mathbf{r} - \mathbf{r}_1|^3}. \end{aligned}$$

Therefore on the surface $z = 0$, writing out the dependence on $x$, $y$ and $z$ explicitly, we have

$$-\left.\frac{\partial G(\mathbf{r}, \mathbf{r}_0)}{\partial z}\right|_{z=0} = \frac{2z_0}{4\pi [(x - x_0)^2 + (y - y_0)^2 + z_0^2]^{3/2}}.$$

Inserting this expression into (17.96) we obtain the solution

$$u(x_0, y_0, z_0) = \frac{z_0}{2\pi} \int_{-\infty}^{\infty} \int_{-\infty}^{\infty} \frac{f(x, y)}{[(x - x_0)^2 + (y - y_0)^2 + z_0^2]^{3/2}} \, dx \, dy. \blacktriangleleft$$

An analogous procedure may be applied in two-dimensional problems. For example, in solving Poisson's equation in two dimensions in the half-space $x > 0$, we again require just one image charge, of strength $q_1 = -1$, at a position $\mathbf{r}_1$ that is the reflection of $\mathbf{r}_0$ in the line $x = 0$. Since we require $G(\mathbf{r}, \mathbf{r}_0) = 0$ when $\mathbf{r}$ lies

† Since $|\mathbf{r} - \mathbf{r}_0|^2 = (\mathbf{r} - \mathbf{r}_0) \cdot (\mathbf{r} - \mathbf{r}_0)$ we have $\nabla |\mathbf{r} - \mathbf{r}_0|^2 = 2(\mathbf{r} - \mathbf{r}_0)$, from which we obtain

$$\nabla (|\mathbf{r} - \mathbf{r}_0|^2)^{1/2} = \frac{1}{2} \frac{2(\mathbf{r} - \mathbf{r}_0)}{(|\mathbf{r} - \mathbf{r}_0|^2)^{1/2}} = \frac{\mathbf{r} - \mathbf{r}_0}{|\mathbf{r} - \mathbf{r}_0|}.$$

Note that this result holds in two *and* three dimensions.

on $x = 0$, the constant in (17.94) must equal zero, and so the Dirichlet Green's function is

$$G(\mathbf{r}, \mathbf{r}_0) = \frac{1}{2\pi}\left(\ln|\mathbf{r} - \mathbf{r}_0| - \ln|\mathbf{r} - \mathbf{r}_1|\right).$$

Clearly $G(\mathbf{r}, \mathbf{r}_0)$ tends to zero as $|\mathbf{r}| \to \infty$. If we wish to solve the two-dimensional Poisson equation in the quarter space $x > 0$, $y > 0$, however, more image points are required.

► *A line charge in the z-direction of charge density $\lambda$ is placed at some position $\mathbf{r}_0$ in the quarter-space $x > 0$, $y > 0$. Calculate the force per unit length on the line charge due to the presence of thin earthed plates along $x = 0$ and $y = 0$.*

Here we wish to solve Poisson's equation

$$\nabla^2 u = -\frac{\lambda}{\epsilon_0}\delta(\mathbf{r} - \mathbf{r}_0)$$

in the quarter space $x > 0$, $y > 0$. It is clear that we require three image line charges with positions and strengths as shown in figure 17.13 (all of which lie outside the region in which we seek a solution). The boundary condition that the electrostatic potential $u = 0$ on $x = 0$ and $y = 0$ (shown as the 'curve' $C$ in figure 17.13) is then automatically satisfied, and so this system of image charges is directly equivalent to the original situation of a single line charge in the presence of the earthed plates along $x = 0$ and $y = 0$. Thus the electrostatic potential is simply equal to the Dirichlet Green's function

$$u(\mathbf{r}) = G(\mathbf{r}, \mathbf{r}_0) = -\frac{\lambda}{2\pi\epsilon_0}\left(\ln|\mathbf{r} - \mathbf{r}_0| - \ln|\mathbf{r} - \mathbf{r}_1| + \ln|\mathbf{r} - \mathbf{r}_2| - \ln|\mathbf{r} - \mathbf{r}_3|\right),$$

which equals zero on $C$ and on the 'surface' at infinity.

The force on the line charge at $\mathbf{r}_0$ is therefore simply that due to the three line charges at $\mathbf{r}_1$, $\mathbf{r}_2$ and $\mathbf{r}_3$. The elecrostatic potential due to a line charge at $\mathbf{r}_i$ ($i = 1$, 2 or 3) is given by the fundamental solution

$$u_i(\mathbf{r}) = \mp\frac{\lambda}{2\pi\epsilon_0}\ln|\mathbf{r} - \mathbf{r}_i| + c,$$

depending on whether the line charge is positive or negative respectively. Therefore the force per unit length on the line charge at $\mathbf{r}_0$, due to the one at $\mathbf{r}_i$, is given by

$$-\lambda\nabla u_i(\mathbf{r})\Big|_{\mathbf{r}=\mathbf{r}_0} = \pm\frac{\lambda^2}{2\pi\epsilon_0}\frac{\mathbf{r}_0 - \mathbf{r}_i}{|\mathbf{r}_0 - \mathbf{r}_i|^2}.$$

Adding the contributions from the three image charges shown in figure 17.13, the total force experienced by the line charge at $\mathbf{r}_0$ is

$$\mathbf{F} = \frac{\lambda^2}{2\pi\epsilon_0}\left(-\frac{\mathbf{r}_0 - \mathbf{r}_1}{|\mathbf{r}_0 - \mathbf{r}_1|^2} + \frac{\mathbf{r}_0 - \mathbf{r}_2}{|\mathbf{r}_0 - \mathbf{r}_2|^2} - \frac{\mathbf{r}_0 - \mathbf{r}_3}{|\mathbf{r}_0 - \mathbf{r}_3|^2}\right),$$

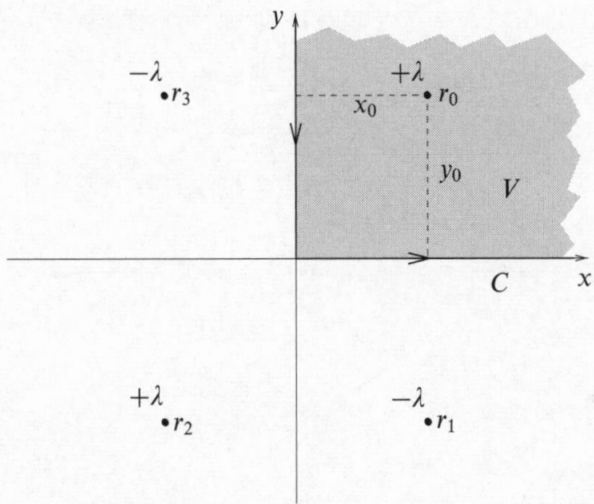


Figure 17.13 The arrangement of images for finding the force on a line charge situated in the (two-dimensional) quarter-space $x > 0$, $y > 0$, when the planes $x = 0$ and $y = 0$ are earthed.

where, from the figure, $\mathbf{r}_0 - \mathbf{r}_1 = 2y_0\mathbf{j}$, $\mathbf{r}_0 - \mathbf{r}_2 = 2x_0\mathbf{i} + 2y_0\mathbf{j}$ and $\mathbf{r}_0 - \mathbf{r}_3 = 2x_0\mathbf{i}$. Thus, in terms of $x_0$ and $y_0$, the total force on the line charge due to the charge induced on the plates is given by

$$\begin{aligned}\mathbf{F} &= \frac{\lambda^2}{2\pi\epsilon_0}\left[-\frac{1}{2y_0}\mathbf{j} + \frac{2x_0\,\mathbf{i} + 2y_0\,\mathbf{j}}{(4x_0^2 + 4y_0^2)} - \frac{1}{2x_0}\mathbf{i}\right] \\ &= -\frac{\lambda^2}{4\pi\epsilon_0(x_0^2 + y_0^2)}\left[\frac{y_0^2}{x_0}\mathbf{i} + \frac{x_0^2}{y_0}\mathbf{j}\right]. \blacktriangleleft\end{aligned}$$

Further generalisations are possible. For instance, solving Poisson's equation in the two-dimensional strip $-\infty < x < \infty$, $0 < y < b$ requires an infinite series of image points.

So far we have considered problems in which the boundary $S$ consists of straight lines (in two dimensions) or planes (in three dimensions), in which simple reflection of the source at $\mathbf{r}_0$ in these boundaries fixes the positions of the image points. For more complicated (curved) boundaries this is no longer possible, and finding the appropriate position of the image source(s) requires further work.

► *Use the method of images to find the Dirichlet Green's function for solving Poisson's equation outside a sphere of radius a centred at the origin.*

We need to find a solution to Poisson's equation valid outside the sphere of radius $a$. Since an image point $\mathbf{r}_1$ cannot lie in this region, it must be located within the sphere. The Green's function for this problem is therefore

$$G(\mathbf{r}, \mathbf{r}_0) = -\frac{1}{4\pi|\mathbf{r} - \mathbf{r}_0|} - \frac{q}{4\pi|\mathbf{r} - \mathbf{r}_1|},$$

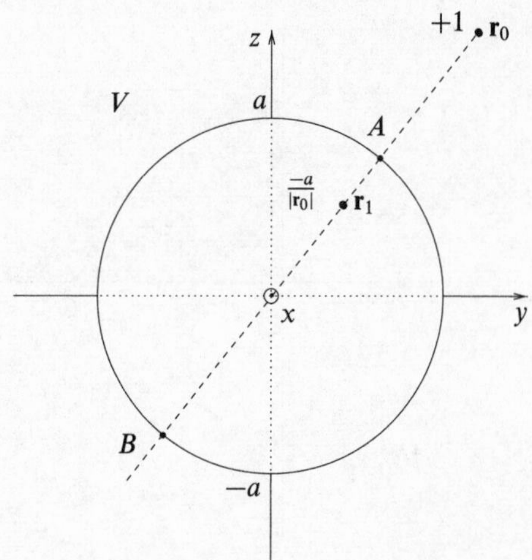


Figure 17.14 The arrangement of images for solving Poisson's equation outside a sphere of radius $a$ centred at the origin.

where $|\mathbf{r}_0| > a$, $|\mathbf{r}_1| < a$ and $q$ is the strength of the image which we have yet to determine. Clearly, $G(\mathbf{r}, \mathbf{r}_0) \to 0$ on the surface at infinity.

By symmetry we expect the image point to lie on the same radial line as the original source, $\mathbf{r}_0$, as shown in figure 17.14, so that $\mathbf{r}_1 = k\mathbf{r}_0$ where $k < 1$. However, for a Dirichlet Green's function we require $G(\mathbf{r} - \mathbf{r}_0) = 0$ on $|\mathbf{r}| = a$, and the form of the Green's function suggests that we need

$$|\mathbf{r} - \mathbf{r}_0| \propto |\mathbf{r} - \mathbf{r}_1| \quad \text{for all } |\mathbf{r}| = a. \tag{17.98}$$

Referring to figure 17.14, if this relationship is to hold over the whole surface of the sphere, then it must certainly hold for the points $A$ and $B$. We thus require

$$\frac{|\mathbf{r}_0| - a}{a - |\mathbf{r}_1|} = \frac{|\mathbf{r}_0| + a}{a + |\mathbf{r}_1|},$$

which reduces to $|\mathbf{r}_1| = a^2/|\mathbf{r}_0|$. Therefore the image point must be located at the position

$$\mathbf{r}_1 = \frac{a^2}{|\mathbf{r}_0|^2}\mathbf{r}_0.$$

It may now be checked that, for this location of the image point, (17.98) is satisfied over the whole sphere. Using the geometrical result

$$\begin{aligned}|\mathbf{r} - \mathbf{r}_1|^2 &= |\mathbf{r}|^2 - \frac{2a^2}{|\mathbf{r}_0|^2}\mathbf{r}\cdot\mathbf{r}_0 + \frac{a^4}{|\mathbf{r}_0|^2} \\ &= \frac{a^2}{|\mathbf{r}_0|^2}\left(|\mathbf{r}_0|^2 - 2\mathbf{r}\cdot\mathbf{r}_0 + a^2\right) \quad \text{for } |\mathbf{r}| = a,\end{aligned} \tag{17.99}$$

we see that, on the surface of the sphere,

$$|\mathbf{r} - \mathbf{r}_1| = \frac{a}{|\mathbf{r}_0|}|\mathbf{r} - \mathbf{r}_0| \quad \text{for } |\mathbf{r}| = a. \tag{17.100}$$

Therefore, in order that $G = 0$ at $|\mathbf{r}| = a$, the strength of the image charge must be $-a/|\mathbf{r}_0|$. Consequently, the Dirichlet Green's function for the exterior of the sphere is

$$G(\mathbf{r}, \mathbf{r}_0) = -\frac{1}{4\pi|\mathbf{r} - \mathbf{r}_0|} + \frac{a/|\mathbf{r}_0|}{4\pi|\mathbf{r} - (a^2/|\mathbf{r}_0|^2)\mathbf{r}_0|}.$$

For a less formal treatment of the same problem see exercise 17.13. ◄

If we seek solutions to Poisson's equation in the *interior* of a sphere then the above analysis still holds, but $\mathbf{r}$ and $\mathbf{r}_0$ are now inside the sphere and the image $\mathbf{r}_1$ lies outside it.

Using similar arguments to those in the previous example, for two-dimensional Dirichlet problems outside the circle $|\mathbf{r}| = a$, we use the same image point as in the three-dimensional case, namely

$$\mathbf{r}_1 = \frac{a^2}{|\mathbf{r}_0|^2}\mathbf{r}_0. \tag{17.101}$$

As illustrated below, however, it is usual to take the image strength as $-1$ in two-dimensional problems.

►*Solve Laplace's equation in the two-dimensional region $|\mathbf{r}| \leq a$, subject to the boundary condition $u = f(\phi)$ on $|\mathbf{r}| = a$.*

In this case we wish to find the Dirichlet Green's function in the interior of a disc of radius $a$, so the image charge must lie outside the disc. Taking the strength of the image to be $-1$, we have

$$G(\mathbf{r}, \mathbf{r}_0) = \frac{1}{2\pi}\ln|\mathbf{r} - \mathbf{r}_0| - \frac{1}{2\pi}\ln|\mathbf{r} - \mathbf{r}_1| + c,$$

where $\mathbf{r}_1 = (a^2/|\mathbf{r}_0|^2)\mathbf{r}_0$ lies outside the disc, and $c$ is a constant that does not necessarily equal zero.

Since we require $G(\mathbf{r}, \mathbf{r}_0) = 0$ when $|\mathbf{r}| = a$, this fixes the value of the constant $c$, and the Dirichlet Green's function for this problem is thus given by

$$G(\mathbf{r}, \mathbf{r}_0) = \frac{1}{2\pi}\left(\ln|\mathbf{r} - \mathbf{r}_0| - \ln\left|\mathbf{r} - \frac{a^2}{|\mathbf{r}_0|^2}\mathbf{r}_0\right| - \ln\frac{|\mathbf{r}_0|}{a}\right), \tag{17.102}$$

where $c$ includes the magnitude of the image charge. Using plane polar coordinates, the solution to the boundary-value problem can be written as a line integral around the circle $\rho = a$:

$$
\begin{aligned}
u(\mathbf{r}_0) &= \int_C f(\mathbf{r}) \frac{\partial G(\mathbf{r}, \mathbf{r}_0)}{\partial n}\, dl \\
&= \int_0^{2\pi} f(\mathbf{r}) \left. \frac{\partial G(\mathbf{r}, \mathbf{r}_0)}{\partial \rho} \right|_{\rho=a} a\, d\phi. \qquad (17.103)
\end{aligned}
$$

The normal derivative of the Green's function (17.102) is given by

$$
\begin{aligned}
\frac{\partial G(\mathbf{r}, \mathbf{r}_0)}{\partial \rho} &= \frac{\mathbf{r}}{|\mathbf{r}|} \cdot \nabla G(\mathbf{r}, \mathbf{r}_0) \\
&= \frac{\mathbf{r}}{2\pi|\mathbf{r}|} \cdot \left( \frac{\mathbf{r} - \mathbf{r}_0}{|\mathbf{r} - \mathbf{r}_0|^2} - \frac{\mathbf{r} - \mathbf{r}_1}{|\mathbf{r} - \mathbf{r}_1|^2} \right). \qquad (17.104)
\end{aligned}
$$

Using the fact that $\mathbf{r}_1 = (a^2/|\mathbf{r}_0|^2)\mathbf{r}_0$ and the geometrical result (17.100), we find that

$$
\left. \frac{\partial G(\mathbf{r}, \mathbf{r}_0)}{\partial \rho} \right|_{\rho=a} = \frac{a^2 - |\mathbf{r}_0|^2}{2\pi a |\mathbf{r} - \mathbf{r}_0|^2}.
$$

In plane polar coordinates, $\mathbf{r} = \rho \cos\phi\, \mathbf{i} + \rho \sin\phi\, \mathbf{j}$ and $\mathbf{r}_0 = \rho_0 \cos\phi_0\, \mathbf{i} + \rho_0 \sin\phi_0\, \mathbf{j}$, and so

$$
\left. \frac{\partial G(\mathbf{r}, \mathbf{r}_0)}{\partial \rho} \right|_{\rho=a} = \left( \frac{1}{2\pi a} \right) \frac{a^2 - \rho_0^2}{a^2 + \rho_0^2 - 2a\rho_0 \cos(\phi - \phi_0)}.
$$

On substituting into (17.103), we obtain

$$
u(\rho_0, \phi_0) = \frac{1}{2\pi} \int_0^{2\pi} \frac{(a^2 - \rho_0^2) f(\phi)\, d\phi}{a^2 + \rho_0^2 - 2a\rho_0 \cos(\phi - \phi_0)}, \qquad (17.105)
$$

which is the solution to the problem. ◄

### 17.5.4 Neumann problems

In a Neumann problem we require the normal derivative of the solution of Poisson's equation to take on specific values on some surface $S$ that bounds $V$, i.e. we require $\partial u(\mathbf{r})/\partial n = f(\mathbf{r})$ on $S$, where $f$ is a given function. As we shall see, much of our discussion of Dirichlet problems can be immediately taken over into the solution of Neumann problems.

As we proved in section 16.7 of the previous chapter, specifying Neumann boundary conditions determines the relevant solution of Poisson's equation to within an (unimportant) additive constant. Unlike Dirichlet conditions, Neumann conditions impose a self-consistency requirement. In order for a solution $u$ to exist, it is necessary that the following consistency condition holds:

$$
\int_S f\, dS = \int_S \nabla u \cdot \hat{\mathbf{n}}\, dS = \int_V \nabla^2 u\, dV = \int_V \rho\, dV, \qquad (17.106)
$$

where we have used the divergence theorem to convert the surface integral into a volume integral. As a physical example, the integral of the normal component of an electric field over a surface bounding a given volume cannot be chosen arbitrarily when the charge inside the volume has already been specified (Gauss's theorem).

Let us again consider (17.86), which is central to our discussion of Green's functions in inhomogeneous problems. It reads

$$u(\mathbf{r}_0) = \int_V G(\mathbf{r}, \mathbf{r}_0)\rho(\mathbf{r})\, dV(\mathbf{r}) + \int_S \left[ u(\mathbf{r}) \frac{\partial G(\mathbf{r}, \mathbf{r}_0)}{\partial n} - G(\mathbf{r}, \mathbf{r}_0) \frac{\partial u(\mathbf{r})}{\partial n} \right] dS(\mathbf{r}).$$

As always, the Green's function must obey

$$\nabla^2 G(\mathbf{r}, \mathbf{r}_0) = \delta(\mathbf{r} - \mathbf{r}_0),$$

where $\mathbf{r}_0$ lies in $V$. In the solution of Dirichlet problems in the previous subsection, we chose the Green's function to obey the boundary condition $G(\mathbf{r}, \mathbf{r}_0) = 0$ on $S$ and, in a similar way, we might wish to choose $\partial G(\mathbf{r}, \mathbf{r}_0)/\partial n = 0$ in the solution of Neumann problems. However, in general this is *not* permitted since the Green's function must obey the consistency condition

$$\int_S \frac{\partial G(\mathbf{r}, \mathbf{r}_0)}{\partial n}\, dS = \int_S \nabla G(\mathbf{r}, \mathbf{r}_0) \cdot \hat{\mathbf{n}}\, dS = \int_V \nabla^2 G(\mathbf{r}, \mathbf{r}_0)\, dV = 1.$$

The simplest permitted boundary condition is therefore

$$\frac{\partial G(\mathbf{r}, \mathbf{r}_0)}{\partial n} = \frac{1}{A} \qquad \text{for } \mathbf{r} \text{ on } S,$$

where $A$ is the area of the surface $S$; this defines a *Neumann Green's function.*

If we require $\partial u(\mathbf{r})/\partial n = f(\mathbf{r})$ on $S$, the solution to Poisson's equation is given by

$$\begin{aligned} u(\mathbf{r}_0) &= \int_V G(\mathbf{r}, \mathbf{r}_0)\rho(\mathbf{r})\, dV(\mathbf{r}) + \frac{1}{A}\int_S u(\mathbf{r})\, dS(\mathbf{r}) - \int_S G(\mathbf{r}, \mathbf{r}_0) f(\mathbf{r})\, dS(\mathbf{r}) \\ &= \int_V G(\mathbf{r}, \mathbf{r}_0)\rho(\mathbf{r})\, dV(\mathbf{r}) + \langle u(\mathbf{r})\rangle_S - \int_S G(\mathbf{r}, \mathbf{r}_0) f(\mathbf{r})\, dS(\mathbf{r}), \end{aligned} \qquad (17.107)$$

where $\langle u(\mathbf{r})\rangle_S$ is the average of $u$ over the surface $S$, and is a freely specifiable constant. For Neumann problems in which the volume $V$ is bounded by a surface $S$ at infinity, we do not need the $\langle u(\mathbf{r})\rangle_S$ term. For example, if we wish to solve a Neumann problem outside the unit sphere centred at the origin, then $r > a$ is the region $V$ throughout which we require the solution; this region may be considered as being bounded by two disconnected surfaces, the surface of the sphere and a surface at infinity. By requiring that $u(\mathbf{r}) \to 0$ as $|\mathbf{r}| \to \infty$, the term $\langle u(\mathbf{r})\rangle_S$ becomes zero.

Much of our discussion of Dirichlet problems can be immediately taken over into the solution of Neumann problems. In particular, we may again use the method of images to find the appropriate Neumann Green's function.

► *Solve Laplace's equation in the two-dimensional region $|\mathbf{r}| \le a$ subject to the boundary condition $\partial u/\partial n = f(\phi)$ on $|\mathbf{r}| = a$ (with $\int_0^{2\pi} f(\phi)\,d\phi = 0$ as required by the consistency condition (17.106)).*

Let us assume, as in Dirichlet problems with this geometry, that a single image charge is placed outside the circle at

$$\mathbf{r}_1 = \frac{a^2}{|\mathbf{r}_0|^2}\mathbf{r}_0,$$

where $\mathbf{r}_0$ is the position of the source inside the circle (see equation (17.101)). Then, from (17.100), we have the useful geometrical result

$$|\mathbf{r} - \mathbf{r}_1| = \frac{a}{|\mathbf{r}_0|}|\mathbf{r} - \mathbf{r}_0| \quad \text{for } |\mathbf{r}| = a. \tag{17.108}$$

Leaving the strength $q$ of the image as a parameter, the Green's function has the form

$$G(\mathbf{r}, \mathbf{r}_0) = \frac{1}{2\pi}\left(\ln|\mathbf{r} - \mathbf{r}_0| + q\ln|\mathbf{r} - \mathbf{r}_1| + c\right). \tag{17.109}$$

Using plane polar coordinates, the radial (normal) derivative of this function is then given by

$$\begin{aligned}\frac{\partial G(\mathbf{r}, \mathbf{r}_0)}{\partial \rho} &= \frac{\mathbf{r}}{|\mathbf{r}|} \cdot \nabla G(\mathbf{r}, \mathbf{r}_0) \\ &= \frac{\mathbf{r}}{2\pi|\mathbf{r}|} \cdot \left[\frac{\mathbf{r} - \mathbf{r}_0}{|\mathbf{r} - \mathbf{r}_0|^2} + \frac{q(\mathbf{r} - \mathbf{r}_1)}{|\mathbf{r} - \mathbf{r}_1|^2}\right].\end{aligned}$$

Using (17.108), on the circumference of the circle $\rho = a$, the radial derivative is

$$\begin{aligned}\left.\frac{\partial G(\mathbf{r}, \mathbf{r}_0)}{\partial \rho}\right|_{\rho=a} &= \frac{1}{2\pi|\mathbf{r}|}\left[\frac{|\mathbf{r}|^2 - \mathbf{r}\cdot\mathbf{r}_0}{|\mathbf{r} - \mathbf{r}_0|^2} + \frac{q|\mathbf{r}|^2 - q(a^2/|\mathbf{r}_0|^2)\mathbf{r}\cdot\mathbf{r}_0}{(a^2/|\mathbf{r}_0|^2)|\mathbf{r} - \mathbf{r}_0|^2}\right] \\ &= \frac{1}{2\pi a}\frac{1}{|\mathbf{r} - \mathbf{r}_0|^2}\left[|\mathbf{r}|^2 + q|\mathbf{r}_0|^2 - (1+q)\mathbf{r}\cdot\mathbf{r}_0\right],\end{aligned}$$

where we have set $|\mathbf{r}|^2 = a^2$ in the second term on the RHS, but not in the first. If we take $q = 1$, the radial derivative simplifies to

$$\left.\frac{\partial G(\mathbf{r}, \mathbf{r}_0)}{\partial \rho}\right|_{\rho=a} = \frac{1}{2\pi a},$$

or $1/L$ where $L$ is the length of the circumference, and so (17.109) with $q = 1$ is the required Neumann Green's function.

Since $\rho(\mathbf{r}) = 0$, the solution to our boundary-value problem is now given by (17.107) as

$$u(\mathbf{r}_0) = \langle u(\mathbf{r})\rangle_C - \int_C G(\mathbf{r}, \mathbf{r}_0)f(\mathbf{r})\,dl(\mathbf{r}),$$

where the integral is around the circumference of the circle $C$. In plane polar coordinates $\mathbf{r} = \rho\cos\phi\,\mathbf{i} + \rho\sin\phi\,\mathbf{j}$ and $\mathbf{r}_0 = \rho_0\cos\phi_0\,\mathbf{i} + \rho_0\sin\phi_0\,\mathbf{j}$, and again using (17.108) we find that on $C$ the Green's function is given by

$$\begin{aligned}
G(\mathbf{r},\mathbf{r}_0)|_{\rho=a} &= \frac{1}{2\pi}\left[\ln|\mathbf{r}-\mathbf{r}_0| + \ln\left(\frac{a}{|\mathbf{r}_0|}|\mathbf{r}-\mathbf{r}_0|\right) + c\right] \\
&= \frac{1}{2\pi}\left(\ln|\mathbf{r}-\mathbf{r}_0|^2 + \ln\frac{a}{|\mathbf{r}_0|} + c\right) \\
&= \frac{1}{2\pi}\left\{\ln\left[a^2+\rho_0^2 - 2a\rho_0\cos(\phi-\phi_0)\right] + \ln\frac{a}{\rho_0} + c\right\}. \qquad (17.110)
\end{aligned}$$

Since $dl = a\,d\phi$ on $C$, the solution to the problem is given by

$$u(\rho_0,\phi_0) = \langle u\rangle_C - \frac{a}{2\pi}\int_0^{2\pi} f(\phi)\ln[a^2+\rho_0^2-2a\rho_0\cos(\phi-\phi_0)]\,d\phi.$$

The contributions of the final two terms terms in the Green's function (17.110) vanish because $\int_0^{2\pi} f(\phi)\,d\phi = 0$. The average value of $u$ around the circumference, $\langle u\rangle_C$, is a freely specifiable constant as we would expect for a Neumann problem. This result should be compared with the result (17.105) for the corresponding Dirichlet problem, but it should be remembered that in the one case $f(\phi)$ is a potential, and in the other the gradient of a potential. ◄

## 17.6 Exercises

17.1 Solve the following first-order partial differential equations by separating the variables.

$$\text{(a)}\ \frac{\partial u}{\partial x} - x\frac{\partial u}{\partial y} = 0, \qquad \text{(b)}\ x\frac{\partial u}{\partial x} - 2y\frac{\partial u}{\partial y} = 0.$$

17.2 The wave equation describing the transverse vibrations of a stretched membrane under tension $T$ and having a uniform surface density $\rho$ is

$$T\left(\frac{\partial^2 u}{\partial x^2} + \frac{\partial^2 u}{\partial y^2}\right) = \rho\frac{\partial^2 u}{\partial t^2}.$$

Find a separable solution appropriate to a membrane stretched on a frame of length $a$ and width $b$, showing that the natural angular frequencies of such a membrane are

$$\omega^2 = \frac{\pi^2 T}{\rho}\left(\frac{n^2}{a^2} + \frac{m^2}{b^2}\right),$$

where $n$ and $m$ are any positive integers.

17.3 Schrödinger's equation for a non-relativistic particle in a constant potential region can be taken as

$$-\frac{\hbar^2}{2m}\left(\frac{\partial^2 u}{\partial x^2} + \frac{\partial^2 u}{\partial y^2} + \frac{\partial^2 u}{\partial z^2}\right) = i\hbar\frac{\partial u}{\partial t}.$$

(a) Find a solution, separable in the four independent variables, that can be written in the form of a plane wave,

$$\psi(x, y, z, t) = A \exp[i(\mathbf{k} \cdot \mathbf{r} - \omega t)].$$

Using the relationships associated with de Broglie ($\mathbf{p} = \hbar\mathbf{k}$) and Einstein ($E = \hbar\omega$), show that the separation constants must be such that

$$p_x^2 + p_y^2 + p_z^2 = 2mE.$$

(b) Obtain a different separable solution describing a particle confined to a box of side $a$ ($\psi$ must vanish at the walls of the box). Show that the energy of the particle can only take the quantised values

$$E = \frac{\hbar^2\pi^2}{2ma^2}(n_x^2 + n_y^2 + n_z^2),$$

where $n_x$, $n_y$, $n_z$ are integers.

17.4 Denoting the three terms of $\nabla^2$ in spherical polars by $\nabla_r^2$, $\nabla_\theta^2$, $\nabla_\phi^2$ in an obvious way, evaluate $\nabla_r^2 u$, etc. for the two functions given below and verify that, in each case, although the individual terms are not necessarily zero their sum $\nabla^2 u$ is zero. Identify the corresponding values of $\ell$ and $m$.

(a) $u(r, \theta, \phi) = \left(Ar^2 + \dfrac{B}{r^3}\right) \dfrac{3\cos^2\theta - 1}{2}$.

(b) $u(r, \theta, \phi) = \left(Ar + \dfrac{B}{r^2}\right) \sin\theta \exp i\phi$.

17.5 Use the expressions at the end of subsection 17.3.2 to verify for $\ell = 0, 1, 2$ that

$$\sum_{m=-\ell}^{\ell} |Y_\ell^m(\theta, \phi)|^2 = \frac{2\ell + 1}{4\pi},$$

and so is independent of the values of $\theta$ and $\phi$. This is true for any $\ell$, but a general proof is more involved. This result helps to reconcile intuition with the apparently arbitrary choice of polar axis in a general quantum mechanical system.

17.6 Express the function

$$f(\theta, \phi) = \sin\theta[\sin^2(\theta/2)\cos\phi + i\cos^2(\theta/2)\sin\phi] + \sin^2(\theta/2)$$

as a sum of spherical harmonics.

17.7 The free transverse vibrations of a thick rod satisfy the equation

$$a^4 \frac{\partial^4 u}{\partial x^4} + \frac{\partial^2 u}{\partial t^2} = 0.$$

Obtain a solution in separated-variable form and, for a rod clamped at one end, $x = 0$, and free at the other, $x = L$, show that the angular frequency of vibration $\omega$ satisfies

$$\cosh\left(\frac{\omega^{1/2}L}{a}\right) = -\sec\left(\frac{\omega^{1/2}L}{a}\right).$$

(At a clamped end both $u$ and $\partial u/\partial x$ vanish, whilst at a free end $\partial^2 u/\partial x^2$ and $\partial^3 u/\partial x^3$ do so.)

17.8 A membrane is stretched between two concentric rings of radii $a$ and $b$ ($b > a$). If the smaller ring is transversely distorted from the planar configuration by an amount $c|\phi|$, $-\pi \le \phi \le \pi$, show that the membrane then has a shape given by

$$u(\rho, \phi) = \frac{c\pi}{2}\frac{\ln(b/\rho)}{\ln(b/a)} - \frac{4c}{\pi}\sum_{m\ \text{odd}} \frac{a^m}{m^2(b^{2m} - a^{2m})}\left(\frac{b^{2m}}{\rho^m} - \rho^m\right)\cos m\phi.$$

17.9 A string of length $L$, fixed at its two ends, is plucked at its mid-point by an amount $A$ and then released. Prove that the subsequent displacement is given by

$$u(x,t) = \sum_{n=0}^{\infty} \frac{8A}{\pi^2(2n+1)^2}\sin\left[\frac{(2n+1)\pi x}{L}\right]\cos\left[\frac{(2n+1)\pi ct}{L}\right],$$

where, in the usual notation, $c^2 = T/\rho$.

Find the total kinetic energy of the string when it passes through its unplucked position, by calculating it in each mode (each $n$) and summing, using the result $\sum_0^\infty (2n+1)^{-2} = \pi^2/8$. Confirm that the total energy is equal to the work done in plucking the string initially.

17.10 Prove that the potential for $\rho < a$ associated with a vertical split cylinder of radius $a$, the two halves of which ($\cos\phi > 0$ and $\cos\phi < 0$) are maintained at equal and opposite potentials $\pm V$, is given by

$$u(\rho, \phi) = \frac{4V}{\pi}\sum_{n=0}^{\infty}\frac{(-1)^n}{2n+1}\left(\frac{\rho}{a}\right)^{2n+1}\cos(2n+1)\phi.$$

17.11 Two identical copper bars are each of length $a$. Initially, one is at 0 °C and the other at 100 °C; they are then joined together end to end and thermally isolated. Obtain in the form of a Fourier series an expression $u(x,t)$ for the temperature at any point a distance $x$ from the join at a later time $t$. (Bear in mind the heat flow conditions at the free ends of the bars.)

Taking $a = 0.5$ m estimate the time it takes for one of the free ends to attain a temperature of 55 °C. The thermal conductivity of copper is $3.8 \times 10^2$ J m$^{-1}$ K$^{-1}$ s$^{-1}$, and its specific heat capacity is $3.4 \times 10^6$ J m$^{-3}$ K$^{-1}$.

17.12 (a) Show that the gravitational potential due to a uniform disc of radius $a$ and mass $M$, centred at the origin, is given for $r < a$ by

$$\frac{2GM}{a}\left[1-\frac{r}{a}P_1(\cos\theta)+\frac{1}{2}\left(\frac{r}{a}\right)^2 P_2(\cos\theta)-\frac{1}{8}\left(\frac{r}{a}\right)^4 P_4(\cos\theta)+\cdots\right],$$

and for $r > a$ by

$$\frac{GM}{r}\left[1-\frac{1}{4}\left(\frac{a}{r}\right)^2 P_2(\cos\theta)+\frac{1}{8}\left(\frac{a}{r}\right)^4 P_4(\cos\theta)-\cdots\right],$$

where the polar axis is normal to the plane of the disc.

(b) Reconcile the presence of a term $P_1(\cos\theta)$, which is odd under $\theta \to \pi - \theta$, with the symmetry with respect to the plane of the disc of the physical system.

(c) Deduce that the gravitational field near an infinite sheet of matter of constant density $\rho$ per unit area is $2\pi G\rho$.

17.13 Point charges $q$ and $-qa/b$ (with $a < b$) are placed respectively at a point $P$, a distance $b$ from the origin $O$, and a point $Q$ between $O$ and $P$, a distance $a^2/b$ from $O$. Show, by considering similar triangles $QOS$ and $SOP$, where $S$ is any point on the surface of the sphere centred at $O$ and of radius $a$, that the net potential anywhere on the sphere due to the two charges is zero.

Use this result (backed up by the uniqueness theorem) to find the force with which a point charge $q$ placed a distance $b$ from the centre of a spherical conductor of radius $a$ ($< b$) is attracted to the sphere (i) if the sphere is earthed, and (ii) if the sphere is uncharged and insulated.

17.14 Consider the PDE $\mathcal{L}u(\mathbf{r}) = \rho(\mathbf{r})$, for which the differential operator $\mathcal{L}$ is given by

$$\mathcal{L} = \nabla \cdot [p(\mathbf{r})\nabla] + q(\mathbf{r}),$$

where $p(\mathbf{r})$ and $q(\mathbf{r})$ are functions of position. By proving the generalised form of Green's theorem,

$$\int_V (\phi\mathcal{L}\psi - \psi\mathcal{L}\phi)\,dV = \oint_S p(\phi\nabla\psi - \psi\nabla\phi)\cdot\hat{\mathbf{n}}\,dS,$$

show that the solution of the PDE is given by

$$u(\mathbf{r}_0) = \int_V G(\mathbf{r},\mathbf{r}_0)\rho(\mathbf{r})\,dV(\mathbf{r}) + \oint_S p(\mathbf{r})\left[u(\mathbf{r})\frac{\partial G(\mathbf{r},\mathbf{r}_0)}{\partial n} - G(\mathbf{r},\mathbf{r}_0)\frac{\partial u(\mathbf{r})}{\partial n}\right] dS(\mathbf{r}),$$

where $G(\mathbf{r},\mathbf{r}_0)$ is the Green's function satisfying $\mathcal{L}G(\mathbf{r},\mathbf{r}_0) = \delta(\mathbf{r}-\mathbf{r}_0)$.

## 17.7 Hints and answers

17.1 (a) $C\exp[\lambda(x^2+2y)]$; (b) $C(x^2y)^\lambda$.

17.2 $u(x,y,t)=\sin(n\pi x/a)\sin(m\pi y/b)(A\sin\omega t+B\cos\omega t)$.

17.3 (a) $-\dfrac{\hbar^2}{2m}\dfrac{X''}{X}=\dfrac{p_x^2}{2m}$, etc., $\dfrac{i\hbar T'}{T}=E$;
(b) As in (a), but with solutions $X=A\sin(p_x x/\hbar)$, etc. with $p_x a/\hbar=n_x\pi$.

17.4 (a) $6u/r^2$, $-6u/r^2$, 0, $\ell=2$, $m=0$. (b) $2u/r^2$, $(\cot^2\theta-1)u/r^2$; $-u/(r^2\sin^2\theta)$, $\ell=1$, $m=1$.

17.6 The first term can only contain $\ell=1,2$ and $m=\pm1$, the second only $\ell=0,1,2$ and $m=0$. $f(\theta,\phi)=(\pi)^{1/2}[Y_0^0-3^{-1/2}Y_1^0-(2/3)^{1/2}Y_1^1-(2/15)^{1/2}Y_2^{-1}]$.

17.7 $(A\cos mx+B\sin mx+C\cosh mx+D\sinh mx)\cos(\omega t+\epsilon)$, with $m^4a^4=\omega^2$.

17.9 $E_n=16\rho A^2c^2/[(2n+1)^2\pi^2L]$; $E=2\rho c^2A^2/L=\int_0^A 2Tv/(L/2)\,dv$.

17.11 Since there is no heat flow at $x=\pm a$, use a series of period $4a$, $u(x,0)=100$ for $0<x\leq 2a$, $u(x,0)=0$ for $-2a\leq x<0$.

$$u(x,t)=50+\frac{200}{\pi}\sum_{n=0}^{\infty}\frac{1}{2n+1}\sin\left[\frac{(2n+1)\pi x}{2a}\right]\exp\left[-\frac{k(2n+1)^2\pi^2 t}{4a^2s}\right].$$

Taking only the $n=0$ term gives $t\approx 2300$ s.

17.12 (a) $u(r=z,0)=2MGa^{-2}[(a^2+z^2)^{1/2}-z]$. (b) For $\theta>\pi/2$, the factor in the square brackets is $(a^2+z^2)^{1/2}+z$. (c) Find $\partial u/\partial r$ at $\theta=0$ for $r<a$, and let $a\to\infty$.

17.13 (i) $q^2ab/[4\pi\epsilon_0(b^2-a^2)^2]$; (ii) $[q^2ab/4\pi\epsilon_0][(b^2-a^2)^{-2}-b^{-4}]$. Obtain (ii) from (i) by adding a further image charge $+qa/b$ at $O$, to give a net zero electrostatic flux from the sphere while maintaining its equipotential property.

# 18

# Complex variables

Throughout this book references have been made to results derived from the theory of complex variables. This theory thus becomes an integral part of the mathematics appropriate to physical applications. The difficulty with it, from the point of view of a book such as the present one, is that although it is of much practical application its underlying basis has a distinctly pure mathematics flavour.

Thus, to adopt a rigorous approach would involve developing a large amount of groundwork in analysis, for example precise definitions of continuity and differentiability, the theory of sets and a detailed study of boundedness. Instead, we will pursue here only those parts of the formal theory that are needed to establish the results used elsewhere in this book and some others of general utility.

In this spirit, the proofs that have been adopted for some of the standard results of complex variable theory have been chosen with an eye to simplicity rather than sophistication. This means that in some cases the imposed conditions are more stringent than would be strictly necessary if more sophisticated proofs were used; where this happens the less restrictive results are usually stated as well. The reader who is interested in a fuller treatment should consult one of the many excellent textbooks on this fascinating subject.†

One further concession to 'hand-waving' has been made in the interests of keeping the treatment to a moderate length. In several places phrases such as 'can be made as small as we like' are used, rather than a careful treatment in terms of 'given $\epsilon > 0$, there exists a $\delta > 0$ such that'. In the authors' experience, some students are more at ease with the former type of statement despite its lack of

† For example, Knopp, *Theory of Functions, Part I* (Dover, 1945); Phillips, *Functions of a Complex Variable* (Oliver and Boyd, 1954); Titchmarsh, *The Theory of Functions* (Oxford, 1952).

precision whilst others, those who would contemplate only the latter, are usually well able to supply it for themselves.

## 18.1 Functions of a complex variable

The quantity $f(z)$ is said to be a function of the complex variable $z$ if to every value of $z$ in a certain domain $R$ (a region of the Argand diagram) there corresponds one or more values of $f(z)$. Stated like this $f(z)$ could be any function consisting of a real and an imaginary part, each of which is, in general, itself a function of $x$ and $y$. If we denote the real and imaginary parts of $f(z)$ by $u$ and $v$ respectively, then

$$f(z) = u(x, y) + iv(x, y).$$

In this chapter, however, we will be primarily concerned with functions that are single-valued, so that for each value of $z$ there corresponds just one value of $f(z)$, and differentiable in a particular sense, which we now discuss.

A function $f(z)$ that is single-valued in some domain $R$ is *differentiable* at the point $z$ in $R$ if the *derivative*

$$f'(z) = \lim_{\Delta z \to 0} \left[ \frac{f(z + \Delta z) - f(z)}{\Delta z} \right], \tag{18.1}$$

exists and is unique, in that its value does not depend upon the direction in the Argand diagram from which $\Delta z$ tends to zero.

►*Show that the function $f(z) = x^2 - y^2 + i2xy$ is differentiable for all values of $z$.*

Considering the definition (18.1), and taking $\Delta z = \Delta x + i\Delta y$, we have

$$\begin{aligned}
\frac{f(z + \Delta z) - f(z)}{\Delta z} \\
&= \frac{(x + \Delta x)^2 - (y + \Delta y)^2 + 2i(x + \Delta x)(y + \Delta y) - x^2 + y^2 - 2ixy}{\Delta x + i\Delta y} \\
&= \frac{2x\Delta x + (\Delta x)^2 - 2y\Delta y - (\Delta y)^2 + 2i(x\Delta y + y\Delta x + \Delta x\Delta y)}{\Delta x + i\Delta y} \\
&= 2x + i2y + \frac{(\Delta x)^2 - (\Delta y)^2 + 2i\Delta x\Delta y}{\Delta x + i\Delta y}.
\end{aligned}$$

Now, in whatever way $\Delta x$ and $\Delta y$ are allowed to tend to zero (e.g. taking $\Delta y = 0$ and letting $\Delta x \to 0$ or vice versa), the last term on the right will tend to zero and the unique limit $2x + i2y$ will be obtained. Since $z$ was arbitrary, $f(z)$ with $u = x^2 - y^2$ and $v = 2xy$ is differentiable at all points in the (finite) complex plane. ◄

We note in the above example that the working can be considerably reduced by recognising that, since $z = x + iy$, we can write

$$f(z) = x^2 - y^2 + 2ixy = (x + iy)^2 = z^2.$$

We then find

$$\begin{aligned} f'(z) = \lim_{\Delta z \to 0} \left[ \frac{(z + \Delta z)^2 - z^2}{\Delta z} \right] &= \lim_{\Delta z \to 0} \left[ \frac{(\Delta z)^2 + 2z\Delta z}{\Delta z} \right] \\ &= \left( \lim_{\Delta z \to 0} \Delta z \right) + 2z = 2z, \end{aligned}$$

from which we see immediately that the limit both exists and is independent of the way in which $\Delta z \to 0$. Thus we have verified that $f(z) = z^2$ is differentiable for all (finite) $z$. We also note that the derivative is analogous to that found for real variables.

Although the definition of a differentiable function clearly includes a wide class of functions, the concept of differentiability is restrictive, and indeed some functions are not differentiable at any point in the complex plane.

►*Show that the function $f(z) = 2y + ix$ is not differentiable anywhere in the complex plane.*

In this case $f(z)$ cannot be written simply in terms of $z$, and so we must consider the limit (18.1) in terms of $x$ and $y$ explicitly. Following the same procedure as in the previous example we find

$$\begin{aligned} \frac{f(z + \Delta z) - f(z)}{\Delta z} &= \frac{2y + 2\Delta y + ix + i\Delta x - 2y - ix}{\Delta x + i\Delta y} \\ &= \frac{2\Delta y + i\Delta x}{\Delta x + i\Delta y}. \end{aligned}$$

In this case the limit will clearly depend on the direction from which $\Delta z \to 0$. Suppose $\Delta z \to 0$ along a line through $z$ of slope $m$, so that $\Delta y = m\Delta x$, then

$$\lim_{\Delta z \to 0} \left[ \frac{f(z + \Delta z) - f(z)}{\Delta z} \right] = \lim_{\Delta x,\, \Delta y \to 0} \left[ \frac{2\Delta y + i\Delta x}{\Delta x + i\Delta y} \right] = \frac{2m + i}{1 + im}.$$

This limit is dependent on $m$ and hence on the direction from which $\Delta z \to 0$. Since this conclusion is independent of the value of $z$, and hence true for all $z$, $f(z) = 2y + ix$ is nowhere differentiable. ◄

A function that is single-valued and differentiable at all points of a domain $R$ is said to be *analytic* (or *regular*) in $R$. A function may be analytic in a domain except at a finite number of points (or an infinite number if the domain is infinite); in this case it is said to be analytic except at these points, which are called the *singularities* of $f(z)$. (In our treatment we will not consider cases in which an infinite number of singularities occur in a finite domain.)

►*Show that the function $f(z) = 1/(1-z)$ is analytic everywhere except at $z = 1$.*

Since $f(z)$ is given explicitly as a function of $z$, evaluation of the limit (18.1) is somewhat easier, and we find

$$\begin{aligned} f'(z) &= \lim_{\Delta z \to 0} \left[ \frac{f(z+\Delta z) - f(z)}{\Delta z} \right] \\ &= \lim_{\Delta z \to 0} \left[ \frac{1}{\Delta z} \left( \frac{1}{1 - z - \Delta z} - \frac{1}{1-z} \right) \right] \\ &= \lim_{\Delta z \to 0} \left[ \frac{1}{(1 - z - \Delta z)(1-z)} \right] = \frac{1}{(1-z)^2}, \end{aligned}$$

independently of the way in which $\Delta z \to 0$, provided $z \neq 1$. Hence $f(z)$ is analytic everywhere except at the singularity $z = 1$. ◄

## 18.2 The Cauchy–Riemann relations

From examining the previous examples, it is apparent that for a function $f(z)$ to be differentiable and hence analytic there must be some particular connection between its real and imaginary parts $u$ and $v$. We next establish what this connection must be, by considering a general function.

If the limit

$$L = \lim_{\Delta z \to 0} \left[ \frac{f(z+\Delta z) - f(z)}{\Delta z} \right], \qquad (18.2)$$

is to exist and be unique, in the way required for differentiability, then two particular ways of letting $\Delta z \to 0$, by moving parallel to the real axis or by moving parallel to the imaginary axis, must produce the same limit. This is certainly a necessary condition, although it may not be sufficient.

If we let $f(z) = u(x, y) + iv(x, y)$ and $\Delta z = \Delta x + i\Delta y$, then we have

$$f(z + \Delta z) = u(x + \Delta x,\ y + \Delta y) + iv(x + \Delta x,\ y + \Delta y),$$

and the limit (18.2) is given by

$$L = \lim_{\Delta x, \Delta y \to 0} \left[ \frac{u(x + \Delta x, y + \Delta y) + iv(x + \Delta x, y + \Delta y) - u(x, y) - iv(x, y)}{\Delta x + i\Delta y} \right].$$

If we first suppose that $\Delta z$ is purely real so that $\Delta y = 0$, we obtain

$$L = \lim_{\Delta x \to 0} \left[ \frac{u(x + \Delta x, y) - u(x, y)}{\Delta x} + i\frac{v(x + \Delta x, y) - v(x, y)}{\Delta x} \right] = \frac{\partial u}{\partial x} + i\frac{\partial v}{\partial x}, \qquad (18.3)$$

provided each limit exists at the point $z$. Similarly, if $\Delta z$ is taken as pure imaginary, so that $\Delta x = 0$, we find

$$L = \lim_{\Delta y \to 0} \left[ \frac{u(x, y + \Delta y) - u(x, y)}{i\Delta y} + i\frac{v(x, y + \Delta y) - v(x, y)}{i\Delta y} \right] = \frac{1}{i}\frac{\partial u}{\partial y} + \frac{\partial v}{\partial y}. \tag{18.4}$$

For $f$ to be differentiable, expressions (18.3) and (18.4) must be identical, and thus equating real and imaginary parts we must have, as a *necessary* condition, that at the point $z$

$$\frac{\partial u}{\partial x} = \frac{\partial v}{\partial y} \quad \text{and} \quad \frac{\partial v}{\partial x} = -\frac{\partial u}{\partial y}. \tag{18.5}$$

These two equations are known as the *Cauchy–Riemann relations.*

We can now see for the earlier examples why (i) $f(z) = x^2 - y^2 + i2xy$ was differentiable, or at least why (ii) $f(z) = 2y + ix$ was not.

(i) $u = x^2 - y^2$, $v = 2xy$:

$$\frac{\partial u}{\partial x} = 2x = \frac{\partial v}{\partial y} \quad \text{and} \quad \frac{\partial v}{\partial x} = 2y = -\frac{\partial u}{\partial y},$$

(ii) $u = 2y$, $v = x$:

$$\frac{\partial u}{\partial x} = 0 = \frac{\partial v}{\partial y} \quad \text{but} \quad \frac{\partial v}{\partial x} = 1 \neq -2 = -\frac{\partial u}{\partial y}.$$

It is apparent that for $f(z)$ to be analytic something more than the existence of the partial derivatives of $u$ and $v$ with respect to $x$ and $y$ is required; this something is that they satisfy the Cauchy–Riemann relations.

We may also enquire as to the *sufficient* conditions for $f(z)$ to be analytic in $R$. It can be shown† that a sufficient condition is that the four partial derivatives exist, *are continuous* and satisfy the Cauchy–Riemann relations. It is the additional requirement of continuity that makes the difference between the necessary conditions and the sufficient conditions.

►*In which domain(s) of the complex plane is $f(z) = |x| - i|y|$ an analytic function?*

Writing $f = u + iv$ it is clear that both $\partial u/\partial y$ and $\partial v/\partial x$ are zero in all four quadrants and hence the second Cauchy–Riemann relation in (18.5) is satisfied everywhere.

Turning to the first Cauchy–Riemann relation, in the first quadrant ($x > 0$, $y > 0$) we have $f(z) = x - iy$ so that

$$\frac{\partial u}{\partial x} = 1, \qquad \frac{\partial v}{\partial y} = -1,$$

† See for example any of the references given earlier.

which clearly violates the first relation in (18.5). Thus $f(z)$ is not analytic in the first quadrant.

Following a similiar argument for the other quadrants, we find

$$\frac{\partial u}{\partial x} = -1 \quad \text{or} \quad +1 \qquad \text{for } x < 0 \text{ and } x > 0 \text{ respectively,}$$

$$\frac{\partial v}{\partial y} = -1 \quad \text{or} \quad +1 \qquad \text{for } y > 0 \text{ and } y < 0 \text{ respectively.}$$

Therefore $\partial u/\partial x$ and $\partial v/\partial y$ are equal, and hence $f(z)$ is analytic, only in the second and fourth quadrants. ◄

Since $x$ and $y$ are related to $z$ and its complex conjugate $z^*$ by

$$x = \frac{1}{2}(z + z^*) \qquad \text{and} \qquad y = \frac{1}{2i}(z - z^*), \tag{18.6}$$

we may formally regard any function $f = u + iv$ as a function of $z$ and $z^*$, rather than $x$ and $y$. If we do this and examine $\partial f/\partial z^*$ we obtain

$$\begin{aligned} \frac{\partial f}{\partial z^*} &= \frac{\partial f}{\partial x}\frac{\partial x}{\partial z^*} + \frac{\partial f}{\partial y}\frac{\partial y}{\partial z^*} \\ &= \left(\frac{\partial u}{\partial x} + i\frac{\partial v}{\partial x}\right)\left(\frac{1}{2}\right) + \left(\frac{\partial u}{\partial y} + i\frac{\partial v}{\partial y}\right)\left(-\frac{1}{2i}\right) \\ &= \frac{1}{2}\left(\frac{\partial u}{\partial x} - \frac{\partial v}{\partial y}\right) + \frac{i}{2}\left(\frac{\partial v}{\partial x} + \frac{\partial u}{\partial y}\right). \end{aligned} \tag{18.7}$$

Now, if $f$ is analytic the Cauchy–Riemann relations (18.5) must be satisfied, and these immediately give that $\partial f/\partial z^*$ is identically zero. Thus we conclude that if $f$ is analytic then $f$ cannot be a function of $z^*$ and any expression representing an analytic function of $z$ can contain $x$ and $y$ only in the combination $x + iy$, *not* in the combination $x - iy$.

We conclude this section by discussing some properties of analytic functions that are of great practical importance in theoretical physics. These can be obtained simply from the requirement that the Cauchy–Riemann relations must be satisfied by the real and imaginary parts of an analytic function.

The most important of these results can be obtained by differentiating the first Cauchy–Riemann relation with respect to one independent variable, and the second with respect to the other independent variable, to obtain

$$\frac{\partial}{\partial x}\left(\frac{\partial u}{\partial x}\right) = \frac{\partial}{\partial x}\left(\frac{\partial v}{\partial y}\right) = \frac{\partial}{\partial y}\left(\frac{\partial v}{\partial x}\right) = -\frac{\partial}{\partial y}\left(\frac{\partial u}{\partial y}\right), \qquad \text{and}$$

$$\frac{\partial}{\partial x}\left(\frac{\partial v}{\partial x}\right) = -\frac{\partial}{\partial x}\left(\frac{\partial u}{\partial y}\right) = -\frac{\partial}{\partial y}\left(\frac{\partial u}{\partial x}\right) = -\frac{\partial}{\partial y}\left(\frac{\partial v}{\partial y}\right).$$

Thus *both* $u$ and $v$ are separately solutions of Laplace's equation in two dimensions, i.e.

$$\frac{\partial^2 u}{\partial x^2} + \frac{\partial^2 u}{\partial y^2} = 0 \qquad \text{and} \qquad \frac{\partial^2 v}{\partial x^2} + \frac{\partial^2 v}{\partial y^2} = 0. \tag{18.8}$$

We shall make use of this result in section 18.9.

A further useful result concerns the two families of curves $u(x, y) =$ constant and $v(x, y) =$ constant, where $u$ and $v$ are the real and imaginary parts of any analytic function $f = u + iv$. As discussed in chapter 8, the vector normal to the curve $u(x, y) =$ constant is given by

$$\nabla u = \frac{\partial u}{\partial x}\mathbf{i} + \frac{\partial u}{\partial y}\mathbf{j}, \tag{18.9}$$

where $\mathbf{i}$ and $\mathbf{j}$ are the unit vectors along the $x$- and $y$- axes respectively. A similar expression exists for $\nabla v$, the normal to the curve $v(x, y) =$ constant. Taking the scalar product of these two normal vectors we obtain

$$\begin{aligned}\nabla u \cdot \nabla v &= \frac{\partial u}{\partial x}\frac{\partial v}{\partial x} + \frac{\partial u}{\partial y}\frac{\partial v}{\partial y} \\ &= -\frac{\partial u}{\partial x}\frac{\partial u}{\partial y} + \frac{\partial u}{\partial y}\frac{\partial u}{\partial x} = 0,\end{aligned}$$

where in the last line we have used the Cauchy–Riemann relations to rewrite the partial derivatives of $v$ as partial derivatives of $u$. Since the scalar product of the normal vectors is zero, they must be orthogonal, and the curves $u(x, y) =$ constant and $v(x, y) =$ constant must therefore intersect at *right angles*.

►*Use the Cauchy–Riemann relations to show that for any analytic function $f = u + iv$, the relation $|\nabla u| = |\nabla v|$ must hold.*

From (18.9) we have

$$|\nabla u|^2 = \nabla u \cdot \nabla u = \left(\frac{\partial u}{\partial x}\right)^2 + \left(\frac{\partial u}{\partial y}\right)^2.$$

Using the Cauchy–Riemann relations to write the partial derivatives of $u$ in terms of those of $v$, we obtain

$$|\nabla u|^2 = \left(\frac{\partial v}{\partial y}\right)^2 + \left(\frac{\partial v}{\partial x}\right)^2 = |\nabla v|^2,$$

from which the result $|\nabla u| = |\nabla v|$ follows immediately. ◄

## 18.3 Power series in a complex variable

The theory of power series in a real variable was considered in chapter 3, which also contained a brief discussion of the natural extension of this theory to a series such as

$$f(z) = \sum_{n=0}^{\infty} a_n z^n, \qquad (18.10)$$

where $z$ is a complex variable and the $a_n$ are in general complex. We now consider complex power series in more detail.

Expression (18.10) is a power series about the origin and may be used for general discussion since a power series about any other point $z_0$ is obtained by a change of variable from $z$ to $z - z_0$. If $z$ were written in its modulus and argument form $z = r \exp i\theta$, expression (18.10) would become

$$f(z) = \sum_{n=0}^{\infty} a_n r^n \exp(in\theta). \qquad (18.11)$$

The series is absolutely convergent if

$$\sum_{n=0}^{\infty} |a_n| r^n, \qquad (18.12)$$

which is a series of positive real terms, is absolutely convergent. Tests for the absolute convergence of real series can thus be used in the present context, and of these the most appropriate form is based on the Cauchy root test. The *radius of convergence* $R$ is defined by

$$1/R = \lim_{n\to\infty} |a_n|^{1/n}. \qquad (18.13)$$

The series (18.10) is absolutely convergent if $|z| < R$ and divergent if $|z| > R$. If $|z| = R$ no particular conclusion may be drawn, and this case must be considered separately, as discussed in subsection 3.5.1.

A circle of radius $R$ and centred on the origin is called the *circle of convergence* of the series $\sum a_n z^n$. The cases $R = 0$ and $R = \infty$ correspond respectively to convergence at the origin only and convergence everywhere. For $R$ finite the convergence occurs in a restricted part of the $z$-plane (the Argand diagram). For a power series about a general point $z_0$, the circle of convergence is of course centred on that point.

►*Find the parts of the z-plane for which the following series are convergent:*

$$(i)\ \sum_{n=0}^{\infty} \frac{z^n}{n!}, \qquad (ii)\ \sum_{n=0}^{\infty} n!z^n, \qquad (iii)\ \sum_{n=1}^{\infty} \frac{z^n}{n}.$$

(i) Since $(n!)^{1/n}$ behaves like $n$ as $n \to \infty$ we find $\lim(1/n!)^{1/n} = 0$. Hence $R = \infty$ and the series is convergent for all $z$.

(ii) Correspondingly, $\lim(n!)^{1/n} = \infty$. Thus $R = 0$ and the series converges only at $z = 0$.

(iii) As $n \to \infty$, $(n)^{1/n}$ has a lower limit of 1 and hence $\lim(1/n)^{1/n} = 1/1 = 1$. Thus the series is absolutely convergent if $|z| < 1$. ◄

Case (iii) of the above example provides a good illustration of the fact that on its circle of convergence a power series may or may not converge. For this particular series the circle of convergence is $|z| = 1$, so let us consider the convergence of the series at two different points on this circle. Taking $z = 1$, the series becomes

$$\sum_{n=1}^{\infty} \frac{1}{n} = 1 + \frac{1}{2} + \frac{1}{3} + \frac{1}{4} + \cdots,$$

which is easily shown to diverge (by, for example, grouping terms, as discussed in subsection 3.3.2). Taking $z = -1$, however, the series is given by

$$\sum_{n=1}^{\infty} \frac{(-1)^n}{n} = -1 + \frac{1}{2} - \frac{1}{3} + \frac{1}{4} - \cdots,$$

which is an alternating series whose terms decrease in magnitude and which therefore converges.

The ratio test discussed in subsection 3.3.2 may also be employed to investigate the absolute convergence of a complex power series. A series is absolutely convergent if

$$\lim_{n\to\infty} \frac{|a_{n+1}||z|^{n+1}}{|a_n||z|^n} = \lim_{n\to\infty} \frac{|a_{n+1}||z|}{|a_n|} < 1 \tag{18.14}$$

and hence the radius of convergence $R$ of the series is given by

$$\frac{1}{R} = \lim_{n\to\infty} \frac{|a_{n+1}|}{|a_n|}.$$

For instance in case (i) of the previous example,

$$\frac{1}{R} = \lim_{n\to\infty} \frac{n!}{(n+1)!} = \lim_{n\to\infty} \frac{1}{n+1} = 0.$$

Thus the series is absolutely convergent for all (finite) $z$, confirming the previous result.

Before turning to particular power series, we conclude this section by stating the important result† that *the power series* $\sum_0^\infty a_n z^n$ *has a sum that is an analytic function of* $z$ *inside its circle of convergence.*

As a corollary to the above theorem, it may further be shown that if $f(z) = \sum a_n z^n$ then, inside the circle of convergence of the series,

$$f'(z) = \sum_{n=0}^{\infty} n a_n z^{n-1}.$$

† For a proof see, for example, Riley, *Mathematical Methods for the Physical Sciences* (CUP, 1974), p. 446.

By repeated application of this result it can be shown that any power series can be differentiated any number of times inside its circle of convergence.

## 18.4 Some elementary functions

In the example at the end of the previous section it was shown that the function $\exp z$ *defined* by

$$\exp z = \sum_{n=0}^{\infty} \frac{z^n}{n!} \tag{18.15}$$

is convergent for all $z$ of finite modulus and is thus, by the discussion of the previous section, an analytic function over the whole $z$-plane.† Like its real-variable counterpart it is called the *exponential function*; also like its real counterpart it is equal to its own derivative.

The multiplication of two exponential functions results in a further exponential function, in accordance with the corresponding result for real variables.

►*Show that* $\exp z_1 \exp z_2 = \exp(z_1 + z_2)$.

Considering the series expansion (18.15) of $\exp z_1$ and a similar expansion for $\exp z_2$, it is clear that the coefficient of $z_1^r z_2^s$ in the corresponding series expansion of $\exp z_1 \exp z_2$ is simply $1/(r!s!)$

But, from (18.15) we also have

$$\exp(z_1 + z_2) = \sum_{n=0}^{\infty} \frac{(z_1 + z_2)^n}{n!}.$$

In order to find the coefficient of $z_1^r z_2^s$ in this expansion, we must clearly consider the term in which $n = r + s$, namely

$$\frac{(z_1 + z_2)^{r+s}}{(r+s)!} = \frac{1}{(r+s)!}\left({}^{r+s}C_0 z_1^{r+s} + \cdots + {}^{r+s}C_s z_1^r z_2^s + \cdots + {}^{r+s}C_{r+s} z_2^{r+s}\right).$$

The coefficient of $z_1^r z_2^s$ is therefore given by

$${}^{r+s}C_s \frac{1}{(r+s)!} = \frac{(r+s)!}{s!r!} \frac{1}{(r+s)!} = \frac{1}{r!s!}.$$

Thus, since all the series involved are absolutely convergent for all $z$, we see that $\exp z_1 \exp z_2 = \exp(z_1 + z_2)$. ◄

As an extension of (18.15) we may also define the complex exponent of a real number $a > 0$ by the equation

$$a^z = \exp(z \ln a), \tag{18.16}$$

† Functions that are analytic in the *whole* $z$-plane are usually called *integral* or *entire* functions.

where $\ln a$ is the natural logarithm of $a$. The particular case $a = e$ and the fact that $\ln e = 1$ enable us to write $\exp z$ interchangeably with $e^z$. If $z$ is real, the definition agrees with the familiar one.

The result that when $z = iy$,

$$\exp iy = \cos y + i \sin y, \tag{18.17}$$

has already been met in equation (2.23). Its immediate extension is that

$$\exp z = (\exp x)(\cos y + i \sin y). \tag{18.18}$$

As $z$ varies over the complex plane the modulus of $\exp z$ takes all real positive values, except that of 0. However, two values of $z$ that differ by $2\pi ni$, for any integral $n$, produce the same value of $\exp z$, as given by (18.18), and so $\exp z$ is periodic with period $2\pi i$. If we denote $\exp z$ by $t$, then the strip $-\pi < y \leq \pi$ in the $z$-plane corresponds to the whole of the $t$-plane, except for the point $t = 0$.

The sine, cosine, sinh and cosh functions of a complex variable are defined from the exponential function exactly as are those for real variables. The functions derived from them (e.g. tan and tanh), the identities they satisfy, and their derivative properties, are also just as for real variables. In view of this we will not give them further attention here.

The inverse function of $\exp z$ is given by $w$, the solution of

$$\exp w = z. \tag{18.19}$$

This inverse function was discussed in chapter 2, but we mention it again here for completeness. By virtue of the discussion following (18.18), $w$ is not uniquely defined since it is indeterminate to the extent of any integral multiple of $2\pi i$. If we denote $z$ by

$$z = r \exp i\theta,$$

where $r$ is the (real) modulus of $z$, and $\theta$ is its argument ($-\pi < \theta \leq \pi$), then multiplying $z$ by $\exp(2in\pi)$, where $n$ is an integer, will result in the same complex number $z$. Thus we may write

$$z = r \exp[i(\theta + 2n\pi)],$$

where $n$ is an integer. If we denote $w$ in (18.19) by

$$w = \text{Ln}\, z = \ln r + i(\theta + 2n\pi), \tag{18.20}$$

where $\ln r$ is the natural logarithm (to base $e$) of the real positive quantity $r$, then $\text{Ln}\, z$ is an infinitely-multivalued function of $z$. Its *principal value* is obtained by taking $n = 0$, so that $-\pi < \theta \leq \pi$, and is denoted by $\ln z$. Thus

$$\ln z = \ln r + i\theta, \qquad \text{with } -\pi < \theta \leq \pi. \tag{18.21}$$

Now that the logarithm of a complex variable has been defined, definition

(18.16) of a general power can be extended to cases other than those in which $a$ is real and positive. If $t$ ($\neq 0$) and $z$ are both complex, the $z$th power of $t$ is defined by

$$t^z = \exp(z \,\mathrm{Ln}\, t). \tag{18.22}$$

Since $\mathrm{Ln}\, t$ is multivalued, so is this definition. Its principal value is obtained by giving $\mathrm{Ln}\, t$ its principal value, $\ln t$.

If $t$ ($\neq 0$) is complex but $z$ is real and equal to $1/n$, then (18.22) provides a definition of the $n$th root of $t$. Because of the multivaluedness of $\mathrm{Ln}\, t$, there will be more than one $n$th root of any given $t$.

►*Show that there are exactly n distinct nth roots of t.*

From (18.22) the $n$th roots of $t$ are given by

$$t^{1/n} = \exp\left(\frac{1}{n}\mathrm{Ln}\, t\right).$$

On the RHS let us denote $t$ as follows:

$$t = r\exp[i(\theta + 2k\pi)],$$

where $k$ is an integer. We then obtain

$$\begin{aligned} t^{1/n} &= \exp\left[\frac{1}{n}\ln r + i\frac{(\theta + 2k\pi)}{n}\right] \\ &= r^{1/n}\exp\left[i\frac{(\theta + 2k\pi)}{n}\right], \end{aligned}$$

where $k = 0, 1, \ldots, n-1$; for other values of $k$ we simply recover the roots already found. Thus $t$ has $n$ distinct $n$th roots. ◄

## 18.5 Multivalued functions and branch cuts

In the definition of an analytic function, one of the conditions imposed was that the function was single-valued. Now, as shown in the previous section, the logarithmic function, a complex power and a complex root are all multivalued. However, it happens that the properties of analytic functions can still be applied to these and other multivalued functions of a complex variable provided that suitable care is taken. This care amounts to identifying the *branch points* of the multivalued function $f(z)$ in question. If $z$ is varied in such a way that its path in the Argand diagram forms a closed curve that encloses a branch point, then $f(z)$ will not, in general, return to its original value.

For definiteness let us consider the multivalued function $f(z) = z^{1/2}$, and denote $z$ by $z = r\exp i\theta$. From figure 18.1($a$), it is clear that, as the point $z$ traverses any closed contour $C$ that does not enclose the origin, $\theta$ will return to its original

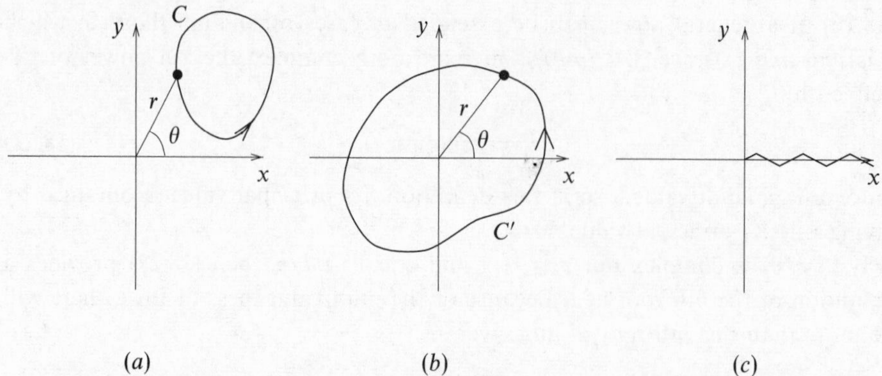


Figure 18.1 (*a*) A closed contour not enclosing the origin; (*b*) a closed contour enclosing the origin; (*c*) a possible branch cut for $f(z) = z^{1/2}$.

value after one complete circuit. However, for any closed contour $C'$ that does enclose the origin, after one circuit $\theta \to \theta + 2\pi$ (see figure 18.1(*b*)). Thus, for the function $f(z) = z^{1/2}$, after one circuit

$$r^{1/2}\exp(i\theta/2) \;\to\; r^{1/2}\exp[i(\theta + 2\pi)/2] = -r^{1/2}\exp(i\theta/2).$$

In other words, the value of $f(z)$ changes around any closed loop enclosing the origin; in this case $f(z) \to -f(z)$. Thus $z = 0$ is a branch point of the function $f(z) = z^{1/2}$.

We note in this case that if any closed contour enclosing the origin is traversed *twice*, then $f(z) = z^{1/2}$ returns to its original value. The number of loops around a branch point required for any given function $f(z)$ to return to its original value depends on the function in question, and for some functions (e.g. $\mathrm{Ln}\, z$, which also has a branch point at the origin) the original value is never recovered.

In order that $f(z)$ may be treated as single-valued we may define a *branch cut* in the Argand diagram. A branch cut is a line (or curve) in the complex plane, which may be thought of as an artificial barrier that we must not cross. Branch cuts are positioned in such a way that we are prevented from making a complete circuit around any one branch point, and the function in question remains single-valued.

For the function $f(z) = z^{1/2}$, we may take as a branch cut any curve starting at the origin $z = 0$ and extending out to $|z| = \infty$ in any direction, since all such curves would equally well prevent us from making a closed loop around the branch point at the origin. It is usual, however, to take the cut along the real or imaginary axis. For example, in figure 18.1(*c*), we take the cut as the positive real axis. By agreeing not to cross this cut, $\theta$ is restricted to lie in the range $0 \le \theta < 2\pi$, and so $f(z)$ will remain single-valued.

These ideas are easily extended to functions with more than one branch point.

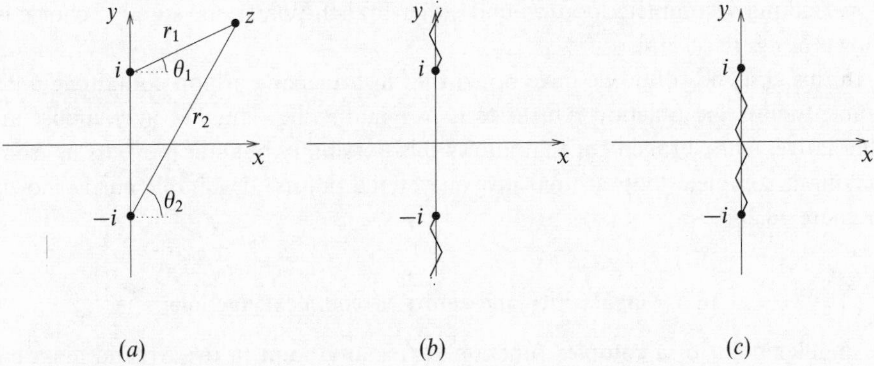


(a) (b) (c)

Figure 18.2 (a) Coordinates used in the analysis of the branch points of $f(z) = (z^2 + 1)^{1/2}$; (b) one possible arrangement of branch cuts; (c) another possible finite branch cut.

►*Find the branch points of $f(z) = \sqrt{z^2 + 1}$, and hence sketch suitable arrangements of branch cuts.*

We begin by writing $f(z)$ as

$$f(z) = \sqrt{z^2 + 1} = \sqrt{(z - i)(z + i)}.$$

As shown above the function $g(z) = z^{1/2}$ has a branch point at $z = 0$. Thus we might expect $f(z)$ to have branch points where the expression under the square root equals zero, i.e. at $z = i$ and $z = -i$.

As shown in figure 18.2(a), we use the notation

$$z - i = r_1 \exp i\theta_1 \qquad \text{and} \qquad z + i = r_2 \exp i\theta_2.$$

We can therefore write $f(z)$ as

$$f(z) = \sqrt{r_1 r_2} \exp(i\theta_1/2) \exp(i\theta_2/2) = \sqrt{r_1 r_2} \exp\left[i(\theta_1 + \theta_2)/2\right].$$

Let us now consider how $f(z)$ changes as we make one complete circuit around various closed loops $C$ in the Argand diagram. If $C$ encloses

(i) neither branch point, then $\theta_1 \to \theta_1$, $\theta_2 \to \theta_2$ and so $f(z) \to f(z)$;
(ii) $z = i$ but not $z = -i$, then $\theta_1 \to \theta_1 + 2\pi$, $\theta_2 \to \theta_2$ and so $f(z) \to -f(z)$;
(iii) $z = -i$ but not $z = i$, then $\theta_1 \to \theta_1$, $\theta_2 \to \theta_2 + 2\pi$ and so $f(z) \to -f(z)$;
(iv) both branch points, then $\theta_1 \to \theta_1 + 2\pi$, $\theta_2 \to \theta_2 + 2\pi$ and so $f(z) \to f(z)$.

Thus, as expected, $f(z)$ changes value around loops containing either $z = i$ or $z = -i$ (but not both). We must therefore choose branch cuts that prevent us from making a complete loop around either branch point; one suitable choice is shown in figure 18.2(*b*).

In this case, however, we have noted that in traversing a loop containing *both* branch points the function returns to its original value. Thus we may choose an alternative, *finite*, branch cut that allows this possibility but still prevents us from making a complete loop around just one of the points. A suitable cut is shown in figure 18.2(*c*). ◄

## 18.6 Singularities and zeroes of complex functions

A singular point of a complex function $f(z)$ is any point in the Argand diagram at which $f(z)$ fails to be analytic. We have already met one sort of singularity, the branch point, and in this section we shall consider other types of singularity, as well discussing the zeroes of complex functions.

If $f(z)$ has a singular point at $z = z_0$ but is analytic at all points in some neighbourhood containing $z_0$ but no other singularities, then $z = z_0$ is called an *isolated singularity*. (Clearly branch points are not isolated singularities.)

The most important type of isolated singularity is the *pole*. If $f(z)$ has the form

$$f(z) = \frac{g(z)}{(z - z_0)^n}, \tag{18.23}$$

where $n$ is a positive integer, $g(z)$ is analytic at all points in some neighbourhood containing $z = z_0$ and $g(z_0) \neq 0$, then $f(z)$ has a *pole of order* $n$ at $z = z_0$. An alternative (though equivalent) definition is that

$$\lim_{z \to z_0} [(z - z_0)^n f(z)] = a, \tag{18.24}$$

where $a$ is a finite, non-zero complex number. (If the limit equals zero then $z = z_0$ is a pole of order less than $n$, or $f(z)$ is analytic there; if the limit is infinite then the pole is of order greater than $n$.) It may also be shown that if $f(z)$ has a pole at $z = z_0$, then $|f(z)| \to \infty$ as $z \to z_0$ from any direction in the Argand diagram.† If no finite value of $n$ can be found such that (18.24) is satisfied then $z = z_0$ is called an *essential singularity*.

►*Find the singularities of the functions*

$$(i)\ f(z) = \frac{1}{1 - z} - \frac{1}{1 + z}, \qquad (ii)\ f(z) = \tanh z.$$

(i) If we write $f(z)$ as

$$f(z) = \frac{1}{1 - z} - \frac{1}{1 + z} = \frac{2z}{(1 - z)(1 + z)},$$

† Although perhaps intuitively obvious this result really requires formal demonstration by analysis.

we see immediately from either (18.23) or (18.24) that $f(z)$ has poles of order 1 (or *simple poles*) at $z = 1$ and $z = -1$.

(ii) In this case we write

$$f(z) = \tanh z = \frac{\sinh z}{\cosh z} = \frac{\exp z - \exp(-z)}{\exp z + \exp(-z)}.$$

Thus $f(z)$ has a singularity when $\exp z = -\exp(-z)$, or equivalently when

$$\exp z = \exp[i(2n+1)\pi] \exp(-z),$$

where $n$ is any integer. Equating the arguments of the exponentials we find $z = (n + \frac{1}{2})\pi i$, for integer $n$.

Furthermore, using l'Hôpital's rule (see chapter 3) we have

$$\lim_{z \to (n+\frac{1}{2})\pi i} \left\{ \frac{[z - (n+1/2)\pi i] \sinh z}{\cosh z} \right\}$$
$$= \lim_{z \to (n+\frac{1}{2})\pi i} \left\{ \frac{[z - (n+1/2)\pi i] \cosh z + \sinh z}{\sinh z} \right\} = 1.$$

Therefore, from (18.24), each singularity is a simple pole. ◄

Another type of singularity exists at points for which the value of $f(z)$ takes an indeterminate form such as $0/0$, but for which $\lim_{z \to z_0} f(z)$ exists and is independent of the direction from which $z_0$ is approached. Such points are called *removable singularities*.

► *Show that $f(z) = (\sin z)/z$ has a removable singularity at $z = 0$.*

It is clear that $f(z)$ takes the indeterminate form $0/0$ at $z = 0$. However, by expanding $\sin z$ as a power series in $z$, we find

$$f(z) = \frac{1}{z}\left( z - \frac{z^3}{3!} + \frac{z^5}{5!} - \cdots \right) = 1 - \frac{z^2}{3!} + \frac{z^4}{5!} - \cdots.$$

Thus $\lim_{z \to 0} f(z) = 1$ independently of the way in which $z \to 0$, and so $f(z)$ has a removable singularity at $z = 0$. ◄

An expression common in mathematics, but which we have so far avoided using explicitly in this chapter, is '$z$ tends to infinity'. For a real variable such as $|z|$ or $R$, 'tending to infinity' has a reasonably well-defined meaning. For a complex variable needing a two-dimensional plane to represent it, the meaning is not well defined. However, it is convenient to have a unique meaning and this is provided by the following *definition*: the behaviour of $f(z)$ *at infinity* is given by that of $f(1/\xi)$ at $\xi = 0$, where $\xi = 1/z$.

►*Find the behaviour at infinity of (i) $f(z) = a + bz^{-2}$, (ii) $f(z) = z(1 + z^2)$, and (iii) $f(z) = \exp z$.*

(i) $f(z) = a + bz^{-2}$; on putting $z = 1/\xi$, $f(1/\xi) = a + b\xi^2$, which is analytic at $\xi = 0$; thus $f$ is analytic at $z = \infty$.

(ii) $f(z) = z(1 + z^2)$: $f(1/\xi) = 1/\xi + 1/\xi^3$; thus $f$ has a pole of order 3 at $z = \infty$.

(iii) $f(z) = \exp z$ : $f(1/\xi) = \sum_0^\infty (n!)^{-1}\xi^{-n}$; thus $f$ has an essential singularity at $z = \infty$. ◄

We conclude this section by briefly mentioning the *zeroes* of a complex function. As the name suggests, if $f(z_0) = 0$ then $z = z_0$ is called a zero of the function $f(z)$. Zeroes are classified in a similar way to poles, in that if

$$f(z) = (z - z_0)^n g(z),$$

where $n$ is a positive integer and $g(z_0) \neq 0$, then $z = z_0$ is called a *zero of order* $n$ of $f(z)$. If $n = 1$ then $z = z_0$ is called a *simple zero*. It may further be shown that if $z = z_0$ is a zero of order $n$ of $f(z)$, then it is also a pole of order $n$ of the function $1/f(z)$.

We will return in section 18.13 to the classification of zeroes and poles in terms of their series expansions.

## 18.7 Complex potentials

Towards the end of section 18.2 it was shown that the real and the imaginary parts of an analytic function of $z$ are separately solutions of Laplace's equation in two dimensions. Analytic functions thus offer a possible way of solving some two-dimensional physical problems describable by a potential satisfying $\nabla^2\phi = 0$. The general method is known as that of *complex potentials*.

We also found that if $f = u + iv$ is an analytic function of $z$ then the curves $u =$ constant and $v =$ constant intersect at right angles. In the context of solutions of Laplace's equation, this result implies that the real and imaginary parts of $f(z)$ have an additional connection between them, for if the set of contours on which one of them is a constant represents the equipotentials of a system then the contours on which the other is constant, being orthogonal to each of the first set, must represent the *corresponding* field lines or stream lines, depending on the context. The analytic function $f$ is the complex potential. It is conventional to use $\phi$ and $\psi$ (rather than $u$ and $v$) to denote the real and imaginary parts of a complex potential, so that $f = \phi + i\psi$.

As an example consider the function

$$f(z) = \frac{-q}{2\pi\epsilon_0} \ln z, \qquad (18.25)$$

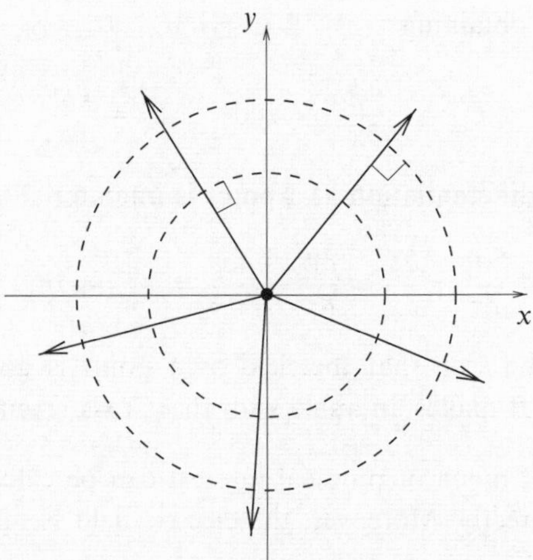


Figure 18.3 The equipotentials (broken) and field lines (solid) for a line charge perpendicular to the $z$-plane.

in connection with the physical situation of a line charge of strength $q$ per unit length passing through the origin, perpendicular to the $z$-plane (figure 18.3). Its real and imaginary parts are

$$\phi = \frac{-q}{2\pi\epsilon_0} \ln|z|, \qquad \psi = \frac{-q}{2\pi\epsilon_0} \arg z. \tag{18.26}$$

The contours in the $z$-plane of $\phi =$ constant are concentric circles and of $\psi =$ constant are radial lines. As expected these are orthogonal sets, but in addition they are respectively the equipotentials and electric field lines appropriate to the field produced by the line charge (the minus sign is needed in (18.25) since the value of $\phi$ must decrease with increasing distance from the origin).

Suppose we make the choice that the real part $\phi$ of the analytic function $f$ gives the conventional potential function; $\psi$ could equally well be selected. Then we may consider how the direction and magnitude of the field are related to $f$.

►*Show that for any complex (electrostatic) potential $f(z)$ the strength of the electric field is given by $E = |f'(z)|$ and that its direction makes an angle of $\pi - \arg[f'(z)]$ with the x-axis.*

Because $\phi =$ constant is an equipotential, the field has components

$$E_x = -\frac{\partial \phi}{\partial x} \qquad \text{and} \qquad E_y = -\frac{\partial \phi}{\partial y}. \tag{18.27}$$

Since $f$ is analytic, (i) we may use the Cauchy–Riemann relations (18.5) to change the second of these, obtaining

$$E_x = -\frac{\partial \phi}{\partial x} \quad \text{and} \quad E_y = \frac{\partial \psi}{\partial x}; \tag{18.28}$$

(ii) the direction of differentiation at a point is immaterial and so

$$\frac{df}{dz} = \frac{\partial f}{\partial x} = \frac{\partial \phi}{\partial x} + i\frac{\partial \psi}{\partial x} = -E_x + iE_y. \tag{18.29}$$

From these it can be seen that the field at a point is given in magnitude by $E = |f'(z)|$ and that it makes an angle with the $x$-axis given by $\pi - \arg[f'(z)]$. ◀

It is apparent that much of physical interest can be calculated by working in terms of $f$ and $z$ directly. Moreover, the electric field **E** may be represented by the quantity

$$\mathcal{E} = E_x + iE_y = -[f'(z)]^*.$$

Complex potentials can be used in two-dimensional fluid mechanics problems in a similar way. If the flow is stationary (i.e. the velocity of the fluid does not depend on time) and irrotational, and the fluid is both incompressible and non-viscous, then the velocity of the fluid can be described by $\mathbf{V} = \nabla\phi$, where $\phi$ is the velocity potential and satisfies $\nabla^2\phi = 0$. If, for a complex potential $f = \phi + i\psi$, the real part $\phi$ is taken to represent the velocity potential then the curves $\psi =$ constant will be the streamlines of the flow. The velocity may be represented in terms of the complex potential by

$$\mathcal{V} = V_x + iV_y = [f'(z)]^*,$$

and the speed of the flow is equal to $|f'(z)|$. Points where $f'(z) = 0$, and so the velocity is zero, are called *stagnation points* of the flow.

In a comparable way to the electrostatic case, a line *source* of fluid at $z = z_0$, perpendicular to the $z$-plane (i.e. a point from which fluid is emerging at a constant rate) is described by the complex potential

$$f(z) = k \ln(z - z_0),$$

where $k$ is the strength of the source. A sink is similarly represented, but with $k$ replaced by $-k$. Other simple examples are as follows:

(i) the flow of a fluid at a constant speed $V_0$ and at an angle $\alpha$ to the $x$-axis is described by $f(z) = V_0(\exp i\alpha)z$;

(ii) vortex flow, in which fluid flows azimuthally in an anticlockwise direction around some point $z_0$, the speed of the flow being inversely proportional to the distance from $z_0$. This flow is described by $f(z) = -ik \ln(z - z_0)$, where $k$ is the strength of the vortex. For a clockwise vortex $k$ is replaced by $-k$.

► *Verify that the complex potential*

$$f(z) = V_0 \left( z + \frac{a^2}{z} \right),$$

*is appropriate to a circular cylinder of radius a placed so that it is perpendicular to a uniform fluid flow of speed $V_0$ parallel to the x-axis.*

Firstly, since $f(z)$ is analytic except at $z = 0$, both its real and imaginary parts satisfy Laplace's equation in the region exterior to the cylinder. Also $f(z) \to V_0 z$ as $z \to \infty$, so that Re $f(z) \to V_0 x$, which is appropriate to a uniform flow of speed $V_0$ in the $x$-direction far from the cylinder.

Writing $z = r \exp i\theta$ and using de Moivre's theorem we have

$$\begin{aligned} f(z) &= V_0 \left[ r \exp i\theta + \frac{a^2}{r} \exp(-i\theta) \right] \\ &= V_0 \left( r + \frac{a^2}{r} \right) \cos\theta + iV_0 \left( r - \frac{a^2}{r} \right) \sin\theta. \end{aligned}$$

Thus we see that the streamlines of the flow described by $f(z)$ are given by

$$\psi = V_0 \left( r - \frac{a^2}{r} \right) \sin\theta = \text{constant}.$$

In particular, $\psi = 0$ on $r = a$, independently of the value of $\theta$, and so $r = a$ must be a streamline. Since there can be no flow of fluid across streamlines, $r = a$ must correspond to a boundary along which the fluid flows tangentially. Thus $f(z)$ is a solution of Laplace's equation that satisfies all the physical boundary conditions of the problem, and so it is the appropriate complex potential. ◄

By a similar argument the complex potential $f(z) = -E(z - a^2/z)$ (note the minus signs) is appropriate to a conducting circular cylinder of radius $a$ placed perpendicular to a uniform electric field $\mathbf{E}$ in the $x$-direction.

The real and imaginary parts of a complex potential $f = \phi + i\psi$ have another interesting relationship in the context of Laplace's equation in electrostatics or fluid mechanics. Let us choose $\phi$ as the conventional potential, so that $\psi$ represents the stream function (or electric field, depending on the application), and consider the difference in the values of $\psi$ at any two points $P$ and $Q$ that are connected

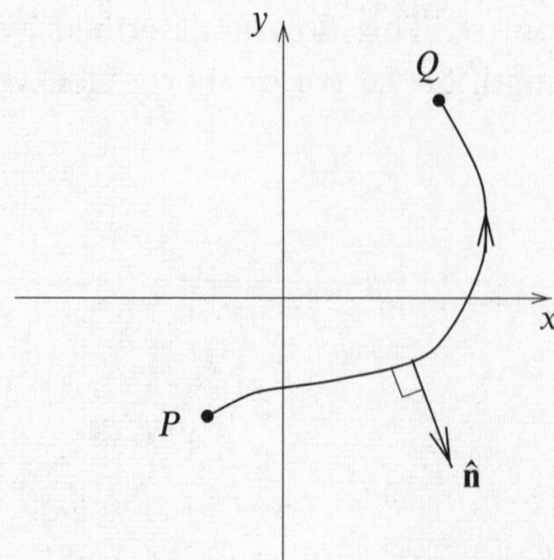


Figure 18.4 A curve joining the points $P$ and $Q$. Also shown is $\hat{\mathbf{n}}$, the unit vector normal to the curve.

by some path $C$, as shown in figure 18.4. This difference is given by

$$\psi(Q) - \psi(P) = \int_P^Q d\psi = \int_P^Q \left( \frac{\partial \psi}{\partial x}\, dx + \frac{\partial \psi}{\partial y}\, dy \right),$$

which, on using the Cauchy–Riemann relations, becomes

$$\begin{aligned} \psi(Q) - \psi(P) &= \int_P^Q \left( -\frac{\partial \phi}{\partial y}\, dx + \frac{\partial \phi}{\partial x}\, dy \right) \\ &= \int_P^Q \nabla\phi \cdot \hat{\mathbf{n}}\, ds \;=\; \int_P^Q \frac{\partial \phi}{\partial n}\, ds, \end{aligned}$$

where $\hat{\mathbf{n}}$ is the vector unit normal to the path $C$, and $s$ is the arc length along the path; the last equality is written in terms of the normal derivative $\partial\phi/\partial n \equiv \nabla\phi \cdot \hat{\mathbf{n}}$.

Now suppose that in an electrostatics application, the path $C$ is the surface of a conductor; then

$$\frac{\partial \phi}{\partial n} = -\frac{\sigma}{\epsilon_0},$$

where $\sigma$ is the surface charge density per unit length normal to the $xy$-plane. Therefore $-\epsilon_0[\psi(Q) - \psi(P)]$ is equal to the charge per unit length normal to the $xy$-plane on the surface of the conductor between the points $P$ and $Q$. Similarly, in fluid mechanics applications, if the density of the fluid is $\rho$ and its velocity $\mathbf{V}$, then

$$\rho[\psi(Q) - \psi(P)] = \rho \int_P^Q \nabla\phi \cdot \hat{\mathbf{n}}\, ds = \rho \int_P^Q \mathbf{V} \cdot \hat{\mathbf{n}}\, ds$$

is equal to the mass flux between $P$ and $Q$ per unit length perpendicular to the $xy$-plane.

► *A conducting circular cylinder of radius a is placed perpendicular to a uniform electric field* **E** *in the x-direction. Find the charge per unit length induced on the half of the cylinder that lies in the region* $x < 0$.

As mentioned after the previous example, the appropriate complex potential for this problem is $f(z) = -E(z - a^2/z)$. Writing $z = r\exp i\theta$ this becomes

$$f(z) = -E\left[r\exp i\theta - \frac{a^2}{r}\exp(-i\theta)\right]$$
$$= -E\left(r - \frac{a^2}{r}\right)\cos\theta - iE\left(r + \frac{a^2}{r}\right)\sin\theta,$$

so that on $r = a$ the imaginary part of $f$ is given by

$$\psi = -2Ea\sin\theta.$$

Therefore the induced charge $q$ per unit length on the left half of the cylinder, between $\theta = \pi/2$ and $\theta = 3\pi/2$, is given by

$$q = 2\epsilon_0 Ea[\sin(3\pi/2) - \sin(\pi/2)] = -4\epsilon_0 Ea. \blacktriangleleft$$

## 18.8 Conformal transformations

We now turn our attention to the subject of transformations, by which we mean a change of coordinates from the complex variable $z = x + iy$ to another, say $w = r + is$, by means of a prescribed formula:

$$w = g(z) = r(x, y) + is(x, y).$$

Under such a transformation, or *mapping*, the Argand diagram for the $z$-variable is transformed into one for the $w$-variable, although the complete $z$-plane might be mapped onto only a part of the $w$-plane, or onto the whole of the $w$-plane, or onto some or all of the $w$-plane covered more than once.

We shall consider only those mappings for which $w$ and $z$ are related by a function $w = g(z)$ and its inverse $z = h(w)$, which are analytic except possibly at a few isolated points; such mappings are called *conformal*. Their important properties are that, except at points at which $g'(z)$ (and hence $h'(z)$) is zero or infinite:

(i) continuous lines in the $z$-plane transform into continuous lines in the $w$-plane;
(ii) the angle between two intersecting curves in the $z$-plane equals the angle between the corresponding curves in the $w$-plane;
(iii) the magnification, as between the $z$- and $w$-plane, of a small line element in the neighbourhood of any particular point is independent of the direction of the element;

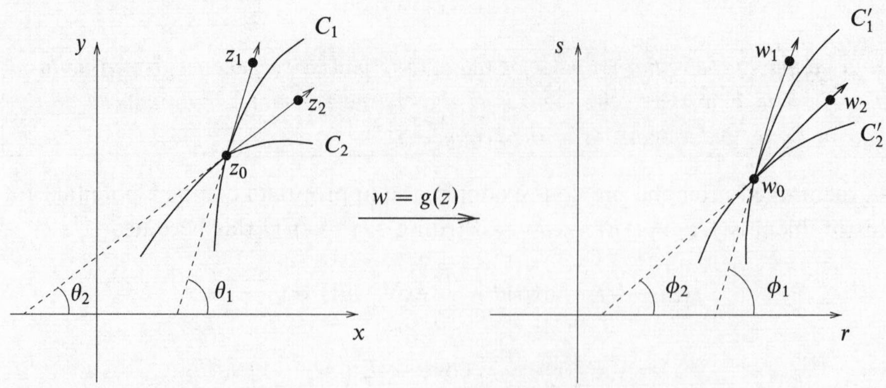


Figure 18.5 Two curves $C_1$ and $C_2$ in the $z$-plane, which are mapped onto $C_1'$ and $C_2'$ in the $w$-plane.

(iv) any analytic function of $z$ transforms to an analytic function of $w$ and vice versa.

Result (i) is immediate, and results (ii) and (iii) can be justified by the following argument. Let two curves $C_1$ and $C_2$ pass through the point $z_0$ in the $z$-plane and $z_1$ and $z_2$ be two points on their respective tangents at $z_0$, each a distance $\rho$ from $z_0$. The same prescription with $w$ replacing $z$ describes the transformed situation; however, the transformed tangents may not be straight lines and the distances of $w_1$ and $w_2$ from $w_0$ have not yet been shown to be equal. This situation is illustrated in figure 18.5.

In the $z$-plane $z_1$ and $z_2$ are given by

$$z_1 - z_0 = \rho \exp i\theta_1 \qquad \text{and} \qquad z_2 - z_0 = \rho \exp i\theta_2.$$

The corresponding descriptions in the $w$-plane are

$$w_1 - w_0 = \rho_1 \exp i\phi_1 \qquad \text{and} \qquad w_2 - w_0 = \rho_2 \exp i\phi_2.$$

The angles $\theta_i$ and $\phi_i$ are clear from figure 18.5. Now since $w = g(z)$, where $g$ is analytic, we have

$$\lim_{z_1 \to z_0} \left( \frac{w_1 - w_0}{z_1 - z_0} \right) = \lim_{z_2 \to z_0} \left( \frac{w_2 - w_0}{z_2 - z_0} \right) = \left. \frac{dg}{dz} \right|_{z=z_0},$$

which may be written as

$$\lim_{\rho \to 0} \left\{ \frac{\rho_1}{\rho} \exp[i(\phi_1 - \theta_1)] \right\} = \lim_{\rho \to 0} \left\{ \frac{\rho_2}{\rho} \exp[i(\phi_2 - \theta_2)] \right\} = g'(z_0). \qquad (18.30)$$

Comparing magnitudes and phases (i.e. arguments) in the equalities (18.30) gives the stated results (ii) and (iii) and adds quantitative information to them, namely that for *small* line elements

$$\rho_1/\rho \approx \rho_2/\rho \approx |g'(z_0)|, \tag{18.31}$$

$$\phi_1 - \theta_1 \approx \phi_2 - \theta_2 \approx \arg g'(z_0). \tag{18.32}$$

For strict comparison with result (ii), (18.32) must be written as $\theta_1 - \theta_2 = \phi_1 - \phi_2$, with an ordinary equality sign, since the angles are only defined in the limit $\rho \to 0$ when (18.32) becomes a true identity. We also see from (18.31) that the linear magnification factor is $|g'(z_0)|$; similarly, small areas are magnified by $|g'(z_0)|^2$.

Since in the neighbourhoods of corresponding points in a transformation angles are preserved and magnifications are independent of direction, it follows that small plane figures are transformed into figures of the same shape, but in general magnified and rotated (but not distorted). We also note, however, that at any point where $g'(z) = 0$, the angle $\arg g'(z)$ through which line elements are rotated is undefined; these are called *critical points* of the transformation.

The final result (iv) is perhaps the most important property of conformal transformations. If $f(z)$ is an analytic function of $z$ and $z = h(w)$ is also analytic, then $F(w) = f(h(w))$ is analytic in $w$. Its importance lies in the fact that the real and imaginary parts of $f = \phi + i\psi$ are, since $f$ is analytic, necessarily solutions of

$$\frac{\partial^2 \phi}{\partial x^2} + \frac{\partial^2 \phi}{\partial y^2} = 0 \qquad \text{and} \qquad \frac{\partial^2 \psi}{\partial x^2} + \frac{\partial^2 \psi}{\partial y^2} = 0. \tag{18.33}$$

Since $F = \Phi + i\Psi$ is also analytic, its real and imaginary parts must also satisfy Laplace's equation in the $w$-plane,

$$\frac{\partial^2 \Phi}{\partial r^2} + \frac{\partial^2 \Phi}{\partial s^2} = 0 \qquad \text{and} \qquad \frac{\partial^2 \Psi}{\partial r^2} + \frac{\partial^2 \Psi}{\partial s^2} = 0. \tag{18.34}$$

Further, suppose that (say) Re $f(z) = \phi$ is constant over a boundary $C$ in the $z$-plane; then Re $F(w) = \Phi$ is constant over $C$ in the $z$-plane. But this is the same as saying that Re $F(w)$ is constant over the boundary $C'$ in the $w$-plane, $C'$ being the curve into which $C$ is transformed by the conformal transformation $w = g(z)$. This is discussed further in the next section.

Examples of useful conformal transformations are numerous. For instance, $w = z + b$, $w = (\exp i\phi)z$ and $w = az$ correspond respectively to a translation by $b$, a rotation through an angle $\phi$, and a stretching (or contraction) in the radial direction (for $a$ real). These three examples can be combined into the general linear transformation $w = az + b$, where $a$ and $b$ are, in general, complex. Another example is the inversion mapping $w = 1/z$, which maps the interior of the unit circle to the exterior and vice versa. Other, more complicated, examples also exist.

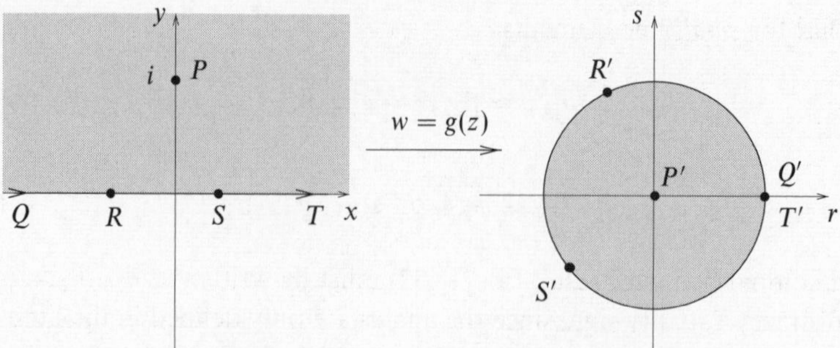


Figure 18.6 Transforming the upper half of the $z$-plane into the interior of the unit circle in the $w$-plane, in such a way that $z = i$ is mapped onto $w = 0$ and the points $x = \pm\infty$ are mapped onto $w = 1$.

►*Show that, if the point $z_0$ lies in the upper half of the z-plane, the transformation*

$$w = (\exp i\phi)\frac{z - z_0}{z - z_0^*}$$

*maps the upper half of the z-plane into the interior of the unit circle in the w-plane. Hence find a similar transformation that maps the point $z = i$ onto $w = 0$ and the points $x = \pm\infty$ onto $w = 1$.*

Taking the modulus of $w$, we have

$$|w| = \left|(\exp i\phi)\frac{z - z_0}{z - z_0^*}\right| = \left|\frac{z - z_0}{z - z_0^*}\right|.$$

However, since the complex conjugate $z_0^*$ is the reflection of $z_0$ in the real axis, if $z$ and $z_0$ both lie in the upper half of the $z$-plane $|z - z_0| \leq |z - z_0^*|$; thus $|w| \leq 1$ as required. We also note that (i) the equality holds only when $z$ lies on the real axis, so this axis is mapped onto the boundary of the unit circle in the $w$-plane; (ii) the point $z_0$ is mapped onto $w = 0$, the origin of the $w$-plane.

By fixing the images of two points in the $z$-plane, the constants $z_0$ and $\phi$ can also be fixed. Since we require the point $z = i$ to be mapped onto $w = 0$, we immediately have $z_0 = i$. By further requiring $z = \pm\infty$ to be mapped onto $w = 1$, we find $1 = w = \exp i\phi$, and so $\phi = 0$. The required transformation is therefore

$$w = \frac{z - i}{z + i},$$

and is illustrated in figure 18.6. ◄

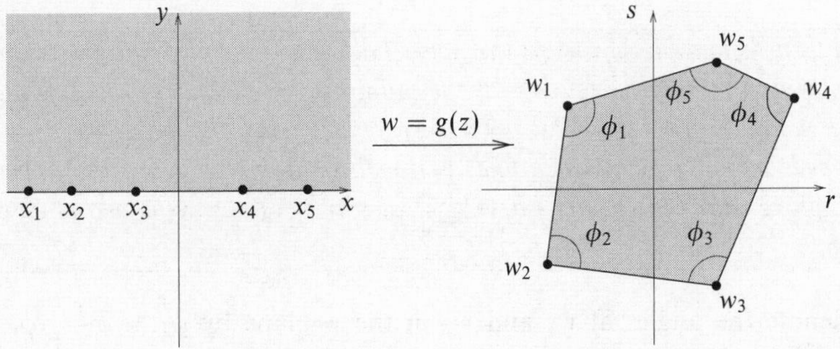


Figure 18.7 Transforming the upper half of the $z$-plane into the interior of a polygon in the $w$-plane, in such a way that the points $x_1, x_2, \ldots, x_n$ are mapped onto the vertices $w_1, w_2, \ldots, w_n$ of the polygon with interior angles $\phi_1, \phi_2, \ldots, \phi_n$.

Finally, we conclude this section by mentioning the rather curious *Schwarz–Christoffel* transformation.† Suppose, as shown in figure 18.7, that we are interested in a (finite) number of points $x_1, x_2, \ldots, x_n$ on the real axis in the $z$-plane. Then by means of the transformation

$$w = \left\{ A \int_0^z (\xi - x_1)^{\phi_1/\pi - 1} (\xi - x_2)^{\phi_2/\pi - 1} \cdots (\xi - x_n)^{\phi_n/\pi - 1} \, d\xi \right\} + B, \tag{18.35}$$

we may map the upper half of the $z$-plane onto the interior of a closed polygon in the $w$-plane having $n$ vertices $w_1, w_2, \ldots, w_n$ (which are the images of $x_1, \ldots, x_n$) with corresponding interior angles $\phi_1, \phi_2, \ldots, \phi_n$, as shown in figure 18.7. The real axis in the $z$-plane is transformed into the boundary of the polygon itself. The constants $A$ and $B$ are in general complex and determine the position, size and orientation of the polygon. It is clear from (18.35) that $dw/dz = 0$ at $x = x_1, \ldots, x_n$, and so the transformation is not conformal at these points.

There are various subtleties associated with the use of the Schwarz–Christoffel transformation. For example, if one of the points on the real axis in the $z$-plane (usually $x_n$) is taken at infinity, then the corresponding factor in (18.35) (i.e. the one involving $x_n$) is not present. In this case, the point(s) $x = \pm\infty$ are considered as one, since they transform to a single vertex of the polygon in the $w$-plane.

We can also map the upper half of the $z$-plane onto an infinite *open* polygon by considering it as the limiting case of some closed polygon.

† Strictly speaking the use of this transformation requires an understanding of complex integrals, which are discussed in section 18.10.

►*Find a transformation that maps the upper half of the z-plane onto the triangular region shown in figure 18.8, such that the points $x_1 = -1$ and $x_2 = 1$ are mapped onto the points $w = -a$ and $w = a$ respectively, and the point(s) $x_3 = \pm\infty$ are mapped onto $w = ib$. Hence find a transformation that maps the upper half of the z-plane into the region $-a < r < a$, $s > 0$ of the w-plane, as shown in figure 18.9.*

Let us denote the angles at $w_1$ and $w_2$ in the $w$-plane by $\phi_1 = \phi_2 = \phi$, where $\phi = \tan^{-1}(b/a)$. Since $x_3$ is taken at infinity we may omit the relevant factor in (18.35) to obtain

$$\begin{aligned} w &= \left\{ A \int_0^z (\xi+1)^{\phi/\pi-1}(\xi-1)^{\phi/\pi-1}\, d\xi \right\} + B \\ &= \left\{ A \int_0^z (\xi^2-1)^{\phi/\pi-1}\, d\xi \right\} + B. \end{aligned} \tag{18.36}$$

The required transformation may then be found be fixing the constants $A$ and $B$ as follows. Since the point $z = 0$ lies on the line segment $x_1x_2$ it will be mapped onto the line segment $w_1w_2$ in the $w$-plane, and by symmetry must be mapped onto the point $w = 0$. Thus setting $z = 0$ and $w = 0$ in (18.36) we obtain $B = 0$. An expression for $A$ can be found in the form of an integral by setting (for example) $z = 1$ and $w = a$ in (18.36).

We may consider the region in the $w$-plane in figure 18.9 to be the limiting case of the triangular region in figure 18.8 with the vertex $w_3$ at infinity. Thus we may use the above, but with the angles at $w_1$ and $w_2$ set to $\phi = \pi/2$. From (18.36), we obtain

$$w = A \int_0^z \frac{d\xi}{\sqrt{\xi^2-1}} = iA \sin^{-1} z.$$

By setting $z = 1$ and $w = a$, we find $iA = 2a/\pi$, so the required transformation is

$$w = \frac{2a}{\pi} \sin^{-1} z. \blacktriangleleft$$

## 18.9 Application of conformal transformations

In the previous section it was shown that, under a conformal transformation $w = g(z)$ from $z = x + iy$ to a new variable $w = r + is$, if a solution of Laplace's equation in some region $R$ of the $xy$-plane can be found as the real or imaginary

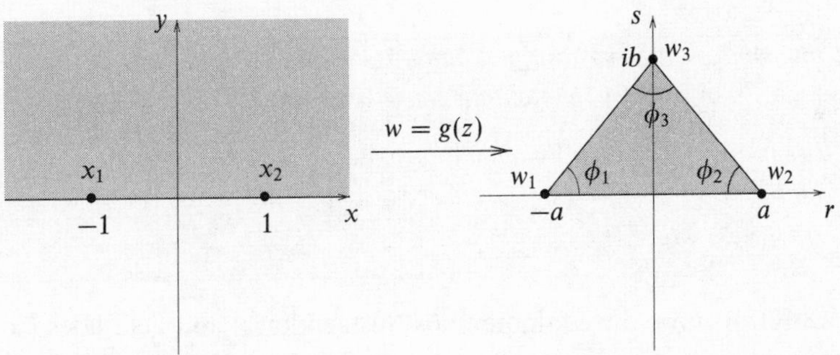


Figure 18.8 Transforming the upper half of the z-plane into the interior of a triangle in the w-plane.

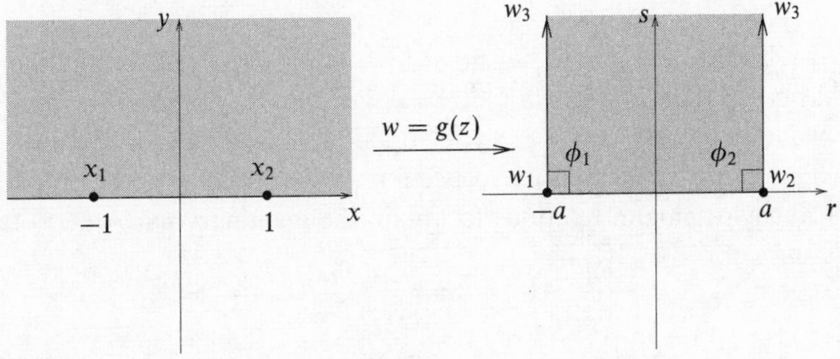


Figure 18.9 Transforming the upper half of the z-plane into the interior of the region $-a < r < a$, $s > 0$ in the w-plane.

part of an analytic function† of $z$, then the same expression put in terms of $r$ and $s$ will be a solution of Laplace's equation in the corresponding region $R'$ of the $w$-plane, and vice versa. In addition, if the solution is constant over the boundary $C$ of the region $R$ in the $xy$-plane, then the solution in the $w$-plane will take the same constant value over the corresponding curve $C'$ that bounds $R'$.

Thus from any two-dimensional solution of Laplace's equation for a particular geometry, further solutions for other geometries can be obtained by making conformal transformations. From the physical point of view the given geometry is usually complicated and the solution is sought by transforming to a simpler one. However, working from simpler to more complicated situations can provide useful experience, so that the reverse procedure can be tackled successfully.

† The original solution in the $xy$-plane need not in fact be explicitly given as the real or imaginary part of an analytic function. Any solution of $\nabla^2\phi = 0$ in the $xy$-plane is carried over into another solution of $\nabla^2\phi = 0$ in the new variables by a conformal transformation, and vice versa.

► *Find the complex electrostatic potential associated with an infinite charged conducting plate ($y = 0$), and thus obtain those associated with*

*(i) a semi-infinite charged conducting plate ($r > 0$, $s = 0$),*
*(ii) the inside of a right-angled charged conducting wedge ($r > 0$, $s = 0$ and $r = 0$, $s > 0$).*

Figure 18.10(*a*) shows the equipotentials (broken lines) and field lines (solid) for the infinite charged conducting plane $y = 0$. Suppose that we elect to make the real part of the complex potential coincide with the conventional electrostatic potential. If the plate is charged to a potential $V$, then clearly

$$\phi(x, y) = V - ky, \tag{18.37}$$

where $k$ is related to the charge density $\sigma$ by $k = \sigma/\epsilon_0$, since physically the electric field $\mathbf{E}$ has components $(0, \sigma/\epsilon_0)$ and $\mathbf{E} = -\nabla\phi$.

Thus what is needed is an analytic function of $z$ of which the real part is $V - ky$. This can be obtained by inspection, but we may proceed formally and use the Cauchy–Riemann relations to obtain the imaginary part $\psi(x, y)$ thus:

$$\frac{\partial \psi}{\partial y} = \frac{\partial \phi}{\partial x} = 0 \qquad \text{and} \qquad \frac{\partial \psi}{\partial x} = -\frac{\partial \phi}{\partial y} = k.$$

Hence $\psi = kx + c$ and, absorbing $c$ into $V$, the required complex potential is

$$f(z) = V - ky + ikx = V + ikz. \tag{18.38}$$

(i) Now consider the transformation

$$w = g(z) = z^2. \tag{18.39}$$

This satisfies the criteria for a conformal mapping (except at $z = 0$) and carries the upper half of the $z$-plane into the entire $w$-plane, with the equipotential plane $y = 0$ going into the half-plane $r > 0$, $s = 0$.

By the general results proved, $f(z)$ when expressed in terms of $r$ and $s$ will give a complex potential of which the real part will be constant on the half-plane in question;

$$F(w) = f(z) = V + ikz = V + ikw^{1/2} \tag{18.40}$$

is thus the required potential. Expressed in terms of $r$, $s$ and $\rho = (r^2 + s^2)^{1/2}$, $w^{1/2}$ is given by

$$w^{1/2} = \rho^{1/2}\left[\left(\frac{\rho + r}{2\rho}\right)^{1/2} + i\left(\frac{\rho - r}{2\rho}\right)^{1/2}\right] \tag{18.41}$$

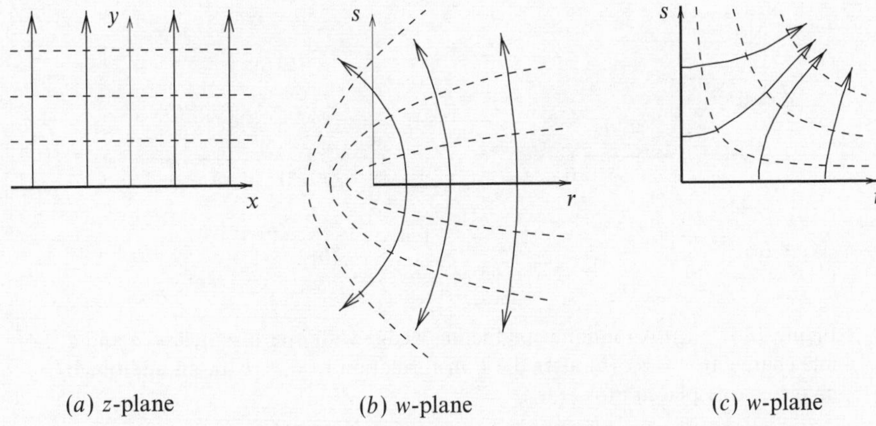


(*a*) *z*-plane (*b*) *w*-plane (*c*) *w*-plane

Figure 18.10 (*a*) The equipotential lines (broken) and field lines (solid) for an infinite charged conducting plane at $y = 0$, where $z = x + iy$; (*b*) after the transformation $w = z^2$; (*c*) after the transformation $w = z^{1/2}$ of the situation shown in (*a*).

and so, for example, the electrostatic potential is given by

$$\Phi(r, s) = \text{Re}\, F(w) = V - \frac{k}{\sqrt{2}} \left[ (r^2 + s^2)^{1/2} - r \right]^{1/2}. \tag{18.42}$$

The corresponding equipotentials and field lines are shown in figure 18.10(*b*). Using results (18.27)–(18.29), the magnitude of the electric field is

$$|\mathbf{E}| = |F'(w)| = |\tfrac{1}{2} ikw^{-1/2}| = \tfrac{1}{2} k(r^2 + s^2)^{-1/4}.$$

(ii) A transformation 'converse' to that used in (i),

$$w = g(z) = z^{1/2},$$

has the effect of mapping the upper half of the $z$-plane into the first quadrant of the $w$-plane and the conducting plane $y = 0$ into the wedge $r > 0$, $s = 0$ and $r = 0$, $s > 0$.

The complex potential now becomes

$$\begin{aligned} F(w) &= V + ikw^2 \\ &= V + ik[(r^2 - s^2) + 2irs], \end{aligned} \tag{18.43}$$

showing that the electrostatic potential is $V - 2krs$, and the electric field has components

$$\mathbf{E} = (2ks, 2kr). \tag{18.44}$$

Figure 18.10(*c*) indicates the approximate equipotentials and field lines. (Note that in both transformations $g'(z)$ is either 0 or $\infty$ at the origin and so neither

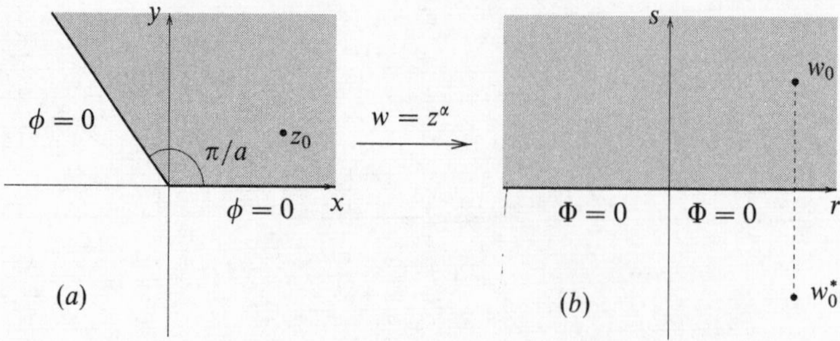


Figure 18.11 (*a*) An infinite conducting wedge with interior angle $\pi/\alpha$ and a line charge at $z = z_0$; (*b*) after the transformation $w = z^\alpha$, with an additional image charge placed at $w = w_0^*$.

transformation is conformal there. Consequently there is no violation of result (ii), given at the start of section 18.8, concerning the angles between intersecting lines.) ◄

The *method of images* discussed in section 17.5 can also be used in conjunction with conformal transformations to solve Laplace's equation in two dimensions.

► *A wedge of angle $\pi/\alpha$ with its vertex at $z = 0$ is formed by two semi-infinite conducting plates as shown in figure 18.11(a). A line charge of strength $q$ per unit length is positioned at $z = z_0$, perpendicular to the z-plane. By considering the transformation $w = z^\alpha$, find the complex electrostatic potential for this situation.*

Let us consider the action of the transformation $w = z^\alpha$ on the lines defining the positions of the conducting plates. The plate that lies along the positive $x$-axis is mapped onto the positive $r$-axis in the $w$-plane, whereas the plate that lies along the direction $\exp(i\pi/\alpha)$ is mapped into the negative $r$-axis, as shown in figure 18.11(*b*). Similarly the line charge at $z_0$ is mapped onto the point $w_0 = z_0^\alpha$.

From figure 18.11(*b*), we see that in the $w$-plane the problem can be solved by introducing a second line charge of opposite sign at the point $w_0^*$, so that the potential $\Phi = 0$ along the $r$-axis. The complex potential for such an arrangement is simply

$$F(w) = -\frac{q}{2\pi\epsilon_0}\ln(w - w_0) + \frac{q}{2\pi\epsilon_0}\ln(w - w_0^*).$$

Substituting $w = z^\alpha$ into the above, it can be shown that the required complex

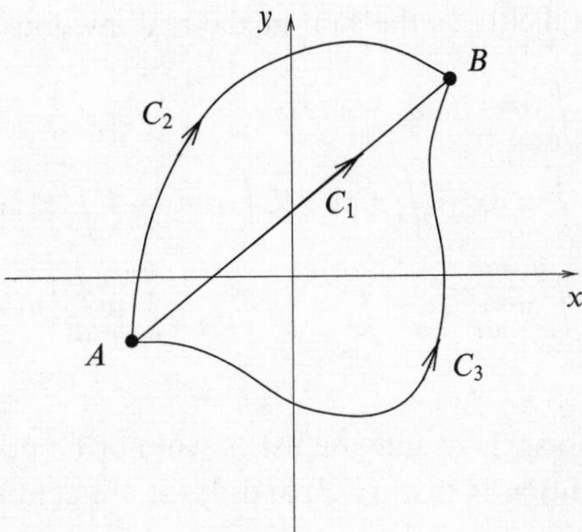


Figure 18.12 Alternative paths for the integral of a function $f(z)$ between $A$ and $B$.

potential in the original $z$-plane is

$$f(z) = \frac{q}{2\pi\epsilon_0} \ln\left(\frac{z^\alpha - z_0^{*\alpha}}{z^\alpha - z_0^\alpha}\right). \blacktriangleleft$$

## 18.10 Complex integrals

Corresponding to integration with respect to a real variable, integration with respect to a complex variable between two complex limits can be defined. Since the $z$-plane is two-dimensional there is clearly greater freedom and hence ambiguity in what is meant by a complex integral. If a complex function $f(z)$ is single-valued and continuous in some region $R$ in the complex plane, then we can define the complex integral of $f(z)$ between two points $A$ and $B$ along some curve in $R$; its value will depend, in general, upon the path taken between $A$ and $B$ (see figure 18.12). However, it will be found that for some paths that are different but bear a particular relationship to each other, the value of the integral does *not* depend upon which of the paths is adopted.

Let a particular path $C$ be described by a continuous (real) parameter $t$ ($\alpha \leq t \leq \beta$) that gives successive positions on $C$ by means of the equations

$$x = x(t), \qquad y = y(t), \tag{18.45}$$

with $t = \alpha$ and $t = \beta$ corresponding to the points $A$ and $B$ respectively. Then the integral along path $C$ of a continuous function $f(z)$ is written

$$\int_C f(z)\, dz \tag{18.46}$$

and is given more explicitly as the sum of the real integrals obtained as follows:

$$\begin{aligned}
\int_C f(z)\,dz &= \int_C (u+iv)(dx+idy) \\
&= \int_C u\,dx - \int_C v\,dy + i\int_C u\,dy + i\int_C v\,dx \\
&= \int_\alpha^\beta u\frac{dx}{dt}\,dt - \int_\alpha^\beta v\frac{dy}{dt}\,dt + i\int_\alpha^\beta u\frac{dy}{dt}\,dt + i\int_\alpha^\beta v\frac{dx}{dt}\,dt.
\end{aligned} \tag{18.47}$$

The question of when such an integral exists will not be pursued, except to state that a sufficient condition is that $dx/dt$ and $dy/dt$ are continuous.

►*Evaluate the complex integral of $f(z) = z^{-1}$ along the circle $|z| = R$, starting and finishing at $z = R$.*

The path $C_1$ is parameterised as follows (figure 18.13($a$)):

$$z(t) = R\cos t + iR\sin t, \qquad 0 \le t \le 2\pi,$$

whilst $f(z)$ is given by

$$f(z) = \frac{1}{x+iy} = \frac{x-iy}{x^2+y^2}.$$

Thus the real and imaginary parts of $f(z)$ are

$$u = \frac{x}{x^2+y^2} = \frac{R\cos t}{R^2} \qquad \text{and} \qquad v = \frac{-y}{x^2+y^2} = -\frac{R\sin t}{R^2}.$$

Hence, using expression (18.47),

$$\begin{aligned}
\int_{C_1} \frac{1}{z}\,dz &= \int_0^{2\pi} \frac{\cos t}{R}(-R\sin t)\,dt - \int_0^{2\pi} \left(\frac{-\sin t}{R}\right) R\cos t\,dt \\
&\quad + i\int_0^{2\pi} \frac{\cos t}{R} R\cos t\,dt + i\int_0^{2\pi} \left(\frac{-\sin t}{R}\right)(-R\sin t)\,dt \\
&= 0 + 0 + i\pi + i\pi = 2\pi i. ◄
\end{aligned} \tag{18.48}$$

With a bit of experience, integrals like the LHS of (18.48) can sometimes be evaluated directly without writing them as four separate real integrals. In the present case,

$$\int_{C_1} \frac{dz}{z} = \int_0^{2\pi} \frac{-R\sin t + iR\cos t}{R\cos t + iR\sin t}\,dt = \int_0^{2\pi} i\,dt = 2\pi i. \tag{18.49}$$

This very important result will be used many times later, and the following should be carefully noted: (i) its value, (ii) that this value is independent of $R$.

In the above example the contour was closed, so that it began and ended at

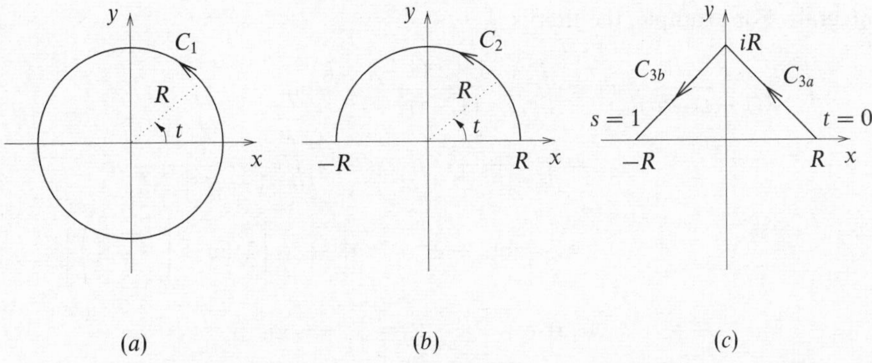


Figure 18.13 Different paths for an integral of $f(z) = z^{-1}$. See text for details.

the same point in the Argand diagram. We can evaluate complex integrals along open paths in a similar way.

►*Evaluate the complex integral of $f(z) = z^{-1}$ along*

*(i) the contour $C_2$ consisting of the semicircle $|z| = R$ in the half-plane $y \geq 0$, (see figure 18.13(b)),*

*(ii) the contour $C_3$ made up of the two straight lines $C_{3a}$ and $C_{3b}$ (see figure 18.13(c)).*

(i) This is just as in the above example, except that now $0 \leq t \leq \pi$. With this change we have from (18.48) or (18.49) that

$$\int_{C_2} \frac{dz}{z} = \pi i. \tag{18.50}$$

(ii) The straight lines that make up the countour $C_3$ may be parameterised as follows:

$$C_{3a},\ z = (1-t)R + itR \quad \text{for } 0 \leq t \leq 1;$$
$$C_{3b},\ z = -sR + i(1-s)R \quad \text{for } 0 \leq s \leq 1.$$

With these parameterisations the required integrals may be written

$$\int_{C_3} \frac{dz}{z} = \int_0^1 \frac{-R+iR}{R+t(-R+iR)}\,dt + \int_0^1 \frac{-R-iR}{iR+s(-R-iR)}\,ds. \tag{18.51}$$

If we could take over from real-variable theory that, for real $t$, $\int (a+bt)^{-1}\,dt = b^{-1}\ln(a+bt)$ even if $a$ and $b$ are complex, then these integrals could be evaluated immediately. However, to do this would be presuming to some extent what we wish to show, and so the evaluation must be made in terms of entirely real integrals. For example, the first is

$$\begin{aligned}\int_0^1 \frac{-R+iR}{R(1-t)+itR}\,dt &= \int_0^1 \frac{(-1+i)(1-t-it)}{(1-t)^2+t^2}\,dt \\ &= \int_0^1 \frac{2t-1}{1-2t+2t^2}\,dt + i\int_0^1 \frac{1}{1-2t+2t^2}\,dt \\ &= \frac{1}{2}\Big[\ln(1-2t+2t^2)\Big]_0^1 + \frac{i}{2}\left[2\tan^{-1}\left(\frac{t-\frac{1}{2}}{\frac{1}{2}}\right)\right]_0^1 \\ &= 0 + \frac{i}{2}\left[\frac{\pi}{2} - \left(-\frac{\pi}{2}\right)\right] = \frac{1}{2}\pi i.\end{aligned}$$

The second integral on the right of (18.51) can also be shown to have the value $\frac{1}{2}\pi i$. Thus

$$\int_{C_3} \frac{dz}{z} = \pi i. \blacktriangleleft$$

Considering the results of the last two examples, which have common integrands, some interesting observations are possible. Firstly, the two integrals from $z = R$ to $z = -R$, along $C_2$ and $C_3$ respectively, have the same value even though the paths taken are different. It also follows that if we took a closed path $C_4$, given by $C_2$ from $R$ to $-R$ and $C_3$ traversed backwards from $-R$ to $R$, then the integral round $C_4$ of $z^{-1}$ would be zero (both parts contributing equal and opposite amounts). This is to be compared with result (18.49), in which closed path $C_1$, beginning and ending at the same place as $C_4$, yields a value $2\pi i$.

It is not true, however, that the integrals along the paths $C_2$ and $C_3$ are equal for any function $f(z)$, or indeed that their values are in general independent of $R$.

►*Evaluate the complex integral of $f(z) = \operatorname{Re} z$ along the paths $C_1$, $C_2$ and $C_3$ as shown in figure 18.13.*

(i) If we take $f(z) = \operatorname{Re} z$ and the contour $C_1$, then

$$\int_{C_1} \operatorname{Re} z\, dz = \int_0^{2\pi} R\cos t(-R\sin t + iR\cos t)\, dt = i\pi R^2.$$

(ii) Using $C_2$ as the contour,

$$\int_{C_2} \operatorname{Re} z\, dz = \int_0^{\pi} R\cos t(-R\sin t + iR\cos t)\, dt = \tfrac{1}{2} i\pi R^2.$$

(iii) Finally the integral along $C_3 = C_{3a} + C_{3b}$ is given by

$$\begin{aligned}\int_{C_3} \operatorname{Re} z\, dz &= \int_0^1 (1-t)R(-R+iR)\, dt + \int_0^1 (-sR)(-R-iR)\, ds \\ &= \tfrac{1}{2}R^2(-1+i) + \tfrac{1}{2}R^2(1+i) = iR^2. \blacktriangleleft\end{aligned}$$

The results of this section demonstrate that the value of an integral between the same two points may depend upon the path that is taken between them, but at the same time suggests that in some circumstances it is independent of the path. The general result is embodied in the result of the next section, namely Cauchy's theorem, which is the cornerstone of the integral calculus of complex variables.

Before discussing Cauchy's theorem, however, we note an important result concerning complex integrals that will be of some use later. Let us consider the integral of a function $f(z)$ along some path $C$. If $M$ is an upper bound on the value of $|f(z)|$ on the path, i.e. $|f(z)| \leq M$ on $C$, and $L$ is the length of the path $C$, then

$$\left| \int_C f(z)\, dz \right| \leq \int_c |f(z)||dz| \leq M \int_C dl = ML. \qquad (18.52)$$

It is straightforward to show that this result does indeed hold for the complex integrals considered in this section.

## 18.11 Cauchy's theorem

*Cauchy's theorem* states that if $f(z)$ is an analytic function, and $f'(z)$ is continuous at each point within and on a closed contour $C$, then

$$\oint_C f(z)\, dz = 0, \qquad (18.53)$$

where from now on we denote the integral around a closed contour by $\oint_C$.

To prove this theorem we will need the two-dimensional form of the divergence theorem, known as Green's theorem in a plane (see section 9.3). This says that if $p$ and $q$ are two functions with continuous first derivatives within and on a closed contour $C$ (bounding a domain $R$) in the $xy$-plane, then

$$\iint_R \left( \frac{\partial p}{\partial x} + \frac{\partial q}{\partial y} \right) dxdy = \oint_C (p\, dy - q\, dx). \qquad (18.54)$$

With $f(z) = u + iv$ and $dz = dx + i\, dy$, this can be applied to

$$I = \oint_C f(z)\, dz = \oint_C (u\, dx - v\, dy) + i \oint_C (v\, dx + u\, dy)$$

to give

$$I = \iint_R \left[ \frac{\partial(-u)}{\partial y} + \frac{\partial(-v)}{\partial x} \right] dx\, dy + i \iint_R \left[ \frac{\partial(-v)}{\partial y} + \frac{\partial u}{\partial x} \right] dx\, dy. \qquad (18.55)$$

Now, recalling that $f(z)$ is analytic and therefore the Cauchy–Riemann relations (18.5) apply, we see that each integrand is identically zero and thus $I$ is also zero; this proves Cauchy's theorem.

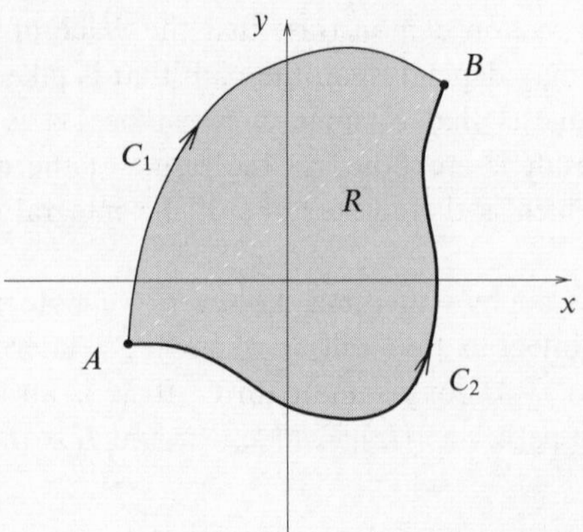


Figure 18.14 Two paths $C_1$ and $C_2$ enclosing a region $R$.

In fact the conditions of the above proof are more stringent than are needed. The continuity of $f'(z)$ is not necessary for the proof of Cauchy's theorem, analyticity of $f(z)$ within and on $C$ being sufficient. However, the proof then becomes more complicated and is too long to be given here.†

The connection between Cauchy's theorem and the zero value of the integral of $z^{-1}$ around the composite path $C_4$ discussed towards the end of the previous section is apparent: the function $z^{-1}$ is analytic in the two regions of the $z$-plane enclosed by contours ($C_2$ and $C_{3a}$) and ($C_2$ and $C_{3b}$).

►*Suppose two points $A$ and $B$ in the complex plane are joined by two different paths $C_1$ and $C_2$. Show that if $f(z)$ is an analytic function on each path and in the region enclosed by the two paths then the integral of $f(z)$ is the same along $C_1$ and $C_2$.*

The situation is shown in figure 18.14. Since $f(z)$ is analytic in $R$, from Cauchy's theorem we have

$$\int_{C_1} f(z)\,dz - \int_{C_2} f(z)\,dz = \oint_{C_1-C_2} f(z)\,dz = 0,$$

since $C_1 - C_2$ forms a closed contour enclosing $R$. Thus we immediately obtain

$$\int_{C_1} f(z)\,dz = \int_{C_2} f(z)\,dz,$$

and so the values of the integrals along $C_1$ and $C_2$ are equal. ◄

An important application of Cauchy's theorem is in proving that in some cases

† The reader may refer to almost any book devoted to complex variables and the theory of functions.

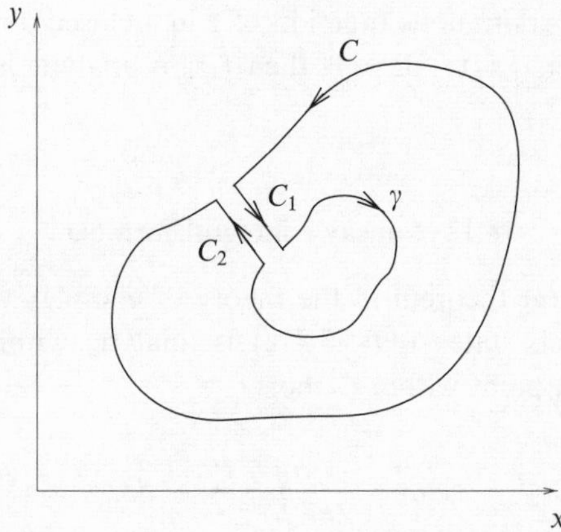


Figure 18.15 The contour used to prove the result (18.56).

it is possible to deform a closed contour $C$ into another contour $\gamma$, in such a way that the integral of some function $f(z)$ around either contour has the same value.

►*Consider two closed contours $C$ and $\gamma$ in the Argand diagram, with $\gamma$ small enough that it lies completely within $C$. Show that if the function $f(z)$ is analytic in the region* between *the two contours, then*

$$\oint_C f(z)\,dz = \oint_\gamma f(z)\,dz. \tag{18.56}$$

To prove this result we consider a contour as shown in figure 18.15. The two close parallel lines $C_1$ and $C_2$ join $\gamma$ and $C$, which are 'cut' to accommodate them. The new contour $\Gamma$ so formed consists of $C$, $C_1$, $\gamma$ and $C_2$.

Within the area bounded by $\Gamma$ the function $f(z)$ is analytic, and therefore by Cauchy's theorem (18.53),

$$\oint_\Gamma f(z)\,dz = 0. \tag{18.57}$$

Now the parts $C_1$ and $C_2$ of $\Gamma$ are traversed in opposite directions and lie (in the limit) on top of each other, and so their contributions to (18.57) cancel. Thus

$$\oint_C f(z)\,dz + \oint_\gamma f(z)\,dz = 0. \tag{18.58}$$

The sense of the integral round $\gamma$ is opposite to the conventional (anticlockwise) one, and so by traversing $\gamma$ in the usual sense, we establish the result (18.56). ◄

A sort of converse of Cauchy's theorem is known as *Morera's theorem*, which states that if $f(z)$ is a continuous function of $z$ in a closed domain $R$ bounded by a curve $C$ and, further, $\oint_C f(z)\,dz = 0$, then $f(z)$ is analytic in $R$.

## 18.12 Cauchy's integral formula

Another very important theorem in the theory of complex variables is *Cauchy's integral formula*, which states that if $f(z)$ is analytic within and on a closed contour $C$ and $z_0$ is a point within $C$ then

$$f(z_0) = \frac{1}{2\pi i} \oint_C \frac{f(z)}{z - z_0}\,dz. \qquad (18.59)$$

This is saying that the value of an analytic function anywhere inside a closed contour is uniquely determined by its values on the contour† and that the specific expression (18.59) can be given for the value at the interior point.

We may prove Cauchy's integral formula by using (18.56) and taking $\gamma$ to be a circle centred on the point $z = z_0$, of small enough radius $\rho$ that it all lies inside $C$. Then, since $f(z)$ is analytic inside $C$, the integrand $f(z)/(z - z_0)$ is analytic in the space between $C$ and $\gamma$. Thus, from (18.56), the integral around $\gamma$ has the same value as that around $C$.

We then use the fact that any point $z$ on $\gamma$ is given by $z = z_0 + \rho \exp i\theta$ (and so $dz = i\rho \exp i\theta\, d\theta$). Thus the value of the integral around $\gamma$ is given by

$$\begin{aligned} I = \oint_\gamma \frac{f(z)}{z - z_0}\,dz &= \int_0^{2\pi} \frac{f(z_0 + \rho \exp i\theta)}{\rho \exp i\theta} i\rho \exp i\theta\, d\theta \\ &= i \int_0^{2\pi} f(z_0 + \rho \exp i\theta)\, d\theta. \end{aligned}$$

If the radius of the circle $\gamma$ is now shrunk to zero, i.e. $\rho \to 0$, the value of the integral $I \to 2\pi i f(z_0)$, thus establishing the result (18.59).

An extension to Cauchy's integral formula can be made, yielding an integral expression for $f'(z_0)$:

$$f'(z_0) = \frac{1}{2\pi i} \int_C \frac{f(z)}{(z - z_0)^2}\,dz, \qquad (18.60)$$

under the same conditions as previously stated.

† The similarity between this and the uniqueness theorem for Dirichlet boundary conditions of chapter 16 is apparent.

►*Prove Cauchy's integral formula for $f'(z_0)$ given in (18.60).*

To show this, we use the definition of a derivative and (18.59) itself to evaluate

$$\begin{aligned} f'(z_0) &= \lim_{h\to 0} \frac{f(z_0+h)-f(z_0)}{h} \\ &= \lim_{h\to 0}\left[\frac{1}{2\pi i}\oint_C \frac{f(z)}{h}\left(\frac{1}{z-z_0-h}-\frac{1}{z-z_0}\right)dz\right] \\ &= \lim_{h\to 0}\left[\frac{1}{2\pi i}\oint_C \frac{f(z)}{(z-z_0-h)(z-z_0)}\,dz\right] \\ &= \frac{1}{2\pi i}\oint_C \frac{f(z)}{(z-z_0)^2}\,dz, \end{aligned}$$

which establishes the result (18.60). ◄

It may be further proved by induction that the $n$th derivative of $f(z)$ is given by a Cauchy integral,

$$f^{(n)}(z_0) = \frac{n!}{2\pi i}\oint_C \frac{f(z)\,dz}{(z-z_0)^{n+1}}. \qquad (18.61)$$

Thus, if the value of the analytic function is known on $C$ then not only may the value of the function at any interior point be calculated, but also the values of *all* its derivatives.

The observant reader will notice that (18.61) may also be obtained by the formal device of differentiating under the integral sign with respect to $z_0$ in Cauchy's integral formula (18.59),

$$\begin{aligned} f^{(n)}(z_0) &= \frac{1}{2\pi i}\oint_C \frac{\partial^n}{\partial z_0^n}\left[\frac{f(z)}{(z-z_0)}\right]dz \\ &= \frac{n!}{2\pi i}\oint_C \frac{f(z)\,dz}{(z-z_0)^{n+1}}. \end{aligned}$$

►*Suppose that $f(z)$ is analytic inside and on a circle $C$ of radius $R$ centred at the point $z = z_0$. If $|f(z)| \le M$ on the circle, where $M$ is some constant, show that*

$$|f^{(n)}(z_0)| \le \frac{Mn!}{R^n}. \qquad (18.62)$$

From (18.61) we have

$$|f^{(n)}(z_0)| = \frac{n!}{2\pi}\left|\oint_C \frac{f(z)\,dz}{(z-z_0)^{n+1}}\right|$$

and using (18.52) this becomes

$$|f^{(n)}(z_0)| \leq \frac{n!}{2\pi}\frac{M}{R^{n+1}}2\pi R = \frac{Mn!}{R^n}.$$

This result is known as *Cauchy's inequality.* ◀

We may use Cauchy's inequality to prove *Liouville's theorem*, which states that if $f(z)$ is analytic and bounded for all $z$ then $f$ is a constant. Setting $n = 1$ in (18.62) and letting $R \to \infty$ we find $|f'(z_0)| = 0$ and hence $f'(z_0) = 0$. Since $f(z)$ is analytic for all $z$ we may take $z_0$ as any point in the $z$-plane and thus $f'(z) = 0$ for all $z$; this implies $f(z) =$ constant. Liouville's theorem may in turn be used to prove the *fundamental theorem of algebra* (see exercise 18.10).

## 18.13 Taylor and Laurent series

Following on from (18.61), we may establish *Taylor's theorem* for functions of a complex variable. If $f(z)$ is analytic inside and on a circle $C$ of radius $R$ centred on the point $z = z_0$, and $z$ is a point inside $C$, then

$$f(z) = \sum_{n=0}^{\infty} a_n(z - z_0)^n, \tag{18.63}$$

where $a_n$ is given by $f^{(n)}(z_0)/n!$. The Taylor expansion is valid inside the region of analyticity and, for any particular $z_0$, can be shown to be unique.

To prove Taylor's theorem (18.63), we note that since $f(z)$ is analytic inside and on $C$, we may use Cauchy's formula to write $f(z)$ as

$$f(z) = \frac{1}{2\pi i}\oint_C \frac{f(\xi)}{\xi - z}\,d\xi, \tag{18.64}$$

where $\xi$ lies on $C$. Now we may expand $(\xi - z)^{-1}$ as a geometric series in $(z - z_0)/(\xi - z_0)$,

$$\frac{1}{\xi - z} = \frac{1}{\xi - z_0}\sum_{n=0}^{\infty}\left(\frac{z - z_0}{\xi - z_0}\right)^n,$$

so (18.64) becomes

$$\begin{aligned} f(z) &= \frac{1}{2\pi i}\oint_C \frac{f(\xi)}{\xi - z_0}\sum_{n=0}^{\infty}\left(\frac{z - z_0}{\xi - z_0}\right)^n d\xi \\ &= \frac{1}{2\pi i}\sum_{n=0}^{\infty}(z - z_0)^n \oint_C \frac{f(\xi)}{(\xi - z_0)^{n+1}}\,d\xi \\ &= \frac{1}{2\pi i}\sum_{n=0}^{\infty}(z - z_0)^n \frac{2\pi i f^{(n)}(z_0)}{n!}, \end{aligned} \tag{18.65}$$

where we have used Cauchy's integral formula (18.61) for derivatives of $f(z)$.

Cancelling the factors of $2\pi i$, we thus establish the result (18.63) with $a_n = f^{(n)}(z_0)/n!$.

►*Show that if $f(z)$ and $g(z)$ are analytic in some region $R$, and $f(z) = g(z)$ within some subregion $S$ of $R$, then $f(z) = g(z)$ throughout $R$.*

It is simpler to consider the (analytic) function $h(z) = f(z) - g(z)$, and show that because $h(z) = 0$ in $S$ it follows that $h(z) = 0$ throughout $R$.

If we choose a point $z = z_0$ in $S$, then we can expand $h(z)$ in a Taylor series about $z_0$,

$$h(z) = h(z_0) + h'(z_0)(z - z_0) + \tfrac{1}{2}h''(z_0)(z - z_0)^2 + \cdots,$$

which will converge inside some circle $C$ that extends at least as far as the boundary of $R$, since $h(z)$ is analytic in $R$. But since $z_0$ lies in $S$, we have

$$h(z_0) = h'(z_0) = h''(z_0) = \cdots = 0,$$

and so $h(z) = 0$ inside $C$. We may now expand about a new point, which can lie anywhere within $C$, and repeat the process. By continuing this procedure we may show that $h(z) = 0$ throughout $R$.

This is called the *identity theorem*, and also applies when $f(z) = g(z)$ just along some curve in $R$, or even just at a countably infinite number of points in $R$. ◄

So far we have assumed that $f(z)$ is analytic inside and on the (circular) contour $C$. If, however, $f(z)$ has a singularity inside $C$ at the point $z = z_0$, then it cannot be expanded as a Taylor series. Nevertheless, suppose, for example, that $f(z)$ has a pole of order $p$ at $z = z_0$ but is analytic at every other point inside $C$ and on $C$ itself. Then the function $g(z) = (z - z_0)^p f(z)$ is analytic at $z = z_0$, and so may be expanded as a Taylor series about $z = z_0$,

$$g(z) = \sum_{n=0}^{\infty} b_n (z - z_0)^n. \qquad (18.66)$$

Thus, for all $z$ inside $C$, $f(z)$ will have a power series representation of the form

$$f(z) = \frac{a_{-p}}{(z - z_0)^p} + \cdots + \frac{a_{-1}}{z - z_0} + a_0 + a_1(z - z_0) + a_2(z - z_0)^2 + \cdots, \qquad (18.67)$$

with $a_{-p} \neq 0$. Such a series, which is an extension of the Taylor expansion, is called a *Laurent series*. By comparing the coefficients in (18.66) and (18.67), we see that $a_n = b_{n+p}$. Now, the coefficients $b_n$ in the Taylor expansion of $g(z)$ are

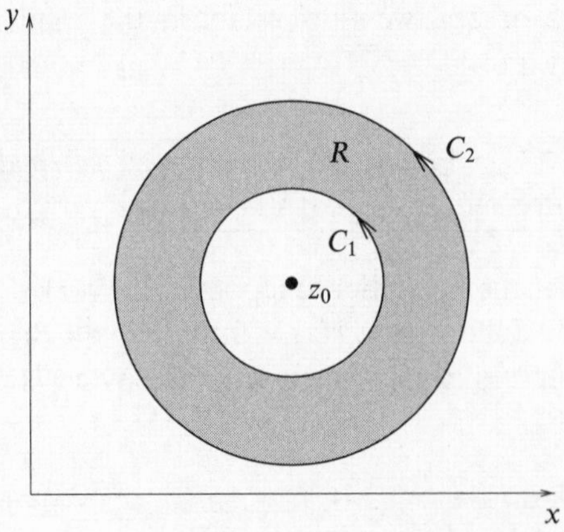


Figure 18.16 The region of convergence $R$ for a Laurent series of $f(z)$ about a point $z = z_0$ where $f(z)$ has a singularity.

seen from (18.65) to be given by

$$b_n = \frac{g^{(n)}(z_0)}{n!} = \frac{1}{2\pi i} \oint \frac{g(z)}{(z - z_0)^{n+1}}\, dz,$$

and so for the coefficients $a_n$ in (18.67) we have

$$a_n = \frac{1}{2\pi i} \oint \frac{g(z)}{(z - z_0)^{n+1+p}}\, dz = \frac{1}{2\pi i} \oint \frac{f(z)}{(z - z_0)^{n+1}}\, dz,$$

an expression that is valid for both positive and negative $n$.

The terms in the Laurent series with $n \geq 0$ are collectively called the *analytic part*, whilst the remainder of the series, consisting of terms in inverse powers of $z$, is called the *principal part*. Depending on the nature of the point $z = z_0$, the principal part may contain an infinite number of terms, so that

$$f(z) = \sum_{n=-\infty}^{+\infty} a_n (z - z_0)^n. \tag{18.68}$$

In this case we would expect the principal part to converge only for $|(z - z_0)^{-1}|$ less than some constant, i.e. *outside* some circle centred on $z_0$. However, the analytic part will converge *inside* some (different) circle also centred on $z_0$. If the latter circle has the greater radius, the Laurent series will converge in the region $R$ *between* the two circles (see figure 18.16); otherwise it does not converge at all.

In fact it may be shown that any function $f(z)$ that is analytic in a region $R$ between any two such circles $C_1$ and $C_2$ centred on $z = z_0$ can be expressed as a Laurent series about $z_0$ that converges in $R$. We note that, depending on the nature of the point $z = z_0$, the inner circle may be a point (when the principal part contain only a finite number of terms) and the outer circle may have an infinite radius.

We may use the Laurent series of a function $f(z)$ about any point $z = z_0$ to classify the nature of that point. If $f(z)$ is actually analytic at $z = z_0$ then in (18.68) all $a_n$ for $n < 0$ must be zero. It may happen that not only are all $a_n$ zero for $n < 0$ but $a_0, a_1, \ldots, a_{m-1}$, are all zero as well. In this case the first non-vanishing term in (18.68) is $a_m(z - z_0)^m$ with $m > 0$, and $f(z)$ is then said to have a *zero of order* $m$ at $z = z_0$.

If $f(z)$ is not analytic at $z = z_0$ then two cases arise, as discussed above ($p$ is here taken as positive).

(i) It is possible to find an integer $p$ such that $a_{-p} \neq 0$ but $a_{-p-k} = 0$ for all integers $k > 0$.

(ii) It is not possible to find such a lowest value of $-p$.

In case (i), $f(z)$ is of the form (18.67) and is described as having a *pole of order* $p$ at $z = z_0$; the value of $a_{-1}$ (not $a_{-p}$) is called the *residue* of $f(z)$ at the pole $z = z_0$, and will play an important part in later applications.

For case (ii), in which the negatively decreasing powers of $z - z_0$ do not terminate, $f(z)$ is said to have an *essential singularity*. These definitions should be compared with those given in section 18.6.

►*Find the Laurent series of*

$$f(z) = \frac{1}{z(z-2)^3}$$

*about the singularities* $z = 0$ *and* $z = 2$ *(separately). Hence verify that* $z = 0$ *is a pole of order* 1 *and* $z = 2$ *is a pole of order* 3*, and find the residue of* $f(z)$ *at each pole.*

To obtain the Laurent series about $z = 0$, we simply write $f(z)$ as

$$\begin{aligned} f(z) &= -\frac{1}{8z(1-z/2)^3} \\ &= -\frac{1}{8z}\left[1 + (-3)\left(-\frac{z}{2}\right) + \frac{(-3)(-4)}{2!}\left(-\frac{z}{2}\right)^2 + \frac{(-3)(-4)(-5)}{3!}\left(-\frac{z}{2}\right)^3 + \cdots\right] \\ &= -\frac{1}{8z} - \frac{3}{16} - \frac{3z}{16} - \frac{5z^2}{32} - \cdots. \end{aligned}$$

Since the lowest power of $z$ is $-1$, the point $z = 0$ is a pole of order 1. The residue of $f(z)$ at $z = 0$ is simply the coefficient of $z^{-1}$ in the Laurent expansion about that point, and is equal to $-1/8$.

The Laurent series about $z = 2$ is most easily found by letting $z = 2 + \xi$ (or

$z-2=\xi$) and substituting into the expression for $f(z)$ to obtain

$$\begin{aligned}
f(z) &= \frac{1}{(2+\xi)\xi^3} = \frac{1}{2\xi^3(1+\xi/2)} \\
&= \frac{1}{2\xi^3}\left[1-\left(\frac{\xi}{2}\right)+\left(\frac{\xi}{2}\right)^2-\left(\frac{\xi}{2}\right)^3+\left(\frac{\xi}{2}\right)^4-\cdots\right] \\
&= \frac{1}{2\xi^3}-\frac{1}{4\xi^2}+\frac{1}{8\xi}-\frac{1}{16}+\frac{\xi}{32}-\cdots \\
&= \frac{1}{2(z-2)^3}-\frac{1}{4(z-2)^2}+\frac{1}{8(z-2)}-\frac{1}{16}+\frac{z-2}{32}-\cdots.
\end{aligned}$$

From this series we see that $z=2$ is a pole of order 3, and that the residue of $f(z)$ at $z=2$ is $1/8$. ◄

As we shall see in the next section, finding the residue of a function at a singularity is of crucial importance in the evaluation of complex integrals. Indeed formulae exist for calculating the residue of a function at a particular (singular) point $z=z_0$, without having to expand the function explicitly as a Laurent series about $z_0$ and identify the coefficient of $(z-z_0)^{-1}$. The type of formula generally depends on the nature of the singularity at which the residue is required.

►*Suppose $f(z)$ has a pole of order $m$ at the point $z=z_0$. By considering the Laurent series of $f(z)$ about $z_0$, derive a general expression for the residue of $f(z)$ at $z=z_0$. Hence evaluate the residue of the function*

$$f(z)=\frac{\exp iz}{(z^2+1)^2}$$

*at the point $z=i$.*

If $f(z)$ has a pole of order $m$ at $z=z_0$ then its Laurent series about this point has the form

$$f(z)=\frac{a_{-m}}{(z-z_0)^m}+\cdots+\frac{a_{-1}}{(z-z_0)}+a_0+a_1(z-z_0)+a_2(z-z_0)^2+\cdots,$$

which, on multiplying both sides of the equation by $(z-z_0)^m$, gives

$$(z-z_0)^m f(z)=a_{-m}+a_{-m+1}(z-z_0)+\cdots+a_{-1}(z-z_0)^{m-1}+\cdots.$$

Differentiating both sides $m-1$ times, we obtain

$$\frac{d^{m-1}}{dz^{m-1}}[(z-z_0)^m f(z)]=(m-1)!\,a_{-1}+\sum_{n=1}^{\infty}b_n(z-z_0)^n,$$

for some coefficients $b_n$. In the limit $z\to z_0$, however, the terms in the sum disappear, and after rearranging we obtain the formula

$$R(z_0)=a_{-1}=\lim_{z\to z_0}\left\{\frac{1}{(m-1)!}\frac{d^{m-1}}{dz^{m-1}}[(z-z_0)^m f(z)]\right\}, \qquad (18.69)$$

which gives the value of the residue of $f(z)$ at the point $z = z_0$.

If we now consider the function

$$f(z) = \frac{\exp iz}{(z^2+1)^2} = \frac{\exp iz}{(z+i)^2(z-i)^2},$$

we see immediately that it has poles of order 2 (or *double* poles) at $z = i$ and $z = -i$. To calculate the residue at (for example) $z = i$, we may apply the formula (18.69) with $m = 2$. Performing the required differentiation we obtain

$$\begin{aligned}\frac{d}{dz}[(z-i)^2 f(z)] &= \frac{d}{dz}\left[\frac{\exp iz}{(z+i)^2}\right] \\ &= \frac{1}{(z+i)^4}[(z+i)^2 i \exp iz - 2(\exp iz)(z+i)].\end{aligned}$$

Setting $z = i$ we find the residue is given by

$$R(i) = \frac{1}{1!}\frac{1}{16}\left(-4ie^{-1} - 4ie^{-1}\right) = -\frac{i}{2e}. \blacktriangleleft$$

An important special case of (18.69) occurs when $f(z)$ has a *simple pole* (a pole of order 1) at $z = z_0$. Then the residue at $z_0$ is given simply by

$$R(z_0) = \lim_{z\to z_0}[(z-z_0)f(z)]. \tag{18.70}$$

If $f(z)$ has a simple pole at $z = z_0$ and, as is often the case, $f(z)$ has the form $g(z)/h(z)$, where $g(z)$ is analytic and non-zero at $z_0$ and $h(z_0) = 0$, then (18.70) becomes

$$\begin{aligned}R(z_0) &= \lim_{z\to z_0}\frac{(z-z_0)g(z)}{h(z)} = g(z_0)\lim_{z\to z_0}\frac{(z-z_0)}{h(z)} \\ &= g(z_0)\lim_{z\to z_0}\frac{1}{h'(z)} = \frac{g(z_0)}{h'(z_0)},\end{aligned} \tag{18.71}$$

where we have used l'Hôpital's rule. This result often provides the simplest way of determining the residue at a simple pole.

## 18.14 Residue theorem

Having seen from Cauchy's theorem that the value of an integral round a closed contour $C$ is zero if the integrand is analytic inside the contour, it is natural to ask what value it takes when the integrand is not analytic inside $C$. The answer to this is contained in the residue theorem, which we now discuss.

Suppose the function $f(z)$ has a pole of order $m$ at the point $z = z_0$, and so can be written as a Laurent series about $z_0$ of the form

$$f(z) = \sum_{n=-m}^{\infty} a_n(z-z_0)^n. \tag{18.72}$$

Now consider the integral $I$ of $f(z)$ around a closed contour $C$ that encloses

$z = z_0$, but no other singular points. Using Cauchy's theorem this integral has the same value as the integral around a circle $\gamma$ of radius $\rho$ centred on $z = z_0$, since $f(z)$ is analytic in the region between $C$ and $\gamma$. On the circle we have $z = z_0 + \rho \exp i\theta$ (and $dz = i\rho \exp i\theta \, d\theta$), and so

$$\begin{aligned} I &= \oint_\gamma f(z)\, dz \\ &= \sum_{n=-m}^{\infty} a_n \oint (z - z_0)^n \, dz \\ &= \sum_{n=-m}^{\infty} a_n \int_0^{2\pi} i\rho^{n+1} \exp[i(n+1)\theta] \, d\theta. \end{aligned}$$

For every term in the series with $n \neq -1$, we have

$$\int_0^{2\pi} i\rho^{n+1} \exp[i(n+1)\theta] \, d\theta = \left[ \frac{i\rho^{n+1} \exp[i(n+1)\theta]}{i(n+1)} \right]_0^{2\pi} = 0,$$

but for the $n = -1$ term we obtain

$$\int_0^{2\pi} i \, d\theta = 2\pi i.$$

Therefore only the term in $(z - z_0)^{-1}$ contributes to the value of the integral around $\gamma$ (and therefore $C$), and $I$ takes the value

$$I = \oint_C f(z)\, dz = 2\pi i a_{-1}. \tag{18.73}$$

Thus the integral around any closed contour containing a single pole of general order $m$ (or, by extension, an essential singularity) is equal to $2\pi i$ times the residue of $f(z)$ at $z = z_0$.

If we extend the above argument to the case where $f(z)$ is continuous within and on a closed contour $C$ and analytic, except for a finite number of poles, within $C$, then we arrive at the *residue theorem*

$$\oint_C f(z)\, dz = 2\pi i \sum_j R_j, \tag{18.74}$$

where $\sum_j R_j$ is the sum of the residues of $f(z)$ at its poles within $C$.

The method of proof is indicated by figure 18.17, in which ($a$) shows the original contour $C$ referred to in (18.74), and ($b$) shows a contour $C'$ giving the same value to the integral – because $f$ is analytic between $C$ and $C'$. Now the contribution to the $C'$ integral from the polygon (a triangle for the case illustrated) joining the small circles is zero, since $f$ is also analytic inside $C'$. Hence the whole value of the integral comes from the circles and, by result (18.73), each of these contributes $2\pi i$ times the residue at the pole it encloses. All the circles are traversed in their positive sense if $C$ is thus traversed and so the residue theorem follows. Formally,

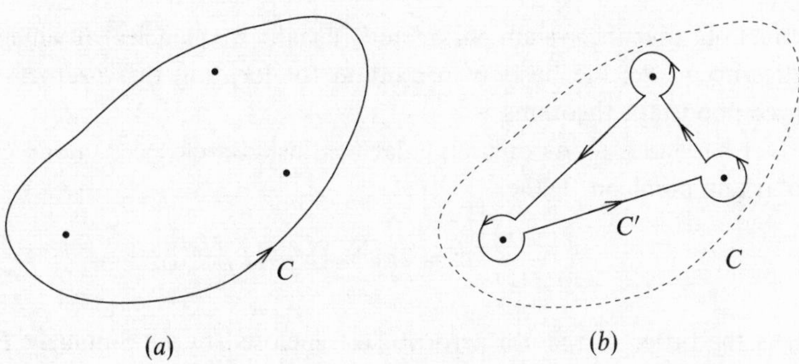


Figure 18.17 The contours used to prove the residue theorem: (*a*) the original contour; (*b*) the contracted contour encircling each of the poles.

Cauchy's theorem (18.53) is a particular case of (18.74) in which $C$ encloses no poles.

Finally we mention another important result, which we will use later. Suppose that $f(z)$ has a simple pole at $z = z_0$, and so may be expanded as the Laurent series

$$f(z) = \phi(z) + a_{-1}(z - z_0)^{-1},$$

where $\phi(z)$ is analytic within some neighbourhood surrounding $z_0$. Let us consider the integral $I$ of $f(z)$ along the *open* contour $C$, which is the arc of a circle of radius $\rho$ centred on $z = z_0$ given by

$$|z - z_0| = \rho, \qquad \theta_1 \leq \arg(z - z_0) \leq \theta_2, \tag{18.75}$$

where $\rho$ is chosen small enough that no singularity of $f$, other than $z = z_0$, lies within the circle. Then $I$ is given by

$$I = \int_C f(z)\,dz = \int_C \phi(z)\,dz + a_{-1}\int_C (z - z_0)^{-1}\,dz.$$

Now let us investigate the value of $I$ as the radius of the arc $C$ tends to zero. As $\rho \to 0$ the first integral tends to zero, since the path becomes of zero length and $\phi$ is analytic and therefore continuous along it. On $C$, $z = \rho e^{i\theta}$ and hence the value of $I$ is given by

$$I = \lim_{\rho\to 0}\int_C f(z)\,dz = \lim_{\rho\to 0}\left(a_{-1}\int_{\theta_1}^{\theta_2}\frac{1}{\rho e^{i\theta}}i\rho e^{i\theta}\,d\theta\right) = ia_{-1}(\theta_2 - \theta_1). \tag{18.76}$$

## 18.15 Location of zeroes

An important use of the residue theorem is in locating the zeroes of functions of a complex variable. The location of such zeroes has a particular application in electrical network and general oscillation theory, since the complex zeroes of certain functions give the system parameters (usually frequencies) at which system instabilities occur. As the basis of a method for locating these zeroes we next prove three important theorems.

(i) If $f(z)$ has poles as its only singularities inside a closed contour $C$ and is not zero at any point on $C$, then

$$\oint_C \frac{f'(z)}{f(z)}\,dz = 2\pi i \sum_j (N_j - P_j). \tag{18.77}$$

Here $N_j$ is the order of the $j$th zero of $f(z)$ enclosed by $C$. Similarly $P_j$ is the order of the $j$th pole of $f(z)$ inside $C$.

To prove this we note that at each position $z_j$, $f(z)$ can be written as

$$f(z) = (z - z_j)^{m_j}\phi(z), \tag{18.78}$$

where $\phi(z)$ is analytic and non-zero at $z = z_j$ and $m_j$ is positive for a zero and negative for a pole. Then the integrand $f'(z)/f(z)$ takes the form

$$\frac{f'(z)}{f(z)} = \frac{m_j}{z - z_j} + \frac{\phi'(z)}{\phi(z)}. \tag{18.79}$$

Since $\phi(z_j) \neq 0$, the second term on the right is analytic; thus the integrand has a simple pole at $z = z_j$, with residue $m_j$. For zeroes $m_j = N_j$, and for poles $m_j = -P_j$, and thus by the residue theorem (18.77) follows.

(ii) If $f(z)$ is analytic inside $C$ and not zero at any point on it, then

$$2\pi \sum_j N_j = \Delta_C[\arg f(z)], \tag{18.80}$$

where $\Delta_C[x]$ denotes the variation in $x$ around the contour $C$.

Since $f$ is analytic there are no $P_j$, and further since

$$\frac{f'(z)}{f(z)} = \frac{d}{dz}[\operatorname{Ln} f(z)], \tag{18.81}$$

equation (18.77) can be written

$$2\pi i \sum N_j = \oint_C \frac{f'(z)}{f(z)}\,dz = \Delta_C[\operatorname{Ln} f(z)]. \tag{18.82}$$

But

$$\Delta_C[\operatorname{Ln} f(z)] = \Delta_C[\ln |f(z)|] + i\Delta_C[\arg f(z)], \tag{18.83}$$

and since $C$ is a closed contour, $\ln|f(z)|$ must return to its original value, and the real term on the right is zero. Comparison of (18.82) and (18.83) then establishes (18.80), which is known as the *principle of the argument*.

(iii) If $f(z)$ and $g(z)$ are analytic within and on a closed contour $C$ and $|g(z)| < |f(z)|$ on $C$, then $f(z)$ and $f(z) + g(z)$ have the same number of zeroes inside $C$; this is *Rouché's theorem*.

With the conditions given, neither $f(z)$ nor $f(z)+g(z)$ can have a zero on $C$. So applying theorem (ii) with an obvious notation,

$$\begin{aligned} 2\pi\textstyle\sum_j N_j(f+g) &= \Delta_C[\arg(f+g)] \\ &= \Delta_C[\arg f] + \Delta_C[\arg(1+g/f)] \\ &= 2\pi\textstyle\sum_k N_k(f) + \Delta_C[\arg(1+g/f)]. \end{aligned} \qquad (18.84)$$

Further, since $|g| < |f|$ on $C$, $1+g/f$ always lies *within* a unit circle centred on $z=1$, thus its argument *always* lies in the range $-\pi/2 < \arg(1+g/f) < \pi/2$ and cannot change by any multiple of $2\pi$. It must therefore return to its original value when $z$ returns to its starting point having traversed $C$. Hence the second term on the right of (18.84) is zero and the theorem is established.

The importance of Rouché's theorem is that for some functions, in particular polynomials, only the behaviour of a single term in the function need be considered if the contour is chosen appropriately. For example, for a polynomial, treated as $f(z)+g(z)$, only the properties of its largest- (smallest-) power, treated as $f(z)$, need be investigated, if a circular contour of large (small) enough radius $R$ is chosen such that, on the contour, the magnitude of the largest (smallest) power term is greater than the sum of the magnitudes of all other terms. Further, if the zeroes of $f(z)+g(z) = \sum_0^N b_n z^n$ are considered as the roots of $f(z)+g(z)=0$, written in the form

$$1 + \frac{g(z)}{f(z)} = 0, \qquad (18.85)$$

then it is apparent that no roots can lie outside (inside) $|z| = R$ and also that $f(z) = b_N z^N$ (or $b_0$) has $N$ (or 0) zeroes inside $|z| = R$; $f+g$ consequently has the same number of zeroes inside the same circle.

A weak form of the *maximum-modulus theorem* may also be deduced. This states that if $f(z)$ is analytic within and on a simple closed contour $C$, then $|f(z)|$ attains its maximum value on the boundary of $C$.

Let $|f(z)| \le M$ on $C$ with equality at one point at least of $C$. Now suppose that there is a point $z=a$ inside $C$ such that $|f(a)| > M$. Then the function $h(z) \equiv f(a)$ is such that $|h(z)| > |-f(z)|$ on $C$, and thus $h(z)$ and $h(z)-f(z)$ have the same number of zeroes inside $C$. But $h(z)$ ($\equiv f(a)$) has no zeroes inside $C$ and by Rouché's theorem this would imply that $f(a)-f(z)$ has no zeroes in $C$. However, $f(a)-f(z)$ clearly has a zero at $z=a$, and so we have a contradiction; the assumption of a point $z=a$ inside $C$ such that $|f(a)| > M$ must be invalid. This establishes the theorem.

The stronger form of the maximum-modulus theorem, which we do not prove, states in addition that the maximum value of $f(z)$ is not attained at any interior point except for the case where $f(z)$ is a constant.

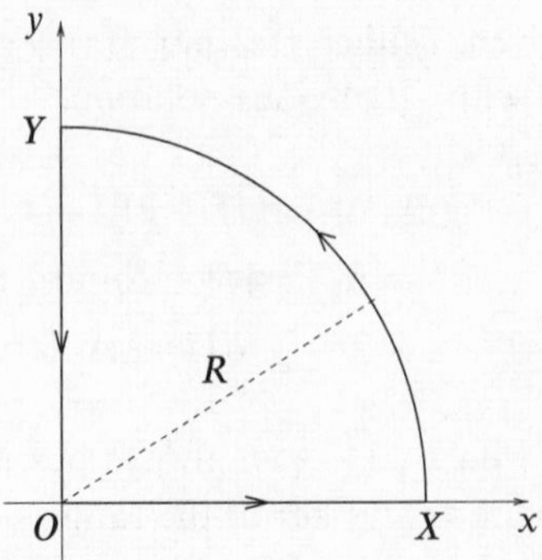


Figure 18.18 A contour for locating the zeroes of a polynomial that occur in the first quadrant of the Argand diagram.

► *Show that the four zeroes of $h(z) = z^4 + z + 1$ occur one in each quadrant of the Argand diagram and that all four lie between the circles $|z| = 2/3$ and $|z| = 3/2$.*

Putting $z = x$ and $z = iy$ shows that no zeroes occur on the real or imaginary axes. They must therefore occur in conjugate pairs (as is shown by taking the complex conjugate of $h(z) = 0$).

Now take $C$ as the contour $OXYO$ shown in figure 18.18 and consider the changes $\Delta[\arg h]$ in the argument of $h(z)$.

(i) $OX$: $\arg h$ is everywhere zero, since $h$ is real, and thus $\Delta_{OX}[\arg h] = 0$.

(ii) $XY$: $z = R \exp i\theta$ and so $\arg h$ changes by an amount

$$\begin{aligned}\Delta_{XY}[\arg h] &= \Delta_{XY}[\arg z^4] + \Delta_{XY}[\arg(1 + z^{-3} + z^{-4})] \\ &= \Delta_{XY}[\arg R^4 e^{4i\theta}] + \Delta_{XY}\left\{\arg[1 + \mathrm{O}(R^{-3})]\right\} \\ &= 2\pi + \mathrm{O}(R^{-3}). \end{aligned} \qquad (18.86)$$

(iii) $YO$: $\arg h = y/(y^4 + 1)$, which starts at $\mathrm{O}(R^{-3})$ and finishes at 0 as $y$ goes from large $R$ to 0. It never reaches $\pi/2$ because $y^4 + 1 = 0$ has no real positive root. Thus $\Delta_{YO}[\arg h] = 0$.

Hence for the complete contour $\Delta_C[\arg h] = 0 + 2\pi + 0 + \mathrm{O}(R^{-3})$ and, if $R$ is allowed to tend to infinity, we deduce from (18.80) that $h(z)$ has one zero in the first quadrant. Furthermore, since the roots occur in conjugate pairs, a second root must lie in the fourth quadrant and the other pair in the second and third quadrants.

To show that the zeroes lie within a given annulus in the $z$-plane we must apply Rouché's theorem.

(i) With $C$ as $|z| = 3/2$, $f = z^4$, $g = z + 1$. Now $|f| = 81/16$ on $C$ and

$|g| \leq 1 + |z| < 5/2 < 81/16$. Thus since $z^4 = 0$ has four roots inside $|z| = 3/2$, so also does $z^4 + z + 1 = 0$.

(ii) With $C$ as $|z| = 2/3$, $f = 1$, $g = z^4 + z$. Now $f = 1$ on $C$ and $|g| \leq |z^4| + |z| = 16/81 + 2/3 = 70/81 < 1$. Thus since $f = 0$ has no roots inside $|z| = 2/3$, neither does $1 + z + z^4 = 0$.

Hence the four zeroes of $h(z) = z^4 + z + 1$ occur one in each quadrant and all lie between the circles $|z| = 2/3$ and $|z| = 3/2$. ◄

A further technique useful in locating function zeroes is explained in exercise 18.13.

## 18.16 Integrals of sinusoidal functions

The remainder of this chapter is devoted to methods of applying contour integration and the residue theorem to various types of definite integral. In each case not much preamble is given since, for this material, the simplest explanation is felt to be via a series of worked examples that can be used as models.

Suppose that an integral of the form

$$\int_0^{2\pi} F(\cos\theta, \sin\theta)\, d\theta \tag{18.87}$$

is to be evaluated. It can be made into a contour integral around the unit circle $C$ by writing $z = \exp i\theta$ and hence

$$\cos\theta = \tfrac{1}{2}(z + z^{-1}), \qquad \sin\theta = -\tfrac{1}{2}i(z - z^{-1}), \qquad d\theta = -iz^{-1}\, dz. \tag{18.88}$$

This contour integral can then be evaluated using the residue theorem, provided the transformed integrand has only a finite number of poles inside the unit circle and none on it.

►*Evaluate*

$$I = \int_0^{2\pi} \frac{\cos 2\theta}{a^2 + b^2 - 2ab\cos\theta}\, d\theta, \qquad b > a > 0. \tag{18.89}$$

By de Moivre's theorem (section 2.4),

$$\cos n\theta = \tfrac{1}{2}(z^n + z^{-n}). \tag{18.90}$$

Using $n = 2$ in (18.90) and straightforward substitution for the other functions of $\theta$ in (18.89) gives

$$I = \frac{i}{2ab} \oint_C \frac{z^4 + 1}{z^2(z - a/b)(z - b/a)}\, dz.$$

Thus there are two poles inside $C$, a double pole at $z = 0$ and a simple pole $z = a/b$ (recall that $b > a$).

We could find the residue of the integrand at $z = 0$ by expanding the integrand as a Laurent series in $z$ and identifying the coefficient of $z^{-1}$. Alternatively, we may use the formula (18.69) with $m = 2$. Denoting the integrand by $f(z)$ we have

$$\frac{d}{dz}[z^2 f(z)] = \frac{d}{dz}\left[\frac{z^4+1}{(z-a/b)(z-b/a)}\right]$$
$$= \frac{(z-a/b)(z-b/a)4z^3 - (z^4+1)[(z-a/b)+(z-b/a)]}{(z-a/b)^2(z-b/a)^2}.$$

Setting $z = 0$ and applying (18.69), we find

$$R(0) = \frac{a}{b} + \frac{b}{a}.$$

For the simple pole at $z = a/b$, using (18.70) the residue is given by

$$R(a/b) = \lim_{z\to(a/b)} \left[(z-a/b)f(z)\right] = \frac{(a/b)^4+1}{(a/b)^2(a/b-b/a)}$$
$$= -\frac{a^4+b^4}{ab(b^2-a^2)}.$$

Therefore by the residue theorem

$$I = 2\pi i \times \frac{i}{2ab}\left[\frac{a^2+b^2}{ab} - \frac{a^4+b^4}{ab(b^2-a^2)}\right] = \frac{2\pi a^2}{b^2(b^2-a^2)}. \blacktriangleleft$$

## 18.17 Some infinite integrals

Suppose we wish to evaluate an integral of the form

$$\int_{-\infty}^{\infty} f(x)\, dx,$$

where $f(z)$ has the following properties.

(i) $f(z)$ is analytic in the upper half-plane, $\operatorname{Im} z \geq 0$, except for a finite number of poles, but with none on the real axis.

(ii) On the semicircle $\Gamma$ (figure 18.19) of radius $R$, $R$ times the maximum of $|f|$ on $\Gamma$ tends to zero as $R \to \infty$ (a sufficient condition is that $zf(z) \to 0$ as $|z| \to \infty$).

(iii) $\int_{-\infty}^{0} f(x)\, dx$ and $\int_{0}^{\infty} f(x)\, dx$ both exist.

The required integral is then given by

$$\int_{-\infty}^{\infty} f(x)\, dx = 2\pi i \times (\text{sum of the residues at poles with } \operatorname{Im} z \geq 0). \tag{18.91}$$

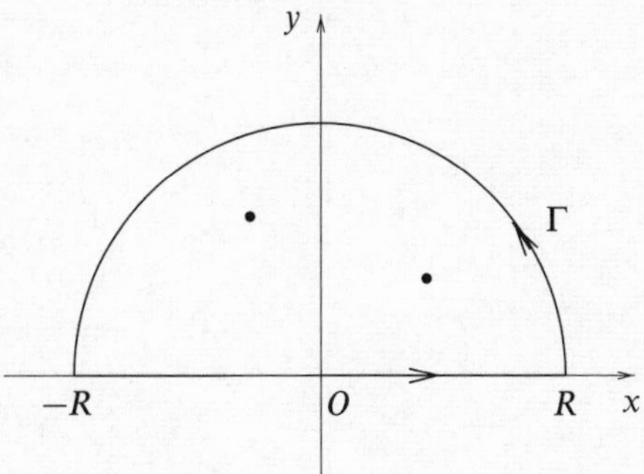


Figure 18.19 A semicircular contour in the upper half-plane.

Condition (ii) ensures that

$$\left| \int_\Gamma f(z)\, dz \right| \le 2\pi R \times (\text{maximum of } |f| \text{ on } \Gamma),$$

which tends to zero as $R \to \infty$, after which (18.91) is obvious from the residue theorem.

►*Evaluate*

$$I = \int_0^\infty \frac{dx}{(x^2+a^2)^4}, \qquad \textit{where } a \textit{ is real.}$$

The complex function $(z^2+a^2)^{-4}$ has poles of order 4 at $z = \pm ai$ of which only $z = ai$ is in the upper half-plane. Conditions (ii) and (iii) are clearly satisfied. For higher-order poles, the formula (18.69) for evaluating residues can be tiresome to apply. Thus we put $z = ai + \xi$ and expand for small $\xi$ to obtain

$$\frac{1}{(z^2+a^2)^4} = \frac{1}{(2ai\xi + \xi^2)^4} = \frac{1}{(2ai\xi)^4}\left(1 - \frac{i\xi}{2a}\right)^{-4}.$$

The coefficient of $\xi^{-1}$ is

$$\frac{1}{(2a)^4}\frac{(-4)(-5)(-6)}{3!}\left(\frac{-i}{2a}\right)^3 = \frac{-5i}{32a^7},$$

and hence by the residue theorem

$$\int_{-\infty}^{\infty} \frac{dx}{(x^2+a^2)^4} = \frac{10\pi}{32a^7},$$

and so $I = 5\pi/32a^7$. ◄

Condition (i) of the previous method required there to be no poles of the integrand on the real axis, but simple poles on the real axis can in fact be

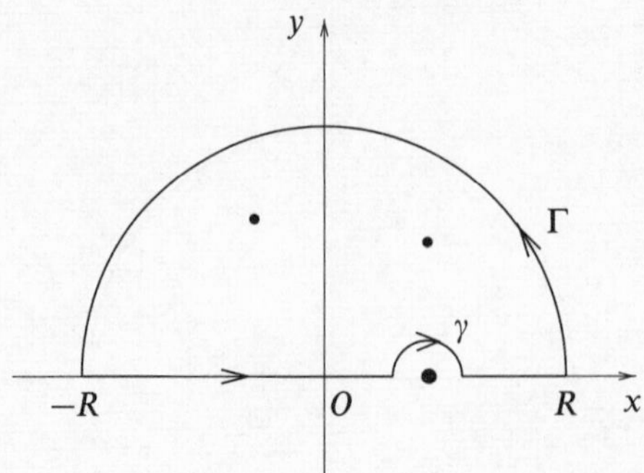


Figure 18.20 An indented contour used when the integrand has a simple pole on the real axis.

accommodated by indenting the contour as shown in figure 18.20. The indentation at the pole $z = z_0$ is in the form of a semicircle $\gamma$ of radius $\rho$ in the upper half-plane, thus excluding the pole from the interior of the contour.

What is then obtained from a contour integration, apart from the contributions for $\Gamma$ and $\gamma$, is called the *principal value of the integral,* defined as $\rho \to 0$ by:

$$P \int_{-R}^{R} f(x)\, dx \equiv \int_{-R}^{z_0-\rho} f(x)\, dx + \int_{z_0+\rho}^{R} f(x)\, dx,$$

The remainder of the calculation goes through as before, but the contribution from the semicircle $\gamma$ must be included. Result (18.76) of section 18.14 shows that since only a simple pole is involved its contribution is

$$-ia_{-1}\pi, \tag{18.92}$$

where $a_{-1}$ is the residue at the pole and the minus sign arises because $\gamma$ is traversed in the clockwise (negative) sense.

We defer giving an example of an indented contour until we have established Jordan's lemma; we will then work through an example illustrating both. *Jordan's lemma* enables infinite integrals involving sinusoidal functions to be evaluated.

**Jordan's lemma.** *For a function $f(z)$ of a complex variable $z$, if*

(i) *$f(z)$ is analytic in the upper half-plane except for a finite number of poles in* $\text{Im}\, z > 0$,

(ii) *the maximum of $|f(z)| \to 0$ as $|z| \to \infty$ in the upper half-plane,*

(iii) $m > 0$,

*then*

$$I_\Gamma = \int_\Gamma e^{imz} f(z)\, dz \to 0 \quad \text{as } R \to \infty, \tag{18.93}$$

*where $\Gamma$ is the same semicircular contour as in figure 18.19.*

Notice that this condition (ii) is less stringent than the earlier condition (ii), since we now only require $M(R) \to 0$ and not $RM(R) \to 0$, where $M$ is the maximum† of $|f(z)|$ on $|z| = R$.

The proof of the lemma is straightforward, once it has been observed that for $0 \le \theta \le \frac{1}{2}\pi$

$$1 \ge \frac{\sin\theta}{\theta} \ge \frac{2}{\pi}. \tag{18.94}$$

Then, since on $\Gamma$ we have $|\exp(imz)| = |\exp(-mR\sin\theta)|$,

$$I_\Gamma \le \int_\Gamma |e^{imz} f(z)|\, |dz| \le MR \int_0^\pi e^{-mR\sin\theta}\, d\theta = 2MR \int_0^{\pi/2} e^{-mR\sin\theta}\, d\theta.$$

Thus, using (18.94),

$$I_\Gamma < 2MR \int_0^{\pi/2} e^{-mR(2\theta/\pi)}\, d\theta = \frac{\pi M}{m}\left(1 - e^{-mR}\right) < \frac{\pi M}{m},$$

and hence tends to zero since $M$ does, as $R \to \infty$.

►*Find the principal value of*

$$\int_{-\infty}^{\infty} \frac{\cos mx}{x-a}\, dx, \qquad \textit{for } a \textit{ real, } m > 0.$$

Consider the function $(z-a)^{-1}\exp(imz)$; although it has no poles in the upper half-plane it does have a simple pole at $z = a$, and further $|(z-a)^{-1}| \to 0$ as $|z| \to \infty$. We use a contour like that shown in figure 18.20 and apply the residue theorem. Symbolically,

$$\int_{-R}^{a-\rho} + \int_\gamma + \int_{a+\rho}^{R} + \int_\Gamma = 0. \tag{18.95}$$

Now as $R \to \infty$ and $\rho \to 0$, $\int_\Gamma \to 0$ by Jordan's lemma, and from (18.91) and (18.92) we obtain

$$P\int_{-\infty}^{\infty} \frac{e^{imx}}{x-a}\, dx - i\pi a_{-1} = 0, \tag{18.96}$$

where $a_{-1}$ is the residue of $(z-a)^{-1}\exp(imz)$ at $z = a$, which is $\exp(ima)$. Then taking the real and imaginary parts of (18.96) gives

$$P\int_{-\infty}^{\infty} \frac{\cos mx}{x-a}\, dx = -\pi \sin ma, \qquad \text{as required,}$$

$$P\int_{-\infty}^{\infty} \frac{\sin mx}{x-a}\, dx = \pi \cos ma, \qquad \text{as a bonus.} ◄$$

† More strictly the least upper bound.

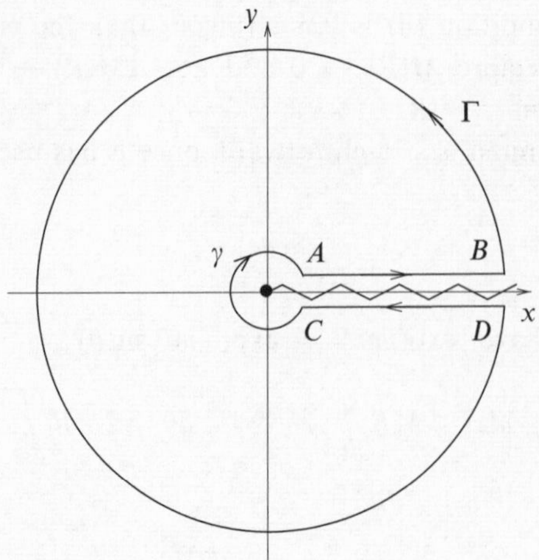


Figure 18.21 A typical cut-plane contour for use with multivalued functions that have a single branch point located at the origin.

## 18.18 Integrals of multivalued functions

We have discussed briefly some of the properties and difficulties associated with certain multivalued functions such as $z^{1/2}$ or $\mathrm{Ln}\, z$. It was mentioned that one method of managing such functions is by means of a 'cut plane'. A similar technique can be used with advantage to evaluate some kinds of infinite integral involving real functions for which the corresponding complex functions are multivalued. A typical contour employed for functions with a single branch point located at the origin is shown in figure 18.21. Here $\Gamma$ is a large circle of radius $R$ and $\gamma$ a small one of radius $\rho$, both centred on the origin. Eventually we will let $R \to \infty$ and $\rho \to 0$.

The value of the method comes from the fact that because the integrand is multivalued, its values along the two lines $AB$ and $CD$ joining $z = \rho$ to $z = R$ are *not* equal and opposite although both are related to the corresponding real integral. Again an example gives the best explanation.

►*Evaluate*

$$I = \int_0^\infty \frac{dx}{(x+a)^3 x^{1/2}}, \qquad a > 0.$$

We consider the integrand $f(z) = (z+a)^{-3} z^{-1/2}$ and note that $|zf(z)| \to 0$ on the two circles as $\rho \to 0$ and $R \to \infty$. Thus the two circles make no contribution to the contour integral.

The only pole of the integrand inside the contour is at $z = -a$ (and is of order 3). To determine its residue we put $z = -a + \xi$ and expand (noting that $(-a)^{1/2}$ equals $a^{1/2}\exp(i\pi/2) = ia^{1/2}$):

$$\frac{1}{(z+a)^3 z^{1/2}} = \frac{1}{\xi^3 i a^{1/2}(1-\xi/a)^{1/2}}$$
$$= \frac{1}{i\xi^3 a^{1/2}}\left(1 + \frac{1}{2}\frac{\xi}{a} + \frac{3}{8}\frac{\xi^2}{a^2} + \cdots\right).$$

The residue is thus $-3i/8a^{5/2}$.

The residue theorem (18.74) now gives

$$\int_{AB} + \int_{\Gamma} + \int_{DC} + \int_{\gamma} = 2\pi i\left(\frac{-3i}{8a^{5/2}}\right).$$

We have seen that $\int_\Gamma$ and $\int_\gamma$ vanish, and if we denote $z$ by $x$ along the line $AB$ then it has the value $z = x\exp 2\pi i$ along the line $DC$ (note that $\exp 2\pi i$ must not be set equal to 1 until after the substitution for $z$ has been made in $\int_{DC}$). Substituting these expressions,

$$\int_0^\infty \frac{dx}{(x+a)^3 x^{1/2}} + \int_\infty^0 \frac{dx}{[x\exp 2\pi i + a]^3 x^{1/2}\exp(\frac{1}{2}2\pi i)} = \frac{3\pi}{4a^{5/2}}.$$

Thus

$$\left(1 - \frac{1}{\exp \pi i}\right)\int_0^\infty \frac{dx}{(x+a)^3 x^{1/2}} = \frac{3\pi}{4a^{5/2}},$$

and

$$I = \frac{1}{2} \times \frac{3\pi}{4a^{5/2}}. \blacktriangleleft$$

## 18.19 Summation of series

Sometimes a real infinite series may be summed if a suitable complex function can be found that has poles on the real axis at positions corresponding to the values of the dummy variable in the summation, and whose residues at these poles are equal to the values of the terms of the series there.

►*By considering*

$$\oint_C \frac{\pi \cot \pi z}{(a+z)^2}\, dz,$$

*where a is not an integer and C is a circle of large radius, evaluate*

$$\sum_{n=-\infty}^{\infty} \frac{1}{(a+n)^2}.$$

The integrand has (i) simple poles at $z =$ integral $n$, $-\infty < n < \infty$, (ii) a double pole at $z = -a$.

(i) To find the residue of $\cot \pi z$, put $z = n + \xi$ for small $\xi$:

$$\cot \pi z = \frac{\cos(n\pi + \xi\pi)}{\sin(n\pi + \xi\pi)} \approx \frac{\cos n\pi}{(\cos n\pi)\xi\pi} = \frac{1}{\xi\pi}.$$

The residue of the integrand at $z = n$ is thus $\pi(a+n)^{-2}\pi^{-1}$.

(ii) Putting $z = -a + \xi$ for small $\xi$ and determining the coefficient of $\xi^{-1}$,†

$$\begin{aligned}\frac{\pi \cot \pi z}{(a+z)^2} &= \frac{\pi}{\xi^2} \cot(-a\pi + \xi\pi) \\ &= \frac{\pi}{\xi^2}\left\{\cot(-a\pi) + \xi\left[\frac{d}{dz}(\cot \pi z)\right]_{z=-a} + \cdots\right\},\end{aligned}$$

so the residue at the double pole $z = -a$ is

$$\pi[-\pi \operatorname{cosec}^2 \pi z]_{z=-a} = -\pi^2 \operatorname{cosec}^2 \pi a.$$

Collecting together these results to express the residue theorem gives

$$I = \oint_C \frac{\pi \cot \pi z}{(a+z)^2}\, dz = 2\pi i \left[\sum_{n=-N}^{N} \frac{1}{(a+n)^2} - \pi^2 \operatorname{cosec}^2 \pi a\right], \qquad (18.97)$$

where $N$ equals the integer part of $R$. But as the radius $R$ of $C$ tends to $\infty$, $\cot \pi z \to \mp i$ (depending on whether Im $z$ is greater or less than zero respectively). Thus

$$I < k \int \frac{dz}{(a+z)^2},$$

which tends to 0 as $R \to \infty$. Thus $I \to 0$ as $R$ (and hence $N$) $\to \infty$ and (18.97) establishes the result

$$\sum_{n=-\infty}^{\infty} \frac{1}{(a+n)^2} = \frac{\pi^2}{\sin^2 \pi a}. \blacktriangleleft$$

Series with alternating signs in the terms, i.e. $(-1)^n$, can also be attempted in this way but using $\operatorname{cosec} \pi z$ instead of $\cot \pi z$, since the former has residue $(-1)^n \pi^{-1}$ at $z = n$; see exercise 18.25.

### 18.20 Inverse Laplace transform

As a final example of contour integration we mention a method whereby the process of Laplace transformation, discussed in chapter 11, can be inverted.

It will be recalled that the Laplace transform $\bar{f}(s)$ of a function $f(x)$, $x \geq 0$, is given by

$$\bar{f}(s) = \int_0^{\infty} e^{-sx} f(x)\, dx, \qquad \text{Re } s > s_0. \qquad (18.98)$$

In chapter 11, functions $f(x)$ were deduced from the transforms by means of a

† This illustrates another useful technique for determining residues.

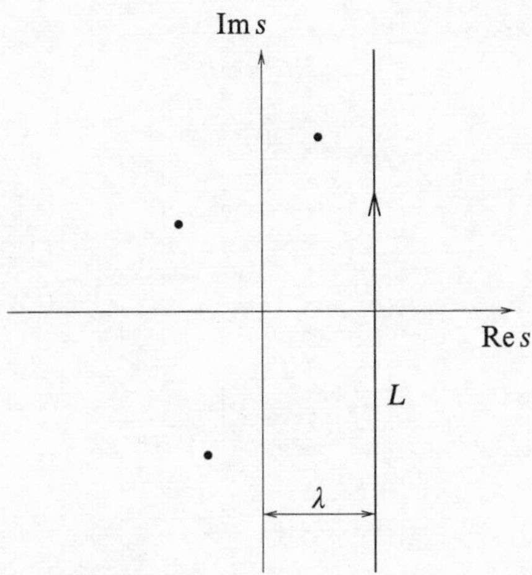


Figure 18.22 The integration path of the inverse Laplace transform is along the infinite line $L$. The quantity $\lambda$ must be positive and large enough for all poles of the integrand to lie to the left of $L$.

prepared dictionary. However, an explicit formula for an unknown inverse may be written in the form of an integral. It is known as the *Bromwich integral*, and is given by

$$f(x) = \frac{1}{2\pi i}\int_{\lambda-i\infty}^{\lambda+i\infty} e^{sx}\bar{f}(s)\,ds, \qquad \lambda > 0, \tag{18.99}$$

where $s$ is treated as a complex variable and the integration is along the line $L$ indicated in figure 18.22. The position of the line is dictated by the requirements that $\lambda$ is positive and that all singularities of $\bar{f}(s)$ lie to the left of the line.

That (18.99) really is the unique inverse of (18.98) is difficult to show for general functions and transforms, but the following verification should at least make it plausible:

$$\begin{aligned}
f(x) &= \frac{1}{2\pi i}\int_{\lambda-i\infty}^{\lambda+i\infty} ds\, e^{sx}\int_0^\infty e^{-su} f(u)\,du, \qquad \text{Re}(s) > 0, \text{ i.e. } \lambda > 0,\\
&= \frac{1}{2\pi i}\int_0^\infty du\, f(u)\int_{\lambda-i\infty}^{\lambda+i\infty} e^{s(x-u)}\,ds\\
&= \frac{1}{2\pi i}\int_0^\infty du\, f(u)\int_{-\infty}^{\infty} e^{\lambda(x-u)}e^{ip(x-u)} i\,dp, \qquad \text{putting } s = \lambda + ip,\\
&= \frac{1}{2\pi}\int_0^\infty f(u)e^{\lambda(x-u)} 2\pi\delta(x-u)\,du\\
&= \begin{cases} f(x) & x \geq 0,\\ 0 & x < 0. \end{cases}
\end{aligned} \tag{18.100}$$

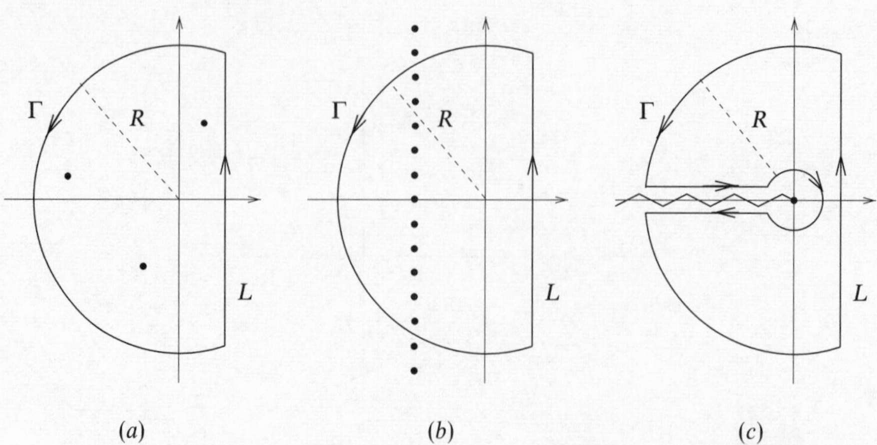


Figure 18.23 Some contour completions for the integration path $L$ of the inverse Laplace transform. For details of when each is appropriate see the main text.

Our main interest here is in the use of contour integration. To employ it to evaluate the line integral in (18.99), the path $L$ must be made into a closed contour in such a way that the contribution from the completion either vanishes or is simply calculable.

A typical completion is shown in figure 18.23($a$) and would be appropriate if $\bar{f}(s)$ had a finite number of poles. For more complicated cases in which $\bar{f}(s)$ has an infinite sequence of poles, but all to the left of $L$ as in figure 18.23($b$), a sequence of circular-arc completions that pass between the poles must be used and $f(x)$ is obtained as a series. If $\bar{f}(s)$ is a multivalued function, a cut plane is needed and a contour such as that shown in figure 18.23($c$) might be appropriate.

We consider here only the simple case in which the contour in figure 18.23($a$) is used; we refer the reader to the exercises at the end of the chapter for others. Ideally, we would like the contribution to the integral from the circular arc $\Gamma$ to tend to zero as its radius $R \to \infty$. Using a modified version of Jordan's lemma, it may be shown that this is indeed the case if there exist constants $M > 0$ and $\alpha > 0$ such that on $\Gamma$

$$|\bar{f}(s)| \leq \frac{M}{R^\alpha}.$$

Moreover, this condition always holds when $\bar{f}(s)$ has the form

$$\bar{f}(s) = \frac{P(s)}{Q(s)},$$

where $P(s)$ and $Q(s)$ are polynomials and the degree of $Q(s)$ is greater than that of $P(s)$.

When the contribution from the part-circle $\Gamma$ tends to zero as $R \to \infty$, we have from the residue theorem that the inverse Laplace transform (18.99) is given simply by

$$f(t) = \sum \left(\text{residues of } \bar{f}(s)e^{sx} \text{ at all poles}\right). \qquad (18.101)$$

►*Find the function $f(x)$ whose Laplace transform is*

$$\bar{f}(s) = \frac{s}{s^2 - k^2},$$

*where $k$ is a constant.*

It is clear that $\bar{f}(s)$ is of the form required for the integral over the circular arc $\Gamma$ to tend to zero as $R \to \infty$, and so we may use the result (18.101). Now

$$\bar{f}(s)e^{sx} = \frac{se^{sx}}{(s-k)(s+k)},$$

and thus has simple poles at $s = k$ and $s = -k$. Using (18.70) the residues at each pole can be easily calculated as

$$R(k) = \frac{ke^{kx}}{2k} \qquad \text{and} \qquad R(-k) = \frac{ke^{-kx}}{2k}.$$

Thus the inverse Laplace transform is given by

$$f(x) = \tfrac{1}{2}\left(e^{kx} + e^{-kx}\right) = \cosh kx.$$

This result may be checked by computing the forward transform of $\cosh kx$. ◄

Sometimes a little more care is required in deciding in which half-plane to close the contour $C$.

►*Find the function $f(x)$ whose Laplace transform is*

$$\bar{f}(s) = \frac{1}{s}(e^{-as} - e^{-bs}),$$

*where $a$ and $b$ are fixed and positive, with $b > a$.*

From (18.99) we have the integral

$$f(x) = \frac{1}{2\pi i}\int_{\lambda - i\infty}^{\lambda + i\infty} \frac{e^{(x-a)s} - e^{(x-b)s}}{s}\, ds. \qquad (18.102)$$

Now, despite appearances to the contrary, the integrand has no poles, as may be confirmed by expanding the exponentials as Taylor series about $s = 0$. Depending on the value of $x$, several cases arise.

(i) For $x < a$ both exponentials in the integrand will tend to zero as $\text{Re}\, s \to \infty$. Thus we may close $L$ with a circular arc $\Gamma$ in the *right* half-plane ($\lambda$ can be as small as desired), and observe that $s \times$ integrand tends to zero everywhere on

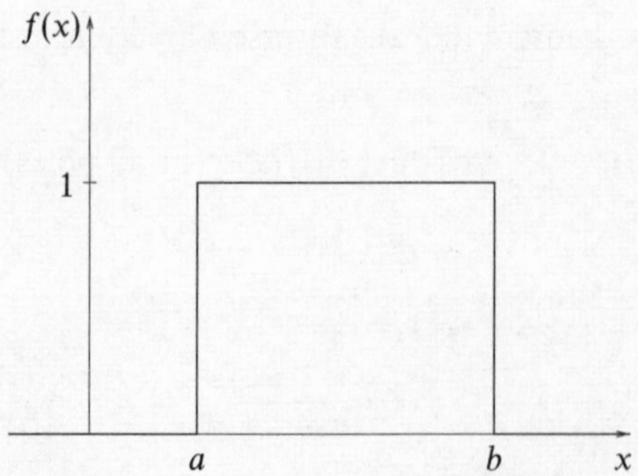


Figure 18.24 The result of the Laplace inversion of $\bar{f}(s) = s^{-1}(e^{-as} - e^{-bs})$ with $b > a$.

$\Gamma$ as $R \to \infty$. With no poles enclosed and no contribution from $\Gamma$, the integral along $L$ must also be zero. Thus

$$f(x) = 0 \qquad \text{for } x < a. \tag{18.103}$$

(ii) For $x > b$ the exponentials in the integrand will tend to zero as Re $s \to -\infty$, and so we may close $L$ in the left half-plane, as in figure 18.23($a$). Again the integral around $\Gamma$ vanishes for infinite $R$ and so, by the residue theorem,

$$f(x) = 0 \qquad \text{for } x > b. \tag{18.104}$$

(iii) For $a < x < b$ the two parts of the integrand behave in different ways and have to be treated separately:

$$I_1 - I_2 \equiv \frac{1}{2\pi i}\int_L \frac{e^{(x-a)s}}{s}\,ds - \frac{1}{2\pi i}\int_L \frac{e^{(x-b)s}}{s}\,ds.$$

The integrand of $I_1$ then vanishes in the far left-hand half-plane, but does now have a (simple) pole at $s = 0$. Closing $L$ in the left half-plane, and using the residue theorem, we obtain

$$I_1 = (\text{residue of } s^{-1}e^{(x-a)s} \text{ at } s = 0) = 1. \tag{18.105}$$

The integrand of $I_2$, however, vanishes in the far right-hand half-plane (and also has a simple pole at $s = 0$) and is evaluated by a circular-arc completion in that half-plane. Such a contour encloses no poles and leads to $I_2 = 0$.

Thus collecting together results (18.103)–(18.105) we obtain

$$f(x) = \begin{cases} 0 & \text{for } x < a, \\ 1 & \text{for } a < x < b, \\ 0 & \text{for } x > b, \end{cases}$$

as shown in figure 18.24. ◄

## 18.21 Exercises

18.1 Find an analytic function of $z = x + iy$ whose imaginary part is

$$(y \cos y + x \sin y) \exp x.$$

18.2 Find a function $f(z)$, analytic in a suitable part of the Argand diagram, for which

$$\operatorname{Re} f = \frac{\sin 2x}{\cosh 2y - \cos 2x}.$$

Where are the singularities of $f(z)$?

18.3 Find the radii of convergence of the following Taylor series:

$$\text{(a)} \sum_{n=2}^{\infty} \frac{z^n}{\ln n}, \quad \text{(b)} \sum_{n=1}^{\infty} \frac{n! z^n}{n^n},$$
$$\text{(c)} \sum_{n=1}^{\infty} z^n n^{\ln n}, \quad \text{(d)} \sum_{n=1}^{\infty} \left(\frac{n+p}{n}\right)^{n^2} z^n, \text{ with } p \text{ real.}$$

18.4 Find the Taylor series expansion about the origin of the function $f(z)$ defined by

$$f(z) = \sum_{r=1}^{\infty} (-1)^{r+1} \sin\left(\frac{pz}{r}\right)$$

where $p$ is a constant. Hence verify that $f(z)$ is a convergent series for all $z$.

18.5 Identify the zeroes, poles and essential singularities of the following functions.

(a) $\tan z$, (b) $[(z-2)/z^2] \sin[1/(1-z)]$, (c) $\exp(1/z)$,
(d) $\tan(1/z)$, (e) $z^{2/3}$.

18.6 For the function

$$f(z) = \ln\left(\frac{z+c}{z-c}\right)$$

where $c$ is real, show that the real part $u$ of $f$ is constant on a circle of radius $c \operatorname{cosech} u$ centred on the point $z = c \coth u$. Use this result to show that the electrical capacitance per unit length of two parallel cylinders of radii $a$, placed with their axes $2d$ apart, is proportional to $[\cosh^{-1}(d/a)]^{-1}$.

18.7 Find a complex potential in the $z$-plane appropriate to a physical situation in which the half-plane $x > 0$, $y = 0$ has zero potential and the half-plane $x < 0$, $y = 0$ has potential $V$.

By making the transformation $w = \frac{1}{2}a(z+z^{-1})$, with $a$ real and positive, find the electrostatic potential associated with the half-plane $r > a$, $s = 0$ and the half-plane $r < -a$, $s = 0$ at potentials 0 and $V$ respectively.

18.8 By considering in turn the transformations

$$z = \tfrac{1}{2}c(w + w^{-1}), \qquad w = \exp\zeta,$$

where $z = x+iy$, $w = r\exp i\theta$, $\zeta = \xi + i\eta$ and $c$ is a real positive constant, show that $z = c\cosh\zeta$ maps the strip $\xi \geq 0$, $0 \leq \eta \leq 2\pi$, onto the whole $z$-plane. Which curves in the $z$-plane correspond to the lines $\xi =$ constant and $\eta =$ constant? Identify those corresponding to $\xi = 0$, $\eta = 0$, and $\eta = 2\pi$.

The electric potential $\phi$ of a charged conducting strip $-c \leq x \leq c$, $y = 0$ satisfies

$$\phi \sim -k\ln(x^2 + y^2)^{1/2} \qquad \text{for large } (x^2 + y^2)^{1/2},$$

with $\phi$ constant on the strip. Show that $\phi = \text{Re}[-k\cosh^{-1}(z/c)]$ and that the magnitude of the electric field near the strip is $k(c^2 - x^2)^{-1/2}$.

18.9 Show that the transformation

$$w = \int_0^z \frac{1}{(\zeta^3 - \zeta)^{1/2}}\, d\zeta$$

transforms the upper half-plane into the interior of a square that has one corner at the origin of the $w$-plane and sides of length $L$, where

$$L = \int_0^{\pi/2} \text{cosec}^{\,1/2}\,\theta \; d\theta.$$

18.10 The *fundamental theorem of algebra* states that a complex polynomial $p_n(z)$ of degree $n$ has precisely $n$ complex roots. By applying Liouville's theorem (see the end of section 18.12) to $f(z) = 1/p_n(z)$ prove that $p_n(z)$ has at least one complex root. Factor out that root to obtain $p_{n-1}(z)$ and by repeating the process prove the above theorem.

18.11 For the equation $8z^3 + z + 1 = 0$:

(a) show that all three roots lie between the circles $|z| = 3/8$ and $|z| = 5/8$;

(b) find the approximate location of the real root, and hence deduce that the complex ones lie in the first and fourth quadrants and have moduli greater than 0.5.

18.12 (a) Prove that $z^8 + 3z^3 + 7z + 5$ has two zeroes in the first quadrant.
(b) Find in which quadrants the zeroes of $2z^3 + 7z^2 + 10z + 6$ lie. Try to locate them.

18.13 The following is a method of determining the number of zeroes of an $n$th-degree polynomial $f(z)$ inside the contour $C$ given by $|z| = R$.

(a) Put $z = R(1 + it)/(1 - it)$ with $t = \tan(\theta/2)$ in $-\infty \le t \le \infty$;
(b) obtain $f(z)$ as
$$\frac{A(t) + iB(t)}{(1 - it)^n}\frac{(1 + it)^n}{(1 + it)^n};$$
(c) show that $\arg f(z) = \tan^{-1}(B/A) + n\tan^{-1} t$;
(d) show that $\Delta_C[\arg f(z)] = \Delta_C[\tan^{-1}(B/A)] + n\pi$;
(e) using inspection or a sketch graph, determine $\Delta_C[\tan^{-1}(B/A)]$ by finding the discontinuities in $B/A$ and evaluating $\tan^{-1}(B/A)$ at $t = \pm\infty$.

Use this method, together with the results of the worked example in section 18.15, to show that the zeroes of $z^4 + z + 1$ in the second and third quadrants have $|z| < 1$.

18.14 By considering the real part of
$$\int \frac{-iz^{n-1}\,dz}{1 - a(z + z^{-1}) + a^2},$$
where $z = \exp i\theta$ and $n$ is a non-negative integer, evaluate
$$\int_0^\pi \frac{\cos n\theta}{1 - 2a\cos\theta + a^2}\,d\theta,$$
for $a$ real and $> 1$.

18.15 Prove that if $f(z)$ has a simple pole at $z_0$, then $1/f(z)$ has residue $1/f'(z_0)$ there. Hence evaluate
$$\int_{-\pi}^{\pi} \frac{\sin\theta}{a - \sin\theta}\,d\theta,$$
where $a$ is real and $> 1$.

18.16 The equation of an ellipse in plane polar coordinates $r, \theta$, with one of its foci at the origin, is
$$\frac{l}{r} = 1 - \epsilon\cos\theta,$$
where $l$ is a length (that of the latus rectum) and $\epsilon$ $(0 < \epsilon < 1)$ is the eccentricity of the ellipse. Express the area of the ellipse as an integral around the unit circle in the complex plane, and show that the only singularity of the integrand inside the circle is a double pole at $z_0 = \epsilon^{-1} - (\epsilon^{-2} - 1)^{1/2}$.

By setting $z = z_0 + \xi$ and expanding the integrand in powers of $\xi$, find the residue at $z_0$ and hence show that the area is equal to $\pi l^2(1-\epsilon^2)^{-3/2}$. (In terms of the semi-axes $a$ and $b$ of the ellipse, $l = b^2/a$ and $\epsilon^2 = (a^2 - b^2)/a^2$.)

18.17 Prove that, for $\alpha > 0$, the integral

$$\int_0^\infty \frac{t \sin \alpha t}{1+t^2}\, dt$$

has the value $(\pi/2)\exp(-\alpha)$.

18.18 Prove

$$\int_0^\infty \frac{\cos mx}{4x^4 + 5x^2 + 1}\, dx = \frac{\pi}{6}\left(4e^{-m/2} - e^{-m}\right) \qquad \text{for } m > 0.$$

18.19 Show that the principal value of the integral

$$\int_{-\infty}^\infty \frac{\cos(x/a)}{x^2 - a^2}\, dx$$

is $-(\pi/a)\sin 1$.

18.20 (a) Prove that the integral of $[\exp(i\pi z^2)]\text{cosec}\,\pi z$ around the parallelogram with corners $\pm 1/2 \pm R\exp(i\pi/4)$ has the value $2i$.

(b) Show that the parts of the contour parallel to the real axis give no contribution when $R \to \infty$.

(c) Evaluate the integrals along the other two sides by putting $z' = r\exp(i\pi/4)$ and working in terms of $z' + 1/2$ and $z' - 1/2$. Hence by letting $R \to \infty$ show that

$$\int_{-\infty}^\infty e^{-\pi r^2}\, dr = 1.$$

18.21 By applying the residue theorem around a wedge-shaped contour of angle $2\pi/n$, with one side along the real axis, prove that the integral

$$\int_0^\infty \frac{dx}{1+x^n},$$

where $n$ is real and $\geq 2$, has the value $(\pi/n)\text{cosec}\,(\pi/n)$.

18.22 Using a suitable cut plane, prove that if $\alpha$ is real and $0 < \alpha < 1$ then

$$\int_0^\infty \frac{x^{-\alpha}}{1+x}\, dx$$

has the value $\pi \,\text{cosec}\, \pi\alpha$.

18.23 Show that

$$\int_0^\infty \frac{\ln x}{x^{3/4}(1+x)}\, dx = -\sqrt{2}\pi^2.$$

18.24 Prove that

$$\sum_{-\infty}^{\infty} \frac{1}{n^2 + \frac{3}{4}n + \frac{1}{8}} = 4\pi.$$

Carry out the summation numerically, say between −4 and 4, and note how much of the sum comes from values near the poles of the contour integration.

18.25 By considering the integral of

$$\left(\frac{\sin \alpha z}{\alpha z}\right)^2 \frac{\pi}{\sin \pi z}, \qquad \alpha < \frac{\pi}{2}$$

around a circle of large radius, prove that

$$\sum_{m=1}^{\infty} (-1)^{m-1} \frac{\sin^2 m\alpha}{(m\alpha)^2} = \frac{1}{2}.$$

18.26 Use the Bromwich inversion and contours such as those shown in figure 18.23(*a*) to find the functions of which the following are the Laplace transforms:

(a) $s(s^2 + b^2)^{-1}$;

(b) $n!(s - a)^{-(n+1)}$, with $n$ a positive integer and $s > a$;

(c) $a(s^2 - a^2)^{-1}$, with $s > |a|$. (Change variable to $t = s - |a|$.)

Compare your answers with those given in table 11.1.

18.27 Use the contour in figure 18.23(*c*) to show that the function with Laplace transform $s^{-1/2}$ is $(\pi x)^{-1/2}$. (For an integrand of the form $r^{-1/2}\exp(-rx)$ change variable to $t = r^{1/2}$.)

## 18.22 Hints and answers

18.1 $\partial u/\partial y = -(\exp x)(y \cos y + x \sin y + \sin y)$; $z \exp z$.

18.2 $f = (\sin 2x - i \sinh 2y)/(\cosh 2y - \cos 2x)$; the special case of $z$ real shows that $f(z) = \cot z$; poles at $z = n\pi$.

18.3 (a) 1; (b) 1; (c) 1; (d) $e^{-p}$.

18.4 The series is given by

$$a_{2n+1} = \frac{(-1)^{n+1} p^{2n+1}}{(2n+1)!} \sum_{r=1}^{\infty} \frac{(-1)^r}{r^{2n+1}},$$

for integer $n \geq 0$, $a_{2n} = 0$; $R^{-1} = \lim\,[p \times 1 \times (2n+1)^{-1}] = 0$, and so series is convergent by the root test.

18.5 (a) zeroes at $z = n\pi$, simple poles at $z = n\pi + \pi/2$, essential singularity at $z = \infty$;
(b) zeroes at $z = \infty$, 2 and $1 - (n\pi)^{-1}$, double pole at $z = 0$, essential singularity at $z = 1$;
(c) zero at $\infty$, essential singularity at $z = 0$;
(d) zeroes at $z = \infty$ and $(n\pi)^{-1}$, simple poles at $z = (n\pi + \pi/2)^{-1}$, essential singularity at $z = 0$;
(e) zero and branch point at the origin, essential singularity at $z = \infty$.

18.6 Set $c \coth u_1 = -d$, $c \coth u_2 = +d$, $|c \operatorname{cosech} u| = a$, and note that the capacity is proportional to $(u_2 - u_1)^{-1}$.

18.7 $f(z) = -i(V/\pi)\ln z$; $-i(V/\pi)\ln\left\{(z/a) \pm [(z/a)^2 - 1]^{1/2}\right\}$.

18.8 $\xi =$ constant, ellipses $x^2(a+1)^{-2} + y^2(a-1)^{-2} = c^2/4a^2$; $\eta =$ constant, hyperbolae $x^2(\cos\alpha)^{-2} - y^2(\sin\alpha)^{-2} = c^2$; cut $-c \le x \le c$, $y = 0$; cut $|x| \ge c$, $y = 0$; the same as for $\eta = 0$.

18.11 (a) $|z| = 3/8$, $|8z^3 + z| \le 51/64 < 1$; $|z| = 5/8$, $|8z^3| = 125/64 > 104/64 \ge |z+1|$ ; (b) write as $8(z-\gamma)(z-\alpha-i\beta)(z-\alpha+i\beta) = 0$, $\gamma < 0$, and then the zero coefficient of $z^2$ shows $\alpha > 0$. Show $-3/8 > \gamma > -1/2$ and use $-8\gamma(\alpha^2 + \beta^2) = 1$.

18.12 (a) For a quarter-circular contour enclosing the first quadrant, the change in the argument of the function is $0 + 8(\pi/2) + 0$ (since $y^8 + 5 = 0$ has no real roots); (b) one negative real zero and a conjugate pair in the second and third quadrants, $-3/2$, $-1 \pm i$.

18.13 $A = 3 - 12t^2 + t^4$, $B = -2t - 2t^3$, $\Delta_C[\tan^{-1}(B/A)] = 0$, $\Delta_C[\arg f(z)] = 4\pi$, hence there are two zeroes inside $|z| = 1$.

18.14 Pole at $z = 1/a$; $\pi a^{-n}(a^2 - 1)^{-1}$.

18.15 The only pole inside the unit circle is at $z = ia - i(a^2 - 1)^{1/2}$, residue is $-(i/2)(a^2 - 1)^{-1/2}$; $2\pi[a(a^2 - 1)^{-1/2} - 1]$.

18.16 The integrand is $2l^2 z(2z - \epsilon z^2 - \epsilon)^{-2}$; residue $= (4\epsilon)^{-1}(\epsilon^{-2} - 1)^{-3/2}$.

18.17 Follow the first example in section 18.17 and use Jordan's lemma, pole at $z = i$.

18.18 Factorise the denominator, showing that the relevant simple poles are at $i/2$ and $i$.

18.19 Use Jordan's lemma and a semicircular contour indented at $z = \pm a$.

18.20 (a) The only pole is at the origin with residue $\pi^{-1}$; (b) each is $\mathrm{O}[\exp(-\pi R^2 - \pi R/\sqrt{2})]$; (c) the sum of the integrals is $2i\int_{-R}^{R}\exp(-\pi r^2)\,dr$.

18.21 The residue at the only pole inside the contour, $z = \exp(i\pi/n)$ is $-n^{-1}\exp(i\pi/n)$. The values of the integrals along the two radii differ by a factor $-\exp(i2\pi/n)$.

18.22 Use a contour like that shown in figure 18.21.

18.23 See previous example.

18.24 Evaluate

$$\int \frac{\pi \cot \pi z}{\left(\frac{1}{2}+z\right)\left(\frac{1}{4}+z\right)}\,dz$$

around a large circle centred on the origin; residue at $z=-\frac{1}{2}$ is 0; residue at $z=-\frac{1}{4}$ is $4\pi\cot\left(-\frac{1}{4}\pi\right)$.

18.25 The behaviour of the integrand for large $|z|$ is $|z|^{-2}\exp[(2\alpha-\pi)|z|]$. The residue at $z=\pm m$, for each integer $m$, is $\sin^2(m\alpha)(-1)^m/(m\alpha)^2$. The contour contributes nothing.
Required summation = [total sum $-$ ($m=0$ term)]/2.

18.26 Poles at (a) $\pm ib$; (b) $t=s-a=0$, of order $n+1$, (c) $t=0$ and $t=-2|a|$.

18.27 $\int_\Gamma$ and $\int_\gamma$ tend to 0 as $R\to\infty$ and $\rho\to 0$. Put $s=r\exp i\pi$ and $s=r\exp(-i\pi)$ on the two sides of the cut and use $\int_0^\infty \exp(-t^2x)\,dt=\frac{1}{2}(\pi/x)^{1/2}$. There are no poles inside the contour.

# 19

# Tensors

It may seem obvious that the quantitative description of physical processes cannot depend on the coordinate system in which they are represented. However, we may turn this argument around: since physical results are indeed independent of the choice of coordinate system, what must this imply about the nature of the quantities involved in the description of physical processes? The study of these implications, and of the classification of physical quantities by means of them, forms the content of the present chapter.

Although the concepts presented here may be applied, with little modification, to more abstract spaces (most notably the four-dimensional space–time of special or general relativity), we shall restrict our attention to our familiar three-dimensional Euclidean space. This removes the need to discuss the properties of differentiable manifolds and their tangent and dual spaces. The reader who is interested in these more technical aspects of tensor calculus in general spaces, and in particular their application to general relativity, should consult one of the many excellent textbooks on the subject.†

Before the presentation of the main development of the subject, we begin by introducing the summation convention, which will prove very useful in writing tensor equations in a more compact form. We then review the effects of a change of basis in a vector space, as discussed in chapter 7. This is followed by an investigation of the rotation of Cartesian coordinate systems, and finally we broaden our discussion to include more general coordinate systems and transformations.

† For example, D'Inverno, *Introducing Einstein's Relativity* (Oxford, 1992); Foster and Nightingale, *A Short Course in General Relativity* (Springer-Verlag, 1994); Schutz, *A First Course in General Relativity* (Cambridge, 1990).

## 19.1 Some notation

Before proceeding further, we introduce the *summation convention* for subscripts, since its use looms large in the work of this chapter. The convention is that any *lower-case* alphabetic subscript that appears *exactly* twice in any term of an expression is to be summed over all the values that a subscript in that position can take (unless the contrary is specifically stated). The subscripted quantities may appear in the numerator and/or the denominator of a term in an expression. This naturally implies that any such pair of repeated subscripts must occur only in subscript positions that have the same range of values. Sometimes the ranges of values have to be specified but usually they are apparent from the context.

The following simple examples illustrate what is meant (in the three-dimensional case):

(i) $a_i x_i$ stands for $a_1 x_1 + a_2 x_2 + a_3 x_3$;

(ii) $a_{ij} b_{jk}$ stands for $a_{i1} b_{1k} + a_{i2} b_{2k} + a_{i3} b_{3k}$;

(iii) $a_{ij} b_{jk} c_k$ stands for $\sum_{j=1}^{3} \sum_{k=1}^{3} a_{ij} b_{jk} c_k$;

(iv) $\dfrac{\partial v_i}{\partial x_i}$ stands for $\dfrac{\partial v_1}{\partial x_1} + \dfrac{\partial v_2}{\partial x_2} + \dfrac{\partial v_3}{\partial x_3}$;

(v) $\dfrac{\partial^2 \phi}{\partial x_i \partial x_i}$ stands for $\dfrac{\partial^2 \phi}{\partial x_1^2} + \dfrac{\partial^2 \phi}{\partial x_2^2} + \dfrac{\partial^2 \phi}{\partial x_3^2}$.

Subscripts that are summed over are called *dummy subscripts* and the others *free subscripts.* It is worth remarking that when introducing a dummy subscript into an expression, care should be taken not to use one that is already present, either as a free or as a dummy subscript. For example, $a_{ij} b_{jk} c_{kl}$ cannot and must not be replaced by $a_{ij} b_{jj} c_{jl}$ or by $a_{il} b_{lk} c_{kl}$, but could be replaced by $a_{im} b_{mk} c_{kl}$ or by $a_{im} b_{mn} c_{nl}$. (Naturally, free subscripts should not be changed at all unless the working calls for it.)

Furthermore, as we have done throughout this book, we will make frequent use of the Kronecker delta $\delta_{ij}$, which is defined by

$$\delta_{ij} = \begin{cases} 1 & \text{if } i = j, \\ 0 & \text{otherwise.} \end{cases}$$

When we have adopted the summation convention, the main use of $\delta_{ij}$ is to replace one subscript by another in certain expressions. For example,

$$a_{ij} \delta_{jk} = a_{ij} \delta_{kj} = a_{ik}; \qquad (19.1)$$

here the dummy index shared by both terms on the left-hand side (namely $j$) has been replaced by the free index carried by the Kronecker delta (namely $k$), and the delta symbol has disappeared. In matrix language, (19.1) can be written as

$\mathsf{AI} = \mathsf{A}$, where $\mathsf{A}$ is the matrix with elements $a_{ij}$ and $\mathsf{I}$ is the unit matrix (of the same dimensions as $\mathsf{A}$).

Similarly, we have

$$b_j\delta_{ij} = b_i,$$

whilst in some expressions we may use the Kronecker delta to replace indices in a number of different ways, such as

$$a_{ij}b_{jk}\delta_{ki} = a_{ij}b_{ji} \quad \text{or} \quad a_{kj}b_{jk},$$

where the two latter expressions are equivalent to one another.

## 19.2 Change of basis

In chapter 7 some attention was given to the subject of changing the basis set (or coordinate system) in a vector space and it was shown that, under such a change, different types of quantities behave in different ways. These results are given in section 7.14, but are summarised below for convenience, using the summation convention. Although throughout this section we shall remind the reader that we are using this convention, it will simply be assumed in the remainder of the chapter.

If we introduce a set of basis vectors $\mathbf{e}_1, \mathbf{e}_2, \mathbf{e}_3$ into our familiar three-dimensional (vector) space, then we can describe any vector $\mathbf{x}$ in terms of its components $x_1, x_2, x_3$ with respect to this basis:

$$\mathbf{x} = x_1\mathbf{e}_1 + x_2\mathbf{e}_2 + x_3\mathbf{e}_3 = x_i\mathbf{e}_i,$$

where we have used the summation convention to write the sum in a more compact form. If we now introduce a new basis $\mathbf{e}'_1, \mathbf{e}'_2, \mathbf{e}'_3$ related to the old one by

$$\mathbf{e}'_j = S_{ij}\mathbf{e}_i \qquad \text{(sum over } i\text{)}, \tag{19.2}$$

where the coefficient $S_{ij}$ is the $i$th component of the vector $\mathbf{e}'_j$ with respect to the unprimed basis, then we may write $\mathbf{x}$ with respect to the new basis as

$$\mathbf{x} = x'_1\mathbf{e}'_1 + x'_2\mathbf{e}'_2 + x'_3\mathbf{e}'_3 = x'_i\mathbf{e}'_i \qquad \text{(sum over } i\text{)}.$$

If we denote the matrix with elements $S_{ij}$ by $\mathsf{S}$, then the components $x'_i$ and $x_i$ in the two bases are related by

$$x'_i = (\mathsf{S}^{-1})_{ij}x_j \qquad \text{(sum over } j\text{)},$$

where, using the summation convention, there is an implicit sum over $j$ from $j = 1$ to $j = 3$. In the special case where the transformation is a rotation of the coordinate axes, the transformation matrix $\mathsf{S}$ is orthogonal, and we have

$$x'_i = (\mathsf{S}^{\mathrm{T}})_{ij}x_j = S_{ji}x_j \qquad \text{(sum over } j\text{)}. \tag{19.3}$$

Scalars behave differently under transformations, however, since they remain unchanged. For example, the value of the scalar product of two vectors $\mathbf{x} \cdot \mathbf{y}$ (which is just a number) is unaffected by the transformation from the unprimed to the primed basis. Different again is the behaviour of linear operators. If a linear operator $\mathcal{A}$ is represented by some matrix $\mathsf{A}$ in a given coordinate system, then in the new (primed) coordinate system it is represented by the new matrix $\mathsf{A}' = \mathsf{S}^{-1}\mathsf{A}\mathsf{S}$.

In this chapter we develop a general formulation to describe and classify these different types of behaviour under a change of basis (or coordinate transformation). In the development, the generic name *tensor* is introduced, and certain scalars, vectors and linear operators are described respectively as tensors of zeroth, first and second order (the *order* – or *rank* – corresponds to the number of subscripts needed to specify a particular element of the tensor). Tensors of third and fourth order will also occupy some of our attention.

## 19.3 Cartesian tensors

We begin our discussion of tensors by considering a particular type of coordinate transformation – namely rotations – and we shall confine our attention strictly to the rotation of Cartesian coordinate systems. Our object is to study the properties of various types of mathematical quantities, and their associated physical interpretations, when they are described in terms of Cartesian coordinates and the axes of the coordinate system are rigidly rotated, whilst keeping the origin fixed, from a basis $\mathbf{e}_1, \mathbf{e}_2, \mathbf{e}_3$ (lying along the $Ox_1$, $Ox_2$ and $Ox_3$ axes) to a new one $\mathbf{e}'_1, \mathbf{e}'_2, \mathbf{e}'_3$ (lying along the $Ox'_1$, $Ox'_2$ and $Ox'_3$ axes).

Since we shall be interested more in how the components of a vector or linear operator are changed by a rotation of the axes, rather than in the relationship between the two sets of basis vectors $\mathbf{e}_i$ and $\mathbf{e}'_i$, let us define the transformation matrix $\mathsf{L}$ as the inverse of the matrix $\mathsf{S}$ in (19.2). Thus, from (19.3), the components of a position vector $\mathbf{x}$, in the old and new bases respectively, are related by

$$x'_i = L_{ij}x_j. \tag{19.4}$$

Because we are considering only rigid rotations of the coordinate axes, the transformation matrix $\mathsf{L}$ will be orthogonal, i.e. such that $\mathsf{L}^{-1} = \mathsf{L}^{\mathrm{T}}$. Therefore the inverse transformation is given by

$$x_i = L_{ji}x'_j. \tag{19.5}$$

The orthogonality of $\mathsf{L}$ also implies relationships among the elements of $\mathsf{L}$, expressing that fact that $\mathsf{L}\mathsf{L}^{\mathrm{T}} = \mathsf{L}^{\mathrm{T}}\mathsf{L} = \mathsf{I}$, which are given in subscript notation by

$$L_{ik}L_{jk} = \delta_{ij} \qquad \text{and} \qquad L_{ki}L_{kj} = \delta_{ij}. \tag{19.6}$$

Furthermore, in terms of the basis vectors of the primed and unprimed Cartesian

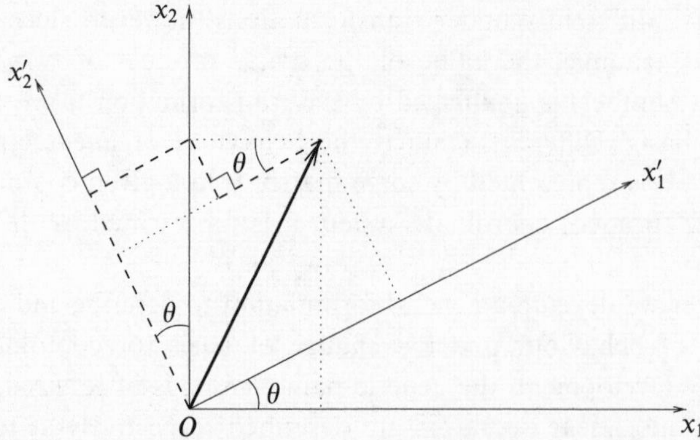


Figure 19.1 Rotation of Cartesian axes by an angle $\theta$ about the $x_3$-axis. The three angles marked $\theta$ and the parallels (broken lines) to the primed axes show how the first two equations of (19.7) are constructed.

coordinate systems, the transformation matrix is given by

$$L_{ij} = \mathbf{e}'_i \cdot \mathbf{e}_j.$$

We note that the product of two rotations is also a rotation. For example, suppose $x'_i = L_{ij}x_j$ and $x''_i = M_{ij}x'_j$, then the composite rotation is described by

$$x''_i = M_{ij}x'_j = M_{ij}L_{jk}x_k = (\mathsf{ML})_{ik}x_k,$$

corresponding to the matrix $\mathsf{ML}$.

►*Find the transformation matrix* $\mathsf{L}$ *corresponding to a rotation of the coordinate axes through an angle* $\theta$ *about the* $\mathbf{e}_3$*-axis (or* $x_3$*-axis), as shown in figure 19.1.*

Taking $\mathbf{x}$ as a position vector, the most obvious choice, we see from the figure that the components of $\mathbf{x}$ with respect to the new (primed) basis are given in terms of the components in the old (unprimed) basis by

$$\begin{aligned} x'_1 &= x_1\cos\theta + x_2\sin\theta \\ x'_2 &= -x_1\sin\theta + x_2\cos\theta \\ x'_3 &= x_3. \end{aligned} \tag{19.7}$$

The (orthogonal) transformation matrix is thus

$$\mathsf{L} = \begin{pmatrix} \cos\theta & \sin\theta & 0 \\ -\sin\theta & \cos\theta & 0 \\ 0 & 0 & 1 \end{pmatrix}.$$

The inverse equations are

$$\begin{aligned} x_1 &= x_1' \cos\theta - x_2' \sin\theta \\ x_2 &= x_1' \sin\theta + x_2' \cos\theta \\ x_3 &= x_3', \end{aligned} \tag{19.8}$$

in line with (19.5). ◄

## 19.4 First- and zero-order Cartesian tensors

Using the above example as a guide, we may consider any set of (three) quantities $v_i$, which are directly or indirectly functions of the coordinates $x_i$ (and possibly some constants), and ask how their values are changed by any rotation of the Cartesian axes. The specific question to be answered is whether the specific forms $v_i'$ in the new variables can be obtained from the old ones $v_i$ using (19.4), i.e.

$$v_i' = L_{ij} v_j. \tag{19.9}$$

If so, the $v_i$ are then said to form the components of a *vector* or *first-order Cartesian tensor*. By definition, the position coordinates are themselves the components of such a tensor. Moreover, since the transformation (19.9) is orthogonal, the components of a first-order Cartesian tensor also obey the inverse relation

$$v_i = L_{ji} v_j'. \tag{19.10}$$

We now consider explicit examples. In order to keep the equations to reasonable proportions, the examples will be restricted to two dimensions. Three-dimensional cases are no different in principle – but much longer to write out.

►*Which of the following pairs of quantities* $(v_1, v_2)$ *are the components of a first-order Cartesian tensor in two dimensions?:*

*(i)* $(x_2, -x_1)$, *(ii)* $(x_2, x_1)$, *(iii)* $(x_1^2, x_2^2)$.

We shall consider the rotation discussed in the previous example, and to save space we denote $\cos\theta$ by $c$ and $\sin\theta$ by $s$.

(i) Here $v_1 = x_2$ and $v_2 = -x_1$, referred to the old axes. In terms of the new coordinates they will be $v_1' = x_2'$ and $v_2' = -x_1'$, i.e.

$$\begin{aligned} v_1' &= x_2' &&= -s x_1 + c x_2 \\ v_2' &= -x_1' &&= -c x_1 - s x_2. \end{aligned} \tag{19.11}$$

Now if we start again and evaluate $v_1'$ and $v_2'$ as given by (19.9) we find that

$$\begin{aligned} v_1' &= L_{11} v_1 + L_{12} v_2 = c x_2 + s(-x_1) \\ v_2' &= L_{21} v_1 + L_{22} v_2 = -s(x_2) + c(-x_1). \end{aligned} \tag{19.12}$$

The expressions in (19.11) and (19.12) for $v_1'$ and for $v_2'$ are respectively the same whatever the values of $\theta$ (i.e. for *all* rotations) and thus by definition (19.9) the pair $(x_2, -x_1)$ *are* components of a first-order Cartesian tensor.

(ii) Here $v_1 = x_2$ and $v_2 = x_1$. Following the same procedure,

$$v_1' = x_2' = -sx_1 + cx_2$$
$$v_2' = x_1' = cx_1 + sx_2.$$

But by (19.9), for a Cartesian tensor we must have

$$\begin{aligned} v_1' &= cv_1 + sv_2 &&= cx_2 + sx_1 \\ v_2' &= (-s)v_1 + cv_2 &&= -sx_2 + cx_1. \end{aligned}$$

These two sets of expressions do not agree and thus the pair $(x_2, x_1)$ do not form the components of a first-order Cartesian tensor.

(iii) $v_1 = x_1^2$ and $v_2 = x_2^2$. The first component alone, as in (ii) above, is sufficient to show that these do *not* form a first-order tensor, since directly

$$v_1' = x_1'^2 = c^2x_1^2 + 2csx_1x_2 + s^2x_2^2,$$

whilst (19.9) requires that

$$v_1' = cv_1 + sv_2 = cx_1^2 + sx_2^2,$$

which is quite different. ◀

There are many physical examples of first-order tensors (i.e. vectors) that will be familiar to the reader. As a straightforward one, we may take the Cartesian components of the momentum of a particle of mass $m$, $(m\dot{x}_1, m\dot{x}_2, m\dot{x}_3)$. These transform in all essentials as do $(x_1, x_2, x_3)$ themselves, since the other operations involved, multiplication by a number and differentiation with respect to time, are quite unaffected by any orthogonal transformation of axes. Similarly, acceleration and force are represented by the components of first-order tensors.

Other more complicated vectors involving the position coordinates more than once, such as the angular momentum of a particle of mass $m$, namely $\mathbf{J} = \mathbf{x} \times \mathbf{p} = m(\mathbf{x} \times \dot{\mathbf{x}})$, are clearly also first-order tensors. That this is so is less obvious in component form than for the above examples, but it may be verified by writing out the components of $\mathbf{J}$ explicitly or by appealing to the quotient law to be discussed in section 19.7 and using the Cartesian tensor $\epsilon_{ijk}$ from section 19.8.

Having considered the effects of rotations on vector-like sets of quantities we may consider quantities that are unchanged by a rotation of axes. In our previous nomenclature these have been called *scalars* but we may also describe them as *tensors of zero order*. They contain only one element (formally, the number of subscripts needed to identify a particular element is zero); the most obvious non-trivial example associated with a rotation of axes is the square of the distance of a point from the origin, $r^2 = x_1^2 + x_2^2 + x_3^2$. In the new coordinate system it will have the form $r'^2 = x_1'^2 + x_2'^2 + x_3'^2$, which for any rotation has the same value as $x_1^2 + x_2^2 + x_3^2$.

In fact any scalar product of two first-order tensors (vectors) is a zero-order tensor (scalar), as might be expected since it can be written in a coordinate-free way as $\mathbf{u} \cdot \mathbf{v}$.

►*By considering the components of the vectors* $\mathbf{u}$ *and* $\mathbf{v}$ *with respect to two Cartesian coordinate systems (related by a rotation), show that the scalar product* $\mathbf{u} \cdot \mathbf{v}$ *is invariant under rotation.*

In the original (unprimed) system the scalar product is given in terms of components by $u_i v_i$ (summed over $i$), and in the rotated (primed) system by

$$u_i' v_i' = L_{ij} u_j L_{ik} v_k = L_{ij} L_{ik} u_j v_k = \delta_{jk} u_j v_k = u_j v_j,$$

where we have used the orthogonality relation (19.6). Since the resulting expression in the rotated system is the same as that in the original system, the scalar product is indeed invariant under rotations. ◄

The above result leads directly to the identification of many physically important quantities as zero-order tensors. Perhaps the most immediate of these is energy, either as potential energy, or as an energy density (e.g. $\mathbf{F} \cdot d\mathbf{r}$, $e\mathbf{E} \cdot d\mathbf{r}$, $\mathbf{D} \cdot \mathbf{E}$, $\mathbf{B} \cdot \mathbf{H}$, $\boldsymbol{\mu} \cdot \mathbf{B}$), but others, such as the angle between two directed quantities, are important. It is the fact that in most analyses of physical situations a scalar quantity (such as energy) is to be found which leads to the situation described in the first paragraph of this chapter. Such quantities are *invariant* under a rotation of axes and so it is possible to work with the most convenient set of axes and still have confidence in the results.

Complementing the way in which a zero-order tensor was obtained from two first-order tensors, so a first-order tensor can be obtained from a scalar. We show this by taking a specific example, that of the electric field $\mathbf{E} = -\nabla\phi$ derived from an electrostatic potential $\phi$, which has components

$$E_i = -\frac{\partial \phi}{\partial x_i}. \tag{19.13}$$

Clearly, $\mathbf{E}$ *is* a first-order tensor, but we may prove this more formally by considering the behaviour of its components (19.13) under a rotation of the coordinate axes, since the components of the electric field $E_i'$ are then given by

$$E_i' = \left(-\frac{\partial \phi}{\partial x_i}\right)' = -\frac{\partial \phi'}{\partial x_i'} = -\frac{\partial x_j}{\partial x_i'}\frac{\partial \phi}{\partial x_j} = L_{ij} E_j, \tag{19.14}$$

where (19.5) has been used to evaluate $\partial x_j / \partial x_i'$. Now (19.14) is in the form (19.9), thus showing that the components of the electric field do behave as the components of a first-order tensor.

►*If $v_i$ are the components of a first-order tensor, show that $\nabla \cdot \mathbf{v} = \partial v_i/\partial x_i$ is a zero-order tensor.*

In the rotated coordinate system $\nabla \cdot \mathbf{v}$ is given by

$$\left(\frac{\partial v_i}{\partial x_i}\right)' = \frac{\partial v_i'}{\partial x_i'} = \frac{\partial x_j}{\partial x_i'}\frac{\partial}{\partial x_j}(L_{ik}v_k) = L_{ij}L_{ik}\frac{\partial v_k}{\partial x_j},$$

since the elements $L_{ij}$ are not functions of position. Using the orthogonality relation (19.6) we then find

$$\frac{\partial v_i'}{\partial x_i'} = L_{ij}L_{ik}\frac{\partial v_k}{\partial x_j} = \delta_{jk}\frac{\partial v_k}{\partial x_j} = \frac{\partial v_j}{\partial x_j}.$$

Hence $\partial v_i/\partial x_i$ is invariant under rotation of the axes, and is thus a zero-order tensor, as was to be expected since it can be written in a coordinate-free way as $\nabla \cdot \mathbf{v}$. ◄

## 19.5 Second- and higher-order Cartesian tensors

Following on from scalars with no subscripts and vectors with one subscript, we turn to sets of quantities that require two subscripts to identify a particular element of the set. Let these quantities by denoted by $T_{ij}$.

Taking (19.9) as a guide we define a *second-order Cartesian tensor* as follows: the $T_{ij}$ form the components of such a tensor if, under the same conditions as for (19.9),

$$T'_{ij} = L_{ik}L_{jl}T_{kl} \qquad (19.15)$$

and

$$T_{ij} = L_{ki}L_{lj}T'_{kl}. \qquad (19.16)$$

We may at the same time define a Cartesian tensor of general order as follows. The set of expressions $T_{ij\cdots k}$ form the components of a Cartesian tensor if, for all rotations of the axes of coordinates given by (19.4) and (19.5), subject to (19.6), the expressions using the new coordinates, $T'_{ij\cdots k}$ are given by

$$T'_{ij\cdots k} = L_{ip}L_{jq}\cdots L_{kr}T_{pq\cdots r} \qquad (19.17)$$

and

$$T_{ij\cdots k} = L_{pi}L_{qj}\cdots L_{rk}T'_{pq\cdots r}. \qquad (19.18)$$

It is apparent that in three dimensions, an $N$th-order Cartesian tensor has $3^N$ components.

Since a second-order tensor has two subscripts, it is natural to display its components in matrix form. The notation $[T_{ij}]$ is used, as well as $\mathsf{T}$, to denote the matrix having $T_{ij}$ as the element in the $i$th row and $j$th column.†

† We can also denote the column matrix containing the elements $v_i$ of a vector by $[v_i]$.

We may think of a second-order tensor **T** as a geometrical entity in a similar way to that in which we viewed linear operators (which transform one vector into another, without reference to any coordinate system), and consider the matrix containing its components as a representation of the tensor with respect to a particular coordinate system. Moreover, the matrix $\mathsf{T} = [T_{ij}]$, containing the components of a second-order tensor, behaves in the same way under orthogonal transformations $\mathsf{T}' = \mathsf{L}\mathsf{T}\mathsf{L}^{\mathrm{T}}$, as the components of a linear operator.

However, not all linear operators are second-order tensors. Specifically, we require the two subscripts in a second-order tensor to refer to the same coordinate system, so that only linear operators that transform a vector to another vector in the same vector space are second-order tensors. Moreover, although the elements $L_{ij}$ of the transformation matrix are written with two subscripts, they cannot be the components of a tensor since the two subscripts each refer to a different coordinate system.

As examples of sets of quantities that are readily shown to be second-order tensors we consider the following.

(i) *The outer product of two vectors.* Let $u_i$ and $v_i$, $i = 1, 2, 3$, be the components of two vectors **u** and **v**, and consider the set of quantities $T_{ij}$ defined by

$$T_{ij} = u_i v_j. \tag{19.19}$$

The set $T_{ij}$ are called the components of the the *outer product* of **u** and **v**. Under rotations the components $T_{ij}$ become

$$T'_{ij} = u'_i v'_j = L_{ik} u_k L_{jl} v_l = L_{ik} L_{jl} u_k v_l = L_{ik} L_{jl} T_{kl}, \tag{19.20}$$

which shows that they do transform as the components of a second-order tensor. Use has been made in (19.20) of the fact that $u_i$ and $v_i$ are the components of first-order tensors.

The outer product of two vectors is often denoted, without reference to any coordinate system, as

$$\mathbf{T} = \mathbf{u} \otimes \mathbf{v}. \tag{19.21}$$

(This is not to be confused with the vector product of two vectors, which is itself a vector, and is discussed in chapter 6.) The expression (19.21) also suggests the basis to which the components $T_{ij}$ of the second-order tensor refer. Since $\mathbf{u} = u_i \mathbf{e}_i$ and $\mathbf{v} = v_i \mathbf{e}_i$, we may write the tensor **T** as

$$\mathbf{T} = u_i \mathbf{e}_i \otimes v_j \mathbf{e}_j = u_i v_j \mathbf{e}_i \otimes \mathbf{e}_j = T_{ij} \mathbf{e}_i \otimes \mathbf{e}_j. \tag{19.22}$$

Moreover, it should be noted that the quantities $T'_{ij}$ are the components of the *same* tensor **T**, but referred to a different coordinate system, i.e.

$$\mathbf{T} = T_{ij} \mathbf{e}_i \otimes \mathbf{e}_j = T'_{ij} \mathbf{e}'_i \otimes \mathbf{e}'_j.$$

These concepts can be straightforwardly extended to higher-order tensors.

(ii) *The gradient of a vector*. Suppose $v_i$ represents the components of a vector; let us consider the quantities generated by forming the derivatives of each $v_i$, $i = 1, 2, 3$, with respect to each $x_j$, $j = 1, 2, 3$, i.e.

$$T_{ij} = \frac{\partial v_i}{\partial x_j}.$$

These nine quantities form the components of a second-order tensor, as can be seen from the fact that

$$T'_{ij} = \frac{\partial v'_i}{\partial x'_j} = \frac{\partial (L_{ik} v_k)}{\partial x_l} \frac{\partial x_l}{\partial x'_j} = L_{ik} \frac{\partial v_k}{\partial x_l} L_{jl} = L_{ik} L_{jl} T_{kl}.$$

In coordinate-free language the tensor $\mathbf{T}$ may be written as $\mathbf{T} = \nabla \mathbf{v}$, and hence gives meaning to the concept of the gradient of a vector, a quantity that was not discussed in chapter 8.

A test of whether any given set of quantities forms the components of a second-order tensor can always be made by direct substitution of the $x'_i$ in terms of the $x_i$, followed by comparison with the right-hand side of (19.15). This procedure is extremely laborious, however, and it is almost always better to try to recognise the set as being one of the forms just developed, or to make alternative tests based on the quotient law of section 19.7.

►*Show that the $T_{ij}$ given by*

$$\mathsf{T} = [T_{ij}] = \begin{pmatrix} x_2^2 & -x_1 x_2 \\ -x_1 x_2 & x_1^2 \end{pmatrix}, \qquad (19.23)$$

*are the components of a second-order tensor.*

We again consider a rotation $\theta$ about the $\mathbf{e}_3$-axis. Carrying out first the direct evaluation we obtain, using (19.7),

$$\begin{aligned}
T'_{11} &= x_2'^2 &&= s^2 x_1^2 - 2sc x_1 x_2 + c^2 x_2^2, \\
T'_{12} &= -x'_1 x'_2 &&= sc x_1^2 + (s^2 - c^2) x_1 x_2 - sc x_2^2, \\
T'_{21} &= -x'_1 x'_2 &&= sc x_1^2 + (s^2 - c^2) x_1 x_2 - sc x_2^2, \\
T'_{22} &= x_1'^2 &&= c^2 x_1^2 + 2sc x_1 x_2 + s^2 x_2^2.
\end{aligned}$$

Now, evaluating the right-hand side of (19.15),

$$\begin{aligned}
T'_{11} &= cc x_2^2 + cs(-x_1 x_2) + sc(-x_1 x_2) + ss x_1^2, \\
T'_{12} &= c(-s) x_2^2 + cc(-x_1 x_2) + s(-s)(-x_1 x_2) + sc x_1^2, \\
T'_{21} &= (-s) c x_2^2 + (-s) s(-x_1 x_2) + cc(-x_1 x_2) + cs x_1^2, \\
T'_{22} &= (-s)(-s) x_2^2 + (-s) c(-x_1 x_2) + c(-s)(-x_1 x_2) + cc x_1^2.
\end{aligned}$$

The corresponding expressions are seen to be the same, showing (as required) that the $T_{ij}$ are the components of a second-order tensor.

The same result could be inferred much more easily by noting that the $T_{ij}$ are in fact the components of the outer product of the vector $(x_2, -x_1)$ with itself. That $(x_2, -x_1)$ are indeed the components of a vector was established by (19.11) and (19.12). ◄

Physical examples involving second-order tensors will be discussed in the later sections of this chapter, but we might note here that, for example, the magnetic susceptibility and electrical conductivity of materials are described by second-order tensors.

## 19.6 The algebra of tensors

Because of the similarity of first- and second-order tensors to column vectors and matrices, it would be expected that similar types of algebraic operation can be carried out with them, and so provide ways of constructing new tensors from old ones. In the remainder of this chapter, instead of referring to the $T_{ij}$ (say) as the *components* of a second-order tensor **T**, we may sometimes simply refer to $T_{ij}$ as the tensor. It should always be remembered, however, that the $T_{ij}$ are in fact just the components of **T** in a given coordinate system, and, similarly, that $T'_{ij}$ refers to the components of the *same* tensor **T** in a different coordinate system.

The addition and subtraction of tensors follows an obvious definition; namely that if $V_{ij\cdots k}$ and $W_{ij\cdots k}$ are (the components of) tensors of the same order, then their sum and difference, $S_{ij\cdots k}$ and $D_{ij\cdots k}$ respectively, are given by

$$\begin{aligned} S_{ij\cdots k} &= V_{ij\cdots k} + W_{ij\cdots k}, \\ D_{ij\cdots k} &= V_{ij\cdots k} - W_{ij\cdots k}, \end{aligned}$$

for each set of values $i, j, \ldots, k$. That $S_{ij\cdots k}$ and $D_{ij\cdots k}$ are the components of tensors follows immediately from the linearity of a rotation of coordinates.

It is equally straightforward to show that if the $T_{ij\cdots k}$ are the components of a tensor, then so is the set of quantities formed by interchanging the order of (a pair of) indices, e.g. $T_{ji\cdots k}$.

If $T_{ji\cdots k}$ is found to be identical with $T_{ij\cdots k}$, then $T_{ij\cdots k}$ is said to be *symmetric* with respect to its first two subscripts (or simply 'symmetric', for second-order tensors). If, however, $T_{ji\cdots k} = -T_{ij\cdots k}$ for every element then it is an *antisymmetric* tensor. An arbitrary tensor is neither symmetric nor antisymmetric but can always be written as the sum of a symmetric tensor $S_{ij\cdots k}$ and an antisymmetric tensor $A_{ij\cdots k}$:

$$\begin{aligned} T_{ij\cdots k} &= \tfrac{1}{2}(T_{ij\cdots k} + T_{ji\cdots k}) + \tfrac{1}{2}(T_{ij\cdots k} - T_{ji\cdots k}) \\ &= S_{ij\cdots k} + A_{ij\cdots k}. \end{aligned}$$

Of course these properties are valid for any pair of subscripts.

In (19.19) in the previous section we had an example of a kind of 'multiplication' of two tensors, thereby producing a tensor of higher order – in that case two first-order tensors were multiplied to give a second-order tensor. Inspection of (19.20) shows that there is nothing particular about the actual orders involved and thus in general that the outer product of an $N$th-order tensor with an $M$th-order tensor will produce an $(M+N)$th-order tensor.

An operation that produces the opposite effect – namely, generates a tensor of smaller rather than larger order – is known as *contraction* and consists of making two of the subscripts equal and summing over all values of the equalised subscripts.

►*Show that the process of contraction of a tensor produces another tensor, but with an order reduced by* 2.

Let $T_{ij\cdots l\cdots m\cdots k}$ be the components of an $N$th-order tensor, then

$$T'_{ij\cdots l\cdots m\cdots k} = \underbrace{L_{ip}L_{jq}\cdots L_{lr}\cdots L_{ms}\cdots L_{kn}}_{N \text{ factors}}\, T_{pq\cdots r\cdots s\cdots n}.$$

Thus if, for example, we make the two subscripts $l$ and $m$ equal, and sum over all values of these subscripts, we obtain

$$\begin{aligned} T'_{ij\cdots l\cdots l\cdots k} &= L_{ip}L_{jq}\cdots L_{lr}\cdots L_{ls}\cdots L_{kn}T_{pq\cdots r\cdots s\cdots n} \\ &= L_{ip}L_{jq}\cdots \delta_{rs}\cdots L_{kn}T_{pq\cdots r\cdots s\cdots n} \\ &= \underbrace{L_{ip}L_{jq}\cdots L_{kn}}_{(N-2) \text{ factors}}\, T_{pq\cdots r\cdots r\cdots n}, \end{aligned}$$

showing that $T_{ij\cdots l\cdots l\cdots k}$ are the components of a (different) Cartesian tensor of order $N-2$. ◄

For a second-rank tensor, the process of contraction is the same as taking the trace of the corresponding matrix. The trace $T_{ii}$ itself is thus a zero-order tensor (or scalar) and hence invariant under rotations, as has been noted in chapter 7.

The process of taking the scalar product of two vectors can be recast into tensor language as forming the outer product $T_{ij} = u_iv_j$ of two first-order tensors $\mathbf{u}$ and $\mathbf{v}$ and then contracting the second-order tensor $\mathbf{T}$ so formed, to give $T_{ii} = u_iv_i$, a scalar (invariant under a rotation of axes).

As yet another example of a familiar operation that is a particular case of a contraction, we may note that the multiplication of a column vector $[u_i]$ by a matrix $[B_{ij}]$ to produce another column vector $[v_i]$,

$$B_{ij}u_j = v_i,$$

can be looked upon as the contraction $T_{ijj}$ of the third-order tensor $T_{ijk}$ formed from the outer product of $B_{ij}$ and $u_k$.

## 19.7 The quotient law

The previous paragraph appears to give a heavy-handed way of describing a familiar operation, but it leads us to ask whether it has a converse. To put the question in more general terms: if we know that **B** and **C** are tensors and also that

$$A_{pq\cdots k\cdots m}B_{ij\cdots k\cdots n} = C_{pq\cdots mij\cdots n}, \tag{19.24}$$

does this imply that the $A_{pq\cdots k\cdots m}$ also form the components of a tensor **A**? Here **A**, **B** and **C** are respectively of $M$th, $N$th and $(M+N-2)$th order and it should be noted that the subscript $k$ which has been contracted may be any of the subscripts in **A** and **B** independently.

The *quotient law* for tensors states that if (19.24) holds in all rotated coordinate frames then the $A_{pq\cdots k\cdots m}$ do indeed form the components of a tensor **A**. To prove it for general $M$ and $N$ is no more difficult regarding the ideas involved than to show it for specific $M$ and $N$, but does involve the introduction of a large number of subscript symbols. We will therefore take the case $M = N = 2$, but it will be readily apparent that the principle of the proof holds for general $M$ and $N$.

We thus start with (say)

$$A_{pk}B_{ik} = C_{pi}, \tag{19.25}$$

where $B_{ik}$ and $C_{pi}$ are arbitrary second-order tensors. Under a rotation of coordinates the set $A_{pk}$ (tensor or not) transforms into a new set of quantities that we will denote by $A'_{pk}$. We thus obtain in succession the following steps, using (19.15), (19.16) and (19.6):

$$\begin{aligned}
A'_{pk}B'_{ik} &= C'_{pi} && \text{(transforming (19.25))},\\
&= L_{pq}L_{ij}C_{qj} && \text{(since } \mathbf{C} \text{ is a tensor)},\\
&= L_{pq}L_{ij}A_{ql}B_{jl} && \text{(from (19.25))},\\
&= L_{pq}L_{ij}A_{ql}L_{mj}L_{nl}B'_{mn} && \text{(since } \mathbf{B} \text{ is a tensor)},\\
&= L_{pq}L_{nl}A_{ql}B'_{in} && \text{(since } L_{ij}L_{mj} = \delta_{im}).
\end{aligned}$$

Now $k$ on the left and $n$ on the right are dummy subscripts and thus we may write

$$(A'_{pk} - L_{pq}L_{kl}A_{ql})B'_{ik} = 0. \tag{19.26}$$

Since $B_{ik}$, and hence $B'_{ik}$, is an arbitrary tensor, we must have

$$A'_{pk} = L_{pq}L_{kl}A_{ql},$$

showing that the $A'_{pk}$ are given by the general formula (19.17) and hence that the set $A_{pk}$ are the components of a second-order tensor. By following an analogous argument, the same result (19.26) and deduction could be obtained if (19.25) were replaced by

$$A_{pk}B_{ki} = C_{pi},$$

i.e. contraction being with respect to a different pair of indices.

Use of the quotient law to test whether a given set of quantities is a tensor is generally much more convenient than making a direct substitution. A particular way in which it is applied is by contracting the given set of quantities, having $N$ subscripts, with an arbitrary $N$th-order tensor (i.e. one having independently variable components) and determining whether the result is a scalar.

►*Use the quotient law to show that the elements of* $\mathsf{T}$, *equation (19.23), are the components of a second-order tensor.*

The outer product $x_i x_j$ is a second-order tensor. Contracting this with the $T_{ij}$ given in (19.23) we obtain

$$T_{ij}x_i x_j = x_2^2 x_1^2 - x_1 x_2 x_1 x_2 - x_1 x_2 x_2 x_1 + x_1^2 x_2^2 = 0,$$

which is clearly invariant (a zeroth-order tensor). Hence by the quotient theorem $T_{ij}$ must also be a tensor. ◄

## 19.8 The tensors $\delta_{ij}$ and $\epsilon_{ijk}$

Throughout this book we have encountered the two-subscript quantity $\delta_{ij}$ defined by

$$\delta_{ij} = \begin{cases} 1 & \text{if } i = j, \\ 0 & \text{otherwise.} \end{cases}$$

Let us now also introduce the three-subscript *Levi–Civita symbol* $\epsilon_{ijk}$, the value of which is given by

$$\epsilon_{ijk} = \begin{cases} +1 & \text{if } i, j, k \text{ is an even permutation of } 1, 2, 3, \\ -1 & \text{if } i, j, k \text{ is an odd permutation of } 1, 2, 3, \\ 0 & \text{otherwise.} \end{cases}$$

We will now show that $\delta_{ij}$ and $\epsilon_{ijk}$ are respectively the components of a second- and a third-order Cartesian tensor. Notice that the coordinates $x_i$ do not appear explicitly in the components of these tensors, their components consisting entirely of 0 and 1.

In passing, we also note that $\epsilon_{ijk}$ is totally antisymmetric, i.e. it changes sign under the interchange of any pair of subscripts. In fact $\epsilon_{ijk}$, or any scalar multiple of it, is the *only* three-subscript quantity with this property.

Treating first $\delta_{ij}$, the proof is straightforward, since if we consider the quantities

$$\delta'_{kl} = L_{ki}L_{lj}\delta_{ij} = L_{ki}L_{li} = \delta_{kl},$$

we have exactly the same expression in the new and old coordinates. Thus $\delta_{ij}$ is a second-order tensor.

Turning now to $\epsilon_{ijk}$, we have to consider the quantity

$$\epsilon'_{lmn} = L_{li}L_{mj}L_{nk}\epsilon_{ijk}.$$

Let us begin, however, by noting that we may use the Levi–Civita symbol to write an expression for the determinant of a $3 \times 3$ matrix $\mathsf{A}$,

$$|\mathsf{A}|\epsilon_{lmn} = A_{li}A_{mj}A_{nk}\epsilon_{ijk}, \qquad (19.27)$$

which may be shown to be equivalent to the Laplace expansion (see chapter 7).† Indeed many of the properties of determinants discussed in chapter 7 can be proved very efficiently using this expression.

►*Evaluate the determinant of the matrix*

$$\mathsf{A} = \begin{pmatrix} 2 & 1 & -3 \\ 3 & 4 & 0 \\ 1 & -2 & 1 \end{pmatrix}.$$

Setting $l = 1$, $m = 2$ and $n = 3$ in (19.27) we find

$$\begin{aligned} |\mathsf{A}| &= \epsilon_{ijk}A_{1i}A_{2j}A_{3k} \\ &= (2)(4)(1) - (2)(0)(-2) - (1)(3)(1) + (-3)(3)(-2) \\ &\quad + (1)(0)(1) - (-3)(4)(1) = 35, \end{aligned}$$

which may be verified using the Laplace expansion method. ◄

Using (19.27) we may now show that the $\epsilon_{ijk}$ are in fact the components of a third-rank tensor, since

$$\epsilon'_{lmn} = L_{li}L_{mj}L_{nk}\epsilon_{ijk} = |\mathsf{L}|\epsilon_{lmn},$$

and as $\mathsf{L}$ is orthogonal, its determinant has the value unity, so that $\epsilon'_{lmn} = \epsilon_{lmn}$. Thus we see that $\epsilon'_{lmn}$ has exactly the properties of $\epsilon_{ijk}$ but with $i, j, k$ replaced by $l, m, n$, i.e. it is the same as the expression $\epsilon_{ijk}$ written using the new coordinates. This shows that $\epsilon_{ijk}$ is a third-order Cartesian tensor.

In addition to providing a convenient notation for the determinant of a matrix, many of the familiar expressions of vector algebra and calculus can also be written as contracted tensors involving $\delta_{ij}$ and $\epsilon_{ijk}$. For example, provided we are using right-handed Cartesian coordinates, the vector product $\mathbf{a} = \mathbf{b} \times \mathbf{c}$ has as

† This may be readily extended to an $N \times N$ matrix $\mathsf{A}$, i.e.

$$|\mathsf{A}|\epsilon_{i_1 i_2 \cdots i_N} = A_{i_1 j_1}A_{i_2 j_2}\cdots A_{i_N j_N}\epsilon_{j_1 j_2 \cdots j_N},$$

where $\epsilon_{i_1 i_2 \cdots i_N}$ equals 1 if $i_1 i_2 \cdots i_N$ is an even permutation of $1, 2, \ldots, N$, and equals $-1$ if it is an odd permutation; otherwise it equals zero.

its $i$th component $a_i = \epsilon_{ijk} b_j c_k$; this should be contrasted with the outer product $\mathbf{T} = \mathbf{b} \otimes \mathbf{c}$, which is a second-order tensor having the components $T_{ij} = b_i c_j$.

►*Write the following as contracted Cartesian tensors: (i)* $\mathbf{a} \cdot \mathbf{b}$*; (ii)* $\nabla^2\phi$*; (iii)* $\nabla \times \mathbf{v}$*; (iv)* $\nabla(\nabla \cdot \mathbf{v})$*; (v)* $\nabla \times (\nabla \times \mathbf{v})$*; (vi)* $(\mathbf{a} \times \mathbf{b}) \cdot \mathbf{c}$.

The corresponding (contracted) tensor expressions are easily found as follows,

$$\begin{aligned}
\mathbf{a} \cdot \mathbf{b} = a_i b_i &= \delta_{ij} a_i b_j, \\
\nabla^2 \phi = \frac{\partial^2 \phi}{\partial x_i \partial x_i} &= \delta_{ij} \frac{\partial^2 \phi}{\partial x_i \partial x_j}, \\
(\nabla \times \mathbf{v})_i = \epsilon_{ijk} \frac{\partial v_k}{\partial x_j}&, \\
[\nabla(\nabla \cdot \mathbf{v})]_i = \frac{\partial}{\partial x_i}\left(\frac{\partial v_j}{\partial x_j}\right) &= \delta_{jk} \frac{\partial^2 v_j}{\partial x_i \partial x_k}, \\
[\nabla \times (\nabla \times \mathbf{v})]_i = \epsilon_{ijk} \frac{\partial}{\partial x_j}\left(\epsilon_{klm} \frac{\partial v_m}{\partial x_l}\right) &= \epsilon_{ijk} \epsilon_{klm} \frac{\partial^2 v_m}{\partial x_j \partial x_l}, \\
(\mathbf{a} \times \mathbf{b}) \cdot \mathbf{c} = \delta_{ij} c_i \epsilon_{jkl} a_k b_l &= \epsilon_{ikl} c_i a_k b_l. \quad ◄
\end{aligned}$$

An important relationship between the $\epsilon$- and $\delta$- tensors is expressed by the identity

$$\epsilon_{ijk} \epsilon_{klm} = \delta_{il} \delta_{jm} - \delta_{im} \delta_{jl}. \tag{19.28}$$

To establish the validity of this identity between two fourth-order tensors (the LHS is a once-contracted sixth-order tensor) we consider the various possibilities that arise.

The RHS of (19.28) has the values

$$+1 \text{ if } i = l \text{ and } j = m \neq i, \tag{19.29}$$

$$-1 \text{ if } i = m \text{ and } j = l \neq i, \tag{19.30}$$

$$0 \text{ for any other set of subscript values } i, j, l, m. \tag{19.31}$$

In each product on the LHS $k$ has the same value in both factors and for a non-zero contribution none of $i, l, j, m$ can have the same value as $k$. Since there are only three values, 1, 2 and 3, that any of the subscripts may take, the only non-zero possibilities are $i = l$ and $j = m$ or vice versa, but not all four subscripts equal (since then each $\epsilon$ factor is zero, as it would be if $i = j$ or $l = m$). This reproduces (19.31) for the LHS of (19.28) and also the conditions (19.29) and (19.30). The values in (19.29) and (19.30) are also reproduced in the LHS of (19.28) since

(i) if $i = l$ and $j = m$, $\epsilon_{ijk} = \epsilon_{lmk} = \epsilon_{klm}$ and, whether $\epsilon_{ijk}$ is $+1$ or $-1$, the product of the two factors is $+1$; and

(ii) if $i = m$ and $j = l$, $\epsilon_{ijk} = \epsilon_{mlk} = -\epsilon_{klm}$, and thus the product $\epsilon_{ijk}\epsilon_{klm}$ (no summation) has the value $-1$.

This concludes the establishment of identity (19.28).

A useful application of (19.28) is in obtaining alternative expressions for the vector quantities resulting from the vector product of a vector product.

►*Obtain an alternative expression for* $\nabla \times (\nabla \times \mathbf{v})$.

As shown in the previous example, $\nabla \times (\nabla \times \mathbf{v})$ can be expressed in tensor form as

$$\begin{aligned}
[\nabla \times (\nabla \times \mathbf{v})]_i &= \epsilon_{ijk}\epsilon_{klm}\frac{\partial^2 v_m}{\partial x_j \partial x_l} \\
&= (\delta_{il}\delta_{jm} - \delta_{im}\delta_{jl})\frac{\partial^2 v_m}{\partial x_j \partial x_l} \\
&= \frac{\partial}{\partial x_i}\left(\frac{\partial v_j}{\partial x_j}\right) - \frac{\partial^2 v_i}{\partial x_j \partial x_j} \\
&= [\nabla(\nabla \cdot \mathbf{v})]_i - \nabla^2 v_i,
\end{aligned}$$

where in the second line we have used the identity (19.28). This result has already been mentioned in chapter 8 and the reader is referred there for a discussion of its applicability. ◄

By examining the various possibilities, it is straightforward to verify that, more generally,

$$\epsilon_{ijk}\epsilon_{pqr} = \begin{vmatrix} \delta_{ip} & \delta_{iq} & \delta_{ir} \\ \delta_{jp} & \delta_{jq} & \delta_{jr} \\ \delta_{kp} & \delta_{kq} & \delta_{kr} \end{vmatrix}, \qquad (19.32)$$

and it is easily seen that (19.28) is a special case of this result. From (19.32) we can derive alternative forms of (19.28), for example,

$$\epsilon_{ijk}\epsilon_{ilm} = \delta_{jl}\delta_{km} - \delta_{jm}\delta_{kl}. \qquad (19.33)$$

The pattern of subscripts in these identities is most easily remembered by noting that the subscripts on the first $\delta$ on the RHS are those which immediately follow (cyclically, if necessary) the common subscript, here $i$, in each $\epsilon$-term on the LHS; the remaining combinations of $j, k, l, m$ as subscripts in the other $\delta$-terms on the RHS can then be filled in automatically.

Contracting (19.33), by setting $j = l$ (say) we obtain, since $\delta_{kk} = 3$ when using the summation convention,

$$\epsilon_{ijk}\epsilon_{ijm} = 3\delta_{km} - \delta_{km} = 2\delta_{km},$$

and by contracting once more, setting $k = m$, we further find that

$$\epsilon_{ijk}\epsilon_{ijk} = 6. \qquad (19.34)$$

## 19.9 Isotropic tensors

It will have been noticed that, unlike most of the tensors discussed (except for scalars), $\delta_{ij}$ and $\epsilon_{ijk}$ have the property that all their components have values that are the same whatever rotation of axes is made, i.e. the component values are independent of the transformation $L_{ij}$. Specifically $\delta_{11}$ has the value 1 in all coordinate frames, whereas for a general second-order tensor **T** all we know is that if $T_{11} = f_{11}(x_1, x_2, x_3)$ then $T'_{11} = f_{11}(x'_1, x'_2, x'_3)$. Tensors with the former property are called *isotropic* (or *invariant*) tensors.

It is important to know how general a tensor can be and still be isotropic, since the description of the physical properties, e.g. conductivity, magnetic susceptibility or tensile strength, of an isotropic medium (i.e. one which has the same properties whichever way it is orientated) involves an isotropic tensor. In the previous section it was shown that $\delta_{ij}$ and $\epsilon_{ijk}$ are second- and third-order isotropic tensors; we will now show that, to within a scalar multiple, they are the only such isotropic tensors.

Let us begin with isotropic second-order tensors. Suppose $T_{ij}$ is an isotropic tensor; then for *any* rotation of the axes we must, by definition, have that

$$T_{ij} = T'_{ij} = L_{ik}L_{jl}T_{kl} \tag{19.35}$$

for each of the nine components.

First consider the rotation of the axes about the $(1, 1, 1)$ direction, which takes $Ox_1$, $Ox_2$, $Ox_3$ into $Ox'_2$, $Ox'_3$, $Ox'_1$ respectively. For this rotation $L_{13} = 1$, $L_{21} = 1$, $L_{32} = 1$ and all other $L_{ij} = 0$. This requires that $T_{11} = T'_{11} = T_{33}$. Similarly $T_{12} = T'_{12} = T_{31}$. Continuing in this way, we find

$$\begin{aligned}
&\text{(a)}\ T_{11} = T_{22} = T_{33},\\
&\text{(b)}\ T_{12} = T_{23} = T_{31},\\
&\text{(c)}\ T_{21} = T_{32} = T_{13}.
\end{aligned}$$

Next consider the rotation of the axes (from their original position) by $\pi/2$ about the $Ox_3$-axis. In this case $L_{12} = -1$, $L_{21} = 1$, $L_{33} = 1$ and all other $L_{ij} = 0$. Amongst other relationships, we must have from (19.35) that

$$\begin{aligned}
T_{13} &= (-1) \times 1 \times T_{23},\\
T_{23} &= 1 \times 1 \times T_{13}.
\end{aligned}$$

Hence $T_{13} = T_{23} = 0$ and therefore, by parts (b) and (c) above, each element $T_{ij} = 0$ except for $T_{11}$, $T_{22}$ and $T_{33}$, which are all the same. This shows that $T_{ij} = \lambda\delta_{ij}$.

►*Show that $\lambda\epsilon_{ijk}$ is the only isotropic third-order Cartesian tensor.*

The general line of attack is as above and so only a minimum of explanation will be given.

$$T_{ijk} = T'_{ijk} = L_{il}L_{jm}L_{kn}T_{lmn} \qquad \text{(in all, there are 27 elements).}$$

Rotate about the $(1, 1, 1)$ direction: this is equivalent to making subscript permutations $1 \to 2 \to 3 \to 1$. We find

(a) $T_{111} = T_{222} = T_{333}$,
(b) $T_{112} = T_{223} = T_{331}$ (and two similar sets),
(c) $T_{123} = T_{231} = T_{312}$ (and a set involving odd permutations of 1, 2, 3).

Rotate by $\pi/2$ about the $Ox_3$-axis: $L_{12} = -1$, $L_{21} = 1$, $L_{33} = 1$, other $L_{ij} = 0$.

(d) $T_{111} = (-1) \times (-1) \times (-1) \times T_{222} = -T_{222}$,
(e) $T_{112} = (-1) \times (-1) \times 1 \times T_{221}$,
(f) $T_{221} = 1 \times 1 \times (-1) \times T_{112}$,
(g) $T_{123} = (-1) \times 1 \times 1 \times T_{213}$.

Relations (a) and (d) show that elements with all subscripts the same are zero. Relations (e), (f) and (b) show that all elements with repeated subscripts are zero. Relations (g) and (c) show that $T_{123} = T_{231} = T_{312} = -T_{213} = -T_{321} = -T_{132}$.

In total, $T_{ijk}$ differs from $\epsilon_{ijk}$ by at most a scalar factor, but since $\epsilon_{ijk}$ (and hence $\lambda\epsilon_{ijk}$) has already been explicitly shown to be an isotropic tensor, $T_{ijk}$ must be the most general third-order isotropic Cartesian tensor. ◄

Using exactly the same procedures employed for $\delta_{ij}$ and $\epsilon_{ijk}$, it may be shown that the only isotropic first-order tensor is the trivial one with all elements zero.

## 19.10 Improper rotations and pseudotensors

So far we have considered rigid rotations of the coordinate axes described by an orthogonal matrix $\mathsf{L}$ with $|\mathsf{L}| = +1$, (19.4). Strictly speaking such transformations are called *proper rotations.* We now broaden our discussion to include transformations that are still described by an orthogonal matrix $\mathsf{L}$, but for which $|\mathsf{L}| = -1$; these are called *improper rotations.*

This kind of transformation can always be considered as an *inversion* of the coordinate axes through the origin represented by the equation

$$x'_i = -x_i, \tag{19.36}$$

combined with a proper rotation. The transformation may alternatively be looked upon as one that changes an initially right-handed coordinate system into a left-handed one; any prior or subsequent proper rotation will not change this state

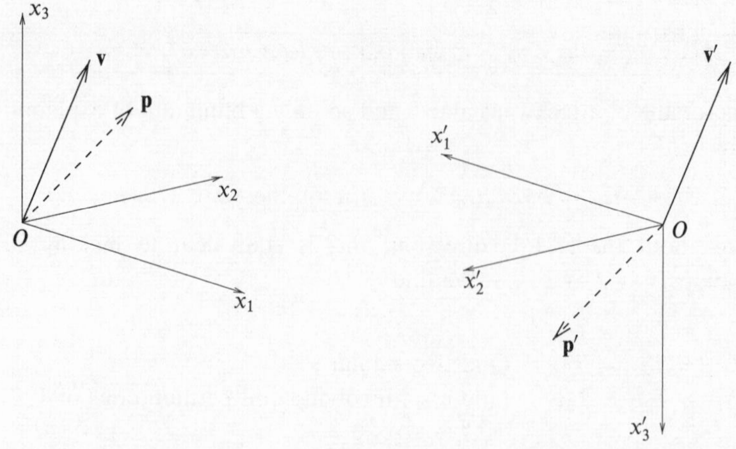


Figure 19.2 The behaviour of a vector **v** and a pseudovector **p** under a reflection through the origin of the coordinate system $x_1, x_2, x_3$ giving the new system $x'_1, x'_2, x'_3$.

of affairs. (Clearly, the matrix corresponding to (19.36) itself is the most obvious example of a transformation with $|\mathsf{L}| = -1$; in this case $L_{ij} = -\delta_{ij}$.)

As we have emphasised in earlier chapters, any real physical vector **v** may be considered as a geometrical object (i.e. an arrow in space), which can be referred to independently of any coordinate system and whose direction and magnitude cannot be altered merely by describing it in terms of a different coordinate system. Thus the components of **v** transform as $v'_i = L_{ij}v_j$ under *all* rotations (proper and improper).

We can define another type of object, however, whose components may also be labelled by a single subscript but which transforms as $v'_i = L_{ij}v_j$ under proper rotations, and as $v'_i = -L_{ij}v_j$ (note the minus sign) under improper rotations. In this case, the $v_i$ are not strictly the components of a true first-order Cartesian tensor but instead are said to form the components of a first-order Cartesian *pseudotensor*, or *pseudovector*.

It is important to realise that a pseudovector (as its name suggests) is not a geometrical object in the usual sense. In particular, it should *not* be considered as a real physical arrow in space, since its direction is reversed by an *improper* transformation of the coordinate axes (such as an inversion through the origin). This is illustrated in figure 19.2, in which the pseudovector **p** is shown as a dotted line to indicate that it is not a real physical vector.

Corresponding to vectors and pseudovectors, zeroth-order objects may be divided into scalars and pseudoscalars – the latter being invariant under rotation but changing sign on reflection. We may also extend the notion of scalars and pseudoscalars, vectors and pseudovectors, to objects with two or more subscripts.

For two subcripts, as defined previously, any quantity with components that transform as $T'_{ij} = L_{ik}L_{jl}T_{kl}$ under *all* rotations (proper and improper) is called a second-order Cartesian tensor. If, however, $T'_{ij} = L_{ik}L_{jl}T_{kl}$ under proper rotations, but $T'_{ij} = -L_{ik}L_{jl}T_{kl}$ under improper ones (which include reflections), then the $T_{ij}$ are the components of a second-order Cartesian pseudotensor. In general the components of Cartesian pseudotensors of arbitary order transform as

$$T'_{ij\cdots k} = |\mathsf{L}|L_{il}L_{jm}\cdots L_{kn}T_{lm\cdots n}, \tag{19.37}$$

where $|\mathsf{L}|$ is the determinant of the transformation matrix.

For example, from (19.27) we have that

$$|\mathsf{L}|\epsilon_{ijk} = L_{il}L_{jm}L_{kn}\epsilon_{lmn},$$

but since $|\mathsf{L}| = \pm 1$ we may rewrite this as

$$\epsilon_{ijk} = |\mathsf{L}|L_{il}L_{jm}L_{kn}\epsilon_{lmn}.$$

From this expression, we see that although $\epsilon_{ijk}$ behaves as a tensor under proper rotations, as discussed in section 19.8, it should properly be regarded as a third-order Cartesian *pseudo*tensor.

►*If $b_j$ and $c_k$ are the components of vectors, show that the quantities $a_i = \epsilon_{ijk}b_jc_k$ form the components of a pseudovector.*

In a new coordinate system we have

$$\begin{aligned} a'_i &= \epsilon'_{ijk}b'_jc'_k \\ &= |\mathsf{L}|L_{il}L_{jm}L_{kn}\epsilon_{lmn}L_{jp}b_pL_{kq}c_q \\ &= |\mathsf{L}|L_{il}\epsilon_{lmn}\delta_{mp}\delta_{nq}b_pc_q \\ &= |\mathsf{L}|L_{il}\epsilon_{lmn}b_mc_n \\ &= |\mathsf{L}|L_{il}a_l, \end{aligned}$$

from which we see immediately that the quantities $a_i$ form the components of a pseudovector. ◄

The above example is worth some further comment. If we denote the vectors with components $b_j$ and $c_k$ by $\mathbf{b}$ and $\mathbf{c}$ respectively, then, as mentioned in section 19.8, the quantities $a_i = \epsilon_{ijk}b_jc_k$ are the components of the real vector $\mathbf{a} = \mathbf{b} \times \mathbf{c}$, *provided we are using a right-handed Cartesian coordinate system.* However, in a different coordinate system, which is left-handed, the quantitites $a'_i = \epsilon'_{ijk}b'_jc'_k$ are *not* the components of the physical vector $\mathbf{a} = \mathbf{b} \times \mathbf{c}$, which has, instead, the components $-a'_i$. It is therefore important to note the handedness of a coordinate system before attempting to write in component form the vector relation $\mathbf{a} = \mathbf{b} \times \mathbf{c}$ (which is true without reference to any coordinate system).

It is worth noting that, although pseudotensors can be useful mathematical objects, the description of the real physical world must usually be in terms of tensors (i.e. scalars, vectors, etc.).† For example, the temperature or density of a gas must be a scalar quantity (rather than a pseudoscalar), since its value does not change when the coordinate system used to describe it is inverted through the origin. Similarly, velocity, magnetic field strength or angular momentum can only be described by a vector, and not by a pseudovector.

At this point, it may be useful to make a brief comment on the distinction between *active* and *passive* transformations of a physical system, as this difference often causes confusion. In this chapter, we are concerned solely with passive transformations, for which the physical system of interest is left unaltered, and only the coordinate system used to describe it is changed. In an active transformation, however, the system itself is altered.

As an example, let us consider a particle of mass $m$ that is located at a position $\mathbf{x}$ relative to the origin $O$ and hence has velocity $\dot{\mathbf{x}}$. The angular momentum of the particle about $O$ is thus $\mathbf{J} = m(\mathbf{x} \times \dot{\mathbf{x}})$. If we merely invert through $O$ the Cartesian coordinates used to describe this system, neither the magnitude nor direction of any these vectors will be changed, since they may be considered simply as arrows in space that are independent of the coordinates used to describe them. If, however, we perform the analogous active transformation on the system, by inverting the position vector of particle through $O$, then it is clear that the direction of particle's velocity will also be reversed, since it is simply the time derivative of the position vector, but the direction of its angular momentum vector remains unaltered. This suggests that vectors can be divided into two categories as follows: *polar* vectors (such as position and velocity) that reverse direction under an active inversion of the physical system through the origin, and *axial* vectors (such as angular momentum) that remain unchanged. It should be emphasised that at no point in this discussion have we introduced the concept of a pseudovector to describe a real physical quantity.

## 19.11 Dual tensors

Although pseudotensors are not themselves appropriate for the description of physical phenomena, they are sometimes useful, since we may use the pseudotensor $\epsilon_{ijk}$ to associate with every *antisymmetric* second-order tensor $A_{ij}$ (in three dimensions) a pseudovector $p_i$ given by

$$p_i = \tfrac{1}{2}\epsilon_{ijk}A_{jk}, \qquad (19.38)$$

† In fact the quantum-mechanical description of elementary particles, such as electrons, protons and neutrons, requires the introduction of a new kind of mathematical object called a *spinor*, which is not a scalar, vector, or more general tensor. The study of spinors, however, falls beyond the scope of this book.

which is called the *dual* of $A_{ij}$. Thus if we denote the antisymmetric tensor **A** by the matrix

$$\mathsf{A} = [A_{ij}] = \begin{pmatrix} 0 & A_{12} & -A_{31} \\ -A_{12} & 0 & A_{23} \\ A_{31} & -A_{23} & 0 \end{pmatrix},$$

then the components of its dual pseudovector are $(p_1, p_2, p_3) = (A_{23}, A_{31}, A_{12})$.

►*Using (19.38) show that* $A_{ij} = \epsilon_{ijk}p_k$.

By contracting both sides of (19.38) with $\epsilon_{ijk}$, we find

$$\epsilon_{ijk}p_k = \tfrac{1}{2}\epsilon_{ijk}\epsilon_{klm}A_{lm}.$$

Using the identity (19.28) we obtain

$$\begin{aligned}\epsilon_{ijk}p_k &= \tfrac{1}{2}(\delta_{il}\delta_{jm} - \delta_{im}\delta_{jl})A_{lm} \\ &= \tfrac{1}{2}(A_{ij} - A_{ji}) = \tfrac{1}{2}(A_{ij} + A_{ij}) = A_{ij},\end{aligned}$$

where in the last line we use the fact that $A_{ij} = -A_{ji}$. ◄

By a simple extension we may associate with every totally antisymmetric third-rank tensor $A_{ijk}$ (i.e. antisymmetric with respect to the interchange of any two of its three subscripts) a dual pseudoscalar $s$ given by

$$s = \frac{1}{3!}\epsilon_{ijk}A_{ijk}. \tag{19.39}$$

Since $A_{ijk}$ is a totally antisymmetric three-subscript quantity, we would expect it to equal some multiple of $\epsilon_{ijk}$ (since this is the only such quantity). In fact $A_{ijk} = s\epsilon_{ijk}$, as can be easily proved by substituting this expression into (19.39) and using (19.34).

## 19.12 Physical applications of tensors

In this section some physical applications of tensors will be given. First-order tensors are familiar as vectors, and so we will concentrate on second-order tensors, starting with a mechanics example.

Consider a collection of rigidly connected point particles of which the $\alpha$th, with mass $m^{(\alpha)}$, is typical, and is positioned at $\mathbf{r}^{(\alpha)}$ with respect to an origin $O$. Suppose that the rigid assembly is rotating about an axis through $O$ with angular velocity $\boldsymbol{\omega}$.

The angular momentum **J** about $O$ of the assembly is given by

$$\mathbf{J} = \sum_{\alpha} \left(\mathbf{r}^{(\alpha)} \times \mathbf{p}^{(\alpha)}\right).$$

But $\mathbf{p}^{(\alpha)} = m^{(\alpha)}\dot{\mathbf{r}}^{(\alpha)}$ and $\dot{\mathbf{r}}^{(\alpha)} = \boldsymbol{\omega}\times\mathbf{r}^{(\alpha)}$, for any $\alpha$, so in subscript form the components of $\mathbf{J}$ are given by

$$\begin{aligned} J_i &= \sum_\alpha m^{(\alpha)}\epsilon_{ijk}x_j^{(\alpha)}\dot{x}_k^{(\alpha)} \\ &= \sum_\alpha m^{(\alpha)}\epsilon_{ijk}x_j^{(\alpha)}\epsilon_{klm}\omega_l x_m^{(\alpha)} \\ &= \sum_\alpha m^{(\alpha)}(\delta_{il}\delta_{jm} - \delta_{im}\delta_{jl})x_j^{(\alpha)}x_m^{(\alpha)}\omega_l \\ &= \sum_\alpha m^{(\alpha)}\left[\left(r^{(\alpha)}\right)^2\delta_{il} - x_i^{(\alpha)}x_l^{(\alpha)}\right]\omega_l \equiv I_{il}\omega_l, \end{aligned} \tag{19.40}$$

where $I_{il}$ is a symmetric second-order Cartesian tensor (by the quotient rule, see section 19.7, since $\mathbf{J}$ and $\boldsymbol{\omega}$ are vectors), which depends only on the distribution of masses in the assembly and not upon the direction or magnitude of $\boldsymbol{\omega}$. The tensor is called the *inertia tensor* at $O$ of the assembly.

A more realistic situation obtains if a continuous rigid body is considered. In this case $m^{(\alpha)}$ must be replaced everywhere by $\rho(\mathbf{r})\,dx\,dy\,dz$, and all summations by integrations over the volume of the body. Written out in full in Cartesians, for a continuous body we would have

$$\mathsf{I} = [I_{ij}] = \begin{pmatrix} \int(y^2+z^2)\rho\,dV & -\int xy\rho\,dV & -\int xz\rho\,dV \\ -\int xy\rho\,dV & \int(z^2+x^2)\rho\,dV & -\int yz\rho\,dV \\ -\int xz\rho\,dV & -\int yz\rho\,dV & \int(x^2+y^2)\rho\,dV \end{pmatrix},$$

where $\rho = \rho(x,y,z)$ is the mass distribution and $dV$ stands for $dx\,dy\,dz$, the integrals being over the whole body. The diagonal elements of this tensor are called the *moments of inertia* and the off-diagonal elements without the minus signs are known as the *products of inertia*.

► *Show that the kinetic energy of the rotating system is given by* $T = \frac{1}{2}I_{jl}\omega_j\omega_l$.

By an argument parallel to that already made for $\mathbf{J}$, the kinetic energy is given by

$$\begin{aligned} T &= \tfrac{1}{2}\sum_\alpha m^{(\alpha)}\left(\dot{\mathbf{r}}^{(\alpha)}\cdot\dot{\mathbf{r}}^{(\alpha)}\right) \\ &= \tfrac{1}{2}\sum_\alpha m^{(\alpha)}\epsilon_{ijk}\omega_j x_k^{(\alpha)}\epsilon_{ilm}\omega_l x_m^{(\alpha)} \\ &= \tfrac{1}{2}\sum_\alpha m^{(\alpha)}(\delta_{jl}\delta_{km} - \delta_{jm}\delta_{kl})x_k^{(\alpha)}x_m^{(\alpha)}\omega_j\omega_l \\ &= \tfrac{1}{2}\sum_\alpha m^{(\alpha)}\left[\delta_{jl}\left(r^{(\alpha)}\right)^2 - x_j^{(\alpha)}x_l^{(\alpha)}\right]\omega_j\omega_l \\ &= \tfrac{1}{2}I_{jl}\omega_j\omega_l. \end{aligned}$$

Alternatively, since $J_j = I_{jl}\omega_l$ we may write the kinetic energy of the rotating system as $T = \frac{1}{2}J_j\omega_j$. ◀

The above example shows that the kinetic energy of the rotating body can be expressed as a scalar obtained by twice contracting $\boldsymbol{\omega}$ with the inertia tensor. It also shows that the moment of inertia of the body about a line given by the unit vector $\hat{\mathbf{n}}$ is $I_{jl}\hat{n}_j\hat{n}_l$ (or $\hat{\mathsf{n}}^{\mathrm{T}}\mathsf{I}\hat{\mathsf{n}}$ in matrix form).

Since $\mathbf{I}$ ($\equiv I_{jl}$) is a real symmetric second-order tensor, it has associated with it three mutually perpendicular directions that are its *principal axes* and have the properties (proved in chapter 7):

(i) with each axis is associated a principal moment of inertia $\lambda_\mu$, $\mu = 1, 2, 3$;

(ii) when the rotation of the body is about one of these axes, the angular velocity and the angular momentum are parallel and given by

$$\mathbf{J} = \mathbf{I}\boldsymbol{\omega} = \lambda_\mu \boldsymbol{\omega},$$

i.e. $\boldsymbol{\omega}$ is an eigenvector of $\mathbf{I}$ with eigenvalue $\lambda_\mu$;

(iii) referred to these axes as coordinate axes, the inertia tensor is diagonal, with diagonal entries $\lambda_1, \lambda_2, \lambda_3$.

Two further examples of physical quantities represented by second-order tensors are magnetic susceptibility and electrical conductivity. In the first case we have (in standard notation)

$$M_i = \chi_{ij}H_j, \tag{19.41}$$

and in the second case

$$j_i = \sigma_{ij}E_j. \tag{19.42}$$

Here $\mathbf{M}$ is the magnetic moment per unit volume and $\mathbf{j}$ the current density (current per unit area). In both cases we have on the left-hand side a vector and on the right-hand side the contraction of a set of quantities with another vector. Each set of quantities must therefore form the components of a second-order tensor.

For isotropic media $\mathbf{M} \propto \mathbf{H}$ and $\mathbf{j} \propto \mathbf{E}$, but for anisotropic materials such as crystals the susceptibility and conductivity may be different along different crystal axes, making $\chi_{ij}$ and $\sigma_{ij}$ general second-order tensors, although they are usually symmetric.

► *The electrical conductivity $\boldsymbol{\sigma}$ in a crystal is measured by an observer to have the components*

$$[\sigma_{ij}] = \begin{pmatrix} 1 & \sqrt{2} & 0 \\ \sqrt{2} & 3 & 1 \\ 0 & 1 & 1 \end{pmatrix}. \tag{19.43}$$

*Show that there is one direction in the crystal along which no current can flow. Does the current flow equally easily in the two perpendicular directions?*

The current density in the crystal is given by $j_i = \sigma_{ij}E_j$, where $\sigma_{ij}$, relative to the observer's coordinate system, is given by (19.43). Since $[\sigma_{ij}]$ is a symmetric matrix, it possess three mutually perpendicular eigenvectors (or principal axes), with respect to which the conductivity tensor is diagonal, with diagonal entries $\lambda_1, \lambda_2, \lambda_3$, the eigenvalues of $[\sigma_{ij}]$.

As discussed in chapter 7, the eigenvalues of $[\sigma_{ij}]$ are given by $|\sigma - \lambda\mathsf{I}| = 0$. Thus we require

$$\begin{vmatrix} 1-\lambda & \sqrt{2} & 0 \\ \sqrt{2} & 3-\lambda & 1 \\ 0 & 1 & 1-\lambda \end{vmatrix} = 0,$$

from which we find

$$(1-\lambda)[(3-\lambda)(1-\lambda)-1] - 2(1-\lambda) = 0.$$

This simplifies to give $\lambda = 0, 1, 4$, so that with respect to its principal axes, the conductivity tensor has components $\sigma'_{ij}$ given by

$$[\sigma'_{ij}] = \begin{pmatrix} 4 & 0 & 0 \\ 0 & 1 & 0 \\ 0 & 0 & 0 \end{pmatrix}.$$

Since $j'_i = \sigma'_{ij}E'_j$ we see immediately that along one of the principal axes there is no current flow and along the two perpendicular directions the current flows are not equal. ◄

We can extend the idea of a second-order tensor that relates two vectors to the case where two physical second-order tensors are related by a fourth-order tensor. The most common occurrence of such relationships is in the theory of elasticity. This is not the place to give a detailed account of elasticity theory, but suffice it to say that the local deformation of an elastic body at any interior point $P$ can be described by a second-order symmetric tensor $e_{ij}$ called the *strain tensor*. It is given by

$$e_{ij} = \frac{1}{2}\left(\frac{\partial u_i}{\partial x_j} + \frac{\partial u_j}{\partial x_i}\right),$$

where $\mathbf{u}$ is the displacement vector under the strain of a small volume element whose unstrained position relative to the origin is $\mathbf{x}$. Similarly we can describe the stress in the body at $P$ by the second-order symmetric *stress tensor* $p_{ij}$ (the quantity $p_{ij}$ is the $x_j$-component of the stress vector acting across a plane through $P$ whose normal lies in the $x_i$-direction). A generalisation of Hooke's law then relates the stress and strain tensors by

$$p_{ij} = c_{ijkl}e_{kl} \qquad (19.44)$$

where $c_{ijkl}$ is a fourth-order Cartesian tensor.

►*Assuming that the most general fourth-order isotropic tensor is*

$$c_{ijkl} = \lambda\delta_{ij}\delta_{kl} + \eta\delta_{ik}\delta_{jl} + \nu\delta_{il}\delta_{jk}, \qquad (19.45)$$

*find the form of (19.44) for an isotropic medium of Young's modulus $E$ and Poisson's ratio $\sigma$.*

For an isotropic medium we must have an isotropic tensor for $c_{ijkl}$, and so we assume the form (19.45). Substituting this into (19.44) yields

$$p_{ij} = \lambda\delta_{ij}e_{kk} + \eta e_{ij} + \nu e_{ji}.$$

But $e_{ij}$ is symmetric, and if we write $\eta + \nu = 2\mu$ this takes the form

$$p_{ij} = \lambda e_{kk}\delta_{ij} + 2\mu e_{ij}.$$

($\lambda$ and $\mu$ are called *Lamé constants*. It will be noted that if $e_{ij} = 0$ for $i \neq j$, then the same is true of $p_{ij}$, i.e. the principal axes of the stress and strain tensors coincide.)

Now consider a simple tension in the $x_1$-direction, i.e. $p_{11} = S$, all other $p_{ij} = 0$. Then denoting $e_{kk}$ (summed over $k$) by $\theta$ we have, in addition to $e_{ij} = 0$ for $i \neq j$, the three equations

$$S = \lambda\theta + 2\mu e_{11},$$
$$0 = \lambda\theta + 2\mu e_{22},$$
$$0 = \lambda\theta + 2\mu e_{33}.$$

Adding them gives

$$S = \theta(3\lambda + 2\mu).$$

Substituting for $\theta$ from this into the first of the three, and recalling that Young's modulus is defined by $S = Ee_{11}$, gives $E$ as

$$E = \frac{\mu(3\lambda + 2\mu)}{\lambda + \mu}. \qquad (19.46)$$

Further, Poisson's ratio is defined as $\sigma = -e_{22}/e_{11}$ (or $-e_{33}/e_{11}$), which is thus

$$\sigma = \left(\frac{1}{e_{11}}\right)\frac{\lambda\theta}{2\mu} = \left(\frac{1}{e_{11}}\right)\left(\frac{\lambda}{2\mu}\right)\frac{Ee_{11}}{3\lambda+2\mu} = \frac{\lambda}{2(\lambda+\mu)}. \tag{19.47}$$

Solving (19.46) and (19.47) for $\lambda$ and $\mu$ gives finally

$$p_{ij} = \frac{\sigma E}{(1+\sigma)(1-2\sigma)}e_{kk}\delta_{ij} + \frac{E}{(1+\sigma)}e_{ij}. \blacktriangleleft$$

## 19.13 Integral theorems for tensors

In chapter 9, we discussed various integral theorems involving vector and scalar fields. Most notably we considered the divergence theorem, which states that, for any vector field **a**,

$$\int_V \nabla\cdot\mathbf{a}\,dV = \oint_S \mathbf{a}\cdot\hat{\mathbf{n}}\,dS, \tag{19.48}$$

where $S$ is the surface enclosing the volume $V$ and $\hat{\mathbf{n}}$ is the outward-pointing unit normal to $S$ at each point.

Writing (19.48) in subscript notation, we have

$$\int_V \frac{\partial a_k}{\partial x_k}\,dV = \oint_S a_k\hat{n}_k\,dS. \tag{19.49}$$

Although we shall not prove it rigorously, (19.49) can be extended in an obvious manner to relate integrals of *tensor fields*, rather than just vector fields, over volumes and surfaces, with the result

$$\int_V \frac{\partial T_{ij\cdots k\cdots m}}{\partial x_k}\,dV = \oint_S T_{ij\cdots k\cdots m}\hat{n}_k\,dS.$$

This form of the divergence theorem for general tensors can be very useful in vector calculus manipulations.

►*A vector field* **a** *satisfies* $\nabla\cdot\mathbf{a} = 0$ *inside some volume* $V$ *and* $\mathbf{a}\cdot\hat{\mathbf{n}} = 0$ *on the boundary surface* $S$. *By considering the divergence theorem applied to* $T_{ij} = x_i a_j$, *show that* $\int_V \mathbf{a}\,dV = \mathbf{0}$.

Applying the divergence theorem to $T_{ij} = x_i a_j$ we find

$$\int_V \frac{\partial T_{ij}}{\partial x_j}\,dV = \int_V \frac{\partial(x_i a_j)}{\partial x_j}\,dV = \oint_S x_i a_j \hat{n}_j\,dS = 0,$$

since $a_j \hat{n}_j = 0$. By expanding the volume integral we thus obtain

$$\begin{aligned}\int_V \frac{\partial(x_i a_j)}{\partial x_j}\, dV &= \int_V \frac{\partial x_i}{\partial x_j} a_j\, dV + \int_V x_i \frac{\partial a_j}{\partial x_j}\, dV \\ &= \int_V \delta_{ij} a_j\, dV \\ &= \int_V a_i\, dV = 0,\end{aligned}$$

where in going from the first to the second line we used the fact that $\partial x_i/\partial x_j = \delta_{ij}$ and $\partial a_j/\partial x_j = 0$. ◄

The other integral theorems discussed in chapter 9 can be extended in a similar way. For example, written in tensor notation, Stokes' theorem states that for a vector field $a_i$,

$$\int_S \epsilon_{ijk} \frac{\partial a_k}{\partial x_j} \hat{n}_i\, dS = \oint_C a_k\, dx_k,$$

which, for a general tensor field, has the straightforward extension

$$\int_S \epsilon_{ijk} \frac{\partial T_{lm\cdots k\cdots n}}{\partial x_j} \hat{n}_i\, dS = \oint_C T_{lm\cdots k\cdots n}\, dx_k.$$

## 19.14 Non-Cartesian coordinates

So far we have restricted our attention to the study of tensors when they are described in terms of Cartesian coordinates and the axes of coordinates are rigidly rotated (sometimes together with an inversion of axes through the origin). In the remainder of this chapter we shall extend the concepts discussed in the previous sections by considering arbitrary coordinate transformations from one general coordinate system to another. Although this generalisation brings with it several complications, many of the properties of Cartesian tensors are still valid for more general tensors. Before considering general coordinate transformations, however, we begin by reminding ourselves of some properties of general curvilinear coordinates, as discussed in chapter 8.

The position of an arbitrary point $P$ in space may be expressed in terms of the three curvilinear coordinates $u_1, u_2, u_3$. If $\mathbf{r}(u_1, u_2, u_3)$ is the position vector of the point $P$, then we saw in chapter 8 that at every such point there exist two sets of basis vectors

$$\mathbf{e}_i = \frac{\partial \mathbf{r}}{\partial u_i} \qquad \text{and} \qquad \boldsymbol{\epsilon}_i = \nabla u_i, \tag{19.50}$$

where $i = 1, 2, 3$. In general, the vectors in each set neither are of unit length nor form an orthogonal basis. The sets $\mathbf{e}_i$ and $\boldsymbol{\epsilon}_i$ are, however, reciprocal systems of vectors, so that

$$\mathbf{e}_i \cdot \boldsymbol{\epsilon}_j = \delta_{ij}. \tag{19.51}$$

In the context of general tensor analysis, it is more usual to denote the second set of vectors $\boldsymbol{\epsilon}_i$ in (19.50) by $\mathbf{e}^i$, with the index placed as a superscript to distinguish it from the (different) vector $\mathbf{e}_i$, which is a member of the first set in (19.50). Although this positioning of the index may seem odd (not least because of the possibility of confusion with powers), it forms part of a slight modification to the summation convention, which we will adopt for the remainder of this chapter. This is as follows: any lower-case alphabetic index which appears exactly twice in any term of an expression, *once as a subscript and once as a superscript*, is to be summed over all the values that an index in that position can take (unless the contrary is specifically stated). All other aspects of the summation convention remain unchanged.

With the introduction of superscripts, the reciprocity relation (19.51) should be rewritten as

$$\mathbf{e}_i \cdot \mathbf{e}^j = \delta_i^{\ j}. \tag{19.52}$$

where the alternative form of the Kronecker delta is defined in a similar way to previously (i.e. it equals unity if $i = j$ and is zero otherwise), but is rewritten so that both sides of (19.52) have one subscript and one superscript.

For similar reasons it is usual to denote the curvilinear coordinates themselves by $u^1, u^2, u^3$, with the index raised, so that

$$\mathbf{e}_i = \frac{\partial \mathbf{r}}{\partial u^i} \qquad \text{and} \qquad \mathbf{e}^i = \nabla u^i. \tag{19.53}$$

From the first equality we see that we may consider a superscript that appears in the denominator of a partial derivative as a subscript.

Given the two bases $\mathbf{e}_i$ and $\mathbf{e}^i$, we may write a general vector $\mathbf{a}$ equally well in terms of either basis as follows:

$$\mathbf{a} = a^1\mathbf{e}_1 + a^2\mathbf{e}_2 + a^3\mathbf{e}_3 = a^i\mathbf{e}_i,$$
$$\mathbf{a} = a_1\mathbf{e}^1 + a_2\mathbf{e}^2 + a_3\mathbf{e}^3 = a_i\mathbf{e}^i,$$

where the $a^i$ are called the *contravariant* components of the vector $\mathbf{a}$ and the $a_i$ the *covariant* components, the position of the index (either as a subscript or superscript) serving to distinguish between them. Similarly, we may call the $\mathbf{e}_i$ the covariant basis vectors and the $\mathbf{e}^i$ the contravariant ones.

►*Show that the contravariant and covariant components of a vector* $\mathbf{a}$ *are given by* $a^i = \mathbf{a} \cdot \mathbf{e}^i$ *and* $a_i = \mathbf{a} \cdot \mathbf{e}_i$ *respectively.*

Firstly, for the contravariant components, we find

$$\mathbf{a} \cdot \mathbf{e}^i = a^j\mathbf{e}_j \cdot \mathbf{e}^i = a^j\delta_j^{\ i} = a^i,$$

where we have used the reciprocity relation (19.52). Similarly, for the covariant components,

$$\mathbf{a} \cdot \mathbf{e}_i = a_j \mathbf{e}^j \cdot \mathbf{e}_i = a_j \delta_i^j = a_i. \blacktriangleleft$$

The reason that the notion of contravariant and covariant components of a vector (and the resulting superscript notation) was not introduced earlier, is that for Cartesian coordinate systems the two sets of vectors $\mathbf{e}_i$ and $\mathbf{e}^i$ are identical, and hence so are the components of a vector with respect to either basis. Thus, for Cartesian coordinates, we may speak simply of the components of the vector, and there is no need to differentiate between contravariance and covariance, or to introduce superscripts to make a distinction between them.

If we consider the components of higher-order tensors in non-Cartesian coordinates, there are even more possibilities. As an example, let us consider a second-order tensor **T**. Using the outer product notation in (19.22), we may write **T** in three different ways:

$$\mathbf{T} = T^{ij} \mathbf{e}_i \otimes \mathbf{e}_j = T^i{}_j \mathbf{e}_i \otimes \mathbf{e}^j = T_{ij} \mathbf{e}^i \otimes \mathbf{e}^j,$$

where $T^{ij}$, $T^i{}_j$ and $T_{ij}$ are called the *contravariant, mixed and covariant* components of **T** respectively. It is important to remember that these three sets of quantities form the components of the *same* tensor **T**, but refer to different (tensor) bases made up from the basis vectors of the coordinate system. Again, if we are using Cartesian coordinates, all three sets of components are identical.

We may generalise straightforwardly to higher-order tensors, for which the components carrying only superscripts or only subscripts are referred to as the contravariant and covariant components respectively, and all others are called mixed components.

## 19.15 The metric tensor

Any particular curvilinear coordinate system is completely characterised (at each point in space) by the nine quantities

$$g_{ij} = \mathbf{e}_i \cdot \mathbf{e}_j, \tag{19.54}$$

which, as we will show, are the covariant components of a symmetric second-order tensor **g** called the *metric tensor*.

Since an infinitesimal vector displacement can be written as $d\mathbf{r} = du^i \mathbf{e}_i$, we find that the square of the infinitesimal arc length $(ds)^2$ can be written in terms of the metric tensor as

$$(ds)^2 = d\mathbf{r} \cdot d\mathbf{r} = du^i \mathbf{e}_i \cdot du^j \mathbf{e}_j = g_{ij}\, du^i du^j. \tag{19.55}$$

It may further be shown that the volume element $dV$ is given by

$$dV = \sqrt{g}\, du^1\, du^2\, du^3, \tag{19.56}$$

where $g$ is the determinant of the matrix $[g_{ij}]$, which has the covariant components of the metric tensor as its elements.

Equations (19.55) and (19.56) should be compared with the analogous ones in section 8.10, from which we see that in the special case where the coordinate system is orthogonal (so that $\mathbf{e}_i \cdot \mathbf{e}_j = 0$ for $i \neq j$), the metric tensor can be written in terms of the coordinate-system scale factors $h_i$, $i = 1, 2, 3$ (see the start of section 8.10) as

$$g_{ij} = \begin{cases} h_i^2 & i = j, \\ 0 & i \neq j, \end{cases}$$

and its determinant is given by $g = h_1^2 h_2^2 h_3^2$.

►*Calculate the elements $g_{ij}$ of the metric tensor for cylindrical polar coordinates. Hence find the infinitesimal arc length $(ds)^2$ and volume $dV$ for this coordinate system.*

In cylindrical polar coordinates, as discussed in section 8.9, $(u^1, u^2, u^3) = (\rho, \phi, z)$, so that the position vector $\mathbf{r}$ of any point $P$ may be written

$$\mathbf{r} = \rho \cos\phi \, \mathbf{i} + \rho \sin\phi \, \mathbf{j} + z \, \mathbf{k},$$

from which we obtain the (covariant) basis vectors

$$\begin{aligned} \mathbf{e}_1 &= \frac{\partial \mathbf{r}}{\partial \rho} &&= \cos\phi \, \mathbf{i} + \sin\phi \, \mathbf{j}, \\ \mathbf{e}_2 &= \frac{\partial \mathbf{r}}{\partial \phi} &&= -\rho \sin\phi \, \mathbf{i} + \rho \cos\phi \, \mathbf{j}, \\ \mathbf{e}_3 &= \frac{\partial \mathbf{r}}{\partial z} &&= \mathbf{k}. \end{aligned} \tag{19.57}$$

Thus the components of the metric tensor $[g_{ij}] = [\mathbf{e}_i \cdot \mathbf{e}_j]$ are found to be

$$\mathsf{G} = [g_{ij}] = \begin{pmatrix} 1 & 0 & 0 \\ 0 & \rho^2 & 0 \\ 0 & 0 & 1 \end{pmatrix}, \tag{19.58}$$

from which we see that, as expected for an orthogonal coordinate system, the metric tensor is diagonal, with the diagonal elements equal to the squares of the scale factors of the coordinate system.

From (19.55), the square of the infinitesimal arc length in this coordinate system is given by

$$(ds)^2 = g_{ij} \, du^i \, du^j = (d\rho)^2 + \rho^2 (d\phi)^2 + (dz)^2,$$

and using (19.56) the volume element is found to be

$$dV = \sqrt{g} \, du^1 \, du^2 \, du^3 = \rho \, d\rho \, d\phi \, dz.$$

These expressions should be compared with those derived in section 8.9, with which they are identical. ◄

We may also express the scalar product of two vectors in terms of the metric tensor,

$$\mathbf{a} \cdot \mathbf{b} = a^i \mathbf{e}_i \cdot b^j \mathbf{e}_j = g_{ij} a^i b^j, \tag{19.59}$$

where we have used the contravariant components of the two vectors. Similarly, using the covariant components, we can write the scalar product as

$$\mathbf{a} \cdot \mathbf{b} = a_i \mathbf{e}^i \cdot b_j \mathbf{e}^j = g^{ij} a_i b_j, \tag{19.60}$$

where we have defined the nine quantities $g^{ij} = \mathbf{e}^i \cdot \mathbf{e}^j$, which (as we shall show) form the contravariant components of the metric tensor $\mathbf{g}$, and are, in general, different from the quantities $g_{ij}$. Finally, we could express the scalar product in terms of the contravariant components of one vector and the covariant components of the other,

$$\mathbf{a} \cdot \mathbf{b} = a_i \mathbf{e}^i \cdot b^j \mathbf{e}_j = a_i b^j \delta^i_j = a_i b^i, \tag{19.61}$$

where we have used the reciprocity relation (19.52). Similarly, we could write

$$\mathbf{a} \cdot \mathbf{b} = a^i \mathbf{e}_i \cdot b_j \mathbf{e}^j = a^i b_j \delta^j_i = a^i b_i. \tag{19.62}$$

By comparing the four alternative expressions (19.59)–(19.62) for the scalar product of two vectors, we can deduce one of the most useful properties of the quantities $g_{ij}$ and $g^{ij}$. Since $g_{ij} a^i b^j = a^i b_i$ holds for any arbitrary vector components $a^i$, it follows that

$$g_{ij} b^j = b_i,$$

which illustrates the fact that the covariant components $g_{ij}$ of the metric tensor can be used to *lower an index*. In other words, we can obtain the covariant components of a vector from its contravariant components. By a similar argument, we have

$$g^{ij} b_j = b^i,$$

so that the contravariant components $g^{ij}$ can be used to perform the reverse operation of *raising an index*.

It is straightforward to show that the contravariant and covariant basis vectors, $\mathbf{e}^i$ and $\mathbf{e}_i$ respectively, are related in an analogous way by

$$\mathbf{e}^i = g^{ij} \mathbf{e}_j \qquad \text{and} \qquad \mathbf{e}_i = g_{ij} \mathbf{e}^j.$$

We also note that, since $\mathbf{e}_i$ and $\mathbf{e}^i$ are reciprocal systems of vectors in three-dimensional space (see chapter 6), we may write

$$\mathbf{e}^i = \frac{\mathbf{e}_j \times \mathbf{e}_k}{\mathbf{e}_i \cdot (\mathbf{e}_j \times \mathbf{e}_k)},$$

for $i, j, k = 1, 2, 3$ and cyclic permutations. A similar expression holds for $\mathbf{e}_i$ in terms of the $\mathbf{e}^i$-basis. Moreover, it may be shown that $|\mathbf{e}_1 \cdot (\mathbf{e}_2 \times \mathbf{e}_3)| = \sqrt{g}$.

►*Show that the matrix $[g^{ij}]$ is the inverse of the matrix $[g_{ij}]$. Hence calculate the contravariant components $g^{ij}$ of the metric tensor in cylindrical polar coordinates.*

Using the index-lowering and index-raising properties of $g_{ij}$ and $g^{ij}$, for an arbitrary vector **a** we find

$$\delta^i_k a^k = a^i = g^{ij} a_j = g^{ij} g_{jk} a^k.$$

But since **a** is arbitrary, we must have

$$g^{ij} g_{jk} = \delta^i_k. \tag{19.63}$$

Denoting the matrix $[g_{ij}]$ by $\mathsf{G}$ and $[g^{ij}]$ by $\hat{\mathsf{G}}$, equation (19.63) can be written in matrix form as $\hat{\mathsf{G}}\mathsf{G} = \mathsf{I}$, where $\mathsf{I}$ is the unit matrix. Hence $\mathsf{G}$ and $\hat{\mathsf{G}}$ are inverse matrices.

Thus, in cylindrical polar coordinates, by inverting the matrix $\mathsf{G}$ in (19.58) we find the elements $g^{ij}$ are given by

$$\hat{\mathsf{G}} = [g^{ij}] = \begin{pmatrix} 1 & 0 & 0 \\ 0 & 1/\rho^2 & 0 \\ 0 & 0 & 1 \end{pmatrix}. \blacktriangleleft$$

So far we have not considered the components of the metric tensor with one subscript and one superscript $g^i_j$. By analogy with (19.54), these mixed components are given by

$$g^i_j = \mathbf{e}^i \cdot \mathbf{e}_j = \delta^j_i,$$

so the components $g^i_j$ are identical to $\delta^i_j$. We may therefore consider the $\delta^i_j$ to be the mixed components of the metric tensor **g**.

## 19.16 General coordinate transformations and tensors

We now discuss the concept of general transformations from one coordinate system, $u^1, u^2, u^3$, to another, $u'^1, u'^2, u'^3$. We can describe the coordinate transform using the three equations

$$u'^i = u'^i(u^1, u^2, u^3),$$

for $i = 1, 2, 3$, in which the new coordinates $u'^i$ can in general be arbitrary functions of the old ones $u^i$, rather than just represent linear orthogonal transformations (rotations) of the coordinate axes. We shall also assume that the transformation can be inverted, so that we can write the old coordinates in terms of the new ones as

$$u^i = u^i(u'^1, u'^2, u'^3),$$

As an example, we may consider the transformation from spherical polar to Cartesian coordinates, given by

$$x = r\sin\theta\cos\phi,$$
$$y = r\sin\theta\sin\phi,$$
$$z = r\cos\theta,$$

which is clearly not a linear transformation.

The two sets of basis vectors in the new coordinate system, $u'^1, u'^2, u'^3$, are given as in (19.53) by

$$\mathbf{e}'_i = \frac{\partial \mathbf{r}}{\partial u'^i} \quad \text{and} \quad \mathbf{e}'^i = \nabla u'^i. \tag{19.64}$$

Considering the first set, we have from the chain rule that

$$\frac{\partial \mathbf{r}}{\partial u^j} = \frac{\partial u'^i}{\partial u^j}\frac{\partial \mathbf{r}}{\partial u'^i},$$

so that the basis vectors in the old and new coordinate systems are related by

$$\mathbf{e}_j = \frac{\partial u'^i}{\partial u^j}\mathbf{e}'_i. \tag{19.65}$$

Now, since we can write any arbitrary vector $\mathbf{a}$ in terms of either basis as

$$\mathbf{a} = a'^i\mathbf{e}'_i = a^j\mathbf{e}_j = a^j\frac{\partial u'^i}{\partial u^j}\mathbf{e}'_i$$

it follows that the contravariant components of a vector must transform as

$$a'^i = \frac{\partial u'^i}{\partial u^j}a^j. \tag{19.66}$$

In fact, we use this relation as the defining property that a set of quantities $a^i$ must have if they are to form the contravariant components of a vector.

►*Find an expression analogous to (19.65) relating the basis vectors* $\mathbf{e}^i$ *and* $\mathbf{e}'^i$ *in the two coordinate systems. Hence deduce the way in which the covariant components of a vector change under a coordinate transformation.*

If we consider the second set of basis vectors in (19.64), $\mathbf{e}'^i = \nabla u'^i$, we have from the chain rule that

$$\frac{\partial u^j}{\partial x} = \frac{\partial u^j}{\partial u'^i}\frac{\partial u'^i}{\partial x},$$

and similarly for $\partial u^j/\partial y$ and $\partial u^j/\partial z$. So the basis vectors in the old and new coordinate systems are related by

$$\mathbf{e}^j = \frac{\partial u^j}{\partial u'^i}\mathbf{e}'^i. \tag{19.67}$$

For any arbitrary vector $\mathbf{a}$,

$$\mathbf{a} = a'_i \mathbf{e}'^i = a_j \mathbf{e}^j = a_j \frac{\partial u^j}{\partial u'^i} \mathbf{e}'^i,$$

and so the covariant components of a vector must transform as

$$a'_i = \frac{\partial u^j}{\partial u'^i} a_j. \tag{19.68}$$

In a similar way to that used in the contravariant case, we take this result as the defining property of the covariant components of a vector. ◄

We may compare the transformation laws (19.66) and (19.68) with those for a first-order Cartesian tensor under a rigid rotation of axes. Let us consider a rotation of Cartesian axes $x^i$ through an angle $\theta$ about the 3-axis to a new set $x'^i$, $i = 1, 2, 3$, as given by (19.7) and the inverse transformation (19.8). It is straightforward to show that

$$\frac{\partial x^j}{\partial x'^i} = \frac{\partial x'^i}{\partial x^j} = L_{ij},$$

where the elements $L_{ij}$ are given by

$$\mathsf{L} = \begin{pmatrix} \cos\theta & \sin\theta & 0 \\ -\sin\theta & \cos\theta & 0 \\ 0 & 0 & 1 \end{pmatrix}.$$

Thus (19.66) and (19.68) agree with our earlier definition in the special case of a rigid rotation of Cartesian axes.

Following on from (19.66) and (19.68), we proceed in a similar way to define general tensors of higher rank. For example, the contravariant, mixed and covariant components, respectively, of a second-order tensor must transform as follows:

$$\text{contravariant components,} \quad T'^{ij} = \frac{\partial u'^i}{\partial u^k} \frac{\partial u'^j}{\partial u^l} T^{kl} \;;$$

$$\text{mixed components,} \quad T'^i{}_j = \frac{\partial u'^i}{\partial u^k} \frac{\partial u^l}{\partial u'^j} T^k{}_l \;;$$

$$\text{covariant components,} \quad T'_{ij} = \frac{\partial u^k}{\partial u'^i} \frac{\partial u^l}{\partial u'^j} T_{kl}.$$

It is important to remember that these quantities form the components of the *same* tensor $\mathbf{T}$ but refer to different tensor bases made up from the basis vectors of the different coordinate systems. For example, in terms of the contravariant components we may write

$$\mathbf{T} = T^{ij} \mathbf{e}_i \otimes \mathbf{e}_j = T'^{ij} \mathbf{e}'_i \otimes \mathbf{e}'_j.$$

We can clearly go on to define tensors of higher order, with arbitrary numbers of covariant (subscript) and contravariant (superscript) indices, by demanding that their components transform as follows:

$$T'^{ij\cdots k}{}_{lm\cdots n} = \frac{\partial u'^i}{\partial u^a}\frac{\partial u'^j}{\partial u^b}\cdots\frac{\partial u'^k}{\partial u^c}\frac{\partial u^d}{\partial u'^l}\frac{\partial u^e}{\partial u'^m}\cdots\frac{\partial u^f}{\partial u'^n}T^{ab\cdots c}{}_{de\cdots f}. \qquad (19.69)$$

Using the revised summation convention described in section 19.14, the algebra of general tensors is completely analogous to that of the Cartesian tensors discussed earlier. For example, as with Cartesian coordinates, the Kronecker delta is a tensor, provided it is written as the mixed tensor $\delta^i_j$, since

$$\delta'^i{}_j = \frac{\partial u'^i}{\partial u^k}\frac{\partial u^l}{\partial u'^j}\delta^k_l = \frac{\partial u'^i}{\partial u^k}\frac{\partial u^k}{\partial u'^j} = \frac{\partial u'^i}{\partial u'^j} = \delta^i_j,$$

where we have used the chain rule to prove the third equality. This also shows that $\delta^i_j$ is isotropic. As discussed at the end of section 19.15, $\delta^i_j$ can be considered as the mixed components of the metric tensor **g**.

► *Show that the quantities $g_{ij} = \mathbf{e}_i \cdot \mathbf{e}_j$ form the covariant components of a second-order tensor.*

In the new (primed) coordinate system we have

$$g'_{ij} = \mathbf{e}'_i \cdot \mathbf{e}'_j,$$

but applying result (19.65) to the inverse transformation, we have

$$\mathbf{e}'_i = \frac{\partial u^k}{\partial u'^i}\mathbf{e}_k,$$

and similarly for $\mathbf{e}'_j$. Thus we may write

$$g'_{ij} = \frac{\partial u^k}{\partial u'^i}\frac{\partial u^l}{\partial u'^j}\,\mathbf{e}_k \cdot \mathbf{e}_l = \frac{\partial u^k}{\partial u'^i}\frac{\partial u^l}{\partial u'^j}g_{kl},$$

which shows that the $g_{ij}$ are indeed the covariant components of a second-order tensor (the metric tensor **g**). ◄

A similar argument to that used in the above example shows that the quantities $g^{ij}$ form the contravariant components of a second-order tensor, such that

$$g'^{ij} = \frac{\partial u'^i}{\partial u^k}\frac{\partial u'^j}{\partial u^l}g^{kl}.$$

In the previous section we discussed the use of the components $g_{ij}$ and $g^{ij}$ in the raising and lowering of indices in contravariant and covariant vectors. This can be extended to tensors of arbitrary rank. In general, contraction of a tensor with $g_{ij}$ will convert the contracted index from being contravariant (superscript) to covariant (subscript), i.e. it is lowered. This can be repeated for as many indices are required. For example,

$$T_{ij} = g_{ik} T^k{}_j = g_{ik} g_{jl} T^{kl}. \qquad (19.70)$$

Similarly contraction with $g^{ij}$ raises an index, i.e.

$$T^{ij} = g^{ik} T_k{}^j = g^{ik} g^{jl} T_{kl}. \qquad (19.71)$$

That (19.70) and (19.71) are mutually consistent may be shown by using the fact that $g^{ik} g_{kj} = \delta^i_j$.

## 19.17 Relative tensors

In section 19.10 we introduced the concept of pseudotensors in the context of the rotation (proper or improper) of a set of Cartesian axes. Generalising to arbitrary coordinate transformations leads to the notion of a *relative tensor*.

For an arbitrary coordinate transformation from one general coordinate system $u^i$ to another $u'^i$, we may define the Jacobian of the transformation (see chapter 5) as the determinant of the transformation matrix $[\partial u'^i/\partial u^j]$, which is usually denoted by

$$J = \left| \frac{\partial u'}{\partial u} \right|.$$

Alternatively, we may interchange the primed and unprimed coordinates to obtain $|\partial u/\partial u'| = 1/J$, which is also often referred to as the Jacobian of the transformation.

Using the Jacobian $J$, we define a relative tensor of weight $w$ as one whose components transform as follows:

$$T'^{ij\cdots k}{}_{lm\cdots n} = \frac{\partial u'^i}{\partial u^a} \frac{\partial u'^j}{\partial u^b} \cdots \frac{\partial u'^k}{\partial u^c} \frac{\partial u^d}{\partial u'^l} \frac{\partial u^e}{\partial u'^m} \cdots \frac{\partial u^f}{\partial u'^n} T^{ab\cdots c}{}_{de\cdots f} \left| \frac{\partial u}{\partial u'} \right|^w. \qquad (19.72)$$

Comparing this expression with (19.69), we see that a true (or *absolute*) general tensor may be considered as a relative tensor of weight $w = 0$. If $w = -1$, on the other hand, the relative tensor is called a general *pseudotensor*, and if $w = 1$ it is called a *tensor density*.

It is worth comparing (19.72) with the definition (19.37) of a Cartesian pseudotensor. For the latter, we are concerned only with its behaviour under a rotation (proper or improper) of Cartesian axes, for which the Jacobian $J = \pm 1$. Thus, general relative tensors of weight $w = -1$ and $w = 1$ would both satisfy the definition (19.37) of a Cartesian pseudotensor.

► *If the $g_{ij}$ are the covariant components of the metric tensor, show that the determinant $g$ of the matrix $[g_{ij}]$ is a relative scalar of weight $w = 2$.*

The components $g_{ij}$ transform as

$$g'_{ij} = \frac{\partial u^k}{\partial u'^i}\frac{\partial u^l}{\partial u'^j} g_{kl}.$$

Defining the matrices $\mathsf{U} = [\partial u^i/\partial u'^j]$, $\mathsf{G} = [g_{ij}]$ and $\mathsf{G}' = [g'_{ij}]$, we may write this expression as

$$\mathsf{G}' = \mathsf{U}^{\mathrm{T}}\mathsf{G}\mathsf{U}.$$

Taking the determinant of both sides, we obtain

$$g' = |\mathsf{U}|^2 g = \left|\frac{\partial u}{\partial u'}\right|^2 g,$$

which shows that $g$ is a relative scalar of weight $w = 2$. ◄

From the discussion in section 19.8, it can be seen that $\epsilon_{ijk}$ is a covariant relative tensor of weight $-1$. We may also define the contravariant tensor $\epsilon^{ijk}$, which is numerically equal to $\epsilon_{ijk}$, but is a relative tensor of weight $+1$.

If two relative tensors have weights $w_1$ and $w_2$ respectively then, from (19.72), the outer product of the two tensors, or any contraction of them, is a relative tensor of weight $w_1 + w_2$. As a special case, we may use $\epsilon_{ijk}$ and $\epsilon^{ijk}$ to construct pseudovectors from antisymmetric tensors and vice versa, in an analogous way to that discussed in section 19.11.

For example, if the $A^{ij}$ are the contravariant components of an antisymmetric tensor ($w = 0$), then

$$p_i = \tfrac{1}{2}\epsilon_{ijk}A^{jk}$$

are the covariant components of a pseudovector ($w = -1$), since $\epsilon_{ijk}$ has weight $w = -1$. Similarly, we may show that

$$A^{ij} = \epsilon^{ijk}p_k.$$

## 19.18 Derivatives of basis vectors and Christoffel symbols

In Cartesian coordinates, the basis vectors $\mathbf{e}_i$ are constant and so their derivatives with respect to the coordinates vanish. In a general coordinate system, however, the basis vectors $\mathbf{e}_i$ and $\mathbf{e}^i$ are functions of the coordinates. In order that we may differentiate general tensors, we must therefore first consider the derivatives of the basis vectors.

Let us consider the derivative $\partial\mathbf{e}_i/\partial u^j$. Since this is itself a vector, it can be written as a linear combination of the basis vectors $\mathbf{e}_k$, $k = 1, 2, 3$. If we introduce the symbol $\Gamma^k{}_{ij}$ to denote the coefficients in this combination, we have

$$\frac{\partial \mathbf{e}_i}{\partial u^j} = \Gamma^k{}_{ij} \mathbf{e}_k. \tag{19.73}$$

The coefficient $\Gamma^k{}_{ij}$ is simply the $k$th component of the vector $\partial \mathbf{e}_i / \partial u^j$. Using the reciprocity relation $\mathbf{e}^i \cdot \mathbf{e}_j = \delta^i_j$, these 27 numbers are given (at each point in space) by

$$\Gamma^k{}_{ij} = \mathbf{e}^k \cdot \frac{\partial \mathbf{e}_i}{\partial u^j}. \tag{19.74}$$

Furthermore, by differentiating the reciprocity relation $\mathbf{e}^i \cdot \mathbf{e}_j = \delta^i_j$ with respect to the coordinates, and using (19.74), it is straightforward to show that the derivatives of the contravariant basis vectors are given by

$$\frac{\partial \mathbf{e}^i}{\partial u^j} = -\Gamma^i{}_{kj} \mathbf{e}^k. \tag{19.75}$$

The symbol $\Gamma^k{}_{ij}$ is called a *Christoffel symbol* (of the second kind), but, despite appearances to the contrary, these quantities do *not* form the components of a third-order tensor. We note that in Cartesian coordinates it is clear from (19.74) that $\Gamma^k{}_{ij} = 0$ for all values of the indices $i$, $j$ and $k$.

►*Using (19.74), deduce the way in which the quantities $\Gamma^k{}_{ij}$ transform under a general coordinate transformation, and hence show that they do not form the components of a third-order tensor.*

In a new coordinate system

$$\Gamma'^k{}_{ij} = \mathbf{e}'^k \cdot \frac{\partial \mathbf{e}'_i}{\partial u'^j},$$

but from (19.67) and (19.65) respectively we have, on reversing primed and unprimed variables,

$$\mathbf{e}'^k = \frac{\partial u'^k}{\partial u^n} \mathbf{e}^n \qquad \text{and} \qquad \mathbf{e}'_i = \frac{\partial u^l}{\partial u'^i} \mathbf{e}_l.$$

Therefore in the new coordinate system the quantities $\Gamma'^k{}_{ij}$ are given by

$$\begin{aligned}
\Gamma'^k{}_{ij} &= \frac{\partial u'^k}{\partial u^n} \mathbf{e}^n \cdot \frac{\partial}{\partial u'^j} \left( \frac{\partial u^l}{\partial u'^i} \mathbf{e}_l \right) \\
&= \frac{\partial u'^k}{\partial u^n} \mathbf{e}^n \cdot \left( \frac{\partial^2 u^l}{\partial u'^j \partial u'^i} \mathbf{e}_l + \frac{\partial u^l}{\partial u'^i} \frac{\partial \mathbf{e}_l}{\partial u'^j} \right) \\
&= \frac{\partial u'^k}{\partial u^n} \frac{\partial^2 u^l}{\partial u'^j \partial u'^i} \mathbf{e}^n \cdot \mathbf{e}_l + \frac{\partial u'^k}{\partial u^n} \frac{\partial u^l}{\partial u'^i} \frac{\partial u^m}{\partial u'^j} \mathbf{e}^n \cdot \frac{\partial \mathbf{e}_l}{\partial u^m} \\
&= \frac{\partial u'^k}{\partial u^l} \frac{\partial^2 u^l}{\partial u'^j \partial u'^i} + \frac{\partial u'^k}{\partial u^n} \frac{\partial u^l}{\partial u'^i} \frac{\partial u^m}{\partial u'^j} \Gamma^n{}_{lm},
\end{aligned} \tag{19.76}$$

where in the last line we have used (19.74) and the reciprocity relation $\mathbf{e}^n \cdot \mathbf{e}_l = \delta^n_l$. From (19.76) we see immediately that the ${\Gamma^k}_{ij}$ do not form the components of a third-order tensor, because of the presence of the first term on the right-hand side. ◄

In a given coordinate system we may, in principle, calculate the ${\Gamma^k}_{ij}$ using (19.74). In practice, however, it is often quicker to use an alternative expression, which we now derive, for the Christoffel symbol in terms of the metric tensor $g_{ij}$ and its derivatives with respect to the coordinates.

First we note that the Christoffel symbol ${\Gamma^k}_{ij}$ is symmetric with respect to the interchange of its two subscripts $i$ and $j$. This is easily shown, since

$$\frac{\partial \mathbf{e}_i}{\partial u^j} = \frac{\partial^2 \mathbf{r}}{\partial u^j \partial u^i} = \frac{\partial^2 \mathbf{r}}{\partial u^i \partial u^j} = \frac{\partial \mathbf{e}_j}{\partial u^i},$$

so that from (19.73) we find ${\Gamma^k}_{ij}\mathbf{e}_k = {\Gamma^k}_{ji}\mathbf{e}_k$. Taking the scalar product with $\mathbf{e}^l$ and using the reciprocity relation $\mathbf{e}_k \cdot \mathbf{e}^l = \delta^l_k$, this immediately gives

$${\Gamma^l}_{ij} = {\Gamma^l}_{ji}.$$

To obtain an expression for ${\Gamma^k}_{ij}$, we then use $g_{ij} = \mathbf{e}_i \cdot \mathbf{e}_j$, and consider the derivative

$$\begin{aligned}\frac{\partial g_{ij}}{\partial u^k} &= \frac{\partial \mathbf{e}_i}{\partial u^k} \cdot \mathbf{e}_j + \mathbf{e}_i \cdot \frac{\partial \mathbf{e}_j}{\partial u^k} \\ &= {\Gamma^l}_{ik}\mathbf{e}_l \cdot \mathbf{e}_j + \mathbf{e}_i \cdot {\Gamma^l}_{jk}\mathbf{e}_l, \\ &= {\Gamma^l}_{ik} g_{lj} + {\Gamma^l}_{jk} g_{il},\end{aligned} \tag{19.77}$$

where we have used the definition (19.73). By cyclically permuting the free indices $i, j, k$ in (19.77), we obtain two further equivalent relations,

$$\frac{\partial g_{jk}}{\partial u^i} = {\Gamma^l}_{ji} g_{lk} + {\Gamma^l}_{ki} g_{jl}, \tag{19.78}$$

and

$$\frac{\partial g_{ki}}{\partial u^j} = {\Gamma^l}_{kj} g_{li} + {\Gamma^l}_{ij} g_{kl}. \tag{19.79}$$

If we now add (19.78) and (19.79) together, and subtract (19.77) from the result, we find

$$\begin{aligned}\frac{\partial g_{jk}}{\partial u^i} + \frac{\partial g_{ki}}{\partial u^j} - \frac{\partial g_{ij}}{\partial u^k} &= {\Gamma^l}_{ji} g_{lk} + {\Gamma^l}_{ki} g_{jl} + {\Gamma^l}_{kj} g_{li} + {\Gamma^l}_{ij} g_{kl} - {\Gamma^l}_{ik} g_{lj} - {\Gamma^l}_{jk} g_{il} \\ &= 2{\Gamma^l}_{ij} g_{kl},\end{aligned}$$

where we have used the symmetry properties of both ${\Gamma^l}_{ij}$ and $g_{ij}$. Contracting both sides with $g^{mk}$ leads to the required expression for the Christoffel symbol in terms of the metric tensor and its derivatives, namely

$${\Gamma^m}_{ij} = \tfrac{1}{2} g^{mk} \left( \frac{\partial g_{jk}}{\partial u^i} + \frac{\partial g_{ki}}{\partial u^j} - \frac{\partial g_{ij}}{\partial u^k} \right). \tag{19.80}$$

►*Calculate the Christoffel symbols $\Gamma^m{}_{ij}$ for cylindrical polar coordinates.*

We may use either (19.73) or (19.80) to calculate the $\Gamma^m{}_{ij}$ for this simple coordinate system. In cylindrical polar coordinates $(u^1, u^2, u^3) = (\rho, \phi, z)$, the basis vectors $\mathbf{e}_i$ are given by (19.57). It is straightforward to show that the only derivatives of these vectors with respect to the coordinates that are non-zero are

$$\frac{\partial \mathbf{e}_\rho}{\partial \phi} = \frac{1}{\rho}\mathbf{e}_\phi, \qquad \frac{\partial \mathbf{e}_\phi}{\partial \rho} = \frac{1}{\rho}\mathbf{e}_\phi, \qquad \frac{\partial \mathbf{e}_\phi}{\partial \phi} = -\rho \mathbf{e}_\rho.$$

Thus, from (19.73), we have immediately that

$$\Gamma^2{}_{12} = \Gamma^2{}_{21} = \frac{1}{\rho} \qquad \text{and} \qquad \Gamma^1{}_{22} = -\rho. \tag{19.81}$$

Alternatively, using (19.80), and the fact that $g_{11} = 1$, $g_{22} = \rho^2$, $g_{33} = 1$ and the other components are zero, we see that the only three non-zero Christoffel symbols are indeed $\Gamma^2{}_{12} = \Gamma^2{}_{21}$ and $\Gamma^1{}_{22}$. These are given by

$$\Gamma^2{}_{12} = \Gamma^2{}_{21} = \frac{1}{2g_{22}}\frac{\partial g_{22}}{\partial u^1} = \frac{1}{2\rho^2}\frac{\partial}{\partial \rho}(\rho^2) = \frac{1}{\rho},$$
$$\Gamma^1{}_{22} = -\frac{1}{2g_{11}}\frac{\partial g_{22}}{\partial u^1} = -\frac{1}{2}\frac{\partial}{\partial \rho}(\rho^2) = -\rho,$$

which agrees with (19.81), found directly from (19.73). ◄

## 19.19 Covariant differentiation

For Cartesian tensors, we noted that the derivative of a scalar is a (covariant) vector. This is also true for general tensors, as may be shown by considering the differential of a scalar

$$d\phi = \frac{\partial \phi}{\partial u^i}\, du^i.$$

Since the $du^i$ are the components of a contravariant vector, and $d\phi$ is a scalar, we have by the quotient rule, discussed in section 19.7, that the quantities $\partial\phi/\partial u^i$ must form the components of a covariant vector.

It is straightforward to show, however, that (unlike in Cartesian coordinates) the differentiation of the components of a general tensor, other than a scalar, with respect to the coordinates does *not* in general result in the components of another tensor. For example, in Cartesian coordinates, if the $v^i$ are the contravariant components of a vector, then the quantites $\partial v^i/\partial x^j$ form the components of a second-order tensor. In general coordinates, however, this is not the case.

►*Show that, in general coordinates, the quantities $\partial v^i/\partial u^j$ do not form the components of a tensor.*

We may show this directly by considering

$$\left(\frac{\partial v^i}{\partial u^j}\right)' = \frac{\partial v'^i}{\partial u'^j} = \frac{\partial u^k}{\partial u'^j}\frac{\partial v'^i}{\partial u^k}$$

$$= \frac{\partial u^k}{\partial u'^j}\frac{\partial}{\partial u^k}\left(\frac{\partial u'^i}{\partial u^l}v^l\right)$$

$$= \frac{\partial u^k}{\partial u'^j}\frac{\partial u'^i}{\partial u^l}\frac{\partial v^l}{\partial u^k} + \frac{\partial u^k}{\partial u'^j}\frac{\partial^2 u'^i}{\partial u^k \partial u^l}v^l. \tag{19.82}$$

The presence of the second term on the right-hand side of (19.82) shows that the $\partial v^i/\partial x^j$ do not form the components of a second-order tensor. This term arises because the 'transformation matrix' $[\partial u'^i/\partial u^j]$ changes with position in space. This is not true in Cartesian coordinates, for which the second term vanishes, and $\partial v^i/\partial x^j$ is a second-order tensor. ◄

We may, however, use the Christoffel symbols discussed in the previous section to define a new *covariant* derivative of the components of a tensor, which does result in the components of another tensor.

Let us first consider the derivative of a vector $\mathbf{v}$ with respect to the coordinates. Writing the vector in terms of its contravariant components $\mathbf{v} = v^i\mathbf{e}_i$, we find

$$\frac{\partial \mathbf{v}}{\partial u^j} = \frac{\partial v^i}{\partial u^j}\mathbf{e}_i + v^i\frac{\partial \mathbf{e}_i}{\partial u^j}, \tag{19.83}$$

where the second term arises because, in general, the basis vectors $\mathbf{e}_i$ are not constant (this term vanishes in Cartesian coordinates). Using (19.73) we may write

$$\frac{\partial \mathbf{v}}{\partial u^j} = \frac{\partial v^i}{\partial u^j}\mathbf{e}_i + v^i\Gamma^k{}_{ij}\mathbf{e}_k.$$

Since $i$ and $k$ are dummy indices in the last term on the right-hand side, we may interchange them to obtain

$$\frac{\partial \mathbf{v}}{\partial u^j} = \frac{\partial v^i}{\partial u^j}\mathbf{e}_i + v^k\Gamma^i{}_{kj}\mathbf{e}_i = \left(\frac{\partial v^i}{\partial u^j} + v^k\Gamma^i{}_{kj}\right)\mathbf{e}_i. \tag{19.84}$$

The reason for the interchanging the dummy indices, as shown in (19.84), is that we may then factor out $\mathbf{e}_i$. The quantity in brackets is called the *covariant derivative*, for which the standard notation is

$$v^i{}_{;j} \equiv \frac{\partial v^i}{\partial u^j} + \Gamma^i{}_{kj}v^k, \tag{19.85}$$

where the semicolon denotes covariant differentiation; a similar short-hand notation also exists for simple partial derivatives, in which a comma is used instead of a semicolon. For example, $\partial v^i/\partial u^j$ is denoted by $v^i{}_{,j}$.

In Cartesian coordinates all the $\Gamma^i{}_{kj}$ are zero, and so the covariant derivative reduces to the simple partial derivative $\partial v^i/\partial u^j$.

Using the short-hand semicolon notation, the derivative of a vector may be written in the very compact form

$$\frac{\partial \mathbf{v}}{\partial u^j} = v^i{}_{;j}\mathbf{e}_i$$

and, by the quotient rule (section 19.7), it is clear that the $v^i{}_{;j}$ are the (mixed) components of a second-order tensor. This may also be verified directly, using the transformation properties of $\partial v^i/\partial u^j$ and $\Gamma^i{}_{kj}$ given in (19.82) and (19.76) respectively.

In general, we may regard the $v^i{}_{;j}$ as the mixed components of a second-order tensor called the covariant derivative of $\mathbf{v}$, which is denoted by $\nabla\mathbf{v}$. In Cartesian coordinates, the components of this tensor are just $\partial v^i/\partial x^j$.

►*Calculate $v^i{}_{;i}$ in cylindrical polar coordinates.*

Contracting (19.85) we obtain

$$v^i{}_{;i} = \frac{\partial v^i}{\partial u^i} + \Gamma^i{}_{ki}v^k.$$

Now from (19.81) we have

$$\begin{aligned}
\Gamma^i{}_{1i} &= \Gamma^1{}_{11} + \Gamma^2{}_{12} + \Gamma^3{}_{13} = 1/\rho,\\
\Gamma^i{}_{2i} &= \Gamma^1{}_{21} + \Gamma^2{}_{22} + \Gamma^3{}_{23} = 0,\\
\Gamma^i{}_{3i} &= \Gamma^1{}_{31} + \Gamma^2{}_{32} + \Gamma^3{}_{33} = 0,
\end{aligned}$$

and so

$$\begin{aligned}
v^i{}_{;i} &= \frac{\partial v^\rho}{\partial \rho} + \frac{\partial v^\phi}{\partial \phi} + \frac{\partial v^z}{\partial z} + \frac{1}{\rho}v^\rho\\
&= \frac{1}{\rho}\frac{\partial}{\partial \rho}(\rho v^\rho) + \frac{\partial v^\phi}{\partial \phi} + \frac{\partial v^z}{\partial z}.
\end{aligned}$$

This result is identical to the expression for the divergence of a vector field in cylindrical polar coordinates given in section 8.9. This is discussed further in section 19.20. ◄

So far we have considered only the covariant derivative of the contravariant components $v^i$ of a vector. The corresponding result for the covariant components $v_i$ may be found in a similar way, by considering the derivative of $\mathbf{v} = v_i\mathbf{e}^i$, and using (19.75), to obtain

$$v_{i;j} = \frac{\partial v_i}{\partial u^j} - \Gamma^k{}_{ij}v_k. \qquad (19.86)$$

Comparing the expressions (19.85) and (19.86) for the covariant derivative of the contravariant and covariant components of a vector respectively, we see that there are some similarities and some differences. It may help to remember that the index with respect to which the covariant derivative is taken ($j$ in this case), is also the last subscript on the Christoffel symbol; the remaining indices can then only be arranged in one way without raising or lowering them. It then remains to remember the sign difference, i.e. that for a covariant index (subscript) the Christoffel symbol carries a minus sign, whereas for a contravariant index (superscript) the sign is positive.

Following a similar procedure that led to equation (19.85), we may obtain expressions for the covariant derivatives of higher-order tensors.

►*By considering the derivative of the second-order tensor* $\mathbf{T}$ *with respect to the coordinate* $u^k$*, find an expression for the covariant derivative* $T^{ij}{}_{;k}$ *of its contravariant components.*

Expressing $\mathbf{T}$ in terms of its contravariant components, we have

$$\begin{aligned}\frac{\partial \mathbf{T}}{\partial u^k} &= \frac{\partial}{\partial u^k}(T^{ij}\mathbf{e}_i \otimes \mathbf{e}_j)\\ &= \frac{\partial T^{ij}}{\partial u^k}\mathbf{e}_i \otimes \mathbf{e}_j + T^{ij}\frac{\partial \mathbf{e}_i}{\partial u^k} \otimes \mathbf{e}_j + T^{ij}\mathbf{e}_i \otimes \frac{\partial \mathbf{e}_j}{\partial u^k}.\end{aligned}$$

Using (19.73), we can rewrite the derivatives of the basis vectors in terms of Christoffel symbols, to obtain

$$\frac{\partial \mathbf{T}}{\partial u^k} = \frac{\partial T^{ij}}{\partial u^k}\mathbf{e}_i \otimes \mathbf{e}_j + T^{ij}\Gamma^l{}_{ik}\mathbf{e}_l \otimes \mathbf{e}_j + T^{ij}\mathbf{e}_i \otimes \Gamma^l{}_{jk}\mathbf{e}_l.$$

Interchanging the dummy indices $i$ and $l$ in the second term and $j$ and $l$ in the third term on the right-hand side, this becomes

$$\frac{\partial \mathbf{T}}{\partial u^k} = \left(\frac{\partial T^{ij}}{\partial u^k} + \Gamma^i{}_{lk}T^{lj} + \Gamma^j{}_{lk}T^{il}\right)\mathbf{e}_i \otimes \mathbf{e}_j,$$

where the expression in brackets is the required covariant derivative

$$T^{ij}{}_{;k} = \frac{\partial T^{ij}}{\partial u^k} + \Gamma^i{}_{lk}T^{lj} + \Gamma^j{}_{lk}T^{il}. \qquad (19.87)$$

Using (19.87), the derivative of the tensor $\mathbf{T}$ with respect to $u^k$ can be written in terms of its contravariant components as

$$\frac{\partial \mathbf{T}}{\partial u^k} = T^{ij}{}_{;k}\mathbf{e}_i \otimes \mathbf{e}_j. \quad ◄$$

Similar results may be obtained for the the covariant derivatives of the mixed and covariant components of a second-order tensor. Collecting these results together, we have

$$T^{ij}{}_{;k} = T^{ij}{}_{,k} + \Gamma^i{}_{lk} T^{lj} + \Gamma^j{}_{lk} T^{il},$$
$$T^i{}_{j;k} = T^i{}_{j,k} + \Gamma^i{}_{lk} T^l{}_j - \Gamma^l{}_{jk} T^i{}_l,$$
$$T_{ij;k} = T_{ij,k} - \Gamma^l{}_{ik} T_{lj} - \Gamma^l{}_{jk} T_{il},$$

where we have used the comma notation for partial derivatives. The position of the indices in these expressions is very systematic: for each contravariant index (superscript) on the LHS we add a term on the RHS containing a Christoffel symbol with a plus sign, and for every covariant index (subscript) we add a corresponding term with a minus sign. This is straightforwardly extended to tensors with an arbitrary number of contravariant and covariant indices.

We note that the quantities $T^{ij}{}_{;k}$, $T^i{}_{j;k}$ and $T_{ij;k}$ are the components of the *same* third-order tensor $\nabla \mathbf{T}$ with respect to different tensor bases, i.e.

$$\nabla \mathbf{T} = T^{ij}{}_{;k} \mathbf{e}_i \otimes \mathbf{e}_j \otimes \mathbf{e}^k = T^i{}_{j;k} \mathbf{e}_i \otimes \mathbf{e}^j \otimes \mathbf{e}^k = T_{ij;k} \mathbf{e}^i \otimes \mathbf{e}^j \otimes \mathbf{e}^k.$$

We conclude this section by considering the covariant derivative of a scalar. The covariant derivative differs from the simple partial derivative with respect to the coordinates only because the basis vectors of the coordinate system change with position in space (hence for Cartesian coordinates there is no difference). However, a scalar $\phi$ does not depend on the basis vectors at all, so its covariant derivative must be the same as its partial derivative, i.e.

$$\phi_{;j} = \frac{\partial \phi}{\partial u^j} = \phi_{,j}. \tag{19.88}$$

## 19.20 Vector operators in tensor form

In section 8.10 we used vector calculus methods to find expressions for vector differential operators, such as grad, div, curl and the Laplacian, in general *orthogonal* curvilinear coordinates (and we considered cylindrical and spherical polars as particular examples). In this section we use the framework of general tensors that we have developed to obtain expressions for these operators in tensor form, which are valid in *all* coordinate systems, whether orthogonal or not.

In order to compare the results obtained here with those given in section 8.10 for orthogonal coordinates, it is necessary to remember that here we are working with the (in general non-unit) basis vectors $\mathbf{e}_i = \partial \mathbf{r}/\partial u^i$ or $\mathbf{e}^i = \nabla u^i$. Thus the components of a vector $\mathbf{v} = v^i \mathbf{e}_i$ are not the same as the components $\hat{v}^i$ in the corresponding unit basis $\hat{\mathbf{e}}_i$. In fact, if the scale factors of the coordinate system are $h_i$, $i = 1, 2, 3$, then $v^i = \hat{v}^i / h_i$ (no sum on $i$).

As mentioned in section 19.15, for an orthogonal coordinate system with scale factors $h_i$, we have

$$g_{ij} = \begin{cases} h_i^2 & \text{if } i = j, \\ 0 & \text{otherwise} \end{cases} \qquad \text{and} \qquad g^{ij} = \begin{cases} 1/h_i^2 & \text{if } i = j, \\ 0 & \text{otherwise,} \end{cases}$$

so that the determinant $g$ of the matrix $[g_{ij}]$ is given by $g = h_1^2 h_2^2 h_3^2$.

*Gradient*. The gradient of a scalar $\phi$ is simply given by

$$\nabla\phi = \phi_{;i}\mathbf{e}^i = \frac{\partial\phi}{\partial u^i}\mathbf{e}^i, \tag{19.89}$$

since the covariant derivative of a scalar is the same as its partial derivative.

*Divergence*. Replacing the partial derivatives that occur in Cartesian coordinates with covariant derivatives, the divergence of a vector field $\mathbf{v}$ in a general coordinate system is given by

$$\nabla \cdot \mathbf{v} = v^i{}_{;i} = \frac{\partial v^i}{\partial u^i} + \Gamma^i{}_{ki} v^k.$$

Using the expression (19.80) for the Christoffel symbol in terms of the metric tensor, we find

$$\Gamma^i{}_{ki} = \tfrac{1}{2} g^{il}\left(\frac{\partial g_{il}}{\partial u^k} + \frac{\partial g_{kl}}{\partial u^i} - \frac{\partial g_{ki}}{\partial u^l}\right) = \tfrac{1}{2} g^{il}\frac{\partial g_{il}}{\partial u^k}. \tag{19.90}$$

The last two terms have cancelled because

$$g^{il}\frac{\partial g_{kl}}{\partial u^i} = g^{li}\frac{\partial g_{ki}}{\partial u^l} = g^{il}\frac{\partial g_{ki}}{\partial u^l},$$

where in the first equality we have interchanged the dummy indices $i$ and $l$, and in the second equality we have used the symmetry of the metric tensor.

We may simplify (19.90) still further by using a result concerning the derivative of the determinant of a matrix whose elements are functions of the coordinates.

►*Suppose* $\mathsf{A} = [a_{ij}]$, $\mathsf{B} = [b^{ij}]$ *and that* $\mathsf{B} = \mathsf{A}^{-1}$. *By considering the determinant* $a = |\mathsf{A}|$, *show that*

$$\frac{\partial a}{\partial u^k} = ab^{ji}\frac{\partial a_{ij}}{\partial u^k}.$$

If we denote the cofactor of the element $a_{ij}$ by $\Delta^{ij}$, then the elements of the inverse matrix are given by (see chapter 7)

$$b^{ij} = \frac{1}{a}\Delta^{ji}. \tag{19.91}$$

However, the determinant of $\mathsf{A}$ is given by

$$a = \sum_j a_{ij}\Delta^{ij},$$

in which we have *fixed* $i$ and explicitly written the sum over $j$ for clarity. Partially differentiating both sides with respect to $a_{ij}$, we then obtain

$$\frac{\partial a}{\partial a_{ij}} = \Delta^{ij}, \tag{19.92}$$

since $a_{ij}$ does not occur in any of the cofactors $\Delta^{ij}$.

Now, if we suppose that the $a_{ij}$ are functions of the coordinates, then so also will be the determinant $a$, and by the chain rule we have

$$\frac{\partial a}{\partial u^k} = \frac{\partial a}{\partial a_{ij}}\frac{\partial a_{ij}}{\partial u^k} = \Delta^{ij}\frac{\partial a_{ij}}{\partial u^k} = ab^{ji}\frac{\partial a_{ij}}{\partial u^k} \tag{19.93}$$

where we have used (19.91) and (19.92). ◄

Applying the result (19.93) to the determinant $g$ of the metric tensor, and remembering that $g^{ik}g_{kj} = \delta^i_j$ and that $g^{ij}$ is symmetric, we obtain

$$\frac{\partial g}{\partial u^k} = gg^{ij}\frac{\partial g_{ij}}{\partial u^k}. \tag{19.94}$$

Substituting (19.94) into (19.90) we find that the expression for the Christoffel symbol can be much simplified to give

$$\Gamma^i{}_{ki} = \frac{1}{2g}\frac{\partial g}{\partial u^k} = \frac{1}{\sqrt{g}}\frac{\partial \sqrt{g}}{\partial u^k}.$$

Thus we finally obtain the expression for the divergence of a vector field in a general coordinate system as

$$\nabla \cdot \mathbf{v} = v^i{}_{;i} = \frac{1}{\sqrt{g}}\frac{\partial}{\partial u^j}(\sqrt{g}v^j). \tag{19.95}$$

*Laplacian.* If we replace $\mathbf{v}$ by $\nabla\phi$ in $\nabla \cdot \mathbf{v}$, then we obtain the Laplacian $\nabla^2\phi$. But, from (19.89), we have

$$\mathbf{v} = v_i\mathbf{e}^i = \frac{\partial \phi}{\partial u^i}\mathbf{e}^i,$$

so the covariant components are given by $v_i = \partial\phi/\partial u^i$. In (19.95), however, we require the contravariant components $v^i$. These may be obtained by raising the index using the metric tensor, to give

$$v^j = g^{jk}v_k = g^{jk}\frac{\partial \phi}{\partial u^k}$$

Substituting this into (19.95) we obtain

$$\nabla^2\phi = \frac{1}{\sqrt{g}}\frac{\partial}{\partial u^j}\left(\sqrt{g}g^{jk}\frac{\partial \phi}{\partial u^k}\right). \tag{19.96}$$

►*Use (19.96) to find the expression for $\nabla^2\phi$ in an orthogonal coordinate system with scale factors $h_i$, $i = 1, 2, 3$.*

For an orthogonal coordinate system $\sqrt{g} = h_1h_2h_3$ and $g^{ij} = 1/h_i^2$ if $i = j$ and $g^{ij} = 0$ otherwise. Therefore, from (19.96) we have

$$\nabla^2\phi = \frac{1}{h_1h_2h_3}\frac{\partial}{\partial u^j}\left(\frac{h_1h_2h_3}{h_j^2}\frac{\partial\phi}{\partial u^j}\right),$$

which agrees with the results of section 8.10. ◄

*Curl.* The special vector form of the curl of a vector field exists only in three dimensions. We therefore consider its more general form, which is also valid in higher-dimensional spaces. In a general space the operation curl $\mathbf{v}$ is defined by

$$(\text{curl }\mathbf{v})_{ij} = v_{i;j} - v_{j;i},$$

which is an antisymmetric covariant tensor.

In fact the difference of derivatives can be simplified, since

$$\begin{aligned}v_{i;j} - v_{j;i} &= \frac{\partial v_i}{\partial u^j} - \Gamma^l{}_{ij}v_l - \frac{\partial v_j}{\partial u^i} + \Gamma^l{}_{ji}v_l \\ &= \frac{\partial v_i}{\partial u^j} - \frac{\partial v_j}{\partial u^i},\end{aligned}$$

where the Christoffel symbols have cancelled because of their symmetry properties. Thus curl $\mathbf{v}$ can be written in terms of partial derivatives as

$$(\text{curl }\mathbf{v})_{ij} = \frac{\partial v_i}{\partial u^j} - \frac{\partial v_j}{\partial u^i}.$$

Generalising slightly the discussion of section 19.17, in three dimensions we may associate with this antisymmetric second-order tensor a vector with contravariant components

$$\begin{aligned}(\nabla\times\mathbf{v})^i &= -\frac{1}{2\sqrt{g}}\epsilon^{ijk}(\text{curl }\mathbf{v})_{jk} \\ &= -\frac{1}{2\sqrt{g}}\epsilon^{ijk}\left(\frac{\partial v_j}{\partial u^k} - \frac{\partial v_k}{\partial u^j}\right) = \frac{1}{\sqrt{g}}\epsilon^{ijk}\frac{\partial v_k}{\partial u^j},\end{aligned}$$

which is analogous to the expression in Cartesian coordinates discussed in section 19.8.

## 19.21 Absolute derivatives along curves

In section 19.19 we discussed how to differentiate a general tensor with respect to the coordinates and introduced the covariant derivative. In this section, we consider the slightly different problem of calculating the derivative of a tensor along a curve $\mathbf{r}(t)$, parameterised by some variable $t$.

Let us begin by considering the derivative of a vector $\mathbf{v}$ along the curve. If we introduce an arbitrary coordinate system $u^i$ with basis vectors $\mathbf{e}_i$, $i = 1, 2, 3$, then we may write $\mathbf{v} = v^i \mathbf{e}_i$, and we have

$$\begin{aligned}\frac{d\mathbf{v}}{dt} &= \frac{dv^i}{dt}\mathbf{e}_i + v^i \frac{d\mathbf{e}_i}{dt} \\ &= \frac{dv^i}{dt}\mathbf{e}_i + v^i \frac{\partial \mathbf{e}_i}{\partial u^k}\frac{du^k}{dt},\end{aligned}$$

where we have used the chain rule to rewrite the last term on the right-hand side. Using (19.73) to write the derivatives of the basis vectors in terms of Christoffel symbols, we obtain

$$\frac{d\mathbf{v}}{dt} = \frac{dv^i}{dt}\mathbf{e}_i + \Gamma^j{}_{ik} v^i \frac{du^k}{dt}\mathbf{e}_j.$$

Interchanging the dummy indices $i$ and $j$ in the last term, we may factor out the basis vector, and we find

$$\frac{d\mathbf{v}}{dt} = \left(\frac{dv^i}{dt} + \Gamma^i{}_{jk} v^j \frac{du^k}{dt}\right)\mathbf{e}_i.$$

The term in parentheses is called the *absolute* (or *intrinsic*) derivative of the components $v^i$ along the curve $\mathbf{r}(t)$, and is usually denoted by

$$\frac{\delta v^i}{\delta t} \equiv \frac{dv^i}{dt} + \Gamma^i{}_{jk} v^j \frac{du^k}{dt} = v^i{}_{;k}\frac{du^k}{dt}.$$

With this notation, we may write

$$\frac{d\mathbf{v}}{dt} = \frac{\delta v^i}{\delta t}\mathbf{e}_i = v^i{}_{;k}\frac{du^k}{dt}\mathbf{e}_i. \tag{19.97}$$

Using the same method, the absolute derivative of the covariant components $v_i$ of a vector is given by

$$\frac{\delta v_i}{\delta t} \equiv v_{i;k}\frac{du^k}{dt}.$$

Similarly, the absolute derivatives of the contravariant, mixed and covariant components of a second-order tensor $\mathbf{T}$ are

$$\begin{aligned}\frac{\delta T^{ij}}{\delta t} &\equiv T^{ij}{}_{;k}\frac{du^k}{dt}, \\ \frac{\delta T^i{}_j}{\delta t} &\equiv T^i{}_{j;k}\frac{du^k}{dt}, \\ \frac{\delta T_{ij}}{\delta t} &\equiv T_{ij;k}\frac{du^k}{dt}.\end{aligned}$$

The derivative of $\mathbf{T}$ along the curve $\mathbf{r}(t)$ may then be written in terms of, for example, its contravariant components as

$$\frac{d\mathbf{T}}{dt} = \frac{\delta T^{ij}}{\delta t}\mathbf{e}_i \otimes \mathbf{e}_j = T^{ij}{}_{;k}\frac{du^k}{dt}\mathbf{e}_i \otimes \mathbf{e}_j.$$

## 19.22 Geodesics

As an example of the use of the absolute derivative, we conclude this chapter with a brief discussion of geodesics. A geodesic in real three-dimensional space is a straight line, which has two equivalent defining properties. Firstly, it is the curve of shortest length between two points and, secondly, its tangent vector always points in the same direction (along the line). Although we have explicitly considered only our familiar three-dimensional space in this chapter, much of the mathematical formalism developed can easily be generalised to more abstract spaces of higher dimensionality in which the familiar ideas of Euclidean geometry are no longer valid. It is often of interest to find geodesic curves in such spaces by using the properties of straight lines in Euclidean space that define a geodesic.

We shall not consider these more complicated spaces explicitly here, but instead derive the equation that a geodesic in Euclidean three-dimensional space (i.e. a straight line) must satisfy, in a sufficiently general way that it may be applied, with little modification, to find the equations satisfied by geodesics in more abstract spaces.

Let us consider a curve $\mathbf{r}(s)$, parameterised by the arc length $s$ from some point on the curve, and choose as our defining property for a geodesic that its tangent vector $\mathbf{t} = d\mathbf{r}/ds$ always points in the same direction everywhere on the curve, i.e.

$$\frac{d\mathbf{t}}{ds} = \mathbf{0}. \tag{19.98}$$

(We could alternatively use the property that the distance between two points is a minimum along a geodesic; using the calculus of variations (see chapter 20), this leads to the same final result (19.99).)

If we now introduce an arbitrary coordinate system $u^i$ with basis vectors $\mathbf{e}_i$, $i = 1, 2, 3$, then we may write $\mathbf{t} = t^i\mathbf{e}_i$, and from (19.97) we find

$$\frac{d\mathbf{t}}{ds} = t^i{}_{;k}\frac{du^k}{ds}\mathbf{e}_i = \mathbf{0}.$$

Writing out the covariant derivative, we obtain

$$\left(\frac{dt^i}{ds} + \Gamma^i{}_{jk}t^j\frac{du^k}{ds}\right)\mathbf{e}_i = \mathbf{0}.$$

But since $t^j = du^j/ds$, we find that the equation satisfied by a geodesic is

$$\frac{d^2u^i}{ds^2} + \Gamma^i{}_{jk}\frac{du^j}{ds}\frac{du^k}{ds} = 0. \tag{19.99}$$

►*Find the equations satisfied by a geodesic (straight line) in cylindrical polar coordinates.*

From (19.81), the only non-zero Christoffel symbols are ${\Gamma^1}_{22} = -\rho$ and ${\Gamma^2}_{12} = {\Gamma^2}_{21} = 1/\rho$. Thus the required geodesic equations are

$$\frac{d^2u^1}{ds^2} + {\Gamma^1}_{22}\frac{du^2}{ds}\frac{du^2}{ds} = 0 \Rightarrow \frac{d^2\rho}{ds^2} - \rho\left(\frac{d\phi}{ds}\right)^2 = 0,$$

$$\frac{d^2u^2}{ds^2} + 2{\Gamma^2}_{12}\frac{du^1}{ds}\frac{du^2}{ds} = 0 \Rightarrow \frac{d^2\phi}{ds^2} + \frac{2}{\rho}\frac{d\rho}{ds}\frac{d\phi}{ds} = 0,$$

$$\frac{d^2u^3}{ds^2} = 0 \Rightarrow \frac{d^2z}{ds^2} = 0. \blacktriangleleft$$

## 19.23 Exercises

19.1 (a) Show that for any general (but fixed) $\phi$, $(u_1, u_2) = (x_1\cos\phi - x_2\sin\phi, x_1\sin\phi + x_2\cos\phi)$ are the components of a first-order tensor in two dimensions.

(b) Show that

$$\begin{pmatrix} x_2^2 & x_1x_2 \\ x_1x_2 & x_1^2 \end{pmatrix}$$

is not a (Cartesian) tensor of order 2. (To establish that a single element does not transform correctly is sufficient.)

19.2 Show how to decompose the tensor $T_{ij}$ into three tensors,

$$T_{ij} = U_{ij} + V_{ij} + S_{ij},$$

where $U_{ij}$ is symmetric and has zero trace, $V_{ij}$ is isotropic and $S_{ij}$ has only three independent components.

19.3 Use the quotient law, discussed in section 19.7, to show that the array

$$\begin{pmatrix} y^2+z^2-x^2 & -2xy & -2xz \\ -2yx & x^2+z^2-y^2 & -2yz \\ -2zx & -2zy & x^2+y^2-z^2 \end{pmatrix}$$

forms a second-order tensor.

19.4 Use tensor methods to establish the following vector identities:

(a) $(\mathbf{u}\times\mathbf{v})\times\mathbf{w} = (\mathbf{u}\cdot\mathbf{w})\mathbf{v} - (\mathbf{v}\cdot\mathbf{w})\mathbf{u}$;

(b) $\operatorname{curl}(\phi\mathbf{u}) = \phi\operatorname{curl}\mathbf{u} + (\operatorname{grad}\phi)\times\mathbf{u}$;

(c) $\operatorname{div}(\mathbf{u}\times\mathbf{v}) = \mathbf{v}\cdot\operatorname{curl}\mathbf{u} - \mathbf{u}\cdot\operatorname{curl}\mathbf{v}$;

(d) $\operatorname{curl}(\mathbf{u}\times\mathbf{v}) = (\mathbf{v}\cdot\operatorname{grad})\mathbf{u} - (\mathbf{u}\cdot\operatorname{grad})\mathbf{v} + \mathbf{u}\operatorname{div}\mathbf{v} - \mathbf{v}\operatorname{div}\mathbf{u}$;

(e) $\operatorname{grad}\frac{1}{2}(\mathbf{u}\cdot\mathbf{u}) = \mathbf{u}\times\operatorname{curl}\mathbf{u} + (\mathbf{u}\cdot\operatorname{grad})\mathbf{u}$.

19.5 Use result (e) of the previous question and the general divergence theorem for tensors to show that

$$\int_S [\mathbf{A}(\mathbf{A}\cdot d\mathbf{S}) - \tfrac{1}{2}A^2 d\mathbf{S}] = \int_V [\mathbf{A}\,\mathrm{div}\mathbf{A} - \mathbf{A}\times \mathrm{curl}\,\mathbf{A}]\, dV.$$

19.6 In four dimensions define second-order antisymmetric tensors $F_{ij}$ and $Q_{ij}$ and a first-order tensor $S_i$ as follows:

(a) $F_{23} = H_1$, $Q_{23} = B_1$ and their cyclic permutations;
(b) $F_{i4} = -D_i$, $Q_{i4} = E_i$ for $i = 1, 2, 3$;
(c) $S_4 = \rho$, $S_i = J_i$ for $i = 1, 2, 3$.

Then, taking $x_4$ as $t$ and the other symbols to have their usual meanings in electromagnetic theory, show that the equations $\sum_j \partial F_{ij}/\partial x_j = S_i$ and $\partial Q_{jk}/\partial x_i + \partial Q_{ki}/\partial x_j + \partial Q_{ij}/\partial x_k = 0$ reproduce Maxwell's equations. Here $i, j, k$ are any set chosen, in such a way that they are all different, from 1, 2, 3, 4.

19.7 A rigid body consists of four particles of masses $m$, $2m$, $3m$, $4m$, respectively situated at the points $(a, a, a)$, $(a, -a, -a)$, $(-a, a, -a)$, $(-a, -a, a)$ and connected together by a light framework.

(a) Find the inertia tensor at the origin and show that the principal moments of inertia are $20ma^2$, $(20 \pm 2\sqrt{5})ma^2$.
(b) Find the principal axes and verify that they are orthogonal.

19.8 A rigid body consists of eight particles each of mass $m$, held together by light rods. In a certain coordinate frame the particles are at

$$\pm a(3, 1, -1), \qquad \pm a(1, -1, 3), \qquad a(1, 3, -1), \qquad a(-1, 1, 3).$$

Show that when the body rotates about an axis through the origin, if the angular velocity and angular momentum vectors are parallel then they must be in one of the ratios $40ma^2$, $64ma^2$ or $72ma^2$.

19.9 The paramagnetic tensor $\chi_{ij}$ of a body placed in a magnetic field, in which its energy density is $-\frac{1}{2}\mu_0 \mathbf{M}\cdot\mathbf{H}$ with $M_i = \sum_j \chi_{ij} H_j$, is

$$\begin{pmatrix} 2k & 0 & 0 \\ 0 & 3k & k \\ 0 & k & 3k \end{pmatrix}.$$

Assuming depolarizing effects are negligible, find how the body will orientate itself if the field is horizontal, in the following circumstances.

(a) The body can rotate freely.
(b) The body is suspended with the $(1, 0, 0)$ axis vertical.
(c) The body is suspended with the $(0, 1, 0)$ axis vertical.

19.10 A block of wood contains a number of thin soft iron nails (of constant permeability). A unit magnetic field directed eastwards induces a magnetic momement with components $(3, 1, -2)$ in the block, and similar fields directed northwards and vertically upwards induce moments $(1, 3, -2)$ and $(-2, 2, 2)$ respectively. Show that all the nails lie parallel to a particular plane.

19.11 For tin the conductivity tensor is diagonal, with entries $a$, $a$, $b$ when referred to its crystal axes. A single crystal is grown in the shape of a long wire of length $L$ and radius $r$, the axis of the wire making polar angle $\theta$ with respect to the crystal's 3-axis. Show that the resistance of the wire is $L(\pi r^2 ab)^{-1}\left(a\cos^2\theta + b\sin^2\theta\right)$.

19.12 By considering an isotropic body subjected to a uniform hydrostatic pressure (no shearing stress) show that the bulk modulus $k$, defined by the ratio of the pressure to the fractional decrease in volume, is given by $k = E/[3(1-2\sigma)]$, where $E$ is Young's modulus and $\sigma$ Poisson's ratio.

19.13 Working in cylindrical polar coordinates $\rho, \phi, z$, parameterise the straight line (geodesic) joining $(1, 0, 0)$ to $(1, \pi/2, 1)$ in terms of $s$, the distance along the line. Show by substitution that the geodesic equations derived at the end of section 19.22 are satisfied.

19.14 In a general coordinate system $u^i$, $i = 1, 2, 3$, in three-dimensional Euclidean space, a volume element is given by

$$dV = |\mathbf{e}_1\,du^1 \cdot (\mathbf{e}_2\,du^2 \times \mathbf{e}_3\,du^3)|.$$

Show that an alternative form for this expression, in terms of the determinant $g$ of the metric tensor, is given by

$$dV = \sqrt{g}\,du^1\,du^2\,du^3.$$

Show that, under a general coordinate transformation to a new coordinate system $u'^i$, the volume element $dV$ remains unchanged, i.e. show that it is a scalar quantity.

19.15 By writing down the expression for the infinitesimal arc length squared, $(ds)^2$, in spherical polar coordinates, find the components $g_{ij}$ of the metric tensor in this coordinate system. Hence, using (19.95), find the expression for the divergence of a vector field $\mathbf{v}$ in spherical polars. Calculate the Christoffel symbols (of the second kind) $\Gamma^i{}_{jk}$ in this coordinate system.

19.16 Find an expression for the second covariant derivative $v_{i;jk} \equiv (v_{i;j})_{;k}$ of a vector $v_i$. By interchanging the order of differentiation and subtracting, we may define the components of the *Riemann tensor* by

$$v_{i;jk} - v_{i;kj} \equiv R^l{}_{ijk} v_l.$$

Show that, in a general coordinate system $u^i$, these components are given by

$$R^l{}_{ijk} = \frac{\partial \Gamma^l{}_{ik}}{\partial u^j} - \frac{\partial \Gamma^l{}_{ij}}{\partial u^k} + \Gamma^m{}_{ik}\Gamma^l{}_{mj} - \Gamma^m{}_{ij}\Gamma^l{}_{mk}.$$

By first considering Cartesian coordinates, show that all the components $R^l{}_{ijk} \equiv 0$ for *any* coordinate system in three-dimensional Euclidean space. (In such a space, we may therefore change the order of the covariant derivatives without changing the resulting expression.)

19.17 A curve $\mathbf{r}(t)$ is parameterised by some scalar variable $t$. Show that the length of the curve between two points $A$ and $B$ is given by

$$L = \int_A^B \sqrt{g_{ij}\frac{du^i}{dt}\frac{du^j}{dt}}\, dt.$$

Using the calculus of variations (see chapter 20), show that the curve $\mathbf{r}(t)$ that minimises $L$ satisfies the equation

$$\frac{d^2u^i}{dt^2} + \Gamma^i{}_{jk}\frac{du^j}{dt}\frac{du^k}{dt} = \frac{\ddot{s}}{\dot{s}}\frac{du^i}{dt},$$

where $s$ is the arc length along the curve, $\dot{s} = ds/dt$ and $\ddot{s} = d^2s/dt^2$. Hence, show that, if the parameter $t$ is of the form $t = as + b$, where $a$ and $b$ are constants, then we recover the equation for a geodesic (19.99) (such a parameter is called an *affine* parameter).

19.18 We may define Christoffel symbols of the first kind by

$$\Gamma_{ijk} = g_{il}\Gamma^l{}_{jk}.$$

Show that these are given by

$$\Gamma_{ijk} = \frac{1}{2}\left(\frac{\partial g_{ik}}{\partial u^j} + \frac{\partial g_{jk}}{\partial u^i} - \frac{\partial g_{ij}}{\partial u^k}\right).$$

By permuting indices, verify that

$$\frac{\partial g_{ij}}{\partial u^k} = \Gamma_{ijk} + \Gamma_{jik}.$$

Using the fact that $\Gamma^l{}_{jk} = \Gamma^l{}_{kj}$, show that

$$g_{ij;k} \equiv 0,$$

i.e. that the covariant derivative of the metric tensor is identically zero in all coordinate systems.

## 19.24 Hints and answers

19.1 (a) $u_1' = x_1\cos(\phi-\theta) - x_2\sin(\phi-\theta)$, etc.;
(b) $u_{11}' = s^2x_1^2 - 2scx_1x_2 + c^2x_2^2 \neq c^2x_2^2 + csx_1x_2 + scx_1x_2 + s^2x_1^2$.

19.2 If $T_0$ is Tr $T_{ij}$, $U_{ij} = \frac{1}{2}(T_{ij}+T_{ji}) - \frac{1}{3}T_0\delta_{ij}$, $V_{ij} = \frac{1}{3}T_0\delta_{ij}$, $S_{ij} = \frac{1}{2}(T_{ij}-T_{ji})$.

19.3 Twice contract the array with the outer product of $(x, y, z)$ with itself to obtain $-(x^2 + y^2 + z^2)^2$, i.e. an invariant.

19.4 (a) $\epsilon_{ijk}\epsilon_{jlm}u_l v_m w_k$ and use (19.28); (b) $\epsilon_{ijk}\partial(\phi u_k)/\partial x_j$; (c) $\partial(\epsilon_{ijk}u_j v_k)/\partial x_i$; (d) $\epsilon_{ijk}\epsilon_{klm}\partial(u_l v_m)/\partial x_j$ and use (19.28); (e) start with $\mathbf{u} \times \operatorname{curl} \mathbf{u}$:

$$\epsilon_{ijk}u_j\epsilon_{klm}\frac{\partial u_m}{\partial x_l} = \cdots = u_j\left(\frac{\partial u_j}{\partial x_i}\right) - u_j\left(\frac{\partial u_i}{\partial x_j}\right).$$

19.5 Write $A_j(\partial A_i/\partial x_j)$ as $\partial(A_iA_j)/\partial x_j - A_i(\partial A_j/\partial x_j)$.

19.6 $\nabla \times \mathbf{H} = \mathbf{J} + \dot{\mathbf{D}}$; $\nabla \cdot \mathbf{D} = \rho$; $\nabla \times \mathbf{E} + \dot{\mathbf{B}} = \mathbf{0}$; $\nabla \cdot \mathbf{B} = 0$.

19.7 (b) $\mathsf{x}_1^{\mathrm{T}} = (2 \;\; -1 \;\; 0)$, $\mathsf{x}_2^{\mathrm{T}} = (1 \;\; 2 \;\; \sqrt{5})$, $\mathsf{x}_3^{\mathrm{T}} = (1 \;\; 2 \;\; -\sqrt{5})$.

19.8 The principal moments give the required ratios.

19.9 Principal susceptibilities and (unnormalised) axes are $\lambda = 4$, $\pm(0, 1, 1)$; $\lambda = 2$, $\pm(c_i, 1, -1)$ with $c_1c_2 = -2$. (a) Lowest energy when $(0, 1, 1)$ axis is parallel to the field; (b) permitted values of orientation are $(0, n_2, n_3)$, hence as in (a); (c) permitted values of orientation are $(n_1, 0, n_3)$, subject to $n_1^2 + n_3^2 = 1$. Energy $= -\frac{1}{2}\mu_0 kH^2V(2n_1^2 + 3n_3^2)$, which minimises when $(0, 0, 1)$ is along the field.

19.10 The principal permeability, in direction $(1, 1, 2)$, has value 0. Thus all the nails lie in the plane to which this is the normal.

19.11 $j_i = \sigma_{ik}E_k$ gives $I \sin\theta \cos\phi = a\pi r^2 E_1$, $I \sin\theta \sin\phi = a\pi r^2 E_2$, $I \cos\theta = b\pi r^2 E_3$. Also $V/L = E_1 \sin\theta\cos\phi + E_2 \sin\theta \sin\phi + E_3 \cos\theta$. The current must flow along the wire; $\mathbf{E}$ is not parallel to the wire.

19.12 Take $p_{11} = p_{22} = p_{33} = -p$, and $p_{ij} = e_{ij} = 0$ for $i \neq j$, leading to $-p = (\lambda + 2\mu/3)e_{ii}$. The fractional volume change is $e_{ii}$; $\lambda$ and $\mu$ are as defined in (19.44) and following.

19.13 $\rho = (1 - 2s/\sqrt{3} + 2s^2/3)^{1/2}$, $\phi = \tan^{-1}[s/(\sqrt{3} - s)]$, $z = s/\sqrt{3}$.

19.14 Use $|\mathbf{e}_1 \cdot (\mathbf{e}_2 \times \mathbf{e}_3)| = \sqrt{g}$.
Recall that $\sqrt{g'} = |\partial u/\partial u'|\sqrt{g}$ and $du'^1\, du'^2\, du'^3 = |\partial u'/\partial u|\, du^1\, du^2\, du^3$.

19.15 $g = r^4 \sin^2\theta$; recall that, for each $i$, $v^i = \hat{v}_i/h_i$, e.g. $v^3 = v_\phi/(r\sin\theta)$; ${\Gamma^1}_{22} = -r$, ${\Gamma^1}_{33} = -r\sin^2\theta$, ${\Gamma^2}_{12} = r^{-1}$, ${\Gamma^2}_{32} = -\sin\theta\cos\theta$, ${\Gamma^3}_{13} = r^{-1}$, ${\Gamma^3}_{23} = \cot\theta$.

19.16 $(v_{i;\,j})_{;k} = (v_{i;\,j})_{,k} - {\Gamma^l}_{ik}v_{l;\,j} - {\Gamma^l}_{jk}v_{i;\,l}$ and $v_{i;\,j} = v_{i,\,j} - {\Gamma^m}_{ij}v_m$. If all components of a tensor equal zero in one coordinate system, they are zero in all coordinate systems.

19.17 Using $\dot{s} = \sqrt{g_{ij}\dot{u}^i\dot{u}^j}$, the Euler–Lagrange equation is

$$\frac{d}{dt}\left(\frac{g_{ik}\dot{u}^i}{\dot{s}}\right) - \frac{1}{2\dot{s}}\frac{\partial g_{ij}}{\partial u^k}\dot{u}^i\dot{u}^j = 0.$$

Perform the $t$-derivative, write

$$\frac{\partial g_{ik}}{\partial u^j} = \frac{1}{2}\left(\frac{\partial g_{ik}}{\partial u^j} + \frac{\partial g_{jk}}{\partial u^i}\right)$$

and multiply through by $g^{lk}$. If $t = as + b$, $\ddot{s} = 0$.

# 20

# Calculus of variations

In chapters 1 and 4 we discussed how to find stationary values of functions of a single variable $f(x)$, of several variables $f(x, y, \ldots)$ and of constrained variables, where $f(x, y, \ldots)$ is subject to the $n$ constraints $g_i(x, y, \ldots) = 0$, $i = 1, \ldots, n$. In all these cases the forms of the functions $f$ and $g_i$ were known, and the problem was one of finding suitable values of the variables $x$, $y$ etc.

We now turn to a different kind of problem in which we are interested in bringing about a particular condition for a given expression (usually maximising or minimising it) by varying the *functions* on which the expression depends. For instance, we might want to know in what shape a fixed length of rope should be arranged so as to enclose the largest possible area, or in what shape it will hang when suspended from two fixed points. In each case we are concerned with a general principle by which the function $y(x)$ that satisfies the given problem may be found.

The calculus of variations provides a method for finding the function $y(x)$ that satisfies a problem of this type. In order to obtain an explicit solution $y(x)$, the problem must first be expressed in a mathematical form, and the form most commonly applicable to such problems is an *integral*. In each of the above questions, the quantity that has to be maximised or minimised by an appropriate choice of the function $y(x)$ may be expressed as an integral involving $y(x)$ and the variables describing the geometry of the situation.

In our example of the rope hanging from two fixed points, we can find its shape by finding the shape function $y(x)$ that minimises the gravitational potential energy of the rope. Each elementary piece of the rope has a gravitational potential energy proportional both to its vertical height above an arbitrary zero level and to the length of the piece. Therefore the total potential energy is given by an integral for the whole rope of such elementary contributions. The particular function $y(x)$ for which the value of this integral is a minimum will give the shape assumed by the hanging rope.

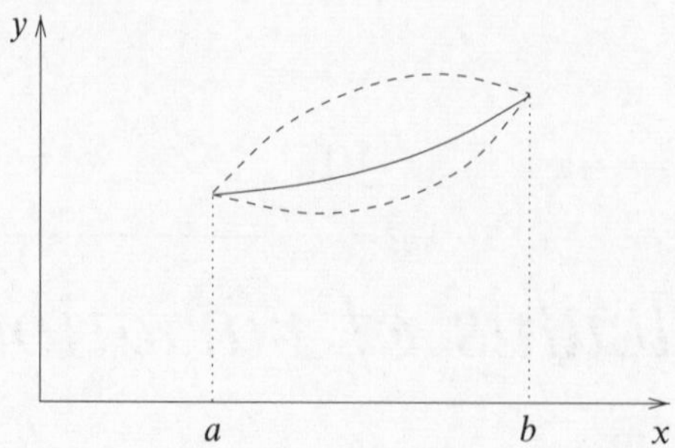


Figure 20.1 Possible paths for the integral (20.1). The solid line is the curve along which the integral is assumed stationary. The broken curves represent small variations from this path.

So in general we are led by this type of question to study the value of an integral whose integrand has a specified form in terms of a certain function and its derivatives, and to study how that value changes when the form of the function is varied. Specifically, we aim to find the function that makes the integral *stationary*, i.e. the function that makes the value of the integral a local maximum or minimum. Note that, unless stated otherwise, $y'$ is used to denote $dy/dx$ throughout this chapter. We also assume that all the functions we need to deal with are sufficiently smooth and differentiable.

## 20.1 The Euler–Lagrange equation

Let us consider the integral

$$I = \int_a^b F(y, y', x)\, dx, \tag{20.1}$$

where $a$, $b$ and the form of the function $F$ are fixed by given considerations, e.g. the physics of the problem, but the curve $y(x)$ has to be chosen so as to make stationary the value of $I$, which is clearly a function (or more accurately a *functional*) of this curve, i.e. $I = I[y(x)]$. Referring to figure 20.1, we wish to find the function $y(x)$ (given, say, by the solid line) such that first-order small changes in it (for example the two broken lines) will make only second-order changes in the value of $I$.

Writing this in a more mathematical form, let us suppose that $y(x)$ is the function required to make $I$ stationary and consider making the replacement

$$y(x) \to y(x) + \alpha\eta(x), \tag{20.2}$$

where the parameter $\alpha$ is small and $\eta(x)$ is an arbitrary function with sufficiently amenable mathematical properties. For the value of $I$ to be stationary with respect to these variations, we require

$$\left.\frac{dI}{d\alpha}\right|_{\alpha=0} = 0 \quad \text{for all } \eta(x). \tag{20.3}$$

Substituting (20.2) into (20.1) and expanding as a Taylor series in $\alpha$ we obtain

$$\begin{aligned} I(y, \alpha) &= \int_a^b F(y + \alpha\eta, y' + \alpha\eta', x)\, dx \\ &= \int_a^b F(y, y', x)\, dx + \int_a^b \left( \frac{\partial F}{\partial y}\alpha\eta + \frac{\partial F}{\partial y'}\alpha\eta' \right) dx + \mathrm{O}(\alpha^2). \end{aligned}$$

With this form for $I(y, \alpha)$ the condition (20.3) implies that for all $\eta(x)$ we require

$$\delta I = \int_a^b \left( \frac{\partial F}{\partial y}\eta + \frac{\partial F}{\partial y'}\eta' \right) dx = 0,$$

where $\delta I$ denotes the first-order variation in the value of $I$ due to the variation (20.2) in the function $y(x)$. Integrating the second term by parts this becomes

$$\left[ \eta \frac{\partial F}{\partial y'} \right]_a^b + \int_a^b \left[ \frac{\partial F}{\partial y} - \frac{d}{dx}\left( \frac{\partial F}{\partial y'} \right) \right] \eta(x)\, dx = 0. \tag{20.4}$$

In order to simplify the result we will, for the moment, assume that the end-points are fixed, i.e. not only $a$ and $b$ are given but also $y(a)$ and $y(b)$. This restriction means that we require $\eta(a) = \eta(b) = 0$, in which case the first term on the LHS of (20.4) equals zero at both end-points. Since (20.4) must be satisfied for arbitrary $\eta(x)$, it is easy to see that we require

$$\frac{\partial F}{\partial y} = \frac{d}{dx}\left( \frac{\partial F}{\partial y'} \right). \tag{20.5}$$

This is known as the *Euler–Lagrange* (EL) equation, and is a differential equation for $y(x)$, since the function $F$ is known.

## 20.2 Special cases

In certain special cases a first integral of the EL equation can be obtained for a general form of $F$.

### *20.2.1 F does not contain y explicitly*

In this case $\partial F/\partial y = 0$, and (20.5) can be integrated immediately giving

$$\frac{\partial F}{\partial y'} = \text{constant}. \tag{20.6}$$

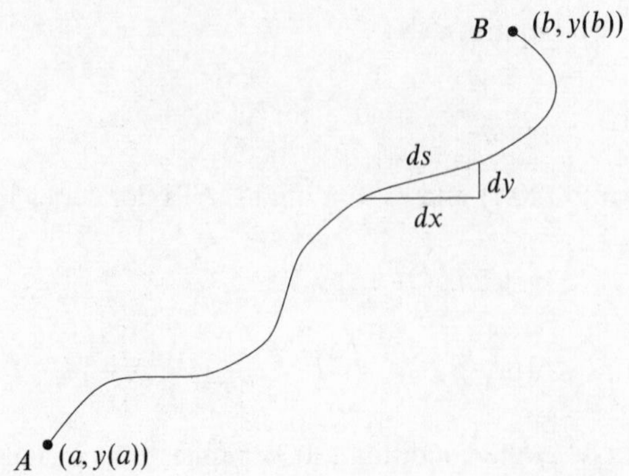


Figure 20.2 An arbitrary path between two fixed points.

► *Show that the shortest curve joining two points is a straight line.*

Let the two points be labelled $A$ and $B$, and have coordinates $(a, y(a))$ and $(b, y(b))$ respectively (see figure 20.2). Whatever the shape of the curve joining $A$ to $B$, the length of an element of path $ds$ is given by

$$ds = \left[(dx)^2 + (dy)^2\right]^{1/2} = (1 + y')^{1/2} dx,$$

and hence the total path length along the curve is given by

$$L = \int_a^b (1 + y'^2)^{1/2}\, dx. \qquad (20.7)$$

We must now apply the results of the previous section to determine that path which makes $L$ stationary (clearly a minimum in this case). Since the integral does not contain $y$ (or indeed $x$) explicitly, we may use (20.6) to obtain

$$k = \frac{\partial F}{\partial y'} = \frac{y'}{(1 + y'^2)^{1/2}}.$$

where $k$ is a constant. This is easily rearranged and integrated to give

$$y = \frac{k}{(1 - k^2)^{1/2}} x + c,$$

which is, as expected, the equation of a straight line in the form $y = mx + c$ with $m = k/(1 - k^2)^{1/2}$. The value of $m$ (or $k$) can be found by demanding that the straight line passes through the points $A$ and $B$ and is given by $m = [y(b) - y(a)]/(b - a)$. Substituting the equation of the straight line into (20.7) we find that the total path length is, as expected, given by

$$L^2 = [y(b) - y(a)]^2 + (b - a)^2. \blacktriangleleft$$

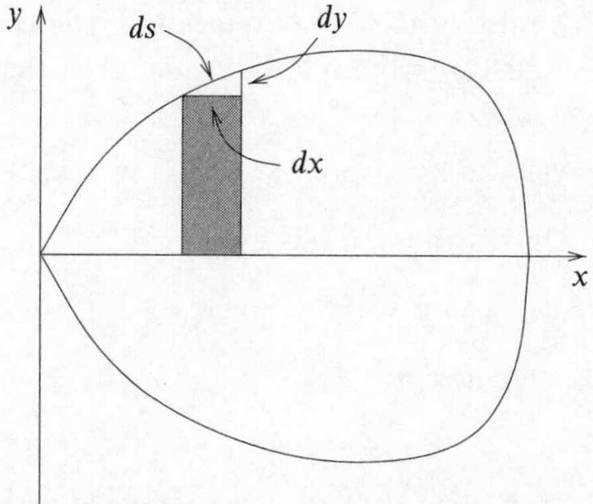


Figure 20.3 A convex closed curve that is symmetrical about the $x$-axis.

### 20.2.2 *F* does not contain *x* explicitly

In this case, multiplying the EL equation (20.5) by $y'$ and using

$$\frac{d}{dx}\left(y'\frac{\partial F}{\partial y'}\right) = y'\frac{d}{dx}\left(\frac{\partial F}{\partial y'}\right) + y''\frac{\partial F}{\partial y'}$$

we obtain

$$y'\frac{\partial F}{\partial y} + y''\frac{\partial F}{\partial y'} = \frac{d}{dx}\left(y'\frac{\partial F}{\partial y'}\right).$$

But since $F$ is a function of $y$ and $y'$ only, and not explicitly of $x$, the LHS of this equation is just the total derivative of $F$, namely $dF/dx$. Hence, integrating we obtain

$$F - y'\frac{\partial F}{\partial y'} = \text{constant.} \qquad (20.8)$$

► *Find the closed convex curve of length l that encloses the greatest possible area.*

Without any loss of generality we can assume that the curve passes through the origin, and can further suppose that it is symmetric with respect to the $x$-axis; this assumption is not essential. Using the distance $s$ along the curve, measured from the origin, as the independent variable and $y$ as the dependent one, we have the boundary conditions $y(0) = y(l/2) = 0$. The element of area shown in figure 20.3 is then given by

$$dA = y\,dx = y\left[(ds)^2 - (dy)^2\right]^{1/2},$$

and the total area by

$$A = 2\int_0^{l/2} y(1-y'^2)^{1/2}\,ds; \qquad (20.9)$$

here $y'$ stands for $dy/ds$ rather than $dy/dx$. Since the integrand does not contain $s$ explicitly, we can use (20.8) to obtain a first integral of the EL equation for $y$, namely

$$y(1 - y'^2)^{1/2} + yy'^2(1 - y'^2)^{-1/2} = k,$$

where $k$ is a constant. On rearranging this gives

$$ky' = \pm(k^2 - y^2)^{1/2},$$

which, using $y(0) = 0$, integrates to

$$y/k = \sin(s/k). \tag{20.10}$$

The other end-point, $y(l/2) = 0$, fixes the value of $k$ as $l/2\pi$ to yield

$$y = \frac{l}{2\pi} \sin \frac{2\pi s}{l}.$$

From this we obtain $dy = \cos(2\pi s/l)\, ds$ and since $(ds)^2 = (dx)^2 + (dy)^2$ we also find that $dx = \pm \sin(2\pi s/l)\, ds$. This in turn can be integrated and, using $x(0) = 0$, gives $x$ in terms of $s$ as

$$x - \frac{l}{2\pi} = -\frac{l}{2\pi} \cos \frac{2\pi s}{l}.$$

We thus obtain the expected result that $x$ and $y$ lie on the circle of radius $l/2\pi$ given by

$$\left(x - \frac{l}{2\pi}\right)^2 + y^2 = \frac{l^2}{4\pi^2}.$$

Substituting the solution (20.10) into the expression for the total area (20.9), it is easily verified that $A = l^2/4\pi$. A much quicker derivation of this result is possible using plane polar coordinates. ◂

The previous two examples have been carried out in some detail, even though the answers are more easily obtained in other ways, expressly so that the method is transparent and the way it works can be filled in mentally at almost every step. The next example, however, does not have such an intuitively obvious solution.

► *Two rings, each of radius $a$, are placed parallel with their centres $2b$ apart and on a common normal. An axially symmetric soap film is formed between them but does not cover the ends of the rings (see figure 20.4). Find the shape assumed by the film.*

Creating the soap film requires an energy $\gamma$ per unit area (numerically equal to the surface tension of the soap solution). So the stable shape of the soap film, i.e. the one that minimises the energy, will also be the one that minimises the surface area (neglecting gravitational effects).

It is obvious that any convex surface shaped such as that shown as the broken

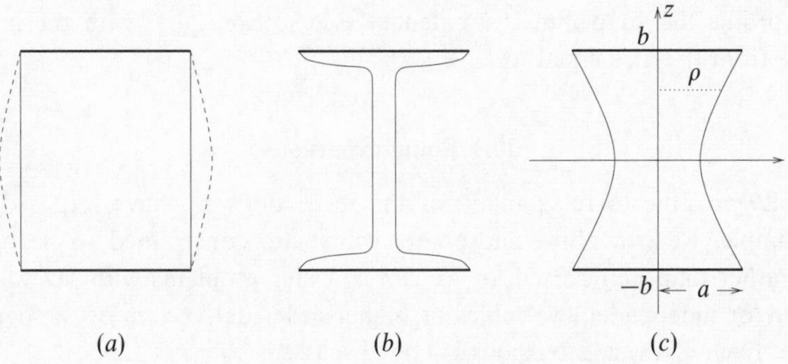


Figure 20.4 Possible soap films between two parallel circular rings.

line in figure 20.4($a$) cannot be a minimum, but it is not clear whether some shape intermediate between the solid cylindrical curve in ($a$), with area $4\pi ab$ (or twice this for the double surface of the film), and the form shown in ($b$), with area $2\pi a^2$, will produce a lower total area than both of these extremes. If there is such a shape (e.g. that in figure 20.4($c$)), then it will be that which best compromises between the criteria of the minimum ring-to-ring distance of the film surface ($a$), and the minimum waist measurement of the surface ($b$).

We take cylindrical polar coordinates as in figure 20.4($c$), and let the radius of the soap film at height $z$ be $\rho(z)$ with $\rho(\pm b) = a$. Counting only one side of the film, the element of surface area between $z$ and $z + dz$ is

$$dS = 2\pi\rho \left[(dz)^2 + (d\rho)^2\right]^{1/2},$$

so the total surface area is given by

$$S = 2\pi \int_{-b}^{b} \rho(1 + \rho'^2)^{1/2}\, dz. \tag{20.11}$$

Since the integrand does not contain $z$ explicitly, we can use (20.8) to obtain an equation for $\rho$ that minimises $S$, i.e.

$$\rho(1 + \rho'^2)^{1/2} - \rho\rho'^2(1 + \rho'^2)^{-1/2} = k,$$

where $k$ is a constant. Multiplying through by $(1 + \rho'^2)^{1/2}$, rearranging to find an explicit expression for $\rho'$ and integrating we find

$$\cosh^{-1}\frac{\rho}{k} = \frac{z}{k} + c.$$

where $c$ is the constant of integration. Using the boundary conditions $\rho(\pm b) = a$, we require $c = 0$ and $k$ such that $a/k = \cosh b/k$ (if $b/a$ is too large, no such $k$ can be found). Thus the curve that minimises the surface area is

$$\rho/k = \cosh(z/k),$$

and in profile the soap film is a catenary (see section 20.4) with the minimum distance from the axis equal to $k$. ◄

## 20.3 Some extensions

It is quite possible to relax many of the restrictions we have imposed so far. For example, we can allow end-points which are constrained to lie on given curves rather than being fixed, or we can consider problems with several dependent and/or independent variables or higher-order derivatives of the dependent variable. Each of these extensions is now discussed.

### *20.3.1 Several dependent variables*

Here we have $F = F(y_1, y_1', y_2, y_2', \ldots, y_n, y_n', x)$ where each $y_i = y_i(x)$. The analysis in this case proceeds as before, leading to $n$ separate but simultaneous equations for the $y_i(x)$,

$$\frac{\partial F}{\partial y_i} = \frac{d}{dx}\left(\frac{\partial F}{\partial y_i'}\right), \qquad i = 1, \ldots, n. \tag{20.12}$$

### *20.3.2 Several independent variables*

With $n$ independent variables, we need to extremise multiple integrals of the form

$$I = \int \cdots \int F\left(y, \frac{\partial y}{\partial x_1}, \ldots, \frac{\partial y}{\partial x_n}, x_1, \ldots, x_n\right) dx_1 \cdots dx_n.$$

Using the same kind of analysis as before, we find that the extremising function $y = y(x_1, \ldots, x_n)$ must satisfy

$$\frac{\partial F}{\partial y} = \sum_{i=1}^{n} \frac{\partial}{\partial x_i}\left(\frac{\partial F}{\partial y_{x_i}}\right), \tag{20.13}$$

where $y_{x_i}$ stands for $\partial y/\partial x_i$.

### *20.3.3 Higher-order derivatives*

If in (20.1) $F = F(y, y', y'', \ldots, y^{(n)}, x)$, then using the same method as before and performing repeated integration by parts, it can be shown that the required extremising function $y(x)$ satisfies

$$\frac{\partial F}{\partial y} - \frac{d}{dx}\left(\frac{\partial F}{\partial y'}\right) + \frac{d^2}{dx^2}\left(\frac{\partial F}{\partial y''}\right) - \cdots + (-1)^n \frac{d^n}{dx^n}\left(\frac{\partial F}{\partial y^{(n)}}\right) = 0, \tag{20.14}$$

provided that $y = y' = \cdots = y^{(n)} = 0$ at both end-points. If $y$ or any of its derivatives are not zero at the end-points then contributions from them will appear on the RHS of (20.14).

### 20.3.4 Variable end-points

We now discuss the very important generalisation to variable end-points. Suppose, as before, we wish to find the function $y(x)$ that extremises the integral

$$I = \int_a^b F(y, y', x)\, dx,$$

but this time we demand only that the lower end-point is fixed, while we allow $y(b)$ to be arbitrary. Repeating the analysis of section 20.1, we find from (20.4) that we require

$$\left[\eta \frac{\partial F}{\partial y'}\right]_a^b + \int_a^b \left[\frac{\partial F}{\partial y} - \frac{d}{dx}\left(\frac{\partial F}{\partial y'}\right)\right] \eta(x)\, dx = 0. \tag{20.15}$$

Obviously the EL equation (20.5) must still hold for the second term on the LHS to vanish. Also, since the lower end-point is fixed, i.e. $\eta(a) = 0$, the first term on the LHS automatically vanishes at the lower limit. However, in order that it also vanishes at the upper limit, we require in addition that

$$\left.\frac{\partial F}{\partial y'}\right|_{x=b} = 0. \tag{20.16}$$

Clearly if both end-points may vary, then $\partial F/\partial y'$ must vanish at both ends.

An interesting, more general, case is where the lower end-point is again fixed at $x = a$, but the upper end-point is free to lie anywhere on the curve $h(x, y) = 0$. Now in this case, the variation in the value of $I$ due to the arbitrary variation (20.2) is given by

$$\delta I = \left[\frac{\partial F}{\partial y'}\eta\right]_a^b + \int_a^b \left(\frac{\partial F}{\partial y} - \frac{d}{dx}\frac{\partial F}{\partial y'}\right) \eta\, dx + F(b)\Delta x, \tag{20.17}$$

where $\Delta x$ is the displacement of the upper end-point, as indicated in figure 20.5, and $F(b)$ is the value of $F$ at $x = b$. In order for (20.17) to be valid, we naturally require the displacement $\Delta x$ to be small. From the figure we see that $\Delta y = \eta(b) + y'(b)\Delta x$. Since the upper end-point must lie on $h(x, y) = 0$ we require that at $x = b$ we have

$$\frac{\partial h}{\partial x}\Delta x + \frac{\partial h}{\partial y}\Delta y = 0,$$

which on substituting our expression for $\Delta y$ and rearranging becomes

$$\left(\frac{\partial h}{\partial x} + y'\frac{\partial h}{\partial y}\right)\Delta x + \frac{\partial h}{\partial y}\eta = 0. \tag{20.18}$$

Also from (20.17) the condition $\delta I = 0$ requires, besides the EL equation, that at $x = b$

$$F\Delta x + \frac{\partial F}{\partial y'}\eta = 0. \tag{20.19}$$

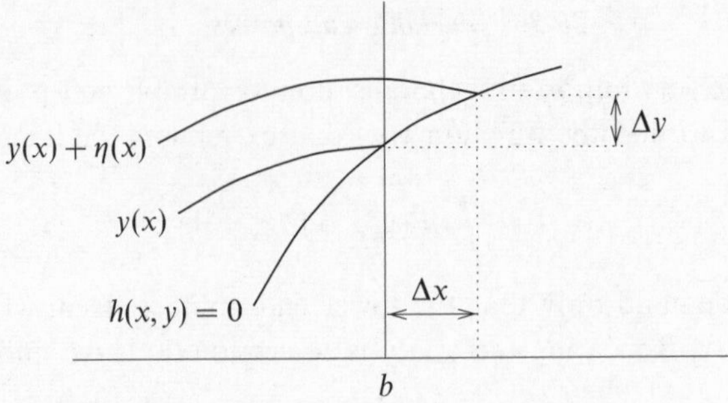


Figure 20.5 Variation of the end-point $b$ along the curve $h(x,y)=0$.

Eliminating $\Delta x$ and $\eta$ between (20.18) and (20.19) leads to the condition that at the end-point

$$\left(F - y'\frac{\partial F}{\partial y'}\right)\frac{\partial h}{\partial y} - \frac{\partial F}{\partial y'}\frac{\partial h}{\partial x} = 0. \tag{20.20}$$

In the special case where the end-point is free to lie anywhere on the vertical line $x=b$, we have $\partial h/\partial x = 1$ and $\partial h/\partial y = 0$. Substituting these values into (20.20), we recover the end-point condition given in (20.16).

►*A frictionless wire in a vertical plane connects two points A and B, A being higher than B. Let the position of A be fixed at the origin of an xy-coordinate system, but allow B to lie anywhere on the vertical line $x = x_0$ (see figure 20.6). Find the shape of the wire such that a bead of mass m placed on it at A will slide under gravity to B in the shortest possible time.*

This is a variant of the famous brachistochrone problem, which is often used to illustrate the calculus of variations. Conservation of energy tells us that the particle speed is given by

$$v = \frac{ds}{dt} = \sqrt{2gy},$$

where $s$ is the path length along the wire and $g$ is the acceleration due to gravity. Since the element of path length is $ds = (1+y'^2)^{1/2}dx$, the total time taken to travel to the line $x = x_0$ is given by

$$t = \int_{x=0}^{x=x_0}\frac{ds}{v} = \frac{1}{\sqrt{2g}}\int_0^{x_0}\sqrt{\frac{1+y'^2}{y}}\,dx.$$

Since the integrand does not contain $x$ explicitly, we can use (20.8) with $F = \sqrt{1+y'^2}/\sqrt{y}$ to find a first integral, which simplifies to give

$$\left[y(1+y'^2)\right]^{1/2} = k,$$

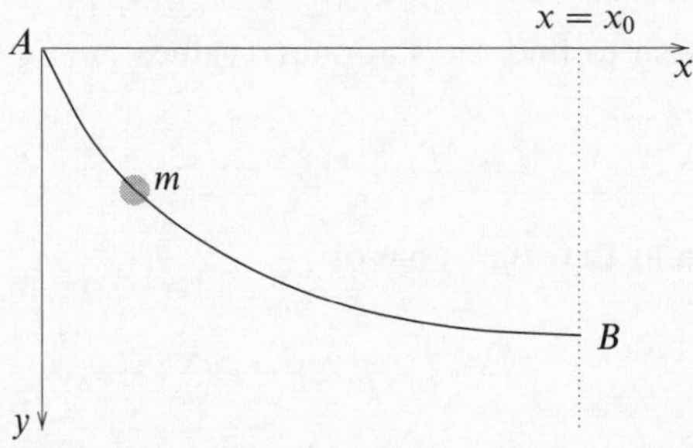


Figure 20.6 A frictionless wire along which a small bead of mass $m$ slides. We seek the shape of the wire that allows the bead to travel from the origin $O$ to the line $x = x_0$ in the least possible time.

where $k$ is a constant. Letting $a = k^2$ and solving for $y'$ we find

$$y' = \frac{dy}{dx} = \sqrt{\frac{a-y}{y}},$$

which on substituting $y = a\sin^2\theta$ integrates to give

$$x = \frac{a}{2}(2\theta - \sin 2\theta) + c.$$

Thus the parametric equations of the curve are given by

$$x = b(\phi - \sin\phi) + c, \qquad y = b(1 - \cos\phi),$$

where $b = a/2$ and $\phi = 2\theta$, and define a cycloid, the curve traced out by a point on the rim of a wheel of radius $b$ rolling along the $x$-axis. We must now use the end-point conditions to determine the constants $b$ and $c$. Since the curve passes through the origin, we see immediately that $c = 0$. Now since $y(x_0)$ is arbitrary, i.e. the upper end-point can lie anywhere on the curve $x = x_0$, the condition (20.20) reduces to (20.16), so that we also require

$$\left.\frac{\partial F}{\partial y'}\right|_{x=x_0} = \left.\frac{y'}{\sqrt{y(1+y'^2)}}\right|_{x=x_0} = 0,$$

which implies that $y' = 0$ at $x = x_0$, or that the tangent to the cycloid at $B$ must be parallel to the $x$-axis; this requires $\pi b = x_0$. ◀

## 20.4 Constrained variation

Just as the problem of finding stationary values of a function $f(x, y)$ subject to the constraint $g(x, y) =$ constant is solved by means of Lagrange's undetermined multipliers (see chapter 4), so the corresponding problem in the calculus of variations is solved by an analogous method.

Suppose that we wish to find the stationary values of

$$I = \int_a^b F(y, y', x)\, dx,$$

subject to the constraint that the value of

$$J = \int_a^b G(y, y', x)\, dx$$

is held constant. Following the method of Lagrange undetermined multipliers let us define a new functional

$$K = I + \lambda J = \int_a^b (F + \lambda G)\, dx,$$

and find its *unconstrained* stationary values. Repeating the analysis of section 20.1 we find that we require

$$\frac{\partial F}{\partial y} - \frac{d}{dx}\left(\frac{\partial F}{\partial y'}\right) + \lambda\left[\frac{\partial G}{\partial y} - \frac{d}{dx}\left(\frac{\partial G}{\partial y'}\right)\right] = 0,$$

which, together with the original constraint $J =$ constant, will yield the required solution $y(x)$.

This method is easily generalised to cases with more than one constraint by the introduction of more Lagrange multipliers. If we wish to find the stationary values of an integral $I$ subject to the multiple constraints that the values of the integrals $J_i$ be held constant for $i = 1, \ldots, n$, then we simply find the unconstrained stationary values of the new integral

$$K = I + \sum_1^n \lambda_i J_i.$$

► *Find the shape assumed by a uniform rope when suspended by its ends from two points at equal heights.*

We will solve this problem using $x$ (see figure 20.7) as the independent variable. Let the rope of length $2L$ be suspended between the points $x = \pm a$, $y = 0$ $(L > a)$, and have uniform density $\rho$ per unit length. We then need to find the stationary value of the rope's gravitational potential energy

$$I = -\rho g \int y\, ds = -\rho g \int_{-a}^{a} y(1 + y'^2)^{1/2}\, dx,$$

with respect to small changes in the form of the rope, but subject to the constraint that the total length of the rope remains constant, i.e.

$$J = \int ds = \int_{-a}^{a} (1 + y'^2)^{1/2}\, dx = 2L.$$

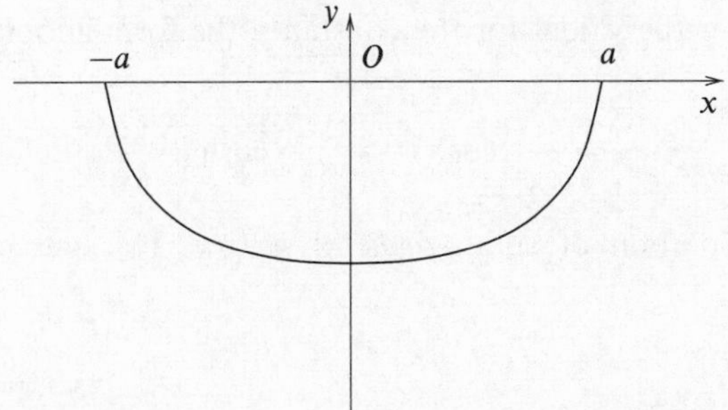


Figure 20.7 A uniform rope with fixed end-points suspended under gravity.

We thus define a new integral (omitting the factor −1 from $I$ for brevity)

$$K = I + \lambda J = \int_{-a}^{a} (\rho g y + \lambda)(1 + y'^2)^{1/2}\, dx,$$

and find its stationary values. Since the integrand does not contain the independent variable $x$ explicitly, we can use (20.8) to find the first integral:

$$(\rho g y + \lambda)\left(1 + y'^2\right)^{1/2} - (\rho g y + \lambda)\left(1 + y'^2\right)^{-1/2} y'^2 = k,$$

where $k$ is a constant; this reduces to

$$y'^2 = \left(\frac{\rho g y + \lambda}{k}\right)^2 - 1.$$

Making the substitution $\rho g y + \lambda = k \cosh z$, this can be integrated easily to give

$$\frac{k}{\rho g} \cosh^{-1}\left(\frac{\rho g y + \lambda}{k}\right) = x + c,$$

where $c$ is the constant of integration.

We now have the three unknowns $\lambda$, $k$ and $c$, which must be evaluated using the two end conditions $y(\pm a) = 0$ and the constraint $J = 2L$. The end conditions give

$$\cosh \frac{\rho g(a + c)}{k} = \frac{\lambda}{k} = \cosh \frac{\rho g(-a + c)}{k},$$

and since $a \neq 0$, these imply $c = 0$ and $\lambda/k = \cosh(\rho g a/k)$. Putting $c = 0$ into the constraint, in which $y' = \sinh(\rho g x/k)$, we obtain

$$\begin{aligned} 2L &= \int_{-a}^{a} [1 + \sinh^2(\rho g x/k)]^{1/2}\, dx \\ &= \frac{2k}{\rho g} \sinh(\rho g a/k). \end{aligned}$$

Collecting together the values for the constants, the form adopted by the rope is therefore

$$y(x) = \frac{k}{\rho g}\left[\cosh\left(\frac{\rho g x}{k}\right) - \cosh\left(\frac{\rho g a}{k}\right)\right],$$

where $k$ is the solution of $\sinh(\rho g a/k) = \rho g L/k$. This curve is known as a catenary. ◄

## 20.5 Physical variational principles

Many results in both classical and quantum physics can be expressed as variational principles, and it is often when expressed in this form that their physical meaning is most clearly understood. Moreover, once a physical phenomenon has been written as a variational principle, we can use all the results derived in this chapter to investigate its behaviour. It is usually possible to identify conserved quantities, or symmetries of the system of interest, which otherwise might only been found with considerable effort. From the wide range of physical variational principles we will select two examples from familiar areas of classical physics, namely geometric optics and mechanics.

### *20.5.1 Fermat's principle in optics*

Fermat's principle in geometrical optics states that a ray of light travelling in a region of variable refractive index follows a path such that the total optical path length (physical length × refractive index) is stationary.

► *From Fermat's principle deduce Snell's law of refraction at an interface.*

Let the interface be at $y =$ constant (see figure 20.8) and let it separate two regions with refractive indices $n_1$ and $n_2$ respectively. For a ray that passes through the points $A$ and $B$, its element of physical path length is $ds = (1 + y'^2)^{1/2}dx$, so its total optical path length is

$$P = \int_A^B n(y)(1 + y'^2)^{1/2}\,dx.$$

Since the integrand does not explicitly contain the independent variable $x$, we use (20.8) to obtain a first integral, which, after some rearrangement, reads

$$n(y)\left(1 + y'^2\right)^{-1/2} = k,$$

where $k$ is a constant. Recalling that $y'$ is the tangent of the angle $\phi$ between the instantaneous direction of the ray and the $x$-axis, this *general* result, which is not

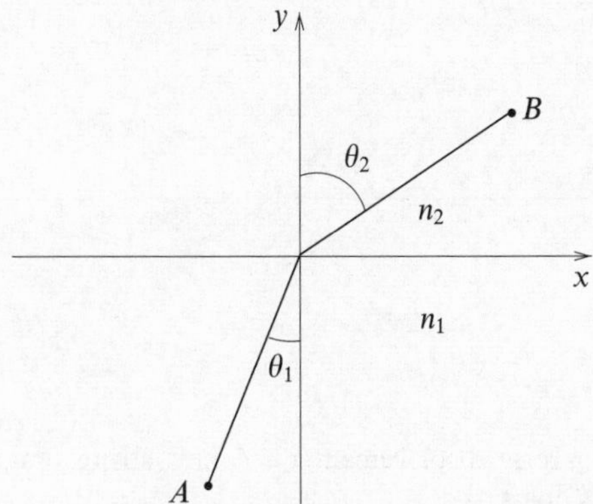


Figure 20.8 Path of a light ray at the plane interface between media with refractive indices $n_1$ and $n_2$, where $n_2 < n_1$.

dependent on the configuration presently under consideration, can be put in the form

$$n \cos \phi = \text{ constant}$$

along a ray, even though $n$ and $\phi$ vary individually.

For our particular configuration $n$ is constant in each medium and therefore so is $y'$. Thus the rays travel in straight lines in each medium (as anticipated in figure 20.8, but not assumed in our analysis), and since $k$ is constant along the *whole* path we have $n_1 \cos \phi_1 = n_2 \cos \phi_2$, or in terms of the conventional angles in the figure

$$n_1 \sin \theta_1 = n_2 \sin \theta_2. \blacktriangleleft$$

### *20.5.2 Hamilton's principle in mechanics*

Consider a mechanical system whose configuration can be uniquely defined by a number of coordinates $q_i$ (usually distances and angles) together with time $t$, and which only experiences forces derivable from a potential. Hamilton's principle states that in moving from one configuration at time $t_0$ to another at time $t_1$ the motion of such a system is such as to make

$$\mathcal{L} = \int_{t_0}^{t_1} L(q_1, \ldots, q_n, \dot{q}_1, \ldots, \dot{q}_n, t)\, dt \qquad (20.21)$$

stationary. The *Lagrangian* $L$ is defined in terms of the kinetic energy $T$ and the potential energy $V$ (with respect to some reference situation) by $L = T - V$. Here $V$ is a function of the $q_i$ only, not of the $\dot{q}_i$. Applying the EL equation to $\mathcal{L}$ we

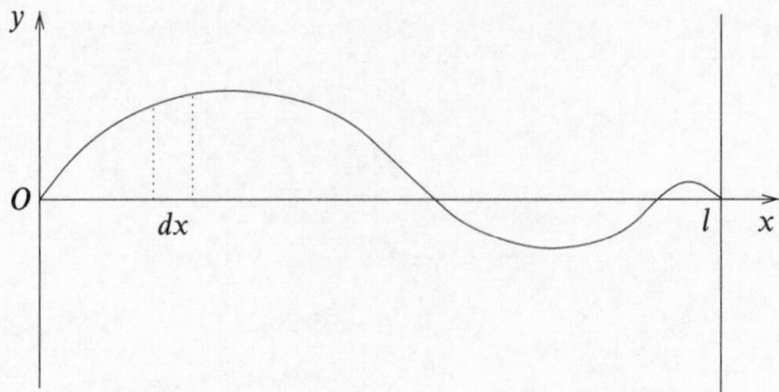


Figure 20.9 Transverse displacement on a taut string that is fixed at two points a distance $l$ apart.

obtain *Lagrange's equations*,

$$\frac{\partial L}{\partial q_i} = \frac{d}{dt}\left(\frac{\partial L}{\partial \dot{q}_i}\right), \qquad i = 1, \ldots, n.$$

► *Using Hamilton's principle derive the wave equation for small transverse oscillations of a taut string.*

In this example we are in fact considering a generalisation of (20.21) to a case involving one isolated independent coordinate $t$, together with a *continuum* in which the $q_i$ become the continuous variable $x$. The expressions for $T$ and $V$ therefore become integrals over $x$ rather than sums over $i$.

If $\rho$ and $\tau$ are the local density and tension of the string, both of which may depend on $x$, then referring to figure 20.9 the kinetic and potential energies of the string are given by

$$T = \int_0^l \frac{\rho}{2}\left(\frac{\partial y}{\partial t}\right)^2 dx, \qquad V = \int_0^l \frac{\tau}{2}\left(\frac{\partial y}{\partial x}\right)^2 dx,$$

and (20.21) becomes

$$\mathcal{L} = \frac{1}{2}\int_{t_0}^{t_1} dt \int_0^l \left[\rho\left(\frac{\partial y}{\partial t}\right)^2 - \tau\left(\frac{\partial y}{\partial x}\right)^2\right] dx.$$

Using (20.13) and the fact that $y$ does not appear explicitly, we obtain

$$\frac{\partial}{\partial t}\left(\rho\frac{\partial y}{\partial t}\right) - \frac{\partial}{\partial x}\left(\tau\frac{\partial y}{\partial x}\right) = 0.$$

If, in addition, $\rho$ and $\tau$ do not depend on $x$ or $t$ then

$$\frac{\partial^2 y}{\partial x^2} = \frac{1}{c^2}\frac{\partial^2 y}{\partial t^2},$$

where $c^2 = \tau/\rho$. This is the wave equation for small transverse oscillations of a taut string. ◄

## 20.6 General eigenvalue problems

We have seen in this chapter that the problem of finding a curve that makes the value of a given integral stationary when the integral is taken along the curve results, in each case, in a differential equation for the curve. It is not a great extension to ask whether this may be used to solve differential equations, by setting up a suitable variational problem and then seeking ways other than the Euler equation of finding or estimating stationary solutions.

We shall be concerned with differential equations of the form $\mathcal{L}y = \lambda\rho(x)y$, where the differential operator $\mathcal{L}$ is self-adjoint, so that $\mathcal{L} = \mathcal{L}^\dagger$ (with appropriate boundary conditions on the solution $y$), and $\rho(x)$ is some weight function, as discussed in chapter 15. In particular, we will concentrate on the Sturm–Liouville equation as an explicit example, but much of what follows can be applied to other equations of this type.

We have already discussed the solution of Sturm–Liouville equations in chapter 15 and the same notation will be used here. In this section, however, we will adopt a variational approach to estimating the eigenvalues of such equations.

Suppose we search for stationary values of the integral

$$I = \int_a^b \left[p(x)y'^2(x) - q(x)y^2(x)\right] dx, \tag{20.22}$$

with $y(a) = y(b) = 0$ and $p$ and $q$ any sufficiently smooth and differentiable functions of $x$. However, in addition we impose a normalisation condition

$$J = \int_a^b \rho(x)y^2(x)\,dx = \text{constant}. \tag{20.23}$$

Here $\rho(x)$ is a positive weight function defined in $a \le x \le b$, but which may in particular cases be a constant.

Then, as in section 20.4, we use undetermined Lagrange multipliers,† and consider $K = I - \lambda J$ given by

$$K = \int_a^b \left[py'^2 - (q + \lambda\rho)y^2\right] dx.$$

On application of the EL equation (20.5) this yields

$$\frac{d}{dx}\left(p\frac{dy}{dx}\right) + qy + \lambda\rho y = 0, \tag{20.24}$$

† We use $-\lambda$, rather than $\lambda$, so that the final equation (20.24) appears in the conventional Sturm–Liouville form.

which is exactly the Sturm–Liouville equation (15.35), with eigenvalue $\lambda$. Now, since both $I$ and $J$ are quadratic in $y$ and its derivative, finding stationary values of $K$ is equivalent to finding stationary values of $I/J$. This may also be shown by considering the functional $\Lambda = I/J$, for which

$$\begin{aligned}\delta\Lambda &= (\delta I/J) - (I/J^2)\,\delta J \\ &= (\delta I - \Lambda\delta J)/J \\ &= \delta K/J.\end{aligned}$$

Hence, extremising $\Lambda$ is equivalent to extremising $K$. Thus we have the important result that *finding functions $y$ that minimise $I/J$ is equivalent to finding functions $y$ that are solutions of the Sturm–Liouville equation and the resulting value of $I/J$ equals the corresponding eigenvalue of the equation.*

Of course this does not tell us how to find such a function $y$ and, naturally, to have to do it by solving (20.24) directly defeats the purpose of the exercise. We will see in the next section how some progress can be made. It is worth recalling that the functions $p(x)$, $q(x)$ and $\rho(x)$ can have many different forms, and so (20.24) represents quite a wide variety of equations.

We now recall some properties of the solutions of the Sturm–Liouville equation. The eigenvalues $\lambda_i$ of (20.24) are real, and will be assumed non-degenerate (for simplicity). We also assume that the corresponding eigenfunctions have been made real, so that normalised eigenfunctions $y_i(x)$ satisfy the orthogonality relation (as in (15.27))

$$\int_a^b y_i y_j \rho\, dx = \delta_{ij}. \qquad (20.25)$$

We further assume the boundary condition

$$\left[y_i p y_j'\right]_{x=a}^{x=b} = 0, \qquad (20.26)$$

which can be satisfied by $y(a) = y(b) = 0$, but also by many other sets of boundary conditions.

►*Show that*

$$\int_a^b \left(y_j' p y_i' - y_j q y_i\right)\, dx = \lambda_i \delta_{ij}, \qquad (20.27)$$

Let $y_i$ be an eigenfunction of (20.24), corresponding to a particular eigenvalue $\lambda_i$, so that

$$(p y_i')' + (q + \lambda_i \rho) y_i = 0.$$

Multiplying this through by $y_j$ and integrating from $a$ to $b$ (the first term by parts) we obtain

$$\left[y_j\,(py_i')\right]_a^b - \int_a^b y_j'(py_i')\,dx + \int_a^b y_j(q+\lambda_i\rho)y_i\,dx = 0. \qquad (20.28)$$

The first term vanishes by virtue of (20.26), and on rearranging the other terms, and using (20.25), we find the result (20.27). ◄

We see at once that, if the function $y(x)$ minimises $I/J$, i.e. satisfies the Sturm–Liouville equation, then putting $y_i = y_j = y$ in (20.25) and (20.27) yields $J$ and $I$ respectively on the left-hand sides; thus, as mentioned above, the minimised value of $I/J$ is just the eigenvalue $\lambda$, introduced originally as the undetermined multiplier.

► *For a function $y$ satisfying the Sturm–Liouville equation, verify that provided (20.26) is satisfied, $\lambda = I/J$.*

We first multiply (20.24) through by $y$ to give

$$y(py')' + qy^2 + \lambda\rho y^2 = 0.$$

Now integrating this expression by parts we have

$$\left[ypy'\right]_a^b - \int_a^b \left(py'^2 - qy^2\right)\,dx + \lambda\int_a^b \rho y^2\,dx = 0.$$

The first term on the LHS is zero, the second is simply $-I$ and the third is $\lambda J$. Thus $\lambda = I/J$. ◄

## 20.7 Estimation of eigenvalues and eigenfunctions

Since the eigenvalues $\lambda_i$ of the Sturm–Liouville equation are the stationary values of $I/J$, it follows that any evaluation of $I/J$ must yield a value that lies between the lowest and highest eigenvalues of the corresponding Sturm–Liouville equation, i.e.

$$\lambda_{\rm min} \le \frac{I}{J} \le \lambda_{\rm max},$$

where, depending on the equation under consideration, either $\lambda_{\rm min} = -\infty$ and $\lambda_{\rm max}$ is finite, or $\lambda_{\rm max} = \infty$ and $\lambda_{\rm min}$ is finite. Notice that here we have departed from direct consideration of the minimising problem and made a statement about a calculation in which no actual minimisation is necessary.

Thus, as an example, for an equation with a finite lowest eigenvalue $\lambda_0$ any evaluation of $I/J$ provides an upper bound on $\lambda_0$. Further, we will now show that the estimate $\lambda$ obtained is a better estimate of $\lambda_0$ that the estimated (guessed) function $y$ is of $y_0$, the true eigenfunction corresponding to $\lambda_0$. The sense in which 'better' is used here will be clear from the final result.

We first expand the estimated or *trial function* $y$ in terms of the complete set $y_i$:

$$y = y_0 + c_1 y_1 + c_2 y_2 + \cdots ,$$

where, if a good trial function has been guessed, the $c_i$ will be small. Using (20.25) we have immediately that $J = 1 + \sum_i |c_i|^2$. The other required integral is

$$I = \int_a^b \left[ p\left( y_0' + \sum_i c_i y_i' \right)^2 - q\left( y_0 + \sum_i c_i y_i \right)^2 \right] dx.$$

On multiplying out the squared terms, all the cross terms vanish because of (20.27) to leave

$$\begin{aligned} \lambda = \frac{I}{J} &= \frac{\lambda_0 + \sum_i |c_i|^2 \lambda_i}{1 + \sum_j |c_j|^2} \\ &= \lambda_0 + \sum_i |c_i|^2 (\lambda_i - \lambda_0) + \mathrm{O}(c^4). \end{aligned}$$

Hence $\lambda$ differs from $\lambda_0$ by a term second order in the $c_i$, even though $y$ differed from $y_0$ by a term first order in the $c_i$. We notice incidentally that, since $\lambda_0 < \lambda_i$ for all $i$, $\lambda$ is shown to be necessarily $\geq \lambda_0$ with equality only if all $c_i = 0$, i.e. if $y \equiv y_0$.

The method can be extended to the second and higher eigenvalues by imposing, in addition to the original constraints and boundary conditions, a restriction of the trial functions to only those that are orthogonal to the eigenfunctions corresponding to lower eigenvalues. (This of course then requires complete or nearly complete knowledge of these latter eigenfunctions.) An example is given at the end of the chapter (exercise 20.13).

We now illustrate the method we have discussed by considering a simple example, one for which, as on previous occasions, the answer is obvious.

►*Solve*

$$-\frac{d^2 y}{dx^2} = \lambda y, \qquad 0 \leq x \leq 1, \tag{20.29}$$

*with boundary conditions*

$$y(0) = 0, \qquad y'(1) = 0. \tag{20.30}$$

In particular we wish to find the lowest value $\lambda_0$ of $\lambda$ for which (20.29) has a solution satisfying (20.30). The exact answer is of course $y = A \sin(x\pi/2)$ and $\lambda_0 = \pi^2/4 \approx 2.47$.

We first note that the Sturm–Liouville equation reduces to (20.29) if we take

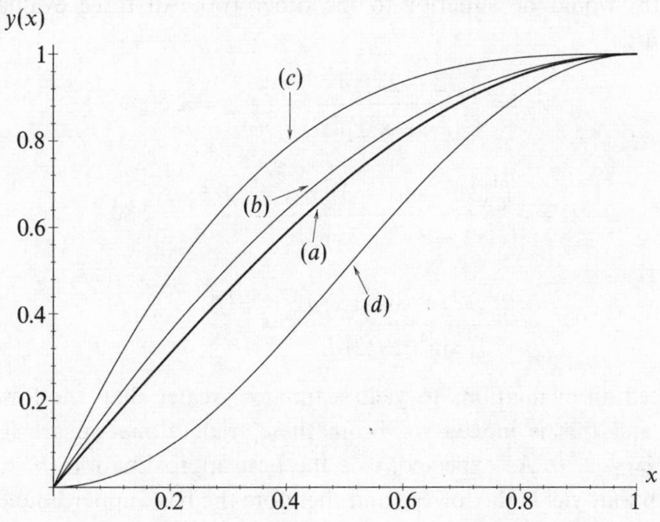


Figure 20.10 Trial solutions used to estimate the lowest eigenvalue $\lambda$ of $-y'' = \lambda y$ with $y(0) = y'(1) = 0$. They are: (*a*) $y = \sin(\pi x/2)$, the exact result; (*b*) $y = 2x - x^2$; (*c*) $y = x^3 - 3x^2 + 3x$; (*d*) $y = \sin^2(\pi x/2)$.

$p(x) = 1$, $q(x) = 0$ and $\rho(x) = 1$ and that the boundary conditions satisfy (20.26). Thus we are able to apply the previous theory.

We will use three trial functions so that the effect on the estimate of $\lambda_0$ of making better or worse 'guesses' can be seen. One further preliminary remark is relevant, namely that the estimate is independent of any constant multiplicative factor in the function used. This is easily verified by looking at the form of $I/J$. We normalise each trial function so that $y(1) = 1$, purely in order to facilitate comparison of the various function shapes.

Figure 20.10 illustrates the trial functions used, curve (*a*) being the exact solution $y = \sin(\pi x/2)$. The other curves are: (*b*) $y(x) = 2x - x^2$, (*c*) $y(x) = x^3 - 3x^2 + 3x$, (*d*) $y(x) = \sin^2(\pi x/2)$. The choice of trial wavefunction is governed by the following considerations.

(i) The boundary conditions (20.30) *must* be satisfied.

(ii) A 'good' trial function ought to mimic the correct solution as far as possible, but it may not be easy to guess even the general shape of the correct solution in some cases.

(iii) The evaluation of $I/J$ should be as simple as possible.

It is easily verified that functions (*b*), (*c*) and (*d*) all satisfy (20.30) but so far as mimicking the correct solution is concerned, we would expect from the figure that (*b*) would be superior to the other two. All three evaluations are straightforward:

$$\lambda_b = \frac{\int_0^1 (2-2x)^2\,dx}{\int_0^1 (2x-x^2)^2\,dx} = \frac{4/3}{8/15} = 2.50,$$

$$\lambda_c = \frac{\int_0^1 (3x^2-6x+3)^2\,dx}{\int_0^1 (x^3-3x^2+3x)^2\,dx} = \frac{9/5}{9/14} = 2.80,$$

$$\lambda_d = \frac{\int_0^1 (\pi^2/4)\sin^2(\pi x)\,dx}{\int_0^1 \sin^4(\pi x/2)\,dx} = \frac{\pi^2/8}{3/8} = 3.29.$$

We expected all evaluations to yield estimates greater than the lowest eigenvalue, 2.47, and this is indeed so. From these trials alone we are (only) able to say that $\lambda_0 \leq 2.50$. As expected, it is the best approximation (*b*) to the true eigenfunction that yields the lowest, and therefore the best, upper bound on $\lambda_0$. ◄

We may generalise the work of this section to other differential equations of the form $\mathcal{L}y = \lambda\rho y$, where $\mathcal{L} = \mathcal{L}^\dagger$. In particular, one finds

$$\lambda_{\min} \leq \frac{I}{J} \leq \lambda_{\max},$$

where $I$ and $J$ are now given by

$$I = \int_a^b y^*(\mathcal{L}y)\,dx \qquad \text{and} \qquad J = \int_a^b \rho y^* y\,dx. \tag{20.31}$$

It is straightforward to show that, for the special case of the Sturm–Liouville equation, for which

$$\mathcal{L}y = -(py')' - qy,$$

the expression for $I$ in (20.31) leads to (20.22).

## 20.8 Adjustment of parameters

Instead of trying to estimate $\lambda_0$ by selecting a large number of different trial functions, we may also use trial functions that include one or more parameters which themselves may be adjusted to give the lowest value to $\lambda = I/J$ and hence the best estimate of $\lambda_0$. The justification for this method comes from the knowledge that no matter what form of function is chosen, nor what values are assigned to the parameters, provided the boundary conditions are satisfied, $\lambda$ can never be less than the required $\lambda_0$.

To illustrate this method an example from quantum mechanics will be used. The time-independent Schrödinger equation is formally written $H\psi = E\psi$, where $H$ is a linear operator, $\psi$ the wavefunction describing a quantum mechanical system and $E$ the energy of the system. The operator $H$ is called the Hamiltonian and for a particle of mass $m$ moving in a one-dimensional harmonic oscillator potential is given by

$$H = -\frac{\hbar^2}{2m}\frac{d^2}{dx^2} + \frac{k}{2}x^2, \tag{20.32}$$

where $\hbar$ is Planck's constant divided by $2\pi$.

►*Estimate the ground-state energy of a quantum harmonic oscillator.*

Using (20.32) in $H\psi = E\psi$, the Schrödinger equation is

$$-\frac{\hbar^2}{2m}\frac{d^2\psi}{dx^2} + \frac{k}{2}x^2\psi = E\psi, \qquad -\infty < x < \infty. \tag{20.33}$$

The boundary conditions are that $\psi$ should vanish as $x \to \pm\infty$. Equation (20.33) is a form of the Sturm–Liouville equation in which $p = \hbar^2/2m$, $q = -kx^2/2$, $\rho = 1$ and $\lambda = E$, and can be solved as previously.

As a trial wavefunction we take $\psi = \exp(-\alpha x^2)$, where $\alpha$ is a positive parameter whose value we will choose later. This function certainly $\to 0$ as $x \to \pm\infty$ and is convenient for calculations. Whether it approximates the true wave function is unknown, but if it does not our estimate will still be valid, although the upper bound will be a poor one.

With $y = \exp(-\alpha x^2)$ and therefore $y' = -2\alpha x \exp(-\alpha x^2)$, the required estimate is

$$E = \lambda = \frac{\int_{-\infty}^{\infty}[(\hbar^2/2m)4\alpha^2x^2 + (k/2)x^2]e^{-2\alpha x^2}\,dx}{\int_{-\infty}^{\infty}e^{-2\alpha x^2}\,dx} = \frac{\hbar^2\alpha}{2m} + \frac{k}{8\alpha}. \tag{20.34}$$

This evaluation is easily performed using the reduction formula

$$I_n = \frac{n-1}{4\alpha}I_{n-2}, \qquad \text{where} \qquad I_n = \int_{-\infty}^{\infty} x^n e^{-2\alpha x^2}\,dx. \tag{20.35}$$

So we have obtained an estimate, given by (20.34), for the ground-state energy, the lowest eigenvalue of $H$, of the oscillator, the estimate involving the parameter $\alpha$. In line with our previous discussion we now minimise $\lambda$ with respect to $\alpha$. Putting $d\lambda/d\alpha = 0$ (clearly a minimum), gives $\alpha = (km)^{1/2}/2\hbar$, which in turn gives as the minimum value for $\lambda$

$$E = \frac{\hbar}{2}\left(\frac{k}{m}\right)^{1/2} = \frac{\hbar\omega}{2}, \tag{20.36}$$

where we have put $(k/m)^{1/2}$ equal to the classical angular frequency $\omega$.

The method thus leads to the conclusion that the ground-state energy $E_0$ is $\leq \frac{1}{2}\hbar\omega$. In fact, as is well known, the equality sign holds, $\frac{1}{2}\hbar\omega$ being just the zero-point energy of a quantum mechanical oscillator. Our estimate gives the exact value because $\exp(-\alpha x^2)$ is exactly the ground state wavefunction if $\alpha$ is as we have determined. ◀

An alternative but equivalent approach to this is developed in the exercises that follow, as is an extension of this particular problem to estimating the second-lowest eigenvalue (see exercise 20.13).

## 20.9 Exercises

20.1 A surface of revolution, whose equation in cylindrical polar coordinates is $\rho = \rho(z)$, is bounded by the circles $\rho = a$, $z = \pm c$ $(a > c)$. Show that the function that makes the surface integral $I = \int \rho^{-1/2}\, dS$ stationary with respect to small variations is given by $\rho(z) = k + z^2/4k$, where $k = \frac{1}{2}[a \pm (a^2 - c^2)^{1/2}]$.

20.2 Show that the lowest value of the integral

$$\int_A^B \frac{(1 + y'^2)^{1/2}}{y}\, dx,$$

where $A$ is $(-1, 1)$ and $B$ is $(1, 1)$, is $2\ln(1 + \sqrt{2})$. Assume that the Euler–Lagrange equation gives a minimising curve.

20.3 The refractive index $n$ of a medium is a function only of the distance $r$ from a fixed point $O$. Prove that the equation of a light ray, assumed to lie in a plane through $O$, travelling in the medium satisfies (in polar coordinates)

$$\frac{1}{r^2}\left(\frac{dr}{d\theta}\right)^2 = \frac{r^2}{a^2}\frac{n^2(r)}{n^2(a)} - 1,$$

where $a$ is the distance of the ray from $O$ at the point at which $dr/d\theta = 0$.

If $n = [1 + (\alpha^2/r^2)]^{1/2}$ and the ray starts and ends far from $O$, find the angle through which the ray is turned if its minimum distance from $O$ is $a$.

20.4 The Lagrangian for a $\pi$-meson is given by

$$L(\mathbf{x}, t) = \tfrac{1}{2}(\dot{\phi}^2 - |\nabla\phi|^2 - \mu^2\phi^2),$$

where $\mu$ is the meson mass and $\phi(\mathbf{x}, t)$ is its wavefunction. Assuming Hamilton's principle find the wave equation satisfied by $\phi$.

20.5 (a) For a system described in terms of coordinates $q_i$ and $t$, show that if $t$ does not appear explicitly in the expressions for $x$, $y$ and $z$ $(x = x(q_i, t)$, etc.), then the kinetic energy $T$ is a homogeneous quadratic function of the $\dot{q}_i$ (it may also involve the $q_i$). Deduce that $\sum_i \dot{q}_i(\partial T/\partial \dot{q}_i) = 2T$.

(b) Assuming that the forces acting on the system are derivable from a potential $V$, show, by expressing $dT/dt$ in terms of $q_i$ and $\dot{q}_i$, that $d(T + V)/dt = 0$.

20.6 For a system specified by the coordinates $q$ and $t$, show that the equation of motion is unchanged if the Lagrangian $L(q, \dot{q}, t)$ is replaced by

$$L_1 = L + \frac{d\phi(q,t)}{dt},$$

where $\phi$ is an arbitrary function. Deduce that the equation of motion of a particle that moves in one dimension subject to a force $-dV(x)/dx$ ($x$ being measured from a point $O$) is unchanged if $O$ is forced to move with a constant velocity $v$ ($x$ still being measured from $O$).

20.7 Derive the differential equations for the polar coordinates $r$, $\theta$ of a particle of unit mass moving in a field of potential $V(r)$. Find the form of $V$ if the path of the particle is given by $r = a \sin\theta$.

20.8 You are provided with a light line of length $\pi a/2$ and some lead shot of total mass $M$. Use a variational method to determine how the lead shot must be distributed along the line if the loaded line is to hang in a circular arc of radius $a$ when its ends are attached to two points at the same height. (Measure the distance $s$ along the line from its centre.)

20.9 In the brachistochrone problem of subsection 20.3.4 show that if the upper end-point can lie anywhere on the curve $h(x, y) = 0$, then the curve of quickest descent $y(x)$ intersects $h(x, y) = 0$ at right angles.

20.10 Show that $y'' - xy + \lambda x^2 y = 0$ has a solution for which $y(0) = y(1) = 0$ and $\lambda \leq 147/4$.

20.11 A drumskin is stretched across a fixed circular rim of radius $a$. Small transverse vibrations of the skin have an amplitude $z(\rho, \phi, t)$ that satisfies

$$\nabla^2 z = \frac{1}{c^2}\frac{\partial^2 z}{\partial t^2}$$

in plane polar coordinates. For a normal mode independent of azimuth, $z = Z(\rho)\cos\omega t$, find the differential equation satisfied by $Z(\rho)$. Using a trial function of the form $a^\nu - \rho^\nu$, obtain an estimate for the lowest normal mode frequency. (The exact answer is $(5.78)^{1/2}c/a$.)

20.12 This is an alternative approach to the example in section 20.8. Using the notation of that section, $H\psi = E\psi$ and the expectation value of the energy of the state $\psi$ is given by $\int \psi^* E\psi\, dv = \int \psi^* H\psi\, dv$. Denote the eigenfunctions of $H$ by $\psi_i$, so that $H\psi_i = E_i\psi_i$, and, since $H$ is self-adjoint, $\int \psi_j^* \psi_i\, dv = \delta_{ij}$.

(a) By writing any function $\psi$ as $\sum c_j \psi_j$ and following an argument similar to that in section 20.7, show that

$$E = \frac{\int \psi^* H\psi\, dv}{\int \psi^* \psi\, dv} \geq E_0,$$

the energy of the lowest state. (This is the Rayleigh–Ritz principle.)

(b) Using the same trial function as in section 20.8 show that the same result is obtained.

20.13 This is an extension to section 20.8 and the previous question. With the ground-state (lowest-energy) wavefunction as $\exp(-\alpha x^2)$, take as a trial function the orthogonal wave function $x^{2n+1}\exp(-\alpha x^2)$ using the integer $n$ as a variable parameter. Use either Sturm–Liouville theory or the Rayleigh–Ritz principle to show that the energy of the second lowest state of a quantum harmonic oscillator is $\leq 3\hbar\omega/2$.

20.14 The Hamiltonian for the hydrogen atom is of the form

$$-\frac{\hbar^2}{2m}\nabla^2 - \frac{q}{4\pi\epsilon_0 r}.$$

For a spherically symmetric state, as may be assumed for the ground state, the only relevant part of $\nabla^2$ is that involving differentiation with respect to $r$.

(a) Define the integrals $J_n$ by

$$J_n = \int_0^\infty r^n e^{-2\beta r}\,dr$$

and show that, for a trial wavefunction of the form $\exp(-\beta r)$ with $\beta > 0$, $\int \psi^* H\psi\,dv$ and $\int \psi^*\psi\,dv$ (see exercise 20.12(a)) can be expressed as $aJ_1 - bJ_2$ and $cJ_2$ respectively, where $a$, $b$, $c$ are factors which you should determine.

(b) Show that the estimate of $E$ is minimised when $\beta = mq^2/(4\pi\epsilon_0\hbar^2)$.

(c) Hence find an upper limit for the ground-state energy of the hydrogen atom. In fact, $\exp(-\beta r)$ is the correct form for the wavefunction and the limit gives the actual value.

20.15 The Sturm–Liouville equation can be extended to two independent variables, $x$ and $z$, with little modification. In equation (20.22) $y'^2$ is replaced by $(\nabla y)^2$ and the integrals of the various functions of $y(x,z)$ become two-dimensional, i.e. the infinitesimal is $dx\,dz$.

The vibrations of a trampoline 4 units long and 1 unit wide satisfy the equation $\nabla^2 y + k^2 y = 0$. By taking the simplest possible permissible polynomial as a trial function, show that the lowest mode of vibration has $k^2 \leq 10.63$ and, by direct solution, that the actual value is 10.49.

## 20.10 Hints and answers

20.2 The minimising curve is $x^2 + y^2 = 2$.

20.3 $I = \int n(r)[r^2 + (dr/d\theta)^2]^{1/2}\, d\theta$. Take axes such that $\theta = 0$ when $r = \infty$. If $\phi = (\pi - \text{deviation angle})/2$ then $\phi = \theta$ at $r = a$, and the equation reduces to

$$\frac{\phi}{(a^2+\alpha^2)^{1/2}} = \int_{-\infty}^{\infty} \frac{dr}{r(r^2-a^2)^{1/2}},$$

which can be evaluated by putting $r = a(y + y^{-1})/2$, or successively $r = a\cosh\psi$, $y = \exp\psi$ to yield a deviation $\pi[(a^2+\alpha^2)^{1/2} - a]/a$.

20.4 $\nabla^2\phi - \partial^2\phi/\partial t^2 = \mu^2\phi$.

20.5 (a) $\partial x/\partial t = 0$ and so $\dot{x} = \sum_i \dot{q}_i \partial x/\partial q_i$; (b) use

$$\sum_i \dot{q}_i \frac{d}{dt}\left(\frac{\partial T}{\partial \dot{q}_i}\right) = \frac{d}{dt}(2T) - \sum_i \ddot{q}_i \frac{\partial T}{\partial \dot{q}_i}.$$

20.6 $\phi(x,t) = m(vx + v^2t/2)$.

20.7 $r^2\dot{\theta} = k$, $\ddot{r} - r\dot{\theta}^2 + dV/dr = 0$, $V(r) = -k^2a^2/2r^4 + \text{constant}$.

20.8 $-\lambda y'(1-y'^2)^{-1/2} = 2gP(s)$, $y = y(s)$, $P(s) = \int_0^s \rho(s')\,ds'$. Solution $y = -a\cos(s/a)$ and $2P(\pi a/4) = M$ give $\lambda = -gM$. Required $\rho(s) = [M/(2a)]\sec^2(s/a)$.

20.10 The equation is of SL form with $p = 1$, $q = -x$ and weight function $x^2$. Try $y = x(1-x)$. The integrals have values 21/60 and 2/210.

20.11 $Z'' + \rho^{-1}Z' + (\omega/c)^2 Z = 0$, with $Z(a) = 0$ and $Z'(0) = 0$, an SL equation with $p = \rho$, $q = 0$, and weight function $\rho/c^2$. Estimate of $\omega^2 = [c^2\nu/(2a^2)][0.5 - 2(\nu+2)^{-1} + (2\nu+2)^{-1}]^{-1}$, which minimises to $c^2(2+\sqrt{2})^2/(2a^2) = 5.83c^2/a^2$ when $\nu = \sqrt{2}$.

20.12 The estimate is $\hbar^2\alpha/(2m) + k/(8\alpha)$ and the minimum occurs at the value of $\alpha$ which makes the two terms equal.

20.13 $E_1 \leq (\hbar\omega/2)(8n^2 + 12n + 3)/(4n+1)$, which has a minimum value of $3\hbar\omega/2$ when integer $n = 0$.

20.14 (a) $a = 4\pi\hbar^2\beta/m - q^2/\epsilon_0$, $b = 2\pi\hbar^2\beta^2/m$, $c = 4\pi$; (c) $-mq^4/[2(4\pi\epsilon_0\hbar)^2]$.

20.15 The SL equation has $p = 1$, $q = 0$, and $\rho = 1$. Use as a trial function $u(x,y) = x(4-x)y(1-y)$. Numerator $= 1088/90$, denominator $= 512/450$. Direct solution $k^2 = 17\pi^2/16$.

# 21

# Integral equations

It is not unusual in the analysis of a physical system to encounter an equation in which an unknown but required function $y(x)$, say, appears under an integral sign. Such an equation is called an *integral equation*, and in this chapter we discuss several methods for solving the more straightforward examples of such equations.

Before embarking on our discussion of methods for solving various integral equations, we begin with a warning that many of the integral equations met in practice cannot be solved by the elementary methods presented here, but must instead be solved numerically, usually on a computer. Nevertheless, the regular occurrence of several simple types of integral equation that may be solved analytically is sufficient reason to explore these equations more fully.

We shall begin this chapter by discussing how a differential equation can be transformed into an integral equation and considering the most common types of linear integral equation. After introducing the operator notation and considering the existence of solutions for various types of equation, we go on to discuss elementary methods of obtaining closed-form solutions of simple integral equations. We then consider the solution of integral equations in terms of infinite series, and conclude by discussing the properties of integral equations with Hermitian kernels, i.e. those in which the integrands have particular symmetry properties.

## 21.1 Obtaining an integral equation from a differential equation

Integral equations occur in many situations, partly because we may always rewrite a differential equation as an integral equation. It is sometimes advantageous to make this transformation, since questions concerning the existence of a solution are more easily answered for integral equations (see section 21.3), and, furthermore, an integral equation can automatically incorporate any boundary conditions on the solution.

We shall illustrate the principles involved by considering the differential equation

$$y''(x) = f(x, y), \tag{21.1}$$

where $f(x, y)$ can be any function of $x$ and $y$, but not of $y'(x)$. Equation (21.1) thus represents a large class of linear and non-linear second-order differential equations.

We can convert (21.1) into the corresponding integral equation by first integrating with respect to $x$ to obtain

$$y'(x) = \int_0^x f(z, y(z))\, dz \; + \; c_1.$$

Integrating once more, we find

$$y(x) = \int_0^x du \int_0^u dz\, f(z, y(z)) \; + \; c_1 x \; + \; c_2.$$

Provided we do not change the region in the $uz$-plane over which the double integral is taken, we can reverse the order of the two integrations. Changing the integration limits appropriately, we find

$$y(x) = \int_0^x dz\, f(z, y(z)) \int_z^x du \; + \; c_1 x \; + \; c_2 \tag{21.2}$$

$$= \int_0^x (x - z) f(z, y(z))\, dz \; + \; c_1 x \; + \; c_2; \tag{21.3}$$

this is a non-linear (for a general $f(x, y)$) *Volterra* integral equation.

It is straightforward to incorporate any boundary conditions on the solution $y(x)$ by fixing the constants $c_1$ and $c_2$ in (21.3). For example, we might have the one-point boundary condition $y(0) = a$ and $y'(0) = b$, for which is it clear that we must set $c_1 = b$ and $c_2 = a$.

## 21.2 Types of integral equation

From (21.3), it is clear that even a relatively simple differential equation such as (21.1) can lead to a corresponding integral equation that is non-linear. In this chapter, however, we will restrict our attention to *linear* integral equations, which have the general form

$$g(x)y(x) = f(x) + \lambda \int_a^b K(x, z) y(z)\, dz. \tag{21.4}$$

In (21.4), $y(x)$ is the unknown function, while the functions $f(x)$, $g(x)$ and $K(x, z)$ are assumed known. $K(x, z)$ is called the *kernel* of the integral equation. The integration limits $a$ and $b$ are also assumed known, and may be constants or functions of $x$, and $\lambda$ is a known constant or parameter.

In fact, we shall be concerned with various special cases of (21.4), which are known by particular names. Firstly, if $g(x) = 0$ then the unknown function $y(x)$ appears only under the integral sign, and (21.4) is called a linear integral equation *of the first kind*. Alternatively, if $g(x) = 1$, so that $y(x)$ appears twice, once inside the integral and once outside, then (21.4) is called a linear integral equation *of the second kind*. In either case, if $f(x) = 0$ the equation is called *homogeneous*, otherwise it is *inhomogeneous*.

We can further distinguish between different types of integral equation by the form of the integration limits $a$ and $b$. If these limits are fixed constants, then the equation is called a *Fredholm* equation. If, however, the upper limit $b = x$ (i.e. it is variable), then the equation is called a *Volterra* equation; such an equation is analogous to one with fixed limits, but for which the kernel $K(x, z) = 0$ for $z > x$. Finally, we note that any equation for which either (or both) of the integration limits is infinite, or for which $K(x, z)$ becomes infinite in the range of integration, is called a *singular* integral equation.

## 21.3 Operator notation and the existence of solutions

There is a close correspondence between linear integral equations and the matrix equations discussed in chapter 7. However, the former involve linear, integral relations between functions in an infinite-dimensional function space (see chapter 15), whereas the latter specify linear relations among vectors in a finite-dimensional vector space.

Since we are restricting our attention to linear integral equations, it will be convenient to introduce the linear integral operator $\mathcal{K}$, whose action on an arbitrary function $y$ is given by

$$\mathcal{K}y = \int_a^b K(x, z)y(z)\,dz. \qquad (21.5)$$

This is analogous to the introduction in chapters 14 and 15 of the notation $\mathcal{L}$ to describe a linear differential operator. Furthermore, we may define the Hermitian conjugate $\mathcal{K}^\dagger$ by

$$\mathcal{K}^\dagger y = \int_a^b K^*(z, x)y(z)\,dz,$$

where the asterisk denotes complex conjugation, and we have reversed the order of the arguments in the kernel.

It is clear from (21.5) that $\mathcal{K}$ is indeed linear. Moreover, since $\mathcal{K}$ operates on the infinite-dimensional space of (reasonable) functions, we may make an obvious analogy with matrix equations, and consider the action of $\mathcal{K}$ on a function $f$ as that of multiplication of a column vector by a matrix (both of infinite dimension).

When written in operator form, the integral equations discussed in the previous section resemble equations familiar from linear algebra. For example, the inhomogeneous Fredholm equation of the first kind may be written as

$$0 = f + \lambda \mathcal{K} y,$$

which has the unique solution $y = -\mathcal{K}^{-1} f/\lambda$, provided $f \neq 0$ and the inverse operator $\mathcal{K}^{-1}$ exists.

Similarly, we may write the corresponding Fredholm equation of the second kind as

$$y = f + \lambda \mathcal{K} y. \tag{21.6}$$

In the homogeneous case, where $f = 0$, this reduces to $y = \lambda \mathcal{K} y$, which is reminiscent of an eigenvalue problem in linear algebra (except that $\lambda$ appears on the other side of the equation) and, similarly, only has solutions for at most a countably infinite set of *eigenvalues* $\lambda = \lambda_1, \lambda_2$, etc. The corresponding solutions $y_i$ are called the eigenfunctions.

In the inhomogeneous case ($f \neq 0$), the solution to (21.6) can be written symbolically as

$$y = (1 - \lambda \mathcal{K})^{-1} f,$$

provided again that the inverse operator exists. It may be shown that, in general, (21.6) does possess a unique solution if $\lambda \neq \lambda_i$, i.e. when $\lambda$ does not equal one of the eigenvalues of the corresponding homogeneous equation.

When $\lambda$ does equal one of these eigenvalues, (21.6) may have either many solutions or no solution at all, depending on the form of $f$. If the function $f$ is orthogonal to *every* eigenfunction of the equation

$$g = \lambda^* \mathcal{K}^\dagger g \tag{21.7}$$

that belongs to the eigenvalue $\lambda^*$, i.e.

$$\langle g|f \rangle = \int_a^b g^*(x) f(x)\, dx = 0$$

for every function $g$ obeying (21.7), then it can be shown that (21.6) has many solutions. Otherwise the equation has no solution. These statements are discussed further in section 21.7, for the special case of integral equations with Hermitian kernels, i.e. those for which $\mathcal{K} = \mathcal{K}^\dagger$.

## 21.4 Closed-form solutions

In certain very special cases, it may be possible to obtain a closed-form solution of an integral equation. The reader should realise, however, when faced with an integral equation, that in general it will not be soluble by the simple methods presented in this section but must instead be solved using (numerical) iterative methods, such as those outlined in section 21.5.

### 21.4.1 Separable kernels

The most straightforward integral equations to solve are Fredholm equations with *separable* (or *degenerate*) kernels. A kernel is separable if it has the form

$$K(x,z) = \sum_{i=1}^{n} \phi_i(x)\psi_i(z), \tag{21.8}$$

where $\phi_i(x)$ are $\psi_i(z)$ are respectively functions of $x$ only and of $z$ only, and the number of terms in the sum, $n$, is finite.

Let us consider the solution of the (inhomogeneous) Fredholm equation of the second kind,

$$y(x) = f(x) + \lambda \int_a^b K(x,z)y(z)\,dz, \tag{21.9}$$

which has a separable kernel of the form (21.8). Writing the kernel in its separated form, the functions $\phi_i(x)$ may be taken outside the integral over $z$ to obtain

$$y(x) = f(x) + \lambda \sum_{i=1}^{n} \phi_i(x) \int_a^b \psi_i(z)y(z)\,dz.$$

Since the integration limits $a$ and $b$ are constant for a Fredholm equation, the integral over $z$ in each term of the sum is just a constant. Denoting these constants by

$$c_i = \int_a^b \psi_i(z)y(z)\,dz, \tag{21.10}$$

the solution to (21.9) is found to be

$$y(x) = f(x) + \lambda \sum_{i=1}^{n} c_i\phi_i(x), \tag{21.11}$$

where the constants $c_i$ can be evalutated by substituting (21.11) into (21.10).

►*Solve the integral equation*

$$y(x) = x + \lambda \int_0^1 (xz + z^2)y(z)\,dz. \tag{21.12}$$

The kernel for this equation is $K(x,z) = xz + z^2$, which is clearly separable, and using the notation in (21.8) we have $\phi_1(x) = x$, $\phi_2(x) = 1$, $\psi_1(z) = z$ and $\psi_2(z) = z^2$. From (21.11) the solution to (21.12) has the form

$$y(x) = x + \lambda(c_1x + c_2),$$

where the constants $c_1$ and $c_2$ are given by (21.10) as

$$c_1 = \int_0^1 z[z + \lambda(c_1 z + c_2)]\, dz = \tfrac{1}{3} + \tfrac{1}{3}\lambda c_1 + \tfrac{1}{2}\lambda c_2,$$

$$c_2 = \int_0^1 z^2[z + \lambda(c_1 z + c_2)]\, dz = \tfrac{1}{4} + \tfrac{1}{4}\lambda c_1 + \tfrac{1}{3}\lambda c_2.$$

These two simultaneous linear equations may be straightforwardly solved for $c_1$ and $c_2$ to give

$$c_1 = \frac{24 + \lambda}{72 - 48\lambda - \lambda^2} \qquad \text{and} \qquad c_2 = \frac{18}{72 - 48\lambda - \lambda^2},$$

so that the solution to (21.12) is

$$y(x) = \frac{(72 - 24\lambda)x + 18\lambda}{72 - 48\lambda - \lambda^2}. \blacktriangleleft$$

In the above example, we see that (21.12) has a (finite) unique solution provided that $\lambda$ is not equal to either root of the quadratic in the denominator of $y(x)$. The roots of this quadratic are in fact the *eigenvalues* of the corresponding homogeneous equation, as mentioned in the previous section. In general, if the separable kernel contains $n$ terms, as in (21.8), there will be $n$ such eigenvalues, although they need not all be different.

Kernels consisting of trigonometric (or similar) functions of sums or differences of $x$ and $z$ are also often separable.

►*Find the eigenvalues and corresponding eigenfunctions of the homogeneous Fredholm equation*

$$y(x) = \lambda \int_0^\pi \sin(x + z)\, y(z)\, dz. \qquad (21.13)$$

The kernel of this integral equation can be written in separated form as

$$K(x, z) = \sin(x + z) = \sin x \cos z + \cos x \sin z,$$

so, comparing with (21.8), we have $\phi_1(x) = \sin x$, $\phi_2(x) = \cos x$, $\psi_1(z) = \cos z$ and $\psi_2(z) = \sin z$.

Thus, from (21.11), the solution to (21.13) has the form

$$y(x) = \lambda(c_1 \sin x + c_2 \cos x),$$

where the constants $c_1$ and $c_2$ are given by

$$c_1 = \lambda \int_0^\pi \cos z\, (c_1 \sin z + c_2 \cos z)\, dz = \frac{\lambda\pi}{2} c_2, \qquad (21.14)$$

$$c_2 = \lambda \int_0^\pi \sin z\, (c_1 \sin z + c_2 \cos z)\, dz = \frac{\lambda\pi}{2} c_1. \qquad (21.15)$$

Combining these two equations we find $c_1 = (\lambda\pi/2)^2 c_1$, and, assuming that $c_1 \neq 0$, this gives $\lambda = \pm 2/\pi$, the two eigenvalues of the integral equation (21.13).

By substituting each of the eigenvalues back into (21.14) and (21.15), we find that the eigenfunctions corresponding to the eigenvalues $\lambda_1 = 2/\pi$ and $\lambda_2 = -2/\pi$ are given respectively by

$$y_1(x) = A(\sin x + \cos x) \quad \text{and} \quad y_2(x) = B(\sin x - \cos x), \tag{21.16}$$

where $A$ and $B$ are arbitrary constants. ◀

### 21.4.2 *Integral transform methods*

If the kernel of an integral equation can be written as a function of the difference $x - z$ of its two arguments, then it is called a *displacement* kernel. An integral equation having such a kernel, and which also has the integration limits $-\infty$ to $\infty$, may be solved by the use of Fourier transforms.

If we consider the following integral equation with a displacement kernel,

$$y(x) = f(x) + \lambda \int_{-\infty}^{\infty} K(x - z)y(z)\,dz, \tag{21.17}$$

the integral over $z$ clearly takes the form of a convolution (see chapter 11). Therefore, Fourier-transforming (21.17) and using the convolution theorem, we obtain

$$\tilde{y}(k) = \tilde{f}(k) + \sqrt{2\pi}\lambda\tilde{K}(k)\tilde{y}(k),$$

which may be rearranged to give

$$\tilde{y}(k) = \frac{\tilde{f}(k)}{1 - \sqrt{2\pi}\lambda\tilde{K}(k)}. \tag{21.18}$$

Taking the inverse Fourier transform, the solution to (21.17) is given by

$$y(x) = \frac{1}{\sqrt{2\pi}} \int_{-\infty}^{\infty} \frac{\tilde{f}(k)\exp(ikx)}{1 - \sqrt{2\pi}\lambda\tilde{K}(k)}\,dk.$$

If we can perform this inverse Fourier transformation, then the solution can be found explicitly; otherwise it must be left in the form of an integral.

►*Find the Fourier transform of the function*

$$g(x) = \begin{cases} 1 & \text{if } |x| \leq a, \\ 0 & \text{if } |x| > a. \end{cases}$$

*Hence find an explicit expression for the solution of the integral equation*

$$y(x) = f(x) + \lambda \int_{-\infty}^{\infty} \frac{\sin(x-z)}{x-z} y(z)\, dz. \qquad (21.19)$$

*Find the solution for the special case* $f(x) = (\sin x)/x$.

The Fourier transform of $g(x)$ is given directly by

$$\tilde{g}(k) = \frac{1}{\sqrt{2\pi}} \int_{-a}^{a} \exp(-ikx)\, dx = \left[ \frac{1}{\sqrt{2\pi}} \frac{\exp(-ikx)}{(-ik)} \right]_{-a}^{a} = \sqrt{\frac{2}{\pi}} \frac{\sin ka}{k}. \qquad (21.20)$$

The kernel of the integral equation (21.19) is $K(x-z) = [\sin(x-z)]/(x-z)$. Using (21.20), it is straightforward to show that the Fourier transform of the kernel is

$$\tilde{K}(k) = \begin{cases} \sqrt{\pi/2} & \text{if } |k| \leq 1, \\ 0 & \text{if } |k| > 1. \end{cases} \qquad (21.21)$$

Thus, using (21.18), we find the Fourier transform of the solution to be

$$\tilde{y}(k) = \begin{cases} \tilde{f}(k)/(1-\pi\lambda) & \text{if } |k| \leq 1, \\ \tilde{f}(k) & \text{if } |k| > 1. \end{cases} \qquad (21.22)$$

Inverse Fourier-transforming, and writing the result in a slightly more convenient form, the solution to (21.19) is given by

$$\begin{aligned} y(x) &= f(x) + \left( \frac{1}{1-\pi\lambda} - 1 \right) \frac{1}{\sqrt{2\pi}} \int_{-1}^{1} \tilde{f}(k) \exp(ikx)\, dk \\ &= f(x) + \frac{\pi\lambda}{1-\pi\lambda} \frac{1}{\sqrt{2\pi}} \int_{-1}^{1} \tilde{f}(k) \exp(ikx)\, dk. \qquad (21.23) \end{aligned}$$

It is clear from (21.22) that when $\lambda = 1/\pi$, which is the only eigenvalue of the corresponding homogeneous equation to (21.19), the solution becomes infinite, as we would expect.

For the special case $f(x) = (\sin x)/x$, the Fourier transform $\tilde{f}(k)$ is identical to that in (21.21), and the solution (21.23) becomes

$$\begin{aligned}
y(x) &= \frac{\sin x}{x} + \left(\frac{\pi\lambda}{1-\pi\lambda}\right)\frac{1}{\sqrt{2\pi}}\int_{-1}^{1}\sqrt{\frac{\pi}{2}}\exp(ikx)\,dk \\
&= \frac{\sin x}{x} + \left(\frac{\pi\lambda}{1-\pi\lambda}\right)\frac{1}{2}\left[\frac{\exp(ikx)}{ix}\right]_{k=-1}^{k=1} \\
&= \frac{\sin x}{x} + \left(\frac{\pi\lambda}{1-\pi\lambda}\right)\frac{\sin x}{x} = \left(\frac{1}{1-\pi\lambda}\right)\frac{\sin x}{x}. \blacktriangleleft
\end{aligned}$$

If the integral equation (21.17) had instead the integration limits 0 and $x$ (so making it a Volterra equation), then its solution could be found, in a similar way, by using the convolution theorem for Laplace transforms (see chapter 11). We would find

$$\bar{y}(s) = \frac{\bar{f}(s)}{1-\lambda\bar{K}(s)},$$

where $s$ is the Laplace transform variable. Often one may use the dictionary of Laplace transforms given in table 11.1 to invert this equation and find the solution $y(x)$. In general, however, the evaluation of inverse Laplace transform integrals is difficult, since (in principle) it requires a contour integration; see chapter 18.

As a final example of the use of Fourier transforms in solving integral equations, we mention equations that have integration limits $-\infty$ and $\infty$, and a kernel of the form

$$K(x,z) = \exp(-ixz).$$

Consider, for example, the inhomogeneous Fredholm equation

$$y(x) = f(x) + \lambda\int_{-\infty}^{\infty}\exp(-ixz)\,y(z)\,dz. \tag{21.24}$$

The integral over $z$ is clearly just (a multiple of) the Fourier transform of $y(z)$, so we can write

$$y(x) = f(x) + \sqrt{2\pi}\lambda\tilde{y}(x). \tag{21.25}$$

If we now take the Fourier transform of (21.25), but continue to denote the independent variable by $x$ (i.e. rather than $k$, for example), we obtain

$$\tilde{y}(x) = \tilde{f}(x) + \sqrt{2\pi}\lambda y(-x). \tag{21.26}$$

Substituting (21.26) into (21.25) we find

$$y(x) = f(x) + \sqrt{2\pi}\lambda\left[\tilde{f}(x) + \sqrt{2\pi}\lambda y(-x)\right],$$

but on making the change $x \to -x$ and substituting back in for $y(-x)$, this gives

$$y(x) = f(x) + \sqrt{2\pi}\lambda\tilde{f}(x) + 2\pi\lambda^2\left[f(-x) + \sqrt{2\pi}\lambda\tilde{f}(-x) + 2\pi\lambda^2 y(x)\right].$$

Thus the solution to (21.24) is given by

$$y(x) = \frac{1}{1-(2\pi)^2\lambda^4}\left[f(x) + (2\pi)^{1/2}\lambda\tilde{f}(x) + 2\pi\lambda^2 f(-x) + (2\pi)^{3/2}\lambda^3\tilde{f}(-x)\right]. \tag{21.27}$$

Clearly, (21.24) possesses a unique solution provided $\lambda \neq \pm 1/\sqrt{2\pi}$ or $\pm i/\sqrt{2\pi}$; these are easily shown to be the eigenvalues of the corresponding homogeneous equation (with $f(x) \equiv 0$).

►*Solve the integral equation*

$$y(x) = \exp(-x^2/2) + \lambda \int_{-\infty}^{\infty} \exp(-ixz)\, y(z)\, dz, \tag{21.28}$$

*where $\lambda$ is a real constant. Show that the solution is unique unless $\lambda$ has one of two particular values. Does a solution exist for either of these two values of $\lambda$?*

Following the argument given above, the solution to (21.28) is given by (21.27) with $f(x) = \exp(-x^2/2)$. In order to write the solution explicitly, however, we must calculate the Fourier transform of $f(x)$. Using equation (11.7), we find $\tilde{f}(k) = \exp(-k^2/2)$, from which we note that $f(x)$ has the special property that its functional form is identical to that of its Fourier transform. Thus, the solution to (21.28) is given by

$$y(x) = \frac{1}{1-(2\pi)^2\lambda^4}\left[1 + (2\pi)^{1/2}\lambda + 2\pi\lambda^2 + (2\pi)^{3/2}\lambda^3\right]\exp\left(-\tfrac{1}{2}x^2\right). \tag{21.29}$$

Since $\lambda$ is restricted to be real, the solution to (21.28) will be unique unless $\lambda = \pm 1/\sqrt{2\pi}$, at which points (21.29) becomes infinite. In order to find whether solutions exist for either of these values of $\lambda$ we must return to equations (21.25) and (21.26).

Let us first consider the case $\lambda = +1/\sqrt{2\pi}$. Putting this value into (21.25) and (21.26), we obtain

$$y(x) = f(x) + \tilde{y}(x), \tag{21.30}$$

$$\tilde{y}(x) = \tilde{f}(x) + y(-x). \tag{21.31}$$

Substituting (21.31) into (21.30) we find

$$y(x) = f(x) + \tilde{f}(x) + y(-x),$$

but on changing $x$ to $-x$ and substituting back in for $y(-x)$, this gives

$$y(x) = f(x) + \tilde{f}(x) + f(-x) + \tilde{f}(-x) + y(x).$$

Thus, in order for a solution to exist, we require that the function $f(x)$ obeys

$$f(x) + \tilde{f}(x) + f(-x) + \tilde{f}(-x) = 0.$$

This is satisfied if $f(x) = -\tilde{f}(x)$, i.e. if the functional form of $f(x)$ is minus the form of its Fourier transform. We may repeat this analysis for the case $\lambda = -1/\sqrt{2\pi}$, and, in a similar way, we find that this time we require $f(x) = \tilde{f}(x)$.

In our case $f(x) = \exp(-x^2/2)$, for which, as we mentioned above, $f(x) = \tilde{f}(x)$. Therefore, (21.28) possesses no solution when $\lambda = +1/\sqrt{2\pi}$, but has many solutions when $\lambda = -1/\sqrt{2\pi}$. ◄

A similar approach to the above may be taken to solve equations with kernels of the form $K(x, y) = \cos xy$ or $\sin xy$, either by considering the integral over $y$ in each case as the real or imaginary part of the corresponding Fourier transform, or by using Fourier cosine or sine transforms directly.

### *21.4.3 Differentiation*

A closed-form solution to a Volterra equation may sometimes be obtained by differentiating the equation to obtain the corresponding differential equation, which may be easier to solve.

►*Solve the integral equation*

$$y(x) = x - \int_0^x xz^2 y(z)\,dz. \tag{21.32}$$

Dividing through by $x$, we obtain

$$\frac{y(x)}{x} = 1 - \int_0^x z^2 y(z)\,dz,$$

which may be differentiated with respect to $x$ to give

$$\frac{d}{dx}\left[\frac{y(x)}{x}\right] = -x^2 y(x) = -x^3\left[\frac{y(x)}{x}\right].$$

This equation may be integrated straightforwardly, and we find

$$\ln\left[\frac{y(x)}{x}\right] = -\frac{x^4}{4} + c,$$

where $c$ is a constant of integration. Thus the solution to (21.32) has the form

$$y(x) = Ax\exp(-x^4/4), \tag{21.33}$$

where $A$ is an arbitrary constant.

Since the original integral equation (21.32) contains no arbitrary constants, neither should its solution. We may calculate the value of the constant, $A$, by substituting the solution (21.33) back into (21.32), from which we find $A = 1$. ◄

## 21.5 Neumann series

As mentioned above, most integral equations met in practice will not be of the simple forms discussed in the last section and so, in general, it is not possible to find closed-form solutions. In such cases, we might try to obtain a solution in the form of an infinite series, as we did for differential equations (see chapter 14).

Let us consider the equation

$$y(x) = f(x) + \lambda \int_a^b K(x,z)y(z)\,dz, \tag{21.34}$$

where either both integration limits are constants (for a Fredholm equation) or the upper limit is variable (for a Volterra equation). Clearly, if $\lambda$ were small, then a crude (but reasonable) approximation to the solution would be

$$y(x) \approx y_0(x) = f(x),$$

where $y_0(x)$ stands for our 'zeroth-order' approximation to the solution (and is not to be confused with an eigenfunction).

Substituting this crude guess under the integral sign in the original equation, we obtain what should be a better approximation:

$$y_1(x) = f(x) + \lambda \int_a^b K(x,z)y_0(z)\,dz = f(x) + \lambda \int_a^b K(x,z)f(z)\,dz,$$

which is first order in $\lambda$. Repeating the procedure once more results in the second-order approximation

$$\begin{aligned} y_2(x) &= f(x) + \lambda \int_a^b K(x,z)y_1(z)\,dz \\ &= f(x) + \lambda \int_a^b dz_1\, K(x,z_1)f(z_1) + \lambda^2 \int_a^b dz_1 \int_a^b dz_2\, K(x,z_1)K(z_1,z_2)f(z_2). \end{aligned}$$

It is clear that we may continue this process to obtain progressively higher-order approximations to the solution. Introducing the functions

$$\begin{aligned} K_1(x,z) &= K(x,z), \\ K_2(x,z) &= \int_a^b dz_1\, K(x,z_1)K(z_1,z), \\ K_3(x,z) &= \int_a^b dz_1 \int_a^b dz_2\, K(x,z_1)K(z_1,z_2)K(z_2,z), \end{aligned}$$

and so on, which obey the recurrence relation

$$K_n(x,z) = \int_a^b K(x,z_1)K_{n-1}(z_1,z)\,dz_1,$$

we may write the $n$th-order approximation as

$$y_n(x) = f(x) + \sum_{m=1}^{n} \lambda^m \int_a^b K_m(x,z) f(z)\, dz. \tag{21.35}$$

The solution to the original integral equation is then given by the limit $y(x) = \lim_{n\to\infty} y_n(x)$, *provided the infinite series converges.* Using (21.35), this solution may be written as

$$y(x) = f(x) + \lambda \int_a^b R(x,z;\lambda) f(z)\, dz, \tag{21.36}$$

where the *resolvent kernel* $R(x,z;\lambda)$ is given by

$$R(x,z;\lambda) = \sum_{m=0}^{\infty} \lambda^m K_{m+1}(x,z). \tag{21.37}$$

Clearly, the resolvent kernel, and hence the series solution, will converge provided $\lambda$ is sufficiently small. In fact, it may be shown that the series converges in some domain of $|\lambda|$ provided the original kernel $K(x,z)$ is bounded in such a way that

$$|\lambda|^2 \int_a^b dx \int_a^b dz\, |K(x,z)|^2 < 1. \tag{21.38}$$

►*Use the Neumann series method to solve the integral equation*

$$y(x) = x + \lambda \int_0^1 xzy(z)\, dz. \tag{21.39}$$

Following the method outlined above, we begin with the crude approximation $y(x) \approx y_0(x) = x$. Substituting this under the integral sign in (21.39), we obtain the next approximation

$$y_1(x) = x + \lambda \int_0^1 xzy_0(z)\, dz = x + \lambda \int_0^1 xz^2 dz = x + \frac{\lambda x}{3},$$

Repeating the procedure once more, we obtain

$$\begin{aligned} y_2(x) &= x + \lambda \int_0^1 xzy_1(z)\, dz \\ &= x + \lambda \int_0^1 xz\left(z + \frac{\lambda z}{3}\right) dz = x + \left(\frac{\lambda}{3} + \frac{\lambda^2}{9}\right) x. \end{aligned}$$

For this simple example, it is easy to see that by continuing this process the solution to (21.39) is obtained as

$$y(x) = x + \left[\frac{\lambda}{3} + \left(\frac{\lambda}{3}\right)^2 + \left(\frac{\lambda}{3}\right)^3 + \cdots\right] x.$$

Clearly the expression in brackets is an infinite geometric series with first term $\lambda/3$ and common ratio $\lambda/3$. Thus, *provided* $|\lambda| < 3$, this infinite series converges to the value $\lambda/(3-\lambda)$, and the solution to (21.39) is

$$y(x) = x + \frac{\lambda x}{3-\lambda} = \frac{3x}{3-\lambda}. \tag{21.40}$$

Finally, we note that the requirement that $|\lambda| < 3$ may also be derived very easily from the condition (21.38). ◄

## 21.6 Fredholm theory

In the previous section, we found that a solution to the integral equation (21.34) can be found as a Neumann series of the form (21.36), where the resolvent kernel $R(x, z; \lambda)$ is written as an infinite power series in $\lambda$. This solution is valid provided the infinite series converges.

A related, but more elegant, approach to the solution of integral equations using infinite series was found by Fredholm. We will not reproduce Fredholm's analysis here, but merely state the useful results. Essentially, *Fredholm theory* provides a formula for the resolvent kernel $R(x, z; \lambda)$ in (21.36) in terms of the ratio of two infinite series

$$R(x, z; \lambda) = \frac{D(x, z; \lambda)}{d(\lambda)}. \tag{21.41}$$

The numerator and denominator in (21.41) are given by

$$D(x, z; \lambda) = \sum_{n=0}^{\infty} \frac{(-1)^n}{n!} D_n(x, z)\lambda^n, \tag{21.42}$$

$$d(\lambda) = \sum_{n=0}^{\infty} \frac{(-1)^n}{n!} d_n\lambda^n, \tag{21.43}$$

where the functions $D_n(x, z)$ and the constants $d_n$ are found from recurrence relations as follows. We start with

$$D_0(x, z) = K(x, z) \qquad \text{and} \qquad d_0 = 1, \tag{21.44}$$

where $K(x, z)$ is the kernel of the original integral equation (21.34). The higher-order coefficients of $\lambda$ in (21.43) and (21.42) are then obtained from the two recurrence relations

$$d_n = \int_a^b D_{n-1}(x, x)\, dx, \tag{21.45}$$

$$D_n(x, z) = K(x, z)d_n - n\int_a^b K(x, z_1)D_{n-1}(z_1, z)\, dz_1. \tag{21.46}$$

Although the formulae for the resolvent kernel appear complicated, they are often simple to apply. Moreover, for the Fredholm solution the power series (21.42) and (21.43) are both guaranteed to converge for all values of $\lambda$, unlike Neumann series, which converge only if the condition (21.38) is satisfied. Thus the Fredholm method leads to a unique, non-singular solution, provided $d(\lambda) \neq 0$. In fact, as we might suspect, the solutions of $d(\lambda) = 0$ give the eigenvalues of the homogeneous equation corresponding to (21.34), i.e. with $f(x) \equiv 0$.

►*Use Fredholm theory to solve the integral equation (21.39).*

Using (21.36) and (21.41), the solution to (21.39) can be written in the form

$$y(x) = x + \lambda \int_0^1 R(x, z; \lambda) z \, dz = x + \lambda \int_0^1 \frac{D(x, z; \lambda)}{d(\lambda)} z \, dz. \qquad (21.47)$$

In order to find the form of the resolvent kernel $R(x, z; \lambda)$, we begin by setting

$$D_0(x, z) = K(x, z) = xz \qquad \text{and} \qquad d_0 = 1,$$

and use the recurrence relations (21.45) and (21.46) to obtain

$$d_1 = \int_0^1 D_0(x, x) \, dx = \int_0^1 x^2 \, dx = \frac{1}{3},$$

$$D_1(x, z) = \frac{xz}{3} - \int_0^1 xz_1^2 z \, dz_1 = \frac{xz}{3} - xz \left[ \frac{z_1^3}{3} \right]_0^1 = 0.$$

Applying the recurrence relations again we find that $d_n = 0$ and $D_n(x, z) = 0$ for $n > 1$. Thus, from (21.42) and (21.43), the numerator and denominator of the resolvent respectively are given by

$$D(x, z; \lambda) = xz \qquad \text{and} \qquad d(\lambda) = 1 - \frac{\lambda}{3}.$$

Substituting these expressions into (21.47), we find that the solution to (21.39) is given by

$$y(x) = x + \lambda \int_0^1 \frac{xz^2}{1 - \lambda/3} \, dz$$

$$= x + \lambda \left[ \frac{x}{1 - \lambda/3} \frac{z^3}{3} \right]_0^1 = x + \frac{\lambda x}{3 - \lambda} = \frac{3x}{3 - \lambda},$$

which, as expected, is the same as the solution (21.40) found by constructing a Neumann series. ◄

## 21.7 Schmidt–Hilbert theory

The Schmidt–Hilbert (SH) theory of integral equations may be considered as analogous to the Sturm–Liouville (SL) theory of differential equations, discussed in chapter 15, and is concerned with the properties of integral equations with *Hermitian* kernels. An Hermitian kernel enjoys the property

$$K(x,z) = K^*(z,x), \tag{21.48}$$

and it is clear that a special case of (21.48) is a real kernel which is symmetric with respect to its two arguments.

Let us begin by considering the homogeneous integral equation

$$y = \lambda \mathcal{K} y,$$

where the integral operator $\mathcal{K}$ has an Hermitian kernel. As discussed in section 21.3, this equation will, in general, have solutions only for particular values of $\lambda = \lambda_i$, the eigenvalues of the integral equation, the corresponding solutions $y_i$ being the eigenfunctions of the equation.

By following similar arguments to those presented in chapter 15 for SL theory, it may be shown that the eigenvalues $\lambda_i$ of an Hermitian kernel are real, and that the corresponding eigenfunctions $y_i$ belonging to different eigenvalues are orthogonal and form a complete set. If the eigenfunctions are suitable normalised, we have

$$\langle y_i | y_j \rangle = \int_a^b y_i^*(x) y_j(x)\, dx = \delta_{ij}. \tag{21.49}$$

If an eigenvalue is degenerate, the eigenfunctions corresponding to that eigenvalue can be made orthogonal by the Gram–Schmidt procedure, in a similar way to that discussed in chapter 15 in the context of SL theory.

Like SL theory, SH theory does not provide a method of obtaining the eigenvalues and eigenfunctions of any particular homogeneous integral equation with an Hermitian kernel; for this we have to turn to the methods discussed in the previous sections. Rather, SH theory is concerned with the general properties of the solutions to such equations. Where SH theory does become useful, however, is in the solution of inhomogeneous integral equations with Hermitian kernels, for which the eigenvalues and eigenfunctions of the corresponding homogeneous equation are already known.

Let us consider the inhomogeneous equation

$$y = f + \lambda \mathcal{K} y, \tag{21.50}$$

where $\mathcal{K} = \mathcal{K}^\dagger$ and for which we know the eigenvalues $\lambda_i$ and normalised eigenfunctions $y_i$ of the corresponding homogeneous problem. Since the eigenfunctions of the Hermitian operator form a complete set, we may expand our unknown function $y$ in terms of them as $y = \sum_i a_i y_i$, where the $a_i$ are the coefficients in the expansion. Substituting this into (21.50) we obtain

$$\sum_i a_i y_i = f + \lambda \mathcal{K} \sum_i a_i y_i = f + \lambda \sum_i \frac{a_i}{\lambda_i} y_i, \tag{21.51}$$

where we have used the fact that $y_i = \lambda_i \mathcal{K} y_i$. Forming the inner product of both sides of (21.51) with $y_j$, we find

$$\sum_i a_i \langle y_j | y_i \rangle = \langle y_j | f \rangle + \lambda \sum_i \frac{a_i}{\lambda_i} \langle y_j | y_i \rangle. \tag{21.52}$$

Since the normalised eigenfunctions of the Hermitian operator $\mathcal{K}$ are orthonormal, i.e. $\langle y_i | y_j \rangle = \delta_{ij}$, we have that the coefficients $a_j$ are given by

$$a_j = \frac{\langle y_j | f \rangle}{1 - \lambda/\lambda_j}. \tag{21.53}$$

Therefore the solution of the inhomogeneous equation (21.50) is found to be

$$y = \sum_j a_j y_j = \sum_j \frac{\langle y_j | f \rangle \; y_j}{1 - \lambda/\lambda_j}. \tag{21.54}$$

We see from (21.54) that the inhomogeneous equation (21.50) has a unique solution provided $\lambda \neq \lambda_i$, i.e. when $\lambda$ is not equal to one of the eigenvalues of the corresponding homogeneous equation. However, if $\lambda$ does equal one of the eigenvalues $\lambda_j$ then, in general, the coefficients $a_j$ become singular and no (finite) solution exists.

Returning to (21.53), however, we notice that even if $\lambda = \lambda_j$ a non-singular solution to the integral equation is still possible provided the function $f$ is orthogonal to every eigenfunction corresponding to the eigenvalue $\lambda_j$, i.e. provided that

$$\langle y_j | f \rangle = \int_a^b y_j^*(x) f(x)\, dx = 0.$$

►*Use Schmidt–Hilbert theory to solve the integral equation*

$$y(x) = \sin(x + \alpha) + \lambda \int_0^{\pi} \sin(x + z) y(z)\, dz. \tag{21.55}$$

It is clear that the kernel $K(x, z) = \sin(x + z)$ is real and symmetric in $x$ and $z$, and is thus Hermitian. In order to solve this inhomogeneous equation using SH theory, however, we must first find the eigenvalues and eigenfunctions of the corresponding homogeneous equation.

In fact, we have already considered the solution of the corresponding homogeneous equation (21.13) in subsection 21.4.1, where we found that it has two eigenvalues $\lambda_1 = 2/\pi$ and $\lambda_2 = -2/\pi$, with eigenfunctions given by (21.16). The normalised eigenfunctions are

$$y_1(x) = \frac{1}{\sqrt{\pi}}(\sin x + \cos x) \quad \text{and} \quad y_2(x) = \frac{1}{\sqrt{\pi}}(\sin x - \cos x), \tag{21.56}$$

and are easily shown to obey the orthonormality condition (21.49).

Using (21.54), the solution to the inhomogeneous equation (21.55) has the form

$$y(x) = a_1 y_1(x) + a_2 y_2(x), \tag{21.57}$$

where the coefficients $a_1$ and $a_2$ are given by (21.53) with $f(x) = \sin(x + \alpha)$. Therefore, using (21.56),

$$a_1 = \frac{1}{1 - \pi\lambda/2} \int_0^\pi \frac{1}{\sqrt{\pi}}(\sin z + \cos z)\sin(z + \alpha)\, dz \;=\; \frac{\sqrt{\pi}}{2 - \pi\lambda}(\cos\alpha + \sin\alpha),$$

$$a_2 = \frac{1}{1 + \pi\lambda/2} \int_0^\pi \frac{1}{\sqrt{\pi}}(\sin z - \cos z)\sin(z + \alpha)\, dz \;=\; \frac{\sqrt{\pi}}{2 + \pi\lambda}(\cos\alpha - \sin\alpha).$$

Substituting these expressions for $a_1$ and $a_2$ into (21.57) and simplifying, we find that the solution to (21.55) is given by

$$y(x) = \frac{1}{1 - (\pi\lambda/2)^2}\left[\sin(x + \alpha) + (\pi\lambda/2)\cos(x - \alpha)\right]. \blacktriangleleft$$

## 21.8 Exercises

21.1 Solve the integral equation

$$\int_0^\infty \cos(uv) y(v)\, dv = \exp(-u^2/2),$$

for the function $y = y(x)$ for $x > 0$. Note that for $x < 0$, $y(x)$ can be chosen as is most convenient.

21.2 Solve

$$\int_0^\infty f(t)\exp(-st)\, dt = \frac{a}{a^2 + s^2}.$$

21.3 Convert

$$f(x) = \exp x + \int_0^x (x - y) f(y)\, dy$$

into a differential equation, and hence show that its solution is

$$(\alpha + \beta x)\exp x + \gamma \exp(-x),$$

where $\alpha$, $\beta$, $\gamma$ are constants that should be determined.

21.4 (a) Consider the inhomogeneous integral equation

$$f(x) = g(x) + \lambda \int_a^b K(x,y) f(y)\, dy;$$

its kernel $K(x,y)$ is real, symmetric and continuous in $a \le x \le b$, $a \le y \le b$. If $\lambda$ is one of the eigenvalues $\lambda_i$ of the homogeneous equation

$$f_i(x) = \lambda_i \int_a^b K(x,y) f_i(y)\, dy,$$

prove that the inhomogeneous equation can only a have non-trivial solution if $g(x)$ is orthogonal to the corresponding eigenfunction $f_i(x)$.

(b) Show that the only values of $\lambda$ for which

$$f(x) = \lambda \int_0^1 xy(x+y) f(y)\, dy$$

has a non-trivial solution are the roots of the equation

$$\lambda^2 + 120\lambda - 240 = 0.$$

(c) Solve

$$f(x) = \mu x^2 + \int_0^1 2xy(x+y) f(y)\, dy.$$

21.5 (a) If the kernel of the integral equation

$$\psi(x) = \lambda \int_a^b K(x,y) \psi(y)\, dy$$

has the form

$$K(x,y) = \sum_{n=0}^{\infty} h_n(x) g_n(y),$$

where the $h_n(x)$ form a complete orthonormal set of functions over the interval $[a,b]$, show that the eigenvalues $\lambda_i$ are given by

$$|\mathsf{M} - \lambda^{-1}\mathsf{I}| = 0,$$

where $\mathsf{M}$ is the matrix with elements

$$M_{kj} = \int_a^b g_k(u) h_j(u)\, du.$$

If the corresponding solutions are $\psi^{(i)}(x) = \sum_{n=0}^{\infty} a_n^{(i)} h_n(x)$, find an expression for $a_n^{(i)}$.

(b) Obtain the eigenvalues and eigenfunctions over the interval $[0, 2\pi]$ if

$$K(x, y) = \sum_{n=1}^{\infty} \frac{1}{n} \cos nx \cos ny.$$

21.6 By considering functions of the form $h(x) = \int_0^x (x - y) f(y)\, dy$, show that the solution $f(x)$ of the integral equation

$$f(x) = x + \frac{1}{2} \int_0^1 |x - y| f(y)\, dy$$

satisfies the equation $f''(x) = f(x)$.

By examining the special cases $x = 0$ and $x = 1$, show that

$$f(x) = \frac{2}{(e + 3)(e + 1)} [(e + 2) \exp x - e \exp(-x)].$$

21.7 For the integral equation

$$y(x) = x^{-3} + \lambda \int_a^b x^2 z^2 y(z)\, dz,$$

show that the resolvent kernel is $5x^2z^2/[5 - \lambda(b^5 - a^5)]$ and hence solve the equation. For what range of $\lambda$ is the solution valid?

21.8 Use Fredholm theory to show that, for the kernel

$$K(x, z) = (x + z) \exp(x - z)$$

over the interval $[0, 1]$, the resolvent kernel is

$$R(x, z; \lambda) = \frac{\exp(x - z)[(x + z) - \lambda(\frac{1}{2}x + \frac{1}{2}z - xz - \frac{1}{3})]}{1 - \lambda - \frac{1}{12}\lambda^2},$$

and hence solve

$$y(x) = x^2 + 2 \int_0^1 (x + z) \exp(x - z)\, y(z)\, dz,$$

expressing your answer in terms of the quantities $I_n$, where $I_n = \int_0^1 u^n \exp(-u)\, du$.

## 21.9 Hints and answers

21.1 Define $y(-x) = y(x)$ and use the cosine Fourier transform inversion theorem; $y(x) = (2/\pi)^{1/2} \exp(-x^2/2)$.

21.2 Use Laplace transform; $f(t) = \sin at$.

21.3 $f''(x) - f(x) = \exp x$; $\alpha = 3/4$, $\beta = 1/2$, $\gamma = 1/4$.

21.4 (b) Set $f(x) = a_1 x^2 + a_2 x$ and obtain $a_1 = (\lambda/4)a_1 + (\lambda/3)a_2$, $a_2 = (\lambda/5)a_1 + (\lambda/4)a_2$; (c) set $f(x) = (\mu + a_1)x^2 + a_2 x$; $f(x) = -6\mu x(5x + 4)$.

21.5 (a) $a_n^{(i)} = \int_a^b h_n(x)\psi(x)\,dx$; (b) use $(1/\sqrt{\pi})\cos nx$ and $(1/\sqrt{\pi})\sin nx$; M is diagonal; eigenvalues $\lambda_k = k/\pi$ with $\psi^{(k)}(x) = (1/\sqrt{\pi})\cos kx$.

21.6 Write $\int_0^1 |x-y|f(y)\,dy$ as $\int_0^x (x-y)f(y)\,dy + \int_x^1 (y-x)f(y)\,dy$.

21.7 $y(x) = x^{-3} + [5x^2\lambda\ln(b/a)]/[5-\lambda(b^5-a^5)]$; $|\lambda| < 5/|b^5-a^5|$.

21.8 $y(x) = x^2 - (3I_3x + I_2)\exp x$.

# 22

# *Group theory*

For systems that have some degree of symmetry, full exploitation of that symmetry is desirable. Significant physical results can sometimes be deduced simply by a study of the symmetry properties of the system under investigation. Consequently it becomes important, for such a system, to identify all those operations (rotations, reflections, inversions) that carry the system into a physically indistiguishable copy of itself.

The study of the properties of the complete set of such operations forms one application of *group theory*. Though this is the aspect of most interest to the physical scientist, group theory itself is a much larger subject and of great importance in its own right. Consequently we leave until the next chapter any direct applications of group theoretical results and concentrate on building up the general mathematical properties of groups.

## 22.1 Groups

As an example of symmetry properties, let us consider the sets of operations, such as rotations, reflections, and inversions, which transform physical objects, for example molecules, into physically indistinguishable copies of themselves, so that only the labelling of identical components of the system (the atoms) changes in the process. For differently shaped molecules there are different sets of operations, but in each case it is a well-defined set, and with a little practice all members of each set can be identified.

As simple examples, consider (*a*) the hydrogen molecule, and (*b*) the ammonia molecule illustrated in figure 22.1. The hydrogen molecule consists of two atoms H of hydrogen, and is carried into itself by any of the following operations:

(i) any rotation about its long axis;
(ii) rotation through $\pi$ about an axis perpendicular to the long axis and passing through the point $M$ that lies midway between the atoms;

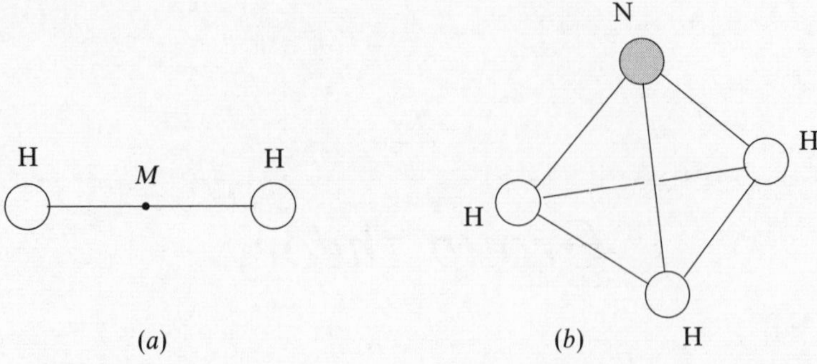


Figure 22.1 (*a*) The hydrogen molecule, and (*b*) the ammonia molecule.

(iii) inversion through the point $M$;

(iv) reflection in the plane that passes through $M$ and has its normal parallel to the long axis.

These operations collectively form the set of symmetry operations for the hydrogen molecule.

The somewhat more complex ammonia molecule consists of a tetrahedron with an equilateral triangular base at the three corners of which lie hydrogen atoms H, whilst a nitrogen atom N is sited at the fourth vertex of the tetrahedron. The set of symmetry operations on this molecule is limited to rotations of $\pi/3$ and $2\pi/3$ about the axis joining the centroid of the equilateral triangle to the nitrogen atom, and reflections in the three planes containing that axis and each of the hydrogen atoms in turn. However, if the nitrogen atom could be replaced by a fourth hydrogen atom, and all interatomic distances equalised in the process, the number of symmetry operations would be greatly increased.

Once *all* the possible operations in any particular set have been identified, it must follow that the result of applying two such operations in succession will be identical to that obtained by the sole application of some third (usually different) operation in the set – for if it were not, a new member of the set would have been found, contradicting the assumption that all members have been identified.

Such observations introduce two of the main considerations relevant to deciding whether a set of objects (here the rotation, reflection and inversion operations) qualifies as a *group* in the mathematically tightly defined sense. These two considerations are (i) whether there is some law for combining two members of the set, and (ii) whether the result of the combination is also a member of the set. The obvious rule of combination has to be that the second operation is carried out on the system that results from application of the first operation, and we have already seen that the second requirement is satisfied by the inclusion of all such operations in the set. However, for a set to qualify as a group, more than these two conditions have to be satisfied, as will now be made clear.

### 22.1.1 Definition of a group

A group $\mathcal{G}$ is a set of elements $\{X, Y, \dots\}$, together with a rule for combining them that associates with each ordered pair $X$, $Y$ a 'product' or combination law $X \bullet Y$ for which the following conditions must be satisfied.

(i) For *every* pair of elements $X$, $Y$ that belongs to $\mathcal{G}$, the product $X \bullet Y$ also belongs to $\mathcal{G}$. (This is known as the *closure property* of the group.)

(ii) For all triples $X$, $Y$, $Z$ the *associative law* holds; in symbols,

$$X \bullet (Y \bullet Z) = (X \bullet Y) \bullet Z. \qquad (22.1)$$

(iii) There exists a unique element $I$, belonging to $\mathcal{G}$, with the property that

$$I \bullet X = X = X \bullet I \qquad (22.2)$$

for *all* $X$ belonging to $\mathcal{G}$. This element $I$ is known as the *identity element* of the group.

(iv) For every element $X$ of $\mathcal{G}$, there exists an element $X^{-1}$, also belonging to $\mathcal{G}$, such that

$$X^{-1} \bullet X = I = X \bullet X^{-1}. \qquad (22.3)$$

$X^{-1}$ is called the *inverse* of $X$.

An alternative notation in common use is to write the elements of a group $\mathcal{G}$ as the set $\{G_1, G_2, \dots\}$ or, more briefly, as $\{G_i\}$, a typical element being denoted by $G_i$.

It should be noticed that, as given, the nature of the operation $\bullet$ is not stated. It should also be noticed that the more general term *element*, rather than *operation*, has been used in this definition. We will see that the general definition of a group allows as elements not only sets of operations on an object but also sets of numbers, of functions and of other objects, provided that the interpretation of $\bullet$ is appropriately defined.

In one of the simplest examples of a group, namely the group of all integers under addition, the operation $\bullet$ is taken to be ordinary addition. In this group the role of the identity $I$ is played by the integer 0, and the inverse of an integer $X$ is $-X$. That requirements (i) and (ii) are satisfied by the integers under addition is trivially obvious. A second simple group, under ordinary multiplication, is formed by the two numbers 1 and $-1$; in this group, closure is obvious, 1 is the identity element, and each element is its own inverse.

It will be apparent from these two examples that the number of elements in a group can be either finite or infinite. In the former case the group is called a *finite group* and the number of elements it contains is called the *order* of the group, which we will denote by $g$; an alternative notation is $|\mathcal{G}|$ but has obvious dangers if matrices are involved. In the notation in which $\mathcal{G} = \{G_1, G_2, \ldots, G_n\}$ the order of the group is clearly $n$.

As we have noted, for the integers under addition zero is the identity. For the group of rotations and reflections, the operation of doing nothing, i.e. the null operation, plays this role. This latter identification may seem artificial, but it is an operation, albeit trivial, which does leave the system in a physically indistinguishable state, and needs to be included. One might add that without it the set of operations would not form a group and none of the powerful results we will derive later in this and the next chapter could be justifiably applied to give deductions of physical significance.

In the examples of rotations and reflections mentioned earlier, $\bullet$ has been taken to mean that the left-hand operation is carried out on the system that results from application of the right-hand operation. Thus

$$Z = X \bullet Y \tag{22.4}$$

means that the effect on the system of carrying out $Z$ is the same as would be obtained by first carrying out $Y$ and then carrying out $X$. The order of the operations should be noted; it is arbitrary in the first instance, but, once chosen, must be adhered to. The choice we have made is dictated by the fact that most of our applications involve the effect of rotations and reflections on functions of space coordinates, and it is usual, and our practice in the rest of this book, to write operators acting on functions to the left of the functions.

It will be apparent that for the above-mentioned group, integers under ordinary addition, it is true that

$$Y \bullet X = X \bullet Y \tag{22.5}$$

for all pairs of integers $X$, $Y$. If any two particular elements of a group satisfy (22.5), they are said to *commute* under the operation $\bullet$; if all pairs of elements in a group satisfy (22.5), then the group is said to be *Abelian*. The set of all integers forms an infinite Abelian group under (ordinary) addition.

As we show below, requirements (iii) and (iv) of the definition of a group are over-demanding (but self-consistent), since in each of equations (22.2) and (22.3) the second equality can be deduced from the first by using the associativity required by (22.1). The mathematical steps in the following arguments are all very simple, but care has to be taken to make sure that nothing that has not yet been proved is used to justify a step. For this reason, and to act as a model in logical deduction, a reference in Roman numerals to the previous result, or to the group definition used, is given over each equality sign. Such explicit detailed referencing soon becomes tiresome, but it should always be available if needed.

►*Using only the first equalities in (22.2) and (22.3), deduce the second ones.*

Consider the expression $X^{-1} \bullet (X \bullet X^{-1})$;

$$X^{-1} \bullet (X \bullet X^{-1}) \overset{\text{(ii)}}{=} (X^{-1} \bullet X) \bullet X^{-1} \overset{\text{(iv)}}{=} I \bullet X^{-1}$$

$$\overset{\text{(iii)}}{=} X^{-1}. \qquad (22.6)$$

But $X^{-1}$ belongs to $\mathcal{G}$, and so from (iv) there is an element $U$ in $\mathcal{G}$ such that

$$U \bullet X^{-1} = I. \qquad \text{(v)}$$

Form the product of $U$ with the two extremes of (22.6) to give

$$U \bullet (X^{-1} \bullet (X \bullet X^{-1})) = U \bullet X^{-1} \overset{\text{(v)}}{=} I. \qquad (22.7)$$

Transforming the left-hand side of this equation gives

$$U \bullet (X^{-1} \bullet (X \bullet X^{-1})) \overset{\text{(ii)}}{=} (U \bullet X^{-1}) \bullet (X \bullet X^{-1})$$

$$\overset{\text{(v)}}{=} I \bullet (X \bullet X^{-1})$$

$$\overset{\text{(iii)}}{=} X \bullet X^{-1}. \qquad (22.8)$$

Comparing (22.7), (22.8) shows that

$$X \bullet X^{-1} = I, \qquad \text{(iv)}'$$

i.e. the second equality in group definition (iv). Similarly

$$X \bullet I \overset{\text{(iv)}}{=} X \bullet (X^{-1} \bullet X) \overset{\text{(ii)}}{=} (X \bullet X^{-1}) \bullet X$$

$$\overset{\text{(iv)}'}{=} I \bullet X$$

$$\overset{\text{(iii)}}{=} X. \qquad \text{(iii}')$$

i.e. the second equality in group definition (iii). ◄

The uniqueness of the identity element $I$ can also be demonstrated rather than assumed. Suppose that $I'$, belonging to $\mathcal{G}$, also has the property

$$I' \bullet X = X = X \bullet I' \qquad \text{for all } X \text{ belonging to } \mathcal{G}.$$

Take $X$ as $I$, then

$$I' \bullet I = I. \qquad (22.9)$$

Further, from (iii′),

$$X = X \bullet I \qquad \text{for all } X \text{ belonging to } \mathcal{G},$$

and setting $X = I'$ gives

$$I' = I' \bullet I. \tag{22.10}$$

It then follows from (22.9), (22.10) that $I = I'$, showing that in any particular group the identity element is unique.

In a similar way it can be shown that the inverse of any particular element is unique. If $U$ and $V$ are two postulated inverses of an element $X$ of $\mathcal{G}$, by considering the product

$$U \bullet (X \bullet V) = (U \bullet X) \bullet V,$$

it can be shown that $U = V$. The proof is left to the reader.

Given the uniqueness of the inverse of any particular group element, it follows that

$$\begin{aligned}(U \bullet V \bullet \cdots \bullet Y \bullet Z) &\bullet (Z^{-1} \bullet Y^{-1} \bullet \cdots \bullet V^{-1} \bullet U^{-1}) \\ &= (U \bullet V \bullet \cdots \bullet Y) \bullet (Z \bullet Z^{-1}) \bullet (Y^{-1} \bullet \cdots \bullet V^{-1} \bullet U^{-1}) \\ &= (U \bullet V \bullet \cdots \bullet Y) \bullet (Y^{-1} \bullet \cdots \bullet V^{-1} \bullet U^{-1}) \\ &\;\;\vdots \\ &= I,\end{aligned}$$

where use has been made of the associativity and of the equations $Z \bullet Z^{-1} = I$ and $I \bullet X = X$. Thus the inverse of a product is the product of the inverses in reverse order, i.e.

$$(U \bullet V \bullet \cdots \bullet Y \bullet Z)^{-1} = (Z^{-1} \bullet Y^{-1} \bullet \cdots \bullet V^{-1} \bullet U^{-1}). \tag{22.11}$$

Further elementary results that can be obtained by arguments similar to those above are as follows.

(i) Given any pair of elements $X, Y$ belonging to $\mathcal{G}$, there exist unique elements $U, V$, also belonging to $\mathcal{G}$, such that

$$X \bullet U = Y \qquad \text{and} \qquad V \bullet X = Y.$$

Clearly $U = X^{-1} \bullet Y$, and $V = Y \bullet X^{-1}$, and they can be shown to be unique. This result is sometimes called the *division axiom.*

(ii) The *cancellation law* can be stated as follows. If

$$X \bullet Y = X \bullet Z$$

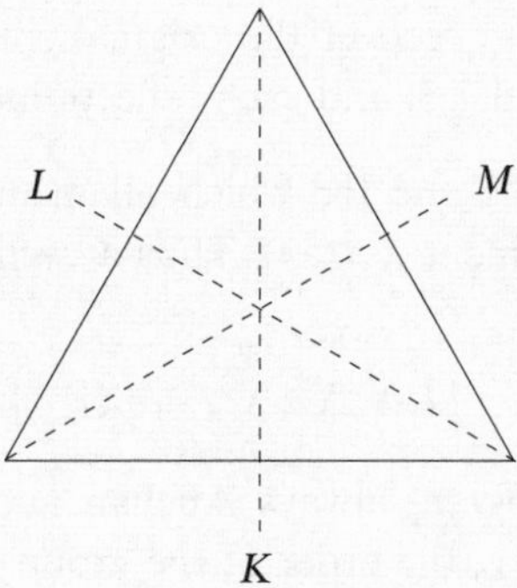


Figure 22.2 Reflections in the three perpendicular bisectors of the sides of an equilateral triangle take the triangle into itself.

for some $X$ belonging to $\mathcal{G}$, then $Y = Z$. Similarly,

$$Y \bullet X = Z \bullet X$$

implies the same conclusion.

(iii) Forming the product of each element of $\mathcal{G}$ with a fixed element $X$ of $\mathcal{G}$ simply permutes the elements of $\mathcal{G}$; this is often written symbolically as $\mathcal{G} \bullet X = \mathcal{G}$. If this were not so, and $X \bullet Y$ and $X \bullet Z$ were not different even though $Y$ and $Z$ were, application of the cancellation law would lead to a contradiction. This result is called the *permutation law*.

In any finite group of order $g$, any element $X$ when combined with itself to form successively $X^2 = X \bullet X$, $X^3 = X \bullet X^2$, ... will, after at most $g - 1$ such combinations, produce the group identity $I$. Of course $X^2$, $X^3$, ... are some of the original elements of the group, and not new ones. If the actual number of combinations needed is $m-1$, i.e. $X^m = I$, then $m$ is called the *order of the element* $X$ in $\mathcal{G}$. The order of the identity of a group is always 1, and that of any other element of a group that is its own inverse is always 2.

►*Determine the order of the group of (two-dimensional) rotations and reflections that take a plane equilateral triangle into itself, and the order of each of the elements. The group is usually known as 3m (to physicists and crystallographers) or $C_{3v}$ (to chemists).*

There are two (clockwise) rotations, by $2\pi/3$ and $4\pi/3$, about an axis perpendicular to the plane of the triangle. In addition, reflections in the perpendicular bisectors of the three sides (see figure 22.2) have the defining property. To these must be added the identity operation. Thus in total there are six distinct operations and $g = 6$ for this group. To reproduce the identity operation either of the rotations has to be applied three times, whilst any of the reflections has to be applied just twice in order to recover the original situation. Thus each rotation element of the group has order 3, and each reflection element has order 2. ◄

A so-called *cyclic group* is one for which all members of the group can be generated from just one element $X$ (say). Thus a cyclic group of order $g$ can be written as

$$\mathcal{G} = \{I, X, X^2, X^3, \ldots, X^{g-1}\}.$$

It is clear that cyclic groups are always Abelian and that each element, apart from the identity, has order $g$, the order of the group itself.

### 22.1.2 Further examples of groups

In this section we consider some sets of objects, each set together with a law of combination, and investigate whether they qualify as groups and, if not, why not.

We have already seen that the integers form a group under ordinary addition, but it is immediately apparent that (even if zero is excluded) they do *not* do so under ordinary multiplication. Unity must be the identity of the set, but the requisite inverse of any integer $n$, namely $1/n$, does not belong to the set of integers for any $n$ other than unity.

Other infinite sets of quantities that do form groups are the sets of all real numbers, or of all complex numbers, under addition, and of the same two sets excluding 0 under multiplication. All of these groups are Abelian.

Although subtraction and division are normally considered the obvious counterparts of the operations of (ordinary) addition and multiplication, they are not acceptable operations for use within groups since the associative law, (22.1), does not hold. Explicitly,

$$X - (Y - Z) \neq (X - Y) - Z,$$
$$X \div (Y \div Z) \neq (X \div Y) \div Z.$$

From within the field of all non-zero complex numbers we can select just those that have unit modulus, i.e. are of the form $e^{i\theta}$ where $0 \leq \theta < 2\pi$, to form a group under multiplication, as can easily be verified:

$$\begin{aligned} e^{i\theta_1} \times e^{i\theta_2} &= e^{i(\theta_1+\theta_2)} && \text{(closure)},\\ e^{i0} &= 1 && \text{(identity)},\\ e^{i(2\pi-\theta)} \times e^{i\theta} &= e^{i2\pi} \equiv e^{i0} = 1 && \text{(inverse)}. \end{aligned}$$

Closely related to the above group is the set of $2 \times 2$ rotation matrices that take the form

$$\mathsf{M}(\theta) = \begin{pmatrix} \cos\theta & -\sin\theta \\ \sin\theta & \cos\theta \end{pmatrix}$$

where, as before, $0 \leq \theta < 2\pi$. These form a group when the law of combination is that of matrix multiplication. The reader can easily verify that

$$\begin{aligned} \mathsf{M}(\theta)\mathsf{M}(\phi) &= \mathsf{M}(\theta+\phi) && \text{(closure)},\\ \mathsf{M}(0) &= \mathsf{I}_2 && \text{(identity)},\\ \mathsf{M}(2\pi-\theta) &= \mathsf{M}^{-1}(\theta) && \text{(inverse)}. \end{aligned}$$

Here $\mathsf{I}_2$ is the unit $2 \times 2$ matrix.

## 22.2 Finite groups

Whilst many properties of physical systems (e.g. angular momentum) are related to the properties of infinite, and, in particular, continuous groups, the symmetry properties of crystals and molecules are more intimately connected with those of finite groups. We therefore concentrate in this section on finite sets of objects that can be combined in a way satisfying the group postulates.

Although it is clear that the set of all integers does not form a group under ordinary multiplication, restricted sets can do so if the operation involved is multiplication (mod $N$) for suitable values of $N$; this operation will be explained below.

As a simple example of a group with only four members, consider the set $S$ defined as follows:

$$S = \{1, 3, 5, 7\} \qquad \text{under multiplication (mod 8).}$$

To find the product (mod 8) of any two elements, we multiply them together in the ordinary way, and then divide the answer by 8, treating the remainder after doing so as the product of the two elements. For example, $5 \times 7 = 35$, which on dividing by 8 gives a remainder of 3. Clearly, since $Y \times Z = Z \times Y$, the full set of different products is

$$\begin{array}{llll} 1\times 1 = 1, & 1\times 3 = 3, & 1\times 5 = 5, & 1\times 7 = 7,\\ 3\times 3 = 1, & 3\times 5 = 7, & 3\times 7 = 5, & \\ 5\times 5 = 1, & 5\times 7 = 3, & & \\ 7\times 7 = 1. & & & \end{array}$$

The first thing to notice is that each multiplication produces a member of the original set, i.e. the set is closed. Obviously the element 1 takes the role of the identity, i.e. $1 \times Y = Y$ for all members $Y$ of the set. Further for each element $Y$ of the set there is an element $Z$ (equal to $Y$, as it happens in this case) such that $Y \times Z = 1$, i.e. each element has an inverse. These observations, together with the associativity of multiplication (mod 8), show that the set $S$ is an Abelian group of order 4.

It is convenient to present the results of combining any two elements of a group in the form of multiplication tables – akin to those which used to appear in elementary arithmetic books before electronic calculators were invented! Written

| | 1 | 3 | 5 | 7 |
|---|---|---|---|---|
| 1 | 1 | 3 | 5 | 7 |
| 3 | 3 | 1 | **7** | 5 |
| 5 | 5 | 7 | 1 | 3 |
| 7 | 7 | 5 | 3 | 1 |

Table 22.1 The table of products for the elements of the group $\mathcal{S} = \{1, 3, 5, 7\}$ under multiplication (mod 8).

in this much more compact form the above example is expressed by table 22.1. Although the order of the two elements being combined does not matter here because the group is Abelian, we adopt the convention that if the product in a general multiplication table is written $X \bullet Y$ then $X$ is taken from the left-hand column and $Y$ is taken from the top row. Thus the bold '**7**' in the table is the result of $3 \times 5$, rather than of $5 \times 3$.

Whilst it would make no difference to the basic information content in a table to present the rows and columns with their headings in random orders, it is usual to list the elements in the same order in both the vertical and horizontal headings in any one table. The actual order of the elements in the common list, whilst arbitrary, is normally chosen to make the table have as much symmetry as possible. This is initially a matter of convenience, but, as we shall see later, some of the more subtle properties of groups are revealed by putting next to each other elements of the group that are alike in certain ways.

Some simple general properties of group multiplication tables can be deduced immediately from the fact that each row or column constitutes the elements of the group.

(i) Each element appears once and only once in each row or column of the table; this must be so since $\mathcal{G} \bullet X = \mathcal{G}$ (i.e. the permutation law).

(ii) The inverse of any element $Y$ can be found by looking along the row in which $Y$ appears in the left-hand column (the $Y$th row), and noting the element $Z$ at the head of the column (the $Z$th column) in which the identity appears as the table entry. An immediate corollary is that whenever the identity appears on the leading diagonal, it indicates that the corresponding header element is of order 2 (unless it happens to be the identity itself).

(iii) For any Abelian group the multiplication table is symmetric about the leading diagonal.

To get used to the ideas involved in using group multiplication tables, we now consider two more sets of integers under multiplication (mod $N$):

$$\mathcal{S}' = \{1, 5, 7, 11\} \quad \text{under multiplication (mod 24), and}$$
$$\mathcal{S}'' = \{1, 2, 3, 4\} \quad \text{under multiplication (mod 5).}$$

(a)

| | 1 | 5 | 7 | 11 |
|---|---|---|---|---|
| 1 | 1 | 5 | 7 | 11 |
| 5 | 5 | 1 | 11 | 7 |
| 7 | 7 | 11 | 1 | 5 |
| 11 | 11 | 7 | 5 | 1 |

(b)

| | 1 | 2 | 3 | 4 |
|---|---|---|---|---|
| 1 | 1 | 2 | 3 | 4 |
| 2 | 2 | 4 | 1 | 3 |
| 3 | 3 | 1 | 4 | 2 |
| 4 | 4 | 3 | 2 | 1 |

Table 22.2 On the left, table 22.2(*a*), the multiplication table for the group $\mathcal{S}' = \{1, 5, 7, 11\}$ under multiplication (mod 24). On the right, table 22.2(*b*), the multiplication table for the group $\mathcal{S}'' = \{1, 2, 3, 4\}$ under multiplication (mod 5).

| | $I$ | $A$ | $B$ | $C$ |
|---|---|---|---|---|
| $I$ | $I$ | $A$ | $B$ | $C$ |
| $A$ | $A$ | $I$ | $C$ | $B$ |
| $B$ | $B$ | $C$ | $I$ | $A$ |
| $C$ | $C$ | $B$ | $A$ | $I$ |

Table 22.3 The common structure exemplified by tables 22.1 and 22.2(*a*).

These have group multiplication tables 22.2(*a*) and (*b*) respectively, as the reader should verify.

If tables 22.1 and 22.2(*a*) for the groups $\mathcal{S}$ and $\mathcal{S}'$ are compared, it will be seen that they have essentially the same structure, i.e if the elements are written as $\{I, A, B, C\}$ in both cases, then the two tables are each equivalent to table 22.3.

For $\mathcal{S}$, $I = 1$, $A = 3$, $B = 5$, $C = 7$ and the law of combination is multiplication (mod 8), whilst for $\mathcal{S}'$, $I = 1$, $A = 5$, $B = 7$, $C = 11$ and the law of combination is multiplication (mod 24). However, the really important point is that the two groups $\mathcal{S}$ and $\mathcal{S}'$ have equivalent group multiplication tables – they are said to be *isomorphic*, a matter to which we will return more formally in section 22.5.

►*Determine the behaviour of the set of four elements*

$$\{1, i, -1, -i\}$$

*under the ordinary multiplication of complex numbers. Show that they form a group and determine whether the group is isomorphic to either of the groups* $\mathcal{S}$ *(itself isomorphic to* $\mathcal{S}'$*) and* $\mathcal{S}''$ *defined above.*

That the elements form a group under the associative operation of complex multiplication is immediate; there is an identity (1), each possible product generates a member of the set, and each element has an inverse (1, $-i$, $-1$, $i$, respectively). The group table has the form shown in table 22.4.

We now ask whether this table can be made to look like table 22.3, which

| | $1$ | $i$ | $-1$ | $-i$ |
|---|---|---|---|---|
| $1$ | $1$ | $i$ | $-1$ | $-i$ |
| $i$ | $i$ | $-1$ | $-i$ | $1$ |
| $-1$ | $-1$ | $-i$ | $1$ | $i$ |
| $-i$ | $-i$ | $1$ | $i$ | $-1$ |

Table 22.4 The group table for the set $\{1, i, -1, -i\}$ under ordinary multiplication of complex numbers.

| | $1$ | $i$ | $-1$ | $-i$ |
|---|---|---|---|---|
| $1$ | $1$ | $i$ | $-1$ | $-i$ |
| $i$ | $i$ | $-1$ | $-i$ | $1$ |
| $-1$ | $-1$ | $-i$ | $1$ | $i$ |
| $-i$ | $-i$ | $1$ | $i$ | $-1$ |

| | 1 | 2 | 4 | 3 |
|---|---|---|---|---|
| 1 | 1 | 2 | 4 | 3 |
| 2 | 2 | 4 | 3 | 1 |
| 4 | 4 | 3 | 1 | 2 |
| 3 | 3 | 1 | 2 | 4 |

Table 22.5 A comparison between tables 22.4 and 22.2($b$), the latter with its columns reordered.

is the standardised form of the tables for $\mathcal{S}$ and $\mathcal{S}'$. Since the identity element of the group (1) will have to be represented by $I$, and '1' only appears on the leading diagonal twice whereas $I$ appears on the leading diagonal four times in table 22.3, it is clear that no amount of relabelling (or, equivalently, no allocation of the symbols $A$, $B$, $C$, amongst $i$, $-1$, $-i$) can bring table 22.4 into the form of table 22.3. We conclude that the group $\{1, i, -1, -i\}$ is not isomorphic to $\mathcal{S}$ or $\mathcal{S}'$. An alternative way of stating the observation is to say that the group contains only one element of order 2 whilst a group corresponding to table 22.3 contains three such elements.

However, if the rows and columns of table 22.2($b$) – in which the identity does appear twice on the diagonal and which therefore has the potential to be equivalent to table 22.4 – are rearranged by making the heading order 1, 2, 4, 3, then the two tables can be compared in the forms shown in table 22.5.

They can thus be seen to have the same structure, namely that shown in table 22.6.

We therefore conclude that the group of four elements $\{1, i, -1, -i\}$ under ordinary multiplication of complex numbers is isomorphic to the group $\{1, 2, 3, 4\}$ under multiplication (mod 5). ◄

What we have done does not prove it, but the two tables 22.3 and 22.6 are in fact the only possible tables for a group of order 4, i.e. a group containing exactly four elements.

| | $I$ | $A$ | $B$ | $C$ |
|---|---|---|---|---|
| $I$ | $I$ | $A$ | $B$ | $C$ |
| $A$ | $A$ | $B$ | $C$ | $I$ |
| $B$ | $B$ | $C$ | $I$ | $A$ |
| $C$ | $C$ | $I$ | $A$ | $B$ |

Table 22.6 The common structure exemplified by tables 22.4 and 22.2($b$), the latter with its columns reordered.

## 22.3 Non-Abelian groups

So far, all the groups for which we have constructed multiplication tables have been based on some form of arithmetic multiplication, a commutative operation, with the result that the groups have been Abelian and the tables symmetric about the leading diagonal. We now turn to examples of groups in which some non-commutation occurs. It should be noted, in passing, that non-commutation *cannot* occur *throughout* a group, as the identity always commutes with any element in its group.

As the first example we consider again as elements of a group the two-dimensional operations which transform an equilateral triangle into itself (see the end of subsection 22.1.1). It has already been shown that there are six such operations; the null operation, two rotations (by $2\pi/3$ and $4\pi/3$ about an axis perpendicular to the plane of the triangle) and three reflections in the perpendicular bisectors of the three sides. To abbreviate we will denote these operations by symbols as follows.

(i) $I$ is the null operation.

(ii) $R$ is a (clockwise) rotation by $2\pi/3$, and $R'$ that by $4\pi/3$.

(iii) $K$, $L$, $M$ are reflections in the three lines indicated in figure 22.2.

Some products of the operations of the form $X \bullet Y$ (where it will be recalled that the symbol $\bullet$ means that the second operation $X$ is carried out on the system resulting from the application of the first operation $Y$) are easily calculated:

$$R \bullet R = R', \qquad R' \bullet R' = R, \qquad R \bullet R' = I = R' \bullet R$$
$$K \bullet K = L \bullet L = M \bullet M = I. \tag{22.12}$$

Others, such as $K \bullet M$, are more difficult, but can be found by a little thought, or by making a model triangle or by drawing a sequence of diagrams such as those following.

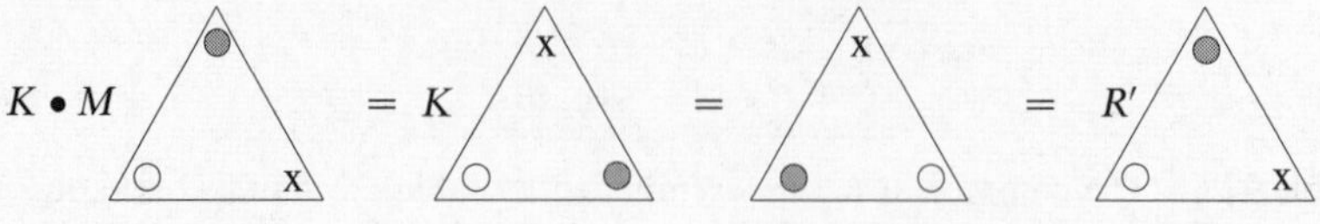

$K \bullet M$ $= K$ $=$ $= R'$

showing that $K \bullet M = R'$. In the same way,

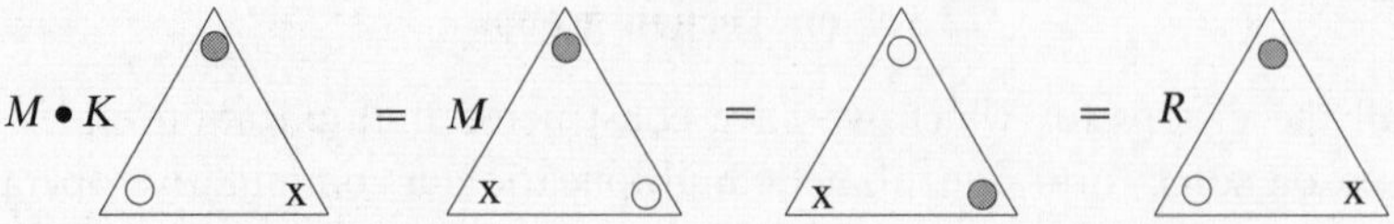

$M \bullet K$ $= M$ $=$ $= R$

shows that $M \bullet K = R$, and

$R \bullet L$ $= R$ $=$ $= K$

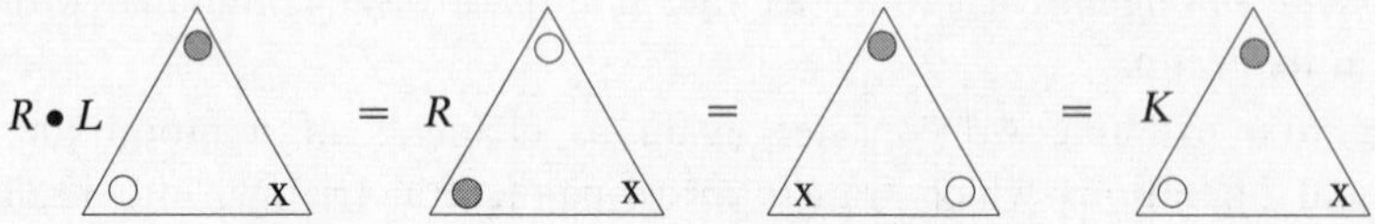

shows that $R \bullet L = K$.

Proceeding in this way we can build up the complete multiplication table (table 22.7). In fact, it is not necessary to draw any more diagrams, as all remaining products can be deduced algebraically from the three found above and the more obvious results (22.12). A number of things may be noticed about this table.

(i) It is *not* symmetric about the leading diagonal, indicating that some pairs of elements in the group do not commute.

(ii) There is some symmetry within the $3 \times 3$ blocks that form the four quarters of the table. This occurs because we have elected to put similar operations close to each other when choosing the order of table headings – the two rotations (or three if $I$ is viewed as a rotation by $0\pi/3$) are next to each other, and the three reflections also occupy adjacent columns and rows. We will return to this later.

That two groups of the same order may be isomorphic carries over to non-Abelian groups. The next two examples are each concerned with sets of six objects; they will be shown to form groups that, although very different in nature from the rotation–reflection group just considered, are isomorphic to it.

| | $I$ | $R$ | $R'$ | $K$ | $L$ | $M$ |
|---|---|---|---|---|---|---|
| $I$ | $I$ | $R$ | $R'$ | $K$ | $L$ | $M$ |
| $R$ | $R$ | $R'$ | $I$ | $M$ | $K$ | $L$ |
| $R'$ | $R'$ | $I$ | $R$ | $L$ | $M$ | $K$ |
| $K$ | $K$ | $L$ | $M$ | $I$ | $R$ | $R'$ |
| $L$ | $L$ | $M$ | $K$ | $R'$ | $I$ | $R$ |
| $M$ | $M$ | $K$ | $L$ | $R$ | $R'$ | $I$ |

Table 22.7 The group table for the two-dimensional symmetry operations on an equilateral triangle.

| | $I$ | $A$ | $B$ | $C$ | $D$ | $E$ |
|---|---|---|---|---|---|---|
| $I$ | $I$ | $A$ | $B$ | $C$ | $D$ | $E$ |
| $A$ | $A$ | $B$ | $I$ | $E$ | $C$ | $D$ |
| $B$ | $B$ | $I$ | $A$ | $D$ | $E$ | $C$ |
| $C$ | $C$ | $D$ | $E$ | $I$ | $A$ | $B$ |
| $D$ | $D$ | $E$ | $C$ | $B$ | $I$ | $A$ |
| $E$ | $E$ | $C$ | $D$ | $A$ | $B$ | $I$ |

Table 22.8 The group table, under matrix multiplication, for the set $\mathcal{M}$ of six orthogonal $2 \times 2$ matrices given by (22.13).

We consider first the set $\mathcal{M}$ of six orthogonal $2 \times 2$ matrices given by

$$I = \begin{pmatrix} 1 & 0 \\ 0 & 1 \end{pmatrix} \quad A = \begin{pmatrix} -\frac{1}{2} & \frac{\sqrt{3}}{2} \\ -\frac{\sqrt{3}}{2} & -\frac{1}{2} \end{pmatrix} \quad B = \begin{pmatrix} -\frac{1}{2} & \frac{-\sqrt{3}}{2} \\ \frac{\sqrt{3}}{2} & -\frac{1}{2} \end{pmatrix}$$
$$C = \begin{pmatrix} -1 & 0 \\ 0 & 1 \end{pmatrix} \quad D = \begin{pmatrix} \frac{1}{2} & -\frac{\sqrt{3}}{2} \\ -\frac{\sqrt{3}}{2} & -\frac{1}{2} \end{pmatrix} \quad E = \begin{pmatrix} \frac{1}{2} & \frac{\sqrt{3}}{2} \\ \frac{\sqrt{3}}{2} & -\frac{1}{2} \end{pmatrix} \tag{22.13}$$

the combination law being that of ordinary matrix multiplication. Here we use italic, rather than the sans serif used for matrices elsewhere, to emphasise that the matrices are group elements.

Although it is tedious to do so, it can be checked that the product of any two of these matrices, in either order, is also in the set. However, the result is generally different in the two cases, as matrix multiplication is non-commutative. The matrix $I$ clearly acts as the identity element of the set, and during the checking for closure it is found that the inverse of each matrix is contained in the set, with $I$, $C$, $D$ and $E$ being their own inverses. The group table is shown in table 22.8.

The similarity to table 22.7 is striking. If $\{R, R', K, L, M\}$ of that table are replaced by $\{A, B, C, D, E\}$ respectively, the two tables are identical, without even the need to reshuffle the rows and columns. The two groups, one of reflections and rotations of an equilateral triangle, the other of matrices, are isomorphic.

Our second example of a group isomorphic to the same rotation–reflection group is provided by a set of functions of an undetermined variable $x$. The functions are as follows:

$$f_1(x) = x, \qquad f_2(x) = 1/(1-x), \qquad f_3(x) = (x-1)/x,$$

$$f_4(x) = 1/x, \qquad f_5(x) = 1-x, \qquad f_6(x) = x/(x-1),$$

and the law of combination is

$$f_i(x) \bullet f_j(x) = f_i(f_j(x)),$$

i.e. the function on the right acts as the argument of the function on the left to produce a new function of $x$. It should be emphasised that it is the functions that are the elements of the group. The variable $x$ is the 'system' on which they act, and plays much the same role as the triangle does in our first example of a non-Abelian group.

To show an explicit example, we calculate the product $f_6 \bullet f_3$. The product will be the function of $x$ obtained by evaluating $y/(y-1)$, when $y$ is set equal to $(x-1)/x$. Explicitly

$$f_6(f_3) = \frac{(x-1)/x}{(x-1)/x \;-\; 1} = 1 - x = f_5(x).$$

Thus $f_6 \bullet f_3 = f_5$. Further examples are

$$f_2 \bullet f_2 = \frac{1}{1 - 1/(1-x)} = \frac{x-1}{x} = f_3,$$

and

$$f_6 \bullet f_6 = \frac{x/(x-1)}{x/(x-1) \;-\; 1} = x = f_1. \tag{22.14}$$

The multiplication table for this set of six functions has all the necessary properties to show that they form a group. Further, if the symbols $f_1, f_2, f_3, f_4, f_5, f_6$ are replaced by $I, A, B, C, D, E$ respectively the table becomes identical to table 22.8. This justifies our earlier claim that this group of functions, with argument substitution as the law of combination, is isomorphic to the group of reflections and rotations of an equilateral triangle.

## 22.4 Permutation groups

The operation of rearranging $n$ distinct objects amongst themselves is called a *permutation* of degree $n$, and since many symmetry operations on physical systems can be viewed in that light, the properties of permutations are of interest. For example, the symmetry operations on an equilateral triangle, to which we have already given much attention, can be considered as the six possible rearrangements of the marked corners of the triangle amongst three fixed points in space, much as in the diagrams used to compute table 22.7. In the same way, the symmetry operations on a cube can be viewed as a rearrangement of its corners amongst eight points in space, albeit with many constraints, or, with fewer complications, as a rearrangement of its body diagonals in space. The details will be left until we review the possible finite groups more systematically.

The notations and conventions used in the literature to describe permutations are very varied and can easily lead to confusion. We will try to avoid this by using letters $a, b, c, \ldots$ (rather than numbers) for the objects that are rearranged by a permutation and by adopting, before long, a 'cycle notation' for the permutations themselves. It is worth emphasising that it is the *permutations*, i.e. the acts of rearranging, and not the objects (represented by letters), that form the elements of permutation groups. The complete group of all permutations of degree $n$ is usually denoted by $S_n$ or $\Sigma_n$. The number of possible permutations of degree $n$ is $n!$, and so this is the order of $S_n$.

Suppose the ordered set of six distinct objects $\{a\ b\ c\ d\ e\ f\}$ is rearranged by some process into $\{b\ e\ f\ a\ d\ c\}$; then we can represent this mathematically as

$$\theta\{a\,b\,c\,d\,e\,f\} = \{b\ e\ f\ a\ d\ c\},$$

where $\theta$ is a permutation of degree 6. The permutation $\theta$ can be denoted by [2 5 6 1 4 3], since the first object, $a$, is replaced by the second, $b$, the second object, $b$, is replaced by the fifth, $e$, the third by the sixth, $f$, etc. The equation can then be written more explicitly as

$$\theta\{a\,b\,c\,d\,e\,f\} = [2\ 5\ 6\ 1\ 4\ 3]\{a\,b\,c\,d\,e\,f\} = \{b\ e\ f\ a\ d\ c\}.$$

If $\phi$ is a second permutation, also of degree 6, then the obvious interpretation of the product $\phi \bullet \theta$ of the two permutations is

$$\phi \bullet \theta\{a\,b\,c\,d\,e\,f\} = \phi(\theta\{a\,b\,c\,d\,e\,f\}).$$

Suppose that $\phi$ is the permutation [4 5 3 6 2 1]; then

$$\begin{aligned}
\phi \bullet \theta\{a\,b\,c\,d\,e\,f\} &= [4\,5\,3\,6\,2\,1][2\,5\,6\,1\,4\,3]\{a\,b\,c\,d\,e\,f\} \\
&= [4\,5\,3\,6\,2\,1]\{b\ e\ f\ a\ d\ c\} \\
&= \{a\ d\ f\ c\ e\ b\} \\
&= [1\,4\,6\,3\,5\,2]\{a\,b\,c\,d\,e\,f\}.
\end{aligned}$$

Written in terms of the permutation notation this result is

$$[4\,5\,3\,6\,2\,1][2\,5\,6\,1\,4\,3] = [1\,4\,6\,3\,5\,2].$$

A concept which is very useful for working with permutations is that of decomposition into cycles. The cycle notation is most easily explained by example. For the permutation $\theta$:

> the 1st object, $a$, has been replaced by the 2nd, $b$;
> the 2nd object, $b$, has been replaced by the 5th, $e$;
> the 5th object, $e$, has been replaced by the 4th, $d$;
> the 4th object, $d$, has been replaced by the 1st, $a$.

This brings us back to the beginning of a closed cycle, which is conveniently represented by the notation (1 2 5 4), in which the successive replacement positions are enclosed, in sequence, in parentheses. Thus (1 2 5 4) means 2nd $\to$ 1st, 5th $\to$ 2nd, 4th $\to$ 5th, 1st $\to$ 4th. It should be noted that the object initially in the first listed position replaces that in the final position indicated in the bracket – here '$a$' is put into the fourth position by the permutation. Clearly the cycle (5 4 1 2), or any other which involved the same numbers in the same relative order, would have exactly the same meaning and effect. The remaining two objects, $c$ and $f$, are interchanged by $\theta$, or, more formally, are rearranged according to a cycle of length 2, or *transposition*, represented by (3 6). Thus the complete representation (specification) of $\theta$ is

$$\theta = (1\ 2\ 5\ 4)(3\ 6).$$

The positions of objects that are unaltered by a permutation are either placed by themselves in a pair of parentheses or omitted altogether. The former is recommended as it helps to indicate how many objects are involved – important when the object in the last position is unchanged, or the permutation is the identity, which leaves all objects unaltered in position! Thus the identity permutation of degree 6 is

$$I = (1)(2)(3)(4)(5)(6),$$

though in practice it is often shortened to (1).

It will be clear that the cycle representation is unique, to within the internal absolute ordering of the numbers in each bracket as already noted, and that each number appears once and only once in the representation of any particular permutation.

The *order of any permutation* of degree $n$ within the group $S_n$ can be read off from the cyclic representation, and is given by the lowest common multiple (LCM) of the lengths of the cycles. Thus $I$ has order 1, as it must, and the permutation $\theta$ discussed above has order 4 (the LCM of 4 and 2).

Expressed in cycle notation our second permutation $\phi$ is (3)(1 4 6)(2 5), and the product $\phi \bullet \theta$ is calculated as

$$\begin{aligned}(3)(1\ 4\ 6)(2\ 5) \bullet (1\ 2\ 5\ 4)(3\ 6)\{a\ b\ c\ d\ e\ f\} &= (3)(1\ 4\ 6)(2\ 5)\{b\ e\ f\ a\ d\ c\} \\ &= \{a\ d\ f\ c\ e\ b\} \\ &= (1)(5)(2\ 4\ 3\ 6)\{a\ b\ c\ d\ e\ f\}.\end{aligned}$$

i.e. expressed as a relationship amongst elements of the group of permutations of degree 6 (not yet proved as a group, but reasonably anticipated), this result reads

$$(3)(1\ 4\ 6)(2\ 5) \bullet (1\ 2\ 5\ 4)(3\ 6) = (1)(5)(2\ 4\ 3\ 6).$$

We note, for practice, that $\phi$ has order 6 (the LCM of 1, 3, and 2), and that the product $\phi \bullet \theta$ has order 4.

The number of elements in the group $S_n$ of all permutations of degree $n$ is $n!$ and clearly increases very rapidly as $n$ increases. Fortunately, to illustrate the essential features of permutation groups it is sufficient to consider the case $n = 3$, which involves only six elements. They are as follows (with labelling which the reader will by now recognise as anticipatory):

$$\begin{array}{lll} I = (1)(2)(3) & A = (1\ 2\ 3) & B = (1\ 3\ 2) \\ C = (1)(2\ 3) & D = (3)(1\ 2) & E = (2)(1\ 3) \end{array}$$

It will be noted that $A$ and $B$ have order 3, whilst $C$, $D$ and $E$ have order 2. As perhaps anticipated, their combination products are exactly those corresponding to table 22.8, with $I$, $C$, $D$ and $E$ being their own inverses. For example, putting in all steps explicitly,

$$\begin{aligned} D \bullet C\{a\ b\ c\} &= (3)(1\ 2) \bullet (1)(2\ 3)\{a\ b\ c\} \\ &= (3)(12)\{a\ c\ b\} \\ &= \{c\ a\ b\} \\ &= (3\ 2\ 1)\{a\ b\ c\} \\ &= (1\ 3\ 2)\{a\ b\ c\} \\ &= B\{a\ b\ c\}. \end{aligned}$$

In brief, the six permutations belonging to $S_3$ form yet another non-Abelian group isomorphic to the rotation–reflection symmetry group of an equilateral triangle.

## 22.5 Mappings between groups

Now that we have available a range of groups that can be used as examples, we return to the study of more general group properties. From here on, when there is no ambiguity we will write the product of two elements, $X \bullet Y$, simply as $XY$, omitting the explicit combination symbol. We will also continue to use 'multiplication' as a loose generic name for the combination process between elements of a group.

If $\mathcal{G}$ and $\mathcal{G}'$ are two groups, we can study the effect of a *mapping*

$$\Phi : \mathcal{G} \to \mathcal{G}'$$

of $\mathcal{G}$ onto $\mathcal{G}'$. If $X$ is an element of $\mathcal{G}$ we denote its *image* in $\mathcal{G}'$ under the mapping $\Phi$ by $X' = \Phi(X)$.

A technical term that we have already used is *isomorphic*. We will now define it formally. Two groups $\mathcal{G} = \{X, Y, \ldots\}$ and $\mathcal{G}' = \{X', Y', \ldots\}$ are said to be *isomorphic* if there is a one-to-one correspondence

$$X \leftrightarrow X', \; Y \leftrightarrow Y', \; \cdots$$

between their elements such that

$$XY = Z \qquad \text{implies} \qquad X'Y' = Z'$$

and vice versa.

In other words, isomorphic groups have the same (multiplication) structure, although they may differ in the nature of their elements, combination law and notation. Clearly if groups $\mathcal{G}$ and $\mathcal{G}'$ are isomorphic, and $\mathcal{G}$ and $\mathcal{G}''$ are isomorphic, then it follows that $\mathcal{G}'$ and $\mathcal{G}''$ are isomorphic. We have already seen an example of four groups (of functions of $x$, of orthogonal matrices, of permutations and of the symmetries of an equilateral triangle) that are isomorphic, all having table 22.8 as their multiplication table.

Although our main interest is in isomorphic relationships between groups, the wider question of mappings of one set of elements onto another is of some importance, and we start with the more general notion of a homomorphism.

*Let $\mathcal{G}$ and $\mathcal{G}'$ be two groups and $\Phi$ a mapping of $\mathcal{G} \to \mathcal{G}'$. If for any pair of elements $X$ and $Y$ in $\mathcal{G}$*

$$(XY)' = X'Y'$$

*then $\Phi$ is called a homomorphism, and $\mathcal{G}'$ is said to be a homomorphic image of $\mathcal{G}$.*

The essential defining relationship, expressed by $(XY)' = X'Y'$, is that the same result is obtained whether the product of two elements is formed first and the image then taken, or the images are taken first and the product then formed.

Three immediate consequences of the above definition are proved as follows.

(i) If $I$ is the identity of $\mathcal{G}$, then $IX = X$ for all $X$ in $\mathcal{G}$. Consequently

$$X' = (IX)' = I'X',$$

for all $X'$ in $\mathcal{G}'$. Thus $I'$ is the identity in $\mathcal{G}'$. In words, the identity element of $\mathcal{G}$ maps into the identity element of $\mathcal{G}'$.

(ii) Further,

$$I' = (XX^{-1})' = X'(X^{-1})'.$$

That is, $(X^{-1})' = (X')^{-1}$. In words, the image of an inverse is the same element in $\mathcal{G}'$ as the inverse of the image.

(iii) If element $X$ in $\mathcal{G}$ is of order $m$, i.e. $I = X^m$, then

$$I' = (X^m)' = (XX^{m-1})' = X'(X^{m-1})' = \cdots = \underbrace{X'X'\cdots X'}_{m \text{ factors}}.$$

In words, the image of an element has the same order as the element.

What distinguishes an isomorphism from the more general homomorphism are the requirements that in an isomorphism:

(I) different elements in $\mathcal{G}$ must map into different elements in $\mathcal{G}'$ (whereas in a homomorphism several elements in $\mathcal{G}$ may have the same image in $\mathcal{G}'$), that is, $x' = y'$ must imply $x = y$;

(II) any element in $\mathcal{G}'$ must be the image of some element in $\mathcal{G}$.

An immediate consequence of (I) and result (iii) for homomorphisms is that groups that are isomorphic each have the same number of elements of any given order.

For a general homomorphism, the set of elements of $\mathcal{G}$ whose image in $\mathcal{G}'$ is $I'$ is called the *kernel* of the homomorphism, and is discussed further in the next section. In an isomorphism the kernel consists of the identity $I$ alone. To illustrate both this point and the general notion of a homomorphism, consider a mapping between the additive group of real numbers $\Re$ and the multiplicative group of complex numbers with unit modulus, $U(1)$. Suppose that the mapping $\Re \to U(1)$ is

$$\Phi : x \to e^{ix};$$

then this is a homomorphism since

$$(x + y)' \to e^{i(x+y)} = e^{ix}e^{iy} = x'y'.$$

However, it is not an isomorphism because many (an infinite number) of the elements of $\Re$ have the same image in $U(1)$. For example, $\pi, 3\pi, 5\pi, \ldots$ in $\Re$ all have the image $-1$ in $U(1)$ and, furthermore, all elements of $\Re$ of the form $2\pi n$, where $n$ is an integer, map onto the identity element in $U(1)$. The latter set forms the kernel of the homomorphism.

For the sake of completeness, we add that a homomorphism for which (I) above holds is said to be a *monomorphism* (or an isomorphism *into*), whilst a homomorphism for which (II) holds is called an *epimorphism* (or an isomorphism *onto*).

Finally, if the initial and final groups are the same, $\mathcal{G} = \mathcal{G}'$, then the isomorphism $\mathcal{G} \to \mathcal{G}'$ is termed an *automorphism*.

(a)

| | $I$ | $A$ | $B$ | $C$ | $D$ | $E$ |
|---|---|---|---|---|---|---|
| $I$ | $\mathbf{I}$ | $\mathbf{A}$ | $\mathbf{B}$ | $C$ | $D$ | $E$ |
| $A$ | $\mathbf{A}$ | $\mathbf{B}$ | $\mathbf{I}$ | $E$ | $C$ | $D$ |
| $B$ | $\mathbf{B}$ | $\mathbf{I}$ | $\mathbf{A}$ | $D$ | $E$ | $C$ |
| $C$ | $C$ | $D$ | $E$ | $I$ | $A$ | $B$ |
| $D$ | $D$ | $E$ | $C$ | $B$ | $I$ | $A$ |
| $E$ | $E$ | $C$ | $D$ | $A$ | $B$ | $I$ |

(b)

| | $I$ | $A$ | $B$ | $C$ |
|---|---|---|---|---|
| $I$ | $\mathbf{I}$ | $\mathbf{A}$ | $B$ | $C$ |
| $A$ | $\mathbf{A}$ | $\mathbf{I}$ | $C$ | $B$ |
| $B$ | $B$ | $C$ | $I$ | $A$ |
| $C$ | $C$ | $B$ | $A$ | $I$ |

Table 22.9 Reproduction of (*a*) table 22.8 and (*b*) table 22.3 with the relevent subgroups shown in bold.

## 22.6 Subgroups

More detailed inspection of tables 22.8 and 22.3 shows that not only do the complete tables have the properties associated with a group multiplication table (see section 22.2), but so do the upper left corners of each table taken on their own. The relevant parts are shown in bold in the tables 22.9(*a*) and (*b*).

This observation immediately prompts the notion of a *subgroup*. A subgroup of a group $\mathcal{G}$ can be formally defined as any non-empty subset $\mathcal{H} = \{H_i\}$ of $\mathcal{G}$, the elements of which themselves behave as a group under the same rule of combination as applies in $\mathcal{G}$ itself. The order of the subgroup is, as for all groups, equal to the number of elements it contains; we will denote it by $h$ or $|\mathcal{H}|$.

All groups $\mathcal{G}$ contain two trivial subgroups;

(i) $\mathcal{G}$ itself,
(ii) the set $\mathcal{I}$ consisting of the identity element alone.

All other subgroups are termed *proper subgroups*. In a group with multiplication table 22.8 the elements $\{I, A, B\}$ form a proper subgroup, as do $\{I, A\}$ in a group with table 22.3 as its group table.

Some groups have no proper subgroups. For example, the so-called *cyclic groups*, mentioned at the end of subsection 22.1.1, have no subgroups other than the whole group or the identity alone. Tables 22.10(*a*) and (*b*) show the multiplication tables for two of these groups. Table 22.6 is also the group table for a cyclic group, that of order 4.

It will be clear that for a cyclic group $\mathcal{G}$ repeated combination of any element with itself generates all other elements of $\mathcal{G}$, before finally reproducing itself. So, for example, in table 22.10(*b*), starting with (say) $D$, repeated combination with itself produces, in turn, $C$, $B$, $A$, $I$ and finally $D$ again. As noted earlier, in any cyclic group $\mathcal{G}$ every element, apart from the identity, is of order $g$, the order of the group itself.

The two tables shown are for groups of orders 3 and 5. It will be proved in subsection 22.7.2 that the order of any group is a multiple of the order of any of

(a)

| | $I$ | $A$ | $B$ |
|---|---|---|---|
| $I$ | $I$ | $A$ | $B$ |
| $A$ | $A$ | $B$ | $I$ |
| $B$ | $B$ | $I$ | $A$ |

(b)

| | $I$ | $A$ | $B$ | $C$ | $D$ |
|---|---|---|---|---|---|
| $I$ | $I$ | $A$ | $B$ | $C$ | $D$ |
| $A$ | $A$ | $B$ | $C$ | $D$ | $I$ |
| $B$ | $B$ | $C$ | $D$ | $I$ | $A$ |
| $C$ | $C$ | $D$ | $I$ | $A$ | $B$ |
| $D$ | $D$ | $I$ | $A$ | $B$ | $C$ |

Table 22.10 The group tables of two cyclic groups, of orders 3 and 5. They have no proper subgroups.

its subgroups (Lagrange's theorem), i.e. in our general notation, $g$ is a multiple of $h$. It thus follows that a group of order $p$, where $p$ is any prime, must be cyclic and cannot have any proper subgroups. The groups for which tables 22.10($a$) and ($b$) are the group tables are two such examples. Groups of non-prime order may (table 22.3) or may not (table 22.6) have proper subgroups.

As we have seen, repeated multiplication of an element $X$ (not the identity) by itself will generate a subgroup $\{X, X^2, X^3, \dots\}$. The subgroup will clearly be Abelian, and if $X$ is of order $m$, i.e. $X^m = I$, the subgroup will have $m$ distinct members. If $m$ is less than $g$ – though, in view of Lagrange's theorem, $m$ must be a factor of $g$ – the subgroup will be a proper subgroup. We can deduce, in passing, that the order of any element of a group is an exact divisor of the order of the group.

Some obvious properties of the subgroups of a group $\mathcal{G}$, which can be listed without formal proof, are as follows.

(i) The identity element of $\mathcal{G}$ belongs to every subgroup $\mathcal{H}$.

(ii) If element $X$ belongs to a subgroup $\mathcal{H}$, so does $X^{-1}$.

(iii) The set of elements in $\mathcal{G}$ that belong to every subgroup of $\mathcal{G}$, themselves form a subgroup, though it may consist of the identity alone.

Properties of subgroups that need more explicit proof are given in the following sections, though some need the development of new concepts before they can be established. However, we can begin with a theorem, applicable to all homomorphisms, not just isomorphisms, that requires no new concepts.

Let $\Phi : \mathcal{G} \to \mathcal{G}'$ be a homomorphism of $\mathcal{G}$ into $\mathcal{G}'$, then

(i) the set of elements $\mathcal{H}'$ in $\mathcal{G}'$ that are images of the elements of $\mathcal{G}$ form a subgroup of $\mathcal{G}'$;

(ii) the set of elements $\mathcal{K}$ in $\mathcal{G}$ that are mapped onto the identity $I'$ in $\mathcal{G}'$ form a subgroup of $\mathcal{G}$.

As indicated in the previous section, the subgroup $\mathcal{K}$ is called the *kernel* of the homomorphism.

To prove (i), suppose $Z$ and $W$ belong to $\mathcal{H}'$, with $Z = X'$ and $W = Y'$, where $X$ and $Y$ belong to $\mathcal{G}$. Then

$$ZW = X'Y' = (XY)'$$

and therefore belongs to $\mathcal{H}'$, and

$$Z^{-1} = (X')^{-1} = (X^{-1})'$$

and therefore belongs to $\mathcal{H}'$. These two results, together with the fact that $I'$ belongs to $\mathcal{H}'$, are enough to establish result (i).

To prove (ii), suppose $X$ and $Y$ belong to $\mathcal{K}$, then

$$(XY)' = X'Y' = I'I' = I' \qquad \text{(closure)},$$

$$I' = (XX^{-1})' = X'(X^{-1})' = I'(X^{-1})' = (X^{-1})'$$

and therefore $X^{-1}$ belongs to $\mathcal{K}$. These two results, together with the fact that $I$ belongs to $\mathcal{K}$, are enough to establish result (ii).

As noted previously, the kernel of the mapping $\Re \to U(1)$ consists of the set of real numbers of the form $2\pi n$ where $n$ is an integer.

In fact the kernel $\mathcal{K}$ of a homomorphism is a *normal* subgroup of $\mathcal{G}$. The defining property of such a subgroup is that for every element $X$ in $\mathcal{G}$ and every element $Y$ in the subgroup, $XYX^{-1}$ belongs to the subgroup. This property is easily verified for the kernel $\mathcal{K}$, since

$$(XYX^{-1})' = X'Y'(X^{-1})' = X'I'(X^{-1})' = X'(X^{-1})' = I'.$$

Anticipating the discussion of subsection 22.7.2, the cosets of a normal subgroup themselves form a group (see exercise 22.7).

## 22.7 Subdividing a group

We have already noted, when looking at the (arbitrary) order of headings in a group table, that some choices appear to make the table more orderly than do others. In the following subsections we will identify ways in which the elements of a group can be divided up into sets with the property that the members of any one set are more like the other members of the set, in some particular regard, than they are like any element that does not belong to the set. We will find that these divisions will be such that the group is *partitioned*, i.e. the elements will be divided into sets in such a way that each element of the group belongs to one, and only one, such set.

We note in passing that the subgroups of a group do *not* form such a partition, not least because the identity element is in every subgroup, rather than being in precisely one. In other words, despite the nomenclature, a group is not simply the aggregate of its proper subgroups.

### 22.7.1 Equivalence relations and classes

We now specify in a more mathematical manner what it means for two elements of a group to be 'more like' one another than like a third element, as mentioned in section 22.2. Our introduction will apply to any set, whether a group or not, but our main interest will ultimately be in two particular applications to groups. We start with the formal definition of an equivalence relation.

An *equivalence relation* on a set $\mathcal{S}$ is a relationship $X \sim Y$ between two elements $X$ and $Y$ belonging to $\mathcal{S}$, in which the definition of $\sim$ must satisfy the requirements of

(i) reflexivity, $X \sim X$;
(ii) symmetry, $X \sim Y$ implies $Y \sim X$;
(iii) transitivity, $X \sim Y$ and $Y \sim Z$ imply $X \sim Z$.

Any particular pair of elements either satisfy or do not satisfy the relationship.

The general notion of an equivalence relation is very straightforward, and the requirements on $\sim$ seem undemanding; but not all relationships qualify. As an example within the topic of groups, if $\sim$ meant 'has the same order as' then clearly all of the requirements would be satisfied. However, if $\sim$ meant 'commutes with' then $\sim$ would not be an equivalence relation, since although $A$ commutes with $I$, and $I$ commutes with $C$, this does not necessarily imply that $A$ commutes with $C$, as is obvious from table 22.8.

It may be shown that an equivalence relation on $\mathcal{S}$ divides up $\mathcal{S}$ into *classes* $\mathcal{C}_i$ such that:

(i) $X$ and $Y$ belong to the same class if, and only if, $X \sim Y$;
(ii) every element $W$ of $\mathcal{S}$ belongs to exactly one class.

This may be shown as follows. Let $X$ belong to $\mathcal{S}$, and define the subset $\mathcal{S}_X$ of $\mathcal{S}$ to be the set of all elements $U$ of $\mathcal{S}$ such that $X \sim U$. Clearly by reflexivity $X$ belongs to $\mathcal{S}_X$. Suppose first that $X \sim Y$, and let $Z$ be any element of $\mathcal{S}_Y$. Then $Y \sim Z$, and hence by transitivity $X \sim Z$, which means that $Z$ belongs to $\mathcal{S}_X$. Conversely, since the symmetry law gives $Y \sim X$, if $Z$ belongs to $\mathcal{S}_X$ then this implies that $Z$ belongs to $\mathcal{S}_Y$. These two results together mean that the two subsets $\mathcal{S}_X$ and $\mathcal{S}_Y$ have the same members and hence are equal.

Now suppose that $\mathcal{S}_X$ equals $\mathcal{S}_Y$. Since $Y$ belongs to $\mathcal{S}_Y$ it also belongs to $\mathcal{S}_X$ and hence $X \sim Y$. This completes the proof of (i), once the distinct subsets of type $\mathcal{S}_X$ are identified as the classes $\mathcal{C}_i$. Statement (ii) is an immediate corollary, the class in question being identified as $\mathcal{S}_W$.

The most important property of an equivalence relation is as follows.

*Two different subsets $\mathcal{S}_X$ and $\mathcal{S}_Y$ can have no element in common, and the collection of all the classes $\mathcal{C}_i$ is a 'partition' of $\mathcal{S}$, i.e. every element in $\mathcal{S}$ belongs to one, and only one, of the classes.*

To prove this, suppose $\mathcal{S}_X$ and $\mathcal{S}_Y$ have an element $Z$ in common; then $X \sim Z$ and $Y \sim Z$ and so by the symmetry and transitivity laws $X \sim Y$. By the above theorem this implies $\mathcal{S}_X$ equals $\mathcal{S}_Y$. But this contradicts the fact that $\mathcal{S}_X$ and $\mathcal{S}_Y$ are different subsets. Hence $\mathcal{S}_X$ and $\mathcal{S}_Y$ can have no element in common.

Finally, if the elements of $\mathcal{S}$ are used in turn to define subsets and hence classes in $\mathcal{S}$, every element $U$ is in the subset $\mathcal{S}_U$ that is either a class already found or constitutes a new one. It follows that the classes exhaust $\mathcal{S}$, i.e. every element is in some class.

Having established the general properties of equivalence relations, we now turn to two specific examples of such relationships, in which the general set $\mathcal{S}$ has the more specialised properties of a group $\mathcal{G}$, and the equivalence relation $\sim$ is chosen in such a way that the relatively transparent general results for equivalence relations can be used to derive powerful, but less obvious, results about the properties of groups.

### 22.7.2 Congruence and cosets

As the first application of equivalence relations we now prove *Lagrange's theorem* which is stated as follows.

*If $\mathcal{G}$ is a finite group of order g, and $\mathcal{H}$ is a subgroup of $\mathcal{G}$ of order h, then g is a multiple of h.*

We take as the definition of $\sim$ that, given $X$ and $Y$ belonging to $\mathcal{G}$, $X \sim Y$ if $X^{-1}Y$ belongs to $\mathcal{H}$. This is the same as saying that $Y = XH_i$ for some element $H_i$ belonging to $\mathcal{H}$; technically $X$ and $Y$ are said to be left-congruent with respect to $\mathcal{H}$.

That this defines an equivalence relation follows from

(i) Reflexivity: $X \sim X$, since $X^{-1}X = I$ and $I$ belongs to any subgroup.
(ii) Symmetry: $X \sim Y$ implies that $X^{-1}Y$ belongs to $\mathcal{H}$, and so, therefore, does its inverse, since $\mathcal{H}$ is a group. But $(X^{-1}Y)^{-1} = Y^{-1}X$ and, as this belongs to $\mathcal{H}$, it follows that $Y \sim X$.
(iii) Transitivity: $X \sim Y$ and $Y \sim Z$ imply that $X^{-1}Y$ and $Y^{-1}Z$ belong to $\mathcal{H}$, and so therefore does their product $(X^{-1}Y)(Y^{-1}Z) = X^{-1}Z$, from which it follows that $X \sim Z$.

With $\sim$ proved as an equivalence relation, we can immediately deduce that it divides $\mathcal{G}$ into disjoint (non-overlapping) classes. For this particular equivalence relation the classes are called the *left cosets* of $\mathcal{H}$. Thus each element of $\mathcal{G}$ is in one and only one left coset of $\mathcal{H}$. The left coset containing any particular $X$ is usually written $X\mathcal{H}$, and denotes the set of elements of the form $XH_i$ (one of which is $X$ itself since $\mathcal{H}$ contains the identity element); it must contain $h$ different elements, since if it did not, and two elements were equal,

$$XH_i = XH_j,$$

we could deduce that $H_i = H_j$ and that $\mathcal{H}$ contained fewer than $h$ elements.

From our general results about equivalence relations it now follows that the left cosets of $\mathcal{H}$ are a 'partition' of $\mathcal{G}$ into a number of sets each containing $h$ members. Since there are $g$ members of $\mathcal{G}$, and each must be in just one of the sets, it follows that $g$ is a multiple of $h$. This concludes the proof of Lagrange's theorem.

The number of left cosets of $\mathcal{H}$ in $\mathcal{G}$ is known as the *index* of $\mathcal{H}$ in $\mathcal{G}$ and written $[\mathcal{G} : \mathcal{H}]$; numerically the index $= g/h$. For the record we note that, for the trivial subgroup $\mathcal{I}$, which contains only the identity element, $[\mathcal{G} : \mathcal{I}] = g$ and that, for a subgroup $\mathcal{J}$ of subgroup $\mathcal{H}$, $[\mathcal{G} : \mathcal{H}][\mathcal{H} : \mathcal{J}] = [\mathcal{G} : \mathcal{J}]$.

As noted earlier, Lagrange's theorem justifies our statement that any group of order $p$, where $p$ is prime, must be cyclic and cannot have any proper subgroups: since any subgroup must have an order that divides $p$, this can only be 1 or $p$, corresponding to the two trivial subgroups $\mathcal{I}$ and the whole group.

It may be helpful to see an example worked through explicitly and we again use the same six-element group.

►*Find the left cosets of the proper subgroup $\mathcal{H}$ of the group $\mathcal{G}$ that has table 22.8 as its multiplication table.*

The subgroup consists of the set of elements $\mathcal{H} = \{I, A, B\}$. We note in passing that it has order 3, which, as required by Lagrange's theorem, is a divisor of 6, the order of $\mathcal{G}$. As in all cases, $\mathcal{H}$ itself provides the first (left) coset, formally the coset

$$I\mathcal{H} = \{II, IA, IB\} = \{I, A, B\}.$$

We continue by choosing an element not already selected, $C$ say, and form

$$C\mathcal{H} = \{CI, CA, CB\} = \{C, D, E\}.$$

These two cosets of $\mathcal{H}$ exhaust $\mathcal{G}$, and are therefore the only cosets, the index of $\mathcal{H}$ in $\mathcal{G}$ being equal to 2.

This completes the example, but it is useful to demonstrate that it would not have mattered if we had taken $D$, say, instead of $I$ to form a first coset

$$D\mathcal{H} = \{DI, DA, DB\} = \{D, E, C\},$$

and then, from previously unselected elements, picked $B$, say:

$$B\mathcal{H} = \{BI, BA, BB\} = \{B, I, A\}.$$

The same two cosets would have resulted. ◄

It will be noticed that the cosets are the same groupings of the elements of $\mathcal{G}$ which we earlier noted as being the choice of adjacent column and row headings that gave the multiplication table its 'neatest' appearance. Furthermore, if $\mathcal{H}$ is a *normal* subgroup of $\mathcal{G}$, then its (left) cosets themselves form a group (see exercise 22.7).

### *22.7.3 Conjugates and classes*

Our second example of an equivalence relation is concerned with those elements $X$ and $Y$ of a group $\mathcal{G}$ that can be connected by a transformation of the form $Y = G_i^{-1}XG_i$, where $G_i$ is an (appropriate) element of $\mathcal{G}$. Thus $X \sim Y$ if there exists an element $G_i$ of $\mathcal{G}$ such that $Y = G_i^{-1}XG_i$. Different pairs of elements $X$ and $Y$ will, in general, require different group elements $G_i$. Elements connected in this way are said to be *conjugates*.

We first need to establish that this does indeed define an equivalence relation, as follows.

(i) Reflexivity: $X \sim X$, since $X = I^{-1}XI$ and $I$ belongs to the group.
(ii) Symmetry: $X \sim Y$ implies $Y = G_i^{-1}XG_i$ and therefore $X = (G_i^{-1})^{-1}YG_i^{-1}$. Since $G_i$ belongs to $\mathcal{G}$, so does $G_i^{-1}$, and it follows that $Y \sim X$.
(iii) Transitivity: $X \sim Y$ and $Y \sim Z$ imply $Y = G_i^{-1}XG_i$ and $Z = G_j^{-1}YG_j$ and therefore $Z = G_j^{-1}G_i^{-1}XG_iG_j = (G_iG_j)^{-1}X(G_iG_j)$. Since $G_i$ and $G_j$ belong to $\mathcal{G}$ so does $G_iG_j$, from which it follows that $X \sim Z$.

These results establish conjugacy as an equivalence relation, and hence show that it divides $\mathcal{G}$ into classes, two elements being in the same class if, and only if, they are conjugate.

Immediate corollaries are:

(i) If $Z$ is in the class containing $I$, then

$$Z = G_i^{-1}IG_i = G_i^{-1}G_i = I.$$

Thus, since any conjugate of $I$ can be shown to be $I$, the identity must be in a class by itself.

(ii) If $X$ is in a class by itself, then

$$Y = G_i^{-1}XG_i$$

must imply that $Y = X$. But

$$X = G_iG_i^{-1}XG_iG_i^{-1}$$

for any $G_i$, and so

$$X = G_i(G_i^{-1}XG_i)G_i^{-1} = G_iYG_i^{-1} = G_iXG_i^{-1},$$

i.e. $XG_i = G_iX$ for all $G_i$.

Thus commutation with all elements of the group is a necessary (and sufficient) condition for any particular group element to be in a class by itself. In an Abelian group each element is in a class by itself.

(iii) In any group $\mathcal{G}$ the set $\mathcal{S}$ of elements in classes by themselves is an Abelian subgroup (known as the *centre* of $\mathcal{G}$). We have shown that $I$ belongs to $\mathcal{S}$, and so if, further, $XG_i = G_iX$ and $YG_i = G_iY$ for all $G_i$ belonging to $\mathcal{G}$, then:

(a) $(XY)G_i = XG_iY = G_i(XY)$, i.e. the closure of $\mathcal{S}$, and

(b) $XG_i = G_iX$ implies $X^{-1}G_i = G_iX^{-1}$, i.e. the inverse of $X$ belongs to $\mathcal{S}$.

Hence $\mathcal{S}$ is a group, and clearly Abelian.

Yet again for illustration purposes we use the six-element group with table 22.8.

►*Find the conjugacy classes of the group $\mathcal{G}$ having table 22.8 as its multiplication table.*

As always, $I$ is in a class by itself, and we need consider it no further.

Consider next the results of forming $X^{-1}AX$, as $X$ runs through the elements of $\mathcal{G}$.

$$\begin{array}{cccccc} I^{-1}AI & A^{-1}AA & B^{-1}AB & C^{-1}AC & D^{-1}AD & E^{-1}AE \\ = IA & = IA & = AI & = CE & = DC & = ED \\ = A & = A & = A & = B & = B & = B \end{array}$$

Only $A$ and $B$ are generated. It is clear that $\{A, B\}$ is one of the conjugacy classes of $\mathcal{G}$. This can be verified by forming all elements $X^{-1}BX$; again only $A$ and $B$ appear.

We now need to pick an element not in the two classes already found. Suppose we pick $C$. Just as for $A$, we compute $X^{-1}CX$, as $X$ runs through the elements of $\mathcal{G}$. The calculations can be done directly using the table and give the following.

$$\begin{array}{llllllll} X & : I & A & B & C & D & E \\ X^{-1}CX & : C & E & D & C & E & D \end{array}$$

Thus $C$, $D$ and $E$ belong to the same class. The group is now exhausted, and so the three conjugacy classes are

$$\{I\}, \qquad \{A, B\}, \qquad \{C, D, E\}. \blacktriangleleft$$

In the case of this small and simple, but non-Abelian, group, only the identity is in a class by itself (i.e. only $I$ commutes with all other elements). It is also the only member of the centre of the group.

Other areas from which examples of conjugacy classes can be taken include permutations and rotations. Two permutations are in the same class if their cycle specifications have the same structure. For example, in $S_5$ the permutations (1 3 5)(2)(4) and (2 5 3)(1)(4) are in the same class as each other but in a different class from that which contains (1 5)(2 4)(3).

In the case of the continuous rotation group, rotations by the same angle $\theta$ about any two axes labelled $i$ and $j$ are in the same class, because the group contains a rotation that takes the first axis into the second. Without going into mathematical details, a rotation about axis $i$ can be represented by the operator $R_i(\theta)$, and the two rotations are connected by a relationship of the form

$$R_j(\theta) = \phi_{ij}^{-1} R_i(\theta) \phi_{ij},$$

where $\phi_{ij}$ is the member of the full continuous rotation group that takes axis $i$ into axis $j$.

## 22.8 Exercises

22.1 For each of the following sets, determine whether they form a group under the operation indicated. Where it is relevant you may assume that matrix multiplication is associative.

(a) The integers (mod 10) under addition.

(b) The integers (mod 10) under multiplication.

(c) The integers $1, 2, 3, 4, 5, 6$ under multiplication (mod 7).

(d) The integers $1, 2, 3, 4, 5$ under multiplication (mod 6).

(e) All matrices of the form

$$\begin{pmatrix} a & a-b \\ 0 & b \end{pmatrix}$$

where $a$ and $b$ are integers (mod 5), and $a \neq 0 \neq b$, under matrix multiplication.

(f) Those elements of the set in (e) which are of order 1 or 2 (taken together).

(g) All matrices of the form

$$\begin{pmatrix} 1 & 0 & 0 \\ a & 1 & 0 \\ b & c & 1 \end{pmatrix} \quad \text{where } a, b, c \text{ are integers,}$$

under matrix multiplication.

22.2 Which of the following relationships between $X$ and $Y$ are equivalence relations? Give a proof of your conclusions in each case.

(a) $X$ and $Y$ are integers and $X - Y$ is odd.
(b) $X$ and $Y$ are integers and $X - Y$ is even.
(c) $X$ and $Y$ are people and have the same postcode.
(d) $X$ and $Y$ are people and have a parent in common.
(e) $X$ and $Y$ are people and have the same mother.
(f) $X$ and $Y$ are $n \times n$ matrices satisfying $Y = PXQ$, where $P$ and $Q$ are elements of a group $\mathcal{G}$ of $n \times n$ matrices.

22.3 Prove that the set $\mathcal{M}$ of matrices

$$A = \begin{pmatrix} a & b \\ 0 & c \end{pmatrix},$$

where $a$, $b$, $c$ are integers (mod 5) and $a \neq 0 \neq c$, form a non-Abelian group under matrix multiplication.

Show that the subset containing elements of $\mathcal{M}$ that are of order 1 or 2 do not form a proper subgroup of $\mathcal{M}$

(a) using Lagrange's theorem,
(b) by direct demonstration that the set is not closed.

22.4 $\mathcal{S}$ is the set of all $2 \times 2$ matrices of the form

$$A = \begin{pmatrix} w & x \\ y & z \end{pmatrix} \qquad \text{where } wz - xy = 1.$$

Show that $\mathcal{S}$ is a group under matrix multiplication. Which element(s) have order 2? Prove that an element $A$ has order 3 if $w + z + 1 = 0$.

22.5 The group of rotations (excluding reflections and inversions), in three dimensions, that take a cube into itself is known as the group 432 (or $O$ in the usual chemical notation). Show that the group has 24 elements, by each of the following methods.

(a) Identify the distinct relevant axes and count the number of qualifying rotations about each.
(b) The orientation of the cube is determined if the directions of two of its body diagonals are given. Consider the number of distinct ways in which one body diagonal can be chosen to be 'vertical' and a second diagonal made to lie along a particular direction.

22.6 Identify the eight symmetry operations on a square. Show that they form a group (known to crystallographers as $4mm$ or to chemists as $C_{4v}$) having one element of order 1, five of order 2, and two of order 4. Find its proper subgroups and the corresponding cosets.

22.7 For the group $\mathcal{G}$ with multiplication table 22.8 and proper subgroup $\mathcal{H} = \{I, A, B\}$, denote the coset $\{I, A, B\}$ by $\mathcal{C}_1$ and the coset $\{C, D, E\}$ by $\mathcal{C}_2$. Form the set of all possible products of a member of $\mathcal{C}_1$ with itself, and denote this by $\mathcal{C}_1\mathcal{C}_1$. Similarly compute $\mathcal{C}_2\mathcal{C}_2$, $\mathcal{C}_1\mathcal{C}_2$ and $\mathcal{C}_2\mathcal{C}_1$. Show that each product coset is equal to $\mathcal{C}_1$ or to $\mathcal{C}_2$, and that a $2 \times 2$ multiplication table can be formed demonstrating that $\mathcal{C}_1$ and $\mathcal{C}_2$ are themselves the elements of a group of order 2. A subgroup like $\mathcal{H}$ whose cosets themselves form a group is a *normal subgroup*.

22.8 In the quaternion group $\mathcal{Q}$ the elements form the set

$$\{1, -1, i, -i, j, -j, k, -k\},$$

with $i^2 = j^2 = k^2 = -1$, $ij = k$ and its cyclic permutations, and $ji = -k$ and its cyclic permutations. Find the proper subgroups of $\mathcal{Q}$ and the corresponding cosets. Show that the subgroup of order 2 is a normal subgroup, but that the other subgroups are not. Show that $\mathcal{Q}$ cannot be isomorphic to the group $4mm$ ($C_{4v}$) considered in question 22.6.

## 22.9 Hints and answers

22.1 † (a) Yes. (b) No, no inverse for 2. (c) Yes. (d) No, $2 \times 3$ is not in the set. (e) Yes. (f) Yes, they form a subgroup of order 4; $[1, 0; 0, 1]$ $[4, 0; 0, 4]$ $[1, 2; 0, 4]$ $[4, 3; 0, 1]$. (g) Yes.

22.2 (a) No, not reflexive. (b) Yes, partition of integers into odd and even. (c) Yes. (d) No, not transitive; $X \to Y \to Z$ if $Y$'s parents both re-marry and $X$ and $Z$ are children of the two second marriages. (e) Yes. (f) Yes.

22.3 † Matrices $[1, 3; 0, 1]$ and $[2, 3; 0, 1]$ do not commute, so group is non-Abelian. (a) 12 elements in set $\{[1, 0; 0, 1]$ $[4, 0; 0, 4]$ $[1, b; 0, 4]$ $[4, b; 0, 1]$ with $b$ arbitrary$\}$. Full group has order $4 \times 4 \times 5 = 80$, which is not divisible by 12. (b) $[1, 0; 0, 4][1, 3; 0, 4] = [1, 3; 0, 1]$, which has order $> 2$.

22.4 † Use $|AB| = |A||B| = 1 \times 1 = 1$ to prove closure. Inverse has $w \leftrightarrow z$, $x \leftrightarrow -x$, $y \leftrightarrow -y$, giving $|A^{-1}| = 1$, i.e. it is in the set. The only element of order 2 is $-I$; $A^2$ can be simplified to $[-(w + 1), -x; -y, -(z + 1)]$.

22.5 (a) Identity $= 1$, three rotations of $\pi$ about face normals, six rotations of $\pm\pi/2$ about face normals, six rotations of $\pi$ about edge diagonals, eight rotations of $\pm 2\pi/3$ about body diagonals. (b) The 'vertical' diagonal can be chosen in $4 \times 2$ ways (either end of each diagonal can be 'up'). There are then three equivalent rotational positions about the vertical. There are thus $4 \times 2 \times 3$ possibilities.

† Where matrix elements are given as a list, the convention used is [row 1; row 2; ...] with individual entries in each row separated by commas.

22.6 With the notation indicated in figure 22.3, $R$ being a rotation of $\pi/2$ about an axis perpendicular to the square: $I$ has order 1; $R^2$, $m_1$, $m_2$, $m_3$, $m_4$ have order 2; $R$, $R^3$ have order 4.

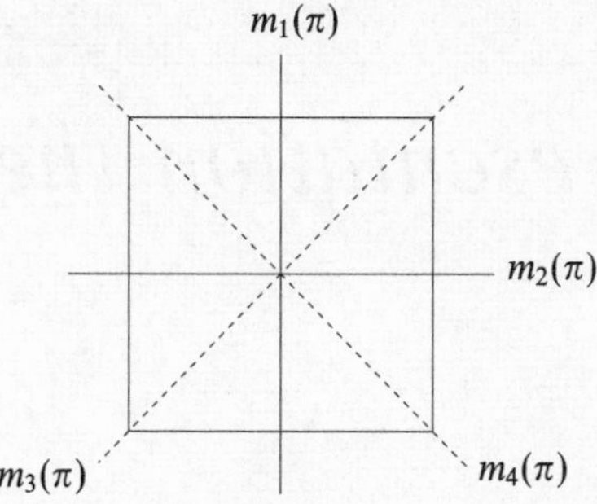


Figure 22.3

subgroup $\{I, R, R^2, R^3\}$ has cosets $\{I, R, R^2, R^3\}$, $\{m_1, m_2, m_3, m_4\}$;
subgroup $\{I, R^2\}$ has cosets $\{I, R^2\}$, $\{R, R^3\}$, $\{m_1, m_2\}$, $\{m_3, m_4\}$;
subgroup $\{I, m_1\}$ has cosets $\{I, m_1\}$, $\{R, m_3\}$, $\{R^2, m_2\}$, $\{R^3, m_4\}$;
subgroup $\{I, m_2\}$ has cosets $\{I, m_2\}$, $\{R, m_4\}$, $\{R^2, m_1\}$, $\{R^3, m_3\}$;
subgroup $\{I, m_3\}$ has cosets $\{I, m_3\}$, $\{R, m_2\}$, $\{R^2, m_4\}$, $\{R^3, m_1\}$;
subgroup $\{I, m_4\}$ has cosets $\{I, m_4\}$, $\{R, m_1\}$, $\{R^2, m_3\}$, $\{R^3, m_2\}$.

22.7 $\mathcal{C}_1\mathcal{C}_1 = \mathcal{C}_2\mathcal{C}_2 = \mathcal{C}_1$, $\mathcal{C}_1\mathcal{C}_2 = \mathcal{C}_2\mathcal{C}_1 = \mathcal{C}_2$.

22.8 Subgroup $\{1, -1\}$ has cosets $\mathcal{C}_1 = \{1, -1\}$, $\mathcal{C}_i = \{i, -i\}$, $\mathcal{C}_j = \{j, -j\}$, $\mathcal{C}_k = \{k, -k\}$. Subgroup $\{1, i, -1, -i\}$ has cosets $\mathcal{D}_i = \{1, i, -1, -i\}$, $\mathcal{D}'_i = \{j, -j, k, -k\}$; corresponding pairs of cosets $\mathcal{D}_j$, $\mathcal{D}_{j'}$ and $\mathcal{D}_k$, $\mathcal{D}_{k'}$ are obtained from subgroups $\{1, j, -1, -j\}$ and $\{1, k, -1, -k\}$ respectively. They can be written down by cyclically permuting $i, j, k$ in $\mathcal{D}_i$, $\mathcal{D}'_i$. The cosets of $\{1, -1\}$ form a group with $\mathcal{C}_1$ as the identity and $\mathcal{C}_i\mathcal{C}_j = \mathcal{C}_k$ etc. The cosets of $\{1, i, -1, -i\}$ do not form a group, since, for example, the product $\mathcal{D}_i\mathcal{D}'_i$ involves all elements of $\mathcal{Q}$. It is sufficient to notice that $4mm$ has six elements of order 2, whilst $\mathcal{Q}$ has only two.

# 23

# *Representation theory*

As indicated at the start of the previous chapter, significant conclusions can often be drawn about a physical system simply from the study of its symmetry properties. That chapter was devoted to setting up a formal mathematical basis with which to describe and classify such properties; the current chapter shows how to implement the consequences of the resulting classifications and obtain concrete physical conclusions about the system under study. The connection between the two chapters is akin to that between working with coordinate-free vectors, each denoted by a single symbol, and working with a coordinate system in which the same vectors are expressed in terms of components, and the analysis is carried out with those components.

The 'coordinate systems' we will choose will be ones that are expressed in terms of matrices; it will be clear that ordinary numbers would not be sufficient, as they make no provision for any non-commutation amongst the elements of a group. Thus, in this chapter the group elements will be *represented* by matrices that have the same commutation relations as the members of the group, whatever the group's original nature (symmetry operations, functional forms, matrices, permutations, etc.). Most of our applications will be concerned with *representations* of the groups that consist of the symmetry operations on molecules containing two or more identical atoms.

In section 23.1, prior to formally developing *representation theory* we demonstrate, with an elementary example, the kind of conclusions that can be reached by arguing purely on symmetry grounds. Sections 23.2–23.10 develop the formal side of representation theory and establish the procedures used in section 23.11 to tackle a variety of problems drawn from across the physical sciences.

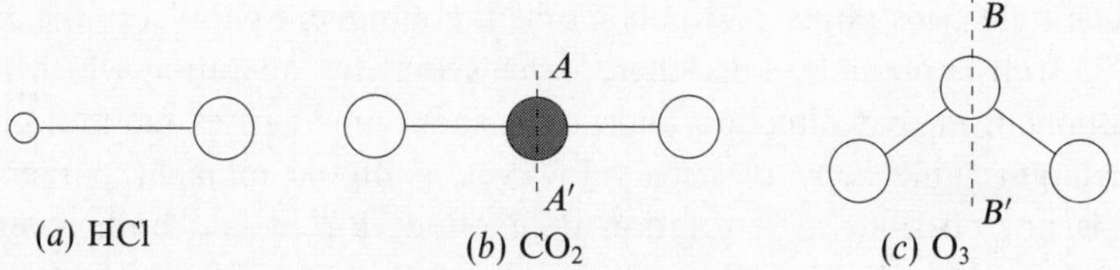


Figure 23.1 Three molecules, (*a*) hydrogen chloride, (*b*) carbon dioxide and (*c*) ozone, for which symmetry considerations impose varying degrees of constraint on their possible electric dipole moments.

## 23.1 Dipole moments of molecules

Some simple consequences of *symmetry* can be demonstrated by considering whether a permanent electric dipole moment can exist in any particular molecule. Three simple molecules, hydrogen chloride, carbon dioxide and ozone, are illustrated in figure 23.1. Although each molecule is electrically neutral, an electric dipole moment will exist in any case in which the centres of gravity of the positive charges (due to protons in the atomic nuclei) and the negative charges (due to the electrons) do not coincide.

For hydrogen chloride there is no reason why they should coincide; indeed, the normal picture of the mechanism causing the atoms to form a molecule is that the electron from the hydrogen atom moves its average position from that of its proton nucleus to somewhere between the hydrogen and chlorine nuclei. There is no compensating movement of positive charge, and a net dipole moment is to be expected – and is found experimentally.

For the linear molecule carbon dioxide it seems obvious that it cannot have a dipole moment, because of its symmetry. Putting this rather more rigorously, we note that any rotation about the long axis of the molecule leaves it totally unchanged; consequently, any component of a permanent electric dipole perpendicular to that axis must be zero. That only leaves the possibility of a component parallel to the axis. However, a rotation of $\pi$ radians about the axis $AA'$ shown in figure 23.1(*b*) carries the molecule into itself, as does a reflection in a plane through the carbon atom and perpendicular to the molecular axis (i.e. one with its normal parallel to the axis). In both cases the two oxygen atoms change places but, as they are identical, the molecule is indistinguishable from the original. Either 'symmetry operation' would reverse the sign of any dipole component directed parallel to the molecular axis; this can only be compatible with the indistinguishability of the original and final systems if this component is zero. Thus on symmetry grounds carbon dioxide cannot have a permanent electric dipole moment.

Finally, for ozone, which is angular rather than linear, symmetry does not place such tight constraints. A dipole moment component parallel to the axis $BB'$ (figure 23.1($c$)) is possible, since there is no symmetry operation which reverses the component in that direction and at the same time carries the molecule into an indistinguishable copy of itself. However, a dipole moment perpendicular to $BB'$ is not possible, as a rotation of $\pi$ about $BB'$ would both reverse any such component and carry the ozone molecule into itself – two contradictory conclusions unless the component is zero.

In summary, symmetry requirements appear in the form that some or all components of permanent electric dipoles in molecules are forbidden; they do not show that the other components do exist, only that they may. The greater the symmetry of the molecule, the tighter the restrictions on potentially non-zero components of its dipole moment.

In section 23.11 other, more complicated, physical situations will be analysed using results derived from representation theory. In anticipation of these results, and since it may help the reader to understand where the developments in the next nine sections are leading, we make here a broad, powerful, but rather formal, statement as follows.

*If a physical system is such that after the application of particular rotations or reflections (or a combination of the two) the final system is indistinguishable from the original, its behaviour, and hence the functions which describe its behaviour, must have the corresponding property of invariance when subjected to the same rotations and reflections.*

## 23.2 Choosing an appropriate formalism

As mentioned in the introduction to this chapter, the elements of a finite group $\mathcal{G}$ can be *represented* by matrices; this is done in the following way. A suitable column matrix $\mathsf{u}$, known as a *basis vector*,† is chosen and written in terms of its components $u_i$ as $\mathsf{u} = (u_1\ u_2\ \cdots\ u_n)^{\mathrm{T}}$. The $u_i$ may be of a variety of natures, e.g. numbers, coordinates, functions, or even a set of labels, though for any one basis vector they will all be of the same kind.

Once chosen, the basis vector can be used to generate an $n$-dimensional *representation* of the group as follows. An element $X$ of the group is selected and its effect on each *basis function* $u_i$ is determined. If the action of $X$ on $u_1$ is to produce $u_1'$, etc., then the set of equations

$$u_i' = Xu_i \tag{23.1}$$

generates a new column matrix $\mathsf{u}' = (u_1'\ u_2'\ \cdots\ u_n')^{\mathrm{T}}$. Having established $\mathsf{u}$ and $\mathsf{u}'$

† This usage of the term *basis vector* is not exactly the same as in chapter 7.

we can determine the $n \times n$ matrix, $\mathsf{M}(X)$ say, that connects them by

$$\mathsf{u}' = \mathsf{M}(X)\mathsf{u}. \tag{23.2}$$

It may seem natural to use the matrix $\mathsf{M}(X)$ so generated as the representative matrix of the element $X$, but in fact, because our choice of convention is that $Z = XY$ implies that the effect of applying element $Z$ is the same as that of first applying $Y$ and then applying $X$ to the result, one further step has to be taken. So that the representative matrices $\mathsf{D}(X)$ may follow the same convention, i.e.

$$\mathsf{D}(Z) = \mathsf{D}(X)\mathsf{D}(Y),$$

and at the same time respect the normal rules of matrix multiplication, it is necessary to take the *transpose* of $\mathsf{M}(X)$ as the representative matrix $\mathsf{D}(X)$. Explicitly,

$$\mathsf{D}(X) = \mathsf{M}^{\mathrm{T}}(X) \tag{23.3}$$

and (23.2) becomes

$$\mathsf{u}' = \mathsf{D}^{\mathrm{T}}(X)\mathsf{u}. \tag{23.4}$$

Thus the procedure for determining the matrix $\mathsf{D}(X)$ that represents the group element $X$ in a representation based on basis vector $\mathsf{u}$ is summarised by equations (23.1)–(23.4).†

This procedure is then repeated for each element $X$ of the group, and the resulting set of matrices $\mathsf{D} = \{\mathsf{D}(X)\}$ is said to be the $n$-dimensional representation of $\mathcal{G}$ having $\mathsf{u}$ as its basis. The need to take the transpose of each matrix $\mathsf{M}(X)$ is not of any fundamental significance, since the only thing that really matters is that the matrices $\mathsf{D}(X)$ have the appropriate multiplication properties – and, as defined, they do.

In cases in which the basis functions are labels the actions of the group elements are such as to cause rearrangements of the labels; correspondingly the matrices $\mathsf{D}(X)$ must contain only '1's and '0's as entries, with each row and each column containing a single '1'.

† An alternative procedure in which a row vector is used as the basis vector is possible. Defining equations of the form $\mathsf{u}^{\mathrm{T}}X = \mathsf{u}^{\mathrm{T}}\mathsf{D}(X)$ are used and no additional transpositions are needed to define the representative matrices. However, row-matrix equations are cumbersome to write out and having the element of the group appear to the right of the basis vector is somewhat unnatural.

►*For the group $S_3$ of permutations on three objects, which has group multiplication table 22.8, with (in cycle notation)*

$$I = (1)(2)(3), \quad A = (1\,2\,3), \quad B = (1\,3\,2),$$

$$C = (1)(2\,3), \quad D = (3)(1\,2), \quad E = (2)(1\,3),$$

*use as the components of a basis vector the ordered letter triplets*

$$u_1 = \{\mathrm{P\,Q\,R}\}, \quad u_2 = \{\mathrm{Q\,R\,P}\}, \quad u_3 = \{\mathrm{R\,P\,Q}\},$$

$$u_4 = \{\mathrm{P\,R\,Q}\}, \quad u_5 = \{\mathrm{Q\,P\,R}\}, \quad u_6 = \{\mathrm{R\,Q\,P}\}.$$

*Generate a six-dimensional representation $\mathsf{D} = \{\mathsf{D}(X)\}$ of the group and confirm that the representative matrices multiply according to table 22.8, e.g.*

$$\mathsf{D}(C)\mathsf{D}(B) = \mathsf{D}(E).$$

It is immediate that the identity permutation $I = (1)(2)(3)$ leaves all $u_i$ unchanged, i.e. $u_i' = u_i$ for all $i$. The representative matrix $\mathsf{D}(I)$ is thus $\mathsf{I}_6$, the $6 \times 6$ unit matrix.

We next take $X$ as the permutation $A = (1\,2\,3)$ and, using (23.1), let it act on each of the components of the basis vector:

$$\begin{aligned} u_1' &= Au_1 = (1\,2\,3)\{\mathrm{P\,Q\,R}\} = \{\mathrm{Q\,R\,P}\} = u_2 \\ u_2' &= Au_2 = (1\,2\,3)\{\mathrm{Q\,R\,P}\} = \{\mathrm{R\,P\,Q}\} = u_3 \\ &\vdots \qquad\qquad\qquad\qquad\qquad\qquad\qquad \vdots \\ u_6' &= Au_6 = (1\,2\,3)\{\mathrm{R\,Q\,P}\} = \{\mathrm{Q\,P\,R}\} = u_5. \end{aligned}$$

The matrix $\mathsf{M}(A)$ has to be such that $\mathsf{u}' = \mathsf{M}(A)\mathsf{u}$ (here dots replace zeroes to aid readability):

$$\mathsf{u}' = \begin{pmatrix} u_2 \\ u_3 \\ u_1 \\ u_6 \\ u_4 \\ u_5 \end{pmatrix} = \begin{pmatrix} \cdot & 1 & \cdot & \cdot & \cdot & \cdot \\ \cdot & \cdot & 1 & \cdot & \cdot & \cdot \\ 1 & \cdot & \cdot & \cdot & \cdot & \cdot \\ \cdot & \cdot & \cdot & \cdot & \cdot & 1 \\ \cdot & \cdot & \cdot & 1 & \cdot & \cdot \\ \cdot & \cdot & \cdot & \cdot & 1 & \cdot \end{pmatrix} \begin{pmatrix} u_1 \\ u_2 \\ u_3 \\ u_4 \\ u_5 \\ u_6 \end{pmatrix} \equiv \mathsf{M}(A)\mathsf{u}.$$

$\mathsf{D}(A)$ is then equal to $\mathsf{M}^{\mathrm{T}}(A)$.

The other $\mathsf{D}(X)$ are calculated in a similar way. In general, if

$$Xu_i = u_j,$$

then $[\mathsf{M}(X)]_{ij} = 1$, leading to $[\mathsf{D}(X)]_{ji} = 1$ and $[\mathsf{D}(X)]_{jk} = 0$ for $k \neq i$. For example,

$$Cu_3 = (1)(23)\{\mathrm{R\,P\,Q}\} = \{\mathrm{R\,Q\,P}\} = u_6$$

implies that $[\mathsf{D}(C)]_{63} = 1$ and $[\mathsf{D}(C)]_{6k} = 0$ for $k = 1, 2, 4, 5, 6$. When calculated in full

$$\mathsf{D}(C) = \begin{pmatrix} \cdot & \cdot & \cdot & 1 & \cdot & \cdot \\ \cdot & \cdot & \cdot & \cdot & 1 & \cdot \\ \cdot & \cdot & \cdot & \cdot & \cdot & 1 \\ 1 & \cdot & \cdot & \cdot & \cdot & \cdot \\ \cdot & 1 & \cdot & \cdot & \cdot & \cdot \\ \cdot & \cdot & 1 & \cdot & \cdot & \cdot \end{pmatrix}, \qquad \mathsf{D}(B) = \begin{pmatrix} \cdot & 1 & \cdot & \cdot & \cdot & \cdot \\ \cdot & \cdot & 1 & \cdot & \cdot & \cdot \\ 1 & \cdot & \cdot & \cdot & \cdot & \cdot \\ \cdot & \cdot & \cdot & \cdot & \cdot & 1 \\ \cdot & \cdot & \cdot & 1 & \cdot & \cdot \\ \cdot & \cdot & \cdot & \cdot & 1 & \cdot \end{pmatrix},$$

$$\mathsf{D}(E) = \begin{pmatrix} \cdot & \cdot & \cdot & \cdot & \cdot & 1 \\ \cdot & \cdot & \cdot & 1 & \cdot & \cdot \\ \cdot & \cdot & \cdot & \cdot & 1 & \cdot \\ \cdot & 1 & \cdot & \cdot & \cdot & \cdot \\ \cdot & \cdot & 1 & \cdot & \cdot & \cdot \\ 1 & \cdot & \cdot & \cdot & \cdot & \cdot \end{pmatrix}$$

from which it can be verified that $\mathsf{D}(C)\mathsf{D}(B) = \mathsf{D}(E)$. ◄

Whilst a representation obtained in this way necessarily has the same dimension as the order of the group it represents, there are, in general, square matrices of both smaller and larger dimensions that can be used to represent the group, though their existence may be less obvious.

One possibility that arises when the group elements are symmetry operations on an object whose position and orientation can be referred to a space coordinate system is called the *natural representation.* In it the representative matrices $\mathsf{D}(X)$ describe, in terms of a fixed coordinate system, what happens to a coordinate system which moves with the object when $X$ is applied. There is usually some redundancy when coordinates are used in this type of representation, since interparticle distances are fixed, and fewer than $3N$ coordinates (where $N$ is the number of identical particles) are needed to specify uniquely the object's position and orientation. Subsection 23.11.1 gives an example that illustrates both the advantages and disadvantages of this type of choice.

We continue here with an example of a natural representation which has no such redundancy.

►*Use the fact that the group considered in the previous worked example is isomorphic to the group of two-dimensional symmetry operations on an equilateral triangle to generate a three-dimensional representation of the group.*

Label the triangle's corners as 1, 2, 3 and three fixed points in space as P, Q, R, so that initially 1 lies at the point P, 2 lies at the point Q, and 3 at the point R. We take P, Q, R as the components of the basis vector.

In figure 23.2, ($a$) shows the initial configuration, and also formally shows the

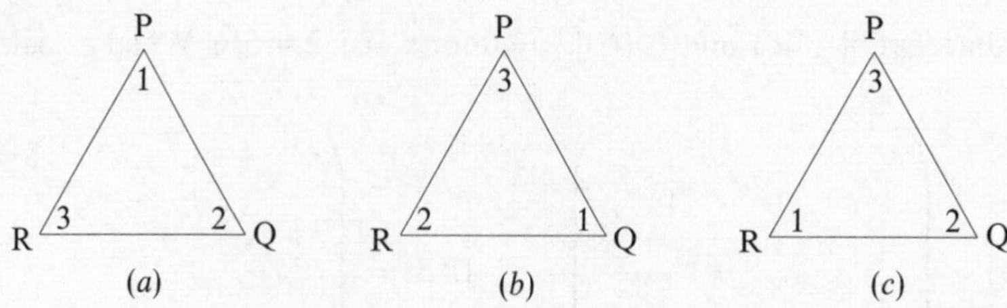


Figure 23.2 Diagram (*a*) shows the definition of the basis vector, (*b*) shows the effect of applying a clockwise rotation of $2\pi/3$, and (*c*) shows the effect of applying a reflection in the mirror axis through Q.

result of applying the identity $I$; it is therefore described by the basis vector, $(\mathrm{P\ Q\ R})^{\mathrm{T}}$.

Diagram (*b*) shows the the effect of a clockwise rotation by $2\pi/3$ (corresponding to element $A$ in the previous example); the new column matrix is $(\mathrm{Q\ R\ P})^{\mathrm{T}}$.

Diagram (*c*) shows the effect of a typical mirror reflection – the one that leaves the corner at point Q unchanged (element $D$ in table 22.8 and the previous example); the new column matrix is now $(\mathrm{R\ Q\ P})^{\mathrm{T}}$.

In similar fashion it can be concluded that the column matrix corresponding to element $B$ (rotation by $4\pi/3$) is $(\mathrm{R\ P\ Q})^{\mathrm{T}}$, and that the other two reflections $C$ and $E$ result in column matrices $(\mathrm{P\ R\ Q})^{\mathrm{T}}$ and $(\mathrm{Q\ P\ R})^{\mathrm{T}}$ respectively. The forms of the representative matrices $\mathsf{D}^{\mathrm{nat}}(X)$ are now determined by equations such as, for element $E$,

$$\begin{pmatrix} \mathrm{Q} \\ \mathrm{P} \\ \mathrm{R} \end{pmatrix} = \begin{pmatrix} 0 & 1 & 0 \\ 1 & 0 & 0 \\ 0 & 0 & 1 \end{pmatrix} \begin{pmatrix} \mathrm{P} \\ \mathrm{Q} \\ \mathrm{R} \end{pmatrix}$$

implying that

$$\mathsf{D}^{\mathrm{nat}}(E) = \begin{pmatrix} 0 & 1 & 0 \\ 1 & 0 & 0 \\ 0 & 0 & 1 \end{pmatrix}^{\mathrm{T}} = \begin{pmatrix} 0 & 1 & 0 \\ 1 & 0 & 0 \\ 0 & 0 & 1 \end{pmatrix}.$$

In this way the full representation is obtained as

$$\mathsf{D}^{\mathrm{nat}}(I) = \begin{pmatrix} 1 & 0 & 0 \\ 0 & 1 & 0 \\ 0 & 0 & 1 \end{pmatrix}, \quad \mathsf{D}^{\mathrm{nat}}(A) = \begin{pmatrix} 0 & 0 & 1 \\ 1 & 0 & 0 \\ 0 & 1 & 0 \end{pmatrix}, \quad \mathsf{D}^{\mathrm{nat}}(B) = \begin{pmatrix} 0 & 1 & 0 \\ 0 & 0 & 1 \\ 1 & 0 & 0 \end{pmatrix},$$

$$\mathsf{D}^{\mathrm{nat}}(C) = \begin{pmatrix} 1 & 0 & 0 \\ 0 & 0 & 1 \\ 0 & 1 & 0 \end{pmatrix}, \quad \mathsf{D}^{\mathrm{nat}}(D) = \begin{pmatrix} 0 & 0 & 1 \\ 0 & 1 & 0 \\ 1 & 0 & 0 \end{pmatrix}, \quad \mathsf{D}^{\mathrm{nat}}(E) = \begin{pmatrix} 0 & 1 & 0 \\ 1 & 0 & 0 \\ 0 & 0 & 1 \end{pmatrix}.$$

It should be emphasised that though the group contains six elements this representation is three-dimensional. ◀

We will concentrate on matrix representations of *finite* groups, particularly rotation and reflection groups (the so-called crystal point groups). The general ideas carry over to infinite groups such as the continuous rotation groups, but in a book such as this, which aims to cover many areas of applicable mathematics, some topics can only be mentioned and not explored. We now give the formal definition of a representation.

**Definition.** *A representation* $\mathsf{D} = \{\mathsf{D}(X)\}$ *of a group* $\mathcal{G}$ *is an assignment of a non-singular square* $n \times n$ *matrix* $\mathsf{D}(X)$ *to each element* $X$ *belonging to* $\mathcal{G}$*, such that*

(i) $\mathsf{D}(I) = \mathsf{I}_n$, *the unit* $n \times n$ *matrix,*
(ii) $\mathsf{D}(X)\mathsf{D}(Y) = \mathsf{D}(XY)$ *for any two elements* $X$ *and* $Y$ *belonging to* $\mathcal{G}$*, i.e. the matrices multiply in the same way as the group elements they represent.*

A representation by $n \times n$ matrices is said to be an *n-dimensional representation* of $\mathcal{G}$. The dimension $n$ is not to be confused with $g$, the order of the group; in fact, $g$ gives the number of matrices needed in the representation, though they might not all be different.

A consequence of the two defining conditions for a representation is that the matrix associated with the inverse of $X$ is the inverse of the matrix associated with $X$. This follows immediately from setting $Y = X^{-1}$ in (ii),

$$\mathsf{D}(X)\mathsf{D}(X^{-1}) = \mathsf{D}(XX^{-1}) = \mathsf{D}(I) = \mathsf{I}_n.$$

Hence

$$\mathsf{D}(X^{-1}) = [\mathsf{D}(X)]^{-1}.$$

As an example, the four-element Abelian group consisting of the set $\{1, i, -1, -i\}$ under ordinary multiplication has a two-dimensional representation based on the column matrix $(1 \quad i)^{\mathrm{T}}$:

$$\mathsf{D}(1) = \begin{pmatrix} 1 & 0 \\ 0 & 1 \end{pmatrix}, \qquad \mathsf{D}(i) = \begin{pmatrix} 0 & -1 \\ 1 & 0 \end{pmatrix},$$

$$\mathsf{D}(-1) = \begin{pmatrix} -1 & 0 \\ 0 & -1 \end{pmatrix}, \qquad \mathsf{D}(-i) = \begin{pmatrix} 0 & 1 \\ -1 & 0 \end{pmatrix}.$$

The reader should check that $\mathsf{D}(i)\mathsf{D}(-i) = \mathsf{D}(1)$, $\mathsf{D}(i)\mathsf{D}(i) = \mathsf{D}(-1)$, etc., i.e. that the matrices have exactly the same multiplication properties as the elements of the group. Having done so, the reader may also wonder why anybody would bother with the representative matrices, when the original elements are so much simpler to handle! As we will see later, once some general properties of matrix representations have been established, the analysis of larger groups, Abelian and non-Abelian, can be reduced to routine, almost cookbook, procedures.

An $n$-dimensional representation of $\mathcal{G}$ is a homomorphism of $\mathcal{G}$ into the set of invertible $n \times n$ matrices (i.e. $n \times n$ matrices that have inverses, or, equivalently, have non-zero determinants); this set is usually known as the general linear group, and denoted by GL($n$). In general the same matrix may represent more than one element of $\mathcal{G}$ but, if all the matrices representing the elements of $\mathcal{G}$ are different, the representation is said to be *faithful*, and the homomorphism becomes an isomorphism onto a subgroup of GL($n$).

A trivial but important representation is $\mathsf{D}(X) = \mathsf{I}_n$ for all elements $X$ of $\mathcal{G}$. Clearly both of the defining relationships are satisfied, and there is no restriction on the value of $n$. However, such a representation is not a faithful one.

To sum up, in the context of a rotation–reflection group, the transposes of the set of $n \times n$ matrices $\mathsf{D}(X)$ that make up a representation $\mathsf{D}$ may be thought of as describing what happens to an $n$-component basis vector of coordinates, $(x\ y\ \cdots)^{\mathrm{T}}$, or of functions, $(\Psi_1\ \Psi_2\ \cdots)^{\mathrm{T}}$, with the $\Psi_i$ themselves functions of coordinates, when the group operation $X$ is carried out on each of the coordinates and functions. For example, to return to the symmetry operations on an equilateral triangle, $R$, the (clockwise) rotation by $2\pi/3$, carries the three-dimensional basis vector $(x\quad y\quad z)^{\mathrm{T}}$ into the column matrix

$$\begin{pmatrix} -\frac{1}{2}x + \frac{\sqrt{3}}{2}y \\ -\frac{\sqrt{3}}{2}x - \frac{1}{2}y \\ z \end{pmatrix}$$

whilst the two-dimensional basis vector of functions $(r^2\quad 3z^2 - r^2)^{\mathrm{T}}$ is unaltered, as neither $r$ nor $z$ is changed by the rotation. The fact that $z$ is unchanged by any of the operations of the group shows that the components $x$, $y$, $z$ actually divide (i.e. are 'reducible', to anticipate a more formal description) into two sets: one comprises $z$, which is unchanged by any of the operations, and the other comprises $x$, $y$, which change as a pair into linear combinations of themselves. This is an important observation to which we return in section 23.4.

## 23.3 Equivalent representations

If $\mathsf{D}$ is an $n$-dimensional representation of a group $\mathcal{G}$, and $\mathsf{Q}$ any fixed invertible $n \times n$ matrix ($|\mathsf{Q}| \neq 0$), then the set of matrices defined by the (*similarity*) transformation

$$\mathsf{D}_{\mathsf{Q}}(X) = \mathsf{Q}^{-1}\mathsf{D}(X)\mathsf{Q} \tag{23.5}$$

also forms a representation $\mathsf{D}_{\mathsf{Q}}$ of $\mathcal{G}$, said to be *equivalent* to $\mathsf{D}$. That they do form a representation is shown by

(i) $\mathsf{D}_{\mathsf{Q}}(I) = \mathsf{Q}^{-1}\mathsf{D}(I)\mathsf{Q} = \mathsf{Q}^{-1}\mathsf{I}_n\mathsf{Q} = \mathsf{I}_n$,

(ii) $\mathsf{D}_{\mathsf{Q}}(X)\mathsf{D}_{\mathsf{Q}}(Y) = \mathsf{Q}^{-1}\mathsf{D}(X)\mathsf{Q}\mathsf{Q}^{-1}\mathsf{D}(Y)\mathsf{Q} = \mathsf{Q}^{-1}\mathsf{D}(X)\mathsf{D}(Y)\mathsf{Q}$
$\qquad = \mathsf{Q}^{-1}\mathsf{D}(XY)\mathsf{Q} = \mathsf{D}_{\mathsf{Q}}(XY)$.

Since we can always transform between equivalent representations using a non-singular matrix $\mathsf{Q}$, we will consider such representations to be one and the same.

Despite the similarity of words and manipulations to those of subsection 22.7.1, that two representations are equivalent does not constitute an 'equivalence relation' – for example, the reflexive property does not hold for a general fixed $\mathsf{Q}$. However, if $\mathsf{Q}$ were not fixed, but simply restricted to belonging to a set of matrices which themselves form a group, then (23.5) would constitute an equivalence relation.

The general invertible matrix $\mathsf{Q}$ that appeared in the definition (23.5) of equivalent matrices describes changes arising from a change in the coordinate system (i.e. in the set of basis functions). As before, suppose that the effect of $X$ on the basis functions is expressed by the action of $\mathsf{M}(X)$ (equal to $\mathsf{D}^{\mathrm{T}}(X)$) on the corresponding basis vector:

$$\mathsf{u}' = \mathsf{M}(X)\mathsf{u} = \mathsf{D}^{\mathrm{T}}(X)\mathsf{u}. \tag{23.6}$$

Then a change of basis would be given by $\mathsf{u}_{\mathsf{Q}} = \mathsf{Q}\mathsf{u}$ and $\mathsf{u}'_{\mathsf{Q}} = \mathsf{Q}\mathsf{u}'$, and we may write

$$\mathsf{u}'_{\mathsf{Q}} = \mathsf{Q}\mathsf{u}' = \mathsf{Q}\mathsf{M}(X)\mathsf{u} = \mathsf{Q}\mathsf{D}^{\mathrm{T}}(X)\mathsf{Q}^{-1}\mathsf{u}_{\mathsf{Q}}. \tag{23.7}$$

This is of the same form as (23.6), i.e.

$$\mathsf{u}'_{\mathsf{Q}} = \mathsf{D}^{\mathrm{T}}{}_{\mathsf{Q}^{\mathrm{T}}}(X)\mathsf{u}_{\mathsf{Q}},$$

where $\mathsf{D}_{\mathsf{Q}^{\mathrm{T}}}(X) = (\mathsf{Q}^{\mathrm{T}})^{-1}\mathsf{D}(X)\mathsf{Q}^{\mathrm{T}}$ is related to $\mathsf{D}(X)$ by a similarity transformation. Thus $\mathsf{D}_{\mathsf{Q}^{\mathrm{T}}}(X)$ is simply the same linear transformation as $\mathsf{D}(X)$, but with respect to a new basis vector $\mathsf{u}_{\mathsf{Q}}$; this supports our contention that representations connected by similarity transformations should be considered as the *same* representation.

►*For the four-element Abelian group consisting of the set $\{1, i, -1, -i\}$ under ordinary multiplication discussed near the end of section 23.2, change the basis vector from $\mathsf{u} = (1 \;\; i)^{\mathrm{T}}$ to $\mathsf{u}_{\mathsf{Q}} = (3-i \;\; 2i-5)^{\mathrm{T}}$. Find the real transformation matrix $\mathsf{Q}$. Show that the transformed representative matrix for the element i, $\mathsf{D}_{\mathsf{Q}^{\mathrm{T}}}(i)$ is*

$$\mathsf{D}_{\mathsf{Q}^{\mathrm{T}}}(i) = \begin{pmatrix} 17 & -29 \\ 10 & -17 \end{pmatrix}$$

*and verify that $\mathsf{D}^{\mathrm{T}}_{\mathsf{Q}^{\mathrm{T}}}(i)\mathsf{u}_{\mathsf{Q}} = i\mathsf{u}_{\mathsf{Q}}$.*

Solving the matrix equation

$$\begin{pmatrix} 3-i \\ 2i-5 \end{pmatrix} = \begin{pmatrix} a & b \\ c & d \end{pmatrix}\begin{pmatrix} 1 \\ i \end{pmatrix},$$

with $a, b, c, d$ real, gives $\mathsf{Q}$ and hence $\mathsf{Q}^{-1}$ as

$$\mathsf{Q} = \begin{pmatrix} 3 & -1 \\ -5 & 2 \end{pmatrix}, \qquad \mathsf{Q}^{-1} = \begin{pmatrix} 2 & 1 \\ 5 & 3 \end{pmatrix}.$$

Following (23.7) we now find the transpose of $\mathsf{D}_{\mathsf{Q}^\mathrm{T}}(i)$ as

$$\mathsf{QD}^\mathrm{T}(i)\mathsf{Q}^{-1} = \begin{pmatrix} 3 & -1 \\ -5 & 2 \end{pmatrix}\begin{pmatrix} 0 & 1 \\ -1 & 0 \end{pmatrix}\begin{pmatrix} 2 & 1 \\ 5 & 3 \end{pmatrix} = \begin{pmatrix} 17 & 10 \\ -29 & -17 \end{pmatrix}$$

and hence $\mathsf{D}_{\mathsf{Q}^\mathrm{T}}(i)$ is as stated.

Finally,

$$\mathsf{D}^\mathrm{T}{}_{\mathsf{Q}^\mathrm{T}}(i)\mathsf{u}_\mathsf{Q} = \begin{pmatrix} 17 & 10 \\ -29 & -17 \end{pmatrix}\begin{pmatrix} 3-i \\ 2i-5 \end{pmatrix} = \begin{pmatrix} 1+3i \\ -2-5i \end{pmatrix}$$

$$= i\begin{pmatrix} 3-i \\ 2i-5 \end{pmatrix} = i\mathsf{u}_\mathsf{Q}. \blacktriangleleft$$

Although we will not prove it, it can be shown that any finite representation of a finite group of linear transformations that preserve spatial length (or, in quantum mechanics, preserve the magnitude of a wavefunction) is equivalent to a representation in which all the matrices are unitary (see chapter 7) and so from now on we will consider only *unitary representations.*

## 23.4 Reducibility of a representation

We have already seen that it is possible to have more than one representation of any particular group. For example, the group $\{1, i, -1, -i\}$ under ordinary multiplication has been shown to have a set of $2 \times 2$ matrices, and a set of four unit $n \times n$ matrices $\mathsf{I}_n$, as two of its possible representations.

Consider two or more representations, $\mathsf{D}^{(1)}, \mathsf{D}^{(2)}, \ldots, \mathsf{D}^{(N)}$ (which may be of different dimensions) of a group $\mathcal{G}$. Combine the matrices $\mathsf{D}^{(1)}(X), \mathsf{D}^{(2)}(X), \ldots, \mathsf{D}^{(N)}(X)$ that correspond to element $X$ of $\mathcal{G}$ into a larger *block-diagonal* matrix:

$$\mathsf{D}(X) = \begin{pmatrix} \boxed{\mathsf{D}^{(1)}(X)} & & & \mathbf{0} \\ & \boxed{\mathsf{D}^{(2)}(X)} & & \\ & & \ddots & \\ \mathbf{0} & & & \boxed{\mathsf{D}^{(N)}(X)} \end{pmatrix} \qquad (23.8)$$

Then $\mathsf{D} = \{\mathsf{D}(X)\}$ is the matrix representation of the group obtained by combining the basis vectors of $\mathsf{D}^{(1)}, \mathsf{D}^{(2)}, \ldots, \mathsf{D}^{(N)}$ into one larger basis vector. If, knowingly or unknowingly, we had started with this larger basis vector and found the matrices of the representation $\mathsf{D}$ to have the form shown in (23.8), or to have a form that can be transformed into this by a similarity transformation (23.5), then we would say that D is *reducible* and that each matrix D($X$) can be written as the *direct sum* of smaller representations:

$$\mathsf{D}(X) = \mathsf{D}^{(1)}(X) \oplus \mathsf{D}^{(2)}(X) \oplus \cdots \oplus \mathsf{D}^{(N)}(X).$$

It may be that $\mathsf{D}^{(1)}(X), \mathsf{D}^{(2)}(X), \ldots, \mathsf{D}^{(N)}$ themselves can be further reduced – i.e. written in block diagonal form. For example, suppose that the representation $\mathsf{D}^{(1)}$, say, has a basis vector $(x \quad y \quad z)^{\mathrm{T}}$; then, for the symmetry group of an equilateral triangle, whilst $x$ and $y$ are mixed together for at least one of the operations $X$, $z$ is never changed. In this case the $3 \times 3$ representative matrix $\mathsf{D}^{(1)}(X)$ can itself be written in block diagonal form as a $2 \times 2$ matrix and a $1 \times 1$ matrix. The direct sum matrix D($X$) can now be written

$$\mathsf{D}(X) = \begin{pmatrix} \begin{array}{cc} a & b \\ c & d \end{array} & & & \mathbf{0} \\ & 1 & & \\ & & \mathsf{D}^{(2)}(X) & \\ & & & \ddots \\ \mathbf{0} & & & & \mathsf{D}^{(N)}(X) \end{pmatrix} \qquad (23.9)$$

but the first two blocks can be reduced no further.

When all the other representations $\mathsf{D}^{(2)}(X)$, ... have been similarly treated, what remains is said to be *irreducible*, and has the characteristic of being block diagonal, with blocks that individually cannot be reduced further. The blocks are known as the *irreducible representations of* $\mathcal{G}$, often abbreviated to the *irreps of* $\mathcal{G}$ and we denote then by $\hat{\mathsf{D}}^{(i)}$. They form the building blocks of representation theory, and it is their properties that are used to analyse any given physical situation which is invariant under the operations that form the elements of $\mathcal{G}$. Any representation can be written as a linear combination of irreps.

If, however, the initial choice u of basis vector for the representation D is arbitrary, as it is in general, then it is unlikely that the matrices D($X$) will assume obviously block diagonal forms, though, as the matrices are square, even a matrix with non-zero entries only in the extreme top right and bottom left positions is technically block diagonal. In general, it will be possible to reduce them to block diagonal matrices with more than one block; this reduction corresponds to a transformation Q to a new basis vector $\mathsf{u}_{\mathsf{Q}}$ as described in section 23.3.

In any particular representation $\mathsf{D}$, each constituent irrep $\hat{\mathsf{D}}^{(i)}$ may appear any number of times, or not at all, subject to the obvious restriction that the sum of all the irrep dimensions must add up to the dimension of $\mathsf{D}$ itself. Let us say that $\hat{\mathsf{D}}^{(i)}$ appears $m_i$ times. The general expansion of $\mathsf{D}$ is then written

$$\mathsf{D} = m_1\hat{\mathsf{D}}^{(1)} \oplus m_2\hat{\mathsf{D}}^{(2)} \oplus \cdots \oplus m_N\hat{\mathsf{D}}^{(N)}, \tag{23.10}$$

where if $\mathcal{G}$ is finite so is $N$.

This is such an important result that we shall now restate the situation in somewhat different language. When the set of matrices that form a representation of a particular group of symmetry operations has been brought to irreducible form, the implications are as follows.

(i) Those components of the basis vector which correspond to rows in the representation matrices with a single-entry block, i.e. a $1 \times 1$ block, are unchanged by the operations of the group. Such a coordinate or function is said to transform according to a one-dimensional irrep of $\mathcal{G}$. In the example given above, that the entry on the third row forms a $1 \times 1$ block implies that the third entry in the basis vector $(x \quad y \quad z \ \cdots)^{\mathrm{T}}$, namely $z$, is invariant under the (two-dimensional) symmetry operations on an equilateral triangle in the $xy$-plane.

(ii) If, in any of the $g$ matrices of the representation, the largest-sized block located on the row or column corresponding to a particular coordinate (or function) in the basis vector is $n \times n$, then that coordinate (or function) is mixed by the symmetry operations with $n - 1$ others and is said to transform according to an $n$-dimensional irrep of $\mathcal{G}$. Thus in the matrix (23.9), $x$ is the first entry in the complete basis vector; the first row of the matrix contains two non-zero entries, as does the first column; and so $x$ is part of a two-component basis vector whose components are mixed by the symmetry operations of $\mathcal{G}$. The other component is $y$.

The result may also be formulated in terms of the more abstract notion of vector spaces (chapter 7). The set of $g$ matrices that form an $n$-dimensional representation $\mathsf{D}$ of the group $\mathcal{G}$ can be thought of as acting on column matrices corresponding to vectors in an $n$-dimensional vector space $V$ spanned by the basis functions of the representation. If there exists a *proper subspace* $W$ of $V$, such that if a vector whose column matrix is $\mathsf{w}$ belongs to $W$ then the vector whose column matrix is $\mathsf{D}(X)\mathsf{w}$ also belongs to $W$, for all $X$ belonging to $\mathcal{G}$, then it follows that $\mathsf{D}$ is reducible. We say that the subspace $W$ is invariant under the actions of the elements of $\mathcal{G}$. With $\mathsf{D}$ unitary, the orthogonal complement $W_{\perp}$ of $W$ is also invariant, and the matrices $\mathsf{D}(X)$ each split into two blocks acting separately on $W$ and $W_{\perp}$. Both $W$ and $W_{\perp}$ may contain further invariant subspaces and be split still further.

As a concrete example of this view we can consider certain proper subspaces of the infinite-dimensional vector space of all functions under the action of the matrices corresponding to the continuous rotation group. Under a rotation of $\alpha$ about an axis perpendicular to the unit circle, a function such as $\sin\theta$ gives $\sin(\theta+\alpha)$ which can be expanded as $\sin\theta\cos\alpha+\cos\theta\sin\alpha$, i.e as a linear combination of $\sin\theta$ and $\cos\theta$; similarly $\cos\theta$ becomes another linear combination of the same two functions. The functions $\sin\theta$ and $\cos\theta$ span an invariant irreducible subspace of the whole space; a corresponding basis vector would be, e.g., $(\sin\theta \ \ \cos\theta)^{\mathrm{T}}$. Similarly, on the surface of the unit sphere the spherical harmonics $Y_{\ell m}(\theta,\phi)$ span an invariant subspace under the action of the full three-dimensional rotation group. This subspace itself contains invariant subspaces corresponding to the different values of $\ell$.

To illustrate further the irreps of a group, we again return to the group $\mathcal{G}$ of two-dimensional rotation and reflection symmetries of an equilateral triangle, or equivalently the permutation group $S_3$, which may be shown, using the methods of section 23.7, to have three irreps.

We have already seen that the set $\mathcal{M}$ of six orthogonal $2\times 2$ matrices given in section (22.3), equation (22.13), is isomorphic to $\mathcal{G}$. These matrices therefore form not only a representation of $\mathcal{G}$, but a faithful one. It should be noticed that, although $\mathcal{G}$ contains six elements, the matrices are only $2\times 2$. However, they contain no invariant $1\times 1$ sub-block (which for $2\times 2$ matrices would require them all to be diagonal) and neither can *all* the matrices be made block diagonal by the *same* similarity transformation; they therefore form a two-dimensional irrep of $\mathcal{G}$.

As previously noted, every group has one (unfaithful) irrep in which every element is represented by the $1\times 1$ matrix $\mathsf{I}_1$, or, more simply, 1.

A third (unfaithful) irrep of $\mathcal{G}$ is given by assignment of the one-dimensional set of 'matrices' $\{1,1,1,-1,-1,-1\}$ to the symmetry operations $\{I,R,R',K,L,M\}$ respectively, or to the group elements $\{I,A,B,C,D,E\}$ respectively (see section 22.3). In terms of the permutation group $S_3$, 1 corresponds to even permutations and $-1$ to odd permutations, 'odd' or 'even' referring to the number of simple pair interchanges to which a permutation is equivalent. That these assignments are in accord with the group multiplication table 22.8 should be checked.

Thus the three irreps of the group $\mathcal{G}$ (i.e. the group $3m$ or $C_{3v}$ or $S_3$), are, using the conventional notation $\mathrm{A}_1$, $\mathrm{A}_2$, E (see section 23.8), as follows:

| | | Element | | | | | |
|---|---|---|---|---|---|---|---|
| | | $I$ | $A$ | $B$ | $C$ | $D$ | $E$ |
| | $\mathrm{A}_1$ | 1 | 1 | 1 | 1 | 1 | 1 |
| Irrep | $\mathrm{A}_2$ | 1 | 1 | 1 | $-1$ | $-1$ | $-1$ |
| | E | $\mathsf{M}_I$ | $\mathsf{M}_A$ | $\mathsf{M}_B$ | $\mathsf{M}_C$ | $\mathsf{M}_D$ | $\mathsf{M}_E$ |

(23.11)

where

$$\mathsf{M}_I = \begin{pmatrix} 1 & 0 \\ 0 & 1 \end{pmatrix}, \quad \mathsf{M}_A = \begin{pmatrix} -\frac{1}{2} & \frac{\sqrt{3}}{2} \\ -\frac{\sqrt{3}}{2} & -\frac{1}{2} \end{pmatrix}, \quad \mathsf{M}_B = \begin{pmatrix} -\frac{1}{2} & -\frac{\sqrt{3}}{2} \\ \frac{\sqrt{3}}{2} & -\frac{1}{2} \end{pmatrix},$$

$$\mathsf{M}_C = \begin{pmatrix} -1 & 0 \\ 0 & 1 \end{pmatrix}, \quad \mathsf{M}_D = \begin{pmatrix} \frac{1}{2} & -\frac{\sqrt{3}}{2} \\ -\frac{\sqrt{3}}{2} & -\frac{1}{2} \end{pmatrix}, \quad \mathsf{M}_E = \begin{pmatrix} \frac{1}{2} & \frac{\sqrt{3}}{2} \\ \frac{\sqrt{3}}{2} & -\frac{1}{2} \end{pmatrix}.$$

## 23.5 The orthogonality theorem for irreducible representations

We now come to the central theorem of representation theory, a theorem that justifies the relatively routine application of certain procedures to determine the restrictions that are inherent in physical systems which have some degree of rotational or reflection symmetry. The development of the theorem is long and quite complex when presented in its entirety, and the reader will have to refer elsewhere for the proof.†

The theorem states that, in a certain sense, the irreps of a group $\mathcal{G}$ are as orthogonal as possible, as follows. If, for each irrep the elements in any one position in each of the $g$ matrices are used to make up $g$-component column matrices, then

(i) any two such column matrices coming from different irreps are orthogonal;
(ii) any two such column matrices coming from different positions in the matrices of the same irrep are orthogonal.

This orthogonality is in addition to the irreps' being in the form of orthogonal (unitary) matrices and thus each comprising mutually orthogonal rows and columns.

More mathematically, if we denote the entry in the $i$th row and $j$th column of a matrix $\mathsf{D}(X)$ by $[\mathsf{D}(X)]_{ij}$, and $\hat{\mathsf{D}}^{(\lambda)}$ and $\hat{\mathsf{D}}^{(\mu)}$ are two irreps of $\mathcal{G}$ having dimensions $n_\lambda$ and $n_\mu$ respectively, then

$$\sum_X \left[\hat{\mathsf{D}}^{(\lambda)}(X)\right]^*_{ij} \left[\hat{\mathsf{D}}^{(\mu)}(X)\right]_{kl} = \frac{g}{n_\lambda}\delta_{ik}\delta_{jl}\delta_{\lambda\mu}. \qquad (23.12)$$

This rather forbidding-looking equation needs some further explanation.

Firstly, the asterisk indicates that the complex conjugate should be taken if necessary, though all our representations so far have involved only real matrix elements. Each Kronecker delta function on the right-hand side has the value 1 if its two subscripts are equal, and has the value 0 otherwise. Thus the right-hand side is only non-zero if $i = k$, $j = l$ and $\lambda = \mu$, all at the same time.

† e.g. *Groups, Representation and Physics*, H. F. Jones (Institute of Physics), *Group Theory in Quantum Mechanics*, J.F. Cornwell (Academic Press), or *Linear Representations of Finite Groups*, J.P. Sore (Springer-Verlag).

Secondly, the summation over the group elements $X$ means that $g$ contributions have to be added together, each contribution being a product of entries drawn from the representative matrices in the two irreps $\hat{\mathsf{D}}^{(\lambda)} = \{\hat{\mathsf{D}}^{(\lambda)}(X)\}$ and $\hat{\mathsf{D}}^{(\mu)} = \{\hat{\mathsf{D}}^{(\mu)}(X)\}$. The $g$ contributions arise as $X$ runs over the $g$ elements of $\mathcal{G}$.

Thus, putting these remarks together, the summation will produce zero if either

(i) the matrix elements are not taken from exactly the same position in every matrix, including cases in which it is not possible to do so because the irreps $\hat{\mathsf{D}}^{(\lambda)}$ and $\hat{\mathsf{D}}^{(\mu)}$ have different dimensions, or

(ii) even if $\hat{\mathsf{D}}^{(\lambda)}$ and $\hat{\mathsf{D}}^{(\mu)}$ do have the same dimensions and the matrix elements are from the same positions in every matrix, they are different irreps, i.e. $\lambda \neq \mu$.

Some numerical illustrations based on the irreps $A_1$, $A_2$ and E of the group $3m$ (or $C_{3v}$ or $S_3$) will probably provide the clearest explanation (see (23.11)).

(a) Take $i = j = k = l = 1$, with $\hat{\mathsf{D}}^{(\lambda)} = A_1$ and $\hat{\mathsf{D}}^{(\mu)} = A_2$. Equation (23.12) then reads

$$1(1) + 1(1) + 1(1) + 1(-1) + 1(-1) + 1(-1) = 0,$$

as expected, since $\lambda \neq \mu$.

(b) Take $(i, j)$ as $(1, 2)$, but $(k, l)$ as $(2, 2)$, corresponding to different matrix positions within the same irrep $\hat{\mathsf{D}}^{(\lambda)} = \hat{\mathsf{D}}^{(\mu)} = \mathrm{E}$. Substituting in (23.12) gives

$$0(1) + \left(-\tfrac{\sqrt{3}}{2}\right)\left(-\tfrac{1}{2}\right) + \left(\tfrac{\sqrt{3}}{2}\right)\left(-\tfrac{1}{2}\right) + 0(1) + \left(-\tfrac{\sqrt{3}}{2}\right)\left(-\tfrac{1}{2}\right) + \left(\tfrac{\sqrt{3}}{2}\right)\left(-\tfrac{1}{2}\right) = 0.$$

(c) Take $(i, j)$ as $(1, 2)$, and $(k, l)$ as $(1, 2)$, corresponding to the same matrix positions within the same irrep $\hat{\mathsf{D}}^{(\lambda)} = \hat{\mathsf{D}}^{(\mu)} = \mathrm{E}$. Substituting in (23.12) gives

$$0(0) + \left(-\tfrac{\sqrt{3}}{2}\right)\left(-\tfrac{\sqrt{3}}{2}\right) + \left(\tfrac{\sqrt{3}}{2}\right)\left(\tfrac{\sqrt{3}}{2}\right) + 0(0) + \left(-\tfrac{\sqrt{3}}{2}\right)\left(-\tfrac{\sqrt{3}}{2}\right) + \left(\tfrac{\sqrt{3}}{2}\right)\left(\tfrac{\sqrt{3}}{2}\right) = \tfrac{6}{2}.$$

(d) No explicit calculation is needed to see that if $i = j = k = l = 1$, with $\hat{\mathsf{D}}^{(\lambda)} = \hat{\mathsf{D}}^{(\mu)} = A_1$ (or $A_2$), then each term in the sum is either $1^2$ or $(-1)^2$ and the total is 6, as predicted by the right-hand side of (23.12) since $g = 6$ and $n_\lambda = 1$.

## 23.6 Characters

The actual matrices of general representations and irreps are cumbersome to work with, and they are not unique since there is always the freedom to change the coordinate system, i.e. the components of the basis vector (see section 23.3), and hence the entries in the matrices. However, one thing that does not change for a matrix under an equivalence (similarity) transformation – i.e. under a change of basis – is the trace of the matrix. This was shown in chapter 7, but is repeated here. The trace of a matrix $\mathsf{A}$ is the sum of its diagonal elements,

$$\mathrm{Tr}\,\mathsf{A} = \sum_{i=1}^{n} \mathsf{A}_{ii}$$

or, using the summation convention, simply $\mathsf{A}_{ii}$. Under a similarity transformation,

$$\begin{aligned}[\mathsf{D}_{\mathsf{Q}}(X)]_{ii} &= [\mathsf{Q}^{-1}]_{ij}[\mathsf{D}(X)]_{jk}[\mathsf{Q}]_{ki} \\ &= [\mathsf{D}(X)]_{jk}[\mathsf{Q}]_{ki}[\mathsf{Q}^{-1}]_{ij} \\ &= [\mathsf{D}(X)]_{jk}[\mathsf{I}]_{kj} \\ &= [\mathsf{D}(X)]_{jj},\end{aligned}$$

showing that the traces of equivalent matrices are equal.

This can be used to simplify greatly work with representations, though with some partial loss of the information content of the full matrices. For example, using trace values alone it is not possible to distinguish between the two groups known as $4mm$ and $\bar{4}2m$ (or as $C_{4v}$ and $D_{2d}$ respectively) even though the two groups are not isomorphic. To make use of these simplifications we now define the characters of a representation.

**Definition.** *The* characters $\chi(\mathsf{D})$ *of a representation* $\mathsf{D}$ *of a group* $\mathcal{G}$ *are defined as the traces of the matrices* $\mathsf{D}(X)$*, where* $X$ *is a typical element of* $\mathcal{G}$.

At this stage there will be $g$ characters, but, as we noted in subsection 22.7.3, elements $A$, $B$ of $\mathcal{G}$ in the same conjugacy class are connected by equations of the form $B = X^{-1}AX$. It follows that their matrix representations are connected by corresponding equations of the form $\mathsf{D}(B) = \mathsf{D}(X^{-1})\mathsf{D}(A)\mathsf{D}(X)$, and so by the argument just given their representations will have equal traces and hence equal characters. Thus *elements in the same conjugacy class have the same characters*, though these will, in general, vary from one representation to another. However, it might also happen that two or more conjugacy classes have the same characters in a representation – indeed, in the trivial irrep $A_1$ (see (23.11)) every element inevitably has the character 1.

For the irrep $A_2$ of the group $3m$, the classes $\{I\}$, $\{A, B\}$ and $\{C, D, E\}$ have characters 1, 1 and $-1$, respectively, whilst they have characters 2, $-1$ and 0, respectively, in irrep E.

We are thus able to draw up a *character table* for the group as shown in table 23.1. This table holds in compact form most of the important information on the behaviour of functions under the two-dimensional rotational and reflection symmetries of an equilateral triangle, i.e. under the elements of group $3m$. The entry under $I$ for any irrep gives the dimension of the irrep, since it is equal to

| $3m$ | $I$ | $A, B$ | $C, D, E$ | |
|---|---|---|---|---|
| $\mathrm{A}_1$ | 1 | 1 | 1 | $z$; $z^2$; $x^2+y^2$ |
| $\mathrm{A}_2$ | 1 | 1 | $-1$ | $R_z$ |
| E | 2 | $-1$ | 0 | $(x, y)$; $(xz, yz)$; $(R_x, R_y)$; $(x^2-y^2, 2xy)$ |

Table 23.1 The character table for the irreps of group $3m$ ($C_{3v}$ or $S_3$). The right-hand column lists some common functions which transform according to the irrep against which each is shown (see text).

the trace of the unit matrix whose dimension is equal to that of the irrep. In other words, for the $\lambda$th irrep $\chi^{(\lambda)}(I) = n_\lambda$ where $n_\lambda$ is its dimension.

In the extreme right-hand column we list some common functions of Cartesian coordinates that transform, under the group $3m$, according to the irrep on whose line they are listed. Thus, as we have seen, $z$, $z^2$, and $x^2 + y^2$ are all unchanged by the group operations (though $x$ and $y$ individually are affected) and are listed against the one-dimensional irrep $\mathrm{A}_1$. Each of the pairs $(x, y)$, $(xz, yz)$, and $(x^2 - y^2, 2xy)$, on the other hand, are mixed as a pair by some of the operations, and are listed against the two-dimensional irrep E; thus each pair forms a basis for this irrep.

The quantities $R_x$, $R_y$ and $R_z$ refer to rotations about the indicated axes; they transform in the same way as the corresponding components of the angular momentum $\mathbf{J}$, and their behaviour can be established by examining how the components of $\mathbf{J} = \mathbf{r} \times \mathbf{p}$ transform under the operations of the group. To do this explicitly is beyond the scope of this book. However, it can be noted that $R_z$, being listed opposite the one-dimensional $\mathrm{A}_2$, is unchanged by $I$ and by the rotations $A$ and $B$ but changes sign under the mirror reflections $C$, $D$, and $E$, as would be expected.

### *23.6.1 Orthogonality property of characters*

Some of the most important properties of characters can be deduced from the orthogonality theorem (23.12),

$$\sum_X \left[\hat{\mathsf{D}}^{(\lambda)}(X)\right]^*_{ij} \left[\hat{\mathsf{D}}^{(\mu)}(X)\right]_{kl} = \frac{g}{n_\lambda}\delta_{ik}\delta_{jl}\delta_{\lambda\mu}.$$

If we set $j = i$ and $l = k$, so that both factors in any particular term in the summation refer to diagonal elements of the representative matrices, and then sum both sides over $i$ and $k$, we obtain

$$\sum_X \sum_{i=1}^{n_\lambda} \sum_{k=1}^{n_\mu} \left[\hat{\mathsf{D}}^{(\lambda)}(X)\right]^*_{ii} \left[\hat{\mathsf{D}}^{(\mu)}(X)\right]_{kk} = \frac{g}{n_\lambda} \sum_{i=1}^{n_\lambda} \sum_{k=1}^{n_\mu} \delta_{ik}\delta_{ik}\delta_{\lambda\mu}.$$

Expressed in term of characters, this reads

$$\sum_X \left[\chi^{(\lambda)}(X)\right]^* \chi^{(\mu)}(X) = \frac{g}{n_\lambda}\sum_{i=1}^{n_\lambda} \delta_{ii}^2 \delta_{\lambda\mu} = \frac{g}{n_\lambda}\sum_{i=1}^{n_\lambda} 1\,\delta_{\lambda\mu} = g\delta_{\lambda\mu}. \tag{23.13}$$

In words, the ($g$-component) 'vectors' formed from the characters of the various irreps of a group are mutually orthogonal, but each one has a squared magnitude (the sum of the squares of its components) equal to the order of the group.

Since, as noted in the previous subsection, group elements in the same class have the same characters, (23.13) can be written as a sum over classes rather than elements. If $c_i$ denotes the number of elements in class $C_i$ and $X_i$ any element of $C_i$, then

$$\sum_i c_i \left[\chi^{(\lambda)}(X_i)\right]^* \chi^{(\mu)}(X_i) = g\delta_{\lambda\mu}. \tag{23.14}$$

For table 23.1 we can verify that these results are valid.

(i) For $\hat{\mathsf{D}}^{(\lambda)} = \hat{\mathsf{D}}^{(\mu)} = \mathrm{A}_1$ or $\mathrm{A}_2$, (23.14) reads

$$1(1) + 2(1) + 3(1) = 6,$$

whilst for $\hat{\mathsf{D}}^{(\lambda)} = \hat{\mathsf{D}}^{(\mu)} = \mathrm{E}$, it gives

$$1(2^2) + 2(1) + 3(0) = 6.$$

(ii) For $\hat{\mathsf{D}}^{(\lambda)} = \mathrm{A}_2$ and $\hat{\mathsf{D}}^{(\mu)} = \mathrm{E}$, say, (23.14) reads

$$1(1)(2) + 2(1)(-1) + 3(-1)(0) = 0.$$

## 23.7 Counting irreps using characters

The expression of a general representation $\mathsf{D} = \{\mathsf{D}(X)\}$ in terms of irreps, as given in (23.10), can be simplified by going from the full matrix form to that of characters. Thus

$$\mathsf{D}(X) = m_1\hat{\mathsf{D}}^{(1)}(X) \oplus m_2\hat{\mathsf{D}}^{(2)}(X) \oplus \cdots \oplus m_N\hat{\mathsf{D}}^{(N)}(X)$$

becomes, on taking the trace of both sides,

$$\chi(X) = \sum_{\lambda=1}^{N} m_\lambda \chi^{(\lambda)}(X). \tag{23.15}$$

Given the characters of the irreps of the group $\mathcal{G}$ to which the elements $X$ belong, and the characters of the representation $\mathsf{D} = \{\mathsf{D}(X)\}$, the $g$ equations (23.15) can be solved as simultaneous equations in the $m_\lambda$, either by inspection or by multiplying both sides by $\left[\chi^{(\mu)}(X)\right]^*$ and summing over $X$, making use of (23.13) and (23.14), to obtain

$$m_\mu = \frac{1}{g}\sum_X \left[\chi^{(\mu)}(X)\right]^* \chi(X) = \frac{1}{g}\sum_i c_i \left[\chi^{(\mu)}(X_i)\right]^* \chi(X_i). \qquad (23.16)$$

That an unambiguous formula can be given for each $m_\lambda$, once the *character set* (the set of characters of each of the group elements or, equivalently, of each of the conjugacy classes) of D is known, shows that, for any particular group, two representations with the same characters are equivalent. This strongly suggests something that can be shown, namely, *the number of irreps = the number of conjugacy classes*. The argument is as follows. Equation (23.15) is a set of simultaneous equations for $N$ unknowns, the $m_\lambda$, some of which may be zero. The value of $N$ is equal to the number of irreps of $\mathcal{G}$. There are $g$ different values of $X$, but the number of *different* equations is only equal to the number of distinct conjugacy classes, since any two elements of $\mathcal{G}$ in the same class have the same character set and therefore generate the same equation. For a unique solution to simultaneous equations in $N$ unknowns, exactly $N$ independent equations are needed. Thus $N$ is also the number of classes, establishing the stated result.

► *Determine the irreps contained in the representation of the group 3m in the vector space spanned by the functions $x^2$, $y^2$, $xy$.*

We first note that, although these functions are not orthogonal, they do form a basis set for a representation. We must establish how they transform under the symmetry operations of group 3*m*. We only need to do so for a representative element of each conjugacy class, and naturally we take the simplest in each case.

The first class contains only $I$ (as always) and clearly $\mathsf{D}(I)$ is the $3 \times 3$ unit matrix.

The second class contains the rotations, $A$ and $B$, and we choose to find $\mathsf{D}(A)$. Since, under $A$, $x \to -\frac{1}{2}x + \frac{\sqrt{3}}{2}y$ and $y \to -\frac{\sqrt{3}}{2}x - \frac{1}{2}y$, it follows that

$$x^2 \to \tfrac{1}{4}x^2 - \tfrac{\sqrt{3}}{2}xy + \tfrac{3}{4}y^2, \qquad y^2 \to \tfrac{3}{4}x^2 + \tfrac{\sqrt{3}}{2}xy + \tfrac{1}{4}y^2 \qquad (23.17)$$

and

$$xy \to \tfrac{\sqrt{3}}{4}x^2 - \tfrac{1}{2}xy - \tfrac{\sqrt{3}}{4}y^2. \qquad (23.18)$$

Hence $\mathsf{D}(A)$ can be deduced and is given below.

| | Classes | | |
|---|---|---|---|
| Irrep | $I$ | $A$ | $C$ |
| $A_1$ | 1 | 1 | 1 |
| $A_2$ | 1 | 1 | −1 |
| E | 2 | −1 | 0 |
| D | 3 | 0 | 1 |

Table 23.2 The characters of the irreps of the group $3m$ and of the representation D, which must be a superposition of some of them.

The third, and final, class contains the reflections, $C$, $D$ and $E$; of these $C$ is much the easiest to deal with. Under $C$, $x \to -x$ and $y \to y$, causing $xy$ to change sign but leaving $x^2$ and $y^2$ unaltered. The three matrices needed are thus $\mathsf{D}(I) = \mathsf{I}_3$,

$$\mathsf{D}(C) = \begin{pmatrix} 1 & 0 & 0 \\ 0 & 1 & 0 \\ 0 & 0 & -1 \end{pmatrix} \quad \text{and} \quad \mathsf{D}(A) = \begin{pmatrix} \frac{1}{4} & \frac{3}{4} & -\frac{\sqrt{3}}{2} \\ \frac{3}{4} & \frac{1}{4} & \frac{\sqrt{3}}{2} \\ \frac{\sqrt{3}}{4} & -\frac{\sqrt{3}}{4} & -\frac{1}{2} \end{pmatrix},$$

and their traces are respectively 3, 1 and 0.

It should be noticed that much more work has been done here than is necessary, since the traces can be computed immediately from the effects of the symmetry operations. All that is needed is the weight of each basis function in the transformed expression for that function; these are clearly 1, 1, 1 for $I$, $\frac{1}{4}$, $\frac{1}{4}$, $-\frac{1}{2}$ for $A$ from (23.17) and (23.18), and 1, 1, −1 for $C$, from the observations made just above the displayed matrices. The traces are then the sums of these weights. The off-diagonal elements of the matrices need not be found, nor need the matrices be written out.

We now need to find a superposition of the characters of the irreps that gives representation D in the bottom line of table 23.2.

By inspection it is obvious that $\mathsf{D} = A_1 \oplus E$, but we can use (23.16) formally:

$$\begin{aligned} m_{A_1} &= \tfrac{1}{6}[1(1)(3) + 2(1)(0) + 3(1)(1)] = 1, \\ m_{A_2} &= \tfrac{1}{6}[1(1)(3) + 2(1)(0) + 3(-1)(1)] = 0, \\ m_E &= \tfrac{1}{6}[1(2)(3) + 2(-1)(0) + 3(0)(1)] = 1. \end{aligned}$$

Thus $A_1$ and E appear once each in the reduction of D, and $A_2$ not at all. Table 23.1 gives the further information, not needed here, that it is the combination $x^2 + y^2$ that transforms as a one-dimensional irrep, and the pair $(x^2 - y^2, 2xy)$ that forms a basis of the two-dimensional irrep, E. ◄

### 23.7.1 Summation rules for irreps

The first summation rule for irreps is a simple restatement of (23.13), with $\mu$ set equal to $\lambda$, which then reads

$$\sum_X \left[\chi^{(\lambda)}(X)\right]^* \chi^{(\lambda)}(X) = g.$$

In words, the sum of the squares (modulus squared if necessary) of the characters of an irrep taken over all elements of the group adds up to the order of the group. For group $3m$ (table 23.1), this takes the following explicit forms:

$$\begin{aligned}
&\text{for } \mathrm{A}_1, \quad 1(1^2) + 2(1^2) + 3(1^2) = 6;\\
&\text{for } \mathrm{A}_2, \quad 1(1^2) + 2(1^2) + 3(-1)^2 = 6;\\
&\text{for } \mathrm{E}, \quad\; 1(2^2) + 2(-1)^2 + 3(0^2) = 6.
\end{aligned}$$

We next prove a theorem that is concerned not with a summation within an irrep but with a summation over irreps.

**Theorem.** *If $n_\mu$ is the dimension of the $\mu$th irrep of a group $\mathcal{G}$, then*

$$\sum_\mu n_\mu^2 = g,$$

*where g is the order of the group.*

*Proof.* Define a representation of the group in the following way. Rearrange the rows of the multiplication table of the group so that whilst the elements in a particular order head the columns, their inverses in the same order head the rows. In this arrangement of the $g \times g$ table, the leading diagonal is entirely occupied by the identity element. Then, for each element $X$ of the group, take as representative matrix the multiplication-table array obtained by replacing $X$ by 1, and all other element symbols by 0. The matrices $\mathsf{D}^{\text{reg}}(X)$ so obtained form the *regular representation* of $\mathcal{G}$; they are each $g \times g$, have one non-zero entry of '1' in each row and column, and (as will be verified by a little experimentation) have the same multiplication structure as the group $\mathcal{G}$ itself, i.e. they form a faithful representation of $\mathcal{G}$.

Although not part of the proof, a simple example may help to make these ideas more transparent. Consider the cyclic group of order 3. Its multiplication table is shown in table 23.3($a$) (a repeat of table 22.10($a$) of the previous chapter), whilst table 23.3($b$) shows the same table reordered so that the columns are labelled in the order $I$, $A$, $B$ but the rows are labelled in the order $I^{-1} = I$, $A^{-1} = B$, $B^{-1} = A$. The three matrices of the regular representation are then

$$\mathsf{D}^{\text{reg}}(I) = \begin{pmatrix} 1 & 0 & 0 \\ 0 & 1 & 0 \\ 0 & 0 & 1 \end{pmatrix}, \quad \mathsf{D}^{\text{reg}}(A) = \begin{pmatrix} 0 & 1 & 0 \\ 0 & 0 & 1 \\ 1 & 0 & 0 \end{pmatrix}, \quad \mathsf{D}^{\text{reg}}(B) = \begin{pmatrix} 0 & 0 & 1 \\ 1 & 0 & 0 \\ 0 & 1 & 0 \end{pmatrix}.$$

(a)

| | $I$ | $A$ | $B$ |
|---|---|---|---|
| $I$ | $I$ | $A$ | $B$ |
| $A$ | $A$ | $B$ | $I$ |
| $B$ | $B$ | $I$ | $A$ |

(b)

| | $I$ | $A$ | $B$ |
|---|---|---|---|
| $I$ | $I$ | $A$ | $B$ |
| $B$ | $B$ | $I$ | $A$ |
| $A$ | $A$ | $B$ | $I$ |

Table 23.3 (*a*) The multiplication table of the cyclic group of order 3, and (*b*) its reordering used to generate the regular representation of the group.

An alternative, more mathematical, definition of the regular representation of a group is

$$\left[\mathsf{D}^{\text{reg}}(G_k)\right]_{ij} = \begin{cases} 1 & \text{if } G_kG_j = G_i, \\ 0 & \text{otherwise.} \end{cases}$$

We now return to the proof. With the construction given, the regular representation has characters as follows:

$$\chi^{\text{reg}}(I) = g, \qquad \chi^{\text{reg}}(X) = 0 \quad \text{if } X \neq I.$$

We now apply (23.16) to $\mathsf{D}^{\text{reg}}$ to obtain for the number $m_\mu$ of times that the irrep $\hat{\mathsf{D}}^{(\mu)}$ appears in $\mathsf{D}^{\text{reg}}$ (see 23.10))

$$m_\mu = \frac{1}{g}\sum_X \left[\chi^{(\mu)}(X)\right]^* \chi^{\text{reg}}(X) = \frac{1}{g}\left[\chi^{(\mu)}(I)\right]^* \chi^{\text{reg}}(I) = \frac{1}{g}n_\mu g = n_\mu.$$

Thus an irrep $\hat{\mathsf{D}}^{(\mu)}$ of dimension $n_\mu$ appears $n_\mu$ times in $\mathsf{D}^{\text{reg}}$, and so by counting the total number of basis functions, or by considering $\chi^{\text{reg}}(I)$, we can conclude that

$$\sum_\mu n_\mu^2 = g. \tag{23.19}$$

This completes the proof.

As it must, our standard demonstration group $3m$ provides an illustration. In this case we have already seen that there are two one-dimensional irreps and one two-dimensional one. This is in accord with (23.19) since

$$1^2 + 1^2 + 2^2 = 6, \qquad \text{which is the order } g \text{ of the group.}$$

Another straightforward application of the relation (23.19), to the group with multiplication table 23.3(*a*), yields immediate results. Since $g = 3$, none of its irreps can have dimension 2 or more, as $2^2 = 4$ is too large for (23.19) to be satisfied. Thus all irreps must be one-dimensional and there must be three of them (consistent with the fact that each element is in a class of its own, and there are therefore three classes). The three irreps are the sets of $1 \times 1$ matrices (numbers)

$$\mathrm{A}_1 = \{1, 1, 1\} \qquad \mathrm{A}_2 = \{1, \omega, \omega^2\} \qquad \mathrm{A}_2^* = \{1, \omega^2, \omega\},$$

where $\omega = \exp(2\pi i/3)$; since the matrices are $1 \times 1$, the same set of nine numbers would, of course, be the entries in the character table for the irreps of the group. The fact that the numbers in each irrep are all cube roots of unity is discussed below. As will be noticed, two of these irreps are complex – an unusual occurrence in most applications – and form a complex conjugate pair of one-dimensional irreps. In practice, they function much as a two-dimensional irrep, but this is to be ignored for formal purposes such as theorems.

A further property of characters can be derived from the fact that all elements in a conjugacy class have the same order. Suppose that the element $X$ has order $m$, i.e. $X^m = I$. This implies for a representation $\mathsf{D}$ of dimension $n$ that

$$[\mathsf{D}(X)]^m = I_n. \tag{23.20}$$

Representations equivalent to $\mathsf{D}$ are generated as before by using similarity transformations of the form

$$\mathsf{D}_{\mathsf{Q}}(X) = \mathsf{Q}^{-1}\mathsf{D}(X)\mathsf{Q}.$$

In particular, if we choose the columns of $\mathsf{Q}$ to be the eigenvectors of $\mathsf{D}(X)$ then, as discussed in chapter 7,

$$\mathsf{D}_{\mathsf{Q}}(X) = \begin{pmatrix} \lambda_1 & & & 0 \\ & \lambda_2 & & \\ & & \ddots & \\ 0 & & & \lambda_n \end{pmatrix}$$

where the $\lambda_i$ are the eigenvalues of $\mathsf{D}(X)$. Therefore, from (23.20), we have that

$$\begin{pmatrix} \lambda_1^m & & & 0 \\ & \lambda_2^m & & \\ & & \ddots & \\ 0 & & & \lambda_n^m \end{pmatrix} = \begin{pmatrix} 1 & & & 0 \\ & 1 & & \\ & & \ddots & \\ 0 & & & 1 \end{pmatrix}.$$

Hence all the eigenvalues $\lambda_i$ are $m$th roots of unity, and $\chi(X)$, the trace of $\mathsf{D}(X)$, is the sum of $n$ $m$th roots of unity. In view of the implications of Lagrange's theorem (section 22.6), the only values of $m$ allowed are the divisors of the order of the group.

## 23.8 Construction of a character table

In order to decompose representations into irreps on a routine basis using characters, it is necessary to have a character table for each group, i.e., for each

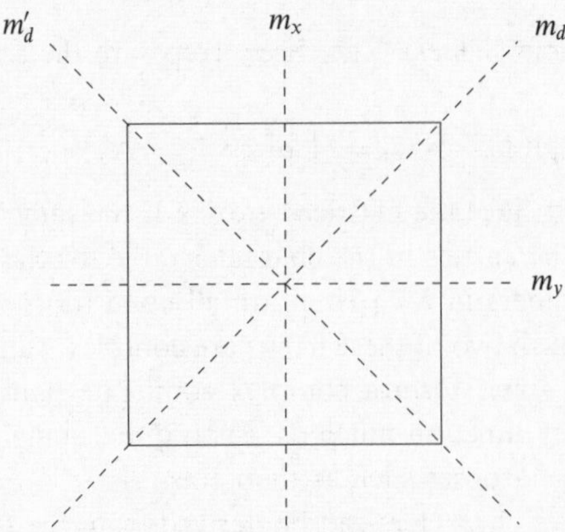


Figure 23.3 The mirror planes associated with 4*mm*, the group of two-dimensional symmetries of a square.

irrep $\mu$ of the group, to know the character $\chi^{(\mu)}(X)$ to be assigned to the class to which $X$ belongs. To construct such a table the following properties may now be used:

(i) the number of classes = the number of irreps;
(ii) the 'vector' formed by the characters from a given irrep is orthogonal to the 'vector' formed by the characters from a different irrep;
(iii) $\sum_\mu n_\mu^2 = g$, where $n_\mu$ is the dimension of the $\mu$th irrep and $g$ the order of the group;
(iv) the identity irrep (one-dimensional with all characters equal to 1) is always present;
(v) $\sum_X \left|\chi^{(\mu)}(X)\right|^2 = g.$
(vi) $\chi^{(\mu)}(X)$ is the sum of $n_\mu$ $m$th roots of unity, where $m$ is the order of $X$.

►*Construct the character table for the group 4mm (or $C_{4v}$) using the properties of classes, irreps and characters so far established.*

The group 4*mm* is the group of two-dimensional symmetries of a square, namely rotations of 0, $\pi/2$, $\pi$ and $3\pi/2$, and reflections in the mirror planes parallel to the coordinate axes and along the main diagonals. These are illustrated in figure 23.3.

For this group there are eight elements:

- the identity, $I$;
- the two rotations $R$ (by $\pi/2$) and $R'$ (by $3\pi/2$);
- the rotation $Q$ (by $\pi$);
- the four mirror reflections $m_x$, $m_y$, $m_d$ and $m_{d'}$.

Requirements (i) to (iv) put tight constraints on the possible character sets, as the following argument shows.

The group is non-Abelian (clearly $Rm_x \neq m_xR$), and so there are fewer than eight classes, and hence fewer than eight irreps. But requirement (iii), with $g = 8$, then implies that at least one irrep has dimension 2 or greater. However, there can be no irrep with dimension 3 or greater, since $3^2 > 8$, nor can there be more than one two-dimensional irrep, since $2^2 + 2^2 = 8$ would rule out a contribution to the sum in (iii) of $1^2$ from the identity irrep, and this must be present. Thus the only possibility is one two-dimensional irrep and, to make the sum in (iii) correct, four one-dimensional irreps.

Using (i) we can now deduce that there are five classes. This same conclusion can be reached by evaluating $X^{-1}YX$ for every pair of elements in $\mathcal{G}$, as in the description of conjugacy classes given in the previous chapter. However, it is tedious to do so and certainly much longer than the above. The five classes are $I$, $Q$, $\{R, R'\}$, $\{m_x, m_y\}$, $\{m_d, m_{d'}\}$.

It is straightforward to show that only $I$ and $Q$ commute with every element of the group, so they are the only ones in classes of their own. Each other class must have at least two members, but, as there are three classes to accommodate $8-2=6$ elements, there must be exactly two in each class. This does not pair up the remaining six, but does say that the five classes have 1, 1, 2, 2, and 2 elements. Of course, if we had started by dividing the group into classes, we would know the number of elements in each class directly.

We cannot entirely ignore the group structure (though it sometimes happens that the results are independent of the group structure – for example, all non-Abelian groups of order 8 have the same character table!), and need to note in the present case that $m_i^2 = I$ for $i = x,\ y,\ d$ or $d'$, and, as can be proved directly, $Rm_i = m_iR'$ for the same four values of label $i$. We also recall that for any pair of elements $X$ and $Y$, $\mathsf{D}(XY) = \mathsf{D}(X)\mathsf{D}(Y)$. We may conclude the following for the one-dimensional irreps.

(a) In view of result (vi), $\chi(m_i) = \mathsf{D}(m_i) = \pm 1$.

(b) Since $R^4 = I$, result (vi) requires that $\chi(R)$ is one of $1, i, -1, -i$. But, since $\mathsf{D}(R)\mathsf{D}(m_i) = \mathsf{D}(m_i)\mathsf{D}(R')$, and the $\mathsf{D}(m_i)$ are just numbers, $\mathsf{D}(R) = \mathsf{D}(R')$. Further $\mathsf{D}(R)\mathsf{D}(R) = \mathsf{D}(R)\mathsf{D}(R') = \mathsf{D}(RR') = \mathsf{D}(I) = 1$, and so $\mathsf{D}(R) = \pm 1 = \mathsf{D}(R')$.

(c) $\mathsf{D}(Q) = \mathsf{D}(RR) = \mathsf{D}(R)\mathsf{D}(R) = 1$.

If we add this to the fact that the characters of the identity irrep $A_1$ are all unity, we can fill in those entries in character table 23.4 that are shown in bold.

Suppose now that the three missing entries in a one-dimensional irrep are $p$, $q$ and $r$, where each can only be $\pm 1$. Then, allowing for the numbers in each class,

| $4mm$ | $I$ | $Q$ | $R, R'$ | $m_x, m_y$ | $m_d, m_{d'}$ |
|---|---|---|---|---|---|
| $A_1$ | **1** | **1** | **1** | **1** | **1** |
| $A_2$ | **1** | **1** | 1 | −1 | −1 |
| $B_1$ | **1** | **1** | −1 | 1 | −1 |
| $B_2$ | **1** | **1** | −1 | −1 | 1 |
| E | **2** | −2 | 0 | 0 | 0 |

Table 23.4 The character table deduced for the group $4mm$. For an explanation of the entries in bold see the text.

orthogonality with the characters of $A_1$ requires that

$$1(1)(1) + 1(1)(1) + 2(1)(p) + 2(1)(q) + 2(1)(r) = 0.$$

The only possibility is that two of $p$, $q$, and $r$ equal $-1$, and the other equals $+1$. This can be achieved in three different ways, corresponding to the need to find three further different one-dimensional irreps. Thus the first four lines of entries in character table 23.4 can be completed. The final line can be completed by requiring it to be orthogonal to the other four. Property (v) has not been used here though it could have replaced part of the argument given. ◀

## 23.9 Group nomenclature

The nomenclature of published character tables, as we have said before, is erratic and sometimes unfortunate; for example, $E$ is often used to represent, not only a two-dimensional irrep, but also the identity operation, where we have used $I$. Thus the symbol $E$ might appear in both the column and row headings of a table, though with quite different meanings in the two cases. In this book we use roman capitals to denote irreps.

One-dimensional irreps are regularly denoted by A and B, with B used if a rotation about the principal axis of $2\pi/n$ has character equal to $-1$. Here $n$ is the highest integer such that a rotation of $2\pi/n$ is a symmetry operation of the system, and the principal axis is the one about which this occurs. For the group of operations on a square, $n = 4$, the axis is the perpendicular to the square and the rotation in question is $R$. The names for the group, $4mm$ and $C_{4v}$, derive from the fact that here $n$ is equal to 4. Similarly, for the operations on an equilateral triangle $n = 3$, the group names are $3m$ and $C_{3v}$, but because the rotation by $2\pi/3$ has character $+1$ in all its one-dimensional irreps (see table 23.1), only A appears in the irrep list.

Two-dimensional irreps are denoted by E, as already noted, and three-dimensional ones by T, although in many cases the symbols are modified by primes and other alphabetic labels to denote variations in behaviour from one irrep to another in respect of mirror reflections and parity inversions. In the study of molecules, alternative names based on molecular angular momentum properties are common. It is beyond the scope of this book to list all these variations, or to give a large selection of character tables; our aim is to demonstrate and justify the use of those found in the literature specifically dedicated to crystal physics or molecular chemistry.

Variations in notation are not restricted to the naming of groups and their irreps, but extend to the symbols used to identify a typical element, and hence all members, of a conjugacy class in a group. In physics these are usually of the types $n_z$ or $\bar{n}_z$ or $m_x$. The first of these denotes a rotation of $2\pi/n$ about the $z$-axis, and the second the same thing followed by parity inversion (all vectors $\mathbf{r}$ go to $-\mathbf{r}$), whilst the third indicates a mirror reflection in a plane, in this case the plane $x = 0$.

Typical chemistry symbols are $NC_n$, $NC_n^2$, $NC_n^x$, $NS_n$, $\sigma_v$, $\sigma^{xy}$. Here the first symbol $N$, where it appears, shows that there are $N$ elements in the class (a useful feature). The subscript $n$ has the same meaning as in the physics notation, but $\sigma$ rather than $m$ is used for a mirror reflection, subscripts $v$, $d$ or $h$, or superscripts $xy$, $xz$ or $yz$, denoting the various orientations of the relevant mirror planes. Symmetries involving parity inversions are denoted by $S$; thus $S_n$ is the chemistry analogue of $\bar{n}$. None of what is said in this and the previous paragraph should be taken as definitive, but merely as a warning of common variations in nomenclature, and as an initial guide to corresponding entities. Before using any set of group character tables, the reader should ensure that he or she understands the precise notation being employed.

## 23.10 Product representations

In quantum mechanical investigations we are often faced with the calculation of what are called matrix elements. These normally take the form of integrals over all space of the product of two or more functions whose analytic forms depend on the microscopic properties (usually angular momentum and its components) of the electrons or nuclei involved. For 'bonding' calculations there are usually two functions involved, whilst for transition probabilities a third function, giving the spatial variation of the interaction Hamiltonian, also appears under the integral sign.

If the environment of the microscopic system under investigation has some symmetry properties, then these can sometimes be used to establish, without detailed evaluation, that the multiple integral must have zero value. We now express the essential content of these ideas in group theoretical language.

Suppose we are given an integral of the form

$$J = \int \Psi\phi \, d\tau \quad \text{or} \quad J = \int \Psi\xi\phi \, d\tau$$

to be evaluated over all space in a situation in which the physical system is invariant under a particular group $\mathcal{G}$ of symmetry operations. To obtain a non-zero answer we must have that the integrand is invariant under each of these operations. In group theoretical language: *the integrand must transform as the identity, the one-dimensional representation* $\mathrm{A}_1$ *of* $\mathcal{G}$. Or more accurately, the non-vanishing part of the integrand must do so.

An alternative way of saying this is that if the integrand transforms, under the symmetry operations of $\mathcal{G}$, according to a representation $\mathsf{D}$, and $\mathsf{D}$ does not contain $\mathrm{A}_1$ amongst its irreps, then the integral $J$ is necessarily zero. It should be noted that the converse is not true; $J$ may be zero even if $\mathrm{A}_1$ is present, since the integral, whilst invariant, may still have the value zero.

It is evident that we need to establish how to find the irreps that go to make up a representation of a double or triple product when we already know, either from observation or from the information provided with character tables, the irreps according to which the components in the product transform. The method is established by the following theorem.

**Theorem.** *The characters of a product representation are the products of the characters of the corresponding constituent elements.*

*Proof.* Suppose that $\{u_i\}$ and $\{v_j\}$ are two sets of basis functions, transforming, under the operations of a group $\mathcal{G}$, according to representations $\mathsf{D}^{(\lambda)}$ and $\mathsf{D}^{(\mu)}$ respectively. Denote by $\mathsf{u}$ and $\mathsf{v}$ the respective corresponding basis vectors, and let $X$ be an element of the group. Then the functions generated from $u_i$ and $v_j$ by the action of $X$ are calculated as follows, using (23.1) and (23.4):

$$u_i' = Xu_i = \left[\left(\mathsf{D}^{(\lambda)}(X)\right)^{\mathrm{T}}\mathsf{u}\right]_i = \left[\mathsf{D}^{(\lambda)}(X)\right]_{ii}u_i + \sum_{l\neq i}\left[\left(\mathsf{D}^{(\lambda)}(X)\right)^{\mathrm{T}}\right]_{il}u_l,$$

$$v_j' = Xv_j = \left[\left(\mathsf{D}^{(\mu)}(X)\right)^{\mathrm{T}}\mathsf{v}\right]_j = \left[\mathsf{D}^{(\mu)}(X)\right]_{jj}v_j + \sum_{m\neq j}\left[\left(\mathsf{D}^{(\mu)}(X)\right)^{\mathrm{T}}\right]_{jm}v_m.$$

Here $[\mathsf{D}(X)]_{ij}$ is just a single element of the matrix $\mathsf{D}(X)$, and $[\mathsf{D}(X)]_{kk} = [\mathsf{D}^{\mathrm{T}}(X)]_{kk}$ is simply a diagonal element from the matrix – the repeated subscript does not indicate summation. Now if we take as basis functions for a product representation, $\mathsf{D}^{\mathrm{prod}}(X)$, the products $w_k = u_iv_j$ (where the various possible pairs of values $i, j$ are labelled by $k$; there are $n_\lambda n_\mu$ of them), we have also that

$$\begin{aligned} w_k' &= Xw_k = Xu_iv_j = (Xu_i)(Xv_j) \\ &= \left[\mathsf{D}^{(\lambda)}(X)\right]_{ii}\left[\mathsf{D}^{(\mu)}(X)\right]_{jj}u_iv_j + \text{terms not involving the product } u_iv_j. \end{aligned}$$

This is to be compared with

$$w_k' = Xw_k = \left[\left(\mathsf{D}^{\mathrm{prod}}(X)\right)^{\mathrm{T}}\mathsf{w}\right]_k = \left[\mathsf{D}^{\mathrm{prod}}(X)\right]_{kk}w_k + \sum_{n\neq k}\left[\left(\mathsf{D}^{\mathrm{prod}}(X)\right)^{\mathrm{T}}\right]_{kn}w_n,$$

showing that

$$\left[\mathsf{D}^{\text{prod}}(X)\right]_{kk} = \left[\mathsf{D}^{(\lambda)}(X)\right]_{ii} \left[\mathsf{D}^{(\mu)}(X)\right]_{jj}.$$

It follows that

$$\begin{aligned}
\chi^{\text{prod}}(X) &= \sum_{k=1}^{n_\lambda n_\mu} \left[\mathsf{D}^{\text{prod}}(X)\right]_{kk} \\
&= \sum_{i=1}^{n_\lambda} \sum_{j=1}^{n_\mu} \left[\mathsf{D}^{(\lambda)}(X)\right]_{ii} \left[\mathsf{D}^{(\mu)}(X)\right]_{jj} \\
&= \left\{ \sum_{i=1}^{n_\lambda} \left[\mathsf{D}^{(\lambda)}(X)\right]_{ii} \right\} \left\{ \sum_{j=1}^{n_\mu} \left[\mathsf{D}^{(\mu)}(X)\right]_{jj} \right\} \\
&= \chi^{(\lambda)}(X)\, \chi^{(\mu)}(X). \qquad (23.21)
\end{aligned}$$

This proves the assertion, and a similar argument leads to the corresponding result for integrands in the form of a product of three or more factors.

An immediate corollary is that *an integral whose integrand is the product of two functions transforming according to two different irreps is necessarily zero*. To see this, we use (23.16) to determine whether irrep $A_1$ appears in the product character set $\chi^{\text{prod}}(X)$.

$$m_{A_1} = \frac{1}{g} \sum_X \left[\chi^{(A_1)}(X)\right]^* \chi^{\text{prod}}(X) = \frac{1}{g} \sum_X \chi^{\text{prod}}(X) = \frac{1}{g} \sum_X \chi^{(\lambda)}(X) \chi^{(\mu)}(X).$$

We have used the fact that $\chi^{(A_1)}(X) = 1$ for all $X$, but now note that, by virtue of (23.13), the expression on the right of this equation is equal to zero unless $\lambda = \mu$. We have ignored any complications due to non-real characters – in practice, they are handled automatically as it is usually $\Psi^*\phi$, rather than $\Psi\phi$, which appears in integrands, though many functions are real in any case, and nearly all characters are.

Equation (23.21) is a general result for integrands but, specifically in the context of chemical bonding, it implies that for the possibility of bonding to exist, the two quantum wavefunctions must transform according to the same irrep. This is discussed further in the next section.

## 23.11 Physical applications of group theory

As we indicated at the start of chapter 22 and discussed in a little more detail at the beginning of the present chapter, some physical systems possess symmetries that allow the results of the present chapter to be used in their analysis. We now consider some of the more common sorts of problem in which the results have ready applications.

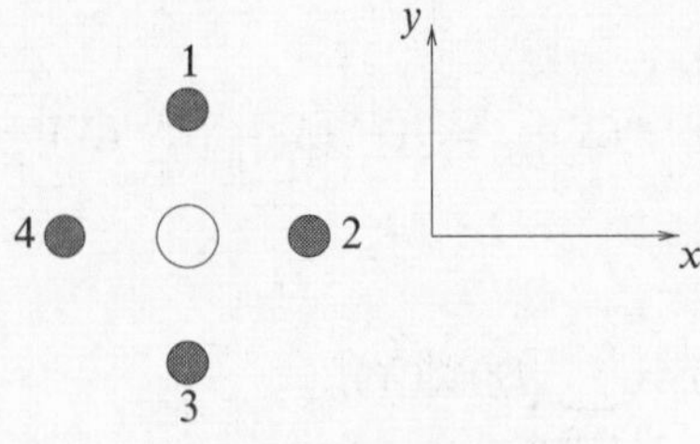


Figure 23.4 A molecule consisting of four atoms of iodine and one of manganese.

### *23.11.1 Bonding in molecules*

We have just seen that whether chemical bonding can take place in a molecule is strongly dependent upon whether the wavefunctions of the two atoms forming a bond transform according to the same irrep. It is thus sometimes useful to be able to find a basis vector that transforms according to a particular irrep of a group of transformations. This can be done if the characters of the irrep are known and a sensible starting point can be guessed. We will not prove it, but, starting from any $n$-dimensional basis vector $\Psi \equiv (\Psi_1 \ \Psi_2 \ \cdots \ \Psi_n)^{\rm T}$, the new vector $\Psi^{(\lambda)} \equiv (\Psi_1^{(\lambda)} \ \Psi_2^{(\lambda)} \ \cdots \ \Psi_n^{(\lambda)})^{\rm T}$ generated by

$$\Psi_i^{(\lambda)} = \sum_X \chi^{(\lambda)^*}(X) X \Psi_i \qquad (23.22)$$

will transform according to the $\lambda$th irrep. Sometimes, if $\Psi$ happens not to contain any part that transforms in the desired way, $\Psi^{(\lambda)}$ is found to be a zero vector and it is necessary to select a new starting vector. An illustration of the use of this 'projection operator' is given in the next example.

►*Consider a molecule made up of four iodine atoms lying at the corners of a square in the xy-plane, with a manganese atom at its centre, as shown in figure 23.4. Investigate whether the molecular orbital on the ring of iodine atoms, given by the superposition of p-state (angular momentum $l = 1$) atomic orbitals*

$$\Psi_1 = \Psi_y(\mathbf{r} - \mathbf{R}_1) + \Psi_x(\mathbf{r} - \mathbf{R}_2) - \Psi_y(\mathbf{r} - \mathbf{R}_3) - \Psi_x(\mathbf{r} - \mathbf{R}_4),$$

*can bond to the d-state atomic orbitals of the manganese atom described by (a) $\phi_1 = (3z^2 - r^2)f(r)$, and (b) $\phi_2 = (x^2 - y^2)f(r)$, where $f(r)$ is a function of r and so is unchanged by any of the symmetry operations of the molecule.*

We have eight basis functions, the atomic orbitals $\Psi_x(N)$ and $\Psi_y(N)$, where $N = 1, 2, 3, 4$ and indicates the position of the iodine atom. Since the wavefunctions are those of $p$-states they have the forms $xf(r)$ or $yf(r)$, where $f(r)$ is unaffected by all operations. The symmetry group of the system is $4mm$, whose character table is table 23.4.

*Case (a)*. The manganese atomic orbital, $\phi_1 = (3z^2 - r^2)f(r)$, at the centre of the molecule, is not affected by any of the symmetry operations, since $z$ and $r$ are unchanged by them. It clearly transforms according to the identity irrep $A_1$. We need to know which combination of the $\Psi_x(N)$ and $\Psi_y(N)$, if any, also transforms according to $A_1$.

We use the projection operator (23.22). If we choose $\Psi_x(1)$ as the arbitrary one-dimensional starting vector, we unfortunately obtain zero (as the reader may wish to verify), but $\Psi_y(1)$ does generate the required answer. The results of acting on $\Psi_y(1)$ with the various symmetry elements $X$ can be written down by inspection (see the discussion in section 23.2).

$$I : \Psi_y(1), \quad Q : -\Psi_y(3), \quad R : \Psi_x(2), \quad R' : -\Psi_x(4),$$

$$m_x : \Psi_y(1), \quad m_y : -\Psi_y(3), \quad m_d : \Psi_x(2), \quad m_{d'} : -\Psi_x(4).$$

Now $\chi^{(A_1)}(X) = 1$ for all $X$, so (23.22) states that the sum of the above results, all with weight 1, gives a vector (in this case of a one-dimensional irrep, just a wavefunction) that transforms according to $A_1$ and is therefore capable of forming a chemical bond with the wavefunction $\phi_1$. It is

$$\Psi^{(A_1)} = 2[\Psi_y(1) - \Psi_y(3) + \Psi_x(2) - \Psi_x(4)],$$

though, of course, the factor 2 is irrelevant. This is precisely the ring orbital $\Psi_1$ given in the problem, but here it is generated, rather than guessed beforehand.

*Case (b)*. The atomic orbital $\phi_2 = (x^2 - y^2)f(r)$ behaves as follows under the action of typical conjugacy class members.

$$I : \phi_2, \quad Q : \phi_2, \quad R : (y^2 - x^2)f(r) = -\phi_2, \quad m_x : \phi_2, \quad m_d : -\phi_2.$$

From this we see that $\phi_2$ transforms as a one-dimensional irrep, but, from table 23.4, that irrep is $B_1$ not $A_1$ (the irrep of $\Psi_1$, as already shown). Thus $\phi_2$ and $\Psi_1$ cannot form a bond. ◄

The original question did not ask for the the ring orbital to which $\phi_2$ may bond, but it can be easily generated by using the values of $X\Psi_y(1)$ calculated in case (a) and weighting them according to the characters of $B_1$, as follows.

$$\begin{aligned}\Psi^{(B_1)} &= \Psi_y(1) - \Psi_y(3) + (-1)\Psi_x(2) - (-1)\Psi_x(4) \\ &\quad + \Psi_y(1) - \Psi_y(3) + (-1)\Psi_x(2) - (-1)\Psi_x(4) \\ &= 2[\Psi_y(1) - \Psi_x(2) - \Psi_y(3) + \Psi_x(4)].\end{aligned}$$

More generally, we can find out which irreps of $4mm$ are present in the space spanned by the basis functions $\Psi_x(N)$ and $\Psi_y(N)$, and at the same time illustrate

| 4mm | $I$ | $Q$ | $R, R'$ | $m_x, m_y$ | $m_d, m_{d'}$ | |
|---|---|---|---|---|---|---|
| $A_1$ | 1 | 1 | 1 | 1 | 1 | $z$; $z^2$; $x^2 + y^2$ |
| $A_2$ | 1 | 1 | 1 | −1 | −1 | $R_z$ |
| $B_1$ | 1 | 1 | −1 | 1 | −1 | $x^2 - y^2$ |
| $B_2$ | 1 | 1 | −1 | −1 | 1 | $xy$ |
| E | 2 | −2 | 0 | 0 | 0 | $(x, y)$; $(xz, yz)$; $(R_x, R_y)$ |

Table 23.5 The character table for the irreps of group 4*mm* (or $C_{4v}$). The right-hand column lists some common functions, or, for the two-dimensional irrep E, pairs of functions, that transform according to the irrep against which they are shown.

the important point that, since we are working with characters, we are only interested in the diagonal elements of the representative matrices. This means that if we work in the natural representation $D^{nat}$ we need consider only those functions that transform, wholly or partially, into themselves.† Since we have no need to write out the matrices explicitly, their size ($8 \times 8$) is no drawback. We now find all the irreps spanned by the set comprising the $\Psi_x(N)$ and $\Psi_y(N)$.

(i) Under $I$ all eight functions are unchanged, and $\chi(I) = 8$.

(ii) The rotations $R$, $R'$ and $Q$ change the value of $N$ in every case and so all diagonal elements of the natural representation are zero and $\chi(R) = \chi(Q) = 0$.

(iii) $m_x$ takes $x$ into $-x$, $y$ into $y$ and, for $N = 1$ and 3, leaves $N$ unchanged, with the consequences (remember the forms of $\Psi_x(N)$ and $\Psi_y(N)$) that

$$\Psi_x(1) \to -\Psi_x(1), \qquad \Psi_x(3) \to -\Psi_x(3),$$

$$\Psi_y(1) \to \Psi_y(1), \qquad \Psi_y(3) \to \Psi_y(3).$$

Thus $\chi(m_x)$ has four non-zero contributions, −1, −1, 1 and 1, together with four zero contributions. The total is thus zero.

(iv) $m_d$ and $m_{d'}$ leave no atom unchanged and so $\chi(m_d) = 0$.

The character set of the natural representation is thus 8, 0, 0, 0, 0, which, either by inspection, or by applying formula (23.16), shows that

$$D^{nat} = A_1 \oplus A_2 \oplus B_1 \oplus B_2 \oplus 2E,$$

i.e. that all possible irreps are present. We have explicitly constructed the combinations of $\Psi_x(N)$ and $\Psi_y(N)$ that transform according to $A_1$ and $B_1$. The others can be found in the same way.

† See section 23.2.

| Function | Irrep | Classes | | |
|---|---|---|---|---|
| | | $I$ | $2C_3$ | $3\sigma_v$ |
| $xy$ | E | 2 | −1 | 0 |
| $x$ | E | 2 | −1 | 0 |
| $x^2 - y^2$ | E | 2 | −1 | 0 |
| product | | 8 | −1 | 0 |

Table 23.6 The character sets, for the group $C_{3v}$ (or $3mm$), of three functions and of their product $x^2y(x^2 - y^2)$.

### *23.11.2 Matrix elements in quantum mechanics*

In section 23.10 we outlined the procedure for determining whether a matrix element that involves the product of three factors as an integrand is necessarily zero. We now illustrate this with a specific worked example.

►*Determine whether a 'dipole' matrix element of the form*

$$J = \int \Psi_{d_1} x \Psi_{d_2}\, d\tau,$$

*where $\Psi_{d_1}$ and $\Psi_{d_2}$ are d-state wavefunctions of the forms $xyf(r)$ and $(x^2-y^2)g(r)$ respectively, can be non-zero (i) in a molecule with symmetry $C_{3v}$ (or 3m), such as ammonia, and (ii) in a molecule with symmetry $C_{4v}$ (or 4mm), such as the* $MnI_4$ *molecule considered in the previous example.*

We will need to make reference to the character tables of the two groups. The table for $C_{3v}$ is table 23.1 (section 23.6); that for $C_{4v}$ is reproduced as table 23.5 from table 23.4, but with the addition of another column showing how some common functions transform.

We make use of (23.21), extended to the product of three functions. No attention need be paid to $f(r)$ and $g(r)$ as they are unaffected by the group operations.

*Case (a).* From the character table 23.1 for $3m$, we see that each of $xy$, $x$ and $x^2 - y^2$ forms part of a basis set transforming according to the two-dimensional irrep E. Thus we may fill in the array of characters (using chemical notation for the classes, except that we continue to use $I$ rather than $E$) as shown in table 23.6. The last line is obtained by multiplying together the corresponding characters for each of the three elements. Now by inspection, or by applying (23.16), i.e.

$$n_{A_1} = \tfrac{1}{6}[1(1)(8) + 2(1)(-1) + 3(1)(0)] = 1,$$

we see that irrep $A_1$ does appear in the reduced representation of the product, and so $J$ is not necessarily zero.

| Function | Irrep | Classes | | | | |
|---|---|---|---|---|---|---|
| | | $I$ | $C_2$ | $2C_6$ | $2\sigma_v$ | $2\sigma_d$ |
| $xy$ | $B_2$ | 1 | 1 | −1 | −1 | 1 |
| $x$ | E | 2 | −2 | 0 | 0 | 0 |
| $x^2 - y^2$ | $B_1$ | 1 | 1 | −1 | 1 | −1 |
| product | | 2 | −2 | 0 | 0 | 0 |

Table 23.7 The character sets, for the group $C_{4v}$ (or $4mm$), of three functions, and of their product $x^2y(x^2 - y^2)$.

*Case (b).* From table 23.5 we find that, under the group $C_{4v}$, $xy$ and $x^2 - y^2$ transform as irreps $B_2$ and $B_1$ respectively, and that $x$ is part of a basis set transforming as E. Thus the calculation table takes the form of table 23.7 (again a chemical notation for the classes has been used).

Here inspection is sufficient, as the product is exactly that of irrep E, and $A_1$ is certainly not present. Thus $J$ is necessarily zero and the dipole matrix element vanishes. ◄

### 23.11.3 Degeneracy of normal modes

As our final area for illustrating the use of group theoretical results to physical systems, we consider the analysis of the normal modes of a vibrating system, which has applications in both chemistry and engineering. Normal modes of a system that are related by some symmetry operation have the same frequency of vibration; hence the modes are *degenerate*. Such modes span a vector space that transforms according to some irrep of the group $\mathcal{G}$ of symmetry operations of the system. Moreover, the degeneracy of the modes equals the dimension of the irrep. As an illustration, we consider the following example.

►*Investigate the possible vibrational modes of the equilateral triangular arrangement of equal masses and springs shown in figure 23.5. Demonstrate that two are degenerate.*

Clearly the symmetry group is that of the symmetry operations on an equilateral triangle, namely $3m$ (or $C_{3v}$), whose character table is table 23.1. As on a previous occasion, it is most convenient to use the natural representation $\mathsf{D}^{\text{nat}}$ of this group (it almost always saves having to write out matrices explicitly) acting on the six-dimensional vector space $(x_1, y_1, x_2, y_2, x_3, y_3)$. In this example the natural and regular representations coincide, but this is not usually the case.

For the rotational and reflection symmetries of the triangle, the class containing $A$ and $B$ in table 23.1 corresponds to the rotations $R$ and $R'$ (or $3_z$ in

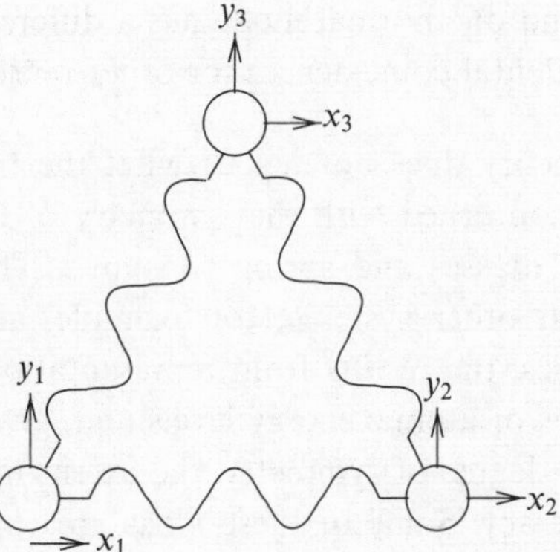


Figure 23.5 An equilateral array of masses and springs.

crystallographic notation, or $C_3$ in the chemical one, as explained in section 23.9). The class containing $C$, $D$, $E$ in table 23.1 corresponds to the three mirror reflections.

Clearly $\chi(I) = 6$, and, since all position labels are changed by a rotation, $\chi(3_z) = 0$. For the mirror reflections the simplest representative class member to choose is the reflection $m_y$ in the plane containing the $y_3$-axis, since then only label 3 is unchanged; under $m_y$, $x_3 \to -x_3$ and $y_3 \to y_3$, leading to the conclusion that $\chi(m_y) = 0$. Thus the character set is 6, 0, 0.

Using (23.16) and the character table 23.1 shows that

$$\mathrm{D}^{\mathrm{nat}} = \mathrm{A}_1 \oplus \mathrm{A}_2 \oplus 2\mathrm{E}.$$

However, we have so far allowed $x_i$, $y_i$ to be completely general, and we must now remove those irreps that do not correspond to vibrations. These will be the irreps corresponding to bodily translations of the triangle and to its rotation without relative motion of the three masses.

Bodily translations are linear motions of the centre of mass (with coordinates $(x_1 + x_2 + x_3)/3$ and $(y_1 + y_2 + y_3)/3$) and they behave under the group operations like $x$ and $y$. Table 23.1 shows that the corresponding irrep is the two-dimensional E; this accounts for one of the two such irreps found in the natural representation.

Table 23.1 also shows (though we have not proved it) that planar bodily rotations of the triangle – rotations about the $z$-axis, denoted by $R_z$ – transform as irrep $\mathrm{A}_2$. When the linear motions of the centre of mass, and pure rotation about it, are removed from our reduced representation, we are left with $\mathrm{A}_1 \oplus \mathrm{E}$. These are thus the irreps corresponding to the internal vibrations of the triangle – one non-degenerate mode and one doubly degenerate mode. The physical interpretation of this, beyond the scope of this book to prove, is that two of the normal modes of the system have the same frequency (and energy quanta in quantum mechanics) and one normal mode has a different frequency and energy quantum (barring accidental coincidences for other reasons). ◄

In general, group theory does not tell us what the frequencies are, since the input to it is entirely concerned with the symmetry of the system and not with the absolute values of masses and spring constants. This limitation extends to analogous situations in other systems; for example, although nothing can be said about their energies, the results from representation theory can be used to predict the degeneracies of atomic energy levels and, given a perturbation whose Hamiltonian has some degree of symmetry, the extent to which the perturbation will resolve the degeneracy. Some of these ideas are explored a little further in the next section and in the exercises.

### 23.11.4 Breaking of degeneracies

If a physical system has a high degree of symmetry, it will normally be the case that some of its eigenvalues (of energy, frequency, angular momentum, etc.) are degenerate. However, if a perturbation that is invariant only under a smaller symmetry group is added, some of the original degeneracies may be broken. The results derived from representation theory can be used to decide the extent of the degeneracy-breaking.

In general we use an $n$-dimensional basis vector, consisting of the $n$ degenerate eigenfunctions, to generate an $n$-dimensional representation of the symmetry group of the perturbation. This representation is then decomposed into irreps. Eigenfunctions that transform according to different irreps need no longer share the same frequency of vibration.

We illustrate this with the following example.

►*A circular drumskin has three equal masses placed on it at the vertices of an equilateral triangle, as shown in figure 23.6. Determine which degenerate normal modes of the drumskin can be split by this perturbation.*

If no masses are present the normal modes of the drum-skin are either non-degenerate or two-fold degenerate. The degenerate eigenfunctions of the $n$th normal mode have the forms (see chapter 17)

$$\Psi = J_n(kr)\cos(n\theta)e^{\pm i\omega t} \qquad \text{or} \qquad J_n(kr)\sin(n\theta)e^{\pm i\omega t}.$$

We therefore consider the two-dimensional vector space space spanned by $\Psi_1 = \sin n\theta$ and $\Psi_2 = \cos n\theta$. This will generate a two-dimensional representation of the group $3m$, the symmetry group of the perturbation. Taking the easiest element

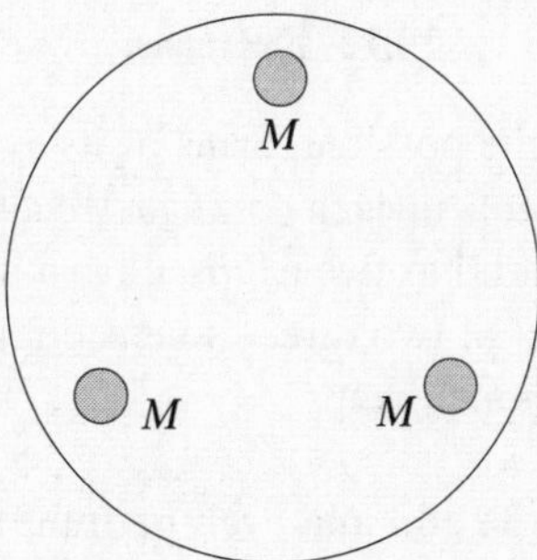


Figure 23.6 A circular drumskin loaded with three symmetrically placed masses.

from each of the three classes (identity, rotations, and reflections) of group $3m$, we have

$$\begin{aligned}
I\Psi_1 &= \Psi_1, \qquad I\Psi_2 = \Psi_2,\\
A\Psi_1 &= \sin\left[n\left(\theta - \tfrac{2}{3}\pi\right)\right] = \left(\cos\tfrac{2}{3}n\pi\right)\Psi_1 - \left(\sin\tfrac{2}{3}n\pi\right)\Psi_2,\\
A\Psi_2 &= \cos\left[n\left(\theta - \tfrac{2}{3}\pi\right)\right] = \left(\cos\tfrac{2}{3}n\pi\right)\Psi_2 + \left(\sin\tfrac{2}{3}n\pi\right)\Psi_1,\\
C\Psi_1 &= \sin[n(\pi-\theta)] = -(\cos n\pi)\Psi_1,\\
C\Psi_2 &= \cos[n(\pi-\theta)] = (\cos n\pi)\Psi_2.
\end{aligned}$$

The three representative matrices are therefore $\mathsf{D}(I) = \mathsf{I}_2$,

$$\mathsf{D}(A) = \begin{pmatrix} \cos\frac{2}{3}n\pi & -\sin\frac{2}{3}n\pi \\ \sin\frac{2}{3}n\pi & \cos\frac{2}{3}n\pi \end{pmatrix} \quad \text{and} \quad \mathsf{D}(C) = \begin{pmatrix} -\cos n\pi & 0 \\ 0 & \cos n\pi \end{pmatrix}.$$

The characters of this representation are $\chi(I) = 2$, $\chi(A) = 2\cos(2n\pi/3)$, and $\chi(C) = 0$. Using (23.16) and table 23.1, we find that

$$\begin{aligned}
m_{\mathrm{A}_1} &= \tfrac{1}{6}\left(2 + 4\cos\tfrac{2}{3}n\pi\right) = m_{\mathrm{A}_2}\\
m_{\mathrm{E}} &= \tfrac{1}{6}\left(4 - 4\cos\tfrac{2}{3}n\pi\right).
\end{aligned}$$

Thus

$$\mathsf{D} = \begin{cases} \mathrm{A}_1 \oplus \mathrm{A}_2 & \text{if } n = 3,\ 6,\ 9,\ \ldots,\\ \mathrm{E} & \text{otherwise.} \end{cases}$$

Hence the normal modes $n = 3,\ 6,\ 9,\ \ldots$ are the sum of two one-dimensional irreps and may therefore be split by the perturbation. For other values of $n$ the representation is irreducible and the degeneracy cannot be split. ◄

## 23.12 Exercises

23.1 (a) By considering the possible forms of its cycle notation, determine the number of elements in each conjugacy class of the permutation group $S_4$ and show that it has five irreps. Give a logical argument that shows they must consist of two three-dimensional, one two-dimensional, and two one-dimensional irreps.

(b) By considering the odd and even permutations in the group establish the characters for one of the one-dimensional irreps.

(c) Form a natural matrix representation of $4 \times 4$ matrices based on a set of objects $\{a, b, c, d\}$, and, by selecting one example from each conjugacy class, show that this natural representation has characters 4, 2, 1, 0, 0. The one-dimensional vector subspace spanned by sets of the form $\{a, a, a, a\}$ is invariant under the permutation group and hence transforms according to the invariant irrep $A_1$. The remaining three-dimensional subspace is irreducible; use this and the characters deduced above to establish the characters of one of the three-dimensional irreps $T_1$.

(d) Complete the character table using orthogonality properties, and check the summation rule for each irrep. You should obtain table 23.8.

| | Typical element and class size | | | | |
|---|---|---|---|---|---|
| | (1) | (12) | (123) | (1234) | (12)(34) |
| Irrep | 1 | 6 | 8 | 6 | 3 |
| $A_1$ | 1 | 1 | 1 | 1 | 1 |
| $A_2$ | 1 | −1 | 1 | −1 | 1 |
| E | 2 | 0 | −1 | 0 | 2 |
| $T_1$ | 3 | 1 | 0 | −1 | −1 |
| $T_2$ | 3 | −1 | 0 | 1 | −1 |

Table 23.8 The character table for group $S_4$.

23.2 In exercise 22.5, the group of pure rotations taking a cube into itself was found to have 24 elements. The group is isomorphic to the permutation group $S_4$, considered in the previous question, and hence has the same character table, once corresponding classes have been established. By counting the number of elements in each class make the correspondences below (the final two cannot be decided purely by counting, and should be taken as given).

| Permutation class type | Symbol (physics) | Action |
|---|---|---|
| (1) | $I$ | none |
| (123) | 3 | rotations about a body diagonal |
| (12)(34) | $2_z$ | rotation of $\pi$ about the normal to a face |
| (12) | $4_z$ | rotations of $\pm\pi/2$ about the normal to a face |
| (1234) | $2_d$ | rotation of $\pi$ about an axis through the centres of opposite edges. |

Reformulate the character table of the previous question in terms of the elements of the rotation symmetry group (432 or $O$) of a cube and use it to answer exercises 23.3 and 23.4.

23.3 In a certain crystalline compound, a thorium atom lies at the centre of a regular octahedron of six sulphur atoms at positions $(\pm a, 0, 0)$, $(0, \pm a, 0)$, $(0, 0, \pm a)$. These can be considered as being positioned at the centres of the faces of a cube of side $2a$. The sulphur atoms produce at the site of the thorium atom an electric field that has the same group of symmetry properties as a cube (432 or $O$).

The five degenerate $d$-electron orbitals of the thorium atom can be expressed, relative to any arbitrary polar axis, as

$$(3\cos^2\theta - 1)f(r), \qquad e^{\pm i\phi}\sin\theta\cos\theta f(r), \qquad e^{\pm 2i\phi}\sin^2\theta f(r).$$

A rotation about that polar axis by an angle $\phi'$ effectively changes $\phi$ to $\phi - \phi'$. Use this to show that the character of the rotation in a representation based on the wavefunctions is given by

$$1 + 2\cos\phi' + 2\cos 2\phi',$$

and hence that the characters of the representation, in the order of the symbols given in exercise 23.2, is 5, −1, 1, −1, 1. Deduce that the five-fold degenerate level is split into two levels, a doublet and a triplet.

23.4 Sulphur hexafluoride is a molecule with the same structure as the crystalline compound in exercise 23.3, except that a sulphur atom is now the central atom. The following are the forms of some of the electronic orbitals of the sulphur atom, together with the irreps according to which they transform under the symmetry group 432 (or $O$).

$$\begin{array}{ll} \Psi_s = f(r), & \mathrm{A}_1 \\ \Psi_{p_1} = zf(r), & \mathrm{T}_1 \\ \Psi_{d_1} = (3z^2 - r^2)f(r), & \mathrm{E} \\ \Psi_{d_2} = (x^2 - y^2)f(r), & \mathrm{E} \\ \Psi_{d_3} = xyf(r), & \mathrm{T}_2. \end{array}$$

The function $x$ transforms according to the irrep $\mathrm{T}_1$. Use the above data to determine whether dipole matrix elements of the form $J = \int \phi_1 x \phi_2 \, d\tau$

can be non-zero in a sulphur hexafluoride molecule for the following pairs $(\phi_1, \phi_2)$: (a) $(\Psi_{d1}, \Psi_s)$; (b) $(\Psi_{d1}, \Psi_{p1})$; (c) $(\Psi_{d2}, \Psi_{d1})$; (d) $(\Psi_s, \Psi_{d3})$; (e) $(\Psi_{p1}, \Psi_s)$.

23.5 The hydrogen atoms in a methane molecule $CH_4$ form a perfect tetrahedron with the carbon atom at its centre. The molecule is most conveniently described mathematically by placing the hydrogen atoms at the points $(1,1,1)$, $(1,-1,-1)$, $(-1,1,-1)$ and $(-1,-1,1)$. The symmetry group to which it belongs, the tetrahedral group ($\bar{4}3m$ or $T_d$) has classes typified by $I$, 3, $2_z$, $m_d$ and $\bar{4}_z$, where the first three are as in exercise 23.2, $m_d$ is a reflection in the mirror plane $x - y = 0$ and $\bar{4}_z$ is a rotation of $\pi/2$ about the $z$-axis followed by an inversion in the origin. A reflection in a mirror plane can be considered as a rotation of $\pi$ about an axis perpendicular to the plane, followed by an inversion in the origin.

The character table for the group $\bar{4}3m$ is very similar to that of the group 432, and has the form shown in table 23.9.

| | Typical element and class size | | | | | |
|---|---|---|---|---|---|---|
| | $I$ | 3 | $2_z$ | $\bar{4}_z$ | $m_d$ | |
| Irreps | 1 | 8 | 3 | 6 | 6 | |
| $A_1$ | 1 | 1 | 1 | 1 | 1 | $x^2 + y^2 + z^2$ |
| $A_2$ | 1 | 1 | 1 | $-1$ | $-1$ | |
| E | 2 | $-1$ | 2 | 0 | 0 | $(x^2 - y^2, 3z^2 - r^2)$ |
| $T_1$ | 3 | 0 | $-1$ | 1 | $-1$ | $(R_x, R_y, R_z)$ |
| $T_2$ | 3 | 0 | $-1$ | $-1$ | 1 | $(x, y, z)$; $(xy, yz, zx)$ |

Table 23.9 The character table for group $\bar{4}3m$.

By following the steps given below, determine how many different internal vibration frequencies the $CH_4$ molecule has.

(a) Consider a representation based on the 12 coordinates $x_i$, $y_i$, $z_i$ for $i = 1, 2, 3, 4$. For those hydrogen atoms that transform into themselves, a rotation through an angle $\theta$ about an axis parallel to one of the coordinate axes gives rise in the natural representation to the diagonal element 1 for the corresponding coordinate and $2\cos\theta$ for the two orthogonal coordinates. If the rotation is followed by an inversion, these entries are multiplied by $-1$. Atoms not transforming into themselves give zero diagonal contribution. Show that the characters of the natural representation are 12, 0, 0, 0, 2 and hence that its expression in terms of irreps is

$$A_1 \oplus E \oplus T_1 \oplus 2T_2.$$

(b) The irreps of the bodily translational and rotational motions are included in this expression and need to be subtracted out. Show that when this is done it can be concluded that there are three different internal vibration frequencies in the $CH_4$ molecule. State their degeneracies and check that they are consistent with the expected number of normal coordinates needed to describe the internal motions of the molecule.

23.6 Demonstrate that (23.22) does indeed generate a set of vectors transforming according to an irrep $\lambda$ by sketching and superposing drawings of an equilateral triangle of springs and masses, based on that shown in figure 23.7.

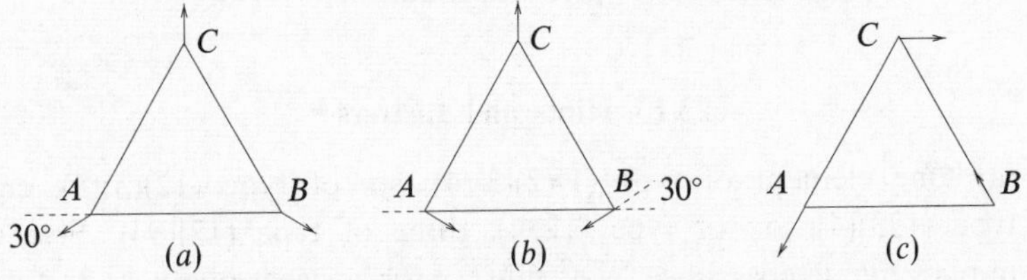


Figure 23.7 The three normal vibration modes of the equilateral array. Mode (*a*) is known as the 'breathing mode'. Modes (*b*) and (*c*) transform according to E and have equal vibrational frequencies.

(a) Starting with an arbitrary displacement of, say, vertex $C$, draw the results of operating on the initial sketch with each of the symmetry elements of the group $3m$ ($C_{3v}$).

(b) Superimpose the results, weighting them according to the characters of irrep $A_1$ (table 23.1 in section 23.6) and verify that the resultant is a symmetrical arrangement in which all three masses move symmetrically towards (or away from) the centroid of the triangle. The mode is illustrated in figure 23.7(*a*).

(c) Start again with a displacement $\delta$ of $C$ parallel to the $x$-axis. Form a similar superposition of sketches weighted according to the characters of irrep E (note that the reflections are not needed). The resultant contains some bodily displacement of the triangle, since this also transforms according to E. Show that the displacement of the centre of mass is $\bar{x} = \delta,\ \bar{y} = 0$. Subtract this out and verify that the remainder is of the form shown in figure 23.7(c).

(d) Using an initial displacement parallel to the $y$-axis, and an analogous procedure, generate the remaining normal mode, degenerate with that in (*c*) and shown in figure 23.7(*b*).

23.7 Further investigation of the crystalline compound considered in exercise 23.3 shows that the octahedron is not quite perfect, but is elongated along the $(1, 1, 1)$ direction with the sulphur atoms at positions $\pm(a + \delta, \delta, \delta)$, $\pm(\delta, a + \delta, \delta)$, $\pm(\delta, \delta, a + \delta)$, where $\delta \ll a$. This structure is invariant under the (crystallographic) symmetry group 32 with three two-fold axes along directions typified by $(1, -1, 0)$. The latter axes, which are perpendicular to the $(1, 1, 1)$ direction, are axes of two-fold symmetry for the perfect octahedron. The group 32 is really the three-dimensional version of the group $3m$ and has the same character table as table 23.1 (section 23.6). Use this to show that, as a result of the distortion of the octahedron, the doublet found in exercise 23.3 is unsplit, but that the triplet breaks up into a singlet and a doublet.

## 23.13 Hints and answers

23.1 (a) One element of type (1)(2)(3)(4), six of type (12)(3)(4), eight of type (123)(4), six of type (1234), three of type (12)(34). Five classes implies five irreps. Since $\sum n_i^2$ must $= 24$, at least one $n_i \geq 3$. Assuming $n_i \geq 4$ leads to a contradiction, and so $n_5$ (say) $= 3$. The inequalities $1^2 + 3(2^2) < 15 < 4(2^2)$ imply that a second $n_i = 3$. $\sum_1^3 n_i^2 = 6$ has only one integer solution. (b) $\mathsf{D}^{(2)}[(12)] = \mathsf{D}^{(2)}[(1234)] = -1$. (c) Characters for $\mathrm{T}_1$ are $(4-1)$, $(2-1)$, $(1-1)$, $(0\ -1)$, $(0-1)$, i.e. 3, 1, 0, $-1$, $-1$.

23.3 The five basis functions of the representation are multiplied by 1, $e^{-i\phi'}$, $e^{+i\phi'}$, $e^{-2i\phi'}$, $e^{+2i\phi'}$ as a result of the rotation. The character is the sum of these for rotations of 0, $2\pi/3$, $\pi$, $\pi/2$, $\pi$. $\mathsf{D}^{\mathrm{rep}} = \mathrm{E} + \mathrm{T}_2$.

23.4 (a) No; (b) yes; (c) no; (d) no; (e) yes.

23.5 (b) The bodily translation has irrep $\mathrm{T}_2$ and the rotation has irrep $\mathrm{T}_1$. The internal vibrations have irreps $\mathrm{A}_1$, E, $\mathrm{T}_2$, with respective degeneracies 1, 2, 3, making six internal coordinates (= 12 total minus three translational minus three rotational).

23.6 (c) $\bar{x} = \frac{1}{3}[2\delta + (-1)(-\frac{1}{2}\delta) + (-1)(-\frac{1}{2}\delta)]$,
$\bar{y} = \frac{1}{3}[0 + (-1)(-\frac{\sqrt{3}}{2}\delta) + (-1)(\frac{\sqrt{3}}{2}\delta)]$.

23.7 The doublet irrep E (characters 2, $-1$, 0) appears in both 432 and 32 and so is unsplit. The triplet $\mathrm{T}_2$ (characters 3, 0, 1) splits under 32 into doublet E (characters 2, $-1$, 0) and singlet $\mathrm{A}_1$ (characters 1, 1, 1).

# 24

# *Probability*

All scientists will know the importance of experiment and observation, and equally be aware that the results of some experiments depend, to a degree, on chance. For example, in an experiment to measure the heights of a random sample of people, we would not be in the least surprised if all the heights were found to be different; but, if the experiment were repeated often enough, we would expect to find some sort of regularity in the results. Statistics, which is the subject of the last part of this chapter, is concerned with the analysis of real experimental data of this sort. First, however, we discuss probability. To a pure mathematician, probability is an entirely theoretical subject based on axioms. Although this axiomatic approach is important, and we will discuss it briefly, we will here adopt an approach to probability more in keeping with its eventual applications in statistics.

We will first discuss the terminology required, with particular reference to the convenient graphical representation of experimental results in Venn diagrams. We then discuss the concept of a random variable and distributions of random variables. It is here that the connection with statistics is made; we assert that the results of many experiments are random variables, and that those results have some sort of regularity, which we call a distribution. Precise definitions of a random variable and a distribution are then given, as are the defining equations for some important distributions. We will also derive some useful quantities associated with these distributions. Application of the theory is left to the end of the chapter.

## 24.1 Venn diagrams

We will call a single performance of an experiment a *trial* and each possible result an *outcome*. The *sample space* $S$ of the experiment is then the set of all possible outcomes of an individual trial. For example, if we throw a six-sided die, there are six possible outcomes that together form the sample space of the experiment.

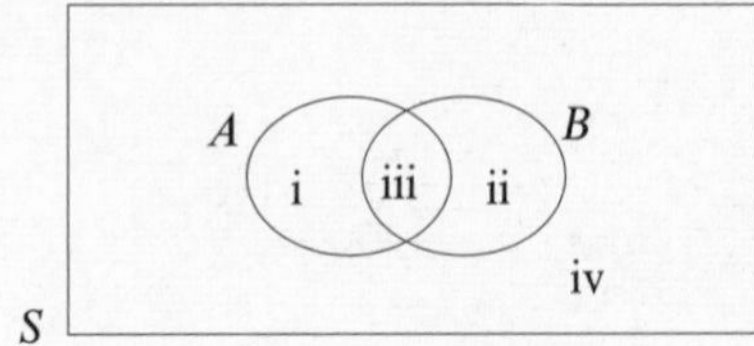


Figure 24.1 A Venn diagram.

At this stage we are not concerned with how likely a particular outcome might be (we will return to the probability of an outcome in due course); for the moment we will concentrate on the classification of possible outcomes. It is clear that some sample spaces are finite (e.g. the outcomes of throwing a die) whilst others are infinite (e.g. the outcomes of measuring people's heights). Most often one is not interested in individual outcomes, but in whether an outcome belongs to a given subset $A$ (say) of the sample space $S$; these subsets are called *events*. For example, we might be interested in whether a person is taller or shorter than 180 cm, in which case we divide the sample space into just two events: namely that the outcome (height measured) is (i) greater than 180 cm or (ii) less than 180 cm.

A common graphical representation of the outcomes of an experiment is the *Venn diagram*. A Venn diagram usually consists of a rectangle, the interior of which represents the sample space, together with one or more closed curves inside it. The interior of each closed curve then represents an event. Figure 24.1 shows a typical Venn diagram representing a sample space $S$ and two events $A$ and $B$. Every possible outcome is then assigned to an appropriate region; in this example there are four regions to consider (marked i to iv in figure 24.1):

(i) outcomes that belong to event $A$ but not to event $B$;
(ii) outcomes that belong to event $B$ but not to event $A$;
(iii) outcomes that belong both to event $A$ and to event $B$;
(iv) outcomes that belong neither to event $A$ nor to event $B$.

►*A six-sided die is thrown. Let event A be 'the number obtained is divisible by 2' and event B be 'the number obtained is divisible by 3'. Draw a Venn diagram to represent these events.*

It is clear that the outcomes 2, 4, 6 belong to event $A$, and that the outcomes 3, 6 belong to event $B$. Of these, 6 belongs to both $A$ and $B$. The remaining outcomes, 1, 5, belong neither to $A$ nor to $B$. The appropriate Venn diagram is shown in figure 24.2. ◄

In the above example, one outcome, 6, is divisible by both 2 and 3 and so belongs to both $A$ and $B$. This outcome is placed in region iii of figure 24.1, which

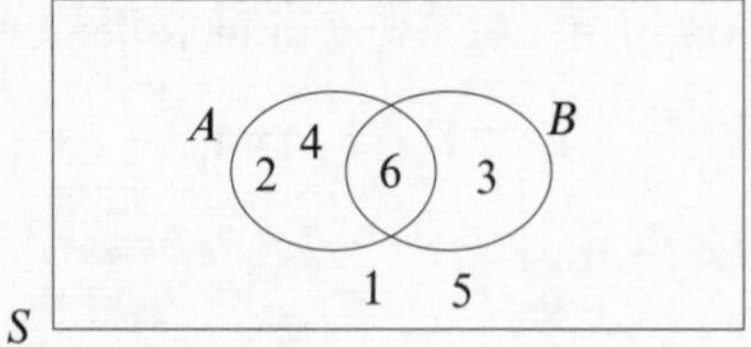


Figure 24.2 The Venn diagram for the outcomes of the die-throwing trials described in the worked example.

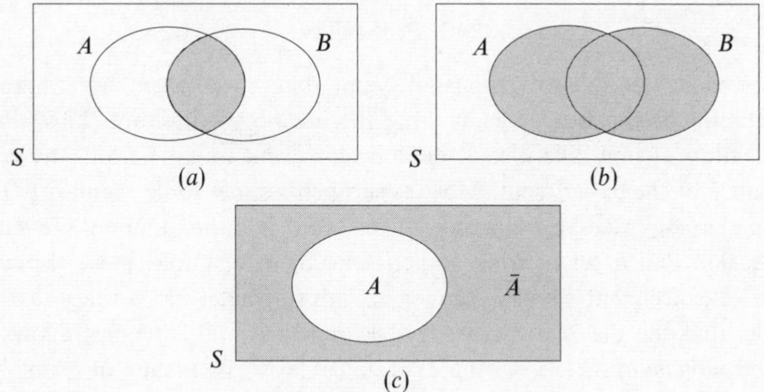


Figure 24.3 Venn diagrams: the shaded regions show (*a*) $A \cap B$, the intersection of two events $A$ and $B$, (*b*) $A \cup B$, the union of events $A$ and $B$, (c) the complement $\bar{A}$ of an event $A$.

is called the *intersection* of $A$ and $B$, and is denoted $A \cap B$ (see figure 24.3(*a*)). If no events lie in the region of intersection, then $A$ and $B$ are said to be *mutually exclusive* or *disjoint*. Similarly, the event comprising all the elements that belong to $A$ or $B$, or both, is called the *union* of $A$ and $B$ and denoted $A \cup B$ (see figure 24.3(*b*)). In the above example, $A \cup B = \{2, 3, 4, 6\}$. An event that contains no outcomes is called the *empty event* and denoted by $\emptyset$. It is sometimes convenient to talk about those outcomes that do *not* belong to a particular event. Outcomes that do not belong to $A$ are called the *complement* of $A$ and are denoted by $\bar{A}$ (see figure 24.3(*c*)).

These ideas are straightforwardly extended to more than two events. If there exist $n$ events $A_1$, $A_2$, ..., $A_n$, in some sample space $S$, then the event consisting of all those outcomes which belong to *one or more* of the $A_i$ is the union of $A_1, A_2, \ldots, A_n$ and is denoted by

$$A_1 \cup A_2 \cup \cdots \cup A_n. \qquad (24.1)$$

Similarly, the event consisting of all the outcomes that belong to *every one* of the

$A_i$ is called the intersection of $A_1, A_2, \ldots, A_n$ and is denoted by

$$A_1 \cap A_2 \cap \cdots \cap A_n. \tag{24.2}$$

If, for *any* pair of values $i, j$ with $i \neq j$

$$A_i \cap A_j = \emptyset, \tag{24.3}$$

then the events $A_i$ and $A_j$ are said to be *mutually exclusive* or *disjoint*.

## 24.2 Probability

In the previous section we discussed Venn diagrams, which are a graphical representation of the outcomes of an experiment. We did not, however, give any indication of how likely each outcome or event might be, on any random performance of the experiment. Most experiments show some regularity. By this we mean that the relative frequency of an event is approximately the same on each occasion that a set of trials is performed. For example, if we throw a die $N$ times we expect that we will throw a six approximately $N/6$ times (assuming, of course, that the die is not biased). The regularity of outcomes allows us to define the *probability*, $P(A)$, as the expected relative frequency of event $A$ in a large number of trials. More quantitatively, if an experiment has a total of $n_S$ outcomes in the sample space $S$, and $n_A$ of these outcomes correspond to the event $A$, then the probability that event $A$ will occur is

$$P(A) = \frac{n_A}{n_S}. \tag{24.4}$$

### *24.2.1 Axioms and theorems*

From (24.4) we may deduce the following properties of the probability $P(A)$.

(i) For any event $A$ in a sample space $S$,

$$0 \leq P(A) \leq 1. \tag{24.5}$$

If $P(A) = 1$, then $A$ is a certainty; if $P(A) = 0$ then $A$ is an impossibility.

(ii) For the entire sample space $S$ we have

$$P(S) = 1, \tag{24.6}$$

which simply states that we are certain to obtain one of the possible outcomes.

(iii) If $\bar{A}$ is the complement of $A$, then from (ii) we have

$$P(\bar{A}) = 1 - P(A). \tag{24.7}$$

This is particularly useful for problems in which evaluating the probability of the complement is easier than evaluating the probability of the event itself.

(iv) If $A$ and $B$ are two events in $S$ then, from the Venn diagrams in figure 24.3, we obtain the *addition rule*

$$P(A \cup B) = P(A) + P(B) - P(A \cap B), \tag{24.8}$$

the final subtraction arising because the probability of the intersection of $A$ and $B$ is counted twice when the probability of $A$ is added to that of $B$. However, if $A$ and $B$ are mutually exclusive ($A \cap B = \emptyset$), then $P(A \cap B) = 0$ and

$$P(A \cup B) = P(A) + P(B). \tag{24.9}$$

►*Calculate the probability of drawing an ace or a spade from a pack of cards.*

Let $A$ be the event that an ace is drawn, and $B$ the event that a spade is drawn. It immediately follows that $P(A) = 4/52 = 1/13$, and $P(B) = 13/52 = 1/4$. The intersection of $A$ and $B$ consists of only the ace of spades and so $P(A \cap B) = 1/52$. Thus, from (24.8)

$$P(A \cup B) = 1/13 + 1/4 - 1/52 = 4/13.$$

In this case it is just as simple to recognise that there are 16 cards in the pack which satisfy the required condition (13 spades plus 3 other aces) and so the probability is 16/52. ◄

Equation (24.9) can easily be extended to more events. If $A_1, A_2, \ldots, A_n$ are mutually exclusive events then

$$P(A_1 \cup A_2 \cup \cdots \cup A_n) = P(A_1) + P(A_2) + \cdots + P(A_n). \tag{24.10}$$

Furthermore, if $A_1, A_2, \ldots, A_n$ (whether mutually exclusive or not) *exhaust* $S$, i.e. are such that

$$A_1 \cup A_2 \cup \cdots \cup A_n = S, \tag{24.11}$$

then

$$P(A_1 \cup A_2 \cup \cdots \cup A_n) = P(S) = 1. \tag{24.12}$$

►*A biased six-sided die has probabilities $p/2$, $p$, $p$, $p$, $p$, $2p$ of showing* 1, 2, 3, 4, 5, 6 *respectively. Calculate $p$.*

Given that the individual events are mutually exclusive, (24.10) can be applied to give

$$P(1 \cup 2 \cup 3 \cup 4 \cup 5 \cup 6) = p/2 + p + p + p + p + 2p = 13p/2.$$

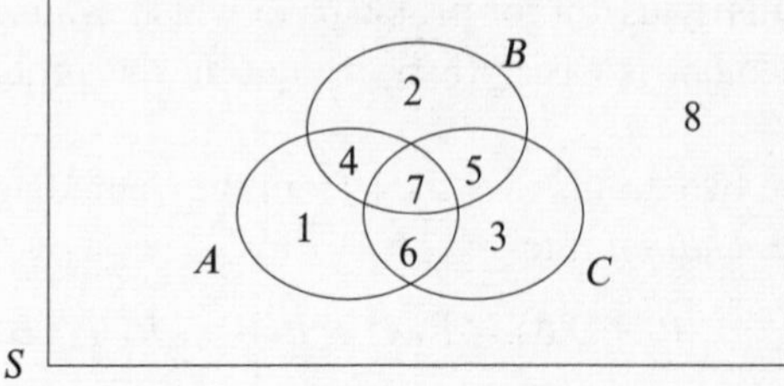


Figure 24.4 The general Venn diagram for three events.

The union of all possible outcomes on the LHS of this equation is clearly the sample space, $S$, and so

$$P(S) = 13p/2.$$

Now using (24.6),

$$13p/2 = P(S) = 1 \quad \Rightarrow \quad p = 2/13. \blacktriangleleft$$

When the possible outcomes of a trial may correspond to more than two events, and those events are not mutually exclusive, the calculation of the probability of the union of a number of events is more complicated than is implied by (24.10), which is valid only for mutually exclusive events.

Consider three events $A$, $B$ and $C$ with a Venn diagram such as is shown in figure 24.4. It will be clear that, in general, the diagram will be divided into eight regions and they will be of four different types. Regions 1, 2 and 3 each correspond to a single event; regions 4, 5 and 6 are each the intersection of exactly two events; region 7 is the three-fold intersection of all three events; and finally region 8 corresponds to none of the events.

For one-event Venn diagrams there are two regions, for the two-event case there are four regions and, as we have just seen, for the three-event case there are eight. In the general $n$-event case there are $2^n$ regions, as is clear from the fact that any particular region $R$ lies either inside or outside the closed curve of any particular event. With two choices (inside or outside) for each of $n$ closed curves, there are $2^n$ different possible combinations with which to characterise $R$. Once $n$ gets beyond three it becomes impossible to draw a simple two-dimensional Venn diagram, but this does not change the results.

The $2^n$ regions will break down into $n+1$ types, with the numbers of each type as follows†

† The symbols ${}^nC_i$, for $i = 0, 1, 2, \ldots, n$, are a convenient notation for combinations; these will be discussed fully in the next section.

$$
\begin{aligned}
{}^nC_0 &= 1, && \text{no events;}\\
{}^nC_1 &= n, && \text{one event but no intersections;}\\
{}^nC_2 &= \tfrac{1}{2}n(n-1), && \text{two-fold intersections;}\\
{}^nC_3 &= \tfrac{1}{3!}n(n-1)(n-2), && \text{three-fold intersections;}\\
&\ \vdots\\
{}^nC_n &= 1, && \text{an } n\text{-fold intersection.}
\end{aligned}
$$

That this makes a total of $2^n$ can be checked by considering the binomial expansion

$$2^n = (1+1)^n = 1 + n + \tfrac{1}{2}n(n-1) + \cdots + 1.$$

When the probabilities of events are combined to calculate the probability of the union of the $n$ events, account must be taken of double, triple etc. counting. The result can be shown (see section 24.3) to be

$$
\begin{aligned}
P(A_1 \cup A_2 \cup \cdots \cup A_n) = \sum_i P(A_i) - \sum_{i,j} P(A_i \cap A_j) + \sum_{i,j,k} P(A_i \cap A_j \cap A_k)\\
- \cdots + (-1)^{n+1} P(A_1 \cap A_2 \cap \cdots \cap A_n). \qquad (24.13)
\end{aligned}
$$

Each summation runs over all possible sets of subscripts, but omitting those in which any two subscripts in a set are the same. Equation (24.8) is a special case of (24.13) in which $n = 2$ and only the first two terms on the RHS survive. We now illustrate these results with extensions of our previous worked example.

►*Calculate the probability of drawing from a pack of cards one that is an ace, or a spade, or shows an even number* (2, 4, 6, 8, 10).

If, as previously, $A$ is the event that an ace is drawn, $P(A) = 4/52$. Similarly the event $B$, that a spade is drawn, has $P(B) = 13/52$. The additional event $C$, that the card is even (but not a picture card) has $P(C) = 20/52$. The two-fold intersections have probabilities

$$
\begin{aligned}
P(A \cap B) &= 1/52\\
P(A \cap C) &= 0\\
P(B \cap C) &= 5/52.
\end{aligned}
$$

There is no three-fold intersection as events $A$ and $C$ are mutually exclusive. Hence

$$P(A \cup B \cup C) = \frac{1}{52}\,[(4 + 13 + 20) - (1 + 0 + 5) + (0)] = \frac{31}{52}.$$

The reader should identify the 31 cards involved. ◄

Our final example has $n = 4$ and includes a four-fold intersection.

►*Find the probability of drawing from a pack a card that has at least one of the following properties:*

*A, it is an ace;*

*B, it is a spade;*

*C, it is a black honour card (ace, king, queen, jack or 10);*

*D, it is a black ace.*

Measuring all probabilities in units of $1/52$, the individual ones are

$$P(A) = 4, \qquad P(B) = 13, \qquad P(C) = 10, \qquad P(D) = 2.$$

The two-fold intersection probabilities, measured in the same units, are

$$\begin{aligned} &P(A \cap B) = 1, \qquad P(A \cap C) = 2, \qquad P(A \cap D) = 2, \\ &P(B \cap C) = 5, \qquad P(B \cap D) = 1, \qquad P(C \cap D) = 2. \end{aligned}$$

The three-fold intersections have probabilities

$$P(A \cap B \cap C) = 1, \quad P(A \cap B \cap D) = 1, \quad P(A \cap C \cap D) = 2, \quad P(B \cap C \cap D) = 1.$$

Finally, the four-fold intersection, requiring all four conditions to hold, is satisfied only by the ace of spades, and hence

$$P(A \cap B \cap C \cap D) = 1 \qquad \text{(in units of } 1/52\text{).}$$

Substituting in (24.13) gives

$$\begin{aligned} P &= \frac{1}{52}\,[(4 + 13 + 10 + 2) - (1 + 2 + 2 + 5 + 1 + 2) + (1 + 1 + 2 + 1) - (1)] \\ &= \frac{20}{52}. \blacktriangleleft \end{aligned}$$

### 24.2.2 Conditional probability

So far we have defined only probabilities of the form 'what is the probability that event $A$ happens'. In this section we turn to *conditional probability*, the probability that a particular event occurs *given* the occurrence of another, possibly related, event. For example, we may wish to know the probability of event $A$, drawing a second ace from a pack of cards, given that event $B$, removal of a first ace, has occurred.

We denote this probability by $P(A|B)$ and may obtain a formula for it by considering the total probability $P(A \cap B) = P(B \cap A)$ that both $A$ and $B$ will occur. This may be written in two ways, i.e.

$$\begin{aligned} P(A \cap B) &= P(A)P(B|A) \\ &= P(B)P(A|B). \end{aligned}$$

From this we obtain

$$P(A|B) = \frac{P(A \cap B)}{P(B)}, \tag{24.14}$$

and

$$P(B|A) = \frac{P(B \cap A)}{P(A)}. \tag{24.15}$$

In terms of Venn diagrams, we may think of $P(A|B)$ as the probability of $A$ in the reduced sample space $B$.

The events $A$ and $B$ are *statistically independent* if $P(A|B) = P(A)$ (or equivalently if $P(B|A) = P(B)$). In words, the probability of $A$ given $B$ is then the same as the probability of $A$ regardless of whether $B$ occurs. For example, if we throw a coin and a die at the same time, we would normally expect that the probability of throwing a six was independent of whether or not a head was thrown. If $A$ and $B$ are statistically independent, it then follows that

$$P(A \cap B) = P(A)P(B). \tag{24.16}$$

This is easily extended to more than two statistically independent events $A_1$, $A_2$, $\ldots$, $A_n$ as follows,

$$P(A_1 \cap A_2 \cap \cdots \cap A_n) = P(A_1)P(A_2)\cdots P(A_n). \tag{24.17}$$

If two events, $A$ and $B$, are mutually exclusive then

$$P(A|B) = 0 = P(B|A). \tag{24.18}$$

We also take this opportunity to distinguish between two different ways in which a *sampling experiment* may be performed. When an experiment consists of drawing objects at random from a given set of such objects, it is termed *sampling a population.* This may be done in two ways: when an object is drawn at random from the set it may either be put aside or returned to the set before the next object is randomly drawn. The former is termed 'sampling without replacement', the latter 'sampling with replacement'.

►*Find the probability of drawing two aces from a pack of cards if (i) the first card drawn is replaced before the second card is drawn, and (ii) the first card is put aside after being drawn.*

Let $A$ be the event that the first card is an ace, and $B$ the event that the second card is an ace. Now

$$P(A \cap B) = P(A)P(B|A),$$

and for both (i) and (ii) we know $P(A) = 4/52 = 1/13$.

(i) If the first card is replaced in the pack before the next is drawn then $P(B|A) = P(B) = 4/52 = 1/13$, since $A$ and $B$ are independent events. We then have

$$P(A \cap B) = P(A)P(B) = \frac{1}{13} \times \frac{1}{13} = \frac{1}{169}.$$

(ii) If the first card is put aside and the second drawn then $A$ and $B$ are not independent and $P(B|A) = 3/51$, so that

$$P(A \cap B) = P(A)P(B|A) = \frac{1}{13} \times \frac{3}{51} = \frac{1}{221}. \blacktriangleleft$$

As a further illustration we now consider a more complicated example based on the card game of poker.

►*From a shuffled deck of 52 playing cards you are dealt five cards. On examining these you find that you have two aces (which we denote by $A$) and three other cards, $X$, $Y$, $Z$, none of which is an ace and which are different from one another. The object is to obtain a 'full house' i.e. three cards of one value and two cards of another value. This is done by discarding either two or three of the cards $X$, $Y$, $Z$ and replacing them with cards from the pack. Are your interests best served by discarding two cards or three cards?*

Let us first consider the case where two of the cards $X$, $Y$, $Z$ are discarded (let us assume $Y$ and $Z$ for definiteness). You are thus left with $A, A, X$ and are dealt two further cards. A full house can then be obtained in three ways, namely drawing $A, X$ or $X, A$ or $X, X$. Remembering that there are now two aces and three cards $X$ left in the 47 remaining cards, the appropriate probabilities are calculated to be

$$P(A \cap X) = P(A)P(X|A) = \frac{2}{47} \times \frac{3}{46}, \qquad P(X \cap A) = P(X)P(A|X) = \frac{3}{47} \times \frac{2}{46},$$

$$P(X \cap X) = P(X)P(X|X) = \frac{3}{47} \times \frac{2}{46}.$$

Thus, in this case

$$P(\text{full house}) = P(A \cap X) + P(X \cap A) + P(X \cap X) = 0.0083.$$

Now, turning to the case where all three of the cards $X$, $Y$, $Z$ are discarded, you are left with $A, A$ and are dealt three further cards. A full house can now be obtained in a large number of ways, namely

$A, u, u$ in any of three orders (i.e. $A, u, u$ or $u, A, u$ or $u, u, A$), where $u = B, X, Y, Z$ with $B \neq$ any of $A, X, Y, Z$,

$v, v, v$ in just one order, where $v = B, X, Y, Z$.

These probabilities can again be calculated:

$$P(A\cap B\cap B) = P(B\cap A\cap B) = P(B\cap B\cap A) = \frac{2}{47}\times\frac{36}{46}\times\frac{3}{45},$$
$$P(A\cap X\cap X) = P(X\cap A\cap X) = P(X\cap X\cap A) = P(A\cap Y\cap Y) = \cdots$$
$$= P(Z\cap Z\cap A) = \frac{2}{47}\times\frac{3}{46}\times\frac{2}{45},$$
$$P(B\cap B\cap B) = \frac{36}{47}\times\frac{3}{46}\times\frac{2}{45},$$
$$P(X\cap X\cap X) = P(Y\cap Y\cap Y) = P(Z\cap Z\cap Z) = \frac{3}{47}\times\frac{2}{46}\times\frac{1}{45}.$$

Summing these probabilities gives

$$P(\text{full house}) = 0.0102.$$

Thus your interests are best served by discarding three cards. ◄

The notion of conditional probability may be straightforwardly extended to several events. For example, in the case of three events $A$, $B$, $C$, we may write $P(A\cap B\cap C)$ in several ways, e.g.

$$\begin{aligned}P(A\cap B\cap C) &= P(C)P(A\cap B|C)\\ &= P(B\cap C)P(A|B\cap C)\\ &= P(C)P(B|C)P(A|B\cap C).\end{aligned}$$

Finally we note without proof a number of useful relationships between conditional probabilities. Let us consider a set of *mutually exclusive* events $A_i$, whose union is $A$. For some other event $B$, we then have

$$P(A|B) = \sum_i P(A_i|B), \tag{24.19}$$

which is the *addition law* for conditional probabilities. Furthermore, if the set of mutually exclusive events $A_i$ exhausts the sample space $S$ then the probability $P(B)$ of some event $B$ in $S$ is given by the *total probability law*

$$P(B) = \sum_i P(B|A_i)P(A_i), \tag{24.20}$$

or, more generally,

$$P(B|C) = \sum_i P(B|A_i)P(A_i|C), \tag{24.21}$$

where $C$ is any other event in $S$.

### 24.2.3 Bayes' theorem

In the previous section we saw that the probability that both an event $A$ and a related event $B$ will occur can be written as

$$P(A)P(B|A) = P(B)P(A|B),$$

from which we obtain *Bayes' theorem*,

$$P(A|B) = \frac{P(A)}{P(B)}P(B|A). \tag{24.22}$$

This theorem clearly shows that $P(B|A) \neq P(A|B)$, unless $P(A) = P(B)$. It is sometimes useful to rewrite $P(B)$, if it is not known directly, as

$$P(B) = P(A)P(B|A) + P(\bar{A})P(B|\bar{A})$$

so that Bayes' theorem becomes

$$P(A|B) = \frac{P(A)P(B|A)}{P(A)P(B|A) + P(\bar{A})P(B|\bar{A})}. \tag{24.23}$$

Finally we note that (24.23) may be written in a more general form if $S$ is not simply made up of $A$ and $\bar{A}$, but rather of any set $A_i$ of mutually exclusive events that exhaust $S$. In this case Bayes' theorem takes the form

$$P(A_k|B) = \frac{P(B|A_k)P(A_k)}{\sum_i P(B|A_i)P(A_i)}. \tag{24.24}$$

►*Suppose that DNA blood matching is 98% reliable, in that a positive match is obtained 98% of the time when the two samples tested are from the same person, but only 2% of the time when the samples are in fact from different people. If one member of a population of 10 000 people commits a crime, find the probability that a person chosen at random, whom the DNA test declares to be guilty, is actually the criminal.*

The probability of a randomly selected individual being guilty of the crime is 1/10 000. Now let $A$ be the event that an individual committed the crime, and $B$ be the event that an individual is implicated by the DNA test, i.e. that a positive match is obtained. Using Bayes' theorem the probability that a person implicated by the test is actually the criminal is

$$P(A|B) = \frac{P(B|A)P(A)}{P(B|A)P(A) + P(B|\bar{A})P(\bar{A})}.$$

Since $P(A) = 1/10000 = 1 - P(\bar{A})$, and we are told that $P(B|A) = 0.98$ and $P(B|\bar{A}) = 0.02$, it follows that

$$P(A|B) = \frac{0.98 \times 1/10000}{(0.98 \times 1/10000) + (0.02 \times 9999/10000)} \approx 0.0049.$$

Thus, although the probability that any random individual committed the crime is 0.0001, the probability that an individual whom is implicated by the test is guilty is 0.0049, i.e. $\sim 50$ times more probable, but still less than $\frac{1}{2}\%$.

This result may, at a first glance, be a little surprising. The problem with the test is that, although it is 98% likely to catch the guilty, the 2% match of the innocent is significant if a large number of people are tested. It is important when dealing with conditional probabilities of this sort to be extremely careful not to introduce invalid assumptions. ◄

Sometimes we are only concerned with the relative probabilities of two events $A$ and $C$, given the occurrence of some other event $B$. From (24.22) we then obtain a different form of Bayes' theorem,

$$\frac{P(A|B)}{P(C|B)} = \frac{P(A)P(B|A)}{P(C)P(B|C)}, \tag{24.25}$$

which does not contain $P(B)$ at all.

## 24.3 Permutations and combinations

In equation (24.4) we defined the probability of an event $A$ in a sample space $S$ as

$$P(A) = \frac{n_A}{n_S},$$

where $n_A$ is the number of outcomes belonging to event $A$, and $n_S$ is the total number of possible outcomes. It is therefore necessary to be able to count the number of possible outcomes in various common situations.

Let us first consider a set of $n$ objects that are all different. We may ask how many ways these $n$ objects may be arranged – i.e. how many *permutations* of these objects exist. This is straightforward to deduce, as follows: the object in the first position may be chosen in $n$ different ways, that in the second in $n-1$ ways, and so on until the final object is positioned. The number of possible arrangements is therefore

$$n(n-1)(n-2)\cdots(1) = n! \tag{24.26}$$

Generalising (24.26) slightly, let us suppose we choose only $k$ $(< n)$ objects from $n$. The number of possible permutations of those $k$ objects selected from $n$ is given by

$$\underbrace{n(n-1)(n-2)\cdots(n-k+1)}_{k \text{ factors}} = \frac{n!}{(n-k)!} \equiv {}^nP_k. \tag{24.27}$$

So far we have assumed that all $n$ objects are different (or *distinguishable*). Let us now consider $n$ objects of which $n_1$ are identical and of type 1, $n_2$ are identical and of type 2, ..., and $n_m$ are identical and of type $m$ (clearly $n = n_1+n_2+\cdots+n_m$). From (24.26) the number of permutations of these $n$ objects is $n!$. However, the number of *distinguishable* permutations is only

$$\frac{n!}{n_1!n_2!\cdots n_m!}, \tag{24.28}$$

since the $i$th group of identical objects can be rearranged in $n_i!$ ways without changing the distinguishable permutation.

►*A set of snooker balls consists of a white, a yellow, a green, a brown, a blue, a pink, a black and 15 reds. How many distinguishable permutations of the balls are there?*

In total there are 22 balls, with the 15 red balls indistinguishable. Thus from (24.28) the number of distinguishable permutations is

$$\frac{22!}{(1!)(1!)(1!)(1!)(1!)(1!)(15!)} = \frac{22!}{15!} = 859\,541\,760. \blacktriangleleft$$

In calculating the number of permutations of various objects we have so far assumed that the objects are sampled *without replacement* – i.e. once an object has been drawn from the set it is put aside. As mentioned previously, however, we may instead replace each object before the next is chosen. The number of permutations of $k$ objects from $n$ *with replacement* may be calculated very easily since the first object can be chosen in $n$ different ways, as can the second, third, etc. Therefore the number of permutations is simply $n^k$. This may also be viewed as the number of permutations of $k$ objects from $n$ where repetitions are allowed, i.e. each object may be used as often as one likes.

► *Find the probability that in a group of k people, at least two have the same birthday (ignoring 29 February).*

There are 365 different possible birthdays for each of $k$ people and so the total number of possible outcomes is $(365)^k$. The number of outcomes for which all the birthdays are different is

$$^{365}P_k = \frac{365!}{(365-k)!},$$

and hence the probability that all the birthdays are different is

$$p = \frac{365!}{(365-k)!\ 365^k}.$$

Now using the complement rule (24.7), the probability that two or more people have the same birthday is

$$1 - p = 1 - \frac{365!}{(365-k)!\ 365^k}.$$

This expression may be conveniently evalutated using Stirling's approximation for $n!$ when $n$ is large, namely

$$n! \sim \sqrt{2\pi n}\left(\frac{n}{e}\right)^n,$$

to give

$$1 - e^{-k}\left(\frac{365}{365-k}\right)^{365-k+0.5}.$$

It is interesting to note that if $k = 23$ the probability is a little greater than a half that at least two people have the same birthday, and if $k = 50$, the probability rises to 0.970. This can prove a good bet at a party of non-mathematicians! ◄

We now consider the number of *combinations* of various objects, where their order is immaterial. From (24.27) we see that the number of permutations of $k$ objects chosen from $n$ is ${}^nP_k = n!/(n-k)!$. Now, since we are no longer concerned with the order of the objects, which can be internally arranged in $k!$ different ways, the number of combinations of $k$ objects from $n$ is

$$\frac{n!}{(n-k)!k!} \equiv {}^nC_k \equiv \binom{n}{k} \quad \text{for } 0 \le k \le n, \tag{24.29}$$

where ${}^nC_k$ is called the *binomial coefficient* since it also appears in the binomial expansion

$$(a+b)^n = \sum_{k=0}^{n} {}^nC_k a^k b^{n-k}. \tag{24.30}$$

The binomial coefficients have the following useful (and easily proved) properties:

$${}^nC_0 = 1, \qquad {}^nC_k = {}^nC_{n-k}, \qquad {}^nC_k + {}^nC_{k+1} = {}^{n+1}C_{k+1},$$

$$\sum_{s=0}^{n-1} {}^{k+s}C_k = {}^{n+k}C_{k+1}, \qquad \sum_{k=0}^{r} {}^pC_k \times {}^qC_{r-k} = {}^{p+q}C_r.$$

As in the case of permutations we might ask how many combinations of $k$ objects may chosen from $n$ *with replacement* (repetition). To calculate this, we note that it is equivalent to the number of ways in which $k$ (indistinguishable) 'balls' can be distributed among $n$ boxes, which may be symbolised as (for example if $k = 7$, $n = 5$)

xxx| |x|xx|x

which denotes three balls in the first box, none in the second, one in the third, two in the fourth and one in the fifth. We therefore need only consider the number of (distinguishable) ways in which $k$ crosses and $n-1$ vertical lines can be arranged, i.e. the number of permutations of $k+n-1$ objects, of which $k$ are identical crosses and $n-1$ are identical lines. This is given by (24.29) as

$$\frac{(k+n-1)!}{k!(n-1)!} = {}^{n+k-1}C_k. \tag{24.31}$$

► *Use the identity*

$$(1-1)^r = 0$$

*to prove the formula given in (24.13) for the probability of an outcome belonging to the union of n events that are not necessarily mutually exclusive.*

For $n$ events $A_1, A_2, \ldots, A_n$, the total region in the Venn diagram $S$ corresponding to the union $A_U \equiv A_1 \cup A_2 \cup \cdots \cup A_n$ will be those parts of the diagram that lie inside any of the $n$ closed curves corresponding to the individual events. The probability $P(A_U)$ will be equal to the sum of the probabilities corresponding to each of the $2^n - 1$ subregions $R_i$ into which the curves divide the non-empty part of $S$. Therefore, in calculating $P(A_U)$, we must ensure that for each $i$ the probability $P(R_i)$ is added in only once.

Let any particular $R_i$ be a subregion of $r$-fold intersection, i.e. it lies inside exactly $r$ of the $n$ closed curves. Without any loss of generality we can suppose these are $A_1, A_2, \ldots, A_r$. The region $R_i$ will then necessarily lie within a subregion of the $s$-fold intersection of each subset (comprising $s$ members) of the events $A_1, A_2, \ldots, A_r$, this being true for each value of $s$ from 1 up to $r$. The number of such intersections will be equal to the number of ways $s$ events can be chosen from the $r$ events $A_1, A_2, \ldots, A_r$, i.e. ${}^rC_s$ ways. This means that in formula (24.13), $P(R_i)$ is included $(-1)^{s+1} \times {}^rC_s$ times (probabilities corresponding to even values of $s$ are subtracted) when the contributions of $s$-fold intersections are evaluated.

The total number of times that $R_i$ is included on the RHS of (24.13) is thus

$$n(R_i) = {}^rC_1 - {}^rC_2 + {}^rC_3 - \cdots + (-1)^{r+1}\,{}^rC_r.$$

But the given identity can be expanded as

$$0 = (1-1)^r = {}^rC_0 - {}^rC_1 + {}^rC_2 - {}^rC_3 + \cdots (-1)^r\,{}^rC_r,$$

and, since ${}^rC_0 = 1$ for any $r$, it follows that $n(R_i) = 1$, whatever the value of $r$. Thus $R_i$, and similarly any other subregion, has its probability counted into $P(A_U)$ only once in total. This establishes (24.13) as a valid prescription for calculating $P(A_1 \cup A_2 \cup \cdots \cup A_n)$. ◄

We conclude this section with an important physical example of the use of permutations and combinations.

►*Consider $k$ balls, placed randomly into $n$ boxes. Calculate the probability of the particular arrangement in which there are $k_1$ balls in the first box, $k_2$ balls in the second box, etc. if the balls are*

*(i) distinguishable with no restrictions on the number in each box,*
*(ii) indistinguishable with no restrictions on the number in each box,*
*(iii) indistinguishable with a maximum of one ball in each box,*
*(iv) distinguishable with a maximum of one ball in each box.*

*Case (i).* If there is no restriction on the number of balls in any given box, then each ball can be placed in any of the $n$ boxes, so that, if the balls are distinguishable, the total number of different arrangements is simply $n^k$. Given that there are $k_i$ balls in the $i$th box, the number of distinguishable arrangements of these balls (without affecting the occupancy numbers) is given by the total number of arrangements of $k$ balls, i.e. $k!$, divided by the product of the number of internal arrangements in each box, i.e.

$$\frac{k!}{k_1!k_2!\cdots k_n!}.$$

So the probability of this particular configuration is

$$\frac{k!}{k_1!k_2!\cdots k_n!}n^{-k}.$$

Such a system of balls and boxes is said to obey Maxwell–Boltzmann statistics. If we replace the balls by (distinguishable) molecules in a gas, for example, then the boxes can be thought of as the energy levels of the system.

*Case (ii).* If, however, the balls are indistinguishable, then from (24.31), the number of possible arrangements of the balls between boxes is

$$\frac{(k+n-1)!}{k!(n-1)!} = {}^{n+k-1}C_k.$$

Given that each arrangement is equally likely the probability of a particular one is

$$\frac{k!(n-1)!}{(k+n-1)!}.$$

Such a system is said to obey Bose–Einstein statistics. If the boxes are again thought of as the energy levels of some physical system, then, in this case, the balls may represent photons.

*Case (iii).* If a maximum of one (indistinguishable) ball can reside in each box, the problem reduces to simply arranging $k$ full boxes and $n-k$ empty boxes. The number of possible arrangements is thus ${}^nC_k$ and so the probability of a particular arrangement is

$$\frac{1}{{}^nC_k} = \frac{(n-k)!k!}{n!}.$$

Such a system is said to obey Fermi–Dirac statistics, and an example is provided by electrons in energy levels.

*Case (iv).* Finally, if the balls are in fact distinguishable, with a maximum of one ball in each box, then the number of arrangements is easily found from the previous result. Since there are $k!$ arrangements of each distribution the number of possible arrangements is $n!/(n-k)!$, and so the probability of a particular arrangement is

$$\frac{(n-k)!}{n!}.$$

This arrangement has the names of no famous scientists attached to it, since it appears never to occur in nature. ◄

## 24.4 Random variables and distributions

Suppose an experiment has an outcome sample space $S$. A real variable $X$ that is defined for all possible outcomes in $S$ (so that a real number – not necessarily unique – is assigned to each possible outcome) is called a *random variable.* The outcome of the experiment may already be a real number and hence a random variable, e.g. the number of heads obtained in ten throws of a coin, or the sum of the values if two dice are thrown. However, more arbitrary assignments are possible, e.g. the assignment of a 'quality' rating to the product of a manufacturing process. Assuming that a probability can be assigned to all possible outcomes in a sample space $S$, it is possible to assign a *probability distribution* to any random variable. Random variables may be divided into two classes, discrete and continuous, and we now examine each of these in turn.

### 24.4.1 Discrete random variables

A random variable $X$ that takes only discrete values $x_1, x_2, \ldots, x_n$, with probabilities $p_1, p_2, \ldots, p_n$, is called a discrete random variable. The number of values $n$ for which $X$ has a non-zero probability is finite or at most countably infinite. As mentioned above, an example of a discrete random variable is the number of heads obtained in 10 throws of a coin. If $X$ is a discrete random variable, we can define a *probability function* (PF) $f(x)$ that assigns probabilities to all the distinct

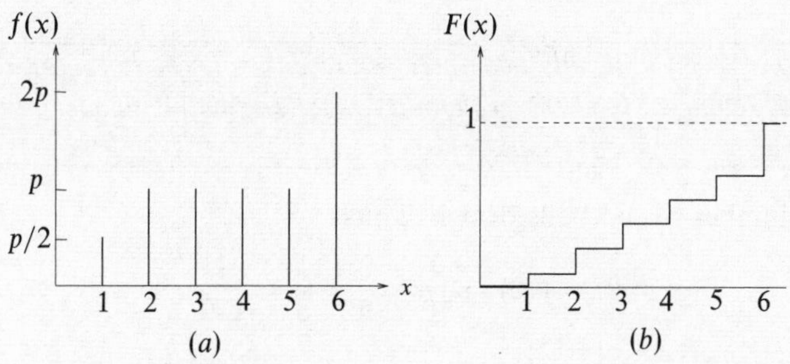


Figure 24.5 (*a*) A typical probability function for a discrete distribution. This particular probability function represents that for the biased die discussed earlier. Since the probabilities must sum to unity we require $p = 2/13$. (*b*) The cumulative probability function for the same discrete distribution. (Note that a different scale has been used for (*b*).)

values that $X$ can take, such that

$$f(x) = P(X = x) = \begin{cases} p_i & \text{if } x = x_i, \\ 0 & \text{otherwise.} \end{cases} \tag{24.32}$$

A typical PF (see figure 24.5) thus consists of spikes, at *valid values* of $X$, whose height at $x$ corresponds to the probability that $X = x$. Since the probabilities must sum to unity, we require

$$\sum_{i=1}^{n} f(x_i) = 1. \tag{24.33}$$

We may also define the *cumulative probability function* (CPF) of $X$, $F(x)$, whose value gives the probability that $X \leq x$, so that

$$F(x) = P(X \leq x) = \sum_{x_i \leq x} f(x_i). \tag{24.34}$$

Hence $F(x)$ is a step function that has upward jumps of $p_i$ at $x = x_i$, and is constant between possible values of $X$. We may also readily calculate the probability that $X$ lies between two limits, $a_1$ and $a_2$ $(a_1 < a_2)$; this is given by

$$P(a_1 < X \leq a_2) = \sum_{a_1 < x_i \leq a_2} f(x_i) = F(a_2) - F(a_1), \tag{24.35}$$

i.e. the sum of all the probabilities for which $x_i$ lies within the relevant interval.

►*A bag contains seven red balls and three white balls. Three balls are drawn at random and not replaced. Find the probability function for the number of red balls drawn.*

Let $X$ be the number of red balls drawn. Thus

$$P(X=0) = f(0) = \frac{3}{10} \times \frac{2}{9} \times \frac{1}{8} = \frac{1}{120},$$
$$P(X=1) = f(1) = \frac{3}{10} \times \frac{2}{9} \times \frac{7}{8} \times 3 = \frac{7}{40},$$
$$P(X=2) = f(2) = \frac{3}{10} \times \frac{7}{9} \times \frac{6}{8} \times 3 = \frac{21}{40},$$
$$P(X=3) = f(3) = \frac{7}{10} \times \frac{6}{9} \times \frac{5}{8} = \frac{7}{24}.$$

It should be noted that $\sum_{i=0}^{3} f(i) = 1$, as expected. ◄

### 24.4.2 Continuous random variables

A random variable $X$ is said to have a *continuous* distribution if $X$ is defined for a continuous range of values between given limits (often $-\infty$ to $\infty$). An example of a continuous random variable is the height of a person drawn from a population, which can take *any* value (within limits!). We can define the *probability density function* (PDF) $f(x)$ of a continuous random variable $X$ such that

$$P(x < X \le x + dx) = f(x)\,dx,$$

i.e. $f(x)\,dx$ is the probability that $X$ lies in the interval $x < X \le x + dx$. Clearly $f(x)$ must be a real function that is everywhere $\ge 0$. If $X$ can only take values between the limits $a_1$ and $a_2$, then in order for the sum of the probabilities of all possible outcomes to be equal to unity, we require

$$\int_{a_1}^{a_2} f(x)\,dx = 1.$$

Often $X$ can take any value between $-\infty$ and $\infty$ and so

$$\int_{-\infty}^{\infty} f(x)\,dx = 1.$$

The probability that $X$ lies in the interval $x_1 < X \le x_2$ is then given by

$$P(x_1 < X \le x_2) = \int_{x_1}^{x_2} f(x)\,dx, \qquad (24.36)$$

i.e. $P(x_1 < X \le x_2)$ is equal to the area under the curve of $f(x)$ between these limits (see figure 24.6).

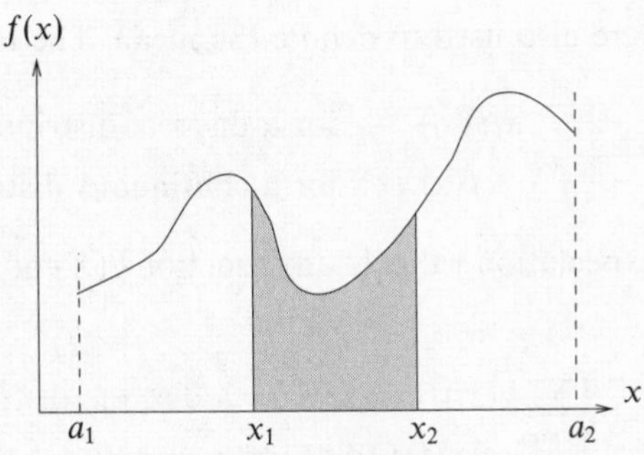


Figure 24.6 The probability density function for a continuous random variable $X$ that can only take values between the limits $a_1$ and $a_2$. The shaded area under the curve gives $P(x_1 < X \leq x_2)$, whereas the total area under the curve, between the limits $a_1$ and $a_2$, is equal to unity.

We may also define the cumulative probability function $F(x)$ for a continuous random variable by

$$F(x) = P(X \leq x) = \int_{a_1}^{x} f(u)\, du, \tag{24.37}$$

where $u$ is a (dummy) integration variable. We can then write

$$P(x_1 < X \leq x_2) = F(x_2) - F(x_1).$$

From (24.37) it is clear that $f(x) = dF(x)/dx$.

►*A random variable $X$ has a PDF $f(x) = e^{-x}$ in the interval $0 < x < \infty$ (and zero elsewhere). Find the probability that $X$ lies in the interval $1 < X \leq 2$.*

As required, the integral of $f(x)$ between 0 and $\infty$ is equal to unity, and from (24.36), we immediately obtain

$$P(1 < X \leq 2) = \int_1^2 f(x)\, dx = \int_1^2 e^{-x}\, dx = -e^{-2} - (-e^{-1}) = 0.23. \blacktriangleleft$$

## 24.5 Properties of distributions

The probability (density) function $f(x)$ contains all the information on the probability distribution of a random variable $X$. It is conventional, however, to characterise $f(x)$ by certain of its properties, which we now discuss.

### *24.5.1 Mean*

The property most commonly used in characterising a probability distribution is the *mean* $E[X]$ (also known as the *expectation* value of $X$). The alternative notations $\mu$ and $\langle x \rangle$ are also used to denote the mean. The mean is defined as

$$E[X] = \begin{cases} \sum_i x_i f(x_i) & \text{for a discrete distribution,} \\ \int_{-\infty}^{\infty} x f(x)\, dx & \text{for a continuous distribution.} \end{cases} \tag{24.38}$$

More generally, the expectation value of any function $g(X)$ of the random variable $X$ is given by

$$E[g(X)] = \begin{cases} \sum_i g(x_i) f(x_i) & \text{for a discrete distribution,} \\ \int_{-\infty}^{\infty} g(x) f(x)\, dx & \text{for a continuous distribution.} \end{cases} \tag{24.39}$$

(It is assumed that the series is absolutely convergent, or that the integral exists, as the case may be. Otherwise we say that the distribution does not have a mean, which is very rare in physical applications.)

►*The probability of finding a 1s electron in a hydrogen atom in a given infinitesimal volume $dV$ is $\psi^*\psi\, dV$, where the wavefunction $\psi$ is given by*

$$\psi = Ae^{-r/a_0}.$$

*Find the value of the real constant $A$ and thereby deduce the mean radius of the electron orbital.*

The value of $A$ is found by requiring the total probability (i.e. the probability that the electron is *somewhere*) to be unity. Since the $1s$ orbital has no $\theta$- or $\phi$-dependence (it is spherically symmetric), $dV$ is simply $4\pi r^2\, dr$, so that

$$\int_{\text{all space}} \psi^*\psi\, dV = A^2 \int_0^{\infty} e^{-2r/a_0} 4\pi r^2\, dr = 1.$$

We note, in passing, that $4\pi r^2 A^2 e^{-2r/a_0}$ is the PDF for the radius of the electron orbit. Integrating by parts we find $A = 1/(\pi a_0^3)^{1/2}$.

If we now consider the random variable

$$R = \text{'radius at which electron is located'},$$

then, using the definition of the mean (24.38),

$$E[R] = \int r\psi^*\psi\, dV.$$

Again using $dV = 4\pi r^2\, dr$, the mean simplifies to

$$E[R] = \int 4\pi r^3 A^2 e^{-2r/a_0}\, dr = \frac{4}{a_0^3} \int_0^{\infty} r^3 e^{-2r/a_0}\, dr.$$

The integral may be evaluated by parts and takes the value $3a_0^4/8$; consequently we find $E[R] = 3a_0/2$. ◄

From its definition it is straightforward to show that the expectation value has the following properties.

(i) If $a$ is a constant then $E[a] = a$.
(ii) If $a$ is a constant then $E[aX] = aE[X]$.
(iii) If $g(X)$ and $h(X)$ are two functions then

$$E[g(X) + h(X)] = E[g(X)] + E[h(X)].$$

Moreover, if $X$ and $Y$ are *any* two random variables with (usually different) probability functions, and $a$ and $b$ are arbitrary constants, then

$$E[aX + bY] = aE[X] + bE[Y].$$

We may use this result to obtain an approximate expression for the expectation $E[f(X,Y)]$ of any arbitrary function $f$. Letting $\mu_X = E[X]$ and $\mu_Y = E[Y]$, and provided $f(X,Y)$ can be reasonably approximated by the linear terms of its Taylor expansion about the point $(\mu_X, \mu_Y)$, we have

$$f(X,Y) \approx f(\mu_X, \mu_Y) + \left(\frac{\partial f}{\partial X}\right)(X - \mu_X) + \left(\frac{\partial f}{\partial Y}\right)(Y - \mu_Y), \qquad (24.40)$$

where the partial derivatives are evaluated at $X = \mu_X$ and $Y = \mu_Y$ respectively. Taking the expectation value of both sides, we find

$$E[f(X,Y)] \approx f(\mu_X, \mu_Y) + \left(\frac{\partial f}{\partial X}\right)(E[X] - \mu_X) + \left(\frac{\partial f}{\partial Y}\right)(E[Y] - \mu_Y) = f(\mu_X, \mu_Y),$$

which gives the approximate result $E[f(X,Y)] \approx f(\mu_X, \mu_Y)$.

### 24.5.2 Mode and median

Although the mean discussed in the last section is the most common measure of the 'average' of a distribution, two other measures are frequently encountered.

The *mode* of a distribution is the value of the random variable $X$ at which the probability (density) function $f(x)$ has its greatest value. If there is more than one value of $X$ for which this is true then each value may equally be called the mode of the distribution.

The *median* $M$ of a distribution is the value of the random variable $X$ at which the cumulative probability function $F(x) = 1/2$, i.e. $F(M) = 1/2$. Related to the median are the lower and upper quartiles $Q_l$ and $Q_u$ of the PDF, which are defined such that

$$F(Q_l) = \tfrac{1}{4}, \qquad F(Q_u) = \tfrac{3}{4}.$$

Thus the median and lower and upper quartiles divide the PDF into four regions each containing 1/4 of the probability. Smaller subdivisions are also possible, e.g. the $n$th percentile, $P_n$, of a PDF is defined by $F(P_n) = n/100$.

### 24.5.3 Variance

The *variance* of a distribution, $V[X]$ (also written $\sigma^2$), is defined by

$$V[X] = E\left[(X-\mu)^2\right] = \begin{cases} \sum_j (x_j-\mu)^2 f(x_j) & \text{for a discrete distribution,} \\ \int_{-\infty}^{\infty} (x-\mu)^2 f(x)\,dx & \text{for a continuous distribution.} \end{cases} \tag{24.41}$$

Here $\mu$ has been written for the mean $E[X]$ of $X$.

The variance of a distribution is always positive; its positive square root is known as the *standard deviation* of the distribution and is denoted by $\sigma$. Roughly speaking, $\sigma$ measures the spread of values (about $x = \mu$) which $X$ can assume. As in the case of the mean, unless the series and the integral in (24.41) converge, the distribution does not have a variance.

From the definition of the variance $V[X]$ we may derive the following properties. If $a$ and $b$ are constants then

(i) $V[a] = 0$,
(ii) $V[aX + b] = a^2 V[X]$.

Furthermore, if $X$ and $Y$ are two random variables that do not depend on one another (i.e. they are independent), and $a$ and $b$ are constants, then

$$V[aX + bY] = a^2 V[X] + b^2 V[Y], \tag{24.42}$$

from which we may also derive

$$V[X+Y] = V[X-Y] = V[X] + V[Y].$$

Provided $X$ and $Y$ are indeed independent random variables, we may obtain an approximate expression for $V[f(X,Y)]$, for any arbitrary function $f(x,y)$, in a similar manner to that used in approximating $E[f(X,Y)]$ in subsection 24.5.1. Taking the variance of both sides of (24.40), and using (24.42), we find

$$V[f(X,Y)] \approx \left(\frac{\partial f}{\partial X}\right)^2 V[X] + \left(\frac{\partial f}{\partial Y}\right)^2 V[Y], \tag{24.43}$$

with the partial derivatives evaluated at $X = \mu_X$ and $Y = \mu_Y$ respectively.

### 24.5.4 Higher moments

The mean or expectation of $X$ is sometimes called the *first moment* of $X$ about zero, since it is defined as the sum or integral of the probability (density) function multiplied by the first power of $x$. By a simple extension the $k$th moment of a distribution is defined by

$$E[X^k] = \begin{cases} \sum_j x_j^k f(x_j) & \text{for a discrete distribution,} \\ \int_{-\infty}^{\infty} x^k f(x)\,dx & \text{for a continuous distribution.} \end{cases} \tag{24.44}$$

A useful result that relates the second moment, the mean and the variance of a distribution is proved using the properties of the expectation operator:

$$\begin{aligned} V[X] &= E\left[(X-\mu)^2\right] \\ &= E\left[X^2 - 2\mu X + \mu^2\right] \\ &= E\left[X^2\right] - 2\mu E[X] + \mu^2 \\ &= E\left[X^2\right] - 2\mu^2 + \mu^2 \\ &= E\left[X^2\right] - \mu^2, \end{aligned} \tag{24.45}$$

or, in an alternative notation,

$$\langle (x-\mu)^2 \rangle = \langle x^2 \rangle - \langle x \rangle^2.$$

►*A biased die has probabilities $p/2$, $p$, $p$, $p$, $p$, $2p$ of showing 1, 2, 3, 4, 5, 6 respectively. Find (i) the mean, (ii) the variance and (iii) the second moment of this probability distribution.*

By demanding that the sum of the probabilities equals unity we require $p = 2/13$. Now, using the definition of the mean (24.38) for a discrete distribution,

$$\begin{aligned} E[X] &= \sum_j x_j f(x_j) \\ &= 1 \times p/2 + 2 \times p + 3 \times p + 4 \times p + 5 \times p + 6 \times 2p \\ &= \frac{53}{2} p = \frac{53}{2} \times \frac{2}{13} = \frac{53}{13}. \end{aligned}$$

Similarly, using the definition of the variance (24.41),

$$\begin{aligned} V[X] &= \sum_j (x_j - \mu)^2 f(x_j) \\ &= \left(1 - \frac{53}{13}\right)^2 \frac{p}{2} + \left(2 - \frac{53}{13}\right)^2 p + \left(3 - \frac{53}{13}\right)^2 p + \left(4 - \frac{53}{13}\right)^2 p \\ &\quad + \left(5 - \frac{53}{13}\right)^2 p + \left(6 - \frac{53}{13}\right)^2 2p \\ &= \left(\frac{3120}{169}\right) p = \frac{480}{169}. \end{aligned}$$

Finally, using the definition of the second moment (24.44),

$$\begin{aligned} E\left[X^2\right] &= \sum_j x_j^2 f(x_j) \\ &= 1^2(p/2) + 2^2 p + 3^2 p + 4^2 p + 5^2 p + 6^2(2p) \\ &= \frac{253}{2} p = \frac{253}{13}. \end{aligned}$$

It is easy to verify that $E\left[X^2\right] - (E[X])^2 = V[X]$. ◄

### *24.5.5 Higher central moments*

The variance $V[X]$ is sometimes called the *second central moment* of the distribution, since it is defined as the sum or integral of the probability (density) function multiplied by the *second* power of $x-\mu$. The origin of the term 'central' is as follows: subtracting $\mu$ from $x$ before squaring effectively transfers the mean of the distribution to the centre of the integral. By a simple extension the $k$th *central* moment of a distribution is defined by

$$E\left[(X-\mu)^k\right] = \begin{cases} \sum_j (x_j-\mu)^k f(x_j) & \text{for a discrete distribution,} \\ \int_{-\infty}^{\infty} (x-\mu)^k f(x)\,dx & \text{for a continuous distribution.} \end{cases} \tag{24.46}$$

Clearly, the first central moment of a distribution is always zero since (for example in the continuous case)

$$\int (x-\mu) f(x)\,dx = \int x f(x)\,dx - \mu \int f(x)\,dx = \mu - (\mu \times 1) = 0.$$

## 24.6 Generating functions

As we saw in chapter 14, when dealing with particular sets of functions $f_n$, each member of the set being characterised by a different non-negative integer $n$, it is sometimes possible to summarise the whole set by a single function of a dummy variable (say $t$), called a generating function. The relationship between the generating function and the $n$th member $f_n$ of the set is that if the generating function is expanded as a power series in $t$, then $f_n$ is the coefficient of $t^n$. For example, in the expansion of the generating function $G(z,t) = (1-2zt+t^2)^{-1/2}$, the coefficient of $t^n$ is the $n$th Legendre polynomial $P_n(z)$, i.e.

$$G(z,t) = (1-2zt+t^2)^{-1/2} = \sum_{n=0}^{\infty} P_n(z) t^n.$$

We found that many useful properties of, and relationships between, the members of a set could be established using the generating function and other functions obtained from it, e.g. its derivatives.

Similar ideas can be used in the area of probability theory, and two types of generating function can be usefully defined, one more generally applicable than the other. The more restricted of the two, applicable only to discrete integral distributions, is called a probability generating function; this is discussed in the next section. The second type, a moment generating function, can be used with both discrete and continuous distributions and is considered in subsection 24.6.2.

### 24.6.1 Probability generating functions

As already indicated, probability generating functions are restricted in applicability to integer distributions, of which the most common (the binomial, the Poisson and the geometric) are considered in this and later sections. In such distributions a random variable may take only non-negative integer values. The actual possible values may be finite or infinite in number, but, for formal purposes, all integers, $0, 1, 2, \ldots$ are considered possible, with all but a finite number having zero probability of occurring, if that is the case.

If, as previously, the probability that the random variable $X$ takes the value $x_n$ is $f(x_n)$, then

$$\sum_n f(x_n) = 1.$$

In the present case, however, only non-negative integral values of $x_n$ are possible, and we can, without ambiguity, write the probability that $X$ takes the value $n$ as $f_n$, with

$$\sum_{n=0}^{\infty} f_n = 1. \qquad (24.47)$$

We may now define the *probability generating function* $\Phi_X(t)$ by

$$\Phi_X(t) \equiv \sum_{n=0}^{\infty} f_n t^n. \qquad (24.48)$$

It is immediately apparent that $\Phi_X(t) = E[t^X]$ and that, by virtue of (24.47), $\Phi_X(1) = 1$.

Probably the simplest example of a probability generating function (PGF) is provided by the random variable $X$ defined by

$$X = \begin{cases} 1 & \text{if the outcome of a single trial is a `success'}, \\ 0 & \text{if the trial ends in `failure'}. \end{cases}$$

If the probability of success is $p$ and that of failure $q$ $(= 1 - p)$, then

$$\Phi_X(t) = qt^0 + pt^1 + 0 + 0 + \cdots = q + pt. \qquad (24.49)$$

This type of random variable is discussed much more fully in subsection 24.7.1 where the binomial distribution is considered.

In a similar but slightly more complicated way, a Poisson-distributed integer variable with mean $\lambda$ (see subsection 24.7.3) has a PGF

$$\Phi_X(t) = \sum_{n=0}^{\infty} \frac{e^{-\lambda}\lambda^n}{n!} t^n = e^{-\lambda} e^{\lambda t}. \qquad (24.50)$$

As must be the case $\Phi_X(1) = 1$.

Useful results will only be obtained from this kind of approach if the summation (24.48) can be carried out explicitly in particular cases, and functions derived from $\Phi_X(t)$ can be shown to be related to meaningful parameters. Two such relationships can be obtained by differentiating (24.48) with respect to $t$. Taking the first derivative we find

$$\frac{d\Phi_X(t)}{dt} = \sum_{n=0}^{\infty} n f_n t^{n-1} \quad \Rightarrow \quad \Phi'_X(1) = \sum_{n=0}^{\infty} n f_n = E[X], \qquad (24.51)$$

and differentiating once more we obtain

$$\frac{d^2\Phi_X(t)}{dt^2} = \sum_{n=0}^{\infty} n(n-1) f_n t^{n-2} \quad \Rightarrow \quad \Phi''_X(1) = \sum_{n=0}^{\infty} n(n-1) f_n = E[X(X-1)]. \qquad (24.52)$$

Equation (24.51) shows that $\Phi'_X(1)$ gives the mean of $X$, and substitution in (24.45) gives

$$\begin{aligned} \Phi''_X(1) + \Phi'_X(1) - \left[\Phi'_X(1)\right]^2 &= E[X(X-1)] + E[X] - (E[X])^2 \\ &= E\left[X^2\right] - E[X] + E[X] - (E[X])^2 \\ &= E\left[X^2\right] - (E[X])^2 \\ &= V[X]. \end{aligned} \qquad (24.53)$$

►*A random variable $X$ is given by the number of trials needed to obtain a first success, when the chance of success at each trial is constant and equal to $p$. Find the probability generating function for $X$ and use it to determine the mean and variance of $X$.*

Clearly, at least one trial is needed, and so $f_0 = 0$. If $n$ ($\geq 1$) trials are needed for the first success, the first $n-1$ trials must have resulted in failure. Thus

$$P(X = n) = q^{n-1}p, \qquad n \geq 1, \qquad (24.54)$$

where $q = 1 - p$ is the probability of failure in each individual trial.

The corresponding probability generating function is thus

$$\begin{aligned} \Phi_X(t) &= \sum_{n=0}^{\infty} f_n t^n = \sum_{n=1}^{\infty} (q^{n-1}p)t^n \\ &= \frac{p}{q}\sum_{n=1}^{\infty}(qt)^n = \frac{p}{q} \times \frac{qt}{1-qt} = \frac{pt}{1-qt}, \end{aligned} \qquad (24.55)$$

where we have used the result for the sum of a geometric series, given in chapter 3, to obtain a closed-form expression for $\Phi_X(t)$. Again, as must be the case, $\Phi_X(1) = 1$.

To find the mean and variance of $X$ we need to evaluate $\Phi'_X(1)$ and $\Phi''_X(1)$. Differentiating (24.55) gives

$$\Phi'_X(t) = \frac{p}{(1-qt)^2} \Rightarrow \Phi'_X(1) = \frac{p}{p^2} = \frac{1}{p},$$
$$\Phi''_X(t) = \frac{2pq}{(1-qt)^3} \Rightarrow \Phi''_X(1) = \frac{2pq}{p^3} = \frac{2q}{p^2}.$$

Thus, using (24.51) and (24.53),

$$E[X] = \Phi'_X(1) = \frac{1}{p},$$
$$\text{and} \quad V[X] = \Phi''_X(1) + \Phi'_X(1) - [\Phi'_X(1)]^2$$
$$= \frac{2q}{p^2} + \frac{1}{p} - \frac{1}{p^2} = \frac{q}{p^2}. \blacktriangleleft$$

A distribution with probabilities of the general form (24.54) is known as a *geometric distribution.* This form of distribution is common in 'waiting time' problems.

*Sums of random variables*

We now turn to considering the sum of two or more random variables, say $X$ and $Y$, and denote by $S_2$ the random variable

$$S_2 = X + Y.$$

If $\Phi_{S_2}(t)$ is the PGF for $S_2$, the coefficient of $t^n$ in its expansion is given by the probability that $X + Y = n$, and is thus equal to the sum of the probabilities that $X = r$ and $Y = n - r$, for all values of $r$ in $0 \leq r \leq n$. Since such outcomes for different values of $r$ are mutually exclusive,

$$P(X+Y=n) = \sum_{r=0}^{\infty} P(X=r)P(Y=n-r). \qquad (24.56)$$

Multiplying both sides of (24.56) by $t^n$, and summing over all values of $n$, enables us to express this relationship in terms of probability generating functions as follows:

$$\Phi_{X+Y}(t) = \sum_{n=0}^{\infty} P(X+Y=n)t^n = \sum_{n=0}^{\infty}\sum_{r=0}^{n} P(X=r)t^r P(Y=n-r)t^{n-r}$$
$$= \sum_{r=0}^{\infty}\sum_{n=r}^{\infty} P(X=r)t^r P(Y=n-r)t^{n-r}.$$

The change in summation order is justified by reference to figure 24.7, which illustrates that the summations are over exactly the same pairs of values of $n$ and

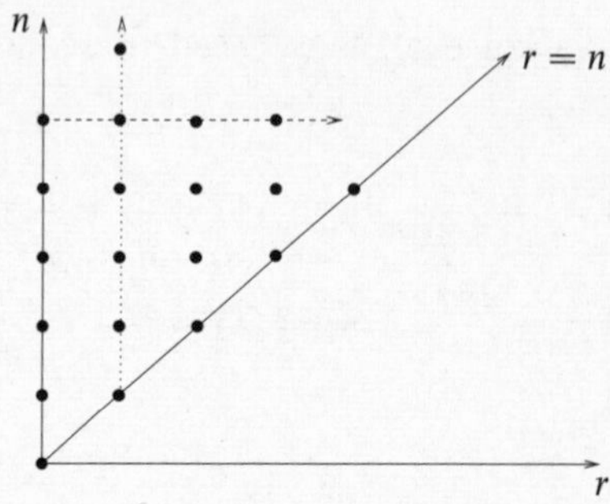


Figure 24.7 The pairs of values of $n$ and $r$ used in the evaluation of $\Phi_{X+Y}(t)$.

$r$, but with the first (inner) summation over 'columns' (vertical line) rather than 'rows' (horizontal line). Now, setting $n = r + s$ gives the final result,

$$\begin{aligned}\Phi_{X+Y}(t) &= \sum_{r=0}^{\infty} P(X = r)t^r \sum_{s=0}^{\infty} P(Y = s)t^s \\ &= \Phi_X(t)\Phi_Y(t),\end{aligned} \tag{24.57}$$

i.e. the PGF of the sum of two independent random variables is equal to the product of their individual PGFs. The same result can be deduced in a less formal way by noting that, if $X$ and $Y$ are independent, then

$$E\left[t^{X+Y}\right] = E\left[t^X\right] E\left[t^Y\right].$$

Clearly result (24.57) can be extended to more than two random variables by writing $S_3 = S_2 + Z$, etc., to give

$$\Phi_{(\sum_{i=1}^n X_i)}(t) = \prod_{i=1}^n \Phi_{X_i}(t), \tag{24.58}$$

and, further, if all the $X_i$ have the same probability distribution,

$$\Phi_{(\sum_{i=1}^n X_i)}(t) = [\Phi_X(t)]^n. \tag{24.59}$$

This latter result has immediate application in the deduction of the PGF for the binomial distribution from that for a single trial, equation (24.49).

*Variable-length sums of random variables*

As a final result in the theory of probability generating functions we show how to calculate the PGF for a sum of $N$ random variables, all with the same probability distribution, when the value of $N$ is itself a random variable, but one with a known probability distribution. In symbols, we wish to find the distribution of

$$S_N = X_1 + X_2 + \cdots + X_N, \tag{24.60}$$

where $N$ is a random variable with $P(N = n) = h_n$ and PGF $\chi_N(t) = \sum h_n t^n$.

The probability $\xi_k$ that $S_N = k$ is given by a sum of conditional probabilities, namely†

$$\begin{aligned}\xi_k &= \sum_{n=0}^{\infty} P(N = n)P(X_0 + X_1 + X_2 + \cdots + X_n = k)\\ &= \sum_{n=0}^{\infty} h_n \times (\text{coefficient of } t^k \text{ in } [\Phi_X(t)]^n).\end{aligned}$$

Multiplying both sides of this equation by $t^k$ and summing over all $k$, we obtain an expression for the PGF $\Xi_S(t)$ of $S_N$:

$$\begin{aligned}\Xi_S(t) = \sum_{k=0}^{\infty} \xi_k t^k &= \sum_{k=0}^{\infty} t^k \sum_{n=0}^{\infty} h_n \times (\text{coefficient of } t^k \text{ in } [\Phi_X(t)]^n)\\ &= \sum_{n=0}^{\infty} h_n \sum_{k=0}^{\infty} (\text{coefficient of } t^k \text{ in } [\Phi_X(t)]^n) t^k\\ &= \sum_{n=0}^{\infty} h_n [\Phi_X(t)]^n\\ &= \chi_N(\Phi_X(t)). \end{aligned} \tag{24.61}$$

In words, the PGF of the sum $S_N$ is given by the compound function $\chi_N(\Phi_X(t))$ obtained by substituting $\Phi_X(t)$ for $t$ in the PGF for the number of terms $N$ in the sum. We illustrate this with the following example.

►*The probability distribution for the number of eggs in a clutch is Poisson-distributed with mean $\lambda$, and the probability of hatching of each egg is $p$ (and is independent of the size of the clutch). Use the results stated in (24.49) and (24.50) to show that the PGF (and hence the probability distribution) for the number of chicks that hatch corresponds to a Poisson distribution having mean $\lambda p$.*

The number of chicks that hatch is given by a sum of the form (24.60) in which $X_i = 1$ if the $i$th chick hatches, and $X_i = 0$ if it does not. As given by (24.49), $\Phi_X(t)$ is thus $(1 - p) + pt$. The value of $N$ is given by a Poisson distribution with mean $\lambda$ and thus from (24.50), in the terminology of our previous discussion,

$$\chi_N(t) = e^{-\lambda} e^{\lambda t}.$$

† Formally $X_0 = 0$ has to be included, since $P(N = 0)$ may be non-zero.

We now substitute these forms into (24.61) to obtain

$$\begin{aligned}\Xi_S(t) &= \exp(-\lambda)\exp(\lambda\Phi_X(t)) \\ &= \exp(-\lambda)\exp\{\lambda[(1-p)+pt]\} \\ &= \exp(-\lambda p)\exp(\lambda pt).\end{aligned}$$

But this is exactly the PGF of a Poisson distribution with mean $\lambda p$.

That this implies that the probability is Poisson-distributed is intuitively obvious since, in the expansion of the PGF as a power series in $t$, every coefficient will be precisely that implied by such a distribution. A solution of the same problem by direct calculation appears in exercise 24.11. ◄

### *24.6.2 Moment generating functions*

As we saw in section 24.5 a probability function is often expressed in terms of its moments. This leads naturally to the second type of generating function, a *moment generating function*. For a random variable $X$, and a real number $t$, the moment generating function (MGF) is defined by

$$\Psi(t) = E\left[e^{tX}\right] = \begin{cases} \sum_i e^{tx_i} f(x_i) & \text{for a discrete distribution,} \\ \int e^{tx} f(x)\,dx & \text{for a continuous distribution.} \end{cases} \qquad (24.62)$$

The MGF will exist for all values of $t$ provided $X$ is bounded, and always exists at the point $t = 0$ where $\Psi(0) = E(1) = 1$.

It will be apparent that the PGF and the MGF for a random variable $X$ are closely related. The former is the expectation of $t^X$ whilst the latter is the expectation of $e^{tX}$:

$$\Phi(t) = E\left[t^X\right], \qquad \Psi(t) = E\left[e^{tX}\right].$$

The MGF can thus be obtained from the PGF by replacing $t$ by $e^t$, and vice versa. The MGF has more general applicability, however, since it can be used with both continuous and discrete distributions, whilst the PGF is restricted to non-negative integer distributions.

Assuming that the MGF exists for all $t$ around the point $t = 0$,

$$\frac{d\Psi(0)}{dt} = \left[\frac{d}{dt}E\left[e^{tX}\right]\right]_{t=0} = E\left[\left(\frac{d}{dt}e^{tX}\right)_{t=0}\right] = E\left[(Xe^{tX})_{t=0}\right] = E[X].$$

By extending this argument it can be shown that the moments of a distribution are given in terms of its MGF by

$$E[X^n] = \Psi^{(n)}(0), \qquad (24.63)$$

and, by substitution in (24.45), that the variance of the distribution is given by

$$V[X] = \Psi''(0) - \left[\Psi'(0)\right]^2, \qquad (24.64)$$

where the prime denotes differentiation with respect to $t$. We may arrive at the same conclusions by writing, for example in the continuous case,

$$\begin{aligned}\Psi(t) = E\left[e^{tX}\right] &= \int_{-\infty}^{\infty} e^{tx} f(x)\, dx \\ &= \int_{-\infty}^{\infty} f(x)\left(1 + tx + \frac{t^2x^2}{2!} + \cdots\right) dx \\ &= 1 + tE[X] + \frac{t^2}{2!}E\left[X^2\right] + \cdots,\end{aligned}$$

and similarly for a discrete distribution. Whilst $\Psi(t)$ is called the moment generating function, it actually generates, not the $n$th moments, but the $n$th moments divided by $n!$.

It is also possible to define the *characteristic function* $\Psi_c$ by

$$\Psi_c(t) = E\left[e^{itX}\right] = 1 + itE[X] - \frac{t^2}{2!}E\left[X^2\right] + \cdots. \tag{24.65}$$

This has the advantage (in the continuous case) of being the Fourier transform of the PDF $f(x)$. However, we consider the additional complications of this definition too numerous to justify its use.

► *The MGF for the Gaussian distribution (see the end of subsection 24.8.1) is given by*

$$\Psi(t) = \exp\left(\mu t + \tfrac{1}{2}\sigma^2 t^2\right).$$

*Find the expectation and variance of this distribution.*

Using (24.63),

$$\begin{aligned}\Psi'(t) &= \left(\mu + \sigma^2 t\right)\exp\left(\mu t + \tfrac{1}{2}\sigma^2 t^2\right) \quad &\Rightarrow \quad E[X] = \Psi'(0) = \mu, \\ \Psi''(t) &= \left[\sigma^2 + (\mu + \sigma^2 t)^2\right]\exp\left(\mu t + \tfrac{1}{2}\sigma^2 t^2\right) \quad &\Rightarrow \quad \Psi''(0) = \sigma^2 + \mu^2.\end{aligned}$$

Thus, using (24.64)

$$V[X] = \sigma^2 + \mu^2 - \mu^2 = \sigma^2.$$

That the mean is found to be $\mu$ and the variance $\sigma^2$ justifies the use of these symbols in the Gaussian distribution. ◄

The following useful theorems follow from the definition of the MGF.

(i) If $\Psi_1$ is the MGF for a random variable $X$, and $\Psi_2$ is the MGF for the random variable $Y = aX + b$, then

$$\Psi_2(t) = E\left[e^{tY}\right] = E\left[e^{t(aX+b)}\right] = e^{bt}E\left[e^{atX}\right] = e^{bt}\Psi_1(at). \tag{24.66}$$

(ii) If $X_1, X_2, \ldots, X_n$ are independent random variables, and $\Psi_i$ for $i = 1, 2, \ldots, n$ is the MGF for $X_i$, then the MGF of $Y = X_1 + X_2 + \cdots + X_n$ is

$$\Psi(t) = E\left[e^{tY}\right] = E\left[e^{t(X_1+X_2+\cdots+X_n)}\right] = E\left[\prod_{i=1}^{n} e^{tX_i}\right].$$

Since the $X_i$ are *independent*,

$$\Psi(t) = \prod_{i=1}^{n} E\left[e^{tX_i}\right],$$

and so

$$\Psi(t) = \prod_{i=1}^{n} \Psi_i(t). \qquad (24.67)$$

In words, the MGF of the sum of $n$ independent random variables is the product of their individual MGFs.

(iii) If the MGF of the random variable $X_1$ is identical to that for $X_2$ then the probability distributions of $X_1$ and $X_2$ are identical. This is intuitively reasonable although a rigorous proof is complicated,† and beyond the scope of this book.

## 24.7 Important discrete distributions

We now consider the more important discrete distributions encountered in physical applications.

### *24.7.1 The binomial distribution*

Perhaps the most important discrete probability distribution is the *binomial distribution*. This distribution describes processes that consist of a number of independent identical *trials* with two possible outcomes, $A$ and $B = \bar{A}$. We may call these outcomes 'success' and 'failure' respectively. Assuming that the probability of a success is $P(A) = p$, then the probability of a failure is $P(B) = q = 1 - p$. If we perform $n$ trials then the discrete random variable

$$X = \text{'number of times } A \text{ occurs'},$$

can take the values $0, 1, 2, \ldots, n$; its distribution amongst these values is described by the *binomial distribution*.

We now calculate the probability that in $n$ trials we obtain $x$ successes (and so

† See, for example, Moran, *An Introduction to Probability Theory* (Oxford Science Publications).

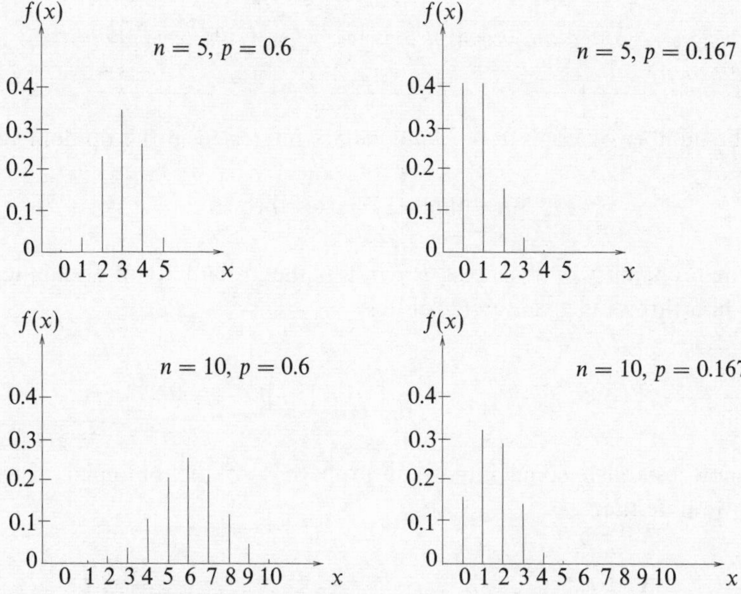


Figure 24.8 Some typical binomial distributions with various combinations of parameters $n$ and $p$.

$n-x$ failures). One way of obtaining such a result is to have $x$ successes followed by $n-x$ failures. Since the trials are assumed independent, the probability of this is

$$\underbrace{pp\cdots p}_{x \text{ times}} \times \underbrace{qq\cdots q}_{n-x \text{ times}} = p^x q^{n-x}.$$

This is, however, just one permutation of $x$ successes and $n-x$ failures. The total number of permutations of $n$ objects, of which $x$ are identical and of type 1, and $n-x$ are identical and of type 2, is given by (24.29) as

$$\frac{n!}{x!(n-x)!} \equiv {}^nC_x.$$

Therefore, the total probability of obtaining $x$ successes from $n$ trials is

$$f(x) = P(X = x) = {}^nC_x\, p^x q^{n-x} = {}^nC_x\, p^x(1-p)^{n-x}, \tag{24.68}$$

which is the *binomial probability distribution formula.* When a random variable $X$ follows the binomial distribution for $n$ trials, with a probability of success $p$, we write $X \sim \text{Bin}(n, p)$. Some typical binomial distributions are shown in figure 24.8.

►*If a single six-sided die is rolled 5 times, what is the probability that a six is thrown exactly 3 times?*

Here the number of 'trials' $n = 5$, and we are interested in the random variable

$$X = \text{number of sixes thrown.}$$

Since the probability of a 'success' is $p = 1/6$, the probability of obtaining exactly 3 sixes in 5 throws is given by (24.68) as

$$P(X = 3) = \frac{5!}{3!(5-3)!}\left(\frac{1}{6}\right)^3\left(\frac{5}{6}\right)^{(5-3)} = 0.032. \blacktriangleleft$$

We now establish some important properties of the binomial distribution. Firstly we note that

$$\sum_{x=0}^{n} f(x) = \sum_{x=0}^{n} {}^nC_x\, p^x(1-p)^{n-x} = [p + (1-p)]^n = 1,$$

as required. Furthermore, from the definitions of $E[X]$ and $V[X]$ for a discrete distribution, we may show that for the binomial distribution

$$E[X] = np, \qquad V[X] = npq.$$

An alternative (and much simpler) method is to use the moment generating function. This is discussed below.

Another useful result is the binomial recurrence formula

$$P(X = x+1) = \frac{p}{q}\left(\frac{n-x}{x+1}\right) P(X = x), \tag{24.69}$$

which enables successive probabilities $P(X = x+k)$, $k = 1, 2, \ldots$, to be calculated once $P(X = x)$ is known, and is often quicker than using (24.68).

►*The random variable $X$ is distributed as $X \sim \mathrm{Bin}(3, \frac{1}{2})$. Evaluate the probability function $f(x)$ using the binomial recurrence formula.*

The probability $P(X = 0)$ may be calculated using (24.68) and is

$$P(X = 0) = {}^3C_0 \left(\tfrac{1}{2}\right)^0 \left(\tfrac{1}{2}\right)^3 = \tfrac{1}{8}.$$

The ratio $p/q = \frac{1}{2}/\frac{1}{2} = 1$ in this case, and so using the binomial recurrence formula (24.69), we find

$$P(X=1) = 1 \times \frac{3-0}{0+1} \times \frac{1}{8} = \frac{3}{8},$$

$$P(X=2) = 1 \times \frac{3-1}{1+1} \times \frac{3}{8} = \frac{3}{8},$$

$$P(X=3) = 1 \times \frac{3-2}{2+1} \times \frac{3}{8} = \frac{1}{8},$$

results which may easily be verified by direct application of (24.68). ◄

*The moment generating function for the binomial distribution*

In order to find the MGF for the binomial distribution we consider the binomial random variable $X$ to be the sum of the random variables $X_i$, $i = 1, 2, \ldots, n$, which are defined by

$$X_i = \begin{cases} 1 & \text{if a `success' occurs on the } i\text{th trial,} \\ 0 & \text{if a `failure' occurs on the } i\text{th trial.} \end{cases}$$

Thus

$$\Psi_i(t) = E\left[e^{tX_i}\right] = 1 \times P(X_i = 0) + e^t P(X_i = 1) = pe^t + q.$$

Now using (24.67), it follows that the MGF for the binomial distribution is given by

$$\Psi(t) = \prod_{i=1}^{n} \Psi_i(t) = (pe^t + q)^n. \tag{24.70}$$

We can now use the moment generating function to derive the mean and variance of the binomial distribution. From (24.70),

$$\Psi'(t) = npe^t(pe^t + q)^{n-1},$$

and from (24.63)

$$E[X] = \Psi'(0) = np(p+q)^{n-1} = np,$$

where the last equality follows from $p + q = 1$.

Differentiating with respect to $t$ once more gives

$$\Psi''(t) = e^t(n-1)np^2(pe^t + q)^{n-2} + e^t np(pe^t + q)^{n-1},$$

and from (24.63)

$$E[X^2] = \Psi''(0) = n^2p^2 - np^2 + np.$$

Thus, using (24.64)

$$V[X] = \Psi''(0) - \left[\Psi'(0)\right]^2 = n^2p^2 - np^2 + np - n^2p^2 = np(1-p) = npq.$$

### 24.7.2 The hypergeometric distribution

In the previous section we saw that the probability of obtaining $x$ successes in $n$ *independent* trials was given by the binomial distribution. Suppose that these $n$ 'trials' actually consist of drawing at random $n$ balls, from a set of $N$ such balls of which $M$ are red and the rest white. Let us consider the random variable $X$ = 'number of red balls drawn'.

On the one hand, if the balls are drawn *with replacement*, then the trials are independent and the probability of drawing a red ball is $p = M/N$ each time. Therefore, the probability of drawing $x$ red balls in $n$ trials is given by the binomial distribution as

$$P(X = x) = \frac{n!}{x!(n-x)!}p^x(1-p)^{n-x}.$$

On the other hand, if the balls are drawn *without replacement* the trials are not independent and the probability of drawing a red ball depends on how many red balls have already been drawn. We can, however, still derive a general formula for the probability of drawing $x$ red balls in $n$ trials, as follows.

The number of ways of drawing $x$ red balls from $M$ is ${}^M C_x$, and the number of ways of drawing $n - x$ white balls from $N - M$ is ${}^{N-M} C_{n-x}$. Therefore, the total number of ways to obtain $x$ red balls in $n$ trials is ${}^M C_x \, {}^{N-M} C_{n-x}$. However, the total number of ways of drawing $n$ objects from $N$ is simply ${}^N C_n$. Hence the probability of $x$ red balls in $n$ trials is

$$\begin{aligned} P(X = x) &= \frac{{}^M C_x \, {}^{N-M} C_{n-x}}{{}^N C_n} \\ &= \frac{M!}{x!(M-x)!} \times \frac{(N-M)!}{(n-x)!(N-M-n+x)!} \times \frac{n!(N-n)!}{N!}, \end{aligned} \qquad (24.71)$$

which is called the *hypergeometric distribution*.

It may be shown that the hypergeometric distribution has mean

$$E[X] = n\frac{M}{N},$$

and variance

$$V[X] = \frac{nM(N-M)(N-n)}{N^2(N-1)}.$$

► *In the UK National Lottery each participant chooses 6 different numbers between 1 and 49. In each weekly draw 6 numbered balls are subsequently chosen. Find the probabilities that a participant chooses* $0, 1, 2, 3, 4, 5, 6$ *numbers correctly.*

The probabilities are given by a hypergeometric distribution with $N$ (the total number of balls) = 49, $M$ (the number of balls drawn) = 6, and $n$ (the number of numbers chosen by each participant) = 6. Thus, substituting in (24.71), we find

$$P(0) = \frac{{}^6C_0\,{}^{43}C_6}{{}^{49}C_6} = \frac{1}{2.29}, \qquad P(1) = \frac{{}^6C_1\,{}^{43}C_5}{{}^{49}C_6} = \frac{1}{2.42},$$

$$P(2) = \frac{{}^6C_2\,{}^{43}C_4}{{}^{49}C_6} = \frac{1}{7.55}, \qquad P(3) = \frac{{}^6C_3\,{}^{43}C_3}{{}^{49}C_6} = \frac{1}{56.6},$$

$$P(4) = \frac{{}^6C_4\,{}^{43}C_2}{{}^{49}C_6} = \frac{1}{1032}, \qquad P(5) = \frac{{}^6C_5\,{}^{43}C_1}{{}^{49}C_6} = \frac{1}{54200},$$

$$P(6) = \frac{{}^6C_6\,{}^{43}C_0}{{}^{49}C_6} = \frac{1}{13.98 \times 10^6}.$$

It can easily be seen that

$$\sum_{i=0}^{6} P(i) = 0.44 + 0.41 + 0.13 + 0.02 + \mathrm{O}(10^{-3}) = 1,$$

as expected. ◄

Note that if the number of trials (balls drawn) is small compared with $N$, $M$ and $N - M$, then not replacing the balls is of little consequence, and we may approximate the hypergeometric distribution by the binomial distribution (with $p = M/N$); this is much easier to evaluate.

### 24.7.3 The Poisson distribution

We have seen that the binomial distribution describes the number of successful outcomes in a certain number of trials $n$. The Poisson distribution also describes the probability of obtaining a given number of successes, but for situations in which the number of 'trials' cannot be enumerated; rather it describes the situation in which discrete events occur in a continuum. Typical examples of discrete random variables $X$ described by a Poisson distribution are the number of telephone calls received by a switchboard in a given interval, or the number of stars above a certain brightness in a particular (small) patch of sky. Given a mean rate of occurrence $\lambda$ of these events in the relevant interval, the Poisson distribution gives the probability $P(X = x)$ that exactly $x$ events will occur.

Suppose we wish to find the probability $P_x(t)$ that a switchboard will receive exactly $x$ telephone calls during a time interval $t$ given that the average numbers of calls received per unit time is $\lambda$. Clearly, in a small time interval $\Delta t$, the probability of receiving a call is $\lambda\Delta t$, provided $\Delta t$ is short enough that the probability of receiving two or more calls in this interval is negligible. Similarly the probability of receiving no call during the same small interval $\Delta t$ is simply $1 - \lambda\Delta t$.

Thus, for $x > 0$, the probabilty of receiving exactly $x$ calls in the total interval $t + \Delta t$ is given by

$$P_x(t + \Delta t) = P_x(t)(1 - \lambda \Delta t) + P_{x-1}(t)\lambda \Delta t.$$

Rearranging the equation, dividing through by $\Delta t$, and letting $\Delta t \to 0$, we obtain the differential recurrence equation

$$\frac{dP_x(t)}{dt} = \lambda P_{x-1}(t) - \lambda P_x(t). \qquad (24.72)$$

For $x = 0$ (i.e. no calls received), however, (24.72) simplifies to

$$\frac{dP_0(t)}{dt} = -\lambda P_0(t),$$

which may be integrated to give $P_0(t) = P_0(0)e^{-\lambda t}$. But since the probability $P_0(0)$ of receiving no calls in a zero time interval must equal unity, we have $P_0(t) = e^{-\lambda t}$. This expression for $P_0(t)$ may then be substituted back into (24.72) with $x = 1$ to obtain a differential equation for $P_1(t)$ that has the solution $P_1(t) = \lambda t e^{-\lambda t}$. We may repeat this process to obtain expressions for $P_2(t), P_3(t), \ldots, P_x(t)$, and we find

$$P_x(t) = \frac{(\lambda t)^x}{x!} e^{-\lambda t}. \qquad (24.73)$$

By setting $t = 1$ in (24.73), we obtain the *Poisson distribution* for obtaining exactly $x$ calls in a unit time interval,

$$f(x) = P(X = x) = \frac{e^{-\lambda}\lambda^x}{x!}. \qquad (24.74)$$

We may alternatively consider the Poisson distribution as the limit of the binomial distribution when the number of trials $n \to \infty$, while the probability of 'success' $p \to 0$, in such a way that $np = \lambda$ remains finite. The probability of $x$ successes in $n$ trials is given by the binomial formula as

$$P(X = x) = \frac{n!}{x!(n-x)!} p^x (1-p)^{n-x}.$$

Under the appropriate limits (with $x$ finite),

$$\lim_{n\to\infty} \frac{n!}{x!(n-x)!} = \frac{n^x}{x!} \qquad \text{and} \qquad \lim_{n\to\infty} \lim_{p\to 0} (1-p)^{n-x} = \lim_{p\to 0} (1-p)^{\lambda/p} = e^{-\lambda},$$

and we again obtain the expression (24.74) where (as we shall show) $\lambda$ is the mean of the distribution.

If a discrete random variable is described by a Poisson distribution of mean $\lambda$, we write $X \sim \text{Po}(\lambda)$. As it must be, the sum of the probabilities is unity:

$$\sum_{x=0}^{\infty} P(X = x) = e^{-\lambda} \sum_{x=0}^{\infty} \frac{\lambda^x}{x!} = e^{-\lambda} e^{\lambda} = 1.$$

From (24.74) we may also derive the *Poisson recurrence formula*,

$$P(X = x + 1) = \frac{\lambda}{x + 1} P(X = x) \qquad \text{for } x = 0, 1, 2, \ldots, \tag{24.75}$$

which enables successive probabilities to be easily calculated once one is known.

►*A person receives on average* 2 *e-mail messages per hour. Assuming that the number of e-mails received has a Poisson distribution, find the probabilities that in any particular hour* 0, 1, 2, 3, 4, 5 *messages are received.*

The Poisson distribution with mean 2 is given by

$$P(X = x) = \frac{2^x}{x!} e^{-2}$$

and so $P(X = 0) = e^{-2} = 0.135$, $P(X = 1) = 2e^{-2} = 0.271$, $P(X = 2) = 2^2 e^{-2}/2! = 0.271$, $P(X = 3) = 2^3 e^{-2}/3! = 0.180$, $P(X = 4) = 2^4 e^{-2}/4! = 0.090$, $P(X = 5) = 2^5 e^{-2}/5! = 0.036$. These results may also be calculated using the recurrence formula (24.75). ◄

The above example illustrates the point that a Poisson distribution typically rises and then falls. It either has a maximum when $x$ is equal to the integer part of $\lambda$ or, if $\lambda$ happens to be an integer, has equal maximal values at $x = \lambda - 1$ and $x = \lambda$. The Poisson distribution always has a long 'tail' towards higher values of $X$. The higher the value of the mean, however, the more symmetric the distribution becomes. Typical Poisson distributions are shown in figure 24.9.

Using the definition (24.38) we may calculate the mean of the Poisson distribution,

$$E[X] = \sum_{x=0}^{\infty} x e^{-\lambda} \frac{\lambda^x}{x!} = \lambda e^{-\lambda} \sum_{x=1}^{\infty} \frac{\lambda^{x-1}}{(x-1)!}.$$

Setting $x' = x - 1$ we obtain

$$E[X] = \lambda e^{-\lambda} \sum_{x'=0}^{\infty} \frac{\lambda^{x'}}{x'!} = \lambda e^{-\lambda} e^{\lambda} = \lambda.$$

The mean of the Poisson distribution is thus $\lambda$, as we asserted earlier.

We may similarly calculate the variance of the Poisson distribution. The simplest method is first to calculate

$$E\left[X^2 - X\right] = E[X(X-1)] = \sum_{x=0}^{\infty} x(x-1) e^{-\lambda} \frac{\lambda^x}{x!}.$$

Dropping the first two terms in the sum, both of which are zero, and letting $x' = x - 2$ we find

$$E\left[X^2 - X\right] = E\left[X^2\right] - E[X] = \lambda^2 e^{-\lambda} \sum_{x'=0}^{\infty} \frac{\lambda^{x'}}{x'!} = \lambda^2 e^{-\lambda} e^{\lambda} = \lambda^2.$$

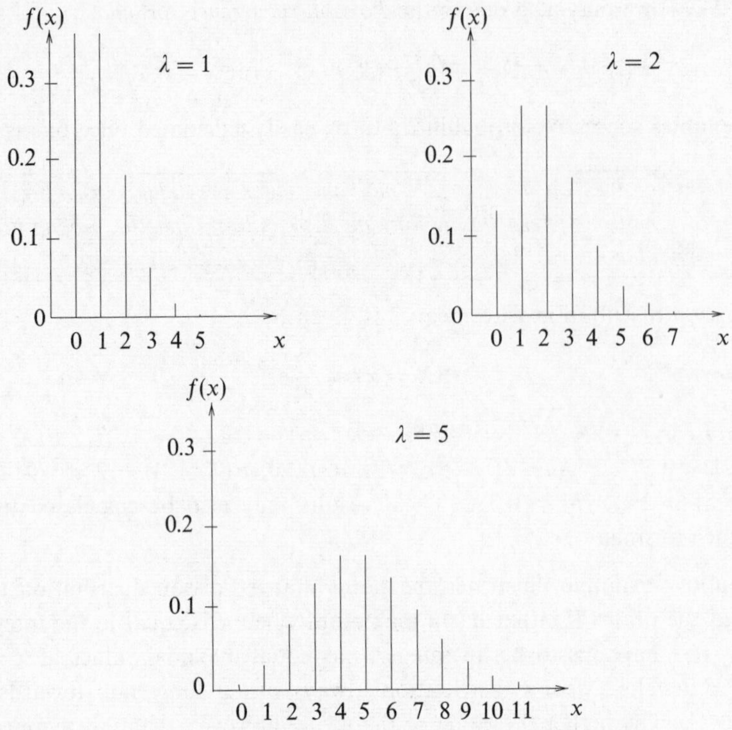


Figure 24.9 Three Poisson distributions for different values of the parameter $\lambda$.

But $E[X] = \lambda$, so $E\left[X^2\right] = \lambda^2 + \lambda$. Hence the variance is given by

$$V[X] = E[X^2] - (E[X])^2 = \lambda^2 + \lambda - \lambda^2 = \lambda.$$

Thus we note that the mean and variance of the Poisson distribution are both equal to $\lambda$.

The moment generating function for the Poisson distribution is given by

$$\Psi(t) = E\left[e^{tX}\right] = \sum_{x=0}^{\infty} \frac{e^{tx}e^{-\lambda}\lambda^x}{x!} = e^{-\lambda} \sum_{x=0}^{\infty} \frac{(\lambda e^t)^x}{x!} = e^{-\lambda}e^{\lambda e^t} = e^{\lambda(e^t-1)}.$$

The MGF can be used, as before, to derive the mean and the variance of the Poisson distribution.

### *24.7.4 The Poisson approximation to the binomial distribution*

Earlier we derived the Poisson distribution as the limit of the binomial distribution when $n \to \infty$ and $p \to 0$ in such a way that $np = \lambda$ remains finite, where $\lambda$ is the mean of the Poisson distribution. It is therefore not surprising that the Poisson distribution is a very good approximation to the binomial distribution for large $n$ ($\geq 50$, say) and small $p$ ($\leq 0.1$, say). Moreover, it is easier to calculate as it involves fewer factorials.

►*In a large batch of light-bulbs, the probability that a bulb is defective is 0.5%. For a sample of 200 bulbs taken at random, find the approximate probabilities that 0, 1 and 2 of the bulbs respectively are defective.*

The random variable $X$ = 'number of defective bulbs in a sample' is distributed as $X \sim \text{Bin}(200, 0.005)$, so that $\lambda = np = 1.0$. Since $n$ is large and $p$ small, we may approximate $X \sim \text{Po}(1)$ so that

$$P(X = x) = e^{-1}\frac{1^x}{x!},$$

from which we find $P(X = 0) = 0.37$, $P(X = 1) = 0.37$, $P(X = 2) = 0.18$. For comparison, it may be noted that the exact values calculated from the binomial distribution are identical to those found here to two decimal places. ◄

### 24.7.5 Multiple Poisson distributions

Suppose $X$ and $Y$ are two *independent* random variables, both of which are described by Poisson distributions with (in general) different means, so that $X \sim \text{Po}(\lambda_1)$ and $Y \sim \text{Po}(\lambda_2)$. Now consider the random variable $Z = X + Y$, which has the probability distribution

$$\begin{aligned}P(Z = z) &= P(X = z)P(Y = 0) + P(X = z - 1)P(Y = 1) + \cdots \\ &\qquad + P(X = 1)P(Y = z - 1) + P(X = 0)P(Y = z) \\ &= \sum_{r=0}^{z} P(X = r)P(Y = z - r).\end{aligned}$$

Since $X \sim \text{Po}(\lambda_1)$ and $Y \sim \text{Po}(\lambda_2)$, this has the form

$$\begin{aligned}P(Z = z) &= e^{-\lambda_1}e^{-\lambda_2}\sum_{r=0}^{z}\frac{\lambda_1^r\lambda_2^{z-r}}{r!(z-r)!} \\ &= e^{-(\lambda_1+\lambda_2)}\frac{(\lambda_1+\lambda_2)^z}{z!}\sum_{r=0}^{z}\frac{z!}{r!(z-r)!}\left(\frac{\lambda_1}{\lambda_1+\lambda_2}\right)^r\left(\frac{\lambda_2}{\lambda_1+\lambda_2}\right)^{z-r}.\end{aligned}$$

But the summation is simply the binomial expansion of

$$\left(\frac{\lambda_1}{\lambda_1+\lambda_2} + \frac{\lambda_2}{\lambda_1+\lambda_2}\right)^z = 1^z = 1.$$

Therefore we conclude that

$$P(Z = z) = e^{-(\lambda_1+\lambda_2)}\frac{(\lambda_1+\lambda_2)^z}{z!},$$

and hence that $Z \sim \text{Po}(\lambda_1 + \lambda_2)$, i.e. $Z$ is also Poisson distributed, and has mean $\lambda_1 + \lambda_2$.

►*Two types of e-mail arrive independently and at random, external ones at a mean rate of 1 every five minutes, and internal ones at a rate of 2 every five minutes. Calculate the probability of receiving 2 or more e-mails in any two-minute interval.*

Let

$$X = \text{number of external e-mails per two-minute interval,}$$
$$Y = \text{number of internal e-mails per two-minute interval.}$$

Since we expect on average 1 external e-mail and 2 internal e-mails every five minutes we have $X \sim \text{Po}(0.4)$ and $Y \sim \text{Po}(0.8)$. Letting $Z = X + Y$ we have $Z \sim \text{Po}(0.4 + 0.8) = \text{Po}(1.2)$. Now

$$P(Z \geq 2) = 1 - P(Z < 2) = 1 - P(Z = 0) - P(Z = 1),$$

and

$$P(Z = 0) = e^{-1.2} = 0.301,$$
$$P(Z = 1) = e^{-1.2}\frac{1.2}{1} = 0.361.$$

Hence $P(Z \geq 2) = 1 - 0.301 - 0.361 = 0.338$. ◄

This can, of course, be extended to any number of Poisson processes, so that if $X_i = \text{Po}(\lambda_i)$, $i = 1, \ldots, n$, then the random variable $Z = X_1 + \cdots + X_n$ is distributed as $Z \sim \text{Po}(\lambda_1 + \cdots + \lambda_n)$.

## 24.8 Important continuous distributions

Having discussed the most commonly encountered discrete probability distributions, we now consider some of the more important continuous probability distributions.

### *24.8.1 The Gaussian distribution*

By far the most important continuous probability distribution is the *Gaussian* or *normal* distribution. The reason for its importance is that a great many random variables of interest, in all areas of the physical sciences and beyond, are described either exactly or approximately by a Gaussian distribution. Moreover, the Gaussian distribution can be used to approximate other, more complicated, probability distributions.

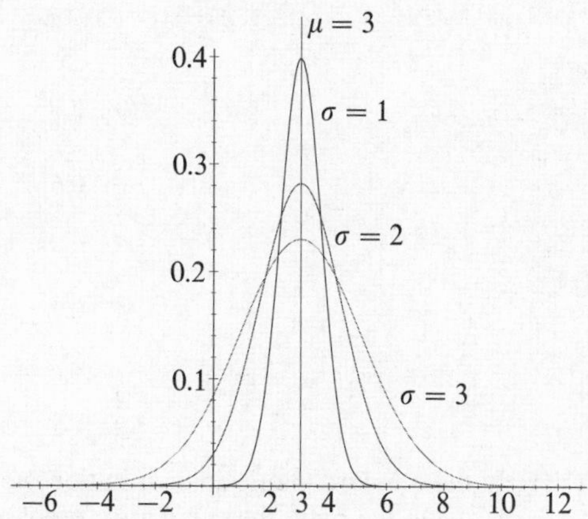


Figure 24.10 The Gaussian or normal distribution for mean $\mu = 3$ and various values of the standard deviation $\sigma$.

The probability density function for a Gaussian distribution of a random variable $X$, with mean $E[X] = \mu$ and variance $V[X] = \sigma^2$, takes the form

$$f(x) = \frac{1}{\sigma\sqrt{2\pi}} \exp\left[-\frac{1}{2}\left(\frac{x-\mu}{\sigma}\right)^2\right]. \tag{24.76}$$

The factor $1/\sqrt{2\pi}$ arises from the normalisation of the distribution,

$$\int_{-\infty}^{\infty} f(x)dx = 1;$$

the evaluation of this integral is discussed in subsection 5.4.2. The Gaussian distribution is symmetric about the point $x = \mu$, and has the characteristic 'bell' shape shown in figure 24.10. The width of the curve is described by the standard deviation $\sigma$: if $\sigma$ is large then the curve is broad, and if $\sigma$ is small the curve is narrow (see figure 24.10). At $x = \mu \pm \sigma$, $f(x)$ falls to $e^{-1/2} \approx 0.61$ of its peak value (at $x = \mu$) – these points are also points of inflection, where $d^2f/dx^2 = 0$. When a random variable $X$ follows a Gaussian distribution with mean $\mu$ and variance $\sigma^2$, we write $X \sim N(\mu, \sigma^2)$.

The effects of changing $\mu$ and $\sigma$ are only to shift the curve along the $x$-axis, or to broaden or narrow it, respectively. Thus all Gaussians are equivalent in that a change of origin and scale can reduce them to a standard form. We therefore consider the random variable $Z = (X - \mu)/\sigma$, for which the PDF takes the form

$$\phi(z) = \frac{1}{\sqrt{2\pi}} \exp\left(-\frac{z^2}{2}\right), \tag{24.77}$$

which is called the *standard Gaussian distribution* and has mean $\mu = 0$ and variance $\sigma^2 = 1$. The random variable $Z$ is called the *standard variable*.

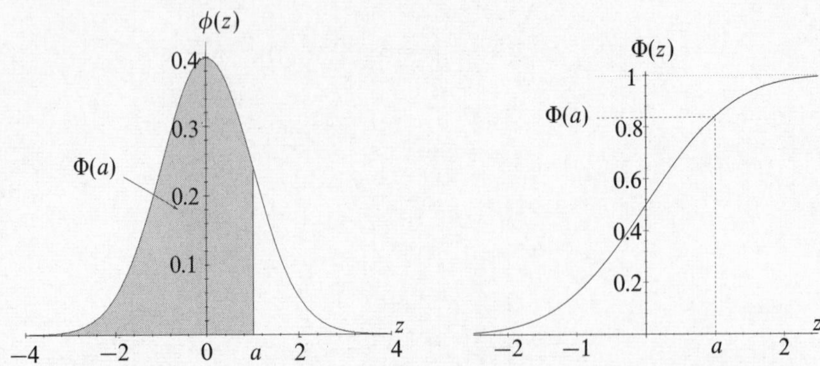


Figure 24.11 On the left, the standard Gaussian distribution $\phi(z)$; the shaded area gives $P(Z < a) = \Phi(a)$. On the right, the cumulative probability function $\Phi(z)$ for a standard Gaussian distribution $\phi(z)$.

From (24.76) we can define the cumulative probability function for a Gaussian distribution as

$$F(x) = P(X < x) = \frac{1}{\sigma\sqrt{2\pi}} \int_{-\infty}^{x} \exp\left[-\frac{1}{2}\left(\frac{u-\mu}{\sigma}\right)^2\right] du, \qquad (24.78)$$

where $u$ is a (dummy) integration variable. Unfortunately, this (indefinite) integral cannot be evaluated analytically. It is therefore standard practice to tabulate values of the cumulative probability function for the standard Gaussian distribution (see figure 24.11), i.e.

$$\Phi(z) = P(Z < z) = \frac{1}{\sqrt{2\pi}} \int_{-\infty}^{z} \exp\left(-\frac{u^2}{2}\right) du. \qquad (24.79)$$

It is usual only to tabulate $\Phi(z)$ for $z > 0$, since it can easily be seen, from figure 24.11 and the symmetry of the Gaussian distribution, that $\Phi(-z) = 1 - \Phi(z)$ (see table 24.1). Using such a table it is then straightforward to evaluate the probability that $Z$ lies in a given range of $z$-values. For example (for $a$ and $b$ constant)

$$P(Z < a) = \Phi(a),$$
$$P(Z > a) = 1 - \Phi(a),$$
$$P(a < Z \leq b) = \Phi(b) - \Phi(a).$$

Remembering that $Z = (X - \mu)/\sigma$, and comparing (24.78) and (24.79), we see that

$$F(x) = \Phi\left(\frac{x-\mu}{\sigma}\right),$$

and so we may also easily calculate the probability that the original random

| $\Phi(z)$ | .00 | .01 | .02 | .03 | .04 | .05 | .06 | .07 | .08 | .09 |
|---|---|---|---|---|---|---|---|---|---|---|
| 0.0 | .5000 | .5040 | .5080 | .5120 | .5160 | .5199 | .5239 | .5279 | .5319 | .5359 |
| 0.1 | .5398 | .5438 | .5478 | .5517 | .5557 | .5596 | .5636 | .5675 | .5714 | .5753 |
| 0.2 | .5793 | .5832 | .5871 | .5910 | .5948 | .5987 | .6026 | .6064 | .6103 | .6141 |
| 0.3 | .6179 | .6217 | .6255 | .6293 | .6331 | .6368 | .6406 | .6443 | .6480 | .6517 |
| 0.4 | .6554 | .6591 | .6628 | .6664 | .6700 | .6736 | .6772 | .6808 | .6844 | .6879 |
| 0.5 | .6915 | .6950 | .6985 | .7019 | .7054 | .7088 | .7123 | .7157 | .7190 | .7224 |
| 0.6 | .7257 | .7291 | .7324 | .7357 | .7389 | .7422 | .7454 | .7486 | .7517 | .7549 |
| 0.7 | .7580 | .7611 | .7642 | .7673 | .7704 | .7734 | .7764 | .7794 | .7823 | .7852 |
| 0.8 | .7881 | .7910 | .7939 | .7967 | .7995 | .8023 | .8051 | .8078 | .8106 | .8133 |
| 0.9 | .8159 | .8186 | .8212 | .8238 | .8264 | .8289 | .8315 | .8340 | .8365 | .8389 |
| 1.0 | .8413 | .8438 | .8461 | .8485 | .8508 | .8531 | .8554 | .8577 | .8599 | .8621 |
| 1.1 | .8643 | .8665 | .8686 | .8708 | .8729 | .8749 | .8770 | .8790 | .8810 | .8830 |
| 1.2 | .8849 | .8869 | .8888 | .8907 | .8925 | .8944 | .8962 | .8980 | .8997 | .9015 |
| 1.3 | .9032 | .9049 | .9066 | .9082 | .9099 | .9115 | .9131 | .9147 | .9162 | .9177 |
| 1.4 | .9192 | .9207 | .9222 | .9236 | .9251 | .9265 | .9279 | .9292 | .9306 | .9319 |
| 1.5 | .9332 | .9345 | .9357 | .9370 | .9382 | .9394 | .9406 | .9418 | .9429 | .9441 |
| 1.6 | .9452 | .9463 | .9474 | .9484 | .9495 | .9505 | .9515 | .9525 | .9535 | .9545 |
| 1.7 | .9554 | .9564 | .9573 | .9582 | .9591 | .9599 | .9608 | .9616 | .9625 | .9633 |
| 1.8 | .9641 | .9649 | .9656 | .9664 | .9671 | .9678 | .9686 | .9693 | .9699 | .9706 |
| 1.9 | .9713 | .9719 | .9726 | .9732 | .9738 | .9744 | .9750 | .9756 | .9761 | .9767 |
| 2.0 | .9772 | .9778 | .9783 | .9788 | .9793 | .9798 | .9803 | .9808 | .9812 | .9817 |
| 2.1 | .9821 | .9826 | .9830 | .9834 | .9838 | .9842 | .9846 | .9850 | .9854 | .9857 |
| 2.2 | .9861 | .9864 | .9868 | .9871 | .9875 | .9878 | .9881 | .9884 | .9887 | .9890 |
| 2.3 | .9893 | .9896 | .9898 | .9901 | .9904 | .9906 | .9909 | .9911 | .9913 | .9916 |
| 2.4 | .9918 | .9920 | .9922 | .9925 | .9927 | .9929 | .9931 | .9932 | .9934 | .9936 |
| 2.5 | .9938 | .9940 | .9941 | .9943 | .9945 | .9946 | .9948 | .9949 | .9951 | .9952 |
| 2.6 | .9953 | .9955 | .9956 | .9957 | .9959 | .9960 | .9961 | .9962 | .9963 | .9964 |
| 2.7 | .9965 | .9966 | .9967 | .9968 | .9969 | .9970 | .9971 | .9972 | .9973 | .9974 |
| 2.8 | .9974 | .9975 | .9976 | .9977 | .9977 | .9978 | .9979 | .9979 | .9980 | .9981 |
| 2.9 | .9981 | .9982 | .9982 | .9983 | .9984 | .9984 | .9985 | .9985 | .9986 | .9986 |
| 3.0 | .9987 | .9987 | .9987 | .9988 | .9988 | .9989 | .9989 | .9989 | .9990 | .9990 |
| 3.1 | .9990 | .9991 | .9991 | .9991 | .9992 | .9992 | .9992 | .9992 | .9993 | .9993 |
| 3.2 | .9993 | .9993 | .9994 | .9994 | .9994 | .9994 | .9994 | .9995 | .9995 | .9995 |
| 3.3 | .9995 | .9995 | .9995 | .9996 | .9996 | .9996 | .9996 | .9996 | .9996 | .9997 |
| 3.4 | .9997 | .9997 | .9997 | .9997 | .9997 | .9997 | .9997 | .9997 | .9997 | .9998 |

Table 24.1 The cumulative probability function $\Phi(z)$ for the standard Gaussian distribution, as given by (24.79). The first decimal place of $z$ is given by the rows, the second by the columns. Thus, for example, $\Phi(1.23) = 0.8907$.

variable $X$ lies in a given $x$-range. For example

$$P(a < X \leq b) = \frac{1}{\sigma\sqrt{2\pi}} \int_a^b \exp\left[-\frac{1}{2}\left(\frac{u-\mu}{\sigma}\right)^2\right] du \tag{24.80}$$

$$= F(b) - F(a) \tag{24.81}$$

$$= \Phi\left(\frac{b-\mu}{\sigma}\right) - \Phi\left(\frac{a-\mu}{\sigma}\right). \tag{24.82}$$

►*If $X$ is described by a Gaussian distribution of mean $\mu$ and variance $\sigma^2$, calculate the probabilities that $X$ lies within $1\sigma$, $2\sigma$ and $3\sigma$ of the mean.*

From (24.82)

$$P(\mu - \sigma < X \leq \mu + \sigma) = \Phi(1) - \Phi(-1) = \Phi(1) - [1 - \Phi(1)],$$

and thus from table 24.1

$$P(\mu - \sigma < X \leq \mu + \sigma) = 2\Phi(1) - 1 = 0.6826 \approx 68.3\%.$$

Similarly,

$$P(\mu - 2\sigma < X \leq \mu + 2\sigma) = 2\Phi(2) - 1 = 0.9544 \approx 95.4\%,$$

and

$$P(\mu - 3\sigma < X \leq \mu + 3\sigma) = 2\Phi(3) - 1 = 0.9974 \approx 99.7\%,$$

Thus we expect $X$ to be distributed in such a way that about $2/3$ of the values will lie between $\mu - \sigma$ and $\mu + \sigma$, 95% will lie within $2\sigma$ of the mean, and 99.7% will lie within $3\sigma$ of the mean. These limits are called the one-, two-, and three-sigma limits respectively. ◄

It is particularly important to note in the above example that the one-, two-, and three-sigma limits are independent of the values of the mean and variance.

There are many other ways in which the Gaussian distribution may be used. We now illustrate some of the uses with more complicated examples.

►*Sawmill A produces boards whose lengths are Gaussian distributed with mean* 209.4 cm *and standard deviation* 5.0 cm. *A board is accepted if it is longer than* 200 cm, *but is rejected otherwise. Show that* 3% *of boards are rejected.*

*Sawmill B produces boards of the same standard deviation, but whose mean length is* 210.1 cm. *Find the proportion of boards rejected if they are drawn at random from the outputs of A and B in the ratio* 3 : 1.

Let $X$ = 'length of boards from $A$' so $X \sim N(209.4,\ 5.0^2)$, and

$$P(X < 200) = \Phi\left(\frac{200-\mu}{\sigma}\right) = \Phi\left(\frac{200-209.4}{5.0}\right) = \Phi(-1.88).$$

But, since $\Phi(-z) = 1 - \Phi(z)$ we have (using table 24.1)

$$P(X < 200) = 1 - \Phi(1.88) = 1 - 0.9699 = 0.0301,$$

i.e. 3.0% of boards are rejected.

Now let $Y$ = 'length of boards from $B$', so that $Y \sim N(210.1,\ 5.0^2)$ and

$$\begin{aligned} P(Y < 200) &= \Phi\left(\frac{200 - 210.1}{5.0}\right) = \Phi(-2.02) \\ &= 1 - \Phi(2.02) \\ &= 1 - 0.9783 = 0.0217. \end{aligned}$$

Therefore, when taken alone, only 2.2% of boards from $B$ are rejected. If, however, boards are drawn at random from $A$ and $B$ in the ratio 3 : 1 then the proportion rejected is

$$\tfrac{1}{4}(3 \times 0.030 + 1 \times 0.022) = 0.028 = 2.8\%. \blacktriangleleft$$

We may sometimes work backwards to derive the mean and standard deviation of a population that is known to be Gaussian distributed.

► *The time taken for a computer 'packet' to travel from Cambridge UK to Cambridge MA is Gaussian distributed. 6.8% of the packets take over* 200 ms *to make the journey, and* 3.0% *take under* 140 ms. *Find the mean and standard deviation of the distribution.*

Let $X$ = 'journey time in ms', and we are told $X \sim N(\mu, \sigma^2)$ where $\mu$ and $\sigma$ are unknown. Since 6.8% of journey times are longer than 200 ms,

$$P(X > 200) = 1 - \Phi\left(\frac{200 - \mu}{\sigma}\right) = 0.068,$$

from which we find

$$\Phi\left(\frac{200 - \mu}{\sigma}\right) = 1 - 0.068 = 0.932.$$

Using table 24.1, we therefore have

$$\frac{200 - \mu}{\sigma} = 1.49. \qquad (24.83)$$

Also, 3.0% of journey times are under 140 ms, so

$$P(X < 140) = \Phi\left(\frac{140 - \mu}{\sigma}\right) = 0.030.$$

Now using $\Phi(-z) = 1 - \Phi(z)$ gives

$$\Phi\left(\frac{\mu - 140}{\sigma}\right) = 1 - 0.030 = 0.970.$$

Using table 24.1 again, we find

$$\frac{\mu - 140}{\sigma} = 1.88. \tag{24.84}$$

Solving the simultaneous equations (24.83) and (24.84) gives $\mu = 173.5$, $\sigma = 17.8$. ◄

*The moment generating function for the Gaussian distribution*

We have already seen how the MGF of the Gaussian distribution can be used to find the mean and variance of the distribution (see subsection 24.6.2), and we here give a derivation of this MGF.

Using the definition of the MGF (24.62)

$$\begin{aligned}\Psi(t) = E\left[e^{tX}\right] &= \int_{-\infty}^{\infty} \frac{1}{\sigma\sqrt{2\pi}} \exp\left[tx - \frac{(x-\mu)^2}{2\sigma^2}\right] dx \\ &= c \exp\left(\mu t + \tfrac{1}{2}\sigma^2 t^2\right),\end{aligned}$$

where the final equality is established by completing the square in the argument of the exponential and writing

$$c = \int_{-\infty}^{\infty} \frac{1}{\sigma\sqrt{2\pi}} \exp\left\{-\frac{[x - (\mu + \sigma^2 t)]^2}{2\sigma^2}\right\} dx.$$

However, the final integral is simply the normalisation integral for the Gaussian distribution, and so $c = 1$ and the MGF is given by

$$\Psi(t) = \exp\left(\mu t + \tfrac{1}{2}\sigma^2 t^2\right). \tag{24.85}$$

### 24.8.2 Gaussian approximation to the binomial distribution

We may consider the Gaussian distribution as the limit of the binomial distribution when the number of trials $n \to \infty$ but the probability of a success $p$ remains finite, so that $np \to \infty$ also. (This contrasts with the Poisson distribution, which corresponds to the limit $n \to \infty$ and $p \to 0$, with $np = \lambda$ remaining finite.) In other words, a Gaussian distribution results when an experiment with a finite probability of success is repeated a large number of times. We now show how this Gaussian limit arises.

The binomial probability function gives the probability of $x$ successes in $n$ trials as

$$f(x) = \frac{n!}{x!(n-x)!} p^x (1-p)^{n-x}.$$

Taking the limit as $n \to \infty$ (and $x \to \infty$) we may approximate the factorials by Stirling's approximation

$$n! \sim \sqrt{2\pi n}\left(\frac{n}{e}\right)^n$$

| $x$ | $f(x)$ (binomial) | $f(x)$ (Gaussian) |
|---|---|---|
| 0 | 0.0001 | 0.0001 |
| 1 | 0.0016 | 0.0014 |
| 2 | 0.0106 | 0.0092 |
| 3 | 0.0425 | 0.0395 |
| 4 | 0.1115 | 0.1119 |
| 5 | 0.2007 | 0.2091 |
| 6 | 0.2508 | 0.2575 |
| 7 | 0.2150 | 0.2091 |
| 8 | 0.1209 | 0.1119 |
| 9 | 0.0403 | 0.0395 |
| 10 | 0.0060 | 0.0092 |

Table 24.2 Comparison of the binomial distribution for $n = 10$ and $p = 0.6$ with its Gaussian approximation.

to obtain

$$\begin{aligned} f(x) &\approx \frac{1}{\sqrt{2\pi n}} \left(\frac{x}{n}\right)^{-x-1/2} \left(\frac{n-x}{n}\right)^{-n+x-1/2} p^x(1-p)^{n-x} \\ &= \frac{1}{\sqrt{2\pi n}} \exp\left[-\left(x+\tfrac{1}{2}\right)\ln\frac{x}{n} - \left(n-x+\tfrac{1}{2}\right)\ln\frac{n-x}{n}\right. \\ &\qquad\qquad \left. + x\ln p + (n-x)\ln(1-p)\right]. \end{aligned}$$

By expanding the argument of the exponential in terms of $y = x - np$, where $1 \ll y \ll np$, and keeping only the dominant terms, it can be shown that

$$f(x) \approx \frac{1}{\sqrt{2\pi n}} \frac{1}{\sqrt{p(1-p)}} \exp\left[-\frac{1}{2}\frac{(x-np)^2}{np(1-p)}\right],$$

which is of Gaussian form with $\mu = np$ and $\sigma = \sqrt{np(1-p)}$.

Thus we see that the *value* of the Gaussian *probability density function* $f(x)$ is a good approximation to the *probability* of obtaining $x$ successes in $n$ trials. This approximation is actually very good even for relatively small $n$. For example, if $n = 10$ and $p = 0.6$, then the Gaussian approximation to the binomial distribution is (24.76) with $\mu = 10 \times 0.6 = 6$ and $\sigma = \sqrt{10 \times 0.6(1-0.6)} = 1.549$. The probability functions $f(x)$ for the binomial and associated Gaussian distributions for these parameters are given in table 24.2, and it can be seen that the Gaussian approximation is a good one.

Strictly speaking, however, since the Gaussian distribution is continuous and the binomial distribution is discrete, we should use the integral of $f(x)$ for the Gaussian distribution in the calculation of approximate binomial probabilities. More specifically, we must apply a *continuity correction* so that the discrete integer $x$ in the binomial distribution becomes the interval $[x-0.5, x+0.5]$ in the Gaussian distribution. Explicitly,

$$P(X = x) \approx \frac{1}{\sigma\sqrt{2\pi}} \int_{x-0.5}^{x+0.5} \exp\left[-\frac{1}{2}\left(\frac{u-\mu}{\sigma}\right)^2\right] du.$$

The Gaussian approximation is particularly useful for estimating the binomial probability that $X$ lies between the (integer) values $x_1$ and $x_2$,

$$P(x_1 < X \le x_2) \approx \frac{1}{\sigma\sqrt{2\pi}} \int_{x_1-0.5}^{x_2+0.5} \exp\left[-\frac{1}{2}\left(\frac{u-\mu}{\sigma}\right)^2\right] du.$$

►*A manufacturer makes computer chips of which* 10% *are defective. For a random sample of* 200 *chips, find the approximate probability that more than* 15 *are defective.*

We first define the random variable

$$X = \text{'number of defective chips in the sample'},$$

which has a binomial distribution $X \sim \text{Bin}(200, 0.1)$. Therefore, the mean and variance of this distribution are

$$E[X] = 200 \times 0.1 = 20 \qquad \text{and} \qquad V[X] = 200 \times 0.1 \times (1 - 0.1) = 18,$$

and we may approximate the binomial distribution with a Gaussian distribution such that $X \sim N(20, 18)$. The standard variable is

$$Z = \frac{X - 20}{\sqrt{18}},$$

and so, using $X = 15.5$ to allow for the continuity correction,

$$\begin{aligned} P(X > 15.5) &= P\left(Z > \frac{15.5 - 20}{\sqrt{18}}\right) = P(Z > -1.06) \\ &= P(Z < 1.06) = 0.86. \end{aligned}$$ ◄

### 24.8.3 Gaussian approximation to the Poisson distribution

We first met the Poisson distribution as the limit of the binomial distribution for $n \to \infty$ and $p \to 0$, taken in such a way that $np = \lambda$ remains finite. Further, in the previous subsection, we considered the Gaussian distribution as the limit of the binomial distribution when $n \to \infty$ but $p$ remains finite, so that $np \to \infty$ also. It should, therefore, come as no surprise that the Gaussian distribution can also be used to approximate the Poisson distribution when the mean $\lambda$ becomes large. The probability function for the Poisson distribution is

$$f(x) = e^{-\lambda} \frac{\lambda^x}{x!},$$

which, on taking the logarithm of both sides, gives

$$\ln f(x) = -\lambda + x \ln \lambda - \ln x!. \tag{24.86}$$

Stirling's approximation for large $x$ gives

$$x! \approx \sqrt{2\pi x} \left(\frac{x}{e}\right)^x$$

so that

$$\ln x! \approx \ln \sqrt{2\pi x} + x \ln x - x,$$

which, on substituting into (24.86), yields

$$\ln f(x) \approx -\lambda + x \ln \lambda - (x \ln x - x) - \ln \sqrt{2\pi x}.$$

Since we expect the Poisson distribution to peak around $x = \lambda$, we substitute $\epsilon = x - \lambda$ to obtain

$$\ln f(x) \approx -\lambda + (\lambda + \epsilon) \left\{ \ln \lambda - \ln \left[ \lambda \left( 1 + \frac{\epsilon}{\lambda} \right) \right] \right\} + (\lambda + \epsilon) - \ln \sqrt{2\pi(\lambda + \epsilon)}.$$

Using the expansion $\ln(1 + z) = z - z^2/2 + \cdots$, we find

$$\begin{aligned}
\ln f(x) &\approx \epsilon - (\lambda + \epsilon) \left( \frac{\epsilon}{\lambda} - \frac{\epsilon^2}{2\lambda^2} \right) - \ln \sqrt{2\pi\lambda} - \left( \frac{\epsilon}{\lambda} - \frac{\epsilon^2}{2\lambda^2} \right) \\
&\approx -\frac{\epsilon^2}{2\lambda} - \ln \sqrt{2\pi\lambda},
\end{aligned}$$

when only the dominant terms are retained after using the fact that $\epsilon$ is of the order of the standard deviation of $x$, i.e. of order $\lambda^{1/2}$. On exponentiating this result we obtain

$$f(x) \approx \frac{1}{\sqrt{2\pi\lambda}} \exp \left[ -\frac{(x - \lambda)^2}{2\lambda} \right],$$

which is the Gaussian distribution with $\mu = \lambda$ and $\sigma^2 = \lambda$.

The larger the value of $\lambda$, the better is the Gaussian approximation to the Poisson distribution; the approximation is reasonable even for $\lambda = 5$, but $\lambda \geq 10$ is safer. As in the case of the Gaussian approximation to the binomial distribution, a continuity correction is necessary since the Poisson distribution is discrete.

►*E-mail messages are received by an author at an average rate of* 1 *per hour. Find the probability that in a day the author receives* 24 *messages or more.*

We first define the random variable

$$X = \text{number of messages received in a day.}$$

Thus $E[X] = 1 \times 24 = 24$, and so $X \sim \text{Po}(24)$. Since $\lambda > 10$ we may approximate the Poisson distribution by $X \sim \text{N}(24, 24)$. Now the standard variable is

$$Z = \frac{X - 24}{\sqrt{24}},$$

and, using the continuity correction, we find

$$\begin{aligned} P(X > 23.5) &= P\left(Z > \frac{23.5 - 24}{\sqrt{24}}\right) \\ &= P(Z > -0.102) = P(Z < 0.102) = 0.54. \end{aligned}$$ ◀

In fact, almost all probability distributions tend towards a Gaussian when the numbers involved become large – that this should happen is required by the central limit theorem, which we discuss in section 24.9.

### *24.8.4 Multiple Gaussian distributions*

Suppose $X$ and $Y$ are *independent* Gaussian-distributed random variables, so that $X \sim N(\mu_1, \sigma_1^2)$ and $Y \sim N(\mu_2, \sigma_2^2)$. Then, the PDFs for the two random variables are

$$f_1(x) = \frac{1}{\sigma_1\sqrt{2\pi}} \exp\left[-\frac{(x-\mu_1)^2}{2\sigma_1^2}\right],$$

$$f_2(y) = \frac{1}{\sigma_2\sqrt{2\pi}} \exp\left[-\frac{(y-\mu_2)^2}{2\sigma_2^2}\right].$$

We now consider the random variable $Z = X + Y$. The PDF for this random variable is (by analogy with the derivation for multiple Poisson distributions)

$$f_3(z) = \int_{-\infty}^{\infty} f_1(x) f_2(z - x)\, dx,$$

which is the convolution of the two individual PDFs (see chapter 11). This integral can be evaluated directly, but the result may be obtained more easily by considering the MGF of each distribution. The MGFs of $f_1$ and $f_2$ are, from (24.85), simply

$$\Phi_1(t) = \exp\left(\mu_1 t + \tfrac{1}{2}\sigma_1^2 t^2\right), \qquad \Phi_2(t) = \exp\left(\mu_2 t + \tfrac{1}{2}\sigma_2^2 t^2\right).$$

But from (24.67) the MGF of the random variable $Z = X + Y$ is then

$$\begin{aligned} \Phi_3(t) = \Phi_1(t)\Phi_2(t) &= \exp\left(\mu_1 t + \tfrac{1}{2}\sigma_1^2 t^2\right) \exp\left(\mu_2 t + \tfrac{1}{2}\sigma_2^2 t^2\right) \\ &= \exp\left[(\mu_1 + \mu_2)t + \tfrac{1}{2}(\sigma_1^2 + \sigma_2^2)t^2\right], \end{aligned}$$

i.e. the MGF for a Gaussian with mean $\mu_1 + \mu_2$ and variance $\sigma_1^2 + \sigma_2^2$. Thus, the PDF for $Z$ is

$$f_3(z) = \frac{1}{\sqrt{\sigma_1^2 + \sigma_2^2}\sqrt{2\pi}} \exp\left\{-\frac{[z - (\mu_1 + \mu_2)]^2}{2(\sigma_1^2 + \sigma_2^2)}\right\},$$

i.e. $Z \sim N(\mu_1 + \mu_2,\ \sigma_1^2 + \sigma_2^2)$.

A similar calculation shows that the random variable $Z = X - Y$ is distributed as $Z \sim N(\mu_1 - \mu_2,\ \sigma_1^2 + \sigma_2^2)$.

►*An executive travels home from her office every evening. Her journey consists of a train ride, followed by a bicycle ride. The time spent on the train is Gaussian distributed with mean 52 minutes and standard deviation 1.8 minutes, while the time for the bicycle journey is Gaussian distributed with mean 8 minutes, and standard deviation 2.6 minutes. Assuming these two factors are independent, estimate the percentage of occasions on which the whole journey takes more than 65 minutes.*

We first define the random variables

$$X = \text{time spent on train}, \qquad Y = \text{time spent on bicycle},$$

so that $X \sim N(52, (1.8)^2)$ and $Y \sim N(8, (2.6)^2)$. Since $X$ and $Y$ are independent, the total journey time $T = X + Y$ is distributed as

$$T \sim N(52 + 8,\ (1.8)^2 + (2.6)^2) = N(60, (3.16)^2).$$

The standard variable is thus

$$Z = \frac{T - 60}{3.16},$$

and the required probability is given by

$$P(T > 65) = P\left(Z > \frac{65 - 60}{3.16}\right) = P(Z > 1.58) = 1 - 0.943 = 0.057,$$

Thus the total journey time exceeds 65 minutes on 5.7% of occasions. ◄

The above results may be easily extended. For example, if the random variables $X_i$, $i = 1, \ldots, n$, are distributed as $X_i \sim N(\mu_i, \sigma_i^2)$, then the random variable $Z = \sum_i X_i$ is distributed as $Z \sim N(\sum_i \mu_i, \sum_i \sigma_i^2)$.

### 24.8.5 The exponential distribution

The exponential distribution with positive parameter $\beta$ is given by

$$f(x) = \begin{cases} \beta e^{-\beta x} & \text{for } x > 0, \\ 0 & \text{for } x \leq 0 \end{cases} \qquad (24.87)$$

and satisfies $\int_{-\infty}^{\infty} f(x)\, dx = 1$ as required. The exponential distribution occurs naturally if we consider the distribution of intervals between successive events in a Poisson process. If the average number of events per unit interval is $\lambda$, then on average there are $\lambda x$ events in interval $x$, so that from the Poisson distribution the probability

$$P(\text{no events in interval } x) = e^{-\lambda x}.$$

But

$$P(\text{no event in interval } x) = 1 - P(\text{an event in interval } x)$$
$$= 1 - \int_0^x f(u)\,du = \int_x^\infty f(u)\,du.$$

Thus

$$\int_x^\infty f(u)\,du = e^{-\lambda x},$$

and differentiating with respect to the lower limit of the integral gives

$$f(x) = \lambda e^{-\lambda x}.$$

This is (24.87) with $\beta$ replaced by $\lambda$. The exponential distribution can also be derived from the hypergeometric distribution.

The expectation and variance of the exponential distribution can be evaluated as $1/\beta$ and $(1/\beta)^2$ respectively. The MGF is given by

$$\Psi(t) = \frac{\beta}{\beta - t}. \tag{24.88}$$

By far the most important occurrence of the exponential distribution in the physical sciences is in thermodynamics, where it is called the Boltzmann distribution and takes the form

$$f(E) = \begin{cases} (1/kT)e^{-E/kT} & \text{for } E > 0, \\ 0 & \text{for } E \leq 0, \end{cases} \tag{24.89}$$

where $E$ is energy, $T$ is temperature and $k$ is Boltzmann's constant. The Boltzmann distribution describes the distribution at a given temperature of the numbers of particles with particular energies. The multiplicative constant $1/kT$ can be obtained by applying the normalisation condition.

### *24.8.6 The uniform distribution*

Finally we mention the very simple, but common, *uniform distribution*, which describes a continuous random variable that has a constant PDF over its allowed range of values. If the limits on $X$ are $a$ and $b$ then

$$f(x) = \begin{cases} (b-a)^{-1} & \text{for } a \leq x \leq b, \\ 0 & \text{otherwise.} \end{cases}$$

The mean and variance are both easily calculated and are

$$E[X] = \frac{a+b}{2}, \qquad V[X] = \frac{(b-a)^2}{12}.$$

## 24.9 The central limit theorem

In subsections 24.8.2 and 24.8.3 we discussed approximating the binomial and Poisson distributions by the Gaussian distribution when the number of trials is large. We now discuss why the Gaussian distribution is so common, and therefore so important.

The *central limit theorem* may be stated as follows.

**Central limit theorem.** *Suppose $X_i$, $i = 1, \ldots, n$, are* independent *random variables, each of which is described by a probability function $f_i(x)$ (these may all be different) with a mean $\mu_i$ and a variance $\sigma_i^2$. The random variable $Z = \left(\sum_i X_i\right)/n$ (i.e. the 'mean' of the $X_i$) has the following properties.*

(i) *Its expectation value is given by $E[Z] = \left(\sum_i \mu_i\right)/n$.*
(ii) *Its variance is given by $V[Z] = \left(\sum_i \sigma_i^2\right)/n^2$.*
(iii) *As $n \to \infty$ the probability function of $Z$ tends to a Gaussian with corresponding mean and variance.*

Properties (i) and (ii) are easily proved, as follows. Firstly

$$E[Z] = \frac{1}{n}(E[X_1] + \cdots + E[X_n]) = \frac{1}{n}(\mu_1 + \cdots + \mu_n) = \frac{\sum_i \mu_i}{n},$$

a result which does *not* require that the $X_i$ are *independent* random variables. If $\mu_i = \mu$ for all $i$, then this becomes

$$E[Z] = \frac{n\mu}{n} = \mu.$$

Secondly, if the $X_i$ *are* independent, it follows from an obvious extension of (24.42) that

$$V[Z] = V\left[\frac{1}{n}(X_1 + \cdots + X_n)\right] = \frac{1}{n^2}(V[X_1] + \cdots + V[X_n]) = \frac{\sum_i \sigma_i^2}{n^2}.$$

Let us now consider property (iii), which is the reason for the ubiquity of the Gaussian distribution, and is most easily proved by considering the moment generating function $G(t)$ of the probability density function $g(z)$ of $Z$. The MGF is given by

$$\begin{aligned}
G(t) &= \int e^{tz} g(z)\,dz \\
&= \int e^{t(x_1+\cdots+x_n)/n} f_1(x_1)\cdots f_n(x_n)\,dx_1\cdots dx_n \\
&= \int e^{tx_1/n} f_1(x_1)\,dx_1 \int e^{tx_2/n} f_2(x_2)\,dx_2 \cdots \int e^{tx_n/n} f_n(x_n)\,dx_n \\
&= \prod_{i=1}^{n} F_i(t/n),
\end{aligned}$$

where $F_i(t)$ is the MGF of $f_i(x_i)$. Now

$$\begin{aligned} F_i\left(\frac{t}{n}\right) &= \int e^{tx_i/n} f_i(x_i)\, dx_i \\ &= 1 + \frac{t}{n}E[X_i] + \tfrac{1}{2}\frac{t^2}{n^2}E[X_i^2] + \cdots \\ &= 1 + \mu_i\frac{t}{n} + \tfrac{1}{2}(\sigma_i^2 + \mu_i^2)\frac{t^2}{n^2} + \cdots, \end{aligned}$$

and as $n$ becomes large

$$F_i\left(\frac{t}{n}\right) \to \exp\left(\frac{\mu_i t}{n} + \tfrac{1}{2}\sigma_i^2\frac{t^2}{n^2}\right),$$

as may be verified by expanding the exponential as a power series in $t/n$. Therefore

$$G(t) = \prod_{i=1}^{n} \exp\left(\frac{\mu_i t}{n} + \tfrac{1}{2}\sigma_i^2\frac{t^2}{n^2}\right) = \exp\left(\frac{\sum_i \mu_i}{n}t + \frac{1}{2}\frac{\sum_i \sigma_i^2}{n^2}t^2\right).$$

Comparing this with the form of the MGF for a Gaussian distribution, (24.85), we can see that $g(z)$ tends to a Gaussian distribution with mean $\sum_i \mu_i/n$ and variance $\sum_i \sigma_i^2/n^2$. In particular, if we consider $Z$ to be the mean of $n$ *independent* measurements of the *same* random variable $X$ (so that $X_i = X$ for $i = 1, \ldots, n$) then, as $n \to \infty$, $Z$ has a Gaussian distribution with mean $\mu$ and variance $\sigma^2/n$.

## 24.10 Transformation of variables

Let $X$ be some random variable for which the probability function $f(x)$ is known. Suppose, however, that we are more interested in the related random variable $Y = F(X)$, where $F(X)$ is some function of $X$. The probability function $g(y)$ for the random variable $Y$ may be straightforwardly calculated by requiring

$$|f(x)\, dx| = |g(y)\, dy| \quad\Rightarrow\quad g(y) = f(x)\left|\frac{dx}{dy}\right|.$$

►*A lighthouse is situated at a distance $L$ from a straight coastline, opposite a point $O$, and sends out a narrow continuous beam of light simultaneously in opposite directions. The beam rotates with constant angular velocity. If the random variable $X$ is the distance along the coastline, measured from $O$, of the spot that the light beam illuminates, find its probability density function.*

The situation is illustrated in figure 24.12. Since the light beam rotates at a constant angular velocity, $\theta$ is distributed uniformly between $-\pi/2$ and $\pi/2$, and so $f(\theta) = 1/\pi$. Now $x = L\tan\theta$, so the PDF for $X$ is given by

$$g(x) = f(\theta)\frac{d\theta}{dx}.$$

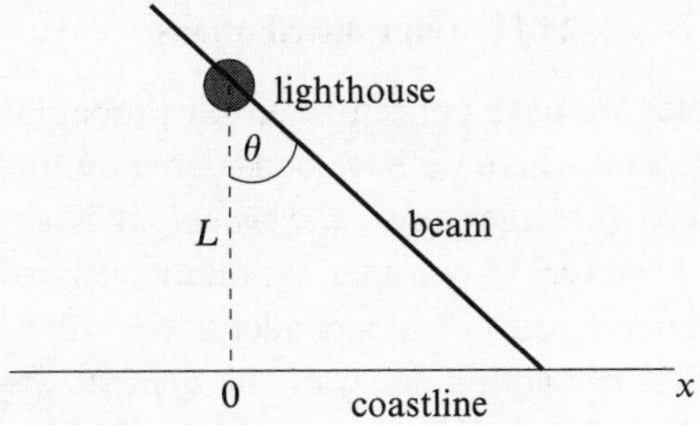


Figure 24.12 The illumination of a coastline by the beam from a lighthouse.

Since $dx/d\theta = L\sec^2\theta = L(1+\tan^2\theta) = L[1+(x/L)^2]$, we find

$$g(x) = \frac{1}{\pi L[1+(x/L)^2]} \qquad \text{for } x = -\infty \text{ to } \infty.$$

A distribution of this form is called a *Cauchy distribution*, and it should be noted that it does not have a variance, as the integral $\int_{-\infty}^{\infty} x^2 g(x)\,dx$ diverges. ◄

Extra care must be taken when the allowed range of values of the new random variable $Y$ is covered more than once as $X$ varies over its own allowed range of values (and vice versa). We note, however, that this will affect only the normalisation of the PDF of the new variable. This normalisation can always be fixed by requiring the integral of the PDF between its allowed limits to equal unity.

►*If the random variable $X$ is Gaussian distributed with mean $\mu$ and variance $\sigma^2$, find the PDF of the new variable $Y = (X-\mu)^2/2\sigma^2$.*

The PDF of $X$ is given by

$$f(x) = \frac{1}{\sigma\sqrt{2\pi}} \exp\left[-\frac{(x-\mu)^2}{2\sigma^2}\right] = \frac{1}{\sigma\sqrt{2\pi}} \exp(-y).$$

Now the probability of a value of $Y$ between $y$ and $y+dy$ is given by

$$g(y)\,dy = 2f(x)\,dx,$$

because as $X$ goes from $-\infty$ to $\infty$, $Y$ goes from $\infty$ to 0 and back again from 0 to $\infty$. Since

$$dy = \frac{x-\mu}{\sigma^2}\,dx = (2y)^{1/2}\sigma^{-1}\,dx,$$

the PDF of $Y$ is

$$g(y) = \frac{2}{\sigma\sqrt{2\pi}} \exp(-y) \frac{\sigma}{(2y)^{1/2}} = \frac{1}{\sqrt{\pi}} y^{-1/2} \exp(-y). \; ◄$$

## 24.11 Joint distributions

Throughout this chapter we have concentrated on probability distributions of a single random variable, and where we have considered multiple random variables we have so far assumed that they are *independent*. It is extremely common in the physical sciences, however, to consider simultaneously two or more random variables, which, in general, are not independent. We will return to the subject of dependence after first presenting some of the general ways of characterising joint distributions. We will concentrate mainly on *bivariate* distributions, i.e. distributions of only two random variables, though the results may readily be extended to multivariate distributions. The subject of multivariate distributions is large and a detailed study is beyond the scope of this book; the interested reader should therefore consult one of the many specialised texts. We do, however, discuss the multinomial and multivariate Gaussian distributions in section 24.13.

The first thing to note when dealing with bivariate distributions is that the distinction between discrete and continuous distributions may not be as clear as for the single variable case; both of the random variables may be discrete, both continuous, or one discrete but the other continuous. In general, for the random variables $X$ and $Y$, the joint distribution will take an infinite number of values unless both $X$ and $Y$ have only a finite number of values.

In this chapter we will consider only the cases where $X$ and $Y$ are either both discrete or both continuous random variables.

### *24.11.1 Discrete bivariate distributions*

In direct analogy with the one-variable (univariate) case, if $X$ is a discrete random variable that takes the values $\{x_i\}$ and $Y$ one that takes the values $\{y_j\}$, then the probability function of the joint distribution is defined as

$$f(x, y) = \begin{cases} P(X = x_i,\ Y = y_j) & \text{for } x = x_i,\ y = y_j, \\ 0 & \text{otherwise.} \end{cases}$$

We may therefore think of $f(x, y)$ as a set of spikes at valid points in the $xy$-plane, whose heights represent the probability of obtaining $X = x_i$ and $Y = y_j$. The normalisation of $f(x, y)$ implies

$$\sum_i \sum_j f(x_i, y_j) = 1, \tag{24.90}$$

where the sums over $i$ and $j$ take all valid values. We can also define the cumulative probability function

$$F(x, y) = \sum_{x_i \le x} \sum_{y_j \le y} f(x_i, y_j), \tag{24.91}$$

from which the probability that $X$ lies in the range $[a_1, a_2]$ and $Y$ lies in the range $[b_1, b_2]$ is given by

$$P(a_1 < X \le a_2,\ b_1 < Y \le b_2) = F(a_2, b_2) - F(a_1, b_2) - F(a_2, b_1) + F(a_1, b_1).$$

Finally, we define $X$ and $Y$ to be *independent* (uncorrelated) if we can write their joint distribution

$$f(x, y) = f_1(x) f_2(y), \tag{24.92}$$

i.e. as the product of two univariate distributions.

### *24.11.2 Continuous bivariate distributions*

In the case where both $X$ and $Y$ are continuous random variables, the PDF of the joint distribution is defined by

$$f(x, y)\,dx\,dy = P(x < X \le x + dx,\ y < Y \le y + dy), \tag{24.93}$$

so $f(x, y)\,dx\,dy$ is the probability that $x$ lies in the range $[x, x + dx]$ and $y$ lies in the range $[y, y + dy]$. It is clear that the two-dimensional function (surface) $f(x, y)$ must be everywhere non-negative, and that normalisation requires

$$\int_{-\infty}^{\infty}\int_{-\infty}^{\infty} f(x, y)\,dx\,dy = 1.$$

It follows further that

$$P(a_1 < X \le a_2,\ b_1 < Y \le b_2) = \int_{b_1}^{b_2}\int_{a_1}^{a_2} f(x, y)\,dx\,dy. \tag{24.94}$$

We can also define the cumulative probability function by

$$F(x, y) = P(X \le x,\ Y \le y) = \int_{-\infty}^{x}\int_{-\infty}^{y} f(u, v)\,du\,dv,$$

from which we see that (as for the discrete case),

$$P(a_1 < X \le a_2,\ b_1 < Y \le b_2) = F(a_2, b_2) - F(a_1, b_2) - F(a_2, b_1) + F(a_1, b_1).$$

Finally we note that the definition of independence (24.92) for discrete bivariate distributions also applies to continuous bivariate distributions.

### *24.11.3 Conditional distributions*

Although a bivariate distribution depends on both $X$ and $Y$, we are sometimes interested only in the probability distribution of $X$ (say) at some given $Y = y_0$ (or vice versa), i.e.

$$P(X = x | Y = y_0),$$

which is called the *conditional probability of X given* $Y = y_0$. The corresponding probability function is

$$g(x) = f(x, y_0),$$

and is called the *conditional* probability function.

### 24.11.4 Marginal distributions

For a bivariate distribution depending on $X$ and $Y$, we may only be interested in the probability function for $X$ *irrespective of the value of* $Y$ (or vice versa). This *marginal* distribution of $X$ is given by

$$g(x) = \begin{cases} \sum_j f(x, y_j) & \text{for a discrete distribution,} \\ \int_{-\infty}^{\infty} f(x, y)\,dy & \text{for a continuous distribution.} \end{cases} \tag{24.95}$$

The corresponding cumulative density functions are

$$G(x) = \begin{cases} \sum_{x_i \le x} g(x_i) & \text{for the discrete case,} \\ \int_{-\infty}^{x} g(u)\,du & \text{for the continuous case.} \end{cases} \tag{24.96}$$

It is clear that analogous definitions exist for the marginal distributions of $Y$, i.e. those which ignore the value of $X$.

## 24.12 Properties of joint distributions

The probability density function $f(x, y)$ contains all the information on the joint probability distribution of two random varibles $X$ and $Y$. In a similar manner to that presented for univariate distributions, however, it is conventional to characterise $f(x, y)$ by certain of its properties, which we now discuss.

### 24.12.1 Expectation values

The expectation values for joint distributions are defined in an exactly analogous way to those for single-variable distributions (24.38), i.e. the expectation value of $X$ is given by

$$E[X] = \mu_X = \begin{cases} \sum_i \sum_j x_i f(x_i, y_j) & \text{for the discrete case,} \\ \int_{-\infty}^{\infty} \int_{-\infty}^{\infty} x f(x, y)\,dx\,dy & \text{for the continuous case.} \end{cases} \tag{24.97}$$

$E[Y]$ is given in a similar manner. In general the expectation value of a function $G(X, Y)$ is given by

$$E[G(X, Y)] = \begin{cases} \sum_i \sum_j g(x_i, y_j) f(x_i, y_j) & \text{for the discrete case,} \\ \int_{-\infty}^{\infty} \int_{-\infty}^{\infty} g(x, y) f(x, y)\,dx\,dy & \text{for the continuous case.} \end{cases}$$

### 24.12.2 Variance

Just as for the mean of a bivariate distribution, the definition of the variance is exactly analogous to that for the single-variable case (24.41), i.e. the variance of $X$ is given by

$$V[X] = \sigma_X^2 = \begin{cases} \sum_i \sum_j (x_i - \mu_X)^2 f(x_i, y_j) & \text{for the discrete case,} \\ \int_{-\infty}^{\infty} \int_{-\infty}^{\infty} (x - \mu_X)^2 f(x, y)\, dx\, dy & \text{for the continuous case.} \end{cases} \tag{24.98}$$

Equivalent definitions exist for the variance of $Y$.

### 24.12.3 Covariance and correlation

Means and variances of joint distributions provide useful information about their marginal distributions, but we have not yet given any indication of how to measure the relationship between the two random variables. Of course, it may be that the two random variables are independent, but often this is not so. For example, if we measure the heights and weights of a sample of people we would not be surprised to find a tendency for tall people to be heavier than short people and vice versa. We will show in this section that two functions, the *covariance* and the *correlation*, can be defined for a bivariate distribution, and that these are useful in characterising the relationship between the two random variables.

The *covariance* of two random variables $X$ and $Y$ is defined by

$$\text{Cov}[X, Y] = E[(X - \mu_X)(Y - \mu_Y)], \tag{24.99}$$

where $\mu_X$ and $\mu_Y$ are the expectation values of $X$ and $Y$ respectively.

Clearly related is the *correlation* of the two random variables, defined by

$$\text{Corr}[X, Y] = \frac{\text{Cov}[X, Y]}{\sigma_X \sigma_Y}, \tag{24.100}$$

where $\sigma_X$ and $\sigma_Y$ are the standard deviations of $X$ and $Y$ respectively. It can be shown that the correlation function lies between $-1$ and $+1$. If the value assumed is negative, $X$ and $Y$ are said to be *negatively correlated*, if it is positive they are said to be *positively correlated* and if it is zero they are said to be *uncorrelated*. We will now justify the use of these terms.

One particularly useful consequence of its definition is that the covariance of two *independent* variables, $X$ and $Y$, is zero. It immediately follows from (24.100) that their correlation is also zero, and this justifies the use of the term 'uncorrelated' for two such variables. To show this extremely important property we first note that

$$\begin{aligned}\text{Cov}[X,Y] &= E[(X-\mu_X)(Y-\mu_Y)] \\ &= E[XY - \mu_X Y - \mu_Y X + \mu_X\mu_Y] \\ &= E[XY] - \mu_X E[Y] - \mu_Y E[X] + \mu_X\mu_Y \\ &= E[XY] - \mu_X\mu_Y. \end{aligned} \tag{24.101}$$

Now, if $X$ and $Y$ are independent, $E[XY] = E[X]E[Y] = \mu_X\mu_Y$ and so $\text{Cov}[X,Y] = 0$. It is important to note that the converse of this result is not necessarily true; two variables dependent on each other can still be uncorrelated.

We have already asserted that if the correlation of two random variables is positive (negative) they are said to be positively (negatively) correlated. We have also stated that the correlation lies between $-1$ and $+1$. The terminology suggests that two identical random variables (i.e. $X = Y$) are completely correlated and that their correlation should be $+1$. Likewise, if $X = -Y$ the functions are completely anticorrelated and their correlation should be $-1$. Values of the correlation function between these extremes show some degree of correlation. In fact it is not necessary that $X = Y$ for $\text{Corr}[X,Y] = 1$; it is sufficient that $Y$ is a linear function of $X$, i.e. $Y = aX + b$ (with $a$ positive). If $a$ is negative then $\text{Corr}[X,Y] = -1$. To show this we first note that $\mu_Y = a\mu_X + b$. Now

$$Y = aX + b = aX + \mu_Y - a\mu_X \quad \Rightarrow \quad Y - \mu_Y = a(X - \mu_X),$$

and so using the definition of the covariance (24.99)

$$\text{Cov}[X,Y] = aE[(X-\mu_X)^2] = a\sigma_X^2.$$

It follows from the definition of the variance that $\sigma_Y = |a|\sigma_X$ and so, using the definition of the correlation (24.100),

$$\text{Corr}[X,Y] = \frac{a\sigma_X^2}{|a|\sigma_X^2} = \frac{a}{|a|},$$

which is the required result.

It should be noted that, even if the possibilities of $X$ and $Y$ being non-zero are mutually exclusive, $\text{Corr}[X,Y]$ need not have value $\pm 1$.

►*A biased die gives probabilities $p/2$, $p$, $p$, $p$, $p$, $2p$ of throwing 1, 2, 3, 4, 5, 6 respectively. If the random variable $X$ is the number shown on the die and the random variable $Y = X^2$, calculate the covariance and correlation of $X$ and $Y$.*

We have already calculated in subsections 24.2.1 and 24.5.3 that $p = 2/13$, $E[X] = 53/13$, $E\left[X^2\right] = 253/13$ and $V[X] = 480/169$. Using (24.101)

$$\text{Cov}[X,Y] = \text{Cov}\left[X, X^2\right] = E\left[X^3\right] - E[X]E\left[X^2\right].$$

Now $E\left[X^3\right]$ is given by

$$E\left[X^3\right] = 1^3 \times \tfrac{1}{2}p + (2^3 + 3^3 + 4^3 + 5^3)p + 6^3 \times 2p$$
$$= \frac{1313}{2}p = 101,$$

and the covariance of $X$ and $Y$ by

$$\mathrm{Cov}[X, Y] = 101 - \frac{53}{13} \times \frac{253}{13} = \frac{3660}{169}.$$

The correlation is defined by $\mathrm{Corr}[X, Y] = \mathrm{Cov}[X, Y]/\sigma_X\sigma_Y$. The standard deviation of $Y$ may be calculated from the definition of the variance, i.e.

$$\sigma_Y^2 = \frac{p}{2}\left(1^2 - \frac{253}{13}\right)^2 + p\left(2^2 - \frac{253}{13}\right)^2 + p\left(3^2 - \frac{253}{13}\right)^2 + p\left(4^2 - \frac{253}{13}\right)^2$$
$$+p\left(5^2 - \frac{253}{13}\right)^2 + 2p\left(6^2 - \frac{253}{13}\right)^2$$
$$= \frac{187356}{169}p = \frac{28824}{169}.$$

Thus

$$\mathrm{Corr}[X, Y] = \frac{3660}{169}\sqrt{\frac{169}{28824}}\sqrt{\frac{169}{480}} \approx 0.984.$$

Thus the random variables $X$ and $Y$ display a strong degree of positive correlation, as we would expect. ◀

We note that the covariance of $X$ and $Y$ occurs in various expressions. For example, if $X$ and $Y$ are *not* independent then

$$V[X + Y] = E\left[(X + Y)^2\right] - [E(X + Y)]^2$$
$$= E\left[X^2\right] + 2E[XY] + E\left[Y^2\right] - \{(E[X])^2 + 2E[X]E[Y] + (E[Y])^2\}$$
$$= V[X] + V[Y] + 2(E[XY] - E[X]E[Y])$$
$$= V[X] + V[Y] + 2\,\mathrm{Cov}[X, Y].$$

More generally, we find (for $a$ and $b$ constant)

$$V[aX + bY] = a^2V[X] + b^2V[Y] + 2ab\,\mathrm{Cov}[X, Y]. \qquad (24.102)$$

Note that if $X$ and $Y$ are in fact independent, then $\mathrm{Cov}[X, Y] = 0$ and we recover the expression (24.42) in subsection 24.5.3.

We may use (24.102) to obtain an approximate expression for $V[f(X, Y)]$ for any arbitrary function $f$, even when the random variables $X$ and $Y$ are correlated. Approximating $f(X, Y)$ by the linear terms of its Taylor expansion about the point $(\mu_X, \mu_Y)$, we have

$$f(X,Y) \approx f(\mu_X,\mu_Y) + \left(\frac{\partial f}{\partial X}\right)(X-\mu_X) + \left(\frac{\partial f}{\partial Y}\right)(Y-\mu_Y),$$

where the partial derivatives are evaluated at $X = \mu_X$ and $Y = \mu_Y$ respectively. Taking the variance of both sides, and using (24.102), we find

$$V[f(X,Y)] \approx \left(\frac{\partial f}{\partial X}\right)^2 V[X] + \left(\frac{\partial f}{\partial Y}\right)^2 V[Y] + 2\left(\frac{\partial f}{\partial X}\right)\left(\frac{\partial f}{\partial Y}\right)\text{Cov}[X,Y].$$

Clearly, if $\text{Cov}[X,Y] = 0$, we recover the result (24.43) derived in subsection 24.5.3.

For several variables $X_i$, $i = 1,\ldots,n$, we can define the symmetric (positive definite) *covariance matrix* whose elements are

$$V_{ij} = \text{Cov}[X_i, X_j], \tag{24.103}$$

and the symmetric (positive definite) *correlation matrix*

$$\rho_{ij} = \text{Corr}[X_i, X_j].$$

The diagonal elements of the covariance matrix are the variances of the variables, whilst those of the correlation matrix are unity.

Our final example in this section is one in which the sense of each correlation is intuitively obvious although, because of the 'scoring system', its magnitude is not.

►*A card is drawn at random from a normal 52-card deck and its identity noted. The card is replaced, the deck shuffled and the process repeated. Random variables $W, X, Y, Z$ are defined as follows:*

$W = 2$ *if the drawn card is a heart; $W = 0$ otherwise.*
$X = 4$ *if the drawn card is an ace, king, or queen; $X = 2$ if the card is a jack or ten; $X = 0$ otherwise.*
$Y = 1$ *if the drawn card is red; $Y = 0$ otherwise.*
$Z = 2$ *if the drawn card is black and an ace, king or queen; $Z = 0$ otherwise.*

*Establish the correlation matrix for $W, X, Y, Z$.*

The means of the variables are given by

$$\mu_W = 2 \times \tfrac{1}{4} = \tfrac{1}{2}, \quad \mu_X = \left(4 \times \tfrac{3}{13}\right) + \left(2 \times \tfrac{2}{13}\right) = \tfrac{16}{13},$$
$$\mu_Y = 1 \times \tfrac{1}{2} = \tfrac{1}{2}, \quad \mu_Z = 2 \times \tfrac{6}{52} = \tfrac{3}{13}.$$

The variances, calculated from $\sigma_U^2 = V[U] = E\left[U^2\right] - (E[U])^2$, where $U = W$,

$X$, $Y$ or $Z$, are

$$\sigma_W^2 = (4 \times \tfrac{1}{4}) - (\tfrac{1}{2})^2 = \tfrac{3}{4}, \quad \sigma_X^2 = (16 \times \tfrac{3}{13}) + (4 \times \tfrac{2}{13}) - (\tfrac{16}{13})^2 = \tfrac{472}{169},$$

$$\sigma_Y^2 = (1 \times \tfrac{1}{2}) - (\tfrac{1}{2})^2 = \tfrac{1}{4}, \quad \sigma_Z^2 = (4 \times \tfrac{6}{52}) - (\tfrac{3}{13})^2 = \tfrac{69}{169}.$$

The covariances are found by first calculating $E[WX]$ etc. and then forming $E[WX] - \mu_W \mu_X$ etc.

$$E[WX] = 2\,(4)\,(\tfrac{3}{52}) + 2\,(2)\,(\tfrac{2}{52}) = \tfrac{8}{13}, \quad \text{Cov}[W,X] = \tfrac{8}{13} - \tfrac{1}{2}\,(\tfrac{16}{13}) = 0,$$

$$E[WY] = 2(1)\,(\tfrac{1}{4}) = \tfrac{1}{2}, \quad \text{Cov}[W,Y] = \tfrac{1}{2} - \tfrac{1}{2}\,(\tfrac{1}{2}) = \tfrac{1}{4},$$

$$E[WZ] = 0, \quad \text{Cov}[W,Z] = 0 - \tfrac{1}{2}\,(\tfrac{3}{13}) = -\tfrac{3}{26},$$

$$E[XY] = 4(1)\,(\tfrac{6}{52}) + 2(1)\,(\tfrac{4}{52}) = \tfrac{8}{13}, \quad \text{Cov}[X,Y] = \tfrac{8}{13} - \tfrac{16}{13}\,(\tfrac{1}{2}) = 0,$$

$$E[XZ] = 4(2)\,(\tfrac{6}{52}) = \tfrac{12}{13}, \quad \text{Cov}[X,Z] = \tfrac{12}{13} - \tfrac{16}{13}\,(\tfrac{3}{13}) = \tfrac{108}{169},$$

$$E[YZ] = 0, \quad \text{Cov}[Y,Z] = 0 - \tfrac{1}{2}\,(\tfrac{3}{13}) = -\tfrac{3}{26}.$$

The correlations $\text{Corr}[W,X]$ and $\text{Corr}[X,Y]$ are clearly zero; the remainder are given by

$$\begin{aligned}
\text{Corr}[W,Y] &= \tfrac{1}{4}\,(\tfrac{3}{4} \times \tfrac{1}{4})^{-1/2} &= 0.577,\\
\text{Corr}[W,Z] &= -\tfrac{3}{26}\,(\tfrac{3}{4} \times \tfrac{69}{169})^{-1/2} &= -0.209,\\
\text{Corr}[X,Z] &= \tfrac{108}{169}\,(\tfrac{472}{169} \times \tfrac{69}{169})^{-1/2} &= 0.598,\\
\text{Corr}[Y,Z] &= -\tfrac{3}{26}\,(\tfrac{1}{4} \times \tfrac{69}{169})^{-1/2} &= -0.361.
\end{aligned}$$

Finally, the correlation matrix is given by

$$\rho = \begin{pmatrix} 1 & 0 & 0.58 & -0.21 \\ 0 & 1 & 0 & 0.60 \\ 0.58 & 0 & 1 & -0.36 \\ -0.21 & 0.60 & -0.36 & 1 \end{pmatrix}.$$

As would be expected, $X$ is uncorrelated with either $W$ or $Y$, colour and face-value being two independent characteristics. Positive correlations are to be expected between $W$ and $Y$, and between $X$ and $Z$; both correlations are fairly strong. Moderate anticorrelations exist between $Z$ and both $W$ and $Y$, reflecting the fact that it is impossible for $W$ and $Y$ to be positive if $Z$ is positive. ◄

## 24.13 Multivariate distributions

So far we have discussed only distributions of one and two random variables, but we have mentioned that multivariate distributions are encountered. In this section we will examine just two of these, the *multinomial distribution*, which is an extension of the binomial distribution, and the *multivariate Gaussian distribution.*

### *24.13.1 The multinomial distribution*

The binomial distribution describes the probability of obtaining $x$ 'successes' from $n$ independent trials, where each trial has only two possible outcomes. This may be generalised to the case where each trial has $k$ possible outcomes with probabilities $p_1, p_2, \ldots, p_k$. If we consider the random variables $X_i$, $i = 1, \ldots, n$, to be the number of outcomes of type $i$ in $n$ trials, then we may calculate their joint probability function

$$f(x_1, x_2, \ldots, x_k) = P(X_1 = x_1,\ X_2 = x_2,\ \ldots,\ X_k = x_k),$$

where we must have $\sum_{i=1}^{k} x_i = n$. In $n$ trials the probability of obtaining $x_1$ outcomes of type 1, followed by $x_2$ outcomes of type 2 etc. is

$$p_1^{x_1} p_2^{x_2} \cdots p_k^{x_k}.$$

However, the number of distinguishable permutations of this result is

$$\frac{n!}{x_1!x_2!\cdots x_k!},$$

and thus

$$f(x_1, x_2, \ldots, x_k) = \frac{n!}{x_1!x_2!\cdots x_k!} p_1^{x_1} p_2^{x_2} \cdots p_k^{x_k}. \qquad (24.104)$$

This is the *multinomial probability distribution.*

If $k = 2$, then the multinomial distribution reduces to the familiar binomial distribution. It must be remembered that, although in this form the binomial distribution appears to be a function of two random variables, in fact, since $p_2 = 1 - p_1$ and $x_2 = n - x_1$, the distribution of $X_1$ is entirely determined by the parameters $p$ and $n$. That $X_1$ has a *binomial* distribution is shown by remembering that it represents the number of objects of a particular type obtained from sampling with replacement, the original definition of the binomial distribution. In fact, any of the random variables $X_i$ has a binomial distribution, i.e. the marginal distribution of each $X_i$ is binomial with parameters $n$ and $p_i$. It immediately follows that

$$E[X_i] = np_i \qquad \text{and} \qquad V[X_i]^2 = np_i(1 - p_i). \qquad (24.105)$$

▶*At a village fête patrons were invited, for a* 10p *entry fee, to pick without looking* 6 *tickets from a drum containing equal large numbers of red, blue and green tickets. If* 5 *or more of the tickets were of the same colour a prize of* 100p *was awarded. A consolation award of* 40p *was made if* 2 *tickets of each colour were picked. Was a good time had by all?*

In this case, all types of outcome (red, blue and green) have the same probabilities. The probability of obtaining any given combination of tickets is given by the multinomial distribution with $n = 6$, $k = 3$, and all $p_i = \frac{1}{3}$ $(i = 1, 2, 3)$.

(i) The probability of picking 6 tickets of the same colour is given by

$$P(6 \text{ of the same colour}) = 3 \times \frac{6!}{6!0!0!}\left(\frac{1}{3}\right)^6\left(\frac{1}{3}\right)^0\left(\frac{1}{3}\right)^0 = \frac{1}{243}.$$

The factor of three is present because three different colours are possible.

(ii) The probability of picking 5 tickets of one colour and 1 ticket of another colour is

$$P(5 \text{ of one colour; 1 of another}) = 3\times2\times\frac{6!}{5!1!0!}\left(\frac{1}{3}\right)^5\left(\frac{1}{3}\right)^1\left(\frac{1}{3}\right)^0 = \frac{4}{81}.$$

The factors of three and two are included because there are three ways to choose the colour of the 5, and then two ways to choose the colour of the remaining ticket.

(iii) Finally, the probability of picking 2 tickets of each colour is

$$P(2 \text{ of each colour}) = \frac{6!}{2!2!2!}\left(\frac{1}{3}\right)^2\left(\frac{1}{3}\right)^2\left(\frac{1}{3}\right)^2 = \frac{10}{81}.$$

Thus the expected return to any patron was, in pence,

$$100\left(\frac{1}{243} + \frac{4}{81}\right) + \left(40 \times \frac{10}{81}\right) = 10.29.$$

A good time was had by all but the stallholder! ◀

### 24.13.2 *The multivariate Gaussian distribution*

A particularly interesting multivariate distribution is the generalisation of the Gaussian distribution to multiple random variables $X_i$, $i = 1, \ldots, n$. If the expectation value of $X_i$ is $E(X_i) = \mu_i$, then the PDF is given by

$$f(x_1, \ldots, x_n) = N \exp\left[-\tfrac{1}{2}\sum_i\sum_j a_{ij}(x_i - \mu_i)(x_j - \mu_j)\right],$$

where $N$ is a normalisation constant that we give below. If we write the column vectors $\mathsf{x} = (x_1\ x_2\ \cdots\ x_n)^{\mathrm{T}}$ and $\mu = (\mu_1\ \mu_2\ \cdots\ \mu_n)^{\mathrm{T}}$, and we denote the matrix with elements $a_{ij}$ by $\mathsf{A}$, then

$$f(\mathsf{x}) = f(x_1, \ldots, x_n) = N \exp\left[-\tfrac{1}{2}(\mathsf{x} - \mu)^{\mathrm{T}}\mathsf{A}(\mathsf{x} - \mu)\right],$$

where $\mathsf{A}$ is symmetric. By considering the MGF of $f(\mathsf{x})$ it may be shown that $\mathsf{A}$ is equal to the inverse of the covariance matrix $\mathsf{V}$ of the $X_i$ (see (24.103)). With the correct normalisation $f(\mathsf{x})$ is given by

$$f(\mathsf{x}) = \frac{1}{(2\pi)^{n/2}\sqrt{|\mathsf{V}|}} \exp\left[-\tfrac{1}{2}(\mathsf{x} - \mu)^{\mathrm{T}}\mathsf{V}^{-1}(\mathsf{x} - \mu)\right]. \qquad (24.106)$$

### *24.13.3 Transformation of variables in multivariate distributions*

Suppose the random variables $X_i$, $i = 1, \ldots, n$, are described by the multivariate PDF $f(x_1, \ldots, x_n)$. If we wish to consider random variables $Y_j$, $j = 1, \ldots, n$, related to the $X_i$ by $Y_j = F_j(X_1, X_2, \ldots, X_n)$ then we may calculate the PDF for the $Y_j$ in a similar way to that in the univariate case by demanding

$$f(x_1, \ldots, x_n)\, dx_1 \cdots dx_n = g(y_1, \ldots, y_n)\, dy_1 \cdots dy_n.$$

From the discussion of changing the variables in multiple integrals given in chapter 5 it follows that

$$g(y_1, \ldots, y_n) = f(x_1, \ldots, x_n) \left| \frac{\partial(x_1, \ldots, x_n)}{\partial(y_1, \ldots, y_n)} \right|,$$

where

$$J = \frac{\partial(x_1, \ldots, x_n)}{\partial(y_1, \ldots, y_n)} = \begin{pmatrix} \dfrac{\partial x_1}{\partial y_1} & \cdots & \dfrac{\partial x_n}{\partial y_1} \\ \vdots & \ddots & \vdots \\ \dfrac{\partial x_1}{\partial y_n} & \cdots & \dfrac{\partial x_n}{\partial y_n} \end{pmatrix},$$

is the Jacobian of the $x_i$ with respect to the $y_j$.

►*Suppose that the random variables $X_i$, $i = 1, 2, \ldots, n$, are independent and Gaussian distributed with means $\mu_i$ and variances $\sigma_i^2$ respectively. Find the PDF for the new variables $Z_i = (X_i - \mu_i)/\sigma_i$, $i = 1, 2, \ldots, n$. By considering an elemental spherical shell in **Z**-space, find the PDF of the random variable $\chi_n^2 = \sum_{i=1}^n Z_i^2$.*

From (24.106) the PDF for the variables $X_i$ is

$$f(x_1, \ldots, x_n) = \frac{1}{(2\pi)^{n/2}\sigma_1 \cdots \sigma_n} \exp\left[-\sum_{i=1}^{n} \frac{(x_i - \mu_i)^2}{2\sigma_i^2}\right].$$

To derive the PDF for the $Z_i$ variables, we require

$$f(x_1,\ldots,x_n)\,dx_1\cdots dx_n = g(z_1,\ldots,z_n)\,dz_1\cdots dz_n,$$

and, noting that $dz_i = dx_i/\sigma_i$, we obtain

$$g(z_1,\ldots,z_n) = \frac{1}{(2\pi)^{n/2}}\exp\left(-\frac{1}{2}\sum_{i=1}^{n} z_i^2\right).$$

Let us now consider the random variable $\chi_n^2 = \sum_{i=1}^n Z_i^2$, which we may regard as the square of distance from the origin in the $n$-dimensional **Z**-space. We now require

$$g(z_1,\ldots,z_n)\,dz_1\cdots dz_n = h(\chi_n^2)d\chi_n^2.$$

If we consider the infinitesimal volume $dV = dz_1\cdots dz_n$ to be that enclosed by the $n$-dimensional spherical shell of radius $\chi_n$ and thickness $d\chi_n$, then we may write $dV = A\chi_n^{n-1}d\chi_n$, for some constant $A$. We thus obtain

$$h(\chi_n^2)d\chi_n^2 \;\;\propto\;\; \exp(-\chi_n^2/2)\chi_n^{n-1}d\chi_n \;\;\propto\;\; \exp(-\chi_n^2/2)\chi_n^{n-2}d\chi_n^2,$$

where we have used the fact that $d\chi_n^2 = 2\chi_n\,d\chi_n$. Thus we see that the PDF for $\chi_n^2$ is given by

$$h(\chi_n^2) = B\exp(-\chi_n^2/2)\chi_n^{n-2},$$

for some constant $B$. This constant may be determined from the normalisation condition

$$\int_0^\infty h(\chi_n^2)\,d\chi_n^2 = 1,$$

and is found to be $B = [2^{n/2}\Gamma(n/2)]^{-1}$.

The PDF $h(\chi_n^2)$ is known as the *chi-squared distribution*, and has numerous applications in statistics. For completeness, we also note that $E[\chi_n^2] = n$ and $V[\chi_n^2] = 2n$. ◄

## 24.14 Statistics: describing data

The discussion in this chapter has, so far, concentrated on probability in relation to theoretical, although commonly occurring, distributions. We now turn briefly to ways in which experimentally obtained data can be characterised in the broadest terms.

In a book of this nature we cannot hope to do justice to a subject as large as statistics; indeed, many would argue that statistics belongs in the realm of experimental science rather than in a mathematics textbook. Nevertheless, physical scientists are regularly called upon to present their data in terms of statistics, and so we will now discuss this aspect of a much more extensive

| 188.7 | 204.7 | 193.2 | 169.0 |
|---|---|---|---|
| 168.1 | 189.8 | 166.3 | 200.0 |

Table 24.3 Experimental data giving eight measurements of the round trip time in milliseconds for a computer 'packet' to travel from Cambridge UK to Cambridge MA in milliseconds.

subject. The reader who requires a fuller treatment is referred to one of the many specialist books.†

Consider an experiment in which a *sample*, a set of $N$ values $x_i$, $i = 1, \ldots, N$, is drawn from a larger set of values, a *population*. It is often useful to describe the sample by simple numbers or *sample statistics* that characterise the data. A sample statistic is any quantity that depends on the sample alone. The ultimate aim is to provide meaningful descriptions of the population from which the sample is drawn. Our discussion of statistics will thus fall into two parts. In this first section we will give the definitions of the most commonly used sample statistics; in the second we will show how they may be used to estimate the parameters of the distribution from which they are drawn.

### *24.14.1 Mean*

The simplest number used to characterise a sample is the mean, which for $N$ values $x_i$, $i = 1, 2, \ldots, N$, is defined by

$$\bar{x} = \frac{1}{N} \sum_{i=1}^{N} x_i. \tag{24.107}$$

In words, the *sample mean* is the sum of the sample values divided by the number of values in the sample.

►*Table 24.3 gives eight values for the round trip time in milliseconds for a computer 'packet' to travel from Cambridge UK to Cambridge MA. Find the sample mean.*

Using (24.107) the sample mean in milliseconds is given by

$$\begin{aligned}\bar{x} &= \tfrac{1}{8}(188.7 + 204.7 + 193.2 + 169.0 + 168.1 + 189.8 + 166.3 + 200.0) \\ &= \frac{1479.8}{8} = 185.0. \blacktriangleleft\end{aligned}$$

Strictly speaking the mean given by (24.107) is the *arithmetic mean* and this

† See, for example, Barlow, *Statistics* (Wiley) or Martin, *Statistics for Physicists* (Academic Press).

is by far the most common definition used for a mean. Other definitions are possible, though less common, and include

(i) the *geometric mean*,

$$\bar{x}_g = \left( \prod_{i=1}^{N} x_i \right)^{1/N}, \tag{24.108}$$

(ii) the *harmonic mean*,

$$\bar{x}_h = \frac{N}{\sum_{i=1}^{N} 1/x_i}, \tag{24.109}$$

(iii) the *root mean square*,

$$\bar{x}_{rms} = \left( \frac{\sum_{i=1}^{N} x_i^2}{N} \right)^{1/2}. \tag{24.110}$$

It should be noted that, whilst $\bar{x}$, $\bar{x}_h$ and $\bar{x}_{rms}$ remain well defined even if some sample values are negative, the value of $\bar{x}_g$ could then become complex; the geometric mean should not be used in such cases.

►*Calculate $\bar{x}_g$, $\bar{x}_h$ and $\bar{x}_{rms}$ for the sample given in table 24.3.*

The geometric mean is given by (24.108) to be

$$\bar{x}_g = (188.7 \times 204.7 \times \cdots \times 200.0)^{1/8} = 184.4.$$

The harmonic mean is given by (24.109) to be

$$\bar{x}_h = \frac{8}{(1/188.7) + (1/204.7) + \cdots + (1/200.0)} = 183.9.$$

Finally, the root mean square is given by (24.110) to be

$$\bar{x}_{rms} = \left[ \tfrac{1}{8}(188.7^2 + 204.7^2 + \cdots + 200.0^2) \right]^{1/2} = 185.5. \blacktriangleleft$$

### 24.14.2 Variance and standard deviation

The variance and standard deviation both give a measure of the spread of values in a sample. The *sample variance* is defined by

$$s^2 = \frac{1}{N} \sum_{i=1}^{N} (x_i - \bar{x})^2, \tag{24.111}$$

and the *sample standard deviation* is the positive square root of the sample variance, i.e.

$$s = \sqrt{\frac{1}{N} \sum_{i=1}^{N} (x_i - \bar{x})^2}. \tag{24.112}$$

►*Find the sample variance and sample standard deviation of the data given in table 24.3.*

We have already found that the sample mean is 185.0 and so, using (24.111),

$$\begin{aligned} s^2 &= \frac{1}{8}\left[(188.7-185.0)^2+(204.7-185.0)^2+\cdots+(200.0-185.0)^2\right] \\ &= \frac{1608.4}{8} = 201.1. \end{aligned}$$

The sample standard deviation is thus

$$s = \sqrt{201.1} = 14.2. \blacktriangleleft$$

Using the definition (24.112), it is clear that in order to calculate the standard deviation of a sample we must first calculate the sample mean. This requirement can be avoided, however, by using an alternative form for $s^2$. From (24.111), we see that

$$\begin{aligned} s^2 &= \frac{1}{N}\sum_{i=1}^{N}(x_i-\bar{x})^2 = \frac{1}{N}\sum_{i=1}^{N}x_i^2 - \frac{1}{N}\sum_{i=1}^{N}2x_i\bar{x} + \frac{1}{N}\sum_{i=1}^{N}\bar{x}^2 \\ &= \overline{x^2} - 2\bar{x}^2 + \bar{x}^2 = \overline{x^2} - \bar{x}^2 \end{aligned}$$

We may therefore write the sample variance $s^2$ as

$$s^2 = \overline{x^2} - \bar{x}^2 = \frac{1}{N}\sum_{i=1}^{N}x_i^2 - \left(\frac{1}{N}\sum_{i=1}^{N}x_i\right)^2, \qquad (24.113)$$

from which the sample standard deviation is found by taking the positive square-root. Thus, by evaluating the quantities $\sum_{i=1}^{N}x_i$ and $\sum_{i=1}^{N}x_i^2$ for our sample, we can calculate the sample mean and sample standard deviation at the same time.

## 24.15 Statistics: estimating parameters

As mentioned previously, the ultimate aim of calculating sample statistics is to provide estimates of the the parameters of the distribution from which the samples were drawn. In this section we introduce some common estimators. In each case we will assume that we are sampling the distribution of a random variable $X$ with mean $E[X] = \mu$ and variance $V[X] = \sigma^2$, and that the sample values $x_1, x_2, \ldots, x_N$ are independent.

### *24.15.1 Mean*

We have already met the sample mean $\bar{x}$. Without any modification this provides an estimate $\hat{\mu}$ of the mean of the population $\mu$ from which the sample was drawn.

This is easily seen by considering the expectation value of the random variable $\bar{X} = \sum_{i=1}^{N} X_i/N$,

$$E\left[\bar{X}\right] = E\left[(\textstyle\sum_{i=1}^{N} X_i)/N\right] = E\left[\textstyle\sum_{i=1}^{N} X_i\right]\Big/ N = N\mu/N = \mu.$$

Thus $\bar{x}$ is indeed an unbiased estimate of the population mean.

### 24.15.2 Variance

A good estimator of the variance of a population is not so straightforward to define as an estimator for the mean. Complications arise because the true mean of the population $\mu$ is unknown (all we have is an estimate $\hat{\mu} = \bar{x}$). If we knew the true mean of the population we could use

$$\frac{1}{N}\sum_{i=1}^{N}(x_i - \mu)^2,$$

as an estimator for the variance, but if the true mean of the population is unknown, a natural alternative is to replace $\mu$ by $\bar{x}$, to obtain the sample variance

$$s^2 = \frac{1}{N}\sum_{i=1}^{N}(x_i - \bar{x})^2. \qquad (24.114)$$

However, the expectation value of the associated random variable $S^2$ is

$$E\left[S^2\right] = E\left[X^2\right] - E\left[\bar{X}^2\right] = V\left[X\right] - V\left[\bar{X}\right],$$

where the last equality follows since $E\left[X\right] = E\left[\bar{X}\right]$ and so

$$E\left[X^2\right] - E\left[\bar{X}^2\right] = E\left[X^2\right] - (E\left[X\right])^2 - \left(E\left[\bar{X}^2\right] - (E\left[\bar{X}\right])^2\right).$$

Now, applying the central limit theorem, $V\left[\bar{X}\right] = V\left[X\right]/N$ and so, denoting $V\left[X\right]$ by $\sigma^2$,

$$E\left[S^2\right] = \frac{N-1}{N}\sigma^2. \qquad (24.115)$$

We can thus see that (24.114) is a biased estimator of the variance of the population, but that it can be made into an unbiased estimator if the multiplicative correction factor $N/(N-1)$, the *Bessel correction*, is applied to (24.114). Thus an unbiased estimator of the variance of the population is defined by

$$\widehat{\sigma^2} = \frac{1}{N-1}\sum_{i=1}^{N}(x_i - \bar{x})^2. \qquad (24.116)$$

The positive square root of the variance estimator gives an estimate $\hat{\sigma}$ of the population standard deviation $\sigma$.

►*Find estimates for the population variance and population standard deviation of the data given in table 24.3.*

We have already found that an estimate of the sample mean is 185.0 and so, using (24.116),

$$\widehat{\sigma^2} = \tfrac{1}{7}\left[(188.7 - 185.0)^2 + (204.7 - 185.0)^2 + \cdots + (200.0 - 185.0)^2\right]$$
$$= \frac{1608.4}{7} = 229.8.$$

An estimate of the standard deviation of the population is thus

$$\hat{\sigma} = \sqrt{229.8} = 15.2. \blacktriangleleft$$

### 24.15.3 Estimating the error in the mean

The sample mean discussed above gives only an approximation to the true mean. It is often important, therefore, to have an estimate of how good the approximation is likely to be. A measure of this is the standard deviation $\sigma_m$ of the estimate of the population mean. Two ways of finding this quantity are now given.

*Method 1: repeating the experiment*

One way to estimate the standard deviation in our estimate of the mean is to repeat the experiment in its entirety a number of times, e.g. in the experiment described above we must find the mean of eight more round trip times, and then another eight, and so on. If we calculate the mean for each of these complete experiments we will obtain a set of values $m_1, m_2, \ldots,$ which themselves form a distribution. This distribution in turn has a standard deviation $\sigma_m$ that may be estimated by

$$\widehat{\sigma_m^2} = \frac{1}{R-1}\sum_{i=1}^{R}(m_i - \bar{m})^2, \qquad (24.117)$$

where $R$ is the number of times the experiment is repeated, and $\bar{m}$ is the mean of the individual means. The difference between $\sigma_m$ and the standard deviation of a single value $\sigma$ is often a source of confusion and so we summarise as follows.

The error on a single measurement of a quantity is given by $\sigma$, which is estimated from an experiment in which $N$ measurements are made. The mean of each set of $N$ values itself has an error $\sigma_m$ that may be estimated by repeating the whole experiment $R$ times ($R$ not usually equalling $N$).

| | | | | | | | | |
|---|---|---|---|---|---|---|---|---|
| trial 1 | 188.8 | 204.7 | 193.2 | 169.0 | 168.1 | 189.8 | 166.3 | 200.0 |
| trial 2 | 203.0 | 177.0 | 179.1 | 193.5 | 172.8 | 181.0 | 174.3 | 197.2 |
| trial 3 | 166.1 | 183.9 | 174.9 | 193.1 | 187.6 | 180.8 | 174.7 | 179.4 |
| trial 4 | 151.0 | 192.1 | 175.0 | 185.4 | 159.6 | 192.3 | 176.4 | 189.9 |
| trial 5 | 183.2 | 172.2 | 190.3 | 181.7 | 175.6 | 202.6 | 185.3 | 189.4 |
| trial 6 | 187.1 | 183.6 | 187.8 | 185.8 | 174.9 | 208.6 | 187.8 | 175.3 |

Table 24.4 Experimental data giving the round trip time in milliseconds for a computer 'packet' to travel from Cambridge UK to Cambridge MA.

| Trial | Mean | Variance | $s$ |
|---|---|---|---|
| 1 | 185.0 | 229.8 | 14.2 |
| 2 | 184.7 | 131.8 | 10.7 |
| 3 | 180.1 | 70.4 | 7.8 |
| 4 | 177.7 | 239.7 | 14.5 |
| 5 | 185.0 | 89.1 | 8.8 |
| 6 | 186.4 | 108.7 | 9.8 |

Table 24.5 The sample mean, sample variance and sample standard deviation of the data given in table 24.4.

►*Table 24.4 gives the results of six repetitions of the experiment described above. Find the standard error in the mean.*

We must first find the sample mean and sample standard deviation for each of the six trials. These are given in table 24.5, and have been calculated using previous results. The mean of the individual means is

$$\bar{m} = \tfrac{1}{6}(185.0 + 184.7 + \cdots + 186.4) = 183.2,$$

and so the variance in our estimation of the population mean is

$$\sigma_m^2 = \tfrac{1}{5}\left[(185 - 183.2)^2 + \cdots + (186.4 - 183.2)^2\right] = 11.5,$$

and the associated standard deviation is $\sqrt{11.5} = 3.4$. ◄

*Method 2: relating $\sigma$ and $\sigma_m$*

In the previous section we showed how to estimate the standard deviation $\sigma_m$ of our estimate of the mean of a population by simply repeating the experiment in its entirety a number of times. In this section we discuss another approach, which does not rely on repeating the experiment.

For a set of $N$ measurements $x_i$, $i = 1, 2, \ldots, N$, the standard deviation $\sigma_m$ on the mean $\bar{x}$ of the measurements may be easily calculated by considering the variance of the associated random variable $\bar{X}$. This is given by

$$V\left[\bar{X}\right] = V\left[(\textstyle\sum_{i=1}^{N} X_i)/N\right] = \frac{1}{N^2}V\left[\textstyle\sum_{i=1}^{N} X_i\right] = \frac{1}{N^2}\textstyle\sum_{i=1}^{N} V\left[X_i\right] = \frac{N\sigma^2}{N^2} = \frac{\sigma^2}{N},$$

and so the standard deviation of the mean is

$$\sigma_m = \frac{\sigma}{\sqrt{N}}. \qquad (24.118)$$

It can be seen that this result gives a similar estimate for the error in the mean to that obtained using method 1.

►*Estimate the error on the mean in trial 1 of the previous example.*

Trial 1 involves eight measurements, and from table 24.5 has a sample standard deviation $s = 14.2$. Thus, applying the Bessel correction, our estimate of the population standard deviation is

$$\widehat{\sigma} = \sqrt{\frac{8}{8-1}}s = 15.2,$$

and our estimate of the error on the mean is $\widehat{\sigma}_m = 15.2/\sqrt{8} = 5.4$. This is in only moderately good agreement with the 3.4 derived using method 1. The discrepancy illustrates the intrinsic imprecision of statistics: had we chosen trial 5, where the sample standard deviation $s = 8.8$, we would have obtained $\widehat{\sigma}_m = 3.3$. ◄

## 24.16 Exercises

24.1 Two duellists, $A$ and $B$, take alternate shots at each other, and the duel is over when a shot (fatal or otherwise!) hits its target. Each shot fired by $A$ has a probability $\alpha$ of hitting $B$, and each shot fired by $B$ has a probability $\beta$ of hitting $A$. Calculate the probabilities $P_1$, $P_2$, $P_3$ that $A$ will win the duel, defined as follows:

(a) $P_1$, $A$ fires the first shot;
(b) $P_2$, $B$ fires the first shot;
(c) $P_3$, they agree to fire simultaneously, rather than alternately.

Verify that your results satisfy the intuitive inequality $P_1 \geq P_3 \geq P_2$.

24.2 (a) Gamblers $A$ and $B$ each roll a fair die, and $B$ wins if his score is strictly greater than $A$'s. Show that the odds are 7 to 5 in $A$'s favour.
(b) Calculate the probabilities of scoring a total $T$ from two rolls of a fair die for $T = 2, 3, \ldots, 12$. Gamblers $C$ and $D$ each roll a fair die twice and score respective totals $T_C$ and $T_D$, with $D$ winning if $T_D > T_C$. Realising that the odds are not equal, $D$ insists that $C$ should increase her stake for each game. $C$ agrees to stake £1.10 per game, as compared to $D$'s £1.00 stake. Who will show a profit?

24.3 An electronics assembly firm buys its microchips from three different suppliers; half of them are bought from firm $X$, whilst firms $Y$ and $Z$ supply 30% and 20% respectively. The suppliers use different quality control procedures and the percentages of defective chips are 2%, 4% and 4% for $X$, $Y$ and $Z$ respectively. The probabilities that a defective chip will fail two or more assembly line tests are 40%, 60% and 80%, again respectively, whilst all defective chips have a 10% chance of escaping detection. An assembler finds a chip that fails only one test. What is the probability that it came from supplier $X$?

24.4 A boy is selected at random from amongst the children of families with $n$ children. It is known that he has at least two sisters. Show that the probability that he has $k-1$ brothers is

$$\frac{(n-1)!}{(2^{n-1}-n)(k-1)!(n-k)!},$$

for $1 \leq k \leq n-2$, and zero for other values of $k$.

24.5 Villages $A$, $B$, $C$ and $D$ are connected by overhead telephone lines joining $AB$, $AC$, $BC$, $BD$ and $CD$. As a result of severe gales, there is a probability $p$ (the same for each link) that any particular link is broken.

(a) Show that the probability that a call can be made from $A$ to $B$ is

$$1-2p^2+p^3.$$

(b) Show that the probability that a call can be made from $D$ to $A$ is

$$1-2p^2-2p^3+5p^4-2p^5.$$

24.6 Kittens from different litters do not get on with each other and fighting breaks out whenever two kittens from different litters are present together. A cage initially contains $x$ kittens from one litter and $y$ from another. To quell the fighting kittens are removed, at random, one at a time, until peace is restored. Show, by induction, that the expected number of kittens finally remaining is

$$N(x,y)=\frac{x}{y+1}+\frac{y}{x+1}.$$

24.7 (*A more difficult question.*)

If the scores in a cup football match are equal at the end of the normal period of play, a 'penalty shoot-out' is held in which each side takes up to five shots (from the penalty spot) alternately, the shoot-out being stopped if one side acquires an unassailable lead (i.e. has a lead greater than its opponents have shots remaining). If the scores are still level after the shoot-out a 'sudden death' competition takes place. Team 1, which takes the first penalty, has a probability $p_1$ (independently of the player involved) of scoring and a probability $q_1$ $(= 1 - p_1)$ of missing; $p_2$ and $q_2$ are defined likewise.

Define $P(i{:}x, y)$ as the probability that team $i$ has scored $x$ goals after $y$ attempts, and let $f(M)$ be the probability that the shoot-out terminates after a *total* of $M$ shots.

(a) Prove that the probability that 'sudden death' will be needed is

$$f(11+) = \sum_{r=0}^{5} ({}^5C_r)^2 (p_1 p_2)^r (q_1 q_2)^{5-r}.$$

(b) Give reasoned arguments (preferably without first looking at the expressions involved) which show that

$$f(M = 2N) = \sum_{r=0}^{2N-6} \left\{ \begin{array}{l} p_2 P(1{:}r, N) P(2{:}5 - N + r, N - 1) \\ + q_2 P(1{:}6 - N + r, N) P(2{:}r, N - 1) \end{array} \right\},$$

for $N = 3, 4, 5$;

$$f(M = 2N + 1) = \sum_{r=0}^{2N-5} \left\{ \begin{array}{l} p_1 P(1{:}5 - N + r, N) P(2{:}r, N) \\ + q_1 P(1{:}r, N) P(2{:}5 - N + r, N) \end{array} \right\},$$

for $N = 3, 4$.

(c) Give an explicit expression for $P(i{:}x, y)$ and hence show that, if the teams are so well matched that $p_1 = p_2 = 1/2$, then

$$f(2N) = \sum_{r=0}^{2N-6} \left( \frac{1}{2^{2N}} \right) \frac{N!(N-1)!6}{r!(N-r)!(6-N+r)!(2N-6-r)!},$$

$$f(2N+1) = \sum_{r=0}^{2N-5} \left( \frac{1}{2^{2N}} \right) \frac{(N!)^2}{r!(N-r)!(5-N+r)!(2N-5-r)!}.$$

(d) Evaluate these expressions to show that, expressing $f(M)$ in units of $2^{-8}$, we have

| $M$ | 6 | 7 | 8 | 9 | 10 | 11+ |
|---|---|---|---|---|---|---|
| $f(M)$ | 8 | 24 | 42 | 56 | 63 | 63 |

Give a simple explanation of why $f(10) = f(11+)$.

24.8 A particle is confined to the one-dimensional space $0 \leq x \leq a$ and classically it can be in any small interval $dx$ in the space with equal probability. However, quantum mechanics gives the result that the probability distribution is proportional to $\sin^2(n\pi x/a)$, where $n$ is an integer. Find the variance in the particle's position in both the classical and quantum mechanical pictures and show that, although they differ, the latter tends to the former in the limit of large $n$, in agreement with the correspondence principle of physics.

24.9 Show, (a) from the corresponding result for the probability generating function and (b) directly, that the moment generating function for the geometric distribution, which gives the probability distribution for the number of trials needed to record a first success, is given by

$$\Psi_1(t) = \frac{p}{e^{-t} - q},$$

where $p$ is the probability of a success in a single trial and $q = 1 - p$. Hence show that the mean and variance of the number of trials needed to record $r$ successes are $r/p$ and $rq/p^2$ respectively.

24.10 The number of errors needing correction on each page of a set of proofs follows a Poisson distribution of mean $\mu$. The cost of the first correction on any page is $\alpha$ and that of subsequent corrections on the same page is $\beta$ each. Prove that the average cost of corrections per page is

$$\alpha + \beta(\mu - 1) - (\alpha - \beta)e^{-\mu}.$$

24.11 The probability distribution for the number of eggs in a clutch is Po($\lambda$), and the probability that each egg will hatch is $p$ (independently of the size of the clutch). Show by direct calculation that the probability distribution for the number of chicks that hatch is Po($\lambda p$) and so justify the assumptions made in the second worked example in subsection 24.6.1.

24.12 A practical-class demonstrator sends his 12 students to the storeroom to collect apparatus for an experiment, but forgets to tell each which type of component to bring. There are three types, $A$, $B$ and $C$, held in the stores (in large numbers) in the ratio 20%, 30% and 50% respectively, and each student picks a component at random. In order to set up one experiment, one unit each of $A$ and $B$, and two units of $C$ are needed. Find an expression for the probability $P(N)$ that at least $N$ experiments can be set up.

(a) Evaluate $P(3)$.

(b) Show that $P(2)$ can be written in the form

$$P(2) = (0.5)^{12} \sum_{i=2}^{6} {}^{12}C_i \,(0.4)^i \sum_{j=2}^{8-i} {}^{12-i}C_j \,(0.6)^j.$$

(c) By considering the conditions under which no experiments can be set up, show that $P(1) = 0.9145$.

24.13 Show that, as the number of trials $n$ becomes large but $np_i = \lambda_i$, $i = 1,\ldots,k-1$, remains finite, the multinomial probability distribution (24.104),

$$M_n(x_1, x_2, \ldots, x_k) = \frac{n!}{x_1!x_2!\cdots x_k!} p_1^{x_1} p_2^{x_2} \cdots p_k^{x_k},$$

can be approximated by a multiple Poisson distribution (with $k-1$ factors)

$$M_n'(x_1, x_2, \ldots, x_{k-1}) = \prod_{i=1}^{k-1} \frac{e^{-\lambda_i} \lambda_i^{x_i}}{x_i!}.$$

(Write $\sum_i^{k-1} p_i = \delta$ and express all terms involving subscript $k$ in terms of $n$ and $\delta$, either exactly or approximately. You will need to use $n! \approx n^{\epsilon}[(n-\epsilon)!]$ and $(1-a/n)^n \approx e^{-a}$ for large $n$.)

(a) Verify that the terms of $M_n'$ when summed over all values of $x_1, x_2, \ldots, x_{k-1}$, add up to unity.

(b) If $k = 7$, and $\lambda_i = 9$ for all $i = 1, 2, \ldots, 6$, estimate, using the appropriate Gaussian approximation, the chance that at least three of $x_1, x_2, \ldots, x_6$ will be 15 or greater.

## 24.17 Hints and answers

24.1 (a) $\alpha(\alpha+\beta-\alpha\beta)^{-1}$; (b) $\alpha(1-\beta)(\alpha+\beta-\alpha\beta)^{-1}$; (c) $\alpha(\alpha+\beta)^{-1}$.

24.2 (b) $P(T_D > T_C) = 0.5\{1-[146/(36)^2]\} = 0.4437$; $C$'s expected return is £2.10(1 − 0.4437) ≈ £1.17 for a £1.10 stake.

24.3 The relative probabilities are $X : Y : Z = 25 : 18 : 4$ (divide by 150 to obtain the absolute probabilities); 25/47.

24.4 Take $A_j$ as the event that a family consists of $j$ boys and $n-j$ girls, and $B$ as the event that the boy has at least two sisters. Apply Bayes' theorem.

24.5 If $q = 1-p$, the probability is $q^3 + 3pq^2 + p^2q$, corresponding to 0, 1, (a particular set of) 2 breaks; (b) similarly, the probability is $q^5 + 5q^4p + (10-2)p^2q^3 + 2p^3q^2$.

24.6 Establish that $N(x, y+1) = [(y+1)N(x,y) + xN(x-1, y+1)]/(x+y+1)$.

24.7 (a) The scores must be equal, at $r$ each, after five attempts each; (b) $M$ can only be even if team 2 gets too far ahead (or drops too far behind) to be caught (or catch up), with conditional probability $p_2$ (or $q_2$). Conversely $M$ can only be odd as a result of a final action by team 1; (c) $P(i : x, y) = {}^yC_x p_i^x q_i^{y-x}$; (d) if the match is still alive at the tenth kick, team 2 is just as likely to lose it as to take it into sudden death.

24.8 $a^2/12$; $a^2/12 - a^2/(2\pi^2 n^2)$.

24.10 Consider 0, 1 and $\geq 2$ errors on a page separately.
24.11 $P(k \text{ chicks hatching}) = \sum_{n=k}^{\infty} \text{Po}(n, \lambda)\ \text{Bin}(n, p)$.
24.12 (a) $[12!(0.5)^6(0.3)^3(0.2)^3]/[6!\,3!\,3!] = 0.063$.
24.13 With the continuity correction $P(x_i \geq 15) = 0.0334$; $7.5 \times 10^{-4}$.

# 25

# *Numerical methods*

It frequently happens that the end product of a calculation or piece of analysis is one or more algebraic or differential equations, or an integral that cannot be evaluated in closed form or in terms of tabulated or pre-programmed functions. From the point of view of the physical scientist or engineer, who needs numerical values for prediction or comparison with experiment, the calculation or analysis is thus incomplete.

With the ready availability of standard packages on powerful computers for the numerical solution of equations, both algebraic and differential, and for the evaluation of integrals, there is, in principle, no need for the investigator to do other than turn to them. However, it should be a part of every engineer's or scientist's competence to have some understanding of the kinds of procedure that are being put into practice within those packages. The present chapter indicates (at a simple level) some of the ways in which analytically intractable problems can be tackled using numerical methods.

In the restricted space available in a book of this nature it is clearly not possible to give anything like a full discussion, even of the elementary points that will be made in this chapter. The limited objective adopted is that of explaining and illustrating by simple examples some of the basic principles involved. The examples used can in many cases be solved in closed form anyway, but this 'obviousness' of the answers should not detract from their illustrative usefulness, and it is hoped that their transparency will help the reader to appreciate some of the inner workings of the methods described.

The student who proposes to study complicated sets of equations or make repeated use of the same procedures by, for example, writing computer programs to carry out the computations, will find it essential to acquire a good understanding of topics hardly mentioned here. Amongst these are the sensitivity of the adopted procedures to the errors introduced by the limited accuracy with which a numerical value can be stored in a computer (rounding errors) and to the errors introduced as a result of approximations made in setting up the numerical procedures (truncation errors). For this scale of application, books specifically devoted to numerical analysis, data analysis and computer programming should also be consulted.

So far as is possible, the method of presentation here is that of indicating and discussing in a qualitative way the main steps in the procedure, and then of following this with an elementary worked example. The examples have been restricted in complexity to a level at which they can be carried out with a pocket calculator. Naturally it will not be possible for the student to check all the numerical values presented unless he or she has a programmable calculator or computer readily available, and even then it might be tedious to do so. However, checking the initial step and at least one step in the middle of each repetitive calculation given is advisable, so that how the symbolic equations are used with actual numbers is understood. There is clearly some advantage in choosing such a step at a point in the calculation where the values involved are changing sufficiently rapidly that whatever calculating device is used will have the accuracy to find any errors.

Where alternative methods for solving the same type of problem are discussed, for example in finding the roots of a polynomial equation, we have usually taken the same example to illustrate each method. This could give the mistaken impression that the methods are very restricted in applicability, but it is felt by the authors that using the same examples repeatedly has sufficient advantages in terms of illustrating the *relative* characteristics of competing methods, as to justify doing so. Once the principles are clear, little is to be gained by using new examples each time and, in fact, having some prior knowledge of the 'correct answer' should allow the reader to judge the efficiency and dangers of particular methods as the successive steps are followed through.

One other point remains to be mentioned. Here, in contrast with every other chapter of this book, the value of a large selection of exercises is not clear cut. The reader with sufficient computing resources to tackle them can easily devise algebraic or differential equations to be solved, or functions to be integrated (which perhaps have arisen in other contexts). Further, the solutions of these problems will be, for the most part, self-checking. Consequently, although a few exercises are included, no attempt has been made to test the full range of ideas treated in this chapter.

## 25.1 Algebraic and transcendental equations

The problem of finding the real roots of an equation of the form $f(x) = 0$, where $f(x)$ is an algebraic or transcendental function of $x$, is one that can sometimes be treated numerically, even if explicit solutions in closed form are not feasible.

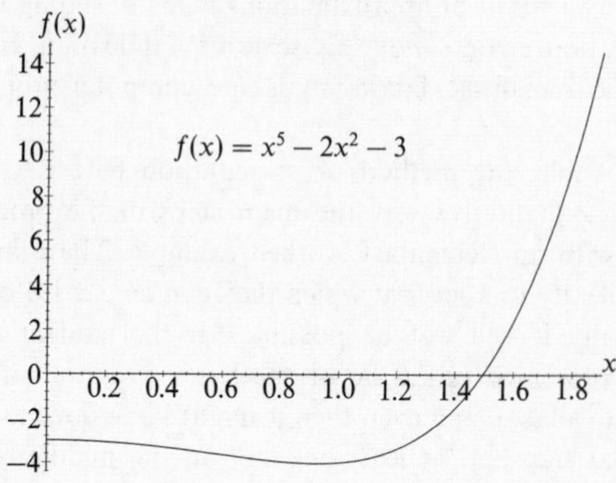


Figure 25.1 A graph of the function $f(x) = x^5 - 2x^2 - 3$ for $x$ in the range $0 \leq x \leq 1.9$.

Examples of the types of equation mentioned are the quartic equation

$$ax^4 + bx + c = 0,$$

and the transcendental equation

$$x - 3\tanh x = 0.$$

The latter type is characterised by the fact that it effectively contains a polynomial of infinite order on the left-hand side.

We will discuss four methods that, in various circumstances, can be used to obtain the real roots of equations of the above types. In all cases we will take as the specific equation to be solved the fifth-order polynomial equation

$$f(x) \equiv x^5 - 2x^2 - 3 = 0. \qquad (25.1)$$

The reasons for using the same equation each time were discussed in the previous section.

For future reference and so that the reader may follow some of the calculations leading to the evaluation of the real root of (25.1), a graph of $f(x)$ in the range $0 \leq x \leq 1.9$ is shown in figure 25.1.

Equation (25.1) is one for which no solution can be found in closed form, that is in the form $x = a$ where $a$ does not explicitly contain $x$. The general scheme to be employed will be an iterative one in which successive approximations to a real root of (25.1) will be obtained, each approximation, it is to be hoped, better than the preceding one; certainly, we require that the approximations converge and that they have as their limit the sought-for root. Let us denote the required root by $\xi$ and the values of successive approximations by $x_1, x_2, \ldots, x_n, \ldots$. Then for any particular method to be successful,

$$\lim_{n\to\infty} x_n = \xi, \qquad \text{where } f(\xi) = 0. \tag{25.2}$$

However, success as defined here is not the only criterion. Since, in practice, only a finite number of iterations will be possible, it is important that the values of $x_n$ be close to that of $\xi$ for all $n > N$, where $N$ is a relatively low number; exactly how low it is naturally depends on the computing resources available and the accuracy required in the final answer.

So that the reader may assess the progress of the calculations that follow, we record that to nine significant figures the real root of (25.1) has the value

$$\xi = 1.495\,106\,40. \tag{25.3}$$

We now consider in turn four methods for determining the value of this root.

### *25.1.1 Rearrangement of the equation*

If the equation can be recast into the form

$$x = \phi(x) \tag{25.4}$$

where $\phi(x)$ is a *slowly* varying function of $x$, then an iteration scheme

$$x_{n+1} = \phi(x_n) \tag{25.5}$$

will often produce a fair approximation to $\xi$ after a few iterations. Clearly $\xi = \phi(\xi)$ since $f(\xi) = 0$, and thus when $x_n$ is close to $x$ the next approximation $x_{n+1}$ will differ little from $x_n$, the actual size of the difference giving an order-of-magnitude indication of the inaccuracy in $x_{n+1}$ (as compared to $\xi$).

In the present case the equation can be written

$$x = (2x^2 + 3)^{1/5}. \tag{25.6}$$

Because of the presence of the $\frac{1}{5}$th power, the RHS is rather insensitive to the value of $x$ used to compute it, and so the form (25.6) fits the general requirements for the method to work satisfactorily. It only remains to choose a starting approximation. It is easy to see from figure 25.1 that the value $x = 1.5$ would be a good starting point, but, so that the behaviour of the procedure at values some way from the actual root can be studied, we will make the poorer choice of $x_1 = 1.7$.

With this starting value and the general recurrence relationship

$$x_{n+1} = (2x_n^2 + 3)^{1/5}, \tag{25.7}$$

successive values can be found. These are recorded in table 25.1. Although not

| $n$ | $x_n$ | $f(x_n)$ |
|---|---|---|
| 1 | 1.7 | 5.42 |
| 2 | 1.544 18 | 1.01 |
| 3 | 1.506 86 | $2.28 \times 10^{-1}$ |
| 4 | 1.497 92 | $5.37 \times 10^{-2}$ |
| 5 | 1.495 78 | $1.28 \times 10^{-2}$ |
| 6 | 1.495 27 | $3.11 \times 10^{-3}$ |
| 7 | 1.495 14 | $7.34 \times 10^{-4}$ |
| 8 | 1.495 12 | $1.76 \times 10^{-4}$ |

Table 25.1 Successive approximations to the root of (25.1) using the rearrangement method.

| $n$ | $A_n$ | $f(A_n)$ | $B_n$ | $f(B_n)$ | $x_n$ | $f(x_n)$ |
|---|---|---|---|---|---|---|
| 1 | 1.0 | −4.0000 | 1.7 | 5.4186 | 1.2973 | −2.6916 |
| 2 | 1.2973 | −2.6916 | 1.7 | 5.4186 | 1.4310 | −1.0957 |
| 3 | 1.4310 | −1.0957 | 1.7 | 5.4186 | 1.4762 | −0.3482 |
| 4 | 1.4762 | −0.3482 | 1.7 | 5.4186 | 1.4897 | −0.1016 |
| 5 | 1.4897 | −0.1016 | 1.7 | 5.4186 | 1.4936 | −0.0289 |
| 6 | 1.4936 | −0.0289 | 1.7 | 5.4186 | 1.4947 | −0.0082 |

Table 25.2 Successive approximations to the root of (25.1) using linear interpolation.

strictly necessary, the value of $f(x_n) \equiv x_n^5 - 2x_n^2 - 3$ is also shown at each stage.

It will be seen that $x_7$ and all later $x_n$ agree with the precise answer (25.3) to within one part in $10^4$. However, $f(x_n)$ and $x_n - \xi$ are both reduced by a factor of only about 4 for each iteration; thus a large number of iterations would be needed to produce a very accurate answer. The factor 4 is of course specific to this particular problem and would be different for a different equation. The successive values of $x_n$ are shown in graph (*a*) of figure 25.2.

### *25.1.2 Linear interpolation*

In this approach two values $A_1$ and $B_1$ of $x$ are chosen with $A_1 < B_1$ and such that $f(A_1)$ and $f(B_1)$ have opposite signs. The chord joining the two points $(A_1, f(A_1))$ and $(B_1, f(B_1))$ is then notionally constructed, as illustrated in graph (*b*) of figure 25.2, and the value $x_1$ at which the chord cuts the $x$-axis is determined by the interpolation formula

$$x_n = \frac{A_n f(B_n) - B_n f(A_n)}{f(B_n) - f(A_n)}, \tag{25.8}$$

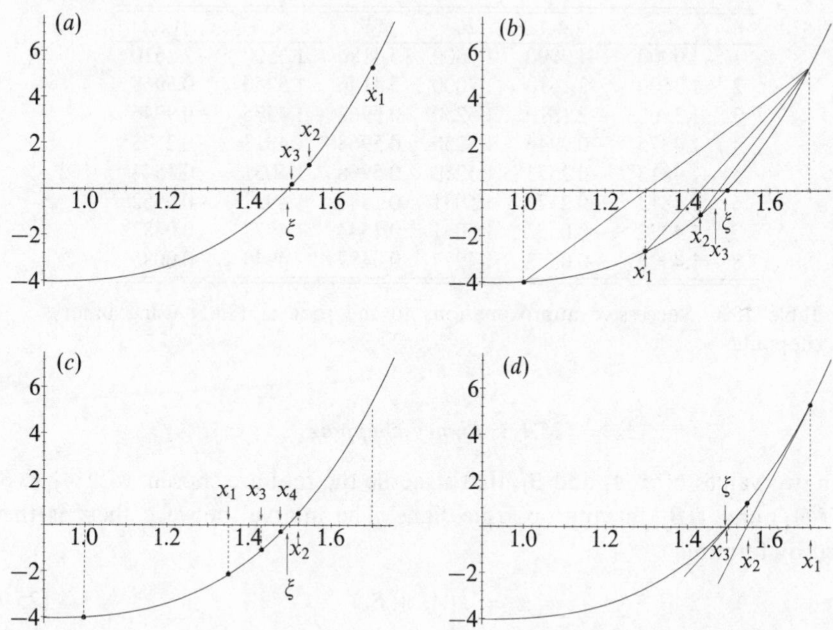


Figure 25.2 Graphical illustrations of the iteration methods discussed in the text: (*a*) rearrangement; (*b*) linear interpolation; (*c*) binary chopping; (*d*) Newton–Raphson.

with $n = 1$. Next $f(x_1)$ is evaluated and the process repeated after replacing by $x_1$ either $A_1$ or $B_1$, according to whether $f(x_1)$ has the same sign as $f(A_1)$ or $f(B_1)$ respectively. In figure 25.2(*b*), $A_1$ is the one replaced.

As in our particular example, there is a tendency (if the curvature of $f(x)$ is of constant sign near the root) for one of the two ends of the successive chords to remain unchanged.

Starting with the initial values $A_1 = 1$ and $B_1 = 1.7$, the results of the first five iterations using (25.8) are given in table 25.2 and indicated in graph (*b*) of figure 25.2.

As with the rearrangement method the improvement in accuracy is a fairly constant factor at each iteration (approximately 3 in this case), and for our particular example there is little to choose between the two. Both tend to their limiting value of $\xi$ monotonically, from either higher or lower values, and this makes it difficult to estimate limits within which $\xi$ can safely be presumed to lie. The next method to be described gives at any stage a range of values within which $\xi$ is *known* to lie.

| $n$ | $A_n$ | $f(A_n)$ | $B_n$ | $f(B_n)$ | $x_n$ | $f(x_n)$ |
|---|---|---|---|---|---|---|
| 1 | 1.0000 | −4.0000 | 1.7000 | 5.4186 | 1.3500 | −2.1610 |
| 2 | 1.3500 | −2.1610 | 1.7000 | 5.4186 | 1.5250 | 0.5968 |
| 3 | 1.3500 | −2.1610 | 1.5250 | 0.5968 | 1.4375 | −0.9946 |
| 4 | 1.4375 | −0.9946 | 1.5250 | 0.5968 | 1.4813 | −0.2573 |
| 5 | 1.4813 | −0.2573 | 1.5250 | 0.5968 | 1.5031 | 0.1544 |
| 6 | 1.4813 | −0.2573 | 1.5031 | 0.1544 | 1.4922 | −0.0552 |
| 7 | 1.4922 | −0.0552 | 1.5031 | 0.1544 | 1.4977 | 0.0487 |
| 8 | 1.4922 | −0.0552 | 1.4977 | 0.0487 | 1.4949 | −0.0085 |

Table 25.3 Successive approximations to the root of (25.1) using binary chopping.

### *25.1.3 Binary chopping*

Again two values of $x$, $A_1$ and $B_1$, that straddle the root are chosen, with $A_1 < B_1$ and $f(A_1)$ and $f(B_1)$ having opposite signs. The interval between them is then halved by forming

$$x_n = \tfrac{1}{2}(A_n + B_n), \tag{25.9}$$

with $n = 1$, and $f(x_1)$ is evaluated. It should be noted that $x_1$ is determined solely by $A_1$ and $B_1$, and not by the values of $f(A_1)$ and $f(B_1)$ as in the linear interpolation method. Now $x_1$ is used to replace either $A_1$ or $B_1$, depending on which of $f(A_1)$ or $f(B_1)$ has the same sign as $f(x_1)$, i.e. if $f(A_1)$ and $f(x_1)$ have the same sign, $x_1$ replaces $A_1$. The process is then repeated to obtain $x_2$, $x_3$ etc.

This has been carried through in table 25.3 for our standard equation (25.1) and is illustrated in figure 25.2($c$). The entries have been rounded to four places of decimals. It is suggested that the reader follows through the sequential replacements of the $A_n$ and $B_n$ in the table and correlates the first few of these with graph ($c$) of figure 25.2.

Clearly the accuracy with which $\xi$ is known in this approach increases by only a factor of 2 at each step, but this accuracy is predictable at the outset of the calculation and (unless $f(x)$ has very violent behaviour near $x = \xi$) a range of $x$ in which $\xi$ lies can be safely stated at any stage. At the stage reached in the last line of table 25.3 it may be stated that $1.4949 < \xi < 1.4977$. Binary chopping thus gives a simple approximation method (it involves less multiplication than linear interpolation for example) that is predictable and relatively safe, although its convergence is slow.

### *25.1.4 Newton–Raphson method*

The Newton–Raphson (NR) procedure is somewhat similar to the interpolation method but, as will be seen, has one distinct advantage over it. Instead of

| $n$ | $x_n$ | $f(x_n)$ |
|---|---|---|
| 1 | 1.7 | 5.42 |
| 2 | 1.545 01 | 1.03 |
| 3 | 1.498 87 | $7.20 \times 10^{-2}$ |
| 4 | 1.495 13 | $4.49 \times 10^{-4}$ |
| 5 | 1.495 106 40 | $2.6 \times 10^{-8}$ |
| 6 | 1.495 106 40 | – |

Table 25.4 Successive approximations to the root of (25.1) using the Newton–Raphson method.

(notionally) constructing the chord between two points on the curve of $f(x)$ against $x$, the tangent to the curve is notionally constructed at each successive value of $x_n$ and the next value $x_{n+1}$ taken as the point at which the tangent cuts the axis $f(x) = 0$. This is illustrated in graph (*d*) of figure 25.2.

If the $n$th value is $x_n$, the tangent to the curve of $f(x)$ at that point has slope $f'(x_n)$ and passes through the point $x = x_n$, $y = f(x_n)$. Its equation is thus

$$y(x) = (x - x_n)f'(x_n) + f(x_n). \tag{25.10}$$

The value of $x$ at which $y = 0$ is then taken as $x_{n+1}$; thus the condition $y(x_{n+1}) = 0$ yields from (25.10) the iteration scheme

$$x_{n+1} = x_n - \frac{f(x_n)}{f'(x_n)}. \tag{25.11}$$

This is the *Newton–Raphson iteration formula.* Clearly, when $x_n$ is close to $\xi$, $x_{n+1}$ is close to $x_n$, as it should be. It is also apparent that if any of the $x_n$ comes close to a stationary point of $f$, so that $f'(x_n)$ is close to zero, the scheme is not going to work well.

For our standard example, (25.11) becomes

$$x_{n+1} = x_n - \frac{x_n^5 - 2x_n^2 - 3}{5x_n^4 - 4x_n} = \frac{4x_n^5 - 2x_n^2 + 3}{5x_n^4 - 4x_n}. \tag{25.12}$$

Again taking a starting value of $x_1 = 1.7$ we obtain in succession the entries in table 25.4. The different values are given to an increasing number of decimal places as the calculation proceeds; $f(x_n)$ is also recorded.

It is apparent that this method is unlike the previous ones in that the increase in accuracy of the answer is not constant throughout the iterations but improves dramatically as the required root is approached. Away from the root the behaviour of the series is less satisfactory and from its geometrical interpretation it can be seen that if, for example, there were a maximum or minimum near the root then the series could oscillate between values on either side of it (instead of 'homing in' on the root). The reason for the good convergence near the root is discussed in the next section.

Of the four methods mentioned, no single one is ideal, and in practice some mixture of them is usually to be preferred. The particular combination of methods selected will depend a great deal on how easily the progress of the calculation may be monitored, but some combination of the first three methods mentioned, followed by the NR scheme if great accuracy were required, would be suitable for most situations.

## 25.2 Convergence of iteration schemes

For iteration schemes in which $x_{n+1}$ can be expressed as a differentiable function of $x_n$, e.g. the rearrangement or NR methods of the previous section, a partial analysis of the conditions necessary for a successful scheme can be made as follows.

Suppose the general iteration formula is expressed as

$$x_{n+1} = F(x_n) \tag{25.13}$$

((25.7) and (25.12) are examples). Then the sequence of values $x_1, x_2, \ldots, x_n, \ldots$ is required to converge to the value $\xi$ which satisfies both

$$f(\xi) = 0 \qquad \text{and} \qquad \xi = F(\xi). \tag{25.14}$$

If the error in the solution at the $n$th stage is $\epsilon_n$, i.e. $x_n = \xi + \epsilon_n$, then

$$\xi + \epsilon_{n+1} = x_{n+1} = F(x_n) = F(\xi + \epsilon_n). \tag{25.15}$$

For the iteration process to converge, a decreasing error is required, i.e. $|\epsilon_{n+1}| < |\epsilon_n|$. To see what this implies about $F$, we expand the right-hand term of (25.15) by means of a Taylor series, and use (25.14) to replace (25.15) by

$$\xi + \epsilon_{n+1} = \xi + \epsilon_n F'(\xi) + \tfrac{1}{2}\epsilon_n^2 F''(\xi) + \cdots. \tag{25.16}$$

This shows that, for small $\epsilon_n$,

$$\epsilon_{n+1} \approx F'(\xi)\epsilon_n$$

and that a necessary (but not sufficient) condition for convergence is that

$$|F'(\xi)| < 1. \tag{25.17}$$

It should be noticed that this is a condition on $F'(\xi)$ and not on $f'(\xi)$, which

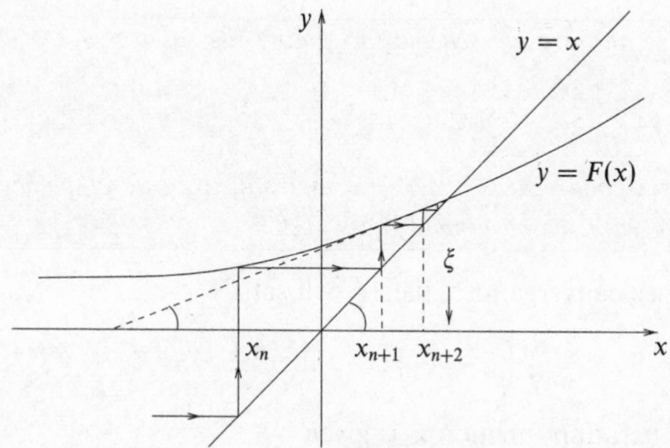


Figure 25.3 Illustration of the convergence of the iteration scheme $x_{n+1} = F(x_n)$ when $0 < F'(\xi) < 1$, where $\xi = F(\xi)$. The line $y = x$ makes an angle $\pi/4$ with the axes. The broken line makes an angle $\tan^{-1} F'(\xi)$ with the $x$-axis.

may have any finite value. Figure 25.3 illustrates in a graphical way how the convergence proceeds for the case $0 < F'(\xi) < 1$.

Equation (25.16) suggests that if $F(x)$ can be chosen so that $F'(\xi) = 0$ then the ratio $|\epsilon_{n+1}/\epsilon_n|$ could be made very small, of order $\epsilon_n$ in fact. To go even further, if it can be arranged that the first few derivatives of $F$ vanish at $x = \xi$, then the convergence, once $x_n$ has become close to $\xi$, could be very rapid indeed. If the first $N - 1$ derivatives of $F$ vanish at $x = \xi$, i.e.

$$F'(\xi) = F''(\xi) = \cdots = F^{(N-1)}(\xi) = 0 \qquad (25.18)$$

and consequently

$$\epsilon_{n+1} = O(\epsilon_n^N), \qquad (25.19)$$

then the scheme is said to have *Nth-order convergence*.

This is the explanation of the significant difference in convergence between the NR scheme and the others discussed (judged by reference to (25.19), so that the differentiability of the function $F$ is not a prerequisite). The NR procedure has second-order convergence as is shown by the following analysis. Since

$$\begin{aligned} F(x) &= x - f(x)/f'(x), \\ F'(x) &= 1 - f'(x)/f'(x) + f(x)f''(x)/[f'(x)]^2 \\ &= f(x)f''(x)/[f'(x)]^2. \end{aligned}$$

Now, provided $f'(\xi) \neq 0$, it follows that $F'(\xi) = 0$ because $f(x) = 0$ at $x = \xi$.

►*The following is an iteration scheme for finding the square root of X:*

$$x_{n+1} = \frac{1}{2}\left(x_n + \frac{X}{x_n}\right). \qquad (25.20)$$

*Show that it has second-order convergence and illustrate its efficiency by finding, say, $\sqrt{16}$ starting with the very poor guess $\sqrt{16} = 1$.*

If this scheme does converge to $\xi$, then $\xi$ will satisfy

$$\xi = \frac{1}{2}\left(\xi + \frac{X}{\xi}\right), \qquad \text{that is, } \xi^2 = X,$$

as required. The iteration function $F$ is given by

$$F(x) = \frac{1}{2}\left(x + \frac{X}{x}\right),$$

and so

$$F'(\xi) = \frac{1}{2}\left(1 - \frac{X}{x^2}\right)_{x=\xi} = 0 \qquad \text{since } \xi^2 = X,$$

whilst

$$F''(\xi) = \left(\frac{X}{x^3}\right)_{x=\xi} = \frac{1}{\xi} \neq 0.$$

Thus the procedure has second-order, but not third-order, convergence.

We now show the procedure in action. Table 25.5 gives successive values of $x_n$ and of $\epsilon_n$, the difference between $x_n$ and the true value of 4.

As we can see the scheme is crude initially, but once $x_n$ gets close to $\xi$, it homes in on the true value extremely rapidly. ◄

## 25.3 Simultaneous linear equations

As we saw in chapter 7, many situations in physical science can be described approximately or exactly by a set of $N$ simultaneous linear equations in $N$ variables (unknowns), $x_i$, $i = 1, 2, \ldots, N$. The equations take the general form

$$\begin{aligned}
A_{11}x_1 + A_{12}x_2 + \cdots + A_{1N}x_N &= b_1, \\
A_{21}x_1 + A_{22}x_2 + \cdots + A_{2N}x_N &= b_2, \\
&\ \vdots \\
A_{N1}x_1 + A_{N2}x_2 + \cdots + A_{NN}x_N &= b_N,
\end{aligned} \qquad (25.21)$$

where the $A_{ij}$ are constants and form the elements of a square matrix $\mathsf{A}$. The $b_i$ are given and form a column vector $\mathsf{b}$. If $\mathsf{A}$ is non-singular, (25.21) can be solved for the $x_i$ using the inverse of $\mathsf{A}$, according to the formula

$$\mathsf{x} = \mathsf{A}^{-1}\mathsf{b}.$$

| $n$ | $x_{n+1}$ | $\epsilon_n$ |
|---|---|---|
| 1 | 8.5 | 4.5 |
| 2 | 5.191 | 1.19 |
| 3 | 4.137 | $1.4 \times 10^{-1}$ |
| 4 | 4.002257 | $2.3 \times 10^{-3}$ |
| 5 | 4.000000637 | $6.4 \times 10^{-7}$ |
| 6 | 4 | – |

Table 25.5 Successive approximations to $\sqrt{16}$ using the iteration scheme (25.20).

This approach was discussed at length in chapter 7 and will not be considered further here.

### *25.3.1 Gaussian elimination*

We follow instead a continuation of one of the earliest techniques acquired by a student of algebra, namely the solving of simultaneous equations (initially only two in number) by the successive elimination of all of the variables but one. This (known as *Gaussian elimination*) is achieved by using, at each stage, one of the equations to obtain an explicit expression for one of the remaining $x_i$ in terms of the others, and then substituting for that $x_i$ in all other remaining equations. Eventually a single linear equation in just one of the unknowns is obtained. This is then solved and the result re-substituted in previously derived equations (in reverse order) to establish values for all the $x_i$.

This method is probably very familiar to the reader and so a specific example to illustrate this alone will not be given. Instead, we will show how a calculation along such lines might be arranged so that the errors due to the inherent lack of precision in any calculating equipment do not become excessive. This can happen if the value of $N$ is large and particularly (and we will merely state this) if the elements $A_{11}, A_{22}, \ldots, A_{NN}$ on the leading diagonal of the matrix in (25.21) are small compared to the off-diagonal elements.

The process to be described is known as *Gaussian elimination with interchange*. The only, but essential, difference from straightforward elimination is that before each variable $x_i$ is eliminated, the equations are reordered to put the largest (in modulus) remaining coefficient of $x_i$ on the leading diagonal.

We will take as an illustration a straightforward three-variable example, which can in fact be solved perfectly well without any interchange since, with simple numbers and only two eliminations to perform, rounding errors do not have a chance to build up. However, the important thing is that the reader should appreciate how this would apply in (say) a computer program for a 1000-variable case, perhaps with unforseeable zeroes or very small numbers appearing on the leading diagonal.

►*Solve the simultaneous equations*

$$\begin{aligned}
&(a) \quad x_1 + 6x_2 - 4x_3 = 8,\\
&(b) \quad 3x_1 - 20x_2 + x_3 = 12,\\
&(c) \quad -x_1 + 3x_2 + 5x_3 = 3.
\end{aligned} \qquad (25.22)$$

Firstly, we interchange rows (a) and (b) to bring the term $3x_1$ onto the leading diagonal. In the following, we label the important equations (I), (II), (III), and the others alphabetically.

$$\begin{aligned}
&(I) \quad 3x_1 - 20x_2 + x_3 = 12,\\
&(d) \quad x_1 + 6x_2 - 4x_3 = 8,\\
&(e) \quad -x_1 + 3x_2 + 5x_3 = 3.
\end{aligned}$$

For (j) = (d) and (e), replace row (j) by

$$\text{row (j)} - \frac{a_{j1}}{3} \times \text{row (I)},$$

where $a_{j1}$ is the coefficient of $x_1$ in row (j), to give the two equations

$$\begin{aligned}
&(II) \quad \left(6 + \tfrac{20}{3}\right) x_2 + \left(-4 - \tfrac{1}{3}\right) x_3 = 8 - \tfrac{12}{3},\\
&(f) \quad \left(3 - \tfrac{20}{3}\right) x_2 + \left(5 + \tfrac{1}{3}\right) x_3 = 3 + \tfrac{12}{3}.
\end{aligned}$$

Now $|6+\frac{20}{3}| > |3-\frac{20}{3}|$ and so no interchange is needed before the next elimination. To eliminate $x_2$, replace row (f) by

$$\text{row (f)} - \frac{\left(-\frac{11}{3}\right)}{\frac{38}{3}} \times \text{row (II)}.$$

This gives

$$(III) \quad \left[\frac{16}{3} + \frac{11}{38} \times \frac{(-13)}{3}\right] x_3 = 7 + \frac{11}{38} \times 4.$$

Collecting together and tidying up the final equations, we have

$$\begin{aligned}
&(I) \quad 3x_1 - 20x_2 + x_3 = 12,\\
&(II) \quad 38x_2 - 13x_3 = 12,\\
&(III) \quad x_3 = 2.
\end{aligned}$$

Starting with (III) and working backwards it is now a simple matter to obtain

$$x_1 = 10, \qquad x_2 = 1, \qquad x_3 = 2. \;◄$$

### *25.3.2 Gauss–Seidel iteration*

In the example considered in the previous subsection an explicit way of solving a set of simultaneous equations was given, the accuracy obtainable being limited only by the rounding errors in the calculating facilities available, and the calculation was planned to minimise these. However, in some situations it may be that only an approximate solution is needed. If, for a large number of variables, this is the case, an iterative method may produce a satisfactory degree of precision with less calculation. Such a method, known as *Gauss–Seidel iteration*, is based upon the following analysis.

The problem is again that of finding the components of the column vector $\mathsf{x}$ that satisfies

$$\mathsf{Ax} = \mathsf{b} \tag{25.23}$$

when $\mathsf{A}$ and $\mathsf{b}$ are a given matrix and column vector respectively.

The steps of the Gauss–Seidel scheme are as follows.

(i) Rearrange the equations (usually by simple division on both sides of each equation) so that all diagonal elements of the new matrix $\mathsf{C}$ are unity, i.e. (25.23) becomes

$$\mathsf{Cx} = \mathsf{d}, \tag{25.24}$$

where $\mathsf{C} = \mathsf{I} - \mathsf{F}$, and $\mathsf{F}$ has zeroes as its diagonal elements.

(ii) Step (i) produces

$$\mathsf{Fx} + \mathsf{d} = \mathsf{Ix} = \mathsf{x}, \tag{25.25}$$

and this forms the basis of an iteration scheme

$$\mathsf{x}_{n+1} = \mathsf{Fx}_n + \mathsf{d}, \tag{25.26}$$

where $\mathsf{x}_n$ is the $n$th approximation to the required solution vector $\boldsymbol{\xi}$.

(iii) To improve the convergence, the matrix $\mathsf{F}$, which has zeroes on its leading diagonal, can be written as the sum of two matrices $\mathsf{L}$ and $\mathsf{U}$ that have non-zero elements only below and above the leading diagonal respectively:

$$\mathsf{L}_{ij} = \begin{cases} \mathsf{F}_{ij} & \text{if } i > j, \\ 0 & \text{otherwise,} \end{cases}$$

$$\mathsf{U}_{ij} = \begin{cases} \mathsf{F}_{ij} & \text{if } i < j, \\ 0 & \text{otherwise.} \end{cases} \tag{25.27}$$

This allows the latest values of the components of $\mathsf{x}$ to be used at each stage and an improved form of (25.26) to be obtained,

$$\mathsf{x}_{n+1} = \mathsf{Lx}_{n+1} + \mathsf{Ux}_n + \mathsf{d}. \tag{25.28}$$

To see why this is possible we note, for example, that when calculating say the fourth component of $\mathsf{x}_{n+1}$, its first three components are already known, and, because of the structure of $\mathsf{L}$, these are the only ones that are needed to evaluate the fourth component of $\mathsf{L}\mathsf{x}_{n+1}$.

►*Obtain an approximate solution to the simultaneous equations*

$$\begin{array}{rrrcl} x_1 & +6x_2 & -4x_3 & = & 8, \\ 3x_1 & -20x_2 & +x_3 & = & 12, \\ -x_1 & +3x_2 & +5x_3 & = & 3. \end{array} \qquad (25.29)$$

*These are the same equations as were solved in subsection 25.3.1.*

Divide the equations by 1, −20 and 5 respectively to give

$$\begin{aligned} x_1 + 6x_2 - 4x_3 &= 8, \\ -0.15x_1 + x_2 - 0.05x_3 &= -0.6, \\ -0.2x_1 + 0.6x_2 + x_3 &= 0.6. \end{aligned}$$

Thus, set out in matrix form, (25.28) is in this case

$$\begin{pmatrix} x_1 \\ x_2 \\ x_3 \end{pmatrix}_{n+1} = \begin{pmatrix} 0 & 0 & 0 \\ 0.15 & 0 & 0 \\ 0.2 & -0.6 & 0 \end{pmatrix} \begin{pmatrix} x_1 \\ x_2 \\ x_3 \end{pmatrix}_{n+1} + \begin{pmatrix} 0 & -6 & 4 \\ 0 & 0 & 0.05 \\ 0 & 0 & 0 \end{pmatrix} \begin{pmatrix} x_1 \\ x_2 \\ x_3 \end{pmatrix}_{n} + \begin{pmatrix} 8 \\ -0.6 \\ 0.6 \end{pmatrix}.$$

Suppose initially ($n = 1$) we guess each component to have the value 2. Then the successive sets of values of the three quantities generated by this scheme are as shown in table 25.6. Even with the rather poor initial guess, a close approximation to the exact result $x_1 = 10$, $x_2 = 1$, $x_3 = 2$ has been obtained in only a few iterations. ◄

### 25.3.3 *Tridiagonal matrices*

Although for the solution of most matrix equations $\mathsf{A}\mathsf{x} = \mathsf{b}$ the number of operations needed increases rapidly with the size $N \times N$ of the matrix (roughly as $N^3$), for one particularly simple kind of matrix the computing required only increases linearly with $N$. This type often occurs in physical situations in which an ordered set of objects interact only with their nearest neighbours, and is one in which only the leading diagonal and the diagonals immediately above and below

| $n$ | $x_1$ | $x_2$ | $x_3$ |
|---|---|---|---|
| 1 | 2 | 2 | 2 |
| 2 | 4 | 0.1 | 1.34 |
| 3 | 12.76 | 1.381 | 2.323 |
| 4 | 9.008 | 0.867 | 1.881 |
| 5 | 10.321 | 1.042 | 2.039 |
| 6 | 9.902 | 0.987 | 1.988 |
| 7 | 10.029 | 1.004 | 2.004 |

Table 25.6 Successive approximations to the solution of simultaneous equations (25.29) using the Gauss–Seidel iteration method.

it contain non-zero entries. Such matrices are known as tridiagonal matrices. They may also be used in numerical approximations to the solutions of certain types of differential equation.

A typical matrix equation involving a tridiagonal matrix is thus

$$\begin{pmatrix} b_1 & c_1 & & & & \mathbf{0} \\ a_2 & b_2 & c_2 & & & \\ & a_3 & b_3 & c_3 & & \\ & & \ddots & \ddots & \ddots & \\ & & & a_{N-1} & b_{N-1} & c_{N-1} \\ \mathbf{0} & & & & a_N & b_N \end{pmatrix} \begin{pmatrix} x_1 \\ x_2 \\ x_3 \\ \vdots \\ x_{N-1} \\ x_N \end{pmatrix} = \begin{pmatrix} y_1 \\ y_2 \\ y_3 \\ \vdots \\ y_{N-1} \\ y_N \end{pmatrix} \qquad (25.30)$$

So as to keep the entries in the matrix as free from subscripts as possible, we have used $a$, $b$ and $c$ to indicate subdiagonal, leading diagonal and superdiagonal elements respectively. As a consequence we have had to change the notation for the column vector on the right-hand side from $\mathsf{b}$ to (say) $\mathsf{y}$.

In such an equation the first and last rows involve $x_1$ and $x_N$ respectively, and so the solution could be found by letting $x_1$ be unknown and then solving in turn each row of the equation in terms of $x_1$, and finally determining $x_1$ by requiring the next-to-last line to generate for $x_N$ an equation compatible with that given by the last line. However, if the matrix is large this becomes a very cumbersome operation and a simpler method is to assume a form of solution

$$x_{i-1} = \theta_{i-1}x_i + \phi_{i-1}. \qquad (25.31)$$

Since the $i$th line of the matrix equation is

$$a_i x_{i-1} + b_i x_i + c_i x_{i+1} = y_i,$$

we must have, by substituting for $x_{i-1}$, that

$$(a_i\theta_{i-1} + b_i)x_i + c_i x_{i+1} = y_i - a_i\phi_{i-1}.$$

This is also in the form of (25.31), but with $i$ replaced by $i+1$. Thus the recurrence formulae for $\theta_i$ and $\phi_i$ are

$$\theta_i = \frac{-c_i}{a_i\theta_{i-1} + b_i}, \qquad \phi_i = \frac{y_i - a_i\phi_{i-1}}{a_i\theta_{i-1} + b_i}, \tag{25.32}$$

provided the denominator does not vanish for any $i$. From the first of the matrix equations it follows that $\theta_1 = -c_1/b_1$ and $\phi_1 = y_1/b_1$. The equations may therefore be solved for the $x_i$ in two stages without carrying through an unknown quantity. First, all the $\theta_i$ and $\phi_i$ are generated using (25.32) and the values of $\theta_1$ and $\phi_1$, and then, as a second stage, (25.31) is used to evaluate the $x_i$, starting with $x_N$ $(= \phi_N)$ and working backwards.

►*Solve the following tridiagonal matrix equation, in which only non-zero elements are shown.*

$$\begin{pmatrix} 1 & 2 & & & & \\ -1 & 2 & 1 & & & \\ & 2 & -1 & 2 & & \\ & & 3 & 1 & 1 & \\ & & & 3 & 4 & 2 \\ & & & & -2 & 2 \end{pmatrix} \begin{pmatrix} x_1 \\ x_2 \\ x_3 \\ x_4 \\ x_5 \\ x_6 \end{pmatrix} = \begin{pmatrix} 4 \\ 3 \\ -3 \\ 10 \\ 7 \\ -2 \end{pmatrix}. \tag{25.33}$$

The solution is set out in table 25.7, in which the arrows indicate the general flow of the calculation. The columns of $a_i$, $b_i$, $c_i$ and $y_i$ are filled in first and then

| | $a_i$ | $b_i$ | $c_i$ | | $a_i\theta_{i-1} + b_i$ | $\theta_i$ | $y_i$ | $a_i\phi_{i-1}$ | $\phi_i$ | $x_i$ | |
|---|---|---|---|---|---|---|---|---|---|---|---|
| ↓ | 0 | 1 | 2 | → | 1 | −2 | 4 | 0 | 4 | 2 | ↑ |
| ↓ | −1 | 2 | 1 | → | 4 | −1/4 | 3 | −4 | 7/4 | 1 | ↑ |
| ↓ | 2 | −1 | 2 | → | −3/2 | 4/3 | −3 | 7/2 | 13/3 | 3 | ↑ |
| ↓ | 3 | 1 | 1 | → | 5 | −1/5 | 10 | 13 | −3/5 | −1 | ↑ |
| ↓ | 3 | 4 | 2 | → | 17/5 | −10/17 | 7 | −9/5 | 44/17 | 2 | ↑ |
| ↓ | −2 | 2 | 0 | → | 54/17 | 0 | −2 | −88/17 | 1 → | 1 | ↑ |

Table 25.7 The solution of tridiagonal matrix equation (25.33). The arrows indicate the general flow of the calculation, as described in the text.

the recurrence relations (25.32) are used to fill in the remainder of the columns to the left of the $x_i$-column line, starting from the top. Finally, the $x_i$-column is completed, working up from the bottom. ◄

## 25.4 Numerical integration

As noted at the start of this chapter, with modern computers and computer packages – some of which will present solutions in algebraic form, where that is possible – the inability to find a closed-form expression for an integral no longer presents a problem. But, just as for the solution of algebraic equations, it is extremely important that scientists and engineers should have some idea of the procedures on which such packages are based. In this section we discuss some of the more elementary methods used to evaluate integrals numerically and at the same time indicate the basis of more sophisticated procedures.

The standard integral evaluation has the form

$$I = \int_a^b f(x)\,dx, \qquad (25.34)$$

where the integrand $f(x)$ may be given in analytic or tabulated form, but for the cases under consideration no closed-form expression for $I$ can be obtained. Numerical evaluations of $I$ are all based on regarding it as the area under the curve of $f(x)$ between the limits $x = a$ and $x = b$, and attempting to estimate that area.

The simplest methods of doing this involve dividing up the interval $a \leq x \leq b$ into $N$ equal sections each of length $h = (b-a)/N$. The dividing points are labelled $x_i$ with $x_0 = a$, $x_N = b$, $i$ running from 0 to $N$. The point $x_i$ is a distance $ih$ from $a$. The central value of $x$ in a strip ($x = x_i + h/2$) is denoted for brevity by $x_{i+1/2}$, and for the same reason $f(x_i)$ is written as $f_i$. This nomenclature is indicated graphically in figure 25.4($a$).

So that we may compare later estimates of the area under the curve with the true value, we next obtain an exact expression for $I$ – even though we cannot evaluate it. To do this we need consider only one strip, say that between $x_i$ and $x_{i+1}$. For this strip the area is, using Taylor's expansion,

$$\begin{aligned}\int_{-h/2}^{h/2} f(x_{i+1/2}+y)\,dy &= \int_{-h/2}^{h/2} \sum_{n=0}^{\infty} f^{(n)}(x_{i+1/2})\frac{y^n}{n!}\,dy \\ &= \sum_{n=0}^{\infty} f_{i+1/2}^{(n)} \int_{-h/2}^{h/2} \frac{y^n}{n!}\,dy \\ &= \sum_{n\ \text{even}}^{\infty} f_{i+1/2}^{(n)} \frac{2}{(n+1)!}\left(\frac{h}{2}\right)^{n+1}. \qquad (25.35)\end{aligned}$$

It should be noticed that in this exact expression, only the even derivatives of $f$ survive the integration and that all derivatives are evaluated at $x_{i+1/2}$. Clearly other exact expressions are possible, e.g. the integral of $f(x_i + y)$ over the range $0 \leq y \leq h$, but we will find (25.35) the most useful for our purposes.

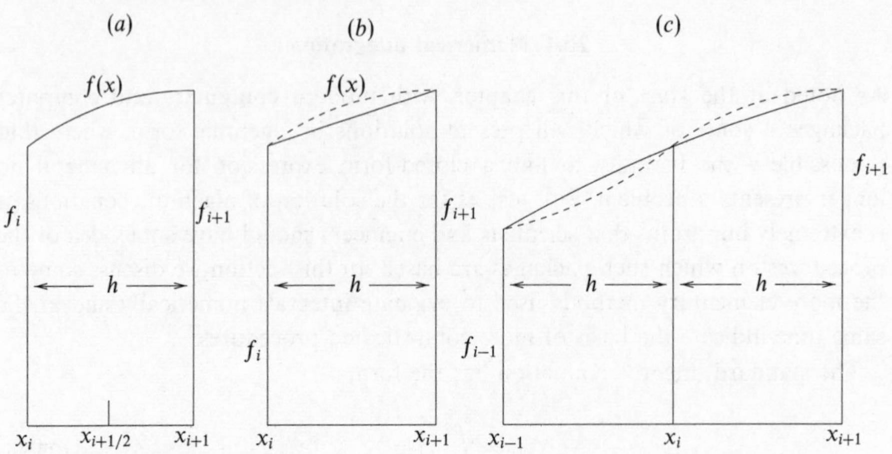


Figure 25.4 (*a*) Definition of nomenclature. (*b*) The approximation in using the trapezium rule; $f(x)$ is indicated by the broken curve. (*c*) Simpson's rule approximation; $f(x)$ is indicated by the broken curve. The solid curve is part of the approximating parabola.

We now turn to practical ways of approximating $I$, given the values of, or a means to calculate, $f_i$, for $i = 0, 1, \ldots, N$.

### 25.4.1 Trapezium rule

In this simple case the area shown in figure 25.4(*a*) is approximated as shown in figure 25.4(*b*), i.e. by a trapezium. The area $A_i$ of the trapezium is

$$A_i = \tfrac{1}{2}(f_i + f_{i+1})h, \tag{25.36}$$

and if such contributions from all strips are added together, the estimate of the total, and hence of $I$, is

$$I(\text{estim.}) = \sum_{i=0}^{N-1} A_i = \frac{h}{2}(f_0 + 2f_1 + 2f_2 + \cdots + 2f_{N-1} + f_N). \tag{25.37}$$

This provides a very simple expression for estimating integral (25.34); its accuracy is limited only by the extent to which $h$ can be made very small (and hence $N$ very large) without making the calculation excessively long. Clearly the estimate provided is only exact if $f(x)$ is a linear function of $x$.

The error made in calculating the area of the strip when the trapezium rule is used may be estimated as follows. The values used are $f_i$ and $f_{i+1}$, as in (25.36).

These can be accurately expressed in terms of $f_{i+1/2}$ and its derivatives by a Taylor series

$$f_{i+1/2\pm 1/2} = f_{i+1/2} \pm \frac{h}{2} f'_{i+1/2} + \frac{1}{2!}\left(\frac{h}{2}\right)^2 f''_{i+1/2} \pm \frac{1}{3!}\left(\frac{h}{2}\right)^3 f^{(3)}_{i+1/2} + \cdots.$$

Thus

$$\begin{aligned} A_i(\text{estim.}) &= \tfrac{1}{2}h(f_i + f_{i+1}), \\ &= h\left[f_{i+1/2} + \frac{1}{2!}\left(\frac{h}{2}\right)^2 f''_{i+1/2} + \mathrm{O}(h^4)\right], \end{aligned}$$

whilst, from the first few terms of the exact result (25.35),

$$A_i(\text{exact}) = hf_{i+1/2} + \frac{2}{3!}\left(\frac{h}{2}\right)^3 f''_{i+1/2} + \mathrm{O}(h^5).$$

Thus the error $\Delta A_i = A_i(\text{estim.}) - A_i(\text{exact})$ is given by

$$\begin{aligned} \Delta A_i &= \left(\tfrac{1}{8} - \tfrac{1}{24}\right) h^3 f''_{i+1/2} + \mathrm{O}(h^5) \\ &\approx \tfrac{1}{12} h^3 f''_{i+1/2}. \end{aligned}$$

The total error in $I(\text{estim.})$ is thus approximately given by

$$\Delta I(\text{estim.}) \approx \tfrac{1}{12} nh^3 \langle f'' \rangle = \tfrac{1}{12}(b-a)h^2\langle f''\rangle, \qquad (25.38)$$

where $\langle f'' \rangle$ represents an average value for the second derivative of $f$ over the interval $a$ to $b$.

►*Use the trapezium rule with $h = 0.5$ to evaluate*

$$I = \int_0^2 (x^2 - 3x + 4)\, dx,$$

*and, by evaluating the integral exactly, examine how well (25.38) estimates the error.*

With $h = 0.5$, we will need five values of $f(x) = x^2 - 3x + 4$ for use in formula (25.37). They are $f(0) = 4$, $f(0.5) = 2.75$, $f(1) = 2$, $f(1.5) = 1.75$ and $f(2) = 2$. Putting these into (25.37) gives

$$I(\text{estim.}) = \frac{0.5}{2}(4 \ + \ 2 \times 2.75 \ + \ 2 \times 2 \ + \ 2 \times 1.75 \ + \ 2) = 4.75.$$

The exact value is

$$I(\text{exact}) = \left[\frac{x^3}{3} - \frac{3x^2}{2} + 4x\right]_0^2 = 4\tfrac{2}{3}.$$

The difference between the estimate of the integral and the exact answer is $1/12$. Equation (25.38) estimates this error as $2 \times 0.25 \times \langle f'' \rangle/12$. Our (deliberately chosen!) integrand is one for which $\langle f'' \rangle$ can be trivially evaluated. Because $f(x)$ is a quadratic function of $x$, its second derivative is constant, and equal to 2 in this case. Thus $\langle f'' \rangle$ has value 2 and (25.38) estimates the error as $1/12$; that the estimate is exactly right should be no surprise since the Taylor expansion for a quadratic polynomial about any point always terminates after three terms and no higher-order terms in $h$ have been ignored in (25.38). ◄

### 25.4.2 *Simpson's rule*

Whereas the trapezium rule makes a linear interpolation of $f$, Simpson's rule effectively mimics the local variation of $f(x)$ using parabolas. The strips are treated two at a time (figure 25.4(c)) and therefore their number, $N$, should be made even.

In the neighbourhood of $x_i$ (where $i$ is odd) it is supposed that $f(x)$ can be adequately represented by a quadratic form

$$f(x_i + y) = f_i + ay + by^2. \tag{25.39}$$

In particular, applying this to $y = \pm h$ yields an expression for $b$,

$$\begin{aligned} f_{i+1} &= f(x_i + h) = f_i + ah + bh^2, \\ f_{i-1} &= f(x_i - h) = f_i - ah + bh^2. \end{aligned}$$

Thus

$$bh^2 = \tfrac{1}{2}(f_{i+1} + f_{i-1} - 2f_i).$$

Now, in the representation (25.39), the area of the double strip from $x_{i-1}$ to $x_{i+1}$ is given by

$$A_i(\text{estim.}) = \int_{-h}^{h} (f_i + ay + by^2)\,dy = 2hf_i + \tfrac{2}{3}bh^3.$$

Substituting for $bh^2$ then yields for the estimated area

$$\begin{aligned} A_i(\text{estim.}) &= 2hf_i + \tfrac{2}{3}h \times \tfrac{1}{2}(f_{i+1} + f_{i-1} - 2f_i) \\ &= \tfrac{1}{3}h(4f_i + f_{i+1} + f_{i-1}), \end{aligned}$$

an expression involving only given quantities. It should be noted that the value of $b$ (or $a$) need never be calculated.

For the full integral

$$I(\text{estim.}) = \frac{h}{3}\left( f_0 + f_N + 4\sum_{m\ \text{odd}} f_m + 2\sum_{m\ \text{even}} f_m \right). \tag{25.40}$$

It can be shown, by following the same procedure as in the trapezium rule case, that the error in the estimated area is approximately

$$\Delta I(\text{estim.}) \approx \frac{(b-a)}{180} h^4 \langle f^{(4)} \rangle.$$

### 25.4.3 Gaussian integration

In the cases considered in the previous two subsections, the function $f$ was mimicked by linear and quadratic functions. These yield exact answers if $f$ itself is a linear or quadratic function (respectively) of $x$. This process could be continued by increasing the order of the polynomial mimicking function so as to increase the accuracy with which more complicated functions $f$ could be numerically integrated; but the same effect can be achieved with less effort by not insisting upon equally spaced points $x_i$.

The detailed analysis of such methods of numerical integration, in which the integration points are not equally spaced and the weightings given to the values at each point do not fall into a few simple groups, is too long to be given here. The reader is referred to books devoted specifically to the theory of numerical analysis, where details of the integration points and weights for many schemes will be found.†

We will content ourselves here with describing Gaussian integration, which is based upon the orthogonality properties, in the interval $-1 \leq x \leq 1$, of the Legendre polynomials $P_\ell(x)$, discussed in subsection 14.6.2. In order to use these properties, the integral between limits $a$ and $b$ in (25.34) has to be changed to one between the limits $-1$ and $+1$. This is easily done with a change of variable from $x$ to $z$ given by

$$z = \frac{2x - b - a}{b - a},$$

so that $I$ becomes

$$I = \frac{(b-a)}{2} \int_{-1}^{1} g(z)\, dz, \tag{25.41}$$

in which $g(z) \equiv f(x)$.

The $n$ integration points $x_i$ for an $n$-point Gaussian integration are given by the zeroes of $P_n(x)$, i.e. the $x_i$ are such that $P_n(x_i) = 0$. The integrand $g(x)$ is mimicked by the $(n-1)$th-degree polynomial

$$G(x) = \sum_{i=1}^{n} \frac{P_n(x)}{(x - x_i)P_n'(x_i)} g(x_i),$$

† The points and weights may be found in, e.g. Abramowitz and Stegun, *Handbook of Mathematical Functions* (Dover, 1965).

which coincides with $g(x)$ at each of the points $x_i$, $i = 1, 2, \ldots, n$. To see this it should be noted that

$$\lim_{x \to x_k} \frac{P_n(x)}{(x - x_i)P_n'(x_i)} = \delta_{ik}.$$

It then follows, to the extent that $g(x)$ is well reproduced by $G(x)$, that

$$\int_{-1}^{1} g(x)\,dx \approx \sum_{i=1}^{n} \frac{g(x_i)}{P_n'(x_i)} \int_{-1}^{1} \frac{P_n(x)}{(x - x_i)}\,dx. \qquad (25.42)$$

The expression

$$w(x_i) \equiv \frac{1}{P_n'(x_i)} \int_{-1}^{1} \frac{P_n(x)}{(x - x_i)}\,dx$$

can be shown, using the properties of Legendre polynomials, to be equal to

$$w_i = \frac{2}{(1 - x_i^2)|P_n'(x_i)|^2},$$

and is the weighting to be attached to the contribution $g(x_i)$ in the sum (25.42), which becomes

$$\int_{-1}^{1} g(x)\,dx \approx \sum_{i=1}^{n} w_i g(x_i). \qquad (25.43)$$

In fact, because of the particular properties of Legendre polynomials, it can be shown that (25.43) integrates exactly any polynomial of degree up to $2n - 1$. The error in the approximate equality is of the order of the $2n$th derivative of $g$, and so, provided $g(x)$ is a reasonably smooth function, the approximation is a good one.

As an example, for a three-point integration, the three $x_i$ are the zeroes of $P_3(x) = \frac{1}{2}(5x^3 - 3x)$, namely 0 and $\pm 0.77460$, and the corresponding weights are

$$2\Big/\left[1 \times \left(-\tfrac{3}{2}\right)^2\right] = \tfrac{8}{9} \qquad \text{and} \qquad 2\Big/\left[(1 - 0.6) \times \left(\tfrac{6}{2}\right)^2\right] = \tfrac{5}{9}.$$

For other forms of integrand, formulae based on other sets of orthogonal functions give better results. For example, integrals over finite ranges involving factors of the forms $(1 - x^2)^{\pm 1/2}$ in the integrand are best treated using formulae based on Chebyshev polynomials, whilst infinite integrals containing $e^{-x}$ $(0 \le x < \infty)$ or $e^{-x^2}$ $(-\infty < x < \infty)$ are best handled using schemes based on Laguerre or Hermite polynomials respectively.

►*Using a three-point formula in each case, evaluate the integral*

$$I = \int_0^1 \frac{1}{1+x^2}\,dx,$$

*(i) using the trapezium rule, (ii) using Simpson's rule, (iii) using Gaussian integration. Also evaluate the integral analytically, and compare the results.*

(i) Using the trapezium rule, we obtain

$$\begin{aligned} I &= \tfrac{1}{2} \times \tfrac{1}{2} \left[ f(0) + 2f\left(\tfrac{1}{2}\right) + f(1) \right] \\ &= \tfrac{1}{4} \left[ 1 + \tfrac{8}{5} + \tfrac{1}{2} \right] = 0.7750. \end{aligned}$$

(ii) Using Simpson's rule, we obtain

$$\begin{aligned} I &= \tfrac{1}{3} \times \tfrac{1}{2} \left[ f(0) + 4f\left(\tfrac{1}{2}\right) + f(1) \right] \\ &= \tfrac{1}{6} \left[ 1 + \tfrac{16}{5} + \tfrac{1}{2} \right] = 0.7833. \end{aligned}$$

(iii) Using Gaussian integration, we obtain

$$\begin{aligned} I &= \frac{1-0}{2} \int_{-1}^{1} \frac{dz}{1 + \frac{1}{4}(z+1)^2} \\ &= \tfrac{1}{2} \left\{ 0.55556\,[f(-0.77460) + f(0.77460)] + 0.88889 f(0) \right\} \\ &= \tfrac{1}{2} \left\{ 0.55556\,[0.987458 + 0.559503] + 0.88889 \times 0.8 \right\} \\ &= 0.78527. \end{aligned}$$

(iv) Exact evaluation gives

$$I = \int_0^1 \frac{dx}{1+x^2} = \left[ \tan^{-1} x \right]_0^1 = \frac{\pi}{4} = 0.78540.$$

In practice a compromise has to be struck between the accuracy of the result achieved and the calculational labour that goes into obtaining it. ◄

## 25.5 Finite differences

It will have been noticed that the previous section included several equations linking sequential values of $f_i$ and the derivatives of $f$ evaluated at one of the $x_i$. In this section, by way of preparation for the numerical treatment of differential equations, we establish these relationships in a more systematic way.

Again we consider a set of values $f_i$ of a function $f(x)$ evaluated at equally spaced points $x_i$, the separation being $h$. The basis for our discussion will again be a Taylor series expansion, but on this occasion about the point $x_i$. It is

$$f_{i\pm1} = f_i \pm h f_i' + \frac{h^2}{2!} f_i'' \pm \frac{h^3}{3!} f_i^{(3)} + \cdots. \qquad (25.44)$$

In this, and subsequently, we denote the $n$th derivative evaluated at $x_i$ by $f_i^{(n)}$.

From (25.44), three different expressions that approximate $f_i^{(1)}$ can be derived. The first of these, obtained by subtracting the $\pm$ equations, is

$$f_i^{(1)} \equiv \left(\frac{df}{dx}\right)_{x_i} = \frac{f_{i+1} - f_{i-1}}{2h} - \frac{h^2}{3!}f_i^{(3)} - \cdots . \tag{25.45}$$

The quantity $(f_{i+1} - f_{i-1})/2h$ is known as the central difference approximation to $f_i^{(1)}$ and can be seen from (25.45) to be in error by approximately $(h^2/6)f_i^{(3)}$.

An alternative approximation, obtained from (25.44+) alone, is given by

$$f_i^{(1)} \equiv \left(\frac{df}{dx}\right)_{x_i} = \frac{f_{i+1} - f_i}{h} - \frac{h}{2!}f_i^{(2)} - \cdots . \tag{25.46}$$

The *forward difference*, $(f_{i+1} - f_i)/h$, is clearly a poorer approximation, since it is in error by approximately $(h/2)f_i^{(2)}$, as compared with $(h^2/6)f_i^{(3)}$. Similarly, the backward difference $(f_i - f_{i-1})/h$ obtained from (25.44−) is not as good as the central difference; the sign of the error is reversed in this case.

This type of differencing approximation can be continued to the higher derivatives of $f$ in an obvious manner. By adding the two equations (25.44±) a central difference approximation to $f_i^{(2)}$ can be obtained:

$$f_i^{(2)} \equiv \left(\frac{d^2f}{dx^2}\right) \approx \frac{f_{i+1} - 2f_i + f_{i-1}}{h^2}. \tag{25.47}$$

The error in this approximation (also known as the second difference of $f$) is easily shown to be about $(h^2/12)f_i^{(4)}$.

Of course, if the function $f(x)$ is a sufficiently simple polynomial in $x$, all derivatives beyond a particular one will vanish and there is no error in taking the differences to obtain the derivatives.

►*The following is copied from the tabulation of a second-degree polynomial $f(x)$ at values of $x$ from* 1 *to* 12 *inclusive,*

2, 2, ?, 8, 14, 22, 32, 46, ?, 74, 92, 112.

*The entries marked* ? *were illegible and in addition one error was made in transcription. Complete and correct the table. Would your procedure have worked if the copying error had been in $f(6)$?*

Write out the entries again in row (a) below, and where possible calculate first differences in row (b) and second differences in row (c). Denote the $j$th entry in row $(n)$ by $(n)_j$.

| | | | | | | | | | | | | |
|---|---|---|---|---|---|---|---|---|---|---|---|---|
| (a) | 2 | 2 | ? | 8 | 14 | 22 | 32 | 46 | ? | 74 | 92 | 112 |
| (b) | 0 | ? | ? | 6 | 8 | 10 | 14 | ? | ? | 18 | 20 | |
| (c) | ? | ? | ? | 2 | 2 | 4 | ? | ? | ? | 2 | | |

Because the polynomial is second-degree the second differences $(c)_j$ (which are proportional to $d^2f/dx^2$) should be constant, and clearly the constant should be 2. That is, $(c)_6$ should equal 2 and $(b)_7$ should equal 12 (not 14). Since all the $(c)_j = 2$, we can conclude that $(b)_2 = 2$, $(b)_3 = 4$, $(b)_8 = 14$, and $(b)_9 = 16$. Working these changes back to row (a) shows that $(a)_3 = 4$, $(a)_8 = 44$ (not 46), and $(a)_9 = 58$.

The entries therefore should read

(a) 2, 2, **4**, 8, 14, 22, 32, **44**, **58**, 74, 92, 112,

where the amended entries are shown in bold type.

It is easily verified that if the error were in $f(6)$ no two computable entries in row (c) would be equal, and it would not be clear what the correct common entry should be. Trial and error might arrive at a self-consistent scheme. ◄

## 25.6 Differential equations

For the remaining sections of this chapter our attention will be on the solution of differential equations by numerical methods. Some of the general difficulties of applying numerical methods to differential equations will be all too apparent. Initially we consider only the simplest kind of equation – one of first order, typically represented by

$$\frac{dy}{dx} = f(x, y), \tag{25.48}$$

where $y$ is taken as the dependent variable and $x$ the independent one. If this equation can be solved analytically then that is the best course to adopt. But sometimes it is not possible to do so and a numerical approach becomes the only one available. Most of the examples we will use can in fact be solved easily by an explicit integration, but, for the purposes of illustration, this is an advantage rather than the reverse, since useful comparisons can then be made between the numerically derived solution and the exact one.

### *25.6.1 Difference equations*

Consider the differential equation

$$\frac{dy}{dx} = -y, \qquad y(0) = 1, \tag{25.49}$$

and the possibility of solving it numerically by approximating $dy/dx$ by a finite difference along the lines indicated in section 25.5. We start with the forward difference

$$\left(\frac{dy}{dx}\right)_{x_i} \approx \frac{y_{i+1} - y_i}{h}, \tag{25.50}$$

| $x$ | $h$ | | | | | | | y(exact) |
|---|---|---|---|---|---|---|---|---|
| | 0.01 | 0.1 | 0.5 | 1.0 | 1.5 | 2 | 3 | |
| 0 | (1) | (1) | (1) | (1) | (1) | (1) | (1) | (1) |
| 0.5 | 0.605 | 0.590 | 0.500 | 0 | −0.500 | −1 | −2 | 0.607 |
| 1.0 | 0.366 | 0.349 | 0.250 | 0 | 0.250 | 1 | 4 | 0.368 |
| 1.5 | 0.221 | 0.206 | 0.125 | 0 | −0.125 | −1 | −8 | 0.223 |
| 2.0 | 0.134 | 0.122 | 0.063 | 0 | 0.063 | 1 | 16 | 0.135 |
| 2.5 | 0.081 | 0.072 | 0.032 | 0 | −0.032 | −1 | −32 | 0.082 |
| 3.0 | 0.049 | 0.042 | 0.016 | 0 | 0.016 | 1 | 64 | 0.050 |

Table 25.8 The solution of differential equation (25.49) using the Euler forward difference method for various values of $h$. The exact solution is also shown.

where we have used the notation of section 25.5, but with $f$ replaced by $y$. In this particular case, it would lead to the recurrence relation,

$$y_{i+1} = y_i + h\left(\frac{dy}{dx}\right)_i = y_i - hy_i = (1-h)y_i. \tag{25.51}$$

Thus, since $y_0 = y(0) = 1$ is given, $y_1 = y(0+h) = y(h)$ can be calculated, and so on (this is the *Euler* method). Table 25.8 shows the values of $y(x)$ obtained if this is done using various values of $h$, for selected values of $x$. The exact solution, $y(x) = \exp(-x)$, is also shown.

It is clear that to maintain anything like a reasonable accuracy only very small steps $h$ can be used. Indeed, if $h$ is taken to be too large, not only is the accuracy bad but, as can be seen, for $h > 1$ the calculated solution oscillates (when it should be monotonic) and for $h > 2$ it diverges. Equation (25.51) is of the form $y_{i+1} = \lambda y_i$, and a necessary condition for non-divergence is $|\lambda| < 1$, i.e. $0 < h < 2$, though this in no way ensures accuracy.

Part of this difficulty arises from the poor approximation (25.50); its right-hand side is a closer approximation to $dy/dx$ evaluated at $x = x_i + h/2$ than to $dy/dx$ at $x = x_i$. This is the result of using a forward difference rather than the more accurate, but of course still approximate, central difference. This more accurate method (*Milne's method*) gives the recurrence relation

$$y_{i+1} = y_{i-1} + 2h\left(\frac{dy}{dx}\right)_i \tag{25.52}$$

in general, and, in this particular case,

$$y_{i+1} = y_{i-1} - 2hy_i. \tag{25.53}$$

An additional difficulty now arises, since two initial values of $y$ are needed. The second must be estimated by other means (e.g. by using a Taylor series, as discussed later) but for illustration purposes we will take the accurate value of

| $x$ | $y$(estim.) | $y$(exact) |
|---|---|---|
| −0.5 | (1.648) | – |
| 0 | (1.000) | (1.000) |
| 0.5 | 0.648 | 0.607 |
| 1.0 | 0.352 | 0.368 |
| 1.5 | 0.296 | 0.223 |
| 2.0 | 0.056 | 0.135 |
| 2.5 | 0.240 | 0.082 |
| 3.0 | −0.184 | 0.050 |

Table 25.9 The solution of differential equation (25.49) using the Milne central difference method with $h = 0.5$ and accurate starting values.

$y(-h) = \exp h$ as the value of $y_{-1}$. If $h$ is taken as, say, 0.5 and (25.53) applied repeatedly, the results shown in table 25.9 are obtained.

Although some improvement in the early values of the calculated $y(x)$ is noticeable, as compared to the corresponding ($h = 0.5$) column of table 25.8, this scheme soon runs into difficulties, as is obvious from the last two lines.

Some part of this poor performance is not really attributable to the approximations made in estimating $dy/dx$ but to the form of the equation itself. *Any* rounding error occurring in the evaluation effectively introduces a small amount of the solution of

$$\frac{dy}{dx} = +y$$

into $y$. This equation has solution $y(x) = \exp x$ and will ultimately dominate the sought-for solution rendering the calculations totally inaccurate.

We have only illustrated, rather than analysed, some of the difficulties associated with simple finite-difference iteration schemes for first-order differential equations, but they may be summarised as (i) insufficiently precise approximations to the derivatives, and (ii) inherent instability due to rounding errors.

### 25.6.2 Taylor series solutions

Since a Taylor series expansion is exact if all its terms are included, and the limits of convergence are not exceeded, we may seek to use one to evaluate $y_1$, $y_2$ etc. for an equation

$$\frac{dy}{dx} = f(x, y), \tag{25.54}$$

when the initial value $y(x_0) = y_0$ is given.

The Taylor series is

$$y(x + h) = y(x) + hy'(x) + \frac{h^2}{2!}y''(x) + \frac{h^3}{3!}y^{(3)}(x) + \cdots . \tag{25.55}$$

In the present notation, at the point $x = x_i$ this is

$$y_{i+1} = y_i + hy_i^{(1)} + \frac{h^2}{2!}y_i^{(2)} + \frac{h^3}{3!}y_i^{(3)} + \cdots. \tag{25.56}$$

But, for the required solution $y(x)$,

$$y_i^{(1)} = \left(\frac{dy}{dx}\right)_{x_i} = f(x_i, y_i) \tag{25.57}$$

is given, and the second derivative can be obtained from it,

$$y_i^{(2)} = \frac{\partial f}{\partial x} + \frac{\partial f}{\partial y}\frac{dy}{dx} = \frac{\partial f}{\partial x} + f\frac{\partial f}{\partial y}, \tag{25.58}$$

evaluated at $x = x_i$, $y = y_i$. This process can be continued for the third and higher derivatives, all of which are to be evaluated at $(x_i, y_i)$.

Having obtained expressions for the derivatives $y_i^{(n)}$ in (25.55), two alternative ways of proceeding are open.

(i) Equation (25.56) is used to evaluate $y_{i+1}$ and then the whole process is repeated to obtain $y_{i+2}$, and so on.

(ii) Equation (25.56) is applied several times, but using a different value of $h$ each time and so the corresponding values of $y(x + h)$ are obtained.

It is clear that, on the one hand, approach (i) does not require so many terms of (25.55) to be kept but, on the other hand, the $y_i(n)$ have to be recalculated at each step. With approach (ii), fairly accurate results for $y$ may be obtained for values of $x$ close to the given starting value, but for large values of $h$ a large number of terms of (25.55) must be retained. As an example of approach (ii) we solve the following problem.

►*Find the numerical solution of the equation*

$$\frac{dy}{dx} = 2y^{3/2}, \qquad y(0) = 1, \tag{25.59}$$

*for* $x = 0.1$ *to* $0.5$ *in steps of* $0.1$. *Compare it with the exact solution obtained analytically.*

Since the right-hand side of the equation does not contain $x$ explicitly, (25.58) is greatly simplified and the calculation becomes a repeated application of

$$y_i^{(n+1)} = \frac{\partial y^{(n)}}{\partial y}\frac{dy}{dx} = f\frac{\partial y^{(n)}}{\partial y}.$$

| $x$ | $y$(estim.) | $y$(exact) |
|---|---|---|
| 0 | 1.0000 | 1.0000 |
| 0.1 | 1.2346 | 1.2346 |
| 0.2 | 1.5619 | 1.5625 |
| 0.3 | 2.0331 | 2.0408 |
| 0.4 | 2.7254 | 2.7778 |
| 0.5 | 3.7500 | 4.0000 |

Table 25.10 The solution of differential equation (25.59) using a Taylor series.

The necessary derivatives and their values at $x = 0$, where $y = 1$, are as follows:

$$\begin{array}{ll}
y(0) = 1 & 1 \\
y' = 2y^{3/2} & 2 \\
y'' = (3/2)(2y^{1/2})(2y^{3/2}) = 6y^2 & 6 \\
y^{(3)} = (12y)2y^{3/2} = 24y^{5/2} & 24 \\
y^{(4)} = (60y^{3/2})2y^{3/2} = 120y^3 & 120 \\
y^{(5)} = (360y^2)2y^{3/2} = 720y^{7/2} & 720.
\end{array}$$

Thus the Taylor expansion of the solution about the origin (thus in fact a Maclaurin series) is

$$y(x) = 1 + 2x + \frac{6}{2!}x^2 + \frac{24}{3!}x^3 + \frac{120}{4!}x^4 + \frac{720}{5!}x^5 + \cdots.$$

Hence, $y(\text{estim.}) = 1 + 2x + 3x^2 + 4x^3 + 5x^4 + 6x^5$. Values calculated from this are given in table 25.10. Comparison with the exact values shows that using the first six terms gives a value which is correct to one part in 100, up to $x = 0.3$. ◄

### 25.6.3 Prediction and correction

An improvement in the accuracy obtainable using difference methods is possible if steps are taken, sometimes retrospectively, to allow for inaccuracies in approximating derivatives by differences. We will describe only the simplest schemes of this kind and begin with a *prediction* method, usually called the *Adams method*.

The forward difference estimate of $y_{i+1}$, namely

$$y_{i+1} = y_i + h\left(\frac{dy}{dx}\right)_i = y_i + hf(x_i, y_i), \qquad (25.60)$$

would give exact results if $y$ were a linear function of $x$ in the range $x_i \le x \le x_i + h$. The idea behind the Adams method is to allow some relaxation of this and suppose that $y$ can be adequately approximated by a parabola over the interval

$x_{i-1} \leq x \leq x_{i+1}$. In the same interval $dy/dx$ can then be approximated by a linear function:

$$f(x, y) = \frac{dy}{dx} \approx a + b(x - x_i), \qquad \text{for } x_i - h \leq x \leq x_i + h.$$

The values of $a$ and $b$ are fixed by the calculated values of $f$ at $x_{i-1}$ and $x_i$, which we may denote by $f_{i-1}$ and $f_i$:

$$a = f_i, \qquad b = \frac{f_i - f_{i-1}}{h}.$$

Thus

$$y_{i+1} - y_i \approx \int_{x_i}^{x_i+h} \left[ f_i + \frac{(f_i - f_{i-1})}{h}(x - x_i) \right] dx,$$

which yields

$$y_{i+1} = y_i + hf_i + \tfrac{1}{2}h(f_i - f_{i-1}). \tag{25.61}$$

The last term of this expression is seen to be a correction to result (25.60). That it is, in some sense, the second-order correction

$$\tfrac{1}{2}h^2 y_{i-1/2}^{(2)}$$

to a first-order formula is apparent.

Such a procedure requires, in addition to a value for $y_0$, a value for either $y_1$ or $y_{-1}$, so that $f_1$ or $f_{-1}$ can be used to initiate the iteration. This has to be obtained by other methods, e.g. a Taylor series expansion.

Improvements to simple difference formulae can also be obtained by using *correction* methods. Here a rough prediction of the value $y_{i+1}$ is first made, and then this is used in a better formula, not originally usable since this formula itself requires a rough value of $y_{i+1}$ for its evaluation. The value of $y_{i+1}$ is then recalculated using this better formula.

Such a scheme based on the forward difference formula might be as follows:

(i) predict $y_{i+1}$ using $y_{i+1} = y_i + hf_i$;
(ii) calculate $f_{i+1}$ using this value;
(iii) recalculate $y_{i+1}$ using $y_{i+1} = y_i + h(f_i + f_{i+1})/2$. Here $(f_i + f_{i+1})/2$ has replaced the $f_i$ used in (i), since it better represents the average value of $dy/dx$ in the interval $x_i \leq x \leq x_i + h$.

Steps (ii) and (iii) can be iterated to improve further the approximation to the average value of $dy/dx$, but will not compensate for higher-order derivatives omitted in the forward difference formula.

Many more complex schemes of prediction and correction, in most cases combining the two in the same process, have been devised, but the reader is referred to more specialist texts for discussions of them. However, because it offers some clear advantages, one group of methods will be set out explicitly.

This is the general class of schemes known as Runge–Kutta methods, discussed in some detail in the next subsection.

### *25.6.4 Runge–Kutta methods*

The Runge–Kutta method of integrating

$$\frac{dy}{dx} = f(x, y), \tag{25.62}$$

is a step-by-step process of obtaining an approximation for $y_{i+1}$ starting from the value of $y_i$. Among its advantages are that no functions other than $f$ are used, no subsidiary differentiation is needed, and no additional starting values need be calculated.

To be set against these advantages is the fact that $f$ is evaluated using somewhat complicated arguments and that this has to be done several times for each increase in the value of $i$. However, once a procedure has been established, for example on a computer, the method usually gives good results.

The basis of the method is to simulate the (accurate) Taylor series for $y(x_i + h)$, not by calculating all the higher derivatives of $y$ at the point $x_i$, but by taking a particular combination of the values of the first derivative of $y$ evaluated at a number of carefully chosen points. Equation (25.62) is used to evaluate these derivatives. The accuracy can be made to be up to whatever power of $h$ is desired but, naturally, the greater the accuracy the more complex the calculation and, in any case, rounding errors cannot ultimately be avoided.

The setting up of the calculational scheme may be illustrated by considering the particular case in which second-order accuracy in $h$ is required. To second order, the Taylor expansion is

$$y_{i+1} = y_i + hf_i + \frac{h^2}{2}\left(\frac{df}{dx}\right)_{x_i}, \tag{25.63}$$

where

$$\left(\frac{df}{dx}\right)_{x_i} = \left(\frac{\partial f}{\partial x} + f\frac{\partial f}{\partial y}\right)_{x_i} \equiv \frac{\partial f_i}{\partial x} + f_i\frac{\partial f_i}{\partial y},$$

the last step being merely the definition of an abbreviated notation.

We assume that this can be simulated by a form

$$y_{i+1} = y_i + \alpha_1 hf_i + \alpha_2 hf(x_i + \beta_1 h,\ y_i + \beta_2 hf_i), \tag{25.64}$$

i.e., by in effect using a weighted mean of the value of $dy/dx$ at $x_i$ and its value at some point yet to be determined. The object is to choose values of $\alpha_1$, $\alpha_2$, $\beta_1$ and $\beta_2$ such that (25.64) coincides with (25.63) up to the coefficient of $h^2$.

Expanding the function $f$ in the last term of (25.64) in a Taylor series of its own we obtain

$$f(x_i + \beta_1 h,\ y_i + \beta_2 h f_i) = f(x_i, y_i) + \beta_1 h \frac{\partial f_i}{\partial x} + \beta_2 h f_i \frac{\partial f_i}{\partial y} + \mathrm{O}(h^2).$$

Putting this result into (25.64) and rearranging in powers of $h$ we obtain

$$y_{i+1} = y_i + (\alpha_1 + \alpha_2) h f_i + \alpha_2 h^2 \left( \beta_1 \frac{\partial f_i}{\partial x} + \beta_2 f_i \frac{\partial f_i}{\partial y} \right). \qquad (25.65)$$

Comparing this with (25.63) shows that there is in fact some freedom remaining in the choice of the $\alpha$'s and $\beta$'s. In terms of an arbitrary $\alpha_1$ ($\neq 1$)

$$\alpha_2 = 1 - \alpha_1, \qquad \beta_1 = \beta_2 = \frac{1}{2(1-\alpha_1)}.$$

One possible choice is $\alpha_1 = 0.5$, and then $\alpha_2 = 0.5$, $\beta_1 = \beta_2 = 1$. In this case the procedure (equation (25.64)) can be summarised by

$$y_{i+1} = y_i + \tfrac{1}{2}(a_1 + a_2), \qquad (25.66)$$

where

$$a_1 = hf(x_i, y_i),$$
$$a_2 = hf(x_i + h, y_i + a_1).$$

Similar schemes giving higher-order accuracy in $h$ can be devised. Two such schemes, given without derivation, are

(i) to order $h^3$,

$$y_{i+1} = y_i + \tfrac{1}{6}(b_1 + 4b_2 + b_3), \qquad (25.67)$$

where

$$b_1 = hf(x_i, y_i),$$
$$b_2 = hf(x_i + \tfrac{1}{2}h, y_i + \tfrac{1}{2}b_1),$$
$$b_3 = hf(x_i + h, y_i + 2b_2 - b_1),$$

(ii) to order $h^4$,

$$y_{i+1} = y_i + \tfrac{1}{6}(c_1 + 2c_2 + 2c_3 + c_4), \qquad (25.68)$$

where

$$c_1 = hf(x_i, y_i),$$
$$c_2 = hf(x_i + \tfrac{1}{2}h, y_i + \tfrac{1}{2}c_1),$$
$$c_3 = hf(x_i + \tfrac{1}{2}h, y_i + \tfrac{1}{2}c_2),$$
$$c_4 = hf(x_i + h, y_i + c_3).$$

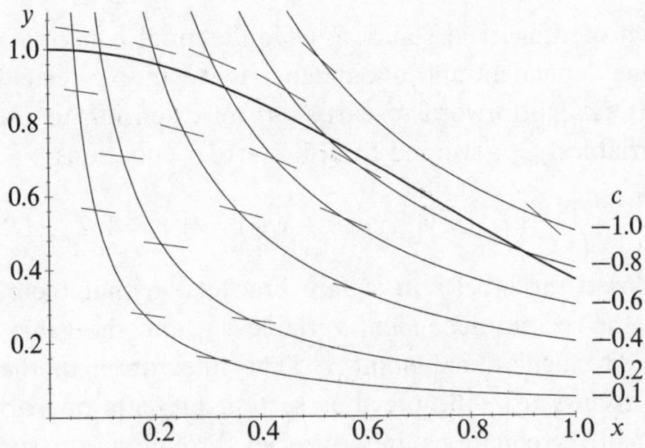


Figure 25.5 The isocline method. The cross lines on each isocline show the slopes that solutions of $dy/dx = -2xy$ must have at the points where they cross the isoclines. The heavy line is the solution with $y(0) = 1$, namely $\exp(-x^2)$.

### 25.6.5 *Isoclines*

The final method to be described for first-order differential equations is not so much numerical as graphical, but since it is sometimes useful it is included here. The method, known as that of *isoclines*, is to sketch for a number of values of a constant $c$ those curves (the isoclines) in the $xy$-plane along which $f(x, y) = c$, i.e. those curves along which $dy/dx$ is a constant of known value. It should be noted that they are not generally straight lines. Since a straight line of slope $dy/dx$ at and through any particular point is a tangent to the curve $y = y(x)$ at that point, small elements of straight lines with slopes appropriate to the isoclines they cut, effectively form the curve $y = y(x)$.

Figure 25.5 illustrates in outline the method as applied to the solution of

$$\frac{dy}{dx} = -2xy. \tag{25.69}$$

The thinner curves (rectangular hyperbolae) are a selection of the isoclines along which $-2xy$ is constant and equal to the corresponding value of $c$. The small cross lines on each curve show the slopes ($= c$) that solutions of (25.69) must have if they cross the curve. The thick line is the solution which has value $y = 1$ at $x = 0$; it takes the slope dictated by the value of $c$ on each isocline it crosses. The analytic solution with these properties is $y(x) = \exp(-x^2)$.

## 25.7 Higher-order equations

The discussion of numerical solutions of differential equations has so far been in terms of one dependent and one independent variable related by a first-order equation. It is straightforward to carry out an extension to the case of several dependent variables $y_{[r]}$ governed by $R$ first-order equations

$$\frac{dy_{[r]}}{dx} = f_{[r]}(x, y_{[1]}, y_{[2]}, \ldots, y_{[R]}), \qquad r = 1, 2, \ldots, R.$$

We have enclosed the label $r$ in square brackets so that there is no confusion between, say, the second dependent variable $y_{[2]}$ and the value $y_2$ of a variable $y$ at the second calculational point $x_2$. The integration of these equations by the methods discussed in the previous section presents no particular difficulty, provided that all the equations are advanced through any particular step before any one of them is taken through the following step.

Higher-order equations in one dependent and one independent variable can be reduced to a set of simultaneous equations, provided that they can be written in the form

$$\frac{d^R y}{dx^R} = f(x, y, y', \ldots, y^{(R-1)}), \qquad (25.70)$$

where $R$ is the order of the equation. To do this, a new set of variables $p_{[r]}$ is defined by

$$p_{[r]} = \frac{d^r y}{dx^r}, \qquad r = 1, 2, \ldots, R-1. \qquad (25.71)$$

Equation (25.70) is then equivalent to the set of simultaneous first-order equations

$$\begin{aligned} \frac{dy}{dx} &= p_{[1]}, \\ \frac{dp_{[r]}}{dx} &= p_{[r+1]}, \qquad r = 1, 2, \ldots, R-2, \\ \frac{dp_{[R-1]}}{dx} &= f(x, y, p_{[1]}, \ldots, p_{[R-1]}). \end{aligned} \qquad (25.72)$$

These can then be treated in the way indicated in the previous paragraph. The extension to more than one dependent variable is straightforward.

In practical problems it often happens that boundary conditions applicable to a higher-order equation do not consist of the function and all its derivatives at one particular point, but of, say, the value of the function at two separate (end-) points. In these cases a solution cannot be found using an explicit step-by-step 'marching' scheme, in which the solution at successive values of the independent variable are calculated using solution values previously found. Other methods have to be tried.

One obvious method is to treat the problem as a 'marching one', but use a number of (intelligently guessed) initial values for the derivatives at the starting point. The aim is then to find, by interpolation or some other form of iteration, those starting values for the derivatives which produce the given value of the function at the finishing point.

In some cases the problem can be reduced by a differencing scheme to a matrix equation. Such a case is that of a second-order equation for $y(x)$ with constant coefficients and given values of $y$ at the two end-points. Consider the second-order equation

$$y'' + 2ky' + \mu y = f(x), \tag{25.73}$$

with the boundary conditions

$$y(0) = A, \qquad y(1) = B.$$

If (25.73) is replaced by a central difference equation,

$$\frac{y_{i+1} - 2y_i + y_{i-1}}{h^2} + 2k\frac{y_{i+1} - y_{i-1}}{2h} + \mu y_i = f(x_i),$$

we obtain from it the recurrence relation

$$(1 + kh)y_{i+1} + (\mu h^2 - 2)y_i + (1 - kh)y_{i-1} = h^2 f(x_i).$$

For $h = 1/(N-1)$ this is in exactly the form of the $N \times N$ tridiagonal matrix equation (25.30), with

$$b_1 = b_N = 1, \qquad c_1 = a_N = 0, \qquad a_i = 1 - kh,$$

$$b_i = \mu h^2 - 2, \qquad c_i = 1 + kh, \qquad i = 2, 3, \ldots, N-1,$$

and the $y_i$ replaced by $y_1 = A$, $y_N = B$ and $y_i = h^2 f(x_i)$ for $i = 2, 3, \ldots, N-1$. The solutions can be obtained as in (25.31) and (25.32).

## 25.8 Partial differential equations

The extension of previous methods to partial differential equations, thus involving two or more independent variables, proceeds in a more or less obvious way. Rather than an interval divided into equal steps by the points at which solutions to the equations are to be found, a mesh of points in two or more dimensions has to be set up and all the variables given an increased number of subscripts.

Considerations of the stability, accuracy and feasibility of particular calculational schemes are in principle the same as for the one-dimensional case, but in practice are too complicated to be discussed here.

Rather than note generalities that we are unable to pursue in any quantitative way, we will conclude this chapter by indicating in outline how two familiar partial differential equations of physical science can be set up for numerical solution. The first of these is Laplace's equation in two dimensions,

$$\frac{\partial^2 \phi}{\partial x^2} + \frac{\partial^2 \phi}{\partial y^2} = 0 \qquad (25.74)$$

with the value of $\phi$ given on the perimeter of a closed domain.

A grid with spacings $\Delta x$ and $\Delta y$ in the two directions is first chosen, so that, for example, $x_i$ stands for the point $x_0 + i\Delta x$ and $\phi_{i,j}$ for the value $\phi(x_i, y_j)$. Next, using a second central difference formula, (25.74) is turned into

$$\frac{\phi_{i+1,j} - 2\phi_{i,j} + \phi_{i-1,j}}{(\Delta x)^2} + \frac{\phi_{i,j+1} - 2\phi_{i,j} + \phi_{i,j-1}}{(\Delta y)^2} = 0, \qquad (25.75)$$

for $i = 0, 1, \ldots, N$ and $j = 0, 1, \ldots, M$. If $(\Delta x)^2 = \lambda(\Delta y)^2$, this becomes the recurrence relationship

$$\phi_{i+1,j} + \phi_{i-1,j} + \lambda(\phi_{i,j+1} + \phi_{i,j-1}) = 2(1+\lambda)\phi_{i,j}. \qquad (25.76)$$

The boundary conditions in their simplest form (for a rectangular domain) mean that

$$\phi_{0,j}, \quad \phi_{N,j}, \quad \phi_{i,0}, \quad \phi_{i,M} \qquad (25.77)$$

have predetermined values. Non-rectangular boundaries can be accommodated, either by more complex boundary value prescriptions or by using non-Cartesian coordinates.

To find a set of values satisfying (25.76), an initial guess of a complete set of values for $\phi_{i,j}$ is made, subject to the requirement that the quantities listed in (25.77) have the given fixed values, and then those values that are not on the boundary are iteratively adjusted in order to try to bring about condition (25.76) everywhere. Clearly one scheme is to set $\lambda = 1$ and recalculate each $\phi_{i,j}$ as the mean of the four current values at neighbouring grid-points, using (25.76) directly, and then iterate this recalculation until no value of $\phi$ changes significantly after a complete cycle through all values of $i$ and $j$. This procedure is the most simplistic of such 'relaxation' methods; the reader is referred to more specialist books for fuller accounts of how this approach can be made faster and more accurate.

Our final example is based upon the one-dimensional diffusion equation for the temperature $\phi$ of a system,

$$\frac{\partial \phi}{\partial t} = \kappa \frac{\partial^2 \phi}{\partial x^2}. \qquad (25.78)$$

If $\phi_{i,j}$ stands for $\phi(x_0 + i\Delta x,\ t_0 + j\Delta t)$, then a forward difference representation of the time derivative and a central difference one for the spatial derivative lead to the following relationship:

$$\frac{\phi_{i,j+1} - \phi_{i,j}}{\Delta t} = \kappa \frac{\phi_{i+1,j} - 2\phi_{i,j} + \phi_{i-1,j}}{(\Delta x)^2}. \qquad (25.79)$$

This allows the construction of an explicit scheme for generating the temperature distribution at later times, given it at an earlier time:

$$\phi_{i,j+1} = \alpha(\phi_{i+1,j} + \phi_{i-1,j}) + (1 - 2\alpha)\phi_{i,j}, \tag{25.80}$$

where $\alpha = \kappa\Delta t/(\Delta x)^2$.

Although this scheme is explicit it is not a good one, because of the asymmetric way in which the differences are formed. However, the effect of this can be minimised if we study and correct for the errors introduced, in the following way. From Taylor's series in time,

$$\phi_{i,j+1} = \phi_{i,j} + \Delta t\frac{\partial \phi_{i,j}}{\partial t} + \tfrac{1}{2}(\Delta t)^2\frac{\partial^2 \phi_{i,j}}{\partial t^2} + \cdots, \tag{25.81}$$

using the same notation as previously. Thus the first correction term to the left-hand side of (25.79) is

$$-\tfrac{1}{2}\Delta t\frac{\partial^2 \phi_{i,j}}{\partial t^2}. \tag{25.82}$$

The first term omitted on the right-hand side of the same equation is, by a similar argument,

$$-\kappa\frac{2(\Delta x)^2}{4!}\frac{\partial^4 \phi_{i,j}}{\partial x^4}. \tag{25.83}$$

But, using the fact that $\phi$ satisfies (25.78) we obtain

$$\frac{\partial^2 \phi}{\partial t^2} = \frac{\partial}{\partial t}\left(\kappa\frac{\partial^2 \phi}{\partial x^2}\right) = \kappa\frac{\partial^2}{\partial x^2}\left(\frac{\partial \phi}{\partial t}\right) = \kappa^2\frac{\partial^4 \phi}{\partial x^4}, \tag{25.84}$$

and so, to this accuracy, the two errors (25.82) and (25.83) can be made to cancel if $\alpha$ is chosen such that

$$-\frac{\kappa^2}{2}\Delta t = -\frac{2\kappa(\Delta x)^2}{4!}, \qquad \text{i.e. } \alpha = \frac{1}{6}.$$

## 25.9 Exercises

25.1 Use an iteration procedure to find to four significant figures the root of the equation $40x = \exp x$.

25.2 Using the Newton–Raphson procedure find, correct to three decimal places, the root nearest to 7 of the equation $4x^3 + 2x^2 - 200x - 50 = 0$.

25.3 (a) Show that if a polynomial equation $g(x) \equiv x^m - f(x) = 0$, where $f(x)$ is a polynomial of degree less than $m$ with $f(0) \neq 0$, is solved using a rearrangement iteration scheme $x_{n+1} = [f(x_n)]^{1/m}$, then the scheme will, in general, have only first-order convergence.

(b) By considering the cubic equation

$$x^3 - ax^2 + 2abx - (b^3 + ab^2) = 0$$

for arbitrary non-zero values of $a$ and $b$, demonstrate that a re-arrangement scheme can, in special cases, give second- (or higher-) order convergence.

25.4 The square root of a number $N$ is to be determined by means of the iteration scheme

$$x_{n+1} = x_n \left[1 - \left(N - x_n^2\right) f(N)\right].$$

Determine how to choose $f(N)$ so that the process has second-order convergence.

Given that $\sqrt{7} \approx 2.65$, calculate $\sqrt{7}$ as accurately as a single application of the formula will allow.

25.5 The following table of values of a polynomial $p(x)$ of low degree contains an error. Identify and correct the erroneous value and extend the table up to $x = 1.2$.

| $x$ | $p(x)$ | $x$ | $p(x)$ |
|---|---|---|---|
| 0.0 | 0.000 | 0.5 | 0.165 |
| 0.1 | 0.011 | 0.6 | 0.216 |
| 0.2 | 0.040 | 0.7 | 0.245 |
| 0.3 | 0.081 | 0.8 | 0.256 |
| 0.4 | 0.128 | 0.9 | 0.243 |

25.6 Use a Taylor series to solve the equation

$$\frac{dy}{dx} + xy = 0, \qquad y(0) = 1,$$

evaluating $y(x)$ for $x = 0.0$ to $0.5$ in steps of 0.1.

25.7 If

$$\frac{dy}{dx} = f(x, y),$$

show that

$$\frac{d^2 f}{dx^2} = \frac{\partial^2 f}{\partial x^2} + 2f\frac{\partial^2 f}{\partial x \partial y} + f^2\frac{\partial^2 f}{\partial y^2} + \frac{\partial f}{\partial x}\frac{\partial f}{\partial y} + f\left(\frac{df}{dy}\right)^2.$$

Hence verify, by substitution and the subsequent expansion of arguments in Taylor series of their own, that the scheme given in (25.67) coincides with the Taylor expansion (25.56), i.e.

$$y_{i+1} = y_i + hy_i^{(1)} + \frac{h^2}{2!}y_i^{(2)} + \frac{h^3}{3!}y_i^{(3)} + \cdots,$$

up to terms in $h^3$.

25.8 Write a computer program that would solve, for a range of values of $\lambda$, the differential equation

$$\frac{dy}{dx} = \frac{1}{\sqrt{x^2 + \lambda y^2}}, \qquad y(0) = 1,$$

using a third-order Runge–Kutta scheme. Consider the difficulties that might arise when $\lambda < 0$.

25.9 Use the isocline approach to sketch the family of curves that satisfy the non-linear first-order differential equation

$$\frac{dy}{dx} = \frac{a}{\sqrt{x^2 + y^2}}.$$

25.10 Laplace's equation

$$\frac{\partial^2 V}{\partial x^2} + \frac{\partial^2 V}{\partial y^2} = 0,$$

is to be solved for the region and boundary conditions shown.

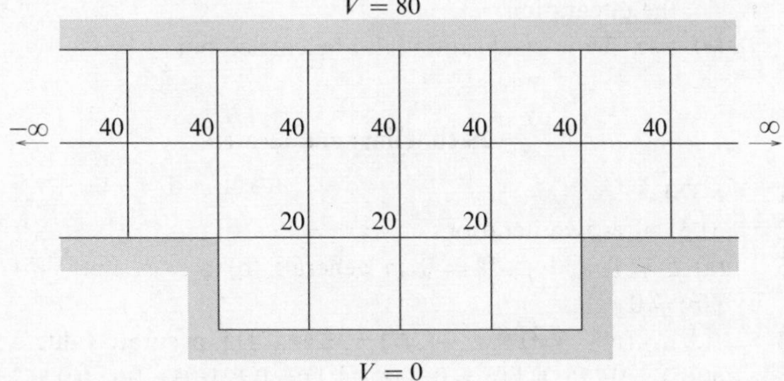


Starting from the given initial guess, and using the simplest possible form of relaxation, obtain a better approximation to the actual solution. Do not aim to be more accurate than $\pm$ 0.5 units and terminate the process when subsequent changes would be no greater than this.

25.11 The Schrödinger equation for a quantum mechanical particle of mass $m$ moving in a one-dimensional harmonic oscillator potential $V(x) = \frac{1}{2}kx^2$ is

$$-\frac{\hbar^2}{2m}\frac{d^2\psi}{dx^2} + \tfrac{1}{2}kx^2\psi = E\psi.$$

For physically acceptable solutions the wavefunction $\psi(x)$ must be finite at $x = 0$, tend to zero as $x \to \pm\infty$, and be normalised, so that $\int |\psi|^2\, dx = 1$. In practice these constraints mean that only certain (quantised) values of $E$, the energy of the particle, are allowed. The allowed values fall into two groups, those for which $y(0) = 0$ and those for which it is non-zero.

Show that if the unit of length is taken as $(\hbar^2/mk)^{1/4}$ and $E$ is measured in units of $\hbar(k/m)^{1/2}$, then the Schrödinger equation takes the form

$$\frac{d^2\psi}{dy^2} + (2E' - y^2)\psi = 0.$$

Devise an outline computerised scheme, using Runge–Kutta integration, which will enable you to:

(a) determine the three lowest allowed values of $E$;
(b) tabulate the normalised wavefunction corresponding to the lowest allowed energy.

You should consider explicitly:

(i) the variables to use in the numerical integration;
(ii) how starting values near $y = 0$ are to be chosen;
(iii) how the condition on $\psi$ as $y \to \pm\infty$ is to be implemented;
(iv) how the required values of $E$ are to be extracted from the results of the integration;
(v) how the normalisation is to be carried out.

## 25.10 Hints and answers

25.1 5.370.

25.2 6.951 after two iterations.

25.3 (a) $\xi \neq 0$ and $f'(\xi) \neq 0$ in general; (b) $\xi = b$, but $f'(b) = 0$ whilst $f(b) \neq 0$.

25.4 $f(N) = (N - 3x^2)^{-1} = -(2N)^{-1}$; 2.6457411, accurate value 2.6457513.

25.5 $p(0.5) = 0.175$, $p(1.0) = 0.200$, $p(1.1) = 0.121$, $p(1.2) = 0.000$.

25.6 $1 - x^2/2 + x^4/8 - x^6/48$; 1.0000, 0.9950, 0.9802, 0.9560, 0.9231, 0.8825; exact solution $y = \exp(-x^2/2)$.

25.9 See figure 25.6.

25.10 See figure 25.7.

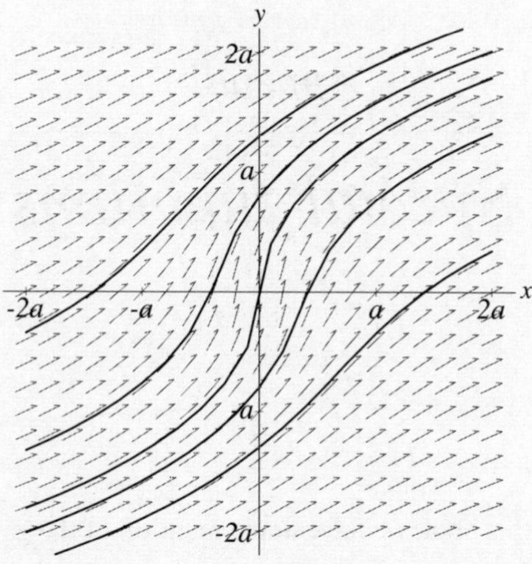


Figure 25.6 Typical solutions $y = y(x)$ (shown by solid lines) of $dy/dx = a(x^2 + y^2)^{-1/2}$. The short arrows give the direction that the tangent to any solution must have at that point.

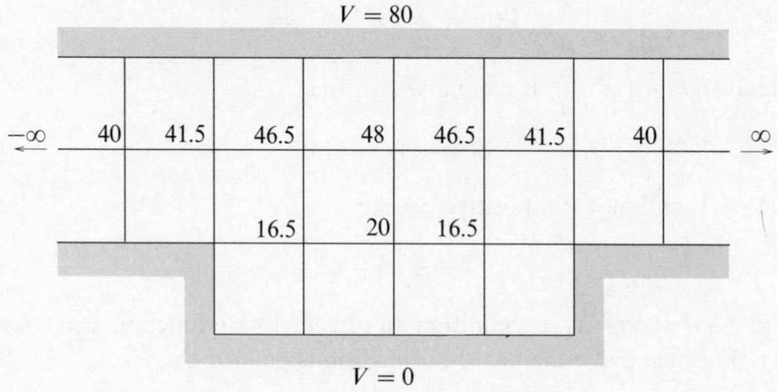


Figure 25.7 The solution to exercise 25.10.

# Appendix

# Special functions

In several places in this book we have made mention of the gamma, beta and error functions. These convenient functions appear in a number of contexts and here we gather together some of their properties. This appendix should be regarded merely as a reference containing some useful relations that are stated without proof.

The *gamma function* $\Gamma(n)$ is defined by

$$\Gamma(n) = \int_0^\infty x^{n-1} e^{-x}\, dx, \tag{A1}$$

which converges for $n > 0$. It can be shown that

$$\Gamma(n+1) = n\Gamma(n), \tag{A2}$$

with $\Gamma(1) = 1$, and, if $n$ is a positive integer,

$$\Gamma(n+1) = n!. \tag{A3}$$

Equation (A3) serves as a definition of the factorial function even for non-integer $n$. For negative $n$ the factorial function is defined by

$$n! = \frac{(n+m)!}{(n+m)(n+m-1)\cdots(n+1)}, \tag{A4}$$

where $m$ is any positive integer that makes $n + m > 0$. Different choices of $m$ ($> -n$) do not lead to different values for $n!$. A plot of the gamma function is given in figure A1.1, where it can be seen that the function is infinite for negative integer values of $n$, in accordance with (A4). A useful result is that $\Gamma\left(\frac{1}{2}\right) = \sqrt{\pi}$, from which $\Gamma(n)$ for half-integral $n$ can be found using (A2). Some immediately derivable factorial values of half integers are

$$\left(-\tfrac{3}{2}\right)! = -2\sqrt{\pi}, \quad \left(-\tfrac{1}{2}\right)! = \sqrt{\pi}, \quad \left(\tfrac{1}{2}\right)! = \tfrac{1}{2}\sqrt{\pi}, \quad \left(\tfrac{3}{2}\right)! = \tfrac{3}{4}\sqrt{\pi}.$$

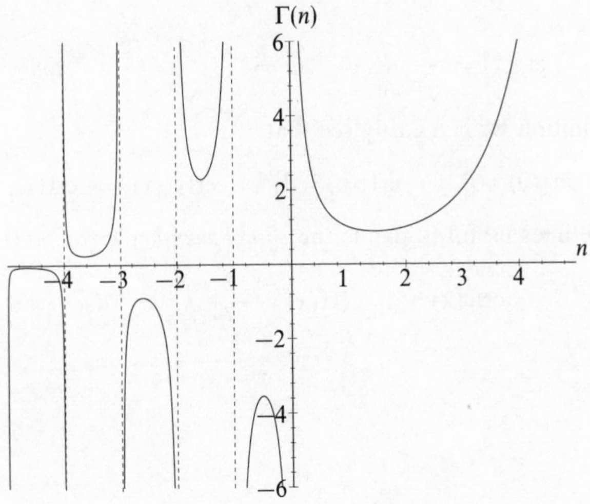


Figure A1.1 The gamma function.

It can also be shown that the gamma function is given by

$$\Gamma(n+1) = \sqrt{2\pi n}\, n^n e^{-n}\left(1 + \frac{1}{12n} + \frac{1}{288n^2} - \frac{139}{51840n^3} + \ldots\right) = n!, \quad \text{(A5)}$$

which is known as *Stirling's asymptotic series*. For large $n$ the first term dominates and so

$$n! \approx \sqrt{2\pi n}\, n^n e^{-n}, \quad \text{(A6)}$$

which is known as *Stirling's approximation*. The gamma function has been encountered in our discussion of Bessel functions (chapter 14). Stirling's approximation is particularly useful in statistical thermodynamics, when arrangements of a large number of particles are to be considered.

The closely related *beta function* is defined by

$$B(m,n) = \int_0^1 x^{m-1}(1-x)^{n-1}\, dx, \quad \text{(A7)}$$

which converges for $m > 0$, $n > 0$. The beta function may be written in terms of the gamma function as

$$B(m,n) = \frac{\Gamma(m)\Gamma(n)}{\Gamma(m+n)}. \quad \text{(A8)}$$

Finally we mention the *error function*, which is encountered in probability theory and in the solutions of some partial differential equations, and which is defined by

$$\operatorname{erf}(x) = \frac{2}{\sqrt{\pi}} \int_0^x e^{-u^2}\, du = 1 - \frac{2}{\sqrt{\pi}} \int_x^\infty e^{-u^2}\, du. \tag{A9}$$

From this definition we can easily see that

$$\operatorname{erf}(0) = 0, \qquad \operatorname{erf}(\infty) = 1, \qquad \operatorname{erf}(-x) = -\operatorname{erf}(x).$$

It is also sometimes useful to define the *complementary error function*

$$\operatorname{erfc}(x) = 1 - \operatorname{erf}(x) = \frac{2}{\sqrt{\pi}} \int_x^\infty e^{-u^2}\, du. \tag{A10}$$

# Index

Where discussion of a topic runs over two consecutive pages, reference is made only to the first of these. For discussions spread over several pages the first and last page numbers are given; these references are usually to the major treatment of the corresponding topic. Some long topics are split, e.g. 'Fourier transforms' and 'Fourier transforms, examples'; in such cases reference to subsequent entries is made in the first entry. The letter 'n' after a page number indicates that the topic is discussed in the footnote on the relevant page.